Engineering Mechanics: Dynamics

Authors

J. L. Meriam • L. G. Kraige

ISBN 9781119213017

Printed in the United States of America 10 9 8 7 6 5 4 3 2 1

List of Titles

Engineering Mechanics Volume 2: Dynamics, 8th edition
by J. L. Meriam and L. G. Kraige
Copyright © 2015, ISBN: 978-1-118-88584-0

Table of Contents

ENGINEERING MECHANICS

VOLUME 2

DYNAMICS

EIGHTH EDITION

ENGINEERING MECHANICS

VOLUME 2
DYNAMICS

EIGHTH EDITION

J.L. MERIAM

L.G. KRAIGE

Virginia Polytechnic Institute and State University

J.N. BOLTON

Bluefield State College

WILEY

On the cover: Liftoff of a Falcon 9 rocket from Vandenberg Air Force Base, California. This SpaceX rocket is a two-stage launch vehicle which generates 1.3 million pounds of thrust at sea level.

Vice President & Executive Publisher	Don Fowley
Executive Marketing Manager	Dan Sayre
Executive Editor	Linda Ratts
Editorial Assistants	Emily Meussner/Francesca Baratta
Content Manager	Karoline Luciano
Production Editor	Ken Santor, Production Management Services provided by Camelot Editorial Services, LLC
Marketing Manager	Christopher Ruel
Senior Designer	Maureen Eide
Cover Design	Wendy Lai
Cover Photo	SPACEX
Electronic Illustrations	Precision Graphics
Senior Photo Editor	Billy Ray
Product Designer	Jennifer Welter
Content Editor	Wendy Ashenberg

This book was set in 9.5/12 New Century Schoolbook Lt Std. by Aptara, Inc., and printed and bound by Quad Graphics Versailles. The cover was printed by Quad Graphics.

This book is printed on acid-free paper. ∞

Founded in 1807, John Wiley & Sons, Inc. has been a valued source of knowledge and understanding for more than 200 years, helping people around the world meet their needs and fulfill their aspirations. Our company is built on a foundation of principles that include responsibility to the communities we serve and where we live and work. In 2008, we launched a Corporate Citizenship Initiative, a global effort to address the environmental, social, economic, and ethical challenges we face in our business. Among the issues we are addressing are carbon impact, paper specifications and procurement, ethical conduct within our business and among our vendors, and community and charitable support. For more information, please visit our website: www.wiley.com/go/citizenship.

Library of Congress Cataloging-in-Publication Data

Meriam, J. L. (James L.)
Dynamics / J. L. Meriam, L. G. Kraige, Virginia Polytechnic Institute and State University, J. N. Bolton, Bluefield State College.—Eighth edition.
pages ; cm—(Engineering mechanics)
Includes bibliographical references and index.
ISBN 978-1-118-88584-0 (cloth)
1. Machinery, Dynamics of. I. Kraige, L. G. (L. Glenn) II. Bolton, J. N. (Jeff N.)
III. Title.
TA352.M45 2015
620.1—dc23
2015016668

ISBN: 978-1-118-88584-0
ISBN: 978-1-119-02253-4 (BRV)

Printed in the United States of America

10 9 8 7 6 5 4 3 2 1

FOREWORD

This series of textbooks was begun in 1951 by the late Dr. James L. Meriam. At that time, the books represented a revolutionary transformation in undergraduate mechanics education. They became the definitive textbooks for the decades that followed as well as models for other engineering mechanics texts that have subsequently appeared. Published under slightly different titles prior to the 1978 First Editions, this textbook series has always been characterized by logical organization, clear and rigorous presentation of the theory, instructive sample problems, and a rich collection of real-life problems, all with a high standard of illustration. In addition to the U.S. versions, the books have appeared in SI versions and have been translated into many foreign languages. These textbooks collectively represent an international standard for undergraduate texts in mechanics.

The innovations and contributions of Dr. Meriam (1917–2000) to the field of engineering mechanics cannot be overstated. He was one of the premier engineering educators of the second half of the twentieth century. Dr. Meriam earned the B.E., M.Eng., and Ph.D. degrees from Yale University. He had early industrial experience with Pratt and Whitney Aircraft and the General Electric Company. During the Second World War he served in the U.S. Coast Guard. He was a member of the faculty of the University of California—Berkeley, Dean of Engineering at Duke University, a faculty member at the California Polytechnic State University, and visiting professor at the University of California—Santa Barbara, finally retiring in 1990. Professor Meriam always placed great emphasis on teaching, and this trait was recognized by his students wherever he taught. He was the recipient of several teaching awards, including the Benjamin Garver Lamme Award, which is the highest annual national award of the American Society of Engineering Education (ASEE).

Dr. L. Glenn Kraige, coauthor of the *Engineering Mechanics* series since the early 1980s, has also made significant contributions to mechanics education. Dr. Kraige earned his B.S., M.S., and Ph.D. degrees at the University of Virginia, principally in aerospace engineering, and he is Professor Emeritus of Engineering Science and Mechanics at Virginia Polytechnic Institute and State University. During the mid-1970s, I had the singular pleasure of chairing Professor Kraige's graduate committee and take particular pride in the fact that he was the first of my fifty Ph.D. graduates. Professor Kraige was invited by Professor Meriam to team with him, thereby ensuring that the Meriam legacy of textbook authorship excellence would be carried forward to future generations of engineers.

In addition to his widely recognized research and publications in the field of spacecraft dynamics, Professor Kraige has devoted his attention to the teaching of mechanics at both introductory and advanced levels. His outstanding teaching has been widely recognized and has earned him teaching awards at the departmental, college, university, state, regional, and national levels. These awards include the Outstanding Educator Award from the State Council of Higher Education for the Commonwealth of Virginia. In 1996, the

Mechanics Division of ASEE bestowed upon him the Archie Higdon Distinguished Educator Award. The Carnegie Foundation for the Advancement of Teaching and the Council for Advancement and Support of Education awarded him the distinction of Virginia Professor of the Year for 1997. In his teaching, Professor Kraige stresses the development of analytical capabilities along with the strengthening of physical insight and engineering judgment. Since the early 1980s, he has worked on personal-computer software designed to enhance the teaching/learning process in statics, dynamics, strength of materials, and higher-level areas of dynamics and vibrations.

Welcomed as a new coauthor for this edition is Dr. Jeffrey N. Bolton, Assistant Professor of Mechanical Engineering Technology at Bluefield State College. Dr. Bolton earned his B.S., M.S., and Ph.D. in Engineering Mechanics from Virginia Polytechnic Institute and State University. His research interests include automatic balancing of six-degree-of-freedom elastically-mounted rotors. He has a wealth of teaching experience, including at Virginia Tech, where he was the 2010 recipient of the Sporn Teaching Award for Engineering Subjects, which is primarily chosen by students. In 2014, Professor Bolton received the Outstanding Faculty Award from Bluefield State College. He has the unusual ability to set high levels of rigor and achievement in the classroom while establishing a high degree of rapport with his students. In addition to maintaining time-tested traditions for future generations of students, Dr. Bolton will bring effective application of technology to this textbook series.

The Eighth Edition of *Engineering Mechanics* continues the same high standards set by previous editions and adds new features of help and interest to students. It contains a vast collection of interesting and instructive problems. The faculty and students privileged to teach or study from the Meriam/Kraige/Bolton *Engineering Mechanics* series will benefit from several decades of investment by three highly accomplished educators. Following the pattern of the previous editions, this textbook stresses the application of theory to actual engineering situations, and at this important task it remains the best.

John L. Junkins
Distinguished Professor of Aerospace Engineering
Holder of the Royce E. Wisebaker '39 Chair in Engineering Innovation
Texas A&M University
College Station, Texas

PREFACE

Engineering mechanics is both a foundation and a framework for most of the branches of engineering. Many of the topics in such areas as civil, mechanical, aerospace, and agricultural engineering, and of course engineering mechanics itself, are based upon the subjects of statics and dynamics. Even in a discipline such as electrical engineering, practitioners, in the course of considering the electrical components of a robotic device or a manufacturing process, may find themselves first having to deal with the mechanics involved.

Thus, the engineering mechanics sequence is critical to the engineering curriculum. Not only is this sequence needed in itself, but courses in engineering mechanics also serve to solidify the student's understanding of other important subjects, including applied mathematics, physics, and graphics. In addition, these courses serve as excellent settings in which to strengthen problem-solving abilities.

PHILOSOPHY

The primary purpose of the study of engineering mechanics is to develop the capacity to predict the effects of force and motion while carrying out the creative design functions of engineering. This capacity requires more than a mere knowledge of the physical and mathematical principles of mechanics; also required is the ability to visualize physical configurations in terms of real materials, actual constraints, and the practical limitations which govern the behavior of machines and structures. One of the primary objectives in a mechanics course is to help the student develop this ability to visualize, which is so vital to problem formulation. Indeed, the construction of a meaningful mathematical model is often a more important experience than its solution. Maximum progress is made when the principles and their limitations are learned together within the context of engineering application.

There is a frequent tendency in the presentation of mechanics to use problems mainly as a vehicle to illustrate theory rather than to develop theory for the purpose of solving problems. When the first view is allowed to predominate, problems tend to become overly idealized and unrelated to engineering with the result that the exercise becomes dull, academic, and uninteresting. This approach deprives the student of valuable experience in formulating problems and thus of discovering the need for and meaning of theory. The second view provides by far the stronger motive for learning theory and leads to a better balance between theory and application. The crucial role played by interest and purpose in providing the strongest possible motive for learning cannot be overemphasized.

Furthermore, as mechanics educators, we should stress the understanding that, at best, theory can only approximate the real world of mechanics rather than the view that the real world approximates the theory. This difference in philosophy is indeed basic and distinguishes the *engineering* of mechanics from the *science* of mechanics.

Over the past several decades, several unfortunate tendencies have occurred in engineering education. First, emphasis on the geometric and physical meanings of prerequisite mathematics appears to have diminished. Second, there has been a significant reduction and even elimination of instruction in graphics, which in the past enhanced the visualization and representation of mechanics problems. Third, in advancing the mathematical level of our treatment of mechanics, there has been a tendency to allow the notational manipulation of vector operations to mask or replace geometric visualization. Mechanics is inherently a subject which depends on geometric and physical perception, and we should increase our efforts to develop this ability.

A special note on the use of computers is in order. The experience of formulating problems, where reason and judgment are developed, is vastly more important for the student than is the manipulative exercise in carrying out the solution. For this reason, computer usage must be carefully controlled. At present, constructing free-body diagrams and formulating governing equations are best done with pencil and paper. On the other hand, there are instances in which the *solution* to the governing equations can best be carried out and displayed using the computer. Computer-oriented problems should be genuine in the sense that there is a condition of design or criticality to be found, rather than "make-work" problems in which some parameter is varied for no apparent reason other than to force artificial use of the computer. These thoughts have been kept in mind during the design of the computer-oriented problems in the Eighth Edition. To conserve adequate time for problem formulation, it is suggested that the student be assigned only a limited number of the computer-oriented problems.

As with previous editions, this Eighth Edition of *Engineering Mechanics* is written with the foregoing philosophy in mind. It is intended primarily for the first engineering course in mechanics, generally taught in the second year of study. *Engineering Mechanics* is written in a style which is both concise and friendly. The major emphasis is on basic principles and methods rather than on a multitude of special cases. Strong effort has been made to show both the cohesiveness of the relatively few fundamental ideas and the great variety of problems which these few ideas will solve.

Pedagogical Features

The basic structure of this textbook consists of an article which rigorously treats the particular subject matter at hand, followed by one or more Sample Problems, followed by a group of Problems. There is a Chapter Review at the end of each chapter which summarizes the main points in that chapter, followed by a Review Problem set.

Problems

The 124 sample problems appear on specially colored pages by themselves. The solutions to typical dynamics problems are presented in detail. In addition, explanatory and cautionary notes (Helpful Hints) in blue type are number-keyed to the main presentation.

There are 1550 homework exercises, of which more than 50 percent are new to the Eighth Edition. The problem sets are divided into *Introductory Problems* and *Representative Problems*. The first section consists of simple, uncomplicated problems designed to help students gain confidence with the new topic, while most of the problems in the second section are of average difficulty and length. The problems are generally arranged in order of increasing difficulty. More difficult exercises appear near the end of the *Representative Problems* and are marked with the symbol ►. *Computer-Oriented Problems*, marked with an asterisk, appear throughout the problem sets as well as in a special section at the end of each chapter. The answers to all problems have been provided in a special section at the end of the textbook.

In recognition of the need for emphasis on SI units, there are approximately two problems in SI units for every one in U.S. customary units. This apportionment between the two sets of units permits anywhere from a 50–50 emphasis to a 100-percent SI treatment.

A notable feature of the Eighth Edition, as with all previous editions, is the wealth of interesting and important problems which apply to engineering design. Whether directly identified as such or not, virtually all of the problems deal with principles and procedures inherent in the design and analysis of engineering structures and mechanical systems.

Illustrations

In order to bring the greatest possible degree of realism and clarity to the illustrations, this textbook series continues to be produced in full color. It is important to note that color is used consistently for the identification of certain quantities:

- *red* for forces and moments
- *green* for velocity and acceleration arrows
- *orange dashes* for selected trajectories of moving points

Subdued colors are used for those parts of an illustration which are not central to the problem at hand. Whenever possible, mechanisms or objects which commonly have a certain color will be portrayed in that color. All of the fundamental elements of technical illustration which have been an essential part of this *Engineering Mechanics* series of textbooks have been retained. The authors wish to restate the conviction that a high standard of illustration is critical to any written work in the field of mechanics.

Special Features

We have retained the following hallmark features of previous editions:

- The main emphasis on the work-energy and impulse-momentum equations is on the time-order form, both for particles in Chapter 3 and rigid bodies in Chapter 6.
- Emphasis has been placed on three-part impulse-momentum diagrams, both for particles and rigid bodies. These diagrams are well integrated with the time-order form of the impulse-momentum equations.
- Within-the-chapter photographs are provided in order to provide additional connection to actual situations in which dynamics has played a major role.
- Approximately 50 percent of the homework problems are new to this Eighth Edition. All new problems have been independently solved in order to ensure a high degree of accuracy.
- All Sample Problems are printed on specially colored pages for quick identification.
- All theory portions have been reexamined in order to maximize rigor, clarity, readability, and level of friendliness.
- Key Concepts areas within the theory presentation have been specially marked and highlighted.
- The Chapter Reviews are highlighted and feature itemized summaries.

Organization

The logical division between particle dynamics (Part I) and rigid-body dynamics (Part II) has been preserved, with each part treating the kinematics prior to the kinetics. This arrangement promotes thorough and rapid progress in rigid-body dynamics with the prior benefit of a comprehensive introduction to particle dynamics.

In Chapter 1, the fundamental concepts necessary for the study of dynamics are established.

Chapter 2 treats the kinematics of particle motion in various coordinate systems, as well as the subjects of relative and constrained motion.

Chapter 3 on particle kinetics focuses on the three basic methods: force-mass-acceleration (Section A), work-energy (Section B), and impulse-momentum (Section C). The special topics of impact, central-force motion, and relative motion are grouped together in a special applications section (Section D) and serve as optional material to be assigned according to instructor preference and available time. With this arrangement, the attention of the student is focused more strongly on the three basic approaches to kinetics.

Chapter 4 on systems of particles is an extension of the principles of motion for a single particle and develops the general relationships which are so basic to the modern comprehension of dynamics. This chapter also includes the topics of steady mass flow and variable mass, which may be considered as optional material.

In Chapter 5 on the kinematics of rigid bodies in plane motion, where the equations of relative velocity and relative acceleration are encountered, emphasis is placed jointly on solution by vector geometry and solution by vector algebra. This dual approach serves to reinforce the meaning of vector mathematics.

In Chapter 6 on the kinetics of rigid bodies, we place great emphasis on the basic equations which govern all categories of plane motion. Special emphasis is also placed on forming the direct equivalence between the actual applied forces and couples and their $m\bar{a}$ and $\bar{I}\alpha$ resultants. In this way the versatility of the moment principle is emphasized, and the student is encouraged to think directly in terms of resultant dynamics effects.

Chapter 7, which may be treated as optional, provides a basic introduction to three-dimensional dynamics which is sufficient to solve many of the more common space-motion problems. For students who later pursue more advanced work in dynamics, Chapter 7 will provide a solid foundation. Gyroscopic motion with steady precession is treated in two ways. The first approach makes use of the analogy between the relation of force and linear-momentum vectors and the relation of moment and angular-momentum vectors. With this treatment, the student can understand the gyroscopic phenomenon of steady precession and can handle most of the engineering problems on gyroscopes without a detailed study of three-dimensional dynamics. The second approach employs the more general momentum equations for three-dimensional rotation where all components of momentum are accounted for.

Chapter 8 is devoted to the topic of vibrations. This full-chapter coverage will be especially useful for engineering students whose only exposure to vibrations is acquired in the basic dynamics course.

Moments and products of inertia of mass are presented in Appendix B. Appendix C contains a summary review of selected topics of elementary mathematics as well as several numerical techniques which the student should be prepared to use in computer-solved problems. Useful tables of physical constants, centroids, and moments of inertia are contained in Appendix D.

Supplements

The following items have been prepared to complement this textbook:

Instructor's Manual

Prepared by the authors and independently checked, fully worked solutions to all problems in the text are available to faculty by contacting their local Wiley representative.

Instructor Lecture Resources

The following resources are available online at www.wiley.com/college/meriam. There may be additional resources not listed.

WileyPLUS is a research-based online environment for effective teaching and learning. WileyPLUS builds students' confidence because it takes the guesswork out of studying by providing students with a clear roadmap: what to do, how to do it, if they did it right. Students will take more initiative so you'll have greater impact on their achievement in the classroom and beyond.

Lecture software specifically designed to aid the lecturer, especially in larger classrooms. Written by the second author and incorporating figures from the textbooks, this software is based on the Macromedia Flash® platform. Major use of animation, concise review of the theory, and numerous sample problems make this tool extremely useful for student self-review of the material.

All ***figures*** in the text are available in electronic format for use in creating lecture presentations.

All ***Sample Problems*** are available as electronic files for display and discussion in the classroom.

ACKNOWLEDGMENTS

Special recognition is due Dr. A. L. Hale, formerly of Bell Telephone Laboratories, for his continuing contribution in the form of invaluable suggestions and accurate checking of the manuscript. Dr. Hale has rendered similar service for all previous versions of this entire series of mechanics books, dating back to the 1950s. He reviews all aspects of the books, including all old and new text and figures. Dr. Hale carries out an independent solution to each new homework exercise and provides the authors with suggestions and needed corrections to the solutions which appear in the *Instructor's Manual*. Dr. Hale is well known for being extremely accurate in his work, and his fine knowledge of the English language is a great asset which aids every user of this textbook.

We would like to thank the faculty members of the Department of Engineering Science and Mechanics at VPI&SU who regularly offer constructive suggestions. These include Saad A. Ragab, Norman E. Dowling, Michael W. Hyer, Michael L. Madigan (now at Texas A&M University), J. Wallace Grant, and Jacob Grohs. Scott L. Hendricks has been particularly effective and accurate in his extensive review of the manuscript and preparation of WileyPlus materials. Nathaniel Greene of Bloomfield State University of Pennsylvania is recognized for his careful reading and suggestions for improvement.

The following individuals (listed in alphabetical order) provided feedback on recent editions, reviewed samples of the Eighth Edition, or otherwise contributed to the Eighth Edition:

Michael Ales, *U.S. Merchant Marine Academy*
Joseph Arumala, *University of Maryland Eastern Shore*
Eric Austin, *Clemson University*
Stephen Bechtel, *Ohio State University*
Peter Birkemoe, *University of Toronto*
Achala Chatterjee, *San Bernardino Valley College*
Jim Shih-Jiun Chen, *Temple University*
Yi-chao Chen, *University of Houston*
Mary Cooper, *Cal Poly San Luis Obispo*
Mukaddes Darwish, *Texas Tech University*
Kurt DeGoede, *Elizabethtown College*
John DesJardins, *Clemson University*
Larry DeVries, *University of Utah*
Craig Downing, *Southeast Missouri State University*
William Drake, *Missouri State University*
Raghu Echempati, *Kettering University*
Amelito Enriquez, *Canada College*
Sven Esche, *Stevens Institute of Technology*
Wallace Franklin, *U.S. Merchant Marine Academy*
Christine Goble, *University of Kentucky*
Barry Goodno, *Georgia Institute of Technology*
Robert Harder, *George Fox University*

Javier Hasbun, *University of West Georgia*
Javad Hashemi, *Texas Tech University*
Robert Hyers, *University of Massachusetts, Amherst*
Matthew Ikle, *Adams State College*
Duane Jardine, *University of New Orleans*
Mariappan Jawaharlal, *California Polytechnic State University, Pomona*
Qing Jiang, *University of California, Riverside*
Jennifer Kadlowec, *Rowan University*
Robert Kern, *Milwaukee School of Engineering*
John Krohn, *Arkansas Tech University*
Keith Lindler, *United States Naval Academy*
Francisco Manzo-Robledo, *Washington State University*
Geraldine Milano, *New Jersey Institute of Technology*
Saeed Niku, *Cal Poly San Luis Obispo*
Wilfrid Nixon, *University of Iowa*
Karim Nohra, *University of South Florida*
Vassilis Panoskaltsis, *Case Western Reserve University*
Chandra Putcha, *California State University, Fullerton*
Blayne Roeder, *Purdue University*
Eileen Rossman, *Cal Poly San Luis Obispo*
Nestor Sanchez, *University of Texas, San Antonio*
Joseph Schaefer, *Iowa State University*
Scott Schiff, *Clemson University*
Sergey Smirnov, *Texas Tech University*
Ertugrul Taciroglu, *UCLA*
Constantine Tarawneh, *University of Texas*
John Turner, *University of Wyoming*
Chris Venters, *Virginia Tech*
Sarah Vigmostad, *University of Iowa*
T. W. Wu, *University of Kentucky*
Mohammed Zikry, *North Carolina State University*

The contributions by the staff of John Wiley & Sons, Inc., including Executive Editor Linda Ratts (recipient of the John Wiley Global Education Editor of the Year Award for 2013), Editorial Assistant Emily Meussner, Production Editor Ken Santor, Senior Designer Maureen Eide, and Photograph Editor Billy Ray, reflect a high degree of professional competence and are duly recognized. We wish to especially acknowledge the critical production efforts of Christine Cervoni of Camelot Editorial Services, LLC. Ms. Cervoni has frequently exceeded expectations over the past several editions. When the inevitable difficulties arise, she is quick to find a solution and regularly helps other production individuals. Helen Walden is recognized for her copy-editing and proofreading of a difficult manuscript. The talented illustrators of Precision Graphics continue to maintain a high standard of illustration excellence.

Finally, we wish to state the extremely significant contribution of our families for their patience and support over the many hours of manuscript preparation. Dale Kraige has managed the preparation of the manuscript for the Eighth Edition and has been a key individual in checking all stages of the proof. In addition, both Stephanie Kokan and David Kraige have contributed problem ideas, illustrations, and solutions to a number of the problems over the past several editions.

We are extremely pleased to participate in extending the time duration of this textbook series well past the sixty-five-year mark. In the interest of providing you with the best possible educational materials over future years, we encourage and welcome all comments and suggestions.

L. Glenn Kraige

Blacksburg, Virginia

Princeton, West Virginia

Conversion Factors

U.S. Customary Units to SI Units

To convert from	To	Multiply by
(Acceleration)		
foot/second^2 (ft/sec^2)	meter/second^2 (m/s^2)	3.048×10^{-1}*
inch/second^2 (in./sec^2)	meter/second^2 (m/s^2)	2.54×10^{-2}*
(Area)		
foot^2 (ft^2)	meter^2 (m^2)	9.2903×10^{-2}
inch^2 (in.^{2})	meter^2 (m^2)	6.4516×10^{-4}*
(Density)		
pound mass/inch^3 (lbm/in.^{3})	kilogram/meter^3 (kg/m^3)	2.7680×10^4
pound mass/foot^3 (lbm/ft^3)	kilogram/meter^3 (kg/m^3)	1.6018×10
(Force)		
kip (1000 lb)	newton (N)	4.4482×10^3
pound force (lb)	newton (N)	4.4482
(Length)		
foot (ft)	meter (m)	3.048×10^{-1}*
inch (in.)	meter (m)	2.54×10^{-2}*
mile (mi), (U.S. statute)	meter (m)	1.6093×10^3
mile (mi), (international nautical)	meter (m)	1.852×10^3*
(Mass)		
pound mass (lbm)	kilogram (kg)	4.5359×10^{-1}
slug (lb-sec^2/ft)	kilogram (kg)	1.4594×10
ton (2000 lbm)	kilogram (kg)	9.0718×10^2
(Moment of force)		
pound-foot (lb-ft)	newton-meter (N·m)	1.3558
pound-inch (lb-in.)	newton-meter (N·m)	0.1129 8
(Moment of inertia, area)		
inch^4	meter^4 (m^4)	41.623×10^{-8}
(Moment of inertia, mass)		
pound-foot-second^2 (lb-ft-sec^2)	kilogram-meter^2 (kg·m^2)	1.3558
(Momentum, linear)		
pound-second (lb-sec)	kilogram-meter/second (kg·m/s)	4.4482
(Momentum, angular)		
pound-foot-second (lb-ft-sec)	newton-meter-second (kg·m^2/s)	1.3558
(Power)		
foot-pound/minute (ft-lb/min)	watt (W)	2.2597×10^{-2}
horsepower (550 ft-lb/sec)	watt (W)	7.4570×10^2
(Pressure, stress)		
atmosphere (std)(14.7 lb/in.^{2})	newton/meter^2 (N/m^2 or Pa)	1.0133×10^5
pound/foot^2 (lb/ft^2)	newton/meter^2 (N/m^2 or Pa)	4.7880×10
pound/inch^2 (lb/in.^{2} or psi)	newton/meter^2 (N/m^2 or Pa)	6.8948×10^3
(Spring constant)		
pound/inch (lb/in.)	newton/meter (N/m)	1.7513×10^2
(Velocity)		
foot/second (ft/sec)	meter/second (m/s)	3.048×10^{-1}*
knot (nautical mi/hr)	meter/second (m/s)	5.1444×10^{-1}
mile/hour (mi/hr)	meter/second (m/s)	4.4704×10^{-1}*
mile/hour (mi/hr)	kilometer/hour (km/h)	1.6093
(Volume)		
foot^3 (ft^3)	meter^3 (m^3)	2.8317×10^{-2}
inch^3 (in.^{3})	meter^3 (m^3)	1.6387×10^{-5}
(Work, Energy)		
British thermal unit (BTU)	joule (J)	1.0551×10^3
foot-pound force (ft-lb)	joule (J)	1.3558
kilowatt-hour (kw-h)	joule (J)	3.60×10^6*

*Exact value

SI Units Used in Mechanics

Quantity	Unit	SI Symbol
(Base Units)		
Length	meter*	m
Mass	kilogram	kg
Time	second	s
(Derived Units)		
Acceleration, linear	meter/second^2	m/s^2
Acceleration, angular	radian/second^2	rad/s^2
Area	meter^2	m^2
Density	kilogram/meter^3	kg/m^3
Force	newton	N (= kg·m/s^2)
Frequency	hertz	Hz (= 1/s)
Impulse, linear	newton-second	N·s
Impulse, angular	newton-meter-second	N·m·s
Moment of force	newton-meter	N·m
Moment of inertia, area	meter^4	m^4
Moment of inertia, mass	kilogram-meter^2	kg·m^2
Momentum, linear	kilogram-meter/second	kg·m/s (= N·s)
Momentum, angular	kilogram-meter^2/second	kg·m^2/s (= N·m·s)
Power	watt	W (= J/s = N·m/s)
Pressure, stress	pascal	Pa (= N/m^2)
Product of inertia, area	meter^4	m^4
Product of inertia, mass	kilogram-meter^2	kg·m^2
Spring constant	newton/meter	N/m
Velocity, linear	meter/second	m/s
Velocity, angular	radian/second	rad/s
Volume	meter^3	m^3
Work, energy	joule	J (= N·m)
(Supplementary and Other Acceptable Units)		
Distance (navigation)	nautical mile	(= 1.852 km)
Mass	ton (metric)	t (= 1000 kg)
Plane angle	degrees (decimal)	°
Plane angle	radian	—
Speed	knot	(1.852 km/h)
Time	day	d
Time	hour	h
Time	minute	min

*Also spelled *metre*.

SI Unit Prefixes

Multiplication Factor	Prefix	Symbol
1 000 000 000 000 = 10^{12}	tera	T
1 000 000 000 = 10^9	giga	G
1 000 000 = 10^6	mega	M
1 000 = 10^3	kilo	k
100 = 10^2	hecto	h
10 = 10	deka	da
0.1 = 10^{-1}	deci	d
0.01 = 10^{-2}	centi	c
0.001 = 10^{-3}	milli	m
0.000 001 = 10^{-6}	micro	μ
0.000 000 001 = 10^{-9}	nano	n
0.000 000 000 001 = 10^{-12}	pico	p

Selected Rules for Writing Metric Quantities

1. (a) Use prefixes to keep numerical values generally between 0.1 and 1000.
   (b) Use of the prefixes hecto, deka, deci, and centi should generally be avoided except for certain areas or volumes where the numbers would be awkward otherwise.
   (c) Use prefixes only in the numerator of unit combinations. The one exception is the base unit kilogram. (*Example:* write kN/m not N/mm; J/kg not mJ/g)
   (d) Avoid double prefixes. (*Example:* write GN not kMN)
2. Unit designations
   (a) Use a dot for multiplication of units. (*Example:* write N·m not Nm)
   (b) Avoid ambiguous double solidus. (*Example:* write N/m^2 not N/m/m)
   (c) Exponents refer to entire unit. (*Example:* mm^2 means (mm)^2)
3. Number grouping
   Use a space rather than a comma to separate numbers in groups of three, counting from the decimal point in both directions. (*Example:* 4 607 321.048 72) Space may be omitted for numbers of four digits. (*Example:* 4296 or 0.0476)

PART I

DYNAMICS OF PARTICLES

Stocktrek Images/Getty Images, Inc.

The International Space Station's Canadarm2 releases the Cygnus spacecraft in October 2013.

1 INTRODUCTION TO DYNAMICS

CHAPTER OUTLINE

1/1 HISTORY AND MODERN APPLICATIONS

Dynamics is that branch of mechanics which deals with the motion of bodies under the action of forces. The study of dynamics in engineering usually follows the study of statics, which deals with the effects of forces on bodies at rest. Dynamics has two distinct parts: *kinematics*, which is the study of motion without reference to the forces which cause motion, and *kinetics*, which relates the action of forces on bodies to their resulting motions. A thorough comprehension of dynamics will provide one of the most useful and powerful tools for analysis in engineering.

History of Dynamics

Dynamics is a relatively recent subject compared with statics. The beginning of a rational understanding of dynamics is credited to Galileo (1564–1642), who made careful observations concerning bodies in free fall, motion on an inclined plane, and motion of the pendulum. He was largely responsible for bringing a scientific approach to the investigation of physical problems. Galileo was continually under severe criticism for refusing to accept the established beliefs of his day, such as the philosophies of Aristotle which held, for example, that heavy bodies fall more rapidly than light bodies. The lack of accurate means for the measurement of time was a severe handicap to Galileo, and

Galileo Galilei
Portrait of Galileo Galilei (1564–1642) (oil on canvas), Sustermans, Justus (1597–1681) (school of)/Galleria Palatina, Florence, Italy/Bridgeman Art Library.

further significant development in dynamics awaited the invention of the pendulum clock by Huygens in 1657.

Newton (1642–1727), guided by Galileo's work, was able to make an accurate formulation of the laws of motion and, thus, to place dynamics on a sound basis. Newton's famous work was published in the first edition of his *Principia*,* which is generally recognized as one of the greatest of all recorded contributions to knowledge. In addition to stating the laws governing the motion of a particle, Newton was the first to correctly formulate the law of universal gravitation. Although his mathematical description was accurate, he felt that the concept of remote transmission of gravitational force without a supporting medium was an absurd notion. Following Newton's time, important contributions to mechanics were made by Euler, D'Alembert, Lagrange, Laplace, Poinsot, Coriolis, Einstein, and others.

Applications of Dynamics

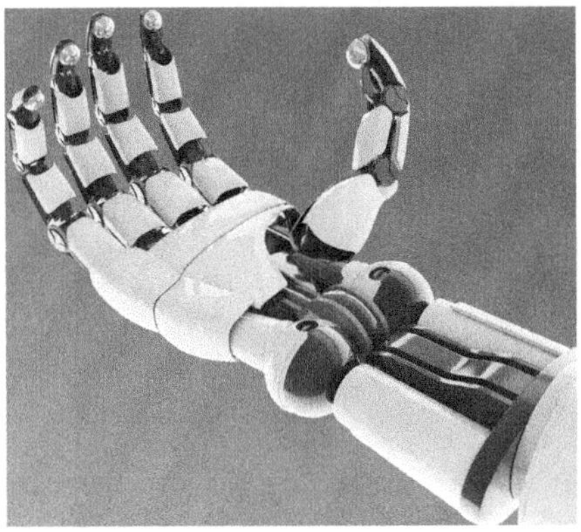

Zhee-Shee/Shutterstock

Artificial hand.

Only since machines and structures have operated with high speeds and appreciable accelerations has it been necessary to make calculations based on the principles of dynamics rather than on the principles of statics. The rapid technological developments of the present day require increasing application of the principles of mechanics, particularly dynamics. These principles are basic to the analysis and design of moving structures, to fixed structures subject to shock loads, to robotic devices, to automatic control systems, to rockets, missiles, and spacecraft, to ground and air transportation vehicles, to electron ballistics of electrical devices, and to machinery of all types such as turbines, pumps, reciprocating engines, hoists, machine tools, etc.

Students with interests in one or more of these and many other activities will constantly need to apply the fundamental principles of dynamics.

1/2 Basic Concepts

The concepts basic to mechanics were set forth in Art. 1/2 of *Vol. 1 Statics*. They are summarized here along with additional comments of special relevance to the study of dynamics.

Space is the geometric region occupied by bodies. Position in space is determined relative to some geometric reference system by means of linear and angular measurements. The basic frame of reference for the laws of Newtonian mechanics is the *primary inertial system* or *astronomical frame of reference*, which is an imaginary set of rectangular axes assumed to have no translation or rotation in space. Measurements show that the laws of Newtonian mechanics are valid for this reference system as long as any velocities involved are negligible compared with the speed of light, which is 300 000 km/s or 186,000 mi/sec. Measurements made with respect to this reference are said to be *absolute*, and this reference system may be considered "fixed" in space.

*The original formulations of Sir Isaac Newton may be found in the translation of his *Principia* (1687), revised by F. Cajori, University of California Press, 1934.

A reference frame attached to the surface of the earth has a somewhat complicated motion in the primary system, and a correction to the basic equations of mechanics must be applied for measurements made relative to the reference frame of the earth. In the calculation of rocket and space-flight trajectories, for example, the absolute motion of the earth becomes an important parameter. For most engineering problems involving machines and structures which remain on the surface of the earth, the corrections are extremely small and may be neglected. For these problems the laws of mechanics may be applied directly with measurements made relative to the earth, and in a practical sense such measurements will be considered *absolute*.

Time is a measure of the succession of events and is considered an absolute quantity in Newtonian mechanics.

Mass is the quantitative measure of the inertia or resistance to change in motion of a body. Mass may also be considered as the quantity of matter in a body as well as the property which gives rise to gravitational attraction.

Force is the vector action of one body on another. The properties of forces have been thoroughly treated in *Vol. 1 Statics*.

A ***particle*** is a body of negligible dimensions. When the dimensions of a body are irrelevant to the description of its motion or the action of forces on it, the body may be treated as a particle. An airplane, for example, may be treated as a particle for the description of its flight path.

A ***rigid body*** is a body whose changes in shape are negligible compared with the overall dimensions of the body or with the changes in position of the body as a whole. As an example of the assumption of rigidity, the small flexural movement of the wing tip of an airplane flying through turbulent air is clearly of no consequence to the description of the motion of the airplane as a whole along its flight path. For this purpose, then, the treatment of the airplane as a rigid body is an acceptable approximation. On the other hand, if we need to examine the internal stresses in the wing structure due to changing dynamic loads, then the deformation characteristics of the structure would have to be examined, and for this purpose the airplane could no longer be considered a rigid body.

Vector and ***scalar*** quantities have been treated extensively in *Vol. 1 Statics*, and their distinction should be perfectly clear by now. Scalar quantities are printed in lightface italic type, and vectors are shown in boldface type. Thus, V denotes the scalar magnitude of the vector $\mathbf{V}$. It is important that we use an identifying mark, such as an underline $\underline{V}$, for all handwritten vectors to take the place of the boldface designation in print. For two nonparallel vectors recall, for example, that $\mathbf{V}_1 + \mathbf{V}_2$ and $V_1 + V_2$ have two entirely different meanings.

We assume that you are familiar with the geometry and algebra of vectors through previous study of statics and mathematics. Students who need to review these topics will find a brief summary of them in Appendix C along with other mathematical relations which find frequent use in mechanics. Experience has shown that the geometry of mechanics is often a source of difficulty for students. Mechanics by its very nature is geometrical, and students should bear this in mind as they review their mathematics. In addition to vector algebra, dynamics requires the use of vector calculus, and the essentials of this topic will be developed in the text as they are needed.

Dynamics involves the frequent use of time derivatives of both vectors and scalars. As a notational shorthand, a dot over a symbol will frequently be used to indicate a derivative with respect to time. Thus, $\dot{x}$ means dx/dt and $\ddot{x}$ stands for d^2x/dt^2.

1/3 Newton's Laws

Newton's three laws of motion, stated in Art. 1/4 of *Vol. 1 Statics*, are restated here because of their special significance to dynamics. In modern terminology they are:

Law I. A particle remains at rest or continues to move with uniform velocity (in a straight line with a constant speed) if there is no unbalanced force acting on it.

Law II. The acceleration of a particle is proportional to the resultant force acting on it and is in the direction of this force.*

Law III. The forces of action and reaction between interacting bodies are equal in magnitude, opposite in direction, and collinear.

These laws have been verified by countless physical measurements. The first two laws hold for measurements made in an absolute frame of reference, but are subject to some correction when the motion is measured relative to a reference system having acceleration, such as one attached to the surface of the earth.

Newton's second law forms the basis for most of the analysis in dynamics. For a particle of mass m subjected to a resultant force $\mathbf{F}$, the law may be stated as

$$\boxed{\mathbf{F} = m\mathbf{a}} \qquad \textbf{(1/1)}$$

where $\mathbf{a}$ is the resulting acceleration measured in a nonaccelerating frame of reference. Newton's first law is a consequence of the second law since there is no acceleration when the force is zero, and so the particle is either at rest or is moving with constant velocity. The third law constitutes the principle of action and reaction with which you should be thoroughly familiar from your work in statics.

1/4 Units

Both the International System of metric units (SI) and the U.S. customary system of units are defined and used in *Vol. 2 Dynamics*, although a stronger emphasis is placed on the metric system because it is replacing the U.S. customary system. However, numerical conversion from one system to the other will often be needed in U.S. engineering

*To some it is preferable to interpret Newton's second law as meaning that the resultant force acting on a particle is proportional to the time rate of change of momentum of the particle and that this change is in the direction of the force. Both formulations are equally correct when applied to a particle of constant mass.

practice for some years to come. To become familiar with each system, it is necessary to think directly in that system. Familiarity with the new system cannot be achieved simply by the conversion of numerical results from the old system.

Tables defining the SI units and giving numerical conversions between U.S. customary and SI units are included inside the front cover of the book. Charts comparing selected quantities in SI and U.S. customary units are included inside the back cover of the book to facilitate conversion and to help establish a feel for the relative size of units in both systems.

The four fundamental quantities of mechanics, and their units and symbols for the two systems, are summarized in the following table:

QUANTITY	DIMENSIONAL SYMBOL	SI UNITS			U.S. CUSTOMARY UNITS		
			UNIT	SYMBOL		UNIT	SYMBOL
Mass	M	Base units	kilogram	kg		slug	—
Length	L	Base units	meter*	m	Base units	foot	ft
Time	T	Base units	second	s	Base units	second	sec
Force	F		newton	N	Base units	pound	lb

*Also spelled *metre*.

As shown in the table, in SI the units for mass, length, and time are taken as base units, and the units for force are derived from Newton's second law of motion, Eq. 1/1. In the U.S. customary system the units for force, length, and time are base units and the units for mass are derived from the second law.

The SI system is termed an *absolute* system because the standard for the base unit kilogram (a platinum-iridium cylinder kept at the International Bureau of Standards near Paris, France) is independent of the gravitational attraction of the earth. On the other hand, the U.S. customary system is termed a *gravitational* system because the standard for the base unit pound (the weight of a standard mass located at sea level and at a latitude of 45°) requires the presence of the gravitational field of the earth. This distinction is a fundamental difference between the two systems of units.

In SI units, by definition, one newton is that force which will give a one-kilogram mass an acceleration of one meter per second squared. In the U.S. customary system a 32.1740-pound mass (1 slug) will have an acceleration of one foot per second squared when acted on by a force of one pound. Thus, for each system we have from Eq. 1/1

Omikron/Photo Researchers, Inc.

The U.S. standard kilogram at the National Bureau of Standards.

SI UNITS	U.S. CUSTOMARY UNITS
$(1\text{ N}) = (1\text{ kg})(1\text{ m/s}^2)$	$(1\text{ lb}) = (1\text{ slug})(1\text{ ft/sec}^2)$
$\text{N} = \text{kg}\cdot\text{m/s}^2$	$\text{slug} = \text{lb}\cdot\text{sec}^2\text{/ft}$

In SI units, the kilogram should be used *exclusively* as a unit of mass and *never* force. Unfortunately, in the MKS (meter, kilogram, second) gravitational system, which has been used in some countries for many years, the kilogram has been commonly used both as a unit of force and as a unit of mass.

In U.S. customary units, the pound is unfortunately used both as a unit of force (lbf) and as a unit of mass (lbm). The use of the unit lbm is especially prevalent in the specification of the thermal properties of liquids and gases. The lbm is the amount of mass which weighs 1 lbf under standard conditions (at a latitude of 45° and at sea level). In order to avoid the confusion which would be caused by the use of two units for mass (slug and lbm), in this textbook we use almost exclusively the unit slug for mass. This practice makes dynamics much simpler than if the lbm were used. In addition, this approach allows us to use the symbol lb to always mean pound force.

Additional quantities used in mechanics and their equivalent base units will be defined as they are introduced in the chapters which follow. However, for convenient reference these quantities are listed in one place in the first table inside the front cover of the book.

Professional organizations have established detailed guidelines for the consistent use of SI units, and these guidelines have been followed throughout this book. The most essential ones are summarized inside the front cover, and you should observe these rules carefully.

1/5 Gravitation

Newton's law of gravitation, which governs the mutual attraction between bodies, is

$$F = G\frac{m_1 m_2}{r^2} \qquad \textbf{(1/2)}$$

where F = the mutual force of attraction between two particles

G = a universal constant called the *constant of gravitation*

m_1, m_2 = the masses of the two particles

r = the distance between the centers of the particles

The value of the gravitational constant obtained from experimental data is $G = 6.673(10^{-11})\ \text{m}^3/(\text{kg}\cdot\text{s}^2)$. Except for some spacecraft applications, the only gravitational force of appreciable magnitude in engineering is the force due to the attraction of the earth. It was shown in *Vol. 1 Statics*, for example, that each of two iron spheres 100 mm in diameter is attracted to the earth with a gravitational force of 37.1 N, which is called its *weight*, but the force of mutual attraction between them if they are just touching is only 0.000 000 095 1 N.

Because the gravitational attraction or weight of a body is a force, it should always be expressed in force units, newtons (N) in SI units and pounds force (lb) in U.S. customary units. To avoid confusion, the word "weight" in this book will be restricted to mean the force of gravitational attraction.

Effect of Altitude

The force of gravitational attraction of the earth on a body depends on the position of the body relative to the earth. If the earth were a perfect homogeneous sphere, a body with a mass of exactly 1 kg would be attracted to the earth by a force of 9.825 N on the surface of the earth, 9.822 N at altitude of 1 km, 9.523 N at an altitude of 100 km, 7.340 N at an altitude of 1000 km, and 2.456 N at an altitude equal to the mean radius of the earth, 6371 km. Thus the variation in gravitational attraction of high-altitude rockets and spacecraft becomes a major consideration.

Every object which falls in a vacuum at a given height near the surface of the earth will have the same acceleration g, regardless of its mass. This result can be obtained by combining Eqs. 1/1 and 1/2 and canceling the term representing the mass of the falling object. This combination gives

$$g = \frac{Gm_e}{R^2}$$

where m_e is the mass of the earth and R is the radius of the earth.* The mass m_e and the mean radius R of the earth have been found through experimental measurements to be $5.976(10^{24})$ kg and $6.371(10^6)$ m, respectively. These values, together with the value of G already cited, when substituted into the expression for g, give a mean value of $g = 9.825\ \text{m/s}^2$.

The variation of g with altitude is easily determined from the gravitational law. If g_0 represents the absolute acceleration due to gravity at sea level, the absolute value at an altitude h is

$$g = g_0 \frac{R^2}{(R + h)^2}$$

where R is the radius of the earth.

Effect of a Rotating Earth

The acceleration due to gravity as determined from the gravitational law is the acceleration which would be measured from a set of axes whose origin is at the center of the earth but which does not rotate with the earth. With respect to these "fixed" axes, then, this value may be termed the *absolute* value of g. Because the earth rotates, the acceleration of a freely falling body as measured from a position attached to the surface of the earth is slightly less than the absolute value.

Accurate values of the gravitational acceleration as measured relative to the surface of the earth account for the fact that the earth is a rotating oblate spheroid with flattening at the poles. These values may

*It can be proved that the earth, when taken as a sphere with a symmetrical distribution of mass about its center, may be considered a particle with its entire mass concentrated at its center.

be calculated to a high degree of accuracy from the 1980 International Gravity Formula, which is

$$g = 9.780\ 327(1 + 0.005\ 279 \sin^2 \gamma + 0.000\ 023 \sin^4 \gamma + \cdots)$$

where γ is the latitude and g is expressed in meters per second squared. The formula is based on an ellipsoidal model of the earth and also accounts for the effect of the rotation of the earth.

The absolute acceleration due to gravity as determined for a nonrotating earth may be computed from the relative values to a close approximation by adding $3.382(10^{-2}) \cos^2\gamma$ m/s^2, which removes the effect of the rotation of the earth. The variation of both the absolute and the relative values of g with latitude is shown in Fig. 1/1 for sea-level conditions.*

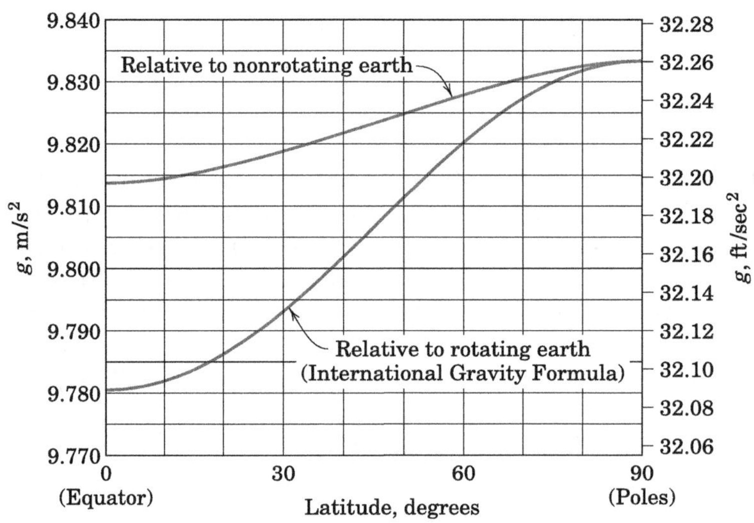

Figure 1/1

Standard Value of g

The standard value which has been adopted internationally for the gravitational acceleration relative to the rotating earth at sea level and at a latitude of 45° is 9.806 65 m/s^2 or 32.1740 ft/sec^2. This value differs very slightly from that obtained by evaluating the International Gravity Formula for $\gamma = 45°$. The reason for the small difference is that the earth is not exactly ellipsoidal, as assumed in the formulation of the International Gravity Formula.

The proximity of large land masses and the variations in the density of the crust of the earth also influence the local value of g by a small but detectable amount. In almost all engineering applications near the surface of the earth, we can neglect the difference between the absolute and relative values of the gravitational acceleration, and the effect of local

*You will be able to derive these relations for a spherical earth after studying relative motion in Chapter 3.

variations. The values of 9.81 m/s^2 in SI units and 32.2 ft/sec^2 in U.S. customary units are used for the sea-level value of g.

Apparent Weight

The gravitational attraction of the earth on a body of mass m may be calculated from the results of a simple gravitational experiment. The body is allowed to fall freely in a vacuum, and its absolute acceleration is measured. If the gravitational force of attraction or true weight of the body is W, then, because the body falls with an absolute acceleration g, Eq. 1/1 gives

$$\mathbf{W} = m\mathbf{g} \qquad \textbf{(1/3)}$$

The *apparent weight* of a body as determined by a spring balance, calibrated to read the correct force and attached to the surface of the earth, will be slightly less than its true weight. The difference is due to the rotation of the earth. The ratio of the apparent weight to the apparent or relative acceleration due to gravity still gives the correct value of mass. The apparent weight and the relative acceleration due to gravity are, of course, the quantities which are measured in experiments conducted on the surface of the earth.

1/6 Dimensions

A given dimension such as length can be expressed in a number of different units such as meters, millimeters, or kilometers. Thus, a *dimension* is different from a *unit*. The *principle of dimensional homogeneity* states that all physical relations must be dimensionally homogeneous; that is, the dimensions of all terms in an equation must be the same. It is customary to use the symbols L, M, T, and F to stand for length, mass, time, and force, respectively. In SI units force is a derived quantity and from Eq. 1/1 has the dimensions of mass times acceleration or

$$F = ML/T^2$$

One important use of the dimensional homogeneity principle is to check the dimensional correctness of some derived physical relation. We can derive the following expression for the velocity v of a body of mass m which is moved from rest a horizontal distance x by a force F:

$$Fx = \tfrac{1}{2}mv^2$$

where the $\frac{1}{2}$ is a dimensionless coefficient resulting from integration. This equation is dimensionally correct because substitution of L, M, and T gives

$$[MLT^{-2}][L] = [M][LT^{-1}]^2$$

Dimensional homogeneity is a necessary condition for correctness of a physical relation, but it is not sufficient, since it is possible to construct an equation which is dimensionally correct but does not represent a correct relation. You should perform a dimensional check on the answer to every problem whose solution is carried out in symbolic form.

1/7 Solving Problems in Dynamics

The study of dynamics concerns the understanding and description of the motions of bodies. This description, which is largely mathematical, enables predictions of dynamical behavior to be made. A dual thought process is necessary in formulating this description. It is necessary to think in terms of both the physical situation and the corresponding mathematical description. This repeated transition of thought between the physical and the mathematical is required in the analysis of every problem.

One of the greatest difficulties encountered by students is the inability to make this transition freely. You should recognize that the mathematical formulation of a physical problem represents an ideal and limiting description, or model, which approximates but never quite matches the actual physical situation.

In Art. 1/8 of *Vol. 1 Statics* we extensively discussed the approach to solving problems in statics. We assume therefore, that you are familiar with this approach, which we summarize here as applied to dynamics.

Approximation in Mathematical Models

Construction of an idealized mathematical model for a given engineering problem always requires approximations to be made. Some of these approximations may be mathematical, whereas others will be physical. For instance, it is often necessary to neglect small distances, angles, or forces compared with large distances, angles, or forces. If the change in velocity of a body with time is nearly uniform, then an assumption of constant acceleration may be justified. An interval of motion which cannot be easily described in its entirety is often divided into small increments, each of which can be approximated.

As another example, the retarding effect of bearing friction on the motion of a machine may often be neglected if the friction forces are small compared with the other applied forces. However, these same friction forces cannot be neglected if the purpose of the inquiry is to determine the decrease in efficiency of the machine due to the friction process. Thus, the type of assumptions you make depends on what information is desired and on the accuracy required.

You should be constantly alert to the various assumptions called for in the formulation of real problems. The ability to understand and make use of the appropriate assumptions when formulating and solving engineering problems is certainly one of the most important characteristics of a successful engineer.

Along with the development of the principles and analytical tools needed for modern dynamics, one of the major aims of this book is to provide many opportunities to develop the ability to formulate good mathematical models. Strong emphasis is placed on a wide range of practical problems which not only require you to apply theory but also force you to make relevant assumptions.

Key Concepts

Method of Attack

An effective method of attack is essential in the solution of dynamics problems, as for all engineering problems. Development of good habits in formulating problems and in representing their solutions will be an invaluable asset. Each solution should proceed with a logical sequence of steps from hypothesis to conclusion. The following sequence of steps is useful in the construction of problem solutions.

1. Formulate the problem:
   (*a*) State the given data.
   (*b*) State the desired result.
   (*c*) State your assumptions and approximations.
2. Develop the solution:
   (*a*) Draw any needed diagrams, and include coordinates which are appropriate for the problem at hand.
   (*b*) State the governing principles to be applied to your solution.
   (*c*) Make your calculations.
   (*d*) Ensure that your calculations are consistent with the accuracy justified by the data.
   (*e*) Be sure that you have used consistent units throughout your calculations.
   (*f*) Ensure that your answers are reasonable in terms of magnitudes, directions, common sense, etc.
   (*g*) Draw conclusions.

The arrangement of your work should be neat and orderly. This will help your thought process and enable others to understand your work. The discipline of doing orderly work will help you to develop skill in problem formulation and analysis. Problems which seem complicated at first often become clear when you approach them with logic and discipline.

Application of Basic Principles

The subject of dynamics is based on a surprisingly few fundamental concepts and principles which, however, can be extended and applied over a wide range of conditions. The study of dynamics is valuable partly because it provides experience in reasoning from fundamentals. This experience cannot be obtained merely by memorizing the kinematic and dynamic equations which describe various motions. It must be obtained through exposure to a wide variety of problem situations which require the choice, use, and extension of basic principles to meet the given conditions.

In describing the relations between forces and the motions they produce, it is essential to define clearly the system to which a principle is to be applied. At times a single particle or a rigid body is the system to be isolated, whereas at other times two or more bodies taken together constitute the system.

The definition of the system to be analyzed is made clear by constructing its ***free-body diagram***. This diagram consists of a closed outline of the external boundary of the system. All bodies which contact and exert forces on the system but are not a part of it are removed and replaced by vectors representing the forces they exert on the isolated system. In this way, we make a clear distinction between the action and reaction of each force, and all forces on and external to the system are accounted for. We assume that you are familiar with the technique of drawing free-body diagrams from your prior work in statics.

Numerical versus Symbolic Solutions

In applying the laws of dynamics, we may use numerical values of the involved quantities, or we may use algebraic symbols and leave the answer as a formula. When numerical values are used, the magnitudes of all quantities expressed in their particular units are evident at each stage of the calculation. This approach is useful when we need to know the magnitude of each term.

The symbolic solution, however, has several advantages over the numerical solution:

1. The use of symbols helps to focus attention on the connection between the physical situation and its related mathematical description.
2. A symbolic solution enables you to make a dimensional check at every step, whereas dimensional homogeneity cannot be checked when only numerical values are used.
3. We can use a symbolic solution repeatedly for obtaining answers to the same problem with different units or different numerical values.

Thus, facility with both forms of solution is essential, and you should practice each in the problem work.

In the case of numerical solutions, we repeat from *Vol. 1 Statics* our convention for the display of results. All given data are taken to be exact, and results are generally displayed to three significant figures, unless the leading digit is a one, in which case four significant figures are displayed. An exception to this rule occurs in the area of orbital mechanics, where answers will generally receive an additional significant figure because of the necessity of increased precision in this discipline.

Solution Methods

Solutions to the various equations of dynamics can be obtained in one of three ways.

1. Obtain a direct mathematical solution by hand calculation, using either algebraic symbols or numerical values. We can solve the large majority of the problems this way.
2. Obtain graphical solutions for certain problems, such as the determination of velocities and accelerations of rigid bodies in two-dimensional relative motion.
3. Solve the problem by computer. A number of problems in *Vol. 2 Dynamics* are designated as *Computer-Oriented Problems*. They appear

at the end of the Review Problem sets and were selected to illustrate the type of problem for which solution by computer offers a distinct advantage.

The choice of the most expedient method of solution is an important aspect of the experience to be gained from the problem work. We emphasize, however, that the most important experience in learning mechanics lies in the formulation of problems, as distinct from their solution per se.

1/8 Chapter Review

This chapter has introduced the concepts, definitions, and units used in dynamics, and has given an overview of the approach used to formulate and solve problems in dynamics. Now that you have finished this chapter, you should be able to do the following:

1. State Newton's laws of motion.
2. Perform calculations using SI and U.S. customary units.
3. Express the law of gravitation and calculate the weight of an object.
4. Discuss the effects of altitude and the rotation of the earth on the acceleration due to gravity.
5. Apply the principle of dimensional homogeneity to a given physical relation.
6. Describe the methodology used to formulate and solve dynamics problems.

stockphoto mania/Shutterstock

Just after liftoff, there is a multitude of critical dynamic events taking place for this jetliner.

Sample Problem 1/1

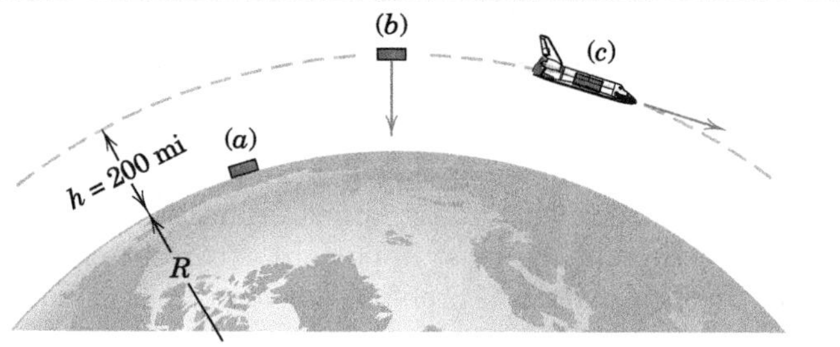

A space-shuttle payload module weighs 100 lb when resting on the surface of the earth at a latitude of 45° north.

(*a*) Determine the mass of the module in both slugs and kilograms, and its surface-level weight in newtons.

(*b*) Now suppose the module is taken to an altitude of 200 miles above the surface of the earth and released there with no velocity relative to the center of the earth. Determine its weight under these conditions in both pounds and newtons.

(*c*) Finally, suppose the module is fixed inside the cargo bay of a space shuttle. The shuttle is in a circular orbit at an altitude of 200 miles above the surface of the earth. Determine the weight of the module in both pounds and newtons under these conditions.

For the surface-level value of the acceleration of gravity relative to a rotating earth, use $g = 32.1740$ ft/sec² (9.80665 m/s²). For the absolute value relative to a nonrotating earth, use $g = 32.234$ ft/sec² (9.825 m/s²). Round off all answers using the rules of this textbook.

Solution. (*a*) From relationship 1/3, we have

① $[W = mg]$ $$m = \frac{W}{g} = \frac{100 \text{ lb}}{32.1740 \text{ ft/sec}^2} = 3.11 \text{ slugs} \qquad \textit{Ans.}$$

Here we have used the acceleration of gravity relative to the rotating earth, because that is the condition of the module in part (*a*). Note that we are using more significant figures in the acceleration of gravity than will normally be required in this textbook (32.2 ft/sec² and 9.81 m/s² will normally suffice).

From the table of conversion factors inside the front cover of the textbook, we see that 1 pound is equal to 4.4482 newtons. Thus, the weight of the module in newtons is

② $$W = 100 \text{ lb}\left[\frac{4.4482 \text{ N}}{1 \text{ lb}}\right] = 445 \text{ N} \qquad \textit{Ans.}$$

Finally, its mass in kilograms is

③ $[W = mg]$ $$m = \frac{W}{g} = \frac{445 \text{ N}}{9.80665 \text{ m/s}^2} = 45.4 \text{ kg} \qquad \textit{Ans.}$$

As another route to the last result, we may convert from pounds mass to kilograms. Again using the table inside the front cover, we have

$$m = 100 \text{ lbm}\left[\frac{0.45359 \text{ kg}}{1 \text{ lbm}}\right] = 45.4 \text{ kg}$$

We recall that 1 lbm is the amount of mass which under standard conditions has a weight of 1 lb of force. We rarely refer to the U.S. mass unit lbm in this textbook series, but rather use the slug for mass. The sole use of slug, rather than the unnecessary use of two units for mass, will prove to be powerful and simple.

Helpful Hints

① Our calculator indicates a result of 3.108099 · · · slugs. Using the rules of significant figure display used in this textbook, we round the written result to three significant figures, or 3.11 slugs. Had the numerical result begun with the digit 1, we would have rounded the displayed answer to four significant figures.

② A good practice with unit conversion is to multiply by a factor such as $\left[\frac{4.4482 \text{ N}}{1 \text{ lb}}\right]$, which has a value of 1, because the numerator and the denominator are equivalent. Be sure that cancellation of the units leaves the units desired—here the units of lb cancel, leaving the desired units of N.

③ Note that we are using a previously calculated result (445 N). We must be sure that when a calculated number is needed in subsequent calculations, it is obtained in the calculator to its full accuracy (444.82 · · ·). If necessary, numbers must be stored in a calculator storage register and then brought out of the register when needed. We must not merely punch 445 into our calculator and proceed to divide by 9.80665—this practice will result in loss of numerical accuracy. Some individuals like to place a small indication of the storage register used in the right margin of the work paper, directly beside the number stored.

Sample Problem 1/1 (Continued)

(*b*) We begin by calculating the absolute acceleration of gravity (relative to the nonrotating earth) at an altitude of 200 miles.

$$\left[g = g_0 \frac{R^2}{(R+h)^2}\right] \qquad g_h = 32.234\left[\frac{3959^2}{(3959+200)^2}\right] = 29.2 \text{ ft/sec}^2$$

The weight at an altitude of 200 miles is then

$$W_h = mg_h = 3.11(29.2) = 90.8 \text{ lb} \qquad \textit{Ans.}$$

We now convert W_h to units of newtons.

$$W_h = 90.8 \text{ lb}\left[\frac{4.4482 \text{ N}}{1 \text{ lb}}\right] = 404 \text{ N} \qquad \textit{Ans.}$$

As an alternative solution to part (*b*), we may use Newton's universal law of gravitation. In U.S. units,

$$\left[F = \frac{Gm_1m_2}{r^2}\right] \qquad W_h = \frac{Gm_em}{(R+h)^2} = \frac{[3.439(10^{-8})][4.095(10^{23})][3.11]}{[(3959+200)(5280)]^2}$$

$$= 90.8 \text{ lb}$$

which agrees with our earlier result. We note that the weight of the module when at an altitude of 200 mi is about 90% of its surface-level weight—it is *not* weightless. We will study the effects of this weight on the motion of the module in Chapter 3.

(*c*) The weight of an object (the force of gravitational attraction) does not depend on the motion of the object. Thus the answers for part (*c*) are the same as those in part (*b*).

$$W_h = 90.8 \text{ lb} \quad \text{or} \quad 404 \text{ N} \qquad \textit{Ans.}$$

This Sample Problem has served to eliminate certain commonly held and persistent misconceptions. First, just because a body is raised to a typical shuttle altitude, it does not become weightless. This is true whether the body is released with no velocity relative to the center of the earth, is inside the orbiting shuttle, or is in its own arbitrary trajectory. And second, the acceleration of gravity is not zero at such altitudes. The only way to reduce both the acceleration of gravity and the corresponding weight of a body to zero is to take the body to an infinite distance from the earth.

PROBLEMS

(Refer to Table D/2 in Appendix D for relevant solar-system values.)

1/1 For the 3500-lb car, determine (*a*) its mass in slugs, (*b*) its weight in newtons, and (*c*) its mass in kilograms.

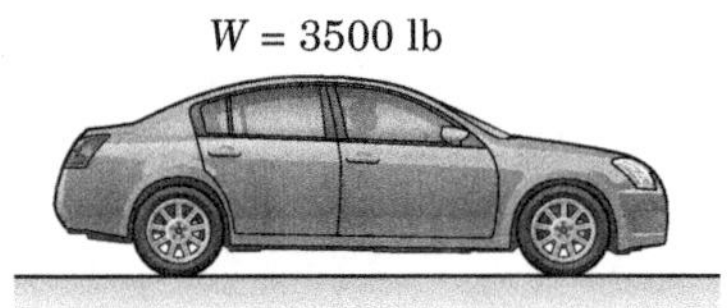

Problem 1/1

1/2 Determine your mass in slugs. Convert your weight to newtons and calculate the corresponding mass in kilograms.

1/3 For the given vectors $\mathbf{V}_1$ and $\mathbf{V}_2$, determine $V_1 + V_2$, $\mathbf{V}_1 + \mathbf{V}_2$, $\mathbf{V}_1 - \mathbf{V}_2$, $\mathbf{V}_1 \times \mathbf{V}_2$, $\mathbf{V}_2 \times \mathbf{V}_1$, and $\mathbf{V}_1 \cdot \mathbf{V}_2$. Consider the vectors to be nondimensional.

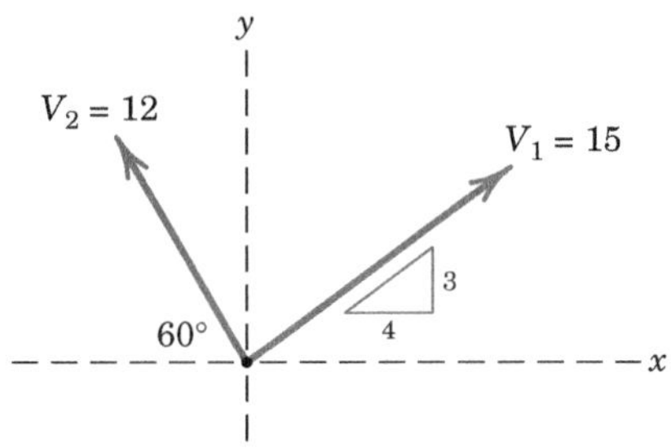

Problem 1/3

1/4 The weight of one dozen apples is 5 lb. Determine the average mass of one apple in both SI and U.S. units and the average weight of one apple in SI units. In the present case, how applicable is the "rule of thumb" that an average apple weighs 1 N?

1/5 Consider two iron spheres, each of diameter 100 mm, which are just touching. At what distance r from the center of the earth will the force of mutual attraction between the contacting spheres be equal to the force exerted by the earth on one of the spheres?

1/6 Two uniform spheres are positioned as shown. Determine the gravitational force which the titanium sphere exerts on the copper sphere. The value of R is 40 mm.

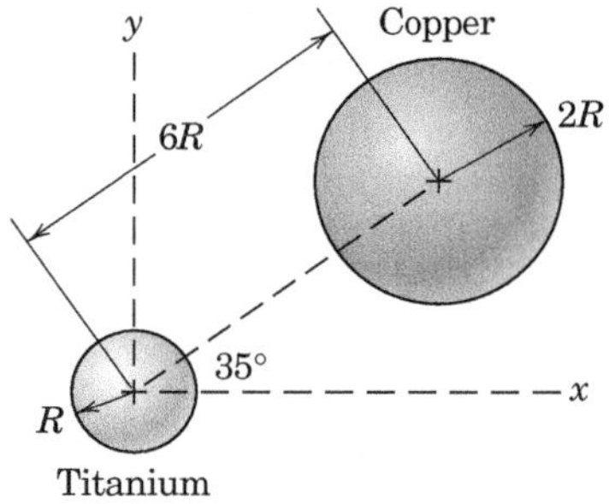

Problem 1/6

1/7 At what altitude h above the north pole is the weight of an object reduced to one-third of its earth-surface value? Assume a spherical earth of radius R and express h in terms of R.

1/8 Determine the absolute weight and the weight relative to the rotating earth of a 60-kg woman if she is standing on the surface of the earth at a latitude of 35°.

1/9 A space shuttle is in a circular orbit at an altitude of 200 mi. Calculate the absolute value of g at this altitude and determine the corresponding weight of a shuttle passenger who weighs 180 lb when standing on the surface of the earth at a latitude of 45°. Are the terms "zero-g" and "weightless," which are sometimes used to describe conditions aboard orbiting spacecraft, correct in the absolute sense?

1/10 Determine the distance h for which the spacecraft S will experience equal attractions from the earth and from the sun. Use Table D/2 of Appendix D as needed.

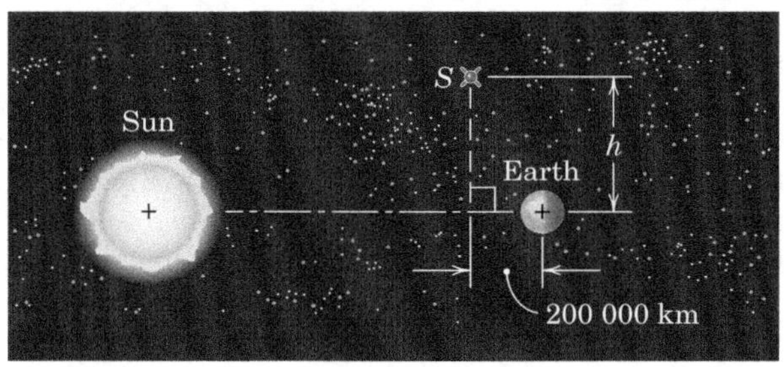

Not to scale

Problem 1/10

1/11 Calculate the distance d from the center of the earth at which a particle experiences equal attractions from the earth and from the moon. The particle is restricted to the line through the centers of the earth and the moon. Justify the two solutions physically. Refer to Table D/2 of Appendix D as needed.

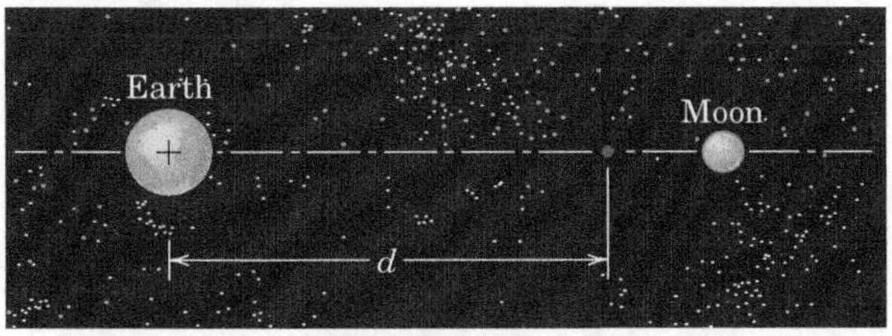

Not to scale

Problem 1/11

1/12 Determine the angle θ at which a particle in Jupiter's circular orbit experiences equal attractions from the sun and from Jupiter. Use Table D/2 of Appendix D as needed.

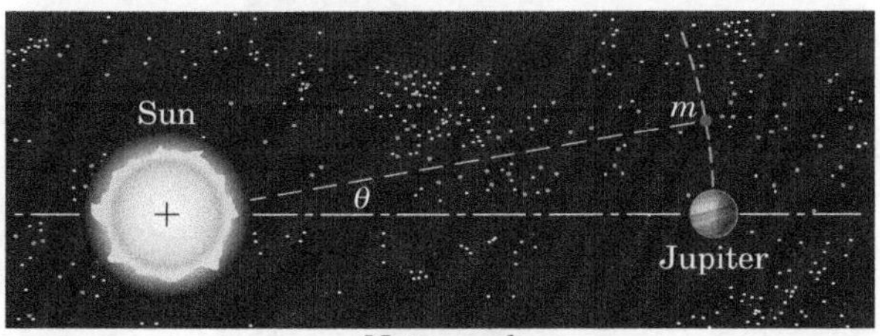

Not to scale

Problem 1/12

1/13 Consider a woman standing on the earth with the sun directly overhead. Determine the ratio R_{es} of the force which the earth exerts on the woman to the force which the sun exerts on her. Neglect the effects of the rotation and oblateness of the earth.

1/14 Determine the ratio R_A of the force exerted by the sun on the moon to that exerted by the earth on the moon for position A of the moon. Repeat for moon position B.

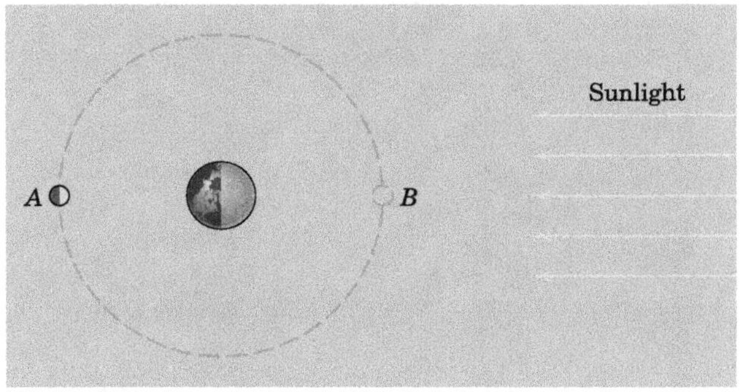

Problem 1/14

1/15 Determine the base units of the expression

$$E = \int_{t_1}^{t_2} mgr\,dt$$

in both SI and U.S. units. The variable m represents mass, g is the acceleration due to gravity, r is distance, and t is time.

1/16 Determine the dimensions of the quantity

$$Q = \frac{1}{2}\rho v^2$$

where ρ is density and v is speed.

Even if this car maintains a constant speed along the winding road, it accelerates laterally, and this acceleration must be considered in the design of the car, its tires, and the roadway itself.

2 Kinematics of Particles

Chapter Outline

2/1 Introduction

Kinematics is the branch of dynamics which describes the motion of bodies without reference to the forces which either cause the motion or are generated as a result of the motion. Kinematics is often described as the "geometry of motion." Some engineering applications of kinematics include the design of cams, gears, linkages, and other machine elements to control or produce certain desired motions, and the calculation of flight trajectories for aircraft, rockets, and spacecraft. A thorough working knowledge of kinematics is a prerequisite to kinetics, which is the study of the relationships between motion and the corresponding forces which cause or accompany the motion.

Particle Motion

We begin our study of kinematics by first discussing in this chapter the motions of points or particles. A particle is a body whose physical dimensions are so small compared with the radius of curvature of its path that we may treat the motion of the particle as that of a point. For example, the wingspan of a jet transport flying between Los Angeles and New York is of no consequence compared with the radius of curvature of

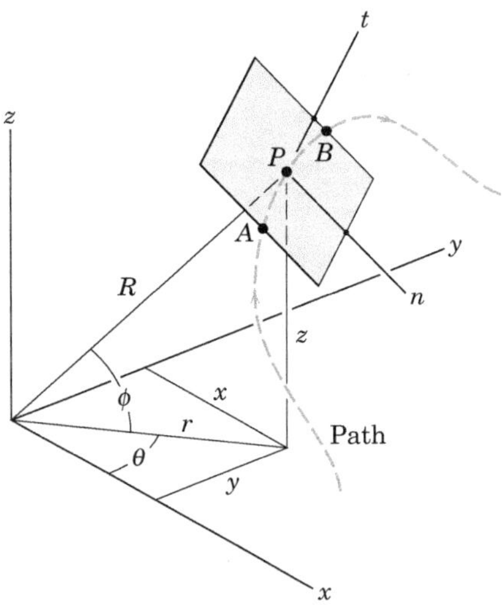

Figure 2/1

its flight path, and thus the treatment of the airplane as a particle or point is an acceptable approximation.

We can describe the motion of a particle in a number of ways, and the choice of the most convenient or appropriate way depends a great deal on experience and on how the data are given. Let us obtain an overview of the several methods developed in this chapter by referring to Fig. 2/1, which shows a particle P moving along some general path in space. If the particle is confined to a specified path, as with a bead sliding along a fixed wire, its motion is said to be *constrained*. If there are no physical guides, the motion is said to be *unconstrained*. A small rock tied to the end of a string and whirled in a circle undergoes constrained motion until the string breaks, after which instant its motion is unconstrained.

Choice of Coordinates

The position of particle P at any time t can be described by specifying its rectangular coordinates* x, y, z, its cylindrical coordinates r, θ, z, or its spherical coordinates R, θ, ϕ. The motion of P can also be described by measurements along the tangent t and normal n to the curve. The direction of n lies in the local plane of the curve.† These last two measurements are called *path variables*.

The motion of particles (or rigid bodies) can be described by using coordinates measured from fixed reference axes (*absolute-motion* analysis) or by using coordinates measured from moving reference axes (*relative-motion* analysis). Both descriptions will be developed and applied in the articles which follow.

With this conceptual picture of the description of particle motion in mind, we restrict our attention in the first part of this chapter to the case of *plane motion* where all movement occurs in or can be represented as occurring in a single plane. A large proportion of the motions of machines and structures in engineering can be represented as plane motion. Later, in Chapter 7, an introduction to three-dimensional motion is presented. We begin our discussion of plane motion with *rectilinear motion*, which is motion along a straight line, and follow it with a description of motion along a plane curve.

2/2 Rectilinear Motion

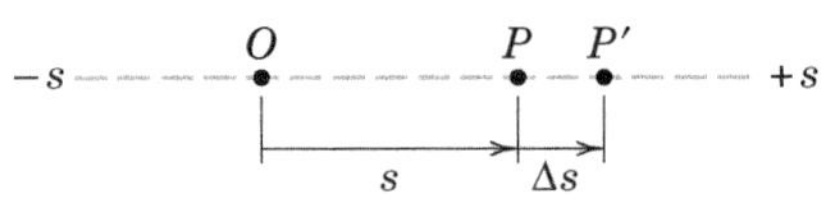

Figure 2/2

Consider a particle P moving along a straight line, Fig. 2/2. The position of P at any instant of time t can be specified by its distance s measured from some convenient reference point O fixed on the line. At time $t + \Delta t$ the particle has moved to P' and its coordinate becomes $s + \Delta s$. The change in the position coordinate during the interval Δt is called the *displacement* Δs of the particle. The displacement would be negative if the particle moved in the negative s-direction.

*Often called *Cartesian* coordinates, named after René Descartes (1596–1650), a French mathematician who was one of the inventors of analytic geometry.

†This plane is called the *osculating* plane, which comes from the Latin word *osculari* meaning "to kiss." The plane which contains P and the two points A and B, one on either side of P, becomes the osculating plane as the distances between the points approach zero.

Velocity and Acceleration

The average velocity of the particle during the interval Δt is the displacement divided by the time interval or $v_{av} = \Delta s/\Delta t$. As Δt becomes smaller and approaches zero in the limit, the average velocity approaches the *instantaneous velocity* of the particle, which is $v = \lim_{\Delta t \to 0} \frac{\Delta s}{\Delta t}$ or

$$\boxed{v = \frac{ds}{dt} = \dot{s}} \qquad \textbf{(2/1)}$$

Thus, the velocity is the time rate of change of the position coordinate s. The velocity is positive or negative depending on whether the corresponding displacement is positive or negative.

The average acceleration of the particle during the interval Δt is the change in its velocity divided by the time interval or $a_{av} = \Delta v/\Delta t$. As Δt becomes smaller and approaches zero in the limit, the average acceleration approaches the *instantaneous acceleration* of the particle, which is $a = \lim_{\Delta t \to 0} \frac{\Delta v}{\Delta t}$ or

$$\boxed{a = \frac{dv}{dt} = \dot{v}} \quad \text{or} \quad \boxed{a = \frac{d^2s}{dt^2} = \ddot{s}} \qquad \textbf{(2/2)}$$

The acceleration is positive or negative depending on whether the velocity is increasing or decreasing. Note that the acceleration would be positive if the particle had a negative velocity which was becoming less negative. If the particle is slowing down, the particle is said to be *decelerating*.

© berc/Fotolia

This sprinter will undergo rectilinear acceleration until he reaches his terminal speed.

Velocity and acceleration are actually vector quantities, as we will see for curvilinear motion beginning with Art. 2/3. For rectilinear motion in the present article, where the direction of the motion is that of the given straight-line path, the sense of the vector along the path is described by a plus or minus sign. In our treatment of curvilinear motion, we will account for the changes in direction of the velocity and acceleration vectors as well as their changes in magnitude.

By eliminating the time dt between Eq. 2/1 and the first of Eqs. 2/2, we obtain a differential equation relating displacement, velocity, and acceleration.* This equation is

$$\boxed{v\,dv = a\,ds} \quad \text{or} \quad \boxed{\dot{s}\,d\dot{s} = \ddot{s}\,ds} \qquad \textbf{(2/3)}$$

Equations 2/1, 2/2, and 2/3 are the differential equations for the rectilinear motion of a particle. Problems in rectilinear motion involving finite changes in the motion variables are solved by integration of these basic differential relations. The position coordinate s, the velocity v, and the acceleration a are algebraic quantities, so that their signs, positive or negative, must be carefully observed. Note that the positive directions for v and a are the same as the positive direction for s.

*Differential quantities can be multiplied and divided in exactly the same way as other algebraic quantities.

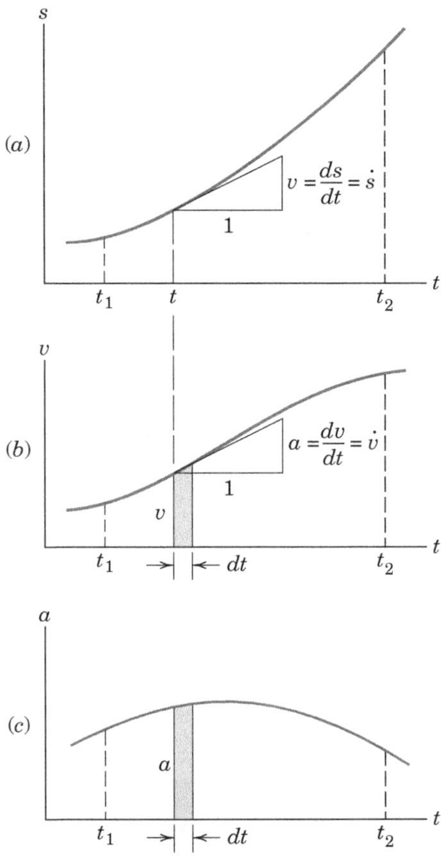

Figure 2/3

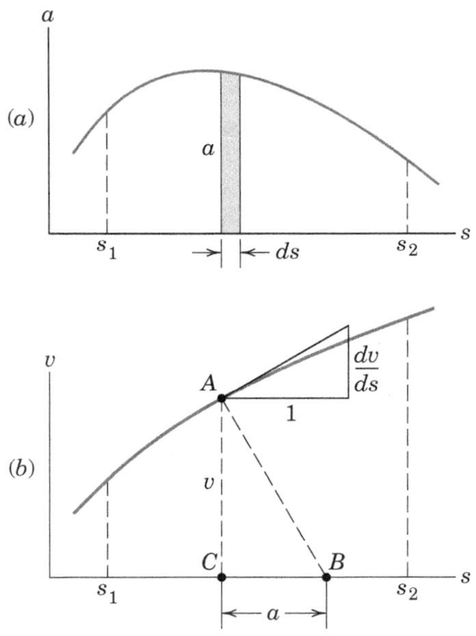

Figure 2/4

Graphical Interpretations

Interpretation of the differential equations governing rectilinear motion is considerably clarified by representing the relationships among s, v, a, and t graphically. Figure 2/3*a* is a schematic plot of the variation of s with t from time t_1 to time t_2 for some given rectilinear motion. By constructing the tangent to the curve at any time t, we obtain the slope, which is the velocity $v = ds/dt$. Thus, the velocity can be determined at all points on the curve and plotted against the corresponding time as shown in Fig. 2/3*b*. Similarly, the slope dv/dt of the v-t curve at any instant gives the acceleration at that instant, and the a-t curve can therefore be plotted as in Fig. 2/3*c*.

We now see from Fig. 2/3*b* that the area under the v-t curve during time dt is $v\ dt$, which from Eq. 2/1 is the displacement ds. Consequently, the net displacement of the particle during the interval from t_1 to t_2 is the corresponding area under the curve, which is

$$\int_{s_1}^{s_2} ds = \int_{t_1}^{t_2} v\ dt \qquad \text{or} \qquad s_2 - s_1 = (\text{area under } v\text{-}t \text{ curve})$$

Similarly, from Fig. 2/3*c* we see that the area under the a-t curve during time dt is $a\ dt$, which, from the first of Eqs. 2/2, is dv. Thus, the net change in velocity between t_1 and t_2 is the corresponding area under the curve, which is

$$\int_{v_1}^{v_2} dv = \int_{t_1}^{t_2} a\ dt \qquad \text{or} \qquad v_2 - v_1 = (\text{area under } a\text{-}t \text{ curve})$$

Note two additional graphical relations. When the acceleration a is plotted as a function of the position coordinate s, Fig. 2/4*a*, the area under the curve during a displacement ds is $a\ ds$, which, from Eq. 2/3, is $v\ dv = d(v^2/2)$. Thus, the net area under the curve between position coordinates s_1 and s_2 is

$$\int_{v_1}^{v_2} v\ dv = \int_{s_1}^{s_2} a\ ds \qquad \text{or} \qquad \tfrac{1}{2}(v_2{}^2 - v_1{}^2) = (\text{area under } a\text{-}s \text{ curve})$$

When the velocity v is plotted as a function of the position coordinate s, Fig. 2/4*b*, the slope of the curve at any point A is dv/ds. By constructing the normal AB to the curve at this point, we see from the similar triangles that $\overline{CB}/v = dv/ds$. Thus, from Eq. 2/3, $\overline{CB} = v(dv/ds) = a$, the acceleration. It is necessary that the velocity and position coordinate axes have the same numerical scales so that the acceleration read on the position coordinate scale in meters (or feet), say, will represent the actual acceleration in meters (or feet) per second squared.

The graphical representations described are useful not only in visualizing the relationships among the several motion quantities but also in obtaining approximate results by graphical integration or differentiation. The latter case occurs when a lack of knowledge of the mathematical relationship prevents its expression as an explicit mathematical function which can be integrated or differentiated. Experimental data and motions which involve discontinuous relationships between the variables are frequently analyzed graphically.

Analytical Integration

If the position coordinate s is known for all values of the time t, then successive mathematical or graphical differentiation with respect to t gives the velocity v and acceleration a. In many problems, however, the functional relationship between position coordinate and time is unknown, and we must determine it by successive integration from the acceleration. Acceleration is determined by the forces which act on moving bodies and is computed from the equations of kinetics discussed in subsequent chapters. Depending on the nature of the forces, the acceleration may be specified as a function of time, velocity, or position coordinate, or as a combined function of these quantities. The procedure for integrating the differential equation in each case is indicated as follows.

(a) Constant Acceleration. When a is constant, the first of Eqs. 2/2 and 2/3 can be integrated directly. For simplicity with $s = s_0$, $v = v_0$, and $t = 0$ designated at the beginning of the interval, then for a time interval t the integrated equations become

$$\int_{v_0}^{v} dv = a\int_0^t dt \qquad \text{or} \qquad v = v_0 + at$$

$$\int_{v_0}^{v} v\,dv = a\int_{s_0}^{s} ds \qquad \text{or} \qquad v^2 = {v_0}^2 + 2a(s - s_0)$$

Substitution of the integrated expression for v into Eq. 2/1 and integration with respect to t give

$$\int_{s_0}^{s} ds = \int_0^t (v_0 + at)\,dt \qquad \text{or} \qquad s = s_0 + v_0 t + \tfrac{1}{2}at^2$$

These relations are necessarily restricted to the special case where the acceleration is constant. The integration limits depend on the initial and final conditions, which for a given problem may be different from those used here. It may be more convenient, for instance, to begin the integration at some specified time t_1 rather than at time $t = 0$.

Caution: The foregoing equations have been integrated for constant acceleration only. A common mistake is to use these equations for problems involving variable acceleration, where they do not apply.

(b) Acceleration Given as a Function of Time, a = f(t). Substitution of the function into the first of Eqs. 2/2 gives $f(t) = dv/dt$. Multiplying by dt separates the variables and permits integration. Thus,

$$\int_{v_0}^{v} dv = \int_0^t f(t)\,dt \qquad \text{or} \qquad v = v_0 + \int_0^t f(t)\,dt$$

From this integrated expression for v as a function of t, the position coordinate s is obtained by integrating Eq. 2/1, which, in form, would be

$$\int_{s_0}^{s} ds = \int_0^t v\,dt \qquad \text{or} \qquad s = s_0 + \int_0^t v\,dt$$

If the indefinite integral is employed, the end conditions are used to establish the constants of integration. The results are identical with those obtained by using the definite integral.

If desired, the displacement s can be obtained by a direct solution of the second-order differential equation $\ddot{s} = f(t)$ obtained by substitution of $f(t)$ into the second of Eqs. 2/2.

(c) Acceleration Given as a Function of Velocity, a = f(v). Substitution of the function into the first of Eqs. 2/2 gives $f(v) = dv/dt$, which permits separating the variables and integrating. Thus,

$$t = \int_0^t dt = \int_{v_0}^{v} \frac{dv}{f(v)}$$

This result gives t as a function of v. Then it would be necessary to solve for v as a function of t so that Eq. 2/1 can be integrated to obtain the position coordinate s as a function of t.

Another approach is to substitute the function $a = f(v)$ into the first of Eqs. 2/3, giving $v\,dv = f(v)\,ds$. The variables can now be separated and the equation integrated in the form

$$\int_{v_0}^{v} \frac{v\,dv}{f(v)} = \int_{s_0}^{s} ds \qquad \text{or} \qquad s = s_0 + \int_{v_0}^{v} \frac{v\,dv}{f(v)}$$

Note that this equation gives s in terms of v without explicit reference to t.

(d) Acceleration Given as a Function of Displacement, a = f(s). Substituting the function into Eq. 2/3 and integrating give the form

$$\int_{v_0}^{v} v\,dv = \int_{s_0}^{s} f(s)\,ds \qquad \text{or} \qquad v^2 = {v_0}^2 + 2\int_{s_0}^{s} f(s)\,ds$$

Next we solve for v to give $v = g(s)$, a function of s. Now we can substitute ds/dt for v, separate variables, and integrate in the form

$$\int_{s_0}^{s} \frac{ds}{g(s)} = \int_0^t dt \qquad \text{or} \qquad t = \int_{s_0}^{s} \frac{ds}{g(s)}$$

which gives t as a function of s. Finally, we can rearrange to obtain s as a function of t.

In each of the foregoing cases when the acceleration varies according to some functional relationship, the possibility of solving the equations by direct mathematical integration will depend on the form of the function. In cases where the integration is excessively awkward or difficult, integration by graphical, numerical, or computer methods can be utilized.

Sample Problem 2/1

The position coordinate of a particle which is confined to move along a straight line is given by $s = 2t^3 - 24t + 6$, where s is measured in meters from a convenient origin and t is in seconds. Determine (a) the time required for the particle to reach a velocity of 72 m/s from its initial condition at $t = 0$, (b) the acceleration of the particle when $v = 30$ m/s, and (c) the net displacement of the particle during the interval from $t = 1$ s to $t = 4$ s.

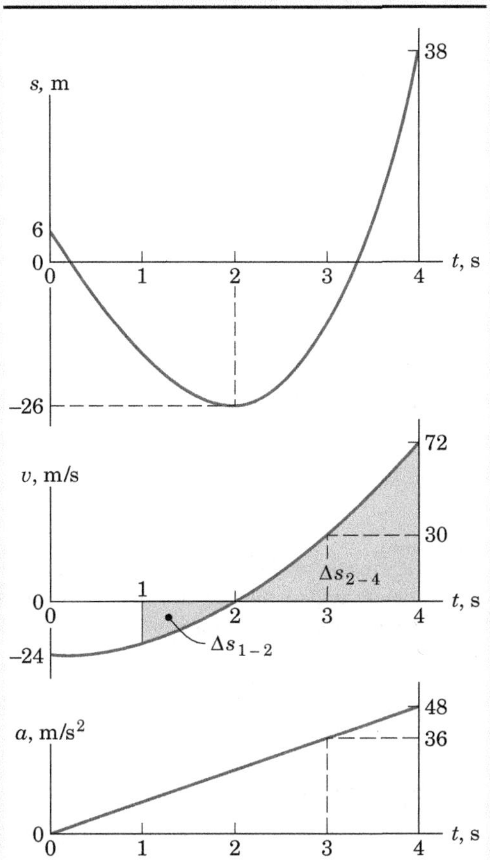

Solution. The velocity and acceleration are obtained by successive differentiation of s with respect to the time. Thus,

$$[v = \dot{s}] \qquad v = 6t^2 - 24 \text{ m/s}$$

$$[a = \dot{v}] \qquad a = 12t \text{ m/s}^2$$

(a) Substituting $v = 72$ m/s into the expression for v gives us $72 = 6t^2 - 24$, from which $t = \pm 4$ s. The negative root describes a mathematical solution for t
① before the initiation of motion, so this root is of no physical interest. Thus, the desired result is

$$t = 4 \text{ s} \qquad \textit{Ans.}$$

(b) Substituting $v = 30$ m/s into the expression for v gives $30 = 6t^2 - 24$, from which the positive root is $t = 3$ s, and the corresponding acceleration is

$$a = 12(3) = 36 \text{ m/s}^2 \qquad \textit{Ans.}$$

(c) The net displacement during the specified interval is

$$\Delta s = s_4 - s_1 \qquad \text{or}$$

$$\Delta s = [2(4^3) - 24(4) + 6] - [2(1^3) - 24(1) + 6]$$

$$= 54 \text{ m} \qquad \textit{Ans.}$$

② which represents the net advancement of the particle along the s-axis from the position it occupied at $t = 1$ s to its position at $t = 4$ s.

To help visualize the motion, the values of s, v, and a are plotted against the time t as shown. Because the area under the v-t curve represents displace-
③ ment, we see that the net displacement from $t = 1$ s to $t = 4$ s is the positive area Δs_{2-4} less the negative area Δs_{1-2}.

Helpful Hints

① Be alert to the proper choice of sign when taking a square root. When the situation calls for only one answer, the positive root is not always the one you may need.

② Note carefully the distinction between italic s for the position coordinate and the vertical s for seconds.

③ Note from the graphs that the values for v are the slopes ($\dot{s}$) of the s-t curve and that the values for a are the slopes ($\dot{v}$) of the v-t curve. *Suggestion:* Integrate $v\,dt$ for each of the two intervals and check the answer for Δs. Show that the total distance traveled during the interval $t = 1$ s to $t = 4$ s is 74 m.

Sample Problem 2/2

A particle moves along the x-axis with an initial velocity $v_x = 50$ ft/sec at the origin when $t = 0$. For the first 4 seconds it has no acceleration, and thereafter it is acted on by a retarding force which gives it a constant acceleration $a_x = -10$ ft/sec^2. Calculate the velocity and the x-coordinate of the particle for the conditions of $t = 8$ sec and $t = 12$ sec and find the maximum positive x-coordinate reached by the particle. ①

Helpful Hints

① Learn to be flexible with symbols. The position coordinate x is just as valid as s.

Solution. The velocity of the particle after $t = 4$ sec is computed from

② $$\left[\int dv = \int a\,dt\right] \qquad \int_{50}^{v_x} dv_x = -10\int_4^t dt \qquad v_x = 90 - 10t \text{ ft/sec}$$

② Note that we integrate to a general time t and then substitute specific values.

and is plotted as shown. At the specified times, the velocities are

$$t = 8 \text{ sec}, \qquad v_x = 90 - 10(8) = 10 \text{ ft/sec}$$

$$t = 12 \text{ sec}, \qquad v_x = 90 - 10(12) = -30 \text{ ft/sec} \qquad \textit{Ans.}$$

The x-coordinate of the particle at any time greater than 4 seconds is the distance traveled during the first 4 seconds plus the distance traveled after the discontinuity in acceleration occurred. Thus,

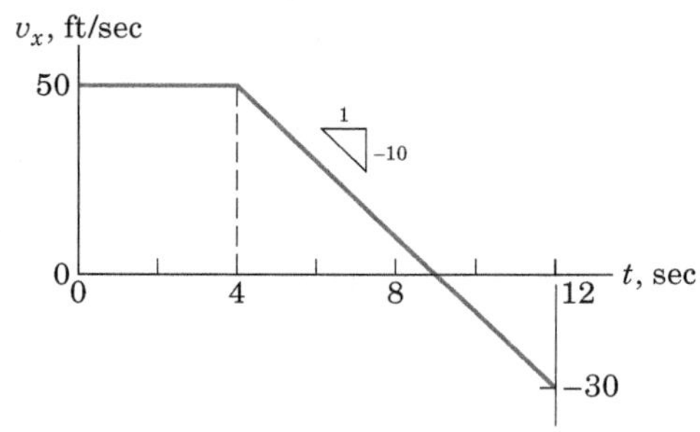

$$\left[\int ds = \int v\,dt\right] \qquad x = 50(4) + \int_4^t (90 - 10t)\,dt = -5t^2 + 90t - 80 \text{ ft}$$

For the two specified times,

$$t = 8 \text{ sec}, \qquad x = -5(8^2) + 90(8) - 80 = 320 \text{ ft}$$

$$t = 12 \text{ sec}, \qquad x = -5(12^2) + 90(12) - 80 = 280 \text{ ft} \qquad \textit{Ans.}$$

The x-coordinate for $t = 12$ sec is less than that for $t = 8$ sec since the motion is in the negative x-direction after $t = 9$ sec. The maximum positive x-coordinate is, then, the value of x for $t = 9$ sec which is

$$x_{\text{max}} = -5(9^2) + 90(9) - 80 = 325 \text{ ft} \qquad \textit{Ans.}$$

③ These displacements are seen to be the net positive areas under the v-t graph up to the values of t in question.

③ Show that the total distance traveled by the particle in the 12 sec is 370 ft.

Sample Problem 2/3

The spring-mounted slider moves in the horizontal guide with negligible friction and has a velocity v_0 in the s-direction as it crosses the mid-position where $s = 0$ and $t = 0$. The two springs together exert a retarding force to the motion of the slider, which gives it an acceleration proportional to the displacement but oppositely directed and equal to $a = -k^2 s$, where k is constant. (The constant is arbitrarily squared for later convenience in the form of the expressions.) Determine the expressions for the displacement s and velocity v as functions of the time t.

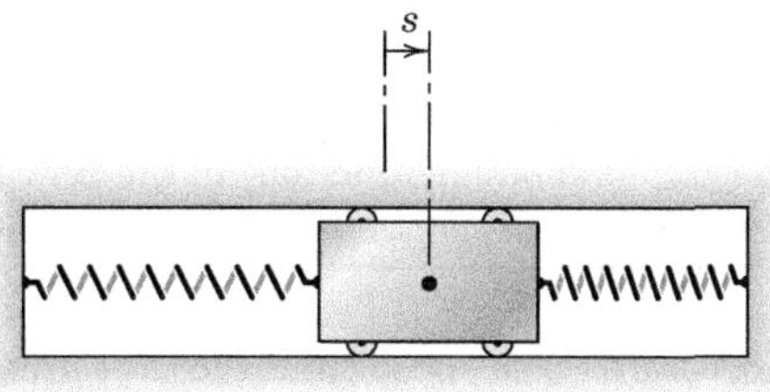

Solution I. Since the acceleration is specified in terms of the displacement, the differential relation $v\,dv = a\,ds$ may be integrated. Thus,

① $$\int v\,dv = \int -k^2 s\,ds + C_1 \text{ a constant,} \quad \text{or} \quad \frac{v^2}{2} = -\frac{k^2 s^2}{2} + C_1$$

When $s = 0$, $v = v_0$, so that $C_1 = v_0{}^2/2$, and the velocity becomes

$$v = +\sqrt{v_0{}^2 - k^2 s^2}$$

The plus sign of the radical is taken when v is positive (in the plus s-direction). This last expression may be integrated by substituting $v = ds/dt$. Thus,

② $$\int \frac{ds}{\sqrt{v_0{}^2 - k^2 s^2}} = \int dt + C_2 \text{ a constant,} \quad \text{or} \quad \frac{1}{k}\sin^{-1}\frac{ks}{v_0} = t + C_2$$

With the requirement of $t = 0$ when $s = 0$, the constant of integration becomes $C_2 = 0$, and we may solve the equation for s so that

$$s = \frac{v_0}{k}\sin kt \qquad \textit{Ans.}$$

The velocity is $v = \dot{s}$, which gives

$$v = v_0 \cos kt \qquad \textit{Ans.}$$

Solution II. Since $a = \ddot{s}$, the given relation may be written at once as

$$\ddot{s} + k^2 s = 0$$

This is an ordinary linear differential equation of second order for which the solution is well known and is

$$s = A \sin Kt + B \cos Kt$$

where A, B, and K are constants. Substitution of this expression into the differential equation shows that it satisfies the equation, provided that $K = k$. The velocity is $v = \dot{s}$, which becomes

$$v = Ak \cos kt - Bk \sin kt$$

The initial condition $v = v_0$ when $t = 0$ requires that $A = v_0/k$, and the condition $s = 0$ when $t = 0$ gives $B = 0$. Thus, the solution is

③ $$s = \frac{v_0}{k}\sin kt \quad \text{and} \quad v = v_0 \cos kt \qquad \textit{Ans.}$$

Helpful Hints

① We have used an indefinite integral here and evaluated the constant of integration. For practice, obtain the same results by using the definite integral with the appropriate limits.

② Again try the definite integral here as above.

③ This motion is called *simple harmonic motion* and is characteristic of all oscillations where the restoring force, and hence the acceleration, is proportional to the displacement but opposite in sign.

Sample Problem 2/4

A freighter is moving at a speed of 8 knots when its engines are suddenly ① stopped. If it takes 10 minutes for the freighter to reduce its speed to 4 knots, determine and plot the distance s in nautical miles moved by the ship and its speed v in knots as functions of the time t during this interval. The deceleration of the ship is proportional to the square of its speed, so that $a = -kv^2$.

Solution. The speeds and the time are given, so we may substitute the expression for acceleration directly into the basic definition $a = dv/dt$ and integrate. Thus,

$$-kv^2 = \frac{dv}{dt} \qquad \frac{dv}{v^2} = -k\,dt \qquad \int_8^v \frac{dv}{v^2} = -k\int_0^t dt$$

② $$-\frac{1}{v} + \frac{1}{8} = -kt \qquad v = \frac{8}{1 + 8kt}$$

Now we substitute the end limits of $v = 4$ knots and $t = \frac{10}{60} = \frac{1}{6}$ hour and get

$$4 = \frac{8}{1 + 8k(1/6)} \qquad k = \frac{3}{4}\text{ mi}^{-1} \qquad v = \frac{8}{1 + 6t} \qquad \textit{Ans.}$$

The speed is plotted against the time as shown.

The distance is obtained by substituting the expression for v into the definition $v = ds/dt$ and integrating. Thus,

$$\frac{8}{1 + 6t} = \frac{ds}{dt} \qquad \int_0^t \frac{8\,dt}{1 + 6t} = \int_0^s ds \qquad s = \frac{4}{3}\ln(1 + 6t) \qquad \textit{Ans.}$$

The distance s is also plotted against the time as shown, and we see that the ship has moved through a distance $s = \frac{4}{3}\ln(1 + \frac{6}{6}) = \frac{4}{3}\ln 2 = 0.924$ mi (nautical) during the 10 minutes.

Helpful Hints

① Recall that one knot is the speed of one nautical mile (6076 ft) per hour. Work directly in the units of nautical miles and hours.

② We choose to integrate to a general value of v and its corresponding time t so that we may obtain the variation of v with t.

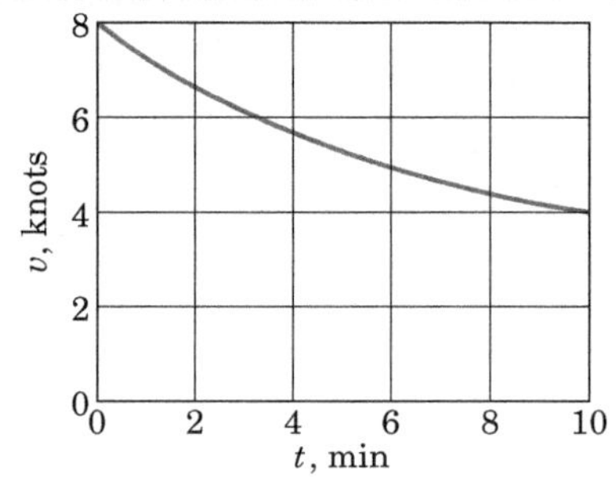

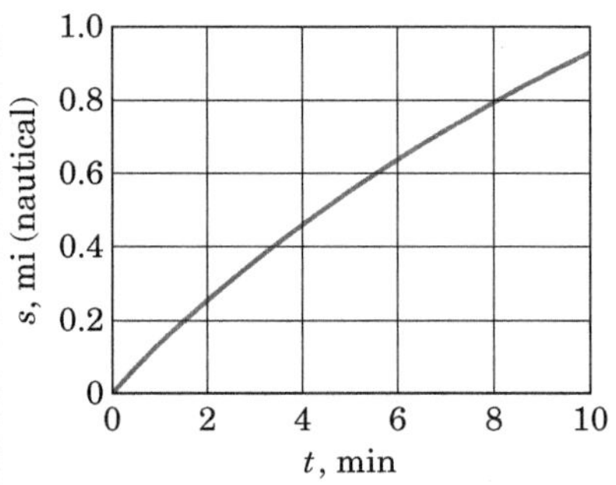

PROBLEMS

Introductory Problems

Problems 2/1 through 2/8 treat the motion of a particle which moves along the s-axis shown in the figure.

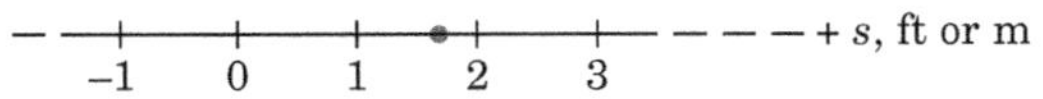

Problems 2/1–2/8

2/1 The velocity of a particle is given by $v = 25t^2 - 80t - 200$, where v is in feet per second and t is in seconds. Plot the velocity v and acceleration a versus time for the first 6 seconds of motion and evaluate the velocity when a is zero.

2/2 The position of a particle is given by $s = 0.27t^3 - 0.65t^2 - 2.35t + 4.4$, where s is in feet and the time t is in seconds. Plot the displacement, velocity, and acceleration as functions of time for the first 5 seconds of motion. Determine the positive time when the particle changes its direction.

2/3 The velocity of a particle which moves along the s-axis is given by $v = 2 - 4t + 5t^{3/2}$, where t is in seconds and v is in meters per second. Evaluate the position s, velocity v, and acceleration a when $t = 3$ s. The particle is at the position $s_0 = 3$ m when $t = 0$.

2/4 The displacement of a particle which moves along the s-axis is given by $s = (-2 + 3t)e^{-0.5t}$, where s is in meters and t is in seconds. Plot the displacement, velocity, and acceleration versus time for the first 20 seconds of motion. Determine the time at which the acceleration is zero.

2/5 The acceleration of a particle is given by $a = 2t - 10$, where a is in meters per second squared and t is in seconds. Determine the velocity and displacement as functions of time. The initial displacement at $t = 0$ is $s_0 = -4$ m, and the initial velocity is $v_0 = 3$ m/s.

2/6 The acceleration of a particle is given by $a = -kt^2$, where a is in meters per second squared and the time t is in seconds. If the initial velocity of the particle at $t = 0$ is $v_0 = 12$ m/s and the particle takes 6 seconds to reverse direction, determine the magnitude and units of the constant k. What is the net displacement of the particle over the same 6-second interval of motion?

2/7 The acceleration of a particle is given by $a = -ks^2$, where a is in meters per second squared, k is a constant, and s is in meters. Determine the velocity of the particle as a function of its position s. Evaluate your expression for $s = 5$ m if $k = 0.1\ \text{m}^{-1}\text{s}^{-2}$ and the initial conditions at time $t = 0$ are $s_0 = 3$ m and $v_0 = 10$ m/s.

2/8 The acceleration of a particle is given by $a = c_1 + c_2 v$, where a is in millimeters per second squared, the velocity v is in millimeters per second, and c_1 and c_2 are constants. If the particle position and velocity at $t = 0$ are s_0 and v_0, respectively, determine expressions for the position s of the particle in terms of the velocity v and time t.

2/9 Calculate the constant acceleration a in g's which the catapult of an aircraft carrier must provide to produce a launch velocity of 180 mi/hr in a distance of 300 ft. Assume that the carrier is at anchor.

2/10 A particle in an experimental apparatus has a velocity given by $v = k\sqrt{s}$, where v is in millimeters per second, the position s is millimeters, and the constant $k = 0.2\ \text{mm}^{1/2}\text{s}^{-1}$. If the particle has a velocity $v_0 = 3$ mm/s at $t = 0$, determine the particle position, velocity, and acceleration as functions of time, and compute the time, position, and acceleration of the particle when the velocity reaches 15 mm/s.

2/11 Ball 1 is launched with an initial vertical velocity $v_1 = 160$ ft/sec. Three seconds later, ball 2 is launched with an initial vertical velocity v_2. Determine v_2 if the balls are to collide at an altitude of 300 ft. At the instant of collision, is ball 1 ascending or descending?

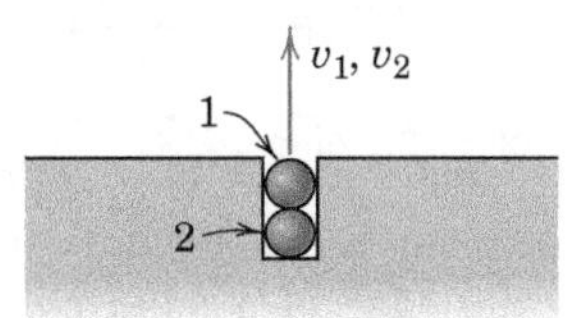

Problem 2/11

2/12 Experimental data for the motion of a particle along a straight line yield measured values of the velocity v for various position coordinates s. A smooth curve is drawn through the points as shown in the graph. Determine the acceleration of the particle when $s = 20$ ft.

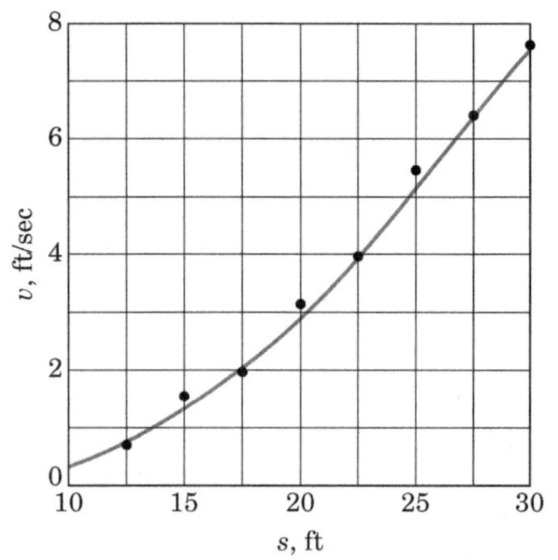

Problem 2/12

2/13 In the pinewood-derby event shown, the car is released from rest at the starting position A and then rolls down the incline and on to the finish line C. If the constant acceleration down the incline is 2.75 m/s^2 and the speed from B to C is essentially constant, determine the time duration t_{AC} for the race. The effects of the small transition area at B can be neglected.

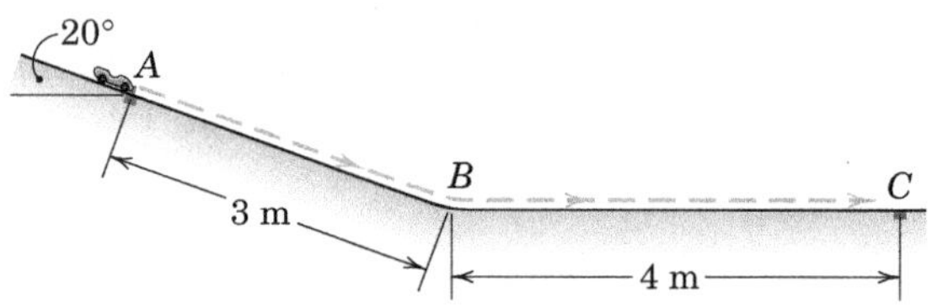

Problem 2/13

2/14 A ball is thrown vertically up with a velocity of 30 m/s at the edge of a 60-m cliff. Calculate the height h to which the ball rises and the total time t after release for the ball to reach the bottom of the cliff. Neglect air resistance and take the downward acceleration to be 9.81 m/s^2.

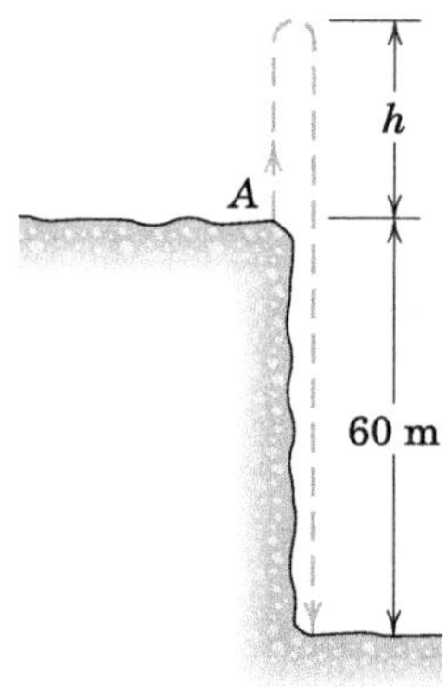

Problem 2/14

2/15 A car comes to a complete stop from an initial speed of 50 mi/hr in a distance of 100 ft. With the same constant acceleration, what would be the stopping distance s from an initial speed of 70 mi/hr?

2/16 The pilot of a jet transport brings the engines to full takeoff power before releasing the brakes as the aircraft is standing on the runway. The jet thrust remains constant, and the aircraft has a near-constant acceleration of $0.4g$. If the takeoff speed is 200 km/h, calculate the distance s and time t from rest to takeoff.

2/17 A game requires that two children each throw a ball upward as high as possible from point O and then run horizontally in opposite directions away from O. The child who travels the greater distance before their thrown ball impacts the ground wins. If child A throws a ball upward with a speed of $v_1 = 70$ ft/sec and immediately runs leftward at a constant speed of $v_A = 16$ ft/sec while child B throws the ball upward with a speed of $v_2 = 64$ ft/sec and immediately runs rightward with a constant speed of $v_B = 18$ ft/sec, which child will win the game?

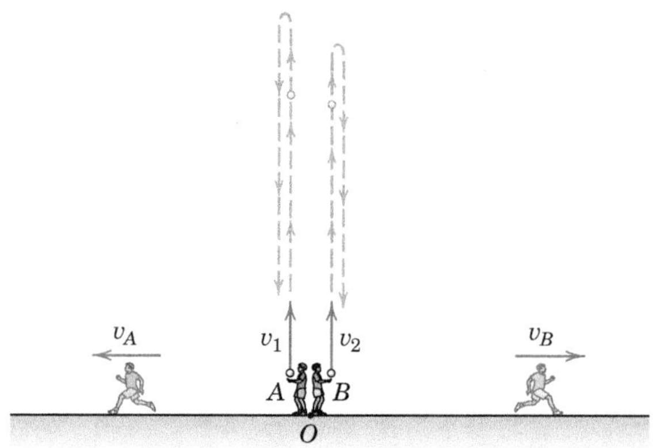

Problem 2/17

Representative Problems

2/18 During an 8-second interval, the velocity of a particle moving in a straight line varies with time as shown. Within reasonable limits of accuracy, determine the amount Δa by which the acceleration at $t = 4$ s exceeds the average acceleration during the interval. What is the displacement Δs during the interval?

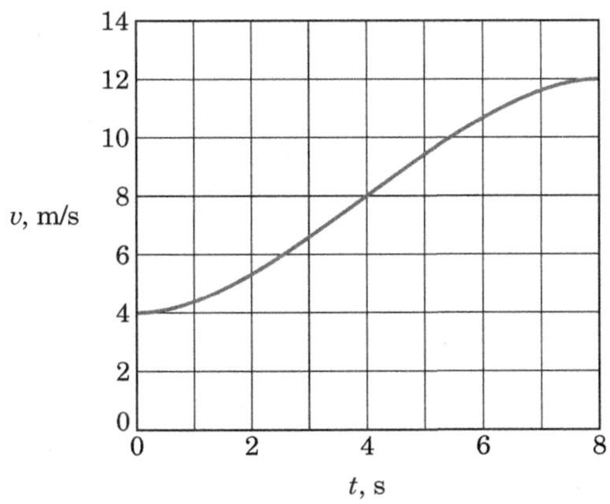

Problem 2/18

2/19 In the final stages of a moon landing, the lunar module descends under retrothrust of its descent engine to within $h = 5$ m of the lunar surface where it has a downward velocity of 2 m/s. If the descent engine is cut off abruptly at this point, compute the impact velocity of the landing gear with the moon. Lunar gravity is $\frac{1}{6}$ of the earth's gravity.

2/20 A girl rolls a ball up an incline and allows it to return to her. For the angle θ and ball involved, the acceleration of the ball along the incline is constant at $0.25g$, directed down the incline. If the ball is released with a speed of 4 m/s, determine the distance s it moves up the incline before reversing its direction and the total time t required for the ball to return to the child's hand.

Problem 2/20

2/21 At a football tryout, a player runs a 40-yard dash in 4.25 seconds. If he reaches his maximum speed at the 16-yard mark with a constant acceleration and then maintains that speed for the remainder of the run, determine his acceleration over the first 16 yards, his maximum speed, and the time duration of the acceleration.

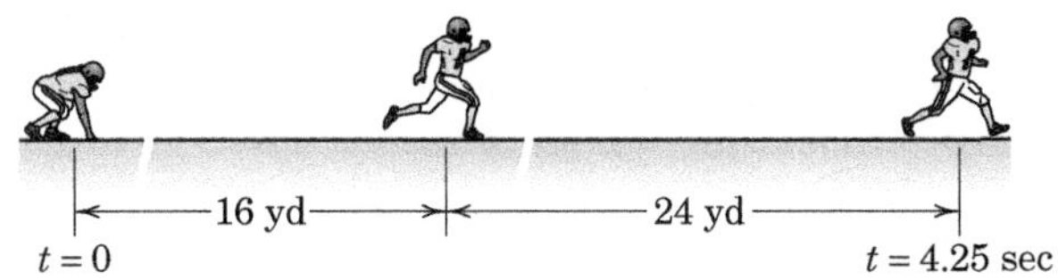

Problem 2/21

2/22 The main elevator A of the CN Tower in Toronto rises about 350 m and for most of its run has a constant speed of 22 km/h. Assume that both the acceleration and deceleration have a constant magnitude of $\frac{1}{4}g$ and determine the time duration t of the elevator run.

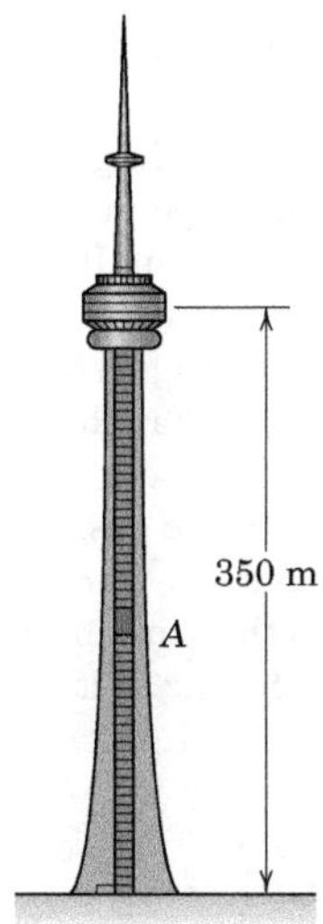

Problem 2/22

2/23 A Scotch-yoke mechanism is used to convert rotary motion into reciprocating motion. As the disk rotates at the constant angular rate ω, a pin A slides in a vertical slot causing the slotted member to displace horizontally according to $x = r\sin(\omega t)$ relative to the fixed disk center O. Determine the expressions for the velocity and acceleration of a point P on the output shaft of the mechanism as functions of time, and determine the maximum velocity and acceleration of point P during one cycle. Use the values $r = 75$ mm and $\omega = \pi$ rad/s.

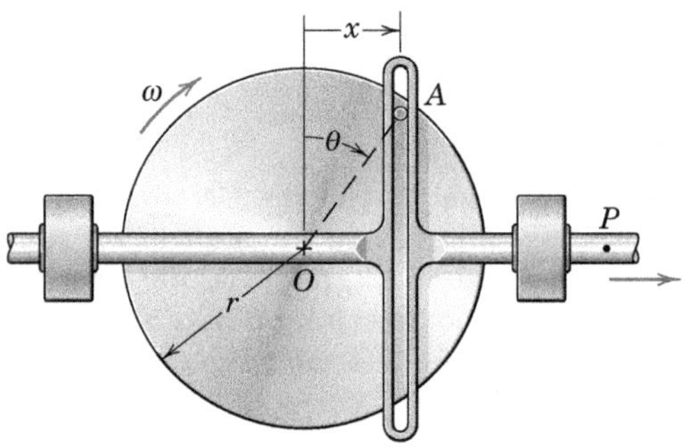

Problem 2/23

2/24 A train which is traveling at 80 mi/hr applies its brakes as it reaches point A and slows down with a constant deceleration. Its decreased velocity is observed to be 60 mi/hr as it passes a point 1/2 mi beyond A. A car moving at 50 mi/hr passes point B at the same instant that the train reaches point A. In an unwise effort to beat the train to the crossing, the driver "steps on the gas." Calculate the constant acceleration a that the car must have in order to beat the train to the crossing by 4 seconds and find the velocity v of the car as it reaches the crossing.

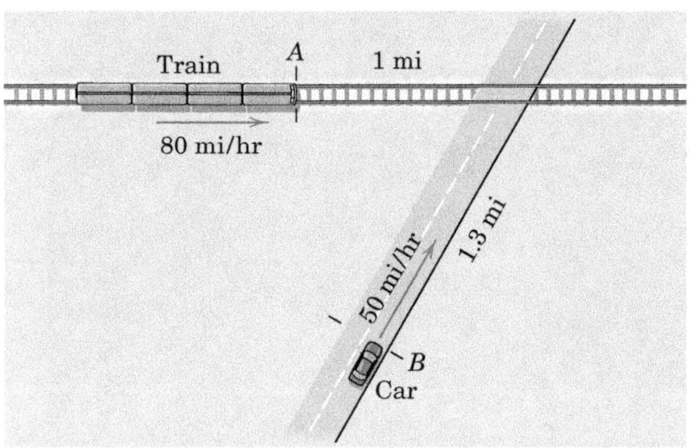

Problem 2/24

2/25 Small steel balls fall from rest through the opening at A at the steady rate of two per second. Find the vertical separation h of two consecutive balls when the lower one has dropped 3 meters. Neglect air resistance.

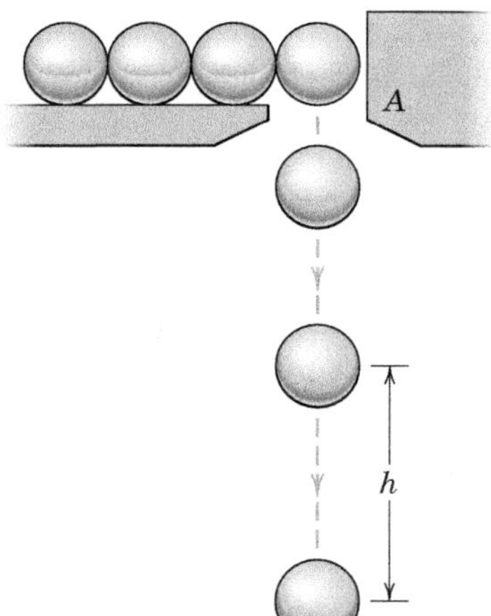

Problem 2/25

2/26 Car A is traveling at a constant speed $v_A = 130$ km/h at a location where the speed limit is 100 km/h. The police officer in car P observes this speed via radar. At the moment when A passes P, the police car begins to accelerate at the constant rate of 6 m/s^2 until a speed of 160 km/h is achieved, and that speed is then maintained. Determine the distance required for the police officer to overtake car A. Neglect any nonrectilinear motion of P.

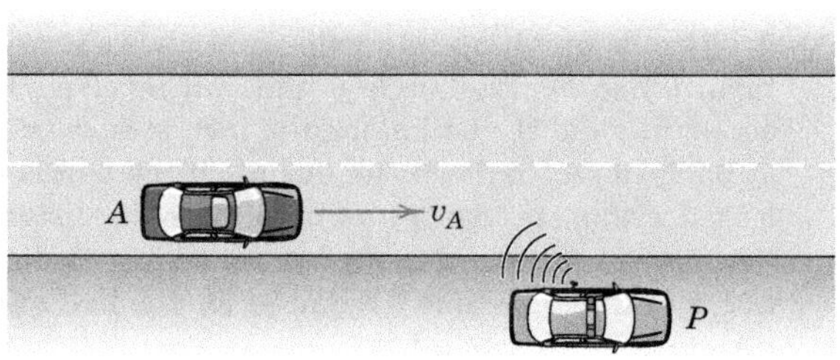

Problem 2/26

2/27 A toy helicopter is flying in a straight line at a constant speed of 4.5 m/s. If a projectile is launched vertically with an initial speed of $v_0 = 28$ m/s, what horizontal distance d should the helicopter be from the launch site S if the projectile is to be traveling downward when it strikes the helicopter? Assume that the projectile travels only in the vertical direction.

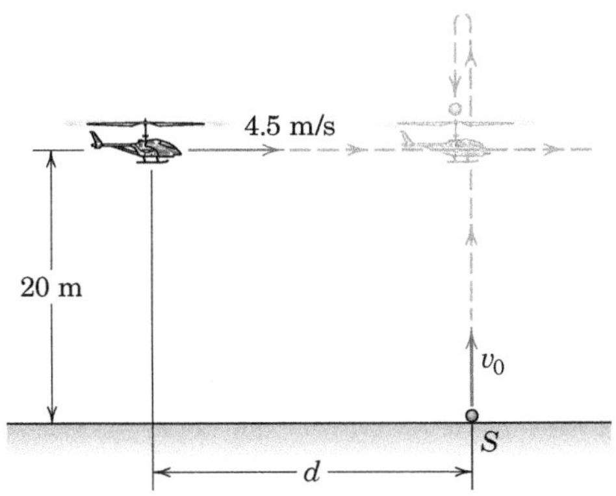

Problem 2/27

2/28 A particle moving along a straight line has an acceleration which varies according to position as shown. If the velocity of the particle at the position $x = -5$ ft is $v = -2$ ft/sec, determine the velocity when $x = 9$ ft.

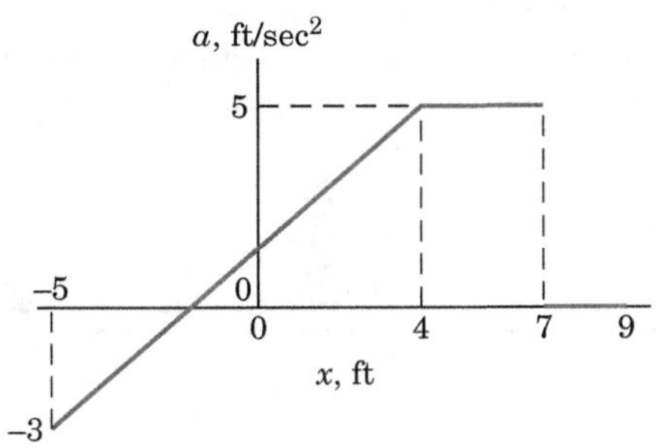

Problem 2/28

2/29 A model rocket is launched from rest with a constant upward acceleration of 3 m/s^2 under the action of a small thruster. The thruster shuts off after 8 seconds, and the rocket continues upward until it reaches its apex. At apex, a small chute opens which ensures that the rocket falls at a constant speed of 0.85 m/s until it impacts the ground. Determine the maximum height h attained by the rocket and the total flight time. Neglect aerodynamic drag during ascent, and assume that the mass of the rocket and the acceleration of gravity are both constant.

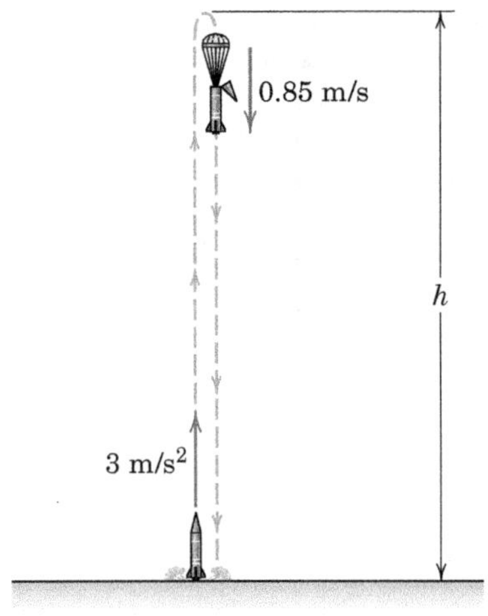

Problem 2/29

2/30 An electric car is subjected to acceleration tests along a straight and level test track. The resulting v-t data are closely modeled over the first 10 seconds by the function $v = 24t - t^2 + 5\sqrt{t}$, where t is the time in seconds and v is the velocity in feet per second. Determine the displacement s as a function of time over the interval $0 \le t \le 10$ sec and specify its value at time $t = 10$ sec.

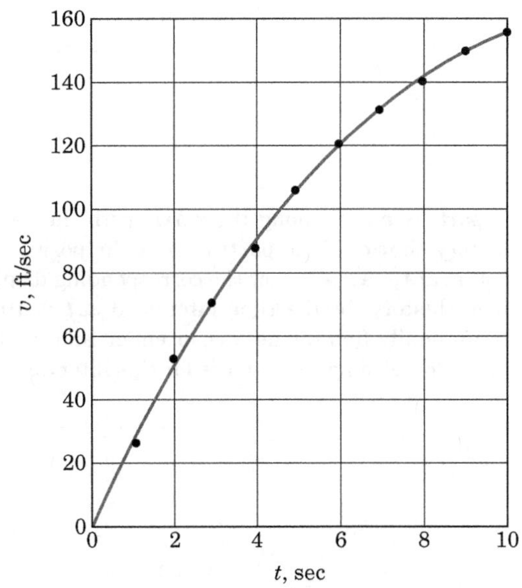

Problem 2/30

2/31 A vacuum-propelled capsule for a high-speed tube transportation system of the future is being designed for operation between two stations A and B, which are 10 km apart. If the acceleration and deceleration are to have a limiting magnitude of $0.6g$ and if velocities are to be limited to 400 km/h, determine the minimum time t for the capsule to make the 10-km trip.

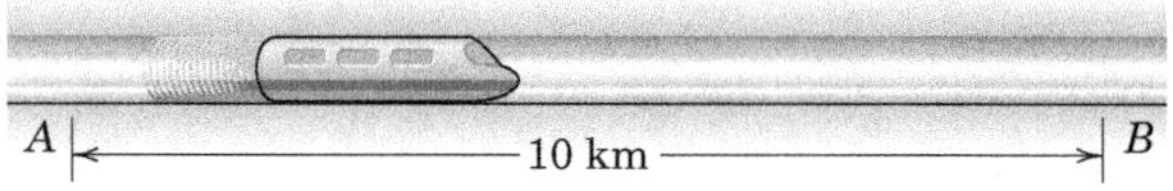

Problem 2/31

2/32 If the velocity v of a particle moving along a straight line decreases linearly with its displacement s from 20 m/s to a value approaching zero at $s = 30$ m, determine the acceleration a of the particle when $s = 15$ m and show that the particle never reaches the 30-m displacement.

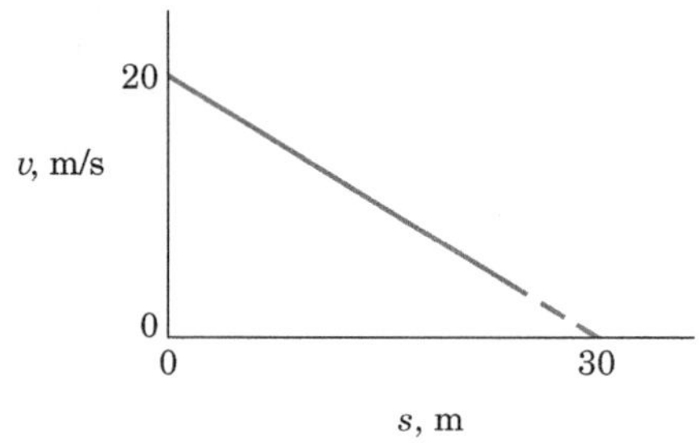

Problem 2/32

2/33 A particle moves along the x-axis with the velocity history shown. If the particle is at the position $x = -4$ in. at time $t = 0$, plot the corresponding displacement history for the time interval $0 \leq t \leq 10$ sec. Additionally, find the net displacement and total distance traveled by the particle for this interval.

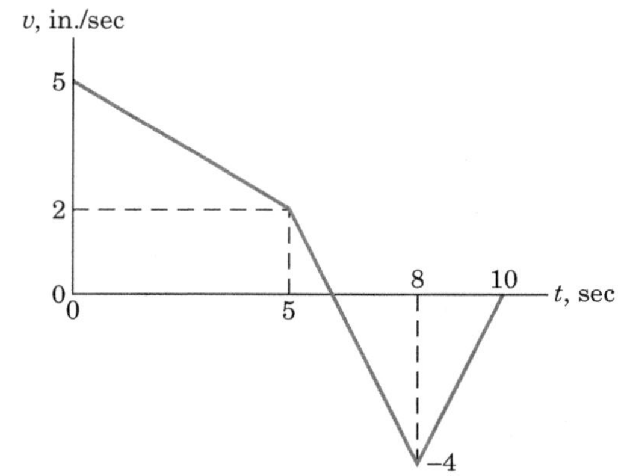

Problem 2/33

2/34 The 230,000-lb space-shuttle orbiter touches down at about 220 mi/hr. At 200 mi/hr its drag parachute deploys. At 35 mi/hr, the chute is jettisoned from the orbiter. If the deceleration in feet per second squared during the time that the chute is deployed is $-0.0003v^2$ (speed v in feet per second), determine the corresponding distance traveled by the orbiter. Assume no braking from its wheel brakes.

Problem 2/34

2/35 Reconsider the rollout of the space-shuttle orbiter of the previous problem. The drag chute is deployed at 200 mi/hr, the wheel brakes are applied at 100 mi/hr until wheelstop, and the drag chute is jettisoned at 35 mi/hr. If the drag chute results in a deceleration of $-0.0003v^2$ (in feet per second squared when the speed v is in feet per second) and the wheel brakes cause a constant deceleration of 5 ft/sec^2, determine the distance traveled from 200 mi/hr to wheelstop.

2/36 The cart impacts the safety barrier with speed $v_0 = 3.25$ m/s and is brought to a stop by the nest of nonlinear springs which provide a deceleration $a = -k_1x - k_2x^3$, where x is the amount of spring deflection from the undeformed position and k_1 and k_2 are positive constants. If the maximum spring deflection is 475 mm and the velocity at half-maximum deflection is 2.85 m/s, determine the values and corresponding units for the constants k_1 and k_2.

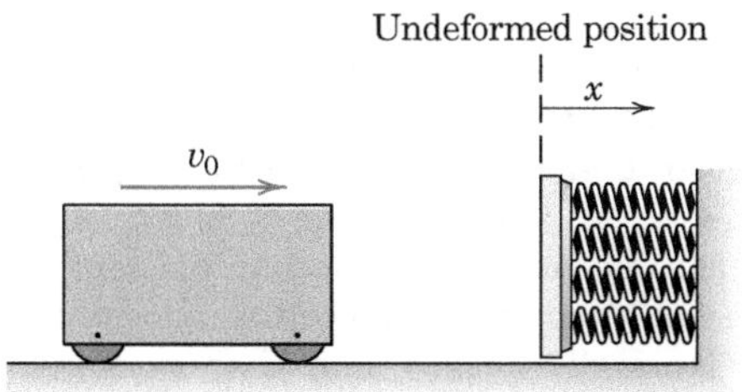

Problem 2/36

2/37 The graph shows the rectilinear acceleration of a particle as a function of time over a 12-second interval. If the particle is at rest at the position $s_0 = 0$ at time $t = 0$, determine the velocity of the particle when (*a*) $t = 4$ s, (*b*) $t = 8$ s, and (*c*) $t = 12$ s.

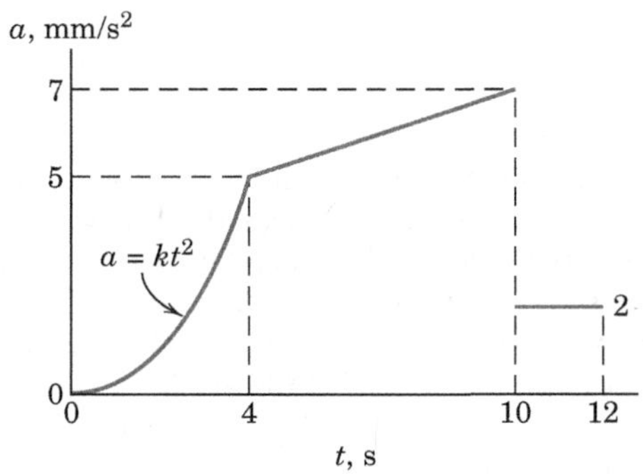

Problem 2/37

2/38 Compute the impact speed of a body released from rest at an altitude $h = 650$ miles above the surface of Mars. (*a*) First assume a constant gravitational acceleration $g_{m_0} = 12.3$ ft/sec^2 (equal to that at the surface) and (*b*) then account for the variation of g with altitude (refer to Art. 1/5). Neglect any effects of atmospheric drag.

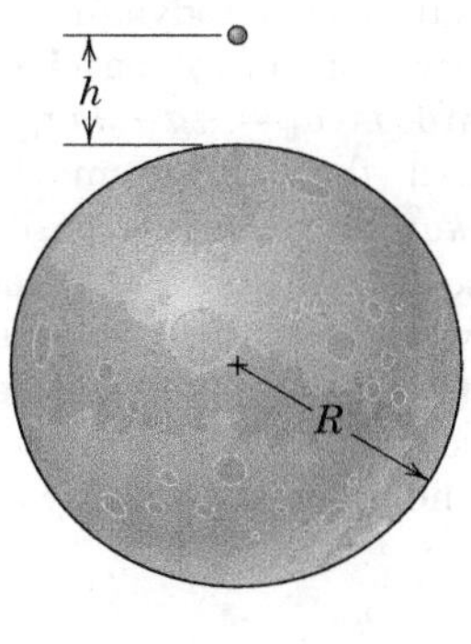

Problem 2/38

2/39 The steel ball A of diameter D slides freely on the horizontal rod which leads to the pole face of the electromagnet. The force of attraction obeys an inverse-square law, and the resulting acceleration of the ball is $a = K/(L - x)^2$, where K is a measure of the strength of the magnetic field. If the ball is released from rest at $x = 0$, determine the velocity v with which it strikes the pole face.

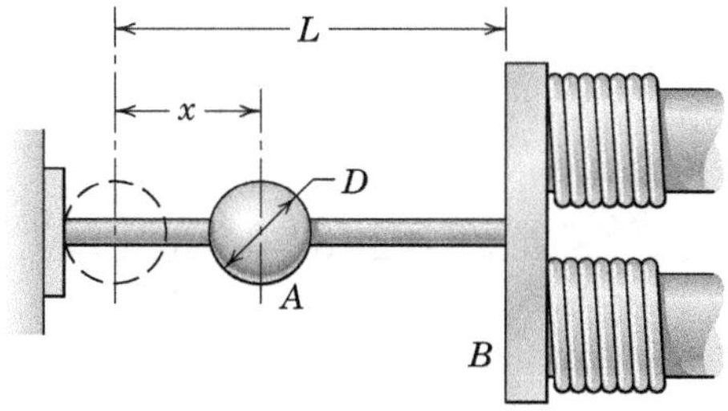

Problem 2/39

***2/40** The falling object has a speed v_0 when it strikes and subsequently deforms the foam arresting material until it comes to rest. The resistance of the foam material to deformation is a function of penetration depth y and object speed v so that the acceleration of the object is $a = g - k_1 v - k_2 y$, where v is the particle speed in inches per second, y is the penetration depth in inches, and k_1 and k_2 are positive constants. Plot the penetration depth y and velocity v of the object as functions of time over the first five seconds for $k_1 = 12$ sec^{-1}, $k_2 = 24$ sec^{-2}, and $v_0 = 25$ in./sec. Determine the time when the penetration depth reaches 95% of its final value.

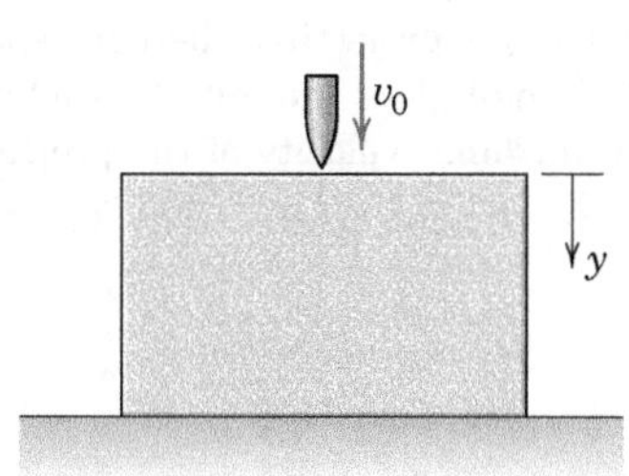

Problem 2/40

2/41 The electronic throttle control of a model train is programmed so that the train speed varies with position as shown in the plot. Determine the time t required for the train to complete one lap.

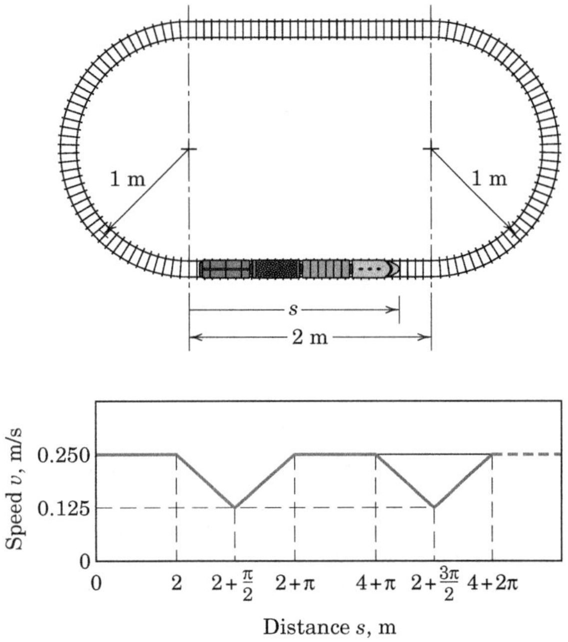

Problem 2/41

2/42 A projectile is fired downward with initial speed v_0 in an experimental fluid and experiences an acceleration $a = \sigma - \eta v^2$, where σ and η are positive constants and v is the projectile speed. Determine the distance traveled by the projectile when its speed has been reduced to one-half of the initial speed v_0. Also, determine the terminal velocity of the projectile. Evaluate for $\sigma = 0.7$ m/s², $\eta = 0.2$ m^{-1}, and $v_0 = 4$ m/s.

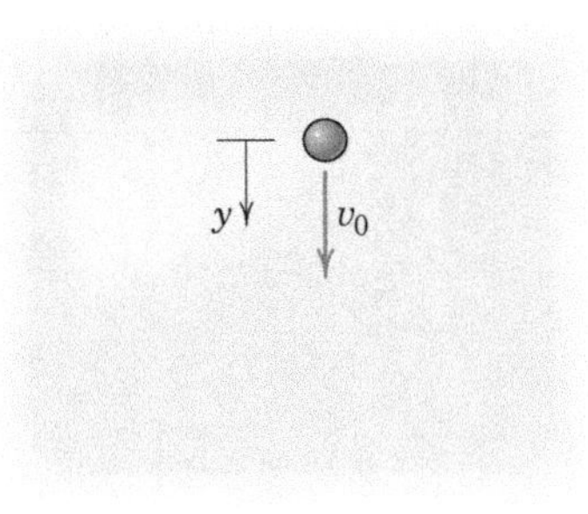

Problem 2/42

2/43 The aerodynamic resistance to motion of a car is nearly proportional to the square of its velocity. Additional frictional resistance is constant, so that the acceleration of the car when coasting may be written $a = -C_1 - C_2v^2$, where C_1 and C_2 are constants which depend on the mechanical configuration of the car. If the car has an initial velocity v_0 when the engine is disengaged, derive an expression for the distance D required for the car to coast to a stop.

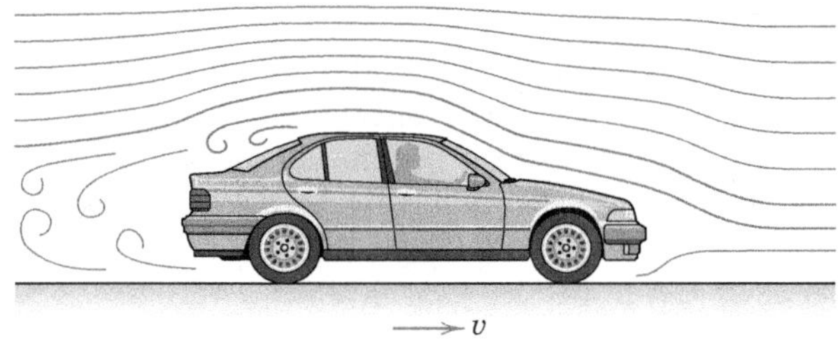

Problem 2/43

2/44 The driver of a car, which is initially at rest at the top A of the grade, releases the brakes and coasts down the grade with an acceleration in feet per second squared given by $a = 3.22 - 0.004v^2$, where v is the velocity in feet per second. Determine the velocity v_B at the bottom B of the grade.

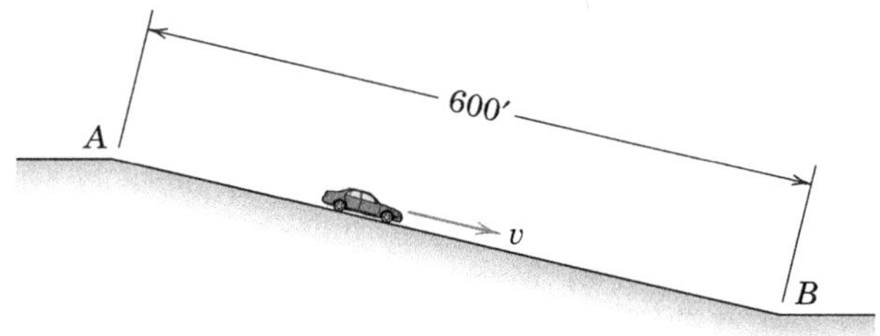

Problem 2/44

2/45 When the effect of aerodynamic drag is included, the y-acceleration of a baseball moving vertically upward is $a_u = -g - kv^2$, while the acceleration when the ball is moving downward is $a_d = -g + kv^2$, where k is a positive constant and v is the speed in meters per second. If the ball is thrown upward at 30 m/s from essentially ground level, compute its maximum height h and its speed v_f upon impact with the ground. Take k to be 0.006 m^{-1} and assume that g is constant.

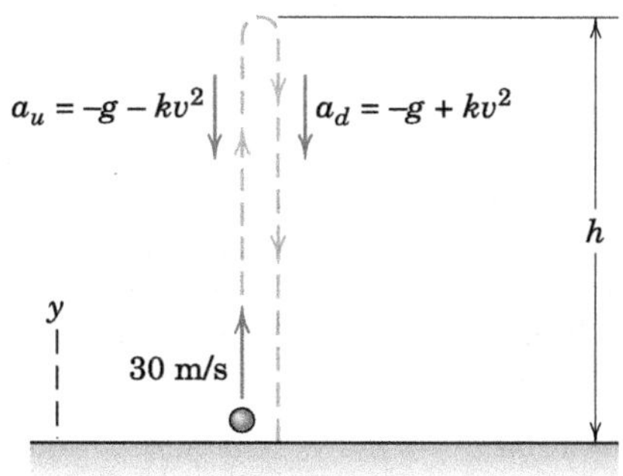

Problem 2/45

2/46 For the baseball of Prob. 2/45 thrown upward with an initial speed of 30 m/s, determine the time t_u from ground to apex and the time t_d from apex to ground.

2/47 The stories of a tall building are uniformly 10 feet in height. A ball A is dropped from the rooftop position shown. Determine the times required for it to pass the 10 feet of the first, tenth, and one-hundredth stories (counted from the top). Neglect aerodynamic drag.

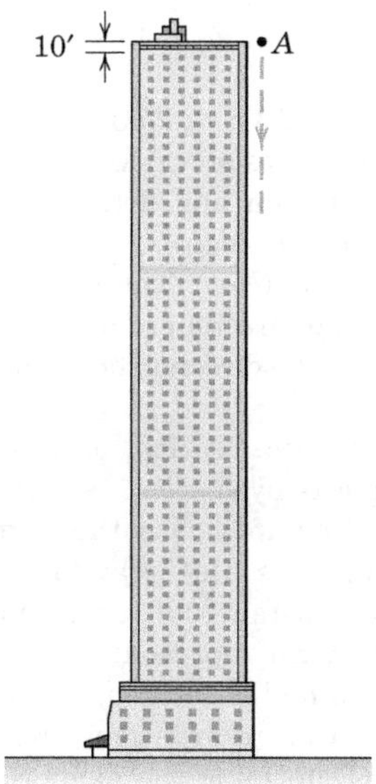

Problem 2/47

2/48 Repeat Prob. 2/47, except now include the effects of aerodynamic drag. The drag force causes an acceleration component in ft/sec^2 of $0.005v^2$ in the direction opposite the velocity vector, where v is in ft/sec.

2/49 On its takeoff roll, the airplane starts from rest and accelerates according to $a = a_0 - kv^2$, where a_0 is the constant acceleration resulting from the engine thrust and $-kv^2$ is the acceleration due to aerodynamic drag. If $a_0 = 2$ m/s^2, $k = 0.00004$ m^{-1}, and v is in meters per second, determine the design length of runway required for the airplane to reach the takeoff speed of 250 km/h if the drag term is (a) excluded and (b) included.

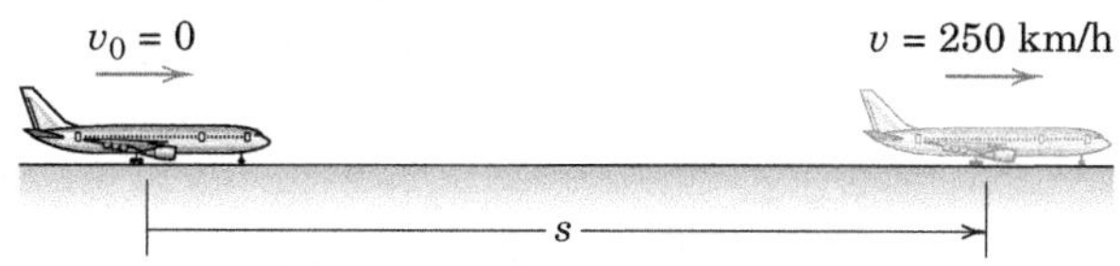

Problem 2/49

2/50 A test projectile is fired horizontally into a viscous liquid with a velocity v_0. The retarding force is proportional to the square of the velocity, so that the acceleration becomes $a = -kv^2$. Derive expressions for the distance D traveled in the liquid and the corresponding time t required to reduce the velocity to $v_0/2$. Neglect any vertical motion.

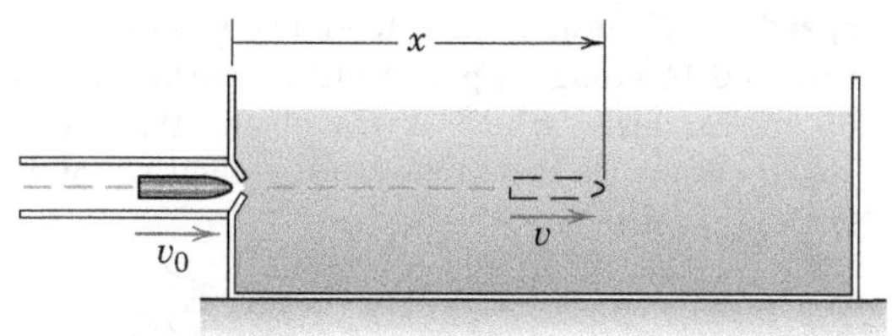

Problem 2/50

2/51 A bumper, consisting of a nest of three springs, is used to arrest the horizontal motion of a large mass which is traveling at 40 ft/sec as it contacts the bumper. The two outer springs cause a deceleration proportional to the spring deformation. The center spring increases the deceleration rate when the compression exceeds 6 in. as shown on the graph. Determine the maximum compression x of the outer springs.

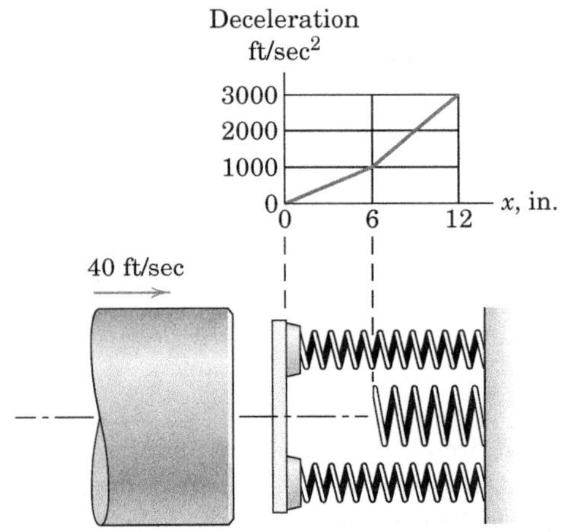

Problem 2/51

2/52 Car A travels at a constant speed of 65 mi/hr. When in the position shown at time $t = 0$, car B has a speed of 25 mi/hr and accelerates at a constant rate of $0.1g$ along its path until it reaches a speed of 65 mi/hr, after which it travels at that constant speed. What is the steady-state position of car A with respect to car B?

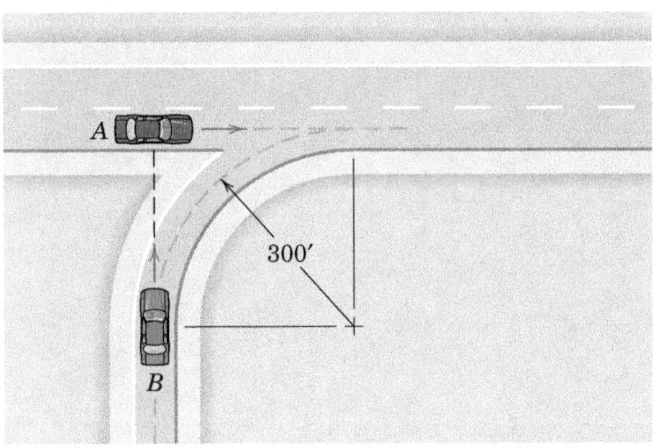

Problem 2/52

2/53 A block of mass m rests on a rough horizontal surface and is attached to a spring of stiffness k. The coefficients of both static and kinetic friction are μ. The block is displaced a distance x_0 to the right of the unstretched position of the spring and released from rest. If the value of x_0 is large enough, the spring force will overcome the maximum available static friction force and the block will slide toward the unstretched position of the spring with an acceleration $a = \mu g - \dfrac{k}{m}x$, where x represents the amount of stretch (or compression) in the spring at any given location in the motion. Use the values $m = 5$ kg, $k = 150$ N/m, $\mu = 0.40$, and $x_0 = 200$ mm and determine the final spring stretch (or compression) x_f when the block comes to a complete stop.

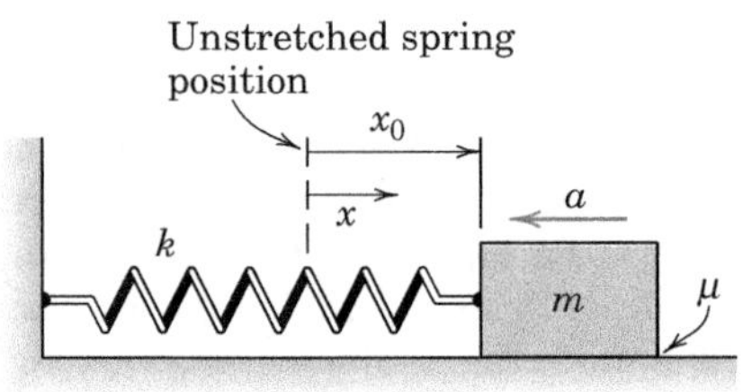

Problem 2/53

▶2/54 The situation of Prob. 2/53 is repeated here. This time, use the values $m = 5$ kg, $k = 150$ N/m, $\mu = 0.40$, and $x_0 = 500$ mm and determine the final spring stretch (or compression) x_f when the block comes to a complete stop. (*Note:* The sign on the μg term is dictated by the direction of motion for the block and always acts in the direction opposite velocity.)

▶2/55 The vertical acceleration of a certain solid-fuel rocket is given by $a = ke^{-bt} - cv - g$, where k, b, and c are constants, v is the vertical velocity acquired, and g is the gravitational acceleration, essentially constant for atmospheric flight. The exponential term represents the effect of a decaying thrust as fuel is burned, and the term $-cv$ approximates the retardation due to atmospheric resistance. Determine the expression for the vertical velocity of the rocket t seconds after firing.

▶**2/56** The preliminary design for a rapid-transit system calls for the train velocity to vary with time as shown in the plot as the train runs the 3.2 km between stations A and B. The slopes of the cubic transition curves (which are of form $a + bt + ct^2 + dt^3$) are zero at the end points. Determine the total run time t between the stations and the maximum acceleration.

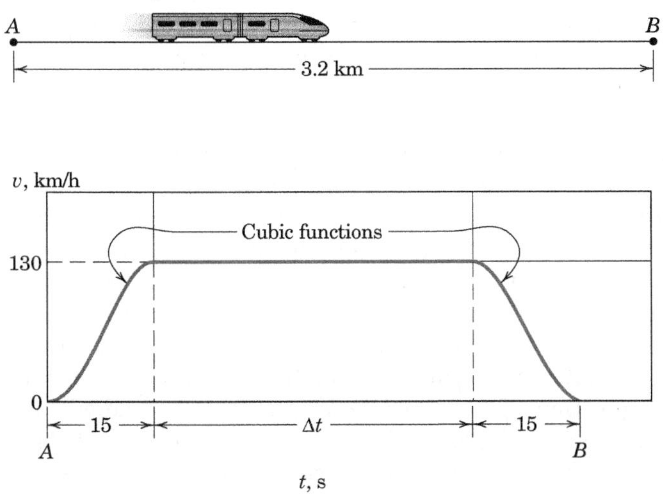

Problem 2/56

2/57 A projectile is fired vertically from point A with an initial speed of 255 ft/sec. Relative to an observer located at B, at what times will the line of sight to the projectile make an angle of 30° with the horizontal? Compute the magnitude of the speed of the projectile at each time, and ignore the effect of aerodynamic drag on the projectile.

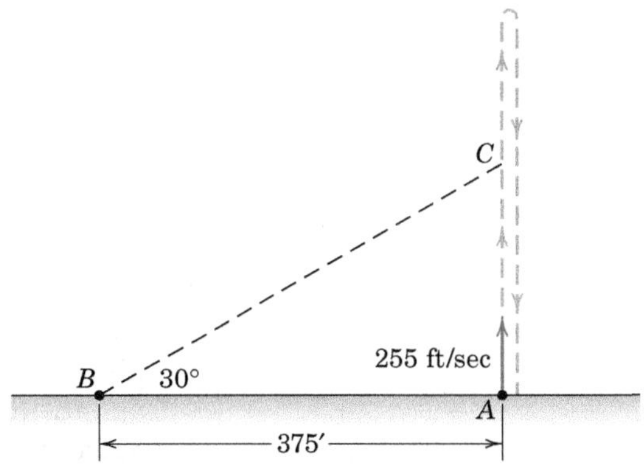

Problem 2/57

▶**2/58** Repeat Prob. 2/57 for the case where aerodynamic drag is included. The magnitude of the drag deceleration is kv^2, where $k = 3.5(10^{-3})$ ft^{-1} and v is the speed in feet per second. The direction of the drag is opposite the motion of the projectile throughout the flight (when the projectile is moving upward, the drag is directed downward, and when the projectile is moving downward, the drag is directed upward).

2/3 Plane Curvilinear Motion

We now treat the motion of a particle along a curved path which lies in a single plane. This motion is a special case of the more general three-dimensional motion introduced in Art. 2/1 and illustrated in Fig. 2/1. If we let the plane of motion be the x-y plane, for instance, then the coordinates z and ϕ of Fig. 2/1 are both zero, and R becomes the same as r. As mentioned previously, the vast majority of the motions of points or particles encountered in engineering practice can be represented as plane motion.

Before pursuing the description of plane curvilinear motion in any specific set of coordinates, we will first use vector analysis to describe the motion, since the results will be independent of any particular coordinate system. What follows in this article constitutes one of the most basic concepts in dynamics, namely, the *time derivative of a vector*. Much analysis in dynamics utilizes the time rates of change of vector quantities. You are therefore well advised to master this topic at the outset because you will have frequent occasion to use it.

Consider now the continuous motion of a particle along a plane curve as represented in Fig. 2/5. At time t the particle is at position A, which is located by the *position vector* $\mathbf{r}$ measured from some convenient fixed origin O. If both the magnitude and direction of $\mathbf{r}$ are known at time t, then the position of the particle is completely specified. At time $t + \Delta t$, the particle is at A', located by the position vector $\mathbf{r} + \Delta\mathbf{r}$. We note, of course, that this combination is vector addition and not scalar addition. The *displacement* of the particle during time Δt is the vector $\Delta\mathbf{r}$ which represents the vector change of position and is clearly independent of the choice of origin. If an origin were chosen at some different location, the position vector $\mathbf{r}$ would be changed, but $\Delta\mathbf{r}$ would be unchanged. The *distance* actually traveled by the particle as it moves along the path from A to A' is the scalar length Δs measured along the path. Thus, we distinguish between the vector displacement $\Delta\mathbf{r}$ and the scalar distance Δs.

Velocity

The *average velocity* of the particle between A and A' is defined as $\mathbf{v}_{av} = \Delta\mathbf{r}/\Delta t$, which is a vector whose direction is that of $\Delta\mathbf{r}$ and whose magnitude is the magnitude of $\Delta\mathbf{r}$ divided by Δt. The average speed of

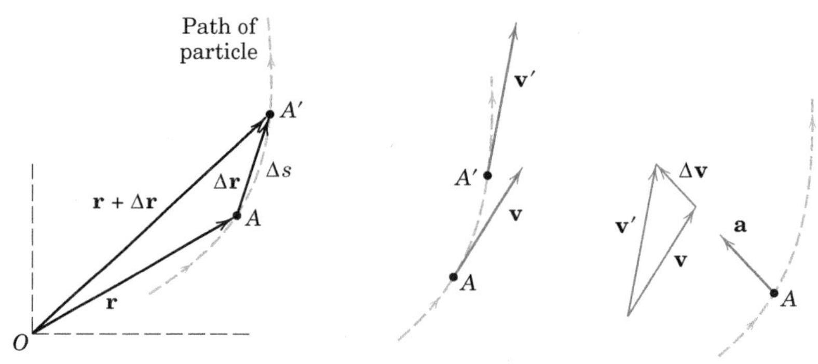

Figure 2/5

the particle between A and A' is the scalar quotient $\Delta s/\Delta t$. Clearly, the magnitude of the average velocity and the speed approach one another as the interval Δt decreases and A and A' become closer together.

The *instantaneous velocity* $\mathbf{v}$ of the particle is defined as the limiting value of the average velocity as the time interval approaches zero. Thus,

$$\mathbf{v} = \lim_{\Delta t \to 0} \frac{\Delta \mathbf{r}}{\Delta t}$$

We observe that the direction of $\Delta \mathbf{r}$ approaches that of the tangent to the path as Δt approaches zero and, thus, the velocity $\mathbf{v}$ is always a vector tangent to the path.

We now extend the basic definition of the derivative of a scalar quantity to include a vector quantity and write

$$\boxed{\mathbf{v} = \frac{d\mathbf{r}}{dt} = \dot{\mathbf{r}}} \qquad \textbf{(2/4)}$$

The derivative of a vector is itself a vector having both a magnitude and a direction. The magnitude of $\mathbf{v}$ is called the *speed* and is the scalar

$$v = |\mathbf{v}| = \frac{ds}{dt} = \dot{s}$$

At this point we make a careful distinction between the *magnitude of the derivative* and the *derivative of the magnitude*. The magnitude of the derivative can be written in any one of the several ways $|d\mathbf{r}/dt| = |\dot{\mathbf{r}}| = \dot{s} = |\mathbf{v}| = v$ and represents the magnitude of the velocity, or the speed, of the particle. On the other hand, the derivative of the magnitude is written $d|\mathbf{r}|/dt = dr/dt = \dot{r}$, and represents the rate at which the length of the position vector $\mathbf{r}$ is changing. Thus, these two derivatives have two entirely different meanings, and we must be extremely careful to distinguish between them in our thinking and in our notation. For this and other reasons, you are urged to adopt a consistent notation for handwritten work for all vector quantities to distinguish them from scalar quantities. For simplicity the underline $\underline{v}$ is recommended. Other handwritten symbols such as $\vec{v}$, $\underset{\sim}{v}$, and $\hat{v}$ are sometimes used.

With the concept of velocity as a vector established, we return to Fig. 2/5 and denote the velocity of the particle at A by the tangent vector $\mathbf{v}$ and the velocity at A' by the tangent $\mathbf{v}'$. Clearly, there is a vector change in the velocity during the time Δt. The velocity $\mathbf{v}$ at A plus (vectorially) the change $\Delta\mathbf{v}$ must equal the velocity at A', so we can write $\mathbf{v}' - \mathbf{v} = \Delta\mathbf{v}$. Inspection of the vector diagram shows that $\Delta\mathbf{v}$ depends both on the change in magnitude (length) of $\mathbf{v}$ and on the change in direction of $\mathbf{v}$. These two changes are fundamental characteristics of the derivative of a vector.

Acceleration

The *average acceleration* of the particle between A and A' is defined as $\Delta\mathbf{v}/\Delta t$, which is a vector whose direction is that of $\Delta\mathbf{v}$. The magnitude of this average acceleration is the magnitude of $\Delta\mathbf{v}$ divided by Δt.

The *instantaneous acceleration* $\mathbf{a}$ of the particle is defined as the limiting value of the average acceleration as the time interval approaches zero. Thus,

$$\mathbf{a} = \lim_{\Delta t \to 0} \frac{\Delta \mathbf{v}}{\Delta t}$$

By definition of the derivative, then, we write

$$\boxed{\mathbf{a} = \frac{d\mathbf{v}}{dt} = \dot{\mathbf{v}}} \qquad \textbf{(2/5)}$$

As the interval Δt becomes smaller and approaches zero, the direction of the change $\Delta\mathbf{v}$ approaches that of the differential change $d\mathbf{v}$ and, thus, of $\mathbf{a}$. The acceleration $\mathbf{a}$, then, includes the effects of both the change in magnitude of $\mathbf{v}$ and the change of direction of $\mathbf{v}$. It is apparent, in general, that the direction of the acceleration of a particle in curvilinear motion is neither tangent to the path nor normal to the path. We do observe, however, that the acceleration component which is normal to the path points toward the center of curvature of the path.

Visualization of Motion

A further approach to the visualization of acceleration is shown in Fig. 2/6, where the position vectors to three arbitrary positions on the path of the particle are shown for illustrative purpose. There is a velocity vector tangent to the path corresponding to each position vector, and the relation is $\mathbf{v} = \dot{\mathbf{r}}$. If these velocity vectors are now plotted from some arbitrary point C, a curve, called the *hodograph*, is formed. The derivatives of these velocity vectors will be the acceleration vectors $\mathbf{a} = \dot{\mathbf{v}}$ which are tangent to the hodograph. We see that the acceleration has the same relation to the velocity as the velocity has to the position vector.

The geometric portrayal of the derivatives of the position vector $\mathbf{r}$ and velocity vector $\mathbf{v}$ in Fig. 2/5 can be used to describe the derivative of any vector quantity with respect to t or with respect to any other scalar variable. Now that we have used the definitions of velocity and acceleration to introduce the concept of the derivative of a vector, it is important to establish the rules for differentiating vector quantities.

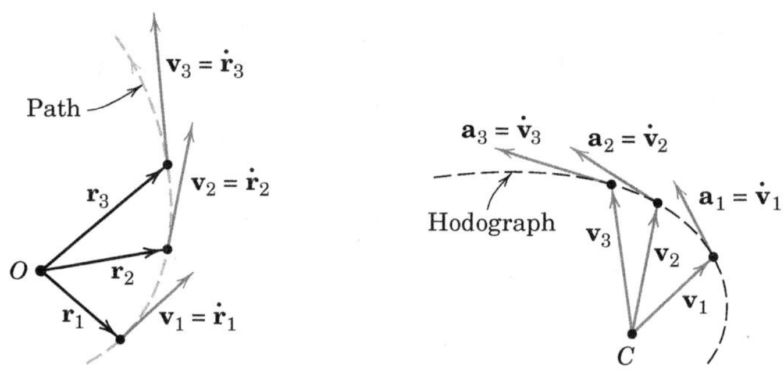

Figure 2/6

These rules are the same as for the differentiation of scalar quantities, except for the case of the cross product where the order of the terms must be preserved. These rules are covered in Art. C/7 of Appendix C and should be reviewed at this point.

Three different coordinate systems are commonly used for describing the vector relationships for curvilinear motion of a particle in a plane: rectangular coordinates, normal and tangential coordinates, and polar coordinates. An important lesson to be learned from the study of these coordinate systems is the proper choice of a reference system for a given problem. This choice is usually revealed by the manner in which the motion is generated or by the form in which the data are specified. Each of the three coordinate systems will now be developed and illustrated.

2/4 Rectangular Coordinates (x-y)

This system of coordinates is particularly useful for describing motions where the x- and y-components of acceleration are independently generated or determined. The resulting curvilinear motion is then obtained by a vector combination of the x- and y-components of the position vector, the velocity, and the acceleration.

Vector Representation

The particle path of Fig. 2/5 is shown again in Fig. 2/7 along with x- and y-axes. The position vector $\mathbf{r}$, the velocity $\mathbf{v}$, and the acceleration $\mathbf{a}$ of the particle as developed in Art. 2/3 are represented in Fig. 2/7 together with their x- and y-components. With the aid of the unit vectors $\mathbf{i}$ and $\mathbf{j}$, we can write the vectors $\mathbf{r}$, $\mathbf{v}$, and $\mathbf{a}$ in terms of their x- and y-components. Thus,

$$\begin{aligned} \mathbf{r} &= x\mathbf{i} + y\mathbf{j} \\ \mathbf{v} = \dot{\mathbf{r}} &= \dot{x}\mathbf{i} + \dot{y}\mathbf{j} \\ \mathbf{a} = \dot{\mathbf{v}} = \ddot{\mathbf{r}} &= \ddot{x}\mathbf{i} + \ddot{y}\mathbf{j} \end{aligned} \qquad \textbf{(2/6)}$$

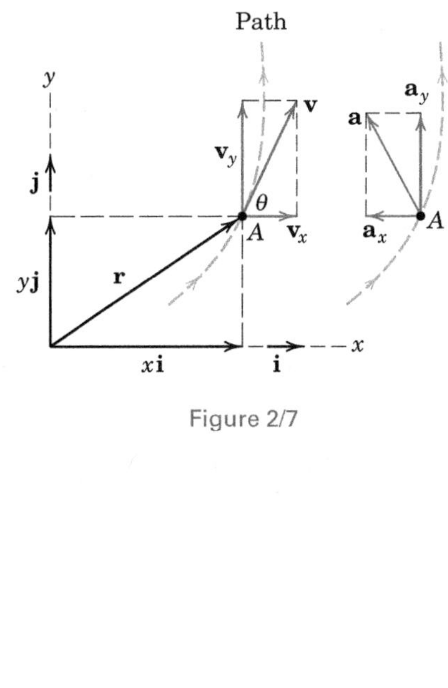

Figure 2/7

As we differentiate with respect to time, we observe that the time derivatives of the unit vectors are zero because their magnitudes and directions remain constant. The scalar values of the components of $\mathbf{v}$ and $\mathbf{a}$ are merely $v_x = \dot{x}$, $v_y = \dot{y}$ and $a_x = \dot{v}_x = \ddot{x}$, $a_y = \dot{v}_y = \ddot{y}$. (As drawn in Fig. 2/7, a_x is in the negative x-direction, so that $\ddot{x}$ would be a negative number.)

As observed previously, the direction of the velocity is always tangent to the path, and from the figure it is clear that

$$v^2 = {v_x}^2 + {v_y}^2 \qquad v = \sqrt{{v_x}^2 + {v_y}^2} \qquad \tan\theta = \frac{v_y}{v_x}$$

$$a^2 = {a_x}^2 + {a_y}^2 \qquad a = \sqrt{{a_x}^2 + {a_y}^2}$$

If the angle θ is measured counterclockwise from the x-axis to $\mathbf{v}$ for the configuration of axes shown, then we can also observe that $dy/dx = \tan\theta = v_y/v_x$.

If the coordinates x and y are known independently as functions of time, $x = f_1(t)$ and $y = f_2(t)$, then for any value of the time we can combine them to obtain $\mathbf{r}$. Similarly, we combine their first derivatives $\dot{x}$ and $\dot{y}$ to obtain $\mathbf{v}$ and their second derivatives $\ddot{x}$ and $\ddot{y}$ to obtain $\mathbf{a}$. On the other hand, if the acceleration components a_x and a_y are given as functions of the time, we can integrate each one separately with respect to time, once to obtain v_x and v_y and again to obtain $x = f_1(t)$ and $y = f_2(t)$. Elimination of the time t between these last two parametric equations gives the equation of the curved path $y = f(x)$.

From the foregoing discussion we can see that the rectangular-coordinate representation of curvilinear motion is merely the superposition of the components of two simultaneous rectilinear motions in the x- and y-directions. Therefore, everything covered in Art. 2/2 on rectilinear motion can be applied separately to the x-motion and to the y-motion.

Projectile Motion

An important application of two-dimensional kinematic theory is the problem of projectile motion. For a first treatment of the subject, we neglect aerodynamic drag and the curvature and rotation of the earth, and we assume that the altitude change is small enough so that the acceleration due to gravity can be considered constant. With these assumptions, rectangular coordinates are useful for the trajectory analysis.

For the axes shown in Fig. 2/8, the acceleration components are

$$a_x = 0 \qquad a_y = -g$$

Integration of these accelerations follows the results obtained previously in Art. 2/2*a* for constant acceleration and yields

$$v_x = (v_x)_0 \qquad v_y = (v_y)_0 - gt$$
$$x = x_0 + (v_x)_0 t \qquad y = y_0 + (v_y)_0 t - \tfrac{1}{2}gt^2$$
$$v_y^2 = (v_y)_0^2 - 2g(y - y_0)$$

In all these expressions, the subscript zero denotes initial conditions, frequently taken as those at launch where, for the case illustrated,

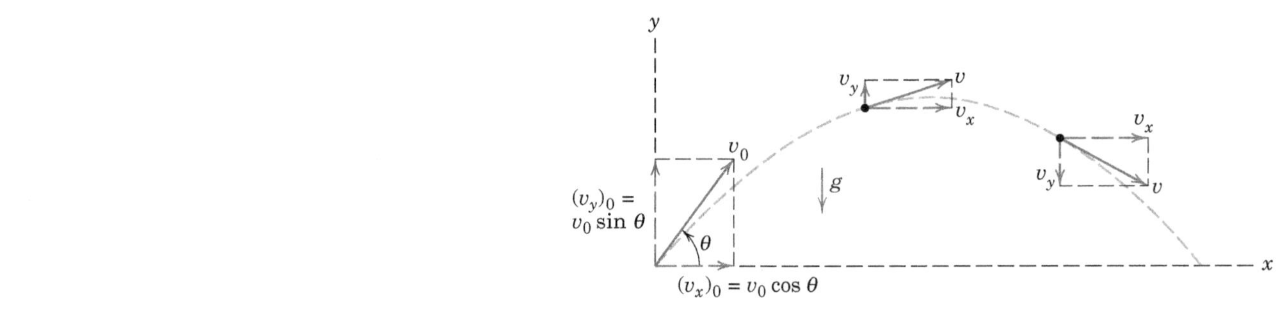

Figure 2/8

$x_0 = y_0 = 0$. Note that the quantity g is taken to be positive throughout this text.

We can see that the x- and y-motions are independent for the simple projectile conditions under consideration. Elimination of the time t between the x- and y-displacement equations shows the path to be parabolic (see Sample Problem 2/6). If we were to introduce a drag force which depends on the speed squared (for example), then the x- and y-motions would be coupled (interdependent), and the trajectory would be nonparabolic.

When the projectile motion involves large velocities and high altitudes, to obtain accurate results we must account for the shape of the projectile, the variation of g with altitude, the variation of the air density with altitude, and the rotation of the earth. These factors introduce considerable complexity into the motion equations, and numerical integration of the acceleration equations is usually necessary.

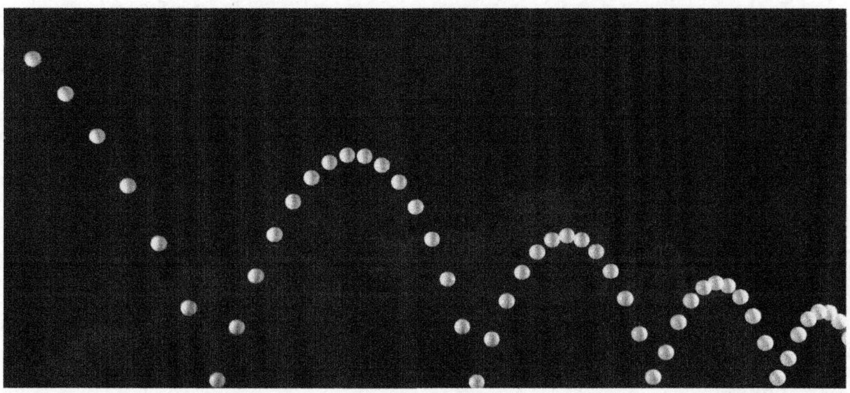

Andrew Davidhazy

This stroboscopic photograph of a bouncing ping-pong ball suggests not only the parabolic nature of the path, but also the fact that the speed is lower near the apex.

Sample Problem 2/5

The curvilinear motion of a particle is defined by $v_x = 50 - 16t$ and $y = 100 - 4t^2$, where v_x is in meters per second, y is in meters, and t is in seconds. It is also known that $x = 0$ when $t = 0$. Plot the path of the particle and determine its velocity and acceleration when the position $y = 0$ is reached.

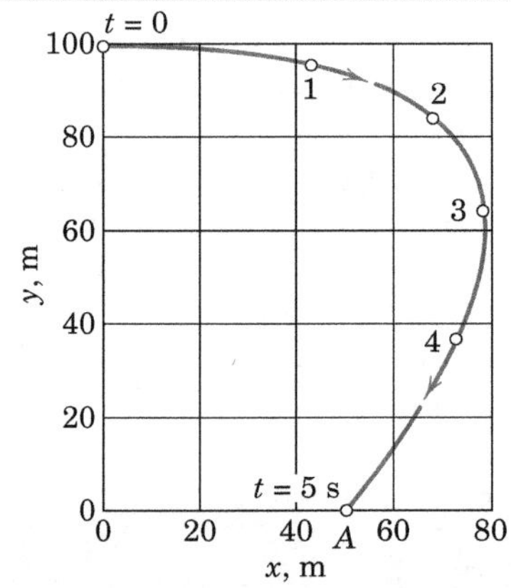

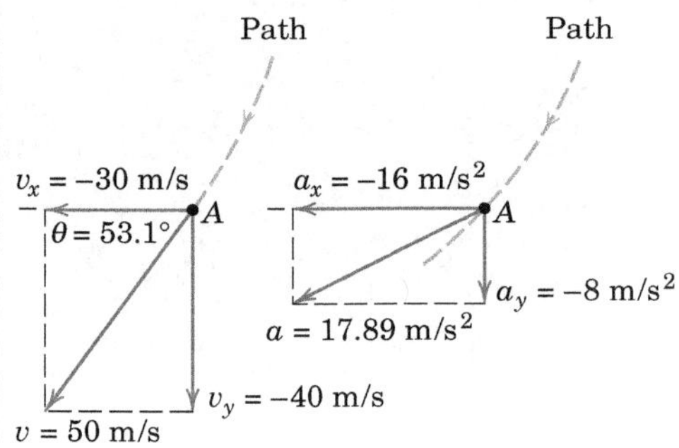

Solution. The x-coordinate is obtained by integrating the expression for v_x, and the x-component of the acceleration is obtained by differentiating v_x. Thus,

$$\left[\int dx = \int v_x\,dt\right] \qquad \int_0^x dx = \int_0^t (50 - 16t)\,dt \qquad x = 50t - 8t^2 \text{ m}$$

$$[a_x = \dot{v}_x] \qquad a_x = \frac{d}{dt}(50 - 16t) \qquad a_x = -16 \text{ m/s}^2$$

The y-components of velocity and acceleration are

$$[v_y = \dot{y}] \qquad v_y = \frac{d}{dt}(100 - 4t^2) \qquad v_y = -8t \text{ m/s}$$

$$[a_y = \dot{v}_y] \qquad a_y = \frac{d}{dt}(-8t) \qquad a_y = -8 \text{ m/s}^2$$

We now calculate corresponding values of x and y for various values of t and plot x against y to obtain the path as shown.

When $y = 0$, $0 = 100 - 4t^2$, so $t = 5$ s. For this value of the time, we have

$$v_x = 50 - 16(5) = -30 \text{ m/s}$$

$$v_y = -8(5) = -40 \text{ m/s}$$

$$v = \sqrt{(-30)^2 + (-40)^2} = 50 \text{ m/s}$$

$$a = \sqrt{(-16)^2 + (-8)^2} = 17.89 \text{ m/s}^2$$

The velocity and acceleration components and their resultants are shown on the separate diagrams for point A, where $y = 0$. Thus, for this condition we may write

$$\mathbf{v} = -30\mathbf{i} - 40\mathbf{j} \text{ m/s} \qquad \textit{Ans.}$$

$$\mathbf{a} = -16\mathbf{i} - 8\mathbf{j} \text{ m/s}^2 \qquad \textit{Ans.}$$

Helpful Hint

We observe that the velocity vector lies along the tangent to the path as it should, but that the acceleration vector is not tangent to the path. Note especially that the acceleration vector has a component that points toward the inside of the curved path. We concluded from our diagram in Fig. 2/5 that it is impossible for the acceleration to have a component that points toward the outside of the curve.

Sample Problem 2/6

A team of engineering students designs a medium-size catapult which launches 8-lb steel spheres. The launch speed is $v_0 = 80$ ft/sec, the launch angle is $\theta = 35°$ above the horizontal, and the launch position is 6 ft above ground level. The students use an athletic field with an adjoining slope topped by an 8-ft fence as shown. Determine:

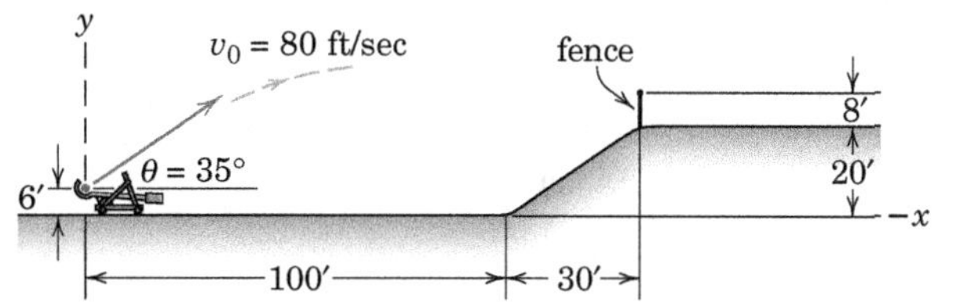

(a) the time duration t_f of the flight

(b) the x-y coordinates of the point of first impact

(c) the maximum height h above the horizontal field attained by the ball

(d) the velocity (expressed as a vector) with which the projectile strikes the ground (or the fence)

Repeat part (*b*) for a launch speed of $v_0 = 75$ ft/sec.

Solution. We make the assumptions of constant gravitational acceleration
① and no aerodynamic drag. With the latter assumption, the 8-lb weight of the projectile is irrelevant. Using the given x-y coordinate system, we begin by checking the y-displacement at the horizontal position of the fence.

$$[x = x_0 + (v_x)_0 t] \qquad 100 + 30 = 0 + (80 \cos 35°)t \qquad t = 1.984 \text{ sec}$$

$$[y = y_0 + (v_y)_0 t - \tfrac{1}{2}gt^2] \quad y = 6 + 80 \sin 35°(1.984) - \tfrac{1}{2}(32.2)(1.984)^2 = 33.7 \text{ ft}$$

(a) Because the y-coordinate of the top of the fence is 20 + 8 = 28 feet, the projectile clears the fence. We now find the flight time by setting $y = 20$ ft:

$$[y = y_0 + (v_y)_0 t - \tfrac{1}{2}gt^2] \quad 20 = 6 + 80 \sin 35°(t_f) - \tfrac{1}{2}(32.2)t_f^2 \quad t_f = 2.50 \text{ sec} \quad \textit{Ans.}$$

$$[x = x_0 + (v_x)_0 t] \quad x = 0 + 80 \cos 35°(2.50) = 164.0 \text{ ft}$$

(b) Thus the point of first impact is $(x, y) = (164.0, 20)$ ft. *Ans.*

(c) For the maximum height:

② $$[v_y{}^2 = (v_y)_0{}^2 - 2g(y - y_0)] \quad 0^2 = (80 \sin 35°)^2 - 2(32.2)(h - 6) \quad h = 38.7 \text{ ft} \quad \textit{Ans.}$$

(d) For the impact velocity:

$$[v_x = (v_x)_0] \qquad v_x = 80 \cos 35° = 65.5 \text{ ft/sec}$$

$$[v_y = (v_y)_0 - gt] \qquad v_y = 80 \sin 35° - 32.2(2.50) = -34.7 \text{ ft/sec}$$

So the impact velocity is $\mathbf{v} = 65.5\mathbf{i} - 34.7\mathbf{j}$ ft/sec. *Ans.*

If $v_0 = 75$ ft/sec, the time from launch to the fence is found by

$$[x = x_0 + (v_x)_0 t] \qquad 100 + 30 = (75 \cos 35°)t \qquad t = 2.12 \text{ sec}$$

and the corresponding value of y is

$$[y = y_0 + (v_y)_0 t - \tfrac{1}{2}gt^2] \quad y = 6 + 80 \sin 35°(2.12) - \tfrac{1}{2}(32.2)(2.12)^2 = 24.9 \text{ ft}$$

For this launch speed, we see that the projectile hits the fence, and the point of impact is

$$(x, y) = (130, 24.9) \text{ ft} \qquad \textit{Ans.}$$

For lower launch speeds, the projectile could land on the slope or even on the level portion of the athletic field.

Helpful Hints

① Neglecting aerodynamic drag is a poor assumption for projectiles with relatively high initial velocities, large sizes, and low weights. In a vacuum, a baseball thrown with an initial speed of 100 ft/sec at 45° above the horizontal will travel about 311 feet over a horizontal range. In sea-level air, the baseball range is about 200 ft, while a typical beachball under the same conditions will travel about 10 ft.

② As an alternative approach, we could find the time at apex where $v_y = 0$, then use that time in the y-displacement equation. Verify that the trajectory apex occurs over the 100-ft horizontal portion of the athletic field.

PROBLEMS

(In the following problems where motion as a projectile in air is involved, neglect air resistance unless otherwise stated and use $g = 9.81$ m/s^2 or $g = 32.2$ ft/sec^2.)

Introductory Problems

2/59 At time $t = 10$ s, the velocity of a particle moving in the x-y plane is $\mathbf{v} = 0.1\mathbf{i} + 2\mathbf{j}$ m/s. By time $t = 10.1$ s, its velocity has become $-0.1\mathbf{i} + 1.8\mathbf{j}$ m/s. Determine the magnitude a_{av} of its average acceleration during this interval and the angle θ made by the average acceleration with the positive x-axis.

2/60 At time $t = 0$, the position vector of a particle moving in the x-y plane is $\mathbf{r} = 5\mathbf{i}$ m. By time $t = 0.02$ s, its position vector has become $5.1\mathbf{i} + 0.4\mathbf{j}$ m. Determine the magnitude v_{av} of its average velocity during this interval and the angle θ made by the average velocity with the positive x-axis.

2/61 At time $t = 0$, a particle is at rest in the x-y plane at the coordinates $(x_0, y_0) = (6, 0)$ in. If the particle is then subjected to the acceleration components $a_x = 0.5 - 0.35t$ in./sec^2 and $a_y = 0.15t - 0.02t^2$ in./sec^2, determine the coordinates of the particle position when $t = 6$ sec. Plot the path of the particle during this time period.

2/62 The rectangular coordinates of a particle which moves with curvilinear motion are given by $x = 10.25t + 1.75t^2 - 0.45t^3$ and $y = 6.32 + 14.65t - 2.48t^2$, where x and y are in millimeters and the time t is in seconds, beginning from $t = 0$. Determine the velocity $\mathbf{v}$ and acceleration $\mathbf{a}$ of the particle when $t = 5$ s. Also, determine the time when the velocity of the particle makes an angle of 45° with the x-axis.

2/63 For a certain interval of motion the pin A is forced to move in the fixed parabolic slot by the horizontal slotted arm which is elevated in the y-direction at the constant rate of 3 in./sec. All measurements are in inches and seconds. Calculate the velocity v and acceleration a of pin A when $x = 6$ in.

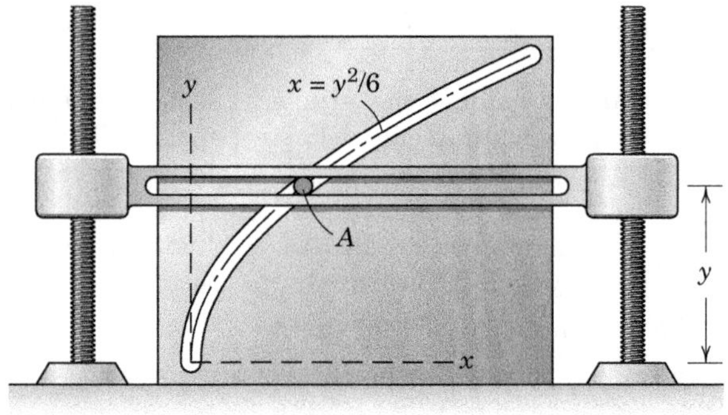

Problem 2/63

2/64 With what minimum horizontal velocity u can a boy throw a rock at A and have it just clear the obstruction at B?

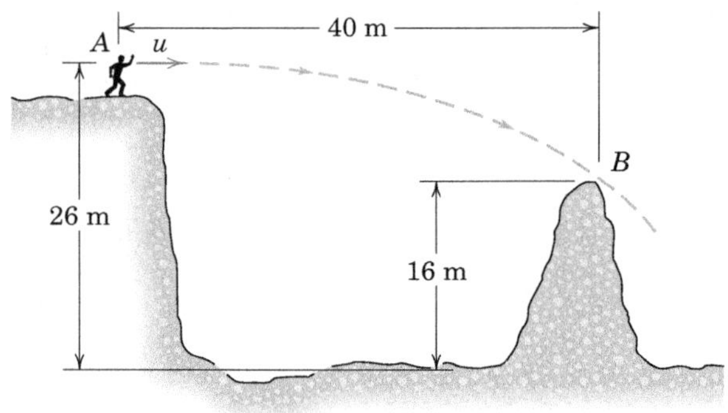

Problem 2/64

2/65 Prove the well-known result that, for a given launch speed v_0, the launch angle $\theta = 45°$ yields the maximum horizontal range R. Determine the maximum range. (Note that this result does not hold when aerodynamic drag is included in the analysis.)

2/66 A placekicker is attempting to make a 64-yard field goal. If the launch angle of the football is 40°, what is the minimum initial speed u which will allow the kicker to succeed?

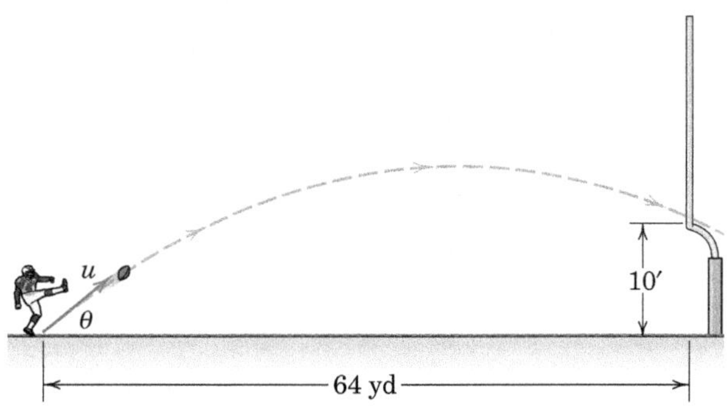

Not to scale

Problem 2/66

2/67 In a basketball game, the point guard A intends to throw a pass to the shooting guard B, who is breaking toward the basket at a constant speed of 12 ft/sec. If the shooting guard is to catch the ball at a height of 7 ft at C while in full stride to execute a layup, determine the speed v_0 and launch angle θ with which the point guard should throw the ball.

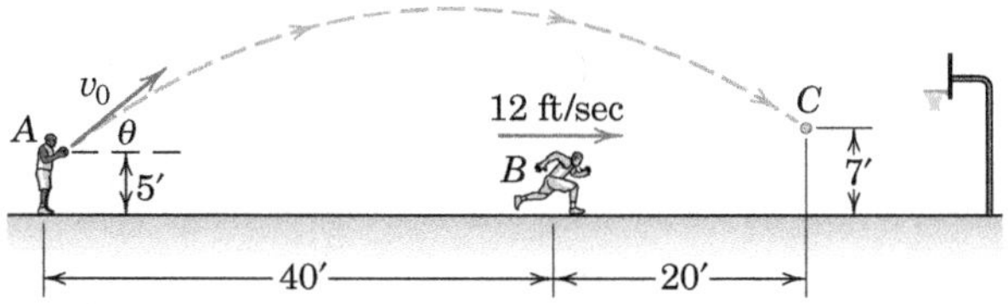

Problem 2/67

2/68 A fireworks show is choreographed to have two shells cross paths at a height of 160 feet and explode at an apex of 200 feet under normal weather conditions. If the shells have a launch angle $\theta = 60°$ above the horizontal, determine the common launch speed v_0 for the shells, the separation distance d between the launch points A and B, and the time from launch at which the shells explode.

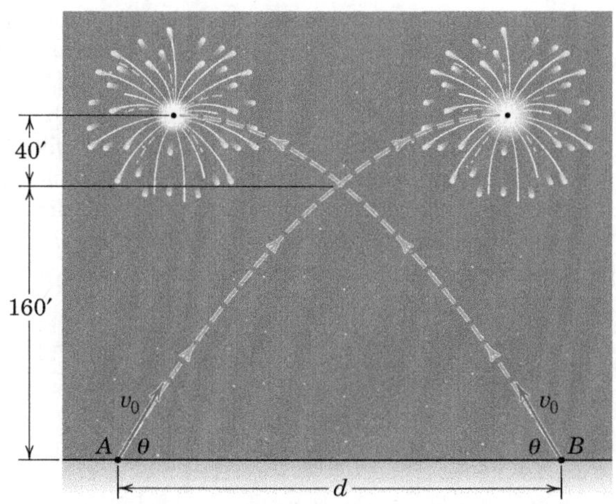

Problem 2/68

2/69 If a strong wind induces a constant rightward acceleration of 16 ft/sec^2 for the fireworks shells of Prob. 2/68, determine the horizontal shift of the crossing point of the shells. Refer to the printed answers for Prob. 2/68 as needed.

2/70 The center of mass G of a high jumper follows the trajectory shown. Determine the component v_0, measured in the vertical plane of the figure, of his takeoff velocity and angle θ if the apex of the trajectory just clears the bar at A. (In general, must the mass center G of the jumper clear the bar during a successful jump?)

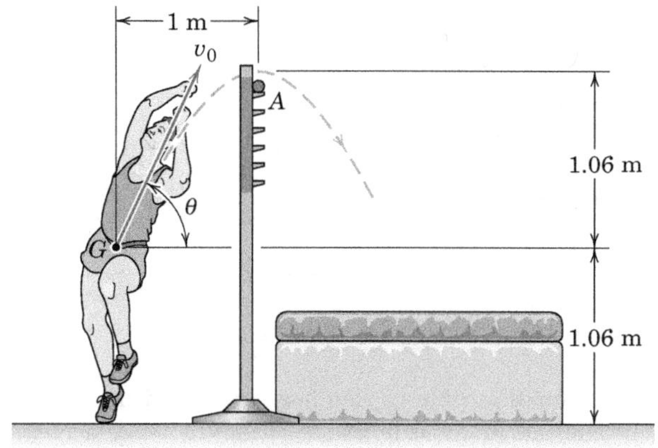

Problem 2/70

Representative Problems

2/71 Electrons are emitted at A with a velocity u at the angle θ into the space between two charged plates. The electric field between the plates is in the direction E and repels the electrons approaching the upper plate. The field produces an acceleration of the electrons in the E-direction of eE/m, where e is the electron charge and m is its mass. Determine the field strength E which will permit the electrons to cross one-half of the gap between the plates. Also find the distance s.

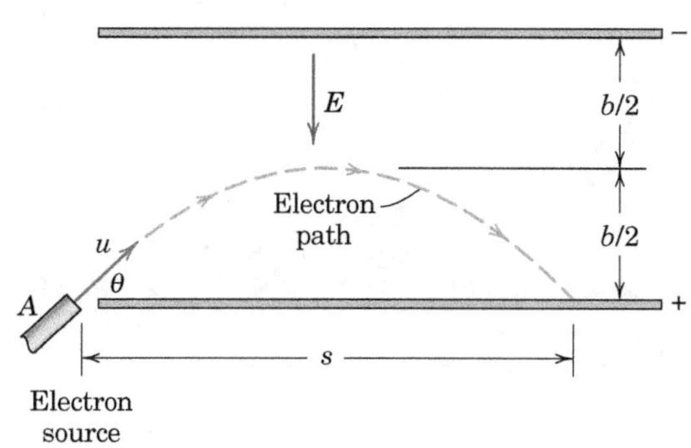

Problem 2/71

2/72 A boy tosses a ball onto the roof of a house. For the launch conditions shown, determine the slant distance s to the point of impact. Also, determine the angle θ which the velocity of the ball makes with the roof at the moment of impact.

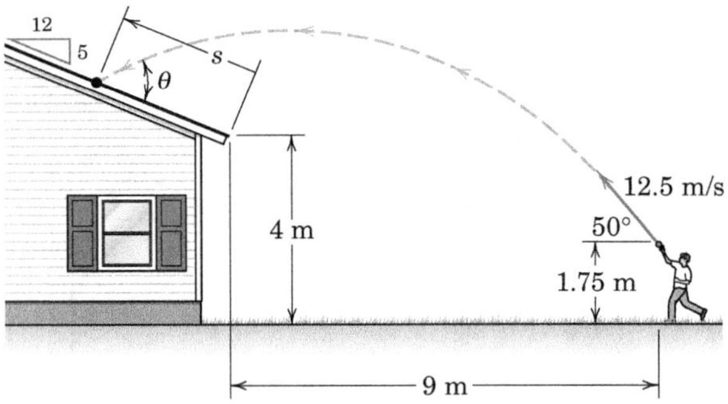

Problem 2/72

2/73 A small airplane flying horizontally with a speed of 180 mi/hr at an altitude of 400 ft above a remote valley drops an emergency medical package at A. The package has a parachute which deploys at B and allows the package to descend vertically at the constant rate of 6 ft/sec. If the drop is designed so that the package is to reach the ground 37 seconds after release at A, determine the horizontal lead L so that the package hits the target. Neglect atmospheric resistance from A to B.

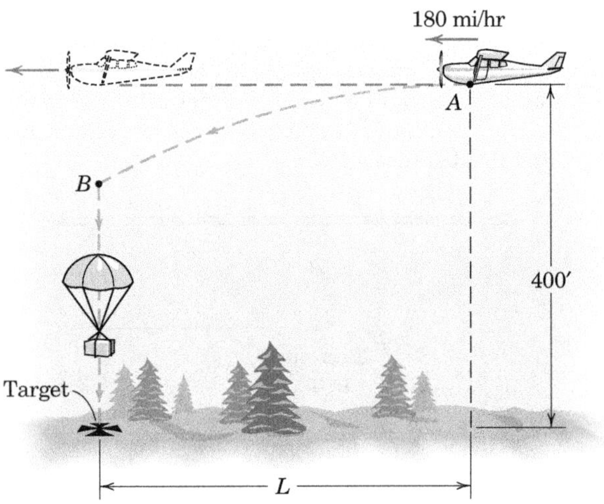

Problem 2/73

2/74 As part of a circus performance, a man is attempting to throw a dart into an apple which is dropped from an overhead platform. Upon release of the apple, the man has a reflex delay of 215 milliseconds before throwing the dart. If the dart is released with a speed $v_0 = 14$ m/s, at what distance d below the platform should the man aim if the dart is to strike the apple before it hits the ground?

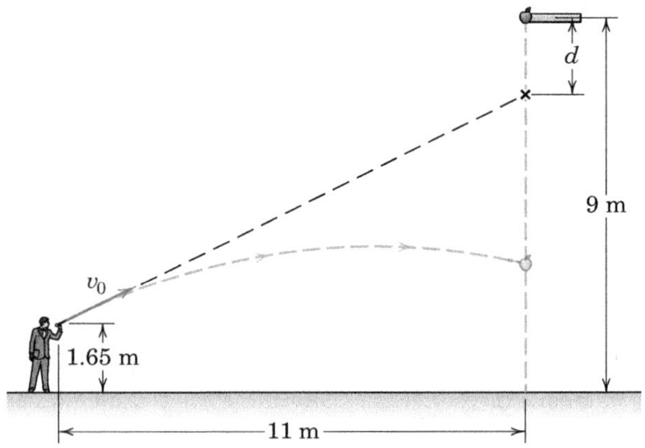

Problem 2/74

2/75 A marksman fires a practice round from A toward a target B. If the target diameter is 160 mm and the target center is at the same altitude as the end of the rifle barrel, determine the range of "shallow" launch angles θ for which the round will strike the target. Neglect aerodynamic drag and assume that the round is directed along the vertical centerline of the target. (*Note:* The word "shallow" indicates a low-flying trajectory for the round.)

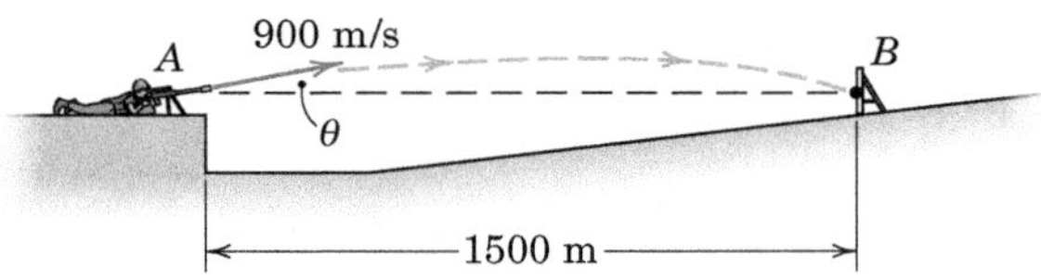

Problem 2/75

2/76 The pilot of an airplane carrying a package of mail to a remote outpost wishes to release the package at the right moment to hit the recovery location A. What angle θ with the horizontal should the pilot's line of sight to the target make at the instant of release? The airplane is flying horizontally at an altitude of 100 m with a velocity of 200 km/h.

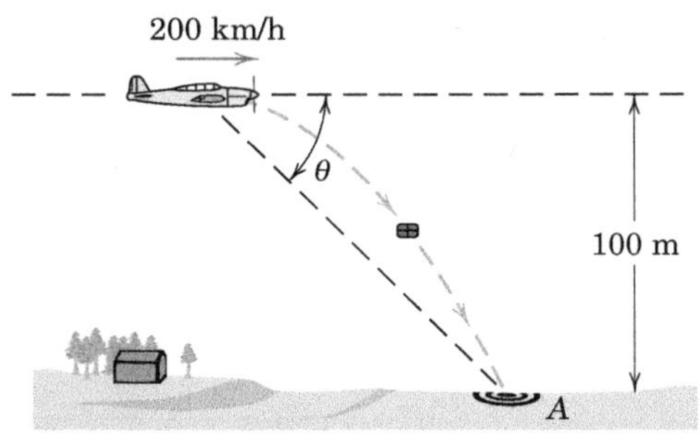

Problem 2/76

2/77 During a baseball practice session, the cutoff man A executes a throw to the third baseman B. If the initial speed of the baseball is $v_0 = 130$ ft/sec, what launch angle θ is best if the ball is to arrive at third base at essentially ground level?

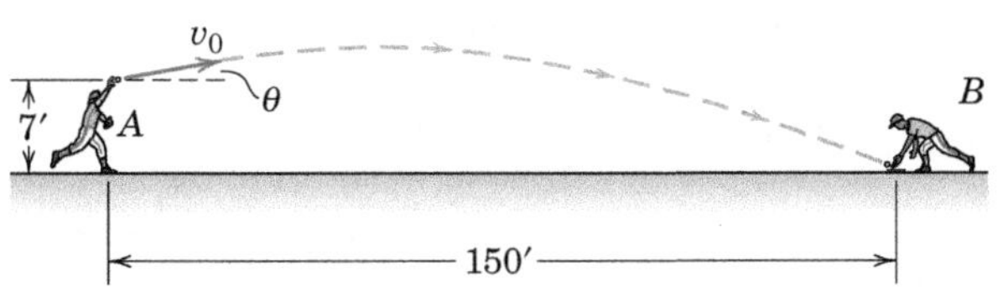

Problem 2/77

2/78 A particle is launched from point A with a horizontal speed u and subsequently passes through a vertical opening of height b as shown. Determine the distance d which will allow the landing zone for the particle to also have a width b. Additionally, determine the range of u which will allow the projectile to pass through the vertical opening for this value of d.

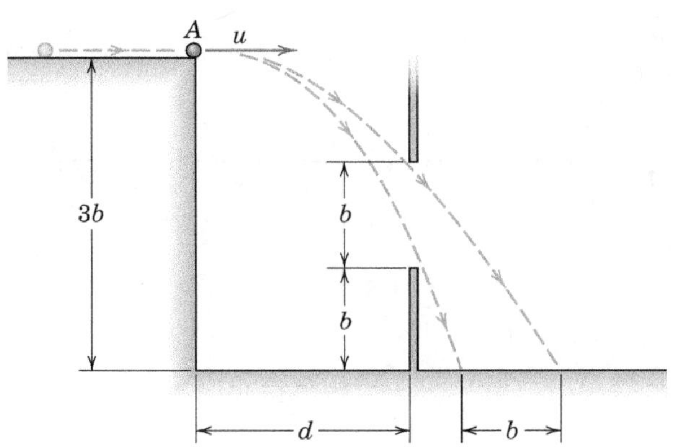

Problem 2/78

2/79 If the tennis player serves the ball horizontally ($\theta = 0$), calculate its velocity v if the center of the ball clears the 0.9-m net by 150 mm. Also find the distance s from the net to the point where the ball hits the court surface. Neglect air resistance and the effect of ball spin.

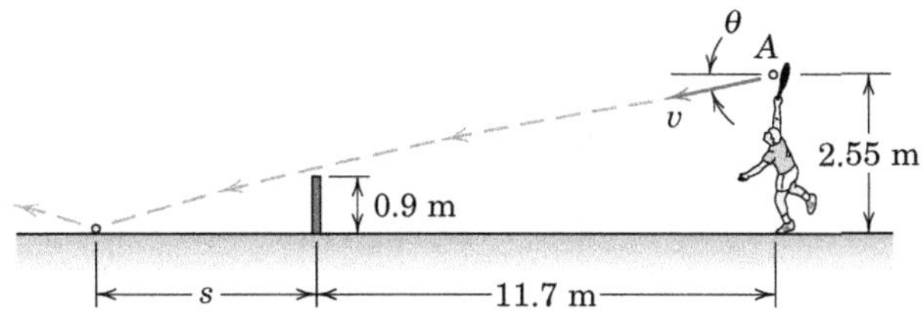

Problem 2/79

2/80 A golfer is attempting to reach the elevated green by hitting his ball under a low-hanging branch in one tree A, but over the top of a second tree B. For $v_0 = 115$ mi/hr and $\theta = 18°$, where does the golf ball land first?

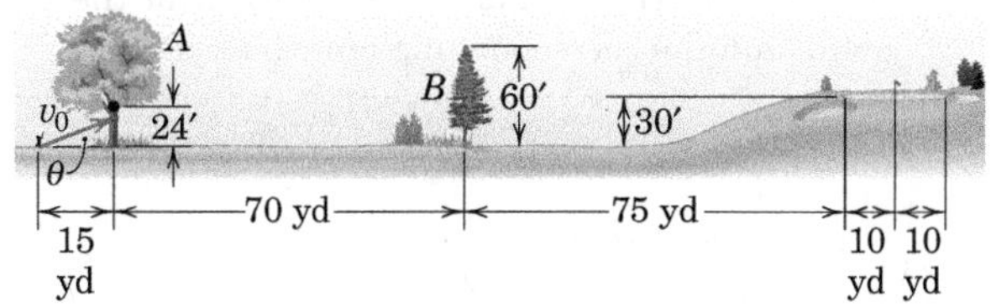

Problem 2/80

2/81 If the launch speed of the golf ball of the previous problem remains $v_0 = 115$ mi/hr, what launch angle θ will put the first impact point of the ball closest to the pin? How far from the pin is this impact point?

2/82 An outfielder experiments with two different trajectories for throwing to home plate from the position shown: (*a*) $v_0 = 42$ m/s with $\theta = 8°$ and (*b*) $v_0 = 36$ m/s with $\theta = 12°$. For each set of initial conditions, determine the time t required for the baseball to reach home plate and the altitude h as the ball crosses the plate.

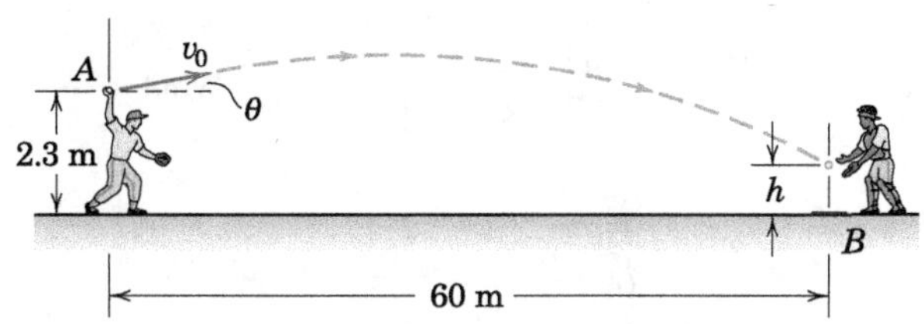

Problem 2/82

2/83 A ski jumper has the takeoff conditions shown. Determine the inclined distance d from the takeoff point A to the location where the skier first touches down in the landing zone, and the total time t_f during which the skier is in the air. For simplicity, assume that the landing zone BC is straight.

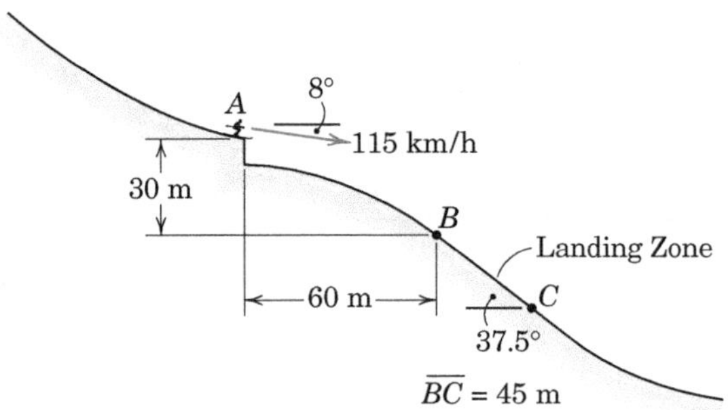

Problem 2/83

2/84 A projectile is launched with a speed $v_0 = 25$ m/s from the floor of a 5-m-high tunnel as shown. Determine the maximum horizontal range R of the projectile and the corresponding launch angle θ.

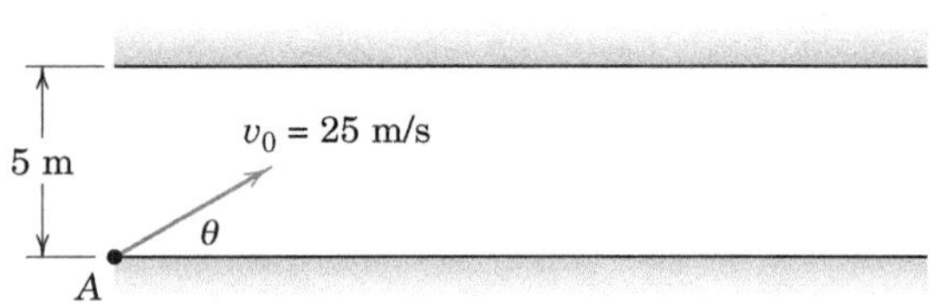

Problem 2/84

2/85 A boy throws a ball upward with a speed $v_0 = 12$ m/s. The wind imparts a horizontal acceleration of 0.4 m/s^2 to the left. At what angle θ must the ball be thrown so that it returns to the point of release? Assume that the wind does not affect the vertical motion.

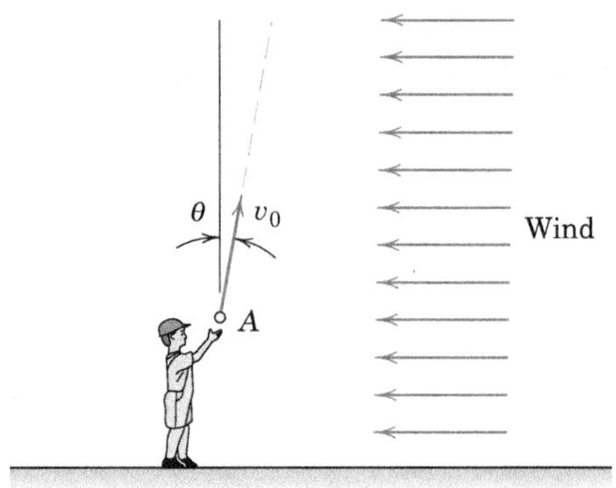

Problem 2/85

2/86 A projectile is launched from point O with the initial conditions shown. Determine the impact coordinates for the projectile if (*a*) $v_0 = 60$ ft/sec and $\theta = 40°$ and (*b*) $v_0 = 85$ ft/sec and $\theta = 15°$.

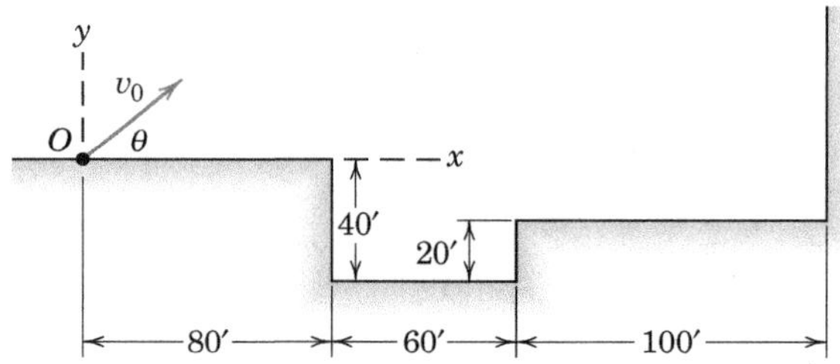

Problem 2/86

2/87 A projectile is launched from point A with the initial conditions shown in the figure. Determine the slant distance s which locates the point B of impact. Calculate the time of flight t.

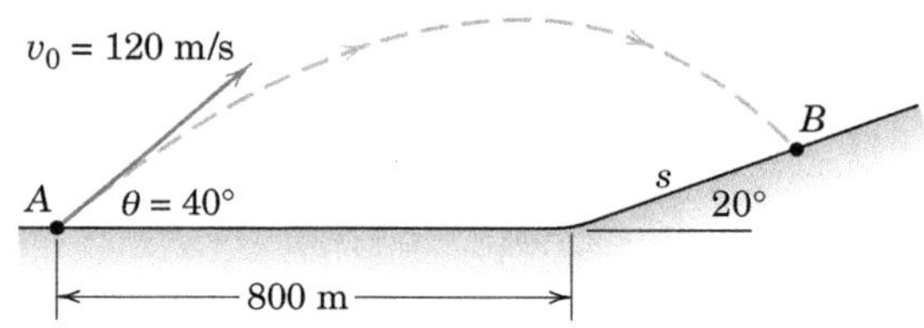

Problem 2/87

2/88 A team of engineering students is designing a catapult to launch a small ball at A so that it lands in the box. If it is known that the initial velocity vector makes a 30° angle with the horizontal, determine the range of launch speeds v_0 for which the ball will land inside the box.

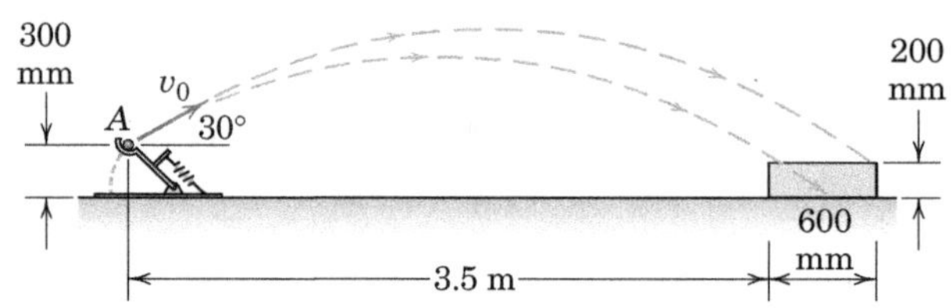

Problem 2/88

2/89 A snow blower travels forward at a constant speed $v_s = 1.4$ ft/sec along the straight and level path shown. The snow is ejected with a speed $v_r = 30$ ft/sec relative to the machine at the 40° angle indicated. Determine the distance d which locates the snow-blower position from which an ejected snow particle ultimately strikes the slender pole P. At what height h above the ground does the snow strike the pole? Points O and Q are at ground level.

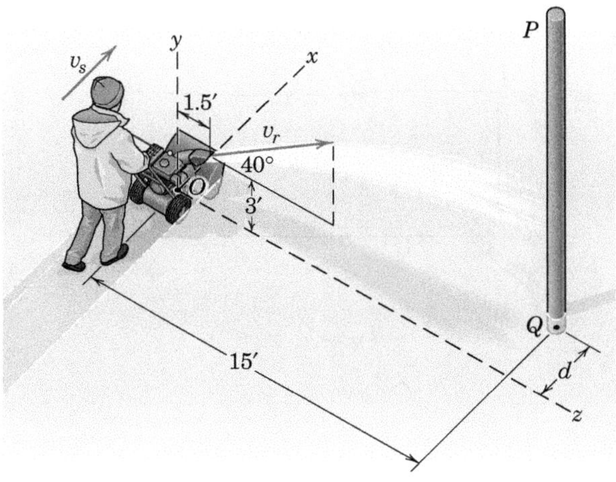

Problem 2/89

2/90 Determine the location h of the spot toward which the pitcher must throw if the ball is to hit the catcher's mitt. The ball is released with a speed of 40 m/s.

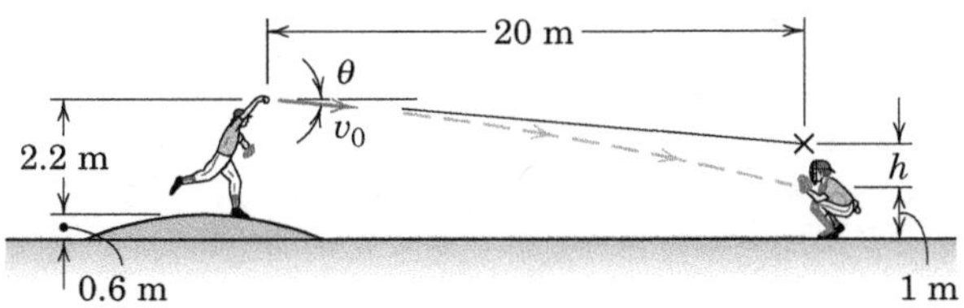

Problem 2/90

2/91 A projectile is launched from point A with $v_0 = 30$ m/s and $\theta = 35°$. Determine the x- and y-coordinates of the point of impact.

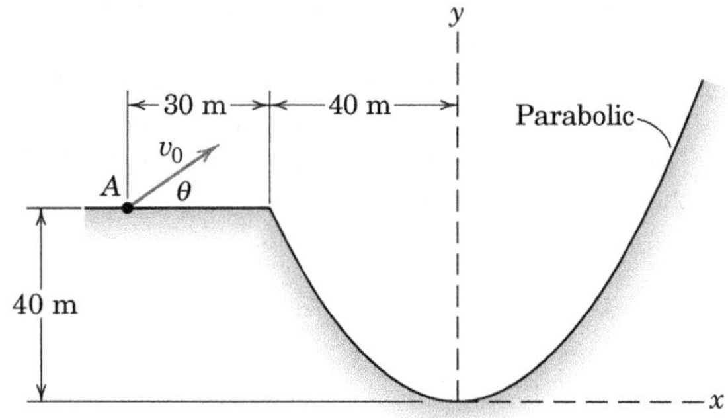

Problem 2/91

2/92 A projectile is fired with a velocity u at right angles to the slope, which is inclined at an angle θ with the horizontal. Derive an expression for the distance R to the point of impact.

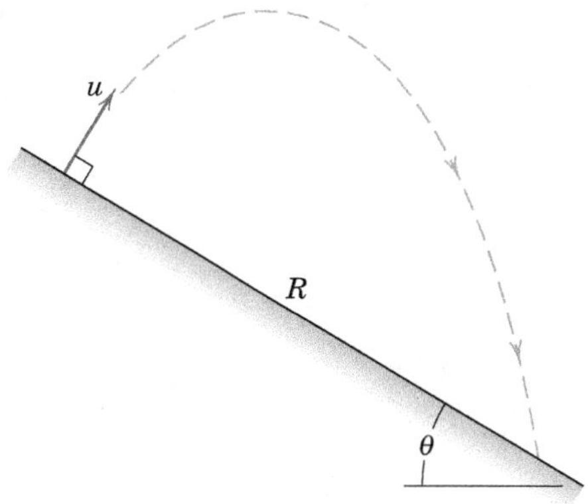

Problem 2/92

2/93 A projectile is launched from point A with an initial speed $v_0 = 100$ ft/sec. Determine the minimum value of the launch angle α for which the projectile will land at point B.

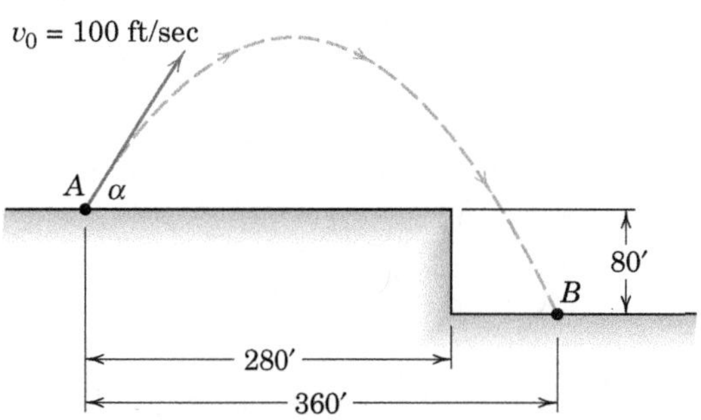

Problem 2/93

2/94 A projectile is launched from point A and lands on the same level at D. Its maximum altitude is h. Determine and plot the fraction f_2 of the total flight time that the projectile is above the level f_1h, where f_1 is a fraction which can vary from zero to 1. State the value of f_2 for $f_1 = \frac{3}{4}$.

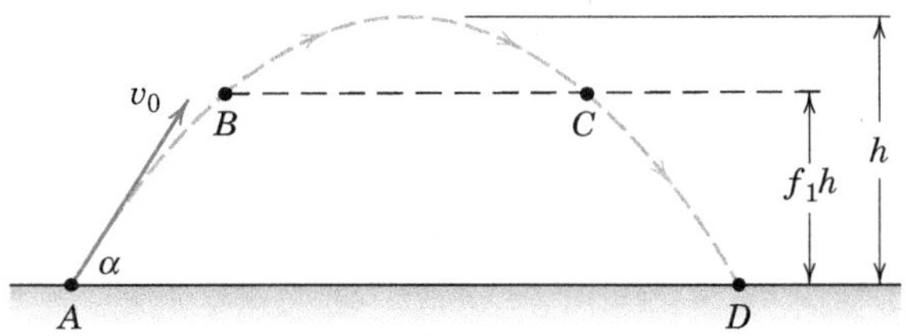

Problem 2/94

▶2/95 A projectile is launched with speed v_0 from point A. Determine the launch angle θ which results in the maximum range R up the incline of angle α (where $0 \le \alpha \le 90°$). Evaluate your results for $\alpha = 0$, 30°, and 45°.

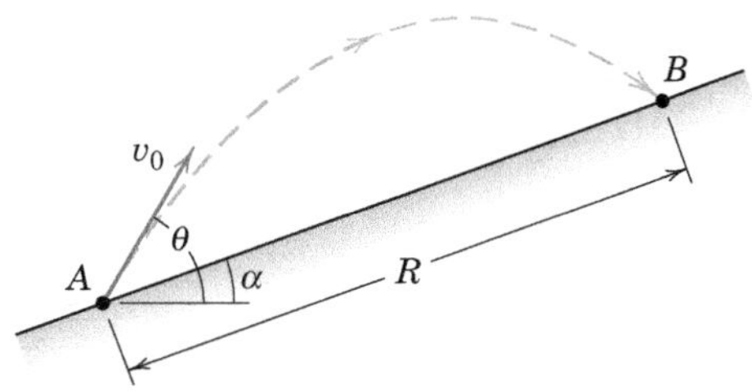

Problem 2/95

▶2/96 A projectile is ejected into an experimental fluid at time $t = 0$. The initial speed is v_0 and the angle to the horizontal is θ. The drag on the projectile results in an acceleration term $\mathbf{a}_D = -k\mathbf{v}$, where k is a constant and $\mathbf{v}$ is the velocity of the projectile. Determine the x- and y-components of both the velocity and displacement as functions of time. What is the terminal velocity? Include the effects of gravitational acceleration.

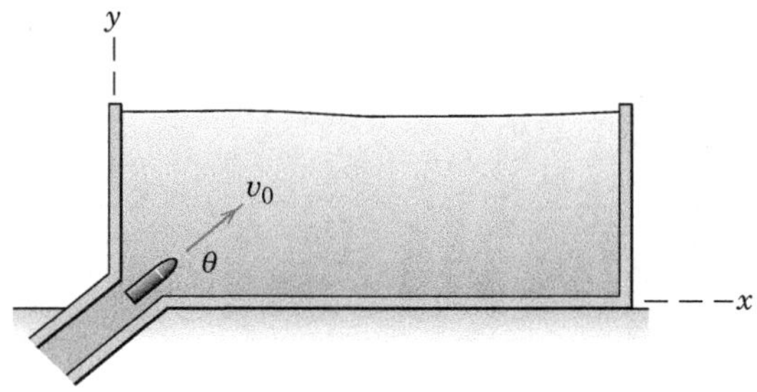

Problem 2/96

2/5 Normal and Tangential Coordinates (n-t)

As we mentioned in Art. 2/1, one of the common descriptions of curvilinear motion uses path variables, which are measurements made along the tangent t and normal n to the path of the particle. These coordinates provide a very natural description for curvilinear motion and are frequently the most direct and convenient coordinates to use. The n- and t-coordinates are considered to move along the path with the particle, as seen in Fig. 2/9 where the particle advances from A to B to C. The positive direction for n at any position is always taken toward the center of curvature of the path. As seen from Fig. 2/9, the positive n-direction will shift from one side of the curve to the other side if the curvature changes direction.

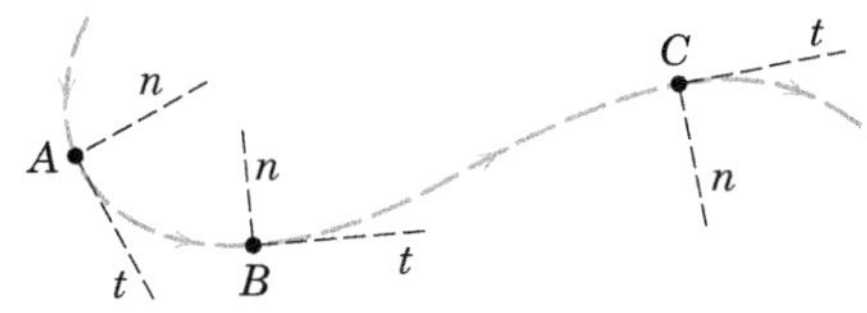

Figure 2/9

Velocity and Acceleration

We now use the coordinates n and t to describe the velocity $\mathbf{v}$ and acceleration $\mathbf{a}$ which were introduced in Art. 2/3 for the curvilinear motion of a particle. For this purpose, we introduce unit vectors $\mathbf{e}_n$ in the n-direction and $\mathbf{e}_t$ in the t-direction, as shown in Fig. 2/10a for the position of the particle at point A on its path. During a differential increment of time dt, the particle moves a differential distance ds along the curve from A to A'. With the radius of curvature of the path at this position designated by ρ, we see that $ds = \rho\, d\beta$, where β is in radians. It is unnecessary to consider the differential change in ρ between A and A' because a higher-order term would be introduced which disappears in the limit. Thus, the magnitude of the velocity can be written $v = ds/dt = \rho\, d\beta/dt$, and we can write the velocity as the vector

$$\mathbf{v} = v\mathbf{e}_t = \rho\dot{\beta}\mathbf{e}_t \tag{2/7}$$

The acceleration $\mathbf{a}$ of the particle was defined in Art. 2/3 as $\mathbf{a} = d\mathbf{v}/dt$, and we observed from Fig. 2/5 that the acceleration is a vector which reflects both the change in magnitude and the change in direction of $\mathbf{v}$. We now differentiate $\mathbf{v}$ in Eq. 2/7 by applying the ordinary rule for the differentiation of the product of a scalar and a vector* and get

$$\mathbf{a} = \frac{d\mathbf{v}}{dt} = \frac{d(v\mathbf{e}_t)}{dt} = v\dot{\mathbf{e}}_t + \dot{v}\mathbf{e}_t \tag{2/8}$$

where the unit vector $\mathbf{e}_t$ now has a nonzero derivative because its direction changes.

To find $\dot{\mathbf{e}}_t$ we analyze the change in $\mathbf{e}_t$ during a differential increment of motion as the particle moves from A to A' in Fig. 2/10a. The unit vector $\mathbf{e}_t$ correspondingly changes to $\mathbf{e}_t'$, and the vector difference $d\mathbf{e}_t$ is shown in part b of the figure. The vector $d\mathbf{e}_t$ in the limit has a magnitude equal to the length of the arc $|\mathbf{e}_t|\, d\beta = d\beta$ obtained by swinging the unit vector $\mathbf{e}_t$ through the angle $d\beta$ expressed in radians.

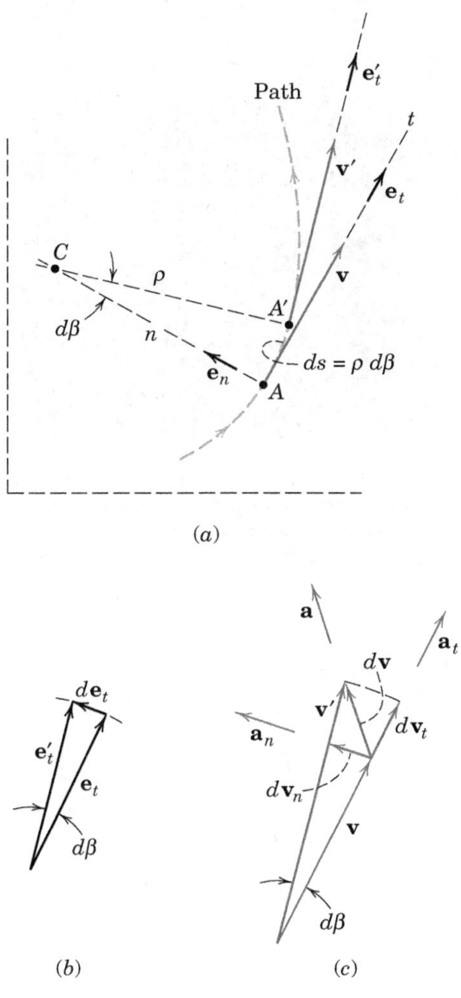

Figure 2/10

*See Art. C/7 of Appendix C.

The direction of $d\mathbf{e}_t$ is given by $\mathbf{e}_n$. Thus, we can write $d\mathbf{e}_t = \mathbf{e}_n \, d\beta$. Dividing by $d\beta$ gives

$$\frac{d\mathbf{e}_t}{d\beta} = \mathbf{e}_n$$

Dividing by dt gives $d\mathbf{e}_t/dt = (d\beta/dt)\mathbf{e}_n$, which can be written

$$\boxed{\dot{\mathbf{e}}_t = \dot{\beta}\mathbf{e}_n} \qquad \textbf{(2/9)}$$

With the substitution of Eq. 2/9 and $\dot{\beta}$ from the relation $v = \rho\dot{\beta}$, Eq. 2/8 for the acceleration becomes

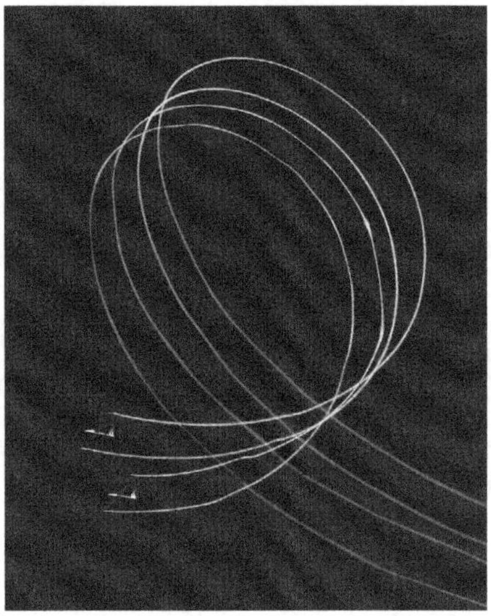

The paths of these two sailplanes strongly suggest the use of path coordinates such as a normal-tangential system.

$$\boxed{\mathbf{a} = \frac{v^2}{\rho}\mathbf{e}_n + \dot{v}\mathbf{e}_t} \qquad \textbf{(2/10)}$$

where

$$a_n = \frac{v^2}{\rho} = \rho\dot{\beta}^2 = v\dot{\beta}$$

$$a_t = \dot{v} = \ddot{s}$$

$$a = \sqrt{a_n{}^2 + a_t{}^2}$$

We stress that $a_t = \dot{v}$ is the time rate of change of the speed v. Finally, we note that $a_t = \dot{v} = d(\rho\dot{\beta})/dt = \rho\ddot{\beta} + \dot{\rho}\dot{\beta}$. This relation, however, finds little use because we seldom have reason to compute $\dot{\rho}$.

Geometric Interpretation

Full understanding of Eq. 2/10 comes only when we clearly see the geometry of the physical changes it describes. Figure 2/10*c* shows the velocity vector $\mathbf{v}$ when the particle is at A and $\mathbf{v}'$ when it is at A'. The vector change in the velocity is $d\mathbf{v}$, which establishes the direction of the acceleration $\mathbf{a}$. The n-component of $d\mathbf{v}$ is labeled $d\mathbf{v}_n$, and in the limit its magnitude equals the length of the arc generated by swinging the vector $\mathbf{v}$ as a radius through the angle $d\beta$. Thus, $|d\mathbf{v}_n| = v \, d\beta$ and the n-component of acceleration is $a_n = |d\mathbf{v}_n|/dt = v(d\beta/dt) = v\dot{\beta}$ as before. The t-component of $d\mathbf{v}$ is labeled $d\mathbf{v}_t$, and its magnitude is simply the change dv in the magnitude or length of the velocity vector. Therefore, the t-component of acceleration is $a_t = dv/dt = \dot{v} = \ddot{s}$ as before. The acceleration vectors resulting from the corresponding vector changes in velocity are shown in Fig. 2/10*c*.

It is especially important to observe that the normal component of acceleration a_n is *always directed toward the center of curvature C*. The tangential component of acceleration, on the other hand, will be in the positive t-direction of motion if the speed v is increasing and in the negative t-direction if the speed is decreasing. In Fig. 2/11 are shown schematic representations of the variation in the acceleration vector for a particle moving from A to B with (a) increasing speed and (b) decreasing speed. At an inflection point on the curve, the normal acceleration v^2/ρ goes to zero because ρ becomes infinite.

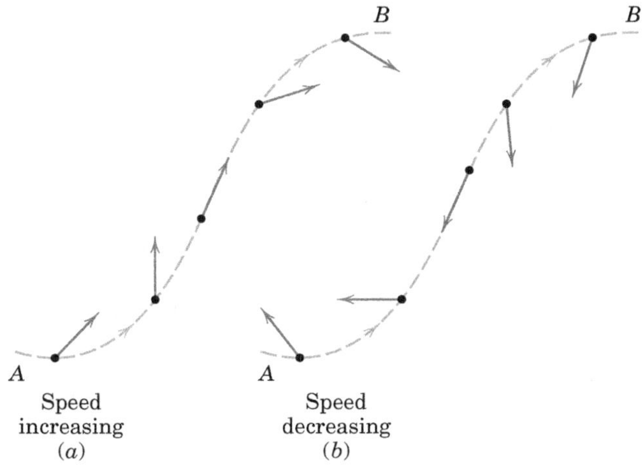

Acceleration vectors for
particle moving from *A* to *B*

Figure 2/11

Circular Motion

Circular motion is an important special case of plane curvilinear motion where the radius of curvature ρ becomes the constant radius r of the circle and the angle β is replaced by the angle θ measured from any convenient radial reference to *OP*, Fig. 2/12. The velocity and the acceleration components for the circular motion of the particle *P* become

$$
\begin{aligned}
v &= r\dot{\theta} \\
a_n &= v^2/r = r\dot{\theta}^2 = v\dot{\theta} \\
a_t &= \dot{v} = r\ddot{\theta}
\end{aligned}
\qquad \textbf{(2/11)}
$$

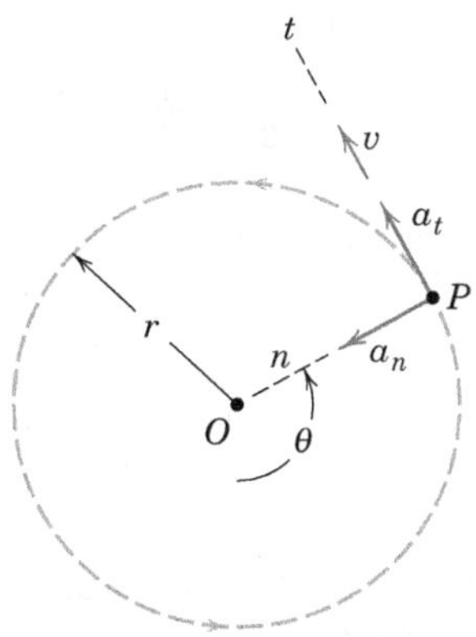

Figure 2/12

We find repeated use for Eqs. 2/10 and 2/11 in dynamics, so these relations and the principles behind them should be mastered.

GaryTramontina/Bloomberg viaGetty Images

An example of uniform circular motion is this car moving with constant speed around a wet skidpad (a circular roadway with a diameter of about 200 feet).

Sample Problem 2/7

To anticipate the dip and hump in the road, the driver of a car applies her brakes to produce a uniform deceleration. Her speed is 100 km/h at the bottom A of the dip and 50 km/h at the top C of the hump, which is 120 m along the road from A. If the passengers experience a total acceleration of 3 m/s^2 at A and if the radius of curvature of the hump at C is 150 m, calculate (a) the radius of curvature ρ at A, (b) the acceleration at the inflection point B, and (c) the total acceleration at C.

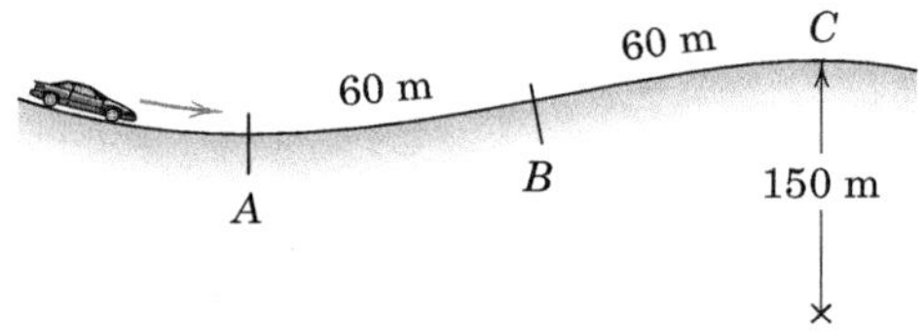

Solution. The dimensions of the car are small compared with those of the
① path, so we will treat the car as a particle. The velocities are

$$v_A = \left(100 \frac{\text{km}}{\text{h}}\right)\left(\frac{1 \text{ h}}{3600 \text{ s}}\right)\left(1000 \frac{\text{m}}{\text{km}}\right) = 27.8 \text{ m/s}$$

$$v_C = 50 \frac{1000}{3600} = 13.89 \text{ m/s}$$

We find the constant deceleration along the path from

$$\left[\int v\, dv = \int a_t\, ds\right] \qquad \int_{v_A}^{v_C} v\, dv = a_t \int_0^s ds$$

$$a_t = \frac{1}{2s}(v_C{}^2 - v_A{}^2) = \frac{(13.89)^2 - (27.8)^2}{2(120)} = -2.41 \text{ m/s}^2$$

(a) Condition at A. With the total acceleration given and a_t determined, we can easily compute a_n and hence ρ from

$$[a^2 = a_n{}^2 + a_t{}^2] \qquad a_n{}^2 = 3^2 - (2.41)^2 = 3.19 \qquad a_n = 1.785 \text{ m/s}^2$$

$$[a_n = v^2/\rho] \qquad \rho = v^2/a_n = (27.8)^2/1.785 = 432 \text{ m} \qquad \textit{Ans.}$$

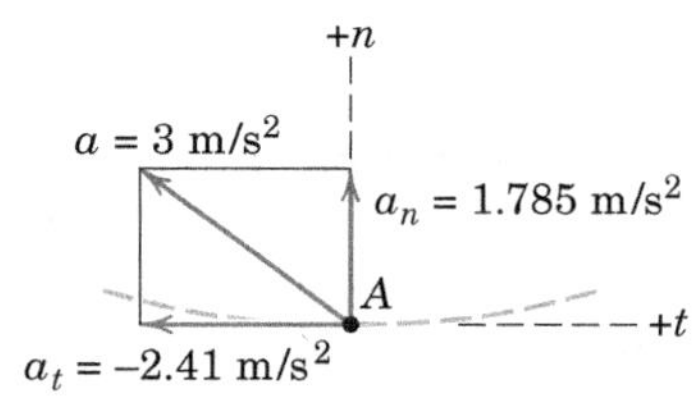

(b) Condition at B. Since the radius of curvature is infinite at the inflection point, $a_n = 0$ and

$$a = a_t = -2.41 \text{ m/s}^2 \qquad \textit{Ans.}$$

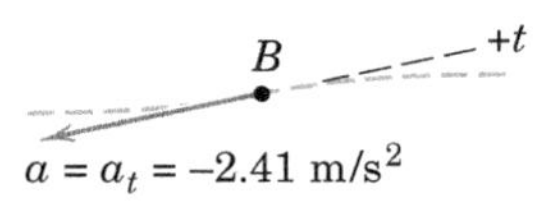

(c) Condition at C. The normal acceleration becomes

$$[a_n = v^2/\rho] \qquad a_n = (13.89)^2/150 = 1.286 \text{ m/s}^2$$

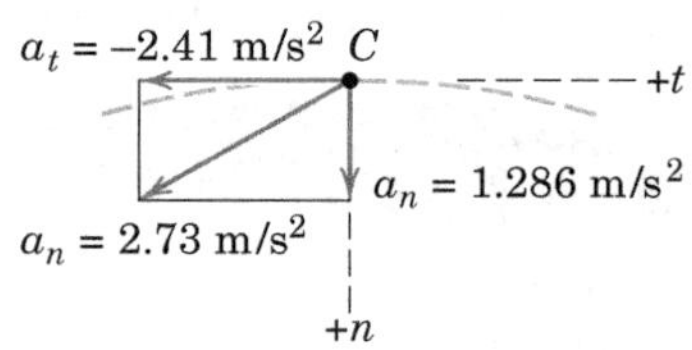

With unit vectors $\mathbf{e}_n$ and $\mathbf{e}_t$ in the n- and t-directions, the acceleration may be written

$$\mathbf{a} = 1.286\mathbf{e}_n - 2.41\mathbf{e}_t \text{ m/s}^2$$

where the magnitude of $\mathbf{a}$ is

$$\left[a = \sqrt{a_n{}^2 + a_t{}^2}\right] \qquad a = \sqrt{(1.286)^2 + (-2.41)^2} = 2.73 \text{ m/s}^2 \qquad \textit{Ans.}$$

The acceleration vectors representing the conditions at each of the three points are shown for clarification.

Helpful Hint

① Actually, the radius of curvature to the road differs by about 1 m from that to the path followed by the center of mass of the passengers, but we have neglected this relatively small difference.

Sample Problem 2/8

A certain rocket maintains a horizontal attitude of its axis during the powered phase of its flight at high altitude. The thrust imparts a horizontal component of acceleration of 20 ft/sec^2, and the downward acceleration component is the acceleration due to gravity at that altitude, which is $g = 30$ ft/sec^2. At the instant represented, the velocity of the mass center G of the rocket along the 15° direction of its trajectory is 12,000 mi/hr. For this position determine (*a*) the radius of curvature of the flight trajectory, (*b*) the rate at which the speed v is increasing, (*c*) the angular rate $\dot{\beta}$ of the radial line from G to the center of curvature C, and (*d*) the vector expression for the total acceleration **a** of the rocket.

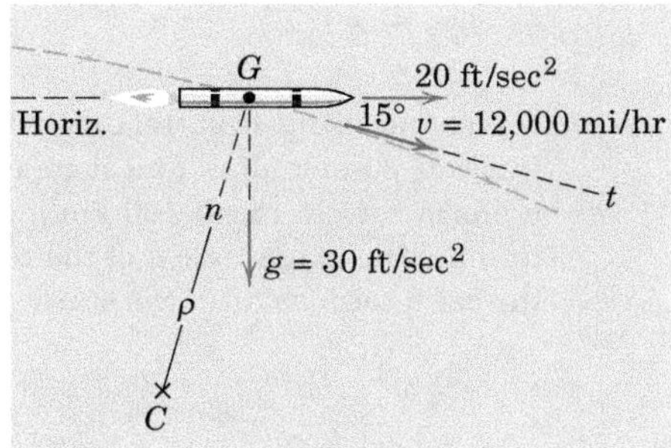

Solution. We observe that the radius of curvature appears in the expression for the normal component of acceleration, so we use n- and t-coordinates to describe the motion of G. The n- and t-components of the total acceleration are ① obtained by resolving the given horizontal and vertical accelerations into their n- and t-components and then combining. From the figure we get

$$a_n = 30 \cos 15° - 20 \sin 15° = 23.8 \text{ ft/sec}^2$$

$$a_t = 30 \sin 15° + 20 \cos 15° = 27.1 \text{ ft/sec}^2$$

(a) We may now compute the radius of curvature from

② $[a_n = v^2/\rho]$ $\qquad \rho = \dfrac{v^2}{a_n} = \dfrac{[(12{,}000)(44/30)]^2}{23.8} = 13.01(10^6) \text{ ft}$ *Ans.*

(b) The rate at which v is increasing is simply the t-component of acceleration.

$[\dot{v} = a_t]$ $\qquad \dot{v} = 27.1 \text{ ft/sec}^2$ *Ans.*

(c) The angular rate $\dot{\beta}$ of line GC depends on v and ρ and is given by

$[v = \rho\dot{\beta}]$ $\qquad \dot{\beta} = v/\rho = \dfrac{12{,}000(44/30)}{13.01(10^6)} = 13.53(10^{-4}) \text{ rad/sec}$ *Ans.*

(d) With unit vectors $\mathbf{e}_n$ and $\mathbf{e}_t$ for the n- and t-directions, respectively, the total acceleration becomes

$$\mathbf{a} = 23.8\mathbf{e}_n + 27.1\mathbf{e}_t \text{ ft/sec}^2$$ *Ans.*

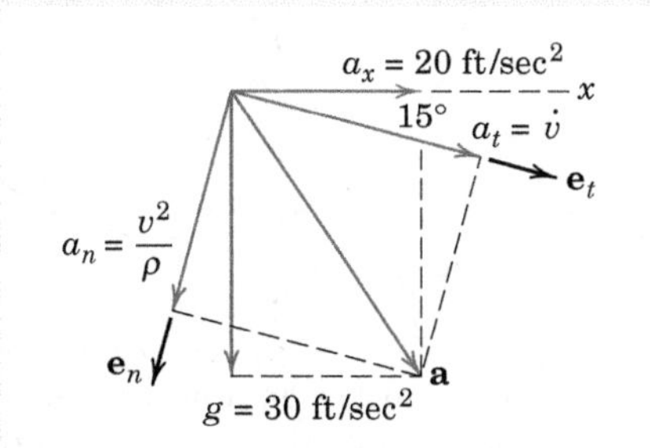

Helpful Hints

① Alternatively, we could find the resultant acceleration and then resolve it into n- and t-components.

② To convert from mi/hr to ft/sec, multiply by $\dfrac{5280 \text{ ft/mi}}{3600 \text{ sec/hr}} = \dfrac{44 \text{ ft/sec}}{30 \text{ mi/hr}}$ which is easily remembered, as 30 mi/hr is the same as 44 ft/sec.

PROBLEMS

Introductory Problems

2/97 A test car starts from rest on a horizontal circular track of 80-m radius and increases its speed at a uniform rate to reach 100 km/h in 10 seconds. Determine the magnitude a of the total acceleration of the car 8 seconds after the start.

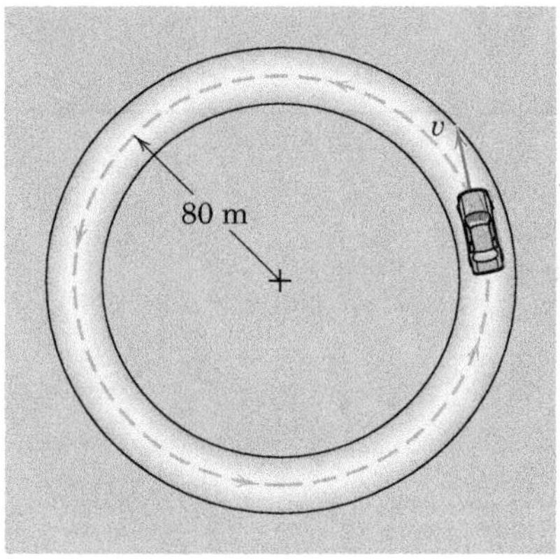

Problem 2/97

2/98 If the compact disc is spinning at a constant angular rate $\dot{\theta} = 360$ rev/min, determine the magnitudes of the accelerations of points A and B at the instant shown.

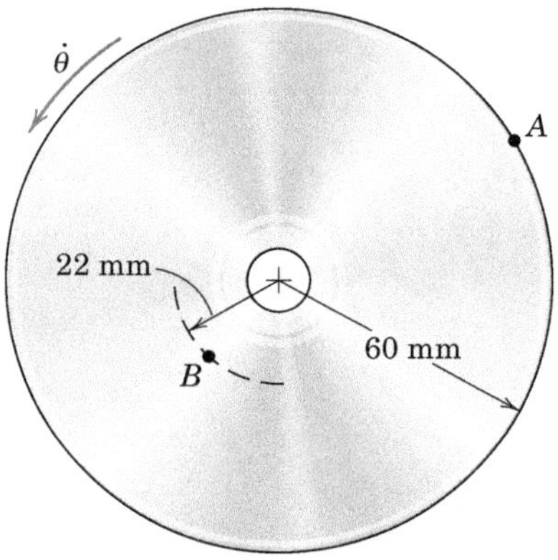

Problem 2/98

2/99 The car moves on a horizontal surface without any slippage of its tires. For each of the eight horizontal acceleration vectors, describe in words the instantaneous motion of the car. The car velocity is directed to the left as shown for all cases.

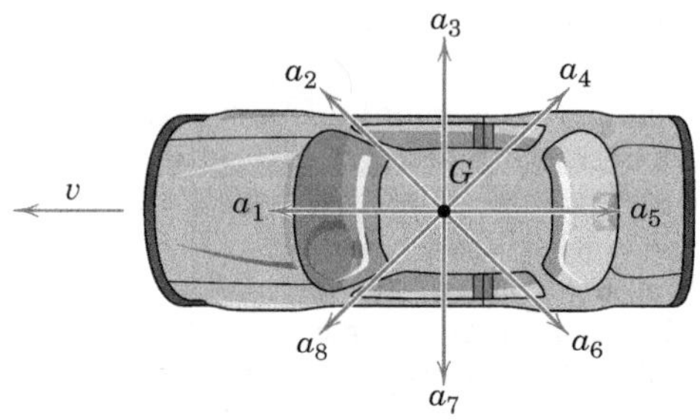

Problem 2/99

2/100 Determine the maximum speed for each car if the normal acceleration is limited to $0.88g$. The roadway is unbanked and level.

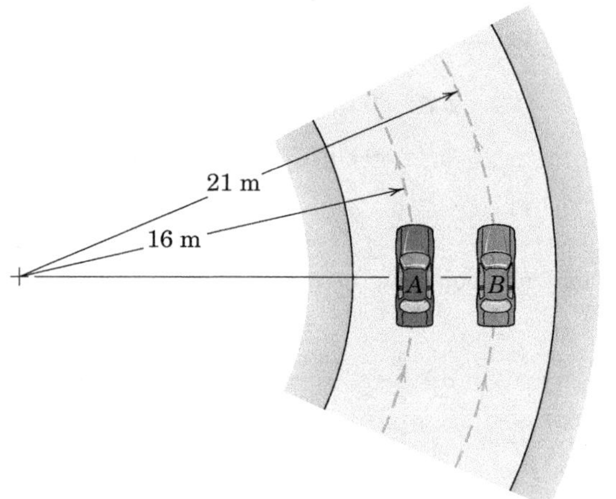

Problem 2/100

2/101 An accelerometer C is mounted to the side of the roller-coaster car and records a total acceleration of $3.5g$ as the empty car passes the bottommost position of the track as shown. If the speed of the car at this position is 215 km/h and is decreasing at the rate of 18 km/h every second, determine the radius of curvature ρ of the track at the position shown.

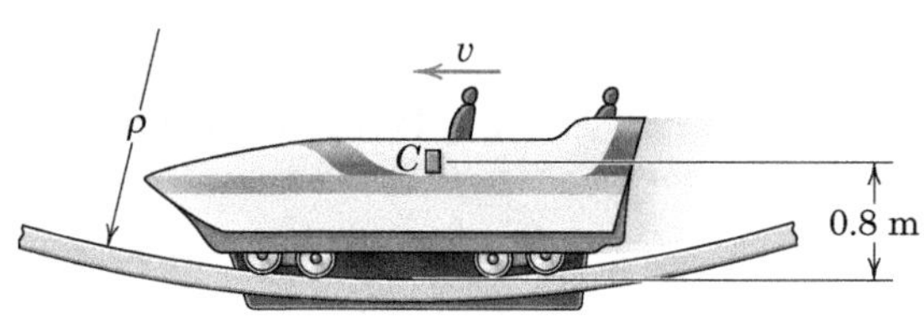

Problem 2/101

2/102 The driver of the truck has an acceleration of $0.4g$ as the truck passes over the top A of the hump in the road at constant speed. The radius of curvature of the road at the top of the hump is 98 m, and the center of mass G of the driver (considered a particle) is 2 m above the road. Calculate the speed v of the truck.

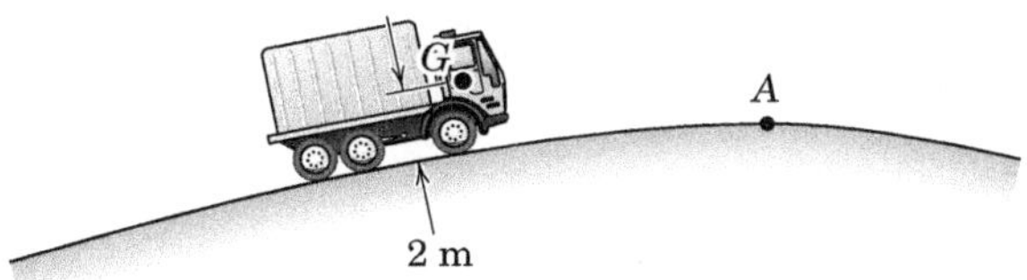

Problem 2/102

2/103 A particle moves along the curved path shown. The particle has a speed $v_A = 12$ ft/sec at time t_A and a speed $v_B = 14$ ft/sec at time t_B. Determine the average values of the normal and tangential accelerations of the particle between points A and B.

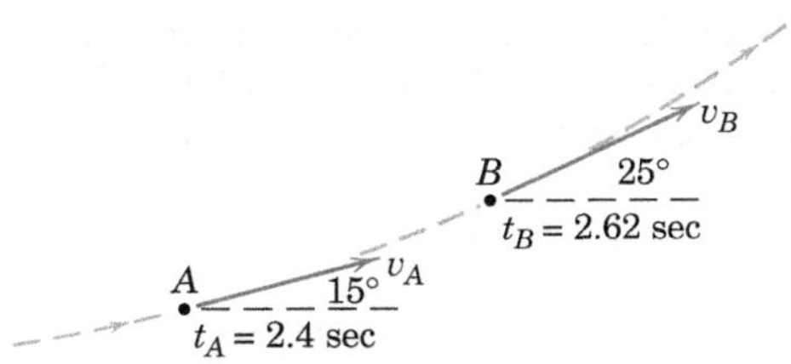

Problem 2/103

2/104 A ship which moves at a steady 20-knot speed (1 knot = 1.852 km/h) executes a turn to port by changing its compass heading at a constant counterclockwise rate. If it requires 60 seconds to alter course 90°, calculate the magnitude of the acceleration **a** of the ship during the turn.

2/105 A sprinter practicing for the 200-m dash accelerates uniformly from rest at A and reaches a top speed of 40 km/h at the 60-m mark. He then maintains this speed for the next 70 meters before uniformly slowing to a final speed of 35 km/h at the finish line. Determine the maximum horizontal acceleration which the sprinter experiences during the run. Where does this maximum acceleration value occur?

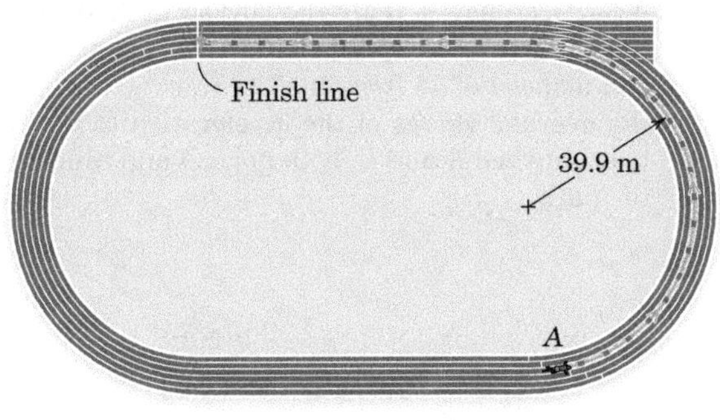

Problem 2/105

2/106 A train enters a curved horizontal section of track at a speed of 100 km/h and slows down with constant deceleration to 50 km/h in 12 seconds. An accelerometer mounted inside the train records a horizontal acceleration of 2 m/s^2 when the train is 6 seconds into the curve. Calculate the radius of curvature ρ of the track for this instant.

2/107 A particle moves on a circular path of radius r = 0.8 m with a constant speed of 2 m/s. The velocity undergoes a vector change $\Delta\mathbf{v}$ from A to B. Express the magnitude of $\Delta\mathbf{v}$ in terms of v and $\Delta\theta$ and divide it by the time interval Δt between A and B to obtain the magnitude of the average acceleration of the particle for (a) $\Delta\theta = 30°$, (b) $\Delta\theta = 15°$, and (c) $\Delta\theta = 5°$. In each case, determine the percentage difference from the instantaneous value of acceleration.

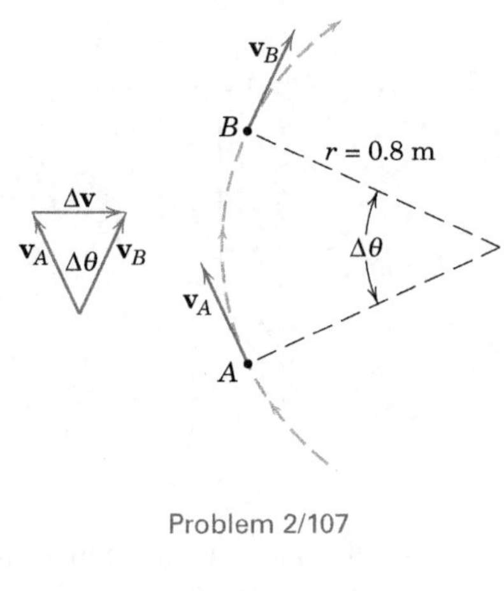

Problem 2/107

2/108 A particle moves along the curved path shown. If the particle has a speed of 40 ft/sec at A at time t_A and a speed of 44 ft/sec at B at time t_B, determine the average values of the acceleration of the particle between A and B, both normal and tangent to the path.

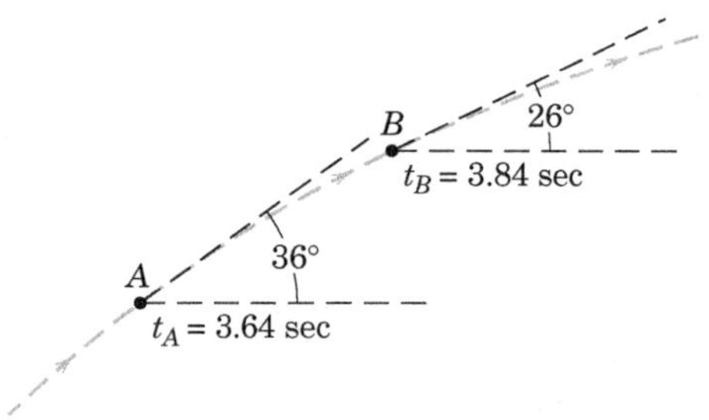

Problem 2/108

Representative Problems

2/109 An overhead view of part of a pinball game is shown. If the plunger imparts a speed of 3 m/s to the ball which travels in the smooth horizontal slot, determine the acceleration **a** of the ball (*a*) just before it exits the curve at C and (*b*) when it is halfway between points D and E. Use the values $r = 150$ mm and $\theta = 60°$.

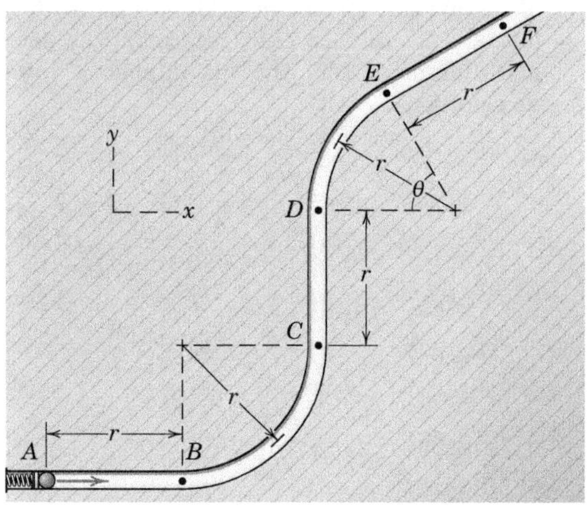

Problem 2/109

2/110 For the pinball game of Prob. 2/109, if the plunger imparts an initial speed of 3 m/s to the ball at time $t = 0$, determine the acceleration **a** of the ball (*a*) at time $t = 0.08$ s and (*b*) at time $t = 0.20$ s. At point F, the speed of the pinball has decreased by 10% from the initial value, and this decrease may be assumed to occur uniformly over the total distance traveled by the pinball. Use the values $r = 150$ mm and $\theta = 60°$.

2/111 The speed of a car increases uniformly with time from 50 km/h at A to 100 km/h at B during 10 seconds. The radius of curvature of the hump at A is 40 m. If the magnitude of the total acceleration of the mass center of the car is the same at B as at A, compute the radius of curvature ρ_B of the dip in the road at B. The mass center of the car is 0.6 m from the road.

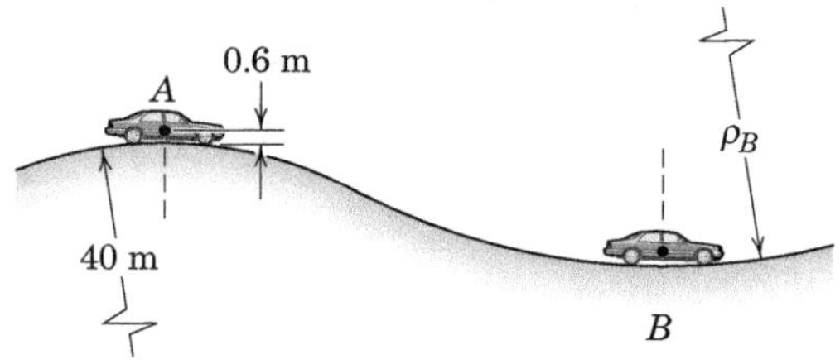

Problem 2/111

2/112 A minivan starts from rest on the road whose constant radius of curvature is 40 m and whose bank angle is 10°. The motion occurs in a horizontal plane. If the constant forward acceleration of the minivan is 1.8 m/s^2, determine the magnitude a of its total acceleration 5 seconds after starting.

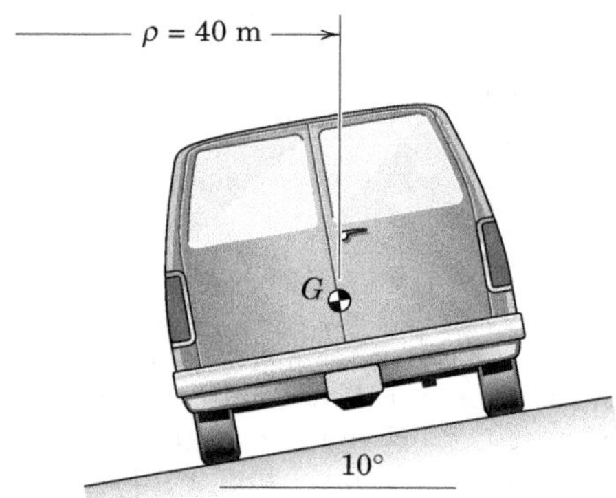

Problem 2/112

2/113 Consider the polar axis of the earth to be fixed in space and compute the magnitudes of the velocity and acceleration of a point P on the earth's surface at latitude 40° north. The mean diameter of the earth is 12 742 km and its angular velocity is $0.7292(10^{-4})$ rad/s.

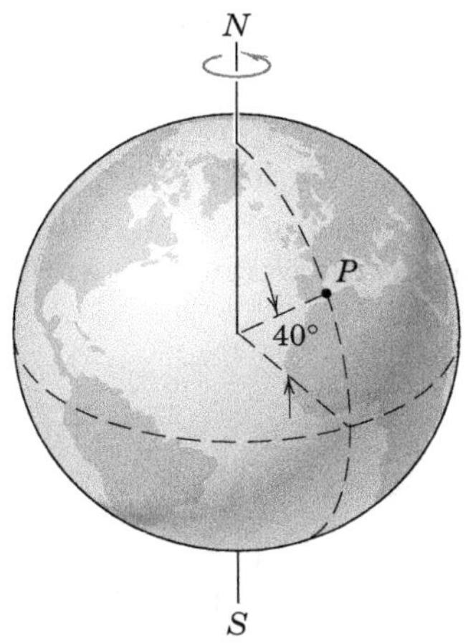

Problem 2/113

2/114 The car C increases its speed at the constant rate of 1.5 m/s^2 as it rounds the curve shown. If the magnitude of the total acceleration of the car is 2.5 m/s^2 at point A where the radius of curvature is 200 m, compute the speed v of the car at this point.

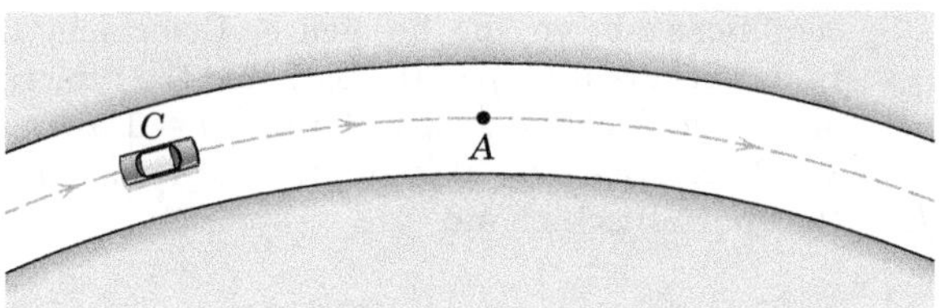

Problem 2/114

2/115 At the bottom A of the vertical inside loop, the magnitude of the total acceleration of the airplane is $3g$. If the airspeed is 800 km/h and is increasing at the rate of 20 km/h per second, calculate the radius of curvature ρ of the path at A.

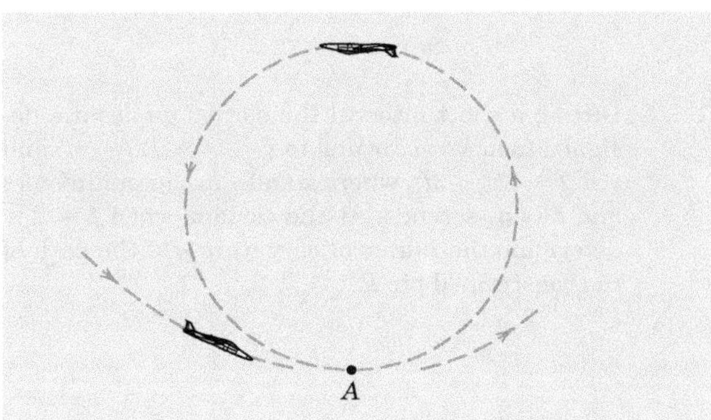

Problem 2/115

2/116 A golf ball is launched with the initial conditions shown in the figure. Determine the radius of curvature of the trajectory and the time rate of change of the speed of the ball (*a*) just after launch and (*b*) at apex. Neglect aerodynamic drag.

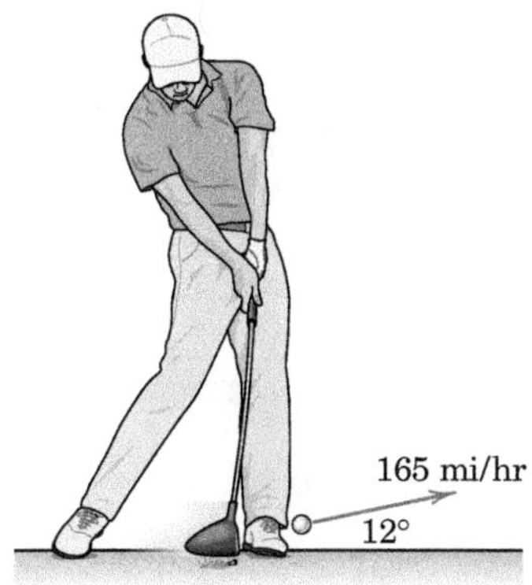

Problem 2/116

***2/117** If the golf ball of Prob. 2/116 is launched at time $t = 0$, determine the two times when the radius of curvature of the trajectory has a value of 1800 ft.

2/118 The preliminary design for a "small" space station to orbit the earth in a circular path consists of a ring (torus) with a circular cross section as shown. The living space within the torus is shown in section A, where the "ground level" is 20 ft from the center of the section. Calculate the angular speed N in revolutions per minute required to simulate standard gravity at the surface of the earth (32.17 ft/sec^2). Recall that you would be unaware of a gravitational field if you were in a nonrotating spacecraft in a circular orbit around the earth.

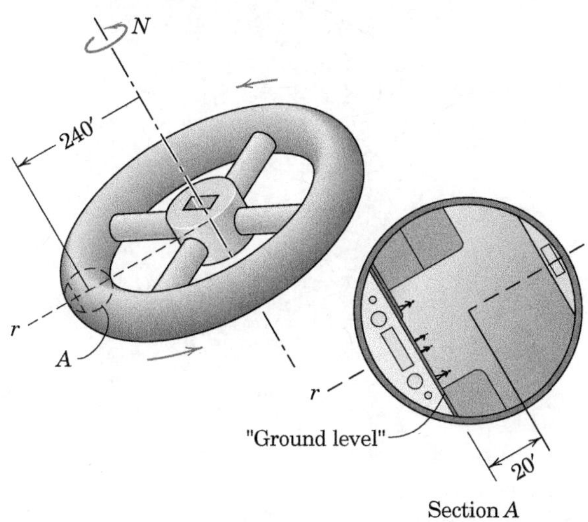

Problem 2/118

2/119 A spacecraft S is orbiting Jupiter in a circular path 1000 km above the surface with a constant speed. Using the gravitational law, calculate the magnitude v of its orbital velocity with respect to Jupiter. Use Table D/2 of Appendix D as needed.

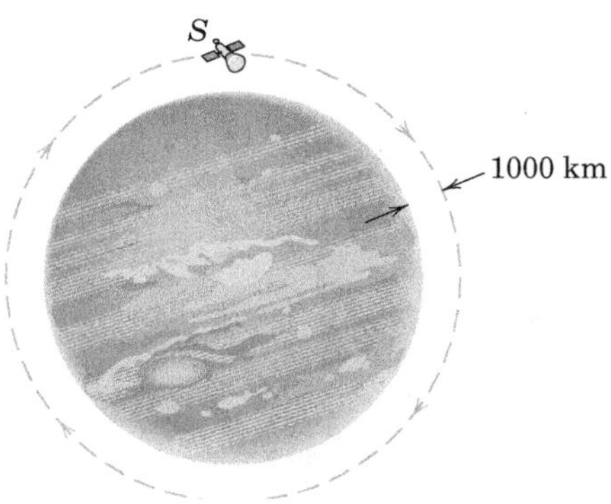

Problem 2/119

2/120 Two cars travel at constant speeds through a curved portion of highway. If the front ends of both cars cross line CC at the same instant, and each driver minimizes his or her time in the curve, determine the distance δ which the second car has yet to go along its own path to reach line DD at the instant the first car reaches there. The maximum horizontal acceleration for car A is $0.60g$ and that for car B is $0.76g$. Which car crosses line DD first?

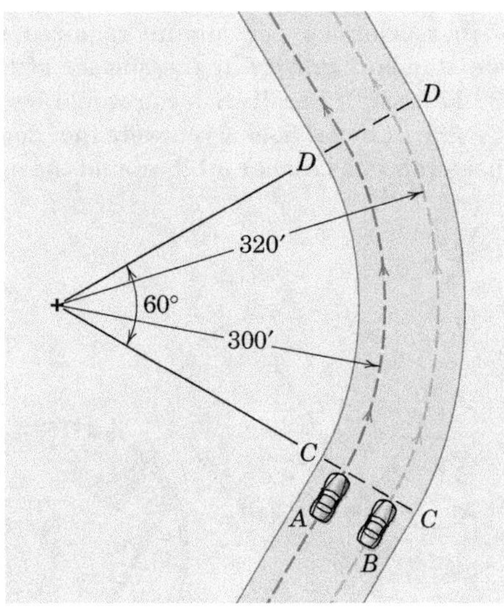

Problem 2/120

2/121 The figure shows a portion of a plate cam used in the design of a control mechanism. The motion of pin P in the fixed slot of the plate cam is controlled by the vertical guide A, which travels horizontally at a constant speed of 6 in./sec over the central sinusoidal portion of the slot. Determine the normal component of acceleration when the pin is at the position $x = 2$ in.

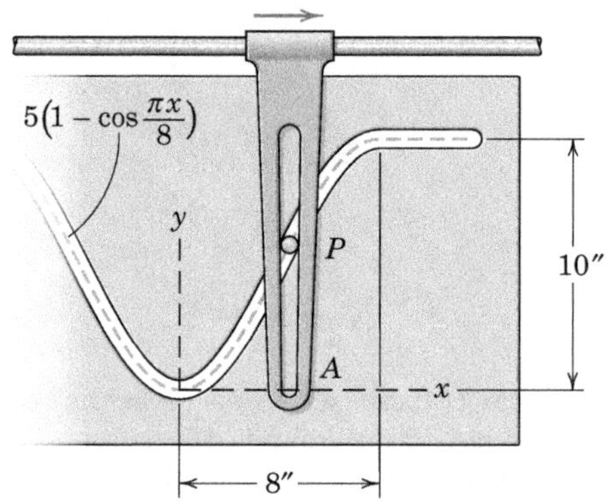

Problem 2/121

2/122 A football player releases a ball with the initial conditions shown in the figure. Determine the radius of curvature ρ of the path and the time rate of change $\dot{v}$ of the speed at times $t = 1$ sec and $t = 2$ sec, where $t = 0$ is the time of release from the quarterback's hand.

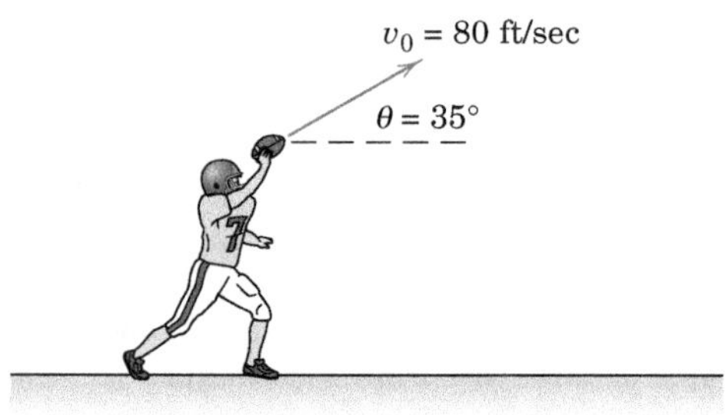

Problem 2/122

2/123 During a short interval the slotted guides are designed to move according to $x = 16 - 12t + 4t^2$ and $y = 2 + 15t - 3t^2$, where x and y are in millimeters and t is in seconds. At the instant when $t = 2$ s, determine the radius of curvature ρ of the path of the constrained pin P.

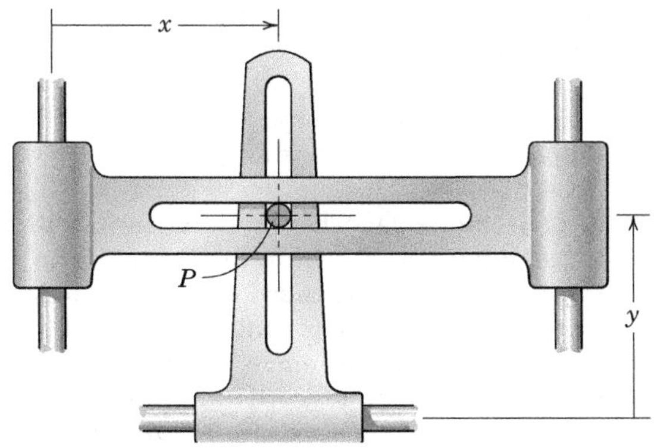

Problem 2/123

2/124 The particle P starts from rest at point A at time $t = 0$ and changes its speed thereafter at a constant rate of $2g$ as it follows the horizontal path shown. Determine the magnitude and direction of its total acceleration (*a*) just before it passes point B, (*b*) just after it passes point B, and (*c*) as it passes point C. State your directions relative to the x-axis shown (CCW positive).

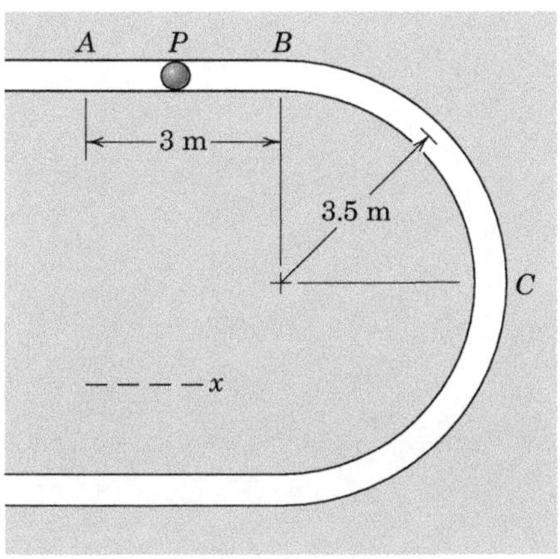

Problem 2/124

2/125 In the design of a timing mechanism, the motion of pin P in the fixed circular slot is controlled by the guide A, which is being elevated by its lead screw. Guide A starts from rest with pin P at the lowest point in the circular slot, and accelerates upward at a constant rate until it reaches a speed of 175 mm/s at the halfway point of its vertical displacement. The guide then decelerates at a constant rate and comes to a stop with pin P at the uppermost point in the circular slot. Determine the n- and t-components of acceleration of pin P once the pin has traveled 30° around the slot from the starting position.

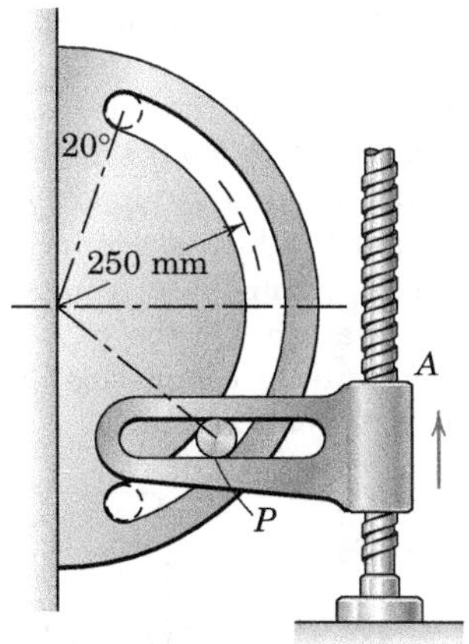

Problem 2/125

2/126 An earth satellite which moves in the elliptical equatorial orbit shown has a velocity v in space of 17 970 km/h when it passes the end of the semi-minor axis at A. The earth has an absolute surface value of g of 9.821 m/s^2 and has a radius of 6371 km. Determine the radius of curvature ρ of the orbit at A.

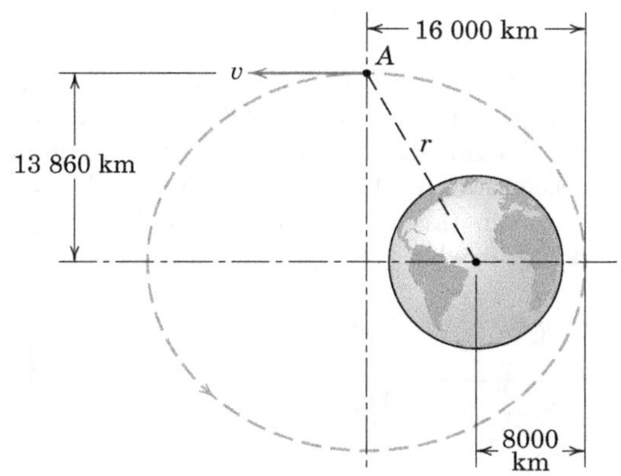

Problem 2/126

2/127 In the design of a control mechanism, the vertical slotted guide is moving with a constant velocity $\dot{x} = 15$ in./sec during the interval of motion from $x = -8$ in. to $x = +8$ in. For the instant when $x = 6$ in., calculate the n- and t-components of acceleration of the pin P, which is confined to move in the parabolic slot. From these results, determine the radius of curvature ρ of the path at this position. Verify your result by computing ρ from the expression cited in Appendix C/10.

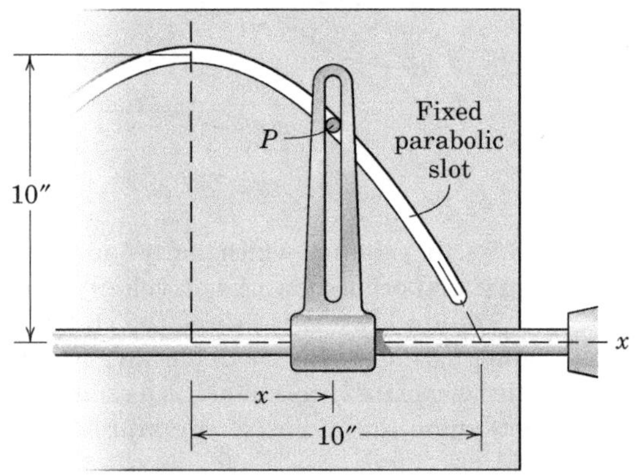

Problem 2/127

▶2/128 In a handling test, a car is driven through the slalom course shown. It is assumed that the car path is sinusoidal and that the maximum lateral acceleration is 0.7g. If the testers wish to design a slalom through which the maximum speed is 80 km/h, what cone spacing L should be used?

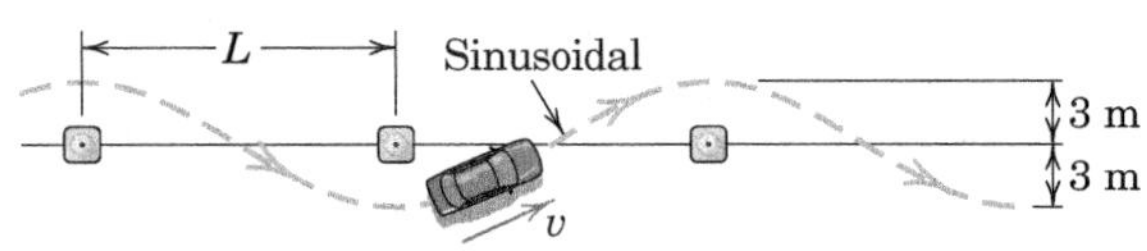

Problem 2/128

▶2/129 A particle which moves with curvilinear motion has coordinates in meters which vary with time t in seconds according to $x = 2t^2 + 3t - 1$ and $y = 5t - 2$. Determine the coordinates of the center of curvature C at time $t = 1$ s.

*2/130 A projectile is launched at time $t = 0$ with the initial conditions shown in the figure. If the wind imparts a constant leftward acceleration of 5 m/s^2, plot the n- and t-components of acceleration and the radius of curvature ρ of the trajectory for the time the projectile is in the air. State the maximum magnitude of each acceleration component along with the time at which it occurs. Additionally, determine the minimum radius of curvature for the trajectory and its corresponding time.

Problem 2/130

2/6 Polar Coordinates (r-θ)

We now consider the third description of plane curvilinear motion, namely, polar coordinates where the particle is located by the radial distance r from a fixed point and by an angular measurement θ to the radial line. Polar coordinates are particularly useful when a motion is constrained through the control of a radial distance and an angular position or when an unconstrained motion is observed by measurements of a radial distance and an angular position.

Figure 2/13a shows the polar coordinates r and θ which locate a particle traveling on a curved path. An arbitrary fixed line, such as the x-axis, is used as a reference for the measurement of θ. Unit vectors $\mathbf{e}_r$ and $\mathbf{e}_\theta$ are established in the positive r- and θ-directions, respectively. The position vector $\mathbf{r}$ to the particle at A has a magnitude equal to the radial distance r and a direction specified by the unit vector $\mathbf{e}_r$. Thus, we express the location of the particle at A by the vector

$$\mathbf{r} = r\mathbf{e}_r$$

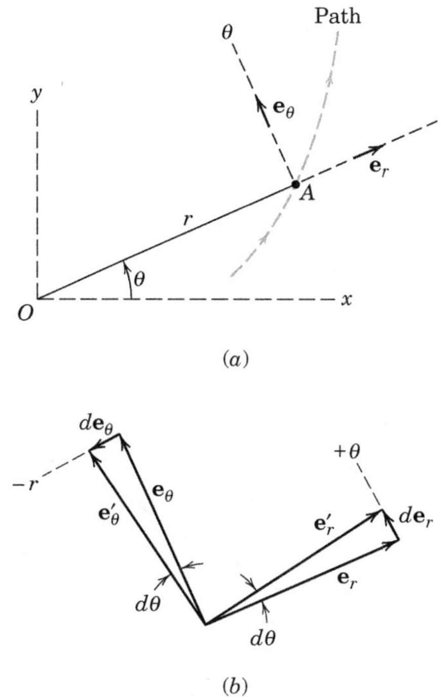

Figure 2/13

Time Derivatives of the Unit Vectors

To differentiate this relation with respect to time to obtain $\mathbf{v} = \dot{\mathbf{r}}$ and $\mathbf{a} = \dot{\mathbf{v}}$, we need expressions for the time derivatives of both unit vectors $\mathbf{e}_r$ and $\mathbf{e}_\theta$. We obtain $\dot{\mathbf{e}}_r$ and $\dot{\mathbf{e}}_\theta$ in exactly the same way we derived $\dot{\mathbf{e}}_t$ in the preceding article. During time dt the coordinate directions rotate through the angle $d\theta$, and the unit vectors also rotate through the same angle from $\mathbf{e}_r$ and $\mathbf{e}_\theta$ to $\mathbf{e}'_r$ and $\mathbf{e}'_\theta$, as shown in Fig. 2/13b. We note that the vector change $d\mathbf{e}_r$ is in the plus θ-direction and that $d\mathbf{e}_\theta$ is in the minus r-direction. Because their magnitudes in the limit are equal to the unit vector as radius times the angle $d\theta$ in radians, we can write them as $d\mathbf{e}_r = \mathbf{e}_\theta\, d\theta$ and $d\mathbf{e}_\theta = -\mathbf{e}_r\, d\theta$. If we divide these equations by $d\theta$, we have

$$\frac{d\mathbf{e}_r}{d\theta} = \mathbf{e}_\theta \qquad \text{and} \qquad \frac{d\mathbf{e}_\theta}{d\theta} = -\mathbf{e}_r$$

If, on the other hand, we divide them by dt, we have $d\mathbf{e}_r/dt = (d\theta/dt)\mathbf{e}_\theta$ and $d\mathbf{e}_\theta/dt = -(d\theta/dt)\mathbf{e}_r$, or simply

$$\boxed{\dot{\mathbf{e}}_r = \dot{\theta}\mathbf{e}_\theta \qquad \text{and} \qquad \dot{\mathbf{e}}_\theta = -\dot{\theta}\mathbf{e}_r} \qquad \textbf{(2/12)}$$

Velocity

We are now ready to differentiate $\mathbf{r} = r\mathbf{e}_r$ with respect to time. Using the rule for differentiating the product of a scalar and a vector gives

$$\mathbf{v} = \dot{\mathbf{r}} = \dot{r}\mathbf{e}_r + r\dot{\mathbf{e}}_r$$

With the substitution of $\dot{\mathbf{e}}_r$ from Eq. 2/12, the vector expression for the velocity becomes

$$\boxed{\mathbf{v} = \dot{r}\mathbf{e}_r + r\dot{\theta}\mathbf{e}_\theta} \qquad \textbf{(2/13)}$$

where $$v_r = \dot{r}$$
$$v_\theta = r\dot{\theta}$$
$$v = \sqrt{v_r^{\,2} + v_\theta^{\,2}}$$

The r-component of $\mathbf{v}$ is merely the rate at which the vector $\mathbf{r}$ stretches. The θ-component of $\mathbf{v}$ is due to the rotation of $\mathbf{r}$.

Acceleration

We now differentiate the expression for $\mathbf{v}$ to obtain the acceleration $\mathbf{a} = \dot{\mathbf{v}}$. Note that the derivative of $r\dot{\theta}\mathbf{e}_\theta$ will produce three terms, since all three factors are variable. Thus,

$$\mathbf{a} = \dot{\mathbf{v}} = (\ddot{r}\mathbf{e}_r + \dot{r}\dot{\mathbf{e}}_r) + (\dot{r}\dot{\theta}\mathbf{e}_\theta + r\ddot{\theta}\mathbf{e}_\theta + r\dot{\theta}\dot{\mathbf{e}}_\theta)$$

Substitution of $\dot{\mathbf{e}}_r$ and $\dot{\mathbf{e}}_\theta$ from Eq. 2/12 and collecting terms give

$$\boxed{\mathbf{a} = (\ddot{r} - r\dot{\theta}^2)\mathbf{e}_r + (r\ddot{\theta} + 2\dot{r}\dot{\theta})\mathbf{e}_\theta} \qquad \textbf{(2/14)}$$

where $$a_r = \ddot{r} - r\dot{\theta}^2$$
$$a_\theta = r\ddot{\theta} + 2\dot{r}\dot{\theta}$$
$$a = \sqrt{a_r^{\,2} + a_\theta^{\,2}}$$

We can write the θ-component alternatively as

$$a_\theta = \frac{1}{r}\frac{d}{dt}(r^2\dot{\theta})$$

which can be verified easily by carrying out the differentiation. This form for a_θ will be useful when we treat the angular momentum of particles in the next chapter.

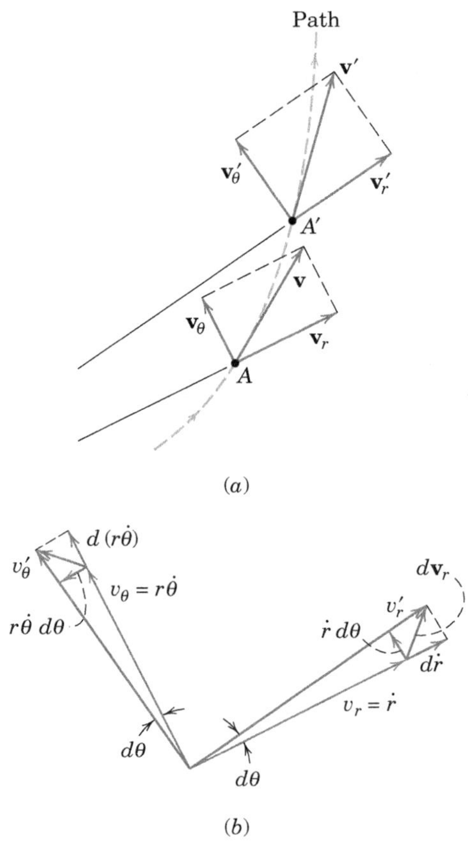

Figure 2/14

Geometric Interpretation

The terms in Eq. 2/14 can be best understood when the geometry of the physical changes can be clearly seen. For this purpose, Fig. 2/14*a* is developed to show the velocity vectors and their r- and θ-components at position A and at position A' after an infinitesimal movement. Each of these components undergoes a change in magnitude and direction as shown in Fig. 2/14*b*. In this figure we see the following changes:

(a) Magnitude Change of $\mathbf{v}_r$. This change is simply the increase in length of v_r or $dv_r = d\dot{r}$, and the corresponding acceleration term is $d\dot{r}/dt = \ddot{r}$ in the positive r-direction.

(b) Direction Change of $\mathbf{v}_r$. The magnitude of this change is seen from the figure to be $v_r\, d\theta = \dot{r}\, d\theta$, and its contribution to the acceleration becomes $\dot{r}\, d\theta/dt = \dot{r}\dot{\theta}$ which is in the positive θ-direction.

(c) Magnitude Change of $\mathbf{v}_\theta$. This term is the change in length of $\mathbf{v}_\theta$ or $d(r\dot{\theta})$, and its contribution to the acceleration is $d(r\dot{\theta})/dt = r\ddot{\theta} + \dot{r}\dot{\theta}$ and is in the positive θ-direction.

(d) Direction Change of $\mathbf{v}_\theta$. The magnitude of this change is $v_\theta \, d\theta = r\dot{\theta}\, d\theta$, and the corresponding acceleration term is observed to be $r\dot{\theta}(d\theta/dt) = r\dot{\theta}^2$ in the negative r-direction.

Collecting terms gives $a_r = \ddot{r} - r\dot{\theta}^2$ and $a_\theta = r\ddot{\theta} + 2\dot{r}\dot{\theta}$ as obtained previously. We see that the term $\ddot{r}$ is the acceleration which the particle would have along the radius in the absence of a change in θ. The term $-r\dot{\theta}^2$ is the normal component of acceleration if r were constant, as in circular motion. The term $r\ddot{\theta}$ is the tangential acceleration which the particle would have if r were constant, but is only a part of the acceleration due to the change in magnitude of $\mathbf{v}_\theta$ when r is variable. Finally, the term $2\dot{r}\dot{\theta}$ is composed of two effects. The first effect comes from that portion of the change in magnitude $d(r\dot{\theta})$ of v_θ due to the change in r, and the second effect comes from the change in direction of $\mathbf{v}_r$. The term $2\dot{r}\dot{\theta}$ represents, therefore, a combination of changes and is not so easily perceived as are the other acceleration terms.

Note the difference between the vector change $d\mathbf{v}_r$ in $\mathbf{v}_r$ and the change dv_r in the magnitude of v_r. Similarly, the vector change $d\mathbf{v}_\theta$ is not the same as the change dv_θ in the magnitude of v_θ. When we divide these changes by dt to obtain expressions for the derivatives, we see clearly that the magnitude of the derivative $|d\mathbf{v}_r/dt|$ and the derivative of the magnitude dv_r/dt are *not* the same. Note also that a_r is not $\dot{v}_r$ and that a_θ is not $\dot{v}_\theta$.

The total acceleration $\mathbf{a}$ and its components are represented in Fig. 2/15. If $\mathbf{a}$ has a component normal to the path, we know from our analysis of n- and t-components in Art. 2/5 that the sense of the n-component *must* be toward the center of curvature.

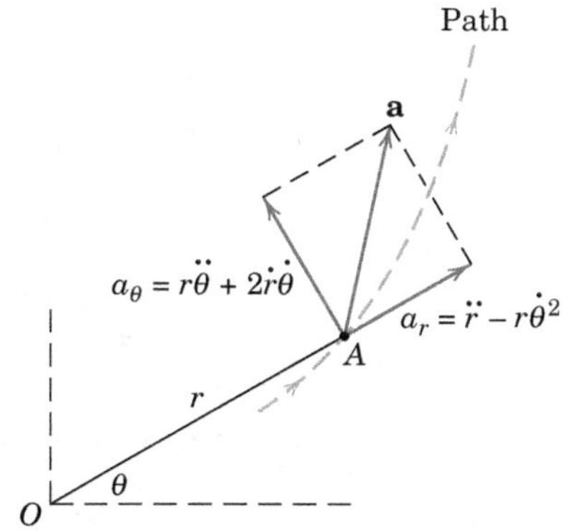

Figure 2/15

Circular Motion

For motion in a circular path with r constant, the components of Eqs. 2/13 and 2/14 become simply

$$v_r = 0 \qquad v_\theta = r\dot{\theta}$$

$$a_r = -r\dot{\theta}^2 \qquad a_\theta = r\ddot{\theta}$$

This description is the same as that obtained with n- and t-components, where the θ- and t-directions coincide but the positive r-direction is in the negative n-direction. Thus, $a_r = -a_n$ for circular motion centered at the origin of the polar coordinates.

The expressions for a_r and a_θ in scalar form can also be obtained by direct differentiation of the coordinate relations $x = r \cos \theta$ and $y = r \sin \theta$ to obtain $a_x = \ddot{x}$ and $a_y = \ddot{y}$. Each of these rectangular components of acceleration can then be resolved into r- and θ-components which, when combined, will yield the expressions of Eq. 2/14.

Sample Problem 2/9

Rotation of the radially slotted arm is governed by $\theta = 0.2t + 0.02t^3$, where θ is in radians and t is in seconds. Simultaneously, the power screw in the arm engages the slider B and controls its distance from O according to $r = 0.2 + 0.04t^2$, where r is in meters and t is in seconds. Calculate the magnitudes of the velocity and acceleration of the slider for the instant when $t = 3$ s.

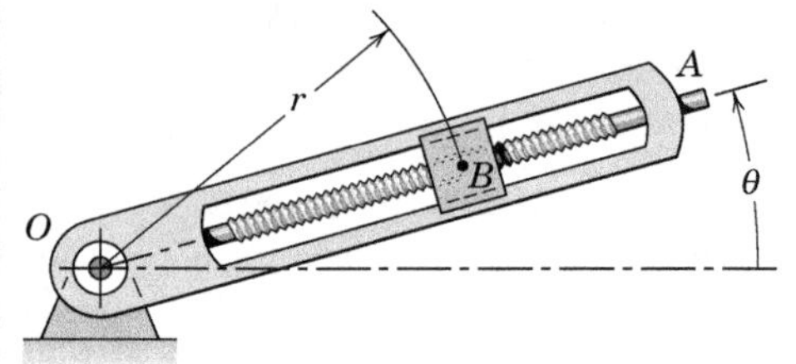

Solution. The coordinates and their time derivatives which appear in the expressions for velocity and acceleration in polar coordinates are obtained first and evaluated for $t = 3$ s. ①

$r = 0.2 + 0.04t^2$ $\qquad r_3 = 0.2 + 0.04(3^2) = 0.56$ m

$\dot{r} = 0.08t$ $\qquad \dot{r}_3 = 0.08(3) = 0.24$ m/s

$\ddot{r} = 0.08$ $\qquad \ddot{r}_3 = 0.08$ m/s^2

$\theta = 0.2t + 0.02t^3$ $\qquad \theta_3 = 0.2(3) + 0.02(3^3) = 1.14$ rad

$\qquad$ or $\theta_3 = 1.14(180/\pi) = 65.3°$

$\dot{\theta} = 0.2 + 0.06t^2$ $\qquad \dot{\theta}_3 = 0.2 + 0.06(3^2) = 0.74$ rad/s

$\ddot{\theta} = 0.12t$ $\qquad \ddot{\theta}_3 = 0.12(3) = 0.36$ rad/s^2

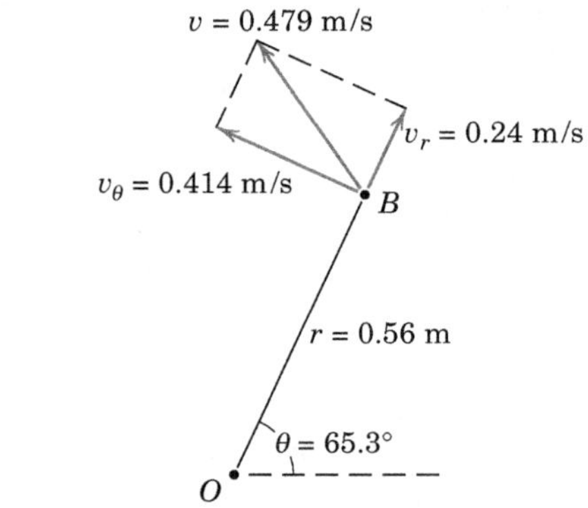

The velocity components are obtained from Eq. 2/13 and for $t = 3$ s are

$[v_\theta = r\dot{\theta}]$ $\qquad v_r = 0.24$ m/s

$[v_\theta = r\dot{\theta}]$ $\qquad v_\theta = 0.56(0.74) = 0.414$ m/s

$[v = \sqrt{{v_r}^2 + {v_\theta}^2}]$ $\qquad v = \sqrt{(0.24)^2 + (0.414)^2} = 0.479$ m/s $\qquad$ *Ans.*

The velocity and its components are shown for the specified position of the arm.

The acceleration components are obtained from Eq. 2/14 and for $t = 3$ s are

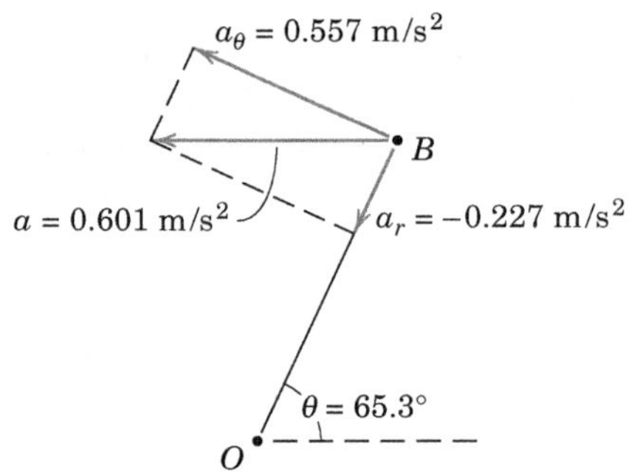

$[a_r = \ddot{r} - r\dot{\theta}^2]$ $\qquad a_r = 0.08 - 0.56(0.74)^2 = -0.227$ m/s^2

$[a_\theta = r\ddot{\theta} + 2\dot{r}\dot{\theta}]$ $\qquad a_\theta = 0.56(0.36) + 2(0.24)(0.74) = 0.557$ m/s^2

$[a = \sqrt{{a_r}^2 + {a_\theta}^2}]$ $\qquad a = \sqrt{(-0.227)^2 + (0.557)^2} = 0.601$ m/s^2 $\qquad$ *Ans.*

The acceleration and its components are also shown for the 65.3° position of the arm.

Plotted in the final figure is the path of the slider B over the time interval $0 \le t \le 5$ s. This plot is generated by varying t in the given expressions for r and θ. Conversion from polar to rectangular coordinates is given by

$$x = r\cos\theta \qquad y = r\sin\theta$$

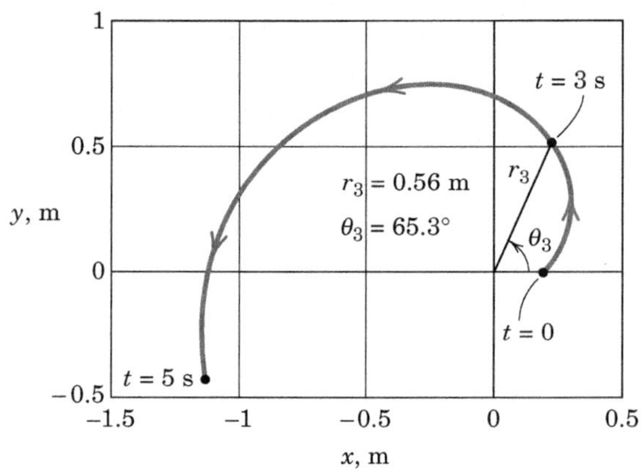

Helpful Hint

① We see that this problem is an example of constrained motion where the center B of the slider is mechanically constrained by the rotation of the slotted arm and by engagement with the turning screw.

Sample Problem 2/10

A tracking radar lies in the vertical plane of the path of a rocket which is coasting in unpowered flight above the atmosphere. For the instant when $\theta = 30°$, the tracking data give $r = 25(10^4)$ ft, $\dot{r} = 4000$ ft/sec, and $\dot{\theta} = 0.80$ deg/sec. The acceleration of the rocket is due only to gravitational attraction and for its particular altitude is 31.4 ft/sec^2 vertically down. For these conditions determine the velocity v of the rocket and the values of $\ddot{r}$ and $\ddot{\theta}$.

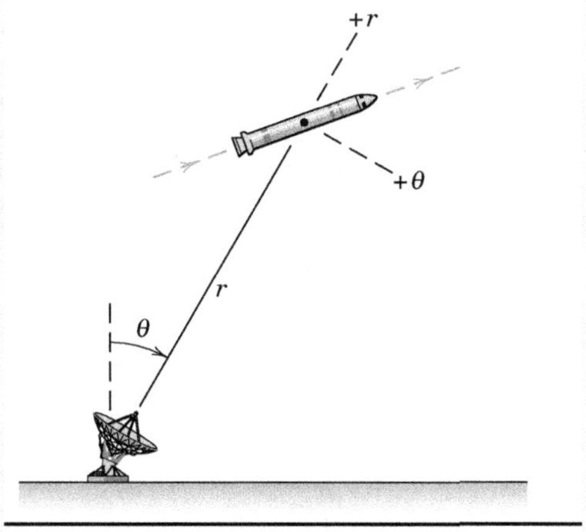

Solution. The components of velocity from Eq. 2/13 are

$$[v_r = \dot{r}] \qquad v_r = 4000 \text{ ft/sec}$$

① $$[v_\theta = r\dot{\theta}] \qquad v_\theta = 25(10^4)(0.80)\left(\frac{\pi}{180}\right) = 3490 \text{ ft/sec}$$

$$[v = \sqrt{v_r^2 + v_\theta^2}] \qquad v = \sqrt{(4000)^2 + (3490)^2} = 5310 \text{ ft/sec} \qquad \textit{Ans.}$$

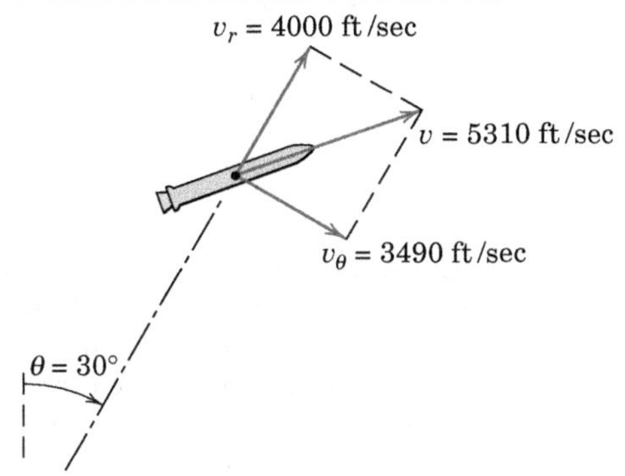

Since the total acceleration of the rocket is $g = 31.4$ ft/sec^2 down, we can easily find its r- and θ-components for the given position. As shown in the figure, they are

② $$a_r = -31.4 \cos 30° = -27.2 \text{ ft/sec}^2$$

$$a_\theta = 31.4 \sin 30° = 15.70 \text{ ft/sec}^2$$

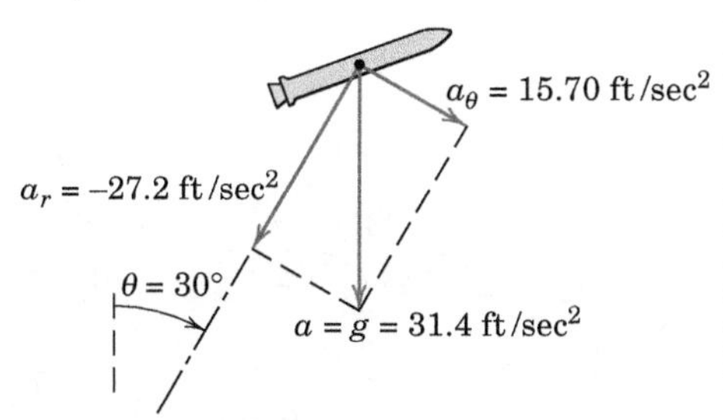

We now equate these values to the polar-coordinate expressions for a_r and a_θ which contain the unknowns $\ddot{r}$ and $\ddot{\theta}$. Thus, from Eq. 2/14

③ $$[a_r = \ddot{r} - r\dot{\theta}^2] \qquad -27.2 = \ddot{r} - 25(10^4)\left(0.80\frac{\pi}{180}\right)^2$$

$$\ddot{r} = 21.5 \text{ ft/sec}^2 \qquad \textit{Ans.}$$

$$[a_\theta = r\ddot{\theta} + 2\dot{r}\dot{\theta}] \qquad 15.70 = 25(10^4)\ddot{\theta} + 2(4000)\left(0.80\frac{\pi}{180}\right)$$

$$\ddot{\theta} = -3.84(10^{-4}) \text{ rad/sec}^2 \qquad \textit{Ans.}$$

Helpful Hints

① We observe that the angle θ in polar coordinates need not always be taken positive in a counterclockwise sense.

② Note that the r-component of acceleration is in the negative r-direction, so it carries a minus sign.

③ We must be careful to convert $\dot{\theta}$ from deg/sec to rad/sec.

PROBLEMS

Introductory Problems

2/131 A car P travels along a straight road with a constant speed $v = 65$ mi/hr. At the instant when the angle $\theta = 60°$, determine the values of $\dot{r}$ in ft/sec and $\dot{\theta}$ in deg/sec.

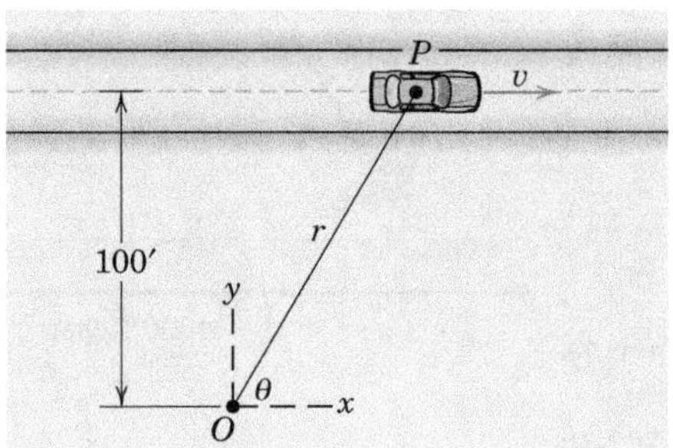

Problem 2/131

2/132 The sprinter begins from rest at position A and accelerates along the track. If the stationary tracking camera at O is rotating counterclockwise at the rate of 12.5 deg/s when the sprinter passes the 60-m mark, determine the speed v of the sprinter and the value of $\dot{r}$.

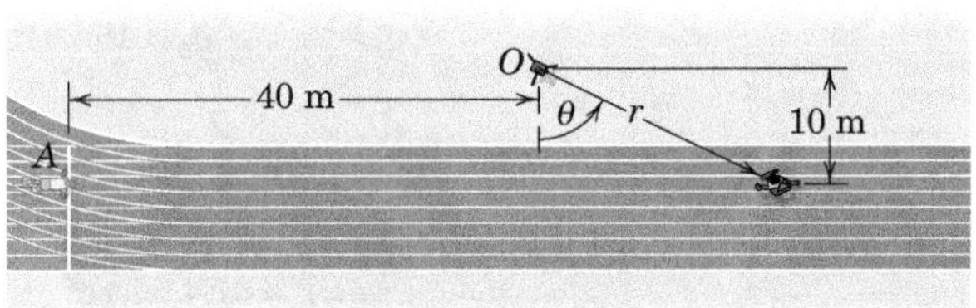

Problem 2/132

2/133 A drone flies over an observer O with constant speed in a straight line as shown. Determine the signs (plus, minus, or zero) for r, $\dot{r}$, $\ddot{r}$, θ, $\dot{\theta}$, and $\ddot{\theta}$ for each of the positions A, B, and C.

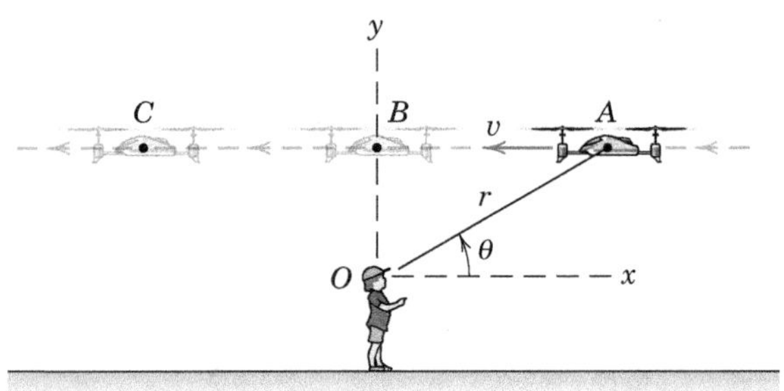

Problem 2/133

2/134 Motion of the sliding block P in the rotating radial slot is controlled by the power screw as shown. For the instant represented, $\dot{\theta} = 0.1$ rad/s, $\ddot{\theta} = -0.04$ rad/s^2, and $r = 300$ mm. Also, the screw turns at a constant speed giving $\dot{r} = 40$ mm/s. For this instant, determine the magnitudes of the velocity $\mathbf{v}$ and acceleration $\mathbf{a}$ of P. Sketch $\mathbf{v}$ and $\mathbf{a}$ if $\theta = 120°$.

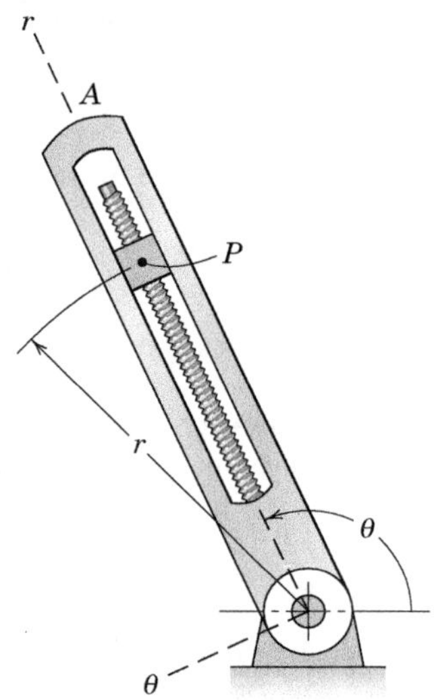

Problem 2/134

2/135 Rotation of bar OA is controlled by the lead screw which imparts a horizontal velocity v to collar C and causes pin P to travel along the smooth slot. Determine the values of $\dot{r}$ and $\dot{\theta}$, where $r = \overline{OP}$, if $h = 160$ mm, $x = 120$ mm, and $v = 25$ mm/s at the instant represented.

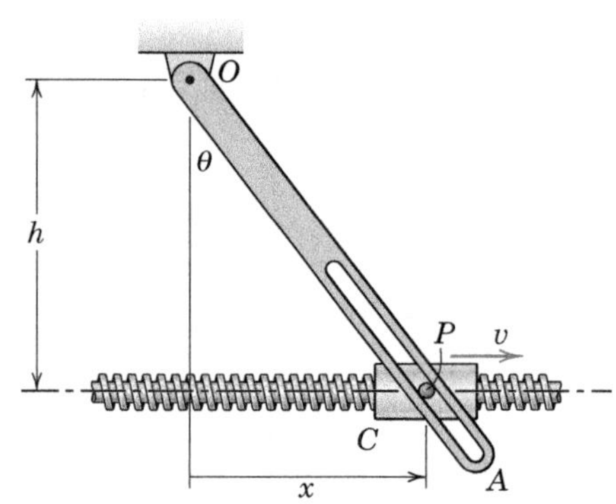

Problem 2/135

2/136 For the bar of Prob. 2/135, determine the values of $\ddot{r}$ and $\ddot{\theta}$ if the velocity of collar C is decreasing at a rate of 5 mm/s^2 at the instant in question. Refer to the printed answers for Prob. 2/135 as needed.

2/137 The boom OAB pivots about point O, while section AB simultaneously extends from within section OA. Determine the velocity and acceleration of the center B of the pulley for the following conditions: $\theta = 20°$, $\dot{\theta} = 5$ deg/sec, $\ddot{\theta} = 2$ deg/sec^2, $l = 7$ ft, $\dot{l} = 1.5$ ft/sec, $\ddot{l} = -4$ ft/sec^2. The quantities $\dot{l}$ and $\ddot{l}$ are the first and second time derivatives, respectively, of the length l of section AB.

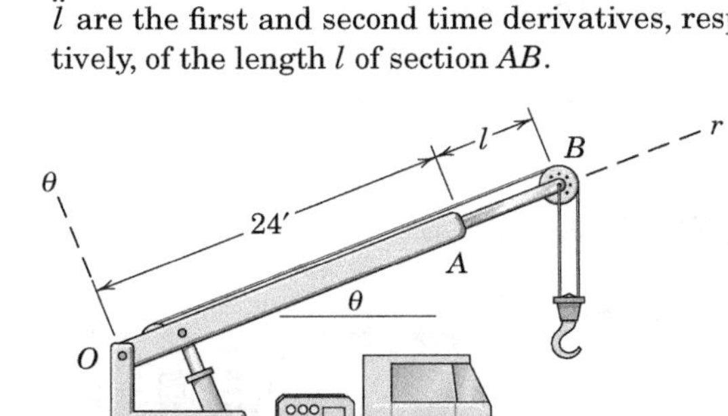

Problem 2/137

2/138 A particle moving along a plane curve has a position vector $\mathbf{r}$, a velocity $\mathbf{v}$, and an acceleration $\mathbf{a}$. Unit vectors in the r- and θ-directions are $\mathbf{e}_r$ and $\mathbf{e}_\theta$, respectively, and both r and θ are changing with time. Explain why each of the following statements is correctly marked as an inequality.

$\dot{\mathbf{r}} \neq v$	$\ddot{\mathbf{r}} \neq a$	$\dot{\mathbf{r}} \neq \dot{r}\mathbf{e}_r$
$\dot{r} \neq v$	$\ddot{r} \neq a$	$\ddot{\mathbf{r}} \neq \ddot{r}\mathbf{e}_r$
$\dot{r} \neq \mathbf{v}$	$\ddot{r} \neq \mathbf{a}$	$\dot{\mathbf{r}} \neq r\dot{\theta}\mathbf{e}_\theta$

2/139 Consider the portion of an excavator shown. At the instant under consideration, the hydraulic cylinder is extending at a rate of 6 in./sec, which is decreasing at the rate of 2 in./sec every second. Simultaneously, the cylinder is rotating about a horizontal axis through O at a constant rate of 10 deg/sec. Determine the velocity $\mathbf{v}$ and acceleration $\mathbf{a}$ of the clevis attachment at B.

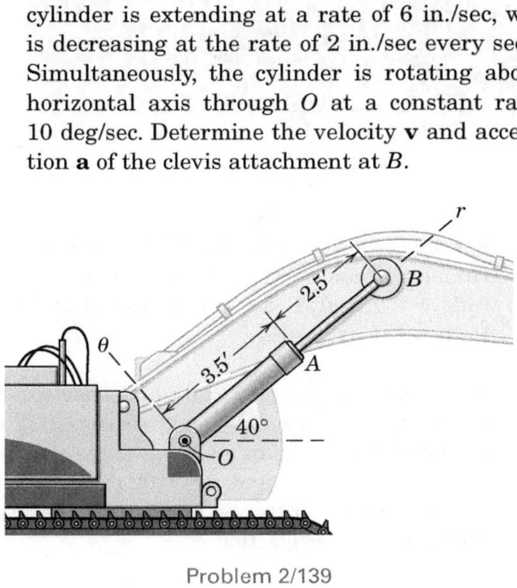

Problem 2/139

2/140 The nozzle shown rotates with constant angular speed Ω about a fixed horizontal axis through point O. Because of the change in diameter by a factor of 2, the water speed relative to the nozzle at A is v, while that at B is $4v$. The water speeds at both A and B are constant. Determine the velocity and acceleration of a water particle as it passes (*a*) point A and (*b*) point B.

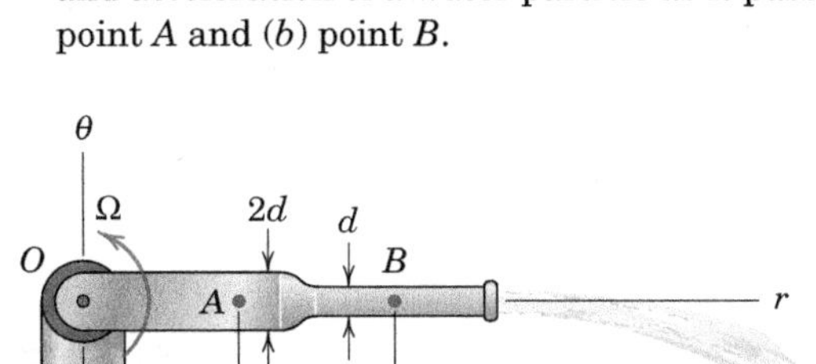

Problem 2/140

2/141 The radial position of a fluid particle P in a certain centrifugal pump with radial vanes is approximated by $r = r_0 \cosh Kt$, where t is time and $K = \dot{\theta}$ is the constant angular rate at which the impeller turns. Determine the expression for the magnitude of the total acceleration of the particle just prior to leaving the vane in terms of r_0, R, and K.

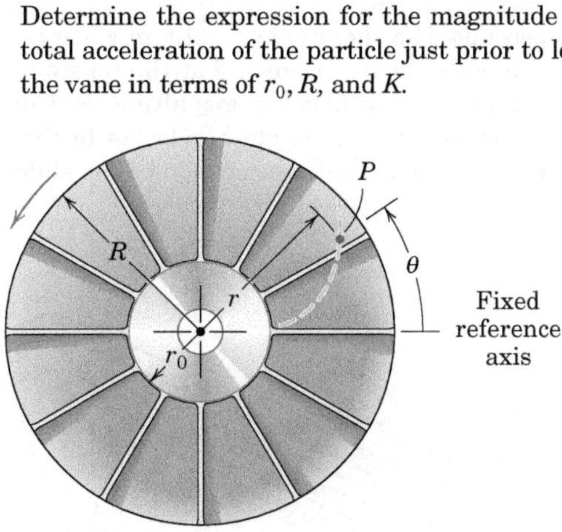

Problem 2/141

2/142 A helicopter starts from rest at point A and travels along the straight-line path with a constant acceleration a. If the speed v = 28 m/s when the altitude of the helicopter is h = 40 m, determine the values of $\dot{r}$, $\ddot{r}$, $\dot{\theta}$, and $\ddot{\theta}$ as measured by the tracking device at O. At this instant, θ = 40°, and the distance d = 160 m. Neglect the small height of the tracking device above the ground.

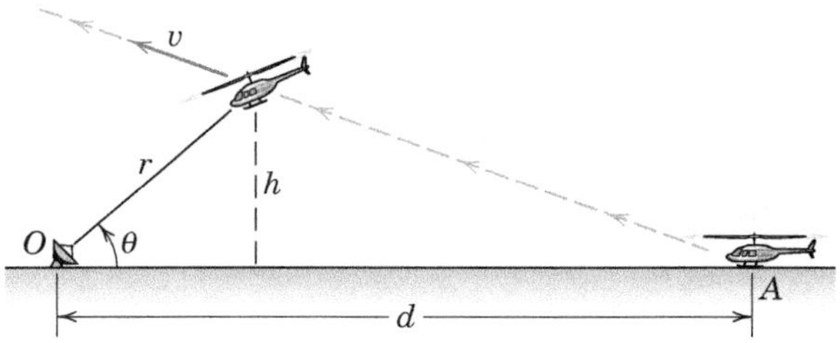

Problem 2/142

2/143 The slider P can be moved inward by means of the string S, while the slotted arm rotates about point O. The angular position of the arm is given by $\theta = 0.8t - \dfrac{t^2}{20}$, where θ is in radians and t is in seconds. The slider is at r = 1.6 m when t = 0 and thereafter is drawn inward at the constant rate of 0.2 m/s. Determine the magnitude and direction (expressed by the angle α relative to the x-axis) of the velocity and acceleration of the slider when t = 4 s.

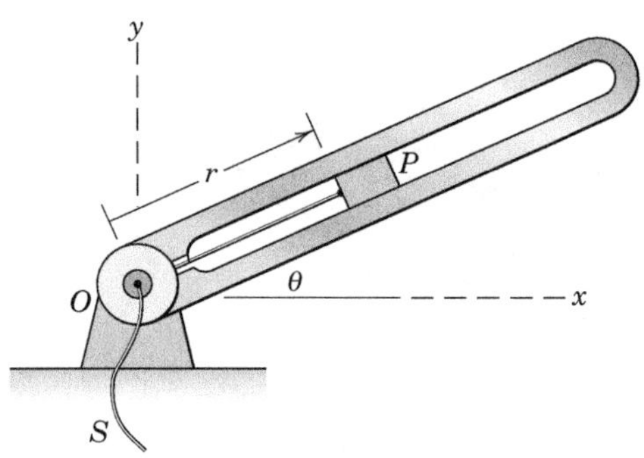

Problem 2/143

Representative Problems

2/144 Cars A and B are both moving with constant speed v on the straight and level highway. They are side-by-side in adjacent lanes as shown. If the radar unit attached to the stationary police car P measures "line-of-sight" velocity, what speed v' will be observed for each car? Use the values v = 70 mi/hr, L = 200 ft, and D = 22 ft.

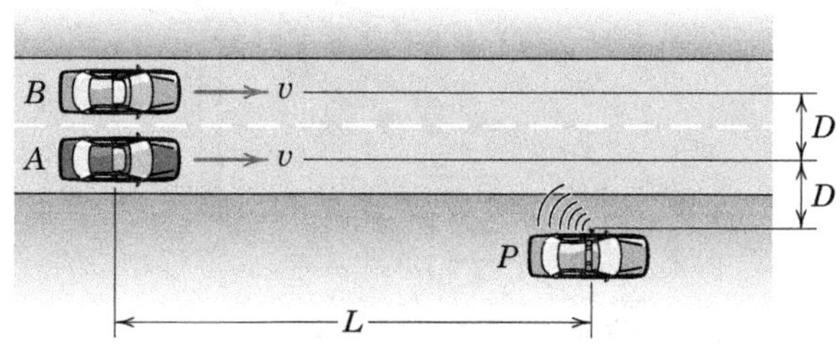

Problem 2/144

2/145 A fireworks shell P is launched upward from point A and explodes at its apex at an altitude of 275 ft. Relative to an observer at O, determine the values of $\dot{r}$ and $\dot{\theta}$ when the shell reaches an altitude y = 175 ft. Neglect aerodynamic drag.

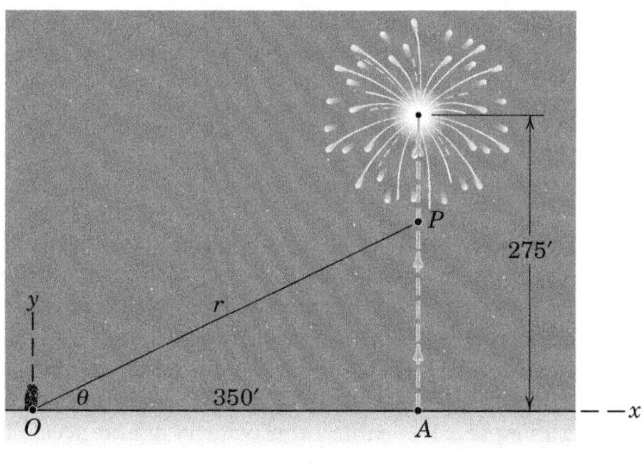

Problem 2/145

2/146 For the fireworks shell of Prob. 2/145, determine the values of $\ddot{r}$ and $\ddot{\theta}$ when the shell reaches an altitude y = 175 ft. Refer to the printed answers for Prob. 2/145 as needed.

2/147 The rocket is fired vertically and tracked by the radar station shown. When θ reaches 60°, other corresponding measurements give the values r = 9 km, $\ddot{r}$ = 21 m/s², and $\dot{\theta}$ = 0.02 rad/s. Calculate the magnitudes of the velocity and acceleration of the rocket at this position.

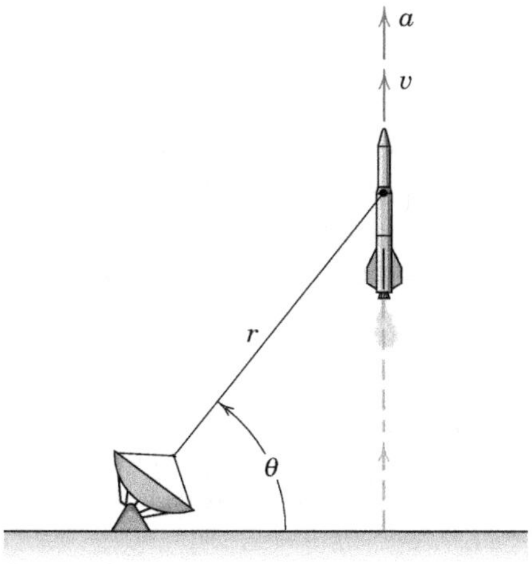

Problem 2/147

2/148 As it passes the position shown, the particle P has a constant speed $v = 100$ m/s along the straight line shown. Determine the corresponding values of $\dot{r}$, $\dot{\theta}$, $\ddot{r}$, and $\ddot{\theta}$.

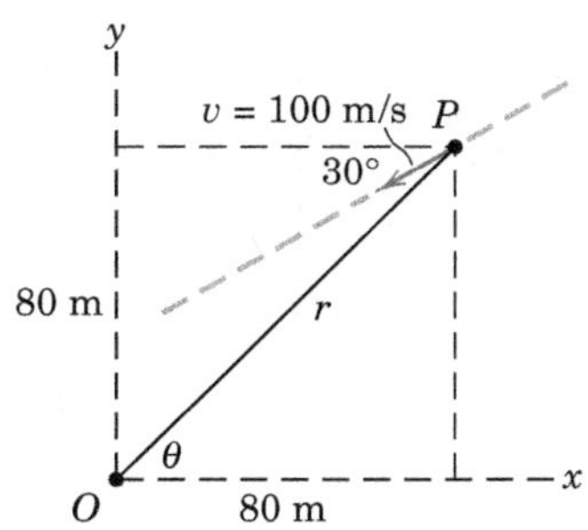

Problem 2/148

2/149 Repeat Prob. 2/148, but now the speed of the particle P is decreasing at the rate of 20 m/s^2 as it moves along the indicated straight path.

***2/150** The diver leaves the platform with an initial upward speed of 2.5 m/s. A stationary camera on the ground is programmed to track the diver throughout the dive by rotating the lens to keep the diver centered in the captured image. Plot $\dot{\theta}$ and $\ddot{\theta}$ as functions of time for the camera over the entire dive and state the values of $\dot{\theta}$ and $\ddot{\theta}$ at the instant the diver enters the water. Treat the diver as a particle which has only vertical motion. Additionally, state the maximum magnitudes of $\dot{\theta}$ and $\ddot{\theta}$ during the dive and the times at which they occur.

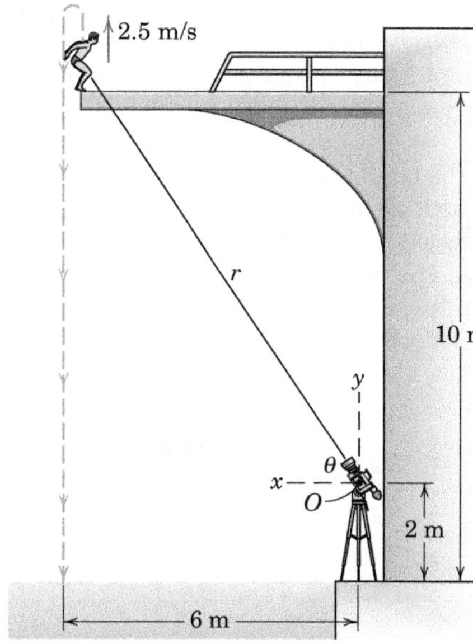

Problem 2/150

2/151 Instruments located at O are part of the ground-traffic control system for a major airport. At a certain instant during the takeoff roll of the aircraft P, the sensors indicate the angle $\theta = 50°$ and the range rate $\dot{r} = 140$ ft/sec. Determine the corresponding speed v of the aircraft and the value of $\dot{\theta}$.

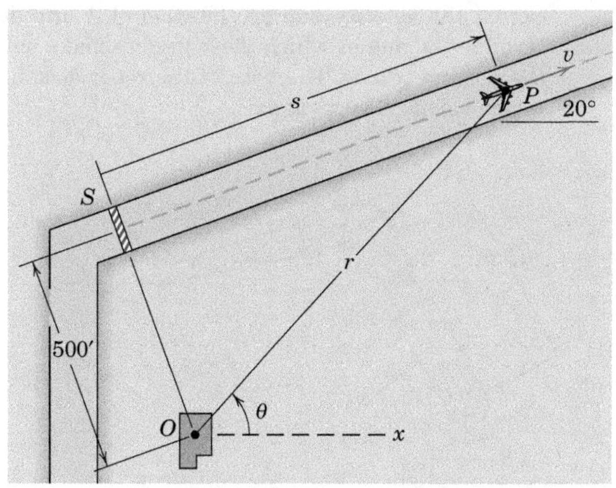

Problem 2/151

2/152 In addition to the information supplied in the previous problem, the sensors at O indicate that $\ddot{r} = 14$ ft/sec^2. Determine the corresponding acceleration a of the aircraft and the value of $\ddot{\theta}$.

2/153 At the bottom of a loop in the vertical (r-θ) plane at an altitude of 400 m, the airplane P has a horizontal velocity of 600 km/h and no horizontal acceleration. The radius of curvature of the loop is 1200 m. For the radar tracking at O, determine the recorded values of $\ddot{r}$ and $\ddot{\theta}$ for this instant.

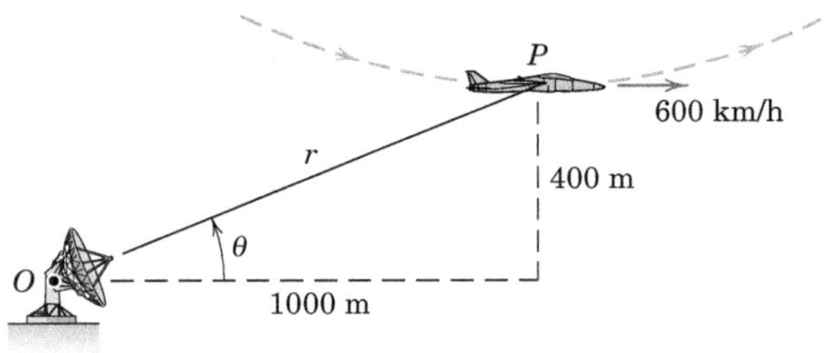

Problem 2/153

2/154 The member OA of the industrial robot telescopes and pivots about the fixed axis at point O. At the instant shown, $\theta = 60°$, $\dot{\theta} = 1.2$ rad/s, $\ddot{\theta} = 0.8$ rad/s^2, $\overline{OA} = 0.9$ m, $\dot{\overline{OA}} = 0.5$ m/s, and $\ddot{\overline{OA}} = -6$ m/s^2. Determine the magnitudes of the velocity and acceleration of joint A of the robot. Also, sketch the velocity and acceleration of A and determine the angles which these vectors make with the positive x-axis. The base of the robot does not revolve about a vertical axis.

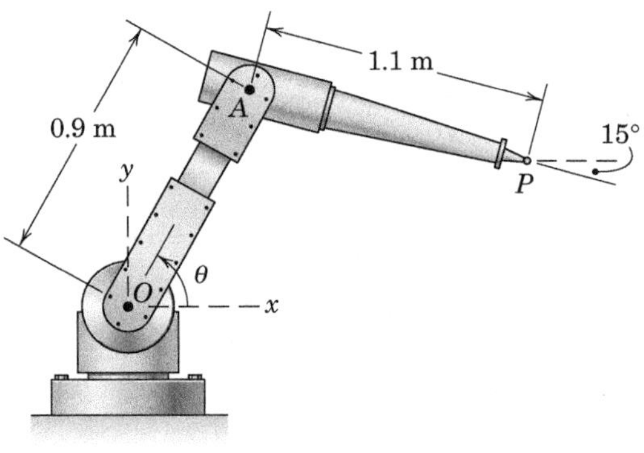

Problem 2/154

2/155 At the instant depicted in the figure, the radar station at O measures the range rate of the space shuttle P to be $\dot{r} = -12{,}272$ ft/sec, with O considered fixed. If it is known that the shuttle is in a circular orbit at an altitude $h = 150$ mi, determine the orbital speed of the shuttle from this information.

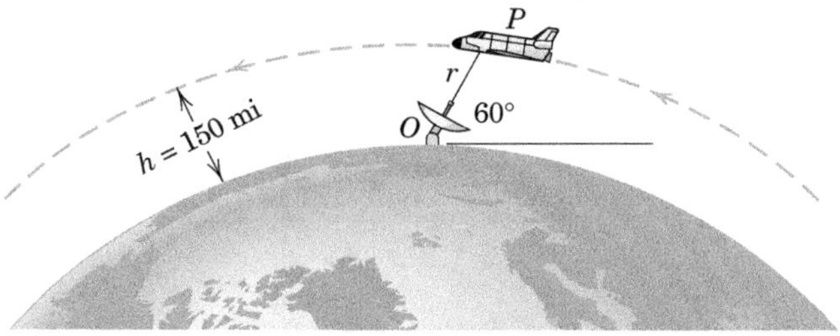

Problem 2/155

2/156 A locomotive is traveling on the straight and level track with a speed $v = 90$ km/h and a deceleration $a = 0.5$ m/s^2 as shown. Relative to the fixed observer at O, determine the quantities $\dot{r}$, $\ddot{r}$, $\dot{\theta}$, and $\ddot{\theta}$ at the instant when $\theta = 60°$ and $r = 400$ m.

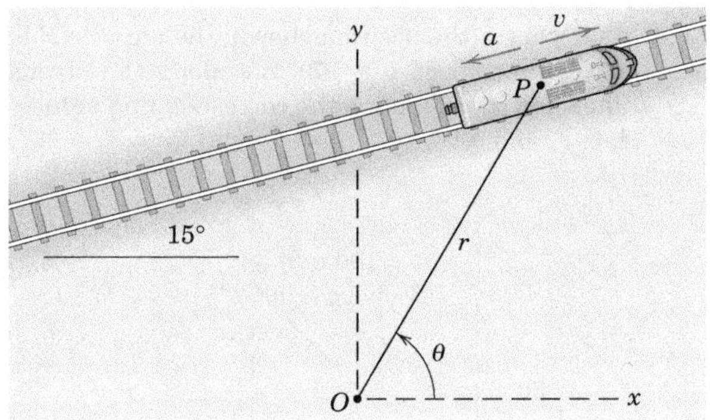

Problem 2/156

2/157 The small block P starts from rest at time $t = 0$ at point A and moves up the incline with constant acceleration a. Determine $\dot{r}$ as a function of time.

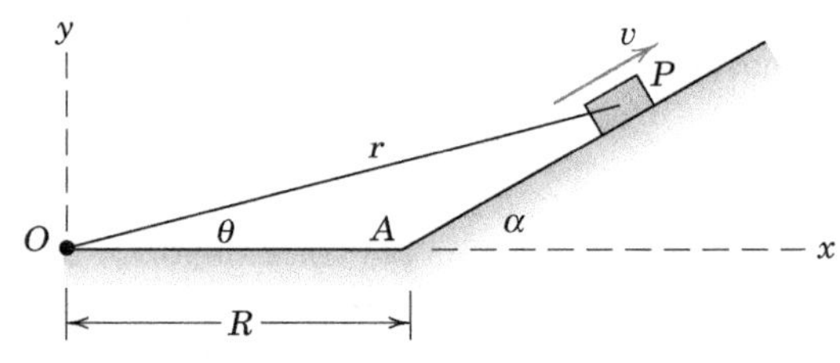

Problem 2/157

2/158 For the conditions of Prob. 2/157, determine $\dot{\theta}$ as a function of time.

2/159 An earth satellite traveling in the elliptical orbit shown has a velocity $v = 12{,}149$ mi/hr as it passes the end of the semiminor axis at A. The acceleration of the satellite at A is due to gravitational attraction and is $32.23[3959/8400]^2 = 7.159$ ft/sec^2 directed from A to O. For position A calculate the values of $\dot{r}$, $\ddot{r}$, $\dot{\theta}$, and $\ddot{\theta}$.

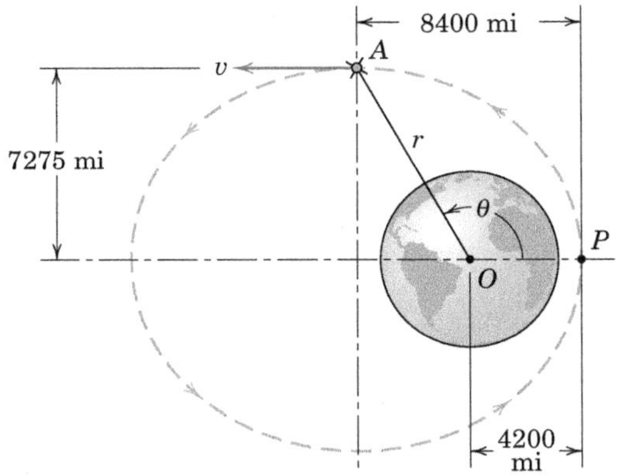

Problem 2/159

2/160 A meteor P is tracked by a radar observatory on the earth at O. When the meteor is directly overhead ($\theta = 90°$), the following observations are recorded: $r = 80$ km, $\dot{r} = -20$ km/s, and $\dot{\theta} = 0.4$ rad/s. (*a*) Determine the speed v of the meteor and the angle β which its velocity vector makes with the horizontal. Neglect any effects due to the earth's rotation. (*b*) Repeat with all given quantities remaining the same, except that $\theta = 75°$.

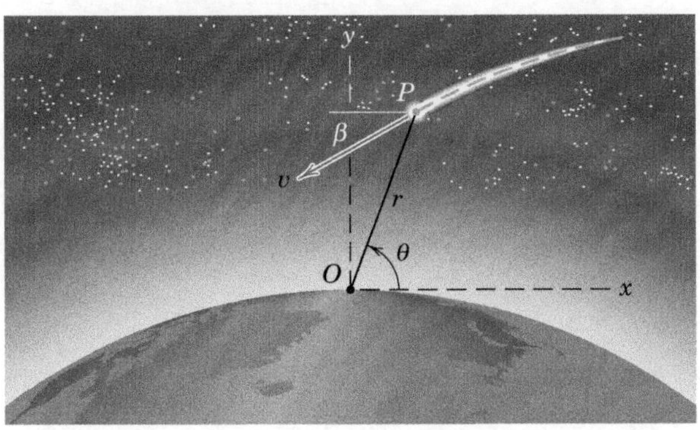

Problem 2/160

2/161 The low-flying aircraft P is traveling at a constant speed of 360 km/h in the holding circle of radius 3 km. For the instant shown, determine the quantities r, $\dot{r}$, $\ddot{r}$, θ, $\dot{\theta}$, and $\ddot{\theta}$ relative to the fixed x-y coordinate system, which has its origin on a mountaintop at O. Treat the system as two-dimensional.

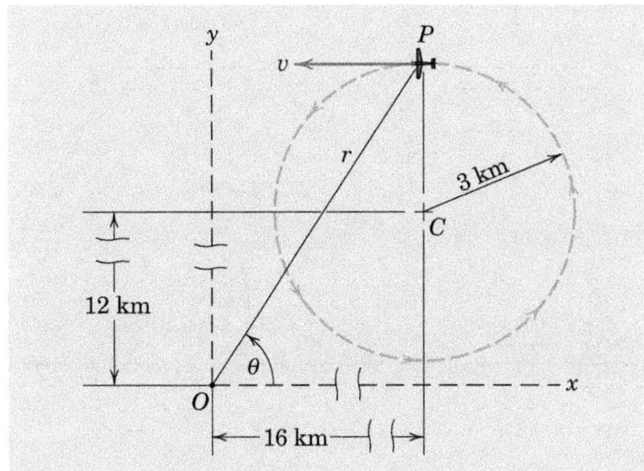

Problem 2/161

▸2/162 At time $t = 0$, the baseball player releases a ball with the initial conditions shown in the figure. Determine the quantities r, $\dot{r}$, $\ddot{r}$, θ, $\dot{\theta}$, and $\ddot{\theta}$, all relative to the x-y coordinate system shown, at time $t = 0.5$ sec.

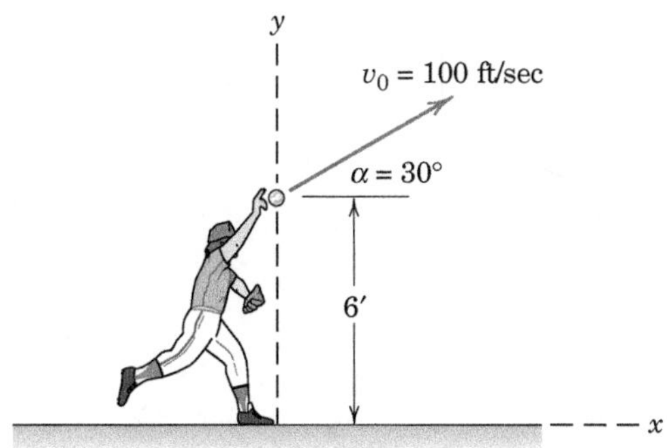

Problem 2/162

▶2/163 The racing airplane is beginning an inside loop in the vertical plane. The tracking station at O records the following data for a particular instant: $r = 90$ m, $\dot{r} = 15.5$ m/s, $\ddot{r} = 74.5$ m/s^2, $\theta = 30°$, $\dot{\theta} = 0.53$ rad/s, and $\ddot{\theta} = -0.29$ rad/s^2. Determine the values of v, $\dot{v}$, ρ, and β at this instant.

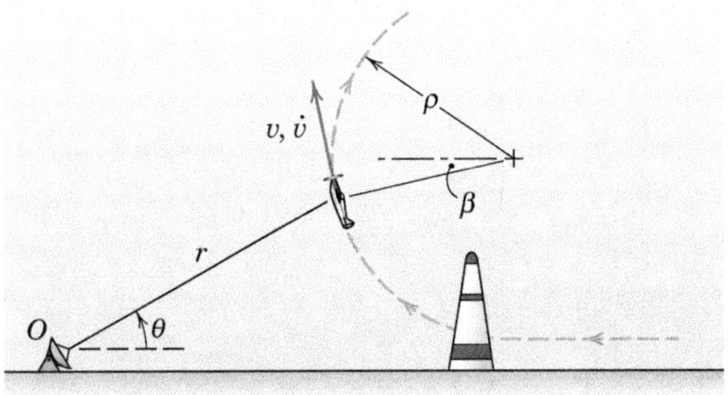

Problem 2/163

▶2/164 A golf ball is driven with the initial conditions shown in the figure. If the wind imparts a constant horizontal deceleration of 4 ft/sec^2, determine the values of r, $\dot{r}$, $\ddot{r}$, θ, $\dot{\theta}$, and $\ddot{\theta}$ when $t = 1.05$ sec. Take the r-coordinate to be measured from the origin.

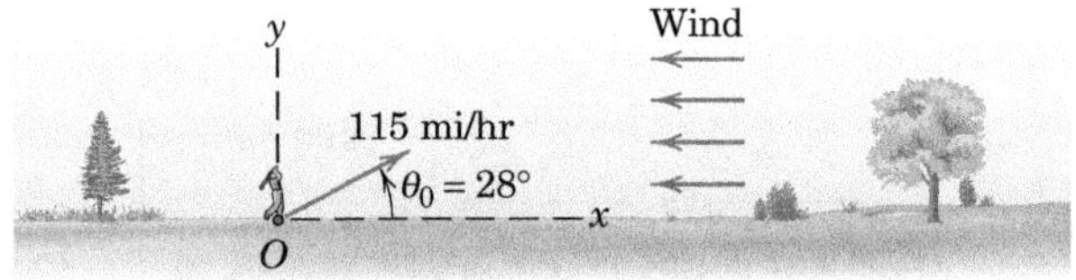

Problem 2/164

2/7 Space Curvilinear Motion

The general case of three-dimensional motion of a particle along a space curve was introduced in Art. 2/1 and illustrated in Fig. 2/1. Three coordinate systems, rectangular (x-y-z), cylindrical (r-θ-z), and spherical (R-θ-ϕ), are commonly used to describe this motion. These systems are indicated in Fig. 2/16, which also shows the unit vectors for the three coordinate systems.*

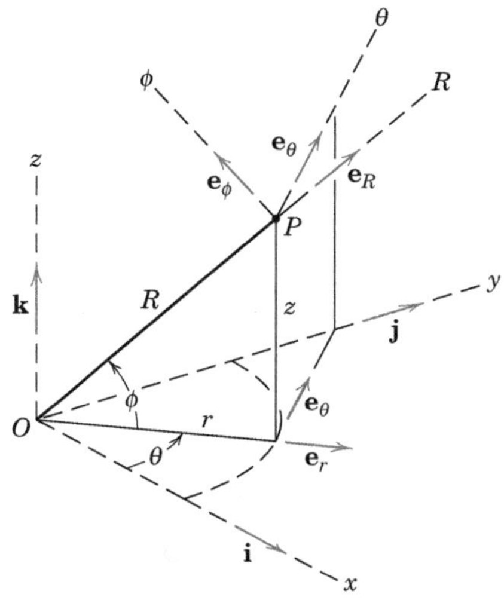

Figure 2/16

Before describing the use of these coordinate systems, we note that a path-variable description, using n- and t-coordinates, which we developed in Art. 2/5, can be applied in the osculating plane shown in Fig. 2/1. We defined this plane as the plane which contains the curve at the location in question. We see that the velocity $\mathbf{v}$, which is along the tangent t to the curve, lies in the osculating plane. The acceleration $\mathbf{a}$ also lies in the osculating plane. As in the case of plane motion, it has a component $a_t = \dot{v}$ tangent to the path due to the change in magnitude of the velocity and a component $a_n = v^2/\rho$ normal to the curve due to the change in direction of the velocity. As before, ρ is the radius of curvature of the path at the point in question and is measured in the osculating plane. This description of motion, which is natural and direct for many plane-motion problems, is awkward to use for space motion because the osculating plane continually shifts its orientation. We will confine our attention, therefore, to the three fixed coordinate systems shown in Fig. 2/16.

Rectangular Coordinates (x-y-z)

The extension from two to three dimensions offers no particular difficulty. We merely add the z-coordinate and its two time derivatives to the two-dimensional expressions of Eqs. 2/6 so that the position vector $\mathbf{R}$, the velocity $\mathbf{v}$, and the acceleration $\mathbf{a}$ become

$$\begin{aligned} \mathbf{R} &= x\mathbf{i} + y\mathbf{j} + z\mathbf{k} \\ \mathbf{v} = \dot{\mathbf{R}} &= \dot{x}\mathbf{i} + \dot{y}\mathbf{j} + \dot{z}\mathbf{k} \\ \mathbf{a} = \dot{\mathbf{v}} = \ddot{\mathbf{R}} &= \ddot{x}\mathbf{i} + \ddot{y}\mathbf{j} + \ddot{z}\mathbf{k} \end{aligned} \quad \textbf{(2/15)}$$

Note that in three dimensions we are using $\mathbf{R}$ in place of $\mathbf{r}$ for the position vector.

Cylindrical Coordinates (r-θ-z)

If we understand the polar-coordinate description of plane motion, then there should be no difficulty with cylindrical coordinates because all that is required is the addition of the z-coordinate and its two time derivatives. The position vector $\mathbf{R}$ to the particle for cylindrical coordinates is simply

$$\mathbf{R} = r\mathbf{e}_r + z\mathbf{k}$$

*In a variation of spherical coordinates commonly used, angle ϕ is replaced by its complement.

In place of Eq. 2/13 for plane motion, we can write the velocity as

$$\boxed{\mathbf{v} = \dot{r}\mathbf{e}_r + r\dot{\theta}\mathbf{e}_\theta + \dot{z}\mathbf{k}} \qquad \textbf{(2/16)}$$

where

$$\begin{aligned} v_r &= \dot{r} \\ v_\theta &= r\dot{\theta} \\ v_z &= \dot{z} \\ v &= \sqrt{{v_r}^2 + {v_\theta}^2 + {v_z}^2} \end{aligned}$$

Similarly, the acceleration is written by adding the z-component to Eq. 2/14, which gives us

$$\boxed{\mathbf{a} = (\ddot{r} - r\dot{\theta}^2)\mathbf{e}_r + (r\ddot{\theta} + 2\dot{r}\dot{\theta})\mathbf{e}_\theta + \ddot{z}\mathbf{k}} \qquad \textbf{(2/17)}$$

where

$$\begin{aligned} a_r &= \ddot{r} - r\dot{\theta}^2 \\ a_\theta &= r\ddot{\theta} + 2\dot{r}\dot{\theta} = \frac{1}{r}\frac{d}{dt}(r^2\dot{\theta}) \\ a_z &= \ddot{z} \\ a &= \sqrt{{a_r}^2 + {a_\theta}^2 + {a_z}^2} \end{aligned}$$

Whereas the unit vectors $\mathbf{e}_r$ and $\mathbf{e}_\theta$ have nonzero time derivatives due to the changes in their directions, we note that the unit vector $\mathbf{k}$ in the z-direction remains fixed in direction and therefore has a zero time derivative.

Nightman 1965/Shutterstock

With base rotation and ladder elevation, spherical coordinates would be a good choice for determining the acceleration of the upper end of the extending ladder.

Spherical Coordinates (R-θ-ϕ)

Spherical coordinates R, θ, ϕ are utilized when a radial distance and two angles are utilized to specify the position of a particle, as in the case of radar measurements, for example. Derivation of the expression for the velocity $\mathbf{v}$ is easily obtained, but the expression for the acceleration $\mathbf{a}$ is more complex because of the added geometry. Consequently, only the results will be cited here.* First we designate unit vectors $\mathbf{e}_R$, $\mathbf{e}_\theta$, $\mathbf{e}_\phi$ as shown in Fig. 2/16. Note that the unit vector $\mathbf{e}_R$ is in the direction in which the particle P would move if R increases but θ and ϕ are held constant. The unit vector $\mathbf{e}_\theta$ is in the direction in which P would move if θ increases while R and ϕ are held constant. Finally, the unit vector $\mathbf{e}_\phi$ is in the direction in which P would move if ϕ increases while R and θ are held constant. The resulting expressions for $\mathbf{v}$ and $\mathbf{a}$ are

$$\boxed{\mathbf{v} = v_R\mathbf{e}_R + v_\theta\mathbf{e}_\theta + v_\phi\mathbf{e}_\phi} \qquad \textbf{(2/18)}$$

where

$$\begin{aligned} v_R &= \dot{R} \\ v_\theta &= R\dot{\theta}\cos\phi \\ v_\phi &= R\dot{\phi} \end{aligned}$$

*For a complete derivation of **v** and **a** in spherical coordinates, see the first author's book *Dynamics*, 2nd edition, 1971, or SI Version, 1975 (John Wiley & Sons, Inc.).

and

$$\boxed{\mathbf{a} = a_R\mathbf{e}_R + a_\theta\mathbf{e}_\theta + a_\phi\mathbf{e}_\phi} \qquad \textbf{(2/19)}$$

where

$$a_R = \ddot{R} - R\dot{\phi}^2 - R\dot{\theta}^2\cos^2\phi$$

$$a_\theta = \frac{\cos\phi}{R}\frac{d}{dt}(R^2\dot{\theta}) - 2R\dot{\theta}\dot{\phi}\sin\phi$$

$$a_\phi = \frac{1}{R}\frac{d}{dt}(R^2\dot{\phi}) + R\dot{\theta}^2\sin\phi\cos\phi$$

Linear algebraic transformations between any two of the three coordinate-system expressions for velocity or acceleration can be developed. These transformations make it possible to express the motion component in rectangular coordinates, for example, if the components are known in spherical coordinates, or vice versa.* These transformations are easily handled with the aid of matrix algebra and a simple computer program.

A portion of the track of this amusement-park ride is in the shape of a helix whose axis is horizontal.

*These coordinate transformations are developed and illustrated in the first author's book *Dynamics*, 2nd edition, 1971, or SI Version, 1975 (John Wiley & Sons, Inc.).

Sample Problem 2/11

The power screw starts from rest and is given a rotational speed $\dot{\theta}$ which increases uniformly with time t according to $\dot{\theta} = kt$, where k is a constant. Determine the expressions for the velocity v and acceleration a of the center of ball A when the screw has turned through one complete revolution from rest. The lead of the screw (advancement per revolution) is L.

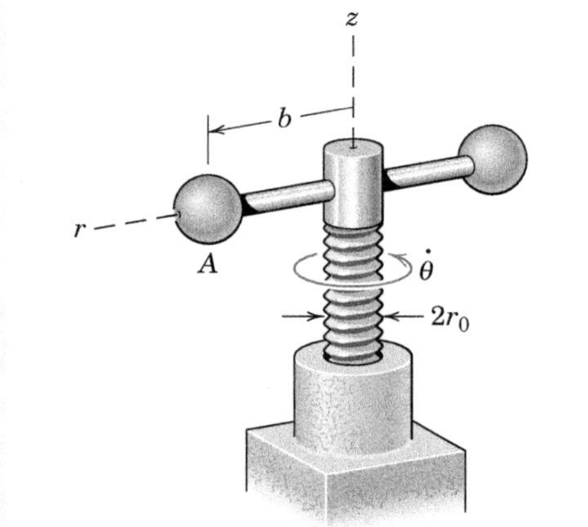

Solution. The center of ball A moves in a helix on the cylindrical surface of radius b, and the cylindrical coordinates r, θ, z are clearly indicated.

Integrating the given relation for $\dot{\theta}$ gives $\theta = \Delta\theta = \int \dot{\theta}\, dt = \frac{1}{2}kt^2$. For one revolution from rest we have

$$2\pi = \frac{1}{2}kt^2$$

giving

$$t = 2\sqrt{\pi/k}$$

Thus, the angular rate at one revolution is

$$\dot{\theta} = kt = k(2\sqrt{\pi/k}) = 2\sqrt{\pi k}$$

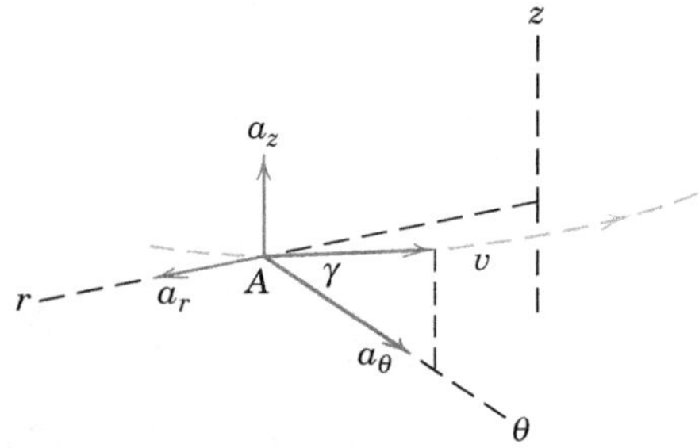

① The helix angle γ of the path followed by the center of the ball governs the relation between the θ- and z-components of velocity and is given by $\tan\gamma = L/(2\pi b)$. Now from the figure we see that $v_\theta = v\cos\gamma$. Substituting $v_\theta = r\dot{\theta} = b\dot{\theta}$
② from Eq. 2/16 gives $v = v_\theta/\cos\gamma = b\dot{\theta}/\cos\gamma$. With $\cos\gamma$ obtained from $\tan\gamma$ and with $\dot{\theta} = 2\sqrt{\pi k}$, we have for the one-revolution position

$$v = 2b\sqrt{\pi k}\,\frac{\sqrt{L^2 + 4\pi^2 b^2}}{2\pi b} = \sqrt{\frac{k}{\pi}}\sqrt{L^2 + 4\pi^2 b^2} \qquad \textit{Ans.}$$

The acceleration components from Eq. 2/17 become

③ $[a_r = \ddot{r} - r\dot{\theta}^2]$ $\qquad a_r = 0 - b(2\sqrt{\pi k})^2 = -4b\pi k$

$[a_\theta = r\ddot{\theta} + 2\dot{r}\dot{\theta}]$ $\qquad a_\theta = bk + 2(0)(2\sqrt{\pi k}) = bk$

$[a_z = \ddot{z} = \dot{v}_z]$ $\qquad a_z = \frac{d}{dt}(v_z) = \frac{d}{dt}(v_\theta \tan\gamma) = \frac{d}{dt}(b\dot{\theta}\tan\gamma)$

$$= (b\tan\gamma)\ddot{\theta} = b\frac{L}{2\pi b}k = \frac{kL}{2\pi}$$

Now we combine the components to give the magnitude of the total acceleration, which becomes

$$a = \sqrt{(-4b\pi k)^2 + (bk)^2 + \left(\frac{kL}{2\pi}\right)^2}$$

$$= bk\sqrt{(1 + 16\pi^2) + L^2/(4\pi^2 b^2)} \qquad \textit{Ans.}$$

Helpful Hints

① We must be careful to divide the lead L by the circumference $2\pi b$ and not the diameter $2b$ to obtain $\tan\gamma$. If in doubt, unwrap one turn of the helix traced by the center of the ball.

② Sketch a right triangle and recall that for $\tan\beta = a/b$ the cosine of β becomes $b/\sqrt{a^2 + b^2}$.

③ The negative sign for a_r is consistent with our previous knowledge that the normal component of acceleration is directed toward the center of curvature.

Sample Problem 2/12

An aircraft P takes off at A with a velocity v_0 of 250 km/h and climbs in the vertical y'-z' plane at the constant 15° angle with an acceleration along its flight path of 0.8 m/s^2. Flight progress is monitored by radar at point O. (*a*) Resolve the velocity of P into cylindrical-coordinate components 60 seconds after takeoff and find $\dot{r}$, $\dot{\theta}$, and $\dot{z}$ for that instant. (*b*) Resolve the velocity of the aircraft P into spherical-coordinate components 60 seconds after takeoff and find $\dot{R}$, $\dot{\theta}$, and $\dot{\phi}$ for that instant.

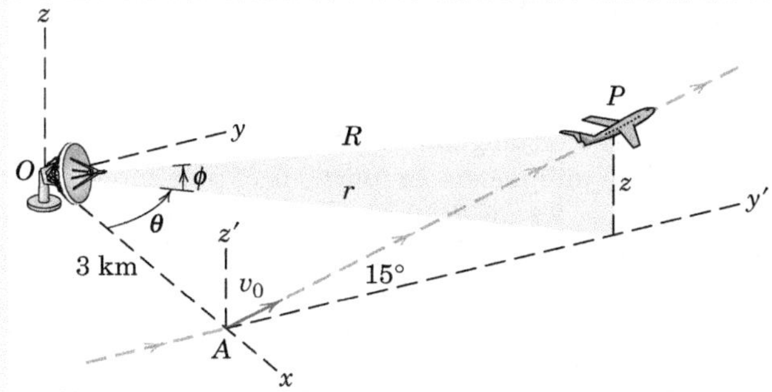

Solution. **(a)** The accompanying figure shows the velocity and acceleration vectors in the y'-z' plane. The takeoff speed is

$$v_0 = \frac{250}{3.6} = 69.4 \text{ m/s}$$

and the speed after 60 seconds is

$$v = v_0 + at = 69.4 + 0.8(60) = 117.4 \text{ m/s}$$

The distance s traveled after takeoff is

$$s = s_0 + v_0 t + \frac{1}{2}at^2 = 0 + 69.4(60) + \frac{1}{2}(0.8)(60)^2 = 5610 \text{ m}$$

The y-coordinate and associated angle θ are

$$y = 5610 \cos 15° = 5420 \text{ m}$$

$$\theta = \tan^{-1}\frac{5420}{3000} = 61.0°$$

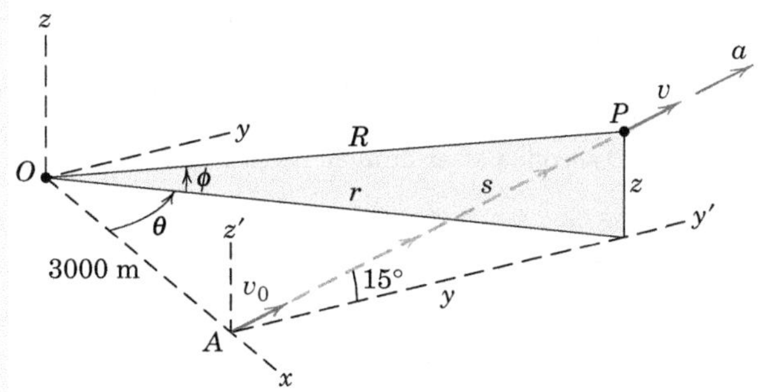

(*a*)

From the figure (*b*) of x-y projections, we have

$$r = \sqrt{3000^2 + 5420^2} = 6190 \text{ m}$$

$$v_{xy} = v \cos 15° = 117.4 \cos 15° = 113.4 \text{ m/s}$$

$$v_r = \dot{r} = v_{xy} \sin \theta = 113.4 \sin 61.0° = 99.2 \text{ m/s} \qquad \textit{Ans.}$$

$$v_\theta = r\dot{\theta} = v_{xy} \cos \theta = 113.4 \cos 61.0° = 55.0 \text{ m/s}$$

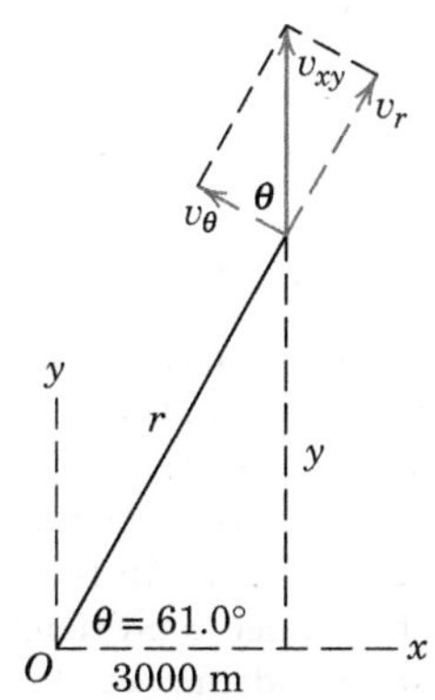

(*b*)

So

$$\dot{\theta} = \frac{55.0}{6190} = 8.88(10^{-3}) \text{ rad/s} \qquad \textit{Ans.}$$

Finally $\dot{z} = v_z = v \sin 15° = 117.4 \sin 15° = 30.4$ m/s *Ans.*

(b) Refer to the accompanying figure (*c*), which shows the x-y plane and various velocity components projected into the vertical plane containing r and R. Note that

$$z = y \tan 15° = 5420 \tan 15° = 1451 \text{ m}$$

$$\phi = \tan^{-1}\frac{z}{r} = \tan^{-1}\frac{1451}{6190} = 13.19°$$

$$R = \sqrt{r^2 + z^2} = \sqrt{6190^2 + 1451^2} = 6360 \text{ m}$$

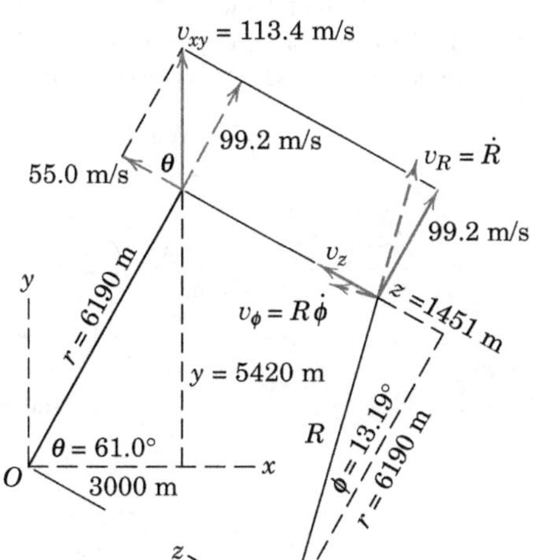

(*c*)

From the figure,

$$v_R = \dot{R} = 99.2 \cos 13.19° + 30.4 \sin 13.19° = 103.6 \text{ m/s} \qquad \textit{Ans.}$$

$$\dot{\theta} = 8.88(10^{-3}) \text{ rad/s, as in part } (a) \qquad \textit{Ans.}$$

$$v_\phi = R\dot{\phi} = 30.4 \cos 13.19° - 99.2 \sin 13.19° = 6.95 \text{ m/s}$$

$$\dot{\phi} = \frac{6.95}{6360} = 1.093(10^{-3}) \text{ rad/s} \qquad \textit{Ans.}$$

PROBLEMS

Introductory Problems

2/165 The rectangular coordinates of a particle are given in millimeters as functions of time t in seconds by $x = 30 \cos 2t$, $y = 40 \sin 2t$, and $z = 20t + 3t^2$. Determine the angle θ_1 between the position vector $\mathbf{r}$ and the velocity $\mathbf{v}$ and the angle θ_2 between the position vector $\mathbf{r}$ and the acceleration $\mathbf{a}$, both at time $t = 2$ s.

2/166 A projectile is launched from point O with an initial velocity of magnitude $v_0 = 600$ ft/sec, directed as shown in the figure. Compute the x-, y-, and z-components of position, velocity, and acceleration 20 seconds after launch. Neglect aerodynamic drag.

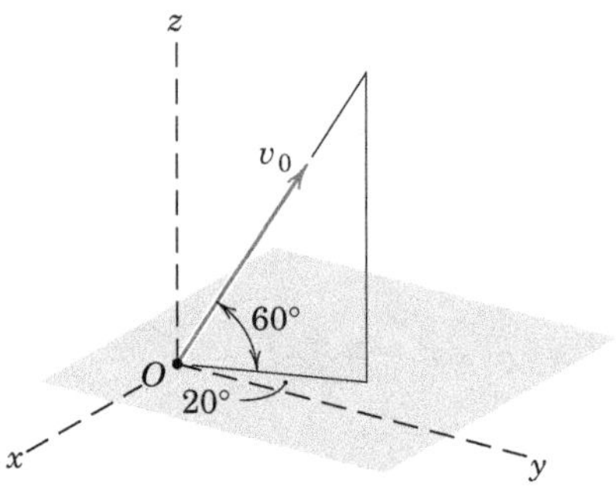

Problem 2/166

2/167 A projectile is launched from point O at a speed $v_0 = 80$ m/s with the goal of hitting the target A. At the launch instant, a strong horizontal wind begins blowing and imparts a constant acceleration of 1.25 m/s^2 to the projectile in the same direction as the wind. If the launch conditions are chosen so that the particle would impact the target in the absence of wind effects, determine the impact coordinates of the projectile. (*Note:* There is more than one solution to this problem, so choose only answers which feature positive angles less than 90°. Select the impact location which is closest to the target.)

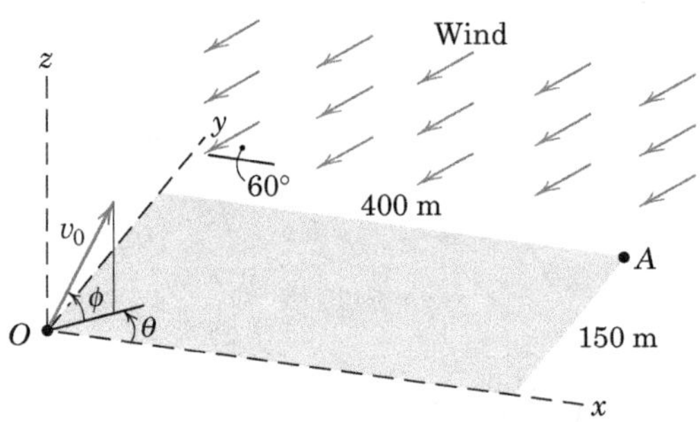

Problem 2/167

2/168 If the launch speed of the projectile in Prob. 2/167 remains unchanged, what values of θ and ϕ (positive and less than 90°) will ensure that the projectile impacts the target at A if the wind conditions are considered. Please list both possible combinations of the angles.

2/169 An amusement ride called the "corkscrew" takes the passengers through the upside-down curve of a horizontal cylindrical helix. The velocity of the cars as they pass position A is 15 m/s, and the component of their acceleration measured along the tangent to the path is $g \cos \gamma$ at this point. The effective radius of the cylindrical helix is 5 m, and the helix angle is $\gamma = 40°$. Compute the magnitude of the acceleration of the passengers as they pass position A.

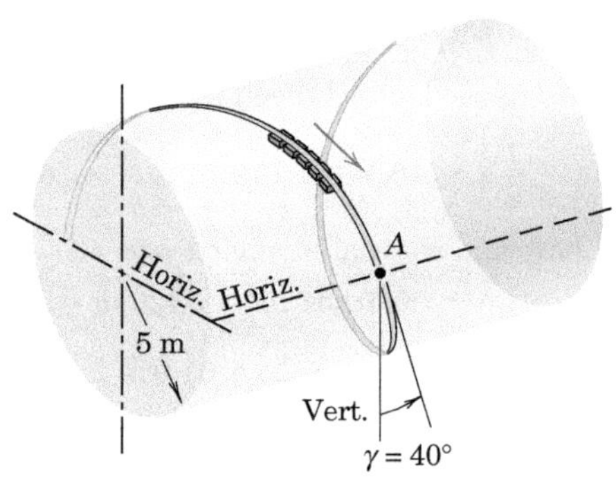

Problem 2/169

2/170 The radar antenna at P tracks the jet aircraft A, which is flying horizontally at a speed u and an altitude h above the level of P. Determine the expressions for the components of the velocity in the spherical coordinates of the antenna motion.

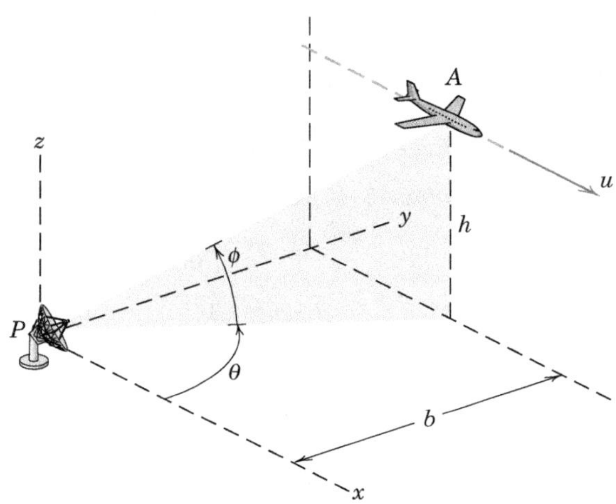

Problem 2/170

2/171 The rotating element in a mixing chamber is given a periodic axial movement $z = z_0 \sin 2\pi nt$ while it is rotating at the constant angular velocity $\dot{\theta} = \omega$. Determine the expression for the maximum magnitude of the acceleration of a point A on the rim of radius r. The frequency n of vertical oscillation is constant.

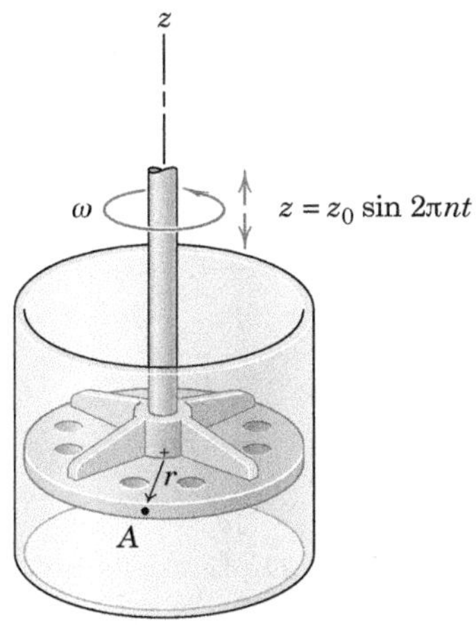

Problem 2/171

Representative Problems

2/172 A helicopter starts from rest at point A and travels along the indicated path with a constant acceleration a. If the helicopter has a speed of 60 m/s when it reaches B, determine the values of $\dot{R}$, $\dot{\theta}$, and $\dot{\phi}$ as measured by the radar tracking device at O at the instant when h = 100 m.

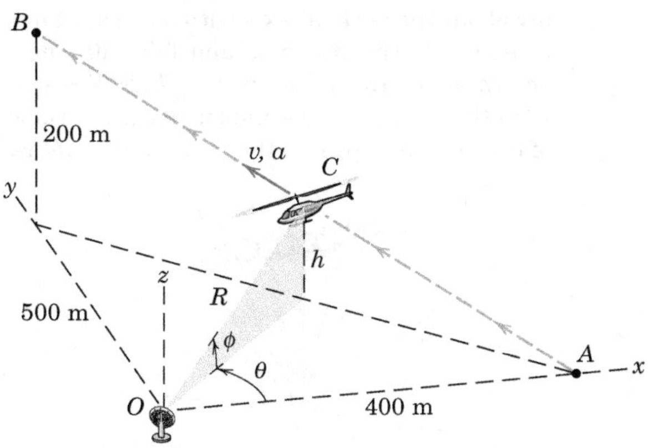

Problem 2/172

2/173 For the helicopter of Prob. 2/172, find the values of $\ddot{R}$, $\ddot{\theta}$, and $\ddot{\phi}$ for the radar tracking device at O at the instant when h = 100 m. Refer to the printed answers for Prob. 2/172 as needed.

2/174 The vertical shaft of the industrial robot rotates at the constant rate ω. The length h of the vertical shaft has a known time history, and this is true of its time derivatives $\dot{h}$ and $\ddot{h}$ as well. Likewise, the values of l, $\dot{l}$, and $\ddot{l}$ are known. Determine the magnitudes of the velocity and acceleration of point P. The lengths h_0 and l_0 are fixed.

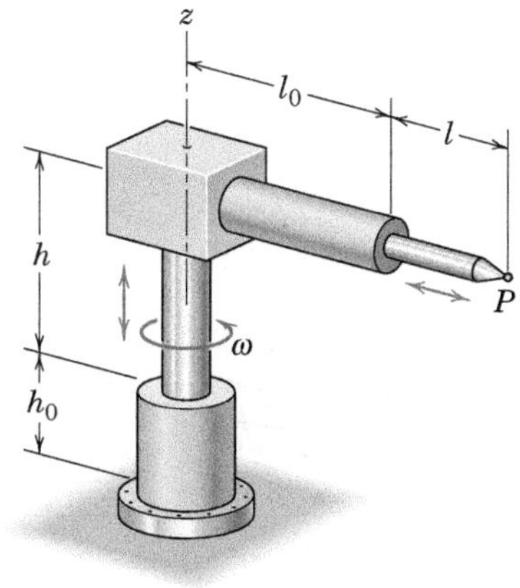

Problem 2/174

2/175 An industrial robot is being used to position a small part P. Calculate the magnitude of the acceleration $\mathbf{a}$ of P for the instant when $\beta = 30°$ if $\dot{\beta}$ = 10 deg/s and $\ddot{\beta}$ = 20 deg/s^2 at this same instant. The base of the robot is revolving at the constant rate ω = 40 deg/s. During the motion arms AO and AP remain perpendicular.

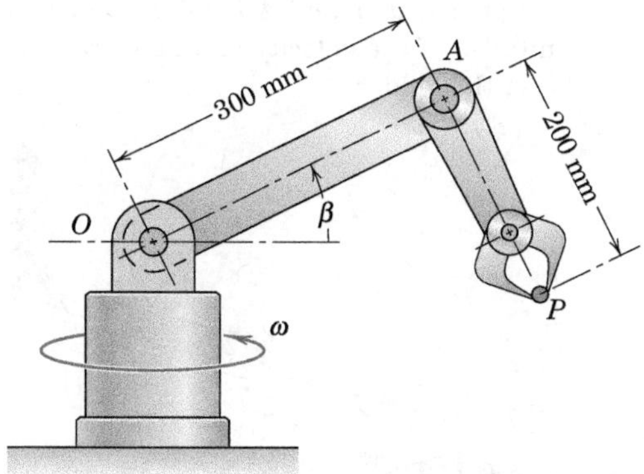

Problem 2/175

2/176 The car A is ascending a parking-garage ramp in the form of a cylindrical helix of 24-ft radius rising 10 ft for each half turn. At the position shown the car has a speed of 15 mi/hr, which is decreasing at the rate of 2 mi/hr per second. Determine the r-, θ-, and z-components of the acceleration of the car.

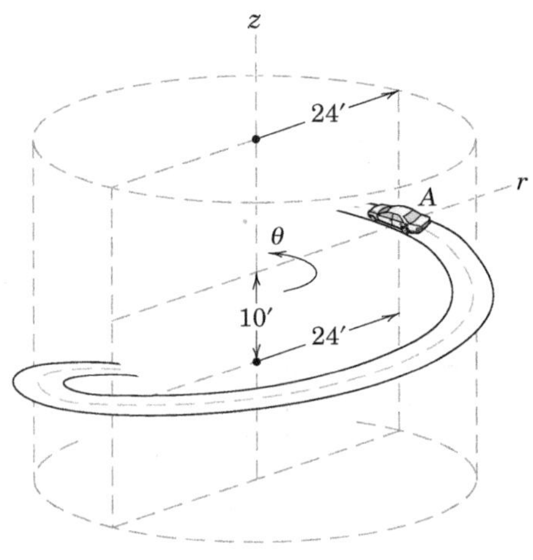

Problem 2/176

2/177 The base structure of the firetruck ladder rotates about a vertical axis through O with a constant angular velocity $\Omega = 10$ deg/s. At the same time, the ladder unit OB elevates at a constant rate $\dot{\phi} = 7$ deg/s, and section AB of the ladder extends from within section OA at the constant rate of 0.5 m/s. At the instant under consideration, $\phi = 30°$, $\overline{OA} = 9$ m, and $\overline{AB} = 6$ m. Determine the magnitudes of the velocity and acceleration of the end B of the ladder.

Problem 2/177

2/178 The rod OA is held at the constant angle $\beta = 30°$ while it rotates about the vertical with a constant angular rate $\dot{\theta} = 120$ rev/min. Simultaneously, the sliding ball P oscillates along the rod with its distance in millimeters from the fixed pivot O given by $R = 200 + 50 \sin 2\pi nt$, where the frequency n of oscillation along the rod is a constant 2 cycles per second and where t is the time in seconds. Calculate the magnitude of the acceleration of P for an instant when its velocity along the rod from O toward A is a maximum.

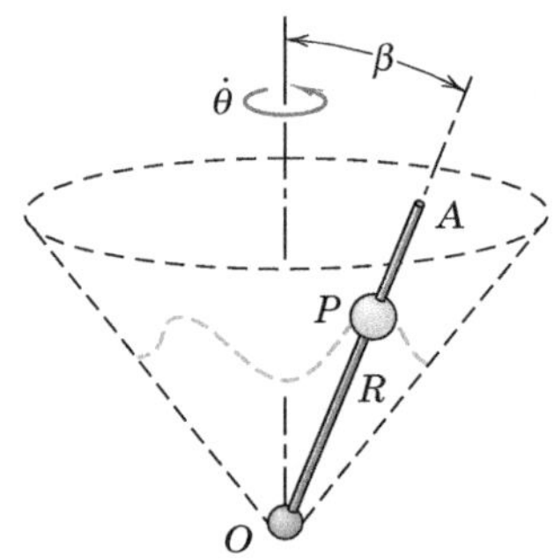

Problem 2/178

►2/179 Beginning with Eq. 2/18, the expression for particle velocity in spherical coordinates, derive the acceleration components in Eq. 2/19. (*Note:* Start by writing the unit vectors for the R-, θ-, and ϕ-coordinates in terms of the fixed unit vectors **i**, **j**, and **k**.)

►2/180 In the design of an amusement-park ride, the cars are attached to arms of length R which are hinged to a central rotating collar which drives the assembly about the vertical axis with a constant angular rate $\omega = \dot{\theta}$. The cars rise and fall with the track according to the relation $z = (h/2)(1 - \cos 2\theta)$. Find the R-, θ-, and ϕ-components of the velocity $\mathbf{v}$ of each car as it passes the position $\theta = \pi/4$ rad.

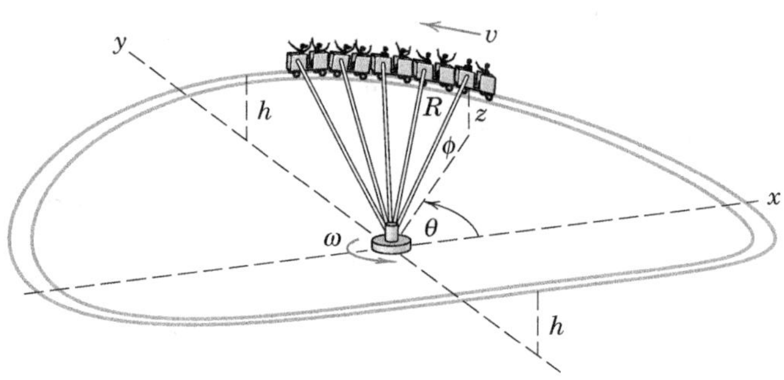

Problem 2/180

▶**2/181** The particle P moves down the spiral path which is wrapped around the surface of a right circular cone of base radius b and altitude h. The angle γ between the tangent to the curve at any point and a horizontal tangent to the cone at this point is constant. Also the motion of the particle is controlled so that $\dot{\theta}$ is constant. Determine the expression for the radial acceleration a_r of the particle for any value of θ.

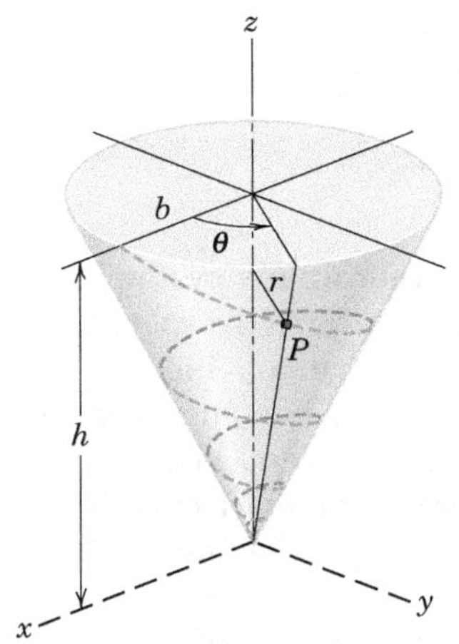

Problem 2/181

▶**2/182** The disk A rotates about the vertical z-axis with a constant speed $\omega = \dot{\theta} = \pi/3$ rad/s. Simultaneously, the hinged arm OB is elevated at the constant rate $\dot{\phi} = 2\pi/3$ rad/s. At time $t = 0$, both $\theta = 0$ and $\phi = 0$. The angle θ is measured from the fixed reference x-axis. The small sphere P slides out along the rod according to $R = 50 + 200t^2$, where R is in millimeters and t is in seconds. Determine the magnitude of the total acceleration $\mathbf{a}$ of P when $t = \frac{1}{2}$ s.

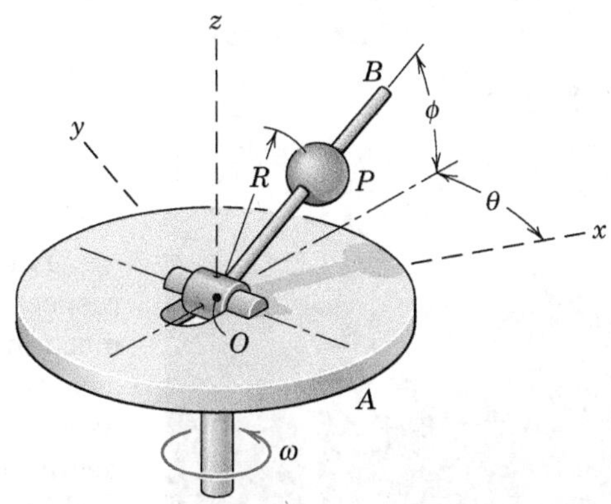

Problem 2/182

2/8 Relative Motion (Translating Axes)

Relative motion is a critical issue in the midair refueling of aircraft.

In the previous articles of this chapter, we have described particle motion using coordinates referred to fixed reference axes. The displacements, velocities, and accelerations so determined are termed *absolute*. It is not always possible or convenient, however, to use a fixed set of axes to describe or to measure motion. In addition, there are many engineering problems for which the analysis of motion is simplified by using measurements made with respect to a moving reference system. These measurements, when combined with the absolute motion of the moving coordinate system, enable us to determine the absolute motion in question. This approach is called a *relative-motion* analysis.

Choice of Coordinate System

The motion of the moving coordinate system is specified with respect to a fixed coordinate system. Strictly speaking, in Newtonian mechanics, this fixed system is the primary inertial system, which is assumed to have no motion in space. For engineering purposes, the fixed system may be taken as any system whose absolute motion is negligible for the problem at hand. For most earthbound engineering problems, it is sufficiently precise to take for the fixed reference system a set of axes attached to the earth, in which case we neglect the motion of the earth. For the motion of satellites around the earth, a nonrotating coordinate system is chosen with its origin on the axis of rotation of the earth. For interplanetary travel, a nonrotating coordinate system fixed to the sun would be used. Thus, the choice of the fixed system depends on the type of problem involved.

We will confine our attention in this article to moving reference systems which translate but do not rotate. Motion measured in rotating systems will be discussed in Art. 5/7 of Chapter 5 on rigid-body kinematics, where this approach finds special but important application. We will also confine our attention here to relative-motion analysis for plane motion.

Vector Representation

Now consider two particles A and B which may have separate curvilinear motions in a given plane or in parallel planes, Fig. 2/17. We will arbitrarily attach the origin of a set of translating (nonrotating) axes x-y to particle B and observe the motion of A from our moving position on B. The position vector of A as measured relative to the frame x-y is $\mathbf{r}_{A/B} = x\mathbf{i} + y\mathbf{j}$, where the subscript notation "A/B" means "A relative to B" or "A with respect to B." The unit vectors along the x- and y-axes are $\mathbf{i}$ and $\mathbf{j}$, and x and y are the coordinates of A measured in the x-y frame. The absolute position of B is defined by the vector $\mathbf{r}_B$ measured from the origin of the fixed axes X-Y. The absolute position of A is seen, therefore, to be determined by the vector

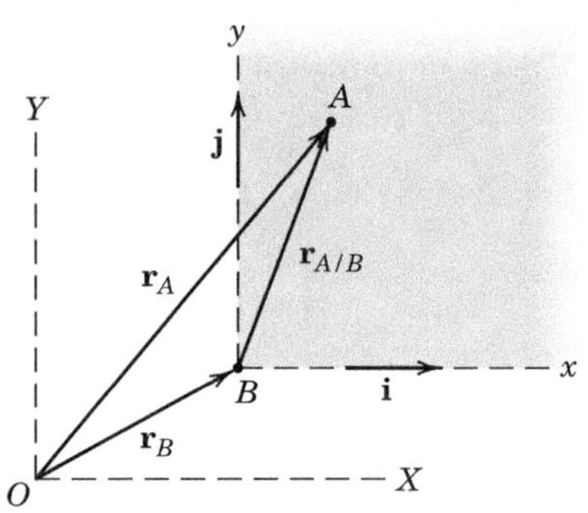

Figure 2/17

$$\mathbf{r}_A = \mathbf{r}_B + \mathbf{r}_{A/B}$$

We now differentiate this vector equation once with respect to time to obtain velocities and twice to obtain accelerations. Thus,

$$\dot{\mathbf{r}}_A = \dot{\mathbf{r}}_B + \dot{\mathbf{r}}_{A/B} \qquad \text{or} \qquad \boxed{\mathbf{v}_A = \mathbf{v}_B + \mathbf{v}_{A/B}} \qquad \textbf{(2/20)}$$

$$\ddot{\mathbf{r}}_A = \ddot{\mathbf{r}}_B + \ddot{\mathbf{r}}_{A/B} \qquad \text{or} \qquad \boxed{\mathbf{a}_A = \mathbf{a}_B + \mathbf{a}_{A/B}} \qquad \textbf{(2/21)}$$

In Eq. 2/20 the velocity which we observe A to have from our position at B attached to the moving axes x-y is $\dot{\mathbf{r}}_{A/B} = \mathbf{v}_{A/B} = \dot{x}\mathbf{i} + \dot{y}\mathbf{j}$. This term is the velocity of A with respect to B. Similarly, in Eq. 2/21 the acceleration which we observe A to have from our nonrotating position on B is $\ddot{\mathbf{r}}_{A/B} = \dot{\mathbf{v}}_{A/B} = \ddot{x}\mathbf{i} + \ddot{y}\mathbf{j}$. This term is the acceleration of A with respect to B. We note that the unit vectors **i** and **j** have zero derivatives because their directions as well as their magnitudes remain unchanged. (Later when we discuss rotating reference axes, we must account for the derivatives of the unit vectors when they change direction.)

Equation 2/20 (or 2/21) states that the absolute velocity (or acceleration) of A equals the absolute velocity (acceleration) of B plus, vectorially, the velocity (or acceleration) of A relative to B. The relative term is the velocity (or acceleration) measurement which an observer attached to the moving coordinate system x-y would make. We can express the relative-motion terms in whatever coordinate system is convenient—rectangular, normal and tangential, or polar—and the formulations in the preceding articles can be used for this purpose. The appropriate fixed system of the previous articles becomes the moving system in the present article.

Additional Considerations

The selection of the moving point B for attachment of the reference coordinate system is arbitrary. As shown in Fig. 2/18, point A could be used just as well for the attachment of the moving system, in which case the three corresponding relative-motion equations for position, velocity, and acceleration are

$$\mathbf{r}_B = \mathbf{r}_A + \mathbf{r}_{B/A} \qquad \mathbf{v}_B = \mathbf{v}_A + \mathbf{v}_{B/A} \qquad \mathbf{a}_B = \mathbf{a}_A + \mathbf{a}_{B/A}$$

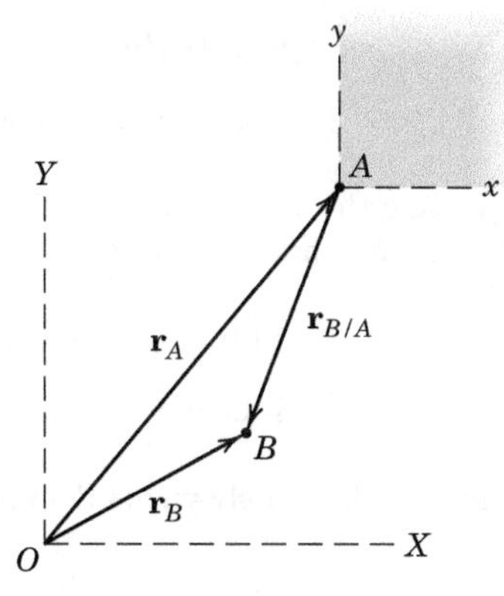

Figure 2/18

It is seen, therefore, that $\mathbf{r}_{B/A} = -\mathbf{r}_{A/B}$, $\mathbf{v}_{B/A} = -\mathbf{v}_{A/B}$, and $\mathbf{a}_{B/A} = -\mathbf{a}_{A/B}$.

In relative-motion analysis, it is important to realize that the acceleration of a particle as observed in a translating system x-y is the same as that observed in a fixed system X-Y if the moving system has a constant velocity. This conclusion broadens the application of Newton's second law of motion (Chapter 3). We conclude, consequently, that a set of axes which has a constant absolute velocity may be used in place of a "fixed" system for the determination of accelerations. A translating reference system which has no acceleration is called an *inertial system*.

Sample Problem 2/13

Passengers in the jet transport A flying east at a speed of 800 km/h observe a second jet plane B that passes under the transport in horizontal flight. Although the nose of B is pointed in the 45° northeast direction, plane B appears to the passengers in A to be moving away from the transport at the 60° angle as shown. Determine the true velocity of B.

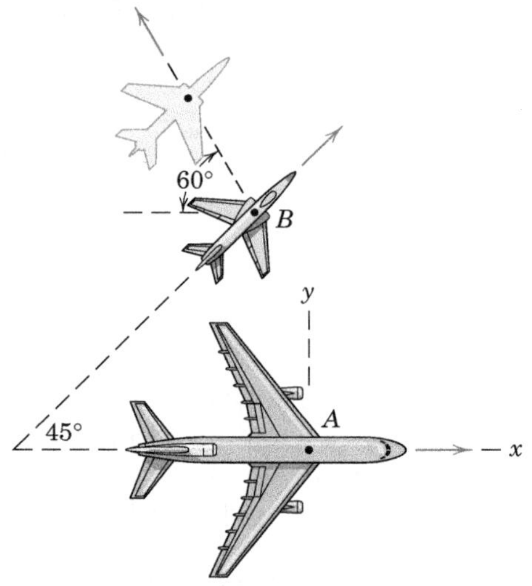

Solution. The moving reference axes x-y are attached to A, from which the relative observations are made. We write, therefore,

① $$\mathbf{v}_B = \mathbf{v}_A + \mathbf{v}_{B/A}$$

Next we identify the knowns and unknowns. The velocity $\mathbf{v}_A$ is given in both magnitude and direction. The 60° direction of $\mathbf{v}_{B/A}$, the velocity which B appears
② to have to the moving observers in A, is known, and the true velocity of B is in the 45° direction in which it is heading. The two remaining unknowns are the magni-
③ tudes of $\mathbf{v}_B$ and $\mathbf{v}_{B/A}$. We may solve the vector equation in any one of three ways.

(I) Graphical. We start the vector sum at some point P by drawing $\mathbf{v}_A$ to a convenient scale and then construct a line through the tip of $\mathbf{v}_A$ with the known direction of $\mathbf{v}_{B/A}$. The known direction of $\mathbf{v}_B$ is then drawn through P, and the intersection C yields the unique solution enabling us to complete the vector triangle and scale off the unknown magnitudes, which are found to be

$$v_{B/A} = 586 \text{ km/h} \qquad \text{and} \qquad v_B = 717 \text{ km/h} \qquad \textit{Ans.}$$

(II) Trigonometric. A sketch of the vector triangle is made to reveal the trigonometry, which gives

④ $$\frac{v_B}{\sin 60°} = \frac{v_A}{\sin 75°} \qquad v_B = 800\frac{\sin 60°}{\sin 75°} = 717 \text{ km/h} \qquad \textit{Ans.}$$

(III) Vector Algebra. Using unit vectors $\mathbf{i}$ and $\mathbf{j}$, we express the velocities in vector form as

$$\mathbf{v}_A = 800\mathbf{i} \text{ km/h} \qquad \mathbf{v}_B = (v_B \cos 45°)\mathbf{i} + (v_B \sin 45°)\mathbf{j}$$

$$\mathbf{v}_{B/A} = (v_{B/A} \cos 60°)(-\mathbf{i}) + (v_{B/A} \sin 60°)\mathbf{j}$$

Substituting these relations into the relative-velocity equation and solving separately for the $\mathbf{i}$ and $\mathbf{j}$ terms give

$$(\mathbf{i}\text{-terms}) \qquad v_B \cos 45° = 800 - v_{B/A} \cos 60°$$

$$(\mathbf{j}\text{-terms}) \qquad v_B \sin 45° = v_{B/A} \sin 60°$$

⑤ Solving simultaneously yields the unknown velocity magnitudes

$$v_{B/A} = 586 \text{ km/h} \qquad \text{and} \qquad v_B = 717 \text{ km/h} \qquad \textit{Ans.}$$

It is worth noting the solution of this problem from the viewpoint of an observer in B. With reference axes attached to B, we would write $\mathbf{v}_A = \mathbf{v}_B + \mathbf{v}_{A/B}$. The apparent velocity of A as observed by B is then $\mathbf{v}_{A/B}$, which is the negative of $\mathbf{v}_{B/A}$.

Helpful Hints

① We treat each airplane as a particle.

② We assume no side slip due to cross wind.

③ Students should become familiar with all three solutions.

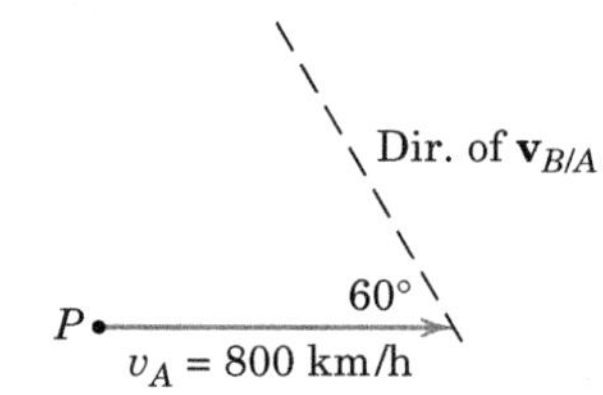

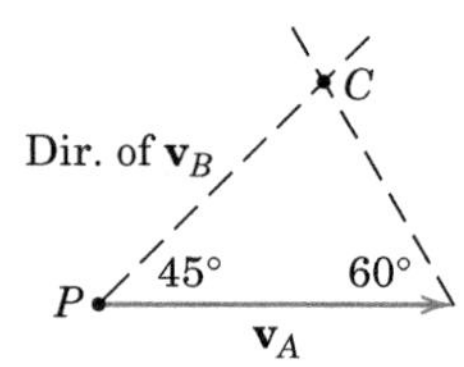

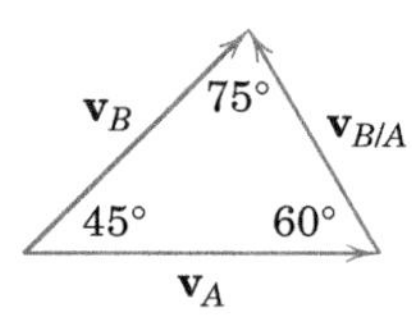

④ We must be prepared to recognize the appropriate trigonometric relation, which here is the law of sines.

⑤ We can see that the graphical or trigonometric solution is shorter than the vector algebra solution in this particular problem.

Sample Problem 2/14

Car A is accelerating in the direction of its motion at the rate of 3 ft/sec^2. Car B is rounding a curve of 440-ft radius at a constant speed of 30 mi/hr. Determine the velocity and acceleration which car B appears to have to an observer in car A if car A has reached a speed of 45 mi/hr for the positions represented.

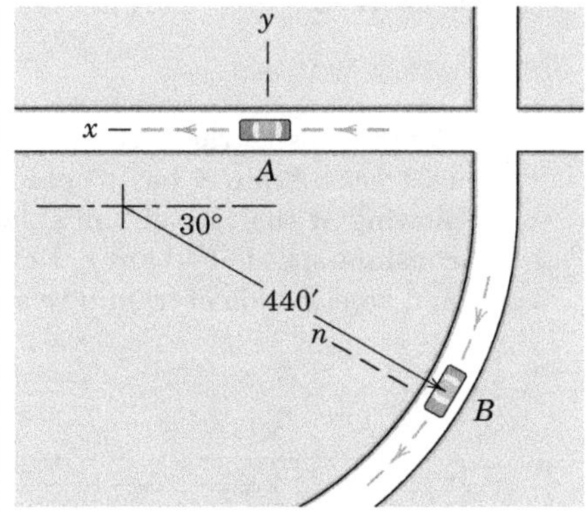

Solution. We choose nonrotating reference axes attached to car A since the motion of B with respect to A is desired.

Velocity. The relative-velocity equation is

$$\mathbf{v}_B = \mathbf{v}_A + \mathbf{v}_{B/A}$$

and the velocities of A and B for the position considered have the magnitudes

$$v_A = 45\frac{5280}{60^2} = 45\frac{44}{30} = 66 \text{ ft/sec} \qquad v_B = 30\frac{44}{30} = 44 \text{ ft/sec}$$

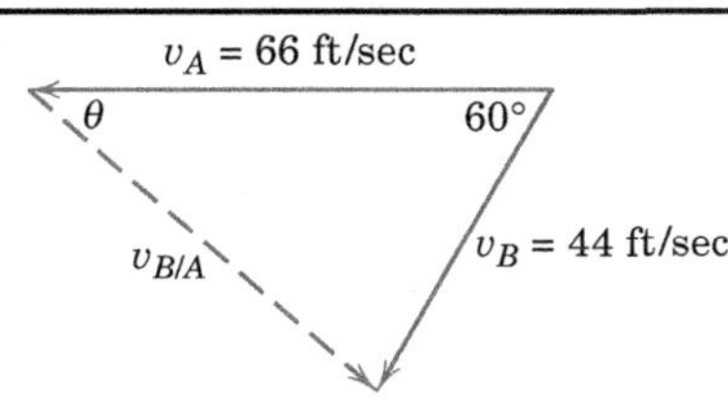

The triangle of velocity vectors is drawn in the sequence required by the equation, and application of the law of cosines and the law of sines gives

① $v_{B/A} = 58.2$ ft/sec $\qquad \theta = 40.9°$ *Ans.*

Acceleration. The relative-acceleration equation is

$$\mathbf{a}_B = \mathbf{a}_A + \mathbf{a}_{B/A}$$

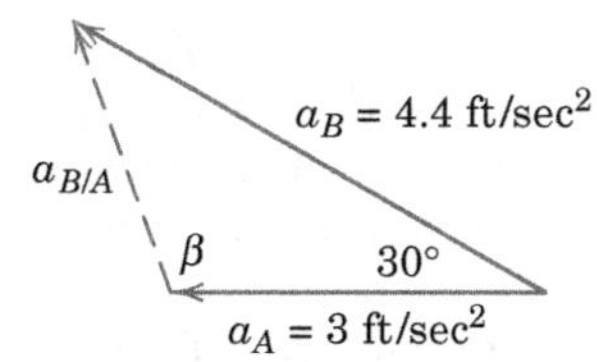

The acceleration of A is given, and the acceleration of B is normal to the curve in the n-direction and has the magnitude

$[a_n = v^2/\rho]$ $\qquad a_B = (44)^2/440 = 4.4$ ft/sec^2

The triangle of acceleration vectors is drawn in the sequence required by the equation as illustrated. Solving for the x- and y-components of $\mathbf{a}_{B/A}$ gives us

$$(a_{B/A})_x = 4.4 \cos 30° - 3 = 0.810 \text{ ft/sec}^2$$

$$(a_{B/A})_y = 4.4 \sin 30° = 2.2 \text{ ft/sec}^2$$

from which $a_{B/A} = \sqrt{(0.810)^2 + (2.2)^2} = 2.34$ ft/sec^2 *Ans.*

The direction of $\mathbf{a}_{B/A}$ may be specified by the angle β which, by the law of sines, becomes

② $\dfrac{4.4}{\sin \beta} = \dfrac{2.34}{\sin 30°} \qquad \beta = \sin^{-1}\left(\dfrac{4.4}{2.34}0.5\right) = 110.2°$ *Ans.*

Helpful Hints

① Alternatively, we could use either a graphical or a vector algebraic solution.

② Be careful to choose between the two values 69.8° and 180 − 69.8 = 110.2°.

Suggestion: To gain familiarity with the manipulation of vector equations, it is suggested that the student rewrite the relative-motion equations in the form $\mathbf{v}_{B/A} = \mathbf{v}_B - \mathbf{v}_A$ and $\mathbf{a}_{B/A} = \mathbf{a}_B - \mathbf{a}_A$ and redraw the vector polygons to conform with these alternative relations.

Caution: So far we are only prepared to handle motion relative to *nonrotating* axes. If we had attached the reference axes rigidly to car B, they would rotate with the car, and we would find that the velocity and acceleration terms relative to the rotating axes are *not* the negative of those measured from the nonrotating axes moving with A. Rotating axes are treated in Art. 5/7.

PROBLEMS

Introductory Problems

2/183 Rapid-transit trains A and B travel on parallel tracks. Train A has a speed of 80 km/h and is slowing at the rate of 2 m/s^2, while train B has a constant speed of 40 km/h. Determine the velocity and acceleration of train B relative to train A.

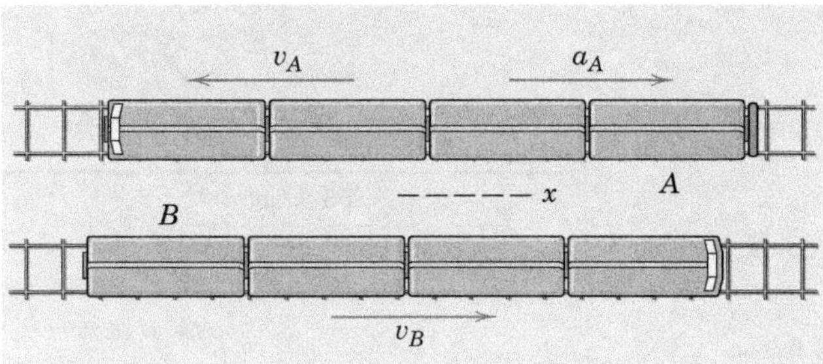

Problem 2/183

2/184 Train A is traveling at a constant speed v_A = 35 mi/hr while car B travels in a straight line along the road as shown at a constant speed v_B. A conductor C in the train begins to walk to the rear of the train car at a constant speed of 4 ft/sec relative to the train. If the conductor perceives car B to move directly westward at 16 ft/sec, how fast is the car traveling?

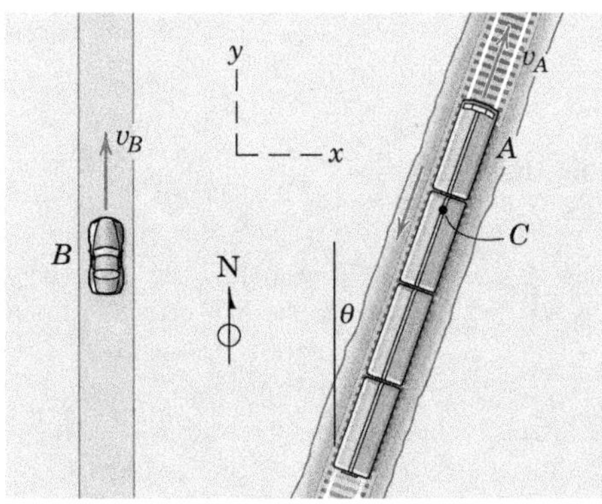

Problem 2/184

2/185 The jet transport B is flying north with a velocity v_B = 600 km/h when a smaller aircraft A passes underneath the transport headed in the 60° direction shown. To passengers in B, however, A appears to be flying sideways and moving east. Determine the actual velocity of A and the velocity which A appears to have relative to B.

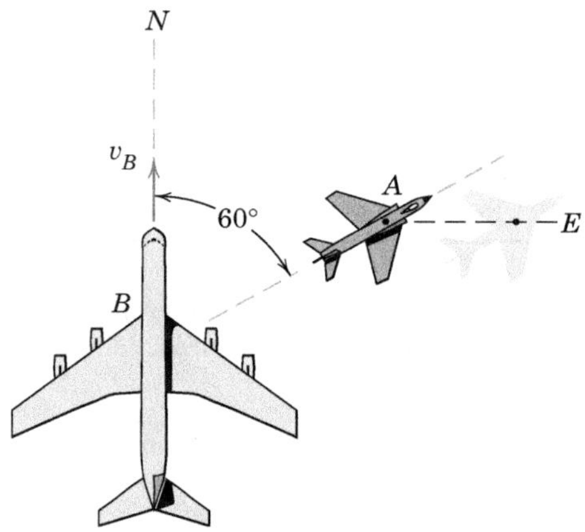

Problem 2/185

2/186 A helicopter approaches a rescue scene. A victim P is drifting along with the river current of speed v_C = 2 m/s. The wind is blowing at a speed v_W = 3 m/s as indicated. Determine the velocity relative to the wind which the helicopter must acquire so that it maintains a steady overhead position relative to the victim.

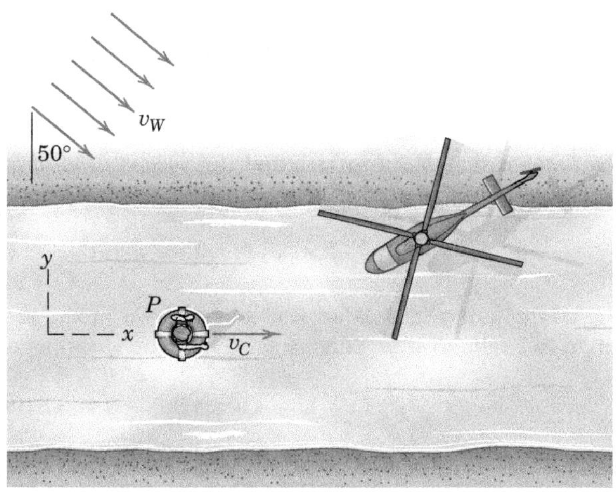

Problem 2/186

2/187 A ship capable of making a speed of 16 knots through still water is to maintain a true course due west while encountering a 3-knot current running from north to south. What should be the heading of the ship (measured clockwise from the north to the nearest degree)? How long does it take the ship to proceed 24 nautical miles due west?

2/188 Train A travels with a constant speed $v_A = 120$ km/h along the straight and level track. The driver of car B, anticipating the railway grade crossing C, decreases the car speed of 90 km/h at the rate of 3 m/s^2. Determine the velocity and acceleration of the train relative to the car.

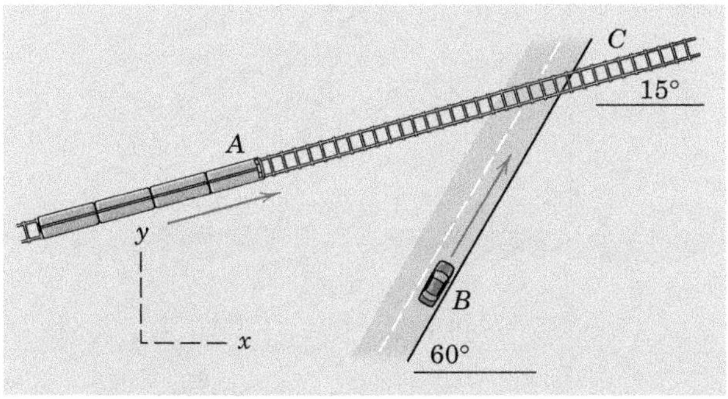

Problem 2/188

2/189 The car A has a forward speed of 18 km/h and is accelerating at 3 m/s^2. Determine the velocity and acceleration of the car relative to observer B, who rides in a nonrotating chair on the Ferris wheel. The angular rate $\Omega = 3$ rev/min of the Ferris wheel is constant.

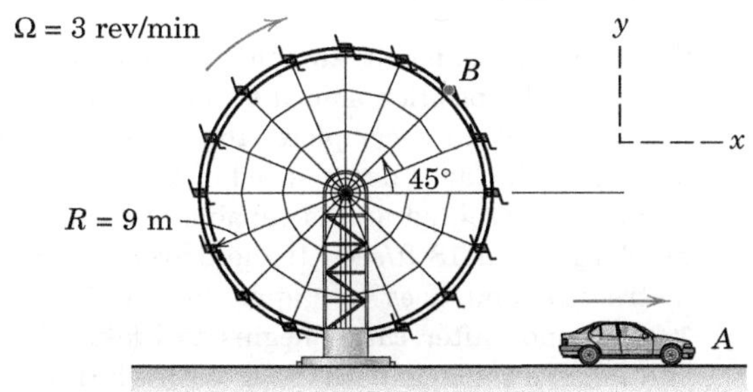

Problem 2/189

2/190 For the instant represented, car A has an acceleration in the direction of its motion, and car B has a speed of 45 mi/hr which is increasing. If the acceleration of B as observed from A is zero for this instant, determine the acceleration of A and the rate at which the speed of B is changing.

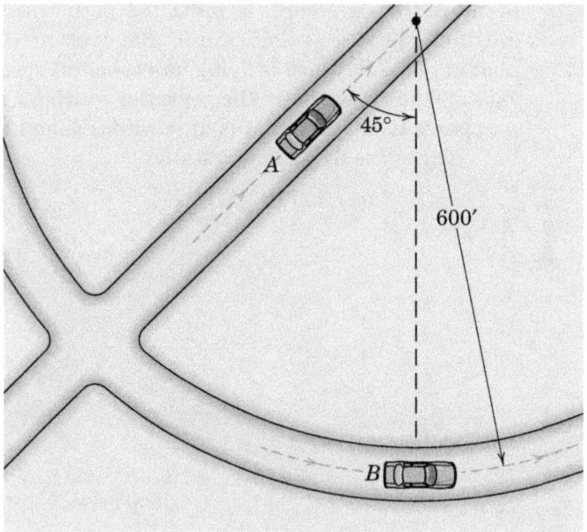

Problem 2/190

Representative Problems

2/191 A drop of water falls with no initial speed from point A of a highway overpass. After dropping 6 m, it strikes the windshield at point B of a car which is traveling at a speed of 100 km/h on the horizontal road. If the windshield is inclined 50° from the vertical as shown, determine the angle θ relative to the normal n to the windshield at which the water drop strikes.

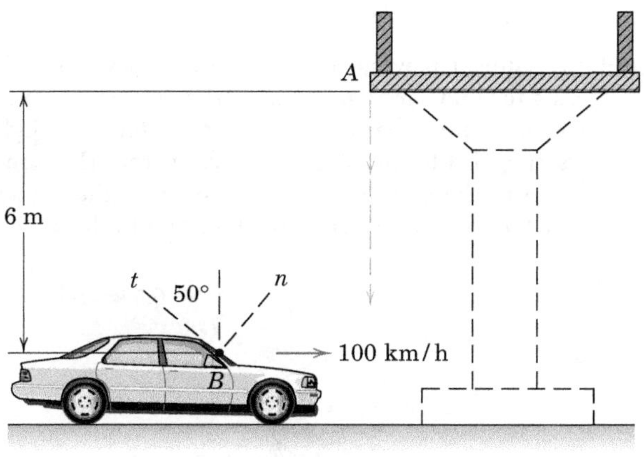

Problem 2/191

2/192 Plane A travels along the indicated path with a constant speed $v_A = 285$ km/h. Relative to the pilot in plane B, which is flying at a constant speed $v_B = 350$ km/h, what are the velocities which plane A appears to have when it is at positions C and E? Both planes are flying horizontally.

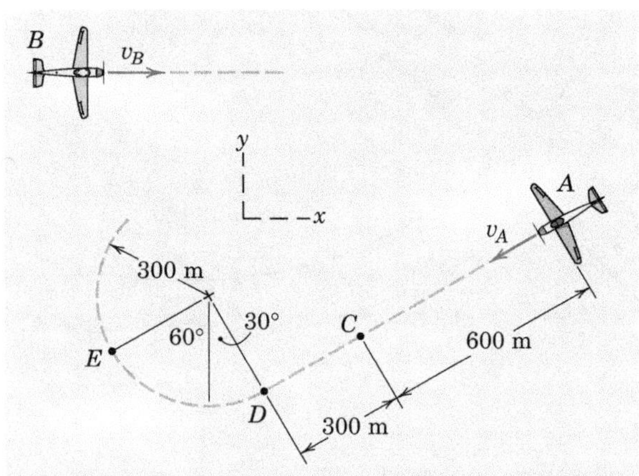

Problem 2/192

2/193 For the planes of Prob. 2/192, beginning at the position shown, plane A increases its speed at a constant rate and acquires a speed of 415 km/h by the time it reaches position E, while plane B experiences a steady deceleration of 1.5 m/s^2. Relative to the pilot in plane B, what are the velocities and accelerations which plane A appears to have when it is at positions C and E?

2/194 A sailboat moving in the direction shown is tacking to windward against a north wind. The log registers a hull speed of 6.5 knots. A "telltale" (light string tied to the rigging) indicates that the direction of the apparent wind is 35° from the centerline of the boat. What is the true wind velocity v_w?

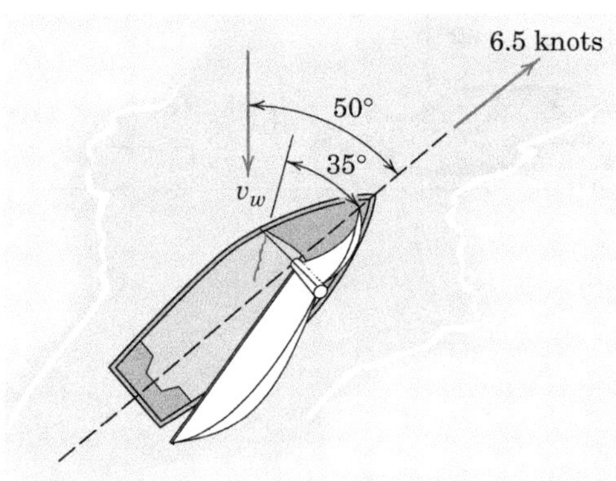

Problem 2/194

2/195 At the instant illustrated, car B has a speed of 30 km/h and car A has a speed of 40 km/h. Determine the values of $\dot{r}$ and $\dot{\theta}$ for this instant where r and θ are measured relative to a longitudinal axis fixed to car B as indicated in the figure.

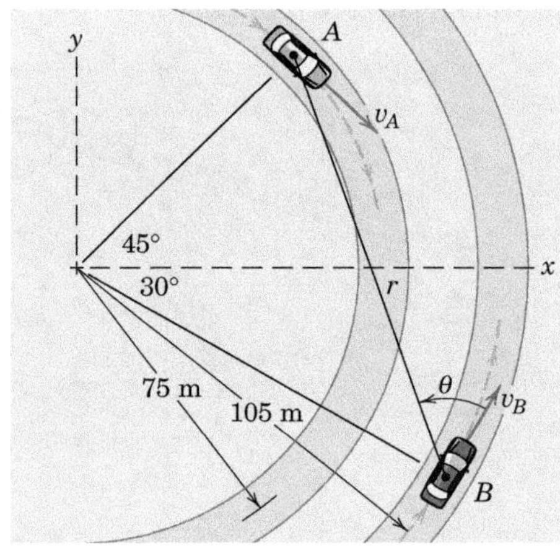

Problem 2/195

2/196 For the cars of Prob. 2/195, determine the instantaneous values of $\ddot{r}$ and $\ddot{\theta}$ if car A is slowing down at a rate of 1.25 m/s^2 and car B is speeding up at a rate of 2.5 m/s^2. Refer to the printed answers for Prob. 2/195 as needed.

2/197 Car A is traveling at 25 mi/hr and applies the brakes at the position shown so as to arrive at the intersection C at a complete stop with a constant deceleration. Car B has a speed of 40 mi/hr at the instant represented and is capable of a maximum deceleration of 18 ft/sec^2. If the driver of car B is distracted, and does not apply his brakes until 1.30 seconds after car A begins to brake, the result being a collision with car A, with what relative speed will car B strike car A? Treat both cars as particles.

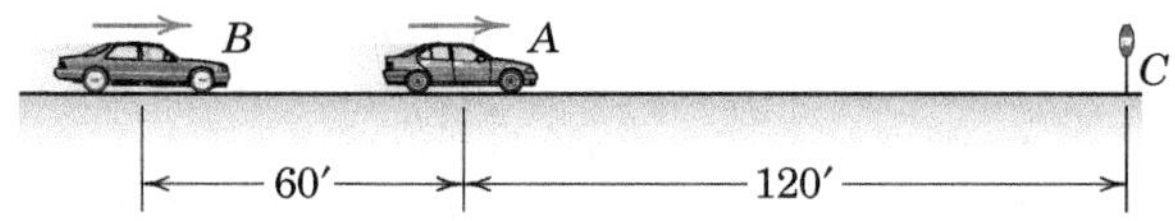

Problem 2/197

2/198 As part of an unmanned-autonomous-vehicle (UAV) demonstration, an unmanned vehicle B launches a projectile A from the position shown while traveling at a constant speed of 30 km/h. The projectile is launched with a speed of 70 m/s relative to the vehicle. At what launch angle α should the projectile be fired to ensure that it strikes a target at C? Compare your answer with that for the case where the vehicle is stationary.

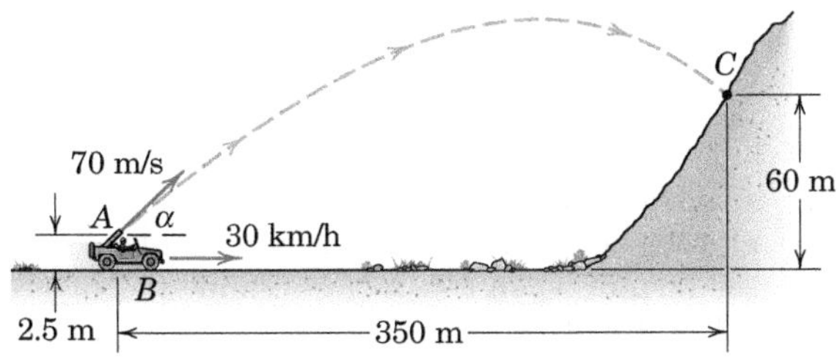

Problem 2/198

2/199 The shuttle orbiter A is in a circular orbit of altitude 200 mi, while spacecraft B is in a geosynchronous circular orbit of altitude 22,300 mi. Determine the acceleration of B relative to a nonrotating observer in shuttle A. Use $g_0 = 32.23$ ft/sec^2 for the surface-level gravitational acceleration and $R = 3959$ mi for the radius of the earth.

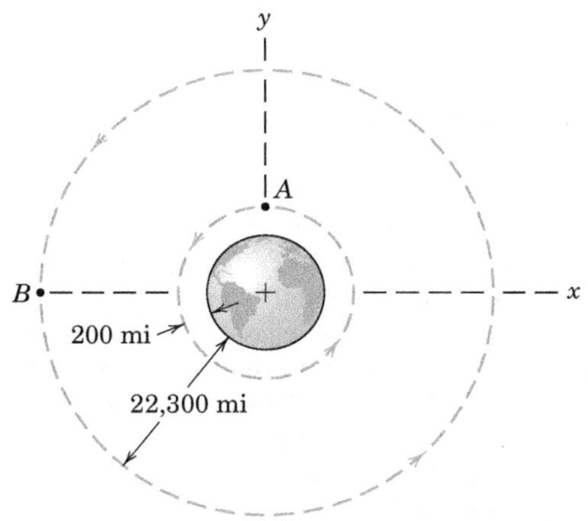

Problem 2/199

2/200 After starting from the position marked with the "x", a football receiver B runs the slant-in pattern shown, making a cut at P and thereafter running with a constant speed $v_B = 7$ yd/sec in the direction shown. The quarterback releases the ball with a horizontal velocity of 100 ft/sec at the instant the receiver passes point P. Determine the angle α at which the quarterback must throw the ball, and the velocity of the ball relative to the receiver when the ball is caught. Neglect any vertical motion of the ball.

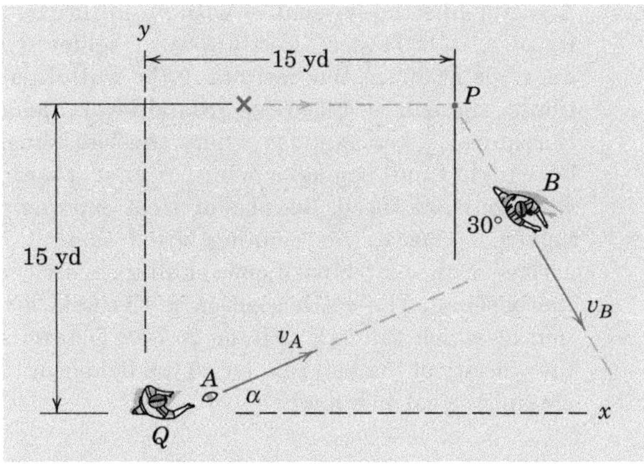

Problem 2/200

2/201 Car A is traveling at the constant speed of 60 km/h as it rounds the circular curve of 300-m radius and at the instant represented is at the position $\theta = 45°$. Car B is traveling at the constant speed of 80 km/h and passes the center of the circle at this same instant. Car A is located with respect to car B by polar coordinates r and θ with the pole moving with B. For this instant determine $v_{A/B}$ and the values of $\dot{r}$ and $\dot{\theta}$ as measured by an observer in car B.

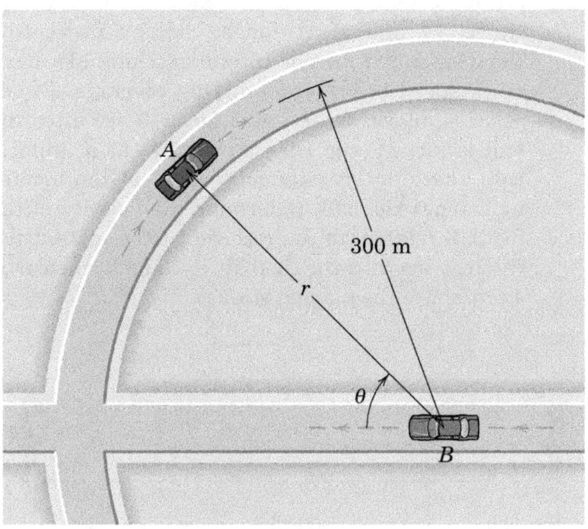

Problem 2/201

2/202 For the conditions of Prob. 2/201, determine the values of $\ddot{r}$ and $\ddot{\theta}$ as measured by an observer in car B at the instant represented. Use the results for $\dot{r}$ and $\dot{\theta}$ cited in the answers for that problem.

2/203 A batter hits the baseball A with an initial velocity of $v_0 = 100$ ft/sec directly toward fielder B at an angle of 30° to the horizontal; the initial position of the ball is 3 ft above ground level. Fielder B requires $\frac{1}{4}$ sec to judge where the ball should be caught and begins moving to that position with constant speed. Because of great experience, fielder B chooses his running speed so that he arrives at the "catch position" simultaneously with the baseball. The catch position is the field location at which the ball altitude is 7 ft. Determine the velocity of the ball relative to the fielder at the instant the catch is made.

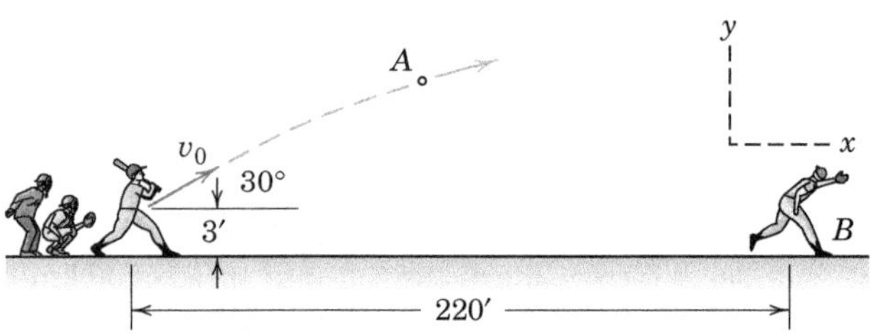

Problem 2/203

2/204 A place kicker A executes a "pooch" kick, which is designed to eliminate a potential return by the receiving team. The "pooch" kick features a high trajectory and short range, thereby preventing the deep kick returner B from reaching his maximum speed before encountering coverage. To offset this, player B hesitates for 1.15 sec after the ball is kicked, and then accelerates at a uniform rate, reaching his maximum speed at the instant he catches the ball. Determine the velocity of the football relative to the receiver at the instant the catch is made if the football is caught when it is 4.5 ft above the playing surface.

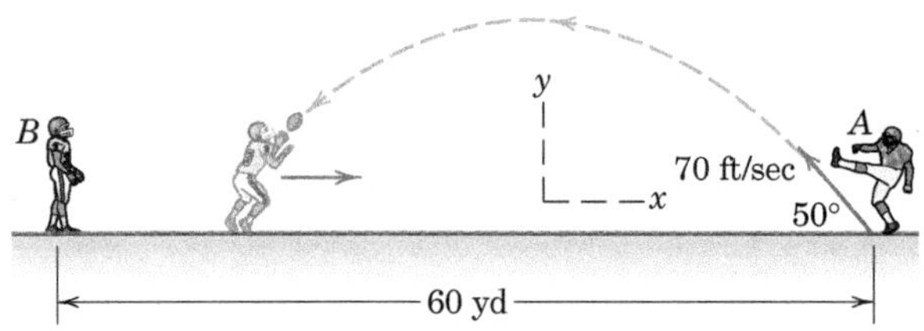

Problem 2/204

▶2/205 The aircraft A with radar detection equipment is flying horizontally at an altitude of 12 km and is increasing its speed at the rate of 1.2 m/s each second. Its radar locks onto an aircraft B flying in the same direction and in the same vertical plane at an altitude of 18 km. If A has a speed of 1000 km/h at the instant when $\theta = 30°$, determine the values of $\ddot{r}$ and $\ddot{\theta}$ at this same instant if B has a constant speed of 1500 km/h.

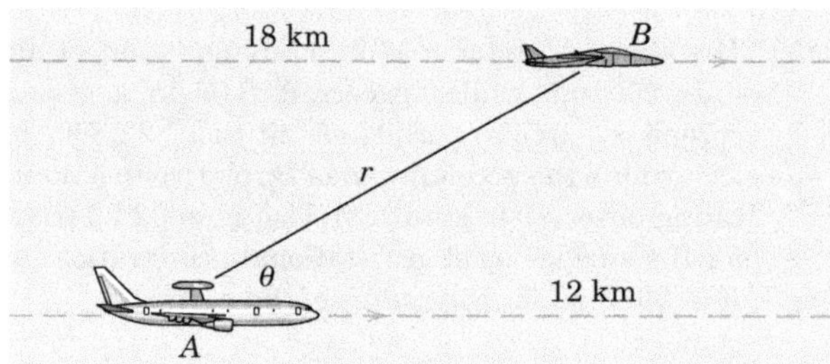

Problem 2/205

▶2/206 At a certain instant after jumping from the airplane A, a skydiver B is in the position shown and has reached a terminal (constant) speed $v_B = 50$ m/s. The airplane has the same constant speed $v_A = 50$ m/s, and after a period of level flight is just beginning to follow the circular path shown of radius $\rho_A = 2000$ m. (*a*) Determine the velocity and acceleration of the airplane relative to the skydiver. (*b*) Determine the time rate of change of the speed v_r of the airplane and the radius of curvature ρ_r of its path, both as observed by the nonrotating skydiver.

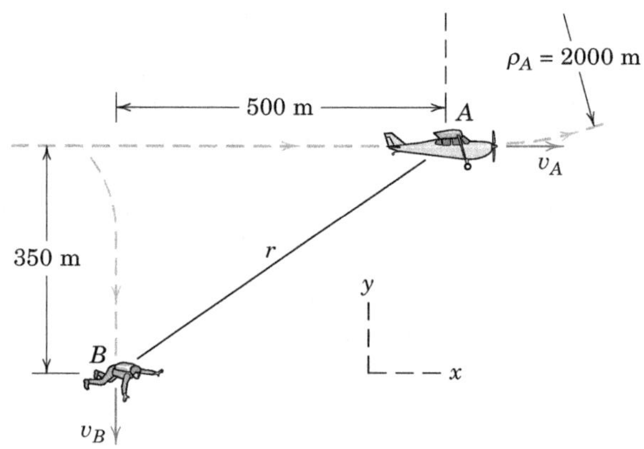

Problem 2/206

2/9 Constrained Motion of Connected Particles

Sometimes the motions of particles are interrelated because of the constraints imposed by interconnecting members. In such cases it is necessary to account for these constraints in order to determine the respective motions of the particles.

One Degree of Freedom

Consider first the very simple system of two interconnected particles A and B shown in Fig. 2/19. It should be quite evident by inspection that the horizontal motion of A is twice the vertical motion of B. Nevertheless we will use this example to illustrate the method of analysis which applies to more complex situations where the results cannot be easily obtained by inspection. The motion of B is clearly the same as that of the center of its pulley, so we establish position coordinates y and x measured from a convenient fixed datum. The total length of the cable is

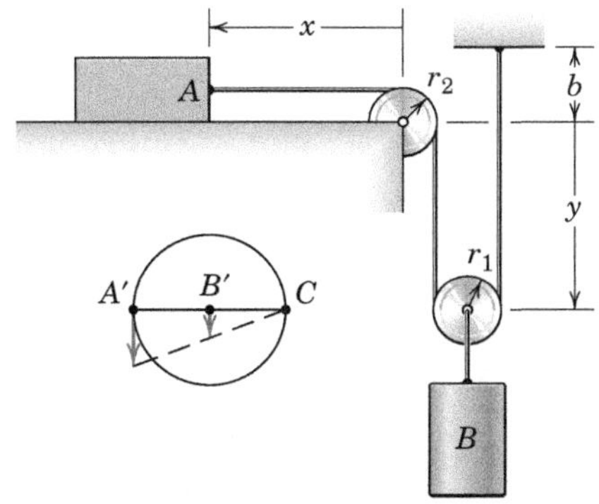

Figure 2/19

$$L = x + \frac{\pi r_2}{2} + 2y + \pi r_1 + b$$

With L, r_2, r_1, and b all constant, the first and second time derivatives of the equation give

$$0 = \dot{x} + 2\dot{y} \qquad \text{or} \qquad 0 = v_A + 2v_B$$
$$0 = \ddot{x} + 2\ddot{y} \qquad \text{or} \qquad 0 = a_A + 2a_B$$

The velocity and acceleration constraint equations indicate that, for the coordinates selected, the velocity of A must have a sign which is opposite to that of the velocity of B, and similarly for the accelerations. The constraint equations are valid for the motion of the system in either direction. We emphasize that $v_A = \dot{x}$ is positive to the left and that $v_B = \dot{y}$ is positive down.

Because the results do not depend on the lengths or pulley radii, we should be able to analyze the motion without considering them. In the lower-left portion of Fig. 2/19 is shown an enlarged view of the horizontal diameter $A'B'C'$ of the lower pulley at an instant of time. Clearly, A' and A have the same motion magnitudes, as do B and B'. During an infinitesimal motion of A', it is easy to see from the triangle that B' moves half as far as A' because point C as a point on the fixed portion of the cable momentarily has no motion. Thus, with differentiation by time in mind, we can obtain the velocity and acceleration magnitude relationships by inspection. The pulley, in effect, is a wheel which rolls on the fixed vertical cable. (The kinematics of a rolling wheel will be treated more extensively in Chapter 5 on rigid-body motion.) The system of Fig. 2/19 is said to have *one degree of freedom* since only one variable, either x or y, is needed to specify the positions of all parts of the system.

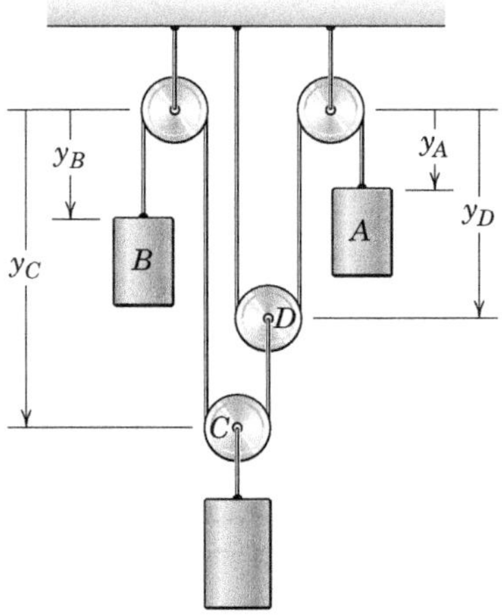

Figure 2/20

Two Degrees of Freedom

The system with *two degrees of freedom* is shown in Fig. 2/20. Here the positions of the lower cylinder and pulley C depend on the separate specifications of the two coordinates y_A and y_B. The lengths of the cables attached to cylinders A and B can be written, respectively, as

$$L_A = y_A + 2y_D + \text{constant}$$

$$L_B = y_B + y_C + (y_C - y_D) + \text{constant}$$

and their time derivatives are

$$0 = \dot{y}_A + 2\dot{y}_D \quad \text{and} \quad 0 = \dot{y}_B + 2\dot{y}_C - \dot{y}_D$$

$$0 = \ddot{y}_A + 2\ddot{y}_D \quad \text{and} \quad 0 = \ddot{y}_B + 2\ddot{y}_C - \ddot{y}_D$$

Eliminating the terms in $\dot{y}_D$ and $\ddot{y}_D$ gives

$$\dot{y}_A + 2\dot{y}_B + 4\dot{y}_C = 0 \quad \text{or} \quad v_A + 2v_B + 4v_C = 0$$

$$\ddot{y}_A + 2\ddot{y}_B + 4\ddot{y}_C = 0 \quad \text{or} \quad a_A + 2a_B + 4a_C = 0$$

It is clearly impossible for the signs of all three terms to be positive simultaneously. So, for example, if both A and B have downward (positive) velocities, then C will have an upward (negative) velocity.

These results can also be found by inspection of the motions of the two pulleys at C and D. For an increment dy_A (with y_B held fixed), the center of D moves up an amount $dy_A/2$, which causes an upward movement $dy_A/4$ of the center of C. For an increment dy_B (with y_A held fixed), the center of C moves up a distance $dy_B/2$. A combination of the two movements gives an upward movement

$$-dy_C = \frac{dy_A}{4} + \frac{dy_B}{2}$$

so that $-v_C = v_A/4 + v_B/2$ as before. Visualization of the actual geometry of the motion is an important ability.

A second type of constraint where the direction of the connecting member changes with the motion is illustrated in the second of the two sample problems which follow.

Sample Problem 2/15

In the pulley configuration shown, cylinder A has a downward velocity of 0.3 m/s. Determine the velocity of B. Solve in two ways.

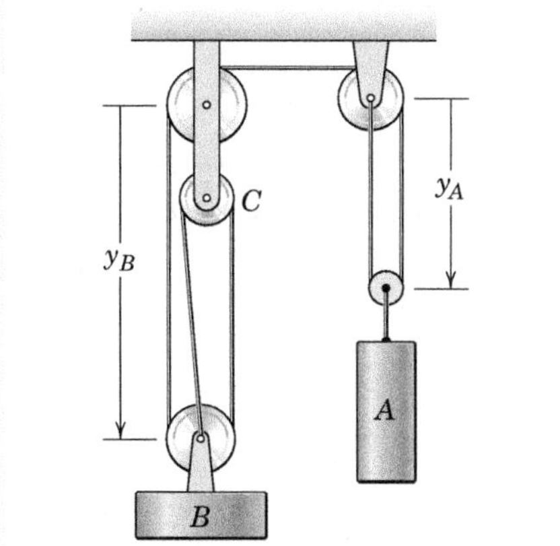

Solution (I). The centers of the pulleys at A and B are located by the coordinates y_A and y_B measured from fixed positions. The total constant length of cable in the pulley system is

$$L = 3y_B + 2y_A + \text{constants}$$

where the constants account for the fixed lengths of cable in contact with the
① circumferences of the pulleys and the constant vertical separation between the two upper left-hand pulleys. Differentiation with time gives

$$0 = 3\dot{y}_B + 2\dot{y}_A$$

Substitution of $v_A = \dot{y}_A = 0.3$ m/s and $v_B = \dot{y}_B$ gives

② $$0 = 3(v_B) + 2(0.3) \qquad \text{or} \qquad v_B = -0.2 \text{ m/s}$$ *Ans.*

Solution (II). An enlarged diagram of the pulleys at A, B, and C is shown. During a differential movement ds_A of the center of pulley A, the left end of its horizontal diameter has no motion since it is attached to the fixed part of the cable. Therefore, the right-hand end has a movement of $2ds_A$ as shown. This movement is transmitted to the left-hand end of the horizontal diameter of the pulley at B. Further, from pulley C with its fixed center, we see that the displacements on each side are equal and opposite. Thus, for pulley B, the right-hand end of the diameter has a downward displacement equal to the upward displacement ds_B of its center. By inspection of the geometry, we conclude that

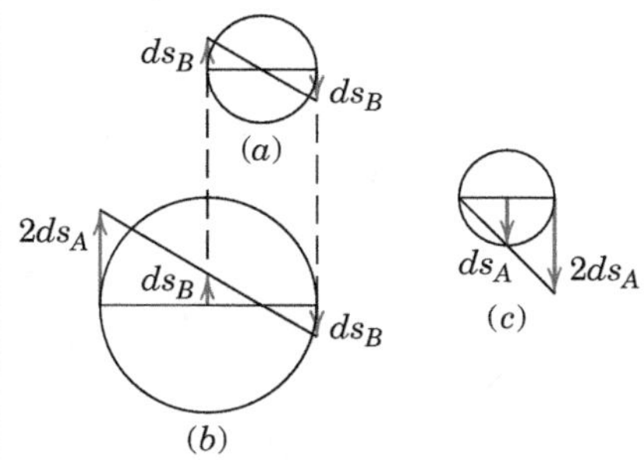

$$2ds_A = 3ds_B \qquad \text{or} \qquad ds_B = \frac{2}{3}ds_A$$

Dividing by dt gives

$$|v_B| = \frac{2}{3}v_A = \frac{2}{3}(0.3) = 0.2 \text{ m/s (upward)}$$ *Ans.*

Helpful Hints

① We neglect the small angularity of the cables between B and C.

② The negative sign indicates that the velocity of B is *upward*.

Sample Problem 2/16

The tractor A is used to hoist the bale B with the pulley arrangement shown. If A has a forward velocity v_A, determine an expression for the upward velocity v_B of the bale in terms of x.

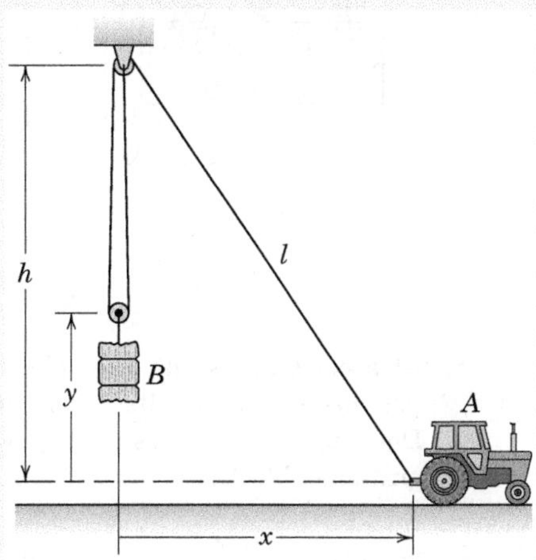

Solution. We designate the position of the tractor by the coordinate x and the position of the bale by the coordinate y, both measured from a fixed reference. The total constant length of the cable is

$$L = 2(h - y) + l = 2(h - y) + \sqrt{h^2 + x^2}$$

① Differentiation with time yields

$$0 = -2\dot{y} + \frac{x\dot{x}}{\sqrt{h^2 + x^2}}$$

Substituting $v_A = \dot{x}$ and $v_B = \dot{y}$ gives

$$v_B = \frac{1}{2}\frac{xv_A}{\sqrt{h^2 + x^2}}$$ *Ans.*

Helpful Hint

① Differentiation of the relation for a right triangle occurs frequently in mechanics.

PROBLEMS

Introductory Problems

2/207 If the velocity $\dot{x}$ of block A up the incline is increasing at the rate of 0.044 m/s each second, determine the acceleration of B.

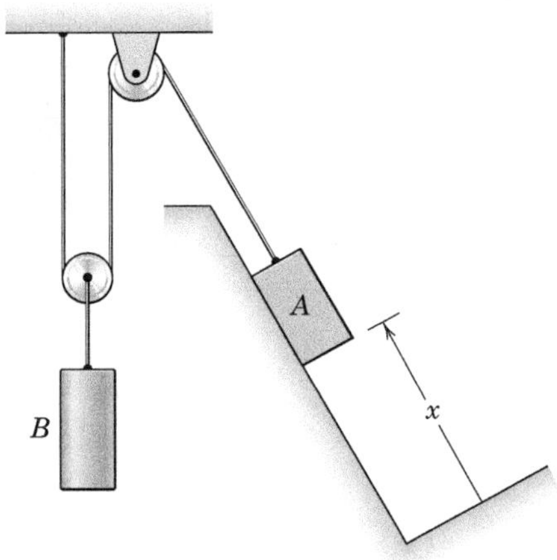

Problem 2/207

2/208 At the instant represented, $\mathbf{v}_{B/A} = 3.5\mathbf{j}$ m/s. Determine the velocity of each body at this instant. Assume that the upper surface of A remains horizontal.

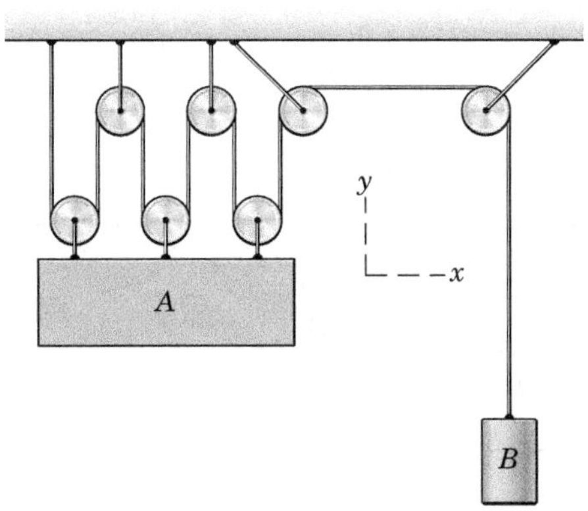

Problem 2/208

2/209 At a certain instant, the velocity of cylinder B is 1.2 m/s down and its acceleration is 2 m/s^2 up. Determine the corresponding velocity and acceleration of block A.

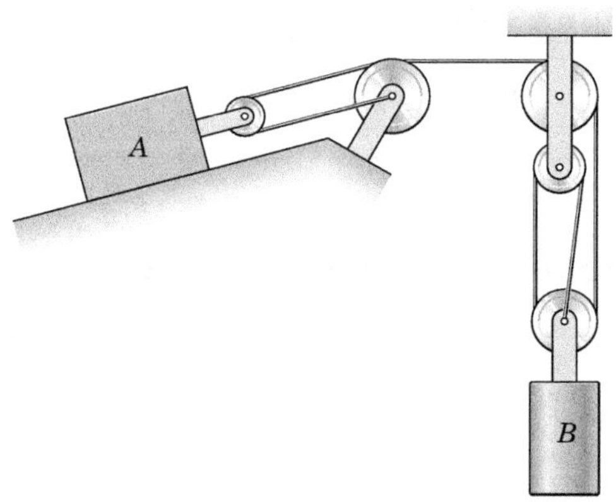

Problem 2/209

2/210 Determine the velocity of cart A if cylinder B has a downward velocity of 2 ft/sec at the instant illustrated. The two pulleys at C are pivoted independently.

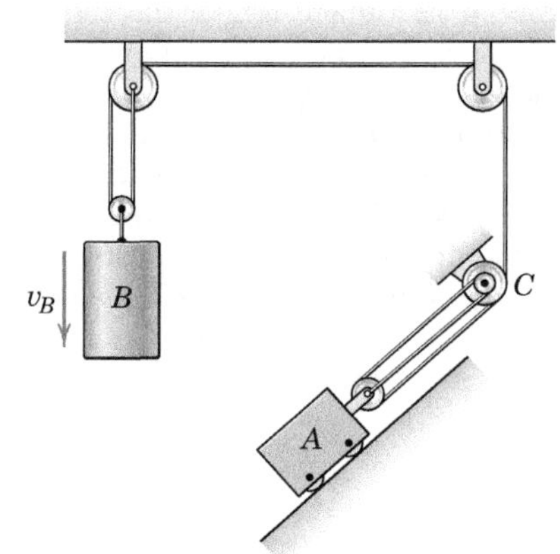

Problem 2/210

2/211 An electric motor M is used to reel in cable and hoist a bicycle into the ceiling space of a garage. Pulleys are fastened to the bicycle frame with hooks at locations A and B, and the motor can reel in cable at a steady rate of 12 in./sec. At this rate, how long will it take to hoist the bicycle 5 feet into the air? Assume that the bicycle remains level.

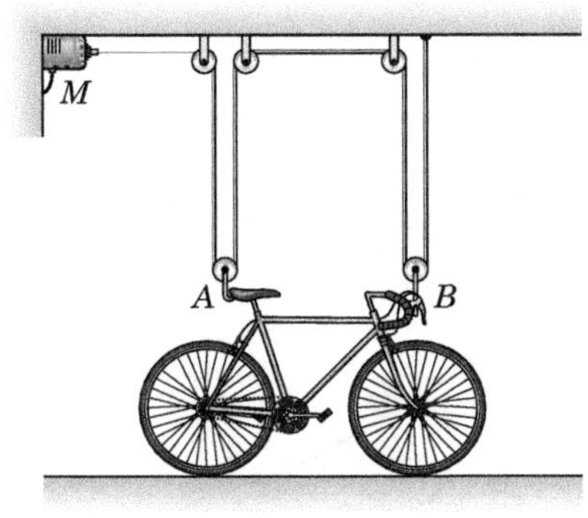

Problem 2/211

2/212 Determine the relation which governs the accelerations of A, B, and C, all measured positive down. Identify the number of degrees of freedom.

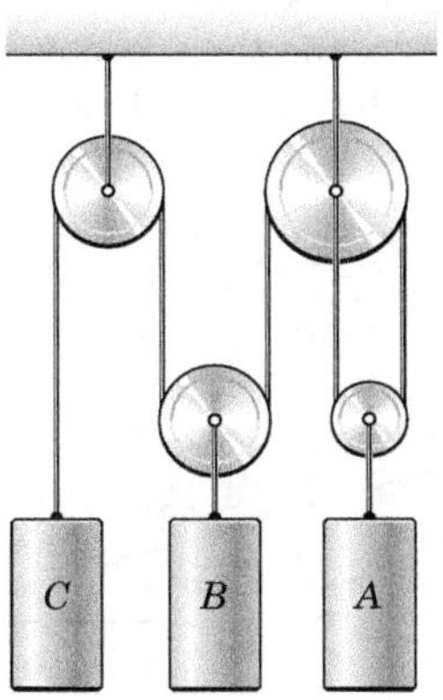

Problem 2/212

Representative Problems

2/213 Determine an expression for the velocity v_A of the cart A down the incline in terms of the upward velocity v_B of cylinder B.

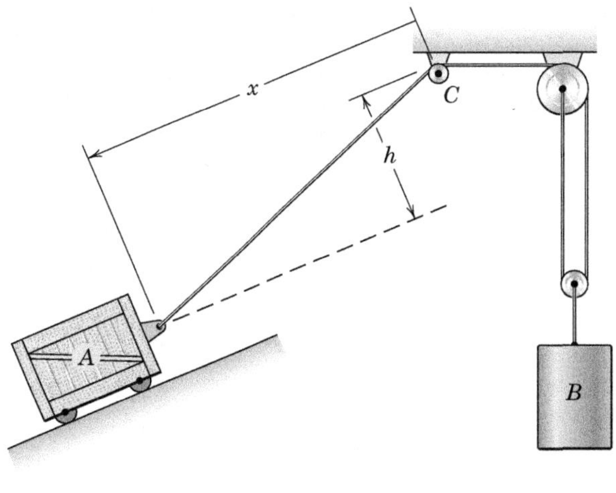

Problem 2/213

2/214 Neglect the diameters of the small pulleys and establish the relationship between the velocity of A and the velocity of B for a given value of y.

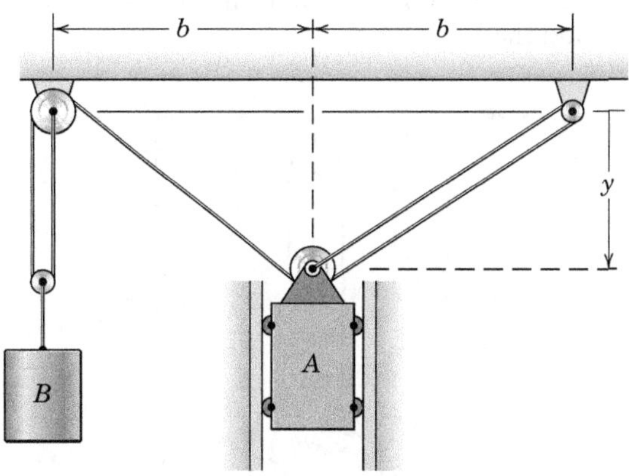

Problem 2/214

2/215 Under the action of force P, the constant acceleration of block B is 6 ft/sec^2 up the incline. For the instant when the velocity of B is 3 ft/sec up the incline, determine the velocity of B relative to A, the acceleration of B relative to A, and the absolute velocity of point C of the cable.

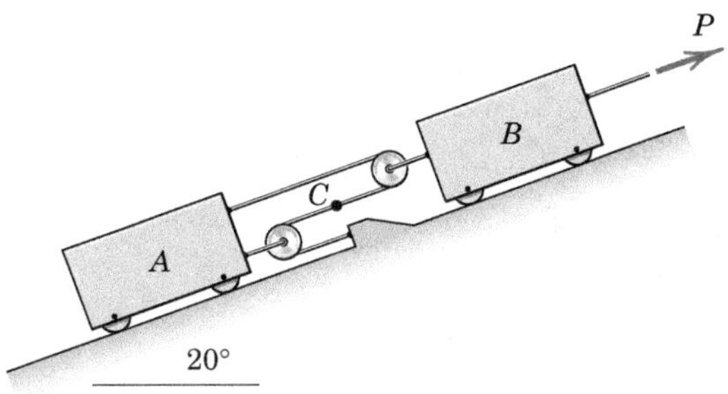

Problem 2/215

2/216 Determine the relationship which governs the velocities of the four cylinders. Express all velocities as positive down. How many degrees of freedom are there?

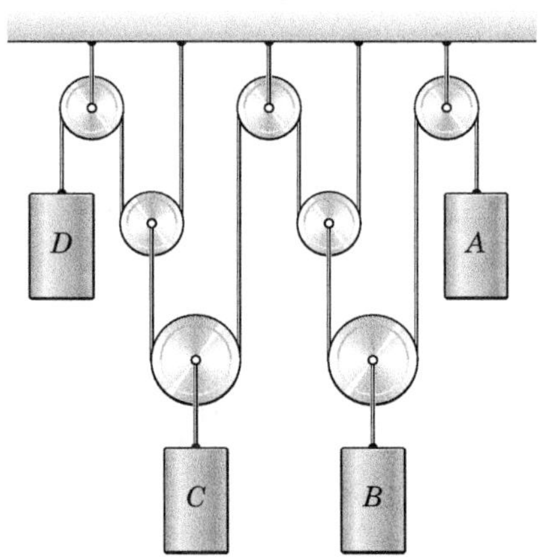

Problem 2/216

2/217 Collars A and B slide along the fixed right-angle rods and are connected by a cord of length L. Determine the acceleration a_x of collar B as a function of y if collar A is given a constant upward velocity v_A.

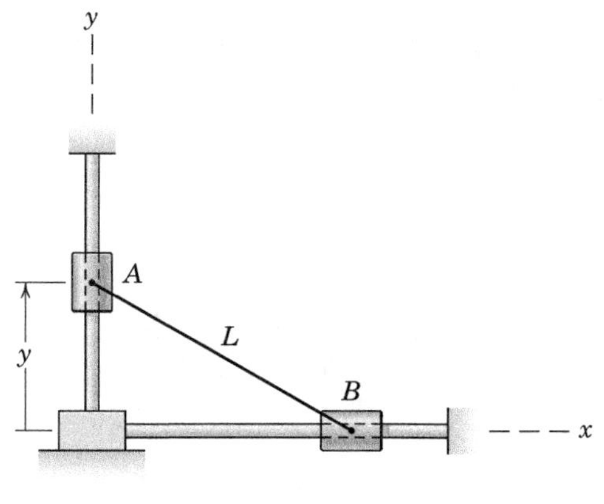

Problem 2/217

2/218 The small sliders A and B are connected by the rigid slender rod. If the velocity of slider B is 2 m/s to the right and is constant over a certain interval of time, determine the speed of slider A when the system is in the position shown.

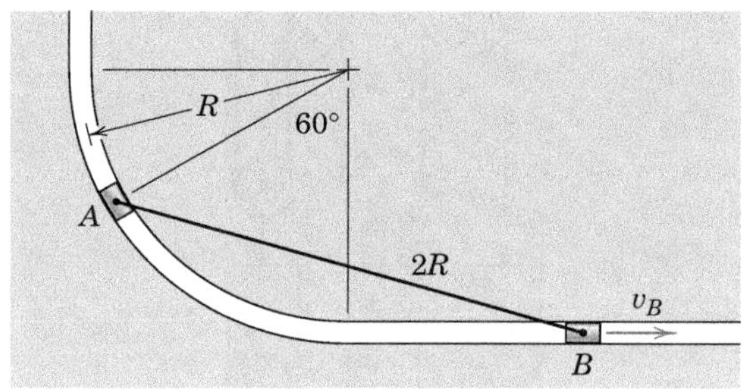

Problem 2/218

2/219 For a given value of y, determine the upward velocity of A in terms of the downward velocity of B. Neglect the diameters of the pulleys.

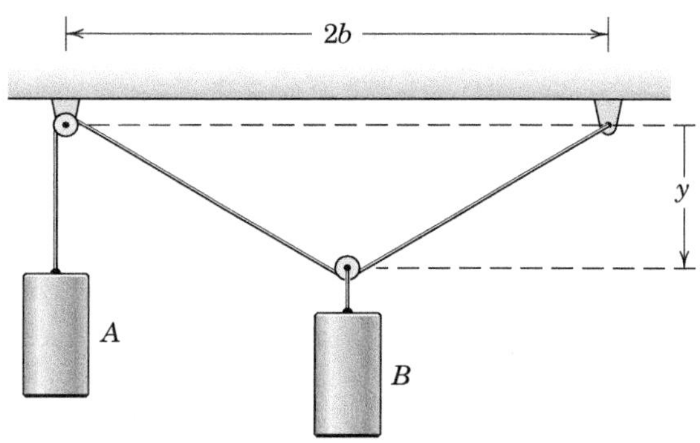

Problem 2/219

2/220 Cart A has a leftward velocity v_A and acceleration a_A at the instant represented. Determine the expressions for the velocity and acceleration of cart B in terms of the position x_A of cart A. Neglect the diameters of the pulleys and assume that there is no mechanical interference. The two pulleys at C are pivoted independently.

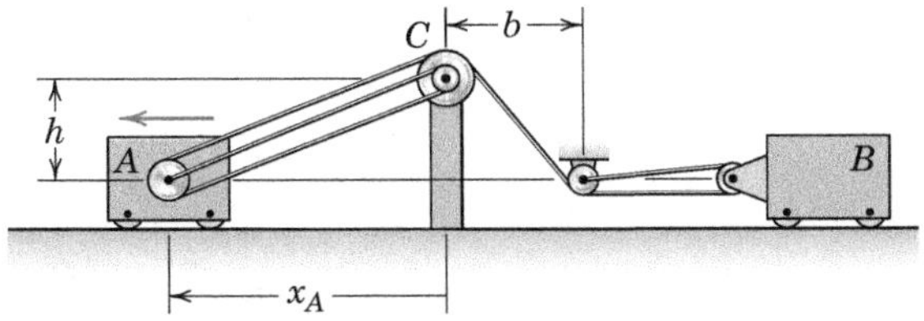

Problem 2/220

2/221 Determine the vertical rise h of the load W during 10 seconds if the hoisting drum draws in cable at the constant rate of 180 mm/s.

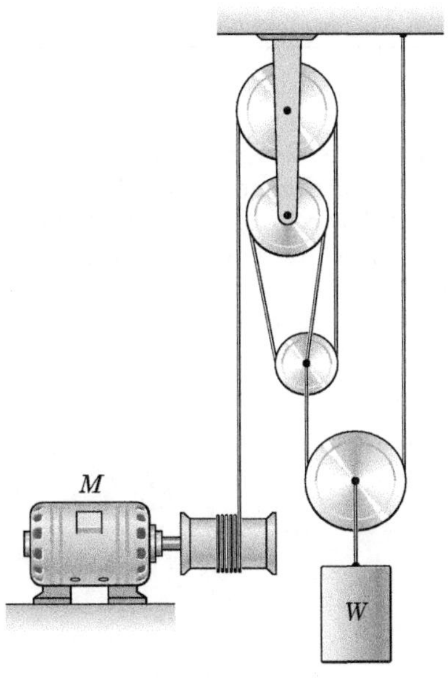

Problem 2/221

2/222 The hoisting system shown is used to easily raise kayaks for overhead storage. Determine expressions for the upward velocity and acceleration of the kayak at any height y if the winch M reels in cable at a constant rate $\dot{l}$. Assume that the kayak remains level.

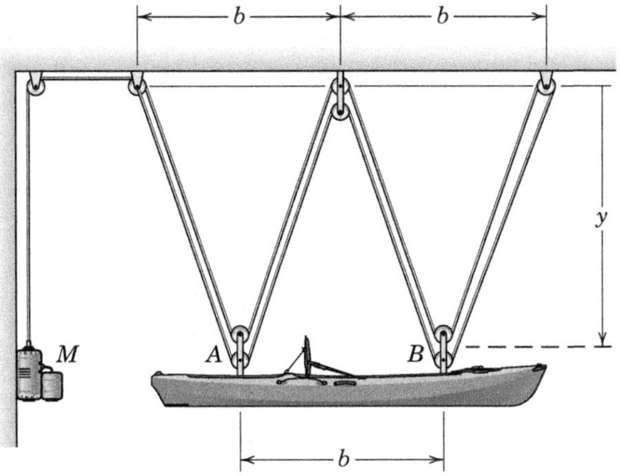

Problem 2/222

2/223 Develop an expression for the upward velocity of cylinder B in terms of the downward velocity of cylinder A. The cylinders are connected by a series of n cables and pulleys in a repeating fashion as shown.

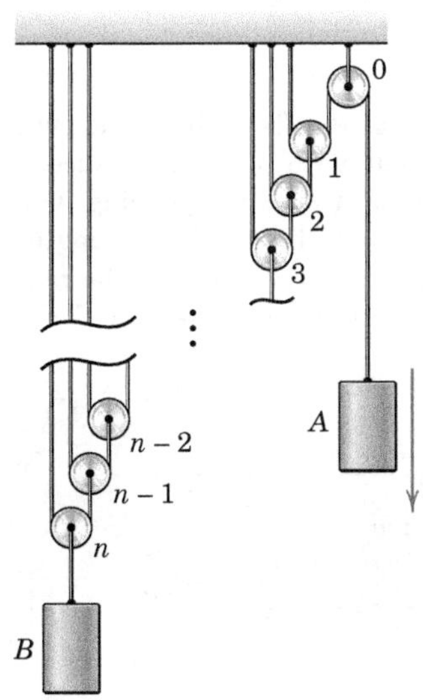

Problem 2/223

2/224 If load B has a downward velocity v_B, determine the upward component $(v_A)_y$ of the velocity of A in terms of b, the boom length l, and the angle θ. Assume that the cable supporting A remains vertical.

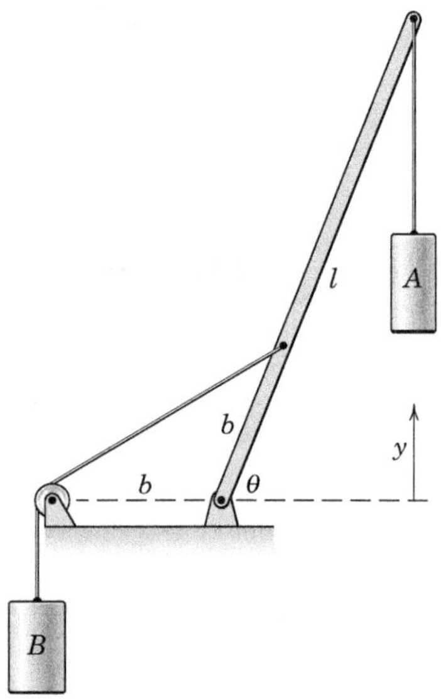

Problem 2/224

2/225 The rod of the fixed hydraulic cylinder is moving to the left with a constant speed $v_A = 25$ mm/s. Determine the corresponding velocity of slider B when $s_A = 425$ mm. The length of the cord is 1050 mm, and the effects of the radius of the small pulley A may be neglected.

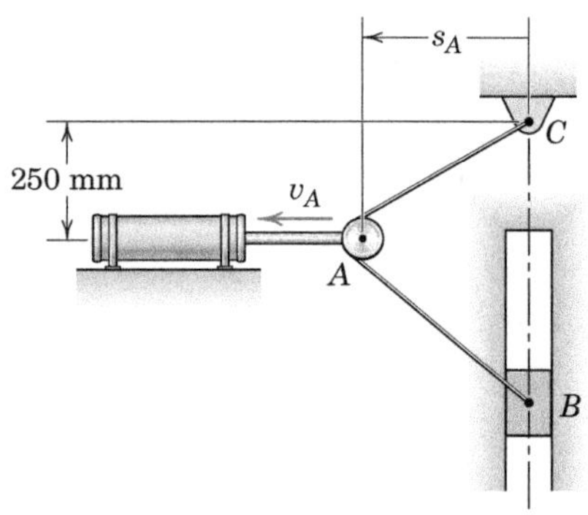

Problem 2/225

▶2/226 With all conditions of Prob. 2/225 remaining the same, determine the acceleration of slider B at the instant when $s_A = 425$ mm.

2/227 The two sliders are connected by the light rigid bar and move in the smooth vertical-plane guide. At the instant illustrated, the speed of slider A is 25 mm/s, $\theta = 45°$, and $\phi = 15°$. Determine the speed of slider B for this instant if $r = 175$ mm.

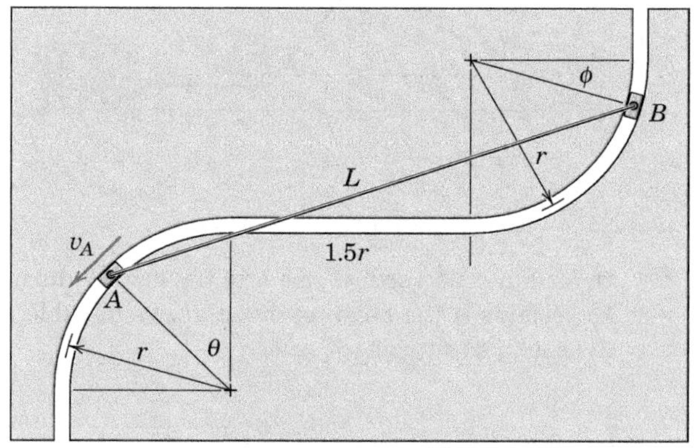

Problem 2/227

▶2/228 For the two sliders of Prob. 2/227, determine the time rate of change of speed for slider B at the location shown if the speed of slider A is constant over a short interval which includes the position shown.

2/10 Chapter Review

In Chapter 2 we have developed and illustrated the basic methods for describing particle motion. The concepts developed in this chapter form the basis for much of dynamics, and it is important to review and master this material before proceeding to the following chapters.

By far the most important concept in Chapter 2 is the time derivative of a vector. The time derivative of a vector depends on direction change as well as magnitude change. As we proceed in our study of dynamics, we will need to examine the time derivatives of vectors other than position and velocity vectors, and the principles and procedures developed in Chapter 2 will be useful for this purpose.

Categories of Motion

The following categories of motion have been examined in this chapter:

1. Rectilinear motion (one coordinate)
2. Plane curvilinear motion (two coordinates)
3. Space curvilinear motion (three coordinates)

In general, the geometry of a given problem enables us to identify the category readily. One exception to this categorization is encountered when only the magnitudes of the motion quantities measured along the path are of interest. In this event, we can use the single distance coordinate measured along the curved path, together with its scalar time derivatives giving the speed $|\dot{s}|$ and the tangential acceleration $\ddot{s}$.

Plane motion is easier to generate and control, particularly in machinery, than space motion, and thus a large fraction of our motion problems come under the plane curvilinear or rectilinear categories.

Use of Fixed Axes

We commonly describe motion or make motion measurements with respect to fixed reference axes (absolute motion) and moving axes (relative motion). The acceptable choice of the fixed axes depends on the problem. Axes attached to the surface of the earth are sufficiently "fixed" for most engineering problems, although important exceptions include earth–satellite and interplanetary motion, accurate projectile trajectories, navigation, and other problems. The equations of relative motion discussed in Chapter 2 are restricted to translating reference axes.

Choice of Coordinates

The choice of coordinates is of prime importance. We have developed the description of motion using the following coordinates:

1. Rectangular (Cartesian) coordinates (x-y) and (x-y-z)
2. Normal and tangential coordinates (n-t)
3. Polar coordinates (r-θ)
4. Cylindrical coordinates (r-θ-z)
5. Spherical coordinates (R-θ-ϕ)

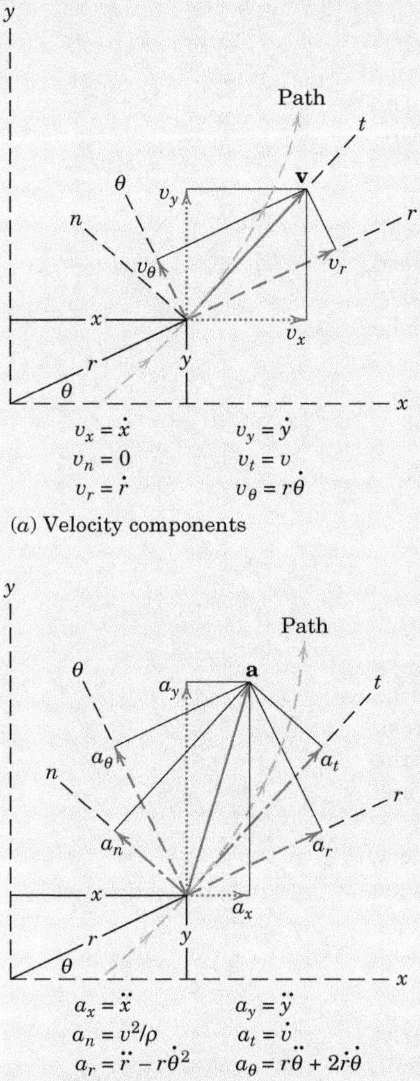

(*a*) Velocity components

(*b*) Acceleration components

Figure 2/21

When the coordinates are not specified, the appropriate choice usually depends on how the motion is generated or measured. Thus, for a particle which slides radially along a rotating rod, polar coordinates are the natural ones to use. Radar tracking calls for polar or spherical coordinates. When measurements are made along a curved path, normal and tangential coordinates are indicated. An x-y plotter clearly involves rectangular coordinates.

Figure 2/21 is a composite representation of the x-y, n-t, and r-θ coordinate descriptions of the velocity **v** and acceleration **a** for curvilinear motion in a plane. It is frequently essential to transpose motion description from one set of coordinates to another, and Fig. 2/21 contains the information necessary for that transition.

Approximations

Making appropriate approximations is one of the most important abilities you can acquire. The assumption of constant acceleration is valid when the forces which cause the acceleration do not vary appreciably. When motion data are acquired experimentally, we must utilize the nonexact data to acquire the best possible description, often with the aid of graphical or numerical approximations.

Choice of Mathematical Method

We frequently have a choice of solution using scalar algebra, vector algebra, trigonometric geometry, or graphical geometry. All of these methods have been illustrated, and all are important to learn. The choice of method will depend on the geometry of the problem, how the motion data are given, and the accuracy desired. Mechanics by its very nature is geometric, so you are encouraged to develop facility in sketching vector relationships, both as an aid to the disclosure of appropriate geometric and trigonometric relations and as a means of solving vector equations graphically. Geometric portrayal is the most direct representation of the vast majority of mechanics problems.

REVIEW PROBLEMS

2/229 In a test of vertical leaping ability, a basketball player (*a*) crouches just before jumping, (*b*) has given his mass center G a vertical velocity v_0 at the instant his feet leave the surface, and (*c*) reaches the maximum height. If the player can raise his mass center 3 feet as shown, estimate the initial velocity v_0 of his mass center in position (*b*).

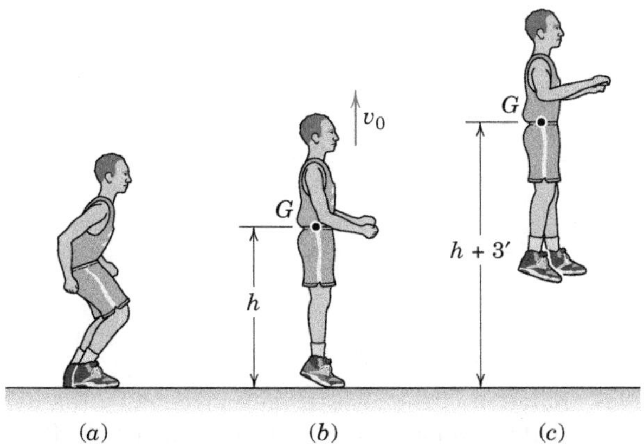

Problem 2/229

2/230 The position s of a particle along a straight line is given by $s = 8e^{-0.4t} - 6t + t^2$, where s is in meters and t is the time in seconds. Determine the velocity v when the acceleration is 3 m/s^2.

2/231 A particle moving in the x-y plane has a velocity $\mathbf{v} = 7.25\mathbf{i} + 3.48\mathbf{j}$ m/s at a certain instant. If the particle then encounters a constant acceleration $\mathbf{a} = 0.85\mathbf{j}$ m/s^2, determine the amount of time which must pass before the direction of the tangent to the trajectory of the particle has been altered by 30°.

2/232 An inexperienced designer of a roadbed for a new high-speed train proposes to join a straight section of track to a circular section of 1000-ft radius as shown. For a train that would travel at a constant speed of 90 mi/hr, plot the magnitude of its acceleration as a function of distance along the track between points A and C and explain why this design is unacceptable.

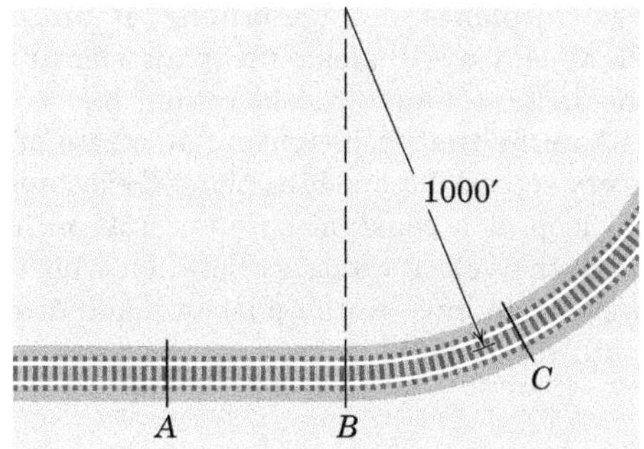

Problem 2/232

2/233 While scrambling directly toward the sideline, the football quarterback Q throws a pass toward the stationary receiver R. At what speed v_Q should the quarterback run if the direction of the velocity of the ball relative to the quarterback is to be directly down the field as indicated? The speed of the ball relative to the quarterback is 60 ft/sec. What is the absolute speed of the ball? Treat the problem as two-dimensional.

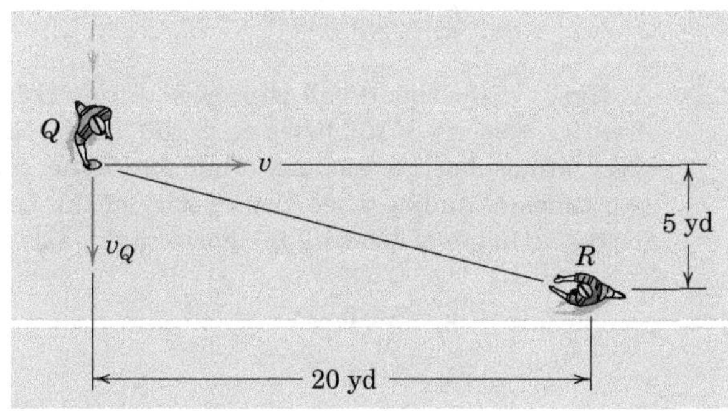

Problem 2/233

2/234 Two airplanes are performing at an air show. Plane A travels along the path shown and, for the instant under consideration, has a speed of 265 mi/hr that is increasing at a rate of 4 mi/hr every second. Meanwhile, plane B executes a vertical loop at a constant speed of 150 mi/hr. Determine the velocity and acceleration which plane B appears to have to the pilot in plane A at the instant represented.

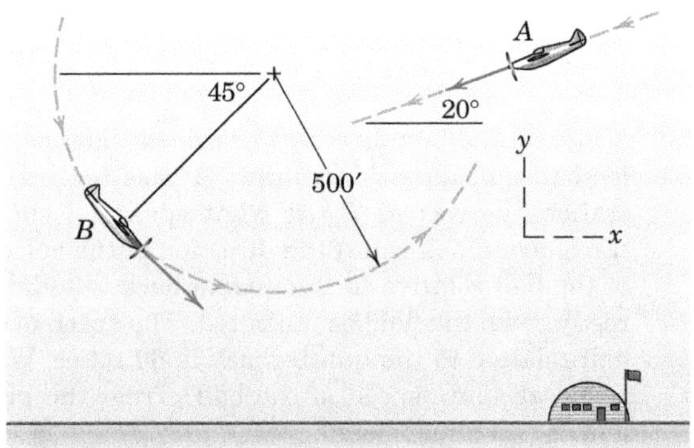

Problem 2/234

2/235 At time $t = 0$ a small ball is projected from point A with a velocity of 200 ft/sec at the 60° angle. Neglect atmospheric resistance and determine the two times t_1 and t_2 when the velocity of the ball makes an angle of 45° with the horizontal x-axis.

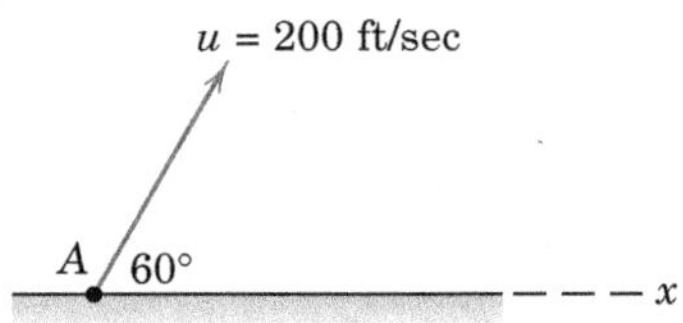

Problem 2/235

2/236 A bicyclist rides along the hard-packed sand beach with a speed $v_B = 16$ mi/hr as indicated. The wind speed is $v_W = 20$ mi/hr. (*a*) Determine the velocity of the wind relative to the bicyclist. (*b*) At what speed v_B would the bicyclist feel the wind coming directly from her left (perpendicular to her path)? What would be this relative speed?

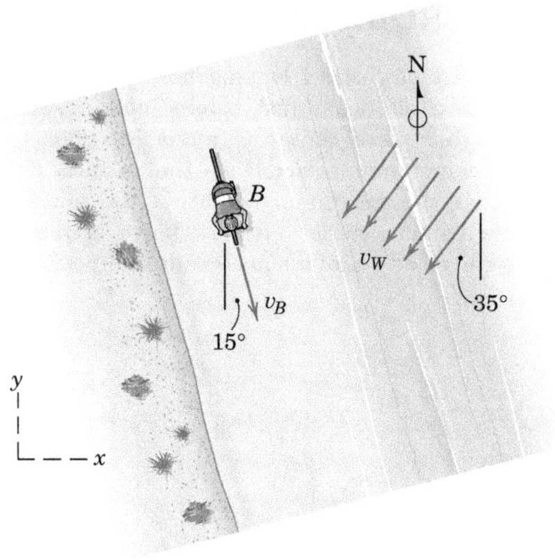

Problem 2/236

2/237 Rotation of the arm OP is controlled by the horizontal motion of the vertical slotted link. If $\dot{x} = 4$ ft/sec and $\ddot{x} = 30$ ft/sec^2 when $x = 2$ in., determine $\dot{\theta}$ and $\ddot{\theta}$ for this instant.

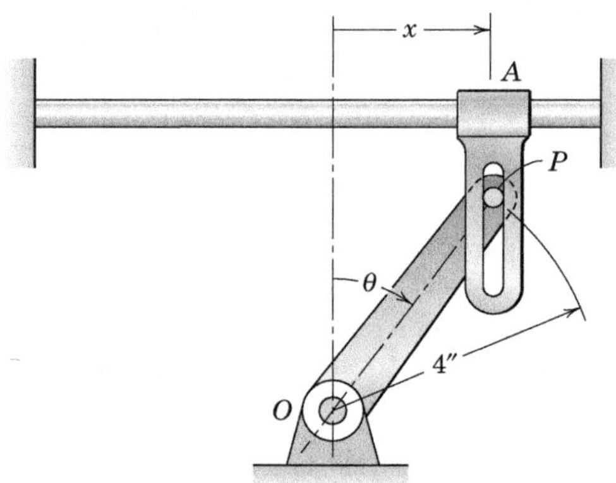

Problem 2/237

2/238 Body A is released from rest in the position shown and moves downward causing body B to lift off the support at C. If motion is controlled such that the magnitude $a_{B/A} = 2.4$ m/s^2 is held constant, determine the amount of time it takes for body B to travel 5 m up the incline and the corresponding speed of body A at the end of that time period. The angle $\theta = 55°$.

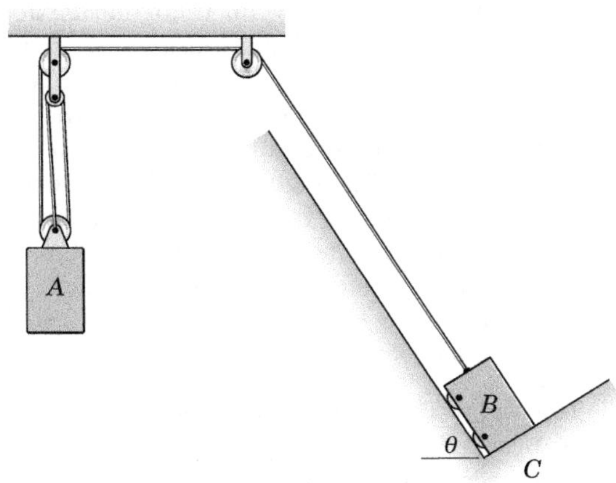

Problem 2/238

2/239 The launching catapult of the aircraft carrier gives the jet fighter a constant acceleration of 50 m/s^2 from rest relative to the flight deck and launches the aircraft in a distance of 100 m measured along the angled takeoff ramp. If the carrier is moving at a steady 30 knots (1 knot = 1.852 km/h), determine the magnitude v of the actual velocity of the fighter when it is launched.

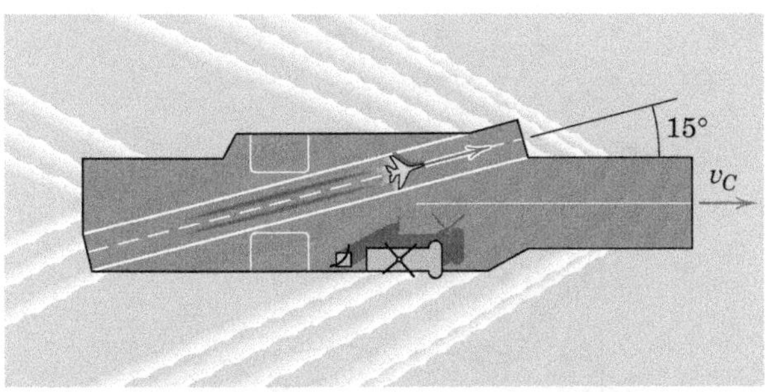

Problem 2/239

2/240 For the instant represented the particle P has a velocity $v = 6$ ft/sec in the direction shown and has acceleration components $a_x = 15$ ft/sec^2 and $a_\theta = -15$ ft/sec^2. Determine a_r, a_y, a_t, a_n, and the radius of curvature ρ of the path for this position. (*Hint:* Draw the related acceleration components of the total acceleration of the particle and take advantage of the simplified geometry for your calculations.)

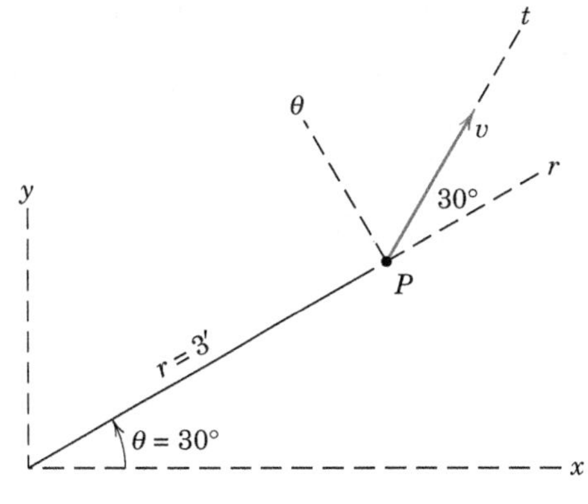

Problem 2/240

2/241 The coordinates of a particle which moves with curvilinear motion are given by $x = 10.25t + 1.75t^2 - 0.45t^3$ and $y = 6.32 + 14.65t - 2.48t^2$, where x and y are in millimeters and the time t is in seconds. Determine the values of v, $\mathbf{v}$, a, $\mathbf{a}$, $\mathbf{e}_t$, $\mathbf{e}_n$, a_t, $\mathbf{a}_t$, a_n, $\mathbf{a}_n$, ρ, and $\dot{\beta}$ (the angular velocity of the normal to the path) when $t = 3.25$ s. Express all vectors in terms of the unit vectors $\mathbf{i}$ and $\mathbf{j}$.

2/242 The coordinates of a particle which moves with curvilinear motion are given by $x = 10.25t + 1.75t^2 - 0.45t^3$ and $y = 6.32 + 14.65t - 2.48t^2$, where x and y are in millimeters and the time t is in seconds. Determine the values of v, $\mathbf{v}$, a, $\mathbf{a}$, $\mathbf{e}_r$, $\mathbf{e}_\theta$, $\mathbf{v}_r$, $\mathbf{v}_r$, v_θ, $\mathbf{v}_\theta$, a_r, $\mathbf{a}_r$, a_θ, $\mathbf{a}_\theta$, r, $\dot{r}$, $\ddot{r}$, θ, $\dot{\theta}$, and $\ddot{\theta}$ when $t = 3.25$ s. Express all vectors in terms of the unit vectors $\mathbf{i}$ and $\mathbf{j}$. Take the r-coordinate to proceed from the origin, and take θ to be measured positive counterclockwise from the positive x-axis.

2/243 As part of a training exercise, the pilot of aircraft A adjusts her airspeed (speed relative to the wind) to 220 km/h while in the level portion of the approach path and thereafter holds her absolute speed constant as she negotiates the 10° glide path. The absolute speed of the aircraft carrier is 30 km/h and that of the wind is 48 km/h. What will be the angle β of the glide path with respect to the horizontal as seen by an observer on the ship?

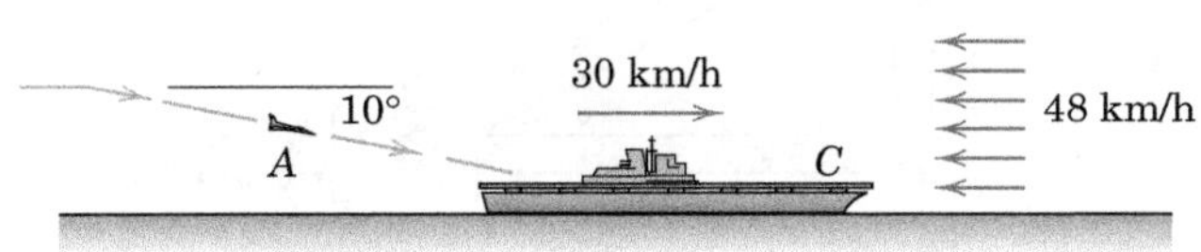

Problem 2/243

2/244 A small aircraft is moving in a horizontal circle with a constant speed of 130 ft/sec. At the instant represented, a small package A is ejected from the right side of the aircraft with a horizontal velocity of 20 ft/sec relative to the aircraft. Neglect aerodynamic effects and calculate the coordinates of the point of impact on the ground.

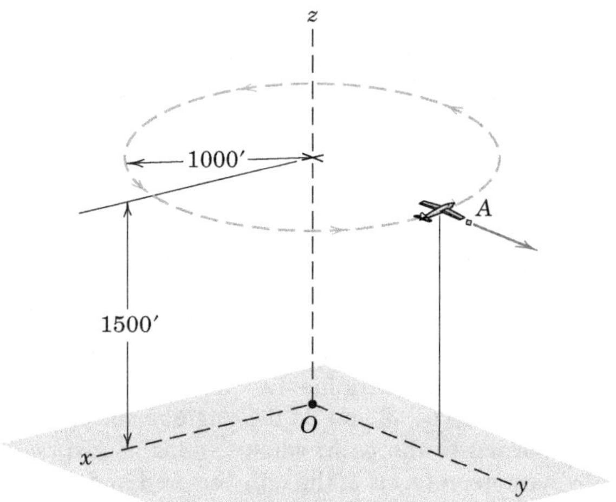

Problem 2/244

2/245 Cylinder A has a constant downward speed of 1 m/s. Compute the velocity of cylinder B for (a) $\theta = 45°$, (b) $\theta = 30°$, and (c) $\theta = 15°$. The spring is in tension throughout the motion range of interest, and the pulleys are connected by the cable of fixed length.

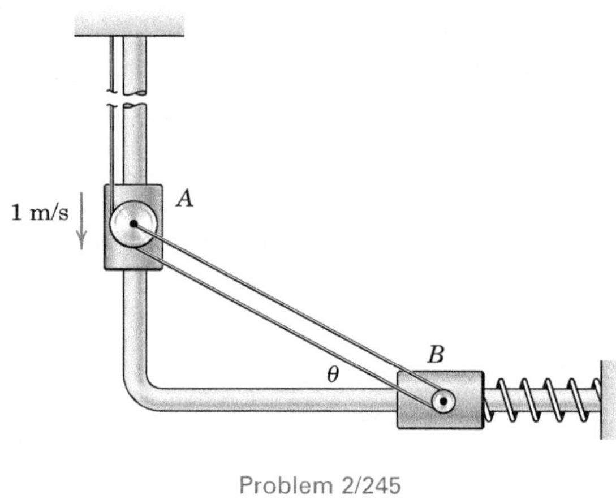

Problem 2/245

2/246 A rocket fired vertically up from the north pole achieves a velocity of 27 000 km/h at an altitude of 350 km when its fuel is exhausted. Calculate the additional vertical height h reached by the rocket before it starts its descent back to the earth. The coasting phase of its flight occurs above the atmosphere. Consult Fig. 1/1 in choosing the appropriate value of gravitational acceleration and use the mean radius of the earth from Table D/2. (*Note:* Launching from the earth's pole avoids considering the effect of the earth's rotation.)

▶2/247 The radar tracking antenna oscillates about its vertical axis according to $\theta = \theta_0 \cos \omega t$, where ω is the constant circular frequency and $2\theta_0$ is the double amplitude of oscillation. Simultaneously, the angle of elevation ϕ is increasing at the constant rate $\dot{\phi} = K$. Determine the expression for the magnitude a of the acceleration of the signal horn (a) as it passes position A and (b) as it passes the top position B, assuming that $\theta = 0$ at this instant.

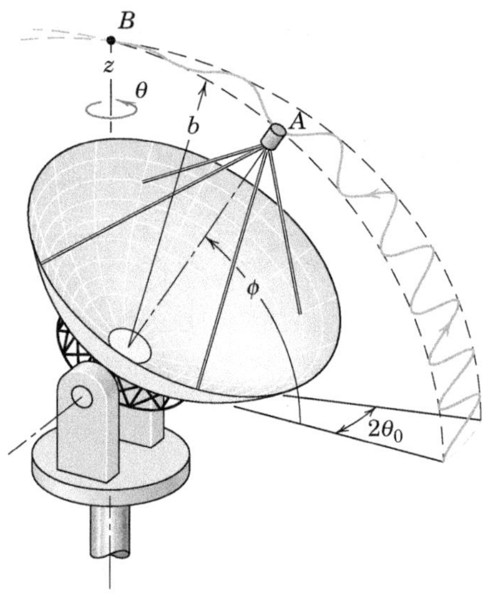

Problem 2/247

*Computer-Oriented Problems

***2/248** A projectile is fired with the given initial conditions. Plot the r- and θ-components of velocity and acceleration as functions of time for the time period during which the particle is in the air. State the value for each component at time $t = 9$ s.

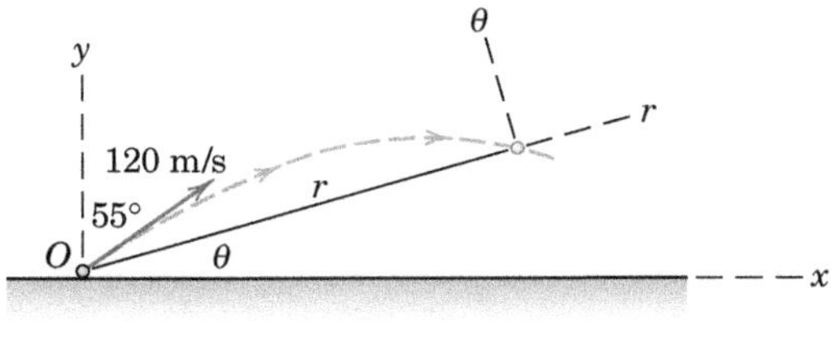

Problem 2/248

***2/249** If all frictional effects are neglected, the expression for the angular acceleration of the simple pendulum is $\ddot{\theta} = \frac{g}{l}\cos\theta$, where g is the acceleration of gravity and l is the length of the rod OA. If the pendulum has a clockwise angular velocity $\dot{\theta} =$ 2 rad/s when $\theta = 0$ at $t = 0$, determine the time t' at which the pendulum passes the vertical position $\theta = 90°$. The pendulum length is $l = 0.6$ m. Also plot the time t versus the angle θ.

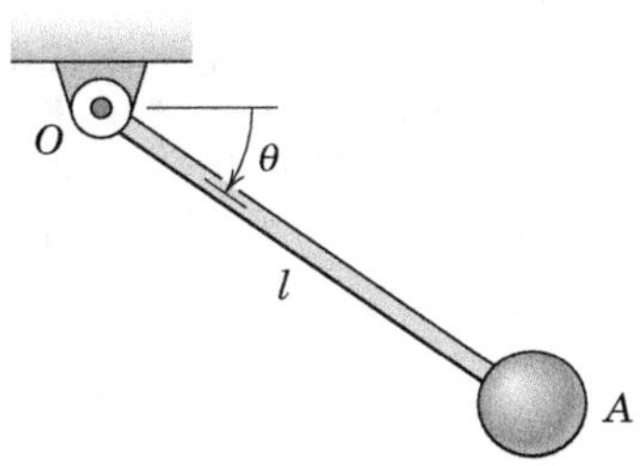

Problem 2/249

***2/250** A baseball is dropped from an altitude $h = 200$ ft and is found to be traveling at 85 ft/sec when it strikes the ground. In addition to gravitational acceleration, which may be assumed constant, air resistance causes a deceleration component of magnitude kv^2, where v is the speed and k is a constant. Determine the value of the coefficient k. Plot the speed of the baseball as a function of altitude y. If the baseball were dropped from a high altitude, but one at which g may still be assumed constant, what would be the terminal velocity v_t? (The *terminal velocity* is that speed at which the acceleration of gravity and that due to air resistance are equal and opposite, so that the baseball drops at a constant speed.) If the baseball were dropped from $h = 200$ ft, at what speed v' would it strike the ground if air resistance were neglected?

***2/251** A ship with a total displacement of 16 000 metric tons (1 metric ton = 1000 kg) starts from rest in still water under a constant propeller thrust $T =$ 250 kN. The ship develops a total resistance to motion through the water given by $R = 4.50v^2$, where R is in kilonewtons and v is in meters per second. The acceleration of the ship is $a = (T - R)/m$, where m equals the mass of the ship in metric tons. Plot the speed v of the ship in knots as a function of the distance s in nautical miles which the ship goes for the first 5 nautical miles from rest. Find the speed after the ship has gone 1 nautical mile. What is the maximum speed which the ship can reach?

***2/252** At time $t = 0$, the 1.8-lb particle P is given an initial velocity $v_0 = 1$ ft/sec at the position $\theta = 0$ and subsequently slides along the circular path of radius $r = 1.5$ ft. Because of the viscous fluid and the effect of gravitational acceleration, the tangential acceleration is $a_t = g\cos\theta - \frac{k}{m}v$, where the constant $k = 0.2$ lb-sec/ft is a drag parameter. Determine and plot both θ and $\dot{\theta}$ as functions of the time t over the range $0 \le t \le 5$ sec. Determine the maximum values of θ and $\dot{\theta}$ and the corresponding values of t. Also determine the first time at which $\theta = 90°$.

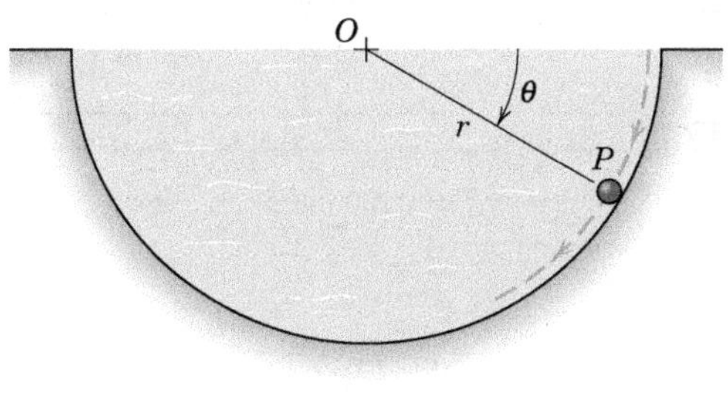

Problem 2/252

***2/253** A projectile is launched from point A with speed $v_0 = 30$ m/s. Determine the value of the launch angle α which maximizes the range R indicated in the figure. Determine the corresponding value of R.

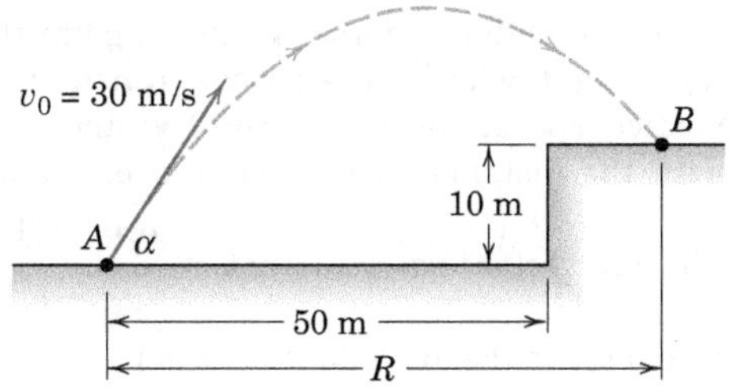

Problem 2/253

***2/254** A low-flying cropduster A is moving with a constant speed of 40 m/s in the horizontal circle of radius 300 m. As it passes the twelve-o'clock position shown at time $t = 0$, car B starts from rest from the position shown and accelerates along the straight road at the constant rate of 3 m/s^2 until it reaches a speed of 30 m/s, after which it maintains that constant speed. Determine the velocity and acceleration of A with respect to B and plot the magnitudes of both these quantities over the time period $0 \le t \le 50$ s as functions of both time and displacement s_B of the car. Determine the maximum and minimum values of both quantities and state the values of the time t and the displacement s_B at which they occur.

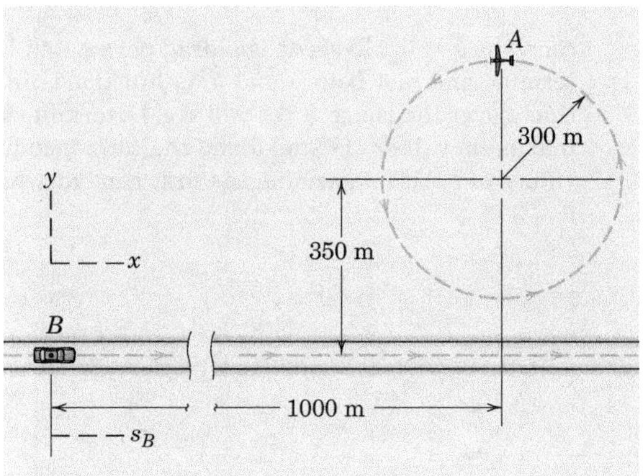

Problem 2/254

***2/255** A particle P is launched from point A with the initial conditions shown. If the particle is subjected to aerodynamic drag, compute the range R of the particle and compare this with the case in which aerodynamic drag is neglected. Plot the trajectories of the particle for both cases. Use the values $v_0 = 65$ m/s, $\theta = 35°$, and $k = 4.0 \times 10^{-3}\,\text{m}^{-1}$. (*Note:* The acceleration due to aerodynamic drag has the form $\mathbf{a}_D = -kv^2\mathbf{t}$, where k is a positive constant, v is the particle speed, and $\mathbf{t}$ is the unit vector associated with the instantaneous velocity $\mathbf{v}$ of the particle. The unit vector $\mathbf{t}$ has the form $\mathbf{t} = \dfrac{v_x\mathbf{i} + v_y\mathbf{j}}{\sqrt{v_x^{\,2} + v_y^{\,2}}}$, where v_x and v_y are the instantaneous x- and y-components of particle velocity, respectively.)

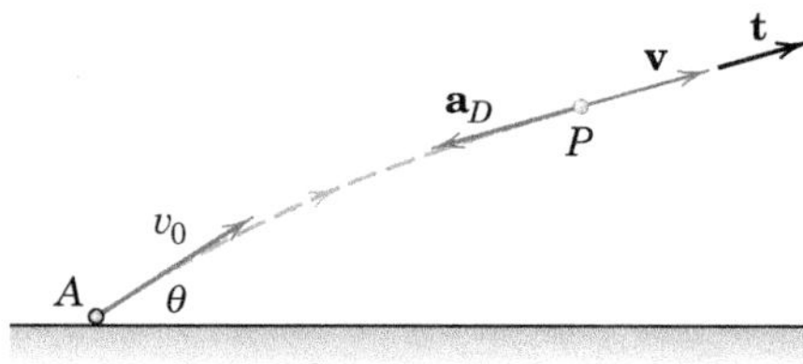

Problem 2/255

***2/256** By means of the control unit M, the pendulum OA is given an oscillatory motion about the vertical given by $\theta = \theta_0 \sin\sqrt{\dfrac{g}{l}}\,t$, where θ_0 is the maximum angular displacement in radians, g is the acceleration of gravity, l is the pendulum length, and t is the time in seconds measured from an instant when OA is vertical. Determine and plot the magnitude a of the acceleration of A as a function of time and as a function of θ over the first quarter cycle of motion. Determine the minimum and maximum values of a and the corresponding values of t and θ. Use the values $\theta_0 = \pi/3$ radians, $l = 0.8$ m, and $g = 9.81$ m/s^2. (*Note:* The prescribed motion is not precisely that of a freely swinging pendulum for large amplitudes.)

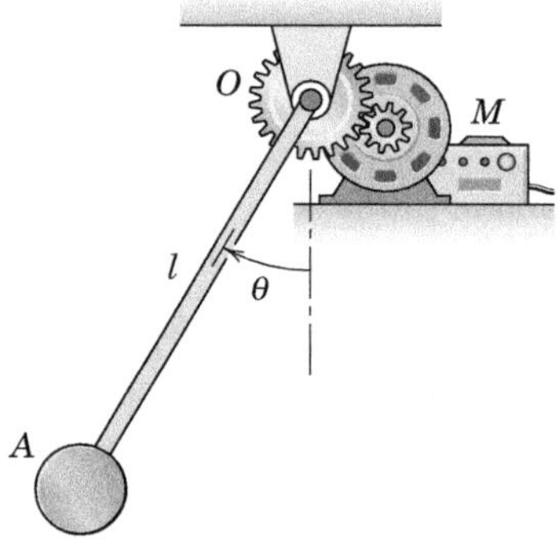

Problem 2/256

Kiev.Victor/Shutterstock

The designers of amusement-park rides such as this roller coaster must not rely upon the principles of equilibrium alone as they develop specifications for the cars and the supporting structure. The particle kinetics of each car must be considered in estimating the involved forces so that a safe system can be designed.

3 Kinetics of Particles

CHAPTER OUTLINE

3/1 Introduction

According to Newton's second law, a particle will accelerate when it is subjected to unbalanced forces. Kinetics is the study of the relations between unbalanced forces and the resulting changes in motion. In Chapter 3 we will study the kinetics of particles. This topic requires that we combine our knowledge of the properties of forces, which we developed in statics, and the kinematics of particle motion just covered in Chapter 2. With the aid of Newton's second law, we can combine these two topics and solve engineering problems involving force, mass, and motion.

The three general approaches to the solution of kinetics problems are: (A) direct application of Newton's second law (called the force-mass-acceleration method), (B) use of work and energy principles, and (C) solution by impulse and momentum methods. Each approach has its special characteristics and advantages, and Chapter 3 is subdivided into Sections A, B, and C, according to these three methods of solution. In addition, a fourth section, Section D, treats special applications and combinations of the three basic approaches. Before proceeding, you should review carefully the definitions and concepts of Chapter 1, because they are fundamental to the developments which follow.

SECTION A FORCE, MASS, AND ACCELERATION

3/2 Newton's Second Law

The basic relation between force and acceleration is found in Newton's second law, Eq. 1/1, the verification of which is entirely experimental. We now describe the fundamental meaning of this law by considering an ideal experiment in which force and acceleration are assumed to be measured without error. We subject a mass particle to the action of a single force $\mathbf{F}_1$, and we measure the acceleration $\mathbf{a}_1$ of the particle in the primary inertial system.* The ratio F_1/a_1 of the magnitudes of the force and the acceleration will be some number C_1 whose value depends on the units used for measurement of force and acceleration. We then repeat the experiment by subjecting the same particle to a different force $\mathbf{F}_2$ and measuring the corresponding acceleration $\mathbf{a}_2$. The ratio F_2/a_2 of the magnitudes will again produce a number C_2. The experiment is repeated as many times as desired.

We draw two important conclusions from the results of these experiments. First, the ratios of applied force to corresponding acceleration all equal the *same* number, provided the units used for measurement are not changed in the experiments. Thus,

$$\frac{F_1}{a_1} = \frac{F_2}{a_2} = \cdots = \frac{F}{a} = C, \qquad \text{a constant}$$

We conclude that the constant C is a measure of some invariable property of the particle. This property is the *inertia* of the particle, which is its *resistance to rate of change of velocity*. For a particle of high inertia (large C), the acceleration will be small for a given force F. On the other hand, if the inertia is small, the acceleration will be large. The mass m is used as a quantitative measure of inertia, and therefore, we may write the expression $C = km$, where k is a constant introduced

*The primary inertial system or astronomical frame of reference is an imaginary set of reference axes which are assumed to have no translation or rotation in space. See Art. 1/2, Chapter 1.

to account for the units used. Thus, we may express the relation obtained from the experiments as

$$F = kma \tag{3/1}$$

where F is the magnitude of the resultant force acting on the particle of mass m, and a is the magnitude of the resulting acceleration of the particle.

The second conclusion we draw from this ideal experiment is that the acceleration is always in the direction of the applied force. Thus, Eq. 3/1 becomes a *vector* relation and may be written

$$\mathbf{F} = km\mathbf{a} \tag{3/2}$$

Although an actual experiment cannot be performed in the ideal manner described, the same conclusions have been drawn from countless accurately performed experiments. One of the most accurate checks is given by the precise prediction of the motions of planets based on Eq. 3/2.

Inertial System

Although the results of the ideal experiment are obtained for measurements made relative to the "fixed" primary inertial system, they are equally valid for measurements made with respect to any nonrotating reference system which translates with a constant velocity with respect to the primary system. From our study of relative motion in Art. 2/8, we know that the acceleration measured in a system translating with no acceleration is the same as that measured in the primary system. Thus, Newton's second law holds equally well in a nonaccelerating system, so that we may define an *inertial system* as any system in which Eq. 3/2 is valid.

If the ideal experiment described were performed on the surface of the earth and all measurements were made relative to a reference system attached to the earth, the measured results would show a slight discrepancy from those predicted by Eq. 3/2, because the measured acceleration would not be the correct absolute acceleration. The discrepancy would disappear when we introduced the correction due to the acceleration components of the earth. These corrections are negligible for most engineering problems which involve the motions of structures and machines on the surface of the earth. In such cases, the accelerations measured with respect to reference axes attached to the surface of the earth may be treated as "absolute," and Eq. 3/2 may be applied with negligible error to experiments made on the surface of the earth.*

*As an example of the magnitude of the error introduced by neglect of the motion of the earth, consider a particle which is allowed to fall from rest (relative to earth) at a height h above the ground. We can show that the rotation of the earth gives rise to an eastward acceleration (Coriolis acceleration) relative to the earth and, neglecting air resistance, that the particle falls to the ground a distance

$$x = \frac{2}{3}\omega\sqrt{\frac{2h^3}{g}}\cos\gamma$$

east of the point on the ground directly under that from which it was dropped. The angular velocity of the earth is $\omega = 0.729(10^{-4})$ rad/s, and the latitude, north or south, is γ. At a latitude of 45° and from a height of 200 m, this eastward deflection would be $x = 43.9$ mm.

An increasing number of problems occur, particularly in the fields of rocket and spacecraft design, where the acceleration components of the earth are of primary concern. For this work it is essential that the fundamental basis of Newton's second law be thoroughly understood and that the appropriate absolute acceleration components be employed.

Before 1905 the laws of Newtonian mechanics had been verified by innumerable physical experiments and were considered the final description of the motion of bodies. The concept of *time*, considered an absolute quantity in the Newtonian theory, received a basically different interpretation in the theory of relativity announced by Einstein in 1905. The new concept called for a complete reformulation of the accepted laws of mechanics. The theory of relativity was subjected to early ridicule, but has been verified by experiment and is now universally accepted by scientists. Although the difference between the mechanics of Newton and that of Einstein is basic, there is a practical difference in the results given by the two theories only when velocities of the order of the speed of light (300×10^6 m/s) are encountered.* Important problems dealing with atomic and nuclear particles, for example, require calculations based on the theory of relativity.

Systems of Units

It is customary to take k equal to unity in Eq. 3/2, thus putting the relation in the usual form of Newton's second law

$$\boxed{\mathbf{F} = m\mathbf{a}} \qquad \textbf{[1/1]}$$

A system of units for which k is unity is known as a *kinetic* system. Thus, for a kinetic system the units of force, mass, and acceleration are not independent. In SI units, as explained in Art. 1/4, the units of force (newtons, N) are derived by Newton's second law from the base units of mass (kilograms, kg) times acceleration (meters per second squared, m/s^2). Thus, N = kg·m/s^2. This system is known as an *absolute* system since the unit for force is dependent on the absolute value of mass.

In U.S. customary units, on the other hand, the units of mass (slugs) are derived from the units of force (pounds force, lb) divided by acceleration (feet per second squared, ft/sec^2). Thus, the mass units are slugs = lb-sec^2/ft. This system is known as a *gravitational* system since mass is derived from force as determined from gravitational attraction.

For measurements made relative to the rotating earth, the relative value of g should be used. The internationally accepted value of g relative to the earth at sea level and at a latitude of 45° is 9.806 65 m/s^2. Except where greater precision is required, the value of 9.81 m/s^2 will be used for g. For measurements relative to a nonrotating earth, the absolute value of g should be used. At a latitude of 45° and at sea level, the absolute value is 9.8236 m/s^2. The sea-level variation in both

*The theory of relativity demonstrates that there is no such thing as a preferred primary inertial system and that measurements of time made in two coordinate systems which have a velocity relative to one another are different. On this basis, for example, the principles of relativity show that a clock carried by the pilot of a spacecraft traveling around the earth in a circular polar orbit of 644 km altitude at a velocity of 27 080 km/h would be slow compared with a clock at the pole by 0.000 001 85 s for each orbit.

the absolute and relative values of g with latitude is shown in Fig. 1/1 of Art. 1/5.

In the U.S. customary system, the standard value of g relative to the rotating earth at sea level and at a latitude of 45° is 32.1740 ft/sec^2. The corresponding value relative to a nonrotating earth is 32.2230 ft/sec^2.

Force and Mass Units

We need to use both SI units and U.S. customary units, so we must have a clear understanding of the correct force and mass units in each system. These units were explained in Art. 1/4, but it will be helpful to illustrate them here using simple numbers before applying Newton's second law. Consider, first, the free-fall experiment as depicted in Fig. 3/1*a* where we release an object from rest near the surface of the earth. We allow it to fall freely under the influence of the force of gravitational attraction W on the body. We call this force the *weight* of the body. In SI units for a mass $m = 1$ kg, the weight is $W = 9.81$ N, and the corresponding downward acceleration a is $g = 9.81$ m/s^2. In U.S. customary units for a mass $m = 1$ lbm (1/32.2 slug), the weight is $W =$ 1 lbf and the resulting gravitational acceleration is $g = 32.2$ ft/sec^2. For a mass $m = 1$ slug (32.2 lbm), the weight is $W = 32.2$ lbf and the acceleration, of course, is also $g = 32.2$ ft/sec^2.

In Fig. 3/1*b* we illustrate the proper units with the simplest example where we accelerate an object of mass m along the horizontal with a force F. In SI units (an absolute system), a force $F = 1$ N causes a mass $m = 1$ kg to accelerate at the rate $a = 1$ m/s^2. Thus, 1 N = 1 kg·m/s^2. In the U.S. customary system (a gravitational system), a force $F = 1$ lbf

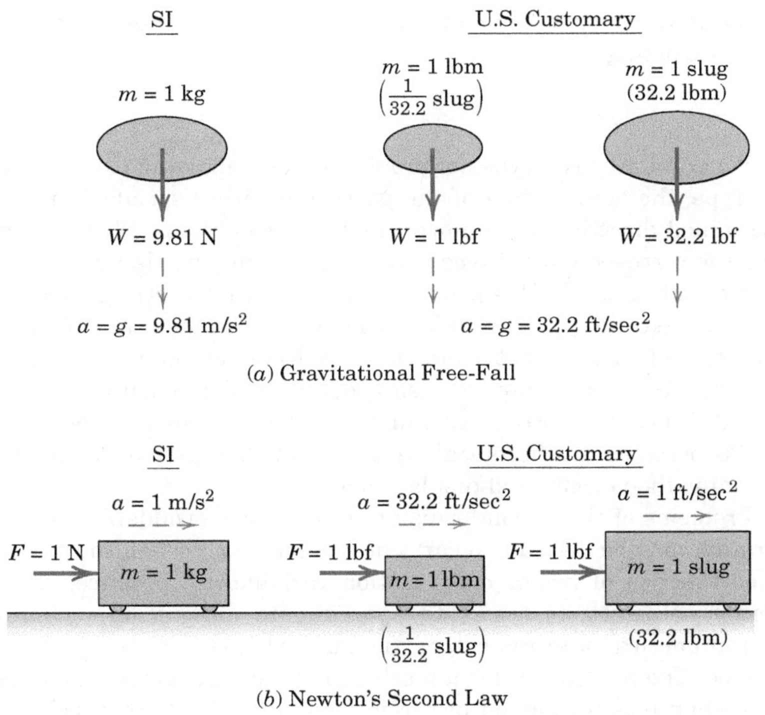

Figure 3/1

causes a mass $m = 1$ lbm (1/32.2 slug) to accelerate at the rate $a = 32.2$ ft/sec^2, whereas a force $F = 1$ lbf causes a mass $m = 1$ slug (32.2 lbm) to accelerate at the rate $a = 1$ ft/sec^2.

We note that in SI units where the mass is expressed in kilograms (kg), the weight W of the body in newtons (N) is given by $W = mg$, where $g = 9.81$ m/s^2. In U.S. customary units, the weight W of a body is expressed in pounds force (lbf), and the mass in slugs (lbf-sec^2/ft) is given by $m = W/g$, where $g = 32.2$ ft/sec^2.

In U.S. customary units, we frequently speak of the weight of a body when we really mean mass. It is entirely proper to specify the mass of a body in pounds (lbm) which must be converted to mass in slugs before substituting into Newton's second law. Unless otherwise stated, the pound (lb) is normally used as the unit of force (lbf).

3/3 Equation of Motion and Solution of Problems

When a particle of mass m is subjected to the action of concurrent forces $\mathbf{F}_1, \mathbf{F}_2, \mathbf{F}_3, \ldots$ whose vector sum is $\Sigma\mathbf{F}$, Eq. 1/1 becomes

$$\boxed{\Sigma\mathbf{F} = m\mathbf{a}} \qquad \textbf{(3/3)}$$

When applying Eq. 3/3 to solve problems, we usually express it in scalar component form with the use of one of the coordinate systems developed in Chapter 2. The choice of an appropriate coordinate system depends on the type of motion involved and is a vital step in the formulation of any problem. Equation 3/3, or any one of the component forms of the force-mass-acceleration equation, is usually called the *equation of motion*. The equation of motion gives the instantaneous value of the acceleration corresponding to the instantaneous values of the forces which are acting.

Two Types of Dynamics Problems

We encounter two types of problems when applying Eq. 3/3. In the first type, the acceleration of the particle is either specified or can be determined directly from known kinematic conditions. We then determine the corresponding forces which act on the particle by direct substitution into Eq. 3/3. This problem is generally quite straightforward.

In the second type of problem, the forces acting on the particle are specified and we must determine the resulting motion. If the forces are constant, the acceleration is also constant and is easily found from Eq. 3/3. When the forces are functions of time, position, or velocity, Eq. 3/3 becomes a differential equation which must be integrated to determine the velocity and displacement.

Problems of this second type are often more formidable, as the integration may be difficult to carry out, particularly when the force is a mixed function of two or more motion variables. In practice, it is frequently necessary to resort to approximate integration techniques, either numerical or graphical, particularly when experimental data are involved. The procedures for a mathematical integration of the acceleration when it is a function of the motion variables were developed in Art. 2/2, and these same procedures apply when the force is a specified function of these same parameters, since force and acceleration differ only by the constant factor of the mass.

Constrained and Unconstrained Motion

There are two physically distinct types of motion, both described by Eq. 3/3. The first type is *unconstrained* motion where the particle is free of mechanical guides and follows a path determined by its initial motion and by the forces which are applied to it from external sources. An airplane or rocket in flight and an electron moving in a charged field are examples of unconstrained motion.

The second type is *constrained* motion where the path of the particle is partially or totally determined by restraining guides. An ice-hockey puck is partially constrained to move in the horizontal plane by the surface of the ice. A train moving along its track and a collar sliding along a fixed shaft are examples of more fully constrained motion. Some of the forces acting on a particle during constrained motion may be applied from outside sources, and others may be the reactions on the particle from the constraining guides. *All forces*, both applied and reactive, which act *on* the particle must be accounted for in applying Eq. 3/3.

The choice of an appropriate coordinate system is frequently indicated by the number and geometry of the constraints. Thus, if a particle is free to move in space, as is the center of mass of the airplane or rocket in free flight, the particle is said to have *three degrees of freedom* since three independent coordinates are required to specify the position of the particle at any instant. All three of the scalar components of the equation of motion would have to be integrated to obtain the space coordinates as a function of time.

If a particle is constrained to move along a surface, as is the hockey puck or a marble sliding on the curved surface of a bowl, only two coordinates are needed to specify its position, and in this case it is said to have *two degrees of freedom*. If a particle is constrained to move along a fixed linear path, as is the collar sliding along a fixed shaft, its position may be specified by the coordinate measured along the shaft. In this case, the particle would have only *one degree of freedom*.

Key Concepts

FREE-BODY DIAGRAM

When applying any of the force-mass-acceleration equations of motion, you must account correctly for *all* forces acting on the particle. The only forces which we may neglect are those whose magnitudes are negligible compared with other forces acting, such as the forces of mutual attraction between two particles compared with their attraction to a celestial body such as the earth. The vector sum $\Sigma \mathbf{F}$ of Eq. 3/3 means the vector sum of *all* forces acting *on* the particle in question. Likewise, the corresponding scalar force summation in any one of the component directions means the sum of the components of *all* forces acting *on* the particle in that particular direction.

The only reliable way to account accurately and consistently for every force is to *isolate* the particle under consideration from *all* contacting and influencing bodies and replace the bodies removed by the forces they exert on the particle isolated. The resulting *free-body diagram* is the means by which every force, known and unknown, which acts on the particle is represented and thus accounted for. Only after this vital step has been completed should you write the appropriate equation or equations of motion.

The free-body diagram serves the same key purpose in dynamics as it does in statics. This purpose is simply to establish a *thoroughly reliable method* for the correct evaluation of the resultant of all actual forces acting on the particle or body in question. In statics this resultant equals zero, whereas in dynamics it is equated to the product of mass and acceleration. When you use the vector form of the equation of motion, remember that it represents several scalar equations and that every equation must be satisfied.

Careful and consistent use of the *free-body method* is the *most important single lesson* to be learned in the study of engineering mechanics. When drawing a free-body diagram, clearly indicate the coordinate axes and their positive directions. When you write the equations of motion, make sure all force summations are consistent with the choice of these positive directions. As an aid to the identification of external forces which act on the body in question, these forces are shown as heavy red vectors in the illustrations in this book. Sample Problems 3/1 through 3/5 in the next article contain five examples of free-body diagrams. You should study these to see how the diagrams are constructed.

In solving problems, you may wonder how to get started and what sequence of steps to follow in arriving at the solution. This difficulty may be minimized by forming the habit of first recognizing some relationship between the desired unknown quantity in the problem and other quantities, known and unknown. Then determine additional relationships between these unknowns and other quantities, known and unknown. Finally, establish the dependence on the original data and develop the procedure for the analysis and computation. A few minutes spent organizing the plan of attack through recognition of the dependence of one quantity on another will be time well spent and will usually prevent groping for the answer with irrelevant calculations.

3/4 Rectilinear Motion

We now apply the concepts discussed in Arts. 3/2 and 3/3 to problems in particle motion, starting with rectilinear motion in this article and treating curvilinear motion in Art. 3/5. In both articles, we will analyze the motions of bodies which can be treated as particles. This simplification is possible as long as we are interested only in the motion of the mass center of the body. In this case we may treat the forces as concurrent through the mass center. We will account for the action of nonconcurrent forces on the motions of bodies when we discuss the kinetics of rigid bodies in Chapter 6.

If we choose the x-direction, for example, as the direction of the rectilinear motion of a particle of mass m, the acceleration in the y- and z-directions will be zero and the scalar components of Eq. 3/3 become

$$\Sigma F_x = ma_x \qquad \Sigma F_y = 0 \qquad \Sigma F_z = 0 \tag{3/4}$$

For cases where we are not free to choose a coordinate direction along the motion, we would have in the general case all three component equations

$$\Sigma F_x = ma_x \qquad \Sigma F_y = ma_y \qquad \Sigma F_z = ma_z \tag{3/5}$$

where the acceleration and resultant force are given by

$$\mathbf{a} = a_x\mathbf{i} + a_y\mathbf{j} + a_z\mathbf{k}$$
$$a = \sqrt{a_x^{\,2} + a_y^{\,2} + a_z^{\,2}}$$
$$\Sigma\mathbf{F} = \Sigma F_x\mathbf{i} + \Sigma F_y\mathbf{j} + \Sigma F_z\mathbf{k}$$
$$|\Sigma\mathbf{F}| = \sqrt{(\Sigma F_x)^2 + (\Sigma F_y)^2 + (\Sigma F_z)^2}$$

Koji Sasahara/AP Images

This view of a car-collision test suggests that very large accelerations and accompanying large forces occur throughout the system of the two cars. The crash dummies are also subjected to large forces, primarily by the shoulder-harness/seat-belt restraints.

Sample Problem 3/1

A 75-kg man stands on a spring scale in an elevator. During the first 3 seconds of motion from rest, the tension T in the hoisting cable is 8300 N. Find the reading R of the scale in newtons during this interval and the upward velocity v of the elevator at the end of the 3 seconds. The total mass of the elevator, man, and scale is 750 kg.

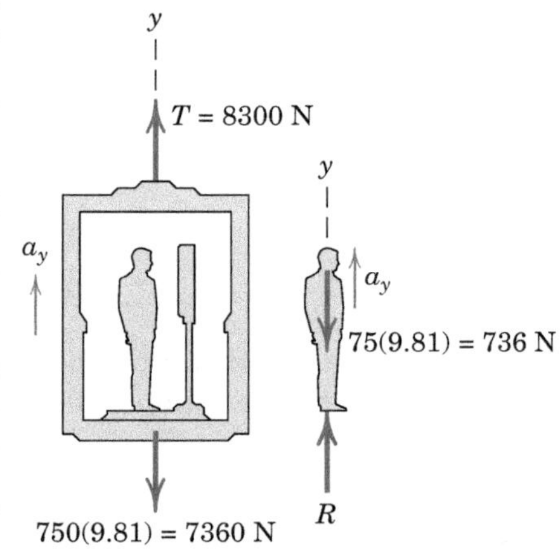

Solution. The force registered by the scale and the velocity both depend on the acceleration of the elevator, which is constant during the interval for which the forces are constant. From the free-body diagram of the elevator, scale, and man taken together, the acceleration is found to be

$$[\Sigma F_y = ma_y] \qquad 8300 - 7360 = 750a_y \qquad a_y = 1.257 \text{ m/s}^2$$

The scale reads the downward force exerted on it by the man's feet. The equal and opposite reaction R to this action is shown on the free-body diagram of the man alone together with his weight, and the equation of motion for him gives

① $$[\Sigma F_y = ma_y] \qquad R - 736 = 75(1.257) \qquad R = 830 \text{ N} \qquad \textit{Ans.}$$

The velocity reached at the end of the 3 seconds is

$$\left[\Delta v = \int a\, dt\right] \qquad v - 0 = \int_0^3 1.257\, dt \qquad v = 3.77 \text{ m/s} \qquad \textit{Ans.}$$

Helpful Hint

① If the scale were calibrated in kilograms, it would read 830/9.81 = 84.6 kg which, of course, is not his true mass since the measurement was made in a noninertial (accelerating) system. *Suggestion:* Rework this problem in U.S. customary units.

Sample Problem 3/2

A small inspection car with a mass of 200 kg runs along the fixed overhead cable and is controlled by the attached cable at A. Determine the acceleration of the car when the control cable is horizontal and under a tension $T = 2.4$ kN. Also find the total force P exerted by the supporting cable on the wheels.

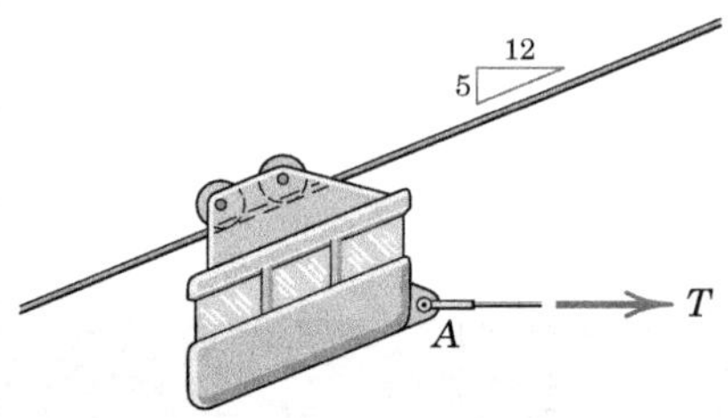

Solution. The free-body diagram of the car and wheels taken together and treated as a particle discloses the 2.4-kN tension T, the weight $W = mg = 200(9.81) = 1962$ N, and the force P exerted on the wheel assembly by the cable.

The car is in equilibrium in the y-direction since there is no acceleration in this direction. Thus,

$$[\Sigma F_y = 0] \qquad P - 2.4(\tfrac{5}{13}) - 1.962(\tfrac{12}{13}) = 0 \qquad P = 2.73 \text{ kN} \qquad \textit{Ans.}$$

① In the x-direction the equation of motion gives

$$[\Sigma F_x = ma_x] \qquad 2400(\tfrac{12}{13}) - 1962(\tfrac{5}{13}) = 200a \qquad a = 7.30 \text{ m/s}^2 \qquad \textit{Ans.}$$

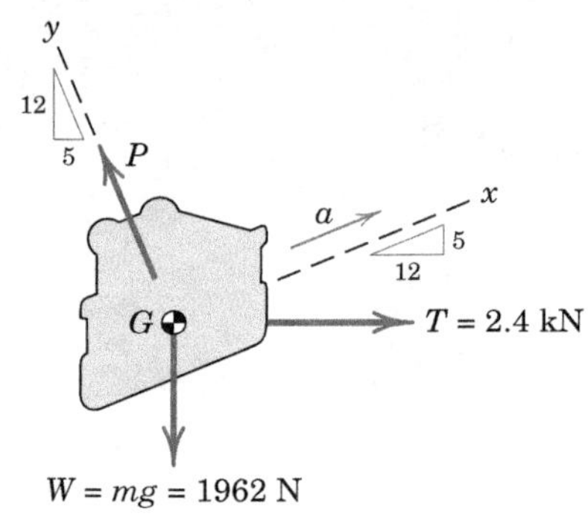

Helpful Hint

① By choosing our coordinate axes along and normal to the direction of the acceleration, we are able to solve the two equations independently. Would this be so if x and y were chosen as horizontal and vertical?

Sample Problem 3/3

The 250-lb concrete block A is released from rest in the position shown and pulls the 400-lb log up the 30° ramp. If the coefficient of kinetic friction between the log and the ramp is 0.5, determine the velocity of the block as it hits the ground at B.

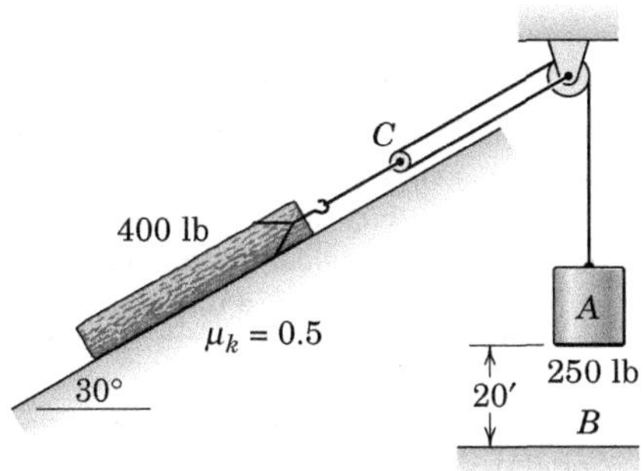

Solution. The motions of the log and the block A are clearly dependent. Although by now it should be evident that the acceleration of the log up the incline is half the downward acceleration of A, we may prove it formally. The constant total length of the cable is $L = 2s_C + s_A +$ constant, where the con-
① stant accounts for the cable portions wrapped around the pulleys. Differentiating twice with respect to time gives $0 = 2\ddot{s}_C + \ddot{s}_A$, or

$$0 = 2a_C + a_A$$

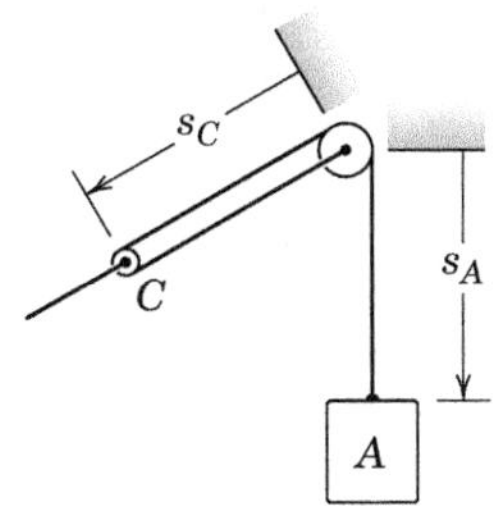

We assume here that the masses of the pulleys are negligible and that they turn with negligible friction. With these assumptions the free-body diagram of the pulley C discloses force and moment equilibrium. Thus, the tension in the cable attached to the log is twice that applied to the block. Note that the accelerations of the log and the center of pulley C are identical.

The free-body diagram of the log shows the friction force $\mu_k N$ for motion up the plane. Equilibrium of the log in the y-direction gives

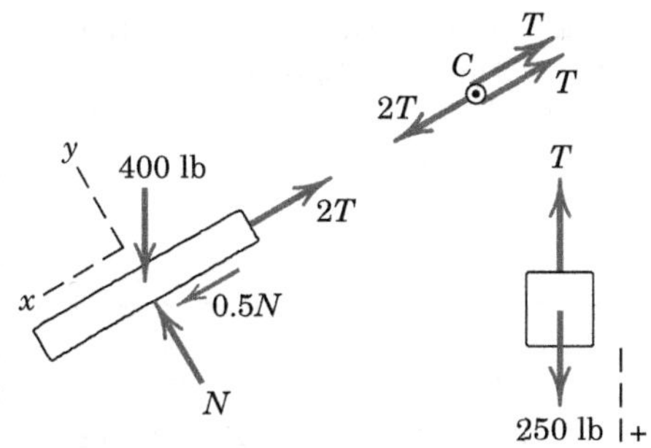

② $[\Sigma F_y = 0]$ $\quad N - 400 \cos 30° = 0 \qquad N = 346$ lb

and its equation of motion in the x-direction gives

$$[\Sigma F_x = ma_x] \qquad 0.5(346) - 2T + 400 \sin 30° = \frac{400}{32.2} a_C$$

For the block in the positive downward direction, we have

③ $[+\downarrow \Sigma F = ma]$ $\qquad 250 - T = \dfrac{250}{32.2} a_A$

Solving the three equations in a_C, a_A, and T gives us

$$a_A = 5.83 \text{ ft/sec}^2 \qquad a_C = -2.92 \text{ ft/sec}^2 \qquad T = 205 \text{ lb}$$

④ For the 20-ft drop with constant acceleration, the block acquires a velocity

$$[v^2 = 2ax] \qquad v_A = \sqrt{2(5.83)(20)} = 15.27 \text{ ft/sec} \qquad \textit{Ans.}$$

Helpful Hints

① The coordinates used in expressing the final kinematic constraint relationship must be consistent with those used for the kinetic equations of motion.

② We can verify that the log will indeed move up the ramp by calculating the force in the cable necessary to initiate motion from the equilibrium condition. This force is $2T = 0.5N + 400 \sin 30° = 373$ lb or $T = 186.5$ lb, which is less than the 250-lb weight of block A. Hence, the log will move up.

③ Note the serious error in assuming that $T = 250$ lb, in which case, block A would not accelerate.

④ Because the forces on this system remain constant, the resulting accelerations also remain constant.

Sample Problem 3/4

The design model for a new ship has a mass of 10 kg and is tested in an experimental towing tank to determine its resistance to motion through the water at various speeds. The test results are plotted on the accompanying graph, and the resistance R may be closely approximated by the dashed parabolic curve shown. If the model is released when it has a speed of 2 m/s, determine the time t required for it to reduce its speed to 1 m/s and the corresponding travel distance x.

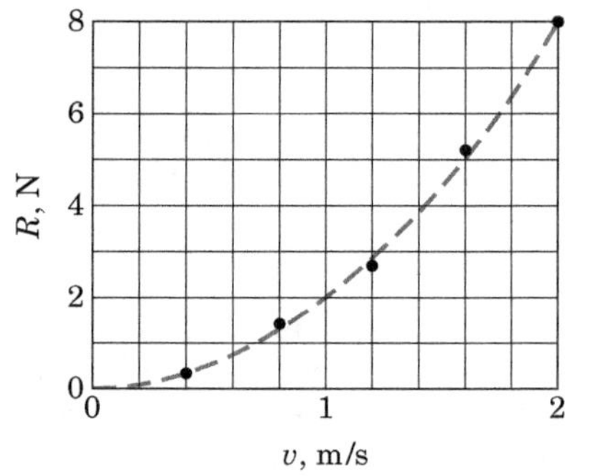

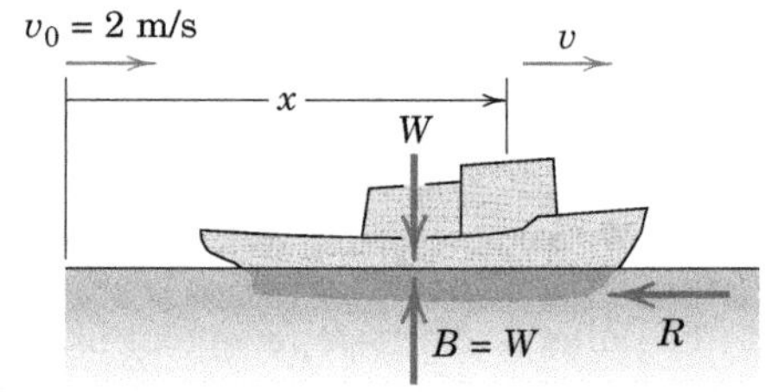

Solution. We approximate the resistance–velocity relation by $R = kv^2$ and find k by substituting $R = 8$ N and $v = 2$ m/s into the equation, which gives $k = 8/2^2 = 2\ \text{N}\cdot\text{s}^2/\text{m}^2$. Thus, $R = 2v^2$.

The only horizontal force on the model is R, so that

① $[\Sigma F_x = ma_x]$ $\qquad -R = ma_x \qquad \text{or} \qquad -2v^2 = 10\dfrac{dv}{dt}$

We separate the variables and integrate to obtain

$$\int_0^t dt = -5\int_2^v \frac{dv}{v^2} \qquad t = 5\left(\frac{1}{v} - \frac{1}{2}\right)\ \text{s}$$

Thus, when $v = v_0/2 = 1$ m/s, the time is $t = 5(\frac{1}{1} - \frac{1}{2}) = 2.5$ s. *Ans.*

The distance traveled during the 2.5 seconds is obtained by integrating $v = dx/dt$. Thus, $v = 10/(5 + 2t)$ so that

② $$\int_0^x dx = \int_0^{2.5} \frac{10}{5 + 2t}\,dt \qquad x = \frac{10}{2}\ln(5 + 2t)\Big|_0^{2.5} = 3.47\ \text{m} \qquad \textit{Ans.}$$

Helpful Hints

① Be careful to observe the minus sign for R.

② *Suggestion:* Express the distance x after release in terms of the velocity v and see if you agree with the resulting relation $x = 5 \ln (v_0/v)$.

Sample Problem 3/5

The collar of mass m slides up the vertical shaft under the action of a force F of constant magnitude but variable direction. If $\theta = kt$ where k is a constant and if the collar starts from rest with $\theta = 0$, determine the magnitude F of the force which will result in the collar coming to rest as θ reaches $\pi/2$. The coefficient of kinetic friction between the collar and shaft is μ_k.

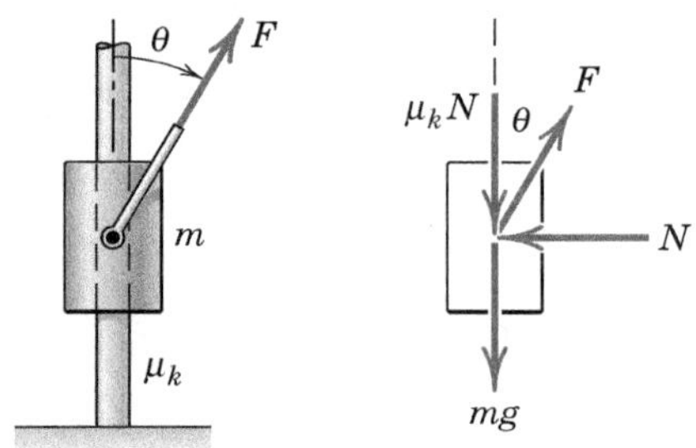

Solution. After drawing the free-body diagram, we apply the equation of motion in the y-direction to get

① $[\Sigma F_y = ma_y]$ $\qquad F\cos\theta - \mu_k N - mg = m\dfrac{dv}{dt}$

where equilibrium in the horizontal direction requires $N = F\sin\theta$. Substituting $\theta = kt$ and integrating first between general limits give

$$\int_0^t (F\cos kt - \mu_k F\sin kt - mg)\,dt = m\int_0^v dv$$

which becomes

$$\frac{F}{k}[\sin kt + \mu_k(\cos kt - 1)] - mgt = mv$$

For $\theta = \pi/2$ the time becomes $t = \pi/2k$, and $v = 0$ so that

② $$\frac{F}{k}[1 + \mu_k(0 - 1)] - \frac{mg\pi}{2k} = 0 \qquad \text{and} \qquad F = \frac{mg\pi}{2(1 - \mu_k)} \qquad \textit{Ans.}$$

Helpful Hints

① If θ were expressed as a function of the vertical displacement y instead of the time t, the acceleration would become a function of the displacement and we would use $v\,dv = a\,dy$.

② We see that the results do not depend on k, the rate at which the force changes direction.

PROBLEMS

Introductory Problems

3/1 The 50-kg crate is projected along the floor with an initial speed of 8 m/s at $x = 0$. The coefficient of kinetic friction is 0.40. Calculate the time required for the crate to come to rest and the corresponding distance x traveled.

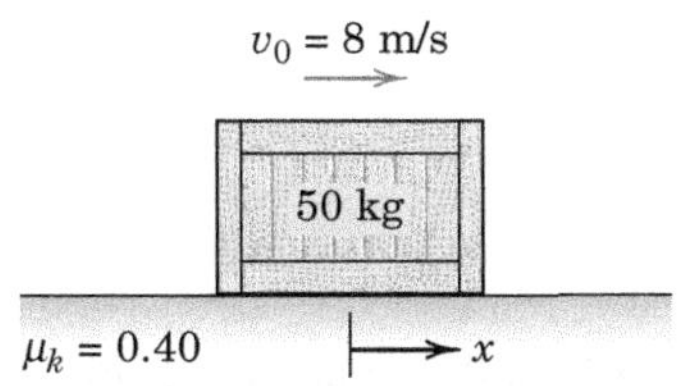

Problem 3/1

3/2 The 50-kg crate is stationary when the force P is applied. Determine the resulting acceleration of the crate if (*a*) $P = 0$, (*b*) $P = 150$ N, and (*c*) $P = 300$ N.

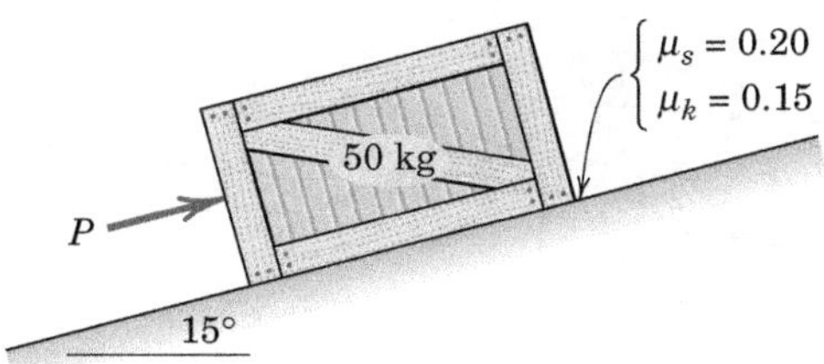

Problem 3/2

3/3 At a certain instant, the 80-lb crate has a velocity of 30 ft/sec up the 20° incline. Calculate the time t required for the crate to come to rest and the corresponding distance d traveled. Also, determine the distance d' traveled when the crate speed has been reduced to 15 ft/sec.

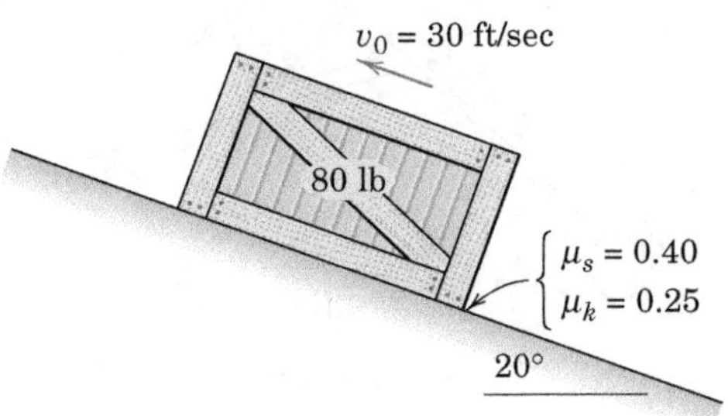

Problem 3/3

3/4 A man pulls himself up the 15° incline by the method shown. If the combined mass of the man and cart is 100 kg, determine the acceleration of the cart if the man exerts a pull of 175 N on the rope. Neglect all friction and the mass of the rope, pulleys, and wheels.

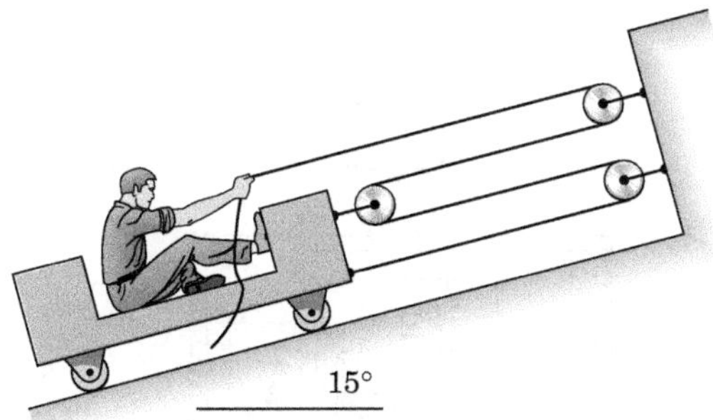

Problem 3/4

3/5 The 10-Mg truck hauls the 20-Mg trailer. If the unit starts from rest on a level road with a tractive force of 20 kN between the driving wheels of the truck and the road, compute the tension T in the horizontal drawbar and the acceleration a of the rig.

Problem 3/5

3/6 A 60-kg woman holds a 9-kg package as she stands within an elevator which briefly accelerates upward at a rate of $g/4$. Determine the force R which the elevator floor exerts on her feet and the lifting force L which she exerts on the package during the acceleration interval. If the elevator support cables suddenly and completely fail, what values would R and L acquire?

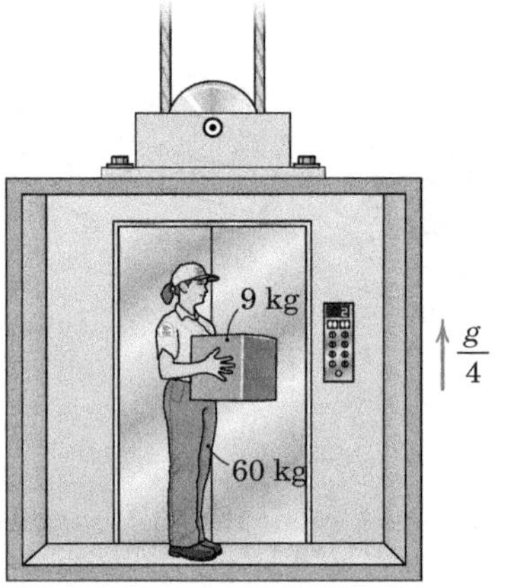

Problem 3/6

3/7 During a brake test, the rear-engine car is stopped from an initial speed of 100 km/h in a distance of 50 m. If it is known that all four wheels contribute equally to the braking force, determine the braking force F at each wheel. Assume a constant deceleration for the 1500-kg car.

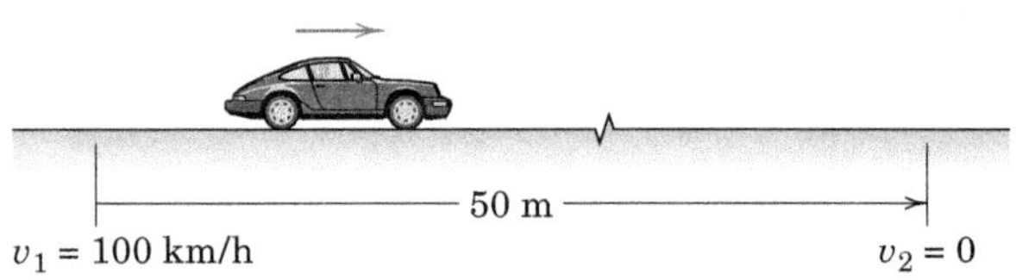

Problem 3/7

3/8 A skier starts from rest on the 40° slope at time $t = 0$ and is clocked at $t = 2.58$ s as he passes a speed checkpoint 20 m down the slope. Determine the coefficient of kinetic friction between the snow and the skis. Neglect wind resistance.

Problem 3/8

3/9 The inexperienced driver of an all-wheel-drive car applies too much throttle as he attempts to accelerate from rest up the slippery 10-percent incline. The result is wheel spin at all four tires, each of which has the same gripping ability. Determine the vehicle acceleration for the conditions of (*a*) light snow, $\mu_k = 0.12$ and (*b*) ice, $\mu_k = 0.05$.

Problem 3/9

3/10 Determine the steady-state angle α if the constant force P is applied to the cart of mass M. The pendulum bob has mass m and the rigid bar of length L has negligible mass. Ignore all friction. Evaluate your expression for $P = 0$.

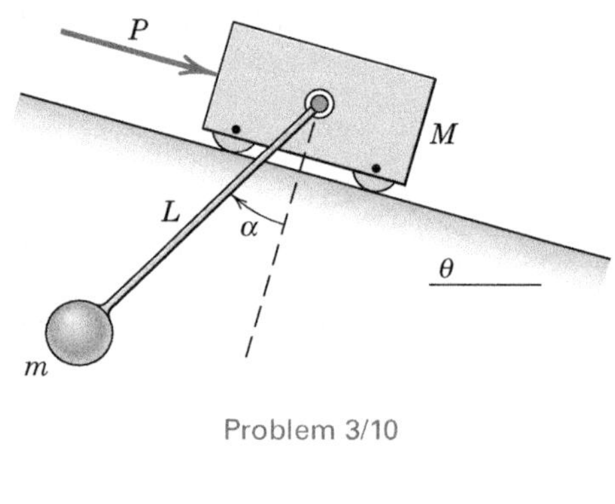

Problem 3/10

3/11 The 300-Mg jet airliner has three engines, each of which produces a nearly constant thrust of 240 kN during the takeoff roll. Determine the length s of runway required if the takeoff speed is 220 km/h. Compute s first for an uphill takeoff direction from A to B and second for a downhill takeoff from B to A on the slightly inclined runway. Neglect air and rolling resistance.

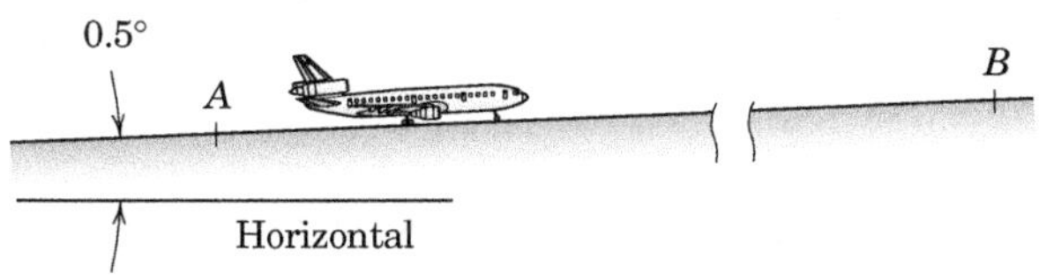

Problem 3/11

3/12 For a given horizontal force P, determine the normal reaction forces at A and B. The mass of the cylinder is m and that of the cart is M. Neglect all friction.

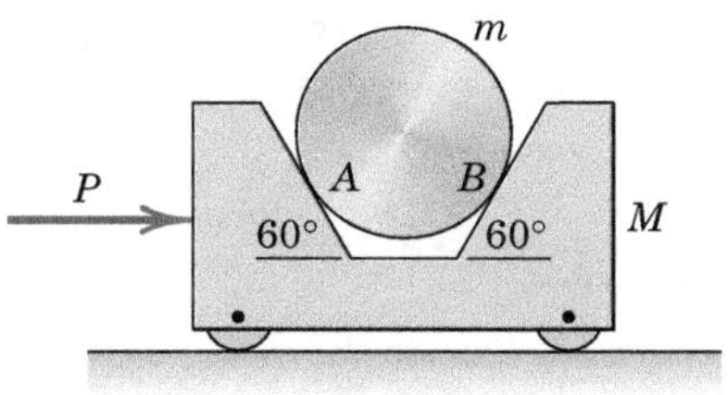

Problem 3/12

3/13 The system of the previous problem is now placed on the 15° incline. What force P will cause the normal reaction force at B to be zero?

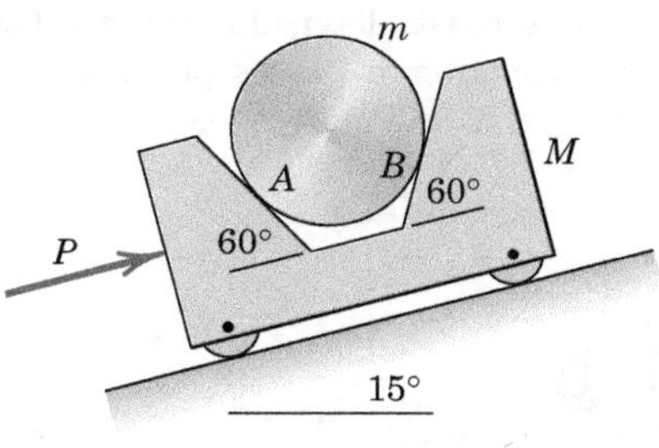

Problem 3/13

3/14 The 750,000-lb jetliner A has four engines, each of which produces a nearly constant thrust of 40,000 lb during the takeoff roll. A small commuter aircraft B taxis toward the end of the runway at a constant speed v_B = 15 mi/hr. Determine the velocity and acceleration which A appears to have relative to an observer in B 10 seconds after A begins its takeoff roll. Neglect air and rolling resistance.

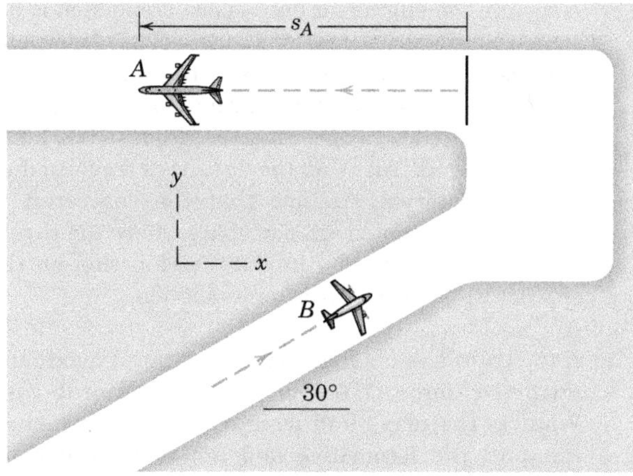

Problem 3/14

Representative Problems

3/15 The drive system of the 350-ton tugboat causes an external thrust P = 7000 lb to be applied as indicated in the figure. If the tugboat pushes an 800-ton coal barge starting from rest, what is the acceleration of the combined unit? Also, determine the force R of interaction between tugboat and barge. Neglect water resistance.

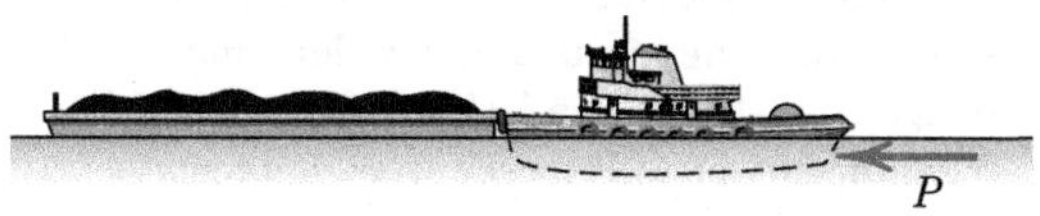

Problem 3/15

3/16 Determine the tension P in the cable which will give the 100-lb block a steady acceleration of 5 ft/sec^2 up the incline.

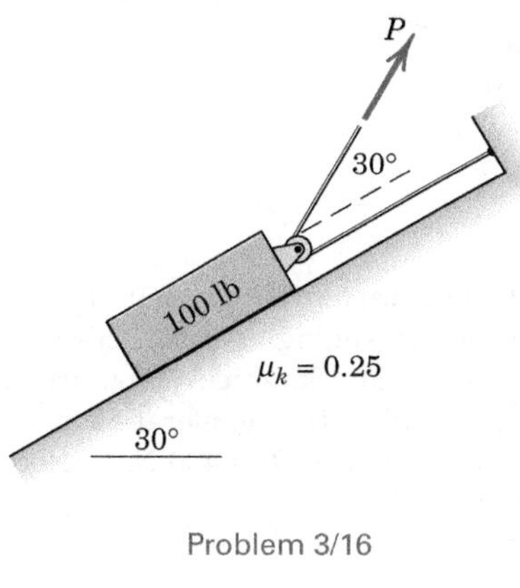

Problem 3/16

3/17 A cesium-ion engine for deep-space propulsion is designed to produce a constant thrust of 2.5 N for long periods of time. If the engine is to propel a 70-Mg spacecraft on an interplanetary mission, compute the time t required for a speed increase from 40 000 km/h to 65 000 km/h. Also find the distance s traveled during this interval. Assume that the spacecraft is moving in a remote region of space where the thrust from its ion engine is the only force acting on the spacecraft in the direction of its motion.

3/18 A toy train has magnetic couplers whose maximum attractive force is 0.2 lb between adjacent cars. What is the maximum force P with which a child can pull the locomotive and not break the train apart at a coupler? If P is slightly exceeded, which coupler fails? Neglect the mass and friction associated with all wheels.

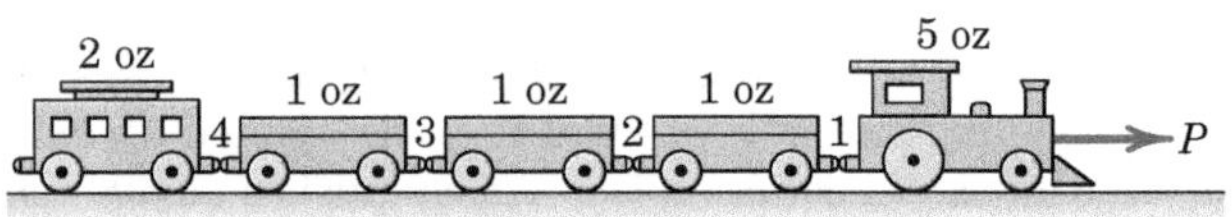

Problem 3/18

3/19 A worker develops a tension T in the cable as he attempts to move the 50-kg cart up the 20° incline. Determine the resulting acceleration of the cart if (*a*) $T = 150$ N and (*b*) $T = 200$ N. Neglect all friction, except that at the worker's feet.

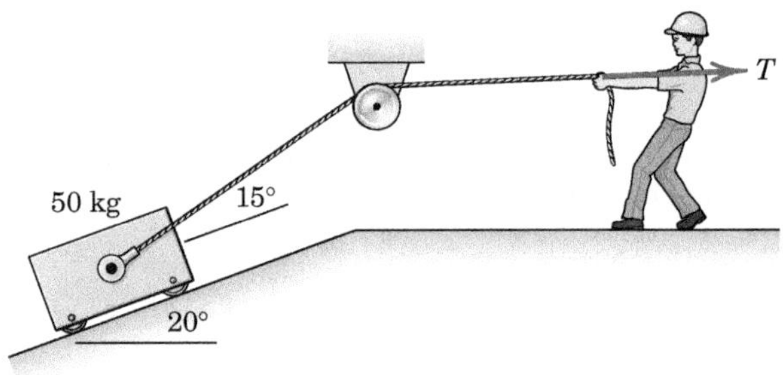

Problem 3/19

3/20 The wheeled cart of Prob. 3/19 is now replaced with a 50-kg sliding wooden crate. The coefficients of static and kinetic friction are given in the figure. Determine the acceleration of the crate if (*a*) $T = 300$ N and (*b*) $T = 400$ N. Neglect friction at the pulley.

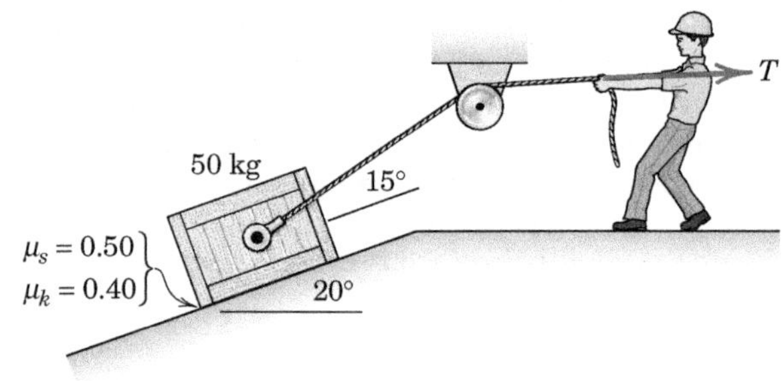

Problem 3/20

3/21 A small package is deposited by the conveyor belt onto the 30° ramp at A with a velocity of 0.8 m/s. Calculate the distance s on the level surface BC at which the package comes to rest. The coefficient of kinetic friction for the package and supporting surface from A to C is 0.30.

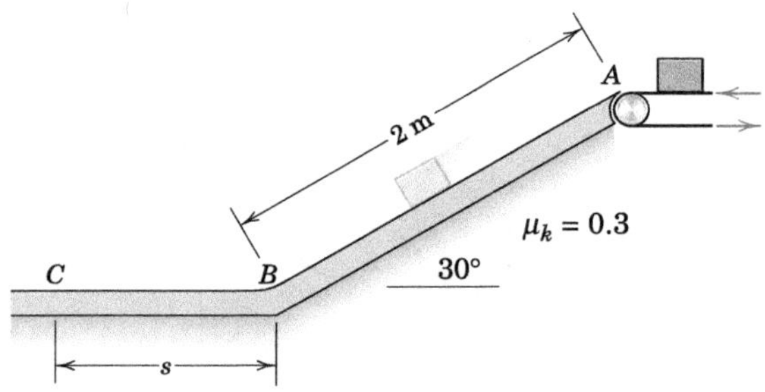

Problem 3/21

3/22 A bicyclist finds that she descends the slope $\theta_1 = 3°$ at a certain constant speed with no braking or pedaling required. The slope changes fairly abruptly to θ_2 at point A. If the bicyclist takes no action but continues to coast, determine the acceleration a of the bike just after it passes point A for the conditions (*a*) $\theta_2 = 5°$ and (*b*) $\theta_2 = 0$.

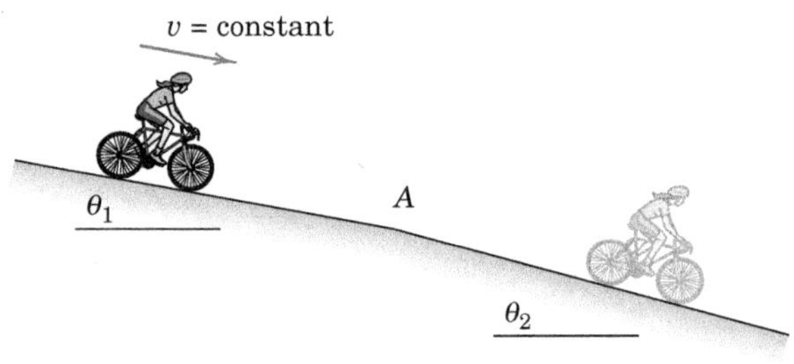

Problem 3/22

3/23 The 5-oz pinewood-derby car is released from rest at the starting line A and crosses the finish line C 2.75 sec later. The transition at B is small and smooth. Assume that the net retarding force is constant throughout the run and find this force.

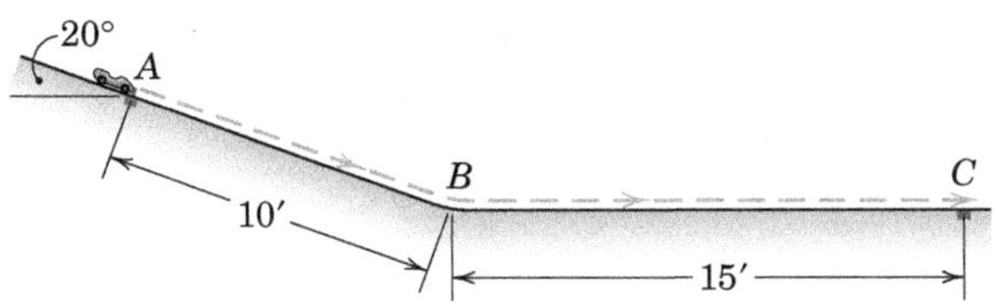

Problem 3/23

3/24 The coefficient of static friction between the flat bed of the truck and the crate it carries is 0.30. Determine the minimum stopping distance s which the truck can have from a speed of 70 km/h with constant deceleration if the crate is not to slip forward.

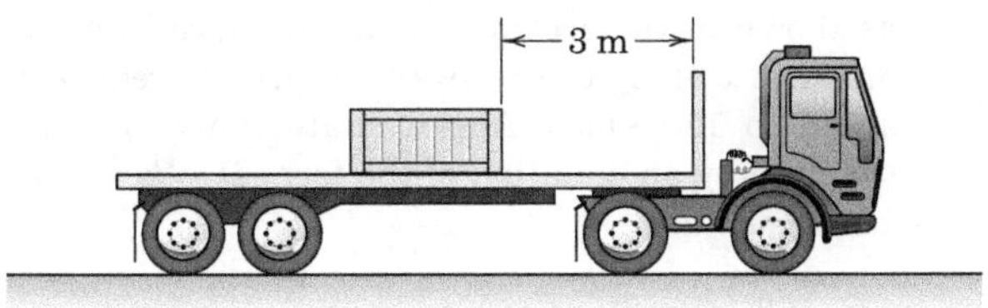

Problem 3/24

3/25 If the truck of Prob. 3/24 comes to a stop from an initial forward speed of 70 km/h in a distance of 50 m with uniform deceleration, determine whether or not the crate strikes the wall at the forward end of the flat bed. If the crate does strike the wall, calculate its speed relative to the truck as the impact occurs. Use the friction coefficients $\mu_s = 0.30$ and $\mu_k = 0.25$.

3/26 The winch takes in cable at the rate of 200 mm/s, and this rate is momentarily increasing at 500 mm/s each second. Determine the tensions in the three cables. Neglect the weights of the pulleys.

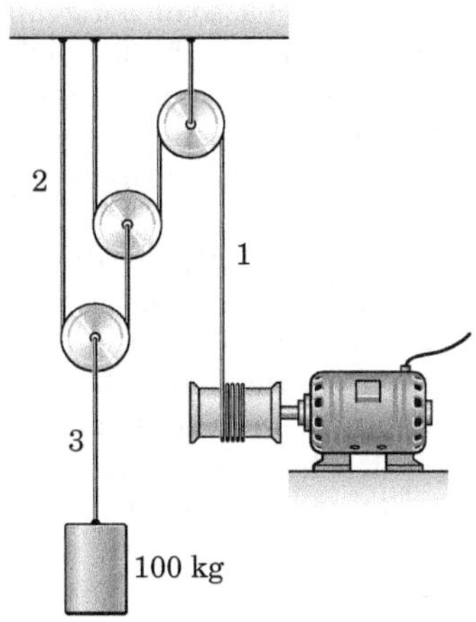

Problem 3/26

3/27 Determine the vertical acceleration of the 60-lb cylinder for each of the two cases. Neglect friction and the mass of the pulleys.

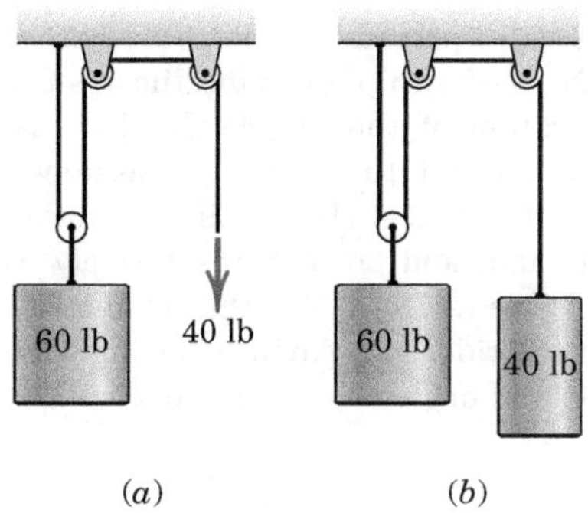

Problem 3/27

3/28 Determine the weight of cylinder B which would cause block A to accelerate (*a*) 5 ft/sec^2 down the incline and (*b*) 5 ft/sec^2 up the incline. Neglect all friction.

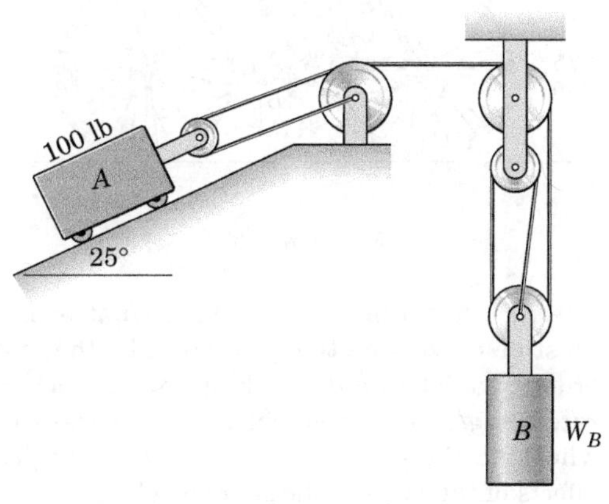

Problem 3/28

3/29 A player pitches a baseball horizontally toward a speed-sensing radar gun. The baseball weighs $5\frac{1}{8}$ oz and has a circumference of $9\frac{1}{8}$ in. If the speed at $x = 0$ is $v_0 = 90$ mi/hr, estimate the speed as a function of x. Assume that the horizontal aerodynamic drag on the baseball is given by $D = C_D(\frac{1}{2}\rho v^2)S$, where C_D is the drag coefficient, ρ is the air density, v is the speed, and S is the cross-sectional area of the baseball. Use a value of 0.3 for C_D. Neglect the vertical component of the motion but comment on the validity of this assumption. Evaluate your answer for $x = 60$ ft, which is the approximate distance between a pitcher's hand and home plate.

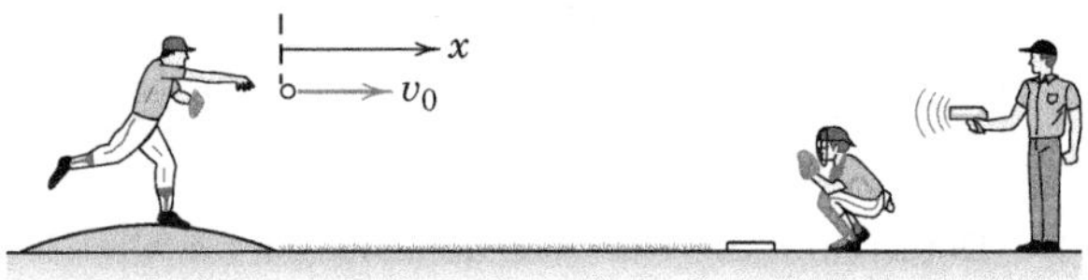

Problem 3/29

3/30 If the rider presses on the pedal with a force $P = 160$ N as shown, determine the resulting forward acceleration of the bicycle. Neglect the effects of the mass of rotating parts, and assume no slippage at the rear wheel. The radii of sprockets A and B are 45 mm and 90 mm, respectively. The mass of the bicycle is 13 kg and that of the rider is 65 kg. Treat the rider as a particle moving with the bicycle frame, and neglect drivetrain friction.

Problem 3/30

3/31 The rack has a mass $m = 50$ kg. What moment M must be exerted on the gear wheel by the motor in order to accelerate the rack up the 60° incline at a rate $a = g/4$? The fixed motor which drives the gear wheel via the shaft at O is not shown. Neglect the effects of the mass of the gear wheel.

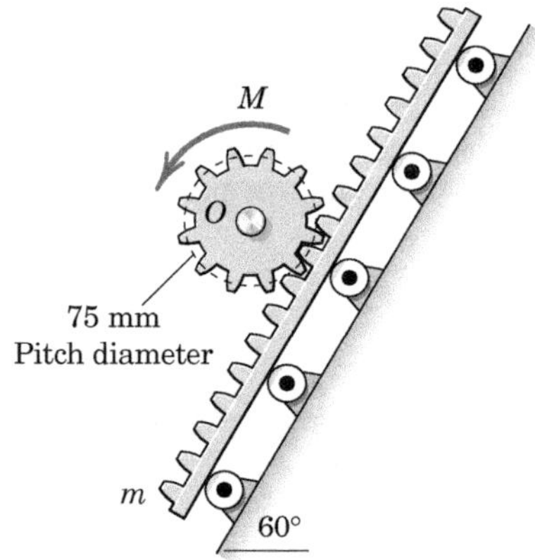

Problem 3/31

3/32 A jet airplane with a mass of 5 Mg has a touchdown speed of 300 km/h, at which instant the braking parachute is deployed and the power shut off. If the total drag on the aircraft varies with velocity as shown in the accompanying graph, calculate the distance x along the runway required to reduce the speed to 150 km/h. Approximate the variation of the drag by an equation of the form $D = kv^2$, where k is a constant.

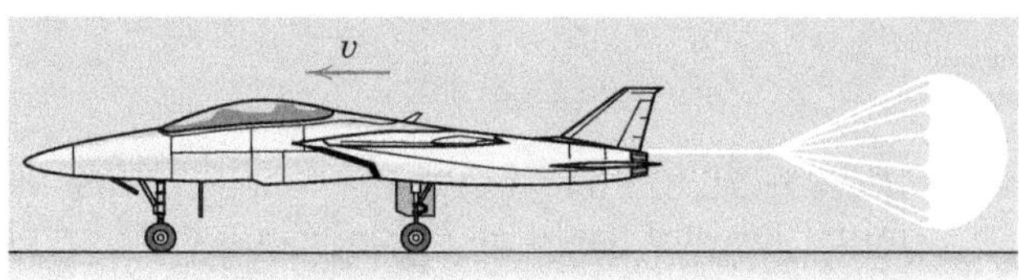

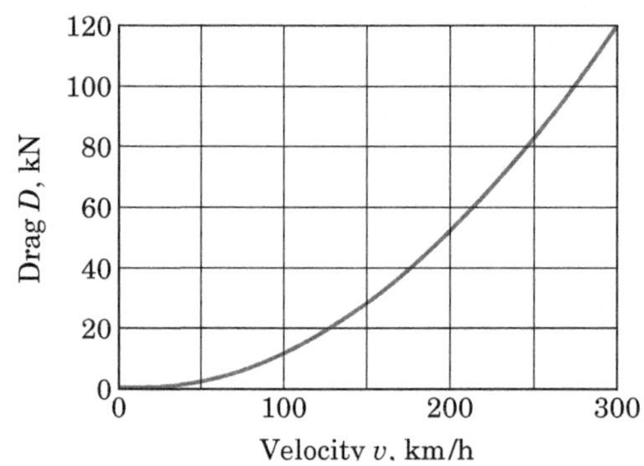

Problem 3/32

3/33 During its final approach to the runway, the aircraft speed is reduced from 300 km/h at A to 200 km/h at B. Determine the net external aerodynamic force R which acts on the 200-Mg aircraft during this interval, and find the components of this force which are parallel to and normal to the flight path.

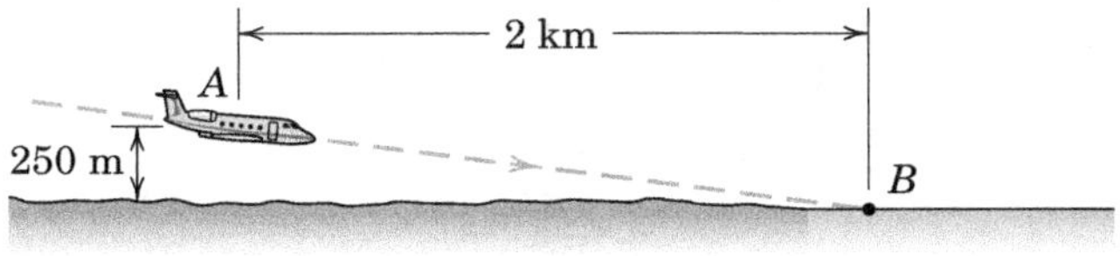

Problem 3/33

3/34 A heavy chain with a mass ρ per unit length is pulled by the constant force P along a horizontal surface consisting of a smooth section and a rough section. The chain is initially at rest on the rough surface with $x = 0$. If the coefficient of kinetic friction between the chain and the rough surface is μ_k, determine the velocity v of the chain when $x = L$. The force P is greater than $\mu_k \rho g L$ in order to initiate motion.

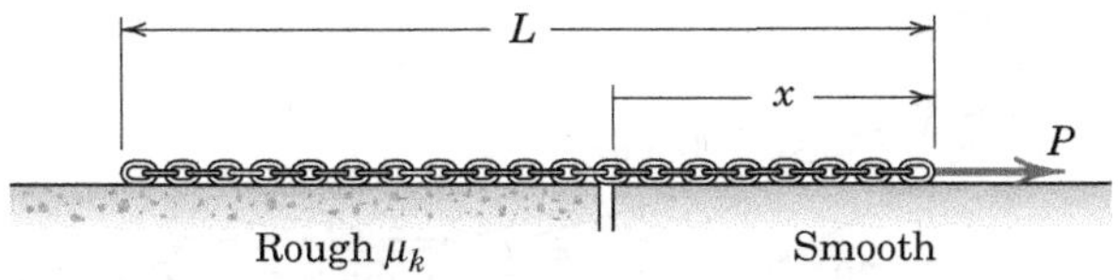

Problem 3/34

3/35 The sliders A and B are connected by a light rigid bar of length $l = 0.5$ m and move with negligible friction in the slots, both of which lie in a horizontal plane. For the position where $x_A = 0.4$ m, the velocity of A is $v_A = 0.9$ m/s to the right. Determine the acceleration of each slider and the force in the bar at this instant.

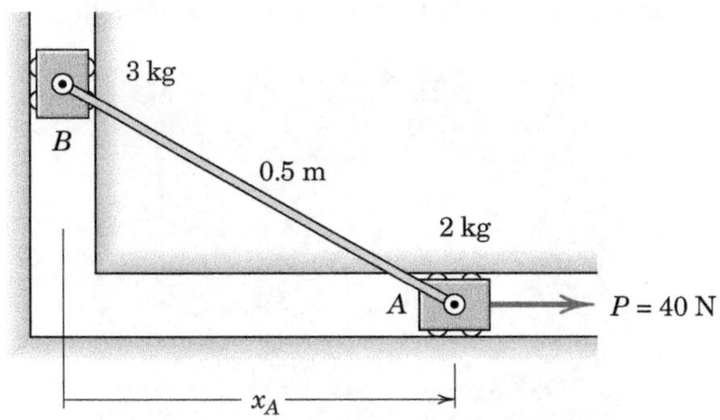

Problem 3/35

3/36 The spring of constant $k = 200$ N/m is attached to both the support and the 2-kg cylinder, which slides freely on the horizontal guide. If a constant 10-N force is applied to the cylinder at time $t = 0$ when the spring is undeformed and the system is at rest, determine the velocity of the cylinder when $x = 40$ mm. Also determine the maximum displacement of the cylinder.

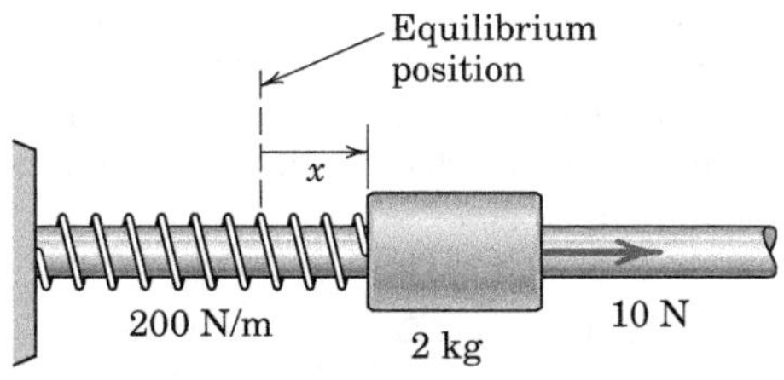

Problem 3/36

3/37 The 4-lb collar is released from rest against the light elastic spring, which has a stiffness of 10 lb/in. and has been compressed a distance of 6 in. Determine the acceleration a of the collar as a function of the vertical displacement x of the collar measured in feet from the point of release. Find the velocity v of the collar when $x = 0.5$ ft. Friction is negligible.

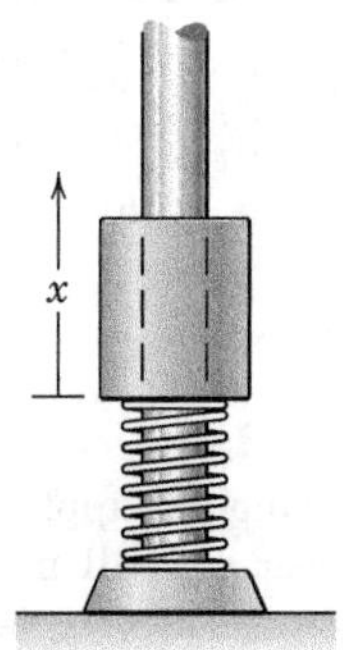

Problem 3/37

3/38 Two configurations for raising an elevator are shown. Elevator A with attached hoisting motor and drum has a total mass of 900 kg. Elevator B without motor and drum also has a mass of 900 kg. If the motor supplies a constant torque of 600 N·m to its 250-mm-diameter drum for 2 s in each case, select the configuration which results in the greater upward acceleration and determine the corresponding velocity v of the elevator 1.2 s after it starts from rest. The mass of the motorized drum is small, thus permitting it to be analyzed as though it were in equilibrium. Neglect the mass of cables and pulleys and all friction.

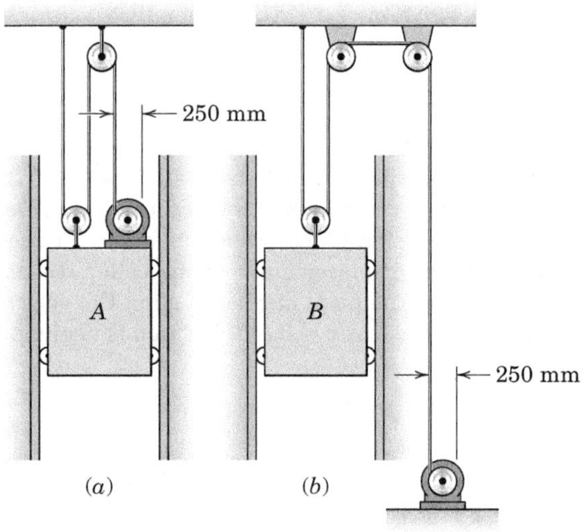

Problem 3/38

3/39 Determine the range of applied force P over which the block of mass m_2 will not slip on the wedge-shaped block of mass m_1. Neglect friction associated with the wheels of the tapered block.

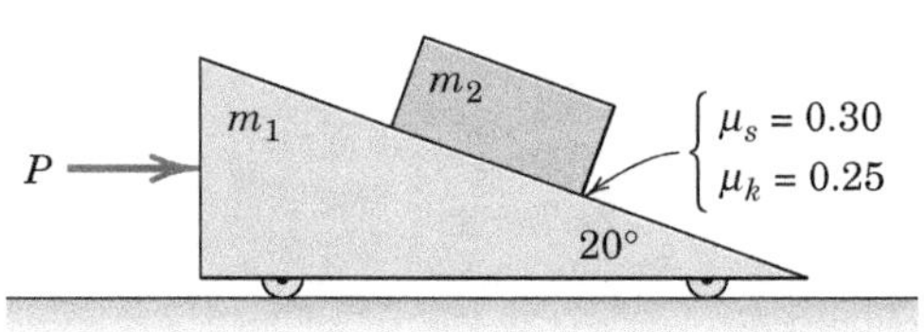

Problem 3/39

3/40 A spring-loaded device imparts an initial vertical velocity of 50 m/s to a 0.15-kg ball. The drag force on the ball is $F_D = 0.002v^2$, where F_D is in newtons when the speed v is in meters per second. Determine the maximum altitude h attained by the ball (*a*) with drag considered and (*b*) with drag neglected.

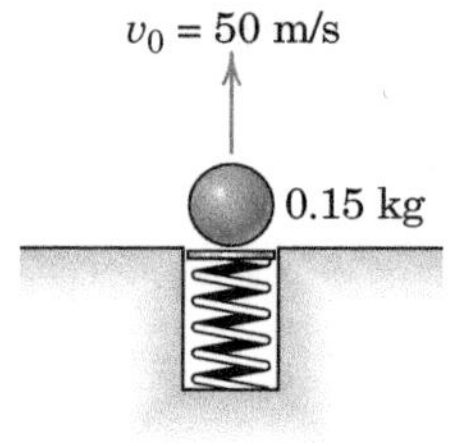

Problem 3/40

3/41 The sliders A and B are connected by a light rigid bar and move with negligible friction in the slots, both of which lie in a horizontal plane. For the position shown, the velocity of A is 0.4 m/s to the right. Determine the acceleration of each slider and the force in the bar at this instant.

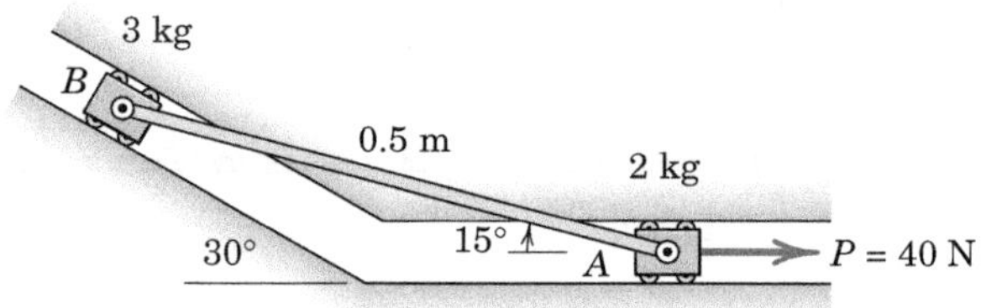

Problem 3/41

3/42 The design of a lunar mission calls for a 1200-kg spacecraft to lift off from the surface of the moon and travel in a straight line from point A and pass point B. If the spacecraft motor has a constant thrust of 2500 N, determine the speed of the spacecraft as it passes point B. Use Table D/2 and the gravitational law from Chapter 1 as needed.

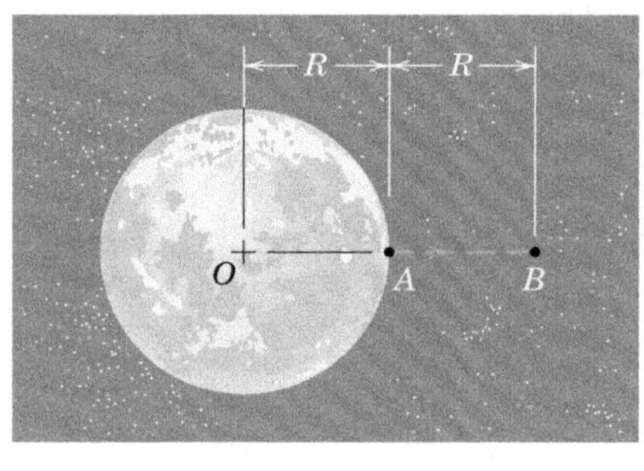

Problem 3/42

3/43 The system is released from rest in the configuration shown at time $t = 0$. Determine the time t when the block of mass m_1 contacts the lower stop of the body of mass m_2. Also, determine the corresponding distance s_2 traveled by m_2. Use the values $m_1 = 0.5$ kg, $m_2 = 2$ kg, $\mu_s = 0.25$, $\mu_k = 0.20$, and $d = 0.4$ m.

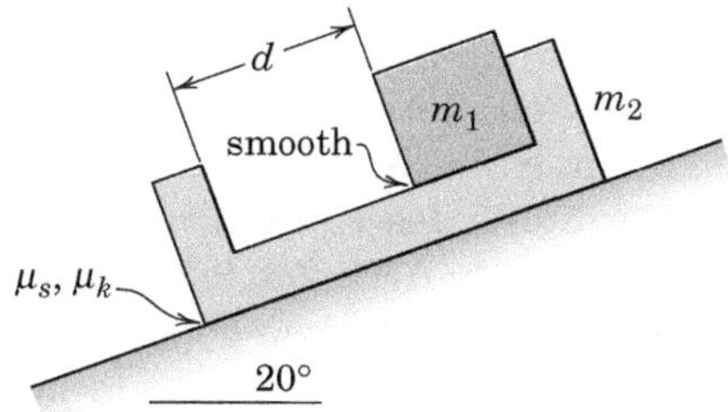

Problem 3/43

3/44 The system of Prob. 3/43 is reconsidered here, only now the interface between the two bodies is not smooth. Use the values $\mu_s = 0.10$ and $\mu_k = 0.08$ between the two bodies. Determine the time t when m_1 contacts the lower stop on m_2 and the corresponding distance s_2 traveled by m_2.

▶3/45 The rod of the fixed hydraulic cylinder is moving to the left with a speed of 100 mm/s, and this speed is momentarily increasing at a rate of 400 mm/s each second at the instant when $s_A = 425$ mm. Determine the tension in the cord at that instant. The mass of slider B is 0.5 kg, the length of the cord is 1050 mm, and the effects of the radius and friction of the small pulley at A are negligible. Find results for cases (*a*) negligible friction at slider B and (*b*) $\mu_k = 0.40$ at slider B. The action is in a vertical plane.

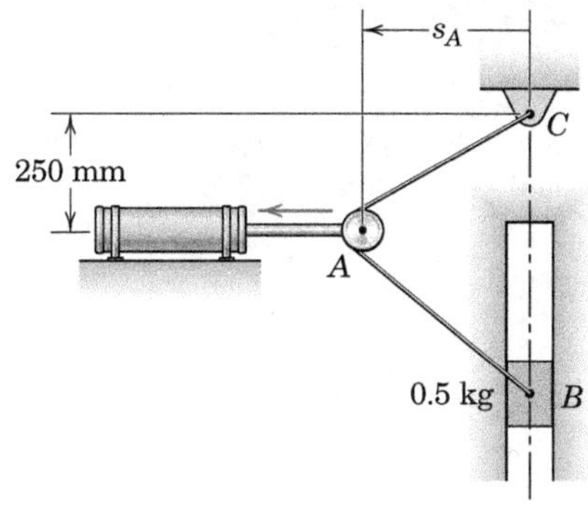

Problem 3/45

▶3/46 Two iron spheres, each of which is 100 mm in diameter, are released from rest with a center-to-center separation of 1 m. Assume an environment in space with no forces other than the force of mutual gravitational attraction and calculate the time t required for the spheres to contact each other and the absolute speed v of each sphere upon contact.

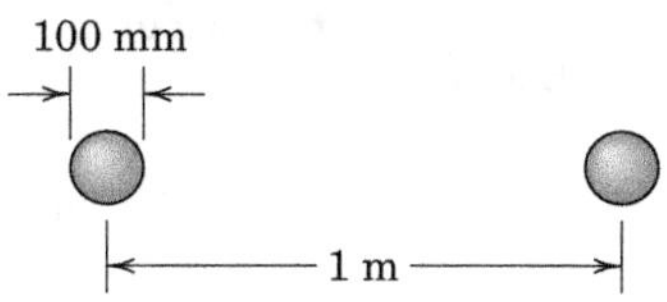

Problem 3/46

Mark Vasey/Getty Images, Inc.

Because of the banking in the turn of this track, the normal reaction force provides most of the normal acceleration of the bobsled.

Noah Clayton/Exactostock-1491/SuperStock

At the highest point of the swing, this child experiences tangential acceleration. An instant later, when she has acquired velocity, she will experience normal acceleration as well.

3/5 Curvilinear Motion

We turn our attention now to the kinetics of particles which move along plane curvilinear paths. In applying Newton's second law, Eq. 3/3, we will make use of the three coordinate descriptions of acceleration in curvilinear motion which we developed in Arts. 2/4, 2/5, and 2/6.

The choice of an appropriate coordinate system depends on the conditions of the problem and is one of the basic decisions to be made in solving curvilinear-motion problems. We now rewrite Eq. 3/3 in three ways, the choice of which depends on which coordinate system is most appropriate.

Rectangular coordinates (Art. 2/4, Fig. 2/7)

$$\boxed{\begin{aligned}\Sigma F_x &= ma_x \\ \Sigma F_y &= ma_y\end{aligned}} \tag{3/6}$$

where $a_x = \ddot{x}$ and $a_y = \ddot{y}$

Normal and tangential coordinates (Art. 2/5, Fig. 2/10)

$$\boxed{\begin{aligned}\Sigma F_n &= ma_n \\ \Sigma F_t &= ma_t\end{aligned}} \tag{3/7}$$

where $a_n = \rho\dot{\beta}^2 = v^2/\rho = v\dot{\beta}$, $a_t = \dot{v}$, and $v = \rho\dot{\beta}$

Polar coordinates (Art. 2/6, Fig. 2/15)

$$\boxed{\begin{aligned}\Sigma F_r &= ma_r \\ \Sigma F_\theta &= ma_\theta\end{aligned}} \tag{3/8}$$

where $a_r = \ddot{r} - r\dot{\theta}^2$ and $a_\theta = r\ddot{\theta} + 2\dot{r}\dot{\theta}$

In applying these motion equations to a body treated as a particle, you should follow the general procedure established in the previous article on rectilinear motion. After you identify the motion and choose the coordinate system, draw the free-body diagram of the body. Then obtain the appropriate force summations from this diagram in the usual way. The free-body diagram should be complete to avoid incorrect force summations.

Once you assign reference axes, you must use the expressions for both the forces and the acceleration which are consistent with that assignment. In the first of Eqs. 3/7, for example, the positive sense of the n-axis is *toward* the center of curvature, and so the positive sense of our force summation ΣF_n must also be *toward* the center of curvature to agree with the positive sense of the acceleration $a_n = v^2/\rho$.

Sample Problem 3/6

Determine the maximum speed v which the sliding block may have as it passes point A without losing contact with the surface.

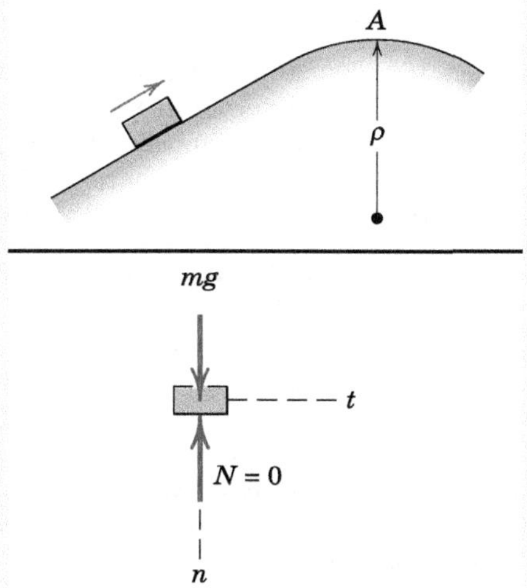

Solution. The condition for loss of contact is that the normal force N which the surface exerts on the block goes to zero. Summing forces in the normal direction gives

$$[\Sigma F_n = ma_n] \qquad mg = m\frac{v^2}{\rho} \qquad v = \sqrt{g\rho} \qquad \textit{Ans.}$$

If the speed at A were less than $\sqrt{g\rho}$, then an upward normal force exerted by the surface on the block would exist. In order for the block to have a speed at A which is greater than $\sqrt{g\rho}$, some type of constraint, such as a second curved surface above the block, would have to be introduced to provide additional downward force.

Sample Problem 3/7

Small objects are released from rest at A and slide down the smooth circular surface of radius R to a conveyor B. Determine the expression for the normal contact force N between the guide and each object in terms of θ and specify the correct angular velocity ω of the conveyor pulley of radius r to prevent any sliding on the belt as the objects transfer to the conveyor.

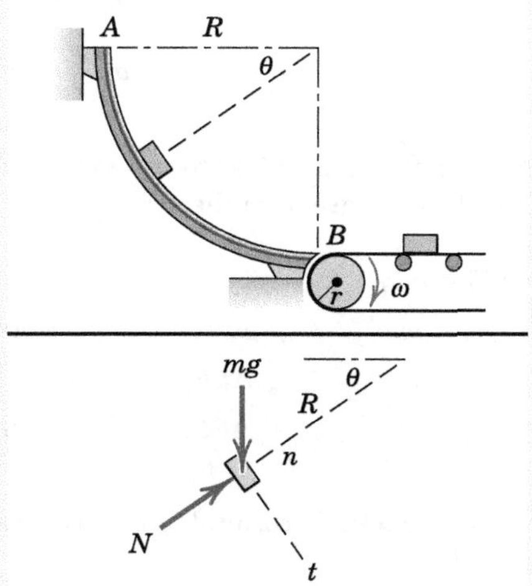

Solution. The free-body diagram of the object is shown together with the coordinate directions n and t. The normal force N depends on the n-component of the acceleration which, in turn, depends on the velocity. The velocity will be cumulative according to the tangential acceleration a_t. Hence, we will find a_t first for any general position.

$$[\Sigma F_t = ma_t] \qquad mg \cos\theta = ma_t \qquad a_t = g\cos\theta$$

① Now we can find the velocity by integrating

$$[v\,dv = a_t\,ds] \qquad \int_0^v v\,dv = \int_0^\theta g\cos\theta\,d(R\theta) \qquad v^2 = 2gR\sin\theta$$

We obtain the normal force by summing forces in the positive n-direction, which is the direction of the n-component of acceleration.

$$[\Sigma F_n = ma_n] \qquad N - mg\sin\theta = m\frac{v^2}{R} \qquad N = 3mg\sin\theta \qquad \textit{Ans.}$$

The conveyor pulley must turn at the rate $v = r\omega$ for $\theta = \pi/2$, so that

$$\omega = \sqrt{2gR}/r \qquad \textit{Ans.}$$

Helpful Hint

① It is essential here that we recognize the need to express the tangential acceleration as a function of position so that v may be found by integrating the kinematical relation $v\,dv = a_t\,ds$, in which all quantities are measured along the path.

Sample Problem 3/8

A 1500-kg car enters a section of curved road in the horizontal plane and slows down at a uniform rate from a speed of 100 km/h at A to a speed of 50 km/h as it passes C. The radius of curvature ρ of the road at A is 400 m and at C is 80 m. Determine the total horizontal force exerted by the road on the tires at positions A, B, and C. Point B is the inflection point where the curvature changes direction.

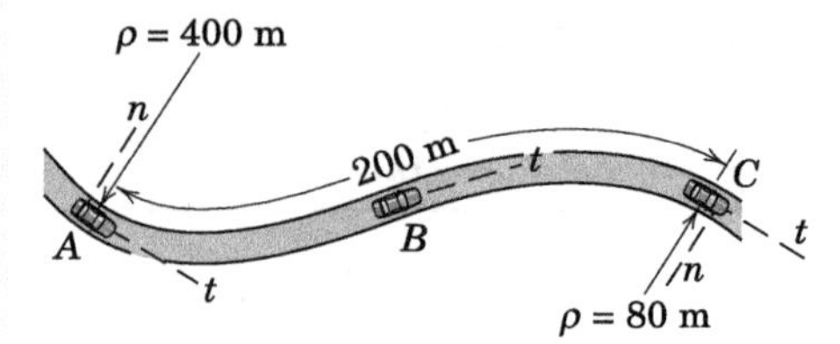

Solution. The car will be treated as a particle so that the effect of all forces exerted by the road on the tires will be treated as a single force. Since the motion is described along the direction of the road, normal and tangential coordinates will be used to specify the acceleration of the car. We will then determine the forces from the accelerations.

The constant tangential acceleration is in the negative t-direction, and its magnitude is given by

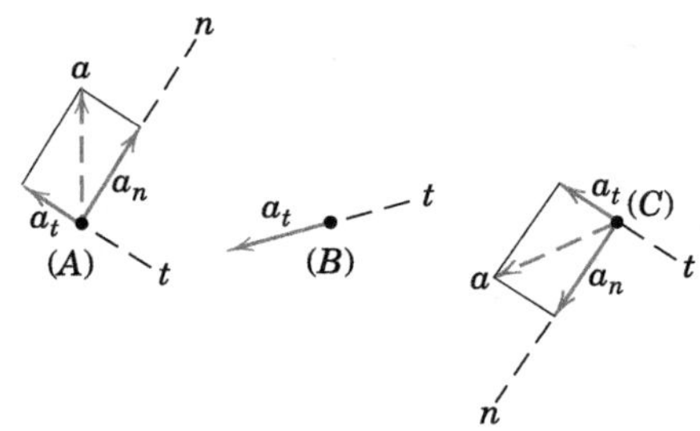

① $[v_C{}^2 = v_A{}^2 + 2a_t\,\Delta s]$ $\quad a_t = \left|\dfrac{(50/3.6)^2 - (100/3.6)^2}{2(200)}\right| = 1.447\ \text{m/s}^2$

The normal components of acceleration at A, B, and C are

② $[a_n = v^2/\rho]$ $\quad$ At A, $\quad a_n = \dfrac{(100/3.6)^2}{400} = 1.929\ \text{m/s}^2$

At B, $\quad a_n = 0$

At C, $\quad a_n = \dfrac{(50/3.6)^2}{80} = 2.41\ \text{m/s}^2$

Application of Newton's second law in both the n- and t-directions to the free-body diagrams of the car gives

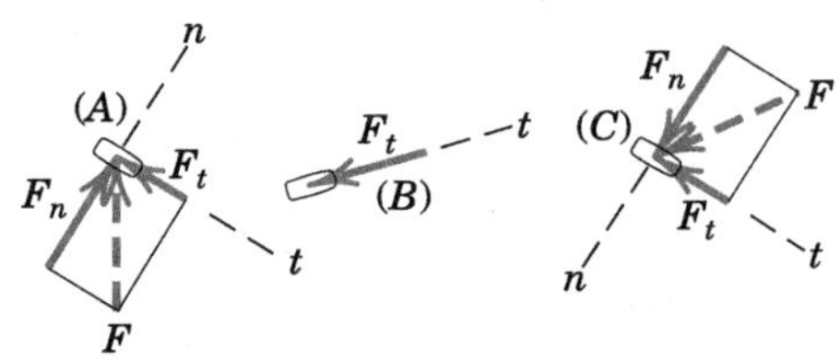

$[\Sigma F_t = ma_t]$ $\quad F_t = 1500(1.447) = 2170\ \text{N}$

③ $[\Sigma F_n = ma_n]$ $\quad$ At A, $\quad F_n = 1500(1.929) = 2890\ \text{N}$

At B, $\quad F_n = 0$

At C, $\quad F_n = 1500(2.41) = 3620\ \text{N}$

Thus, the total horizontal force acting on the tires becomes

At A, $\quad F = \sqrt{F_n{}^2 + F_t{}^2} = \sqrt{(2890)^2 + (2170)^2} = 3620\ \text{N}$ $\quad$ *Ans.*

At B, $\quad F = F_t = 2170\ \text{N}$ $\quad$ *Ans.*

④ At C, $\quad F = \sqrt{F_n{}^2 + F_t{}^2} = \sqrt{(3620)^2 + (2170)^2} = 4220\ \text{N}$ $\quad$ *Ans.*

Helpful Hints

① Recognize the numerical value of the conversion factor from km/h to m/s as 1000/3600 or 1/3.6.

② Note that a_n is always directed toward the center of curvature.

③ Note that the direction of F_n must agree with that of a_n.

④ The angle made by **a** and **F** with the direction of the path can be computed if desired.

Sample Problem 3/9

Compute the magnitude v of the velocity required for the spacecraft S to maintain a circular orbit of altitude 200 mi above the surface of the earth.

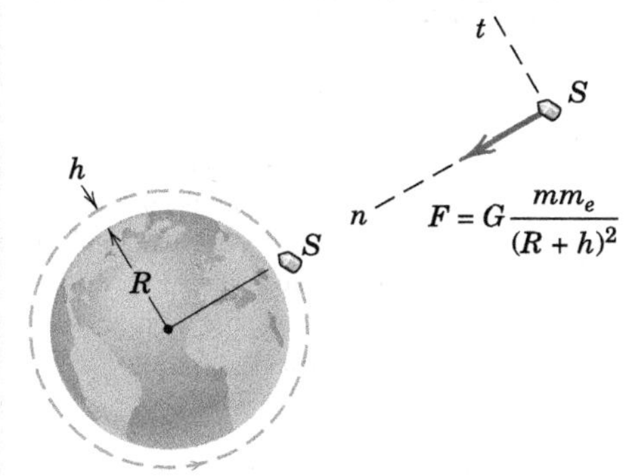

Solution. The only external force acting on the spacecraft is the force of grav-
① itational attraction to the earth (i.e., its weight), as shown in the free-body diagram. Summing forces in the normal direction yields

$$[\Sigma F_n = ma_n] \quad G\frac{mm_e}{(R+h)^2} = m\frac{v^2}{(R+h)}, \quad v = \sqrt{\frac{Gm_e}{(R+h)}} = R\sqrt{\frac{g}{(R+h)}}$$

where the substitution $gR^2 = Gm_e$ has been made. Substitution of numbers gives

$$v = (3959)(5280)\sqrt{\frac{32.234}{(3959+200)(5280)}} = 25{,}326 \text{ ft/sec} \qquad \textit{Ans.}$$

Helpful Hint

① Note that, for observations made within an inertial frame of reference, there is no such quantity as "centrifugal force" acting in the minus n-direction. Note also that neither the spacecraft nor its occupants are "weightless," because the weight in each case is given by Newton's law of gravitation. For this altitude, the weights are only about 10 percent less than the earth-surface values. Finally, the term "zero-g" is also misleading. It is only when we make our observations with respect to a coordinate system which has an acceleration equal to the gravitational acceleration (such as in an orbiting spacecraft) that we appear to be in a "zero-g" environment. The quantity which does go to zero aboard orbiting spacecraft is the familiar normal force associated with, for example, an object in contact with a horizontal surface within the spacecraft.

Sample Problem 3/10

Tube A rotates about the vertical O-axis with a constant angular rate $\dot{\theta} = \omega$ and contains a small cylindrical plug B of mass m whose radial position is controlled by the cord which passes freely through the tube and shaft and is wound around the drum of radius b. Determine the tension T in the cord and the horizontal component F_θ of force exerted by the tube on the plug if the constant angular rate of rotation of the drum is ω_0 first in the direction for case (a) and second in the direction for case (b). Neglect friction.

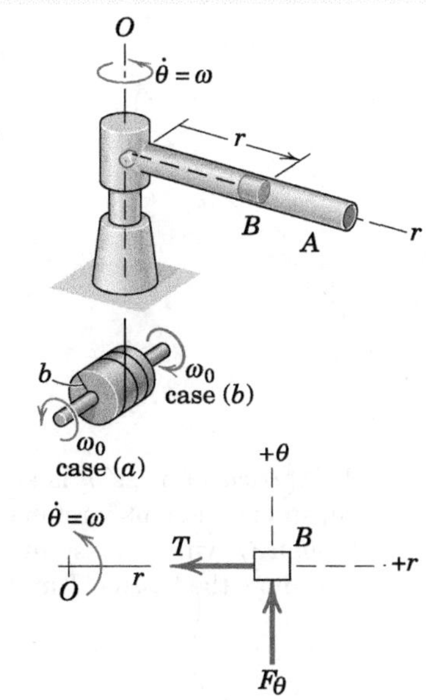

Solution. With r a variable, we use the polar-coordinate form of the equations of motion, Eqs. 3/8. The free-body diagram of B is shown in the horizontal plane and discloses only T and F_θ. The equations of motion are

$$[\Sigma F_r = ma_r] \qquad -T = m(\ddot{r} - r\dot{\theta}^2)$$

$$[\Sigma F_\theta = ma_\theta] \qquad F_\theta = m(r\ddot{\theta} + 2\dot{r}\dot{\theta})$$

Case (a). With $\dot{r} = +b\omega_0$, $\ddot{r} = 0$, and $\ddot{\theta} = 0$, the forces become

$$T = mr\omega^2 \qquad F_\theta = 2mb\omega_0\omega \qquad \textit{Ans.}$$

① ***Case (b).*** With $\dot{r} = -b\omega_0$, $\ddot{r} = 0$, and $\ddot{\theta} = 0$, the forces become

$$T = mr\omega^2 \qquad F_\theta = -2mb\omega_0\omega \qquad \textit{Ans.}$$

Helpful Hint

① The minus sign shows that F_θ is in the direction opposite to that shown on the free-body diagram.

PROBLEMS

Introductory Problems

3/47 The small 2-kg block A slides down the curved path and passes the lowest point B with a speed of 4 m/s. If the radius of curvature of the path at B is 1.5 m, determine the normal force N exerted on the block by the path at this point. Is knowledge of the friction properties necessary?

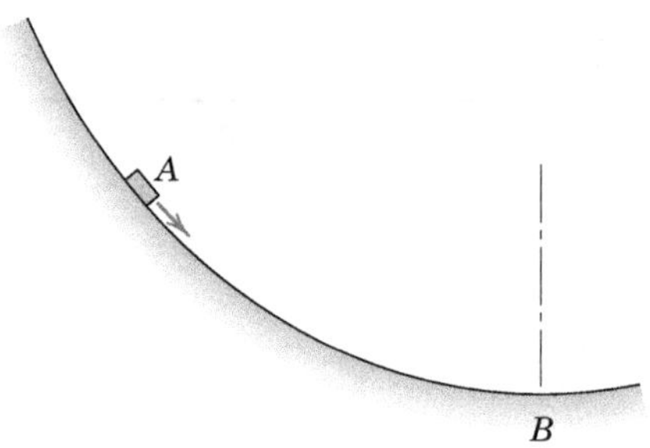

Problem 3/47

3/48 If the 2-kg block passes over the top B of the circular portion of the path with a speed of 3.5 m/s, calculate the magnitude N_B of the normal force exerted by the path on the block. Determine the maximum speed v which the block can have at A without losing contact with the path.

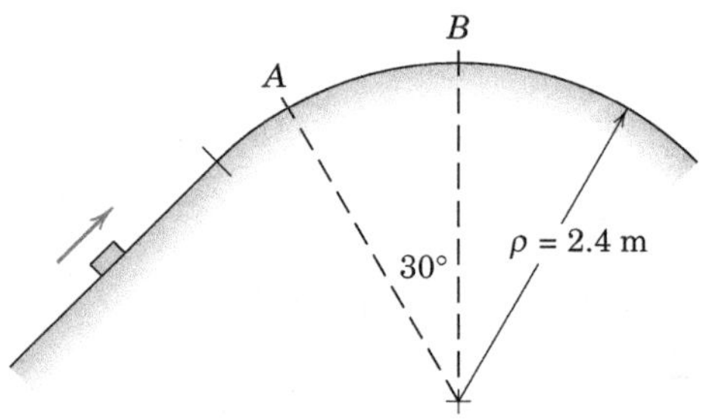

Problem 3/48

3/49 The particle of mass m is attached to the light rigid rod, and the assembly rotates about a horizontal axis through O with a constant angular velocity $\dot{\theta} = \omega$. Determine the tension T in the rod as a function of θ.

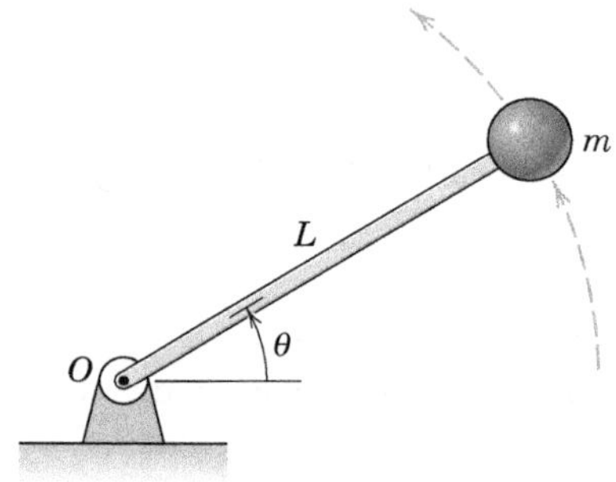

Problem 3/49

3/50 If the 180-lb ski-jumper attains a speed of 80 ft/sec as he approaches the takeoff position, calculate the magnitude N of the normal force exerted by the snow on his skis just before he reaches A.

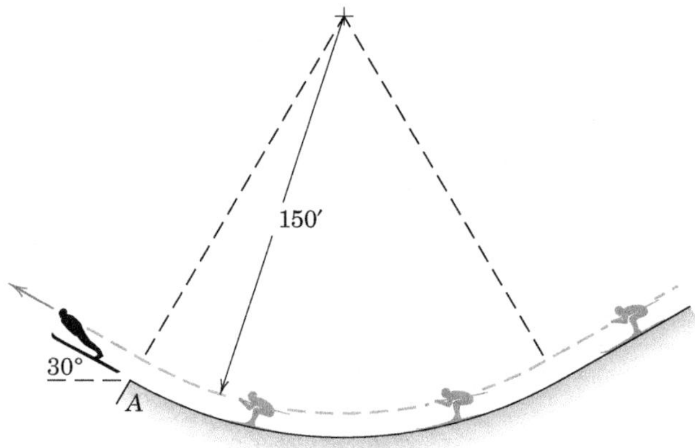

Problem 3/50

3/51 The 4-oz slider has a speed $v = 3$ ft/sec as it passes point A of the smooth guide, which lies in a horizontal plane. Determine the magnitude R of the force which the guide exerts on the slider (a) just before it passes point A of the guide and (b) as it passes point B.

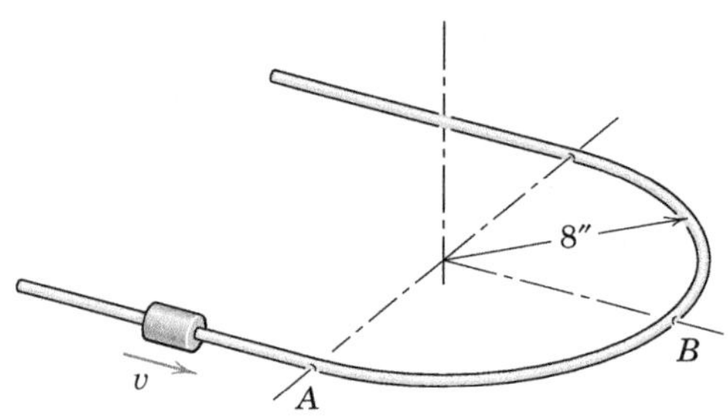

Problem 3/51

3/52 A jet transport plane flies in the trajectory shown in order to allow astronauts to experience the "weightless" condition similar to that aboard orbiting spacecraft. If the speed at the highest point is 600 mi/hr, what is the radius of curvature ρ necessary to exactly simulate the orbital "free-fall" environment?

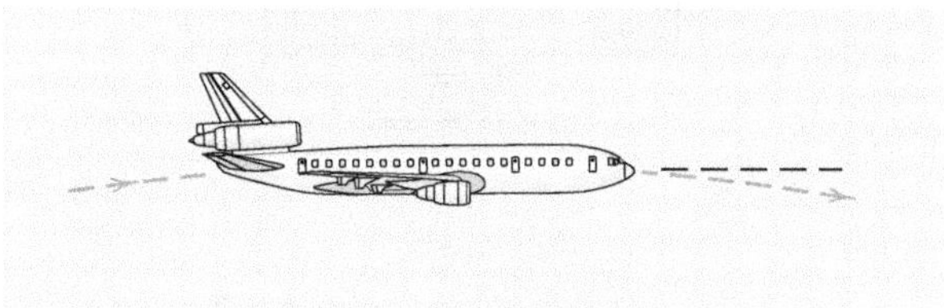

Problem 3/52

3/53 In the design of a space station to operate outside the earth's gravitational field, it is desired to give the structure a rotational speed N which will simulate the effect of the earth's gravity for members of the crew. If the centers of the crew's quarters are to be located 12 m from the axis of rotation, calculate the necessary rotational speed N of the space station in revolutions per minute.

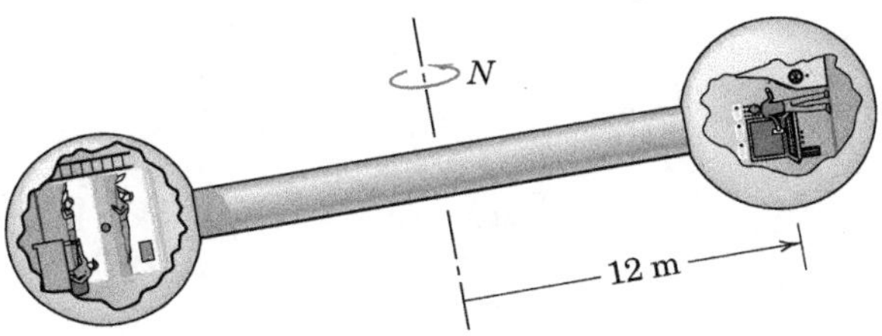

Problem 3/53

3/54 Determine the speed which the 630-kg four-man bobsled must have in order to negotiate the turn without reliance on friction. Also find the net normal force exerted on the bobsled by the track.

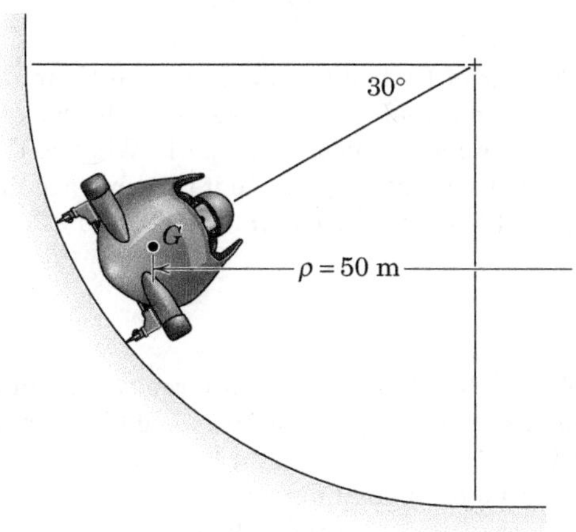

Problem 3/54

3/55 The hollow tube is pivoted about a horizontal axis through point O and is made to rotate in the vertical plane with a constant counterclockwise angular velocity $\dot{\theta} = 3$ rad/sec. If a 0.2-lb particle is sliding in the tube toward O with a velocity of 6 ft/sec relative to the tube when the position $\theta = 30°$ is passed, calculate the magnitude N of the normal force exerted by the wall of the tube on the particle at this instant.

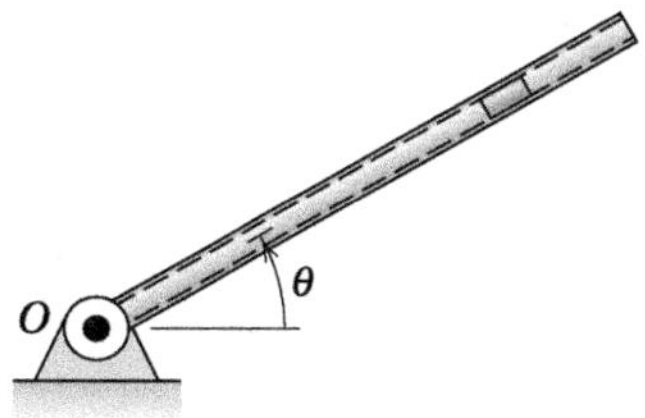

Problem 3/55

3/56 A Formula-1 car encounters a hump which has a circular shape with smooth transitions at both ends. (*a*) What speed v_B will cause the car to lose contact with the road at the topmost point B? (*b*) For a speed $v_A = 190$ km/h, what is the normal force exerted by the road on the 640-kg car as it passes point A?

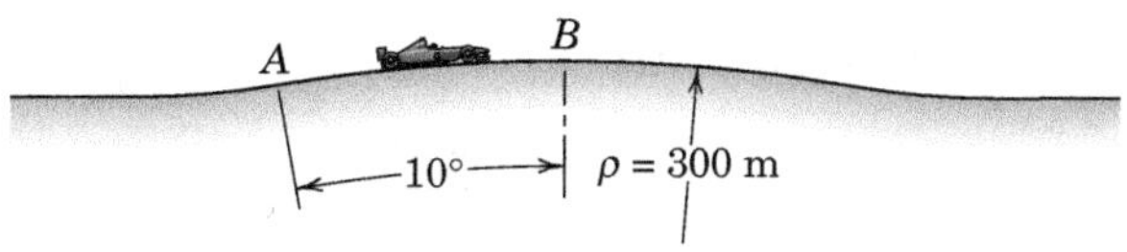

Problem 3/56

3/57 The small spheres are free to move on the inner surface of the rotating spherical chambers shown in section with radius $R = 200$ mm. If the spheres reach a steady-state angular position $\beta = 45°$, determine the angular velocity Ω of the device.

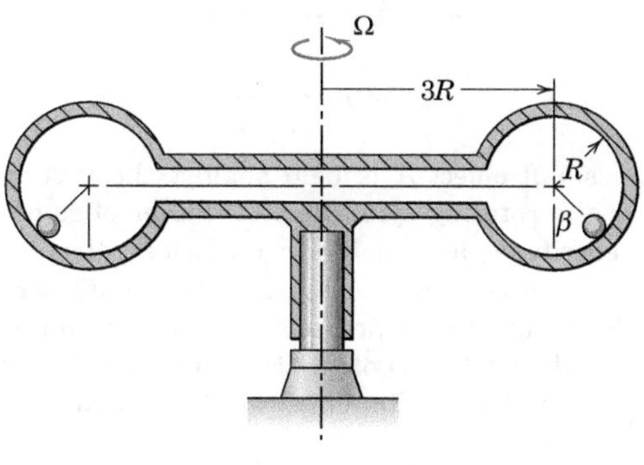

Problem 3/57

3/58 A 180-lb snowboarder has speed $v = 15$ ft/sec when in the position shown on the halfpipe. Determine the normal force on his snowboard and the magnitude of his total acceleration at the instant depicted. Use a value $\mu_k = 0.10$ for the coefficient of kinetic friction between the snowboard and the surface. Neglect the weight of the snowboard and assume that the mass center G of the snowboarder is 3 feet from the surface of the snow.

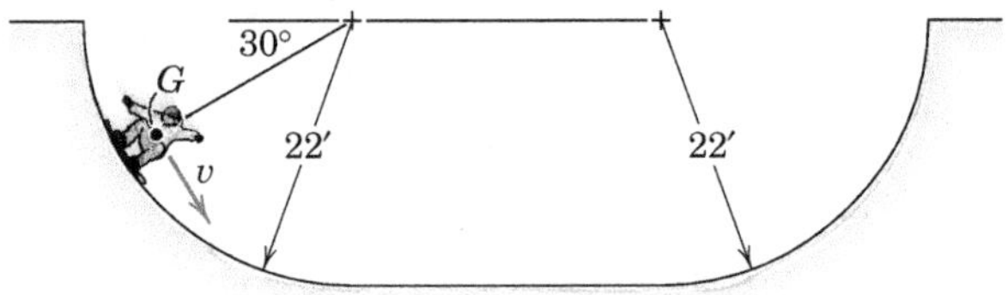

Problem 3/58

3/59 A child twirls a small 50-g ball attached to the end of a 1-m string so that the ball traces a circle in a vertical plane as shown. What is the minimum speed v which the ball must have when in position 1? If this speed is maintained throughout the circle, calculate the tension T in the string when the ball is in position 2. Neglect any small motion of the child's hand.

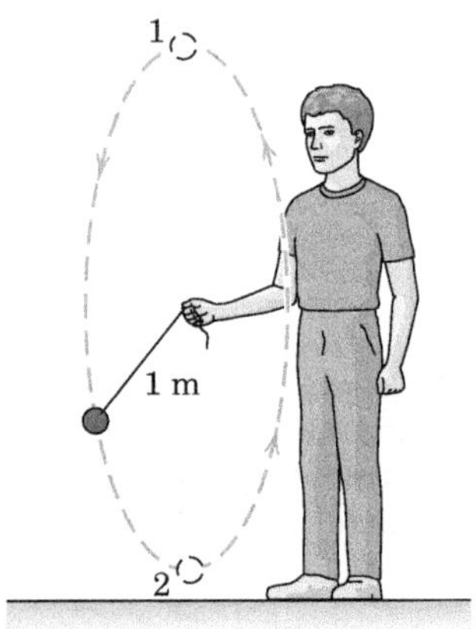

Problem 3/59

3/60 A small object A is held against the vertical side of the rotating cylindrical container of radius r by centrifugal action. If the coefficient of static friction between the object and the container is μ_s, determine the expression for the minimum rotational rate $\dot{\theta} = \omega$ of the container which will keep the object from slipping down the vertical side.

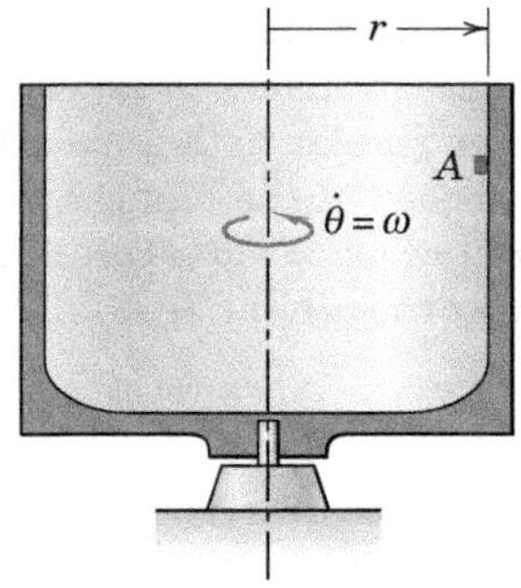

Problem 3/60

3/61 The standard test to determine the maximum lateral acceleration of a car is to drive it around a 200-ft-diameter circle painted on a level asphalt surface. The driver slowly increases the vehicle speed until he is no longer able to keep both wheel pairs straddling the line. If this maximum speed is 35 mi/hr for a 3000-lb car, determine its lateral acceleration capability a_n in g's and compute the magnitude F of the total friction force exerted by the pavement on the car tires.

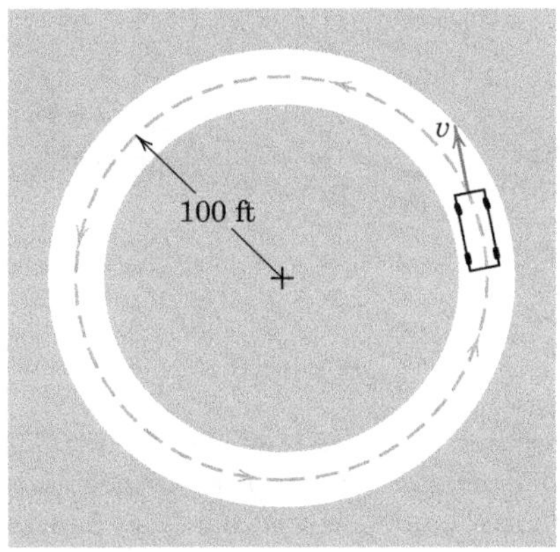

Problem 3/61

3/62 The car of Prob. 3/61 is traveling at 25 mi/hr when the driver applies the brakes, and the car continues to move along the circular path. What is the maximum deceleration possible if the tires are limited to a total horizontal friction force of 2400 lb?

Representative Problems

3/63 The flatbed truck carries a large section of circular pipe secured only by the two fixed blocks A and B of height h. The truck is in a left turn of radius ρ. Determine the maximum speed for which the pipe will be restrained. Use the values $\rho = 60$ m, $h = 0.1$ m, and $R = 0.8$ m.

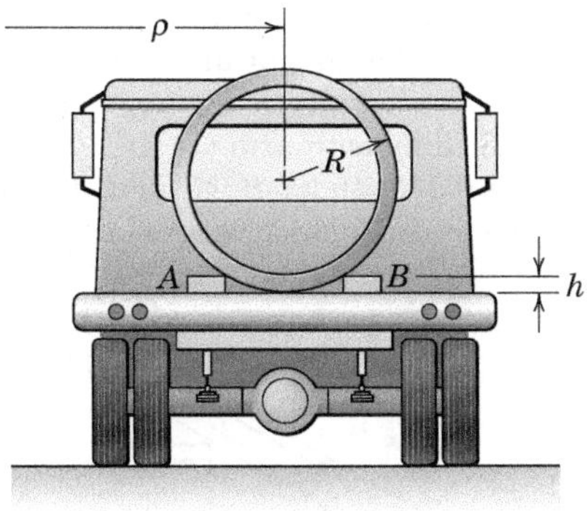

Problem 3/63

3/64 The particle of mass $m = 0.2$ kg travels with constant speed v in a circular path around the conical body. Determine the tension T in the cord. Neglect all friction, and use the values $h = 0.8$ m and $v = 0.6$ m/s. For what value of v does the normal force go to zero?

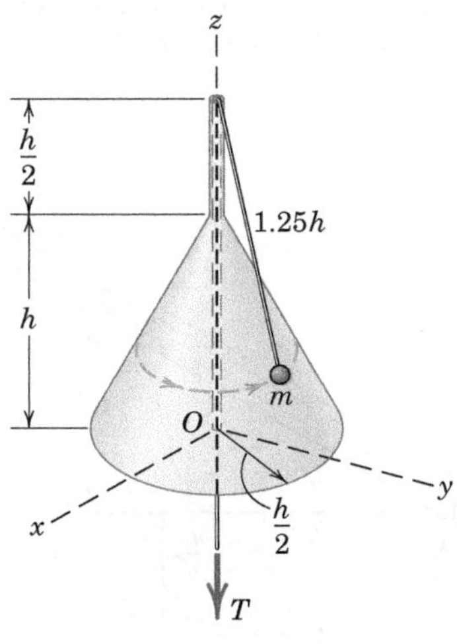

Problem 3/64

3/65 Calculate the necessary rotational speed N for the aerial ride in an amusement park in order that the arms of the gondolas will assume an angle $\theta = 60°$ with the vertical. Neglect the mass of the arms to which the gondolas are attached and treat each gondola as a particle.

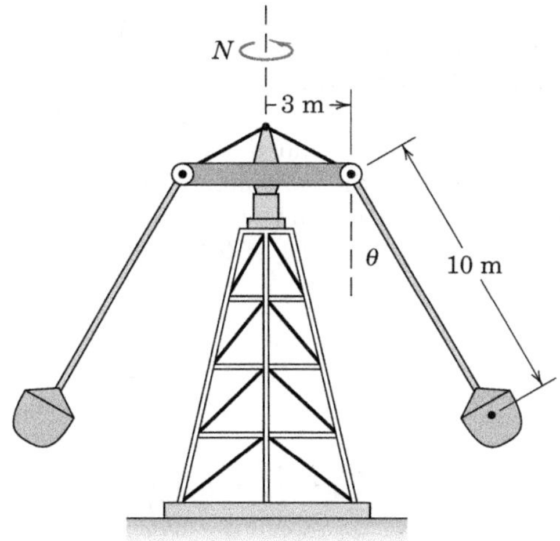

Problem 3/65

3/66 A 0.2-kg particle P is constrained to move along the vertical-plane circular slot of radius $r = 0.5$ m and is confined to the slot of arm OA, which rotates about a horizontal axis through O with a constant angular rate $\Omega = 3$ rad/s. For the instant when $\beta = 20°$, determine the force N exerted on the particle by the circular constraint and the force R exerted on it by the slotted arm.

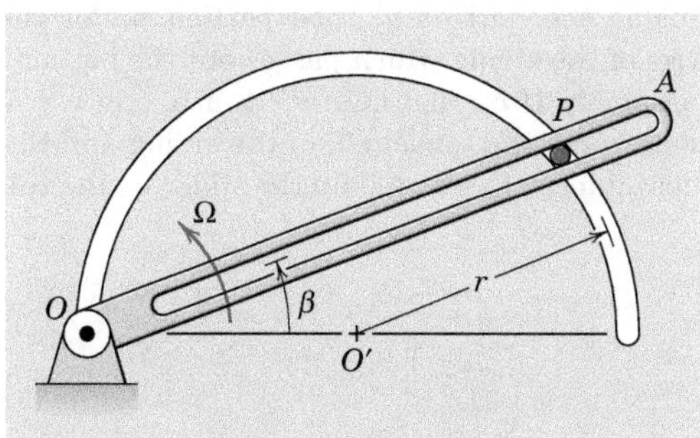

Problem 3/66

3/67 Repeat the previous problem, only now the slotted arm is rotating with angular velocity $\Omega = 3$ rad/s, and this rate is increasing at 5 rad/s^2.

3/68 At the instant under consideration, the cable attached to the cart of mass m_1 is tangent to the circular path of the cart. If the upward speed of the cylinder of mass m_2 is $v_2 = 1.2$ m/s, determine the acceleration of m_1 and the tension T in the cable. What would be the maximum speed of m_2 for which m_1 remains in contact with the surface? Use the values $R = 1.75$ m, $m_1 = 0.4$ kg, $m_2 = 0.6$ kg, and $\beta = 30°$.

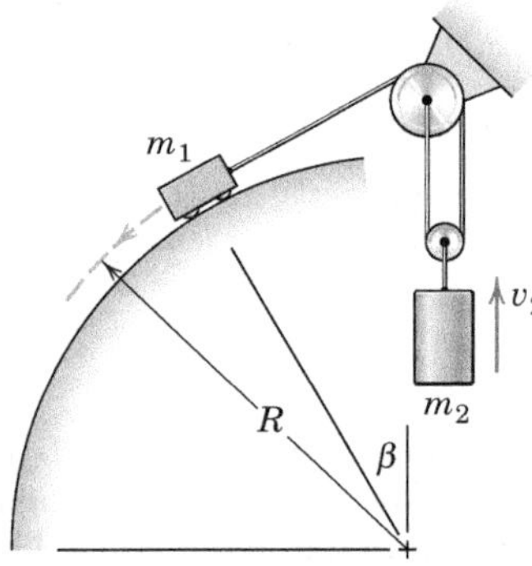

Problem 3/68

3/69 The hollow tube assembly rotates about a vertical axis with angular velocity $\omega = \dot{\theta} = 4$ rad/s and $\dot{\omega} = \ddot{\theta} = -2$ rad/s^2. A small 0.2-kg slider P moves inside the horizontal tube portion under the control of the string which passes out the bottom of the assembly. If $r = 0.8$ m, $\dot{r} = -2$ m/s, and $\ddot{r} = 4$ m/s^2, determine the tension T in the string and the horizontal force F_θ exerted on the slider by the tube.

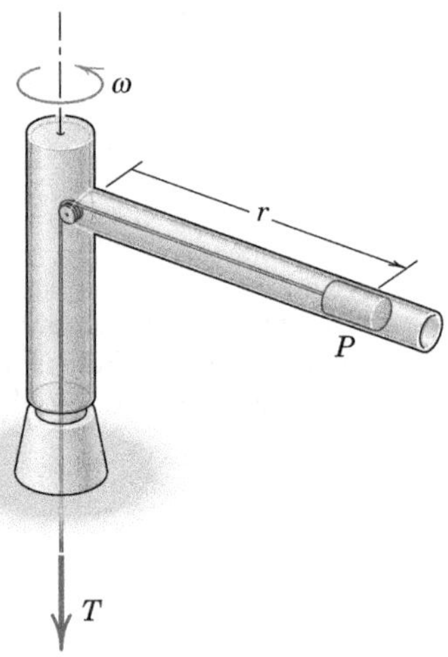

Problem 3/69

3/70 The slotted arm OA rotates about a fixed axis through O. At the instant under consideration, $\theta = 30°$, $\dot{\theta} = 45$ deg/s, and $\ddot{\theta} = 20$ deg/s^2. Determine the forces applied by both arm OA and the sides of the slot to the 0.2-kg slider B. Neglect all friction, and let $L = 0.6$ m. The motion occurs in a vertical plane.

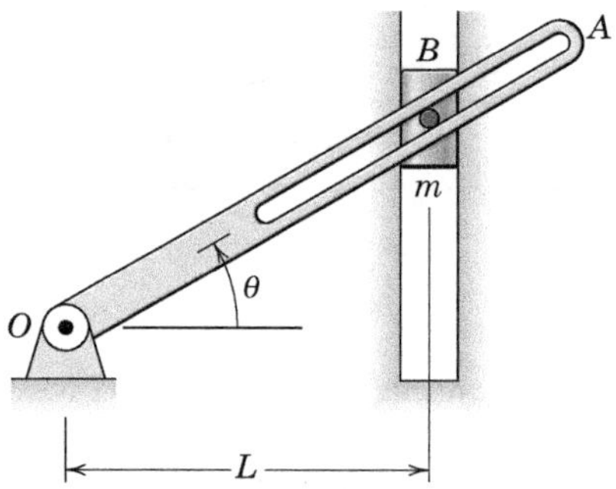

Problem 3/70

3/71 The configuration of Prob. 3/70 is now modified as shown in the figure. Use all the data of Prob. 3/70 and determine the forces applied to the slider B by both arm OA and the sides of the slot. Neglect all friction.

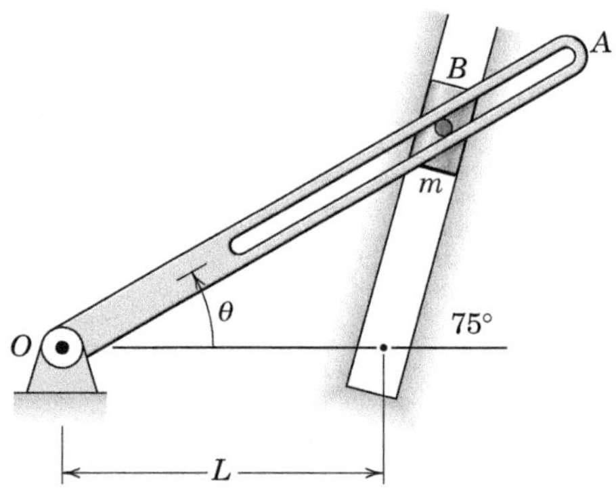

Problem 3/71

3/72 Determine the altitude h (in kilometers) above the surface of the earth at which a satellite in a circular orbit has the same period, 23.9344 h, as the earth's absolute rotation. If such an orbit lies in the equatorial plane of the earth, it is said to be geosynchronous, because the satellite does not appear to move relative to an earth-fixed observer.

3/73 The quarter-circular slotted arm OA is rotating about a horizontal axis through point O with a constant counterclockwise angular velocity Ω = 7 rad/sec. The 0.1-lb particle P is epoxied to the arm at the position β = 60°. Determine the tangential force F parallel to the slot which the epoxy must support so that the particle does not move along the slot. The value of R = 1.4 ft.

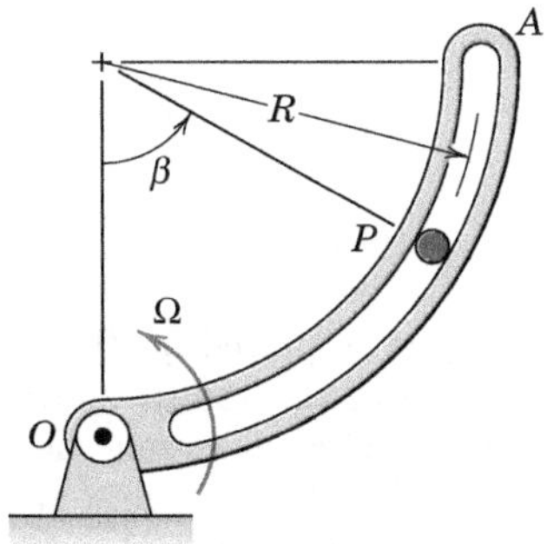

Problem 3/73

3/74 A 2-kg sphere S is being moved in a vertical plane by a robotic arm. When the angle θ is 30°, the angular velocity of the arm about a horizontal axis through O is 50 deg/s clockwise and its angular acceleration is 200 deg/s^2 counterclockwise. In addition, the hydraulic element is being shortened at the constant rate of 500 mm/s. Determine the necessary minimum gripping force P if the coefficient of static friction between the sphere and the gripping surfaces is 0.50. Compare P with the minimum gripping force P_s required to hold the sphere in static equilibrium in the 30° position.

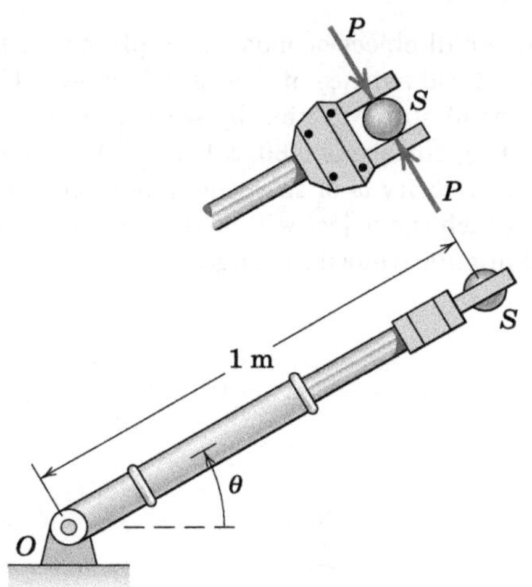

Problem 3/74

3/75 The cars of an amusement park ride have a speed v_A = 22 m/s at A and a speed v_B = 12 m/s at B. If a 75-kg rider sits on a spring scale (which registers the normal force exerted on it), determine the scale readings as the car passes points A and B. Assume that the person's arms and legs do not support appreciable force.

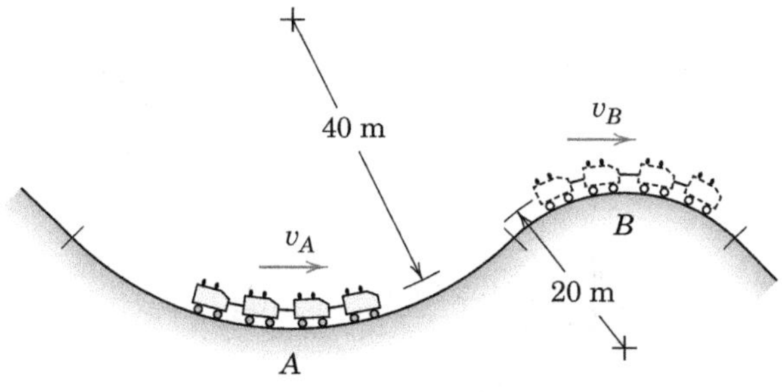

Problem 3/75

3/76 The rocket moves in a vertical plane and is being propelled by a thrust T of 32 kN. It is also subjected to an atmospheric resistance R of 9.6 kN. If the rocket has a velocity of 3 km/s and if the gravitational acceleration is 6 m/s^2 at the altitude of the rocket, calculate the radius of curvature ρ of its path for the position described and the time-rate-of-change of the magnitude v of the velocity of the rocket. The mass of the rocket at the instant considered is 2000 kg.

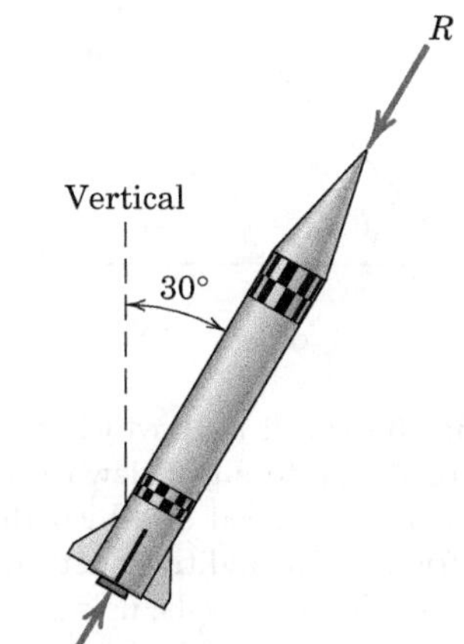

Problem 3/76

3/77 The robot arm is elevating and extending simultaneously. At a given instant, $\theta = 30°$, $\dot{\theta} = 40$ deg/s, $\ddot{\theta} = 120$ deg/s^2, $l = 0.5$ m, $\dot{l} = 0.4$ m/s, and $\ddot{l} = -0.3$ m/s^2. Compute the radial and transverse forces F_r and F_θ that the arm must exert on the gripped part P, which has a mass of 1.2 kg. Compare with the case of static equilibrium in the same position.

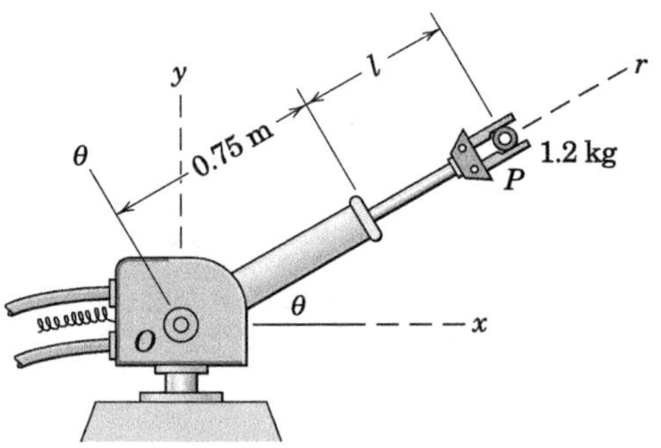

Problem 3/77

3/78 The 0.1-lb projectile A is subjected to a drag force of magnitude kv^2, where the constant $k = 0.0002$ lb-sec^2/ft^2. This drag force always opposes the velocity **v**. At the instant depicted, $v = 100$ ft/sec, $\theta = 45°$, and $r = 400$ ft. Determine the corresponding values of $\ddot{r}$ and $\ddot{\theta}$.

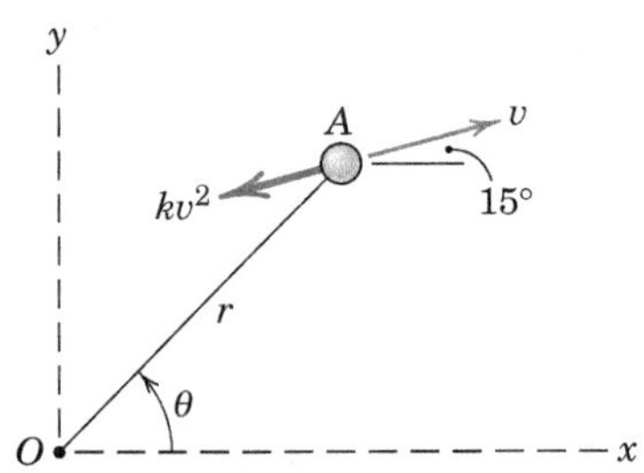

Problem 3/78

3/79 Determine the speed v at which the race car will have no tendency to slip sideways on the banked track, that is, the speed at which there is no reliance on friction. In addition, determine the minimum and maximum speeds, using the coefficient of static friction $\mu_s = 0.90$. State any assumptions.

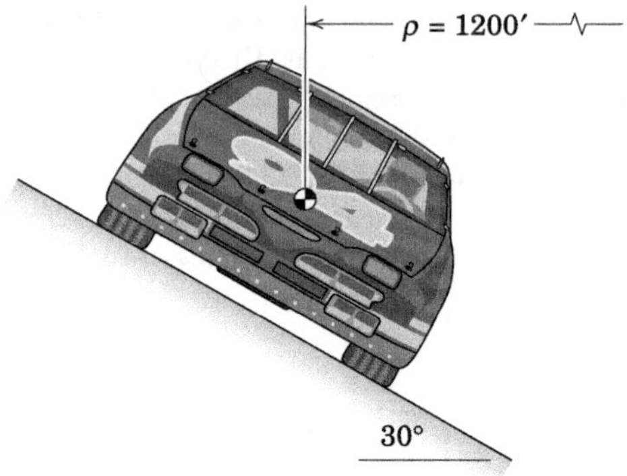

Problem 3/79

3/80 The small object is placed on the inner surface of the conical dish at the radius shown. If the coefficient of static friction between the object and the conical surface is 0.30, for what range of angular velocities ω about the vertical axis will the block remain on the dish without slipping? Assume that speed changes are made slowly so that any angular acceleration may be neglected.

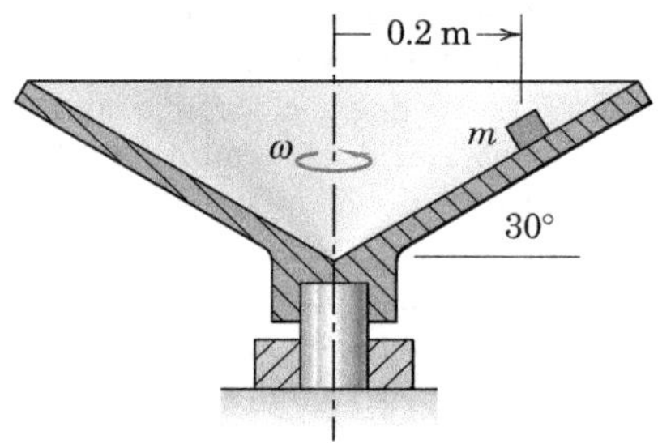

Problem 3/80

3/81 The small object of mass m is placed on the rotating conical surface at the radius shown. If the coefficient of static friction between the object and the rotating surface is 0.80, calculate the maximum angular velocity ω of the cone about the vertical axis for which the object will not slip. Assume very gradual angular-velocity changes.

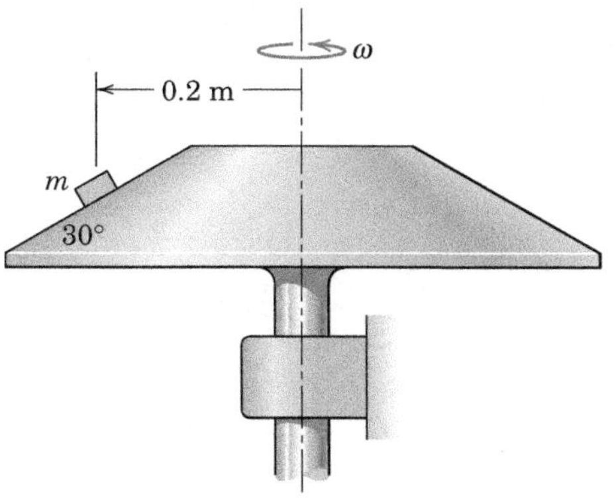

Problem 3/81

3/82 The spring-mounted 0.8-kg collar A oscillates along the horizontal rod, which is rotating at the constant angular rate $\dot{\theta} = 6$ rad/s. At a certain instant, r is increasing at the rate of 800 mm/s. If the coefficient of kinetic friction between the collar and the rod is 0.40, calculate the friction force F exerted by the rod on the collar at this instant.

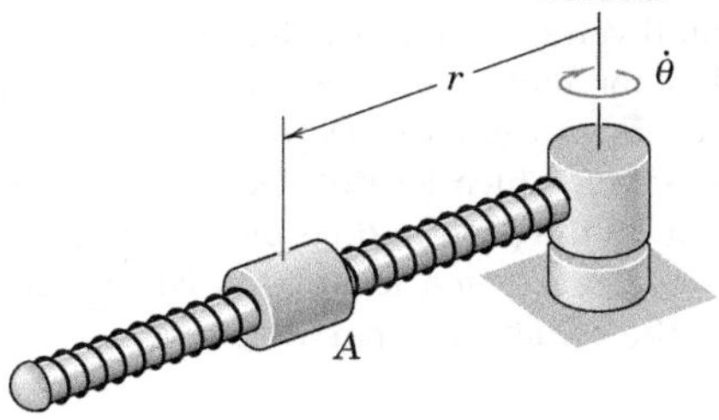

Problem 3/82

3/83 The slotted arm revolves in the horizontal plane about the fixed vertical axis through point O. The 3-lb slider C is drawn toward O at the constant rate of 2 in./sec by pulling the cord S. At the instant for which $r = 9$ in., the arm has a counterclockwise angular velocity $\omega = 6$ rad/sec and is slowing down at the rate of 2 rad/sec^2. For this instant, determine the tension T in the cord and the magnitude N of the force exerted on the slider by the sides of the smooth radial slot. Indicate which side, A or B, of the slot contacts the slider.

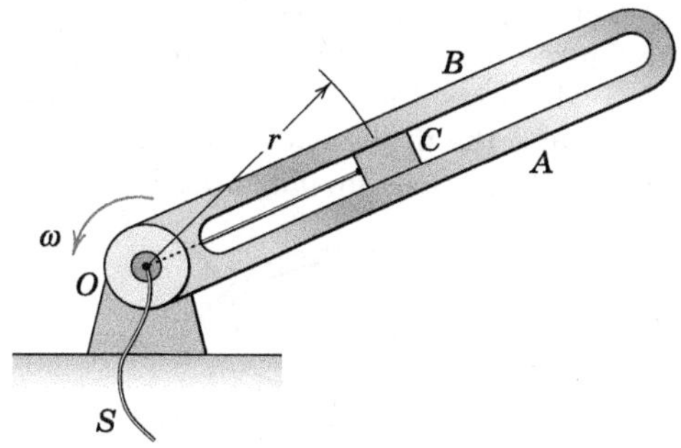

Problem 3/83

3/84 Beginning from rest when $\theta = 20°$, a 35-kg child slides with negligible friction down the sliding board which is in the shape of a 2.5-m circular arc. Determine the tangential acceleration and speed of the child, and the normal force exerted on her (*a*) when $\theta = 30°$ and (*b*) when $\theta = 90°$.

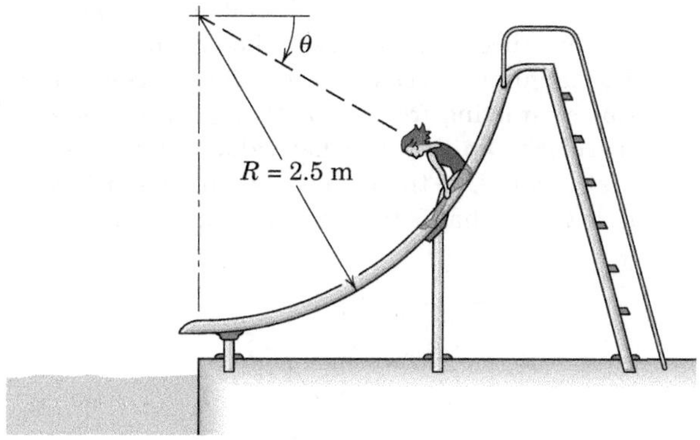

Problem 3/84

3/85 A small coin is placed on the horizontal surface of the rotating disk. If the disk starts from rest and is given a constant angular acceleration $\ddot{\theta} = \alpha$, determine an expression for the number of revolutions N through which the disk turns before the coin slips. The coefficient of static friction between the coin and the disk is μ_s.

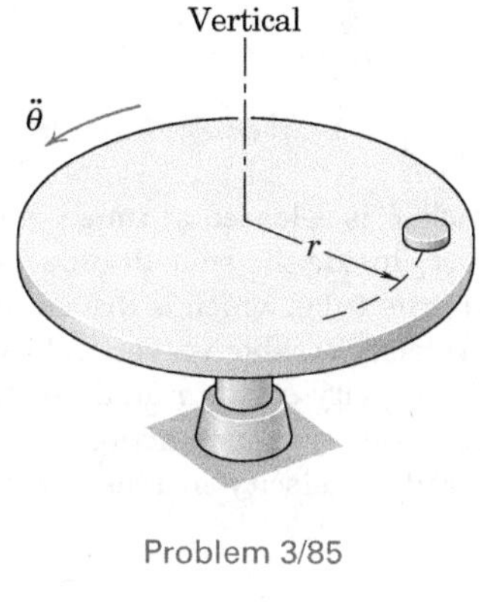

Problem 3/85

3/86 The rotating drum of a clothes dryer is shown in the figure. Determine the angular velocity Ω of the drum which results in loss of contact between the clothes and the drum at $\theta = 50°$. Assume that the small vanes prevent slipping until loss of contact.

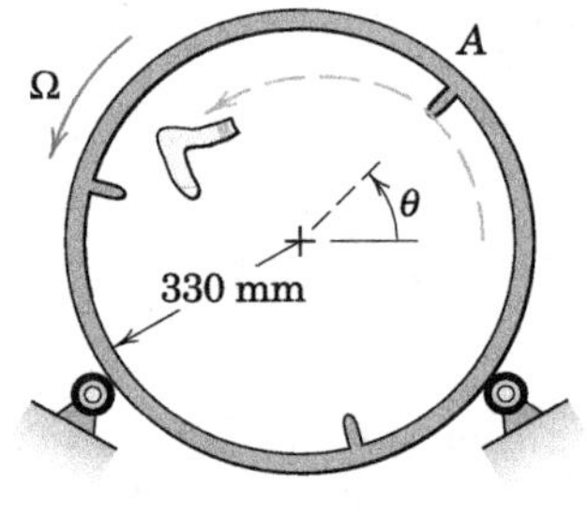

Problem 3/86

3/87 The disk spins about the fixed axis BB, which is inclined at the angle α to the vertical z-axis. A small block A is placed on the disk in its lowest position P at a distance r from the axis when the disk is at rest. The angular velocity $\omega = \dot{\theta}$ is then increased very slowly, starting from zero. At what value of ω will the block slip, and at what value of θ will the slip first occur? Use the values $\alpha = 20°$, $r = 0.4$ m, and $\mu_s = 0.60$. What is the critical value of ω if $\alpha = 0$?

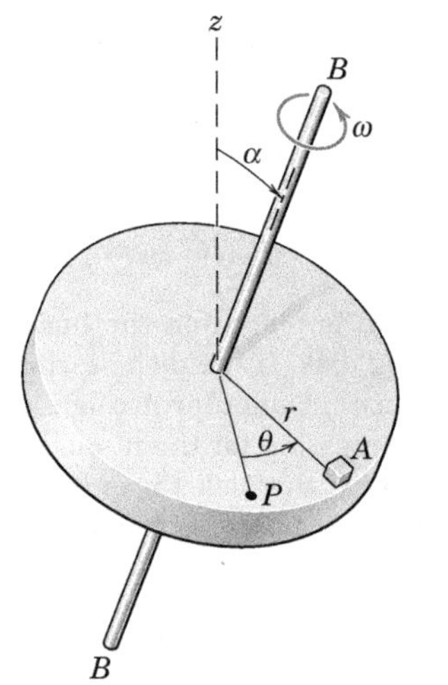

Problem 3/87

3/88 The particle P is released at time $t = 0$ from the position $r = r_0$ inside the smooth tube with no velocity relative to the tube, which is driven at the constant angular velocity ω_0 about a vertical axis. Determine the radial velocity v_r, the radial position r, and the transverse velocity v_θ as functions of time t. Explain why the radial velocity increases with time in the absence of radial forces. Plot the absolute path of the particle during the time it is inside the tube for $r_0 = 0.1$ m, $l = 1$ m, and $\omega_0 = 1$ rad/s.

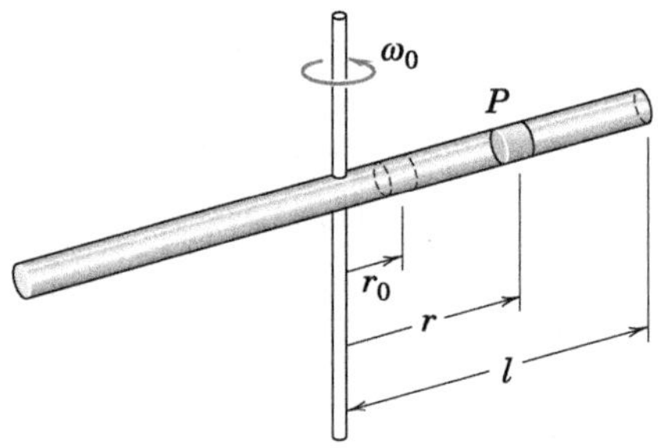

Problem 3/88

3/89 Remove the assumption of smooth surfaces as stated in Prob. 3/88 and assume a coefficient of kinetic friction μ_k between the particle and rotating tube. Determine the radial position r of the particle as a function of time t if it is released with no relative velocity at $r = r_0$ when $t = 0$. Assume that static friction is overcome.

3/90 A small vehicle enters the top A of the circular path with a horizontal velocity v_0 and gathers speed as it moves down the path. Determine an expression for the angle β which locates the point where the vehicle leaves the path and becomes a projectile. Evaluate your expression for $v_0 = 0$. Neglect friction and treat the vehicle as a particle.

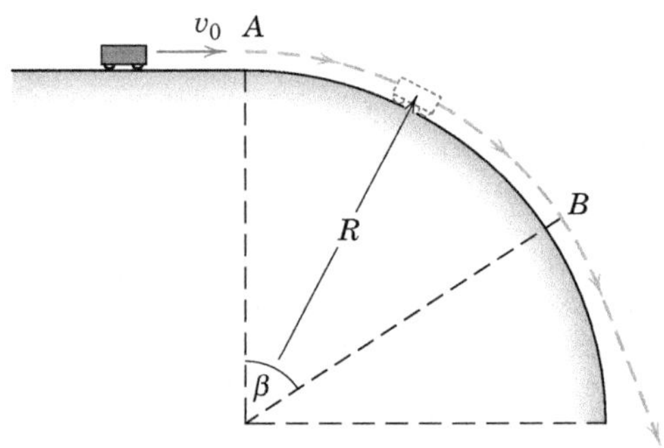

Problem 3/90

3/91 The spacecraft P is in the elliptical orbit shown. At the instant represented, its speed is $v = 13{,}244$ ft/sec. Determine the corresponding values of $\dot{r}$, $\dot{\theta}$, $\ddot{r}$, and $\ddot{\theta}$. Use $g = 32.23$ ft/sec^2 as the acceleration of gravity on the surface of the earth and $R = 3959$ mi as the radius of the earth.

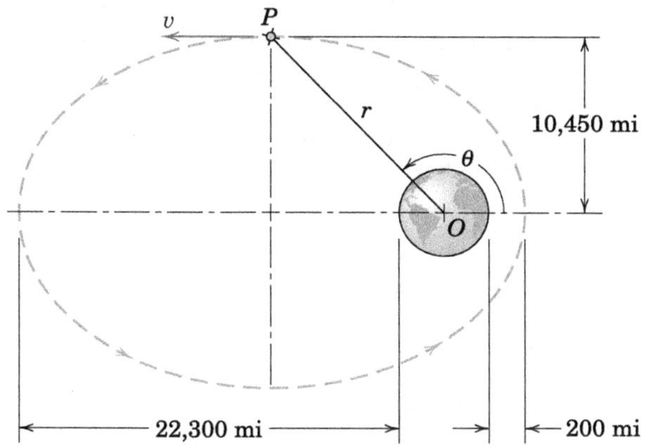

Problem 3/91

3/92 The uniform slender rod of length L, mass m, and cross-sectional area A is rotating in a horizontal plane about the vertical central axis O-O at a constant high angular velocity ω. By analyzing the horizontal forces on the accelerating differential element shown, derive an expression for the tensile stress σ in the rod as a function of r. The stress, commonly referred to as centrifugal stress, equals the tensile force divided by the cross-sectional area A.

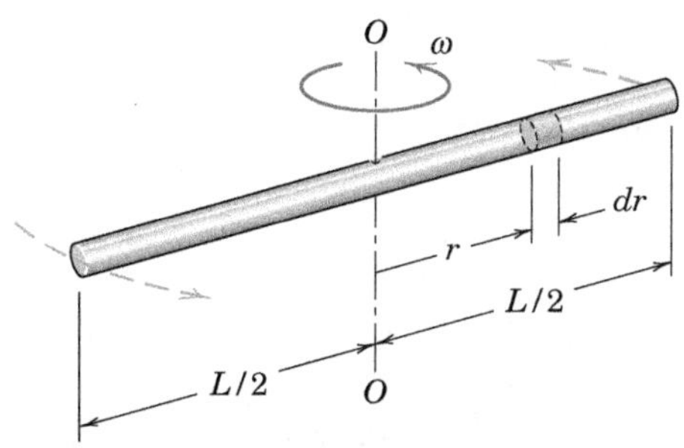

Problem 3/92

▶3/93 A small object is released from rest at A and slides with friction down the circular path. If the coefficient of friction is 0.20, determine the velocity of the object as it passes B. (*Hint:* Write the equations of motion in the n- and t-directions, eliminate N, and substitute $v\,dv = a_t r\,d\theta$. The resulting equation is a linear nonhomogeneous differential equation of the form $dy/dx + f(x)y = g(x)$, the solution of which is well known.)

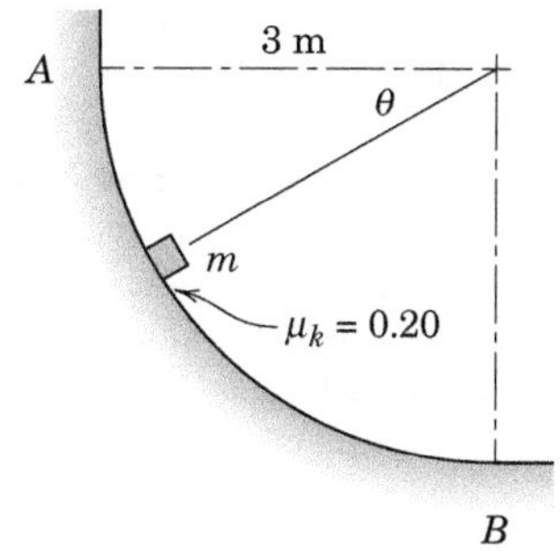

Problem 3/93

▶3/94 The slotted arm OB rotates in a horizontal plane about point O of the fixed circular cam with constant angular velocity $\dot{\theta} = 15$ rad/s. The spring has a stiffness of 5 kN/m and is uncompressed when $\theta = 0$. The smooth roller A has a mass of 0.5 kg. Determine the normal force N which the cam exerts on A and also the force R exerted on A by the sides of the slot when $\theta = 45°$. All surfaces are smooth. Neglect the small diameter of the roller.

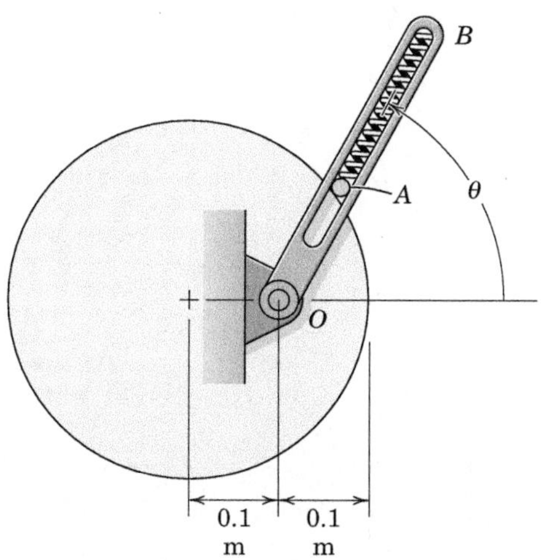

Problem 3/94

▶3/95 Each tire on the 1350-kg car can support a maximum friction force parallel to the road surface of 2500 N. This force limit is nearly constant over all possible rectilinear and curvilinear car motions and is attainable only if the car does not skid. Under this maximum braking, determine the total stopping distance s if the brakes are first applied at point A when the car speed is 25 m/s and if the car follows the centerline of the road.

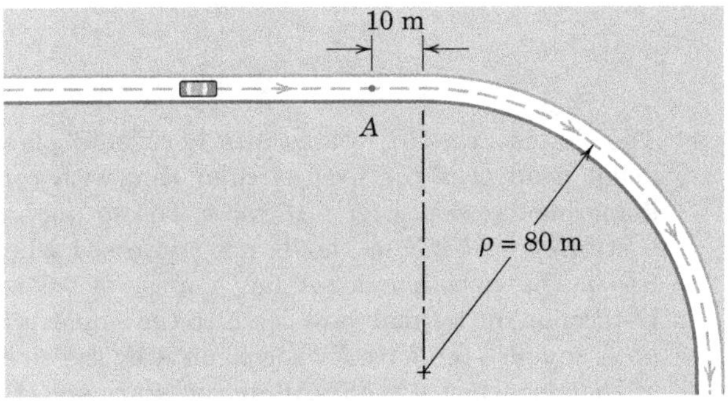

Problem 3/95

▶3/96 A small collar of mass m is given an initial velocity of magnitude v_0 on the horizontal circular track fabricated from a slender rod. If the coefficient of kinetic friction is μ_k, determine the distance traveled before the collar comes to rest. (*Hint:* Recognize that the friction force depends on the net normal force.)

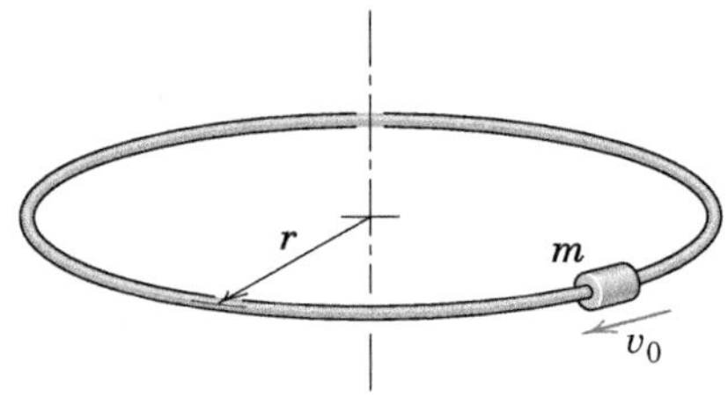

Problem 3/96

SECTION B WORK AND ENERGY

3/6 Work and Kinetic Energy

In the previous two articles, we applied Newton's second law $\mathbf{F} = m\mathbf{a}$ to various problems of particle motion to establish the instantaneous relationship between the net force acting on a particle and the resulting acceleration of the particle. When we needed to determine the change in velocity or the corresponding displacement of the particle, we integrated the computed acceleration by using the appropriate kinematic equations.

There are two general classes of problems in which the cumulative effects of unbalanced forces acting on a particle are of interest to us. These cases involve (1) integration of the forces with respect to the displacement of the particle and (2) integration of the forces with respect to the time they are applied. We may incorporate the results of these integrations directly into the governing equations of motion so that it becomes unnecessary to solve directly for the acceleration. Integration with respect to displacement leads to the equations of work and energy, which are the subject of this article. Integration with respect to time leads to the equations of impulse and momentum, discussed in Section C.

Definition of Work

We now develop the quantitative meaning of the term "work."* Figure 3/2a shows a force $\mathbf{F}$ acting on a particle at A which moves along the path shown. The position vector $\mathbf{r}$ measured from some convenient origin O locates the particle as it passes point A, and $d\mathbf{r}$ is the differential displacement associated with an infinitesimal movement from A to A'. The work done by the force $\mathbf{F}$ during the displacement $d\mathbf{r}$ is defined as

$$dU = \mathbf{F} \cdot d\mathbf{r}$$

The magnitude of this dot product is $dU = F\,ds \cos\alpha$, where α is the angle between $\mathbf{F}$ and $d\mathbf{r}$ and where ds is the magnitude of $d\mathbf{r}$. This expression may be interpreted as the displacement multiplied by the force component $F_t = F\cos\alpha$ in the direction of the displacement, as represented by the dashed lines in Fig. 3/2b. Alternatively, the work dU may be interpreted as the force multiplied by the displacement component $ds \cos\alpha$ in the direction of the force, as represented by the full lines in Fig. 3/2b.

With this definition of work, it should be noted that the component $F_n = F\sin\alpha$ normal to the displacement does no work. Thus, the work dU may be written as

$$dU = F_t\,ds$$

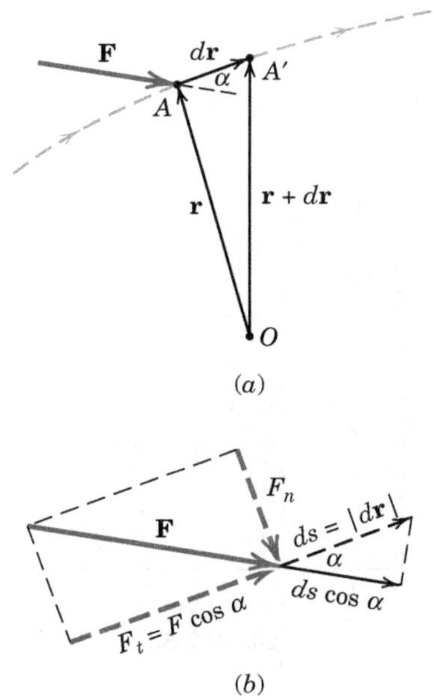

Figure 3/2

*The concept of work was also developed in the study of virtual work in Chapter 7 of *Vol. 1 Statics*.

Work is positive if the working component F_t is in the direction of the displacement and negative if it is in the opposite direction. Forces which do work are termed *active forces*. Constraint forces which do no work are termed *reactive forces*.

Units of Work

The SI units of work are those of force (N) times displacement (m) or N·m. This unit is given the special name *joule* (J), which is defined as the work done by a force of 1 N acting through a distance of 1 m in the direction of the force. Consistent use of the joule for work (and energy) rather than the units N·m will avoid possible ambiguity with the units of moment of a force or torque, which are also written N·m.

In the U.S. customary system, work has the units of ft-lb. Dimensionally, work and moment are the same. In order to distinguish between the two quantities, it is recommended that work be expressed as foot pounds (ft-lb) and moment as pound feet (lb-ft). It should be noted that work is a scalar as given by the dot product and involves the product of a force and a distance, both measured along the same line. Moment, on the other hand, is a vector as given by the cross product and involves the product of force and distance measured at right angles to the force.

Calculation of Work

During a finite movement of the point of application of a force, the force does an amount of work equal to

$$U = \int_1^2 \mathbf{F}\cdot d\mathbf{r} = \int_1^2 (F_x\,dx + F_y\,dy + F_z\,dz)$$

or

$$U = \int_{s_1}^{s_2} F_t\,ds$$

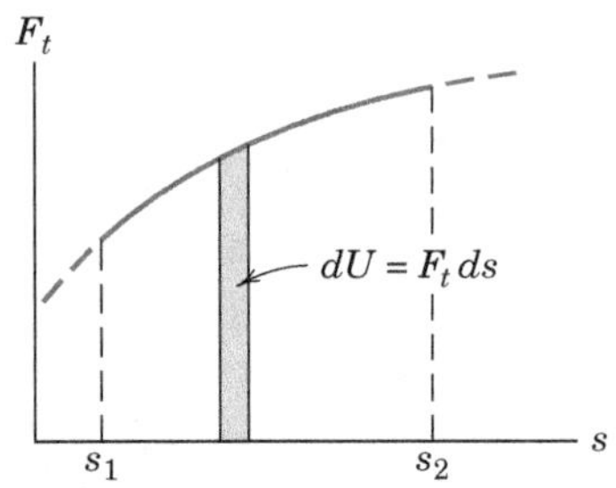

Figure 3/3

In order to carry out this integration, it is necessary to know the relations between the force components and their respective coordinates or the relation between F_t and s. If the functional relationship is not known as a mathematical expression which can be integrated but is specified in the form of approximate or experimental data, then we can compute the work by carrying out a numerical or graphical integration as represented by the area under the curve of F_t versus s, as shown in Fig. 3/3.

Examples of Work

When work must be calculated, we may always begin with the definition of work, $U = \int \mathbf{F}\cdot d\mathbf{r}$, insert appropriate vector expressions for the force $\mathbf{F}$ and the differential displacement vector $d\mathbf{r}$, and carry out the required integration. With some experience, simple work calculations, such as those associated with constant forces, may be performed by inspection. We now formally compute the work associated with three frequently occurring forces: constant forces, spring forces, and weights.

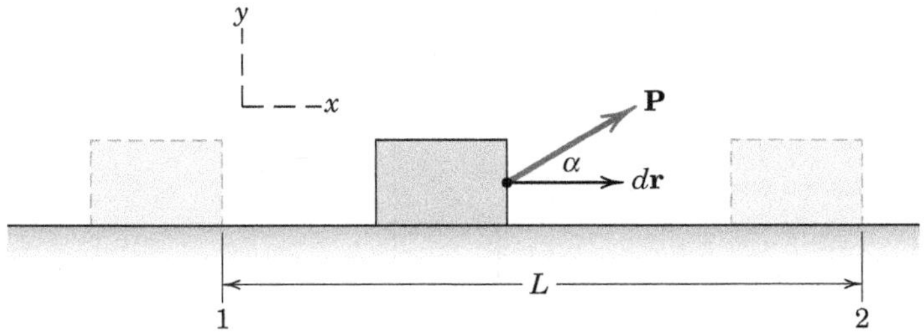

Figure 3/4

(1) Work Associated with a Constant External Force. Consider the constant force **P** applied to the body as it moves from position 1 to position 2, Fig. 3/4. With the force **P** and the differential displacement $d\mathbf{r}$ written as vectors, the work done on the body by the force is

$$U_{1\text{-}2} = \int_1^2 \mathbf{F}\cdot d\mathbf{r} = \int_1^2 [(P\cos\alpha)\mathbf{i} + (P\sin\alpha)\mathbf{j}]\cdot dx\,\mathbf{i}$$

$$= \int_{x_1}^{x_2} P\cos\alpha\, dx = P\cos\alpha(x_2 - x_1) = PL\cos\alpha \qquad \textbf{(3/9)}$$

As previously discussed, this work expression may be interpreted as the force component $P\cos\alpha$ times the distance L traveled. Should α be between 90° and 270°, the work would be negative. The force component $P\sin\alpha$ normal to the displacement does no work.

(2) Work Associated with a Spring Force. We consider here the common linear spring of stiffness k where the force required to stretch or compress the spring is proportional to the deformation x, as shown in Fig. 3/5*a*. We wish to determine the work done on the body by the spring force as the body undergoes an arbitrary displacement from an initial position x_1 to a final position x_2. The force exerted by the spring on the body is $\mathbf{F} = -kx\mathbf{i}$, as shown in Fig. 3/5*b*. From the definition of work, we have

$$U_{1\text{-}2} = \int_1^2 \mathbf{F}\cdot d\mathbf{r} = \int_1^2 (-kx\mathbf{i})\cdot dx\,\mathbf{i} = -\int_{x_1}^{x_2} kx\,dx = \tfrac{1}{2}k(x_1^2 - x_2^2) \qquad \textbf{(3/10)}$$

If the initial position is the position of zero spring deformation so that $x_1 = 0$, then the work is negative for any final position $x_2 \neq 0$. This is verified by recognizing that if the body begins at the undeformed spring position and then moves to the right, the spring force is to the left; if the body begins at $x_1 = 0$ and moves to the left, the spring force is to the right. On the other hand, if we move from an arbitrary initial position $x_1 \neq 0$ to the undeformed final position $x_2 = 0$, we see that the work is positive. In any movement toward the undeformed spring position, the spring force and the displacement are in the same direction.

In the *general* case, of course, neither x_1 nor x_2 is zero. The magnitude of the work is equal to the shaded trapezoidal area of Fig. 3/5*a*. In calculating the work done on a body by a spring force, care must be

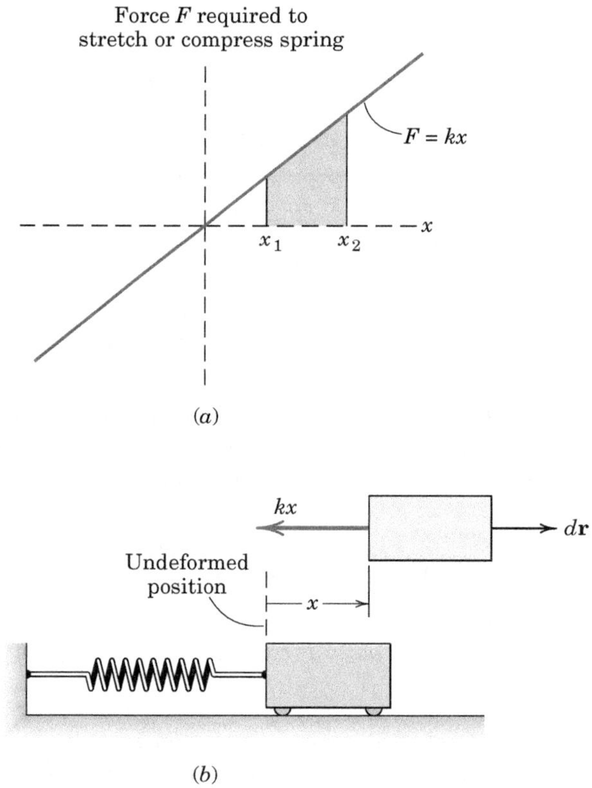

Figure 3/5

taken to ensure that the units of k and x are consistent. If x is in meters (or feet), k must be in N/m (or lb/ft). In addition, be sure to recognize that the variable x represents a deformation from the unstretched spring length and *not* the total length of the spring.

The expression $F = kx$ is actually a static relationship which is true only when elements of the spring have no acceleration. The dynamic behavior of a spring when its mass is accounted for is a fairly complex problem which will not be treated here. We shall assume that the mass of the spring is small compared with the masses of other accelerating parts of the system, in which case the linear static relationship will not involve appreciable error.

(3) Work Associated with Weight. *Case (a) g = constant.* If the altitude variation is sufficiently small so that the acceleration of gravity g may be considered constant, the work done by the weight mg of the body shown in Fig. 3/6*a* as the body is displaced from an arbitrary altitude y_1 to a final altitude y_2 is

$$U_{1\text{-}2} = \int_1^2 \mathbf{F}\cdot d\mathbf{r} = \int_1^2 (-mg\mathbf{j})\cdot(dx\mathbf{i} + dy\mathbf{j})$$

$$= -mg\int_{y_1}^{y_2} dy = -mg(y_2 - y_1) \tag{3/11}$$

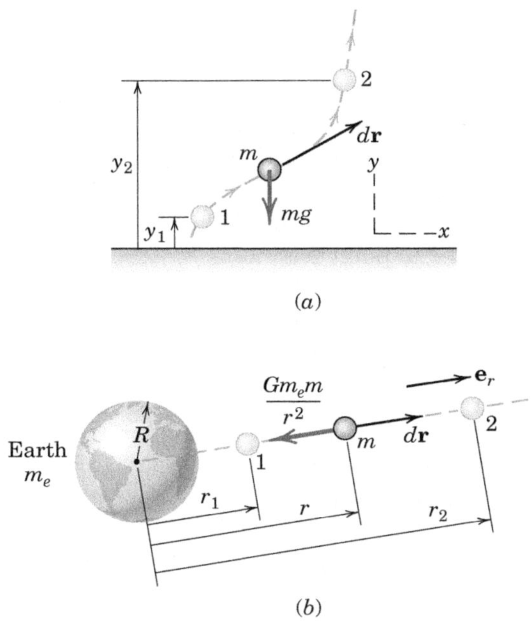

Figure 3/6

We see that horizontal movement does not contribute to this work. We also note that if the body rises (perhaps due to other forces not shown), then $(y_2 - y_1) > 0$ and this work is negative. If the body falls, $(y_2 - y_1) < 0$ and the work is positive.

Case (b) $g \neq$ constant. If large changes in altitude occur, then the weight (gravitational force) is no longer constant. We must therefore use the gravitational law (Eq. 1/2) and express the weight as a variable force of magnitude $F = \dfrac{Gm_em}{r^2}$, as indicated in Fig. 3/6*b*. Using the radial coordinate shown in the figure allows the work to be expressed as

$$U_{1\text{-}2} = \int_1^2 \mathbf{F}\cdot d\mathbf{r} = \int_1^2 \frac{-Gm_em}{r^2}\mathbf{e}_r\cdot dr\,\mathbf{e}_r = -Gm_em\int_{r_1}^{r_2}\frac{dr}{r^2}$$

$$= Gm_em\left(\frac{1}{r_2} - \frac{1}{r_1}\right) = mgR^2\left(\frac{1}{r_2} - \frac{1}{r_1}\right) \qquad \textbf{(3/12)}$$

where the equivalence $Gm_e = gR^2$ was established in Art. 1/5, with g representing the acceleration of gravity at the earth's surface and R representing the radius of the earth. The student should verify that if a body rises to a higher altitude $(r_2 > r_1)$, this work is negative, as it was in case (*a*). If the body falls to a lower altitude $(r_2 < r_1)$, the work is positive. Be sure to realize that r represents a radial distance from the center of the earth and not an altitude $h = r - R$ above the surface of the earth. As in case (*a*), had we considered a transverse displacement in addition to the radial displacement shown in Fig. 3/6*b*, we would have concluded that the transverse displacement, because it is perpendicular to the weight, does not contribute to the work.

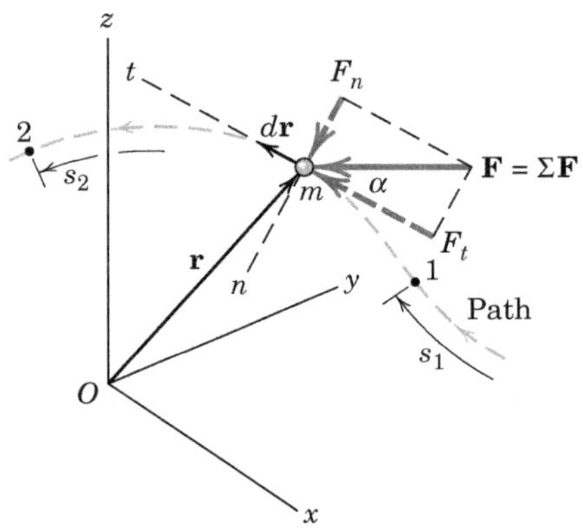

Figure 3/7

Work and Curvilinear Motion

We now consider the work done on a particle of mass m, Fig. 3/7, moving along a curved path under the action of the force $\mathbf{F}$, which stands for the resultant $\Sigma\mathbf{F}$ of all forces acting on the particle. The position of m is specified by the position vector $\mathbf{r}$, and its displacement along its path during the time dt is represented by the change $d\mathbf{r}$ in its position vector. The work done by $\mathbf{F}$ during a finite movement of the particle from point 1 to point 2 is

$$U_{1\text{-}2} = \int_1^2 \mathbf{F}\cdot d\mathbf{r} = \int_{s_1}^{s_2} F_t\, ds$$

where the limits specify the initial and final end points of the motion.

When we substitute Newton's second law $\mathbf{F} = m\mathbf{a}$, the expression for the work of all forces becomes

$$U_{1\text{-}2} = \int_1^2 \mathbf{F}\cdot d\mathbf{r} = \int_1^2 m\mathbf{a}\cdot d\mathbf{r}$$

But $\mathbf{a}\cdot d\mathbf{r} = a_t\, ds$, where a_t is the tangential component of the acceleration of m. In terms of the velocity v of the particle, Eq. 2/3 gives $a_t ds = v\, dv$. Thus, the expression for the work of $\mathbf{F}$ becomes

$$U_{1\text{-}2} = \int_1^2 \mathbf{F}\cdot d\mathbf{r} = \int_{v_1}^{v_2} mv\, dv = \tfrac{1}{2}m(v_2{}^2 - v_1{}^2) \qquad \textbf{(3/13)}$$

where the integration is carried out between points 1 and 2 along the curve, at which points the velocities have the magnitudes v_1 and v_2, respectively.

Principle of Work and Kinetic Energy

The *kinetic energy* T of the particle is defined as

$$\boxed{T = \tfrac{1}{2}mv^2} \qquad \textbf{(3/14)}$$

and is the total work which must be done on the particle to bring it from a state of rest to a velocity v. Kinetic energy T is a scalar quantity with the units of N·m or joules (J) in SI units and ft-lb in U.S. customary units. Kinetic energy is *always* positive, regardless of the direction of the velocity.

Equation 3/13 may be restated as

$$U_{1\text{-}2} = T_2 - T_1 = \Delta T \qquad \textbf{(3/15)}$$

which is the *work-energy equation* for a particle. The equation states that the *total work done* by all forces acting on a particle as it moves from point 1 to point 2 equals the corresponding *change in kinetic energy* of the particle. Although T is always positive, the change ΔT may be positive, negative, or zero. When written in this concise form, Eq. 3/15 tells us that the work always results in a *change* of kinetic energy.

Alternatively, the work-energy relation may be expressed as the initial kinetic energy T_1 plus the work done $U_{1\text{-}2}$ equals the final kinetic energy T_2, or

$$\boxed{T_1 + U_{1\text{-}2} = T_2} \qquad \textbf{(3/15a)}$$

When written in this form, the terms correspond to the natural sequence of events. Clearly, the two forms 3/15 and 3/15*a* are equivalent.

Advantages of the Work-Energy Method

We now see from Eq. 3/15 that a major advantage of the method of work and energy is that it avoids the necessity of computing the acceleration and leads directly to the velocity changes as functions of the forces which do work. Further, the work-energy equation involves only those forces which do work and thus give rise to changes in the magnitude of the velocities.

We consider now a system of two particles joined together by a connection which is frictionless and incapable of any deformation. The forces in the connection are equal and opposite, and their points of application necessarily have identical displacement components in the direction of the forces. Therefore, the net work done by these internal forces is zero during any movement of the system. Thus, Eq. 3/15 is applicable to the entire system, where $U_{1\text{-}2}$ is the total or net work done on the system by forces external to it and ΔT is the change, $T_2 - T_1$, in the total kinetic energy of the system. The total kinetic energy is the sum of the kinetic energies of both elements of the system. We thus see that another advantage of the work-energy method is that it enables us to analyze a system of particles joined in the manner described without dismembering the system.

Application of the work-energy method requires isolation of the particle or system under consideration. For a single particle you should draw a *free-body diagram* showing all externally applied forces. For a system of particles rigidly connected without springs, draw an *active-force diagram* showing only those external forces which do work (active forces) on the entire system.*

Power

The capacity of a machine is measured by the time rate at which it can do work or deliver energy. The total work or energy output is not a measure of this capacity since a motor, no matter how small, can deliver a large amount of energy if given sufficient time. On the other hand, a large and powerful machine is required to deliver a large amount of energy in a short period of time. Thus, the capacity of a machine is rated by its *power*, which is defined as the *time rate of doing work*.

*The active-force diagram was introduced in the method of virtual work in statics. See Chapter 7 of *Vol. 1 Statics*.

The power which must be produced by a bike rider depends on the bicycle speed and the propulsive force which is exerted by the supporting surface on the rear wheel. The driving force depends on the slope being negotiated.

Accordingly, the power P developed by a force $\mathbf{F}$ which does an amount of work U is $P = dU/dt = \mathbf{F}\cdot d\mathbf{r}/dt$. Because $d\mathbf{r}/dt$ is the velocity $\mathbf{v}$ of the point of application of the force, we have

$$\boxed{P = \mathbf{F}\cdot\mathbf{v}} \qquad \textbf{(3/16)}$$

Power is clearly a scalar quantity, and in SI it has the units of N·m/s = J/s. The special unit for power is the *watt* (W), which equals one joule per second (J/s). In U.S. customary units, the unit for mechanical power is the *horsepower* (hp). These units and their numerical equivalences are

$$1 \text{ W} = 1 \text{ J/s}$$

$$1 \text{ hp} = 550 \text{ ft-lb/sec} = 33{,}000 \text{ ft-lb/min}$$

$$1 \text{ hp} = 746 \text{ W} = 0.746 \text{ kW}$$

Efficiency

The ratio of the work done *by* a machine to the work done *on* the machine during the same time interval is called the *mechanical efficiency* e_m of the machine. This definition assumes that the machine operates uniformly so that there is no accumulation or depletion of energy within it. Efficiency is always less than unity since every device operates with some loss of energy and since energy cannot be created within the machine. In mechanical devices which involve moving parts, there will always be some loss of energy due to the negative work of kinetic friction forces. This work is converted to heat energy which, in turn, is dissipated to the surroundings. The mechanical efficiency at any instant of time may be expressed in terms of mechanical power P by

$$e_m = \frac{P_{\text{output}}}{P_{\text{input}}} \qquad \textbf{(3/17)}$$

In addition to energy loss by mechanical friction, there may also be electrical and thermal energy loss, in which case, the *electrical efficiency* e_e and *thermal efficiency* e_t are also involved. The *overall efficiency* e in such instances is

$$e = e_m e_e e_t$$

Sample Problem 3/11

Calculate the velocity v of the 50-kg crate when it reaches the bottom of the chute at B if it is given an initial velocity of 4 m/s down the chute at A. The coefficient of kinetic friction is 0.30.

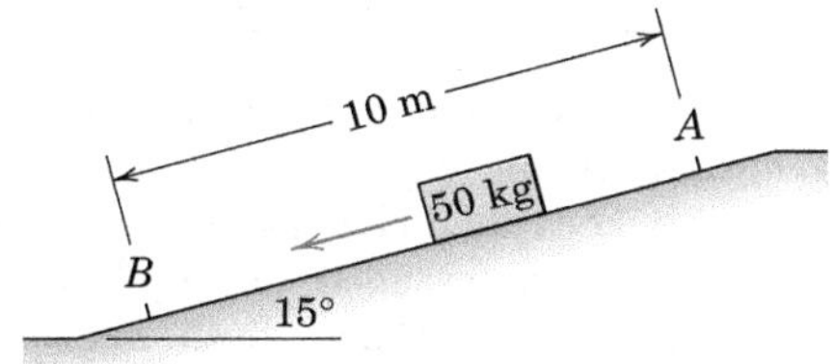

Solution. The free-body diagram of the crate is drawn and includes the normal force R and the kinetic friction force F calculated in the usual manner. The work done by the weight is positive, whereas that done by the friction force is negative. The total work done on the crate during the motion is

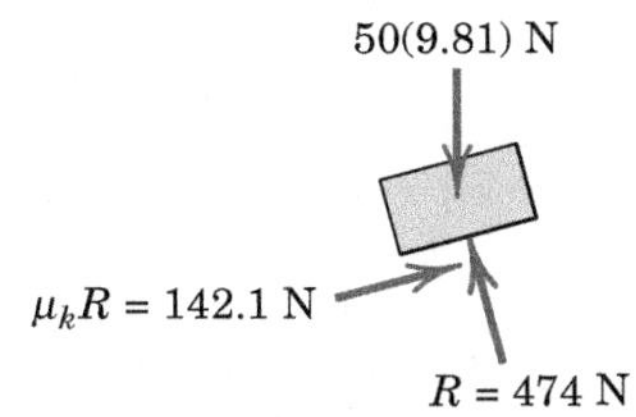

① $[U = Fs]$ $\quad U_{1\text{-}2} = 50(9.81)(10 \sin 15°) - 142.1(10) = -151.9 \text{ J}$

The work-energy equation gives

$$[T_1 + U_{1\text{-}2} = T_2] \qquad \frac{1}{2}mv_1^2 + U_{1\text{-}2} = \frac{1}{2}mv_2^2$$

$$\frac{1}{2}(50)(4)^2 - 151.9 = \frac{1}{2}(50)v_2^2$$

$$v_2 = 3.15 \text{ m/s} \qquad \textit{Ans.}$$

Since the net work done is negative, we obtain a decrease in the kinetic energy.

Helpful Hint

① The work due to the weight depends only on the *vertical* distance traveled.

Sample Problem 3/12

The flatbed truck, which carries an 80-kg crate, starts from rest and attains a speed of 72 km/h in a distance of 75 m on a level road with constant acceleration. Calculate the work done by the friction force acting on the crate during this interval if the static and kinetic coefficients of friction between the crate and the truck bed are (*a*) 0.30 and 0.28, respectively, or (*b*) 0.25 and 0.20, respectively.

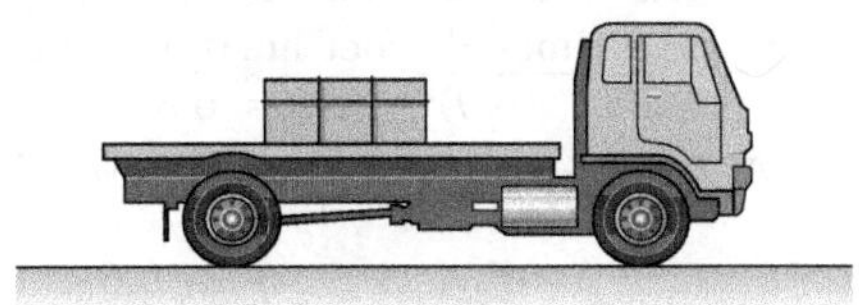

Solution. If the crate does not slip on the bed, its acceleration will be that of the truck, which is

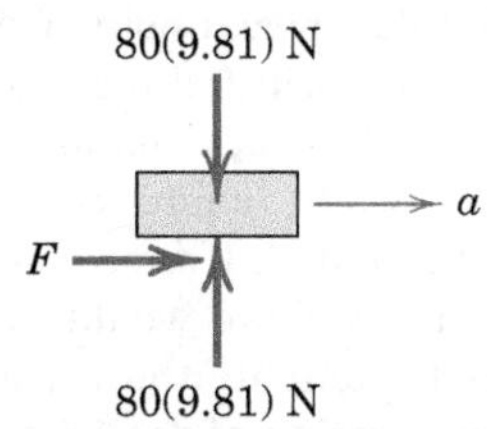

$$[v^2 = 2as] \qquad a = \frac{v^2}{2s} = \frac{(72/3.6)^2}{2(75)} = 2.67 \text{ m/s}^2$$

Case (a). This acceleration requires a friction force on the block of

$$[F = ma] \qquad F = 80(2.67) = 213 \text{ N}$$

which is less than the maximum possible value of $\mu_s N = 0.30(80)(9.81) = 235$ N. Therefore, the crate does not slip and the work done by the actual static friction force of 213 N is

① $[U = Fs]$ $\quad U_{1\text{-}2} = 213(75) = 16\,000 \text{ J} \quad$ or $\quad 16 \text{ kJ} \qquad$ *Ans.*

Case (b). For $\mu_s = 0.25$, the maximum possible friction force is 0.25(80)(9.81) = 196.2 N, which is slightly less than the value of 213 N required for no slipping. Therefore, we conclude that the crate slips, and the friction force is governed by the kinetic coefficient and is $F = 0.20(80)(9.81) = 157.0$ N. The acceleration becomes

$$[F = ma] \qquad a = F/m = 157.0/80 = 1.962 \text{ m/s}^2$$

The distances traveled by the crate and the truck are in proportion to their accelerations. Thus, the crate has a displacement of (1.962/2.67)75 = 55.2 m, and the work done by kinetic friction is

② $[U = Fs]$ $\quad U_{1\text{-}2} = 157.0(55.2) = 8660 \text{ J} \quad$ or $\quad 8.66 \text{ kJ} \qquad$ *Ans.*

Helpful Hints

① We note that static friction forces do no work when the contacting surfaces are both at rest. When they are in motion, however, as in this problem, the static friction force acting on the crate does positive work and that acting on the truck bed does negative work.

② This problem shows that a kinetic friction force can do positive work when the surface which supports the object and generates the friction force is in motion. If the supporting surface is at rest, then the kinetic friction force acting on the moving part always does negative work.

Sample Problem 3/13

The 50-kg block at A is mounted on rollers so that it moves along the fixed horizontal rail with negligible friction under the action of the constant 300-N force in the cable. The block is released from rest at A, with the spring to which it is attached extended an initial amount $x_1 = 0.233$ m. The spring has a stiffness $k = 80$ N/m. Calculate the velocity v of the block as it reaches position B.

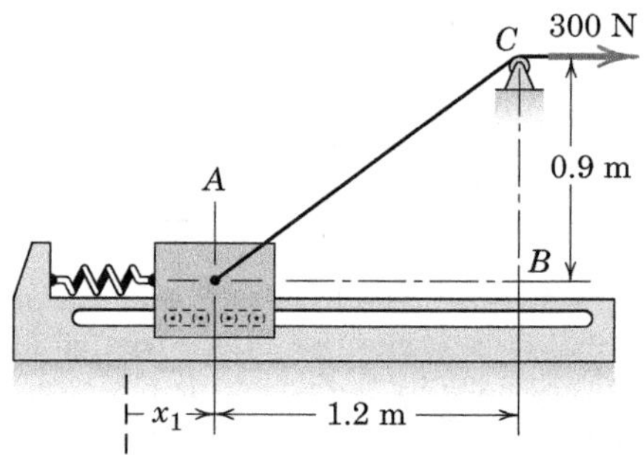

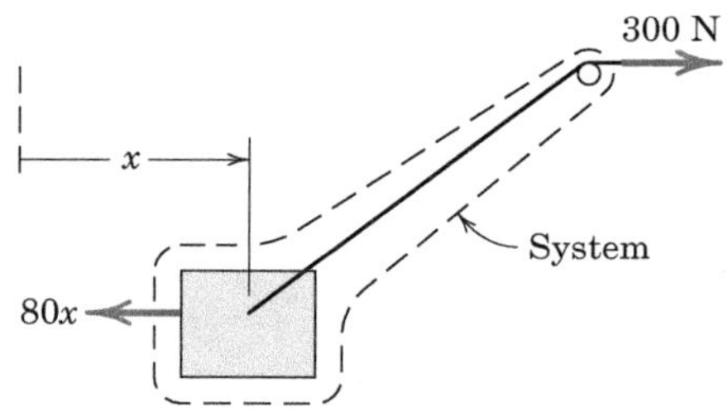

Solution. It will be assumed initially that the stiffness of the spring is small enough to allow the block to reach position B. The active-force diagram for the system composed of both block and cable is shown for a general position. The spring force $80x$ and the 300-N tension are the only forces external to this system which do work on the system. The force exerted on the block by the rail, the weight of the block, and the reaction of the small pulley on the cable do no work on the system and are not included on the active-force diagram.

As the block moves from $x_1 = 0.233$ m to $x_2 = 0.233 + 1.2 = 1.433$ m, the work done by the spring force acting on the block is

① $$[U_{1\text{-}2} = \frac{1}{2}k(x_1{}^2 - x_2{}^2)] \qquad U_{1\text{-}2} = \frac{1}{2}\,80[0.233^2 - (0.233 + 1.2)^2]$$
$$= -80.0 \text{ J}$$

The work done on the system by the constant 300-N force in the cable is the force times the net horizontal movement of the cable over pulley C, which is $\sqrt{(1.2)^2 + (0.9)^2} - 0.9 = 0.6$ m. Thus, the work done is $300(0.6) = 180$ J. We now apply the work-energy equation to the system and get

$$[T_1 + U_{1\text{-}2} = T_2] \qquad 0 - 80.0 + 180 = \frac{1}{2}(50)v^2 \qquad v = 2.00 \text{ m/s} \qquad \textit{Ans.}$$

We take special note of the advantage to our choice of system. If the block alone had constituted the system, the horizontal component of the 300-N cable tension on the block would have to be integrated over the 1.2-m displacement. This step would require considerably more effort than was needed in the solution as presented. If there had been appreciable friction between the block and its guiding rail, we would have found it necessary to isolate the block alone in order to compute the variable normal force and, hence, the variable friction force. Integration of the friction force over the displacement would then be required to evaluate the negative work which it would do.

Helpful Hint

① Recall that this general formula is valid for any initial and final spring deflections x_1 and x_2, positive (spring in tension) or negative (spring in compression). In deriving the spring-work formula, we assumed the spring to be linear, which is the case here.

Sample Problem 3/14

The power winch A hoists the 800-lb log up the 30° incline at a constant speed of 4 ft/sec. If the power output of the winch is 6 hp, compute the coefficient of kinetic friction μ_k between the log and the incline. If the power is suddenly increased to 8 hp, what is the corresponding instantaneous acceleration a of the log?

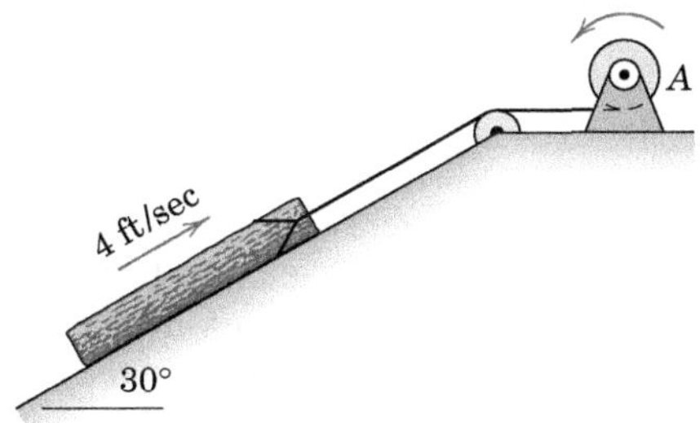

Solution. From the free-body diagram of the log, we get $N = 800 \cos 30° = 693$ lb, and the kinetic friction force becomes $693\mu_k$. For constant speed, the forces are in equilibrium so that

$$[\Sigma F_x = 0] \qquad T - 693\mu_k - 800 \sin 30° = 0 \qquad T = 693\mu_k + 400$$

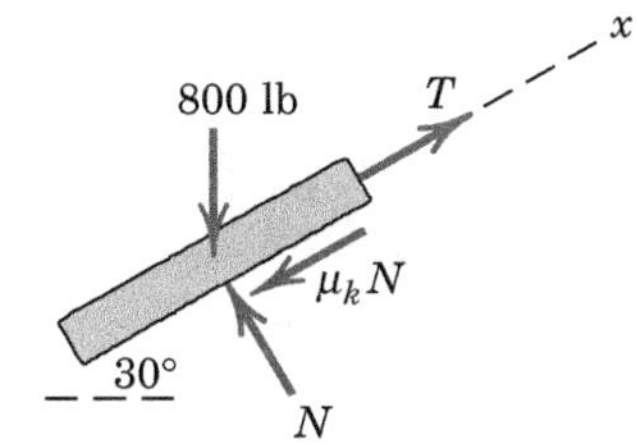

The power output of the winch gives the tension in the cable

① $$[P = Tv] \qquad T = P/v = 6(550)/4 = 825 \text{ lb}$$

Substituting T gives

$$825 = 693\mu_k + 400 \qquad \mu_k = 0.613 \qquad \textit{Ans.}$$

When the power is increased, the tension momentarily becomes

$$[P = Tv] \qquad T = P/v = 8(550)/4 = 1100 \text{ lb}$$

and the corresponding acceleration is given by

$$[\Sigma F_x = ma_x] \qquad 1100 - 693(0.613) - 800 \sin 30° = \frac{800}{32.2}a$$

② $$a = 11.07 \text{ ft/sec}^2 \qquad \textit{Ans.}$$

Helpful Hints

① Note the conversion from horsepower to ft-lb/sec.

② As the speed increases, the acceleration will drop until the speed stabilizes at a value higher than 4 ft/sec.

Sample Problem 3/15

A satellite of mass m is put into an elliptical orbit around the earth. At point A, its distance from the earth is $h_1 = 500$ km and it has a velocity $v_1 = 30\,000$ km/h. Determine the velocity v_2 of the satellite as it reaches point B, a distance $h_2 = 1200$ km from the earth.

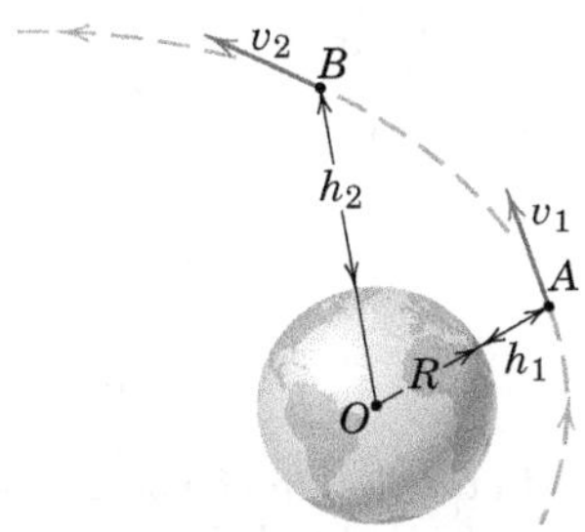

Solution. The satellite is moving outside of the earth's atmosphere so that the only force acting on it is the gravitational attraction of the earth. For the large change in altitude of this problem, we cannot assume that the acceleration due to gravity is constant. Rather, we must use the work expression, derived in this article, which accounts for variation in the gravitational acceleration with altitude. Put another way, the work expression accounts for the variation of the weight $F = \dfrac{Gmm_e}{r^2}$ with altitude. This work expression is

$$U_{1\text{-}2} = mgR^2\left(\frac{1}{r_2} - \frac{1}{r_1}\right)$$

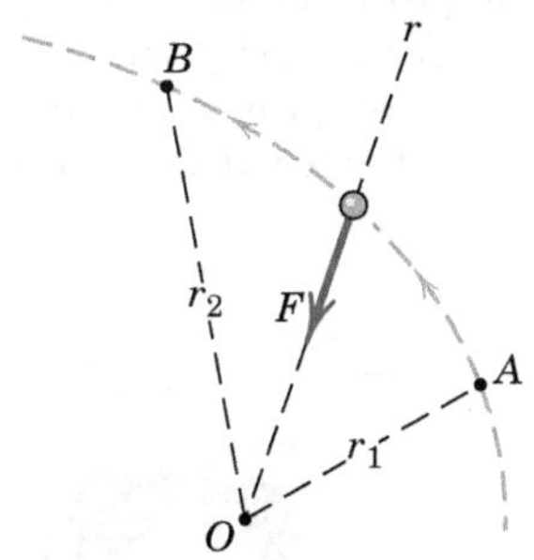

The work-energy equation $T_1 + U_{1\text{-}2} = T_2$ gives

① ② $$\tfrac{1}{2}mv_1^2 + mgR^2\left(\frac{1}{r_2} - \frac{1}{r_1}\right) = \tfrac{1}{2}mv_2^2 \qquad v_2^2 = v_1^2 + 2gR^2\left(\frac{1}{r_2} - \frac{1}{r_1}\right)$$

Substituting the numerical values gives

$$v_2^2 = \left(\frac{30\,000}{3.6}\right)^2 + 2(9.81)[(6371)(10^3)]^2\left(\frac{10^{-3}}{6371 + 1200} - \frac{10^{-3}}{6371 + 500}\right)$$

$$= 69.44(10^6) - 10.72(10^6) = 58.73(10^6)(\text{m/s})^2$$

$$v_2 = 7663 \text{ m/s} \qquad \text{or} \qquad v_2 = 7663(3.6) = 27\,590 \text{ km/h} \qquad \textit{Ans.}$$

Helpful Hints

① Note that the result is independent of the mass m of the satellite.

② Consult Table D/2, Appendix D, to find the radius R of the earth.

PROBLEMS

Introductory Problems

3/97 The spring is unstretched at the position $x = 0$. Under the action of a force P, the cart moves from the initial position $x_1 = -6$ in. to the final position $x_2 = 3$ in. Determine (*a*) the work done on the cart by the spring and (*b*) the work done on the cart by its weight.

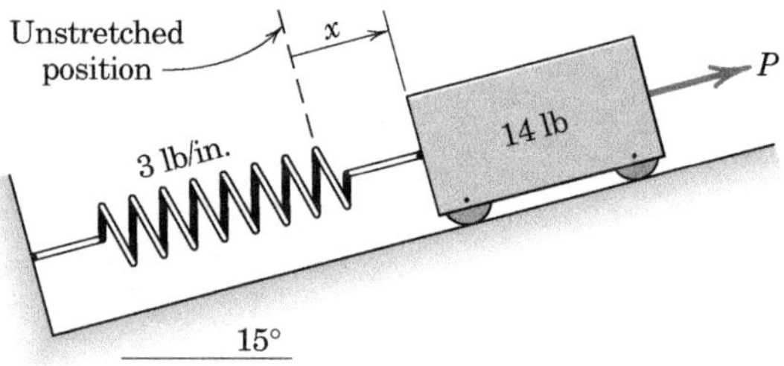

Problem 3/97

3/98 The small cart has a speed $v_A = 4$ m/s as it passes point A. It moves without appreciable friction and passes over the top hump of the track. Determine the cart speed as it passes point B. Is knowledge of the shape of the track necessary?

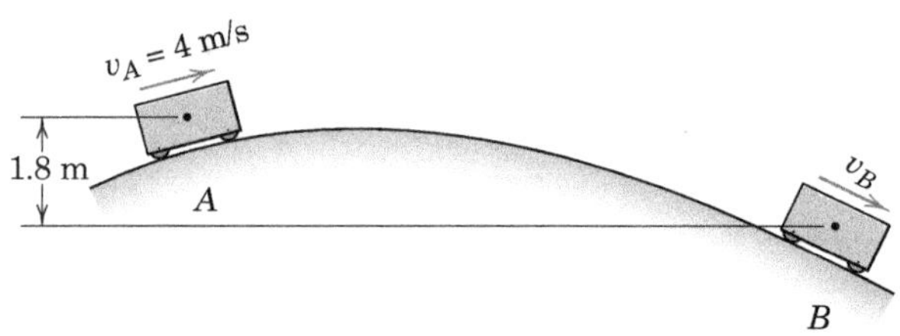

Problem 3/98

3/99 In the design of a spring bumper for a 3500-lb car, it is desired to bring the car to a stop from a speed of 5 mi/hr in a distance equal to 6 in. of spring deformation. Specify the required stiffness k for each of the two springs behind the bumper. The springs are undeformed at the start of impact.

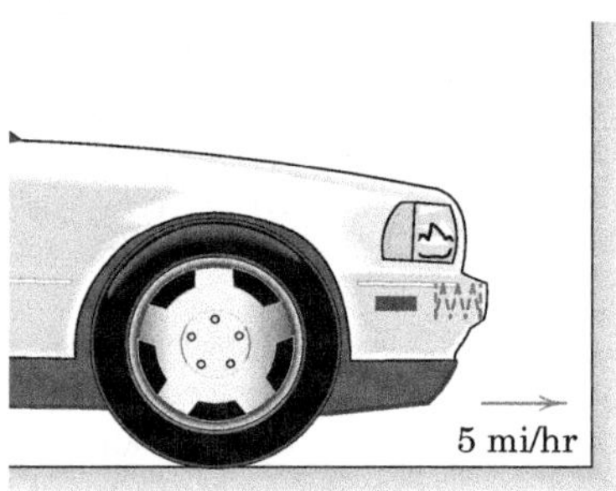

Problem 3/99

3/100 The 2-kg collar is at rest in position A when the constant force P is applied as shown. Determine the speed of the collar as it passes position B if (*a*) $P = 25$ N and (*b*) $P = 40$ N. The curved rod lies in a vertical plane, and friction is negligible.

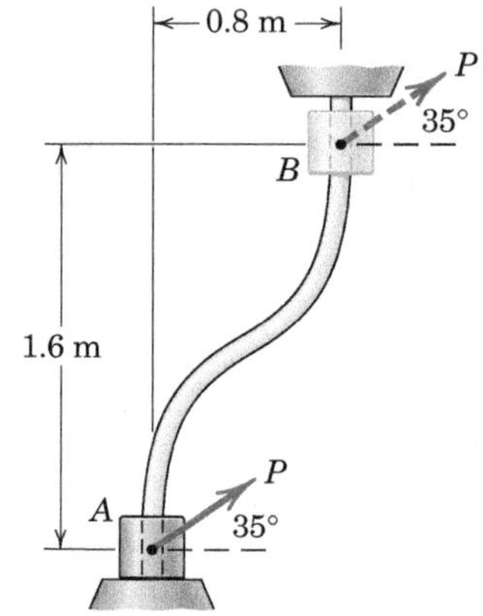

Problem 3/100

3/101 The 0.5-kg collar C starts from rest at A and slides with negligible friction on the fixed rod in the vertical plane. Determine the velocity v with which the collar strikes end B when acted upon by the 5-N force, which is constant in direction. Neglect the small dimensions of the collar.

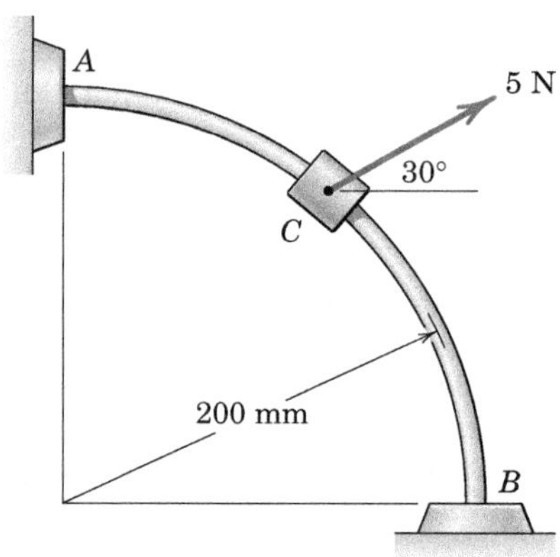

Problem 3/101

3/102 The small 0.1-kg slider enters the "loop-the-loop" with a speed $v_A = 12$ m/s as it passes point A, and it has a speed $v_B = 10$ m/s as it exits at point B. Determine the work done by friction between points A and B. The track lies in a vertical plane. Assume that contact is maintained throughout.

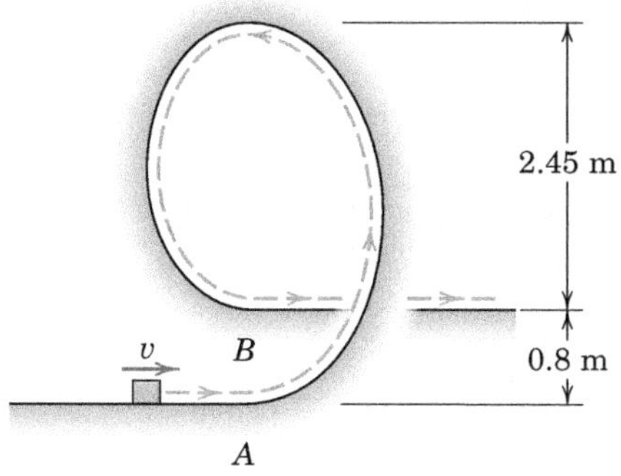

Problem 3/102

3/103 The man and his bicycle together weigh 200 lb. What power P is the man developing in riding up a 5-percent grade at a constant speed of 15 mi/hr?

Problem 3/103

3/104 The car is moving with a speed $v_0 = 65$ mi/hr up the 6-percent grade, and the driver applies the brakes at point A, causing all wheels to skid. The coefficient of kinetic friction for the rain-slicked road is $\mu_k = 0.60$. Determine the stopping distance s_{AB}. Repeat your calculations for the case when the car is moving downhill from B to A.

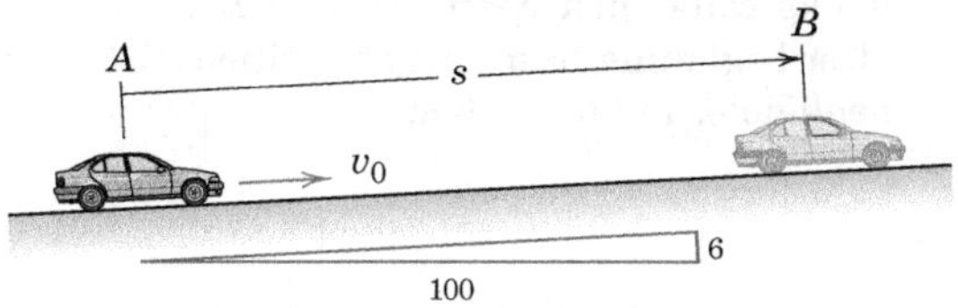

Problem 3/104

3/105 The small 0.2-kg slider is known to move from position A to position B along the vertical-plane slot. Determine (*a*) the work done on the body by its weight and (*b*) the work done on the body by the spring. The distance $R = 0.8$ m, the spring modulus $k = 180$ N/m, and the unstretched length of the spring is 0.6 m.

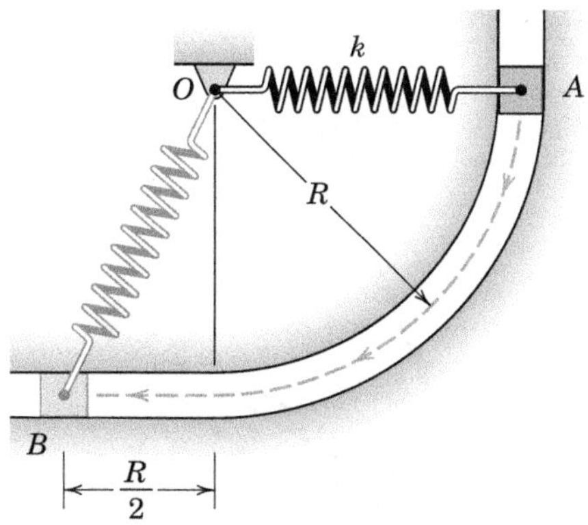

Problem 3/105

3/106 The 2-kg collar is released from rest at A and slides down the inclined fixed rod in the vertical plane. The coefficient of kinetic friction is 0.40. Calculate (*a*) the velocity v of the collar as it strikes the spring and (*b*) the maximum deflection x of the spring.

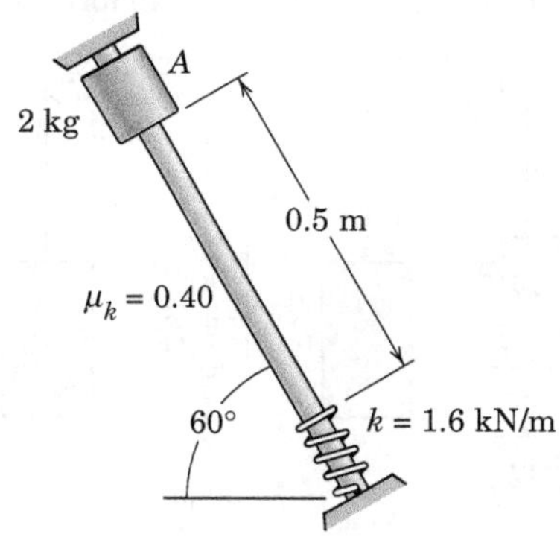

Problem 3/106

3/107 The 30-lb collar A is released from rest in the position shown and slides with negligible friction up the fixed rod inclined 30° from the horizontal under the action of a constant force $P = 50$ lb applied to the cable. Calculate the required stiffness k of the spring so that its maximum deflection equals 6 in. The position of the small pulley at B is fixed.

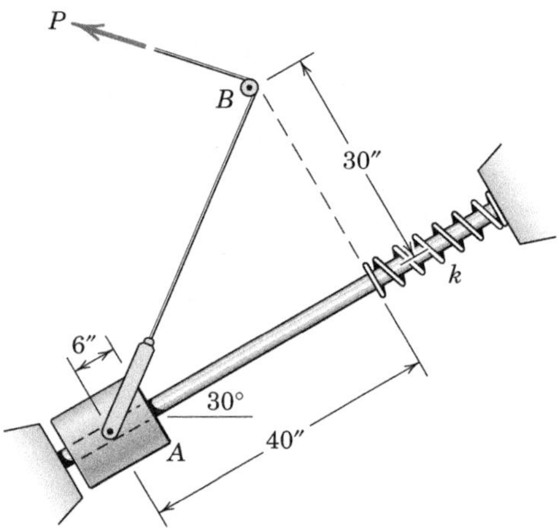

Problem 3/107

3/108 Each of the two systems is released from rest. Calculate the speed v of each 60-lb cylinder after the 40-lb cylinder has dropped 2 ft. The 30-lb cylinder of case (a) is replaced by a 30-lb force in case (b).

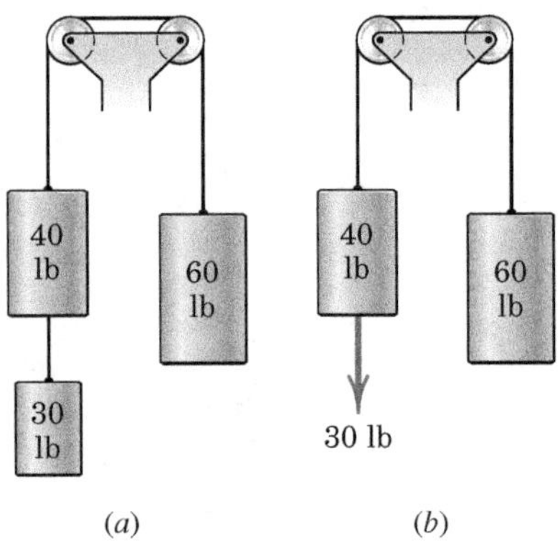

Problem 3/108

3/109 Each of the two systems is released from rest. Calculate the speed v of each 60-lb cylinder after the 40-lb cylinder has dropped 2 ft. The 30-lb cylinder of case (a) is replaced by a 30-lb force in case (b).

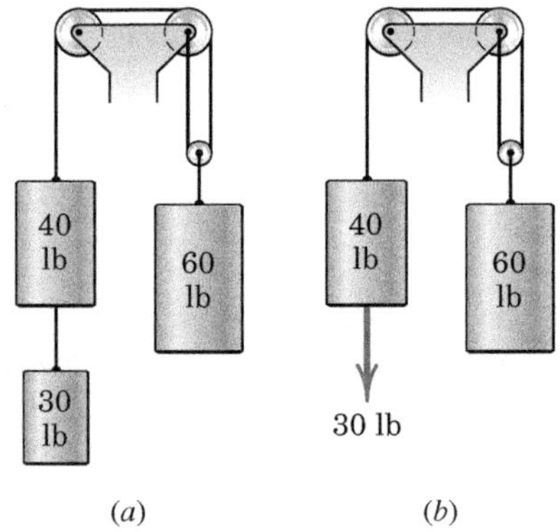

Problem 3/109

3/110 The 0.8-kg collar travels with negligible friction on the vertical rod under the action of the constant force $P = 20$ N. If the collar starts from rest at A, determine its speed as it passes point B. The value of $R = 1.6$ m.

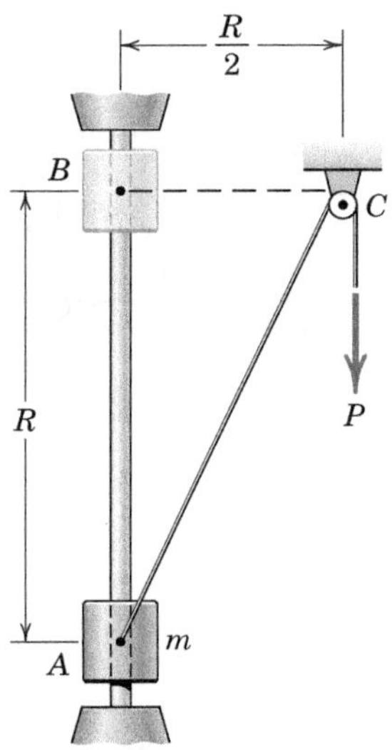

Problem 3/110

3/111 The system of the previous problem is rearranged as shown. For what constant force P will the 0.8-kg collar just reach position B with no speed after beginning from rest at position A? Friction is negligible, and $R = 1.6$ m.

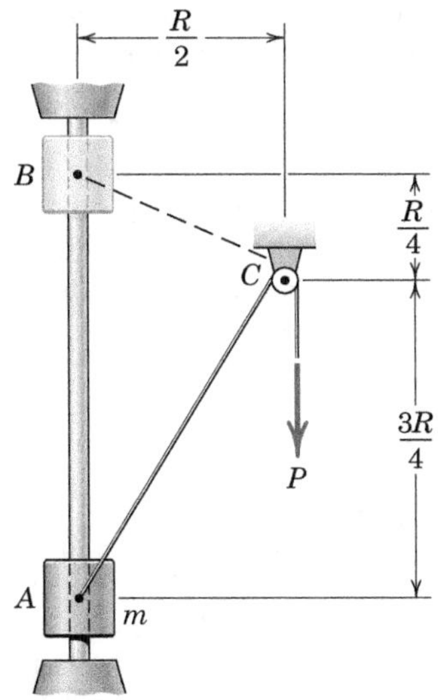

Problem 3/111

3/112 The 120-lb woman jogs up the flight of stairs in 5 seconds. Determine her average power output. Convert all given information to SI units and repeat your calculation.

Problem 3/112

Representative Problems

3/113 The 0.8-kg collar slides freely on the fixed circular rod. Calculate the velocity v of the collar as it hits the stop at B if it is elevated from rest at A by the action of the constant 40-N force in the cord. The cord is guided by the small fixed pulleys.

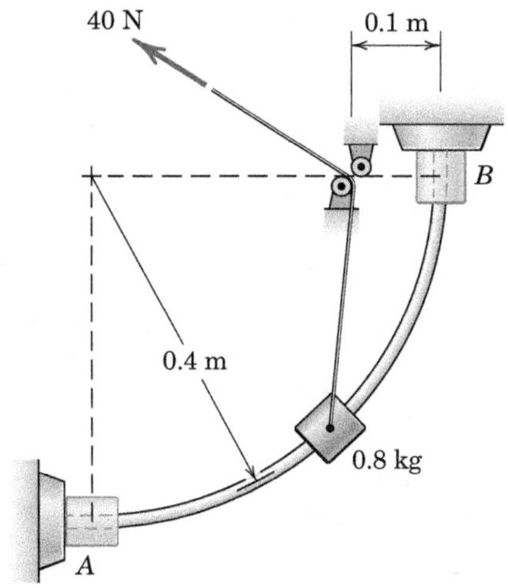

Problem 3/113

3/114 The 4-kg ball and the attached light rod rotate in the vertical plane about the fixed axis at O. If the assembly is released from rest at $\theta = 0$ and moves under the action of the 60-N force, which is maintained normal to the rod, determine the velocity v of the ball as θ approaches 90°. Treat the ball as a particle.

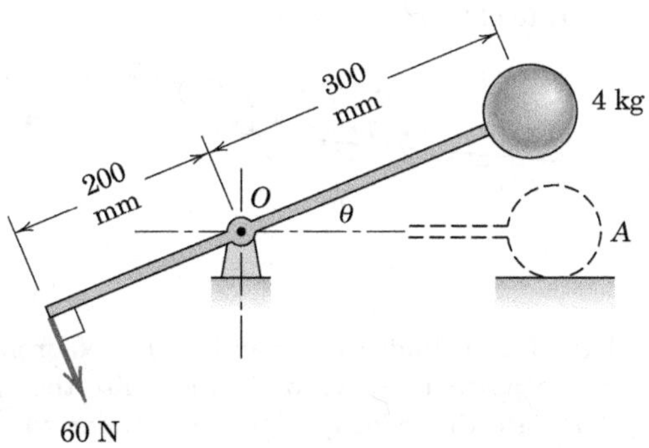

Problem 3/114

3/115 The position vector of a particle is given by $\mathbf{r} = 8t\mathbf{i} + 1.2t^2\mathbf{j} - 0.5(t^3 - 1)\mathbf{k}$, where t is the time in seconds from the start of the motion and where $\mathbf{r}$ is expressed in meters. For the condition when $t = 4$ s, determine the power P developed by the force $\mathbf{F} = 40\mathbf{i} - 20\mathbf{j} - 36\mathbf{k}$ N which acts on the particle.

3/116 An escalator handles a steady load of 30 people per minute in elevating them from the first to the second floor through a vertical rise of 24 ft. The average person weighs 140 lb. If the motor which drives the unit delivers 4 hp, calculate the mechanical efficiency e of the system.

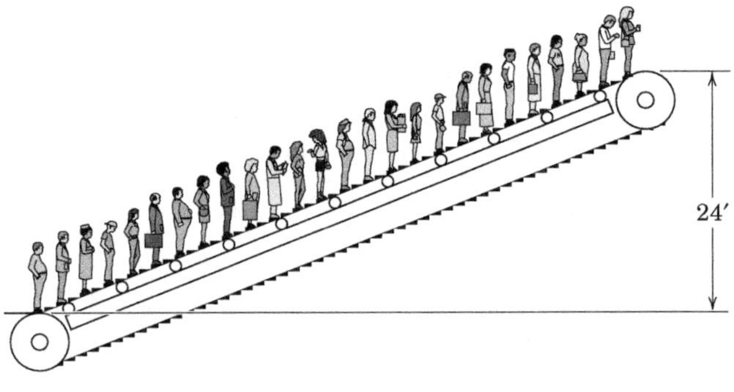

Problem 3/116

3/117 A 3600-lb car travels up the 6-percent incline shown. The car is subjected to a 60-lb aerodynamic drag force and a 50-lb force due to all other factors such as rolling resistance. Determine the power output required at a speed of 65 mi/hr if (*a*) the speed is constant and (*b*) the speed is increasing at the rate of $0.05g$.

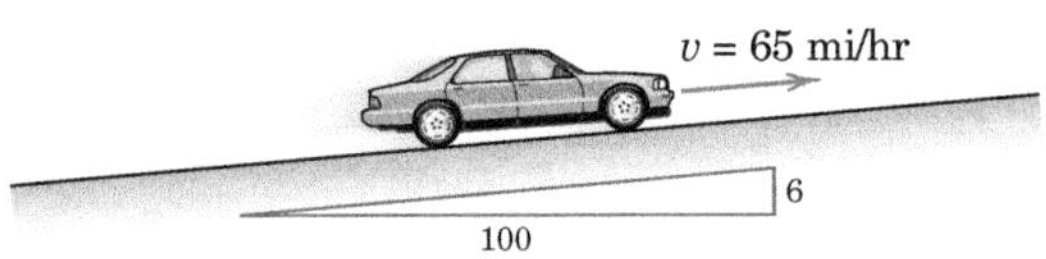

Problem 3/117

3/118 The 15-lb cylindrical collar is released from rest in the position shown and drops onto the spring. Calculate the velocity v of the cylinder when the spring has been compressed 2 in.

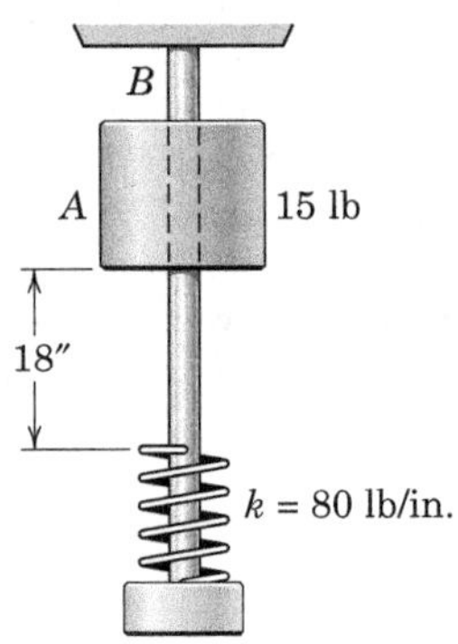

Problem 3/118

3/119 In the design of a conveyor-belt system, small metal blocks are discharged with a velocity of 0.4 m/s onto a ramp by the upper conveyor belt shown. If the coefficient of kinetic friction between the blocks and the ramp is 0.30, calculate the angle θ which the ramp must make with the horizontal so that the blocks will transfer without slipping to the lower conveyor belt moving at the speed of 0.14 m/s.

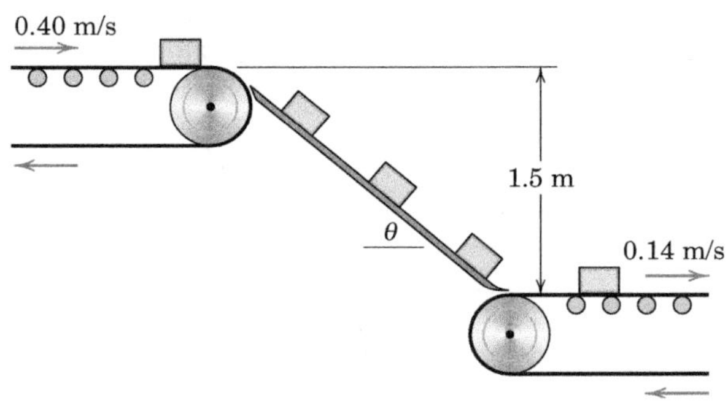

Problem 3/119

3/120 The collar of mass m is released from rest while in position A and subsequently travels with negligible friction along the vertical-plane circular guide. Determine the normal force (magnitude and direction) exerted by the guide on the collar (*a*) just before the collar passes point B, (*b*) just after the collar passes point B (i.e., the collar is now on the curved portion of the guide), (*c*) as the collar passes point C, and (*d*) just before the collar passes point D. Use the values $m = 0.4$ kg, $R = 1.2$ m, and $k = 200$ N/m. The unstretched length of the spring is $0.8R$.

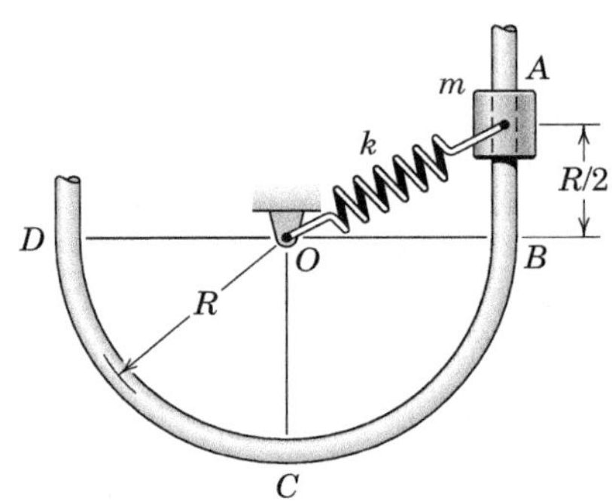

Problem 3/120

3/121 A nonlinear automobile spring is tested by having a 150-lb cylinder impact it with a speed $v_0 =$ 12 ft/sec. The spring resistance is shown in the accompanying graph. Determine the maximum deflection δ of the spring with and without the nonlinear term present. The small platform at the top of the spring has negligible weight.

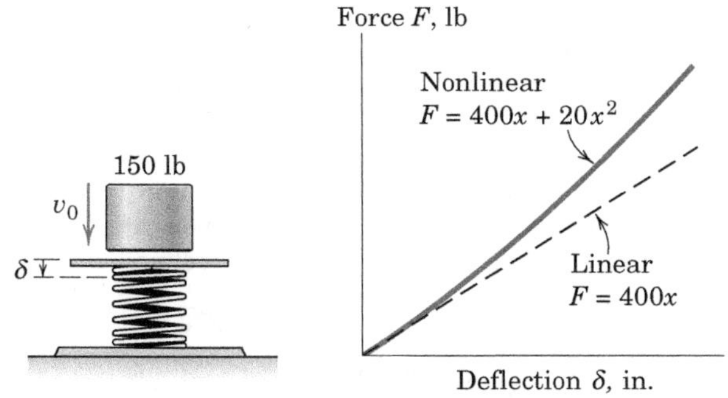

Problem 3/121

3/122 The motor unit A is used to elevate the 300-kg cylinder at a constant rate of 2 m/s. If the power meter B registers an electrical input of 2.20 kW, calculate the combined electrical and mechanical efficiency e of the system.

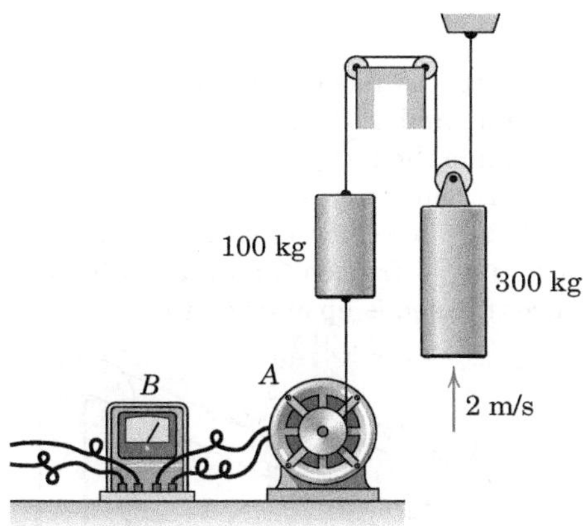

Problem 3/122

3/123 A 90-lb boy starts from rest at the bottom A of a 10-percent incline and increases his speed at a constant rate to 5 mi/hr as he passes B, 50 ft along the incline from A. Determine his power output as he approaches B.

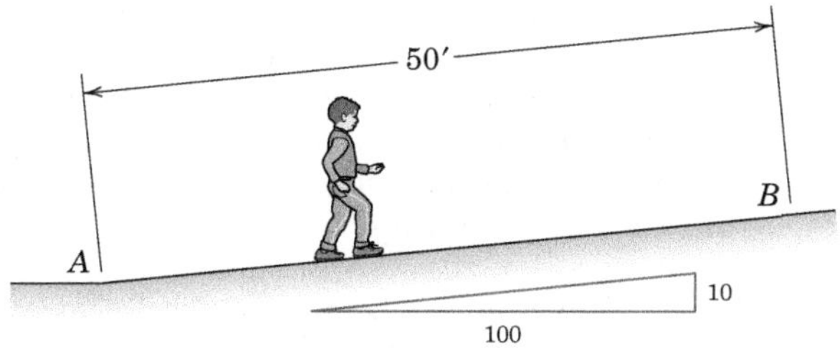

Problem 3/123

3/124 A projectile is launched from the north pole with an initial vertical velocity v_0. What value of v_0 will result in a maximum altitude of $R/3$? Neglect aerodynamic drag and use $g = 9.825$ m/s^2 as the surface-level acceleration due to gravity.

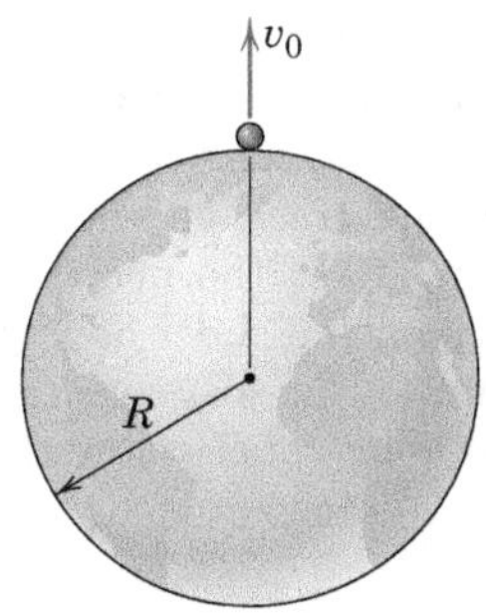

Problem 3/124

3/125 Two 425,000-lb locomotives pull fifty 200,000-lb coal hoppers. The train starts from rest and accelerates uniformly to a speed of 40 mi/hr over a distance of 8000 ft on a level track. The constant rolling resistance of each car is 0.005 times its weight. Neglect all other retarding forces and assume that each locomotive contributes equally to the tractive force. Determine (*a*) the tractive force exerted by each locomotive at 20 mi/hr, (*b*) the power required from each locomotive at 20 mi/hr, (*c*) the power required from each locomotive as the train speed approaches 40 mi/hr, and (*d*) the power required from each locomotive if the train cruises at a steady 40 mi/hr.

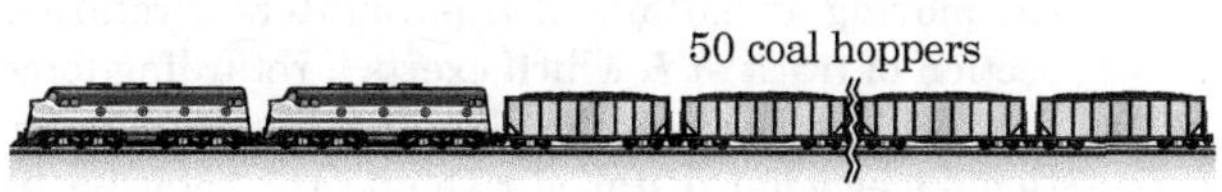

Problem 3/125

3/126 A car with a mass of 1500 kg starts from rest at the bottom of a 10-percent grade and acquires a speed of 50 km/h in a distance of 100 m with constant acceleration up the grade. What is the power P delivered to the drive wheels by the engine when the car reaches this speed?

3/127 The third stage of a rocket fired vertically up over the north pole coasts to a maximum altitude of 500 km following burnout of its rocket motor. Calculate the downward velocity v of the rocket when it has fallen 100 km from its position of maximum altitude. (Use the mean value of 9.825 m/s^2 for g and 6371 km for the mean radius of the earth.)

3/128 The small slider of mass m is released from rest while in position A and then slides along the vertical-plane track. The track is smooth from A to D and rough (coefficient of kinetic friction μ_k) from point D on. Determine (a) the normal force N_B exerted by the track on the slider just after it passes point B, (b) the normal force N_C exerted by the track on the slider as it passes the bottom point C, and (c) the distance s traveled along the incline past point D before the slider stops.

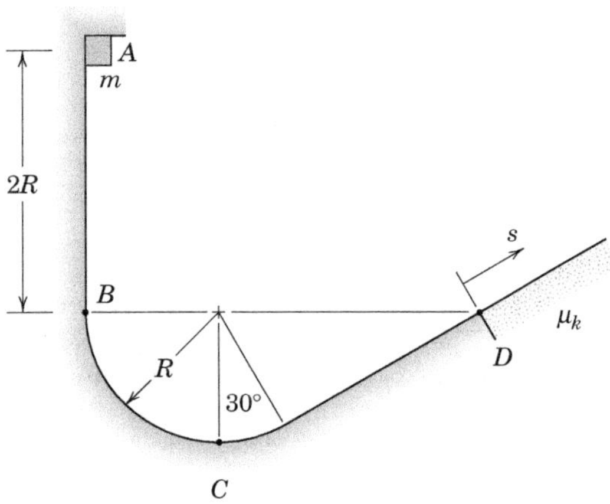

Problem 3/128

3/129 In a railroad classification yard, a 68-Mg freight car moving at 0.5 m/s at A encounters a retarder section of track at B which exerts a retarding force of 32 kN on the car in the direction opposite to motion. Over what distance x should the retarder be activated in order to limit the speed of the car to 3 m/s at C?

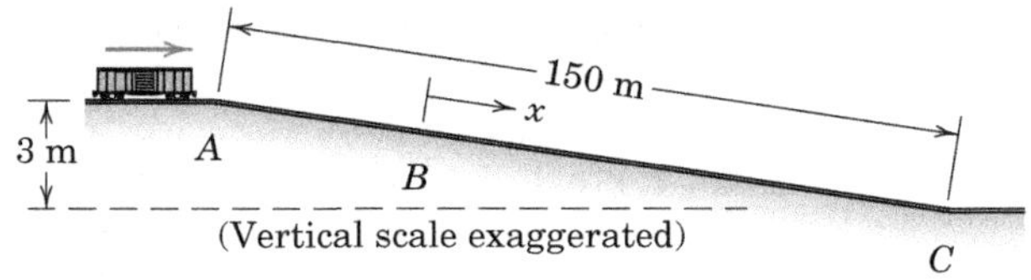

Problem 3/129

3/130 The system is released from rest with no slack in the cable and with the spring unstretched. Determine the distance s traveled by the 4-kg cart before it comes to rest (a) if m approaches zero and (b) if m = 3 kg. Assume no mechanical interference and no friction, and state whether the distance traveled is up or down the incline.

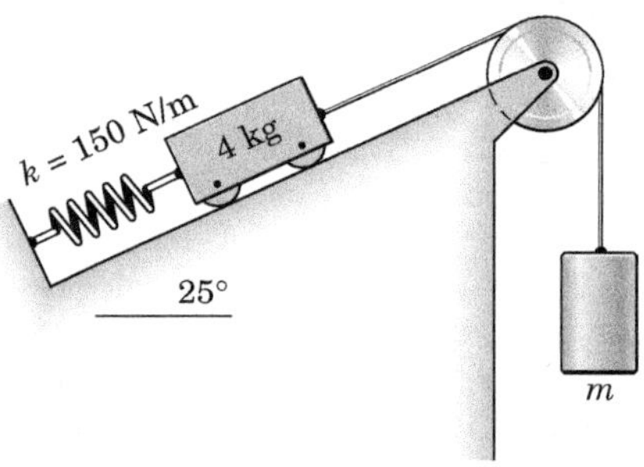

Problem 3/130

3/131 The system is released from rest with no slack in the cable and with the spring stretched 200 mm. Determine the distance s traveled by the 4-kg cart before it comes to rest (a) if m approaches zero and (b) if m = 3 kg. Assume no mechanical interference and no friction, and state whether the distance traveled is up or down the incline.

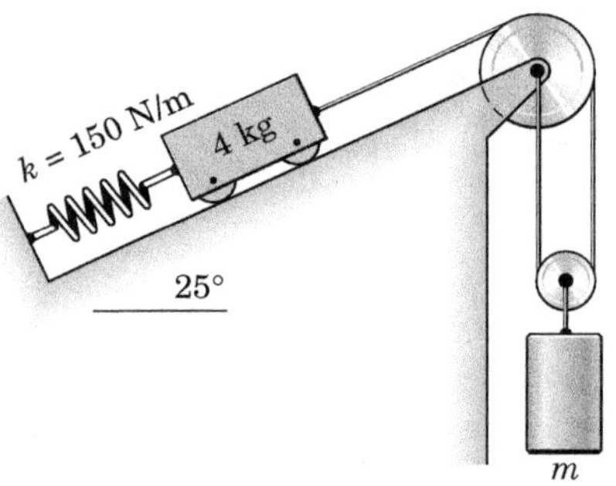

Problem 3/131

3/132 It is experimentally determined that the drive wheels of a car must exert a tractive force of 560 N on the road surface in order to maintain a steady vehicle speed of 90 km/h on a horizontal road. If it is known that the overall drivetrain efficiency is $e_m = 0.70$, determine the required motor power output P.

3/133 Once under way at a steady speed, the 1000-kg elevator A rises at the rate of 1 story (3 m) per second. Determine the power input P_{in} into the motor unit M if the combined mechanical and electrical efficiency of the system is $e = 0.8$.

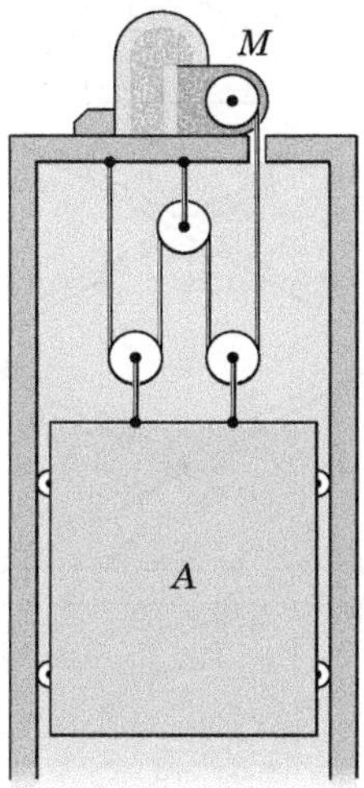

Problem 3/133

3/134 Calculate the horizontal velocity v with which the 48-lb carriage must strike the spring in order to compress it a maximum of 4 in. The spring is known as a "hardening" spring, since its stiffness increases with deflection as shown in the accompanying graph.

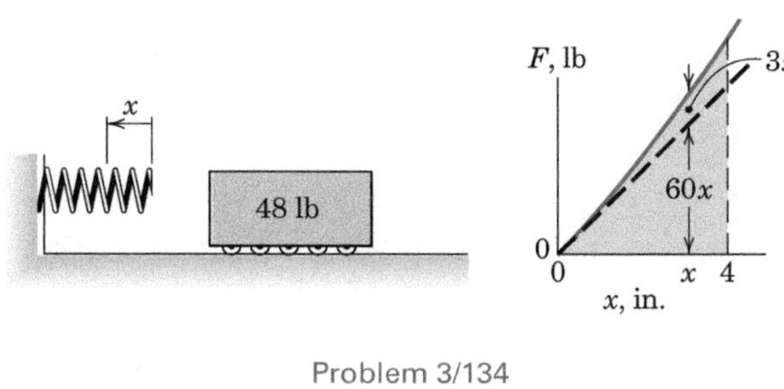

Problem 3/134

3/135 The 6-kg cylinder is released from rest in the position shown and falls on the spring, which has been initially precompressed 50 mm by the light strap and restraining wires. If the stiffness of the spring is 4 kN/m, compute the additional deflection δ of the spring produced by the falling cylinder before it rebounds.

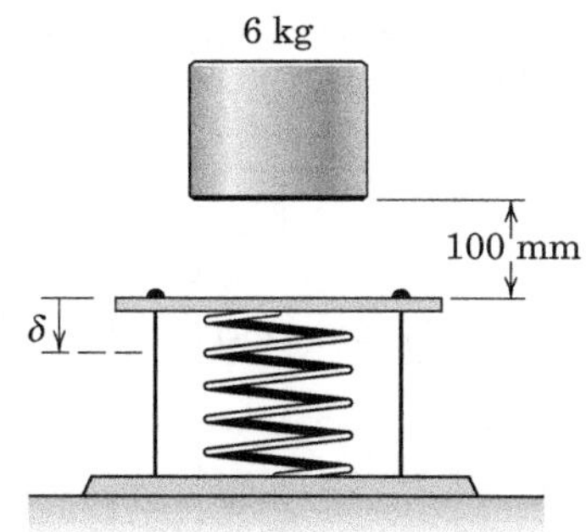

Problem 3/135

3/136 The nest of two springs is used to bring the 0.5-kg plunger A to a stop from a speed of 5 m/s and reverse its direction of motion. The inner spring increases the deceleration, and the adjustment of its position is used to control the exact point at which the reversal takes place. If this point is to correspond to a maximum deflection $\delta = 200$ mm for the outer spring, specify the adjustment of the inner spring by determining the distance s. The outer spring has a stiffness of 300 N/m and the inner one a stiffness of 150 N/m.

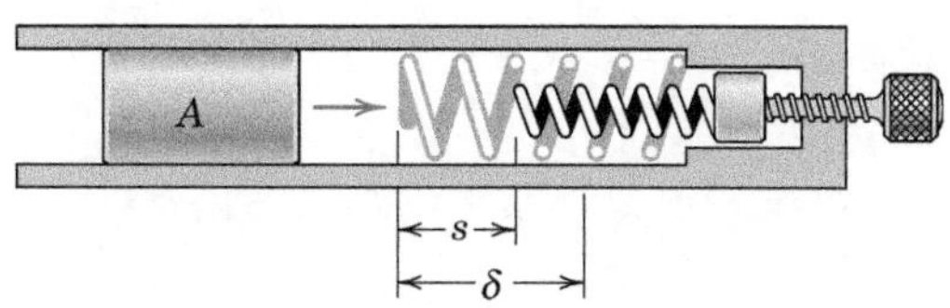

Problem 3/136

3/137 Extensive testing of an experimental 2000-lb automobile reveals the aerodynamic drag force F_D and the total nonaerodynamic rolling-resistance force F_R to be as shown in the plot. Determine (*a*) the power required for steady speeds of 30 and 60 mi/hr on a level road, (*b*) the power required for a steady speed of 60 mi/hr both up and down a 6-percent incline, and (*c*) the steady speed at which no power is required going down the 6-percent incline.

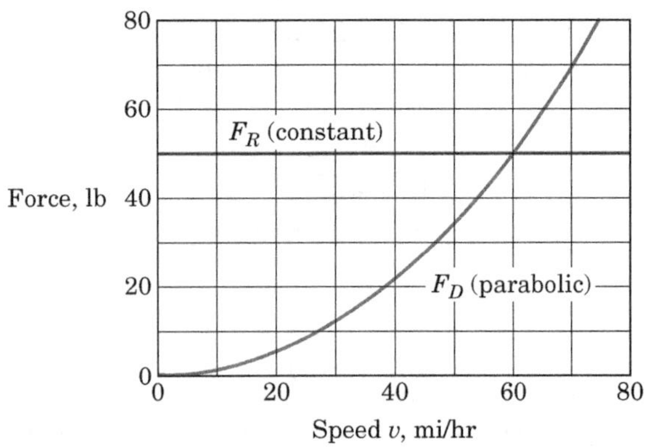

Problem 3/137

3/138 The vertical motion of the 50-lb block is controlled by the two forces P applied to the ends A and B of the linkage, where A and B are constrained to move in the horizontal guide. If forces $P = 250$ lb are applied with the linkage initially at rest with $\theta = 60°$, determine the upward velocity v of the block as θ approaches 180°. Neglect friction and the weight of the links and note that P is greater than its equilibrium value of $(5W/2) \cot 30° = 217$ lb.

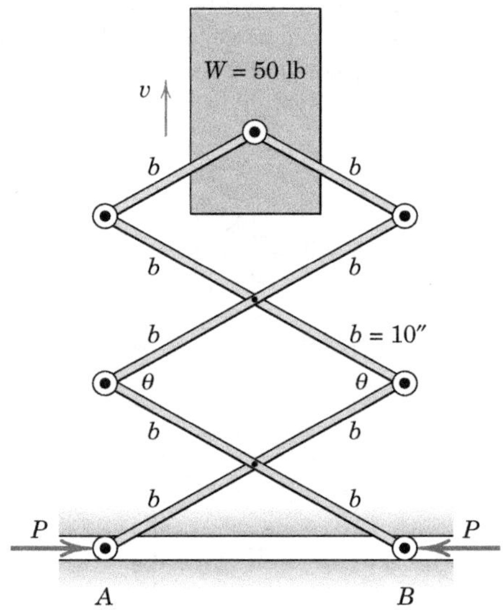

Problem 3/138

3/7 Potential Energy

In the previous article on work and kinetic energy, we isolated a particle or a combination of joined particles and determined the work done by gravity forces, spring forces, and other externally applied forces acting on the particle or system. We did this to evaluate U in the work-energy equation. In the present article we will introduce the concept of *potential energy* to treat the work done by gravity forces and by spring forces. This concept will simplify the analysis of many problems.

Gravitational Potential Energy

We consider first the motion of a particle of mass m in close proximity to the surface of the earth, where the gravitational attraction (weight) mg is essentially constant, Fig. 3/8a. The *gravitational potential energy* V_g of the particle is defined as the work mgh done *against* the gravitational field to elevate the particle a distance h above some arbitrary reference plane (called a *datum*), where V_g is taken to be zero. Thus, we write the potential energy as

$$\boxed{V_g = mgh} \qquad \textbf{(3/18)}$$

This work is called potential energy because it may be converted into energy if the particle is allowed to do work on a supporting body while it returns to its lower original datum plane. In going from one level at $h = h_1$ to a higher level at $h = h_2$, the *change* in potential energy becomes

$$\Delta V_g = mg(h_2 - h_1) = mg\Delta h$$

The corresponding work done *by* the gravitational force on the particle is $-mg\Delta h$. Thus, the work done by the gravitational force is the negative of the change in potential energy.

When large changes in altitude in the field of the earth are encountered, Fig. 3/8b, the gravitational force $Gmm_e/r^2 = mgR^2/r^2$ is no longer constant. The work done *against* this force to change the radial position of the particle from r_1 to r_2 is the change $(V_g)_2 - (V_g)_1$ in gravitational potential energy, which is

$$\int_{r_1}^{r_2} mgR^2 \frac{dr}{r^2} = mgR^2\left(\frac{1}{r_1} - \frac{1}{r_2}\right) = (V_g)_2 - (V_g)_1$$

It is customary to take $(V_g)_2 = 0$ when $r_2 = \infty$, so that with this datum we have

$$\boxed{V_g = -\frac{mgR^2}{r}} \qquad \textbf{(3/19)}$$

In going from r_1 to r_2, the corresponding change in potential energy is

$$\Delta V_g = mgR^2\left(\frac{1}{r_1} - \frac{1}{r_2}\right)$$

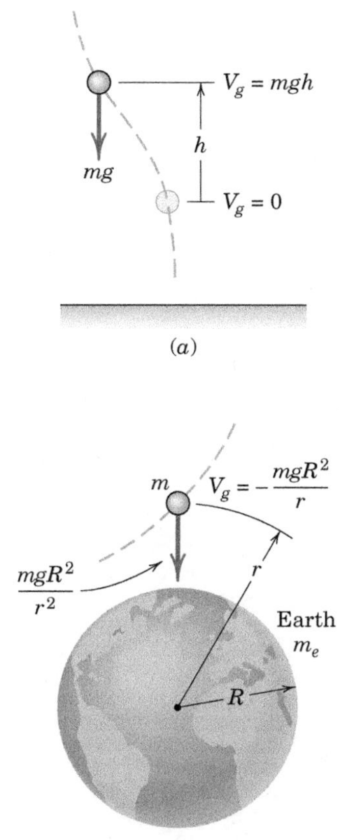

Figure 3/8

which, again, is the *negative* of the work done *by* the gravitational force. We note that the potential energy of a given particle depends only on its position, h or r, and not on the particular path it followed in reaching that position.

Elastic Potential Energy

The second example of potential energy occurs in the deformation of an elastic body, such as a spring. The work which is done on the spring to deform it is stored in the spring and is called its *elastic potential energy* V_e. This energy is recoverable in the form of work done by the spring on the body attached to its movable end during the release of the deformation of the spring. For the one-dimensional linear spring of stiffness k, which we discussed in Art. 3/6 and illustrated in Fig. 3/5, the force supported by the spring at any deformation x, tensile or compressive, from its undeformed position is $F = kx$. Thus, we define the elastic potential energy of the spring as the work done on it to deform it an amount x, and we have

$$V_e = \int_0^x kx\,dx = \frac{1}{2}kx^2 \qquad \textbf{(3/20)}$$

If the deformation, either tensile or compressive, of a spring increases from x_1 to x_2 during the motion, then the change in potential energy of the spring is its final value minus its initial value or

$$\Delta V_e = \frac{1}{2}k({x_2}^2 - {x_1}^2)$$

which is positive. Conversely, if the deformation of a spring decreases during the motion interval, then the change in potential energy of the spring becomes negative. The magnitude of these changes is represented by the shaded trapezoidal area in the F-x diagram of Fig. 3/5*a*.

Because the force exerted *on* the spring *by* the moving body is equal and opposite to the force F exerted *by* the spring *on* the body, it follows that the work done on the spring is the negative of the work done on the body. Therefore, we may replace the work U done by the spring on the body by $-\Delta V_e$, the negative of the potential energy change for the spring, provided the spring is now included within the system.

Work-Energy Equation

With the elastic member included in the system, we now modify the work-energy equation to account for the potential-energy terms. If $U'_{1\text{-}2}$ stands for the work of all external forces *other than* gravitational forces and spring forces, we may write Eq. 3/15 as $U'_{1\text{-}2} + (-\Delta V_g) + (-\Delta V_e) = \Delta T$ or

$$U'_{1\text{-}2} = \Delta T + \Delta V \qquad \textbf{(3/21)}$$

where ΔV is the change in total potential energy, gravitational plus elastic.

This alternative form of the work-energy equation is often far more convenient to use than Eq. 3/15, since the work of both gravity and spring forces is accounted for by focusing attention on the end-point positions of the particle and on the end-point lengths of the elastic spring. The path followed between these end-point positions is of no consequence in the evaluation of ΔV_g and ΔV_e.

Note that Eq. 3/21 may be rewritten in the equivalent form

$$\boxed{T_1 + V_1 + U'_{1\text{-}2} = T_2 + V_2} \qquad \textbf{(3/21}\boldsymbol{a}\textbf{)}$$

To help clarify the difference between the use of Eqs. 3/15 and 3/21, Fig. 3/9 shows schematically a particle of mass m constrained to move along a fixed path under the action of forces F_1 and F_2, the gravitational force $W = mg$, the spring force F, and the normal reaction N. In Fig. 3/9*b*, the particle is isolated with its free-body diagram. The work done by each of the forces F_1, F_2, W, and the spring force $F = kx$ is evaluated, say, from A to B, and equated to the change ΔT in kinetic energy using Eq. 3/15. The constraint reaction N, if normal to the path, will do no work. The alternative approach is shown in Fig. 3/9*c*, where the spring is included as a part of the isolated system. The work done during the interval by F_1 and F_2 is the $U'_{1\text{-}2}$-term of Eq. 3/21 with the changes in elastic and gravitational potential energies included on the energy side of the equation.

We note with the first approach that the work done by $F = kx$ could require a somewhat awkward integration to account for the changes in magnitude and direction of F as the particle moves from A

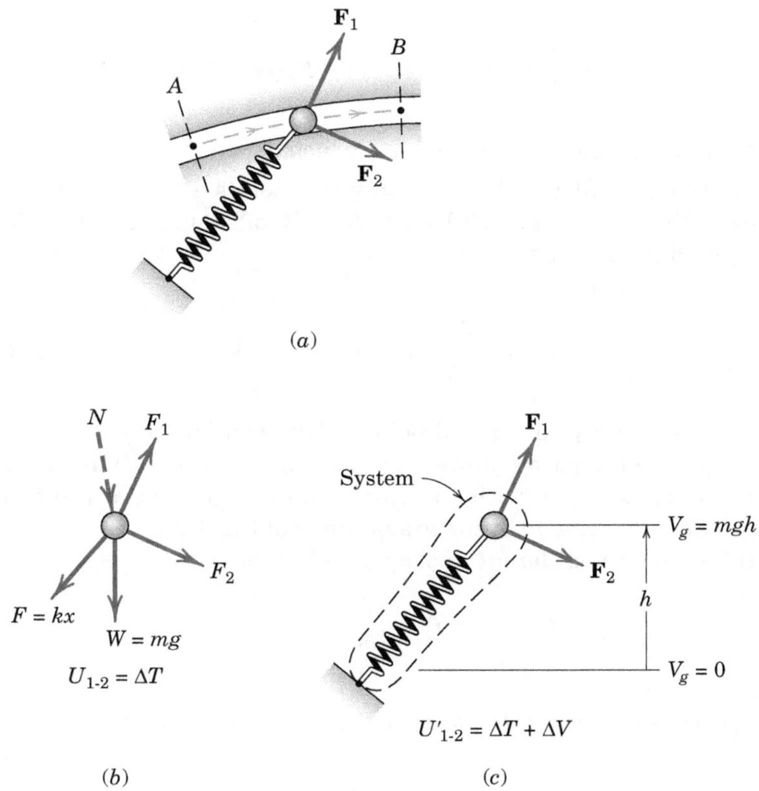

Figure 3/9

to B. With the second approach, however, only the initial and final lengths of the spring are required to evaluate ΔV_e. This greatly simplifies the calculation.

For problems where the only forces are gravitational, elastic, and nonworking constraint forces, the U'-term of Eq. 3/21*a* is zero, and the energy equation becomes

$$T_1 + V_1 = T_2 + V_2 \qquad \text{or} \qquad E_1 = E_2 \tag{3/22}$$

where $E = T + V$ is the total mechanical energy of the particle and its attached spring. When E is constant, we see that transfers of energy between kinetic and potential may take place as long as the total mechanical energy $T + V$ does not change. Equation 3/22 expresses the *law of conservation of dynamical energy*.

Conservative Force Fields*

We have observed that the work done against a gravitational or an elastic force depends only on the net change of position and not on the particular path followed in reaching the new position. Forces with this characteristic are associated with *conservative force fields*, which possess an important mathematical property.

Consider a force field where the force $\mathbf{F}$ is a function of the coordinates, Fig. 3/10. The work done by $\mathbf{F}$ during a displacement $d\mathbf{r}$ of its point of application is $dU = \mathbf{F}\cdot d\mathbf{r}$. The total work done along its path from 1 to 2 is

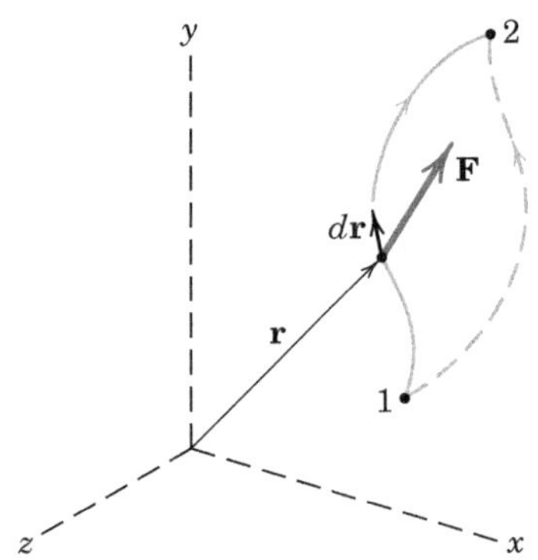

Figure 3/10

$$U = \int \mathbf{F}\cdot d\mathbf{r} = \int (F_x\,dx + F_y\,dy + F_z\,dz)$$

The integral $\int \mathbf{F}\cdot d\mathbf{r}$ is a line integral which depends, in general, on the particular path followed between any two points 1 and 2 in space. If, however, $\mathbf{F}\cdot d\mathbf{r}$ is an *exact differential*† $-dV$ of some scalar function V of the coordinates, then

$$U_{1\text{-}2} = \int_{V_1}^{V_2} -dV = -(V_2 - V_1) \tag{3/23}$$

which depends only on the end points of the motion and which is thus *independent* of the path followed. The minus sign before dV is arbitrary but is chosen to agree with the customary designation of the sign of potential energy change in the gravity field of the earth.

If V exists, the differential change in V becomes

$$dV = \frac{\partial V}{\partial x}dx + \frac{\partial V}{\partial y}dy + \frac{\partial V}{\partial z}dz$$

*Optional.

†Recall that a function $d\Phi = P\,dx + Q\,dy + R\,dz$ is an exact differential in the coordinates x-y-z if

$$\frac{\partial P}{\partial y} = \frac{\partial Q}{\partial x} \qquad \frac{\partial P}{\partial z} = \frac{\partial R}{\partial x} \qquad \frac{\partial Q}{\partial z} = \frac{\partial R}{\partial y}$$

Comparison with $-dV = \mathbf{F}\cdot d\mathbf{r} = F_x\,dx + F_y\,dy + F_z\,dz$ gives us

$$F_x = -\frac{\partial V}{\partial x} \qquad F_y = -\frac{\partial V}{\partial y} \qquad F_z = -\frac{\partial V}{\partial z}$$

The force may also be written as the vector

$$\mathbf{F} = -\boldsymbol{\nabla} V \tag{3/24}$$

where the symbol $\boldsymbol{\nabla}$ stands for the vector operator “del”, which is

$$\boldsymbol{\nabla} = \mathbf{i}\frac{\partial}{\partial x} + \mathbf{j}\frac{\partial}{\partial y} + \mathbf{k}\frac{\partial}{\partial z}$$

The quantity V is known as the *potential function*, and the expression $\boldsymbol{\nabla} V$ is known as the *gradient of the potential function*.

When force components are derivable from a potential as described, the force is said to be *conservative*, and the work done by $\mathbf{F}$ between any two points is independent of the path followed.

Sample Problem 3/16

The 6-lb slider is released from rest at position 1 and slides with negligible friction in a vertical plane along the circular rod. The attached spring has a stiffness of 2 lb/in. and has an unstretched length of 24 in. Determine the velocity of the slider as it passes position 2.

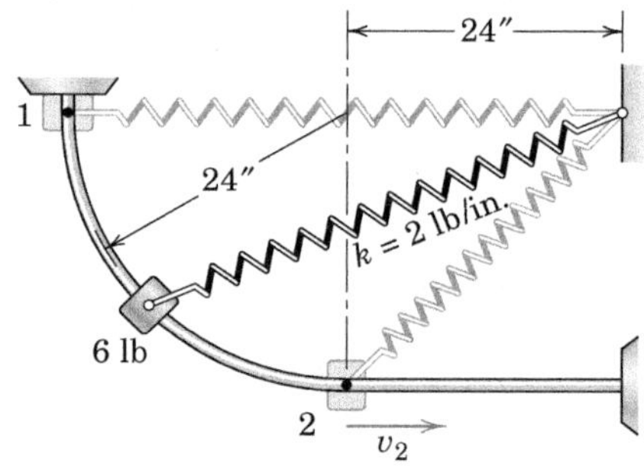

Solution. The work done by the weight and the spring force on the slider will be treated using potential-energy methods. The reaction of the rod on the slider is normal to the motion and does no work. Hence, $U'_{1\text{-}2} = 0$. We define the datum to be at the level of position 1, so that the gravitational potential energies are

① $$V_1 = 0$$

$$V_2 = -mgh = -6\left(\frac{24}{12}\right) = -12 \text{ ft-lb}$$

The initial and final elastic (spring) potential energies are

$$V_1 = \frac{1}{2}kx_1^2 = \frac{1}{2}(2)(12)\left(\frac{24}{12}\right)^2 = 48 \text{ ft-lb}$$

$$V_2 = \frac{1}{2}kx_2^2 = \frac{1}{2}(2)(12)\left(\frac{24\sqrt{2}}{12} - \frac{24}{12}\right)^2 = 8.24 \text{ ft-lb}$$

Substitution into the alternative work-energy equation yields

$$[T_1 + V_1 + U'_{1\text{-}2} = T_2 + V_2] \qquad 0 + 48 + 0 = \frac{1}{2}\left(\frac{6}{32.2}\right)v_2^2 - 12 + 8.24$$

$$v_2 = 23.6 \text{ ft/sec} \qquad \textit{Ans.}$$

Helpful Hint

① Note that if we evaluated the work done by the spring force acting on the slider by means of the integral $\int \mathbf{F}\cdot d\mathbf{r}$, it would necessitate a lengthy computation to account for the change in the magnitude of the force, along with the change in the angle between the force and the tangent to the path. Note further that v_2 depends only on the end conditions of the motion and does not require knowledge of the shape of the path.

Sample Problem 3/17

The 10-kg slider moves with negligible friction up the inclined guide. The attached spring has a stiffness of 60 N/m and is stretched 0.6 m in position A, where the slider is released from rest. The 250-N force is constant and the pulley offers negligible resistance to the motion of the cord. Calculate the velocity v_C of the slider as it passes point C.

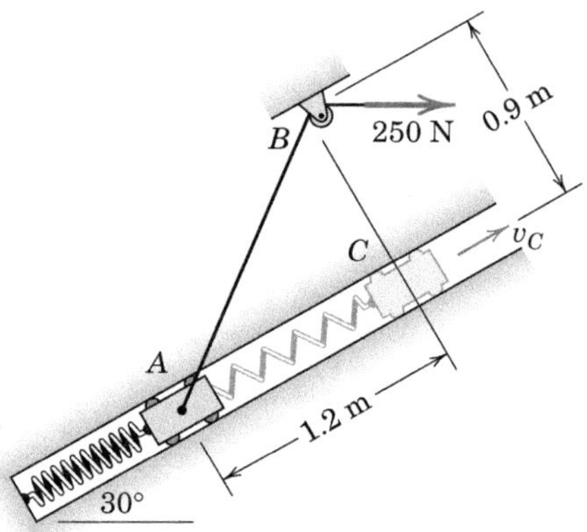

Solution. The slider and inextensible cord together with the attached spring will be analyzed as a system, which permits the use of Eq. 3/21a. The only nonpotential force doing work on this system is the 250-N tension applied to the cord. While the slider moves from A to C, the point of application of the 250-N force moves a distance of $\overline{AB} - \overline{BC}$ or $1.5 - 0.9 = 0.6$ m.

① ② $$U'_{A\text{-}C} = 250(0.6) = 150 \text{ J}$$

We define a datum at position A so that the initial and final gravitational potential energies are

$$V_A = 0 \qquad V_C = mgh = 10(9.81)(1.2 \sin 30°) = 58.9 \text{ J}$$

The initial and final elastic potential energies are

$$V_A = \frac{1}{2}kx_A^2 = \frac{1}{2}(60)(0.6)^2 = 10.8 \text{ J}$$

$$V_C = \frac{1}{2}kx_B^2 = \frac{1}{2}60(0.6 + 1.2)^2 = 97.2 \text{ J}$$

Substitution into the alternative work-energy equation 3/21a gives

$$[T_A + V_A + U'_{A\text{-}C} = T_C + V_C] \qquad 0 + 0 + 10.8 + 150 = \frac{1}{2}(10)v_C^2 + 58.9 + 97.2$$

$$v_C = 0.974 \text{ m/s} \qquad \textit{Ans.}$$

Helpful Hints

① Do not hesitate to use subscripts tailored to the problem at hand. Here we use A and C rather than 1 and 2.

② The reactions of the guides on the slider are normal to the direction of motion and do no work.

Sample Problem 3/18

The system shown is released from rest with the lightweight slender bar OA in the vertical position shown. The torsional spring at O is undeflected in the initial position and exerts a restoring moment of magnitude $k_T\theta$ on the bar, where θ is the counterclockwise angular deflection of the bar. The string S is attached to point C of the bar and slips without friction through a vertical hole in the support surface. For the values $m_A = 2$ kg, $m_B = 4$ kg, $L = 0.5$ m, and $k_T = 13$ N·m/rad:

(*a*) Determine the speed v_A of particle A when θ reaches 90°.

(*b*) Plot v_A as a function of θ over the range $0 \le \theta \le 90°$. Identify the maximum value of v_A and the value of θ at which this maximum occurs.

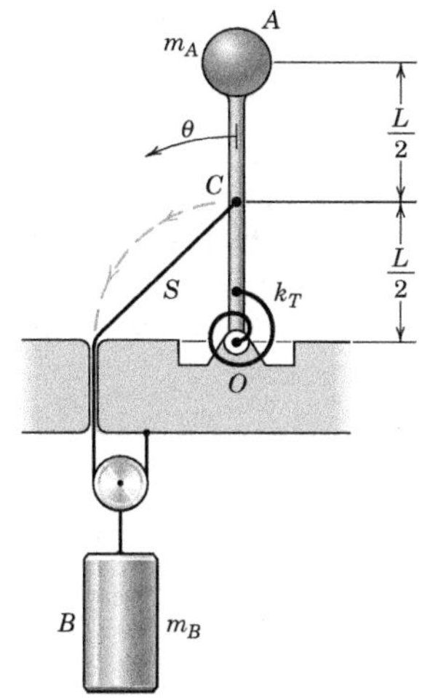

Solution (a). We begin by establishing a general relationship for the potential energy associated with the deflection of a torsional spring. Recalling that the change in potential energy is the work done on the spring to deform it, we write

$$V_e = \int_0^\theta k_T\theta\, d\theta = \frac{1}{2}k_T\theta^2$$

We also need to establish the relationship between v_A and v_B when $\theta = 90°$. Noting that the speed of point C is always $v_A/2$, and further noting that the speed of cylinder B is one-half the speed of point C at $\theta = 90°$, we conclude that at $\theta = 90°$,

$$v_B = \frac{1}{4}v_A$$

Establishing datums at the initial altitudes of bodies A and B, and with state 1 at $\theta = 0$ and state 2 at $\theta = 90°$, we write

$[T_1 + V_1 + U'_{1\text{-}2} = T_2 + V_2]$

① $$0 + 0 + 0 = \frac{1}{2}m_Av_A^2 + \frac{1}{2}m_Bv_B^2 - m_AgL - m_Bg\left(\frac{L\sqrt{2}}{4}\right) + \frac{1}{2}k_T\left(\frac{\pi}{2}\right)^2$$

With numbers:

$$0 = \frac{1}{2}(2)v_A^2 + \frac{1}{2}(4)\left(\frac{v_A}{4}\right)^2 - 2(9.81)(0.5) - 4(9.81)\left(\frac{0.5\sqrt{2}}{4}\right) + \frac{1}{2}(13)\left(\frac{\pi}{2}\right)^2$$

Solving, $\quad v_A = 0.794$ m/s $\quad$ *Ans.*

(b). We leave our definition of the initial state 1 as is, but now redefine state 2 to be associated with an arbitrary value of θ. From the accompanying diagram constructed for an arbitrary value of θ, we see that the speed of cylinder B can be written as

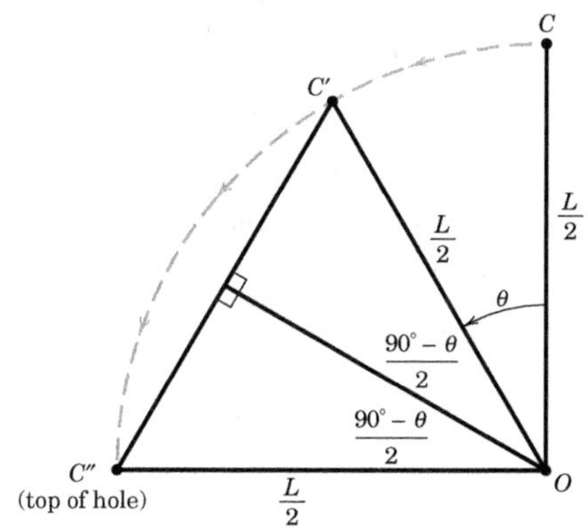

② $$v_B = \frac{1}{2}\left|\frac{d}{dt}(\overline{C'C''})\right| = \frac{1}{2}\left|\frac{d}{dt}\left[2\frac{L}{2}\sin\left(\frac{90° - \theta}{2}\right)\right]\right|$$

$$= \frac{1}{2}\left|L\left(-\frac{\dot\theta}{2}\right)\cos\left(\frac{90° - \theta}{2}\right)\right| = \frac{L\dot\theta}{4}\cos\left(\frac{90° - \theta}{2}\right)$$

Finally, because $v_A = L\dot\theta$, $\quad v_B = \dfrac{v_A}{4}\cos\left(\dfrac{90° - \theta}{2}\right)$

$[T_1 + V_1 + U'_{1\text{-}2} = T_2 + V_2]$

$$0 + 0 + 0 = \frac{1}{2}m_Av_A^2 + \frac{1}{2}m_B\left[\frac{v_A}{4}\cos\left(\frac{90° - \theta}{2}\right)\right]^2 - m_AgL(1 - \cos\theta)$$

$$- m_Bg\left(\frac{1}{2}\right)\left[\frac{L\sqrt{2}}{2} - 2\frac{L}{2}\sin\left(\frac{90° - \theta}{2}\right)\right] + \frac{1}{2}k_T\theta^2$$

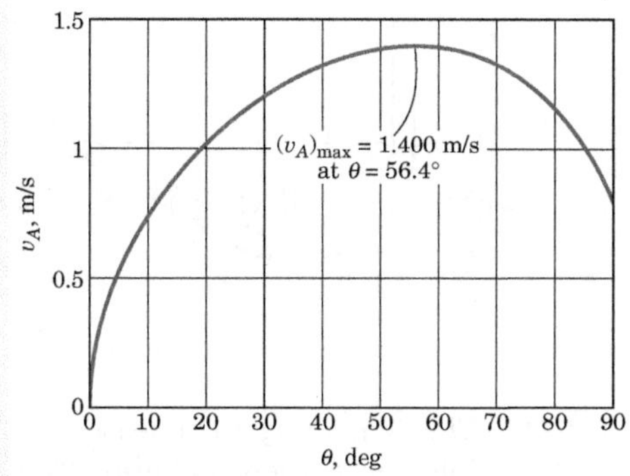

Upon substitution of the given quantities, we vary θ to produce the plot of v_A versus θ. The maximum value of v_A is seen to be

$(v_A)_{max} = 1.400$ m/s at $\theta = 56.4°$ $\quad$ *Ans.*

Helpful Hints

① Note that mass B will move downward by one-half of the length of string initially above the supporting surface. This downward distance is $\frac{1}{2}\left(\frac{L}{2}\sqrt{2}\right) = \frac{L\sqrt{2}}{4}$.

② The absolute-value signs reflect the fact that v_B is known to be positive.

PROBLEMS

Introductory Problems

3/139 The two particles of equal mass are joined by a rod of negligible mass. If they are released from rest in the position shown and slide on the smooth guide in the vertical plane, calculate their velocity v when A reaches B's position and B is at B'.

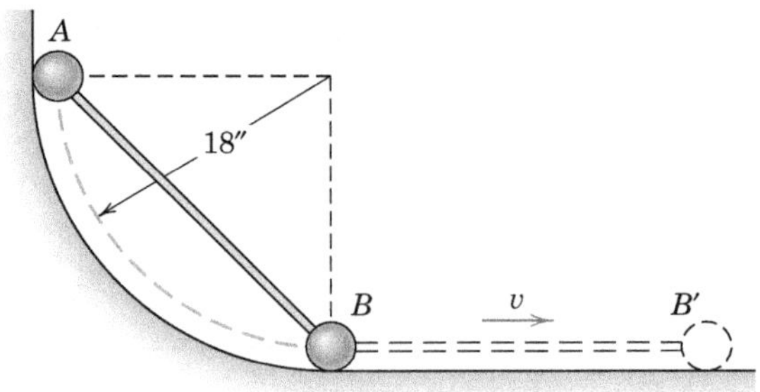

Problem 3/139

3/140 The 1.2-kg slider is released from rest in position A and slides without friction along the vertical-plane guide shown. Determine (*a*) the speed v_B of the slider as it passes position B and (*b*) the maximum deflection δ of the spring.

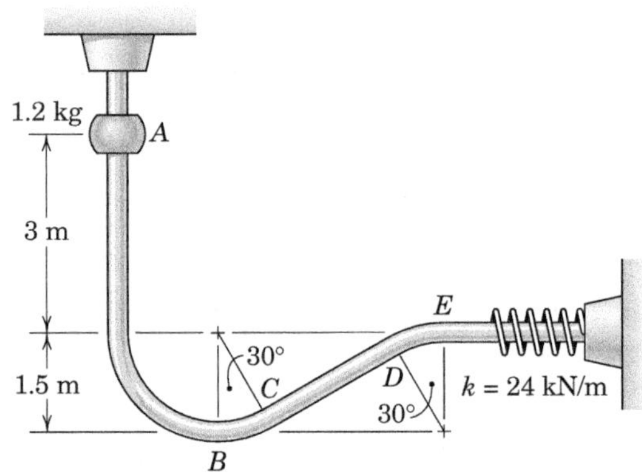

Problem 3/140

3/141 The 2-kg plunger is released from rest in the position shown where the spring of stiffness $k = 500$ N/m has been compressed to one-half its uncompressed length of 200 mm. Calculate the maximum height h above the starting position reached by the plunger.

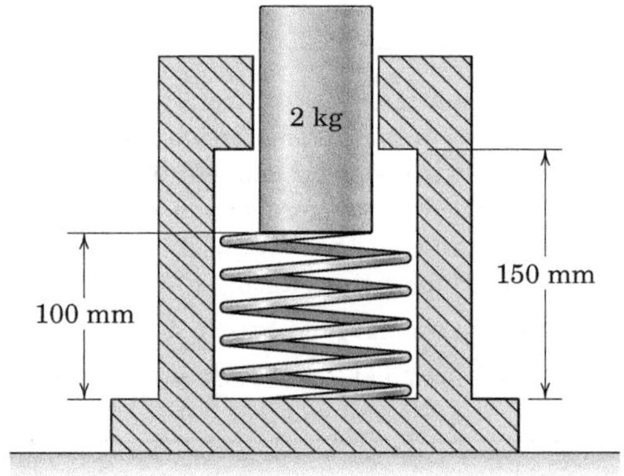

Problem 3/141

3/142 The system is released from rest with the spring initially stretched 3 in. Calculate the velocity v of the cylinder after it has dropped 0.5 in. The spring has a stiffness of 6 lb/in. Neglect the mass of the small pulley.

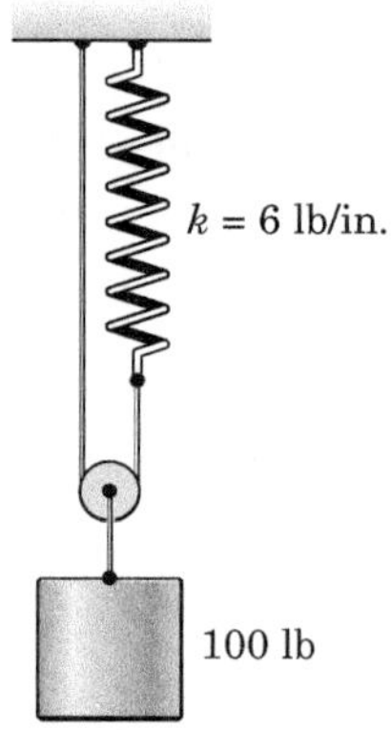

Problem 3/142

3/143 The 3-lb collar is released from rest at A and slides freely down the inclined rod. If the spring constant $k = 4$ lb/ft and the unstretched length of the spring is 50 in., determine the speed of the collar as it passes point B.

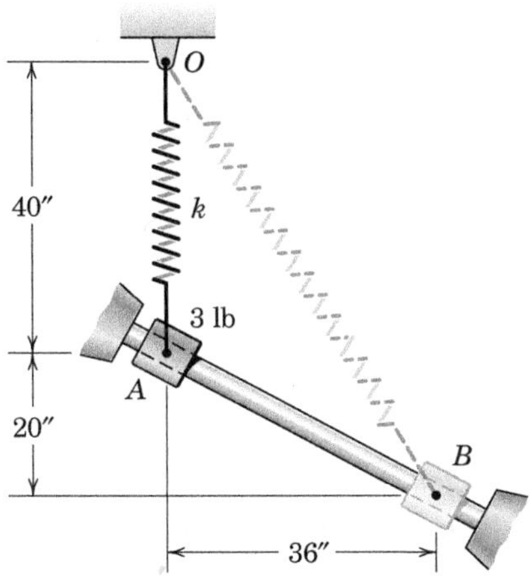

Problem 3/143

3/144 Determine the unstretched spring length which would cause the 3-lb collar of the previous problem to have no speed as it arrives at position B. All other conditions of the previous problem remain the same.

3/145 A bead with a mass of 0.25 kg is released from rest at A and slides down and around the fixed smooth wire. Determine the force N between the wire and the bead as it passes point B.

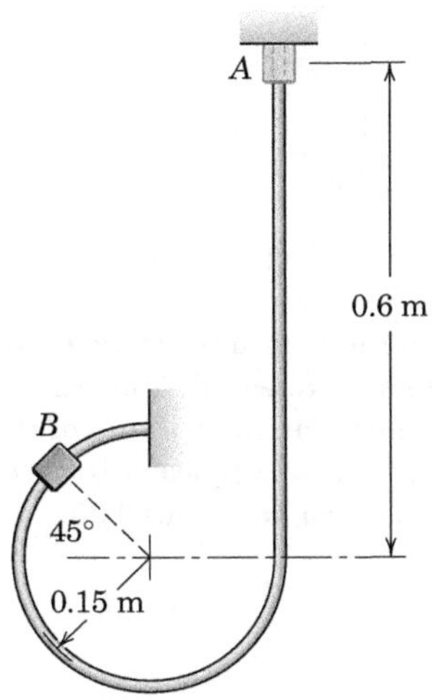

Problem 3/145

3/146 The 0.8-kg particle is attached to the system of two light rigid bars, all of which move in a vertical plane. The spring is compressed an amount $b/2$ when $\theta = 0$, and the length $b = 0.30$ m. The system is released from rest in a position slightly above that for $\theta = 0$. (*a*) If the maximum value of θ is observed to be 50°, determine the spring constant k. (*b*) For $k = 400$ N/m, determine the speed v of the particle when $\theta = 25°$. Also find the corresponding value of $\dot{\theta}$.

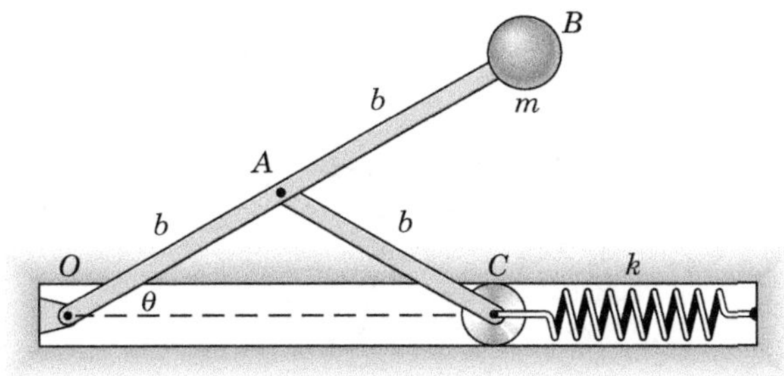

Problem 3/146

3/147 The light rod is pivoted at O and carries the 8- and 10-lb particles. If the rod is released from rest at $\theta = 30°$ and swings in the vertical plane, calculate (*a*) the velocity v of the 8-lb particle just before it hits the spring and (*b*) the maximum compression x of the spring. Assume that x is small so that the position of the rod when the spring is compressed is essentially vertical.

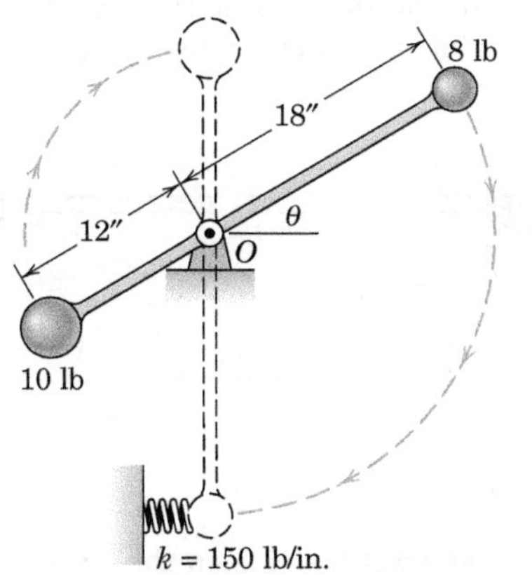

Problem 3/147

3/148 The two springs, each of stiffness $k = 1.2$ kN/m, are of equal length and undeformed when $\theta = 0$. If the mechanism is released from rest in the position $\theta = 20°$, determine its angular velocity $\dot{\theta}$ when $\theta = 0$. The mass m of each sphere is 3 kg. Treat the spheres as particles and neglect the masses of the light rods and springs.

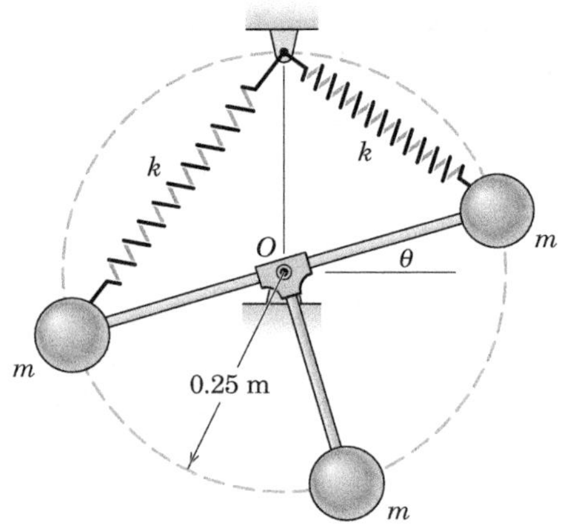

Problem 3/148

3/149 The particle of mass $m = 1.2$ kg is attached to the end of the light rigid bar of length $L = 0.6$ m. The system is released from rest while in the horizontal position shown, at which the torsional spring is undeflected. The bar is then observed to rotate 30° before stopping momentarily. (*a*) Determine the value of the torsional spring constant k_T. (*b*) For this value of k_T, determine the speed v of the particle when $\theta = 15°$.

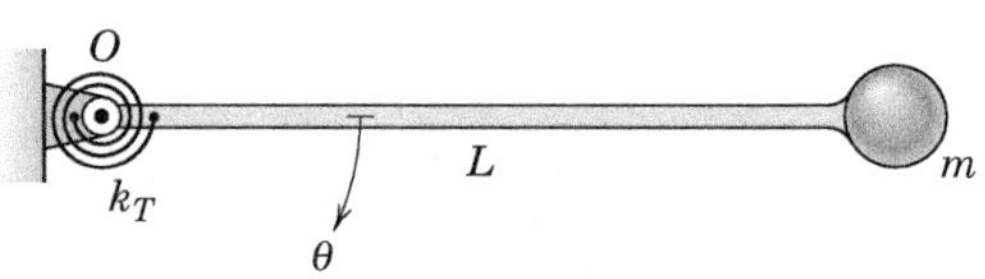

Problem 3/149

Representative Problems

3/150 The 10-kg collar slides on the smooth vertical rod and has a velocity $v_1 = 2$ m/s in position A where each spring is stretched 0.1 m. Calculate the velocity v_2 of the collar as it passes point B.

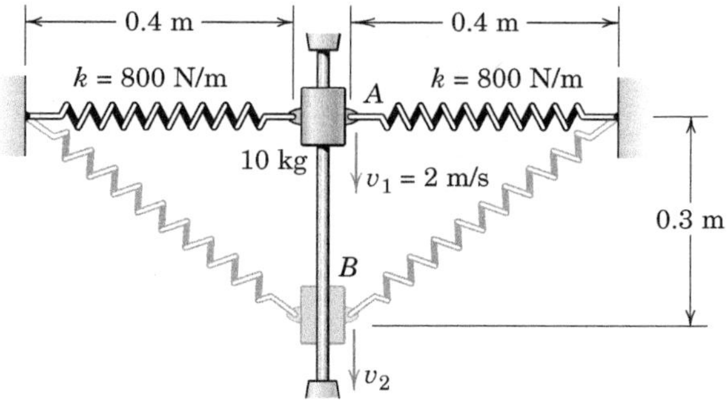

Problem 3/150

3/151 The system is released from rest with the spring initially stretched 2 in. Calculate the velocity of the 100-lb cylinder after it has dropped 6 in. Also determine the maximum drop distance of the cylinder. Neglect the mass and friction of the pulleys.

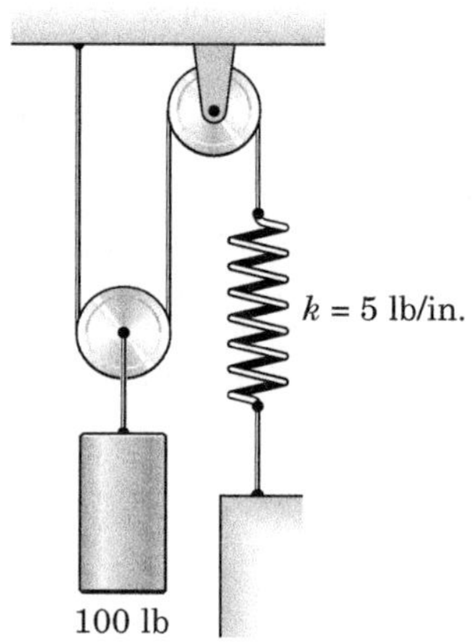

Problem 3/151

3/152 The spring has an unstretched length of 25 in. If the system is released from rest in the position shown, determine the speed v of the ball (*a*) when it has dropped a vertical distance of 10 in. and (*b*) when the rod has rotated 35°.

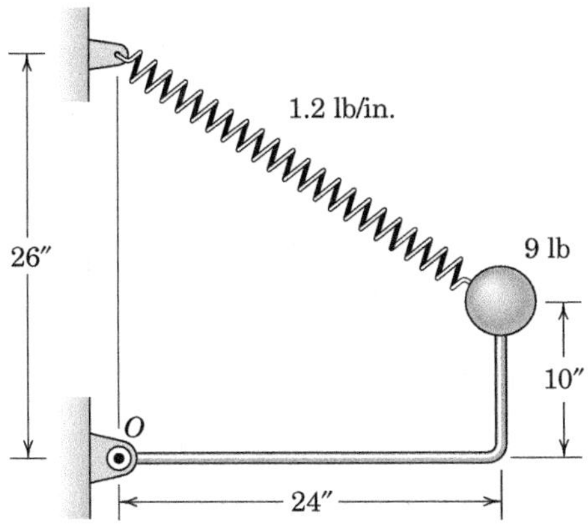

Problem 3/152

3/153 The two wheels consisting of hoops and spokes of negligible mass rotate about their respective centers and are pressed together sufficiently to prevent any slipping. The 3-lb and 2-lb eccentric masses are mounted on the rims of the wheels. If the wheels are given a slight nudge from rest in the equilibrium positions shown, compute the angular velocity $\dot{\theta}$ of the larger of the two wheels when it has revolved through a quarter of a revolution and put the eccentric masses in the dashed positions shown. Note that the angular velocity of the small wheel is twice that of the large wheel. Neglect any friction in the wheel bearings.

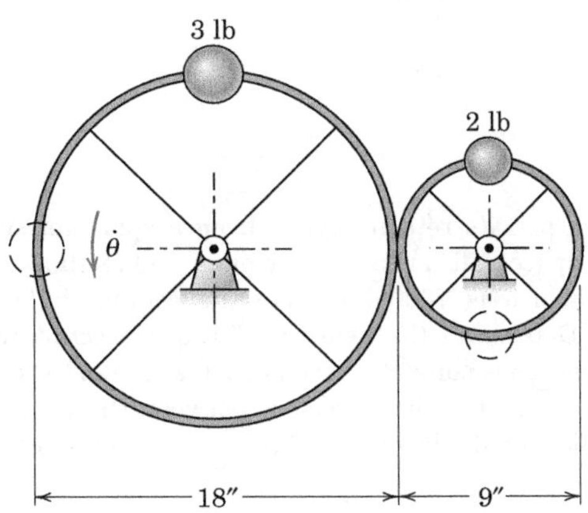

Problem 3/153

3/154 The slider of mass m is released from rest in position A and slides without friction along the vertical-plane guide shown. Determine the height h such that the normal force exerted by the guide on the slider is zero as the slider passes point C. For this value of h, determine the normal force as the slider passes point B.

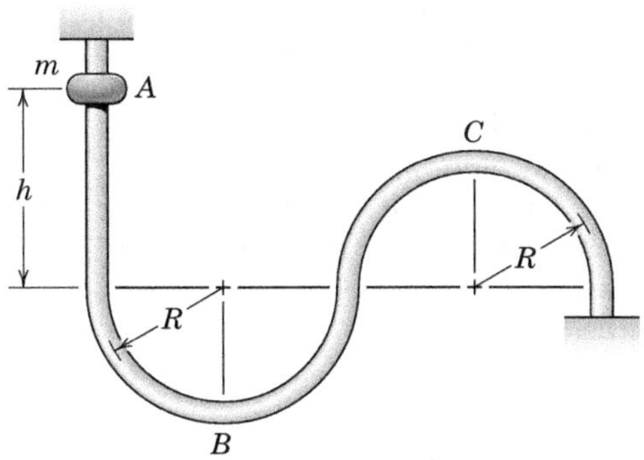

Problem 3/154

3/155 The two 1.5-kg spheres are released from rest and gently nudged outward from the position $\theta = 0$ and then rotate in a vertical plane about the fixed centers of their attached gears, thus maintaining the same angle θ for both rods. Determine the velocity v of each sphere as the rods pass the position $\theta = 30°$. The spring is unstretched when $\theta = 0$, and the masses of the two identical rods and the two gear wheels may be neglected.

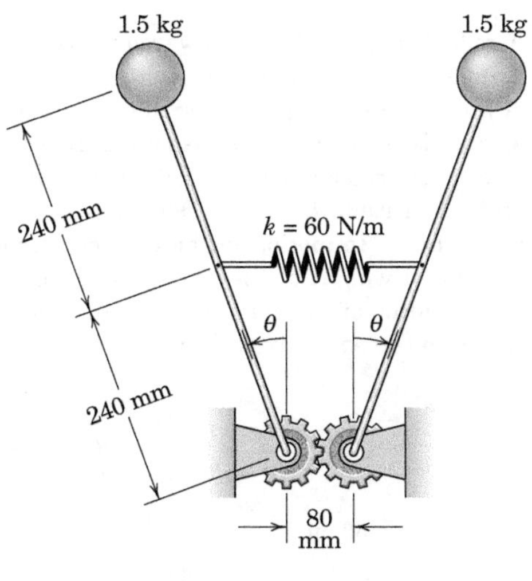

Problem 3/155

3/156 In the design of an inside loop for an amusement-park ride, it is desired to maintain the same centripetal acceleration throughout the loop. Assume negligible loss of energy during the motion and determine the radius of curvature ρ of the path as a function of the height y above the low point A, where the velocity and radius of curvature are v_0 and ρ_0, respectively. For a given value of ρ_0, what is the minimum value of v_0 for which the vehicle will not leave the track at the top of the loop?

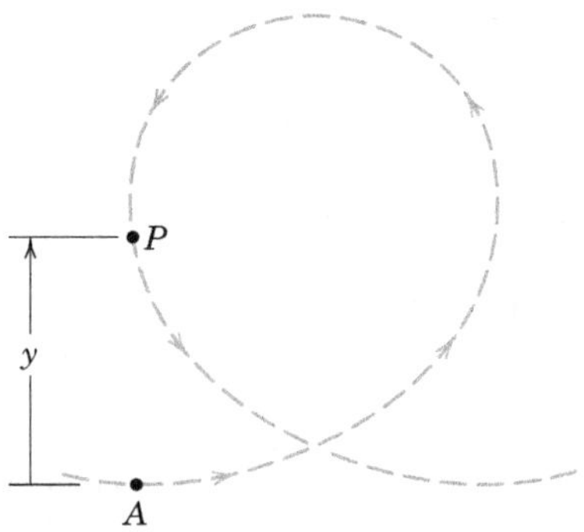

Problem 3/156

3/157 A rocket launches an unpowered space capsule at point A with an absolute velocity $v_A = 8000$ mi/hr at an altitude of 25 mi. After the capsule has traveled a distance of 250 mi measured along its absolute space trajectory, its velocity at B is 7600 mi/hr and its altitude is 50 mi. Determine the average resistance P to motion in the rarified atmosphere. The earth weight of the capsule is 48 lb, and the mean radius of the earth is 3959 mi. Consider the center of the earth fixed in space.

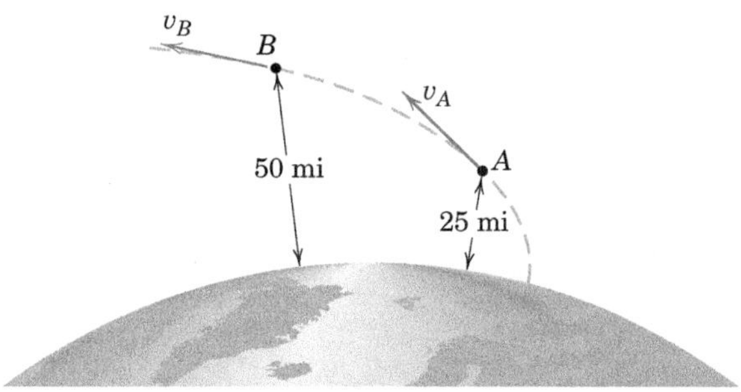

Problem 3/157

3/158 The projectile of Prob. 3/124 is repeated here. By the method of this article, determine the vertical launch velocity v_0 which will result in a maximum altitude of $R/3$. The launch is from the north pole and aerodynamic drag can be neglected. Use $g = 9.825$ m/s^2 as the surface-level acceleration due to gravity.

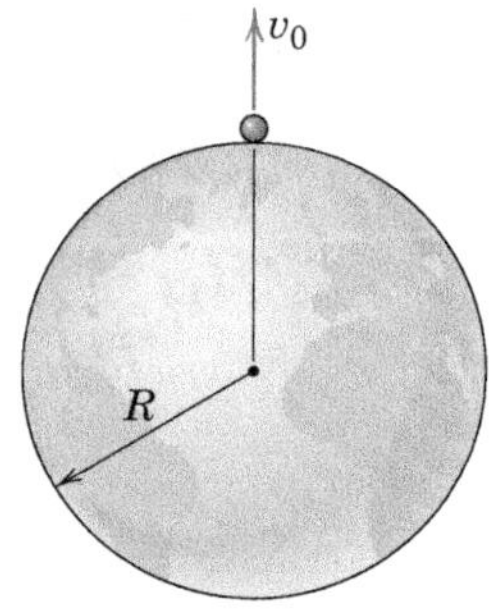

Problem 3/158

3/159 The small bodies A and B each of mass m are connected and supported by the pivoted links of negligible mass. If A is released from rest in the position shown, calculate its velocity v_A as it crosses the vertical centerline. Neglect any friction.

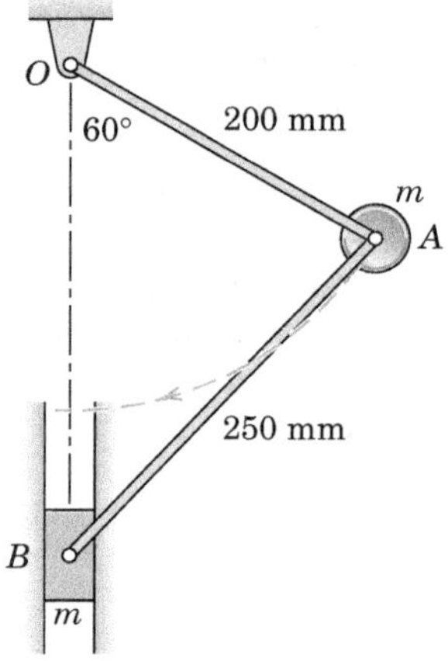

Problem 3/159

3/160 Upon its return voyage from a space mission, the spacecraft has a velocity of 24 000 km/h at point A, which is 7000 km from the center of the earth. Determine the velocity of the spacecraft when it reaches point B, which is 6500 km from the center of the earth. The trajectory between these two points is outside the effect of the earth's atmosphere.

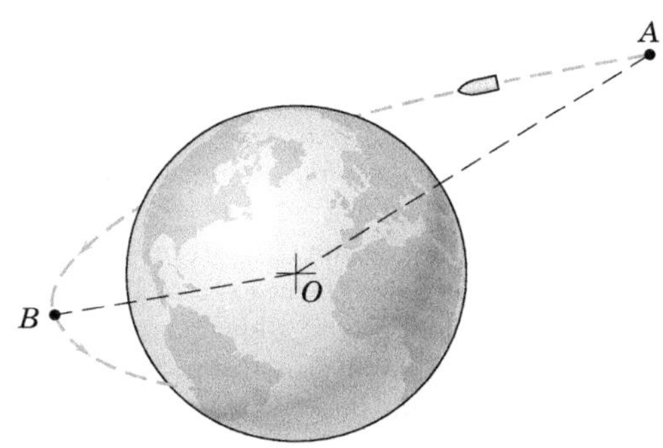

Problem 3/160

3/161 A 175-lb pole vaulter carrying a uniform 16-ft, 10-lb pole approaches the jump with a velocity v and manages to barely clear the bar set at a height of 18 ft. As he clears the bar, his velocity and that of the pole are essentially zero. Calculate the minimum possible value of v required for him to make the jump. Both the horizontal pole and the center of gravity of the vaulter are 42 in. above the ground during the approach.

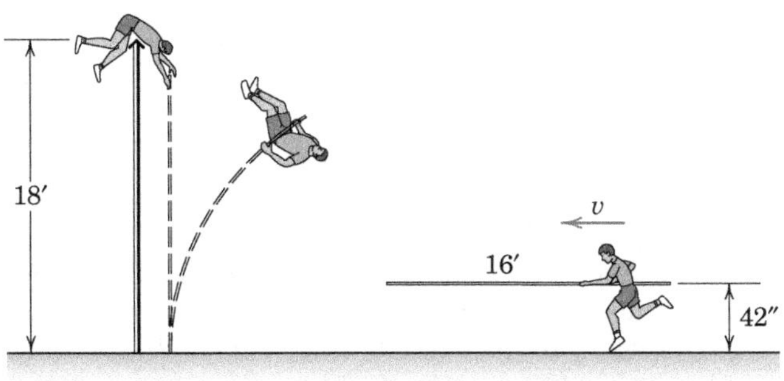

Problem 3/161

3/162 When the mechanism is released from rest in the position where $\theta = 60°$, the 4-kg carriage drops and the 6-kg sphere rises. Determine the velocity v of the sphere when $\theta = 180°$. Neglect the mass of the links and treat the sphere as a particle.

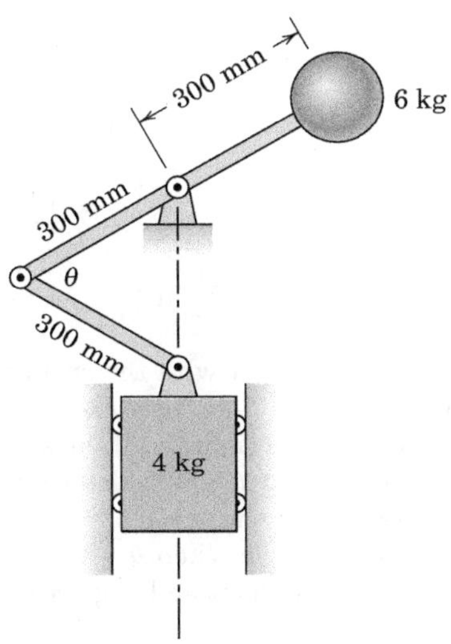

Problem 3/162

3/163 The cars of an amusement-park ride have a speed $v_1 = 90$ km/h at the lowest part of the track. Determine their speed v_2 at the highest part of the track. Neglect energy loss due to friction. (*Caution:* Give careful thought to the change in potential energy of the system of cars.)

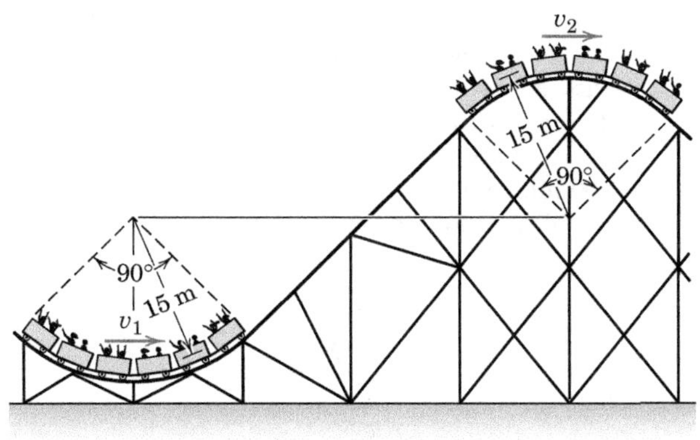

Problem 3/163

3/164 A satellite is put into an elliptical orbit around the earth and has a velocity v_P at the perigee position P. Determine the expression for the velocity v_A at the apogee position A. The radii to A and P are, respectively, r_A and r_P. Note that the total energy remains constant.

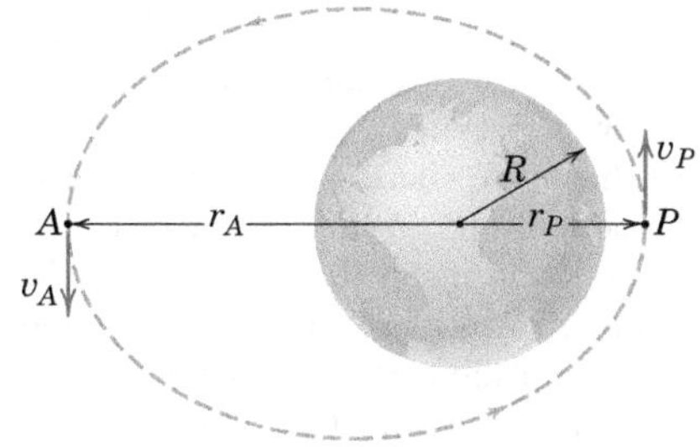

Problem 3/164

3/165 Calculate the maximum velocity of slider B if the system is released from rest with $x = y$. Motion is in the vertical plane. Assume that friction is negligible. The sliders have equal masses, and the motion is restricted to $y \geq 0$.

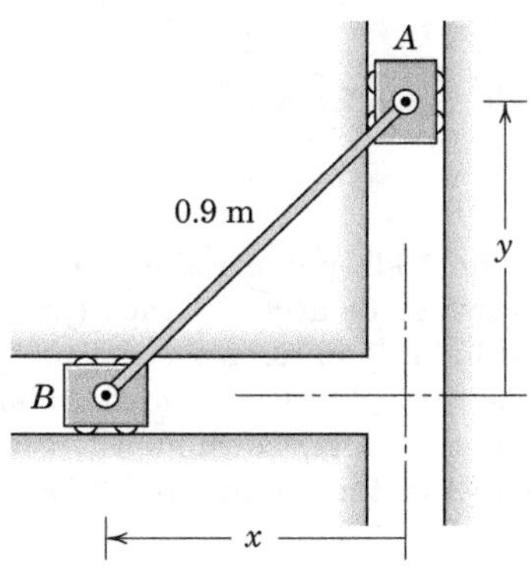

Problem 3/165

3/166 The system is initially moving with the cable taut, the 10-kg block moving down the rough incline with a speed of 0.3 m/s, and the spring stretched 25 mm. By the method of this article, (*a*) determine the velocity v of the block after it has traveled 100 mm, and (*b*) calculate the distance traveled by the block before it comes to rest.

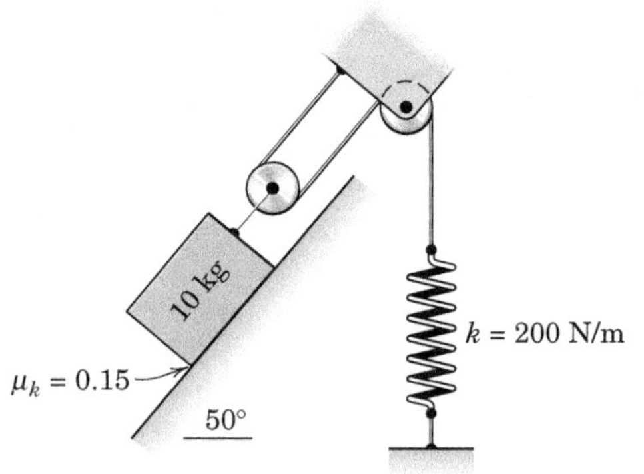

Problem 3/166

3/167 A spacecraft m is heading toward the center of the moon with a velocity of 2000 mi/hr at a distance from the moon's surface equal to the radius R of the moon. Compute the impact velocity v with the surface of the moon if the spacecraft is unable to fire its retro-rockets. Consider the moon fixed in space. The radius R of the moon is 1080 mi, and the acceleration due to gravity at its surface is 5.32 ft/sec^2.

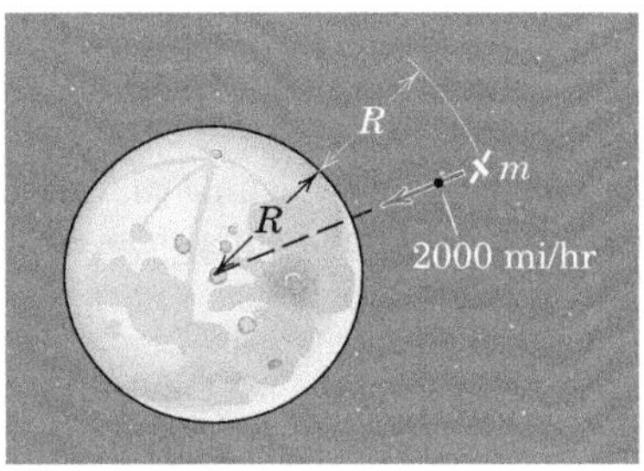

Problem 3/167

3/168 When the 10-lb plunger is released from rest in its vertical guide at $\theta = 0$, each spring of stiffness $k = 20$ lb/in. is uncompressed. The links are free to slide through their pivoted collars and compress their springs. Calculate the velocity v of the plunger when the position $\theta = 30°$ is passed.

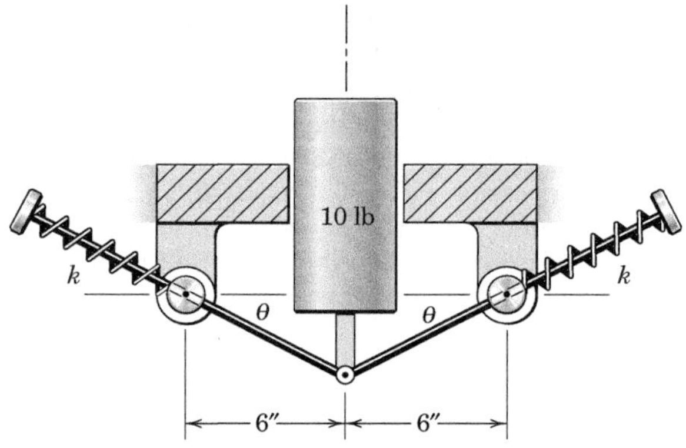

Problem 3/168

3/169 The system is released from rest with the angle $\theta = 90°$. Determine $\dot{\theta}$ when θ reaches 60°. Use the values $m_1 = 1$ kg, $m_2 = 1.25$ kg, and $b = 0.40$ m. Neglect friction and the mass of bar OB, and treat the body B as a particle.

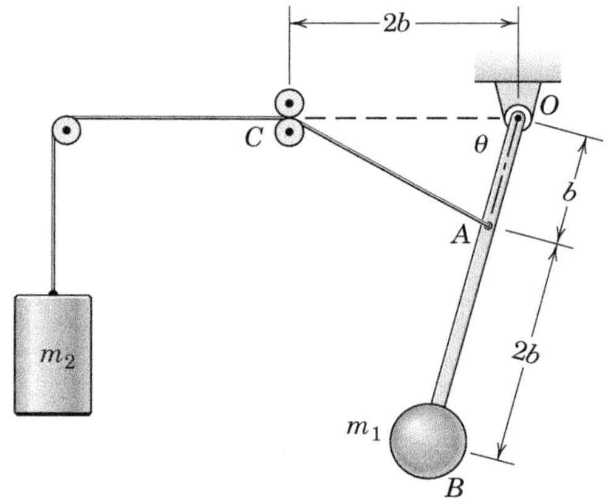

Problem 3/169

3/170 The system is at rest with the spring unstretched when $\theta = 0$. The 3-kg particle is then given a slight nudge to the right. (*a*) If the system comes to momentary rest at $\theta = 40°$, determine the spring constant k. (*b*) For the value $k = 100$ N/m, find the speed of the particle when $\theta = 25°$. Use the value $b = 0.40$ m throughout and neglect friction.

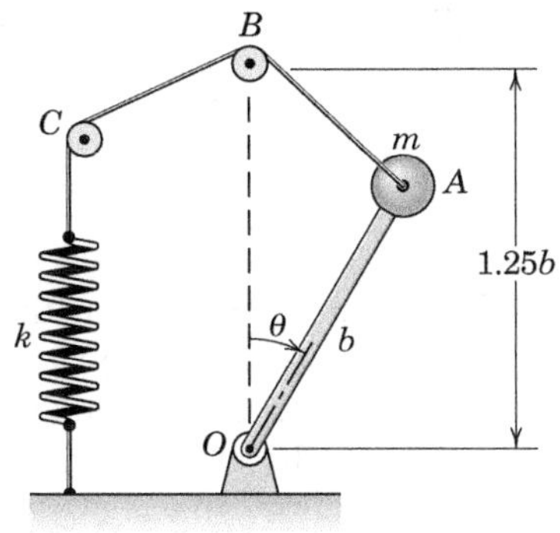

Problem 3/170

3/171 The 0.6-kg slider is released from rest at A and slides down the smooth parabolic guide (which lies in a vertical plane) under the influence of its own weight and of the spring of constant 120 N/m. Determine the speed of the slider as it passes point B and the corresponding normal force exerted on it by the guide. The unstretched length of the spring is 200 mm.

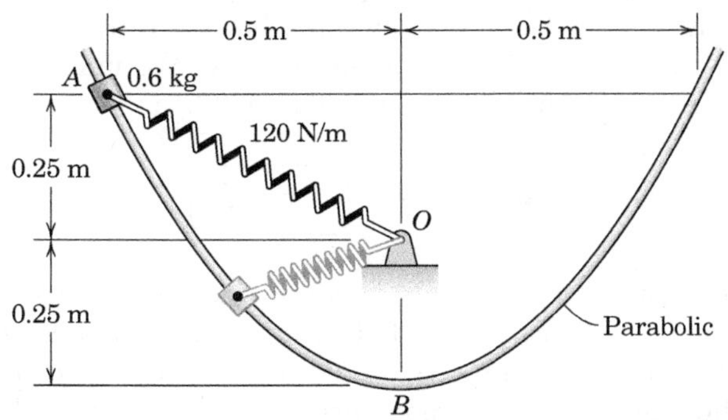

Problem 3/171

►3/172 The two particles of mass m and $2m$, respectively, are connected by a rigid rod of negligible mass and slide with negligible friction in a circular path of radius r on the inside of the vertical circular ring. If the unit is released from rest at $\theta = 0$, determine (a) the velocity v of the particles when the rod passes the horizontal position, (b) the maximum velocity v_{max} of the particles, and (c) the maximum value of θ.

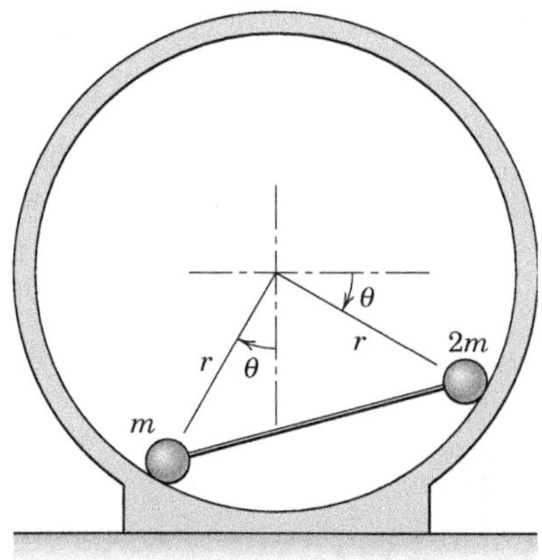

Problem 3/172

SECTION C IMPULSE AND MOMENTUM

3/8 Introduction

In the previous two articles, we focused attention on the equations of work and energy, which are obtained by integrating the equation of motion $\mathbf{F} = m\mathbf{a}$ with respect to the displacement of the particle. We found that the velocity changes could be expressed directly in terms of the work done or in terms of the overall changes in energy. In the next two articles, we will integrate the equation of motion with respect to time rather than displacement. This approach leads to the equations of impulse and momentum. These equations greatly facilitate the solution of many problems in which the applied forces act during extremely short periods of time (as in impact problems) or over specified intervals of time.

3/9 Linear Impulse and Linear Momentum

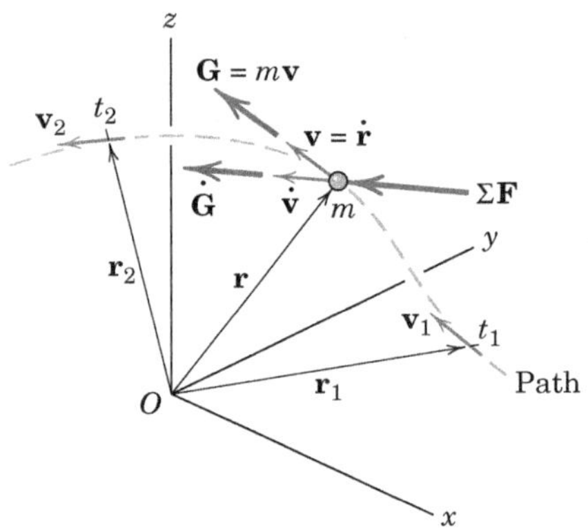

Figure 3/11

Consider again the general curvilinear motion in space of a particle of mass m, Fig. 3/11, where the particle is located by its position vector $\mathbf{r}$ measured from a fixed origin O. The velocity of the particle is $\mathbf{v} = \dot{\mathbf{r}}$ and is tangent to its path (shown as a dashed line). The resultant $\Sigma\mathbf{F}$ of all forces on m is in the direction of its acceleration $\dot{\mathbf{v}}$. We may now write the basic equation of motion for the particle, Eq. 3/3, as

$$\Sigma\mathbf{F} = m\dot{\mathbf{v}} = \frac{d}{dt}(m\mathbf{v}) \qquad \text{or} \qquad \boxed{\Sigma\mathbf{F} = \dot{\mathbf{G}}} \qquad \textbf{(3/25)}$$

where the product of the mass and velocity is defined as the *linear momentum* $\mathbf{G} = m\mathbf{v}$ of the particle. Equation 3/25 states that *the resultant of all forces acting on a particle equals its time rate of change of linear momentum*. In SI the units of linear momentum $m\mathbf{v}$ are seen to be kg·m/s, which also equals N·s. In U.S. customary units, the units of linear momentum $m\mathbf{v}$ are [lb/(ft/sec^2)] [ft/sec] = lb-sec.

Because Eq. 3/25 is a vector equation, we recognize that, in addition to the equality of the magnitudes of $\Sigma\mathbf{F}$ and $\dot{\mathbf{G}}$, the direction of the resultant force coincides with the direction of the rate of change in linear momentum, which is the direction of the rate of change in velocity. Equation 3/25 is one of the most useful and important relationships in dynamics, and it is valid as long as the mass m of the particle is not changing with time. The case where m changes with time is discussed in Art. 4/7 of Chapter 4.

We now write the three scalar components of Eq. 3/25 as

$$\Sigma F_x = \dot{G}_x \qquad \Sigma F_y = \dot{G}_y \qquad \Sigma F_z = \dot{G}_z \qquad \textbf{(3/26)}$$

These equations may be applied independently of one another.

The Linear Impulse-Momentum Principle

All that we have done so far in this article is to rewrite Newton's second law in an alternative form in terms of momentum. But we are now able to describe the effect of the resultant force $\Sigma\mathbf{F}$ on the linear momentum of the particle over a finite period of time simply by integrating Eq. 3/25 with respect to the time t. Multiplying the equation by dt gives $\Sigma\mathbf{F}\,dt = d\mathbf{G}$, which we integrate from time t_1 to time t_2 to obtain

$$\int_{t_1}^{t_2} \Sigma\mathbf{F}\,dt = \mathbf{G}_2 - \mathbf{G}_1 = \Delta\mathbf{G} \qquad \textbf{(3/27)}$$

Here the linear momentum at time t_2 is $\mathbf{G}_2 = m\mathbf{v}_2$ and the linear momentum at time t_1 is $\mathbf{G}_1 = m\mathbf{v}_1$. The product of force and time is defined as the *linear impulse* of the force, and Eq. 3/27 states that *the total linear impulse on m equals the corresponding change in linear momentum of m.*

Alternatively, we may write Eq. 3/27 as

$$\mathbf{G}_1 + \int_{t_1}^{t_2} \Sigma\mathbf{F}\,dt = \mathbf{G}_2 \qquad \textbf{(3/27}\boldsymbol{a}\textbf{)}$$

which says that the initial linear momentum of the body plus the linear impulse applied to it equals its final linear momentum.

The impulse integral is a vector which, in general, may involve changes in both magnitude and direction during the time interval. Under these conditions, it will be necessary to express $\Sigma\mathbf{F}$ and $\mathbf{G}$ in component form and then combine the integrated components. The components of Eq. 3/27*a* are the scalar equations

$$\begin{aligned} m(v_1)_x + \int_{t_1}^{t_2} \Sigma F_x\,dt &= m(v_2)_x \\ m(v_1)_y + \int_{t_1}^{t_2} \Sigma F_y\,dt &= m(v_2)_y \\ m(v_1)_z + \int_{t_1}^{t_2} \Sigma F_z\,dt &= m(v_2)_z \end{aligned} \qquad \textbf{(3/27}\boldsymbol{b}\textbf{)}$$

These three scalar impulse-momentum equations are completely independent.

Whereas Eq. 3/27 clearly stresses that the external linear impulse causes a change in the linear momentum, the order of the terms in Eqs. 3/27*a* and 3/27*b* corresponds to the natural sequence of events. While the form of Eq. 3/27 may be best for the experienced dynamicist, the form of Eqs. 3/27*a* and 3/27*b* is very effective for the beginner.

We now introduce the concept of the *impulse-momentum diagram.* Once the body to be analyzed has been clearly identified and isolated, we construct three drawings of the body as shown in Fig. 3/12. In the first drawing, we show the initial momentum $m\mathbf{v}_1$, or components thereof. In

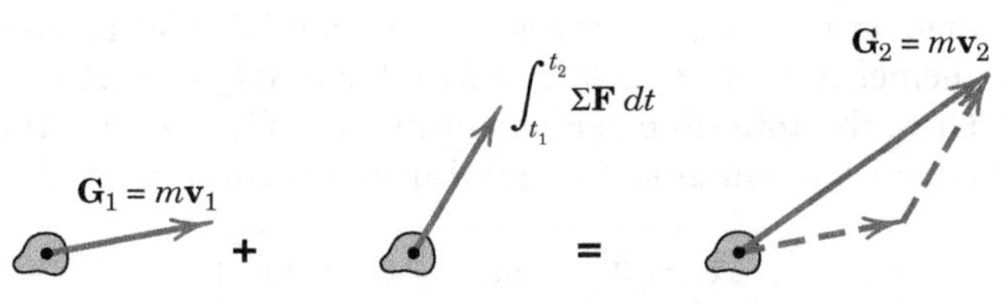

Figure 3/12

The impact force exerted by the bat on this ball will usually be much larger than the weight of the ball.

the second or middle drawing, we show all the external linear impulses (or components thereof). In the final drawing, we show the final linear momentum $m\mathbf{v}_2$ (or its components). The writing of the impulse-momentum equations 3/27*b* then follows directly from these drawings, with a clear one-to-one correspondence between diagrams and equation terms.

We note that the center diagram is very much like a free-body diagram, except that the impulses of the forces appear rather than the forces themselves. As with the free-body diagram, it is necessary to include the effects of *all* forces acting on the body, except those forces whose magnitudes are negligible.

In some cases, certain forces are very large and of short duration. Such forces are called *impulsive forces*. An example is a force of sharp impact. We frequently assume that impulsive forces are constant over their time of duration, so that they can be brought outside the linear-impulse integral. In addition, we frequently assume that *nonimpulsive forces* can be neglected in comparison with impulsive forces. An example of a nonimpulsive force is the weight of a baseball during its collision with a bat—the weight of the ball (about 5 oz) is small compared with the force (which could be several hundred pounds in magnitude) exerted on the ball by the bat.

There are cases where a force acting on a particle varies with the time in a manner determined by experimental measurements or by other approximate means. In this case a graphical or numerical integration must be performed. If, for example, a force F acting on a particle in a given direction varies with the time t as indicated in Fig. 3/13, then the impulse, $\int_{t_1}^{t_2} F\,dt$, of this force from t_1 to t_2 is the shaded area under the curve.

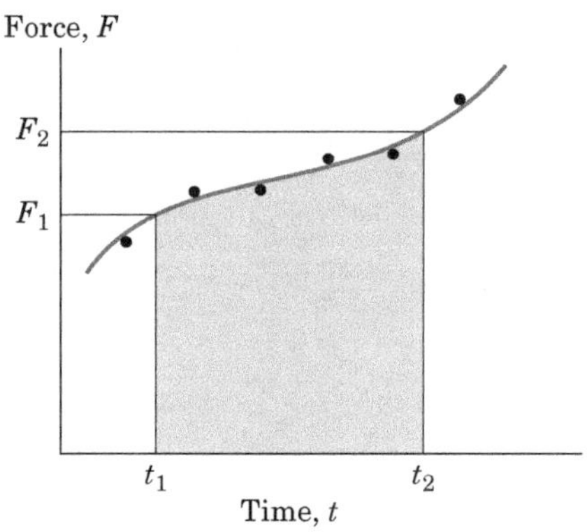

Figure 3/13

Conservation of Linear Momentum

If the resultant force on a particle is zero during an interval of time, we see that Eq. 3/25 requires that its linear momentum $\mathbf{G}$ remain constant. In this case, the linear momentum of the particle is said to be *conserved*. Linear momentum may be conserved in one coordinate direction, such as x, but not necessarily in the y- or z-direction. A careful examination of the impulse-momentum diagram of the particle will disclose whether the total linear impulse on the particle in a particular direction is zero. If it is, the corresponding linear momentum is unchanged (conserved) in that direction.

Consider now the motion of two particles a and b which interact during an interval of time. If the interactive forces $\mathbf{F}$ and $-\mathbf{F}$ between them are the only unbalanced forces acting on the particles during the interval, it follows that the linear impulse on particle a is the negative of the linear impulse on particle b. Therefore, from Eq. 3/27, the change in linear momentum $\Delta\mathbf{G}_a$ of particle a is the negative of the change $\Delta\mathbf{G}_b$ in linear momentum of particle b. So we have $\Delta\mathbf{G}_a = -\Delta\mathbf{G}_b$ or $\Delta(\mathbf{G}_a + \mathbf{G}_b) = \mathbf{0}$. Thus, the total linear momentum $\mathbf{G} = \mathbf{G}_a + \mathbf{G}_b$ for the system of the two particles remains constant during the interval, and we write

$$\boxed{\Delta\mathbf{G} = \mathbf{0} \qquad \text{or} \qquad \mathbf{G}_1 = \mathbf{G}_2} \qquad \textbf{(3/28)}$$

Equation 3/28 expresses the *principle of conservation of linear momentum*.

Sample Problem 3/19

A tennis player strikes the tennis ball with her racket when the ball is at the uppermost point of its trajectory as shown. The horizontal velocity of the ball just before impact with the racket is $v_1 = 50$ ft/sec, and just after impact its velocity is $v_2 = 70$ ft/sec directed at the 15° angle as shown. If the 2-oz ball is in contact with the racket for 0.02 sec, determine the magnitude of the average force **R** exerted by the racket on the ball. Also determine the angle β made by **R** with the horizontal.

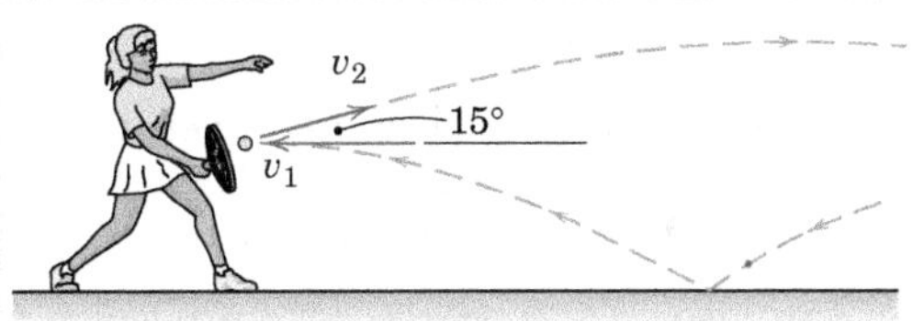

Solution. We construct the impulse-momentum diagrams for the ball as follows:

①

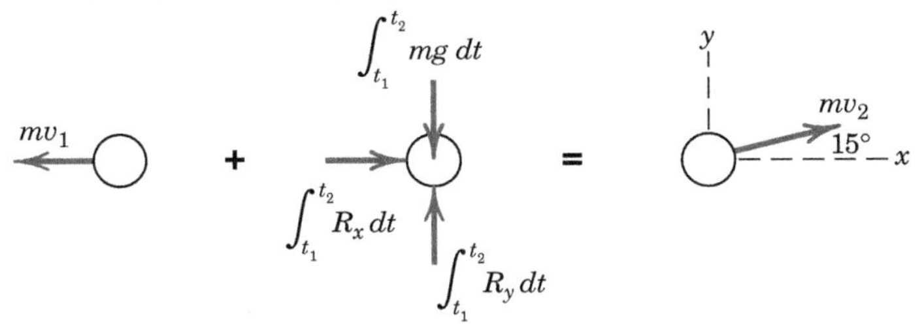

② $$\left[m(v_x)_1 + \int_{t_1}^{t_2} \Sigma F_x dt = m(v_x)_2\right] \qquad -\frac{2/16}{32.2}(50) + R_x(0.02) = \frac{2/16}{32.2}(70 \cos 15°)$$

$$\left[m(v_y)_1 + \int_{t_1}^{t_2} \Sigma F_y dt = m(v_y)_2\right]$$

$$\frac{2/16}{32.2}(0) + R_y(0.02) - (2/16)(0.02) = \frac{2/16}{32.2}(70 \sin 15°)$$

We can now solve for the impact forces as

$$R_x = 22.8 \text{ lb}$$

$$R_y = 3.64 \text{ lb}$$

We note that the impact force $R_y = 3.64$ lb is considerably larger than the 0.125-lb weight of the ball. Thus, the weight mg, a nonimpulsive force, could have been neglected as small in comparison with R_y. Had we neglected the weight, the computed value of R_y would have been 3.52 lb.

We now determine the magnitude and direction of **R** as

$$R = \sqrt{R_x^2 + R_y^2} = \sqrt{22.8^2 + 3.64^2} = 23.1 \text{ lb} \qquad \textit{Ans.}$$

$$\beta = \tan^{-1}\frac{R_y}{R_x} = \tan^{-1}\frac{3.64}{22.8} = 9.06° \qquad \textit{Ans.}$$

Helpful Hints

① Recall that for the impulse-momentum diagrams, initial linear momentum goes in the first diagram, all external linear impulses go in the second diagram, and final linear momentum goes in the third diagram.

② For the linear impulse $\int_{t_1}^{t_2} R_x\,dt$, the average impact force R_x is a constant, so that it can be brought outside the integral sign, resulting in $R_x \int_{t_1}^{t_2} dt = R_x(t_2 - t_1) = R_x \Delta t$. The linear impulse in the y-direction has been similarly treated.

Sample Problem 3/20

A 2-lb particle moves in the vertical y-z plane (z up, y horizontal) under the action of its weight and a force **F** which varies with time. The linear momentum of the particle in pound-seconds is given by the expression $\mathbf{G} = \frac{3}{2}(t^2 + 3)\mathbf{j} - \frac{2}{3}(t^3 - 4)\mathbf{k}$, where t is the time in seconds. Determine **F** and its magnitude for the instant when $t = 2$ sec.

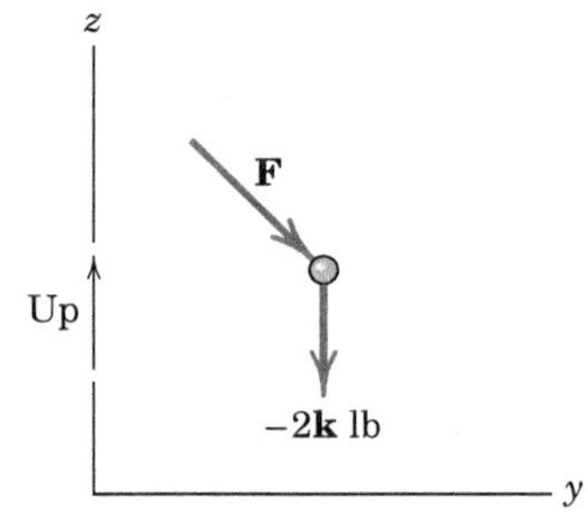

Solution. The weight expressed as a vector is $-2\mathbf{k}$ lb. Thus, the force-momentum equation becomes

① $[\Sigma\mathbf{F} = \dot{\mathbf{G}}]$ $$\mathbf{F} - 2\mathbf{k} = \frac{d}{dt}\left[\tfrac{3}{2}(t^2 + 3)\mathbf{j} - \tfrac{2}{3}(t^3 - 4)\mathbf{k}\right]$$
$$= 3t\mathbf{j} - 2t^2\mathbf{k}$$

For $t = 2$ sec, $$\mathbf{F} = 2\mathbf{k} + 3(2)\mathbf{j} - 2(2^2)\mathbf{k} = 6\mathbf{j} - 6\mathbf{k} \text{ lb}$$ *Ans.*

Thus, $$F = \sqrt{6^2 + 6^2} = 6\sqrt{2} \text{ lb}$$ *Ans.*

Helpful Hint

① Don't forget that $\Sigma\mathbf{F}$ includes *all* external forces acting on the particle, including the weight.

Sample Problem 3/21

A particle with a mass of 0.5 kg has a velocity of 10 m/s in the x-direction at time $t = 0$. Forces $\mathbf{F}_1$ and $\mathbf{F}_2$ act on the particle, and their magnitudes change with time according to the graphical schedule shown. Determine the velocity $\mathbf{v}_2$ of the particle at the end of the 3-s interval. The motion occurs in the horizontal x-y plane.

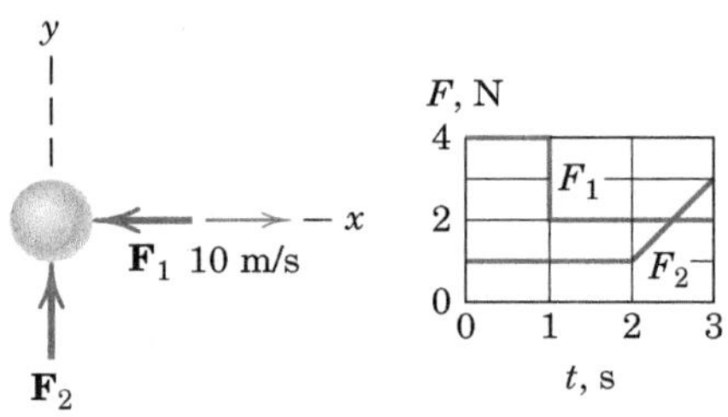

Solution. First, we construct the impulse-momentum diagrams as shown.

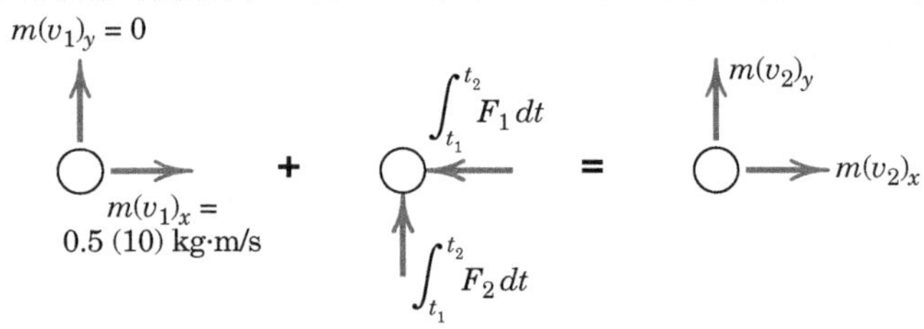

Then the impulse-momentum equations follow as

① $\left[m(v_1)_x + \int_{t_1}^{t_2} \Sigma F_x\, dt = m(v_2)_x\right]$ $$0.5(10) - [4(1) + 2(3 - 1)] = 0.5(v_2)_x$$
$$(v_2)_x = -6 \text{ m/s}$$

$\left[m(v_1)_y + \int_{t_1}^{t_2} \Sigma F_y\, dt = m(v_2)_y\right]$ $$0.5(0) + [1(2) + 2(3 - 2)] = 0.5(v_2)_y$$
$$(v_2)_y = 8 \text{ m/s}$$

Thus,

$$\mathbf{v}_2 = -6\mathbf{i} + 8\mathbf{j} \text{ m/s} \quad \text{and} \quad v_2 = \sqrt{6^2 + 8^2} = 10 \text{ m/s}$$

$$\theta_x = \tan^{-1}\frac{8}{-6} = 126.9°$$ *Ans.*

Although not called for, the path of the particle for the first 3 seconds is plotted in the figure. The velocity at $t = 3$ s is shown together with its components.

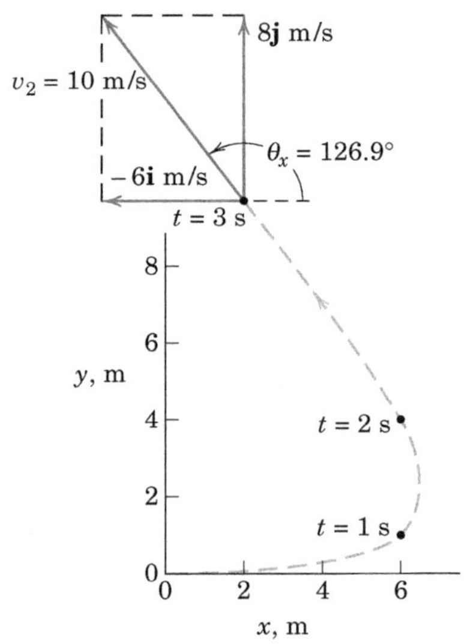

Helpful Hint

① The impulse in each direction is the corresponding area under the force-time graph. Note that F_1 is in the negative x-direction, so its impulse is negative.

Sample Problem 3/22

The loaded 150-kg skip is rolling down the incline at 4 m/s when a force P is applied to the cable as shown at time $t = 0$. The force P is increased uniformly with the time until it reaches 600 N at $t = 4$ s, after which time it remains constant at this value. Calculate (a) the time t' at which the skip reverses its direction and (b) the velocity v of the skip at $t = 8$ s. Treat the skip as a particle.

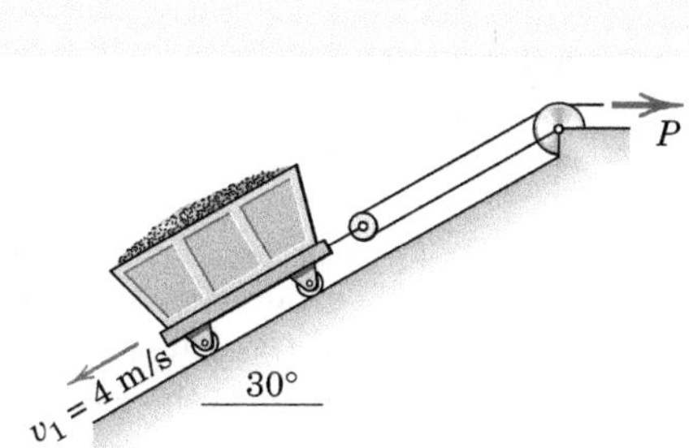

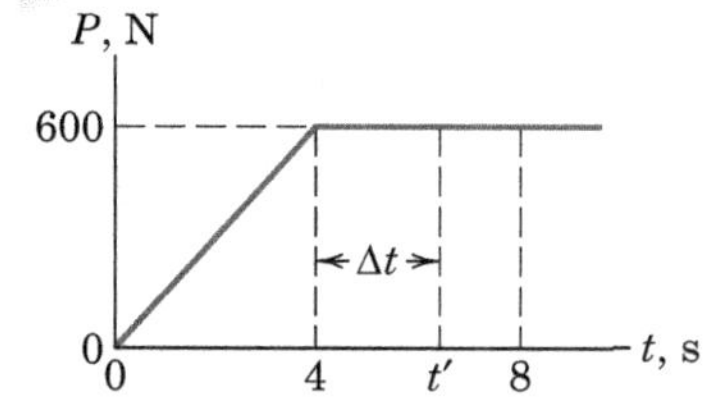

Solution. The stated variation of P with the time is plotted, and the impulse-momentum diagrams of the skip are drawn.

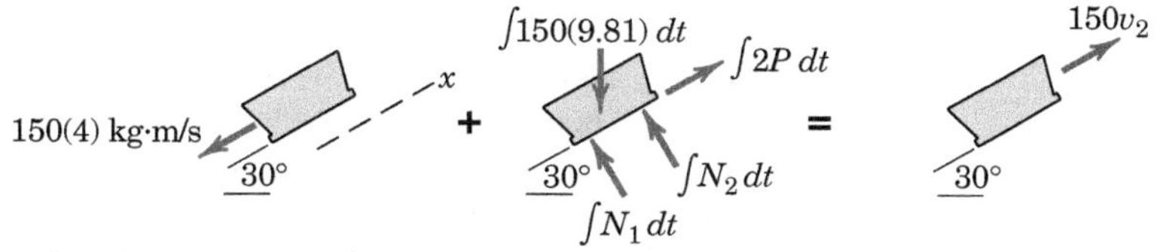

Part (a). The skip reverses direction when its velocity becomes zero. We will assume that this condition occurs at $t = 4 + \Delta t$ s. The impulse-momentum equation applied consistently in the positive x-direction gives

$$\left[m(v_1)_x + \int \Sigma F_x \, dt = m(v_2)_x\right]$$

① $$150(-4) + \frac{1}{2}(4)(2)(600) + 2(600)\Delta t - 150(9.81)\sin 30°(4 + \Delta t) = 150(0)$$

$$\Delta t = 2.46 \text{ s} \qquad t' = 4 + 2.46 = 6.46 \text{ s} \qquad \textit{Ans.}$$

Part (b). Applying the momentum equation to the entire 8-s interval gives

$$\left[m(v_1)_x + \int \Sigma F_x \, dt = m(v_2)_x\right]$$

$$150(-4) + \frac{1}{2}(4)(2)(600) + 4(2)(600) - 150(9.81)\sin 30°(8) = 150(v_2)_x$$

$$(v_2)_x = 4.76 \text{ m/s} \qquad \textit{Ans.}$$

The same result is obtained by analyzing the interval from t' to 8 s.

Helpful Hint

① The impulse-momentum diagram keeps us from making the error of using the impulse of P rather than $2P$ or of forgetting the impulse of the component of the weight. The first term in the linear impulse is the triangular area of the P-t relation for the first 4 s, doubled for the force of $2P$.

Sample Problem 3/23

The 50-g bullet traveling at 600 m/s strikes the 4-kg block centrally and is embedded within it. If the block slides on a smooth horizontal plane with a velocity of 12 m/s in the direction shown prior to impact, determine the velocity $\mathbf{v}_2$ of the block and embedded bullet immediately after impact.

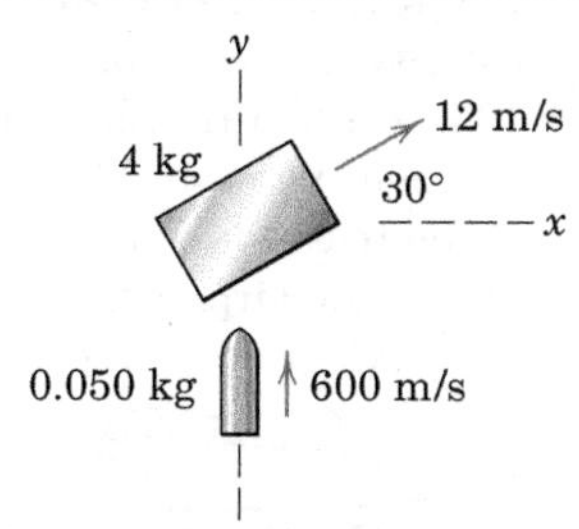

Solution. Since the force of impact is internal to the system composed of the block and bullet and since there are no other external forces acting on the system in the plane of motion, it follows that the linear momentum of the system is conserved. Thus,

① $$[\mathbf{G}_1 = \mathbf{G}_2] \quad 0.050(600\mathbf{j}) + 4(12)(\cos 30°\mathbf{i} + \sin 30°\mathbf{j}) = (4 + 0.050)\mathbf{v}_2$$

$$\mathbf{v}_2 = 10.26\mathbf{i} + 13.33\mathbf{j} \text{ m/s} \qquad \textit{Ans.}$$

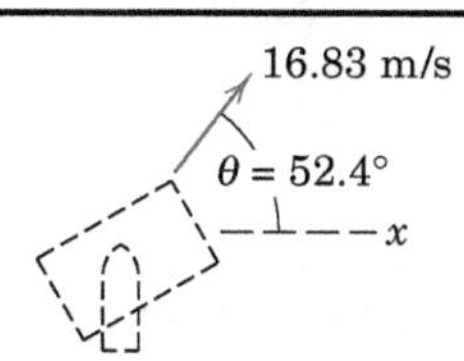

The final velocity and its direction are given by

$$\left[v = \sqrt{v_x^2 + v_y^2}\right] \qquad v_2 = \sqrt{(10.26)^2 + (13.33)^2} = 16.83 \text{ m/s} \qquad \textit{Ans.}$$

$$[\tan\theta = v_y/v_x] \qquad \tan\theta = \frac{13.33}{10.26} = 1.299 \qquad \theta = 52.4° \qquad \textit{Ans.}$$

Helpful Hint

① Working with the vector form of the principle of conservation of linear momentum is clearly equivalent to working with the component form.

PROBLEMS

Introductory Problems

3/173 A 0.2-kg wad of clay is released from rest and drops 2 m to a concrete floor. The clay does not rebound, and the collision lasts 0.04 s. Determine the time average of the force which the floor exerts on the clay during the impact.

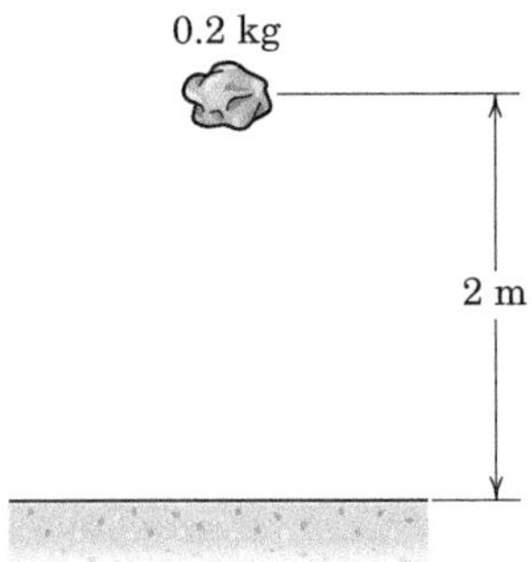

Problem 3/173

3/174 The two orbital maneuvering engines of the space shuttle develop 26 kN of thrust each. If the shuttle is traveling in orbit at a speed of 28 000 km/h, how long would it take to reach a speed of 28 100 km/h after the two engines are fired? The mass of the shuttle is 90 Mg.

3/175 A jet-propelled airplane with a mass of 10 Mg is flying horizontally at a constant speed of 1000 km/h under the action of the engine thrust T and the equal and opposite air resistance R. The pilot ignites two rocket-assist units, each of which develops a forward thrust T_0 of 8 kN for 9 s. If the velocity of the airplane in its horizontal flight is 1050 km/h at the end of the 9 s, calculate the time-average increase ΔR in air resistance. The mass of the rocket fuel used is negligible compared with that of the airplane.

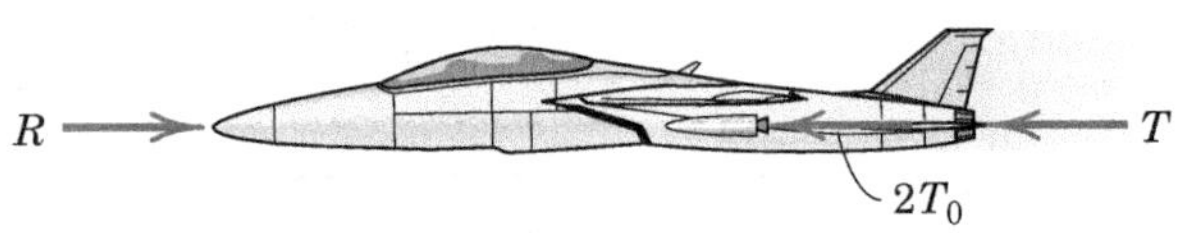

Problem 3/175

3/176 The velocity of a 1.2-kg particle is given by $\mathbf{v} = 1.5t^3\mathbf{i} + (2.4 - 3t^2)\mathbf{j} + 5\mathbf{k}$, where $\mathbf{v}$ is in meters per second and the time t is in seconds. Determine the linear momentum $\mathbf{G}$ of the particle, its magnitude G, and the net force $\mathbf{R}$ which acts on the particle when $t = 2$ s.

3/177 A 75-g projectile traveling at 600 m/s strikes and becomes embedded in the 40-kg block, which is initially stationary. Compute the energy lost during the impact. Express your answer as an absolute value $|\Delta E|$ and as a percentage n of the original system energy E.

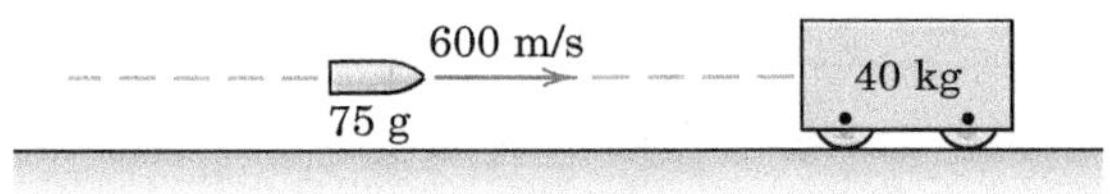

Problem 3/177

3/178 A 60-g bullet is fired horizontally with a velocity $v_1 = 600$ m/s into the 3-kg block of soft wood initially at rest on the horizontal surface. The bullet emerges from the block with the velocity $v_2 = 400$ m/s, and the block is observed to slide a distance of 2.70 m before coming to rest. Determine the coefficient of kinetic friction μ_k between the block and the supporting surface.

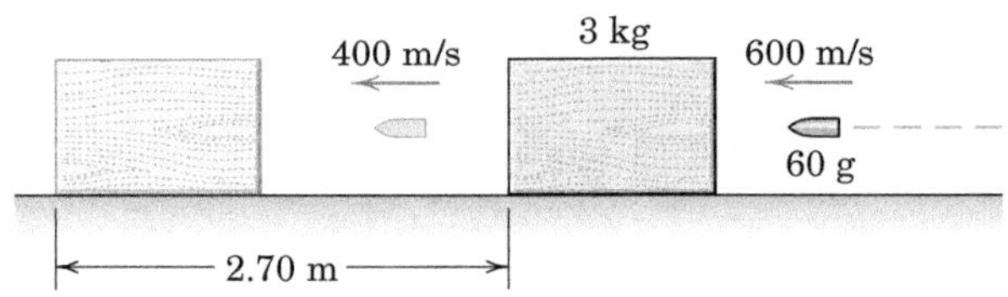

Problem 3/178

3/179 Careful measurements made during the impact of the 200-g metal cylinder with the spring-loaded plate reveal a semielliptical relation between the contact force F and the time t of impact as shown. Determine the rebound velocity v of the cylinder if it strikes the plate with a velocity of 6 m/s.

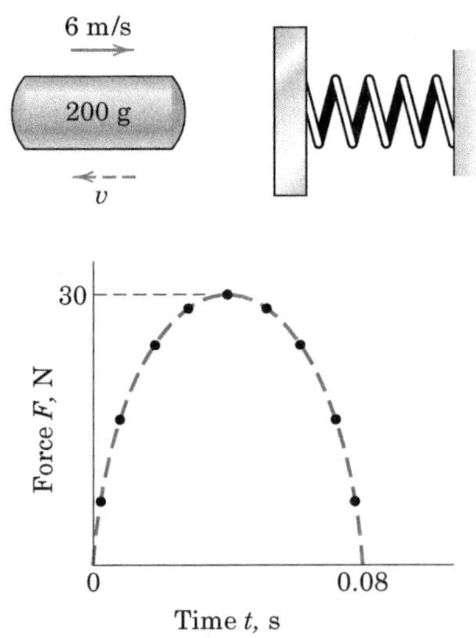

Problem 3/179

3/180 A 0.25-kg particle is moving with a velocity $\mathbf{v}_1 = 2\mathbf{i} + \mathbf{j} - \mathbf{k}$ m/s at time $t_1 = 2$ s. If the single force $\mathbf{F} = (4 + 2t)\mathbf{i} + (t^2 - 2)\mathbf{j} + 5\mathbf{k}$ N acts on the particle, determine its velocity $\mathbf{v}_2$ at time $t_2 = 4$ s.

3/181 The 90-kg man dives from the 40-kg canoe. The velocity indicated in the figure is that of the man relative to the canoe just after loss of contact. If the man, woman, and canoe are initially at rest, determine the horizontal component of the absolute velocity of the canoe just after separation. Neglect drag on the canoe, and assume that the 60-kg woman remains motionless relative to the canoe.

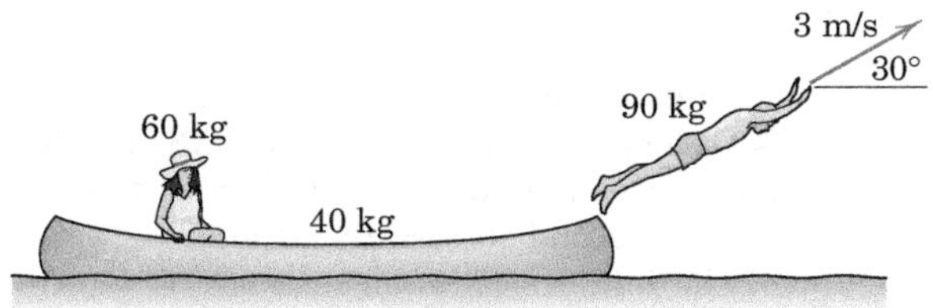

Problem 3/181

3/182 A 4-kg object, which is moving on a smooth horizontal surface with a velocity of 10 m/s in the $-x$-direction, is subjected to a force F_x which varies with time as shown. Approximate the experimental data by the dashed line and determine the velocity of the object (*a*) at $t = 0.6$ s and (*b*) at $t = 0.9$ s.

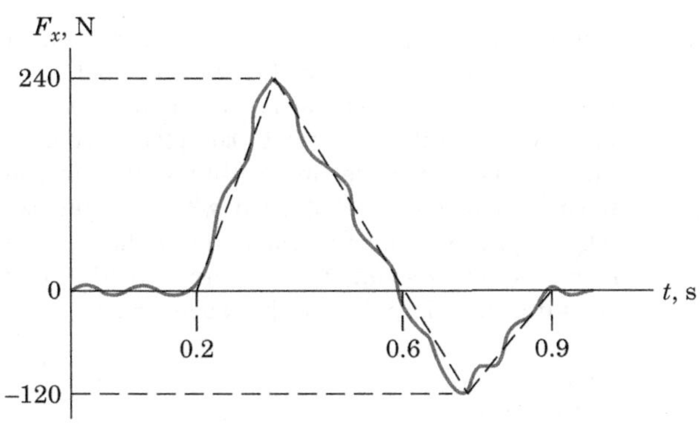

Problem 3/182

3/183 Crate A is traveling down the incline with a speed of 4 m/s when in the position shown. It later strikes and becomes attached to crate B. Determine the distance d moved by the pair after the collision. The coefficient of kinetic friction is $\mu_k = 0.40$ for both crates.

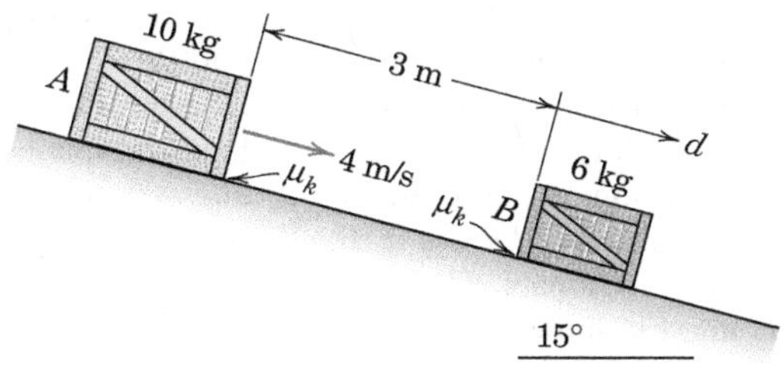

Problem 3/183

3/184 The 15 200-kg lunar lander is descending onto the moon's surface with a velocity of 2 m/s when its retro-engine is fired. If the engine produces a thrust T for 4 s which varies with time as shown and then cuts off, calculate the velocity of the lander when $t = 5$ s, assuming that it has not yet landed. Gravitational acceleration at the moon's surface is 1.62 m/s^2.

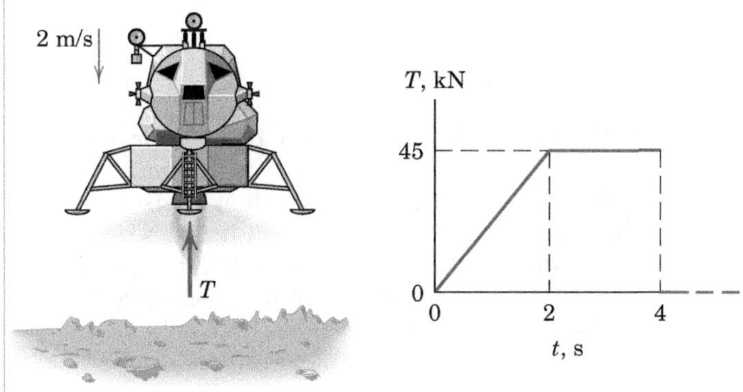

Problem 3/184

3/185 A boy weighing 100 lb runs and jumps on his 20-lb sled with a horizontal velocity of 15 ft/sec. If the sled and boy coast 80 ft on the level snow before coming to rest, compute the coefficient of kinetic friction μ_k between the snow and the runners of the sled.

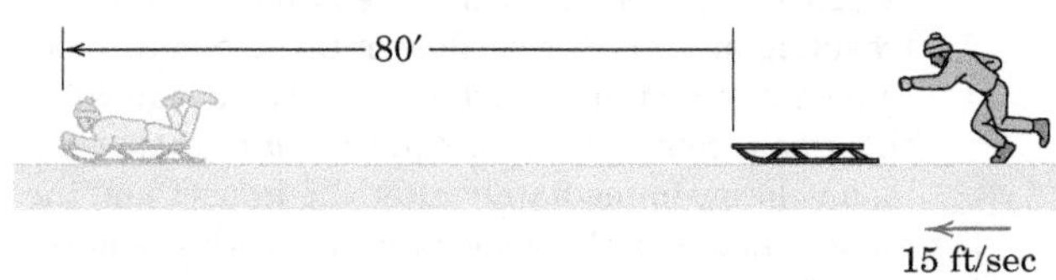

Problem 3/185

3/186 The snowboarder is traveling with a velocity of 6 m/s as shown when he lands on the incline with no rebound. If the impact has a time duration of 0.1 s, determine his speed v along the incline just after impact and the total time-average normal force exerted by the incline on the snowboard during the impact. The combined mass of the athlete and his snowboard is 60 kg.

Problem 3/186

Representative Problems

3/187 The tow truck with attached 1200-kg car accelerates uniformly from 30 km/h to 70 km/h over a 15-s interval. The average rolling resistance for the car over this speed interval is 500 N. Assume that the 60° angle shown represents the time average configuration and determine the average tension in the tow cable.

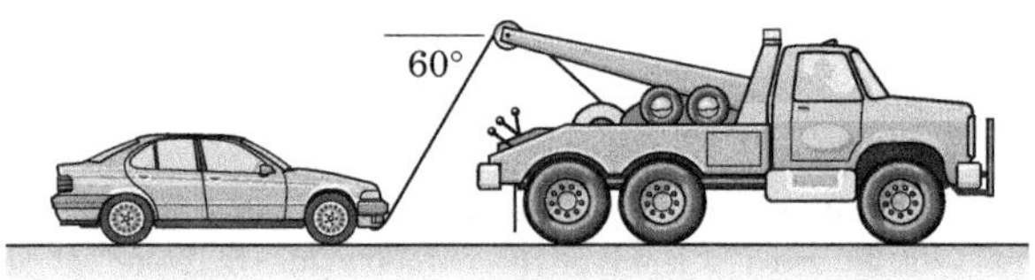

Problem 3/187

3/188 Car A weighing 3200 lb and traveling north at 20 mi/hr collides with car B weighing 3600 lb and traveling at 30 mi/hr as shown. If the two cars become entangled and move together as a unit after the crash, compute the magnitude v of their common velocity immediately after the impact and the angle θ made by the velocity vector with the north direction.

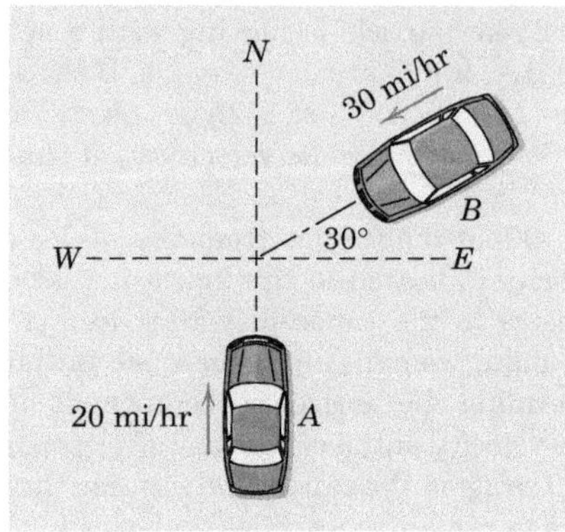

Problem 3/188

3/189 A railroad car of mass m and initial speed v collides with and becomes coupled with the two identical cars. Compute the final speed v' of the group of three cars and the fractional loss n of energy if (a) the initial separation distance $d = 0$ (that is, the two stationary cars are initially coupled together with no slack in the coupling) and (b) the distance $d \neq 0$ so that the cars are uncoupled and slightly separated. Neglect rolling resistance.

Problem 3/189

3/190 The 600,000-lb jet airliner has a touchdown velocity $v = 120$ mi/hr directed $\theta = 0.5°$ below the horizontal. The touchdown process of the eight main wheels takes 0.6 sec to complete. Treat the aircraft as a particle and estimate the average normal reaction force at each wheel during this 0.6-sec process, during which tires deflect, struts compress, etc. Assume that the aircraft lift equals the aircraft weight during the touchdown.

Problem 3/190

3/191 The collar of mass m slides on the rough horizontal shaft under the action of the force F of constant magnitude $F \leq mg$ but variable direction. If $\theta = kt$ where k is a constant, and if the collar has a speed v_1 to the right when $\theta = 0$, determine the velocity v_2 of the collar when θ reaches 90°. Also determine the value of F which renders $v_2 = v_1$.

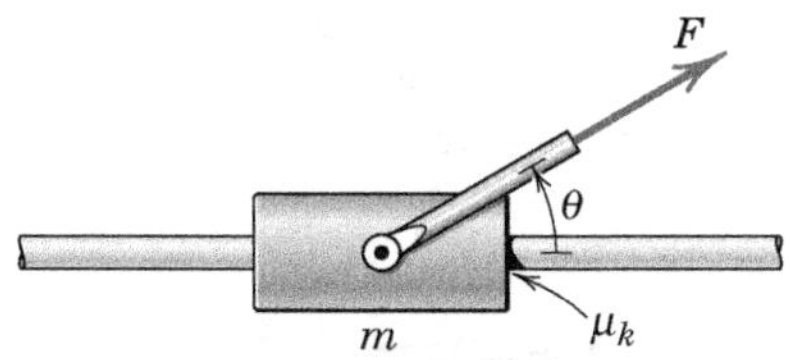

Problem 3/191

3/192 The 140-g projectile is fired with a velocity of 600 m/s and picks up three washers, each with a mass of 100 g. Find the common velocity v of the projectile and washers. Determine also the loss $|\Delta E|$ of energy during the interaction.

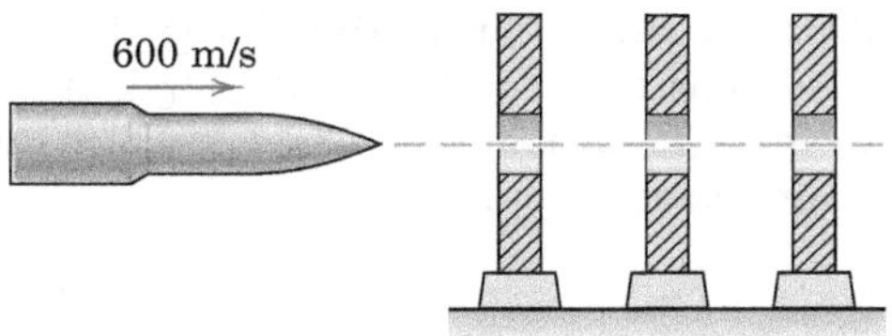

Problem 3/192

3/193 The third and fourth stages of a rocket are coasting in space with a velocity of 18 000 km/h when a small explosive charge between the stages separates them. Immediately after separation the fourth stage has increased its velocity to $v_4 = 18\ 060$ km/h. What is the corresponding velocity v_3 of the third stage? At separation the third and fourth stages have masses of 400 and 200 kg, respectively.

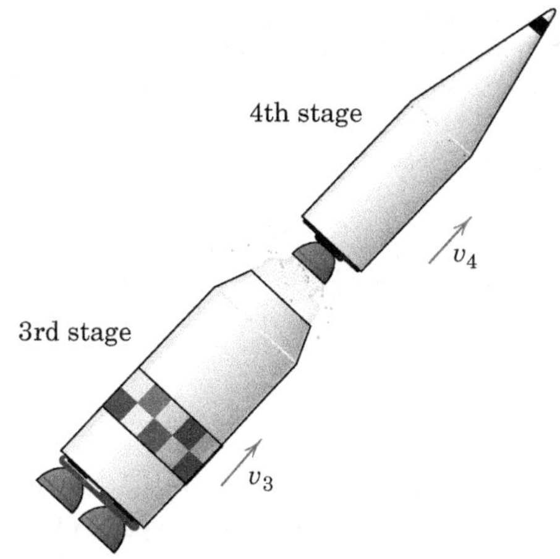

Problem 3/193

3/194 The initially stationary 20-kg block is subjected to the time-varying horizontal force whose magnitude P is shown in the plot. Note that the force is zero for all times greater than 3 s. Determine the time t_s at which the block comes to rest.

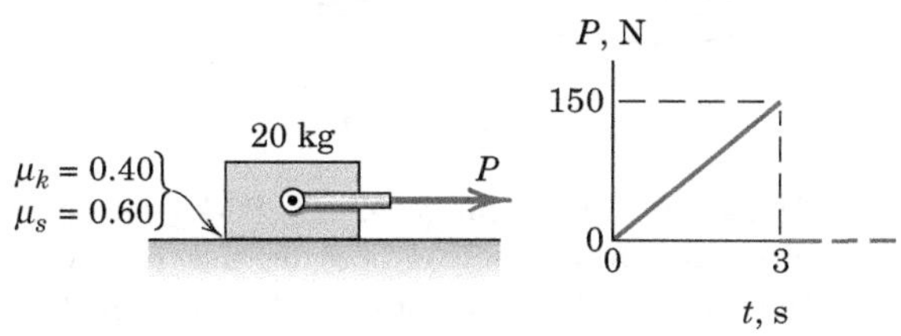

Problem 3/194

3/195 All elements of the previous problem remain unchanged, except that the force P is now held at a constant 30° angle relative to the horizontal. Determine the time t_s at which the initially stationary 20-kg block comes to rest.

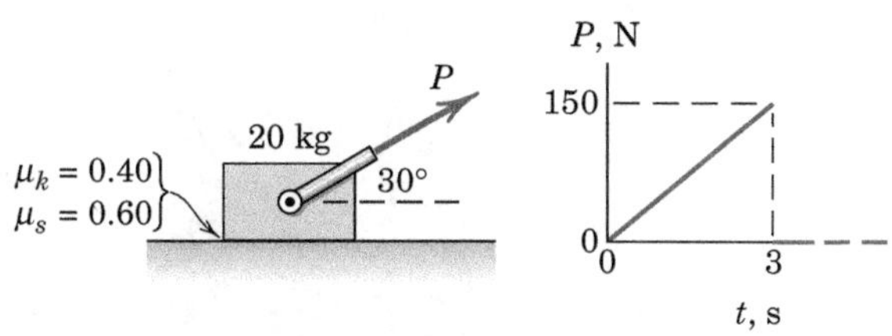

Problem 3/195

3/196 The spring of modulus $k = 200$ N/m is compressed a distance of 300 mm and suddenly released with the system at rest. Determine the absolute velocities of both masses when the spring is unstretched. Neglect friction.

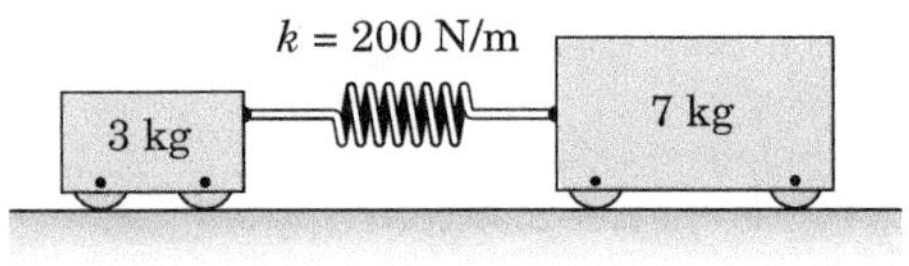

Problem 3/196

3/197 The pilot of a 90,000-lb airplane which is originally flying horizontally at a speed of 400 mi/hr cuts off all engine power and enters a 5° glide path as shown. After 120 seconds the airspeed is 360 mi/hr. Calculate the time-average drag force D (air resistance to motion along the flight path).

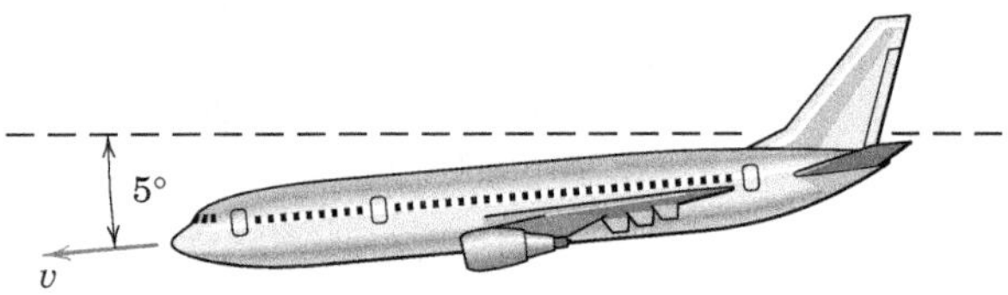

Problem 3/197

3/198 The space shuttle launches an 800-kg satellite by ejecting it from the cargo bay as shown. The ejection mechanism is activated and is in contact with the satellite for 4 s to give it a velocity of 0.3 m/s in the z-direction relative to the shuttle. The mass of the shuttle is 90 Mg. Determine the component of velocity v_f of the shuttle in the minus z-direction resulting from the ejection. Also find the time average F_{av} of the ejection force.

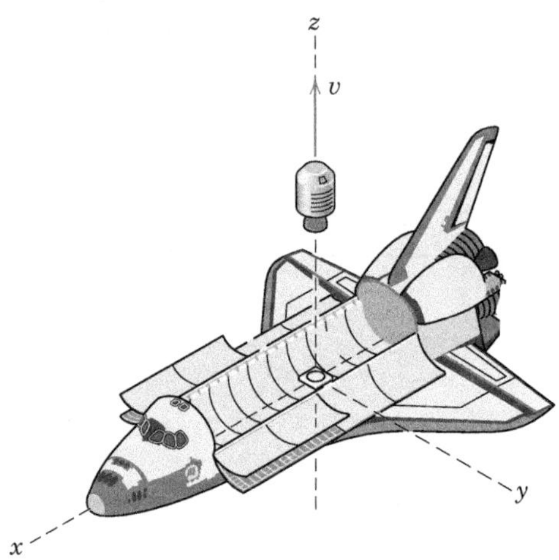

Problem 3/198

3/199 The hydraulic braking system for the truck and trailer is set to produce equal braking forces for the two units. If the brakes are applied uniformly for 5 seconds to bring the rig to a stop from a speed of 20 mi/hr down the 10-percent grade, determine the force P in the coupling between the trailer and the truck. The truck weighs 20,000 lb and the trailer weighs 15,000 lb.

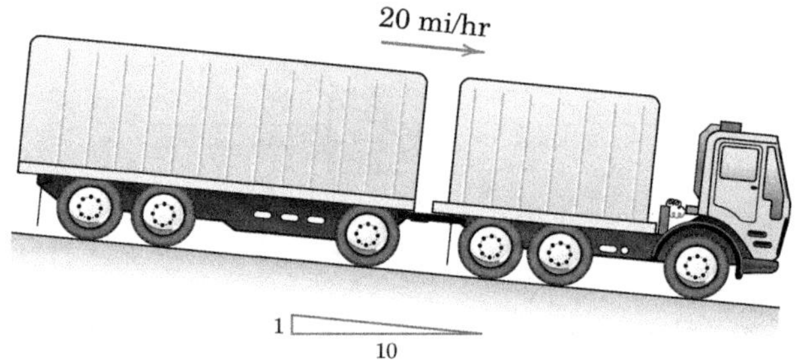

Problem 3/199

3/200 The 100-lb block is stationary at time $t = 0$, and then it is subjected to the force P shown. Note that the force is zero for all times beyond $t = 15$ sec. Determine the velocity v of the block at time $t = 15$ sec. Also calculate the time t at which the block again comes to rest.

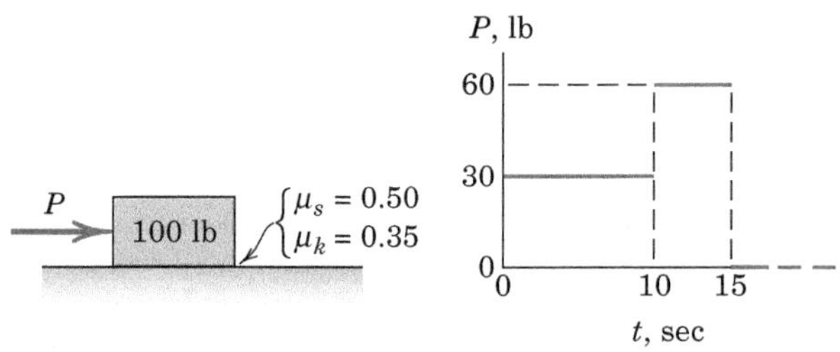

Problem 3/200

3/201 Car B is initially stationary and is struck by car A moving with initial speed $v_1 = 20$ mi/hr. The cars become entangled and move together with speed v' after the collision. If the time duration of the collision is 0.1 sec, determine (*a*) the common final speed v', (*b*) the average acceleration of each car during the collision, and (*c*) the magnitude R of the average force exerted by each car on the other car during the impact. All brakes are released during the collision.

Problem 3/201

3/202 The 2.4-kg particle moves in the horizontal x-y plane and has the velocity shown at time $t = 0$. If the force $F = 2 + 3t^2/4$ newtons, where t is time in seconds, is applied to the particle in the y-direction beginning at time $t = 0$, determine the velocity v of the particle 4 seconds after F is applied and specify the corresponding angle θ measured counterclockwise from the x-axis to the direction of the velocity.

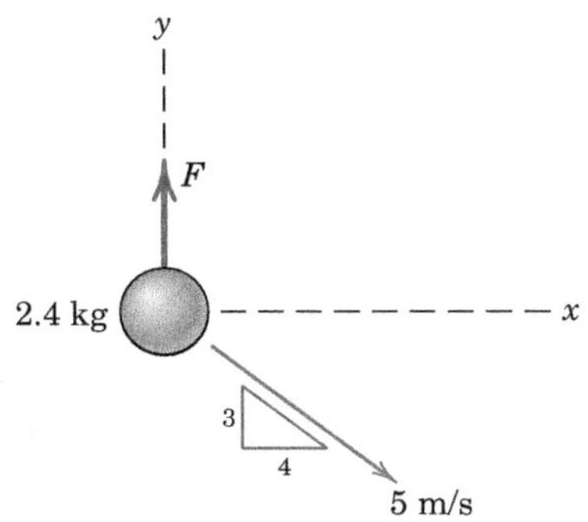

Problem 3/202

3/203 The 1.62-oz golf ball is struck by the five-iron and acquires the velocity shown in a time period of 0.001 sec. Determine the magnitude R of the average force exerted by the club on the ball. What acceleration magnitude a does this force cause, and what is the distance d over which the launch velocity is achieved, assuming constant acceleration?

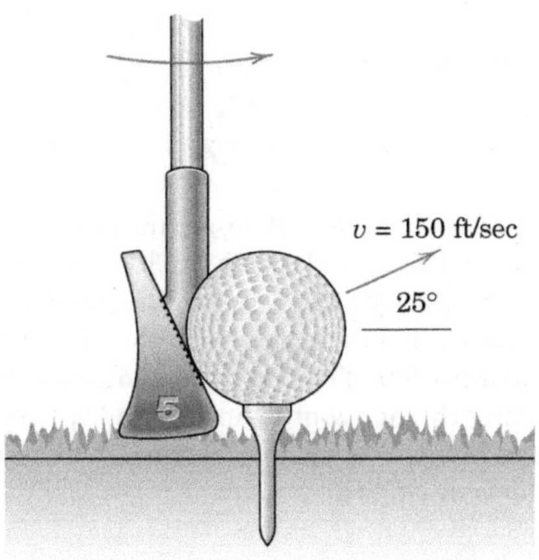

Problem 3/203

3/204 The ice-hockey puck with a mass of 0.20 kg has a velocity of 12 m/s before being struck by the hockey stick. After the impact the puck moves in the new direction shown with a velocity of 18 m/s. If the stick is in contact with the puck for 0.04 s, compute the magnitude of the average force **F** exerted by the stick on the puck during contact, and find the angle β made by **F** with the x-direction.

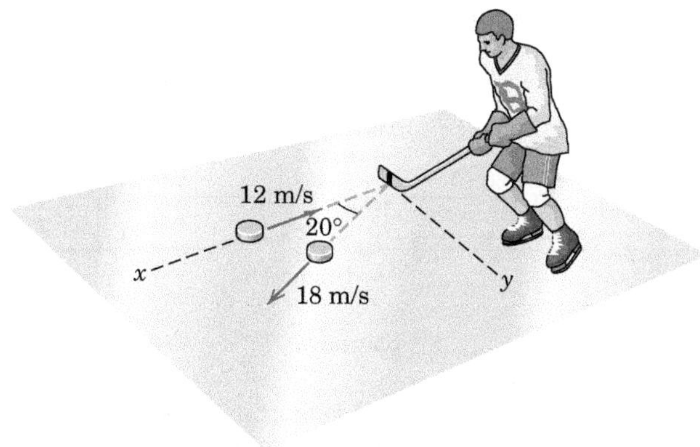

Problem 3/204

3/205 The baseball is traveling with a horizontal velocity of 85 mi/hr just before impact with the bat. Just after the impact, the velocity of the $5\frac{1}{8}$-oz ball is 130 mi/hr directed at 35° to the horizontal as shown. Determine the x- and y-components of the average force **R** exerted by the bat on the baseball during the 0.005-sec impact. Comment on the treatment of the weight of the baseball (a) during the impact and (b) over the first few seconds after impact.

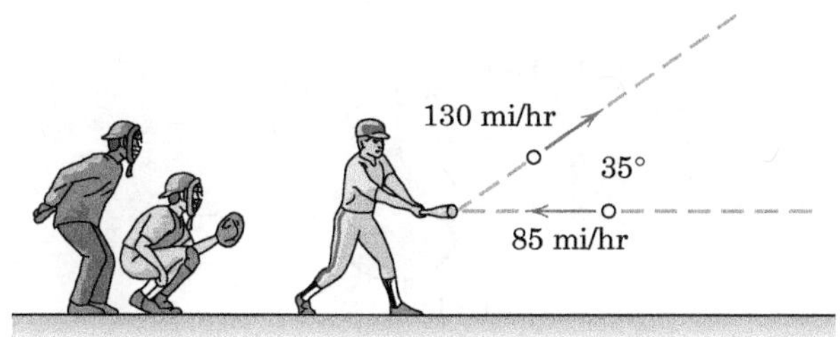

Problem 3/205

3/206 A spacecraft in deep space is programmed to increase its speed by a desired amount Δv by burning its engine for a specified time duration t. Twenty-five percent of the way through the burn, the engine suddenly malfunctions and thereafter produces only half of its normal thrust. What percent n of Δv is achieved if the rocket motor is fired for the planned time t? How much extra time t' would the rocket need to operate in order to compensate for the failure?

3/207 The slider of mass $m_1 = 0.4$ kg moves along the smooth support surface with velocity $v_1 = 5$ m/s when in the position shown. After negotiating the curved portion, it moves onto the inclined face of an initially stationary block of mass $m_2 = 2$ kg. The coefficient of kinetic friction between the slider and the block is $\mu_k = 0.30$. Determine the velocity v' of the system after the slider has come to rest relative to the block. Neglect friction at the small wheels, and neglect any effects associated with the transition.

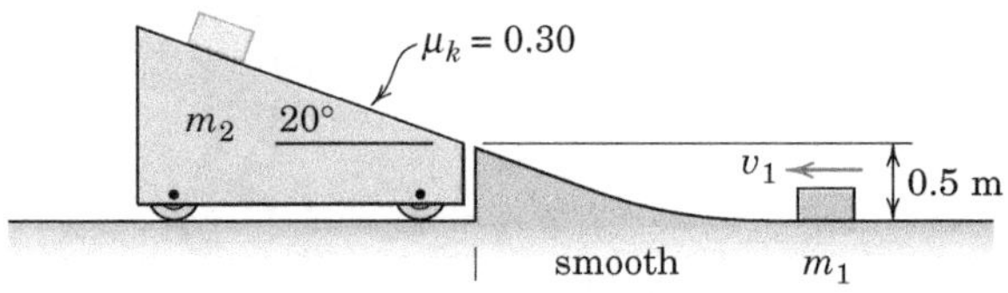

Problem 3/207

3/208 The 1.2-lb sphere is moving in the horizontal *x-y* plane with a velocity of 10 ft/sec in the direction shown and encounters a steady flow of air in the *x*-direction. If the air stream exerts an essentially constant force of 0.2 lb on the sphere in the *x*-direction, determine the time *t* required for the sphere to cross the *y*-axis again.

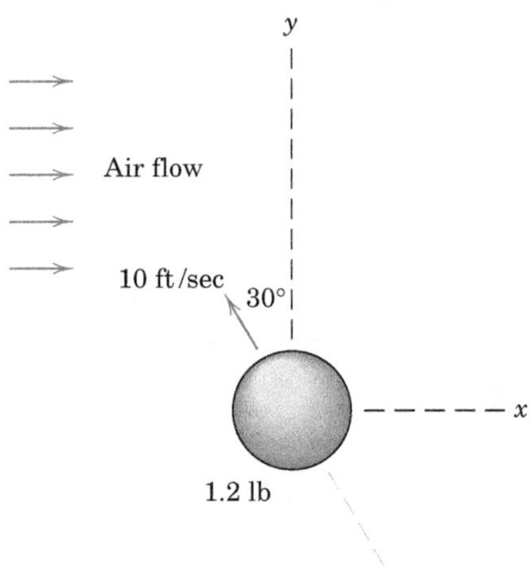

Problem 3/208

3/209 The ballistic pendulum is a simple device to measure projectile velocity v by observing the maximum angle θ to which the box of sand with embedded projectile swings. Calculate the angle θ if the 2-oz projectile is fired horizontally into the suspended 50-lb box of sand with a velocity v = 2000 ft/sec. Also find the percentage of energy lost during impact.

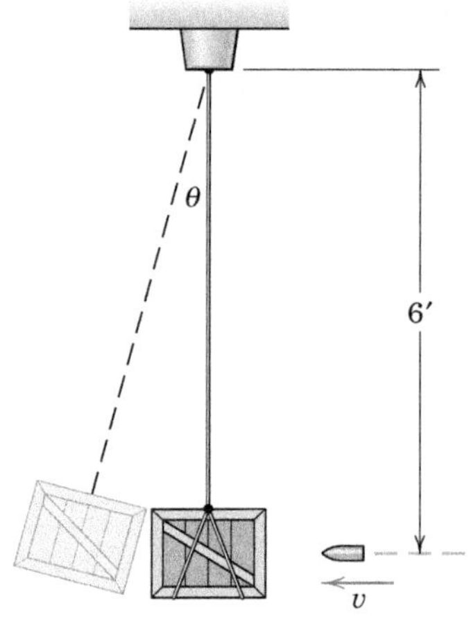

Problem 3/209

3/210 A tennis player strikes the tennis ball with her racket while the ball is still rising. The ball speed before impact with the racket is $v_1 = 15$ m/s and after impact its speed is $v_2 = 22$ m/s, with directions as shown in the figure. If the 60-g ball is in contact with the racket for 0.05 s, determine the magnitude of the average force **R** exerted by the racket on the ball. Find the angle β made by **R** with the horizontal. Comment on the treatment of the ball weight during impact.

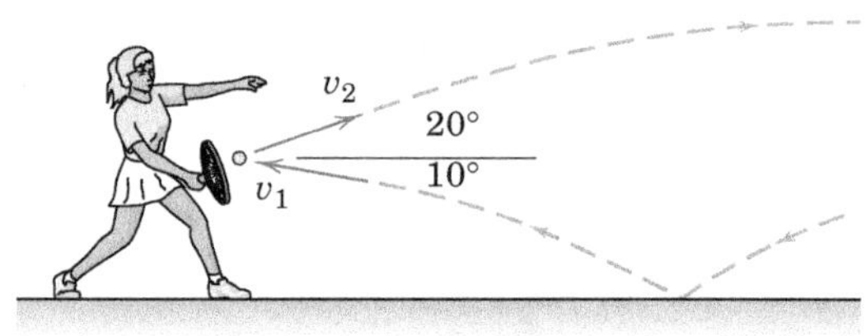

Problem 3/210

3/211 The 80-lb boy has taken a running jump from the upper surface and lands on his 10-lb skateboard with a velocity of 16 ft/sec in the plane of the figure as shown. If his impact with the skateboard has a time duration of 0.05 sec, determine the final speed v along the horizontal surface and the total normal force N exerted by the surface on the skateboard wheels during the impact.

Problem 3/211

3/212 The wad of clay A is projected as shown at the same instant that cylinder B is released. The two bodies collide and stick together at C and then ultimately strike the horizontal surface at D. Determine the horizontal distance d. Use the values $v_0 = 12$ m/s, $\theta = 40°$, $L = 6$ m, $m_A = 0.1$ kg, and $m_B = 0.2$ kg.

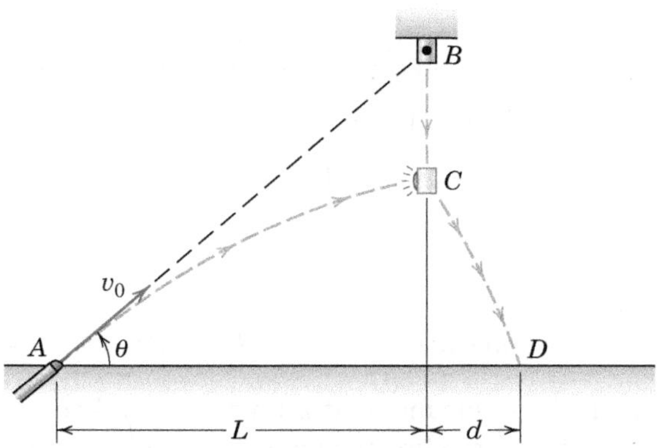

Problem 3/212

3/213 The two mine cars of equal mass are connected by a rope which is initially slack. Car A is given a shove which imparts to it a velocity of 4 ft/sec with car B initially at rest. When the slack is taken up, the rope suffers a tension impact which imparts a velocity to car B and reduces the velocity of car A. (a) If 40 percent of the kinetic energy of car A is lost during the rope impact, calculate the velocity v_B imparted to car B. (b) Following the initial impact, car B overtakes car A and the two are coupled together. Calculate their final common velocity v_C.

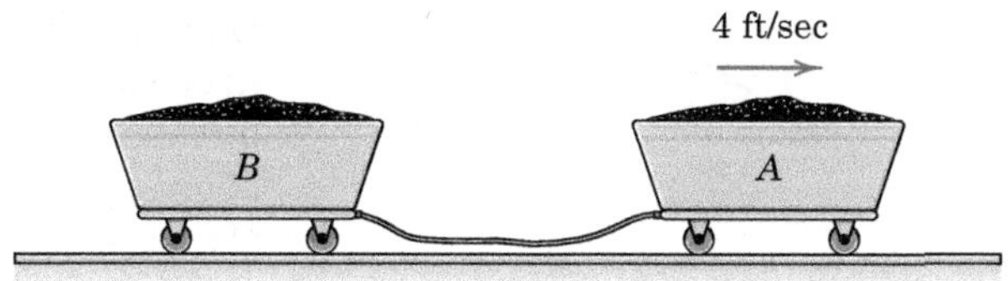

Problem 3/213

▶3/214 Two barges, each with a displacement (mass) of 500 Mg, are loosely moored in calm water. A stunt driver starts his 1500-kg car from rest at A, drives along the deck, and leaves the end of the 15° ramp at a speed of 50 km/h relative to the barge and ramp. The driver successfully jumps the gap and brings his car to rest relative to barge 2 at B. Calculate the velocity v_2 imparted to barge 2 just after the car has come to rest on the barge. Neglect the resistance of the water to motion at the low velocities involved.

Problem 3/214

3/10 Angular Impulse and Angular Momentum

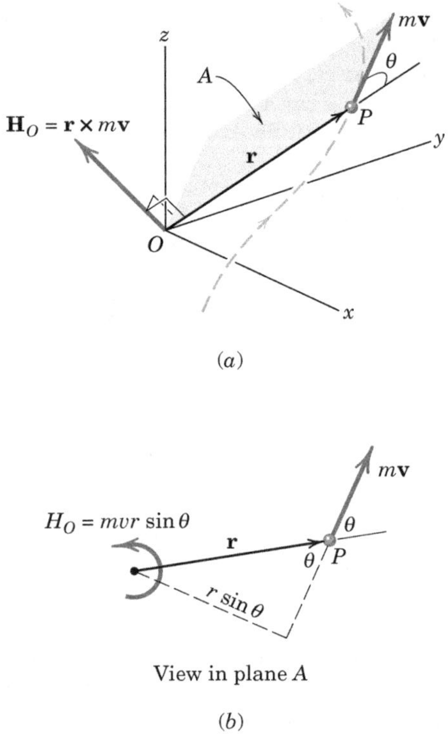

Figure 3/14

In addition to the equations of linear impulse and linear momentum, there exists a parallel set of equations for angular impulse and angular momentum. First, we define the term *angular momentum*. Figure 3/14*a* shows a particle P of mass m moving along a curve in space. The particle is located by its position vector $\mathbf{r}$ with respect to a convenient origin O of fixed coordinates x-y-z. The velocity of the particle is $\mathbf{v} = \dot{\mathbf{r}}$, and its linear momentum is $\mathbf{G} = m\mathbf{v}$. The *moment* of the *linear momentum vector* $m\mathbf{v}$ about the origin O is defined as the *angular momentum* $\mathbf{H}_O$ of P about O and is given by the cross-product relation for the moment of a vector

$$\boxed{\mathbf{H}_O = \mathbf{r} \times m\mathbf{v}} \tag{3/29}$$

The angular momentum then is a vector perpendicular to the plane A defined by $\mathbf{r}$ and $\mathbf{v}$. The sense of $\mathbf{H}_O$ is clearly defined by the right-hand rule for cross products.

The scalar components of angular momentum may be obtained from the expansion

$$\mathbf{H}_O = \mathbf{r} \times m\mathbf{v} = m(v_z y - v_y z)\mathbf{i} + m(v_x z - v_z x)\mathbf{j} + m(v_y x - v_x y)\mathbf{k}$$

$$\mathbf{H}_O = m\begin{vmatrix} \mathbf{i} & \mathbf{j} & \mathbf{k} \\ x & y & z \\ v_x & v_y & v_z \end{vmatrix} \tag{3/30}$$

so that

$$H_x = m(v_z y - v_y z) \qquad H_y = m(v_x z - v_z x) \qquad H_z = m(v_y x - v_x y)$$

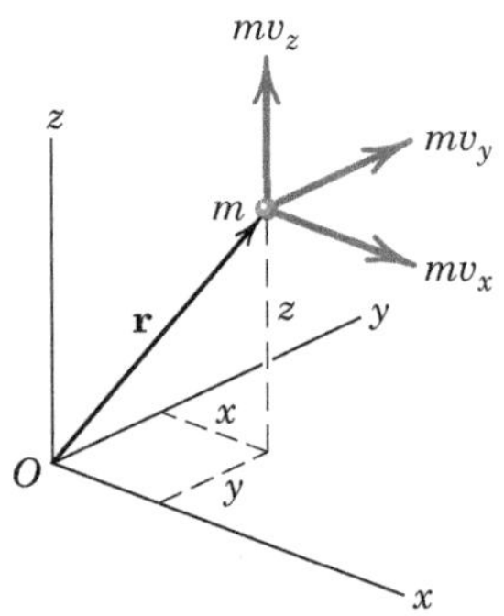

Figure 3/15

Each of these expressions for angular momentum may be checked easily from Fig. 3/15, which shows the three linear-momentum components, by taking the moments of these components about the respective axes.

To help visualize angular momentum, we show in Fig. 3/14*b* a two-dimensional representation in plane A of the vectors shown in part *a* of the figure. The motion is viewed in plane A defined by $\mathbf{r}$ and $\mathbf{v}$. The magnitude of the moment of $m\mathbf{v}$ about O is simply the linear momentum mv times the moment arm $r \sin\theta$ or $mvr \sin\theta$, which is the magnitude of the cross product $\mathbf{H}_O = \mathbf{r} \times m\mathbf{v}$.

Angular momentum is the moment of linear momentum and must not be confused with linear momentum. In SI units, angular momentum has the units $\text{kg}\cdot(\text{m/s})\cdot\text{m} = \text{kg}\cdot\text{m}^2/\text{s} = \text{N}\cdot\text{m}\cdot\text{s}$. In the U.S. customary system, angular momentum has the units $[\text{lb/(ft/sec}^2)][\text{ft/sec}][\text{ft}] =$ lb-ft-sec.

Rate of Change of Angular Momentum

We are now ready to relate the moment of the forces acting on the particle P to its angular momentum. If $\Sigma\mathbf{F}$ represents the resultant of *all* forces acting on the particle P of Fig. 3/14, the moment $\mathbf{M}_O$ about the origin O is the vector cross product

$$\Sigma\mathbf{M}_O = \mathbf{r} \times \Sigma\mathbf{F} = \mathbf{r} \times m\dot{\mathbf{v}}$$

where Newton's second law $\Sigma\mathbf{F} = m\dot{\mathbf{v}}$ has been substituted. We now differentiate Eq. 3/29 with time, using the rule for the differentiation of a cross product (see item 9, Art. C/7, Appendix C) and obtain

$$\dot{\mathbf{H}}_O = \dot{\mathbf{r}} \times m\mathbf{v} + \mathbf{r} \times m\dot{\mathbf{v}} = \mathbf{v} \times m\mathbf{v} + \mathbf{r} \times m\dot{\mathbf{v}}$$

The term $\mathbf{v} \times m\mathbf{v}$ is zero since the cross product of parallel vectors is identically zero. Substitution into the expression for $\Sigma\mathbf{M}_O$ gives

$$\boxed{\Sigma\mathbf{M}_O = \dot{\mathbf{H}}_O} \qquad \textbf{(3/31)}$$

Equation 3/31 states that the *moment about the fixed point O of all forces acting on m equals the time rate of change of angular momentum of m about O*. This relation, particularly when extended to a system of particles, rigid or nonrigid, provides one of the most powerful tools of analysis in dynamics.

Equation 3/31 is a vector equation with scalar components

$$\Sigma M_{O_x} = \dot{H}_{O_x} \qquad \Sigma M_{O_y} = \dot{H}_{O_y} \qquad \Sigma M_{O_z} = \dot{H}_{O_z} \qquad \textbf{(3/32)}$$

The Angular Impulse-Momentum Principle

Equation 3/31 gives the instantaneous relation between the moment and the time rate of change of angular momentum. To obtain the effect of the moment $\Sigma\mathbf{M}_O$ on the angular momentum of the particle over a finite period of time, we integrate Eq. 3/31 from time t_1 to time t_2. Multiplying the equation by dt gives $\Sigma\mathbf{M}_O\,dt = d\mathbf{H}_O$, which we integrate to obtain

$$\int_{t_1}^{t_2} \Sigma\mathbf{M}_O\,dt = (\mathbf{H}_O)_2 - (\mathbf{H}_O)_1 = \Delta\mathbf{H}_O \qquad \textbf{(3/33)}$$

where $(\mathbf{H}_O)_2 = \mathbf{r}_2 \times m\mathbf{v}_2$ and $(\mathbf{H}_O)_1 = \mathbf{r}_1 \times m\mathbf{v}_1$. The product of moment and time is defined as *angular impulse*, and Eq. 3/33 states that the *total angular impulse on m about the fixed point O equals the corresponding change in angular momentum of m about O*.

Alternatively, we may write Eq. 3/33 as

$$\boxed{(\mathbf{H}_O)_1 + \int_{t_1}^{t_2} \Sigma\mathbf{M}_O\,dt = (\mathbf{H}_O)_2} \qquad \textbf{(3/33a)}$$

which states that the initial angular momentum of the particle plus the angular impulse applied to it equals its final angular momentum. The units of angular impulse are clearly those of angular momentum, which are N·m·s or kg·m^2/s in SI units and lb-ft-sec in U.S. customary units.

As in the case of linear impulse and linear momentum, the equation of angular impulse and angular momentum is a vector equation where changes in direction as well as magnitude may occur during the interval of integration. Under these conditions, it is necessary to express $\Sigma\mathbf{M}_O$

and $\mathbf{H}_O$ in component form and then combine the integrated components. The x-component of Eq. 3/33a is

$$(H_{O_x})_1 + \int_{t_1}^{t_2} \Sigma M_{O_x}\,dt = (H_{O_x})_2$$

or

$$m(v_z y - v_y z)_1 + \int_{t_1}^{t_2} \Sigma M_{O_x}\,dt = m(v_z y - v_y z)_2 \qquad \textbf{(3/33b)}$$

where the subscripts 1 and 2 refer to the values of the respective quantities at times t_1 and t_2. Similar expressions exist for the y- and z-components of the angular impulse-momentum equation.

Plane-Motion Applications

The foregoing angular-impulse and angular-momentum relations have been developed in their general three-dimensional forms. Most of the applications of interest to us, however, can be analyzed as plane-motion problems where moments are taken about a single axis normal to the plane of motion. In this case, the angular momentum may change magnitude and sense, but the direction of the vector remains unaltered.

Thus, for a particle of mass m moving along a curved path in the x-y plane, Fig. 3/16, the angular momenta about O at points 1 and 2 have the magnitudes $(H_O)_1 = |\mathbf{r}_1 \times m\mathbf{v}_1| = mv_1d_1$ and $(H_O)_2 = |\mathbf{r}_2 \times m\mathbf{v}_2| = mv_2d_2$, respectively. In the illustration both $(H_O)_1$ and $(H_O)_2$ are represented in the counterclockwise sense in accord with the direction of the moment of the linear momentum. The scalar form of Eq. 3/33a applied to the motion between points 1 and 2 during the time interval t_1 to t_2 becomes

$$(H_O)_1 + \int_{t_1}^{t_2} \Sigma M_O\,dt = (H_O)_2 \quad \text{or} \quad mv_1d_1 + \int_{t_1}^{t_2} \Sigma Fr \sin\theta\,dt = mv_2d_2$$

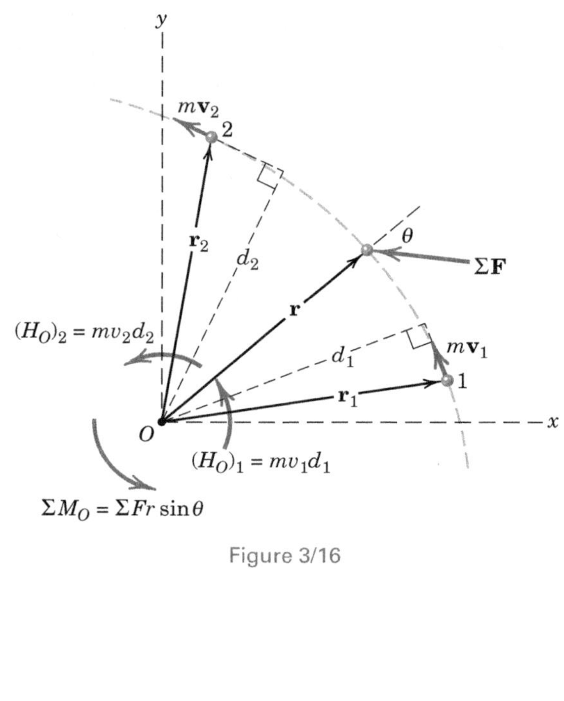

Figure 3/16

This example should help clarify the relation between the scalar and vector forms of the angular impulse-momentum relations.

Whereas Eq. 3/33 clearly stresses that the external angular impulse causes a change in the angular momentum, the order of the terms in Eqs. 3/33*a* and 3/33*b* corresponds to the natural sequence of events. Equation 3/33*a* is analogous to Eq. 3/27*a*, just as Eq. 3/31 is analogous to Eq. 3/25.

As was the case for linear-momentum problems, we encounter *impulsive* (large magnitude, short duration) and *nonimpulsive* forces in angular-momentum problems. The treatment of these forces was discussed in Art. 3/9.

Equations 3/25 and 3/31 add no new basic information since they are merely alternative forms of Newton's second law. We will discover in subsequent chapters, however, that the motion equations expressed in terms of the time rate of change of momentum are applicable to the motion of rigid and nonrigid bodies and provide a very general and powerful approach to many problems. The full generality of Eq. 3/31 is usually not required to describe the motion of a single particle or the plane motion of rigid bodies, but it does have important use in the analysis of the space motion of rigid bodies introduced in Chapter 7.

Conservation of Angular Momentum

If the resultant moment about a fixed point O of all forces acting on a particle is zero during an interval of time, Eq. 3/31 requires that its angular momentum $\mathbf{H}_O$ about that point remain constant. In this case, the angular momentum of the particle is said to be *conserved.* Angular momentum may be conserved about one axis but not about another axis. A careful examination of the free-body diagram of the particle will disclose whether the moment of the resultant force on the particle about a fixed point is zero, in which case, the angular momentum about that point is unchanged (conserved).

Consider now the motion of two particles a and b which interact during an interval of time. If the interactive forces $\mathbf{F}$ and $-\mathbf{F}$ between them are the only unbalanced forces acting on the particles during the interval, it follows that the moments of the equal and opposite forces about any fixed point O not on their line of action are equal and opposite. If we apply Eq. 3/33 to particle a and then to particle b and add the two equations, we obtain $\Delta\mathbf{H}_a + \Delta\mathbf{H}_b = \mathbf{0}$ (where all angular momenta are referred to point O). Thus, the total angular momentum for the system of the two particles remains constant during the interval, and we write

$$\Delta\mathbf{H}_O = \mathbf{0} \quad \text{or} \quad (\mathbf{H}_O)_1 = (\mathbf{H}_O)_2 \qquad \textbf{(3/34)}$$

which expresses the *principle of conservation of angular momentum.*

Sample Problem 3/24

A small sphere has the position and velocity indicated in the figure and is acted upon by the force F. Determine the angular momentum $\mathbf{H}_O$ about point O and the time derivative $\dot{\mathbf{H}}_O$.

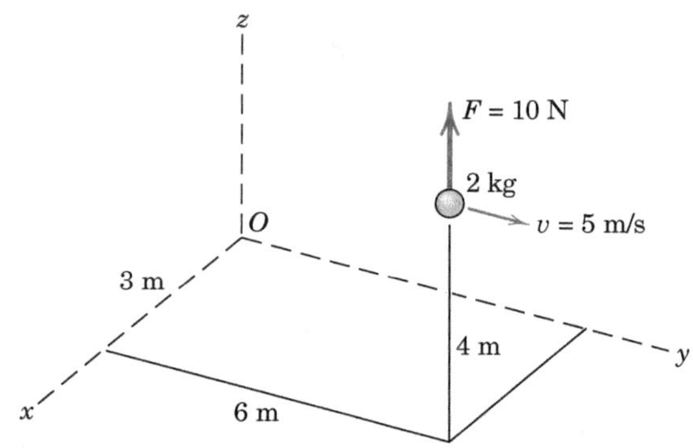

Solution. We begin with the definition of angular momentum and write

$$\mathbf{H}_O = \mathbf{r} \times m\mathbf{v}$$

$$= (3\mathbf{i} + 6\mathbf{j} + 4\mathbf{k}) \times 2(5\mathbf{j})$$

$$= -40\mathbf{i} + 30\mathbf{k} \text{ N}\cdot\text{m/s} \qquad \textit{Ans.}$$

From Eq. 3/31,

$$\dot{\mathbf{H}}_O = \mathbf{M}_O$$

$$= \mathbf{r} \times \mathbf{F}$$

$$= (3\mathbf{i} + 6\mathbf{j} + 4\mathbf{k}) \times 10\mathbf{k}$$

$$= 60\mathbf{i} - 30\mathbf{j} \text{ N}\cdot\text{m} \qquad \textit{Ans.}$$

As with moments of forces, the position vector must run *from* the reference point (O in this case) *to* the line of action of the linear momentum $m\mathbf{v}$. Here $\mathbf{r}$ runs directly to the particle.

Sample Problem 3/25

A comet is in the highly eccentric orbit shown in the figure. Its speed at the most distant point A, which is at the outer edge of the solar system, is $v_A = 740$ m/s. Determine its speed at the point B of closest approach to the sun.

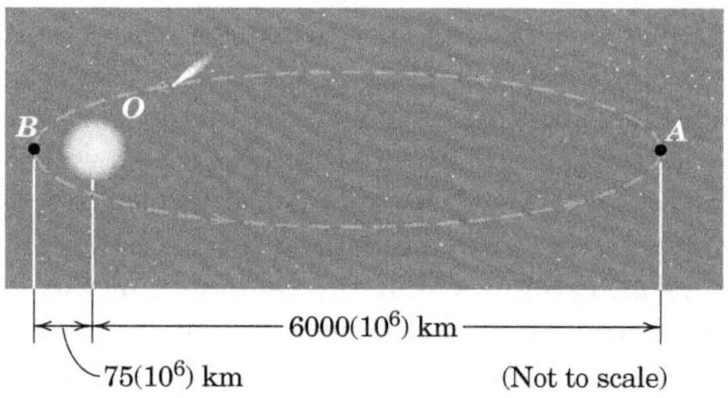

Solution. Because the only significant force acting on the comet, the gravitational force exerted on it by the sun, is central (points to the sun center O), angular momentum about O is conserved.

$$(H_O)_A = (H_O)_B$$

$$mr_Av_A = mr_Bv_B$$

$$v_B = \frac{r_Av_A}{r_B} = \frac{6000(10^6)740}{75(10^6)}$$

$$v_B = 59\,200 \text{ m/s} \qquad \textit{Ans.}$$

Sample Problem 3/26

The assembly of the light rod and two end masses is at rest when it is struck by the falling wad of putty traveling with speed v_1 as shown. The putty adheres to and travels with the right-hand end mass. Determine the angular velocity $\dot{\theta}_2$ of the assembly just after impact. The pivot at O is frictionless, and all three masses may be assumed to be particles.

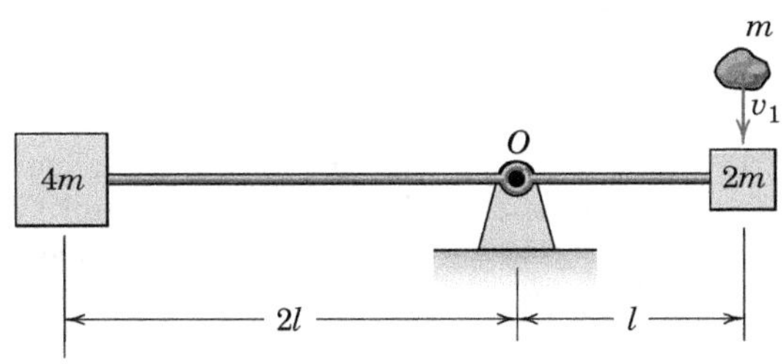

Solution. If we ignore the angular impulses associated with the weights during the collision process, then system angular momentum about O is conserved during the impact.

$$(H_O)_1 = (H_O)_2$$

$$mv_1 l = (m + 2m)(l\dot{\theta}_2)l + 4m(2l\dot{\theta}_2)2l$$

$$\dot{\theta}_2 = \frac{v_1}{19l} \text{ CW} \qquad \textit{Ans.}$$

Note that each angular-momentum term is written in the form mvd, and the final transverse velocities are expressed as radial distances times the common final angular velocity $\dot{\theta}_2$.

Sample Problem 3/27

A small mass particle is given an initial velocity $\mathbf{v}_0$ tangent to the horizontal rim of a smooth hemispherical bowl at a radius r_0 from the vertical centerline, as shown at point A. As the particle slides past point B, a distance h below A and a distance r from the vertical centerline, its velocity $\mathbf{v}$ makes an angle θ with the horizontal tangent to the bowl through B. Determine θ.

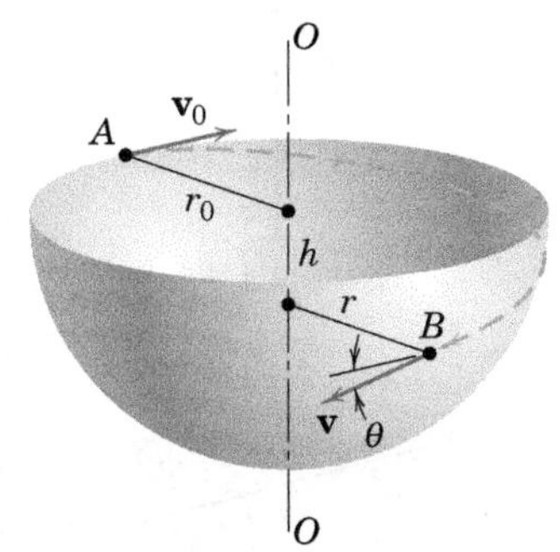

Solution. The forces on the particle are its weight and the normal reaction exerted by the smooth surface of the bowl. Neither force exerts a moment about the axis O-O, so that angular momentum is conserved about that axis. Thus,

① $[(H_O)_1 = (H_O)_2]$ $\qquad mv_0 r_0 = mvr \cos\theta$

Also, energy is conserved so that $E_1 = E_2$. Thus

$[T_1 + V_1 = T_2 + V_2]$ $\qquad \frac{1}{2}mv_0^2 + mgh = \frac{1}{2}mv^2 + 0$

$$v = \sqrt{v_0^2 + 2gh}$$

Eliminating v and substituting $r^2 = r_0^2 - h^2$ give

$$v_0 r_0 = \sqrt{v_0^2 + 2gh}\sqrt{r_0^2 - h^2}\cos\theta$$

$$\theta = \cos^{-1}\frac{1}{\sqrt{1 + \frac{2gh}{v_0^2}}\sqrt{1 - \frac{h^2}{r_0^2}}} \qquad \textit{Ans.}$$

Helpful Hint

① The angle θ is measured in the plane tangent to the hemispherical surface at B.

PROBLEMS

Introductory Problems

3/215 Determine the magnitude H_O of the angular momentum of the 2-kg sphere about point O (*a*) by using the vector definition of angular momentum and (*b*) by using an equivalent scalar approach. The center of the sphere lies in the *x*-*y* plane.

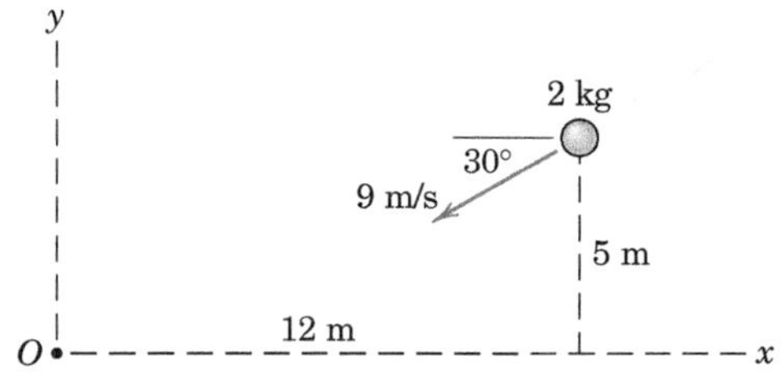

Problem 3/215

3/216 At a certain instant, the particle of mass m has the position and velocity shown in the figure, and it is acted upon by the force **F**. Determine its angular momentum about point O and the time rate of change of this angular momentum.

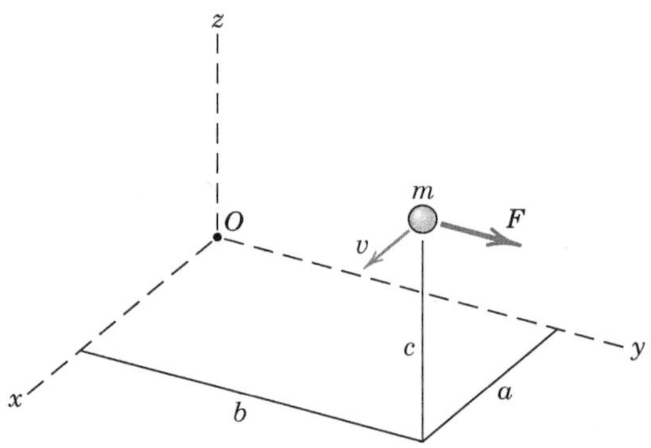

Problem 3/216

3/217 The 3-kg sphere moves in the *x*-*y* plane and has the indicated velocity at a particular instant. Determine its (*a*) linear momentum, (*b*) angular momentum about point O, and (*c*) kinetic energy.

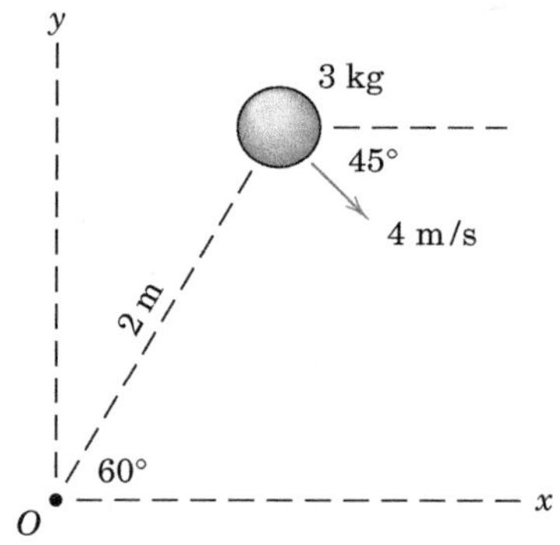

Problem 3/217

3/218 The particle of mass m is gently nudged from the equilibrium position A and subsequently slides along the smooth elliptical path which lies in a vertical plane. Determine the magnitude of its angular momentum about point O as it passes (*a*) point B and (*b*) point C. In each case, determine the time rate of change of H_O.

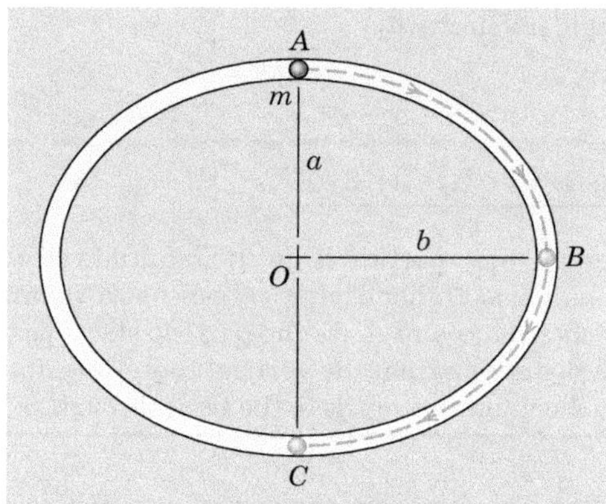

Problem 3/218

3/219 The assembly starts from rest and reaches an angular speed of 150 rev/min under the action of a 20-N force T applied to the string for t seconds. Determine t. Neglect friction and all masses except those of the four 3-kg spheres, which may be treated as particles.

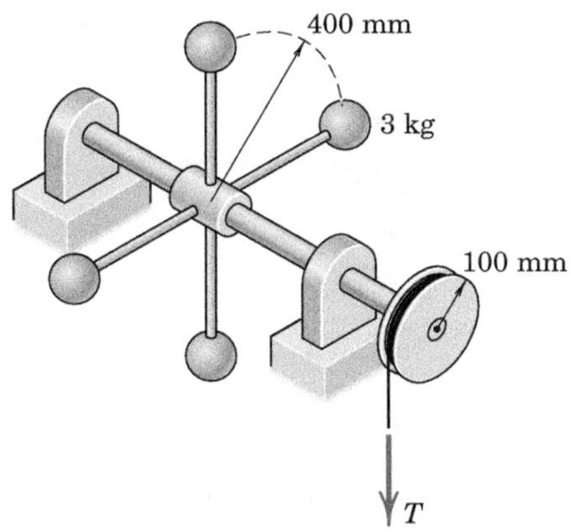

Problem 3/219

3/220 Just after launch from the earth, the space-shuttle orbiter is in the 37 × 137–mi orbit shown. At the apogee point A, its speed is 17,290 mi/hr. If nothing were done to modify the orbit, what would be its speed at the perigee P? Neglect aerodynamic drag. (Note that the normal practice is to add speed at A, which raises the perigee altitude to a value that is well above the bulk of the atmosphere.)

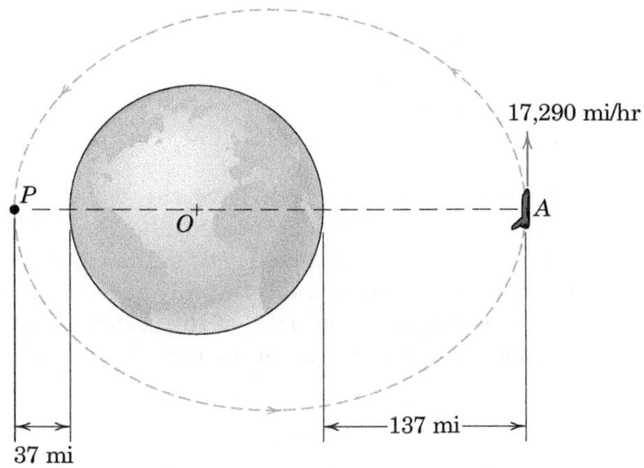

Problem 3/220

3/221 The rigid assembly which consists of light rods and two 1.2-kg spheres rotates freely about a vertical axis. The assembly is initially at rest and then a constant couple $M = 2$ N·m is applied for 5 s. Determine the final angular velocity of the assembly. Treat the small spheres as particles.

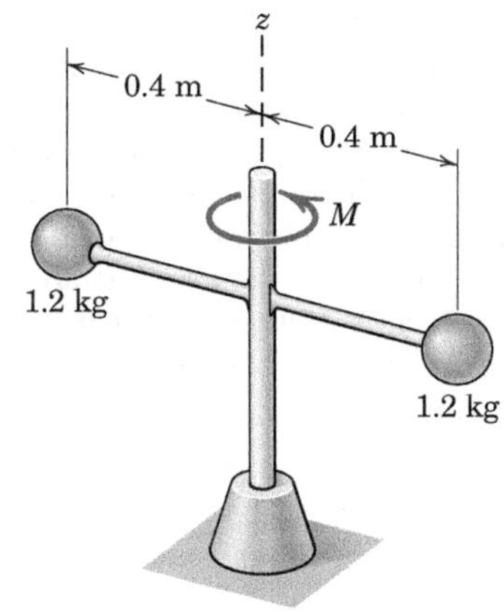

Problem 3/221

3/222 All conditions of the previous problem remain the same, except now the applied couple varies with time according to $M = 2t$, where t is in seconds and M is in newton-meters. Determine the angular velocity of the assembly at time $t = 5$ s.

Representative Problems

3/223 The small particle of mass m and its restraining cord are spinning with an angular velocity ω on the horizontal surface of a smooth disk, shown in section. As the force F is slightly relaxed, r increases and ω changes. Determine the rate of change of ω with respect to r and show that the work done by F during a movement dr equals the change in kinetic energy of the particle.

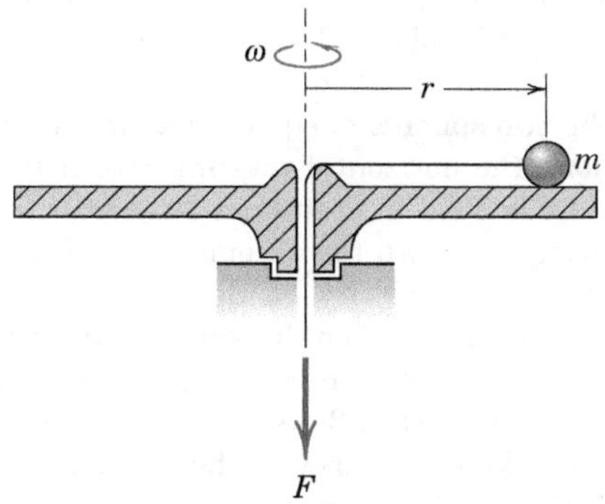

Problem 3/223

3/224 A particle with a mass of 4 kg has a position vector in meters given by $\mathbf{r} = 3t^2\mathbf{i} - 2t\mathbf{j} - 3\mathbf{k}$, where t is the time in seconds. For $t = 5$ s determine the angular momentum of the particle and the moment of all forces on the particle, both about the origin O of coordinates.

3/225 The 6-kg sphere and 4-kg block (shown in section) are secured to the arm of negligible mass which rotates in the vertical plane about a horizontal axis at O. The 2-kg plug is released from rest at A and falls into the recess in the block when the arm has reached the horizontal position. An instant before engagement, the arm has an angular velocity $\omega_0 = 2$ rad/s. Determine the angular velocity ω of the arm immediately after the plug has wedged itself in the block.

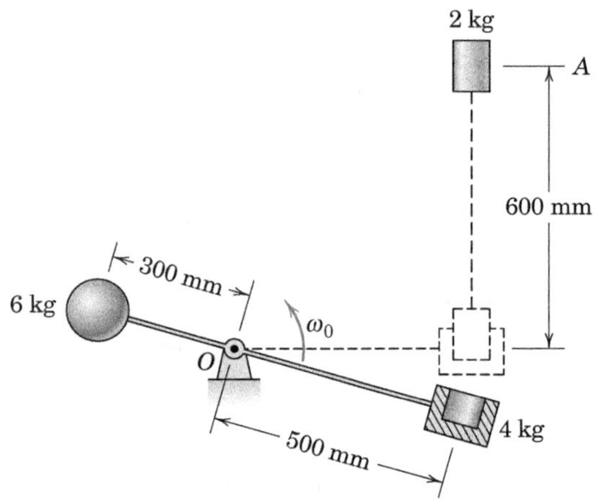

Problem 3/225

3/226 A 0.7-lb particle is located at the position $\mathbf{r}_1 = 2\mathbf{i} + 3\mathbf{j} + \mathbf{k}$ ft and has the velocity $\mathbf{v}_1 = \mathbf{i} + \mathbf{j} + 2\mathbf{k}$ ft/sec at time $t = 0$. If the particle is acted upon by a single force which has the moment $\mathbf{M}_O = (4 + 2t)\mathbf{i} + (3 - t^2)\mathbf{j} + 5\mathbf{k}$ lb-ft about the origin O of the coordinate system in use, determine the angular momentum about O of the particle when $t = 4$ sec.

3/227 The two spheres of equal mass m are able to slide along the horizontal rotating rod. If they are initially latched in position a distance r from the rotating axis with the assembly rotating freely with an angular velocity ω_0, determine the new angular velocity ω after the spheres are released and finally assume positions at the ends of the rod at a radial distance of $2r$. Also find the fraction n of the initial kinetic energy of the system which is lost. Neglect the small mass of the rod and shaft.

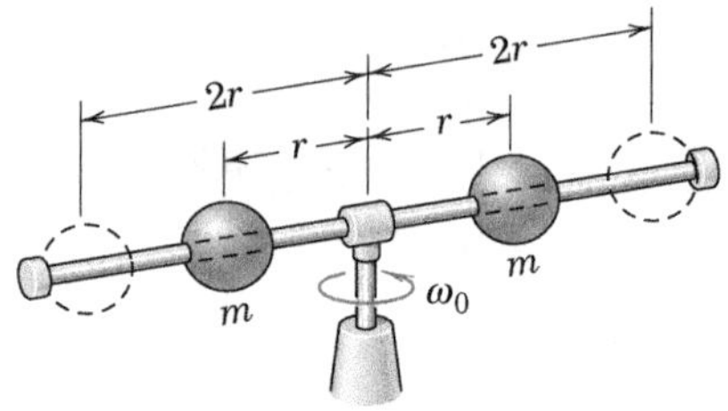

Problem 3/227

3/228 A particle of mass m moves with negligible friction on a horizontal surface and is connected to a light spring fastened at O. At position A the particle has the velocity $v_A = 4$ m/s. Determine the velocity v_B of the particle as it passes position B.

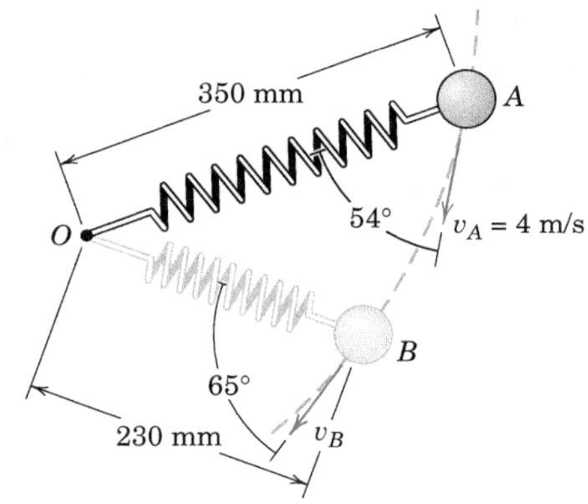

Problem 3/228

3/229 The small spheres, which have the masses and initial velocities shown in the figure, strike and become attached to the spiked ends of the rod, which is freely pivoted at O and is initially at rest. Determine the angular velocity ω of the assembly after impact. Neglect the mass of the rod.

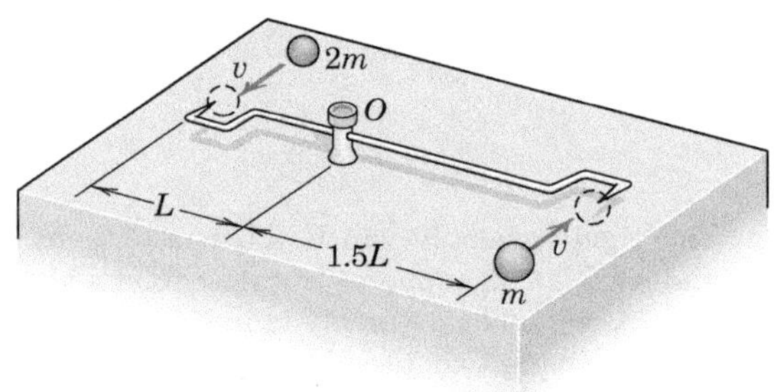

Problem 3/229

3/230 The particle of mass m is launched from point O with a horizontal velocity $\mathbf{u}$ at time $t = 0$. Determine its angular momentum $\mathbf{H}_O$ relative to point O as a function of time.

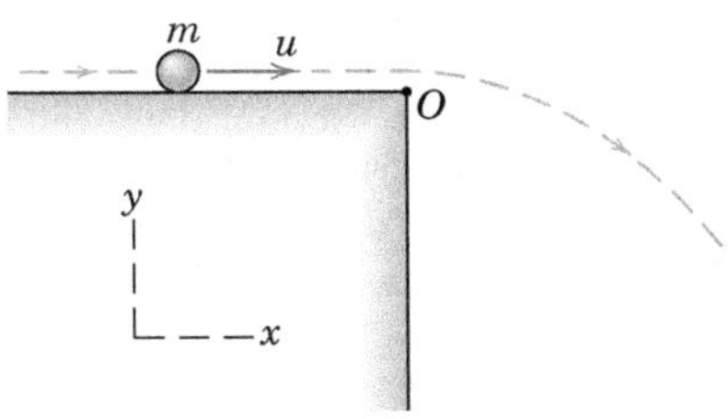

Problem 3/230

3/231 A wad of clay of mass m_1 with an initial horizontal velocity v_1 hits and adheres to the massless rigid bar which supports the body of mass m_2, which can be assumed to be a particle. The pendulum assembly is freely pivoted at O and is initially stationary. Determine the angular velocity $\dot{\theta}$ of the combined body just after impact. Why is linear momentum of the system not conserved?

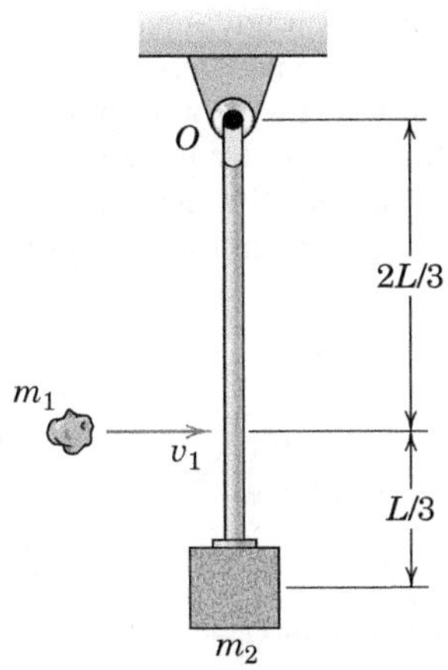

Problem 3/231

3/232 A particle of mass m is released from rest in position A and then slides down the smooth vertical-plane track. Determine its angular momentum about both points A and D (*a*) as it passes position B and (*b*) as it passes position C.

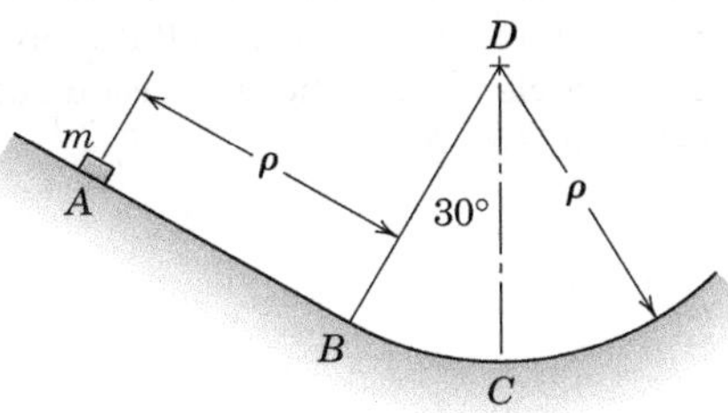

Problem 3/232

3/233 At the point A of closest approach to the sun, a comet has a velocity $v_A = 188{,}500$ ft/sec. Determine the radial and transverse components of its velocity v_B at point B, where the radial distance from the sun is $75(10^6)$ mi.

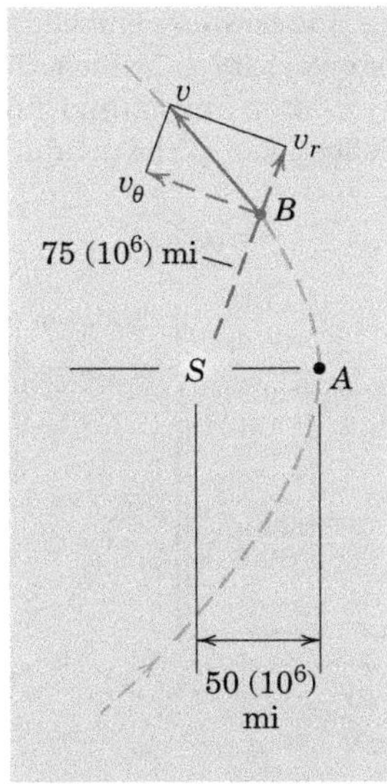

Problem 3/233

3/234 A particle moves on the inside surface of a smooth conical shell and is given an initial velocity $\mathbf{v}_0$ tangent to the horizontal rim of the surface at A. As the particle slides past point B, a distance z below A, its velocity $\mathbf{v}$ makes an angle θ with the horizontal tangent to the surface through B. Determine expressions for θ and the speed v.

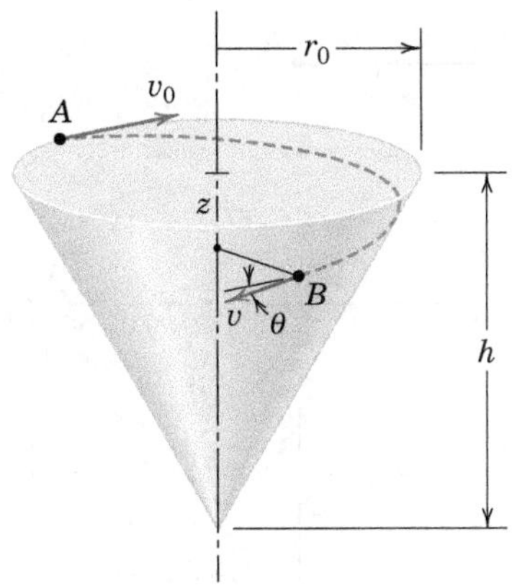

Problem 3/234

3/235 A pendulum consists of two 3.2-kg concentrated masses positioned as shown on a light but rigid bar. The pendulum is swinging through the vertical position with a clockwise angular velocity ω = 6 rad/s when a 50-g bullet traveling with velocity v = 300 m/s in the direction shown strikes the lower mass and becomes embedded in it. Calculate the angular velocity ω' which the pendulum has immediately after impact and find the maximum angular deflection θ of the pendulum.

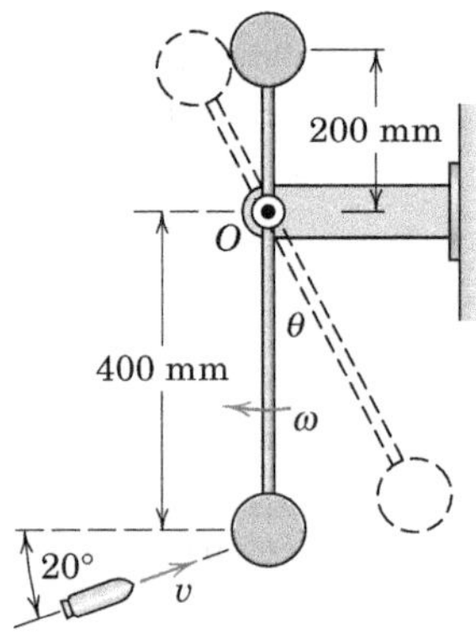

Problem 3/235

3/236 The central attractive force F on an earth satellite can have no moment about the center O of the earth. For the particular elliptical orbit with major and minor axes as shown, a satellite will have a velocity of 33 880 km/h at the perigee altitude of 390 km. Determine the velocity of the satellite at point B and at apogee A. The radius of the earth is 6371 km.

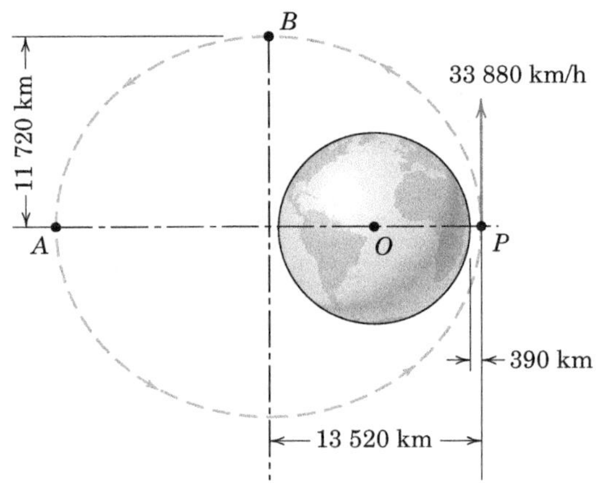

Problem 3/236

3/237 A particle is launched with a horizontal velocity v_0 = 0.55 m/s from the 30° position shown and then slides without friction along the funnel-like surface. Determine the angle θ which its velocity vector makes with the horizontal as the particle passes level O-O. The value of r is 0.9 m.

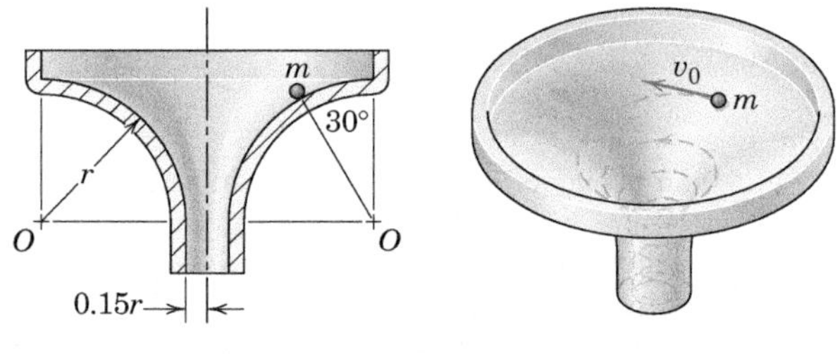

Problem 3/237

3/238 The 0.4-lb ball and its supporting cord are revolving about the vertical axis on the fixed smooth conical surface with an angular velocity of 4 rad/sec. The ball is held in the position b = 14 in. by the tension T in the cord. If the distance b is reduced to the constant value of 9 in. by increasing the tension T in the cord, compute the new angular velocity ω and the work $U'_{1\text{-}2}$ done on the system by T.

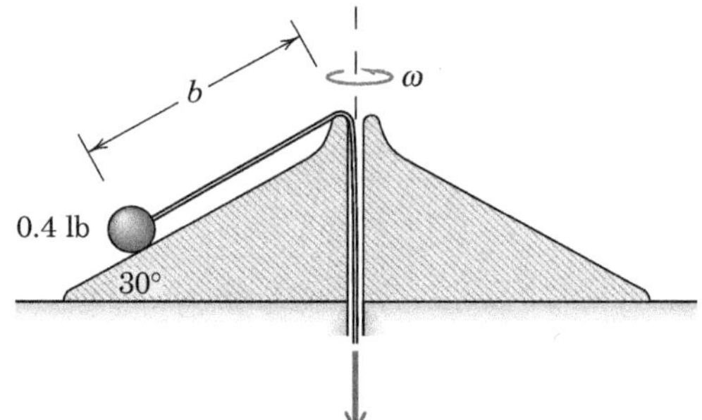

Problem 3/238

3/239 The 0.02-kg particle moves along the dashed trajectory shown and has the indicated velocities at positions A and B. Calculate the time average of the moment about O of the resultant force P acting on the particle during the 0.5 second required for it to go from A to B.

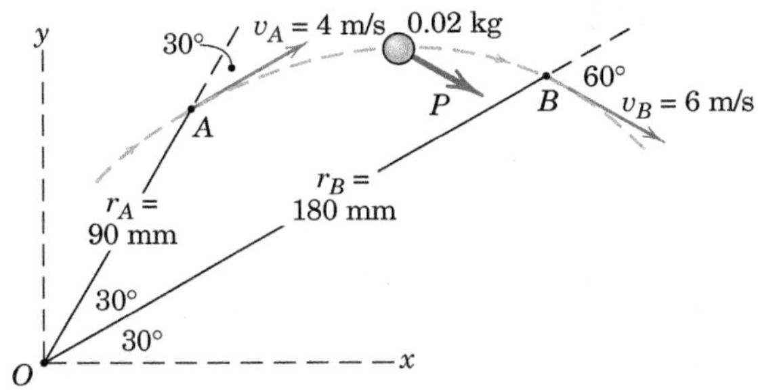

Problem 3/239

▶**3/240** The assembly of two 5-kg spheres is rotating freely about the vertical axis at 40 rev/min with $\theta = 90°$. If the force F which maintains the given position is increased to raise the base collar and reduce θ to 60°, determine the new angular velocity ω. Also determine the work U done by F in changing the configuration of the system. Assume that the mass of the arms and collars is negligible.

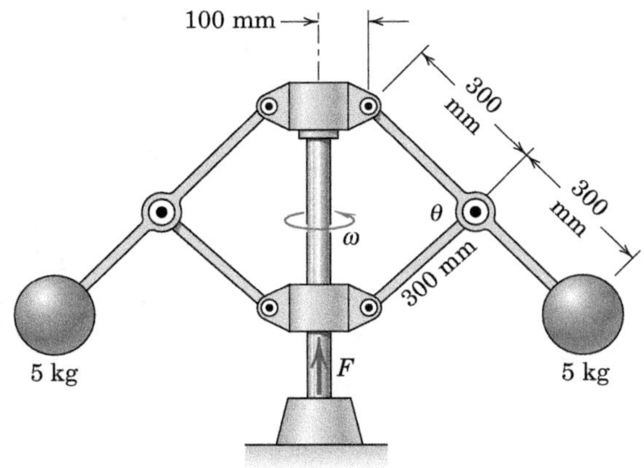

Problem 3/240

SECTION D SPECIAL APPLICATIONS

3/11 Introduction

The basic principles and methods of particle kinetics were developed and illustrated in the first three sections of this chapter. This treatment included the direct use of Newton's second law, the equations of work and energy, and the equations of impulse and momentum. We paid special attention to the kind of problem for which each of the approaches was most appropriate.

Several topics of specialized interest in particle kinetics will be briefly treated in Section D:

1. Impact
2. Central-force motion
3. Relative motion

These topics involve further extension and application of the fundamental principles of dynamics, and their study will help to broaden your background in mechanics.

3/12 Impact

The principles of impulse and momentum have important use in describing the behavior of colliding bodies. *Impact* refers to the collision between two bodies and is characterized by the generation of relatively large contact forces which act over a very short interval of time. It is important to realize that an impact is a very complex event involving material deformation and recovery and the generation of heat and sound. Small changes in the impact conditions may cause large changes in the impact process and thus in the conditions immediately following the impact. Therefore, we must be careful not to rely heavily on the results of impact calculations.

Direct Central Impact

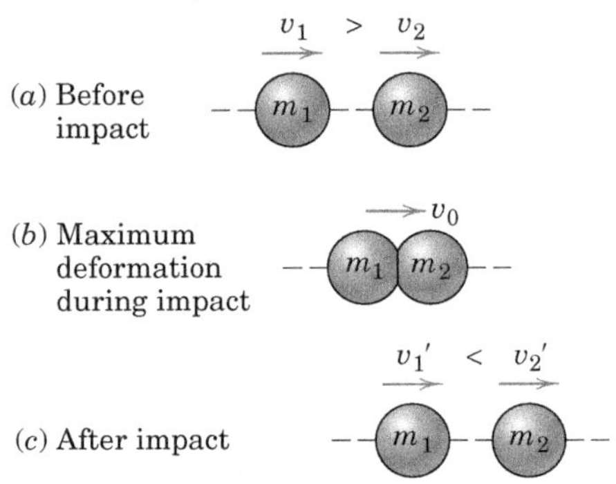

Figure 3/17

As an introduction to impact, we consider the collinear motion of two spheres of masses m_1 and m_2, Fig. 3/17*a*, traveling with velocities v_1 and v_2. If v_1 is greater than v_2, collision occurs with the contact forces directed along the line of centers. This condition is called *direct central impact.*

Following initial contact, a short period of increasing deformation takes place until the contact area between the spheres ceases to increase. At this instant, both spheres, Fig. 3/17*b*, are moving with the same velocity v_0. During the remainder of contact, a period of restoration occurs during which the contact area decreases to zero. In the final condition shown in part *c* of the figure, the spheres now have new velocities $v_1{}'$ and $v_2{}'$, where $v_1{}'$ must be less than $v_2{}'$. All velocities are arbitrarily assumed positive to the right, so that with this scalar notation a velocity to the left would carry a negative sign. If the impact is not overly severe and if the spheres are highly elastic, they will regain their original shape following the restoration. With a more severe impact and with less elastic bodies, a permanent deformation may result.

Because the contact forces are equal and opposite during impact, the linear momentum of the system remains unchanged, as discussed in Art. 3/9. Thus, we apply the law of conservation of linear momentum and write

$$m_1 v_1 + m_2 v_2 = m_1 v_1{}' + m_2 v_2{}' \qquad \textbf{(3/35)}$$

We assume that any forces acting on the spheres during impact, other than the large internal forces of contact, are relatively small and produce negligible impulses compared with the impulse associated with each internal impact force. In addition, we assume that no appreciable change in the positions of the mass centers occurs during the short duration of the impact.

Coefficient of Restitution

For given masses and initial conditions, the momentum equation contains two unknowns, $v_1{}'$ and $v_2{}'$. Clearly, we need an additional relationship to find the final velocities. This relationship must reflect the capacity of the contacting bodies to recover from the impact and can be expressed by the ratio e of the magnitude of the restoration impulse to the magnitude of the deformation impulse. This ratio is called the *coefficient of restitution.*

Let F_r and F_d represent the magnitudes of the contact forces during the restoration and deformation periods, respectively, as shown in Fig. 3/18. For particle 1 the definition of e together with the impulse-momentum equation give us

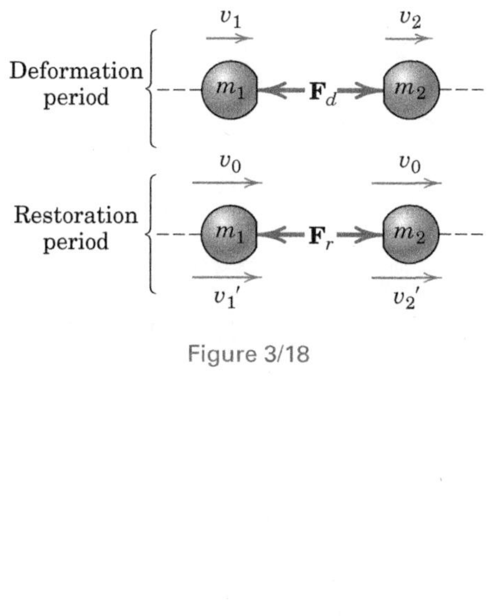

Figure 3/18

$$e = \frac{\displaystyle\int_{t_0}^{t} F_r\,dt}{\displaystyle\int_0^{t_0} F_d\,dt} = \frac{m_1[-v_1{}' - (-v_0)]}{m_1[-v_0 - (-v_1)]} = \frac{v_0 - v_1{}'}{v_1 - v_0}$$

Similarly, for particle 2 we have

$$e = \frac{\displaystyle\int_{t_0}^{t} F_r\,dt}{\displaystyle\int_0^{t_0} F_d\,dt} = \frac{m_2\,(v_2{}' - v_0)}{m_2\,(v_0 - v_2)} = \frac{v_2{}' - v_0}{v_0 - v_2}$$

We are careful in these equations to express the change of momentum (and therefore Δv) in the same direction as the impulse (and thus the force). The time for the deformation is taken as t_0 and the total time of contact is t. Eliminating v_0 between the two expressions for e gives us

$$e = \frac{v_2{}' - v_1{}'}{v_1 - v_2} = \frac{|\text{relative velocity of separation}|}{|\text{relative velocity of approach}|} \qquad \textbf{(3/36)}$$

If the two initial velocities v_1 and v_2 and the coefficient of restitution e are known, then Eqs. 3/35 and 3/36 give us two equations in the two unknown final velocities v_1' and v_2'.

Energy Loss During Impact

Impact phenomena are almost always accompanied by energy loss, which may be calculated by subtracting the kinetic energy of the system just after impact from that just before impact. Energy is lost through the generation of heat during the localized inelastic deformation of the material, through the generation and dissipation of elastic stress waves within the bodies, and through the generation of sound energy.

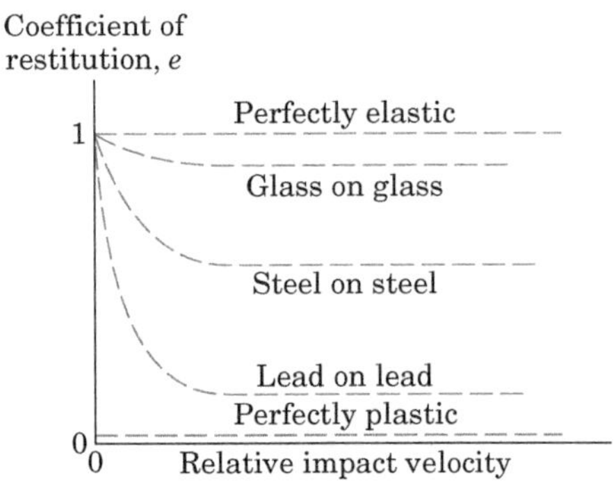

Figure 3/19

According to this classical theory of impact, the value $e = 1$ means that the capacity of the two particles to recover equals their tendency to deform. This condition is one of *elastic impact* with no energy loss. The value $e = 0$, on the other hand, describes *inelastic* or *plastic impact* where the particles cling together after collision and the loss of energy is a maximum. All impact conditions lie somewhere between these two extremes.

Also, it should be noted that a coefficient of restitution must be associated with a *pair* of contacting bodies. The coefficient of restitution is frequently considered a constant for given geometries and a given combination of contacting materials. Actually, it depends on the impact velocity and approaches unity as the impact velocity approaches zero as shown schematically in Fig. 3/19. A handbook value for e is generally unreliable.

Oblique Central Impact

We now extend the relationships developed for direct central impact to the case where the initial and final velocities are not parallel, Fig. 3/20. Here spherical particles of mass m_1 and m_2 have initial velocities $\mathbf{v}_1$ and $\mathbf{v}_2$ in the same plane and approach each other on a collision course, as shown in part a of the figure. The directions of the velocity vectors are measured from the direction tangent to the contacting surfaces, Fig. 3/20b. Thus, the initial velocity components along the t- and n-axes are $(v_1)_n = -v_1 \sin \theta_1$, $(v_1)_t = v_1 \cos \theta_1$, $(v_2)_n = v_2 \sin \theta_2$, and $(v_2)_t = v_2 \cos \theta_2$. Note that

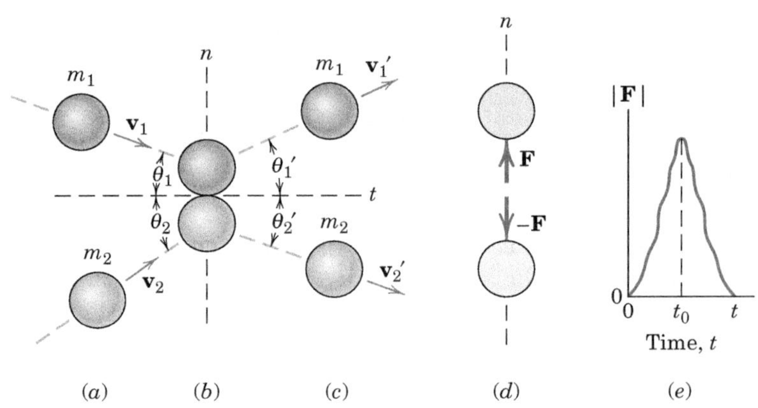

Figure 3/20

$(v_1)_n$ is a negative quantity for the particular coordinate system and initial velocities shown.

The final rebound conditions are shown in part *c* of the figure. The impact forces are **F** and $-\mathbf{F}$, as seen in part *d* of the figure. They vary from zero to their peak value during the deformation portion of the impact and back again to zero during the restoration period, as indicated in part *e* of the figure where *t* is the duration of the impact interval.

For given initial conditions of m_1, m_2, $(v_1)_n$, $(v_1)_t$, $(v_2)_n$, and $(v_2)_t$, there will be four unknowns, namely, $(v_1')_n$, $(v_1')_t$, $(v_2')_n$, and $(v_2')_t$. The four needed equations are obtained as follows:

(1) Momentum of the system is conserved in the *n*-direction. This gives

$$m_1(v_1)_n + m_2(v_2)_n = m_1(v_1')_n + m_2(v_2')_n$$

(2) and (3) The momentum for each particle is conserved in the *t*-direction since there is no impulse on either particle in the *t*-direction. Thus,

$$m_1(v_1)_t = m_1(v_1')_t$$

$$m_2(v_2)_t = m_2(v_2')_t$$

(4) The coefficient of restitution, as in the case of direct central impact, is the positive ratio of the recovery impulse to the deformation impulse. Equation 3/36 applies, then, to the velocity components in the *n*-direction. For the notation adopted with Fig. 3/20, we have

$$e = \frac{(v_2')_n - (v_1')_n}{(v_1)_n - (v_2)_n}$$

Once the four final velocity components are found, the angles θ_1' and θ_2' of Fig. 3/20 may be easily determined.

Kunstgalerie Aquarius/Getty Images, Inc.

The motion of pool balls after impact is easily investigated with the principles of direct and oblique central impact.

Sample Problem 3/28

The ram of a pile driver has a mass of 800 kg and is released from rest 2 m above the top of the 2400-kg pile. If the ram rebounds to a height of 0.1 m after impact with the pile, calculate (*a*) the velocity v_p' of the pile immediately after impact, (*b*) the coefficient of restitution e, and (*c*) the percentage loss of energy due to the impact.

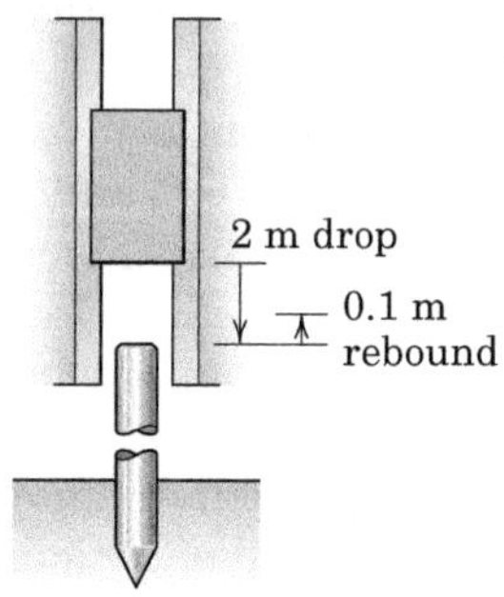

Solution. Conservation of energy during free fall gives the initial and final velocities of the ram from $v = \sqrt{2gh}$. Thus,

$$v_r = \sqrt{2(9.81)(2)} = 6.26 \text{ m/s} \qquad v_r' = \sqrt{2(9.81)(0.1)} = 1.40 \text{ m/s}$$

① ***(a)*** Conservation of momentum ($G_1 = G_2$) for the system of the ram and pile gives

$$800(6.26) + 0 = 800(-1.401) + 2400v_p' \qquad v_p' = 2.55 \text{ m/s} \qquad \textit{Ans.}$$

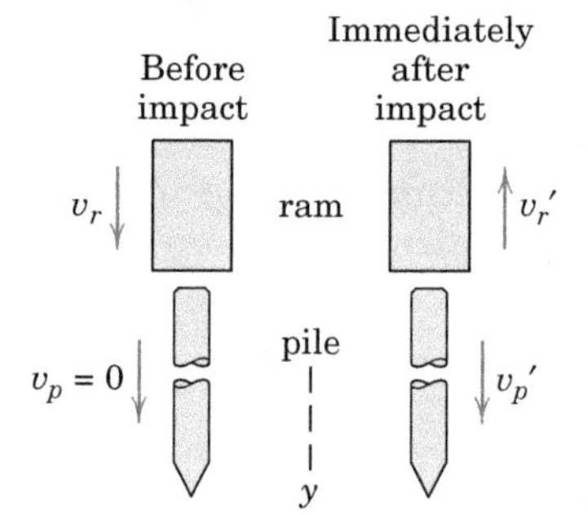

(b) The coefficient of restitution yields

$$e = \frac{|\text{rel. vel. separation}|}{|\text{rel. vel. approach}|} \qquad e = \frac{2.55 + 1.401}{6.26 + 0} = 0.631 \qquad \textit{Ans.}$$

(c) The kinetic energy of the system just before impact is the same as the potential energy of the ram above the pile and is

$$T = V_g = mgh = 800(9.81)(2) = 15\,700 \text{ J}$$

The kinetic energy T' just after impact is

$$T' = \tfrac{1}{2}(800)(1.401)^2 + \tfrac{1}{2}(2400)(2.55)^2 = 8620 \text{ J}$$

The percentage loss of energy is, therefore,

$$\frac{15\,700 - 8620}{15\,700}(100) = 45.1\% \qquad \textit{Ans.}$$

Helpful Hint

① The impulses of the weights of the ram and pile are very small compared with the impulses of the impact forces and thus are neglected during the impact.

Sample Problem 3/29

A ball is projected onto the heavy plate with a velocity of 50 ft/sec at the 30° angle shown. If the effective coefficient of restitution is 0.5, compute the rebound velocity v' and its angle θ'.

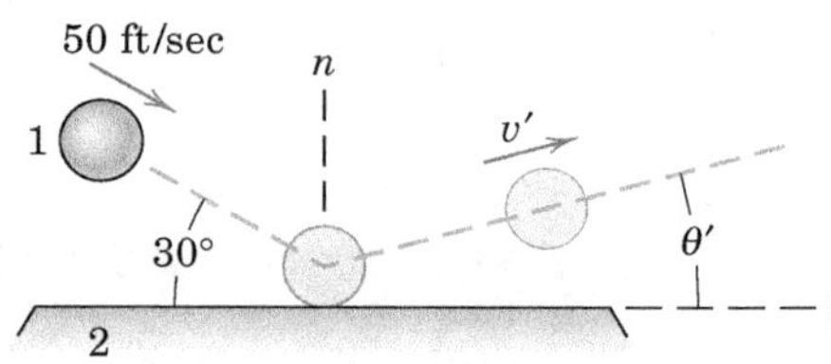

Solution. Let the ball be denoted body 1 and the plate body 2. The mass of the heavy plate may be considered infinite and its corresponding velocity zero after impact. The coefficient of restitution is applied to the velocity components normal to the plate in the direction of the impact force and gives

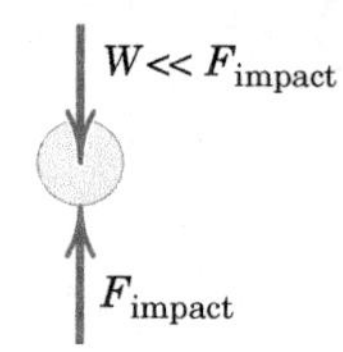

① $$e = \frac{(v_2')_n - (v_1')_n}{(v_1)_n - (v_2)_n} \qquad 0.5 = \frac{0 - (v_1')_n}{-50 \sin 30° - 0} \qquad (v_1')_n = 12.5 \text{ ft/sec}$$

Momentum of the ball in the t-direction is unchanged since, with assumed smooth surfaces, there is no force acting on the ball in that direction. Thus,

$$m(v_1)_t = m(v_1')_t \qquad (v_1')_t = (v_1)_t = 50 \cos 30° = 43.3 \text{ ft/sec}$$

The rebound velocity v' and its angle θ' are then

$$v' = \sqrt{(v_1')_n{}^2 + (v_1')_t{}^2} = \sqrt{12.5^2 + 43.3^2} = 45.1 \text{ ft/sec} \qquad \textit{Ans.}$$

$$\theta' = \tan^{-1}\left(\frac{(v_1')_n}{(v_1')_t}\right) = \tan^{-1}\left(\frac{12.5}{43.3}\right) = 16.10° \qquad \textit{Ans.}$$

Helpful Hint

① We observe here that for infinite mass there is no way of applying the principle of conservation of momentum for the system in the n-direction. From the free-body diagram of the ball during impact, we note that the impulse of the weight W is neglected since W is very small compared with the impact force.

Sample Problem 3/30

Spherical particle 1 has a velocity $v_1 = 6$ m/s in the direction shown and collides with spherical particle 2 of equal mass and diameter and initially at rest. If the coefficient of restitution for these conditions is $e = 0.6$, determine the resulting motion of each particle following impact. Also calculate the percentage loss of energy due to the impact.

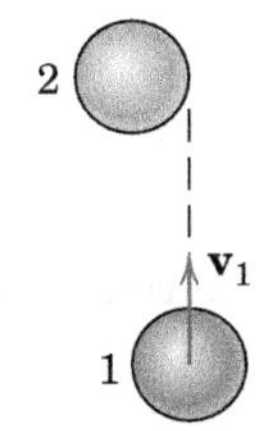

Solution. The geometry at impact indicates that the normal n to the contacting surfaces makes an angle $\theta = 30°$ with the direction of $\mathbf{v}_1$, as indicated ① in the figure. Thus, the initial velocity components are $(v_1)_n = v_1 \cos 30° = 6 \cos 30° = 5.20$ m/s, $(v_1)_t = v_1 \sin 30° = 6 \sin 30° = 3$ m/s, and $(v_2)_n = (v_2)_t = 0$.

Momentum conservation for the two-particle system in the n-direction gives

$$m_1(v_1)_n + m_2(v_2)_n = m_1(v_1')_n + m_2(v_2')_n$$

or, with $m_1 = m_2$,

$$5.20 + 0 = (v_1')_n + (v_2')_n \qquad (a)$$

The coefficient-of-restitution relationship is

$$e = \frac{(v_2')_n - (v_1')_n}{(v_1)_n - (v_2)_n} \qquad 0.6 = \frac{(v_2')_n - (v_1')_n}{5.20 - 0} \qquad (b)$$

② Simultaneous solution of Eqs. a and b yields

$$(v_1')_n = 1.039 \text{ m/s} \qquad (v_2')_n = 4.16 \text{ m/s}$$

Conservation of momentum for each particle holds in the t-direction because, with assumed smooth surfaces, there is no force in the t-direction. Thus for particles 1 and 2, we have

$$m_1(v_1)_t = m_1(v_1')_t \qquad (v_1')_t = (v_1)_t = 3 \text{ m/s}$$

③ $$m_2(v_2)_t = m_2(v_2')_t \qquad (v_2')_t = (v_2)_t = 0$$

The final speeds of the particles are

$$v_1' = \sqrt{(v_1')_n{}^2 + (v_1')_t{}^2} = \sqrt{(1.039)^2 + 3^2} = 3.17 \text{ m/s} \qquad \textit{Ans.}$$

$$v_2' = \sqrt{(v_2')_n{}^2 + (v_2')_t{}^2} = \sqrt{(4.16)^2 + 0^2} = 4.16 \text{ m/s} \qquad \textit{Ans.}$$

The angle θ' which $\mathbf{v}_1'$ makes with the t-direction is

$$\theta' = \tan^{-1}\left(\frac{(v_1')_n}{(v_1')_t}\right) = \tan^{-1}\left(\frac{1.039}{3}\right) = 19.11° \qquad \textit{Ans.}$$

The kinetic energies just before and just after impact, with $m = m_1 = m_2$, are

$$T = \tfrac{1}{2}m_1 v_1{}^2 + \tfrac{1}{2}m_2 v_2{}^2 = \tfrac{1}{2}m(6)^2 + 0 = 18m$$

$$T' = \tfrac{1}{2}m_1 v_1'^2 + \tfrac{1}{2}m_2 v_2'^2 = \tfrac{1}{2}m(3.17)^2 + \tfrac{1}{2}m(4.16)^2 = 13.68m$$

The percentage energy loss is then

$$\frac{|\Delta E|}{E}(100) = \frac{T - T'}{T}(100) = \frac{18m - 13.68m}{18m}(100) = 24.0\% \qquad \textit{Ans.}$$

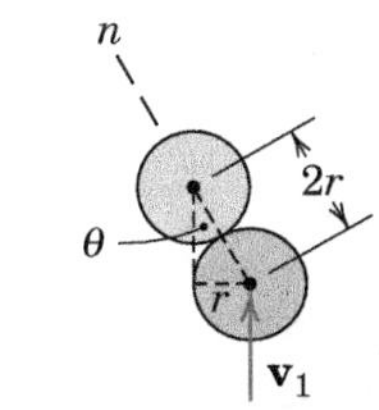

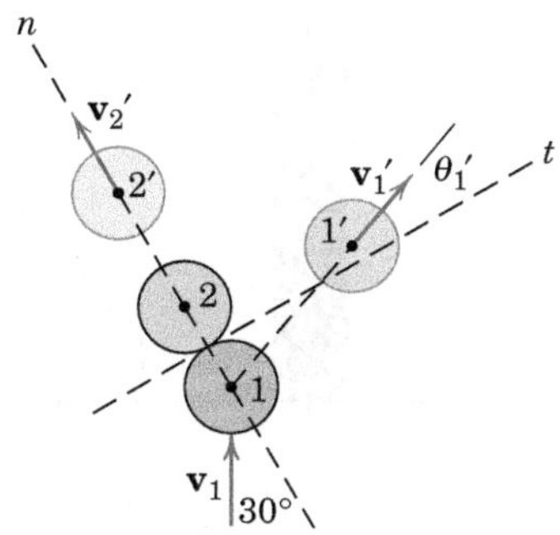

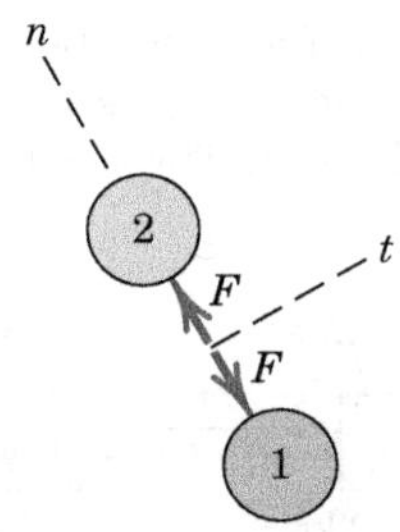

Helpful Hints

① Be sure to set up n- and t-coordinates which are, respectively, normal to and tangent to the contacting surfaces. Calculation of the 30° angle is critical to all that follows.

② Note that, even though there are four equations in four unknowns for the standard problem of oblique central impact, only one pair of the equations is coupled.

③ We note that particle 2 has no initial or final velocity component in the t-direction. Hence, its final velocity $\mathbf{v}_2'$ is restricted to the n-direction.

PROBLEMS

Introductory Problems

3/241 As a check of the basketball before the start of a game, the referee releases the ball from the overhead position shown, and the ball rebounds to about waist level. Determine the coefficient of restitution e and the percentage n of the original energy lost during the impact.

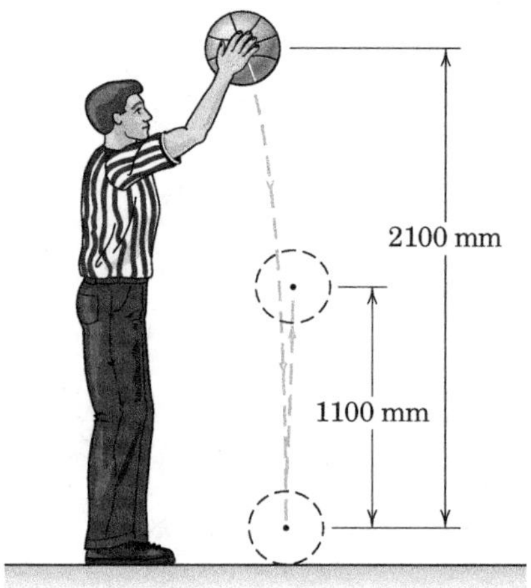

Problem 3/241

3/242 Compute the final velocities v_1' and v_2' after collision of the two cylinders which slide on the smooth horizontal shaft. The coefficient of restitution is $e = 0.8$.

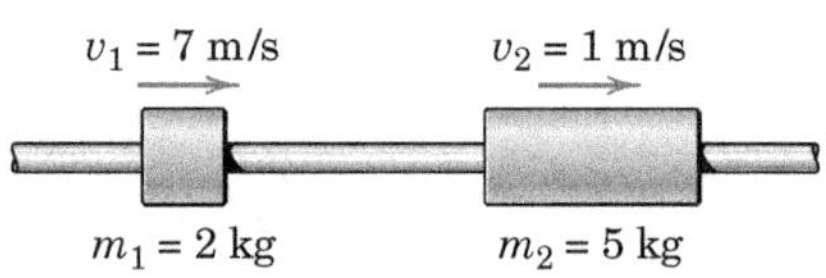

Problem 3/242

3/243 Car B is initially stationary and is struck by car A, which is moving with speed v. The mass of car B is pm, where m is the mass of car A and p is a positive constant. If the coefficient of restitution is $e = 0.1$, express the speeds v_A' and v_B' of the two cars at the end of the impact in terms of p and v. Evaluate your expressions for $p = 0.5$.

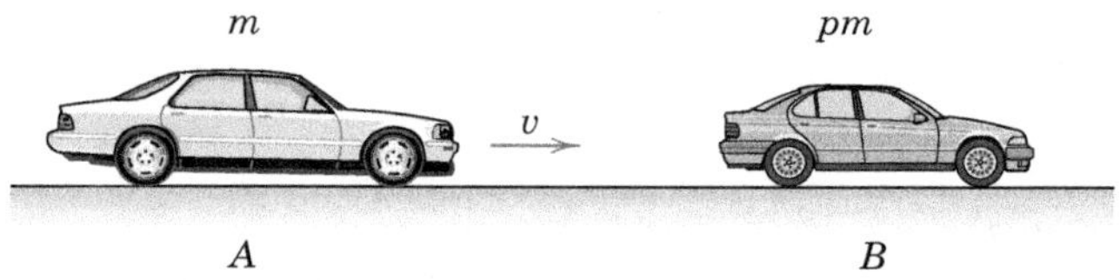

Problem 3/243

3/244 The sphere of mass m_1 travels with an initial velocity v_1 directed as shown and strikes the stationary sphere of mass m_2. For a given coefficient of restitution e, what condition on the mass ratio m_1/m_2 ensures that the final velocity of m_2 is greater than v_1?

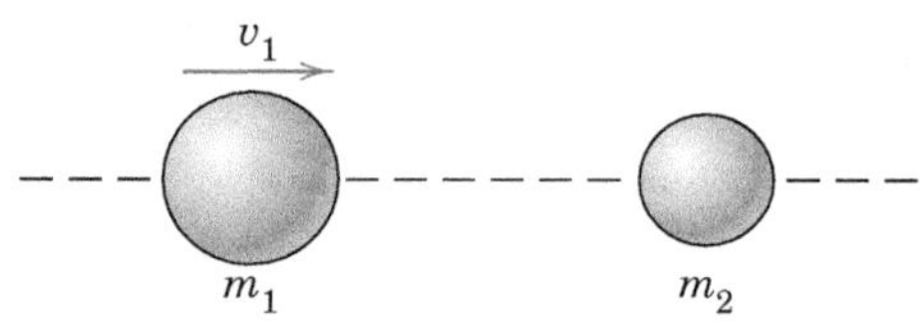

Problem 3/244

3/245 A tennis ball is projected toward a smooth surface with speed v as shown. Determine the rebound angle θ' and the final speed v'. The coefficient of restitution is 0.6.

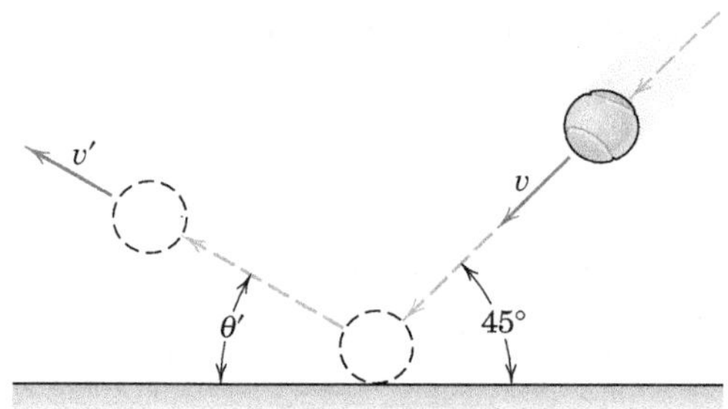

Problem 3/245

3/246 Determine the coefficient of restitution e for a steel ball dropped from rest at a height h above a heavy horizontal steel plate if the height of the second rebound is h_2.

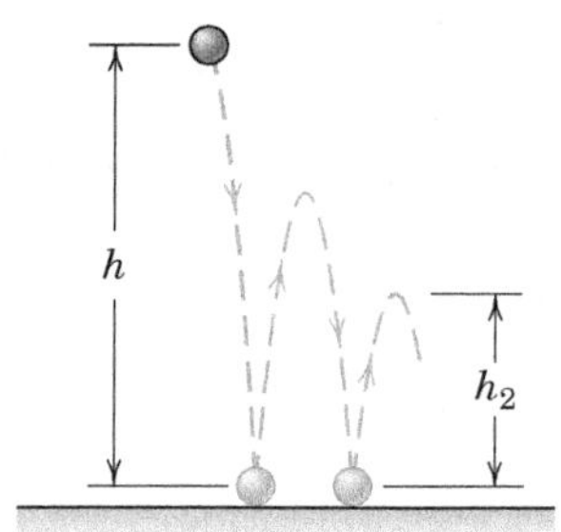

Problem 3/246

3/247 Determine the value of the coefficient of restitution e for which the outgoing angle is one-half of the incoming angle θ as shown. Evaluate your general expression for $\theta = 40°$.

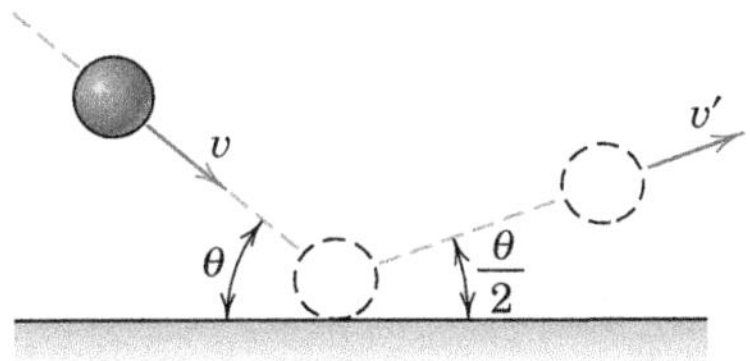

Problem 3/247

3/248 To pass inspection, steel balls designed for use in ball bearings must clear the fixed bar A at the top of their rebound when dropped from rest through the vertical distance $H = 36$ in. onto the heavy inclined steel plate. If balls which have a coefficient of restitution of less than 0.7 with the rebound plate are to be rejected, determine the position of the bar by specifying h and s. Neglect any friction during impact.

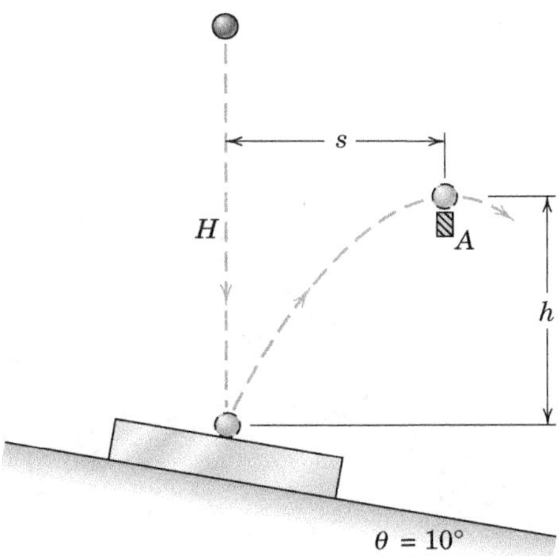

Problem 3/248

3/249 The cart of mass $m_1 = 3$ kg is moving to the right with a speed $v_1 = 6$ m/s when it collides with the initially stationary barrier of mass $m_2 = 5$ kg. The coefficient of restitution for this collision is $e = 0.75$. Determine the maximum deflection δ of the barrier, which is connected to three springs, each of which has a modulus of 4 kN/m and is undeformed before the impact.

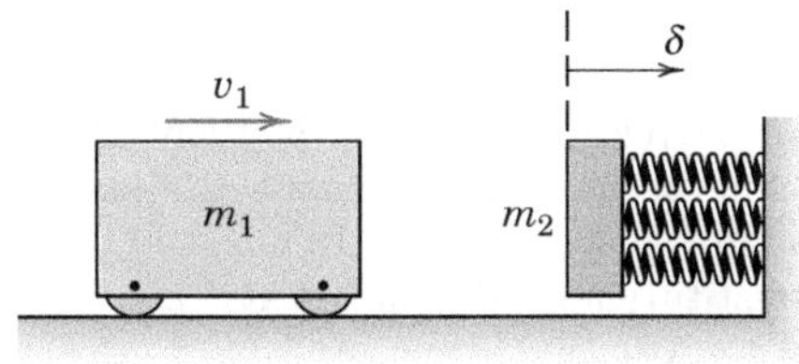

Problem 3/249

3/250 If the center of the ping-pong ball is to clear the net as shown, at what height h should the ball be horizontally served? Also determine h_2. The coefficient of restitution for the impacts between ball and table is $e = 0.9$, and the radius of the ball is $r = 0.75$ in.

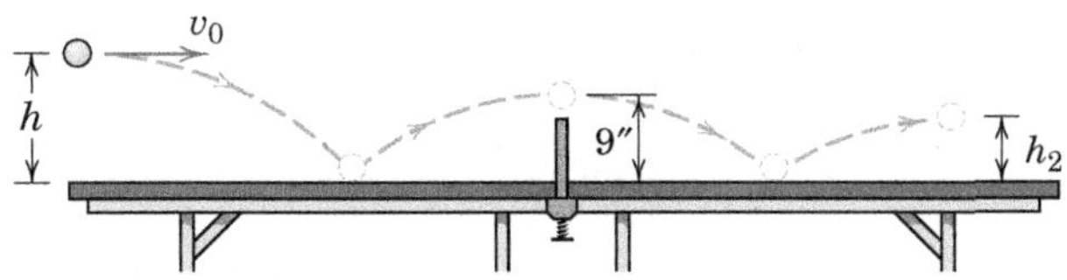

Problem 3/250

3/251 Two steel balls of the same diameter are connected by a rigid bar of negligible mass as shown and are dropped in the horizontal position from a height of 150 mm above the heavy steel and brass base plates. If the coefficient of restitution between the ball and the steel base is 0.6 and that between the other ball and the brass base is 0.4, determine the angular velocity ω of the bar immediately after impact. Assume that the two impacts are simultaneous.

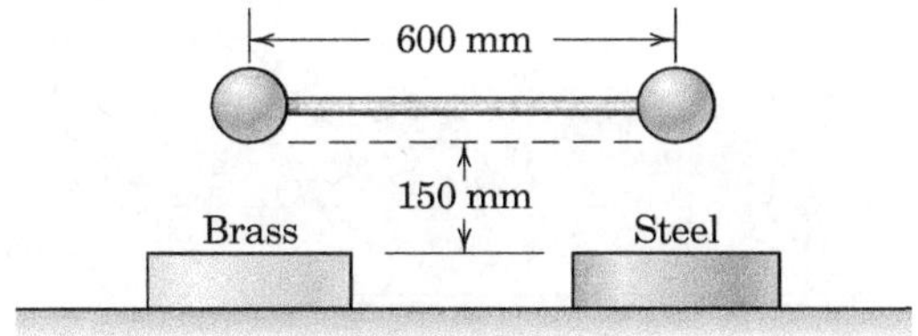

Problem 3/251

3/252 Freight car A of mass m_A is rolling to the right when it collides with freight car B of mass m_B initially at rest. If the two cars are coupled together at impact, show that the fractional loss of energy equals $m_B/(m_A + m_B)$.

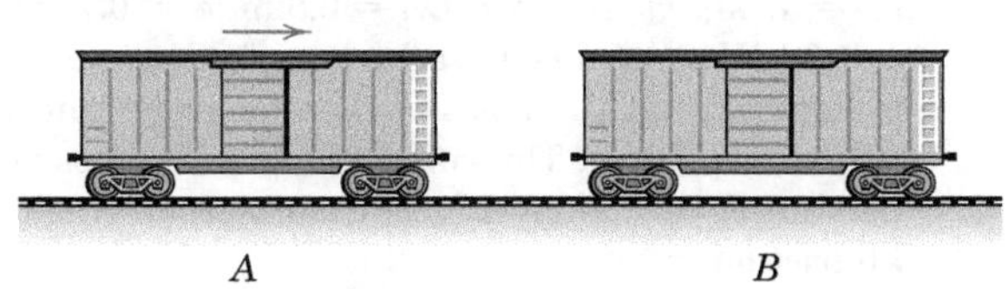

Problem 3/252

Representative Problems

3/253 A small ball is projected horizontally toward an incline as shown. Determine the slant range R. The initial speed is $v_0 = 16$ m/s, and the coefficient of restitution for the impact at A is $e = 0.6$.

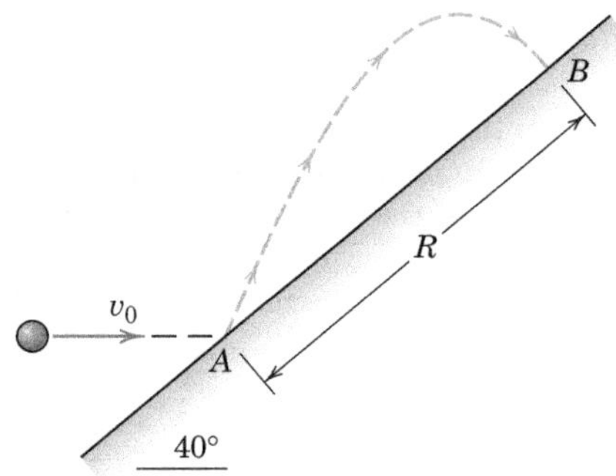

Problem 3/253

3/254 A miniature-golf shot from position A to the hole D is to be accomplished by "banking off" the 45° wall. Using the theory of this article, determine the location x for which the shot can be made. The coefficient of restitution associated with the wall collision is $e = 0.8$.

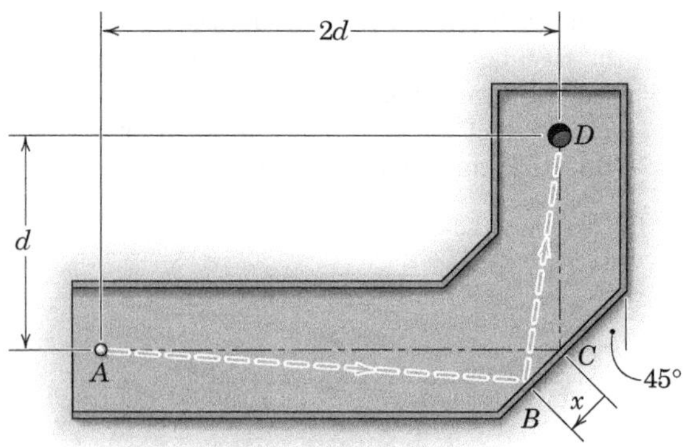

Problem 3/254

3/255 The pendulum is released from the 60° position and then strikes the initially stationary cylinder of mass m_2 when OA is vertical. Determine the maximum spring compression δ. Use the values $m_1 = 3$ kg, $m_2 = 2$ kg, $\overline{OA} = 0.8$ m, $e = 0.7$, and $k = 6$ kN/m. Assume that the bar of the pendulum is light so that the mass m_1 is effectively concentrated at point A. The rubber cushion S stops the pendulum just after the collision is over. Neglect all friction.

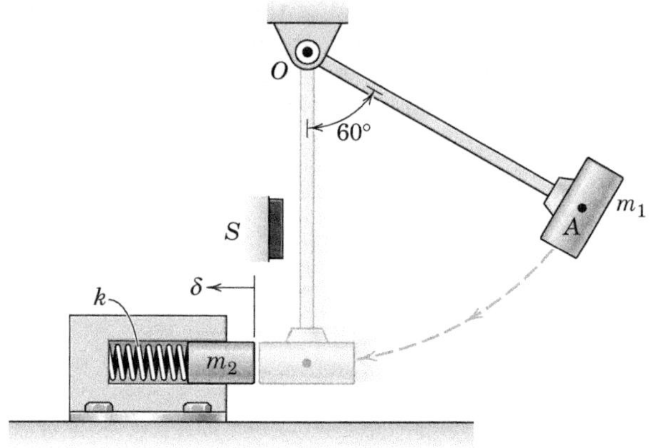

Problem 3/255

3/256 A 0.1-kg meteor and a 1000-kg spacecraft have the indicated absolute velocities just before colliding. The meteor punches a hole entirely through the spacecraft. Instruments indicate that the velocity of the meteor relative to the spacecraft just after the collision is $\mathbf{v}_{m/s}' = -1880\mathbf{i} - 6898\mathbf{j}$ m/s. Determine the direction θ of the absolute velocity of the spacecraft after the collision.

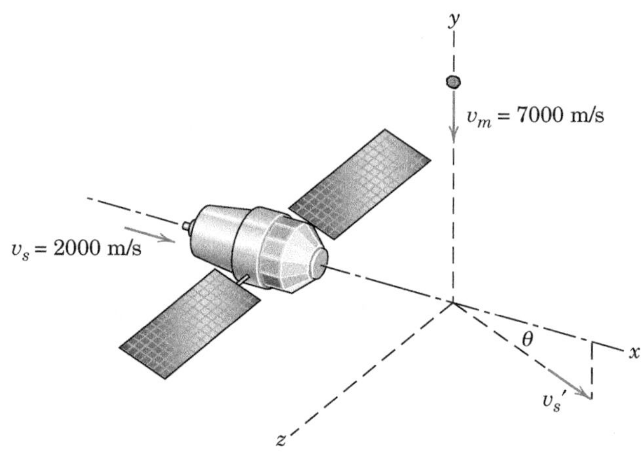

Problem 3/256

3/257 In a pool game the cue ball A must strike the eight ball in the position shown in order to send it to the pocket P with a velocity v_2'. The cue ball has a velocity v_1 before impact and a velocity v_1' after impact. The coefficient of restitution is 0.9. Both balls have the same mass and diameter. Calculate the rebound angle θ and the fraction n of the kinetic energy which is lost during the impact.

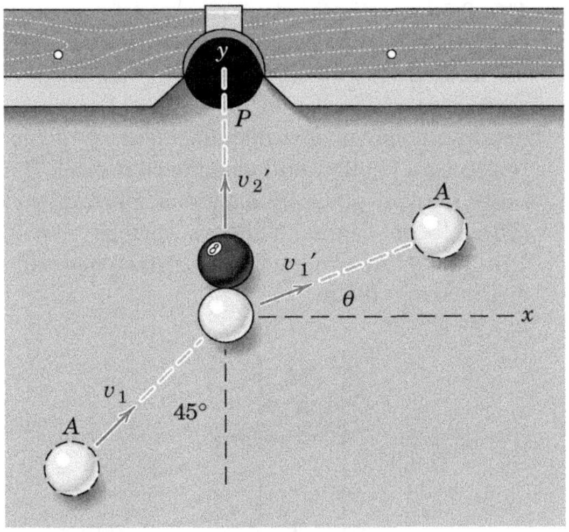

Problem 3/257

3/258 Determine the coefficient of restitution e which will allow the ball to bounce down the steps as shown. The tread and riser dimensions, d and h, respectively, are the same for every step, and the ball bounces the same distance h' above each step. What horizontal velocity v_x is required so that the ball lands in the center of each tread?

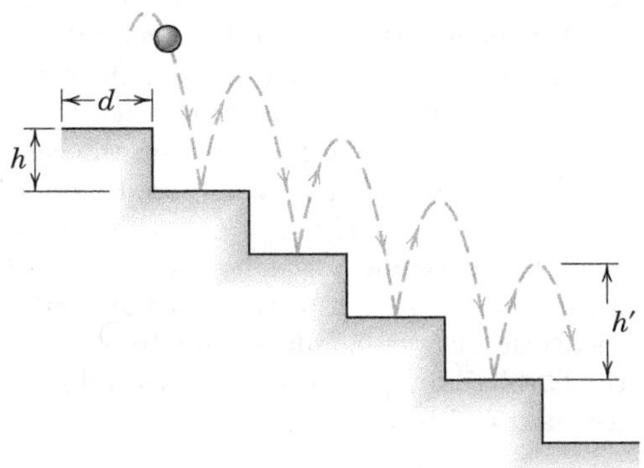

Problem 3/258

3/259 Sphere A has a mass of 23 kg and a radius of 75 mm, while sphere B has a mass of 4 kg and a radius of 50 mm. If the spheres are traveling initially along the parallel paths with the speeds shown, determine the velocities of the spheres immediately after impact. Specify the angles θ_A and θ_B with respect to the x-axis made by the rebound velocity vectors. The coefficient of restitution is 0.4 and friction is neglected.

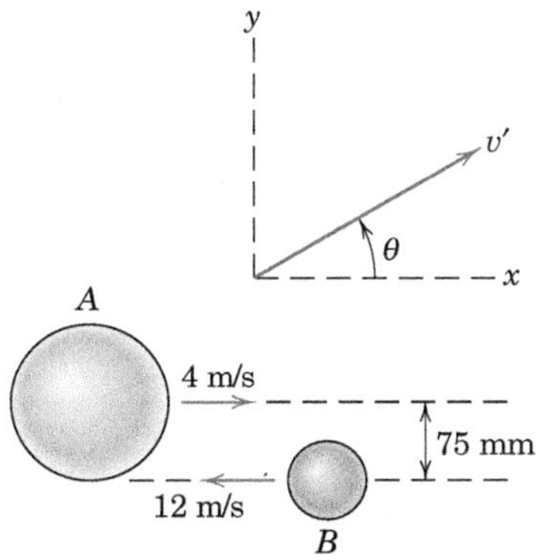

Problem 3/259

3/260 During a pregame warmup period, two basketballs collide above the hoop when in the positions shown. Just before impact, ball 1 has a velocity v_1 which makes a 30° angle with the horizontal. If the velocity v_2 of ball 2 just before impact has the same magnitude as v_1, determine the two possible values of the angle θ, measured from the horizontal, which will cause ball 1 to go directly through the center of the basket. The coefficient of restitution is $e = 0.8$.

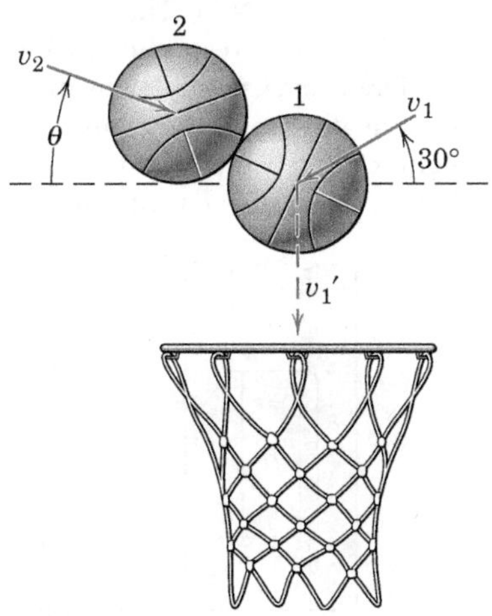

Problem 3/260

3/261 Two identical hockey pucks moving with initial velocities v_A and v_B collide as shown. If the coefficient of restitution is $e = 0.75$, determine the velocity (magnitude and direction θ with respect to the positive x-axis) of each puck just after impact. Also calculate the percentage loss n of system kinetic energy.

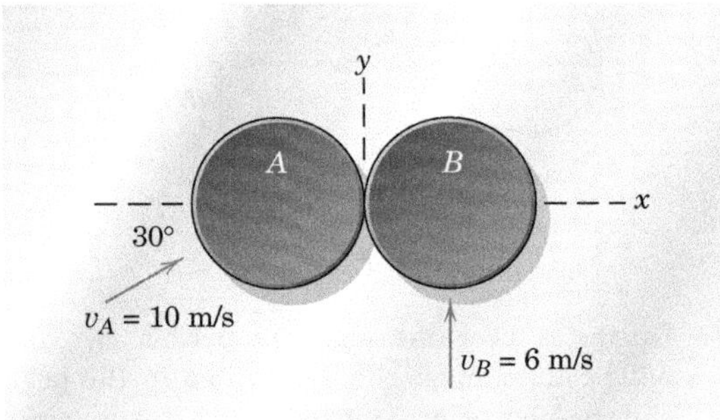

Problem 3/261

3/262 Repeat the previous problem, only now the mass of puck B is twice that of puck A.

3/263 The 3000-kg anvil A of the drop forge is mounted on a nest of heavy coil springs having a combined stiffness of $2.8(10^6)$ N/m. The 600-kg hammer B falls 500 mm from rest and strikes the anvil, which suffers a maximum downward deflection of 24 mm from its equilibrium position. Determine the height h of rebound of the hammer and the coefficient of restitution e which applies.

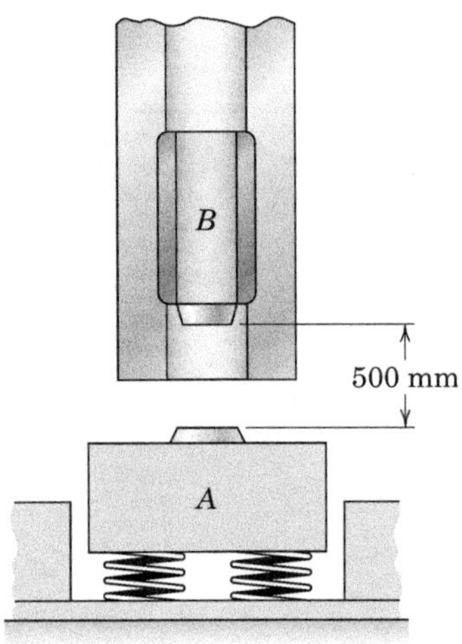

Problem 3/263

3/264 The 0.5-kg cylinder A is released from rest from the position shown and drops the distance $h_1 = 0.6$ m. It then collides with the 0.4-kg block B; the coefficient of restitution is $e = 0.8$. Determine the maximum downward displacement h_2 of block B. Neglect all friction and assume that block B is initially held in place by a hidden mechanism until the collision begins. The two springs of modulus $k = 500$ N/m are initially unstretched, and the distance $d = 0.8$ m.

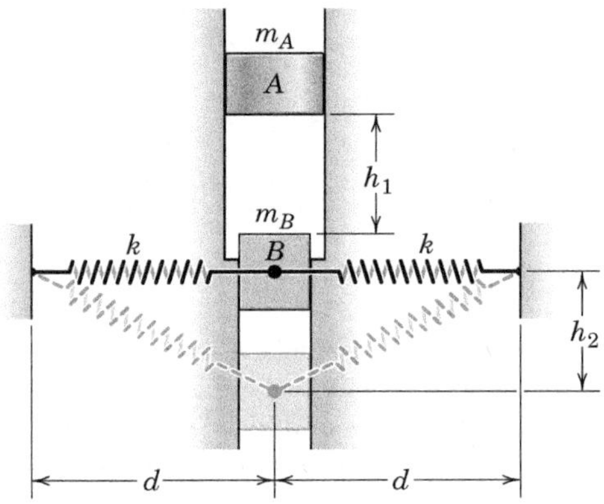

Problem 3/264

▶3/265 The elements of a device designed to measure the coefficient of restitution of bat–baseball collisions are shown. The 1-lb "bat" A is a short length of wood or aluminum which is projected to the right with a speed $v_A = 60$ ft/sec within the confines of the horizontal slot. Just before and after the moment of impact, body A is free to move horizontally. The baseball B weighs 5.125 oz and has an initial speed $v_B = 125$ ft/sec. If the coefficient of restitution is $e = 0.5$, determine the final speed of the baseball and the angle β which its final velocity makes with the horizontal.

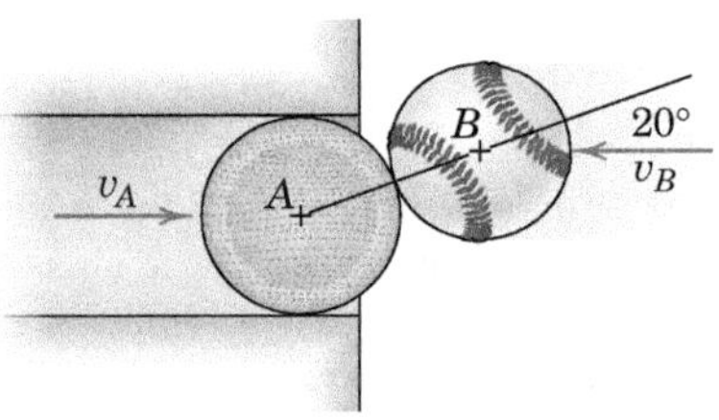

Problem 3/265

▶3/266 A child throws a ball from point A with a speed of 50 ft/sec. It strikes the wall at point B and then returns exactly to point A. Determine the necessary angle α if the coefficient of restitution in the wall impact is $e = 0.5$.

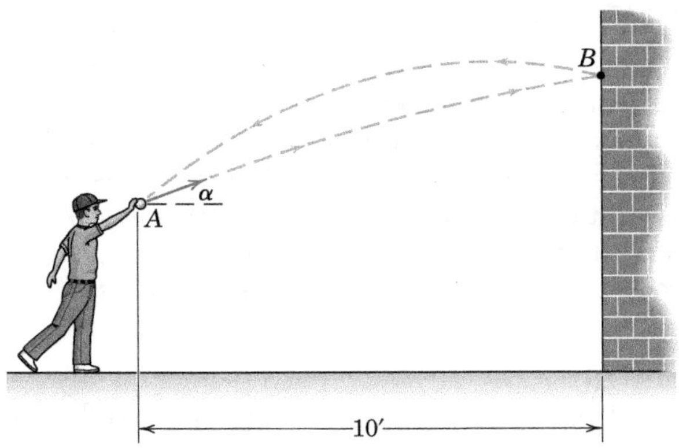

Problem 3/266

▶3/267 The 2-kg sphere is projected horizontally with a velocity of 10 m/s against the 10-kg carriage which is backed up by the spring with stiffness of 1600 N/m. The carriage is initially at rest with the spring uncompressed. If the coefficient of restitution is 0.6, calculate the rebound velocity v', the rebound angle θ, and the maximum travel δ of the carriage after impact.

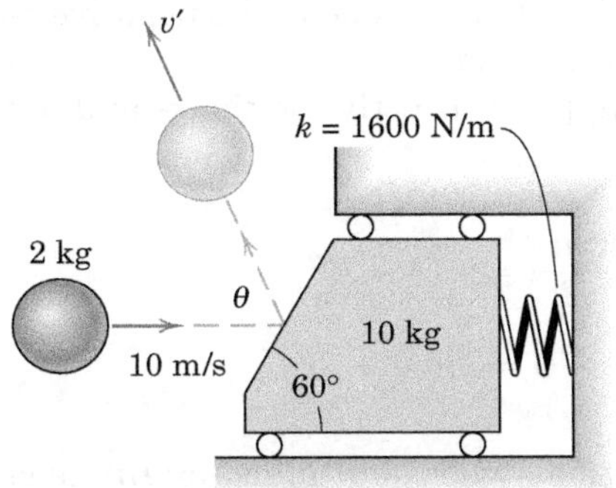

Problem 3/267

▶3/268 A small ball is projected horizontally as shown and bounces at point A. Determine the range of initial speed v_0 for which the ball will ultimately land on the horizontal surface at B. The coefficient of restitution at A is $e = 0.8$ and the distance $d = 4$ m.

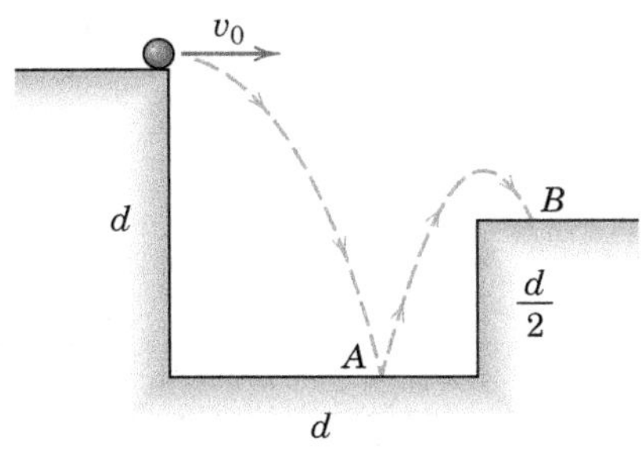

Problem 3/268

3/13 Central-Force Motion

When a particle moves under the influence of a force directed toward a fixed center of attraction, the motion is called *central-force motion*. The most common example of central-force motion is the orbital movement of planets and satellites. The laws which govern this motion were deduced from observation of the motions of the planets by J. Kepler (1571–1630). An understanding of central-force motion is required to design high-altitude rockets, earth satellites, and space vehicles.

Motion of a Single Body

Consider a particle of mass m, Fig. 3/21, moving under the action of the central gravitational attraction

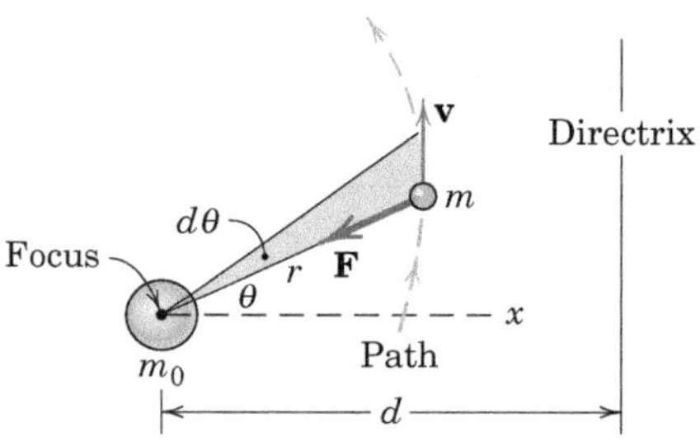

Figure 3/21

$$F = G\frac{mm_0}{r^2}$$

where m_0 is the mass of the attracting body, which is assumed to be fixed, G is the universal gravitational constant, and r is the distance between the centers of the masses. The particle of mass m could represent the earth moving about the sun, the moon moving about the earth, or a satellite in its orbital motion about the earth above the atmosphere.

The most convenient coordinate system to use is polar coordinates in the plane of motion since $\mathbf{F}$ will always be in the negative r-direction and there is no force in the θ-direction.

Equations 3/8 may be applied directly for the r- and θ-directions to give

$$\begin{aligned} -G\frac{mm_0}{r^2} &= m(\ddot{r} - r\dot{\theta}^2) \\ 0 &= m(r\ddot{\theta} + 2\dot{r}\dot{\theta}) \end{aligned} \qquad \textbf{(3/37)}$$

The second of the two equations when multiplied by r/m is seen to be the same as $d(r^2\dot{\theta})/dt = 0$, which is integrated to give

$$r^2\dot{\theta} = h, \qquad \text{a constant} \qquad \textbf{(3/38)}$$

The physical significance of Eq. 3/38 is made clear when we note that the angular momentum $\mathbf{r} \times m\mathbf{v}$ of m about m_0 has the magnitude $mr^2\dot{\theta}$. Thus, Eq. 3/38 merely states that the angular momentum of m about m_0 remains constant (is conserved). This statement is easily deduced from Eq. 3/31, which shows that the angular momentum $\mathbf{H}_O$ remains constant (is conserved) if there is no moment acting on the particle about a fixed point O.

We observe that during time dt, the radius vector sweeps out an area, shaded in Fig. 3/21, equal to $dA = (\frac{1}{2}r)(r\,d\theta)$. Therefore, the rate at which area is swept by the radius vector is $\dot{A} = \frac{1}{2}r^2\dot{\theta}$, which is constant according to Eq. 3/38. This conclusion is expressed in Kepler's *second law* of planetary motion, which states that the areas swept through in equal times are equal.

The shape of the path followed by m may be obtained by solving the first of Eqs. 3/37, with the time t eliminated through combination with Eq. 3/38. To this end the mathematical substitution $r = 1/u$ is useful. Thus, $\dot{r} = -(1/u^2)\dot{u}$, which from Eq. 3/38 becomes $\dot{r} = -h(\dot{u}/\dot{\theta})$ or $\dot{r} = -h(du/d\theta)$. The second time derivative is $\ddot{r} = -h(d^2u/d\theta^2)\dot{\theta}$, which by combining with Eq. 3/38, becomes $\ddot{r} = -h^2u^2(d^2u/d\theta^2)$. Substitution into the first of Eqs. 3/37 now gives

$$-Gm_0u^2 = -h^2u^2\frac{d^2u}{d\theta^2} - \frac{1}{u}h^2u^4$$

or

$$\frac{d^2u}{d\theta^2} + u = \frac{Gm_0}{h^2} \tag{3/39}$$

which is a nonhomogeneous linear differential equation.

The solution of this familiar second-order equation may be verified by direct substitution and is

$$u = \frac{1}{r} = C\cos(\theta + \delta) + \frac{Gm_0}{h^2}$$

where C and δ are the two integration constants. The phase angle δ may be eliminated by choosing the x-axis so that r is a minimum when $\theta = 0$. Thus,

$$\frac{1}{r} = C\cos\theta + \frac{Gm_0}{h^2} \tag{3/40}$$

Conic Sections

The interpretation of Eq. 3/40 requires a knowledge of the equations for conic sections. We recall that a conic section is formed by the locus of a point which moves so that the ratio e of its distance from a point (focus) to a line (directrix) is constant. Thus, from Fig. 3/21, $e = r/(d - r\cos\theta)$, which may be rewritten as

$$\frac{1}{r} = \frac{1}{d}\cos\theta + \frac{1}{ed} \tag{3/41}$$

which is the same form as Eq. 3/40. Thus, we see that the motion of m is along a conic section with $d = 1/C$ and $ed = h^2/(Gm_0)$, or

$$e = \frac{h^2C}{Gm_0} \tag{3/42}$$

The three cases to be investigated correspond to $e < 1$ (ellipse), $e = 1$ (parabola), and $e > 1$ (hyperbola). The trajectory for each of these cases is shown in Fig. 3/22.

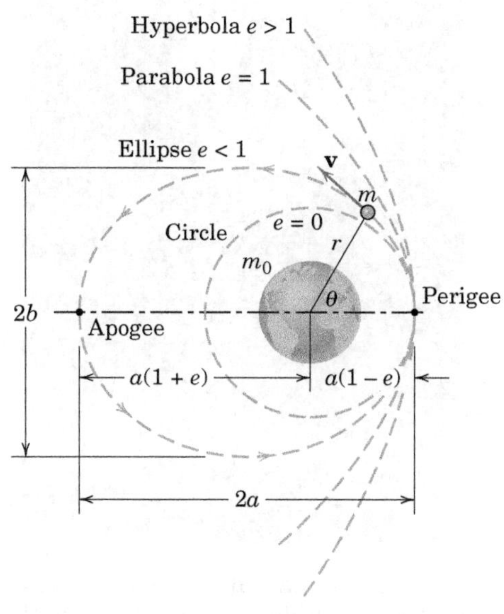

Figure 3/22

Case 1: ellipse ($e < 1$). From Eq. 3/41 we deduce that r is a minimum when $\theta = 0$ and is a maximum when $\theta = \pi$. Thus,

$$2a = r_{\min} + r_{\max} = \frac{ed}{1+e} + \frac{ed}{1-e} \qquad \text{or} \qquad a = \frac{ed}{1-e^2}$$

With the distance d expressed in terms of a, Eq. 3/41 and the maximum and minimum values of r may be written as

$$\frac{1}{r} = \frac{1 + e\cos\theta}{a(1-e^2)}$$
$$r_{\min} = a(1-e) \qquad r_{\max} = a(1+e) \tag{3/43}$$

In addition, the relation $b = a\sqrt{1-e^2}$, which comes from the geometry of the ellipse, gives the expression for the semiminor axis. We see that the ellipse becomes a circle with $r = a$ when $e = 0$. Equation 3/43 is an expression of Kepler's *first law*, which says that the planets move in elliptical orbits around the sun as a focus.

The period τ for the elliptical orbit is the total area A of the ellipse divided by the constant rate $\dot{A}$ at which the area is swept through. Thus, from Eq. 3/38,

$$\tau = \frac{A}{\dot{A}} = \frac{\pi ab}{\frac{1}{2}r^2\dot{\theta}} \qquad \text{or} \qquad \tau = \frac{2\pi ab}{h}$$

We can eliminate reference to $\dot{\theta}$ or h in the expression for τ by substituting Eq. 3/42, the identity $d = 1/C$, the geometric relationships $a = ed/(1-e^2)$ and $b = a\sqrt{1-e^2}$ for the ellipse, and the equivalence $Gm_0 = gR^2$. The result after simplification is

$$\boxed{\tau = 2\pi\frac{a^{3/2}}{R\sqrt{g}}} \tag{3/44}$$

Courtesy NASA/JPL-Caltech

Artist conception of the Mars Reconnaissance Orbiter, which arrived at Mars in March 2006.

In this equation note that R is the mean radius of the central attracting body and g is the absolute value of the acceleration due to gravity at the surface of the attracting body.

Equation 3/44 expresses Kepler's *third law* of planetary motion which states that the square of the period of motion is proportional to the cube of the semimajor axis of the orbit.

Case 2: parabola ($e = 1$). Equations 3/41 and 3/42 become

$$\frac{1}{r} = \frac{1}{d}(1 + \cos\theta) \qquad \text{and} \qquad h^2C = Gm_0$$

The radius vector becomes infinite as θ approaches π, so the dimension a is infinite.

Case 3: hyperbola ($e > 1$). From Eq. 3/41 we see that the radial distance r becomes infinite for the two values of the polar angle θ_1 and

$-\theta_1$ defined by $\cos\theta_1 = -1/e$. Only branch I corresponding to $-\theta_1 < \theta < \theta_1$, Fig. 3/23, represents a physically possible motion. Branch II corresponds to angles in the remaining sector (with r negative). For this branch, positive r's may be used if θ is replaced by $\theta - \pi$ and $-r$ by r. Thus, Eq. 3/41 becomes

$$\frac{1}{-r} = \frac{1}{d}\cos(\theta - \pi) + \frac{1}{ed} \quad \text{or} \quad \frac{1}{r} = -\frac{1}{ed} + \frac{\cos\theta}{d}$$

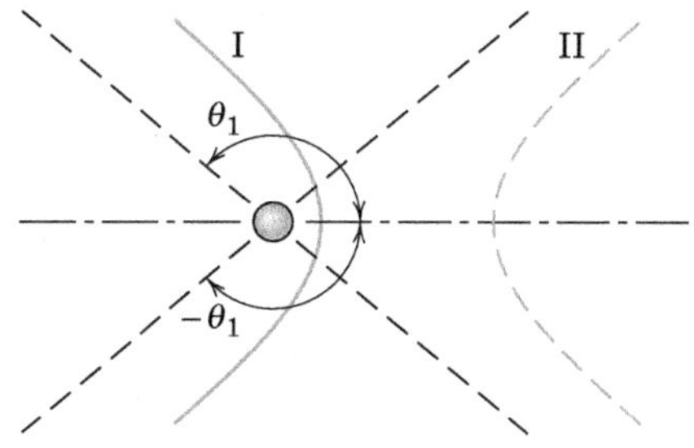

Figure 3/23

But this expression contradicts the form of Eq. 3/40 where Gm_0/h^2 is necessarily positive. Thus branch II does not exist (except for repulsive forces).

Energy Analysis

Now consider the energies of particle m. The system is conservative, and the constant energy E of m is the sum of its kinetic energy T and potential energy V. The kinetic energy is $T = \frac{1}{2}mv^2 = \frac{1}{2}m(\dot{r}^2 + r^2\dot{\theta}^2)$ and the potential energy from Eq. 3/19 is $V = -mgR^2/r$.

Recall that g is the absolute acceleration due to gravity measured at the surface of the attracting body, R is the radius of the attracting body, and $Gm_0 = gR^2$. Thus,

$$E = \frac{1}{2}m(\dot{r}^2 + r^2\dot{\theta}^2) - \frac{mgR^2}{r}$$

This constant value of E can be determined from its value at $\theta = 0$, where $\dot{r} = 0$, $1/r = C + gR^2/h^2$ from Eq. 3/40, and $r\dot{\theta} = h/r$ from Eq. 3/38. Substituting this into the expression for E and simplifying yield

$$\frac{2E}{m} = h^2C^2 - \frac{g^2R^4}{h^2}$$

Now C is eliminated by substitution of Eq. 3/42, which may be written as $h^2C = egR^2$, to obtain

$$e = +\sqrt{1 + \frac{2Eh^2}{mg^2R^4}} \qquad \textbf{(3/45)}$$

The plus value of the radical is mandatory since by definition e is positive. We now see that for the

elliptical orbit $e < 1$, E is negative

parabolic orbit $e = 1$, E is zero

hyperbolic orbit $e > 1$, E is positive

These conclusions, of course, depend on the arbitrary selection of the datum condition for zero potential energy ($V = 0$ when $r = \infty$).

The expression for the velocity v of m may be found from the energy equation, which is

$$\frac{1}{2}mv^2 - \frac{mgR^2}{r} = E$$

The total energy E is obtained from Eq. 3/45 by combining Eq. 3/42 and $1/C = d = a(1 - e^2)/e$ to give for the elliptical orbit

$$E = -\frac{gR^2m}{2a} \tag{3/46}$$

Substitution into the energy equation yields

$$v^2 = 2gR^2\left(\frac{1}{r} - \frac{1}{2a}\right) \tag{3/47}$$

from which the magnitude of the velocity may be computed for a particular orbit in terms of the radial distance r.

Next, combining the expressions for r_{min} and r_{max} corresponding to perigee and apogee, Eq. 3/43, with Eq. 3/47 results in a pair of expressions for the respective velocities at these two positions for the elliptical orbit:

$$\begin{aligned} v_P &= R\sqrt{\frac{g}{a}}\sqrt{\frac{1+e}{1-e}} = R\sqrt{\frac{g}{a}}\sqrt{\frac{r_{max}}{r_{min}}} \\ v_A &= R\sqrt{\frac{g}{a}}\sqrt{\frac{1-e}{1+e}} = R\sqrt{\frac{g}{a}}\sqrt{\frac{r_{min}}{r_{max}}} \end{aligned} \tag{3/48}$$

Selected numerical data pertaining to the solar system are included in Appendix D and are useful in applying the foregoing relationships to problems in planetary motion.

Summary of Assumptions

The foregoing analysis is based on three assumptions:

1. The two bodies possess spherical mass symmetry so that they may be treated as if their masses were concentrated at their centers, that is, as if they were particles.
2. There are no forces present except the gravitational force which each mass exerts on the other.
3. Mass m_0 is fixed in space.

Assumption (1) is excellent for bodies which are distant from the central attracting body, which is the case for most heavenly bodies. A significant class of problems for which assumption (1) is poor is that of artificial satellites in the very near vicinity of oblate planets. As a comment on assumption (2), we note that aerodynamic drag on a low-altitude earth satellite is a force which usually cannot be ignored in the orbital analysis. For an artificial satellite in earth orbit, the error of assumption (3) is negligible because the ratio of the mass of the satellite to that of the earth is very small. On the other hand, for the earth–moon system, a small but significant error is introduced if assumption (3) is invoked—note that the lunar mass is about 1/81 times that of the earth.

Perturbed Two-Body Problem

We now account for the motion of both masses and allow the presence of other forces in addition to those of mutual attraction by considering the *perturbed two-body problem*. Figure 3/24 depicts the major mass m_0, the minor mass m, their respective position vectors $\mathbf{r}_1$ and $\mathbf{r}_2$ measured relative to an inertial frame, the gravitation forces $\mathbf{F}$ and $-\mathbf{F}$, and a non-two-body force $\mathbf{P}$ which is exerted on mass m. The force $\mathbf{P}$ may be due to aerodynamic drag, solar pressure, the presence of a third body, on-board thrusting activities, a nonspherical gravitational field, or a combination of these and other sources.

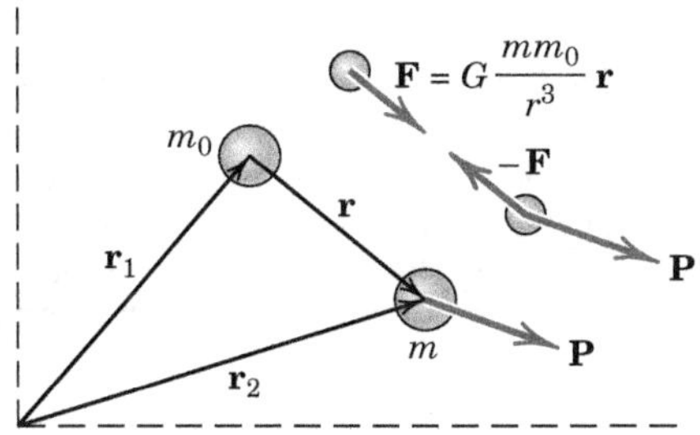

Figure 3/24

Application of Newton's second law to each mass results in

$$G\frac{mm_0}{r^3}\mathbf{r} = m_0\ddot{\mathbf{r}}_1 \qquad \text{and} \qquad -G\frac{mm_0}{r^3}\mathbf{r} + \mathbf{P} = m\ddot{\mathbf{r}}_2$$

Dividing the first equation by m_0, the second equation by m, and subtracting the first equation from the second give

$$-G\frac{(m_0 + m)}{r^3}\mathbf{r} + \frac{\mathbf{P}}{m} = \ddot{\mathbf{r}}_2 - \ddot{\mathbf{r}}_1 = \ddot{\mathbf{r}}$$

or

$$\ddot{\mathbf{r}} + G\frac{(m_0 + m)}{r^3}\mathbf{r} = \frac{\mathbf{P}}{m} \qquad \textbf{(3/49)}$$

Equation 3/49 is a second-order differential equation which, when solved, yields the relative position vector $\mathbf{r}$ as a function of time. Numerical techniques are usually required for the integration of the scalar differential equations which are equivalent to the vector equation 3/49, especially if $\mathbf{P}$ is nonzero.

Restricted Two-Body Problem

If $m_0 >> m$ and $\mathbf{P} = \mathbf{0}$, we have the restricted two-body problem, the equation of motion of which is

$$\ddot{\mathbf{r}} + G\frac{m_0}{r^3}\mathbf{r} = \mathbf{0} \qquad \textbf{(3/49a)}$$

With $\mathbf{r}$ and $\ddot{\mathbf{r}}$ expressed in polar coordinates, Eq. 3/49a becomes

$$(\ddot{r} - r\dot{\theta}^2)\mathbf{e}_r + (r\ddot{\theta} + 2\dot{r}\dot{\theta})\mathbf{e}_\theta + G\frac{m_0}{r^3}(r\mathbf{e}_r) = \mathbf{0}$$

When we equate coefficients of like unit vectors, we recover Eqs. 3/37.

Comparison of Eq. 3/49 (with $\mathbf{P} = \mathbf{0}$) and Eq. 3/49a enables us to relax the assumption that mass m_0 is fixed in space. If we replace m_0 by $(m_0 + m)$ in the expressions derived with the assumption of m_0 fixed, then we obtain expressions which account for the motion of m_0. For example, the corrected expression for the period of elliptical motion of m about m_0 is, from Eq. 3/44,

$$\tau = 2\pi\frac{a^{3/2}}{\sqrt{G(m_0 + m)}} \qquad \textbf{(3/49b)}$$

where the equality $R^2g = Gm_0$ has been used.

Sample Problem 3/31

An artificial satellite is launched from point B on the equator by its carrier rocket and inserted into an elliptical orbit with a perigee altitude of 2000 km. If the apogee altitude is to be 4000 km, compute (*a*) the necessary perigee velocity v_P and the corresponding apogee velocity v_A, (*b*) the velocity at point C where the altitude of the satellite is 2500 km, and (*c*) the period τ for a complete orbit.

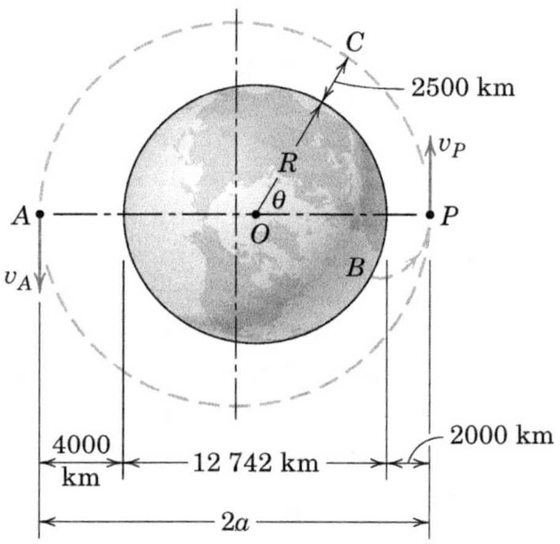

Solution. ***(a)*** The perigee and apogee velocities for specified altitudes are given by Eqs. 3/48, where

① $$r_{max} = 6371 + 4000 = 10\,371 \text{ km}$$

$$r_{min} = 6371 + 2000 = 8371 \text{ km}$$

$$a = (r_{min} + r_{max})/2 = 9371 \text{ km}$$

Thus,

$$v_P = R\sqrt{\frac{g}{a}}\sqrt{\frac{r_{max}}{r_{min}}} = 6371(10^3)\sqrt{\frac{9.825}{9371(10^3)}}\sqrt{\frac{10\,371}{8371}}$$

$$= 7261 \text{ m/s} \quad \text{or} \quad 26\,140 \text{ km/h} \qquad \textit{Ans.}$$

$$v_A = R\sqrt{\frac{g}{a}}\sqrt{\frac{r_{min}}{r_{max}}} = 6371(10^3)\sqrt{\frac{9.825}{9371(10^3)}}\sqrt{\frac{8371}{10\,371}}$$

$$= 5861 \text{ m/s} \quad \text{or} \quad 21\,099 \text{ km/h} \qquad \textit{Ans.}$$

(b) For an altitude of 2500 km the radial distance from the center of the earth is $r = 6371 + 2500 = 8871$ km. From Eq. 3/47 the velocity at point C becomes

② $$v_C{}^2 = 2gR^2\left(\frac{1}{r} - \frac{1}{2a}\right) = 2(9.825)[(6371)(10^3)]^2\left(\frac{1}{8871} - \frac{1}{18\,742}\right)\frac{1}{10^3}$$

$$= 47.353(10^6)(\text{m/s})^2$$

$$v_C = 6881 \text{ m/s} \quad \text{or} \quad 24\,773 \text{ km/h} \qquad \textit{Ans.}$$

(c) The period of the orbit is given by Eq. 3/44, which becomes

③ $$\tau = 2\pi\frac{a^{3/2}}{R\sqrt{g}} = 2\pi\frac{[(9371)(10^3)]^{3/2}}{(6371)(10^3)\sqrt{9.825}} = 9026 \text{ s}$$

$$\text{or} \quad \tau = 2.507 \text{ h} \qquad \textit{Ans.}$$

Helpful Hints

① The mean radius of 12 742/2 = 6371 km from Table D/2 in Appendix D is used. Also the absolute acceleration due to gravity $g = 9.825$ m/s^2 from Art. 1/5 will be used.

② We must be careful with units. It is often safer to work in base units, meters in this case, and convert later.

③ We should observe here that the time interval between successive overhead transits of the satellite as recorded by an observer on the equator is longer than the period calculated here since the observer will have moved in space due to the counterclockwise rotation of the earth, as seen looking down on the north pole.

PROBLEMS

(Unless otherwise indicated, the velocities mentioned in the problems which follow are measured from a nonrotating reference frame moving with the center of the attracting body. Also, aerodynamic drag is to be neglected unless stated otherwise. Use $g = 9.825$ m/s^2 (32.23 ft/sec^2) for the absolute gravitational acceleration at the surface of the earth and treat the earth as a sphere of radius $R = 6371$ km (3959 mi).)

Introductory Problems

3/269 Calculate the velocity of a spacecraft which orbits the moon in a circular path of 80-km altitude.

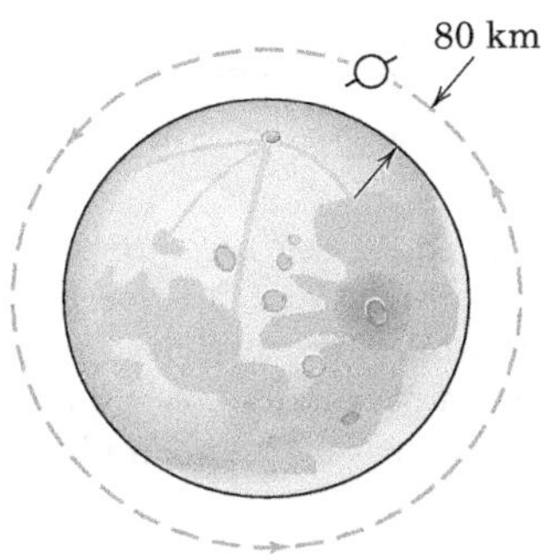

Problem 3/269

3/270 What velocity v must the space shuttle have in order to release the Hubble space telescope in a circular earth orbit 590 km above the surface of the earth?

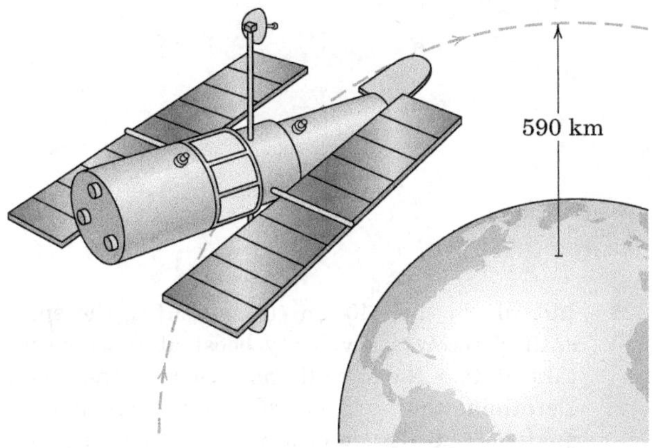

Problem 3/270

3/271 Show that the path of the moon is concave toward the sun at the position shown. Assume that the sun, earth, and moon are in the same line.

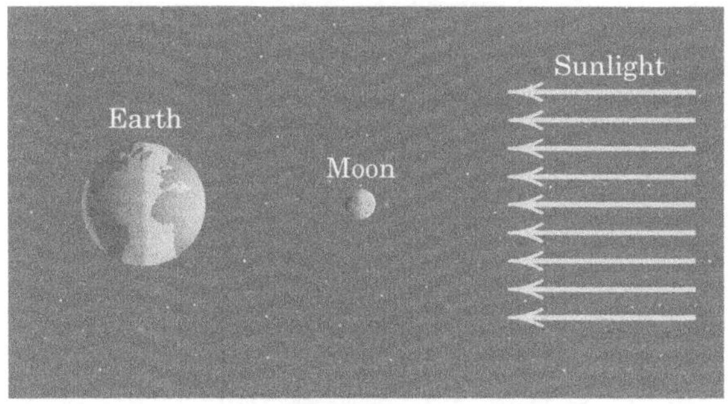

Problem 3/271

3/272 A satellite is in a circular polar orbit of altitude 300 km. Determine the separation d at the equator between the ground tracks (shown dashed) associated with two successive overhead passes of the satellite.

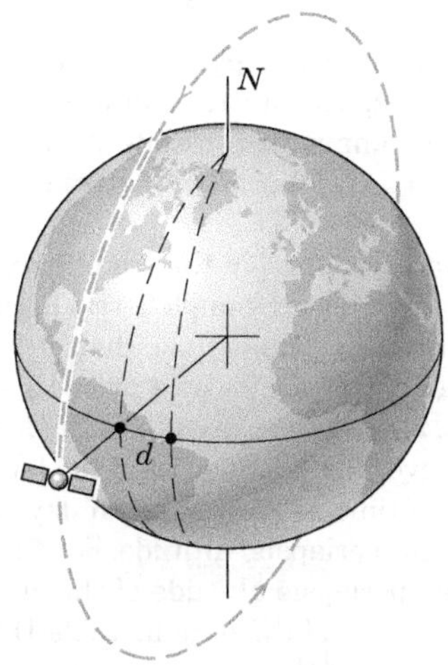

Problem 3/272

3/273 Determine the apparent velocity v_{rel} of a satellite moving in a circular equatorial orbit 200 mi above the earth as measured by an observer on the equator (*a*) for a west-to-east orbit and (*b*) for an east-to-west orbit. Why is the west-to-east orbit more easily achieved?

3/274 A spacecraft is in an initial circular orbit with an altitude of 350 km. As it passes point P, onboard thrusters give it a velocity boost of 25 m/s. Determine the resulting altitude gain Δh at point A.

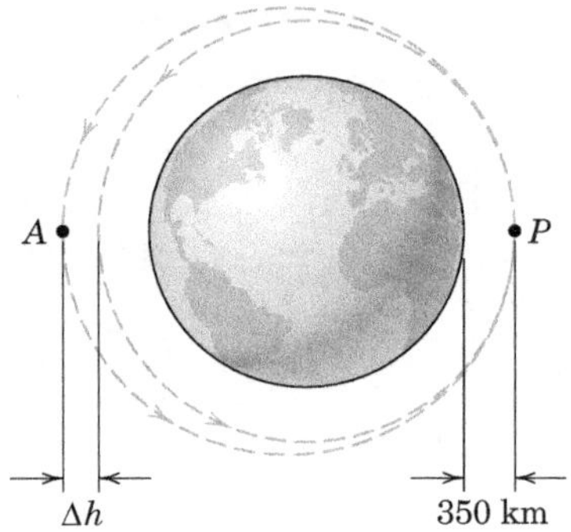

Problem 3/274

3/275 If the perigee altitude of an earth satellite is 240 km and the apogee altitude is 400 km, compute the eccentricity e of the orbit and the period τ of one complete orbit in space.

3/276 Determine the energy difference ΔE between an 80 000-kg space-shuttle orbiter on the launch pad in Cape Canaveral (latitude 28.5°) and the same orbiter in a circular orbit of altitude $h = 300$ km.

3/277 The Mars orbiter for the Viking mission was designed to make one complete trip around the planet in exactly the same time that it takes Mars to revolve once about its own axis. This time is 24 h, 37 min, 23 s. In this way, it is possible for the orbiter to pass over the landing site of the lander capsule at the same time in each Martian day at the orbiter's minimum (periapsis) altitude. For the Viking I mission, the periapsis altitude of the orbiter was 1508 km. Make use of the data in Table D/2 in Appendix D and compute the maximum (apoapsis) altitude h_a for the orbiter in its elliptical path.

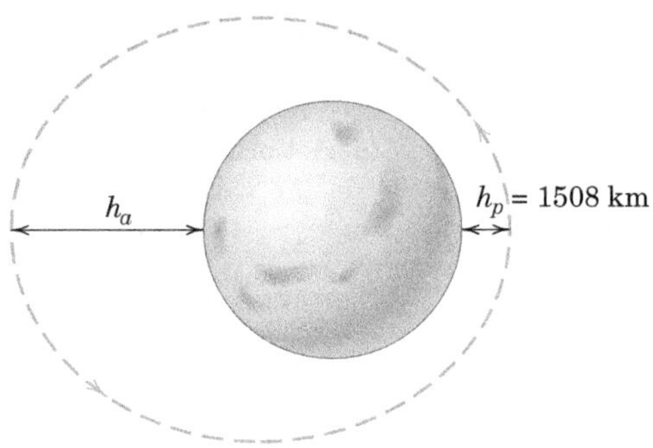

Problem 3/277

3/278 A "drag-free" satellite is one which carries a small mass inside a chamber as shown. If the satellite speed decreases because of drag, the mass speed will not, and so the mass moves relative to the chamber as indicated. Sensors detect this change in the position of the mass within the chamber, and the satellite thruster is periodically fired to re-center the mass. In this manner, compensation is made for drag. If the satellite is in a circular earth orbit of 200-km altitude and a total thruster burn time of 300 seconds occurs during 10 orbits, determine the drag force D acting on the 100-kg satellite. The thruster force T is 2 N.

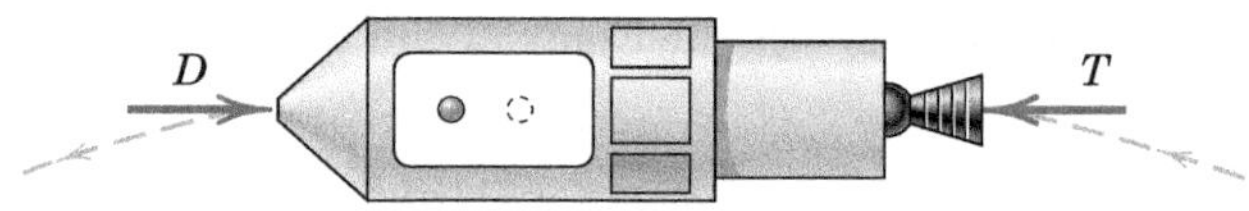

Problem 3/278

3/279 Determine the speed v required of an earth satellite at point A for (*a*) a circular orbit, (*b*) an elliptical orbit of eccentricity $e = 0.1$, (*c*) an elliptical orbit of eccentricity $e = 0.9$, and (*d*) a parabolic orbit. In cases (*b*), (*c*), and (*d*), A is the orbit perigee.

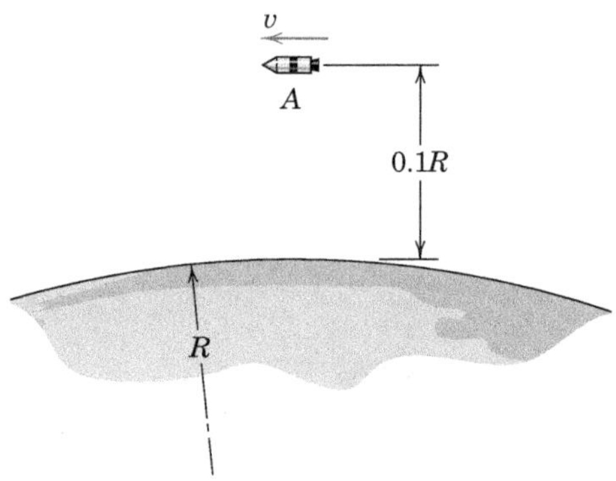

Problem 3/279

Representative Problems

3/280 Initially in the 240-km circular orbit, the spacecraft S receives a velocity boost at P which will take it to $r \rightarrow \infty$ with no speed at that point. Determine the required velocity increment Δv at point P and also determine the speed when $r = 2r_P$. At what value of θ does r become $2r_P$?

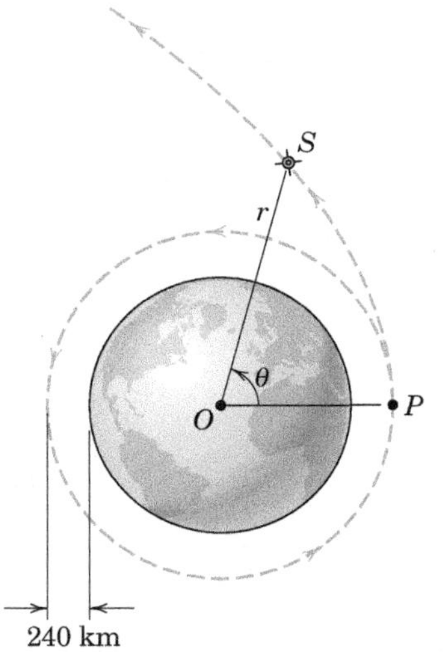

Problem 3/280

3/281 The binary star system consists of stars A and B, both of which orbit about the system mass center. Compare the orbital period τ_f calculated with the assumption of a fixed star A with the period τ_{nf} calculated without this assumption.

Problem 3/281

3/282 If the earth were suddenly deprived of its orbital velocity around the sun, find the time t which it would take for the earth to "fall" to the location of the center of the sun. *(Hint:* The time would be one-half the period of a degenerate elliptical orbit around the sun with the semiminor axis approaching zero.) Refer to Table D/2 for the exact period of the earth around the sun.

3/283 Just after launch from the earth, the space-shuttle orbiter is in the 37 × 137-mi orbit shown. The first time that the orbiter passes the apogee A, its two orbital-maneuvering-system (OMS) engines are fired to circularize the orbit. If the weight of the orbiter is 175,000 lb and the OMS engines have a thrust of 6000 lb each, determine the required time duration Δt of the burn.

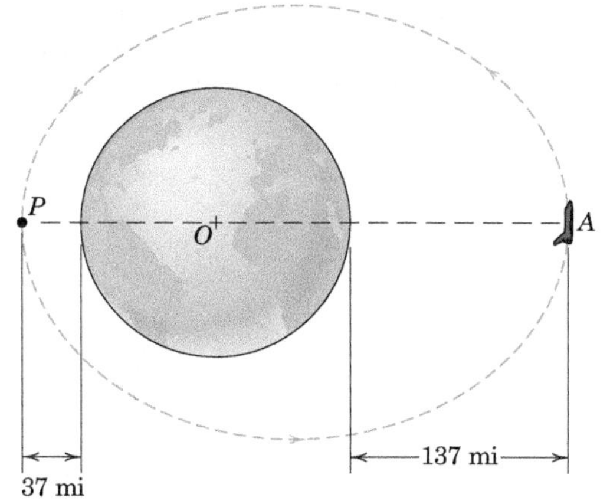

Problem 3/283

3/284 A spacecraft is in a circular orbit of radius $3R$ around the moon. At point A, the spacecraft ejects a probe which is designed to arrive at the surface of the moon at point B. Determine the necessary velocity v_r of the probe relative to the spacecraft just after ejection. Also calculate the position θ of the spacecraft when the probe arrives at point B.

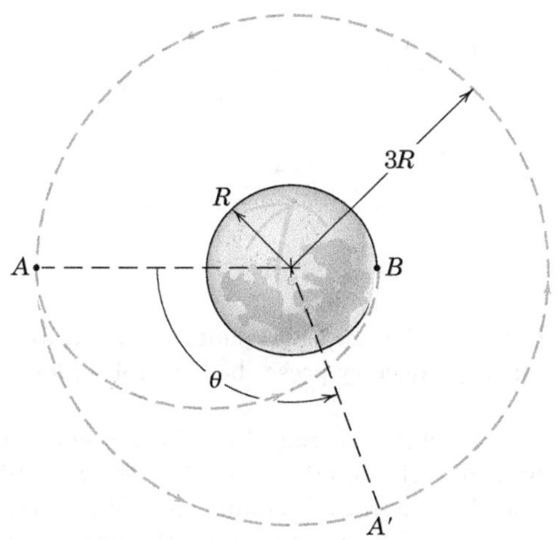

Problem 3/284

3/285 A projectile is launched from B with a speed of 2000 m/s at an angle α of 30° with the horizontal as shown. Determine the maximum altitude h_{max}.

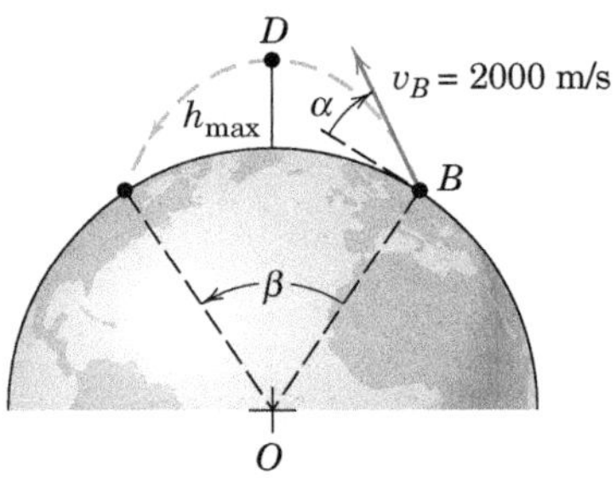

Problem 3/285

3/286 Compute the magnitude of the necessary launch velocity at B if the projectile trajectory is to intersect the earth's surface so that the angle β equals 90°. The altitude at the highest point of the trajectory is $0.5R$.

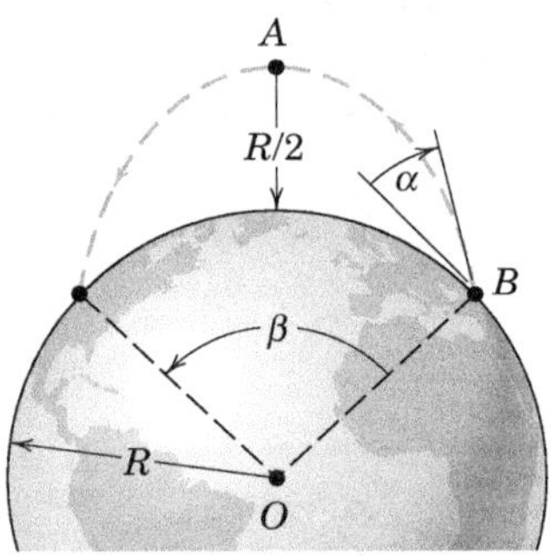

Problem 3/286

3/287 Compute the necessary launch angle α at point B for the trajectory prescribed in Prob. 3/286.

3/288 Two satellites B and C are in the same circular orbit of altitude 500 miles. Satellite B is 1000 mi ahead of satellite C as indicated. Show that C can catch up to B by "putting on the brakes." Specifically, by what amount Δv should the circular-orbit velocity of C be reduced so that it will rendezvous with B after one period in its new elliptical orbit? Check to see that C does not strike the earth in the elliptical orbit.

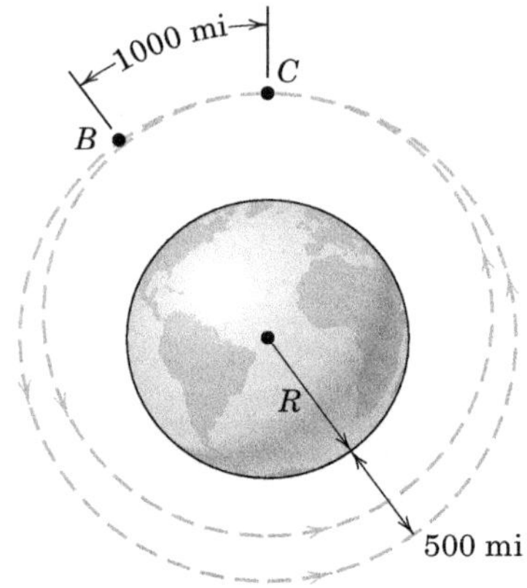

Problem 3/288

3/289 Determine the necessary amount Δv by which the circular-orbit velocity of satellite C should be reduced if the catch-up maneuver of Prob. 3/288 is to be accomplished with not one but two periods in a new elliptical orbit.

3/290 The 175,000-lb space-shuttle orbiter is in a circular orbit of altitude 200 miles. The two orbital-maneuvering-system (OMS) engines, each of which has a thrust of 6000 lb, are fired in retrothrust for 150 seconds. Determine the angle β which locates the intersection of the shuttle trajectory with the earth's surface. Assume that the shuttle position B corresponds to the completion of the OMS burn and that no loss of altitude occurs during the burn.

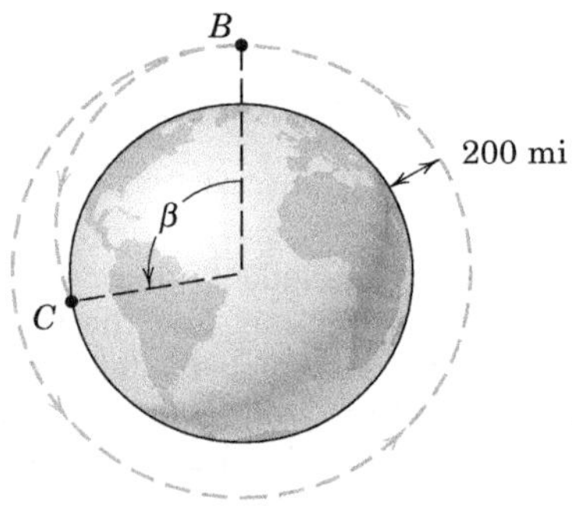

Problem 3/290

3/291 Compare the orbital period of the moon calculated with the assumption of a fixed earth with the period calculated without this assumption.

3/292 A satellite is placed in a circular polar orbit a distance H above the earth. As the satellite goes over the north pole at A, its retro-rocket is activated to produce a burst of negative thrust which reduces its velocity to a value which will ensure an equatorial landing. Derive the expression for the required reduction Δv_A of velocity at A. Note that A is the apogee of the elliptical path.

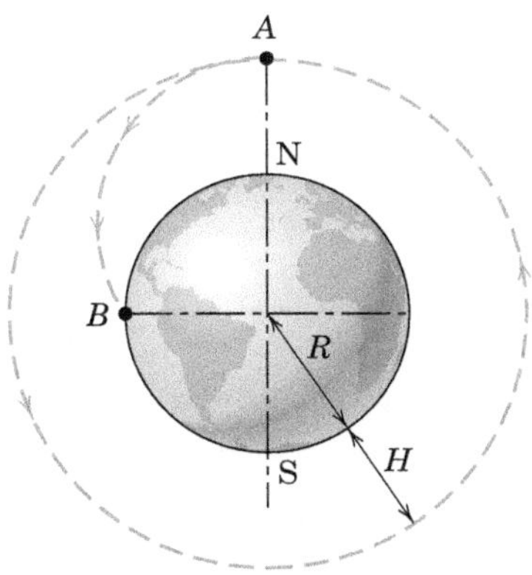

Problem 3/292

3/293 The perigee and apogee altitudes above the surface of the earth of an artificial satellite are h_p and h_a, respectively. Derive the expression for the radius of curvature ρ_p of the orbit at the perigee position. The radius of the earth is R.

3/294 A spacecraft moving in a west-to-east equatorial orbit is observed by a tracking station located on the equator. If the spacecraft has a perigee altitude $H = 150$ km and velocity v when directly over the station and an apogee altitude of 1500 km, determine an expression for the angular rate p (relative to the earth) at which the antenna dish must be rotated when the spacecraft is directly overhead. Compute p. The angular velocity of the earth is $\omega = 0.7292(10^{-4})$ rad/s.

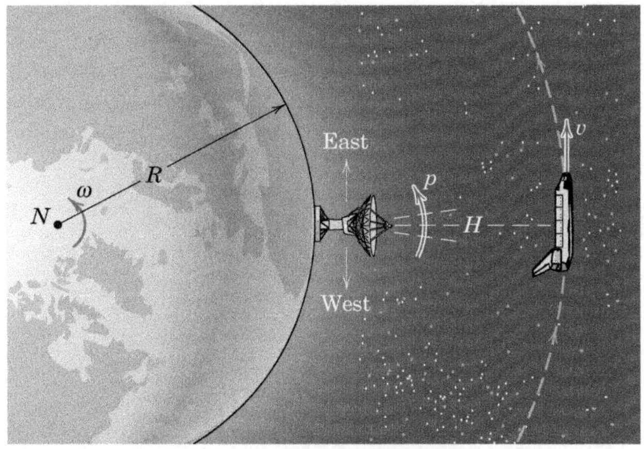

Problem 3/294

3/295 Sometime after launch from the earth, a spacecraft S is in the orbital path of the earth at some distance from the earth at position P. What velocity boost Δv at P is required so that the spacecraft arrives at the orbit of Mars at A as shown?

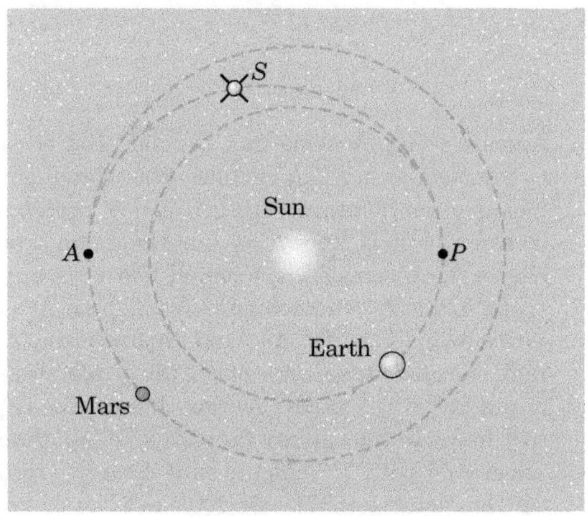

Problem 3/295

*3/296 In 1995 a spacecraft called the Solar and Heliospheric Observatory (SOHO) was placed into a circular orbit about the sun and inside that of the earth as shown. Determine the distance h so that the period of the spacecraft orbit will match that of the earth, with the result that the spacecraft will remain between the earth and the sun in a "halo" orbit.

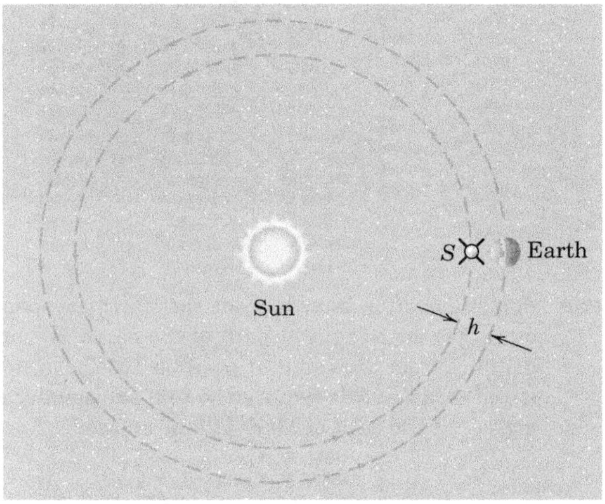

Problem 3/296

▶3/297 A space vehicle moving in a circular orbit of radius r_1 transfers to a larger circular orbit of radius r_2 by means of an elliptical path between A and B. (This transfer path is known as the Hohmann transfer ellipse.) The transfer is accomplished by a burst of speed Δv_A at A and a second burst of speed Δv_B at B. Write expressions for Δv_A and Δv_B in terms of the radii shown and the value of g of the acceleration due to gravity at the earth's surface. If each Δv is positive, how can the velocity for path 2 be less than the velocity for path 1? Compute each Δv if $r_1 = (6371 + 500)$ km and $r_2 = (6371 + 35\,800)$ km. Note that r_2 has been chosen as the radius of a geosynchronous orbit.

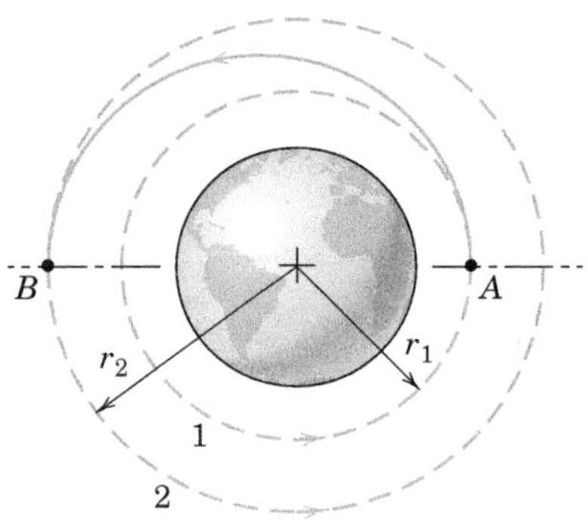

Problem 3/297

▶3/298 At the instant represented in the figure, a small experimental satellite A is ejected from the shuttle orbiter with a velocity $v_r = 100$ m/s relative to the shuttle, directed toward the center of the earth. The shuttle is in a circular orbit of altitude $h = 200$ km. For the resulting elliptical orbit of the satellite, determine the semimajor axis a and its orientation, the period τ, eccentricity e, apogee speed v_a, perigee speed v_p, r_{max}, and r_{min}. Sketch the satellite orbit.

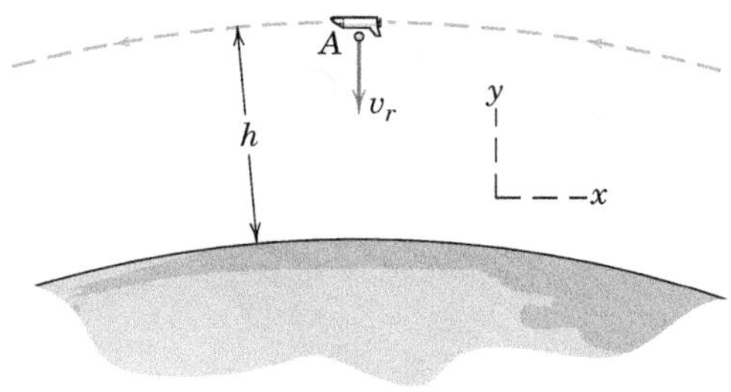

Problem 3/298

▶3/299 A spacecraft in an elliptical orbit has the position and velocity indicated in the figure at a certain instant. Determine the semimajor axis length a of the orbit and find the acute angle α between the semimajor axis and the line l. Does the spacecraft eventually strike the earth?

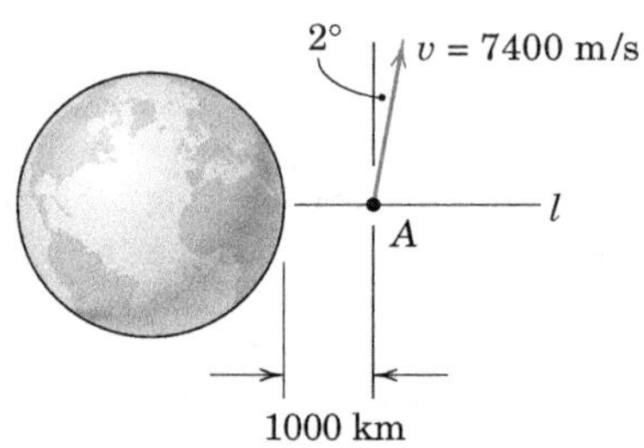

Problem 3/299

▶3/300 The satellite has a velocity at B of 3200 m/s in the direction indicated. Determine the angle β which locates the point C of impact with the earth.

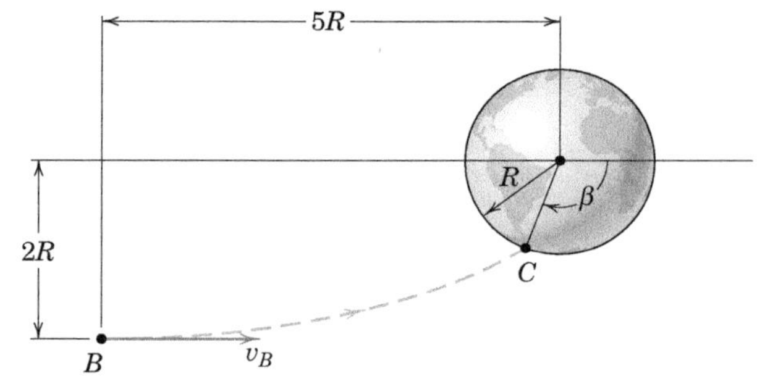

Problem 3/300

3/14 Relative Motion

Up to this point in our development of the kinetics of particle motion, we have applied Newton's second law and the equations of work-energy and impulse-momentum to problems where all measurements of motion were made with respect to a reference system which was considered fixed. The nearest we can come to a "fixed" reference system is the primary inertial system or astronomical frame of reference, which is an imaginary set of axes attached to the fixed stars. All other reference systems then are considered to have motion in space, including any reference system attached to the moving earth.

The acceleration of points attached to the earth as measured in the primary system are quite small, however, and we normally neglect them for most earth-surface measurements. For example, the acceleration of the center of the earth in its near-circular orbit around the sun considered fixed is 0.00593 m/s^2 (or 0.01946 ft/sec^2), and the acceleration of a point on the equator at sea level with respect to the center of the earth considered fixed is 0.0339 m/s^2 (or 0.1113 ft/sec^2). Clearly, these accelerations are small compared with g and with most other significant accelerations in engineering work. Thus, we make only a small error when we assume that our earth-attached reference axes are equivalent to a fixed reference system.

Relative-Motion Equation

We now consider a particle A of mass m, Fig. 3/25, whose motion is observed from a set of axes x-y-z which translate with respect to a fixed reference frame X-Y-Z. Thus, the x-y-z directions always remain parallel to the X-Y-Z directions. We postpone discussion of motion relative to a rotating reference system until Arts. 5/7 and 7/7. The acceleration of the origin B of x-y-z is $\mathbf{a}_B$. The acceleration of A as observed from or relative to x-y-z is $\mathbf{a}_{\text{rel}} = \mathbf{a}_{A/B} = \ddot{\mathbf{r}}_{A/B}$, and by the relative-motion principle of Art. 2/8, the absolute acceleration of A is

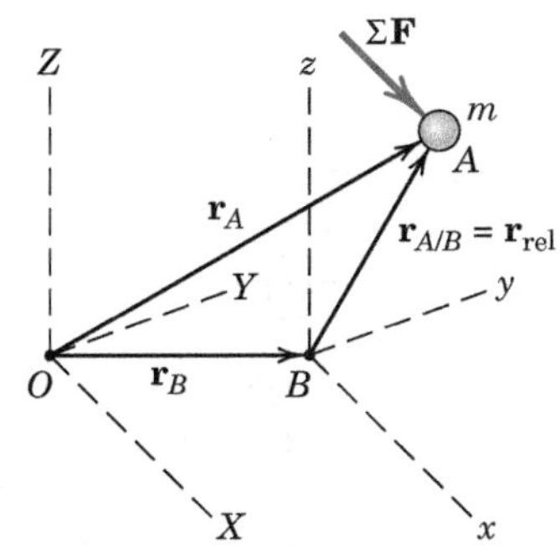

Figure 3/25

$$\mathbf{a}_A = \mathbf{a}_B + \mathbf{a}_{\text{rel}}$$

Thus, Newton's second law $\Sigma\mathbf{F} = m\mathbf{a}_A$ becomes

$$\Sigma\mathbf{F} = m(\mathbf{a}_B + \mathbf{a}_{\text{rel}}) \quad \textbf{(3/50)}$$

We can identify the force sum $\Sigma\mathbf{F}$, as always, by a complete free-body diagram. This diagram will appear the same to an observer in x-y-z or to one in X-Y-Z as long as only the real forces acting on the particle are represented. We can conclude immediately that Newton's second law does not hold with respect to an accelerating system since $\Sigma\mathbf{F} \neq m\mathbf{a}_{\text{rel}}$.

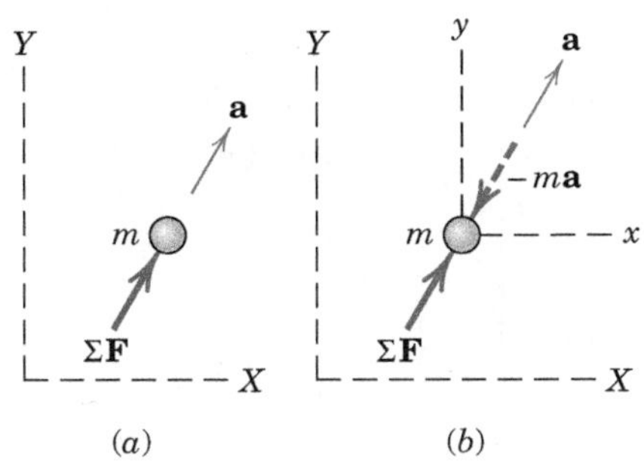

Figure 3/26

D'Alembert's Principle

The particle acceleration we measure from a fixed set of axes X-Y-Z, Fig. 3/26a, is its absolute acceleration $\mathbf{a}$. In this case the familiar relation $\Sigma\mathbf{F} = m\mathbf{a}$ applies. When we observe the particle from a moving system

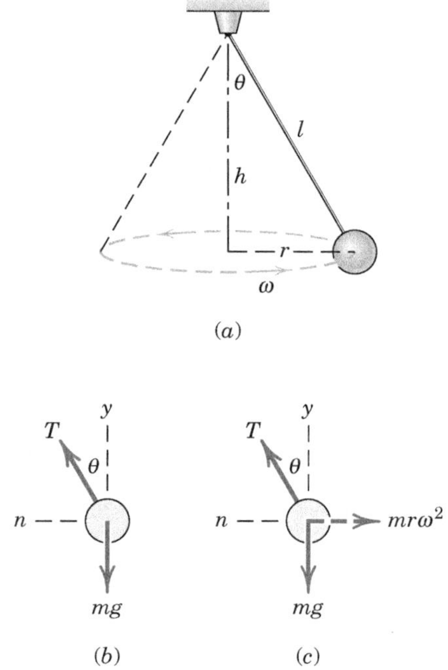

Figure 3/27

x-y-z attached to the particle, Fig. 3/26*b*, the particle necessarily appears to be at rest or in equilibrium in x-y-z. Thus, the observer who is accelerating with x-y-z concludes that a force $-m\mathbf{a}$ acts on the particle to balance $\Sigma\mathbf{F}$. This point of view, which allows the treatment of a dynamics problem by the methods of statics, was an outgrowth of the work of D'Alembert contained in his *Traité de Dynamique* published in 1743.

This approach merely amounts to rewriting the equation of motion as $\Sigma\mathbf{F} - m\mathbf{a} = \mathbf{0}$, which assumes the form of a zero force summation if $-m\mathbf{a}$ is treated as a force. This fictitious force is known as the *inertia force*, and the artificial state of equilibrium created is known as *dynamic equilibrium*. The apparent transformation of a problem in dynamics to one in statics has become known as *D'Alembert's principle*.

Opinion differs concerning the original interpretation of D'Alembert's principle, but the principle in the form in which it is generally known is regarded in this book as being mainly of historical interest. It evolved when understanding and experience with dynamics were extremely limited and was a means of explaining dynamics in terms of the principles of statics, which were more fully understood. This excuse for using an artificial situation to describe a real one is no longer justified, as today a wealth of knowledge and experience with dynamics strongly supports the direct approach of thinking in terms of dynamics rather than statics. It is somewhat difficult to understand the long persistence in the acceptance of statics as a way of understanding dynamics, particularly in view of the continued search for the understanding and description of physical phenomena in their most direct form.

We cite only one simple example of the method known as D'Alembert's principle. The conical pendulum of mass m, Fig. 3/27*a*, is swinging in a horizontal circle, with its radial line r having an angular velocity ω. In the straightforward application of the equation of motion $\Sigma\mathbf{F} = m\mathbf{a}_n$ in the direction n of the acceleration, the free-body diagram in part b of the figure shows that $T \sin\theta = mr\omega^2$. When we apply the equilibrium requirement in the y-direction, $T\cos\theta - mg = 0$, we can find the unknowns T and θ. But if the reference axes are attached to the particle, the particle will appear to be in equilibrium relative to these axes. Accordingly, the inertia force $-m\mathbf{a}$ must be added, which amounts to visualizing the application of $mr\omega^2$ in the direction opposite to the acceleration, as shown in part c of the figure. With this pseudo free-body diagram, a zero force summation in the n-direction gives $T\sin\theta - mr\omega^2 = 0$ which, of course, gives us the same result as before.

We may conclude that no advantage results from this alternative formulation. The authors recommend against using it since it introduces no simplification and adds a nonexistent force to the diagram. In the case of a particle moving in a circular path, this hypothetical inertia force is known as the *centrifugal force* since it is directed away from the center and is opposite to the direction of the acceleration. You are urged to recognize that there is no actual centrifugal force acting on the particle. The only actual force which may properly be called centrifugal is the horizontal component of the tension T exerted *by* the particle *on* the cord.

Constant-Velocity, Nonrotating Systems

In discussing particle motion relative to moving reference systems, we should note the special case where the reference system has a constant velocity and no rotation. If the x-y-z axes of Fig. 3/25 have a constant velocity, then $\mathbf{a}_B = \mathbf{0}$ and the acceleration of the particle is $\mathbf{a}_A = \mathbf{a}_{\text{rel}}$. Therefore, we may write Eq. 3/50 as

$$\boxed{\Sigma\mathbf{F} = m\mathbf{a}_{\text{rel}}} \qquad \textbf{(3/51)}$$

which tells us that Newton's second law holds for measurements made in a system moving with a constant velocity. Such a system is known as an inertial system or as a Newtonian frame of reference. Observers in the moving system and in the fixed system will also agree on the designation of the resultant force acting on the particle from their identical free-body diagrams, provided they avoid the use of any so-called "inertia forces."

We will now examine the parallel question concerning the validity of the work-energy equation and the impulse-momentum equation relative to a constant-velocity, nonrotating system. Again, we take the x-y-z axes of Fig. 3/25 to be moving with a constant velocity $\mathbf{v}_B = \dot{\mathbf{r}}_B$ relative to the fixed axes X-Y-Z. The path of the particle A relative to x-y-z is governed by $\mathbf{r}_{\text{rel}}$ and is represented schematically in Fig. 3/28. The work done by $\Sigma\mathbf{F}$ relative to x-y-z is $dU_{\text{rel}} = \Sigma\mathbf{F}\cdot d\mathbf{r}_{\text{rel}}$. But $\Sigma\mathbf{F} = m\mathbf{a}_A = m\mathbf{a}_{\text{rel}}$ since $\mathbf{a}_B = \mathbf{0}$. Also $\mathbf{a}_{\text{rel}}\cdot d\mathbf{r}_{\text{rel}} = \mathbf{v}_{\text{rel}}\cdot d\mathbf{v}_{\text{rel}}$ for the same reason that $a_t\, ds = v\, dv$ in Art. 2/5 on curvilinear motion. Thus, we have

$$dU_{\text{rel}} = m\mathbf{a}_{\text{rel}}\cdot d\mathbf{r}_{\text{rel}} = mv_{\text{rel}}\, dv_{\text{rel}} = d(\tfrac{1}{2}mv_{\text{rel}}{}^2)$$

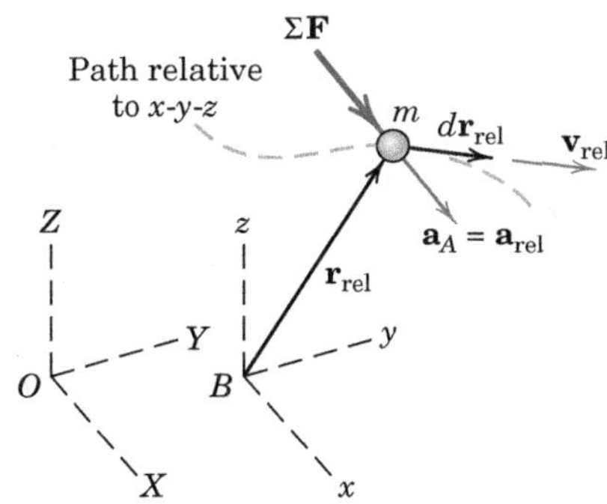

Figure 3/28

We define the kinetic energy relative to x-y-z as $T_{\text{rel}} = \frac{1}{2}mv_{\text{rel}}{}^2$ so that we now have

$$\boxed{dU_{\text{rel}} = dT_{\text{rel}}} \quad \text{or} \quad \boxed{U_{\text{rel}} = \Delta T_{\text{rel}}} \qquad \textbf{(3/52)}$$

which shows that the work-energy equation holds for measurements made relative to a constant-velocity, nonrotating system.

Relative to x-y-z, the impulse on the particle during time dt is $\Sigma\mathbf{F}\, dt = m\mathbf{a}_A\, dt = m\mathbf{a}_{\text{rel}}\, dt$. But $m\mathbf{a}_{\text{rel}}\, dt = m\, d\mathbf{v}_{\text{rel}} = d(m\mathbf{v}_{\text{rel}})$ so

$$\Sigma\mathbf{F}\, dt = d(m\mathbf{v}_{\text{rel}})$$

We define the linear momentum of the particle relative to x-y-z as $\mathbf{G}_{\text{rel}} = m\mathbf{v}_{\text{rel}}$, which gives us $\Sigma\mathbf{F}\, dt = d\mathbf{G}_{\text{rel}}$. Dividing by dt and integrating give

$$\boxed{\Sigma\mathbf{F} = \dot{\mathbf{G}}_{\text{rel}}} \quad \text{and} \quad \boxed{\int \Sigma\mathbf{F}\, dt = \Delta\mathbf{G}_{\text{rel}}} \qquad \textbf{(3/53)}$$

Thus, the impulse-momentum equations for a fixed reference system also hold for measurements made relative to a constant-velocity, nonrotating system.

Finally, we define the relative angular momentum of the particle about a point in x-y-z, such as the origin B, as the moment of the relative linear momentum. Thus, $(\mathbf{H}_B)_{rel} = \mathbf{r}_{rel} \times \mathbf{G}_{rel}$. The time derivative gives $(\dot{\mathbf{H}}_B)_{rel} = \dot{\mathbf{r}}_{rel} \times \mathbf{G}_{rel} + \mathbf{r}_{rel} \times \dot{\mathbf{G}}_{rel}$. The first term is nothing more than $\mathbf{v}_{rel} \times m\mathbf{v}_{rel} = \mathbf{0}$, and the second term becomes $\mathbf{r}_{rel} \times \Sigma\mathbf{F} = \Sigma\mathbf{M}_B$, the sum of the moments about B of all forces on m. Thus, we have

$$\boxed{\Sigma\mathbf{M}_B = (\dot{\mathbf{H}}_B)_{rel}} \qquad \textbf{(3/54)}$$

which shows that the moment-angular momentum relation holds with respect to a constant-velocity, nonrotating system.

Although the work-energy and impulse-momentum equations hold relative to a system translating with a constant velocity, the individual expressions for work, kinetic energy, and momentum differ between the fixed and the moving systems. Thus,

$$(dU = \Sigma\mathbf{F}\cdot d\mathbf{r}_A) \neq (dU_{rel} = \Sigma\mathbf{F}\cdot d\mathbf{r}_{rel})$$
$$(T = \tfrac{1}{2}mv_A{}^2) \neq (T_{rel} = \tfrac{1}{2}mv_{rel}{}^2)$$
$$(\mathbf{G} = m\mathbf{v}_A) \neq (\mathbf{G}_{rel} = m\mathbf{v}_{rel})$$

Equations 3/51 through 3/54 are formal proof of the validity of the Newtonian equations of kinetics in any constant-velocity, nonrotating system. We might have surmised these conclusions from the fact that $\Sigma\mathbf{F} = m\mathbf{a}$ depends on acceleration and not velocity. We are also ready to conclude that there is no experiment which can be conducted in and relative to a constant-velocity, nonrotating system (Newtonian frame of reference) which discloses its absolute velocity. Any mechanical experiment will achieve the same results in any Newtonian system.

Relative motion is a critical issue during aircraft-carrier landings.

Sample Problem 3/32

A simple pendulum of mass m and length r is mounted on the flatcar, which has a constant horizontal acceleration a_0 as shown. If the pendulum is released from rest relative to the flatcar at the position $\theta = 0$, determine the expression for the tension T in the supporting light rod for any value of θ. Also find T for $\theta = \pi/2$ and $\theta = \pi$.

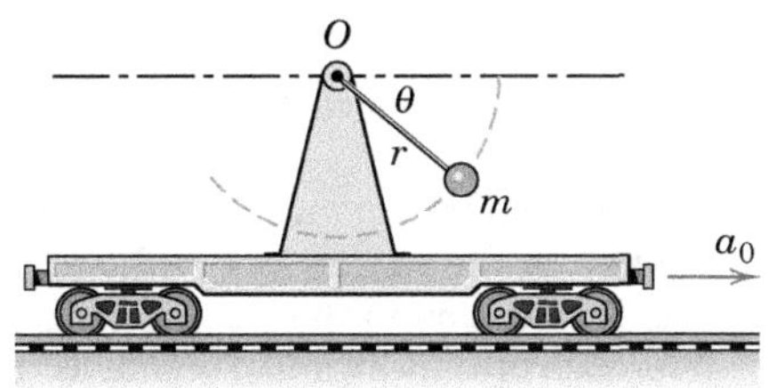

Solution. We attach our moving x-y coordinate system to the translating car with origin at O for convenience. Relative to this system, n- and t-coordinates are the natural ones to use since the motion is circular within x-y. The acceleration of m is given by the relative-acceleration equation

$$\mathbf{a} = \mathbf{a}_0 + \mathbf{a}_{\text{rel}}$$

where $\mathbf{a}_{\text{rel}}$ is the acceleration which would be measured by an observer riding with the car. He would measure an n-component equal to $r\dot{\theta}^2$ and a t-component equal to $r\ddot{\theta}$. The three components of the absolute acceleration of m are shown in the separate view.

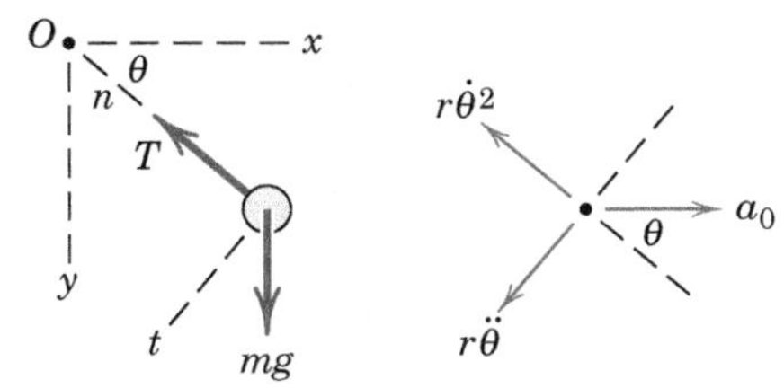

Free-body diagram

Acceleration components

First, we apply Newton's second law to the t-direction and get

① $[\Sigma F_t = ma_t]$ $\quad mg \cos \theta = m(r\ddot{\theta} - a_0 \sin \theta)$

$$r\ddot{\theta} = g \cos \theta + a_0 \sin \theta$$

Integrating to obtain $\dot{\theta}$ as a function of θ yields

$[\dot{\theta}\, d\dot{\theta} = \ddot{\theta}\, d\theta]$ $\quad \displaystyle\int_0^{\dot{\theta}} \dot{\theta}\, d\dot{\theta} = \int_0^{\theta} \frac{1}{r}(g \cos \theta + a_0 \sin \theta)\, d\theta$

$$\frac{\dot{\theta}^2}{2} = \frac{1}{r}[g \sin \theta + a_0(1 - \cos \theta)]$$

We now apply Newton's second law to the n-direction, noting that the n-component of the absolute acceleration is $r\dot{\theta}^2 - a_0 \cos \theta$.

② $[\Sigma F_n = ma_n]$ $\quad T - mg \sin \theta = m(r\dot{\theta}^2 - a_0 \cos \theta)$

$$= m[2g \sin \theta + 2a_0(1 - \cos \theta) - a_0 \cos \theta]$$

$$T = m[3g \sin \theta + a_0(2 - 3 \cos \theta)] \qquad \textit{Ans.}$$

For $\theta = \pi/2$ and $\theta = \pi$, we have

$$T_{\pi/2} = m[3g(1) + a_0\,(2 - 0)] = m(3g + 2a_0) \qquad \textit{Ans.}$$

$$T_{\pi} = m[3g(0) + a_0\,(2 - 3[-1])] = 5ma_0 \qquad \textit{Ans.}$$

Helpful Hints

① We choose the t-direction first since the n-direction equation, which contains the unknown T, will involve $\dot{\theta}^2$, which, in turn, is obtained from an integration of $\ddot{\theta}$.

② Be sure to recognize that $\dot{\theta}\, d\dot{\theta} = \ddot{\theta}\, d\theta$ may be obtained from $v\, dv = a_t\, ds$ by dividing by r^2.

Sample Problem 3/33

The flatcar moves with a constant speed v_0 and carries a winch which produces a constant tension P in the cable attached to the small carriage. The carriage has a mass m and rolls freely on the horizontal surface starting from rest relative to the flatcar at $x = 0$, at which instant $X = x_0 = b$. Apply the work-energy equation to the carriage, first, as an observer moving with the frame of reference of the car and, second, as an observer on the ground. Show the compatibility of the two expressions.

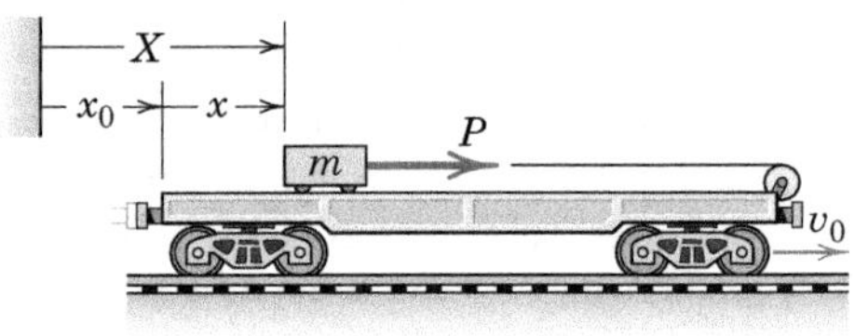

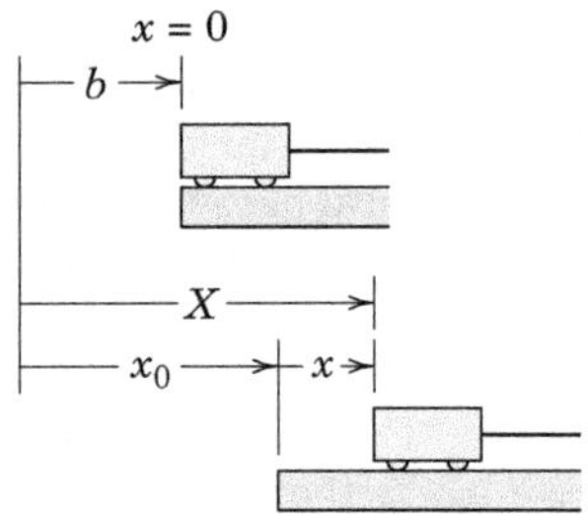

Solution. To the observer on the flatcar, the work done by P is

① $$U_{\text{rel}} = \int_0^x P\,dx = Px \qquad \text{for constant } P$$

The change in kinetic energy relative to the car is

$$\Delta T_{\text{rel}} = \frac{1}{2}m(\dot{x}^2 - 0)$$

The work-energy equation for the moving observer becomes

$$[U_{\text{rel}} = \Delta T_{\text{rel}}] \qquad Px = \frac{1}{2}m\dot{x}^2$$

To the observer on the ground, the work done by P is

$$U = \int_b^X P\,dX = P(X - b)$$

The change in kinetic energy for the ground measurement is

② $$\Delta T = \frac{1}{2}m(\dot{X}^2 - v_0^2)$$

The work-energy equation for the fixed observer gives

$$[U = \Delta T] \qquad P(X - b) = \frac{1}{2}m(\dot{X}^2 - v_0^2)$$

To reconcile this equation with that for the moving observer, we can make the following substitutions:

$$X = x_0 + x, \qquad \dot{X} = v_0 + \dot{x}, \qquad \ddot{X} = \ddot{x}$$

Thus,

$$P(X - b) = Px + P(x_0 - b) = Px + m\ddot{x}\,(x_0 - b)$$

③ $$= Px + m\ddot{x}v_0t = Px + mv_0\dot{x}$$

and

$$\dot{X}^2 - v_0^2 = (v_0^2 + \dot{x}^2 + 2v_0\dot{x} - v_0^2) = \dot{x}^2 + 2v_0\dot{x}$$

The work-energy equation for the fixed observer now becomes

$$Px + mv_0\dot{x} = \frac{1}{2}m\dot{x}^2 + mv_0\dot{x}$$

which is merely $Px = \frac{1}{2}m\dot{x}^2$, as concluded by the moving observer. We see, therefore, that the difference between the two work-energy expressions is

$$U - U_{\text{rel}} = T - T_{\text{rel}} = mv_0\dot{x}$$

Helpful Hints

① The only coordinate which the moving observer can measure is x.

② To the ground observer, the initial velocity of the carriage is v_0, so its initial kinetic energy is $\frac{1}{2}mv_0^2$.

③ The symbol t stands for the time of motion from $x = 0$ to $x = x$. The displacement $x_0 - b$ of the carriage is its velocity v_0 times the time t or $x_0 - b = v_0t$. Also, since the constant acceleration times the time equals the velocity change, $\ddot{x}t = \dot{x}$.

PROBLEMS

Introductory Problems

3/301 The flatbed truck is traveling at the constant speed of 60 km/h up the 15-percent grade when the 100-kg crate which it carries is given a shove which imparts to it an initial relative velocity $\dot{x} = 3$ m/s toward the rear of the truck. If the crate slides a distance $x = 2$ m measured on the truck bed before coming to rest on the bed, compute the coefficient of kinetic friction μ_k between the crate and the truck bed.

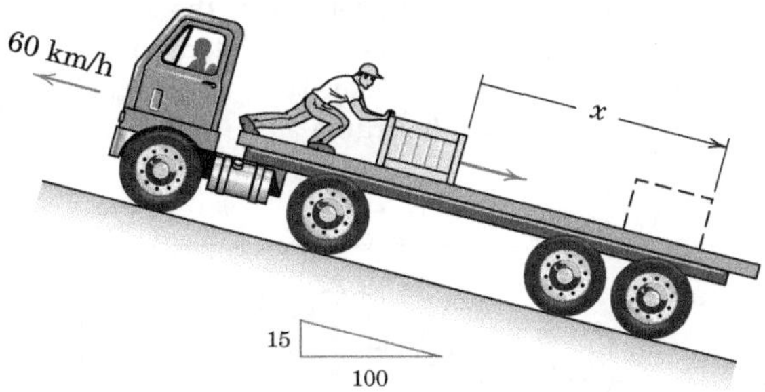

Problem 3/301

3/302 If the spring of constant k is compressed a distance δ as indicated, calculate the acceleration a_{rel} of the block of mass m_1 relative to the frame of mass m_2 upon release of the spring. The system is initially stationary.

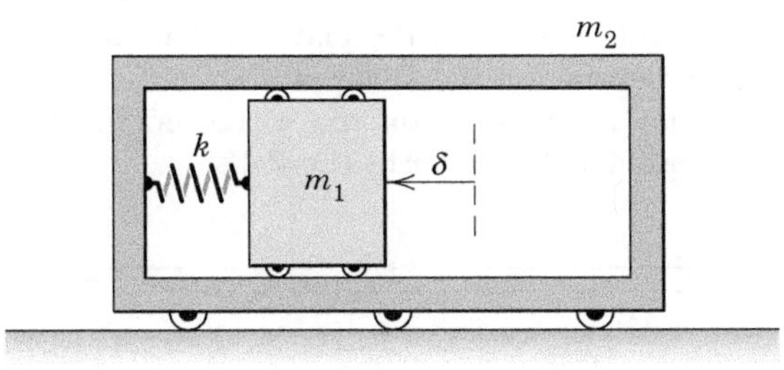

Problem 3/302

3/303 The cart with attached x-y axes moves with an absolute speed $v = 2$ m/s to the right. Simultaneously, the light arm of length $l = 0.5$ m rotates about point B of the cart with angular velocity $\dot{\theta} = 2$ rad/s. The mass of the sphere is $m = 3$ kg. Determine the following quantities for the sphere when $\theta = 0$: $\mathbf{G}$, $\mathbf{G}_{rel}$, T, T_{rel}, $\mathbf{H}_O$, $(\mathbf{H}_B)_{rel}$ where the subscript "rel" indicates measurement relative to the x-y axes. Point O is an inertially fixed point coincident with point B at the instant under consideration.

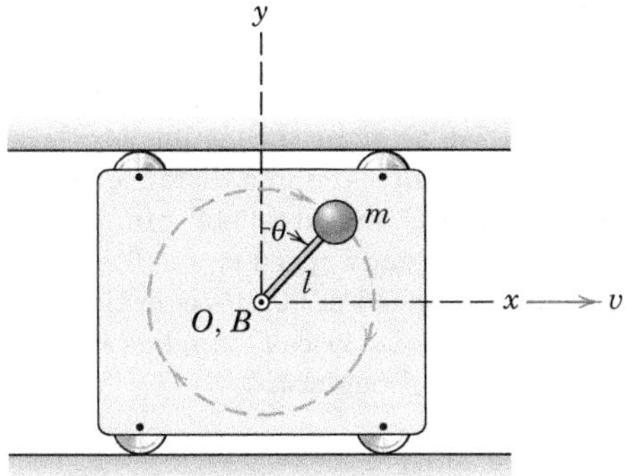

Problem 3/303

3/304 The aircraft carrier is moving at a constant speed and launches a jet plane with a mass of 3 Mg in a distance of 75 m along the deck by means of a steam-driven catapult. If the plane leaves the deck with a velocity of 240 km/h relative to the carrier and if the jet thrust is constant at 22 kN during takeoff, compute the constant force P exerted by the catapult on the airplane during the 75-m travel of the launch carriage.

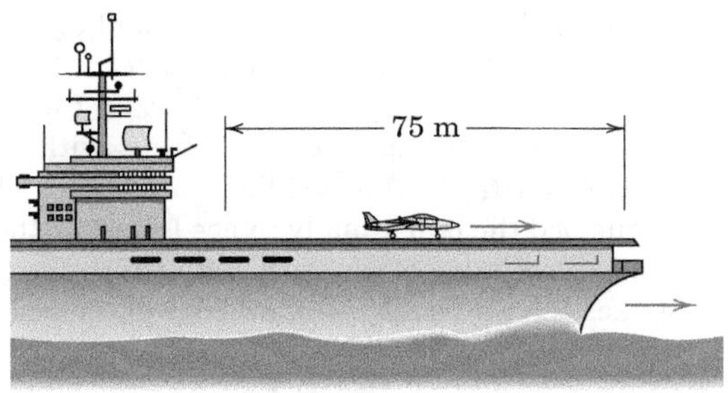

Problem 3/304

3/305 The 4000-lb van is driven from position A to position B on the barge, which is towed at a constant speed $v_0 = 10$ mi/hr. The van starts from rest relative to the barge at A, accelerates to $v = 15$ mi/hr relative to the barge over a distance of 80 ft, and then stops with a deceleration of the same magnitude. Determine the magnitude of the net force F between the tires of the van and the barge during this maneuver.

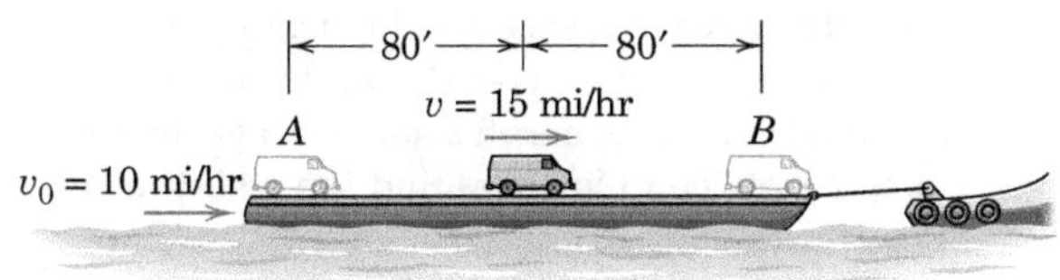

Problem 3/305

Representative Problems

3/306 The launch catapult of the aircraft carrier gives the 7-Mg jet airplane a constant acceleration and launches the airplane in a distance of 100 m measured along the angled takeoff ramp. The carrier is moving at a steady speed $v_C = 16$ m/s. If an absolute aircraft speed of 90 m/s is desired for takeoff, determine the net force F supplied by the catapult and the aircraft engines.

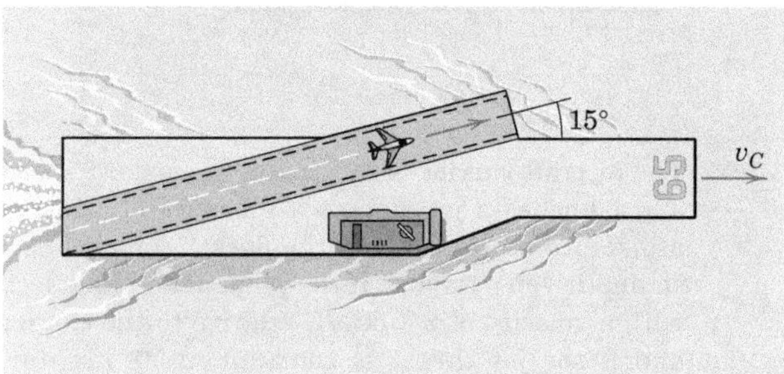

Problem 3/306

3/307 The coefficients of friction between the flatbed of the truck and crate are $\mu_s = 0.80$ and $\mu_k = 0.70$. The coefficient of kinetic friction between the truck tires and the road surface is 0.90. If the truck stops from an initial speed of 15 m/s with maximum braking (wheels skidding), determine where on the bed the crate finally comes to rest or the velocity v_{rel} relative to the truck with which the crate strikes the wall at the forward edge of the bed.

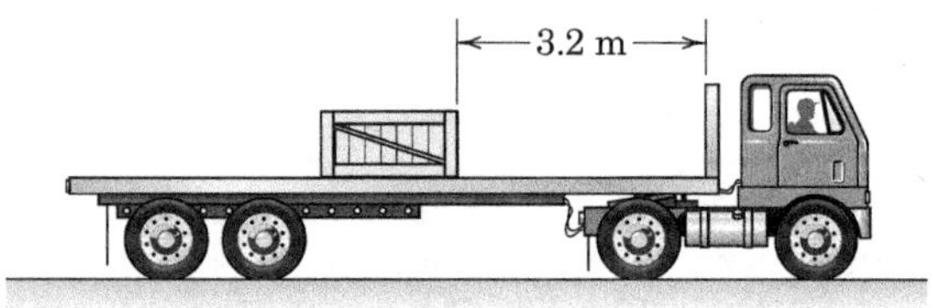

Problem 3/307

3/308 A boy of mass m is standing initially at rest relative to the moving walkway, which has a constant horizontal speed u. He decides to accelerate his progress and starts to walk from point A with a steadily increasing speed and reaches point B with a speed $\dot{x} = v$ relative to the walkway. During his acceleration he generates an average horizontal force F between his shoes and the walkway. Write the work-energy equations for his absolute and relative motions and explain the meaning of the term muv.

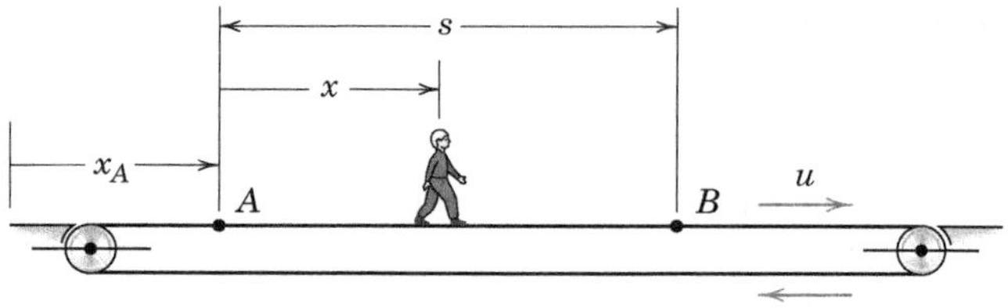

Problem 3/308

3/309 The block of mass m is attached to the frame by the spring of stiffness k and moves horizontally with negligible friction within the frame. The frame and block are initially at rest with $x = x_0$, the uncompressed length of the spring. If the frame is given a constant acceleration a_0, determine the maximum velocity $\dot{x}_{\text{max}} = (v_{\text{rel}})_{\text{max}}$ of the block relative to the frame.

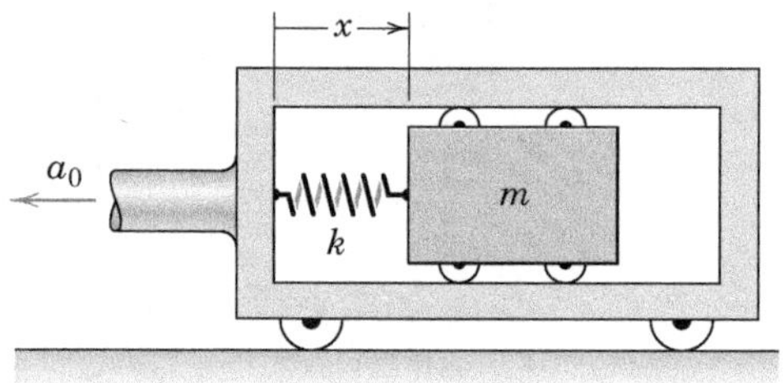

Problem 3/309

3/310 The slider A has a mass of 2 kg and moves with negligible friction in the 30° slot in the vertical sliding plate. What horizontal acceleration a_0 should be given to the plate so that the absolute acceleration of the slider will be vertically down? What is the value of the corresponding force R exerted on the slider by the slot?

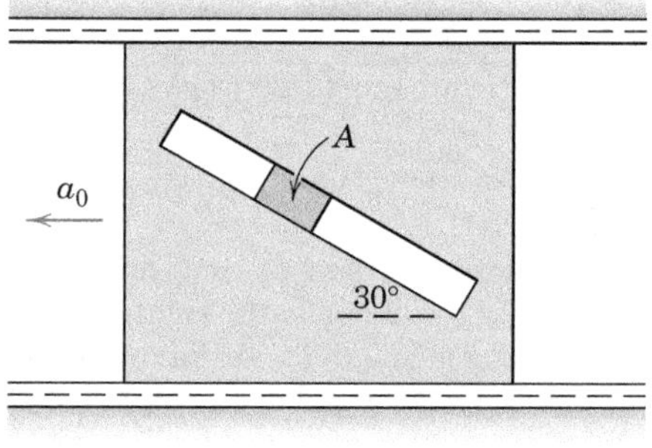

Problem 3/310

3/311 The ball A of mass 10 kg is attached to the light rod of length $l = 0.8$ m. The mass of the carriage alone is 250 kg, and it moves with an acceleration a_O as shown. If $\dot{\theta} = 3$ rad/s when $\theta = 90°$, find the kinetic energy T of the system if the carriage has a velocity of 0.8 m/s (a) in the direction of a_O and (b) in the direction opposite to a_O. Treat the ball as a particle.

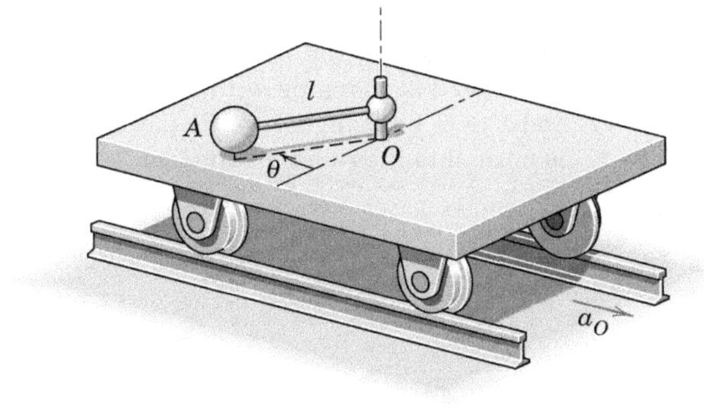

Problem 3/311

3/312 Consider the system of Prob. 3/311 where the mass of the ball is $m = 10$ kg and the length of the light rod is $l = 0.8$ m. The ball–rod assembly is free to rotate about a vertical axis through O. The carriage, rod, and ball are initially at rest with $\theta = 0$ when the carriage is given a constant acceleration $a_O = 3$ m/s^2. Write an expression for the tension T in the rod as a function of θ and calculate T for the position $\theta = \pi/2$.

3/313 A simple pendulum is placed on an elevator, which accelerates upward as shown. If the pendulum is displaced an amount θ_0 and released from rest relative to the elevator, find the tension T_0 in the supporting light rod when $\theta = 0$. Evaluate your result for $\theta_0 = \pi/2$.

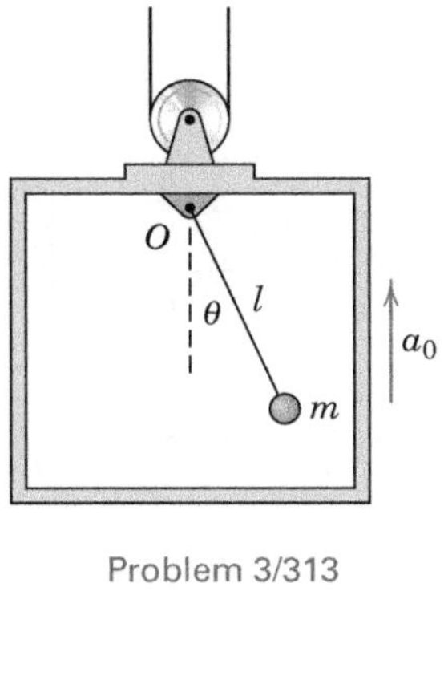

Problem 3/313

3/314 A boy of mass m is standing initially at rest relative to the moving walkway inclined at the angle θ and moving with a constant speed u. He decides to accelerate his progress and starts to walk from point A with a steadily increasing speed and reaches point B with a speed v_r relative to the walkway. During his acceleration he generates a constant average force F tangent to the walkway between his shoes and the walkway surface. Write the work-energy equations for the motion between A and B for his absolute motion and his relative motion and explain the meaning of the term muv_r. If the boy weighs 150 lb and if $u = 2$ ft/sec, $s = 30$ ft, and $\theta = 10°$, calculate the power P_{rel} developed by the boy as he reaches the speed of 2.5 ft/sec relative to the walkway.

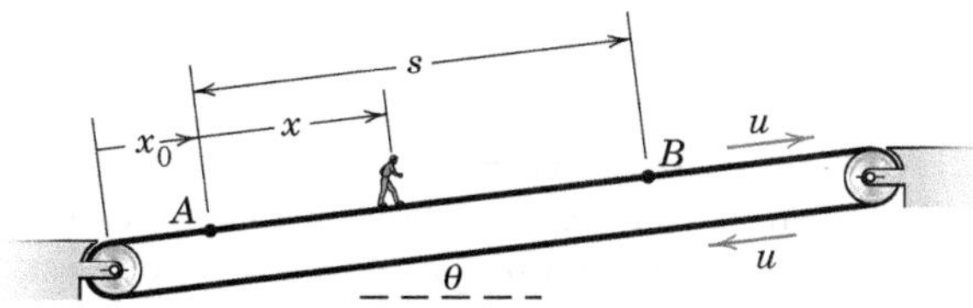

Problem 3/314

▶3/315 A ball is released from rest relative to the elevator at a distance h_1 above the floor. The speed of the elevator at the time of ball release is v_0. Determine the bounce height h_2 of the ball (a) if v_0 is constant and (b) if an upward elevator acceleration $a = g/4$ begins at the instant the ball is released. The coefficient of restitution for the impact is e.

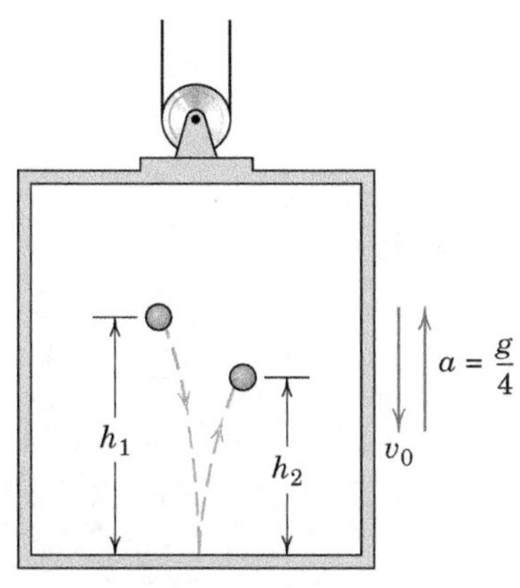

Problem 3/315

▶3/316 The small slider A moves with negligible friction down the tapered block, which moves to the right with constant speed $v = v_0$. Use the principle of work-energy to determine the magnitude v_A of the absolute velocity of the slider as it passes point C if it is released at point B with no velocity relative to the block. Apply the equation, both as an observer fixed to the block and as an observer fixed to the ground, and reconcile the two relations.

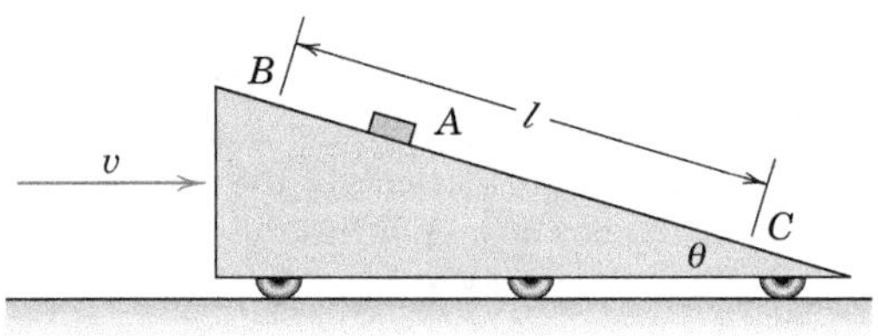

Problem 3/316

▶3/317 When a particle is dropped from rest relative to the surface of the earth at a latitude γ, the initial apparent acceleration is the relative acceleration due to gravity g_{rel}. The absolute acceleration due to gravity g is directed toward the center of the earth. Derive an expression for g_{rel} in terms of g, R, ω, and γ, where R is the radius of the earth treated as a sphere and ω is the constant angular velocity of the earth about the polar axis considered fixed. Although axes x-y-z are attached to the earth and hence rotate, we may use Eq. 3/50 as long as the particle has no velocity relative to x-y-z. (*Hint:* Use the first two terms of the binomial expansion for the approximation.)

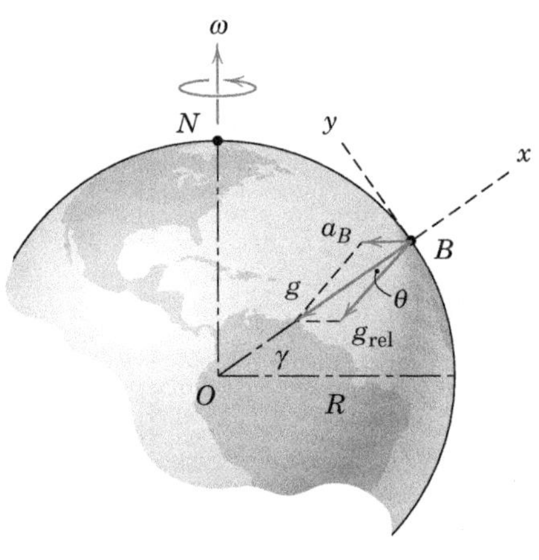

Problem 3/317

▶3/318 The figure represents the space shuttle S, which is (a) in a circular orbit about the earth and (b) in an elliptical orbit where P is its perigee position. The exploded views on the right represent the cabin space with its x-axis oriented in the direction of the orbit. The astronauts conduct an experiment by applying a known force F in the x-direction to a small mass m. Explain why $F = m\ddot{x}$ does or does not hold in each case, where x is measured within the spacecraft. Assume that the shuttle is between perigee and apogee in the elliptical orbit so that the orbital speed is changing with time. Note that the t- and x-axes are tangent to the path, and the θ-axis is normal to the radial r-direction.

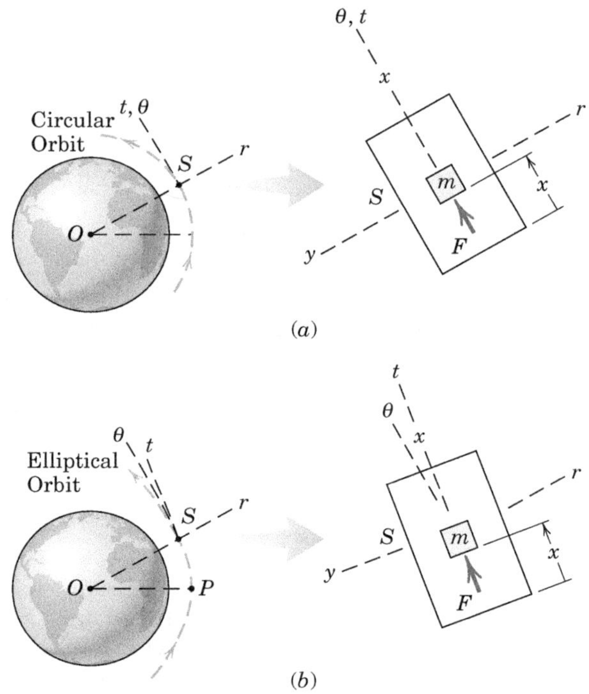

Problem 3/318

3/15 Chapter Review

In Chapter 3 we have developed the three basic methods of solution to problems in particle kinetics. This experience is central to the study of dynamics and lays the foundation for the subsequent study of rigid-body and nonrigid-body dynamics. These three methods are summarized as follows:

1. Direct Application of Newton's Second Law

First, we applied Newton's second law $\Sigma\mathbf{F} = m\mathbf{a}$ to determine the instantaneous relation between forces and the acceleration they produce. With the background of Chapter 2 for identifying the kind of motion and with the aid of our familiar free-body diagram to be certain that all forces are accounted for, we were able to solve a large variety of problems using x-y, n-t, and r-θ coordinates for plane-motion problems and x-y-z, r-θ-z, and R-θ-ϕ coordinates for space problems.

2. Work-Energy Equations

Next, we integrated the basic equation of motion $\Sigma\mathbf{F} = m\mathbf{a}$ with respect to displacement and derived the scalar equations for work and energy. These equations enable us to relate the initial and final velocities to the work done during an interval by forces external to our defined system. We expanded this approach to include potential energy, both elastic and gravitational. With these tools we discovered that the energy approach is especially valuable for conservative systems, that is, systems wherein the loss of energy due to friction or other forms of dissipation is negligible.

3. Impulse-Momentum Equations

Finally, we rewrote Newton's second law in the form of force equals time rate of change of linear momentum and moment equals time rate of change of angular momentum. Then we integrated these relations with respect to time and derived the impulse and momentum equations. These equations were then applied to motion intervals where the forces were functions of time. We also investigated the interactions between particles under conditions where the linear momentum is conserved and where the angular momentum is conserved.

In the final section of Chapter 3, we employed these three basic methods in specific application areas as follows:

1. We noted that the impulse-momentum method is convenient in developing the relations governing particle impacts.
2. We observed that the direct application of Newton's second law enables us to determine the trajectory properties of a particle under central-force attraction.
3. Finally, we saw that all three basic methods may be applied to particle motion relative to a translating frame of reference.

Successful solution of problems in particle kinetics depends on knowledge of the prerequisite particle kinematics. Furthermore, the principles of particle kinetics are required to analyze particle systems and rigid bodies, which are covered in the remainder of *Dynamics*.

REVIEW PROBLEMS

3/319 The block of weight W is given an initial velocity $v_1 = 20$ ft/sec up the 20° incline at point A. Calculate the velocity v_2 with which the block passes A as it slides back down.

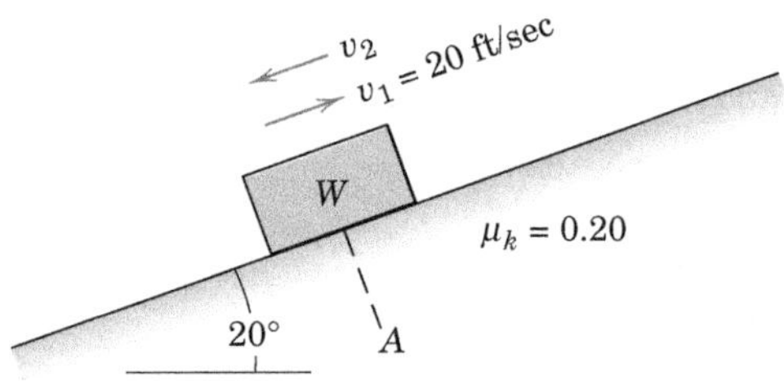

Problem 3/319

3/320 A particle of mass m is attached to the end of the light rigid rod of length L, and the assembly rotates freely about a horizontal axis through the pivot O. The particle is given an initial speed v_0 when the assembly is in the horizontal position $\theta = 0$. Determine the speed v of the particle as a function of θ.

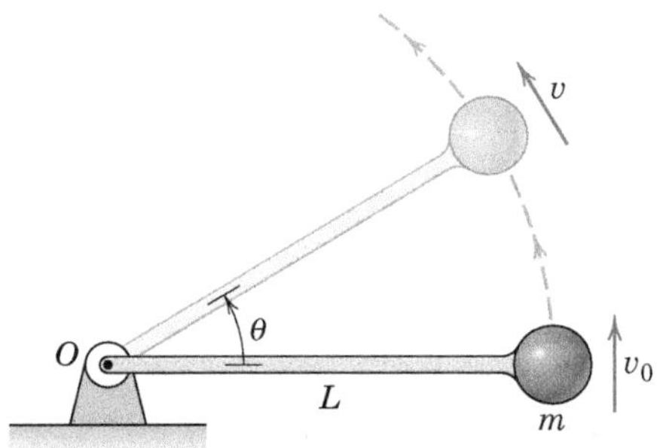

Problem 3/320

3/321 A 30-g tire-balance weight is attached to a vertical surface of the wheel rim by means of an adhesive backing. The tire–wheel unit is then given a final test on the tire-balance machine. If the adhesive can support a maximum shear force of 80 N, determine the maximum rotational speed N for which the weight remains fixed to the wheel. Assume very gradual speed changes.

Problem 3/321

3/322 The simple 2-kg pendulum is released from rest in the horizontal position. As it reaches the bottom position, the cord wraps around the smooth fixed pin at B and continues in the smaller arc in the vertical plane. Calculate the magnitude of the force R supported by the pin at B when the pendulum passes the position $\theta = 30°$.

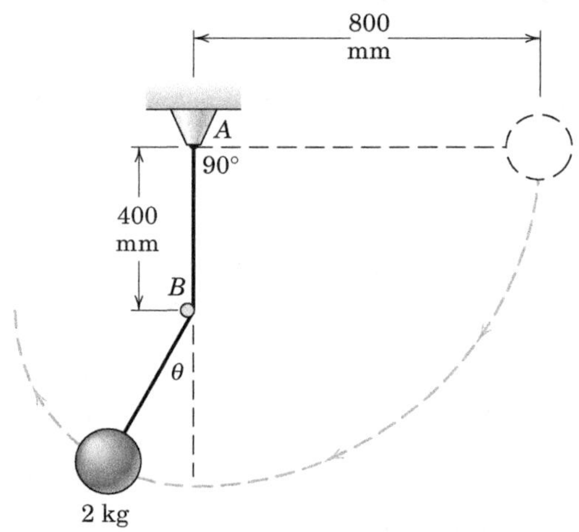

Problem 3/322

3/323 The small 2-kg carriage is moving freely along the horizontal with a speed of 4 m/s at time $t = 0$. A force applied to the carriage in the direction opposite to motion produces two impulse "peaks," one after the other, as shown by the graphical plot of the readings of the instrument that measured the force. Approximate the loading by the dashed lines and determine the velocity v of the carriage for $t = 1.5$ s.

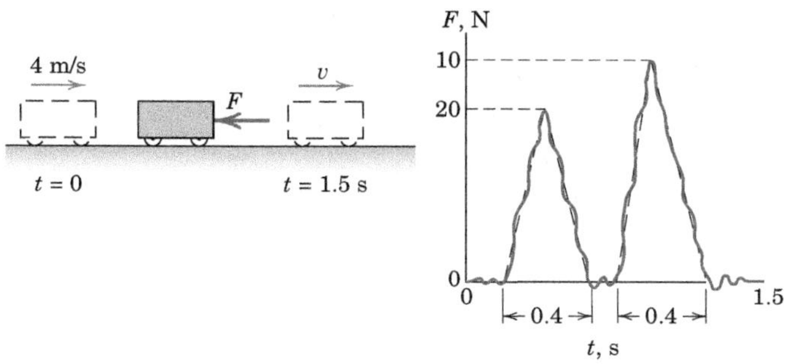

Problem 3/323

3/324 For the elliptical orbit of a spacecraft around the earth, determine the speed v_A at point A which results in a perigee altitude at B of 200 km. What is the eccentricity e of the orbit?

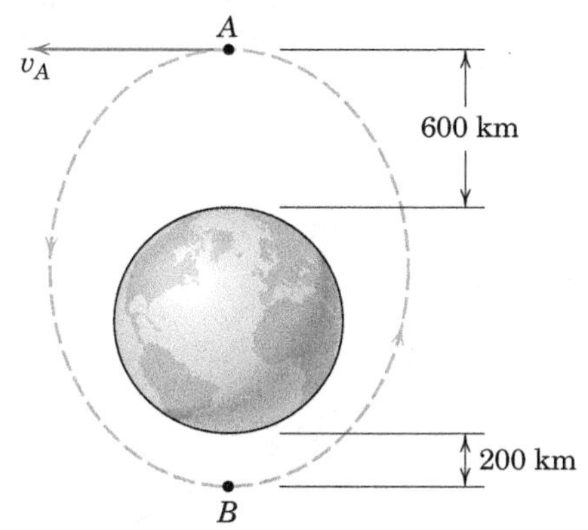

Problem 3/324

3/325 At a steady speed of 200 mi/hr along a level track, the racecar is subjected to an aerodynamic force of 900 lb and an overall rolling resistance of 200 lb. If the drivetrain efficiency is $e = 0.90$, what power P must the motor produce?

Problem 3/325

3/326 The spring of stiffness k is compressed and suddenly released, sending the particle of mass m sliding along the track. Determine the minimum spring compression δ for which the particle will not lose contact with the loop-the-loop track. The sliding surface is smooth except for the rough portion of length s equal to R, where the coefficient of kinetic friction is μ_k.

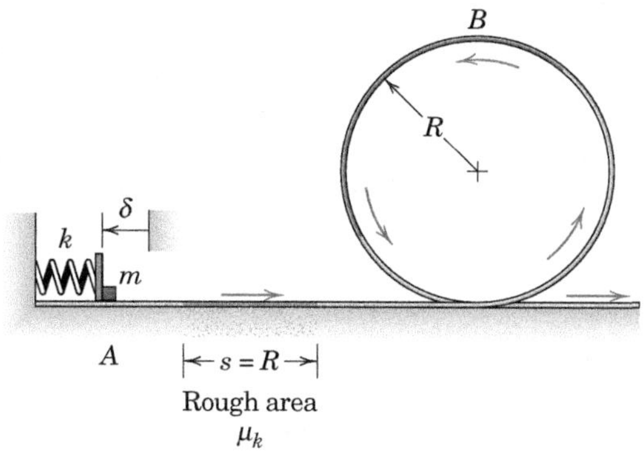

Problem 3/326

3/327 The small slider has a speed $v_A = 15$ ft/sec as it passes point A. Neglecting friction, determine its speed as it passes point B.

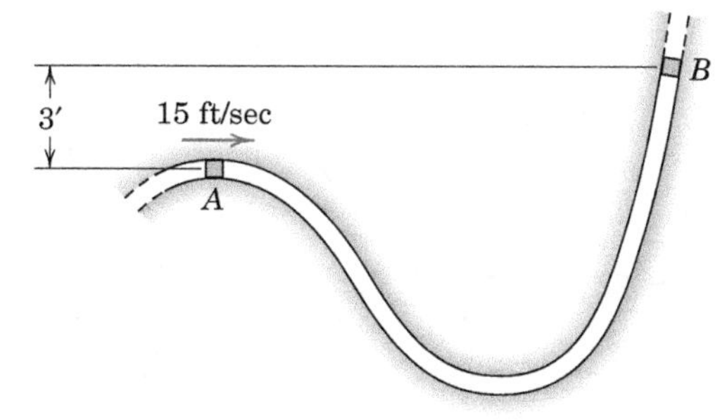

Problem 3/327

3/328 Six identical spheres are arranged as shown in the figure. The two spheres at the left end are released from the displaced positions and strike sphere 3 with speed v_1. Assuming that the common coefficient of restitution is $e = 1$, explain why two spheres leave the right end of the row with speed v_1 instead of one sphere leaving the right end with speed $2v_1$.

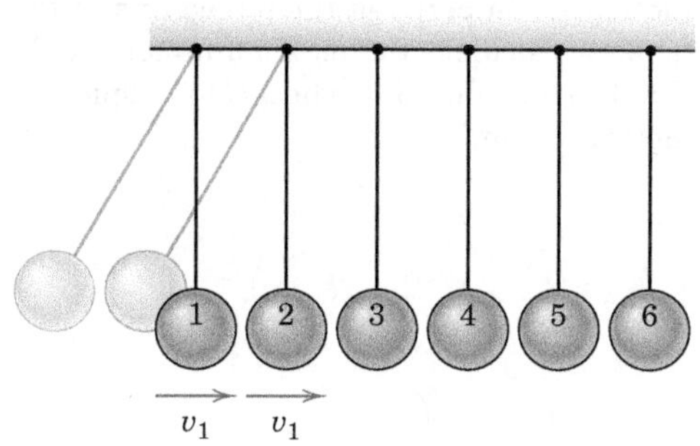

Problem 3/328

3/329 The 180-lb exerciser is beginning to execute a bicep curl. When in the position shown with his right elbow fixed, he causes the 20-lb cylinder to accelerate upward at the rate $g/4$. Neglect the effects of the mass of his lower arm and estimate the normal reaction forces at A and B. Friction is sufficient to prevent slipping.

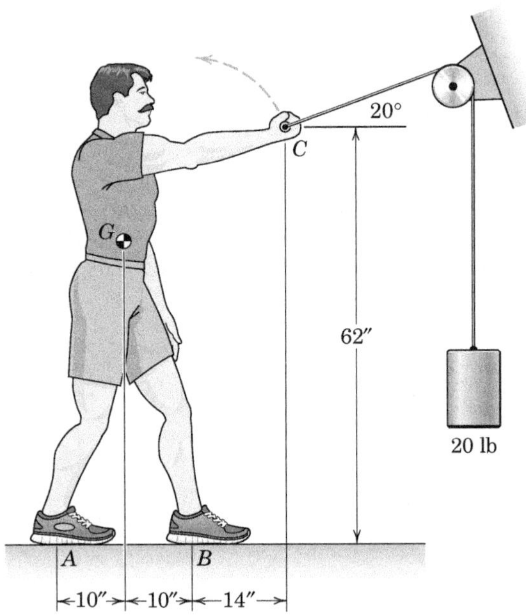

Problem 3/329

3/330 The figure shows a centrifugal clutch consisting in part of a rotating spider A which carries four plungers B. As the spider is made to rotate about its center with a speed ω, the plungers move outward and bear against the interior surface of the rim of wheel C, causing it to rotate. The wheel and spider are independent except for frictional contact. If each plunger has a mass of 2 kg with a center of mass at G, and if the coefficient of kinetic friction between the plungers and the wheel is 0.40, calculate the maximum moment M which can be transmitted to wheel C for a spider speed of 3000 rev/min.

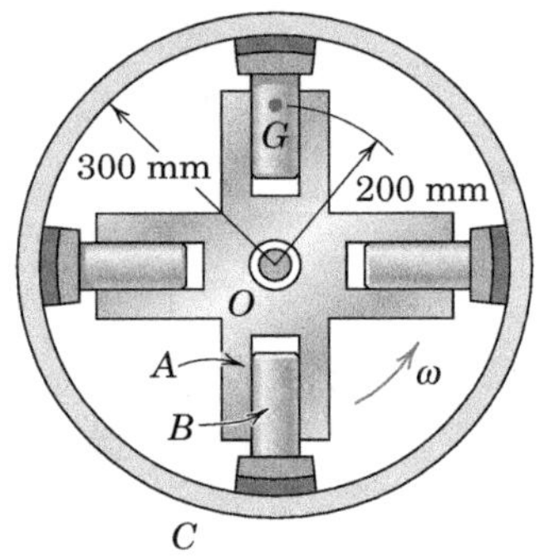

Problem 3/330

3/331 A ball is thrown from point O with a velocity of 30 ft/sec at a 60° angle with the horizontal and bounces on the inclined plane at A. If the coefficient of restitution is 0.6, calculate the magnitude v of the rebound velocity at A. Neglect air resistance.

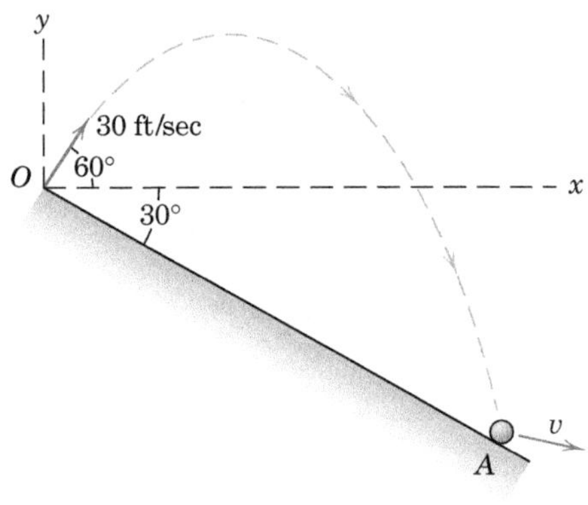

Problem 3/331

3/332 The pickup truck is used to hoist the 40-kg bale of hay as shown. If the truck has reached a constant velocity $v = 5$ m/s when $x = 12$ m, compute the corresponding tension T in the rope.

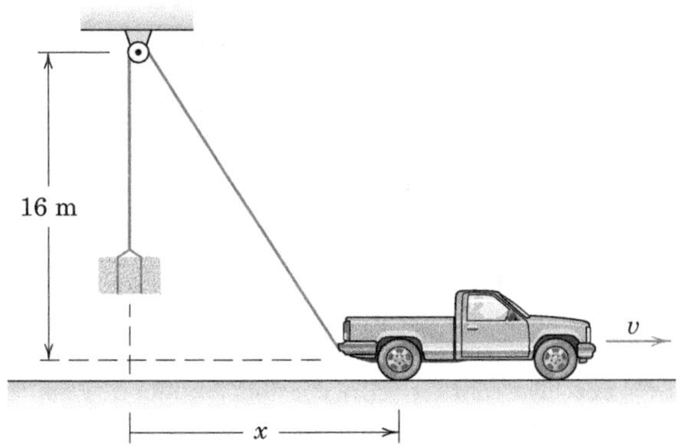

Problem 3/332

3/333 For a given value of the force P, determine the steady-state spring compression δ, which is measured relative to the unstretched length of the spring of modulus k. The mass of the cart is M and that of the slider is m. Neglect all friction. State the values of P and δ associated with the equilibrium condition.

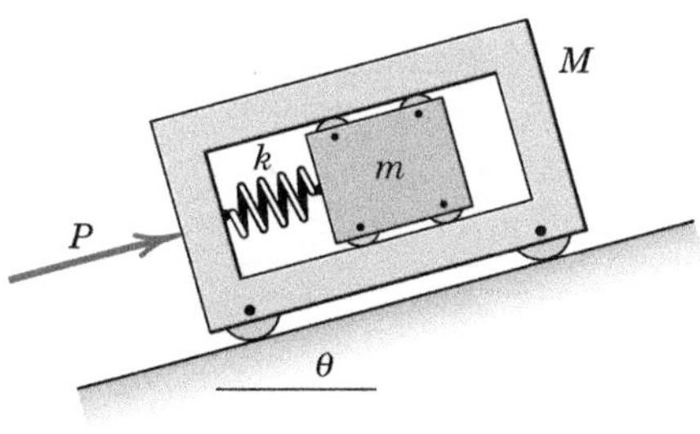

Problem 3/333

3/334 The 200-kg glider B is being towed by airplane A, which is flying horizontally with a constant speed of 220 km/h. The tow cable has a length $r = 60$ m and may be assumed to form a straight line. The glider is gaining altitude and when θ reaches 15°, the angle is increasing at the constant rate $\dot{\theta} =$ 5 deg/s. At the same time the tension in the tow cable is 1520 N for this position. Calculate the aerodynamic lift L and drag D acting on the glider.

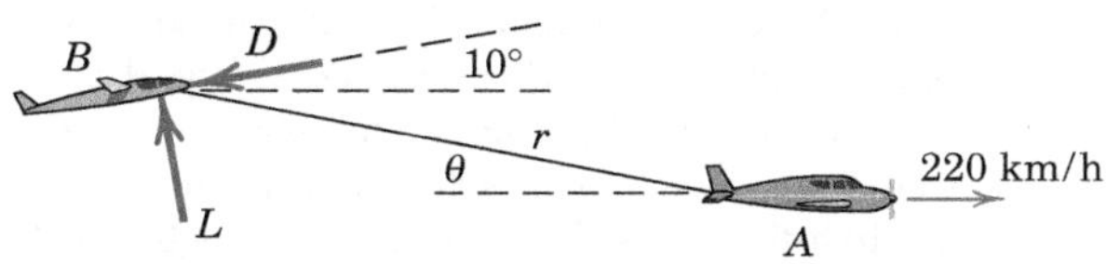

Problem 3/334

3/335 An electromagnetic catapult system is being designed to replace a steam-driven system on an aircraft carrier. The requirements include accelerating a 12 000-kg aircraft from rest to a speed of 70 m/s over a distance of 90 m. What constant force F must the catapult exert on the aircraft?

Problem 3/335

3/336 The 2-lb piece of putty is dropped 6 ft onto the 18-lb block initially at rest on the two springs, each with a stiffness $k = 3$ lb/in. Calculate the additional deflection δ of the springs due to the impact of the putty, which adheres to the block upon contact.

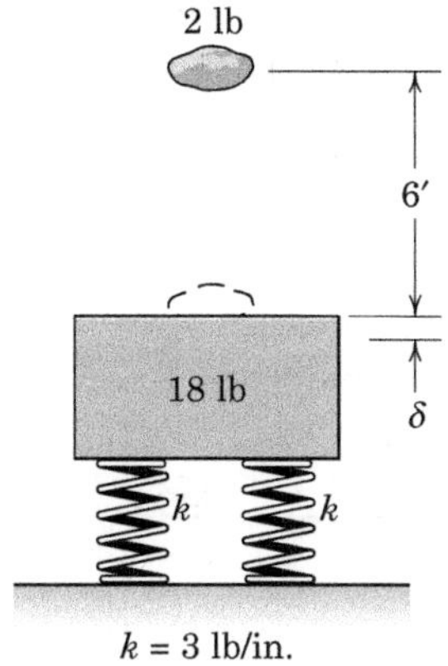

Problem 3/336

3/337 The member OA rotates about a horizontal axis through O with a constant counterclockwise angular velocity $\omega = 3$ rad/sec. As it passes the position $\theta = 0$, a small block of mass m is placed on it at a radial distance $r = 18$ in. If the block is observed to slip at $\theta = 50°$, determine the coefficient of static friction μ_s between the block and the member.

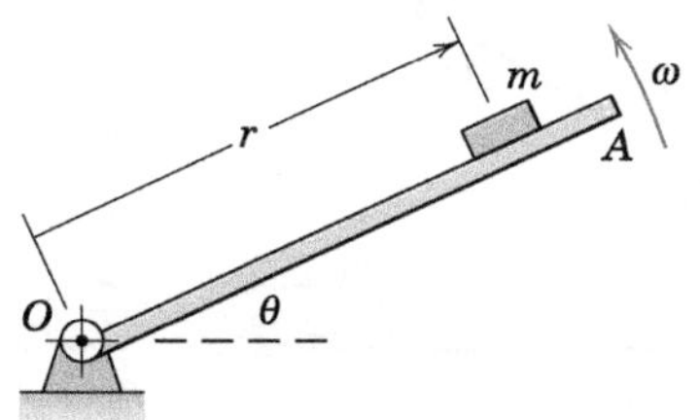

Problem 3/337

3/338 An automobile accident occurs as follows: The driver of a full-size car (vehicle A, 4000 lb) is traveling on a dry, level road and approaches a stationary compact car (vehicle B, 2000 lb). Just 50 feet before collision, he applies the brakes, skidding all wheels. After impact, vehicle A skids an additional 50 ft and vehicle B, whose driver had all brakes fully applied, skids 100 ft. The final positions of the vehicles are shown in the figure. If the coefficient of kinetic friction is 0.90, was the driver of vehicle A exceeding the speed limit of 55 mi/hr before the initial application of his brakes?

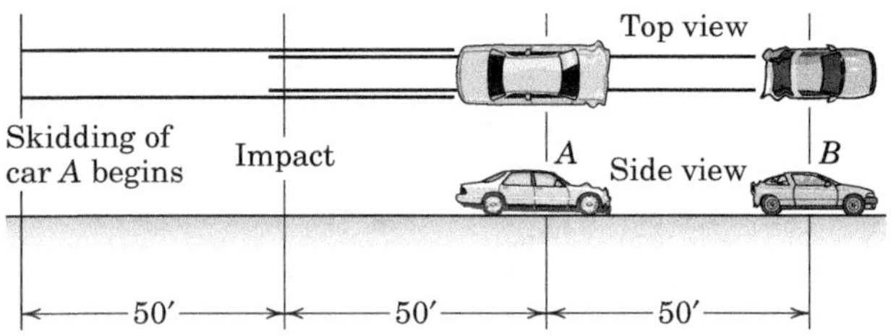

Problem 3/338

3/339 The 3-kg block A is released from rest in the 60° position shown and subsequently strikes the 1-kg cart B. If the coefficient of restitution for the collision is $e = 0.7$, determine the maximum displacement s of cart B beyond point C. Neglect friction.

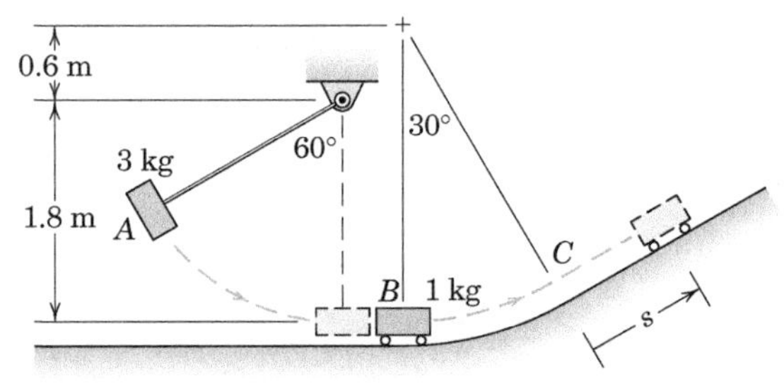

Problem 3/339

3/340 The bungee jumper, an 80-kg man, falls from the bridge at A with the bungee cord secured to his ankles. He falls 20 m before the 17-m length of elastic bungee cord begins to stretch. The 3 m of rope above the elastic cord has no appreciable stretch. The man is observed to drop a total of 44 m before being projected upward. Neglect any energy loss and calculate (a) the stiffness k of the bungee cord (increase in tension per meter of elongation), (b) the maximum velocity v_{max} of the man during his fall, and (c) his maximum acceleration a_{max}. Treat the man as a particle located at the end of the bungee cord.

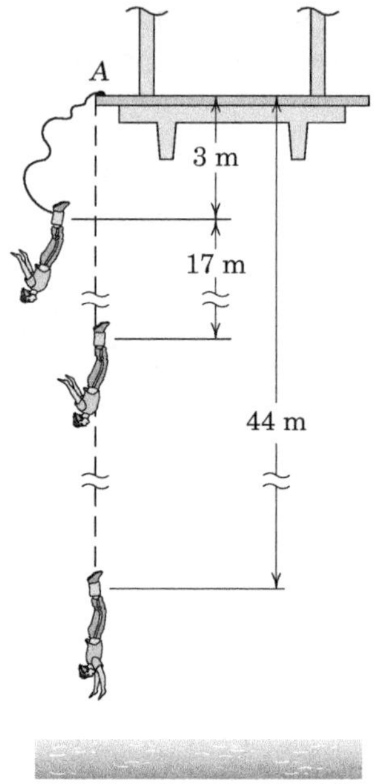

Problem 3/340

3/341 One of the functions of the space shuttle is to release communications satellites at low altitude. A booster rocket is fired at B, placing the satellite in an elliptical transfer orbit, the apogee of which is at the altitude necessary for a geosynchronous orbit. (A geosynchronous orbit is an equatorial-plane circular orbit whose period is equal to the absolute rotational period of the earth. A satellite in such an orbit appears to remain stationary to an earth-fixed observer.) A second booster rocket is then fired at C, and the final circular orbit is achieved. On one of the early space-shuttle missions, a 1500-lb satellite was released from the shuttle at B, where $h_1 = 170$ miles. The booster rocket was to fire for $t = 90$ seconds, forming a transfer orbit with $h_2 = 22{,}300$ miles. The rocket failed during its burn. Radar observations determined the apogee altitude of the transfer orbit to be only 700 miles. Determine the actual time t' which the rocket motor operated before failure. Assume negligible mass change during the booster rocket firing.

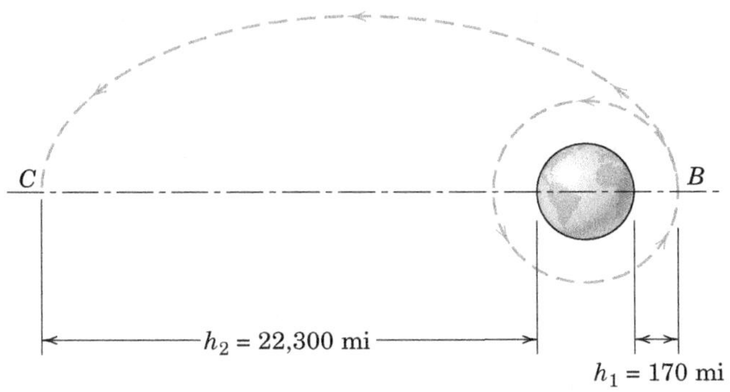

Problem 3/341

3/342 The system is released from rest with the spring initially stretched 25 mm. Calculate the velocity v of the 60-kg cart after it has moved 100 mm down the incline. Also determine the maximum distance traveled by the cart before it stops momentarily. Neglect the mass and friction of the pulleys and wheels.

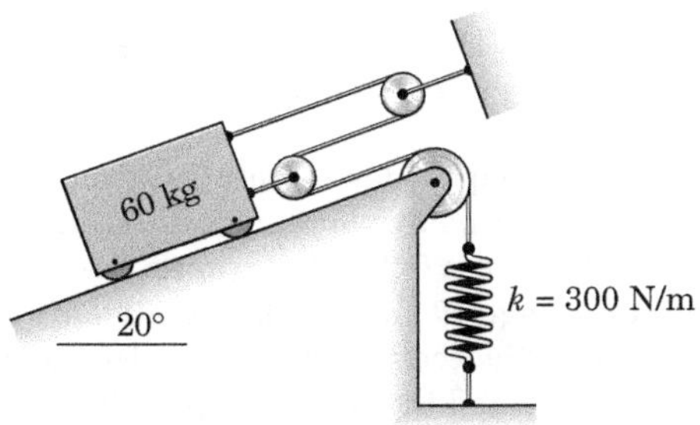

Problem 3/342

3/343 A short train consists of a 400,000-lb locomotive and three 200,000-lb hopper cars. The locomotive exerts a constant friction force of 40,000 lb on the rails as the train starts from rest. (a) If there is 1 ft of slack in each of the three couplers before the train begins moving, estimate the speed v of the train just after car C begins to move. Slack removal is a plastic short-duration impact. Neglect all friction except that of the locomotive tractive force and neglect the tractive force during the short time duration of the impacts associated with the slack removal. (b) If there is no slack in the train couplers, determine the speed v' which is acquired when the train has moved 3 ft.

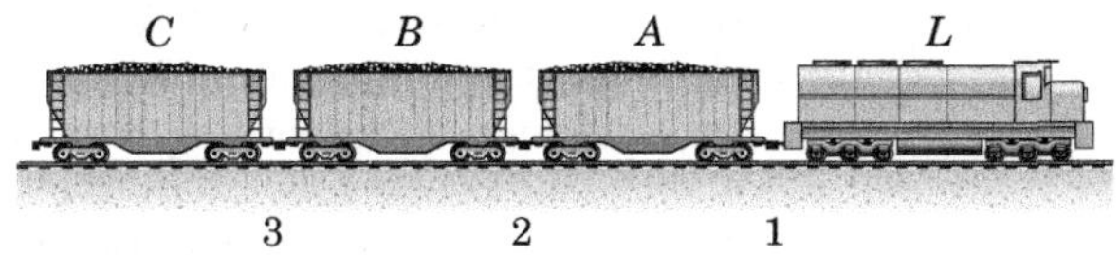

Problem 3/343

3/344 A car of mass m is traveling at a road speed v_r along an equatorial east–west highway at sea level. If the road follows the curvature of the earth, derive an expression for the difference ΔP between the total force exerted by the road on the car for eastward travel and the total force for westward travel. Calculate ΔP for $m = 1500$ kg and $v_r =$ 200 km/h. The angular velocity ω of the earth is $0.7292(10^{-4})$ rad/s. Neglect the motion of the center of the earth.

▶3/345 The retarding forces which act on the race car are the drag force F_D and a nonaerodynamic force F_R. The drag force is $F_D = C_D(\frac{1}{2}\rho v^2)S$, where C_D is the drag coefficient, ρ is the air density, v is the car speed, and $S = 30$ ft^2 is the projected frontal area of the car. The nonaerodynamic force F_R is constant at 200 lb. With its sheet metal in good condition, the race car has a drag coefficient $C_D = 0.3$ and it has a corresponding top speed $v = 200$ mi/hr. After a minor collision, the damaged front-end sheet metal causes the drag coefficient to be $C_D' =$ 0.4. What is the corresponding top speed v' of the race car?

Problem 3/345

3/346 The satellite of Sample Problem 3/31 has a perigee velocity of 26 140 km/h at the perigee altitude of 2000 km. What is the minimum increase Δv in velocity required of its rocket motor at this position to allow the satellite to escape from the earth's gravity field?

3/347 A long fly ball strikes the wall at point A (where $e_1 = 0.5$) and then hits the ground at B (where $e_2 = 0.3$). The outfielder likes to catch the ball when it is 4 ft above the ground and 2 ft in front of him as shown. Determine the distance x from the wall where he can catch the ball as described. Note the two possible solutions.

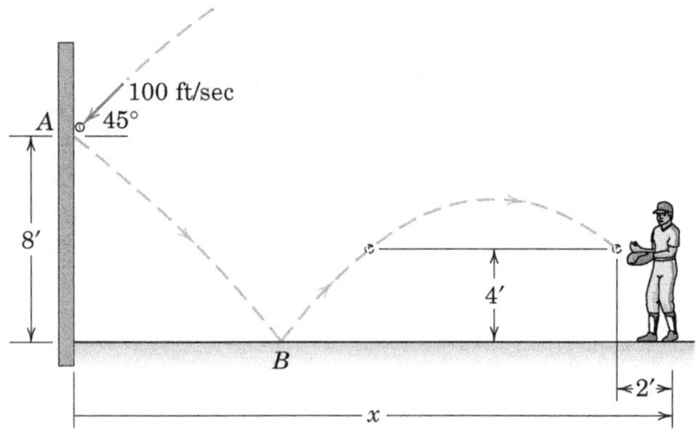

Problem 3/347

Computer-Oriented Problems

***3/348** The system is released from rest while in the position shown with the torsional spring undeflected. The rod has negligible mass, and all friction is negligible. Determine (*a*) the value of $\dot{\theta}$ when θ is 30° and (*b*) the maximum value of θ. Use the values $m = 5$ kg, $M = 8$ kg, $L = 0.8$ m, and $k_T = 100$ N·m/rad.

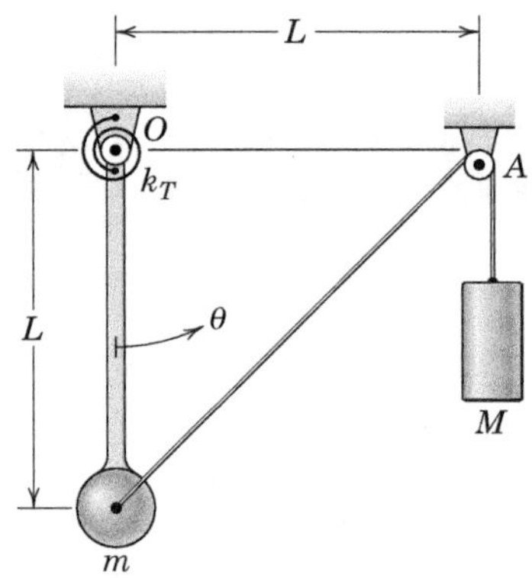

Problem 3/348

***3/349** The cylinder of the previous problem is now replaced by a linear spring of constant k as shown. The system is released from rest in the position shown with the torsional spring undeflected but with the linear spring stretched 500 mm. Determine (*a*) the value of $\dot{\theta}$ when θ is 30° and (*b*) the maximum value of θ. Use the values $m = 5$ kg, $L = 0.8$ m, $k_T = 100$ N·m/rad, and $k = 400$ N/m.

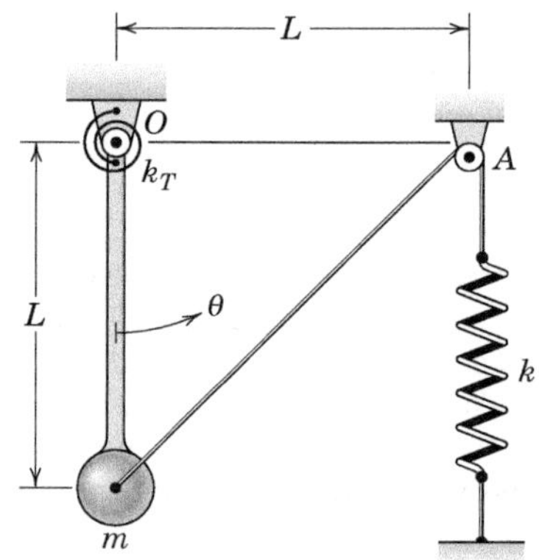

Problem 3/349

***3/350** The bowl-shaped device rotates about a vertical axis with a constant angular velocity $\omega = 6$ rad/s. The value of r is 0.2 m. Determine the range of the position angle θ for which a stationary value is possible if the coefficient of static friction between the particle and the surface is $\mu_s = 0.20$.

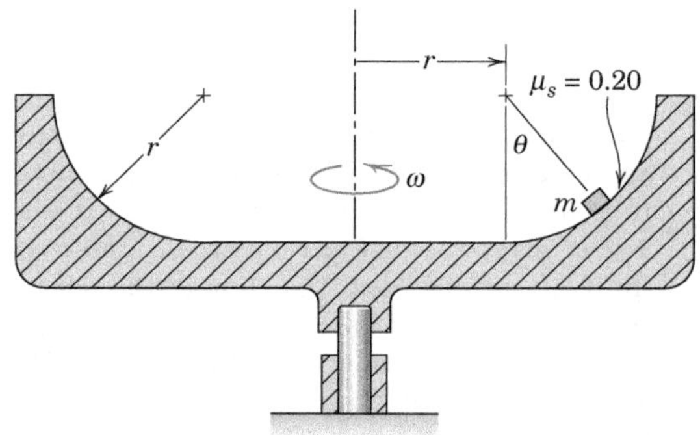

Problem 3/350

***3/351** If the vertical frame starts from rest with a constant acceleration a and the smooth sliding collar A is initially at rest in the bottom position $\theta = 0$, plot $\dot{\theta}$ as a function of θ and find the maximum position angle $\theta_{\max}$ reached by the collar. Use the values $a = g/2$ and $r = 0.3$ m.

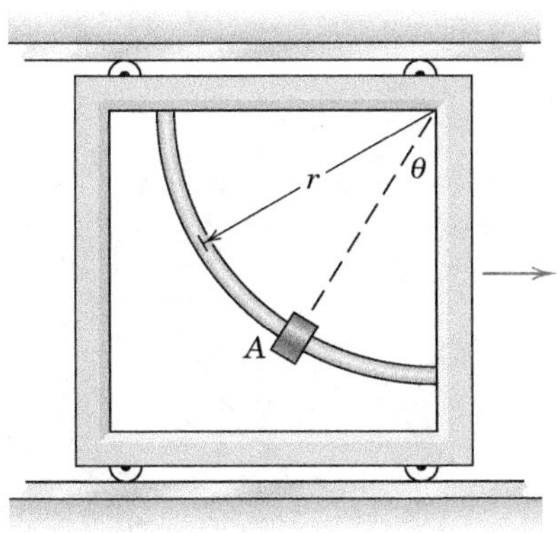

Problem 3/351

***3/352** The tennis player practices by hitting the ball against the wall at A. The ball bounces off the court surface at B and then up to its maximum height at C. For the conditions shown in the figure, plot the location of point C for values of the coefficient of restitution in the range $0.5 \leq e \leq 0.9$. (The value of e is common to both A and B.) For what value of e is $x = 0$ at point C, and what is the corresponding value of y?

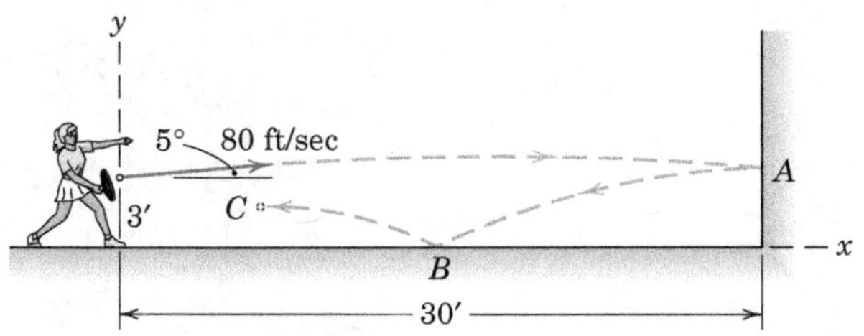

Problem 3/352

***3/353** A particle of mass m is introduced with zero velocity at $r = 0$ when $\theta = 0$. It slides outward through the smooth hollow tube, which is driven at the constant angular velocity ω_0 about a horizontal axis through point O. If the length l of the tube is 1 m and $\omega_0 = 0.5$ rad/s, determine the time t after release and the angular displacement θ for which the particle exits the tube.

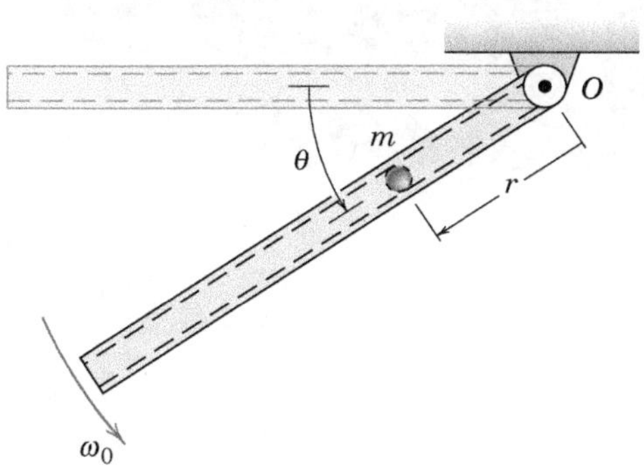

Problem 3/353

***3/354** The elements of a device designed to measure the coefficient of restitution of bat–baseball collisions are repeated here from Prob. 3/265. The 1-lb "bat" A is a length of wood or aluminum which is projected to the right with a speed $v_A = 60$ ft/sec and is confined to move horizontally in the smooth slot. Just before and after the moment of impact, body A is free to move horizontally. The baseball B weighs 5.125 oz and has an initial speed $v_B = 125$ ft/sec. Determine the immediate post-impact speed $v_B{}'$ of the baseball and the resulting horizontal distance R traveled by the baseball over the range $0.4 \leq e \leq 0.6$, where e is the coefficient of restitution. The range is to be calculated assuming that the baseball is initially 3 ft above a horizontal ground.

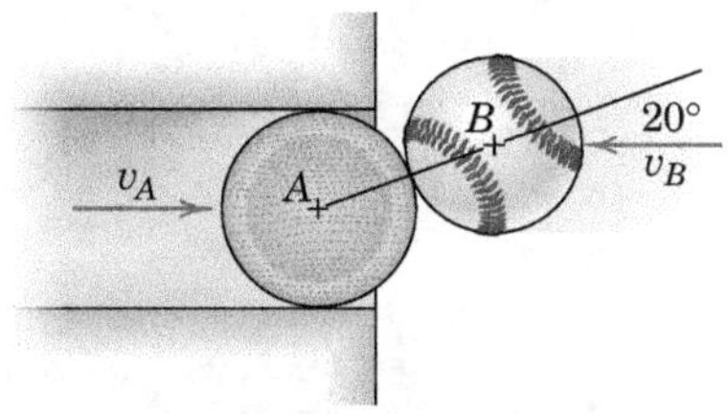

Problem 3/354

***3/355** The system of Prob. 3/166 is repeated here. The system is released from rest at position $x = 0$ with the cable taut at time $t = 0$, with the 10-kg block moving down the rough incline with a speed of 0.3 m/s, and with the spring stretched 25 mm. Plot the velocity of the block as a function of the distance x traveled down the incline and determine the maximum velocity of the block and its corresponding position.

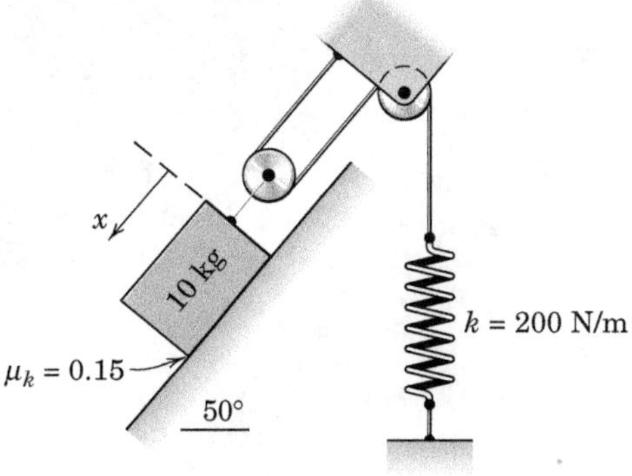

Problem 3/355

***3/356** For the conditions stated in the previous problem, determine the time t at which the velocity of the block is a maximum.

© Ben Lé/Corbis Images

The forces of interaction between the rotating blades of this jet engine and the air which passes over them is a subject which is introduced in this chapter.

4 Kinetics of Systems of Particles

Chapter Outline

4/1 Introduction

In the previous two chapters, we have applied the principles of dynamics to the motion of a particle. Although we focused primarily on the kinetics of a single particle in Chapter 3, we mentioned the motion of two particles, considered together as a system, when we discussed work-energy and impulse-momentum.

Our next major step in the development of dynamics is to extend these principles, which we applied to a single particle, to describe the motion of a general system of particles. This extension will unify the remaining topics of dynamics and enable us to treat the motion of both rigid bodies and nonrigid systems.

Recall that a rigid body is a solid system of particles wherein the distances between particles remain essentially unchanged. The overall motions found with machines, land and air vehicles, rockets and spacecraft, and many moving structures provide examples of rigid-body problems. On the other hand, we may need to study the time-dependent changes in the shape of a nonrigid, but solid, body due to elastic or inelastic deformations. Another example of a nonrigid body is a defined mass of liquid or gaseous particles flowing at a specified rate. Examples are the air and fuel flowing through the turbine of an aircraft engine, the burned gases issuing from the nozzle of a rocket motor, or the water passing through a rotary pump.

Although we can extend the equations for single-particle motion to a general system of particles without much difficulty, it is difficult to understand the generality and significance of these extended principles without considerable problem experience. For this reason, you should frequently review the general results obtained in the following articles during the remainder of your study of dynamics. In this way, you will understand how these broader principles unify dynamics.

Key Concepts

4/2 Generalized Newton's Second Law

We now extend Newton's second law of motion to cover a general mass system which we model by considering n mass particles bounded by a closed surface in space, Fig. 4/1. This bounding envelope, for example, may be the exterior surface of a given rigid body, the bounding surface of an arbitrary portion of the body, the exterior surface of a rocket containing both rigid and flowing particles, or a particular volume of fluid particles. In each case, the system to be considered is the mass within the envelope, and that mass must be clearly defined and isolated.

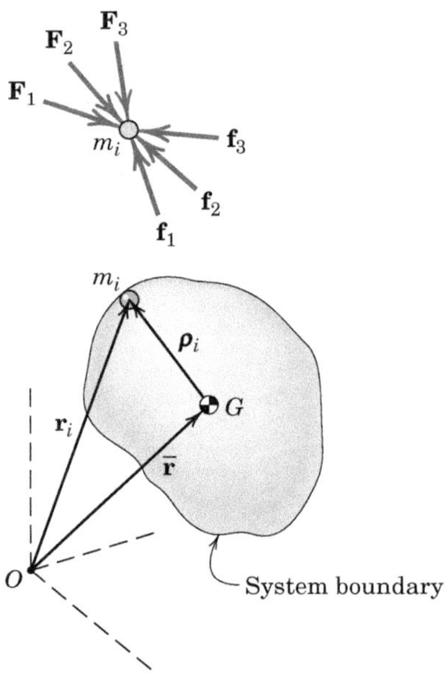

Figure 4/1

Figure 4/1 shows a representative particle of mass m_i of the system isolated with forces $\mathbf{F}_1$, $\mathbf{F}_2$, $\mathbf{F}_3$, . . . acting on m_i from sources *external* to the envelope, and forces $\mathbf{f}_1$, $\mathbf{f}_2$, $\mathbf{f}_3$, . . . acting on m_i from sources *internal* to the system boundary. The external forces are due to contact with external bodies or to external gravitational, electric, or magnetic effects. The internal forces are forces of reaction with other mass particles within the boundary. The particle of mass m_i is located by its position vector $\mathbf{r}_i$ measured from the nonaccelerating origin O of a Newtonian set of reference axes.* The center of mass G of the isolated system of particles is located by the position vector $\bar{\mathbf{r}}$ which, from the definition of the mass center as covered in statics, is given by

$$m\bar{\mathbf{r}} = \Sigma m_i \mathbf{r}_i$$

where the total system mass is $m = \Sigma m_i$. The summation sign Σ represents the summation $\Sigma_{i=1}^{n}$ over all n particles.

Newton's second law, Eq. 3/3, when applied to m_i gives

$$\mathbf{F}_1 + \mathbf{F}_2 + \mathbf{F}_3 + \cdots + \mathbf{f}_1 + \mathbf{f}_2 + \mathbf{f}_3 + \cdots = m_i \ddot{\mathbf{r}}_i$$

where $\ddot{\mathbf{r}}_i$ is the acceleration of m_i. A similar equation may be written for each of the particles of the system. If these equations written for *all* particles of the system are added together, the result is

$$\Sigma\mathbf{F} + \Sigma\mathbf{f} = \Sigma m_i \ddot{\mathbf{r}}_i$$

The term $\Sigma\mathbf{F}$ then becomes the vector sum of *all* forces acting on all particles of the isolated system from sources external to the system, and

*It was shown in Art. 3/14 that any nonrotating and nonaccelerating set of axes constitutes a Newtonian reference system in which the principles of Newtonian mechanics are valid.

$\Sigma\mathbf{f}$ becomes the vector sum of all forces on all particles produced by the internal actions and reactions between particles. This last sum is identically zero since all internal forces occur in pairs of equal and opposite actions and reactions. By differentiating the equation defining $\bar{\mathbf{r}}$ twice with time, we have $m\ddot{\bar{\mathbf{r}}} = \Sigma m_i\ddot{\mathbf{r}}_i$ where m has a zero time derivative as long as mass is not entering or leaving the system.* Substitution into the summation of the equations of motion gives

$$\Sigma\mathbf{F} = m\ddot{\bar{\mathbf{r}}} \qquad \text{or} \qquad \mathbf{F} = m\bar{\mathbf{a}} \tag{4/1}$$

where $\bar{\mathbf{a}}$ is the acceleration $\ddot{\bar{\mathbf{r}}}$ of the center of mass of the system.

Equation 4/1 is the generalized Newton's second law of motion for a mass system and is called the *equation of motion of m*. The equation states that the resultant of the external forces on *any* system of masses equals the total mass of the system times the acceleration of the center of mass. This law expresses the so-called *principle of motion of the mass center*.

Observe that $\bar{\mathbf{a}}$ is the acceleration of the mathematical point which represents instantaneously the position of the mass center for the given n particles. For a nonrigid body, this acceleration need not represent the acceleration of any particular particle. Note also that Eq. 4/1 holds for each instant of time and is therefore an instantaneous relationship. Equation 4/1 for the mass system had to be proved, as it cannot be inferred directly from Eq. 3/3 for a single particle.

Equation 4/1 may be expressed in component form using x-y-z coordinates or whatever coordinate system is most convenient for the problem at hand. Thus,

$$\Sigma F_x = m\bar{a}_x \qquad \Sigma F_y = m\bar{a}_y \qquad \Sigma F_z = m\bar{a}_z \tag{4/1a}$$

Although Eq. 4/1, as a vector equation, requires that the acceleration vector $\bar{\mathbf{a}}$ have the same direction as the resultant external force $\Sigma\mathbf{F}$, it does not follow that $\Sigma\mathbf{F}$ necessarily passes through G. In general, in fact, $\Sigma\mathbf{F}$ does not pass through G, as will be shown later.

4/3 Work-Energy

In Art. 3/6 we developed the work-energy relation for a single particle, and we noted that it applies to a system of two joined particles. Now consider the general system of Fig. 4/1, where the work-energy relation for the representative particle of mass m_i is $(U_{1\text{-}2})_i = \Delta T_i$. Here $(U_{1\text{-}2})_i$ is the work done on m_i during an interval of motion by all forces $\mathbf{F}_1 + \mathbf{F}_2 + \mathbf{F}_3 + \cdots$ applied from sources external to the system and by all forces $\mathbf{f}_1 + \mathbf{f}_2 + \mathbf{f}_3 + \cdots$ applied from sources internal to the system. The kinetic energy of m_i is $T_i = \frac{1}{2}m_i v_i^2$, where v_i is the magnitude of the particle velocity $\mathbf{v}_i = \dot{\mathbf{r}}_i$.

*If m is a function of time, a more complex situation develops; this situation is discussed in Art. 4/7 on variable mass.

Work-Energy Relation

For the entire system, the sum of the work-energy equations written for all particles is $\Sigma(U_{1\text{-}2})_i = \Sigma\Delta T_i$, which may be represented by the same expressions as Eqs. 3/15 and 3/15*a* of Art. 3/6, namely,

$$U_{1\text{-}2} = \Delta T \quad \text{or} \quad T_1 + U_{1\text{-}2} = T_2 \tag{4/2}$$

where $U_{1\text{-}2} = \Sigma(U_{1\text{-}2})_i$, the work done by all forces, external and internal, on all particles, and ΔT is the change in the total kinetic energy $T = \Sigma T_i$ of the system.

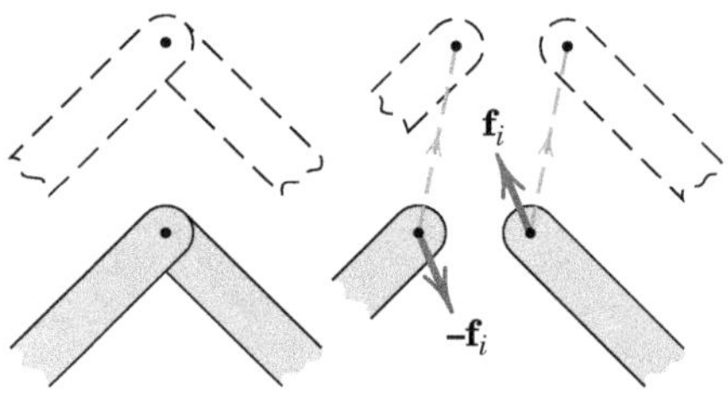

Figure 4/2

For a rigid body or a system of rigid bodies joined by ideal frictionless connections, no net work is done by the internal interacting forces or moments in the connections. We see that the work done by all pairs of internal forces, labeled here as $\mathbf{f}_i$ and $-\mathbf{f}_i$, at a typical connection, Fig. 4/2, in the system is zero since their points of application have identical displacement components while the forces are equal but opposite. For this situation $U_{1\text{-}2}$ becomes the work done on the system by the external forces only.

For a nonrigid mechanical system which includes elastic members capable of storing energy, a part of the work done by the external forces goes into changing the internal elastic potential energy V_e. Also, if the work done by the gravity forces is *excluded* from the work term and is accounted for instead by the changes in gravitational potential energy V_g, then we may equate the work $U'_{1\text{-}2}$ done on the system during an interval of motion to the change ΔE in the total mechanical energy of the system. Thus, $U'_{1\text{-}2} = \Delta E$ or

$$U'_{1\text{-}2} = \Delta T + \Delta V \tag{4/3}$$

or

$$T_1 + V_1 + U'_{1\text{-}2} = T_2 + V_2 \tag{4/3a}$$

which are the same as Eqs. 3/21 and 3/21*a*. Here, as in Chapter 3, $V = V_g + V_e$ represents the total potential energy.

Kinetic Energy Expression

We now examine the expression $T = \Sigma\frac{1}{2}m_i v_i^{\,2}$ for the kinetic energy of the mass system in more detail. By our principle of relative motion discussed in Art. 2/8, we may write the velocity of the representative particle as

$$\mathbf{v}_i = \bar{\mathbf{v}} + \dot{\boldsymbol{\rho}}_i$$

where $\bar{\mathbf{v}}$ is the velocity of the mass center G and $\dot{\boldsymbol{\rho}}_i$ is the velocity of m_i with respect to a translating reference frame moving with the mass center G. We recall the identity $v_i^2 = \mathbf{v}_i \cdot \mathbf{v}_i$ and write the kinetic energy of the system as

$$T = \Sigma \frac{1}{2} m_i \mathbf{v}_i \cdot \mathbf{v}_i = \Sigma \frac{1}{2} m_i (\overline{\mathbf{v}} + \dot{\boldsymbol{\rho}}_i) \cdot (\overline{\mathbf{v}} + \dot{\boldsymbol{\rho}}_i)$$
$$= \Sigma \frac{1}{2} m_i \overline{v}^2 + \Sigma \frac{1}{2} m_i |\dot{\boldsymbol{\rho}}_i|^2 + \Sigma m_i \overline{\mathbf{v}} \cdot \dot{\boldsymbol{\rho}}_i$$

Because $\boldsymbol{\rho}_i$ is measured from the mass center, $\Sigma m_i \boldsymbol{\rho}_i = \mathbf{0}$ and the third term is $\overline{\mathbf{v}} \cdot \Sigma m_i \dot{\boldsymbol{\rho}}_i = \overline{\mathbf{v}} \cdot \frac{d}{dt} \Sigma(m_i \boldsymbol{\rho}_i) = 0$. Also $\Sigma \frac{1}{2} m_i \overline{v}^2 = \frac{1}{2} \overline{v}^2 \Sigma m_i = \frac{1}{2} m \overline{v}^2$. Therefore, the total kinetic energy becomes

$$T = \frac{1}{2} m \overline{v}^2 + \Sigma \frac{1}{2} m_i |\dot{\boldsymbol{\rho}}_i|^2 \qquad \textbf{(4/4)}$$

This equation expresses the fact that the total kinetic energy of a mass system equals the kinetic energy of mass-center translation of the system as a whole plus the kinetic energy due to motion of all particles relative to the mass center.

4/4 Impulse-Momentum

We now develop the concepts of momentum and impulse as applied to a system of particles.

Linear Momentum

From our definition in Art. 3/8, the linear momentum of the representative particle of the system depicted in Fig. 4/1 is $\mathbf{G}_i = m_i \mathbf{v}_i$ where the velocity of m_i is $\mathbf{v}_i = \dot{\mathbf{r}}_i$.

The linear momentum of the system is defined as the vector sum of the linear momenta of all of its particles, or $\mathbf{G} = \Sigma m_i \mathbf{v}_i$. By substituting the relative-velocity relation $\mathbf{v}_i = \overline{\mathbf{v}} + \dot{\boldsymbol{\rho}}_i$ and noting again that $\Sigma m_i \boldsymbol{\rho}_i = m\overline{\boldsymbol{\rho}} = \mathbf{0}$, we obtain

$$\mathbf{G} = \Sigma m_i (\overline{\mathbf{v}} + \dot{\boldsymbol{\rho}}_i) = \Sigma m_i \overline{\mathbf{v}} + \frac{d}{dt} \Sigma m_i \boldsymbol{\rho}_i$$
$$= \overline{\mathbf{v}} \, \Sigma m_i + \frac{d}{dt}(\mathbf{0})$$

or

$$\mathbf{G} = m\overline{\mathbf{v}} \qquad \textbf{(4/5)}$$

Thus, the linear momentum of any system of constant mass is the product of the mass and the velocity of its center of mass.

The time derivative of $\mathbf{G}$ is $m\dot{\overline{\mathbf{v}}} = m\overline{\mathbf{a}}$, which by Eq. 4/1 is the resultant external force acting on the system. Thus, we have

$$\Sigma \mathbf{F} = \dot{\mathbf{G}} \qquad \textbf{(4/6)}$$

which has the same form as Eq. 3/25 for a single particle. Equation 4/6 states that the resultant of the external forces on any mass system equals the time rate of change of the linear momentum of the system. It is an alternative form of the generalized second law of motion, Eq. 4/1. As was noted at the end of the last article, $\Sigma\mathbf{F}$, in general, does not pass through the mass center G. In deriving Eq. 4/6, we differentiated with respect to time and assumed that the total mass is constant. Thus, the equation does not apply to systems whose mass changes with time.

Angular Momentum

We now determine the angular momentum of our general mass system about the fixed point O, about the mass center G, and about an arbitrary point P, shown in Fig. 4/3, which may have an acceleration $\mathbf{a}_P = \ddot{\mathbf{r}}_P$.

About a Fixed Point O. The angular momentum of the mass system about the point O, fixed in the Newtonian reference system, is defined as the vector sum of the moments of the linear momenta about O of all particles of the system and is

$$\mathbf{H}_O = \Sigma(\mathbf{r}_i \times m_i\mathbf{v}_i)$$

The time derivative of the vector product is $\dot{\mathbf{H}}_O = \Sigma(\dot{\mathbf{r}}_i \times m_i\mathbf{v}_i) + \Sigma(\mathbf{r}_i \times m_i\dot{\mathbf{v}}_i)$. The first summation vanishes since the cross product of two parallel vectors $\dot{\mathbf{r}}_i$ and $m_i\mathbf{v}_i$ is zero. The second summation is $\Sigma(\mathbf{r}_i \times m_i\mathbf{a}_i) = \Sigma(\mathbf{r}_i \times \mathbf{F}_i)$, which is the vector sum of the moments about O of all forces acting on all particles of the system. This moment sum $\Sigma\mathbf{M}_O$ represents only the moments of forces external to the system, since the internal forces cancel one another and their moments add up to zero. Thus, the moment sum is

$$\boxed{\Sigma\mathbf{M}_O = \dot{\mathbf{H}}_O} \qquad \textbf{(4/7)}$$

which has the same form as Eq. 3/31 for a single particle.

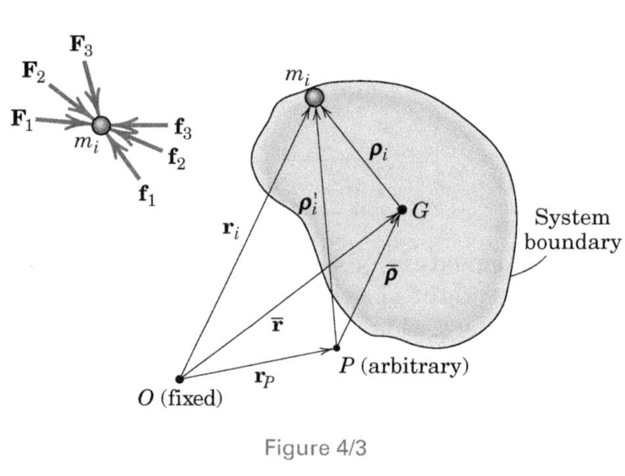

Figure 4/3

Equation 4/7 states that the resultant vector moment about any fixed point of all external forces on any system of mass equals the time rate of change of angular momentum of the system about the fixed point. As in the linear-momentum case, Eq. 4/7 does not apply if the total mass of the system is changing with time.

About the Mass Center G. The angular momentum of the mass system about the mass center G is the sum of the moments of the linear momenta about G of all particles and is

$$\mathbf{H}_G = \Sigma \boldsymbol{\rho}_i \times m_i \dot{\mathbf{r}}_i \tag{4/8}$$

We may write the absolute velocity $\dot{\mathbf{r}}_i$ as $(\dot{\bar{\mathbf{r}}} + \dot{\boldsymbol{\rho}}_i)$ so that $\mathbf{H}_G$ becomes

$$\mathbf{H}_G = \Sigma \boldsymbol{\rho}_i \times m_i(\dot{\bar{\mathbf{r}}} + \dot{\boldsymbol{\rho}}_i) = \Sigma \boldsymbol{\rho}_i \times m_i \dot{\bar{\mathbf{r}}} + \Sigma \boldsymbol{\rho}_i \times m_i \dot{\boldsymbol{\rho}}_i$$

The first term on the right side of this equation may be rewritten as $-\dot{\bar{\mathbf{r}}} \times \Sigma m_i \boldsymbol{\rho}_i$, which is zero because $\Sigma m_i \boldsymbol{\rho}_i = \mathbf{0}$ by definition of the mass center. Thus, we have

$$\mathbf{H}_G = \Sigma \boldsymbol{\rho}_i \times m_i \dot{\boldsymbol{\rho}}_i \tag{4/8a}$$

The expression of Eq. 4/8 is called the *absolute* angular momentum because the absolute velocity $\dot{\mathbf{r}}_i$ is used. The expression of Eq. 4/8*a* is called the *relative* angular momentum because the relative velocity $\dot{\boldsymbol{\rho}}_i$ is used. With the mass center G as a reference, the absolute and relative angular momenta are seen to be identical. We will see that this identity does not hold for an arbitrary reference point P; there is no distinction for a fixed reference point O.

Differentiating Eq. 4/8 with respect to time gives

$$\dot{\mathbf{H}}_G = \Sigma \dot{\boldsymbol{\rho}}_i \times m_i(\dot{\bar{\mathbf{r}}} + \dot{\boldsymbol{\rho}}_i) + \Sigma \boldsymbol{\rho}_i \times m_i \ddot{\mathbf{r}}_i$$

The first summation is expanded as $\Sigma \dot{\boldsymbol{\rho}}_i \times m_i \dot{\bar{\mathbf{r}}} + \Sigma \dot{\boldsymbol{\rho}}_i \times m_i \dot{\boldsymbol{\rho}}_i$. The first term may be rewritten as $-\dot{\bar{\mathbf{r}}} \times \Sigma m_i \dot{\boldsymbol{\rho}}_i = -\dot{\bar{\mathbf{r}}} \times \dfrac{d}{dt} \Sigma m_i \boldsymbol{\rho}_i$, which is zero from the definition of the mass center. The second term is zero because the cross product of parallel vectors is zero. With $\mathbf{F}_i$ representing the sum of all external forces acting on m_i and $\mathbf{f}_i$ the sum of all internal forces acting on m_i, the second summation by Newton's second law becomes $\Sigma \boldsymbol{\rho}_i \times (\mathbf{F}_i + \mathbf{f}_i) = \Sigma \boldsymbol{\rho}_i \times \mathbf{F}_i = \Sigma \mathbf{M}_G$, the sum of all external moments about point G. Recall that the sum of all internal moments $\Sigma \boldsymbol{\rho}_i \times \mathbf{f}_i$ is zero. Thus, we are left with

$$\boxed{\Sigma \mathbf{M}_G = \dot{\mathbf{H}}_G} \tag{4/9}$$

where we may use either the absolute or the relative angular momentum.

Equations 4/7 and 4/9 are among the most powerful of the governing equations in dynamics and apply to any defined system of constant mass—rigid or nonrigid.

About an Arbitrary Point P. The angular momentum about an arbitrary point P (which may have an acceleration $\ddot{\mathbf{r}}_P$) will now be expressed with the notation of Fig. 4/3. Thus,

$$\mathbf{H}_P = \Sigma\boldsymbol{\rho}_i' \times m_i\dot{\mathbf{r}}_i = \Sigma(\bar{\boldsymbol{\rho}} + \boldsymbol{\rho}_i) \times m_i\dot{\mathbf{r}}_i$$

The first term may be written as $\bar{\boldsymbol{\rho}} \times \Sigma m_i\dot{\mathbf{r}}_i = \bar{\boldsymbol{\rho}} \times \Sigma m_i\mathbf{v}_i = \bar{\boldsymbol{\rho}} \times m\bar{\mathbf{v}}$. The second term is $\Sigma\boldsymbol{\rho}_i \times m_i\dot{\mathbf{r}}_i = \mathbf{H}_G$. Thus, rearranging gives

$$\boxed{\mathbf{H}_P = \mathbf{H}_G + \bar{\boldsymbol{\rho}} \times m\bar{\mathbf{v}}} \qquad \textbf{(4/10)}$$

Equation 4/10 states that the absolute angular momentum about any point P equals the angular momentum about G plus the moment about P of the linear momentum $m\bar{\mathbf{v}}$ of the system considered concentrated at G.

We now make use of the principle of moments developed in our study of statics where we represented a force system by a resultant force through any point, such as G, and a corresponding couple. Figure 4/4 represents the resultants of the external forces acting on the system expressed in terms of the resultant force $\Sigma\mathbf{F}$ through G and the corresponding couple $\Sigma\mathbf{M}_G$. We see that the sum of the moments about P of all forces external to the system must equal the moment of their resultants. Therefore, we may write

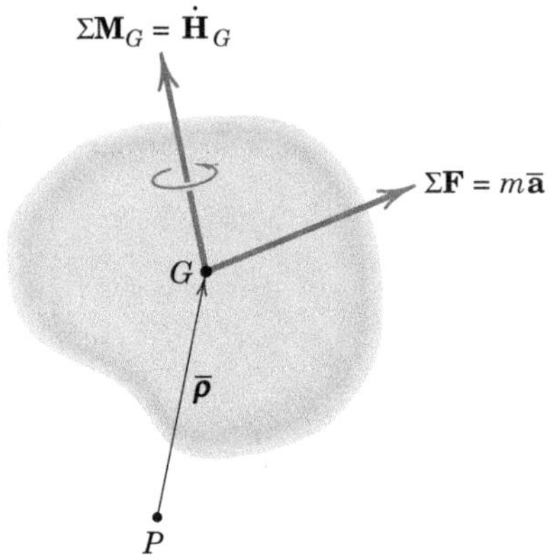

Figure 4/4

$$\Sigma\mathbf{M}_P = \Sigma\mathbf{M}_G + \bar{\boldsymbol{\rho}} \times \Sigma\mathbf{F}$$

which, by Eqs. 4/9 and 4/6, becomes

$$\boxed{\Sigma\mathbf{M}_P = \dot{\mathbf{H}}_G + \bar{\boldsymbol{\rho}} \times m\bar{\mathbf{a}}} \qquad \textbf{(4/11)}$$

Equation 4/11 enables us to write the moment equation about any convenient moment center P and is easily visualized with the aid of Fig. 4/4. This equation forms a rigorous basis for much of our treatment of planar rigid-body kinetics in Chapter 6.

We may also develop similar momentum relationships by using the momentum relative to P. Thus, from Fig. 4/3

$$(\mathbf{H}_P)_{\text{rel}} = \Sigma\boldsymbol{\rho}_i' \times m_i\dot{\boldsymbol{\rho}}_i'$$

where $\dot{\boldsymbol{\rho}}_i'$ is the velocity of m_i relative to P. With the substitution $\boldsymbol{\rho}_i' = \bar{\boldsymbol{\rho}} + \boldsymbol{\rho}_i$ and $\dot{\boldsymbol{\rho}}_i' = \dot{\bar{\boldsymbol{\rho}}} + \dot{\boldsymbol{\rho}}_i$ we may write

$$(\mathbf{H}_P)_{\text{rel}} = \Sigma\bar{\boldsymbol{\rho}} \times m_i\dot{\bar{\boldsymbol{\rho}}} + \Sigma\bar{\boldsymbol{\rho}} \times m_i\dot{\boldsymbol{\rho}}_i + \Sigma\boldsymbol{\rho}_i \times m_i\dot{\bar{\boldsymbol{\rho}}} + \Sigma\boldsymbol{\rho}_i \times m_i\dot{\boldsymbol{\rho}}_i$$

The first summation is $\bar{\boldsymbol{\rho}} \times m\bar{\mathbf{v}}_{\text{rel}}$. The second summation is $\bar{\boldsymbol{\rho}} \times \dfrac{d}{dt}\Sigma m_i\boldsymbol{\rho}_i$ and the third summation is $-\dot{\bar{\boldsymbol{\rho}}} \times \Sigma m_i\boldsymbol{\rho}_i$ where both are zero by definition of the mass center. The fourth summation is $(\mathbf{H}_G)_{\text{rel}}$. Rearranging gives us

$$(\mathbf{H}_P)_{\text{rel}} = (\mathbf{H}_G)_{\text{rel}} + \bar{\boldsymbol{\rho}} \times m\bar{\mathbf{v}}_{\text{rel}} \qquad \textbf{(4/12)}$$

where $(\mathbf{H}_G)_{rel}$ is the same as $\mathbf{H}_G$ (see Eqs. 4/8 and 4/8*a*). Note the similarity of Eqs. 4/12 and 4/10.

The moment equation about P may now be expressed in terms of the angular momentum relative to P. We differentiate the definition $(\mathbf{H}_P)_{rel} = \Sigma\boldsymbol{\rho}_i' \times m_i\dot{\boldsymbol{\rho}}_i'$ with time and make the substitution $\ddot{\mathbf{r}}_i = \ddot{\mathbf{r}}_P + \ddot{\boldsymbol{\rho}}_i'$ to obtain

$$(\dot{\mathbf{H}}_p)_{rel} = \Sigma\dot{\boldsymbol{\rho}}_i' \times m_i\dot{\boldsymbol{\rho}}_i' + \Sigma\boldsymbol{\rho}_i' \times m_i\ddot{\mathbf{r}}_i - \Sigma\boldsymbol{\rho}_i' \times m_i\ddot{\mathbf{r}}_P$$

The first summation is identically zero, and the second summation is the sum $\Sigma\mathbf{M}_P$ of the moments of all external forces about P. The third summation becomes $\Sigma\boldsymbol{\rho}_i' \times m_i\mathbf{a}_P = -\mathbf{a}_P \times \Sigma m_i\boldsymbol{\rho}_i' = -\mathbf{a}_P \times m\bar{\boldsymbol{\rho}} = \bar{\boldsymbol{\rho}} \times m\mathbf{a}_P$. Substituting and rearranging terms give

$$\Sigma\mathbf{M}_P = (\dot{\mathbf{H}}_P)_{rel} + \bar{\boldsymbol{\rho}} \times m\mathbf{a}_P \qquad \textbf{(4/13)}$$

The form of Eq. 4/13 is convenient when a point P whose acceleration is known is used as a moment center. The equation reduces to the simpler form

$$\Sigma\mathbf{M}_P = (\dot{\mathbf{H}}_P)_{rel} \text{ if } \begin{cases} 1.\ \mathbf{a}_P = \mathbf{0} \text{ (equivalent to Eq. 4/7)} \\ 2.\ \bar{\boldsymbol{\rho}} = \mathbf{0} \text{ (equivalent to Eq. 4/9)} \\ 3.\ \bar{\boldsymbol{\rho}} \text{ and } \mathbf{a}_P \text{ are parallel } (\mathbf{a}_P \text{ directed} \\ \quad \text{toward or away from } G) \end{cases}$$

4/5 Conservation of Energy and Momentum

Under certain common conditions, there is no net change in the total mechanical energy of a system during an interval of motion. Under other conditions, there is no net change in the momentum of a system. These conditions are treated separately as follows.

Conservation of Energy

A mass system is said to be *conservative* if it does not lose energy by virtue of internal friction forces which do negative work or by virtue of inelastic members which dissipate energy upon cycling. If no work is done on a conservative system during an interval of motion by external forces (other than gravity or other potential forces), then none of the energy of the system is lost. For this case, $U_{1\text{-}2}' = 0$ and we may write Eq. 4/3 as

$$\Delta T + \Delta V = 0 \qquad \textbf{(4/14)}$$

or

$$T_1 + V_1 = T_2 + V_2 \qquad \textbf{(4/14a)}$$

which expresses the *law of conservation of dynamical energy*. The total energy $E = T + V$ is a constant, so that $E_1 = E_2$. This law holds only in the ideal case where internal kinetic friction is sufficiently small to be neglected.

Conservation of Momentum

If, for a certain interval of time, the resultant external force $\Sigma\mathbf{F}$ acting on a conservative or nonconservative mass system is zero, Eq. 4/6 requires that $\dot{\mathbf{G}} = \mathbf{0}$, so that during this interval

$$\boxed{\mathbf{G}_1 = \mathbf{G}_2} \qquad \textbf{(4/15)}$$

which expresses the *principle of conservation of linear momentum*. Thus, in the absence of an external impulse, the linear momentum of a system remains unchanged.

Similarly, if the resultant moment about a fixed point O or about the mass center G of all external forces on any mass system is zero, Eq. 4/7 or 4/9 requires, respectively, that

$$\boxed{(\mathbf{H}_O)_1 = (\mathbf{H}_O)_2 \qquad \text{or} \qquad (\mathbf{H}_G)_1 = (\mathbf{H}_G)_2} \qquad \textbf{(4/16)}$$

These relations express the *principle of conservation of angular momentum* for a general mass system in the absence of an angular impulse. Thus, if there is no angular impulse about a fixed point (or about the mass center), the angular momentum of the system about the fixed point (or about the mass center) remains unchanged. Either equation may hold without the other.

We proved in Art. 3/14 that the basic laws of Newtonian mechanics hold for measurements made relative to a set of axes which translate with a constant velocity. Thus, Eqs. 4/1 through 4/16 are valid provided all quantities are expressed relative to the translating axes.

Equations 4/1 through 4/16 are among the most important of the basic derived laws of mechanics. In this chapter we have derived these laws for the most general system of constant mass to establish the generality of these laws. Common applications of these laws are specific mass systems such as rigid and nonrigid solids and certain fluid systems, which are discussed in the following articles. Study these laws carefully and compare them with their more restricted forms encountered earlier in Chapter 3.

The principles of particle-system kinetics form the foundation for the study of the forces associated with the water-spraying equipment of this firefighting boat.

Sample Problem 4/1

The system of four particles has the indicated particle masses, positions, velocities, and external forces. Determine $\bar{\mathbf{r}}$, $\dot{\bar{\mathbf{r}}}$, $\ddot{\bar{\mathbf{r}}}$, T, $\mathbf{G}$, $\mathbf{H}_O$, $\dot{\mathbf{H}}_O$, $\mathbf{H}_G$, and $\dot{\mathbf{H}}_G$.

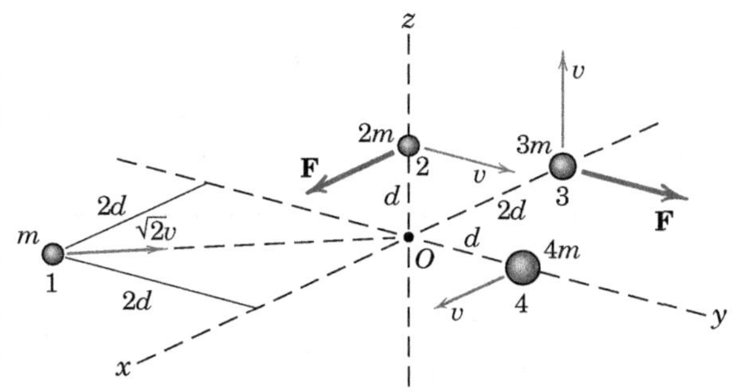

Solution. The position of the mass center of the system is

① $$\bar{\mathbf{r}} = \frac{\Sigma m_i \mathbf{r}_i}{\Sigma m_i} = \frac{m(2d\mathbf{i} - 2d\mathbf{j}) + 2m(d\mathbf{k}) + 3m(-2d\mathbf{i}) + 4m(d\mathbf{j})}{m + 2m + 3m + 4m}$$

$$= d(-0.4\mathbf{i} + 0.2\mathbf{j} + 0.2\mathbf{k}) \qquad \textit{Ans.}$$

$$\dot{\bar{\mathbf{r}}} = \frac{\Sigma m_i \dot{\mathbf{r}}_i}{\Sigma m_i} = \frac{m(-v\mathbf{i} + v\mathbf{j}) + 2m(v\mathbf{j}) + 3m(v\mathbf{k}) + 4m(v\mathbf{i})}{10m}$$

$$= v(0.3\mathbf{i} + 0.3\mathbf{j} + 0.3\mathbf{k}) \qquad \textit{Ans.}$$

$$\ddot{\bar{\mathbf{r}}} = \frac{\Sigma \mathbf{F}}{\Sigma m_i} = \frac{F\mathbf{i} + F\mathbf{j}}{10m} = \frac{F}{10m}(\mathbf{i} + \mathbf{j}) \qquad \textit{Ans.}$$

$$T = \Sigma \frac{1}{2} m_i v_i^2 = \frac{1}{2}[m(\sqrt{2}v)^2 + 2mv^2 + 3mv^2 + 4mv^2] = \frac{11}{2}mv^2 \qquad \textit{Ans.}$$

$$\mathbf{G} = (\Sigma m_i)\dot{\bar{\mathbf{r}}} = 10m(v)(0.3\mathbf{i} + 0.3\mathbf{j} + 0.3\mathbf{k}) = mv(3\mathbf{i} + 3\mathbf{j} + 3\mathbf{k}) \qquad \textit{Ans.}$$

② $$\mathbf{H}_O = \Sigma \mathbf{r}_i \times m_i \dot{\mathbf{r}}_i = \mathbf{0} - 2mvd\mathbf{i} + 3mv(2d)\mathbf{j} - 4mvd\mathbf{k}$$

$$= mvd(-2\mathbf{i} + 6\mathbf{j} - 4\mathbf{k}) \qquad \textit{Ans.}$$

$$\dot{\mathbf{H}}_O = \Sigma \mathbf{M}_O = -2dF\mathbf{k} + Fd\mathbf{j} = Fd(\mathbf{j} - 2\mathbf{k}) \qquad \textit{Ans.}$$

For $\mathbf{H}_G$, we use Eq. 4/10:

③ $$[\mathbf{H}_G = \mathbf{H}_O + \bar{\boldsymbol{\rho}} \times m\bar{\mathbf{v}}]$$

$$\mathbf{H}_G = mvd(-2\mathbf{i} + 6\mathbf{j} - 4\mathbf{k}) - d(-0.4\mathbf{i} + 0.2\mathbf{j} + 0.2\mathbf{k}) \times 10mv(0.3\mathbf{i} + 0.3\mathbf{j} + 0.3\mathbf{k}) = mvd(-2\mathbf{i} + 4.2\mathbf{j} - 2.2\mathbf{k}) \qquad \textit{Ans.}$$

For $\dot{\mathbf{H}}_G$, we could use Eq. 4/9 or Eq. 4/11 with P replaced by O. Using the latter, we have

$$[\dot{\mathbf{H}}_G = \Sigma \mathbf{M}_O - \bar{\boldsymbol{\rho}} \times m\bar{\mathbf{a}}]$$

④ $$\dot{\mathbf{H}}_G = Fd(\mathbf{j} - 2\mathbf{k}) - d(-0.4\mathbf{i} + 0.2\mathbf{j} + 0.2\mathbf{k}) \times 10m\left(\frac{F}{10m}\right)(\mathbf{i} + \mathbf{j})$$

$$= Fd(0.2\mathbf{i} + 0.8\mathbf{j} - 1.4\mathbf{k}) \qquad \textit{Ans.}$$

Helpful Hints

① All summation signs are from $i = 1$ to 4, and all are performed in order of the mass numbers in the given figure.

② Because of the simple geometry, the cross products are performed by inspection.

③ Using Eq. 4/10 with P replaced by O is more expedient than using Eq. 4/8 or 4/8a. The m in Eq. 4/10 is the total mass, which is $10m$ in this example. The quantity $\bar{\boldsymbol{\rho}}$ in Eq. 4/10, with P replaced by O, is $\bar{\mathbf{r}}$.

④ We again recognize that $\bar{\boldsymbol{\rho}} = \bar{\mathbf{r}}$ here and that the mass of this system is $10m$.

Sample Problem 4/2

Each of the three balls has a mass m and is welded to the rigid equiangular frame of negligible mass. The assembly rests on a smooth horizontal surface. If a force **F** is suddenly applied to one bar as shown, determine (*a*) the acceleration of point O and (*b*) the angular acceleration $\ddot{\theta}$ of the frame.

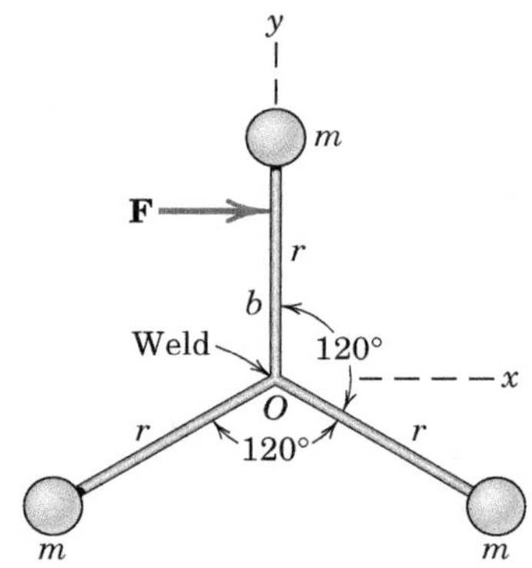

Solution. ***(a)*** Point O is the mass center of the system of the three balls, so that its acceleration is given by Eq. 4/1.

① $[\Sigma \mathbf{F} = m\bar{\mathbf{a}}]$ $\qquad F\mathbf{i} = 3m\bar{\mathbf{a}} \qquad \bar{\mathbf{a}} = \mathbf{a}_O = \dfrac{F}{3m}\mathbf{i}$ *Ans.*

(b) We determine $\ddot{\theta}$ from the moment principle, Eq. 4/9. To find $\mathbf{H}_G$ we note that the velocity of each ball relative to the mass center O as measured in the nonrotating axes x-y is $r\dot{\theta}$, where $\dot{\theta}$ is the common angular velocity of the spokes. The angular momentum of the system about O is the sum of the moments of the relative linear momenta as shown by Eq. 4/8, so it is expressed by

$$H_O = H_G = 3(mr\dot{\theta})r = 3mr^2\dot{\theta}$$

② Equation 4/9 now gives

$[\Sigma \mathbf{M}_G = \dot{\mathbf{H}}_G]$ $\qquad Fb = \dfrac{d}{dt}(3mr^2\dot{\theta}) = 3mr^2\ddot{\theta} \qquad \text{so} \qquad \ddot{\theta} = \dfrac{Fb}{3mr^2}$ *Ans.*

Helpful Hints

① We note that the result depends only on the magnitude and direction of **F** and not on b, which locates the line of action of **F**.

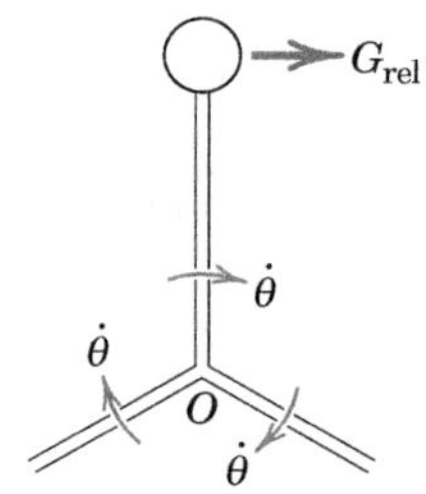

② Although $\dot{\theta}$ is initially zero, we need the expression for $H_O = H_G$ in order to get $\dot{H}_G$. We observe also that $\ddot{\theta}$ is independent of the motion of O.

Sample Problem 4/3

Consider the same conditions as for Sample Problem 4/2, except that the spokes are freely hinged at O and so do not constitute a rigid system. Explain the difference between the two problems.

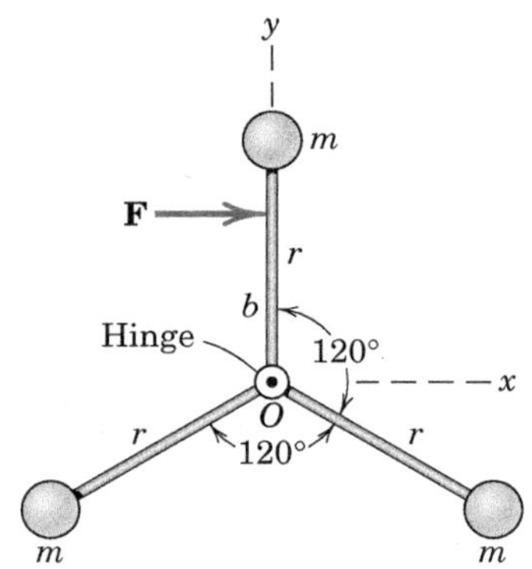

Solution. The generalized Newton's second law holds for any mass system, so that the acceleration $\bar{\mathbf{a}}$ of the mass center G is the same as with Sample Problem 4/2, namely,

$$\bar{\mathbf{a}} = \frac{F}{3m}\mathbf{i}$$ *Ans.*

Although G coincides with O at the instant represented, the motion of the hinge O is not the same as the motion of G since O will not remain the center of mass as the angles between the spokes change.

① Both ΣM_G and $\dot{H}_G$ have the same values for the two problems at the instant represented. However, the angular motions of the spokes in this problem are all different and are not easily determined.

Helpful Hint

① This present system could be dismembered and the motion equations written for each of the parts, with the unknowns eliminated one by one. Or a more sophisticated method using the equations of Lagrange could be employed. (See the first author's *Dynamics, 2nd Edition, SI Version*, 1975, for a discussion of this approach.)

Sample Problem 4/4

A shell with a mass of 20 kg is fired from point O, with a velocity $u = 300$ m/s in the vertical x-z plane at the inclination shown. When it reaches the top of its trajectory at P, it explodes into three fragments A, B, and C. Immediately after the explosion, fragment A is observed to rise vertically a distance of 500 m above P, and fragment B is seen to have a horizontal velocity $\mathbf{v}_B$ and eventually lands at point Q. When recovered, the masses of the fragments A, B, and C are found to be 5, 9, and 6 kg, respectively. Calculate the velocity which fragment C has immediately after the explosion. Neglect atmospheric resistance.

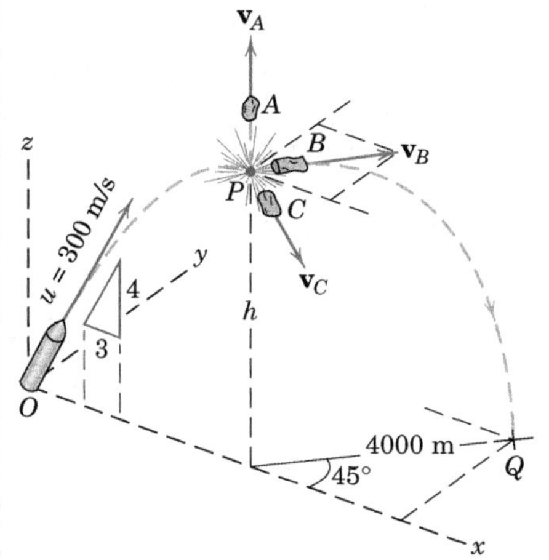

Solution. From our knowledge of projectile motion, the time required for the shell to reach P and its vertical rise are

$$t = u_z/g = 300(4/5)/9.81 = 24.5 \text{ s}$$

$$h = \frac{u_z^2}{2g} = \frac{[(300)(4/5)]^2}{2(9.81)} = 2940 \text{ m}$$

The velocity of A has the magnitude

$$v_A = \sqrt{2gh_A} = \sqrt{2(9.81)(500)} = 99.0 \text{ m/s}$$

With no z-component of velocity initially, fragment B requires 24.5 s to return to the ground. Thus, its horizontal velocity, which remains constant, is

$$v_B = s/t = 4000/24.5 = 163.5 \text{ m/s}$$

Since the force of the explosion is internal to the system of the shell and its three fragments, the linear momentum of the system remains unchanged during the explosion. Thus,

① $[\mathbf{G}_1 = \mathbf{G}_2]$ $\quad m\mathbf{v} = m_A\mathbf{v}_A + m_B\mathbf{v}_B + m_C\mathbf{v}_C$

$$20(300)(\tfrac{3}{5})\mathbf{i} = 5(99.0\mathbf{k}) + 9(163.5)(\mathbf{i}\cos 45° + \mathbf{j}\sin 45°) + 6\mathbf{v}_C$$

$$6\mathbf{v}_C = 2560\mathbf{i} - 1040\mathbf{j} - 495\mathbf{k}$$

② $$\mathbf{v}_C = 427\mathbf{i} - 173.4\mathbf{j} - 82.5\mathbf{k} \text{ m/s}$$

$$v_C = \sqrt{(427)^2 + (173.4)^2 + (82.5)^2} = 468 \text{ m/s} \qquad \textit{Ans.}$$

Helpful Hints

① The velocity $\mathbf{v}$ of the shell at the top of its trajectory is, of course, the constant horizontal component of its initial velocity $\mathbf{u}$, which becomes $u(3/5)$.

② We note that the mass center of the three fragments while still in flight continues to follow the same trajectory which the shell would have followed if it had not exploded.

Sample Problem 4/5

The 32.2-lb carriage A moves horizontally in its guide with a speed of 4 ft/sec and carries two assemblies of balls and light rods which rotate about a shaft at O in the carriage. Each of the four balls weighs 3.22 lb. The assembly on the front face rotates counterclockwise at a speed of 80 rev/min, and the assembly on the back side rotates clockwise at a speed of 100 rev/min. For the entire system, calculate (a) the kinetic energy T, (b) the magnitude G of the linear momentum, and (c) the magnitude H_O of the angular momentum about point O.

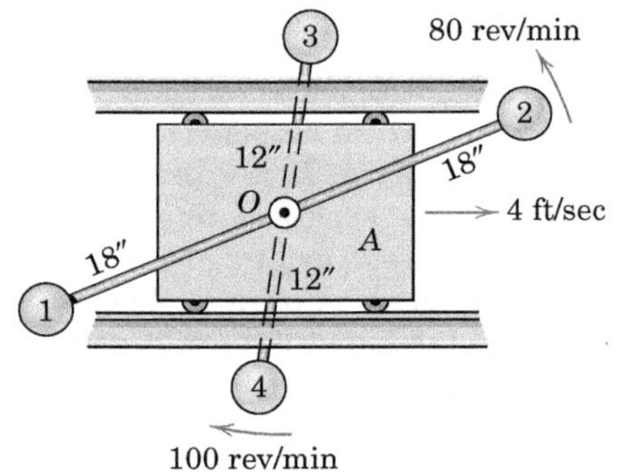

Solution. ***(a) Kinetic energy.*** The velocities of the balls with respect to O are

$$[|\dot{\boldsymbol{\rho}}_i| = v_{\text{rel}} = r\dot{\theta}] \qquad (v_{\text{rel}})_{1,2} = \frac{18}{12}\frac{80(2\pi)}{60} = 12.57 \text{ ft/sec}$$

$$(v_{\text{rel}})_{3,4} = \frac{12}{12}\frac{100(2\pi)}{60} = 10.47 \text{ ft/sec}$$

The kinetic energy of the system is given by Eq. 4/4. The translational part is

① $$\frac{1}{2}m\bar{v}^2 = \frac{1}{2}\left(\frac{32.2}{32.2} + 4\frac{3.22}{32.2}\right)(4^2) = 11.20 \text{ ft-lb}$$

The rotational part of the kinetic energy depends on the squares of the relative velocities and is

② $$\Sigma\frac{1}{2}m_i|\dot{\boldsymbol{\rho}}_i|^2 = 2\left[\frac{1}{2}\frac{3.22}{32.2}(12.57)^2\right]_{(1,2)} + 2\left[\frac{1}{2}\frac{3.22}{32.2}(10.47)^2\right]_{(3,4)}$$

$$= 15.80 + 10.96 = 26.8 \text{ ft-lb}$$

The total kinetic energy is

$$T = \frac{1}{2}m\bar{v}^2 + \Sigma\frac{1}{2}m_i|\dot{\boldsymbol{\rho}}_i|^2 = 11.20 + 26.8 = 38.0 \text{ ft-lb} \qquad \textit{Ans.}$$

(b) Linear momentum. The linear momentum of the system by Eq. 4/5 is the total mass times v_O, the velocity of the center of mass. Thus,

③ $$[\mathbf{G} = m\bar{\mathbf{v}}] \qquad G = \left(\frac{32.2}{32.2} + 4\frac{3.22}{32.2}\right)(4) = 5.6 \text{ lb-sec} \qquad \textit{Ans.}$$

(c) Angular momentum about O. The angular momentum about O is due to the moments of the linear momenta of the balls. Taking counterclockwise as positive, we have

$$H_O = \Sigma|\mathbf{r}_i \times m_i\mathbf{v}_i|$$

④ $$H_O = \left[2\left(\frac{3.22}{32.2}\right)\left(\frac{18}{12}\right)(12.57)\right]_{(1,2)} - \left[2\left(\frac{3.22}{32.2}\right)\left(\frac{12}{12}\right)(10.47)\right]_{(3,4)}$$

$$= 3.77 - 2.09 = 1.676 \text{ ft-lb-sec} \qquad \textit{Ans.}$$

Helpful Hints

① Note that the mass m is the total mass, carriage plus the four balls, and that $\bar{v}$ is the velocity of the mass center O, which is the carriage velocity.

② Note that the direction of rotation, clockwise or counterclockwise, makes no difference in the calculation of kinetic energy, which depends on the square of the velocity.

③ There is a temptation to overlook the contribution of the balls since their linear momenta relative to O in each pair are in opposite directions and cancel. However, each ball also has a velocity component $\bar{\mathbf{v}}$ and hence a momentum component $m_i\bar{\mathbf{v}}$.

④ Contrary to the case of kinetic energy where the direction of rotation was immaterial, angular momentum is a vector quantity and the direction of rotation must be accounted for.

PROBLEMS

Introductory Problems

4/1 The system of three particles has the indicated particle masses, velocities, and external forces. Determine $\bar{\mathbf{r}}$, $\dot{\bar{\mathbf{r}}}$, $\ddot{\bar{\mathbf{r}}}$, T, $\mathbf{H}_O$, and $\dot{\mathbf{H}}_O$ for this two-dimensional system.

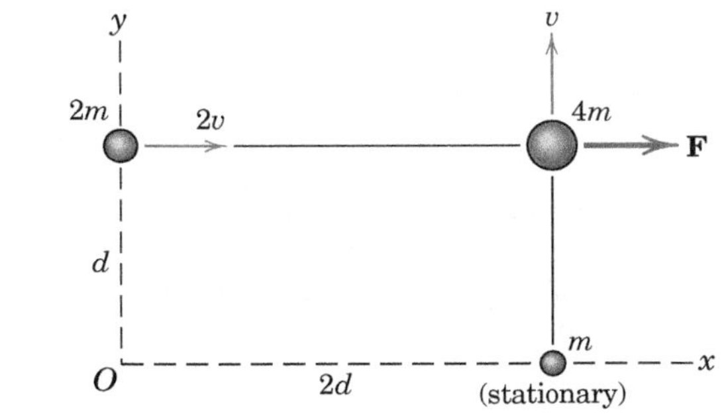

Problem 4/1

4/2 For the particle system of Prob. 4/1, determine $\mathbf{H}_G$ and $\dot{\mathbf{H}}_G$.

4/3 The system of three particles has the indicated particle masses, velocities, and external forces. Determine $\bar{\mathbf{r}}$, $\dot{\bar{\mathbf{r}}}$, $\ddot{\bar{\mathbf{r}}}$, T, $\mathbf{H}_O$, and $\dot{\mathbf{H}}_O$ for this three-dimensional system.

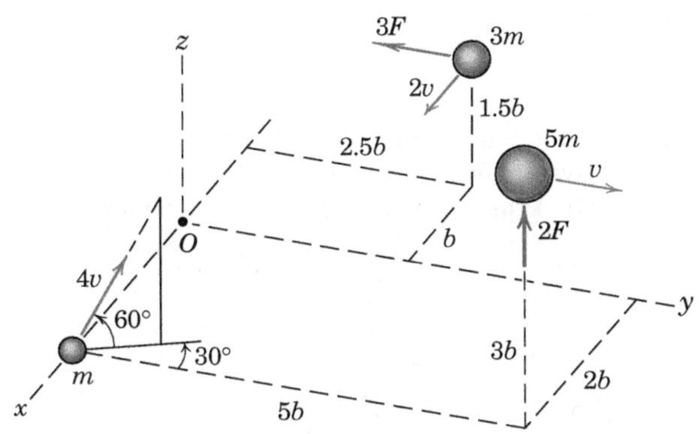

Problem 4/3

4/4 For the particle system of Prob. 4/3, determine $\mathbf{H}_G$ and $\dot{\mathbf{H}}_G$.

4/5 The system consists of the two smooth spheres, each weighing 3 lb and connected by a light spring, and the two bars of negligible weight hinged freely at their ends and hanging in the vertical plane. The spheres are confined to slide in the smooth horizontal guide. If a horizontal force $F = 10$ lb is applied to the one bar at the position shown, what is the acceleration of the center C of the spring? Why does the result not depend on the dimension b?

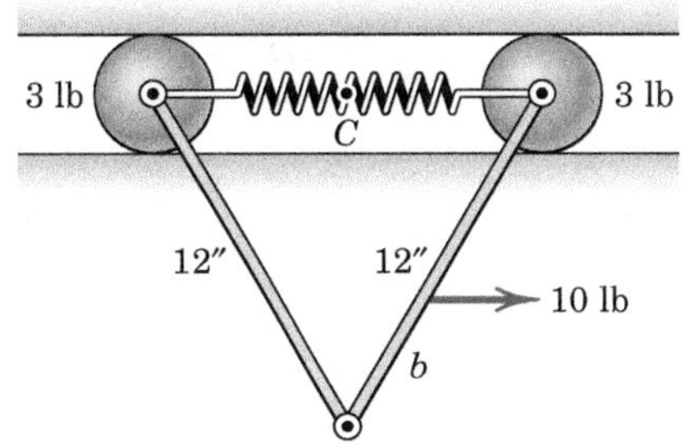

Problem 4/5

4/6 The two small spheres, each of mass m, are rigidly connected by a rod of negligible mass and are released from rest in the position shown and slide down the smooth circular guide in the vertical plane. Determine their common velocity v as they reach the horizontal dashed position. Also find the force N between sphere 1 and the supporting surface an instant before the sphere reaches the bottom position A.

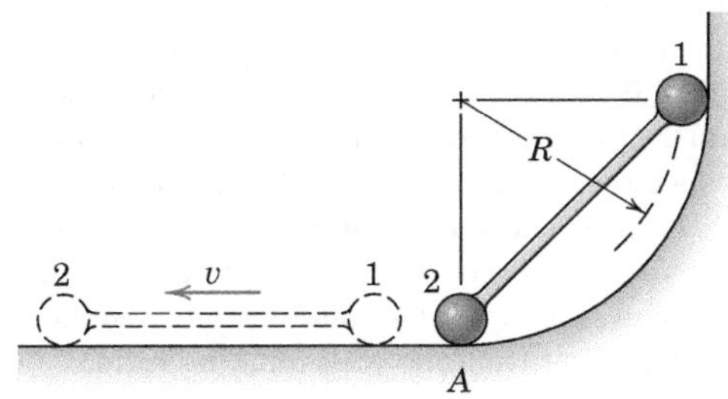

Problem 4/6

4/7 The total linear momentum of a system of five particles at time $t = 2.2$ s is given by $\mathbf{G}_{2.2} = 3.4\mathbf{i} - 2.6\mathbf{j} + 4.6\mathbf{k}$ kg·m/s. At time $t = 2.4$ s, the linear momentum has changed to $\mathbf{G}_{2.4} = 3.7\mathbf{i} - 2.2\mathbf{j} + 4.9\mathbf{k}$ kg·m/s. Calculate the magnitude F of the time average of the resultant of the external forces acting on the system during the interval.

4/8 The angular momentum of a system of six particles about a fixed point O at time $t = 4$ s is $\mathbf{H}_4 = 3.65\mathbf{i} + 4.27\mathbf{j} - 5.36\mathbf{k}$ kg·m²/s. At time $t = 4.1$ s, the angular momentum is $\mathbf{H}_{4.1} = 3.67\mathbf{i} + 4.30\mathbf{j} - 5.20\mathbf{k}$ kg·m²/s. Determine the average value of the resultant moment about point O of all forces acting on all particles during the 0.1-s interval.

4/9 Three monkeys A, B, and C weighing 20, 25, and 15 lb, respectively, are climbing up and down the rope suspended from D. At the instant represented, A is descending the rope with an acceleration of 5 ft/sec^2, and C is pulling himself up with an acceleration of 3 ft/sec^2. Monkey B is climbing up with a constant speed of 2 ft/sec. Treat the rope and monkeys as a complete system and calculate the tension T in the rope at D.

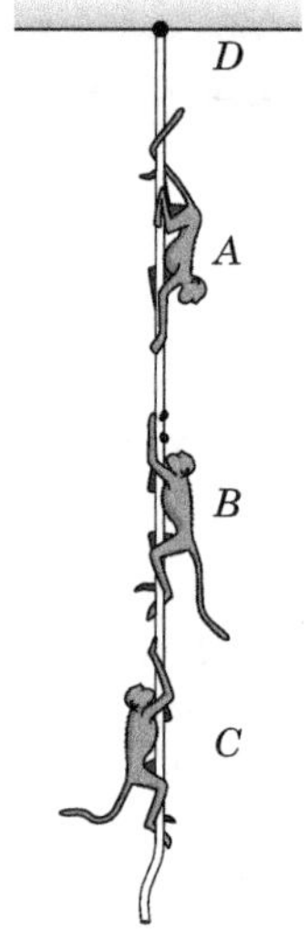

Problem 4/9

4/10 The monkeys of Prob. 4/9 are now climbing along the heavy rope wall suspended from the uniform beam. If monkeys A, B, and C have velocities of 5, 3, and 2 ft/sec, and accelerations of 1.5, 0.5, and 2 ft/sec^2, respectively, determine the changes in the reactions at D and E caused by the motion and weight of the monkeys. The support at E makes contact with only one side of the beam at a time. Assume for this analysis that the rope wall remains rigid.

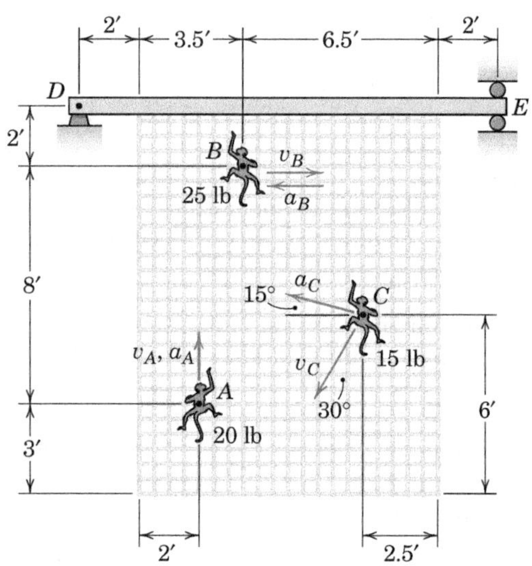

Problem 4/10

4/11 The two spheres, each of mass m, are connected by the spring and hinged bars of negligible mass. The spheres are free to slide in the smooth guide up the incline θ. Determine the acceleration a_C of the center C of the spring.

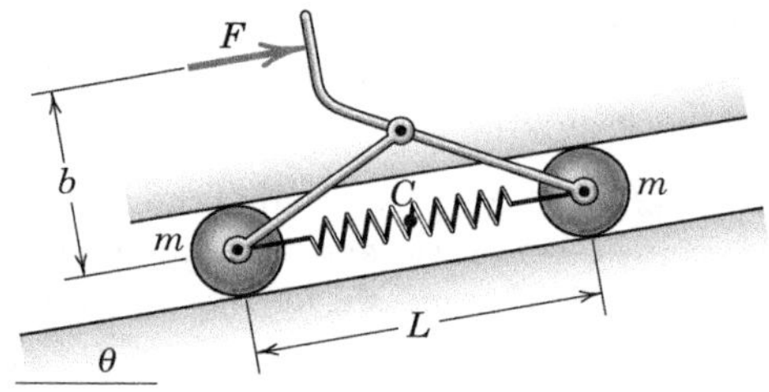

Problem 4/11

4/12 Each of the five connected particles has a mass of 0.5 kg, and G is the mass center of the system. At a certain instant the angular velocity of the body is $\omega = 2$ rad/s and the linear velocity of G is $v_G = 4$ m/s in the direction shown. Determine the linear momentum of the body and its angular momentum about G and about O.

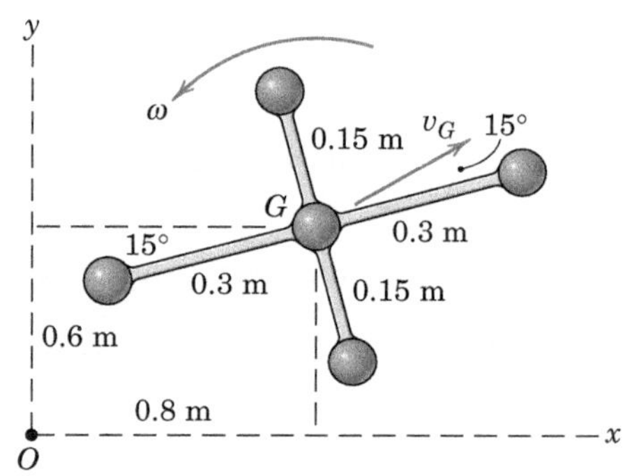

Problem 4/12

4/13 Calculate the acceleration of the center of mass of the system of the four 10-kg cylinders. Neglect friction and the mass of the pulleys and cables.

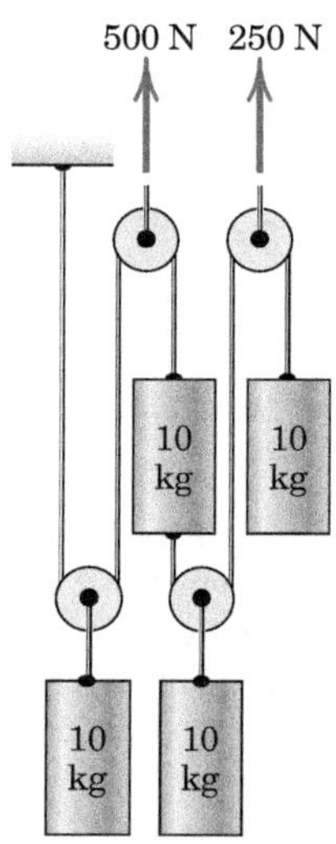

Problem 4/13

4/14 The four systems slide on a smooth horizontal surface and have the same mass m. The configurations of mass in the two pairs are identical. What can be said about the acceleration of the mass center for each system? Explain any difference in the accelerations of the members.

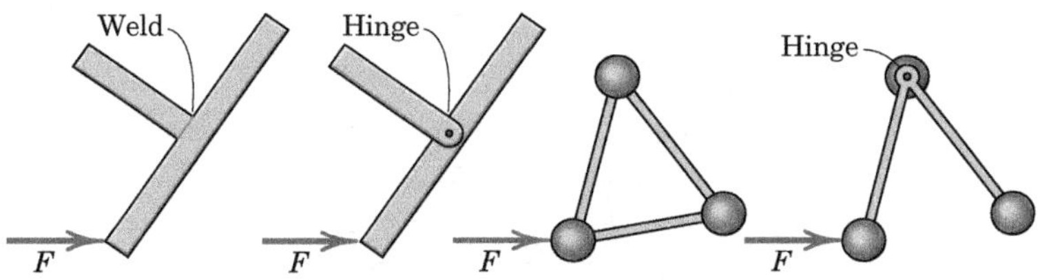

Problem 4/14

Representative Problems

4/15 Calculate the vertical acceleration of the system mass center and the individual vertical accelerations of spheres 1 and 2 for the cases (*a*) $\alpha = \beta$ and (*b*) $\alpha \neq \beta$. Each sphere has a mass of 2 kg, and the 50-N force is applied vertically to the junction O of the two light wires. The spheres are initially at rest. State any assumptions.

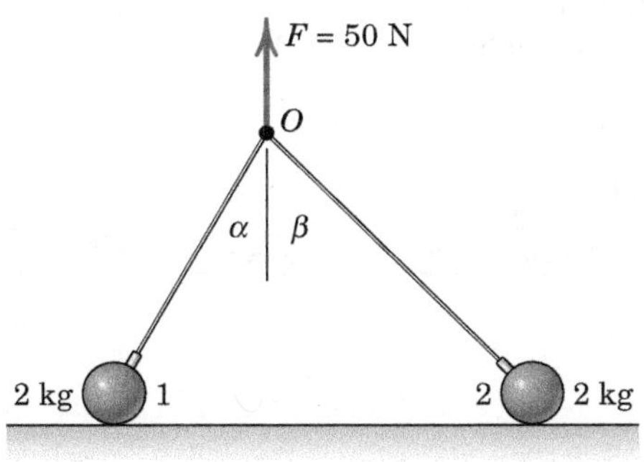

Problem 4/15

4/16 The two small spheres, each of mass m, and their connecting rod of negligible mass are rotating about their mass center G with an angular velocity ω. At the same instant the mass center has a velocity v in the x-direction. Determine the angular momentum $\mathbf{H}_O$ of the assembly at the instant when G has coordinates x and y.

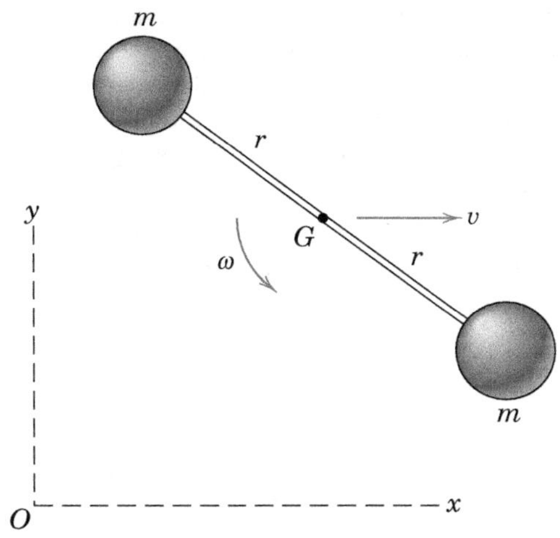

Problem 4/16

4/17 A department-store escalator makes an angle of 30° with the horizontal and takes 40 seconds to transport a person from the first to the second floor with a vertical rise of 20 ft. At a certain instant, there are 10 people on the escalator averaging 150 lb per person and standing at rest relative to the moving steps. Additionally, three boys averaging 120 lb each are running down the escalator at a speed of 2 ft/sec relative to the moving steps. Calculate the power output P of the driving motor to maintain the constant speed of the escalator. The no-load power without passengers is 2.2 hp to overcome friction in the mechanism.

4/18 A centrifuge consists of four cylindrical containers, each of mass m, at a radial distance r from the rotation axis. Determine the time t required to bring the centrifuge to an angular velocity ω from rest under a constant torque M applied to the shaft. The diameter of each container is small compared with r, and the mass of the shaft and supporting arms is small compared with m.

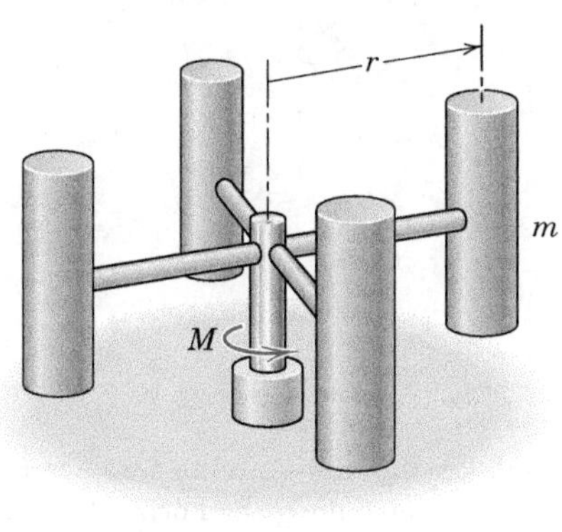

Problem 4/18

4/19 The three small spheres are welded to the light rigid frame which is rotating in a horizontal plane about a vertical axis through O with an angular velocity $\dot{\theta} = 20$ rad/s. If a couple $M_O = 30$ N·m is applied to the frame for 5 seconds, compute the new angular velocity $\dot{\theta}'$.

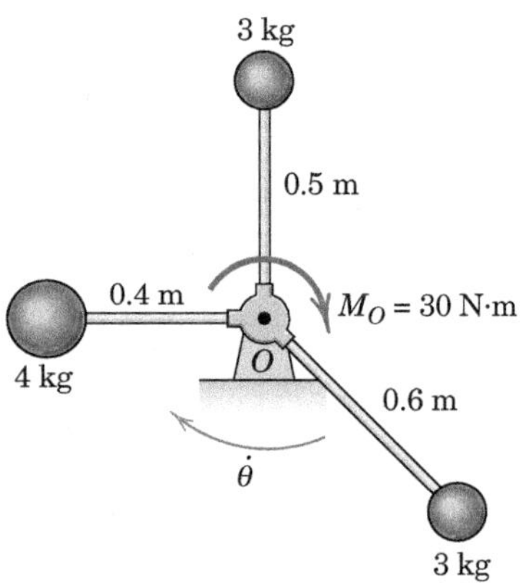

Problem 4/19

4/20 Billiard ball A is moving in the y-direction with a velocity of 2 m/s when it strikes ball B of identical size and mass initially at rest. Following the impact, the balls are observed to move in the directions shown. Calculate the velocities v_A and v_B which the balls have immediately after the impact. Treat the balls as particles and neglect any friction forces acting on the balls compared with the force of impact.

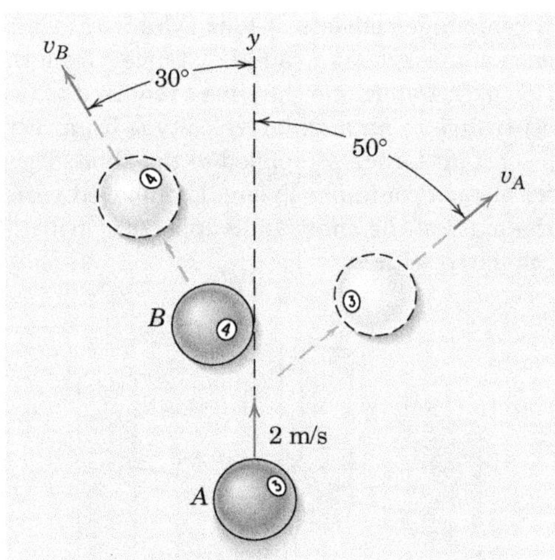

Problem 4/20

4/21 The 300-kg and 400-kg mine cars are rolling in opposite directions along the horizontal track with the respective speeds of 0.6 m/s and 0.3 m/s. Upon impact the cars become coupled together. Just prior to impact, a 100-kg boulder leaves the delivery chute with a velocity of 1.2 m/s in the direction shown and lands in the 300-kg car. Calculate the velocity v of the system after the boulder has come to rest relative to the car. Would the final velocity be the same if the cars were coupled before the boulder dropped?

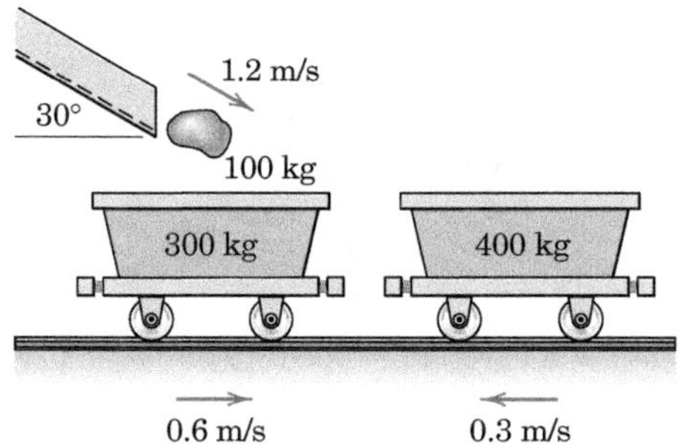

Problem 4/21

4/22 The three freight cars are rolling along the horizontal track with the velocities shown. After the impacts occur, the three cars become coupled together and move with a common velocity v. The weights of the loaded cars A, B, and C are 130,000, 100,000, and 150,000 lb, respectively. Determine v and calculate the percentage loss n of energy of the system due to coupling.

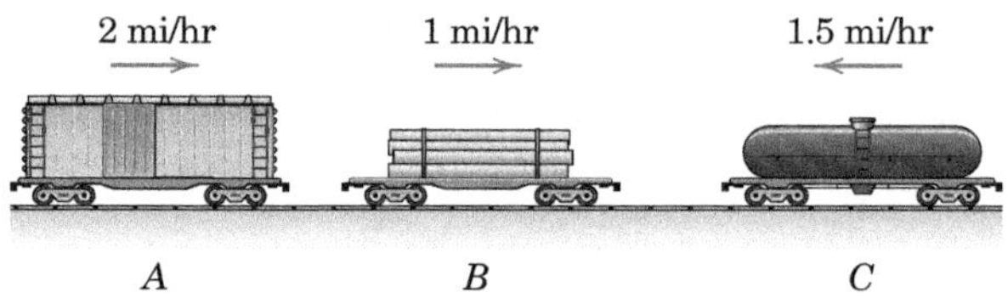

Problem 4/22

4/23 The man of mass m_1 and the woman of mass m_2 are standing on opposite ends of the platform of mass m_0 which moves with negligible friction and is initially at rest with $s = 0$. The man and woman begin to approach each other. Derive an expression for the displacement s of the platform when the two meet in terms of the displacement x_1 of the man relative to the platform.

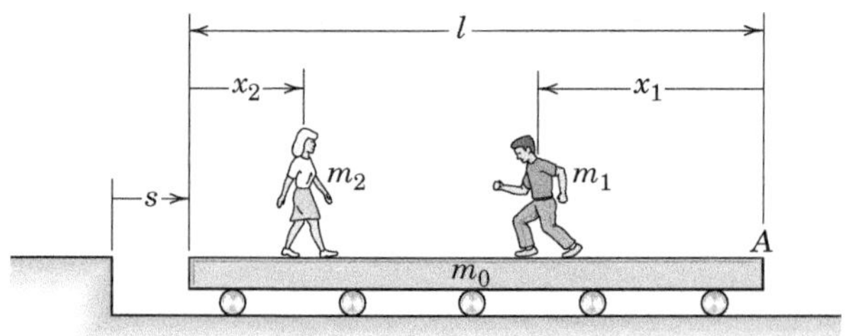

Problem 4/23

4/24 The woman A, the captain B, and the sailor C weigh 120, 180, and 160 lb, respectively, and are sitting in the 300-lb skiff, which is gliding through the water with a speed of 1 knot. If the three people change their positions as shown in the second figure, find the distance x from the skiff to the position where it would have been if the people had not moved. Neglect any resistance to motion afforded by the water. Does the sequence or timing of the change in positions affect the final result?

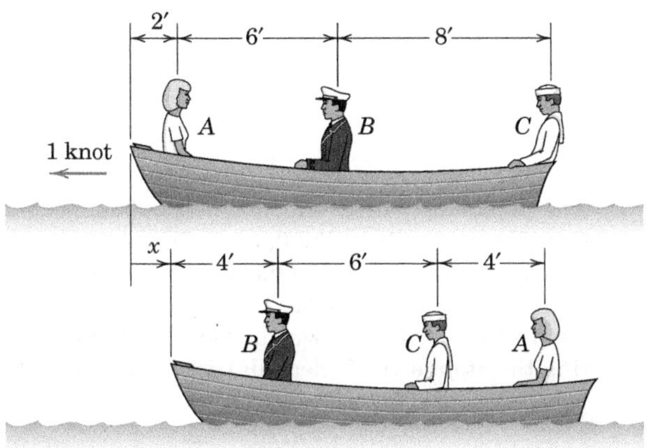

Problem 4/24

4/25 Each of the bars A and B has a mass of 10 kg and slides in its horizontal guideway with negligible friction. Motion is controlled by the lever of negligible mass connected to the bars as shown. Calculate the acceleration of point C on the lever when the 200-N force is applied as indicated. To verify your result, analyze the kinetics of each member separately and determine a_C by kinematic considerations from the calculated accelerations of the two bars.

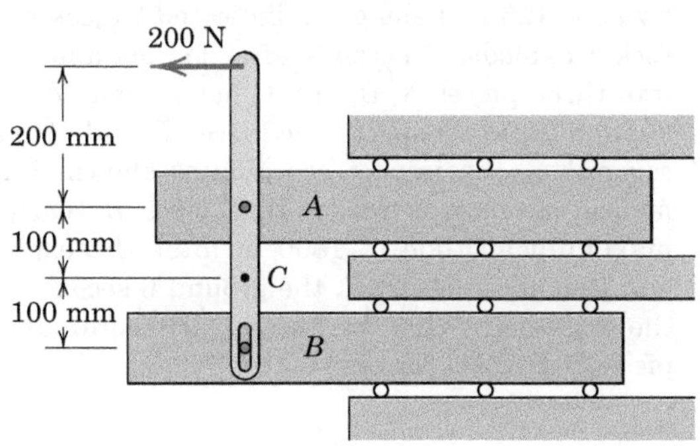

Problem 4/25

4/26 The three small steel balls, each of mass 2.75 kg, are connected by the hinged links of negligible mass and equal length. They are released from rest in the positions shown and slide down the quarter-circular guide in the vertical plane. When the upper sphere reaches the bottom position, the spheres have a horizontal velocity of 1.560 m/s. Calculate the energy loss ΔQ due to friction and the total impulse I_x on the system of three spheres during this interval.

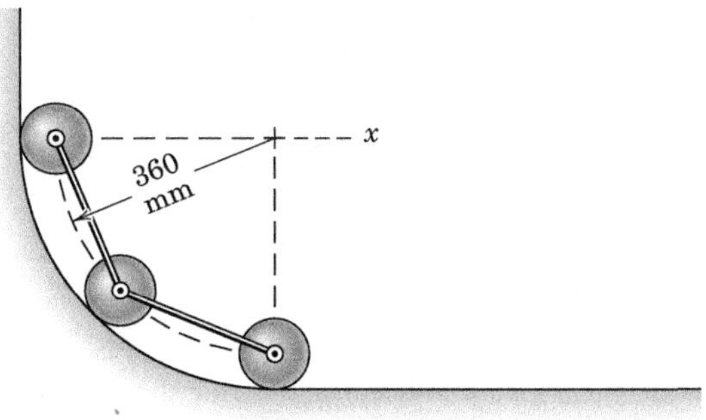

Problem 4/26

4/27 Two steel balls, each of mass m, are welded to a light rod of length L and negligible mass and are initially at rest on a smooth horizontal surface. A horizontal force of magnitude F is suddenly applied to the rod as shown. Determine (a) the instantaneous acceleration $\bar{a}$ of the mass center G and (b) the corresponding rate $\ddot{\theta}$ at which the angular velocity of the assembly about G is changing with time.

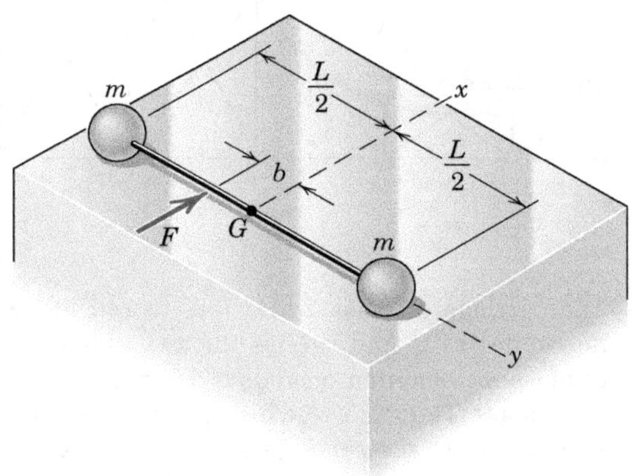

Problem 4/27

4/28 The small car, which has a mass of 20 kg, rolls freely on the horizontal track and carries the 5-kg sphere mounted on the light rotating rod with r = 0.4 m. A geared motor drive maintains a constant angular speed $\dot{\theta}$ = 4 rad/s of the rod. If the car has a velocity v = 0.6 m/s when θ = 0, calculate v when θ = 60°. Neglect the mass of the wheels and any friction.

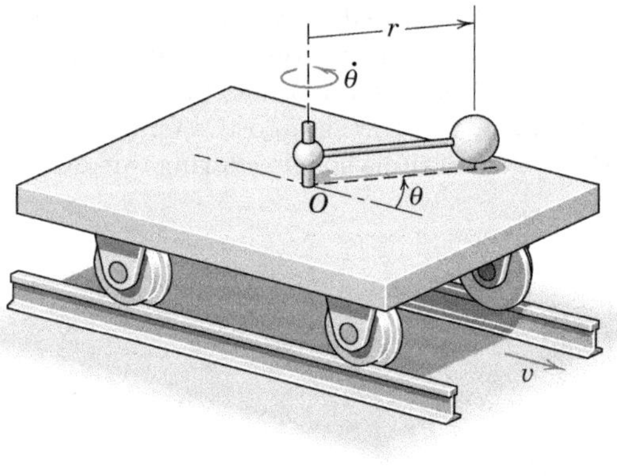

Problem 4/28

4/29 The cars of a roller-coaster ride have a speed of 30 km/h as they pass over the top of the circular track. Neglect any friction and calculate their speed v when they reach the horizontal bottom position. At the top position, the radius of the circular path of their mass centers is 18 m, and all six cars have the same mass.

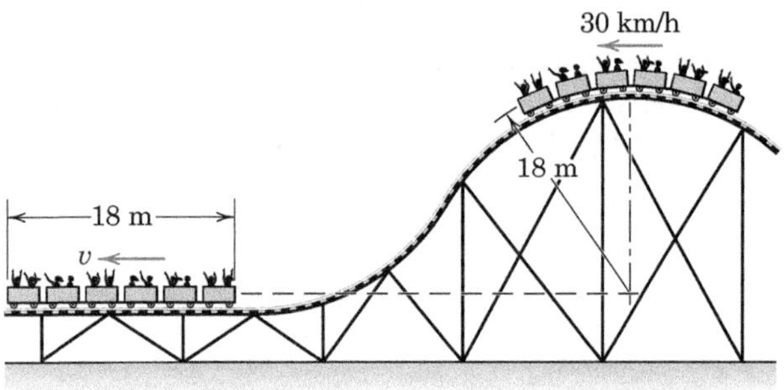

Problem 4/29

4/30 The carriage of mass $2m$ is free to roll along the horizontal rails and carries the two spheres, each of mass m, mounted on rods of length l and negligible mass. The shaft to which the rods are secured is mounted in the carriage and is free to rotate. If the system is released from rest with the rods in the vertical position where $\theta = 0$, determine the velocity v_x of the carriage and the angular velocity $\dot{\theta}$ of the rods for the instant when $\theta = 180°$. Treat the carriage and the spheres as particles and neglect any friction.

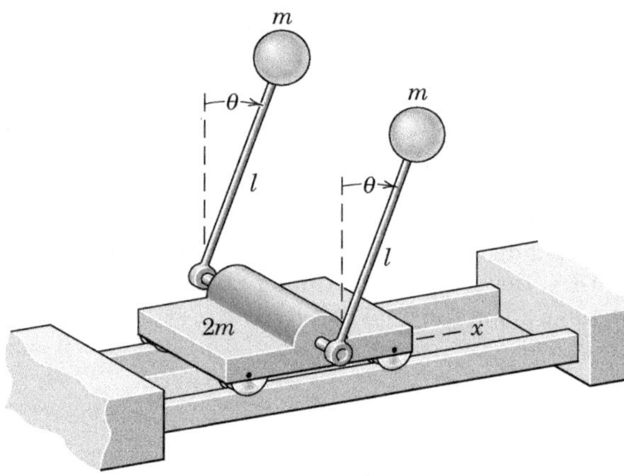

Problem 4/30

▶4/31 The 50,000-lb flatcar supports a 15,000-lb vehicle on a 5° ramp built on the flatcar. If the vehicle is released from rest with the flatcar also at rest, determine the velocity v of the flatcar when the vehicle has rolled $s = 40$ ft down the ramp just before hitting the stop at B. Neglect all friction and treat the vehicle and the flatcar as particles.

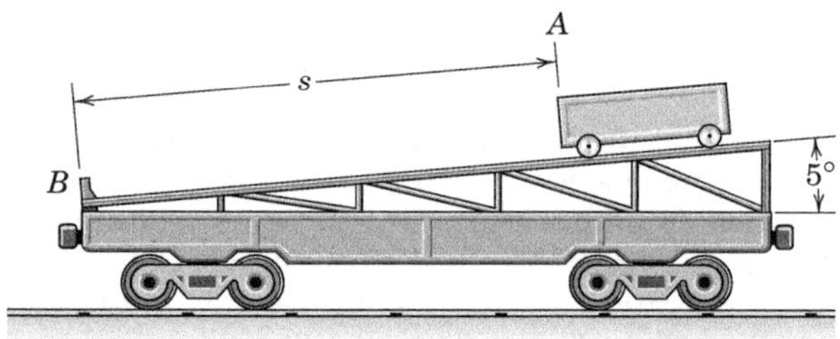

Problem 4/31

▶4/32 A 60-kg rocket is fired from O with an initial velocity $v_0 = 125$ m/s along the indicated trajectory. The rocket explodes 7 seconds after launch and breaks into three pieces A, B, and C having masses of 10, 30, and 20 kg, respectively. Pieces B and C are recovered at the impact coordinates shown. Instrumentation records reveal that piece B reached a maximum altitude of 1500 m after the explosion and that piece C struck the ground 6 seconds after the explosion. What are the impact coordinates for piece A? Neglect air resistance.

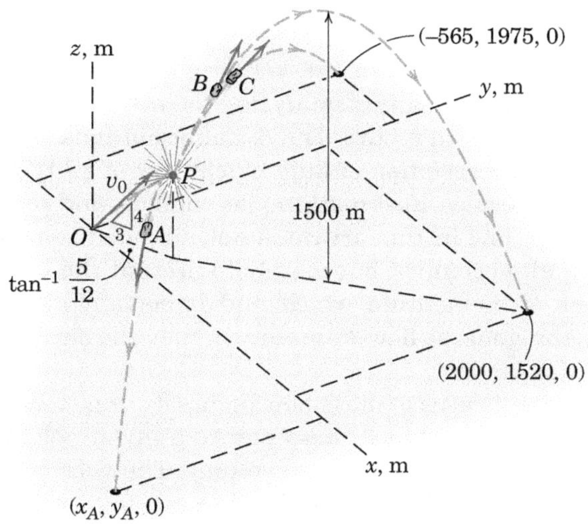

Problem 4/32

▶**4/33** A horizontal bar of mass m_1 and small diameter is suspended by two wires of length l from a carriage of mass m_2 which is free to roll along the horizontal rails. If the bar and carriage are released from rest with the wires making an angle θ with the vertical, determine the velocity $v_{b/c}$ of the bar relative to the carriage and the velocity v_c of the carriage at the instant when $\theta = 0$. Neglect all friction and treat the carriage and the bar as particles in the vertical plane of motion.

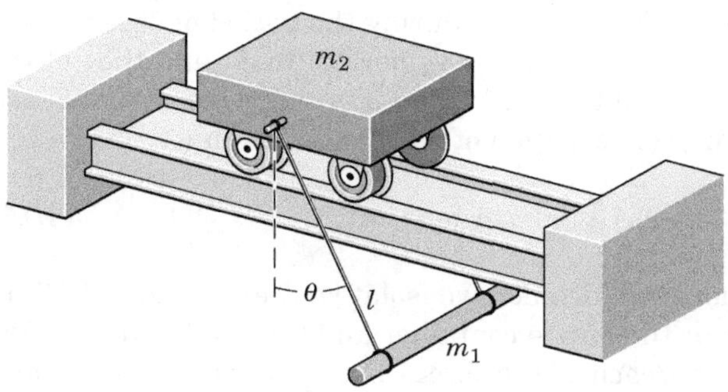

Problem 4/33

▶**4/34** In the unstretched position the coils of the 3-lb spring are just touching one another, as shown in part a of the figure. In the stretched position the force P, proportional to x, equals 200 lb when x = 20 in. If end A of the spring is suddenly released, determine the velocity v_A of the coil end A, measured positive to the left, as it approaches its unstretched position at $x = 0$. What happens to the kinetic energy of the spring?

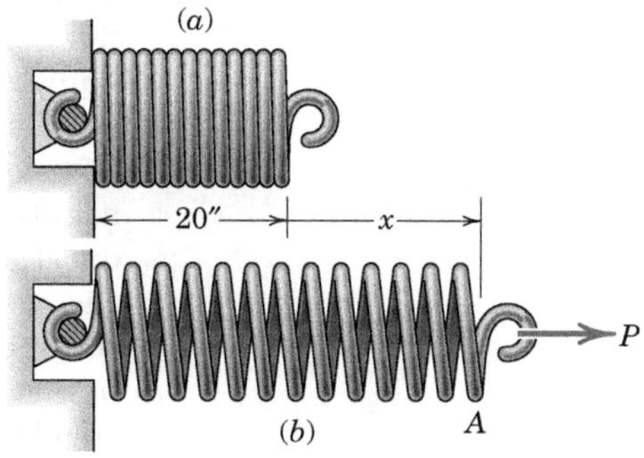

Problem 4/34

4/6 Steady Mass Flow

The momentum relation developed in Art. 4/4 for a general system of mass provides us with a direct means of analyzing the action of mass flow where a change of momentum occurs. The dynamics of mass flow is of great importance in the description of fluid machinery of all types including turbines, pumps, nozzles, air-breathing jet engines, and rockets. The treatment of mass flow in this article is not intended to take the place of a study of fluid mechanics, but merely to present the basic principles and equations of momentum which find important use in fluid mechanics and in the general flow of mass whether the form be liquid, gaseous, or granular.

One of the most important cases of mass flow occurs during steady-flow conditions where the rate at which mass enters a given volume equals the rate at which mass leaves the same volume. The volume in question may be enclosed by a rigid container, fixed or moving, such as the nozzle of a jet aircraft or rocket, the space between blades in a gas turbine, the volume within the casing of a centrifugal pump, or the volume within the bend of a pipe through which a fluid is flowing at a steady rate. The design of such fluid machines depends on the analysis of the forces and moments associated with the corresponding momentum changes of the flowing mass.

Analysis of Flow Through a Rigid Container

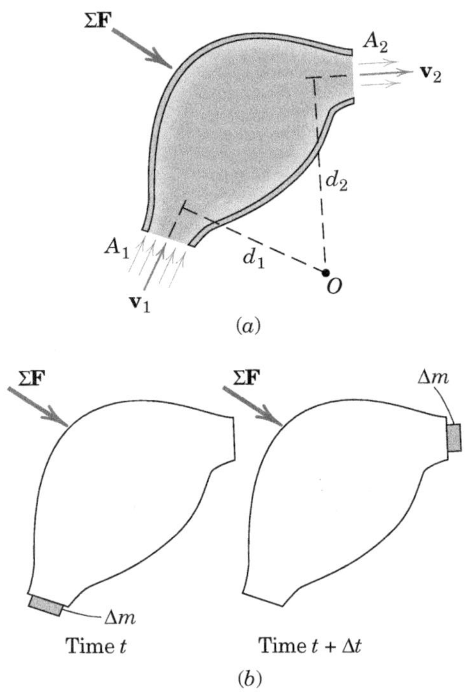

Figure 4/5

Consider a rigid container, shown in section in Fig. 4/5*a*, into which mass flows in a steady stream at the rate m' through the entrance section of area A_1. Mass leaves the container through the exit section of area A_2 at the same rate, so that there is no accumulation or depletion of the total mass within the container during the period of observation. The velocity of the entering stream is $\mathbf{v}_1$ normal to A_1 and that of the leaving stream is $\mathbf{v}_2$ normal to A_2. If ρ_1 and ρ_2 are the respective densities of the two streams, conservation of mass requires that

$$\rho_1 A_1 v_1 = \rho_2 A_2 v_2 = m' \qquad \textbf{(4/17)}$$

To describe the forces which act, we isolate either the mass of fluid within the container or the entire container and the fluid within it. We would use the first approach if the forces between the container and the fluid were to be described, and we would adopt the second approach when the forces external to the container are desired.

The latter situation is our primary interest, in which case, the *system isolated* consists of the fixed structure of the container and the fluid within it at a particular instant of time. This isolation is described by a free-body diagram of the mass within a closed volume defined by the exterior surface of the container and the entrance and exit surfaces. We must account for all forces applied *externally* to this system, and in Fig. 4/5*a* the vector sum of this external force system is denoted by $\Sigma\mathbf{F}$. Included in $\Sigma\mathbf{F}$ are

1. the forces exerted on the container at points of its attachment to other structures, including attachments at A_1 and A_2, if present,

2. the forces acting on the fluid within the container at A_1 and A_2 due to any static pressure which may exist in the fluid at these positions, and
3. the weight of the fluid and structure if appreciable.

The resultant $\Sigma\mathbf{F}$ of all of these external forces must equal $\dot{\mathbf{G}}$, the time rate of change of the linear momentum of the isolated system. This statement follows from Eq. 4/6, which was developed in Art. 4/4 for any systems of constant mass, rigid or nonrigid.

Incremental Analysis

The expression for $\dot{\mathbf{G}}$ may be obtained by an incremental analysis. Figure 4/5*b* illustrates the system at time t when the system mass is that of the container, the mass within it, and an increment Δm about to enter during time Δt. At time $t + \Delta t$ the same total mass is that of the container, the mass within it, and an equal increment Δm which leaves the container in time Δt. The linear momentum of the container and mass within it between the two sections A_1 and A_2 remains unchanged during Δt so that the change in momentum of the system in time Δt is

$$\Delta\mathbf{G} = (\Delta m)\mathbf{v}_2 - (\Delta m)\mathbf{v}_1 = \Delta m(\mathbf{v}_2 - \mathbf{v}_1)$$

Division by Δt and passage to the limit yield $\dot{\mathbf{G}} = m'\Delta\mathbf{v}$, where

$$m' = \lim_{\Delta t \to 0}\left(\frac{\Delta m}{\Delta t}\right) = \frac{dm}{dt}$$

Thus, by Eq. 4/6

$$\boxed{\Sigma\mathbf{F} = m'\Delta\mathbf{v}} \qquad \textbf{(4/18)}$$

© Hayden Verry//age fotostock Spain S.L./Corbis Images

The blades of a helicopter impart downward momentum to a column of air, thereby creating the forces necessary for hovering and maneuvering.

Equation 4/18 establishes the relation between the resultant force on a steady-flow system and the corresponding mass flow rate and vector velocity increment.*

Alternatively, we may note that the time rate of change of linear momentum is the vector difference between the rate at which linear momentum leaves the system and the rate at which linear momentum enters the system. Thus, we may write $\dot{\mathbf{G}} = m'\mathbf{v}_2 - m'\mathbf{v}_1 = m'\Delta\mathbf{v}$, which agrees with the foregoing result.

We can now see one of the powerful applications of our general force-momentum equation which we derived for any mass system. Our system here includes a body which is rigid (the structural container for the mass stream) and particles which are in motion (the flow of mass). By defining the boundary of the system, the mass within which is constant for steady-flow conditions, we are able to utilize the generality of Eq. 4/6. However, we must be very careful to account for *all* external

*We must be careful not to interpret dm/dt as the time derivative of the mass of the isolated system. That derivative is zero since the system mass is constant for a steady-flow process. To help avoid confusion, the symbol m' rather than dm/dt is used to represent the steady mass flow rate.

forces acting *on* the system, and they become clear if our free-body diagram is correct.

Angular Momentum in Steady-Flow Systems

A similar formulation is obtained for the case of angular momentum in steady-flow systems. The resultant moment of all external forces about some fixed point O on or off the system, Fig. 4/5a, equals the time rate of change of angular momentum of the system about O. This fact was established in Eq. 4/7 which, for the case of steady flow in a single plane, becomes

$$\Sigma M_O = m'(v_2 d_2 - v_1 d_1) \qquad \textbf{(4/19)}$$

When the velocities of the incoming and outgoing flows are not in the same plane, the equation may be written in vector form as

$$\boxed{\Sigma \mathbf{M}_O = m'(\mathbf{d}_2 \times \mathbf{v}_2 - \mathbf{d}_1 \times \mathbf{v}_1)} \qquad \textbf{(4/19}\boldsymbol{a}\textbf{)}$$

where $\mathbf{d}_1$ and $\mathbf{d}_2$ are the position vectors to the centers of A_1 and A_2 from the fixed reference O. In both relations, the mass center G may be used alternatively as a moment center by virtue of Eq. 4/9.

Equations 4/18 and 4/19a are very simple relations which find important use in describing relatively complex fluid actions. Note that these equations relate *external* forces to the resultant changes in momentum and are independent of the flow path and momentum changes *internal* to the system.

The foregoing analysis may also be applied to systems which move with constant velocity by noting that the basic relations $\Sigma \mathbf{F} = \dot{\mathbf{G}}$ and $\Sigma \mathbf{M}_O = \dot{\mathbf{H}}_O$ or $\Sigma \mathbf{M}_G = \dot{\mathbf{H}}_G$ apply to systems moving with constant velocity as discussed in Arts. 3/12 and 4/4. The only restriction is that the mass within the system remain constant with respect to time.

Three examples of the analysis of steady mass flow are given in the following sample problems, which illustrate the application of the principles embodied in Eqs. 4/18 and 4/19a.

Dan Barnes/Getty Images, Inc.

The principles of steady mass flow are critical to the design of this hovercraft.

Sample Problem 4/6

The smooth vane shown diverts the open stream of fluid of cross-sectional area A, mass density ρ, and velocity v. (*a*) Determine the force components R and F required to hold the vane in a fixed position. (*b*) Find the forces when the vane is given a constant velocity u less than v and in the direction of v.

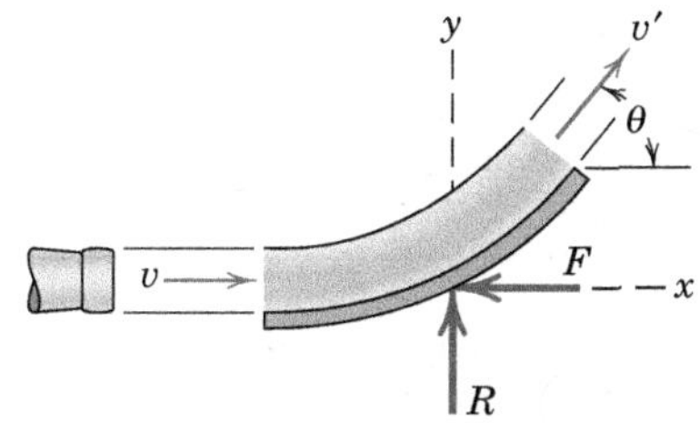

Fixed vane

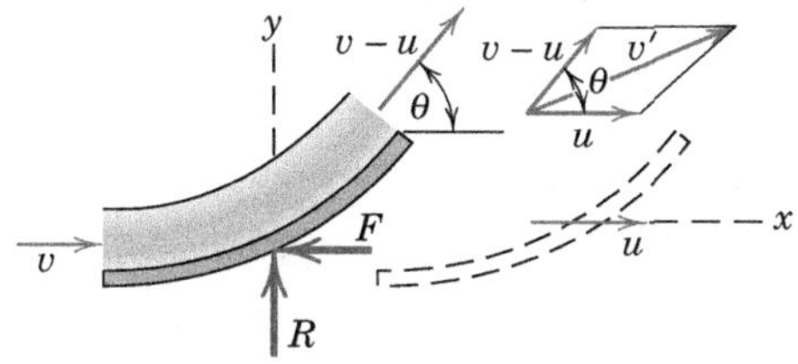

Moving vane

Solution. Part (a). The free-body diagram of the vane together with the fluid portion undergoing the momentum change is shown. The momentum equation may be applied to the isolated system for the change in motion in both the x- and y-directions. With the vane stationary, the magnitude of the exit velocity v' equals that of the entering velocity v with fluid friction neglected. The changes in the velocity components are then

① $$\Delta v_x = v' \cos\theta - v = -v(1 - \cos\theta)$$

and

$$\Delta v_y = v' \sin\theta - 0 = v \sin\theta$$

The mass rate of flow is $m' = \rho A v$, and substitution into Eq. 4/18 gives

$[\Sigma F_x = m'\Delta v_x]$ $$-F = \rho A v[-v(1 - \cos\theta)]$$

$$F = \rho A v^2(1 - \cos\theta) \qquad \textit{Ans.}$$

$[\Sigma F_y = m'\Delta v_y]$ $$R = \rho A v[v \sin\theta]$$

$$R = \rho A v^2 \sin\theta \qquad \textit{Ans.}$$

Part (b). In the case of the moving vane, the final velocity v' of the fluid upon exit is the vector sum of the velocity u of the vane plus the velocity of the fluid relative to the vane $v - u$. This combination is shown in the velocity diagram to the right of the figure for the exit conditions. The x-component of v' is the sum of the components of its two parts, so $v'_x = (v - u)\cos\theta + u$. The change in x-velocity of the stream is

$$\Delta v_x = (v - u)\cos\theta + (u - v) = -(v - u)(1 - \cos\theta)$$

The y-component of v' is $(v - u)\sin\theta$, so that the change in the y-velocity of the stream is $\Delta v_y = (v - u)\sin\theta$.

The mass rate of flow m' is the mass undergoing momentum change per unit of time. This rate is the mass flowing over the vane per unit time and *not* the rate of issuance from the nozzle. Thus,

$$m' = \rho A(v - u)$$

The impulse-momentum principle of Eq. 4/18 applied in the positive coordinate directions gives

② $[\Sigma F_x = m'\Delta v_x]$ $$-F = \rho A(v - u)[-(v - u)(1 - \cos\theta)]$$

$$F = \rho A(v - u)^2(1 - \cos\theta) \qquad \textit{Ans.}$$

$[\Sigma F_y = m'\Delta v_y]$ $$R = \rho A(v - u)^2 \sin\theta \qquad \textit{Ans.}$$

Helpful Hints

① Be careful with algebraic signs when using Eq. 4/18. The change in v_x is the final value minus the initial value measured in the positive x-direction. Also we must be careful to write $-F$ for ΣF_x.

② Observe that for given values of u and v, the angle for maximum force F is $\theta = 180°$.

Sample Problem 4/7

For the moving vane of Sample Problem 4/6, determine the optimum speed u of the vane for the generation of maximum power by the action of the fluid on the vane.

Solution. The force R shown with the figure for Sample Problem 4/6 is normal to the velocity of the vane so it does no work. The work done by the force F shown is negative, but the power developed by the force (reaction to F) exerted by the fluid on the moving vane is

$[P = Fu]$ $$P = \rho A(v - u)^2 u(1 - \cos\theta)$$

The velocity of the vane for maximum power for the one blade in the stream is specified by

$\left[\dfrac{dP}{du} = 0\right]$ $$\rho A(1 - \cos\theta)(v^2 - 4uv + 3u^2) = 0$$

(1) $$(v - 3u)(v - u) = 0 \qquad u = \frac{v}{3}$$ *Ans.*

The second solution $u = v$ gives a minimum condition of zero power. An angle $\theta = 180°$ completely reverses the flow and clearly produces both maximum force and maximum power for any value of u.

Helpful Hint

(1) The result here applies to a single vane only. In the case of multiple vanes, such as the blades on a turbine disk, the rate at which fluid issues from the nozzles is the same rate at which fluid is undergoing momentum change. Thus, $m' = \rho A v$ rather than $\rho A(v - u)$. With this change, the optimum value of u turns out to be $u = v/2$.

Sample Problem 4/8

The offset nozzle has a discharge area A at B and an inlet area A_0 at C. A liquid enters the nozzle at a static gage pressure p through the fixed pipe and issues from the nozzle with a velocity v in the direction shown. If the constant density of the liquid is ρ, write expressions for the tension T, shear Q, and bending moment M in the pipe at C.

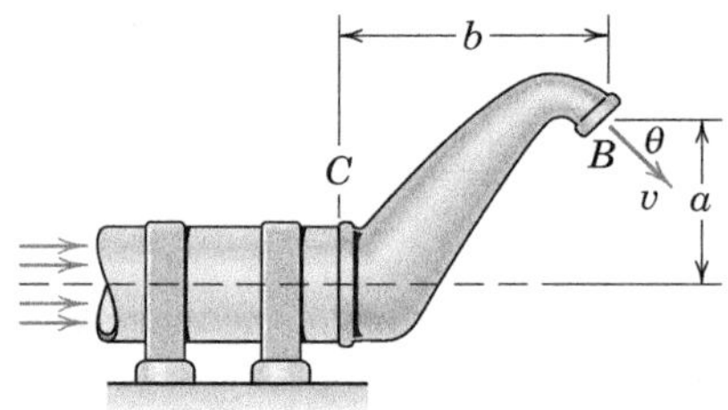

Solution. The free-body diagram of the nozzle and the fluid within it shows the tension T, shear Q, and bending moment M acting on the flange of the nozzle where it attaches to the fixed pipe. The force pA_0 on the fluid within the nozzle due to the static pressure is an additional external force.

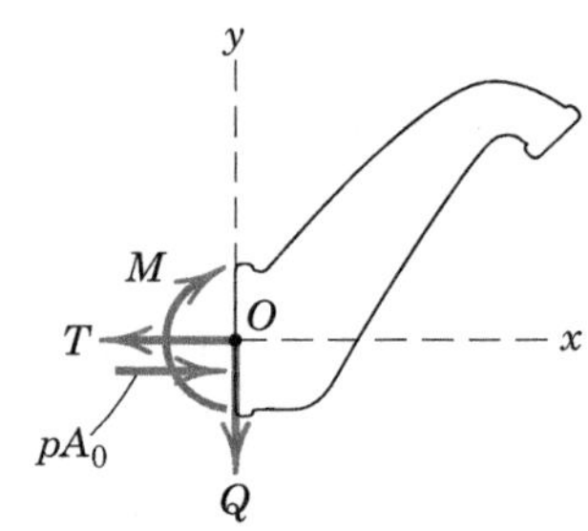

Continuity of flow with constant density requires that

$$Av = A_0 v_0$$

where v_0 is the velocity of the fluid at the entrance to the nozzle. The momentum principle of Eq. 4/18 applied to the system in the two coordinate directions gives

$[\Sigma F_x = m'\Delta v_x]$ $$pA_0 - T = \rho A v(v\cos\theta - v_0)$$

(1) $$T = pA_0 + \rho A v^2\left(\frac{A}{A_0} - \cos\theta\right)$$ *Ans.*

$[\Sigma F_y = m'\Delta v_y]$ $$-Q = \rho A v(-v\sin\theta - 0)$$

$$Q = \rho A v^2 \sin\theta$$ *Ans.*

The moment principle of Eq. 4/19 applied in the clockwise sense gives

(2) $[\Sigma M_O = m'(v_2 d_2 - v_1 d_1)]$ $$M = \rho A v(va\cos\theta + vb\sin\theta - 0)$$

$$M = \rho A v^2(a\cos\theta + b\sin\theta)$$ *Ans.*

Helpful Hints

(1) Again, be careful to observe the correct algebraic signs of the terms on both sides of Eqs. 4/18 and 4/19.

(2) The forces and moment acting on the pipe are equal and opposite to those shown acting on the nozzle.

Sample Problem 4/9

An air-breathing jet aircraft of total mass m flying with a constant speed v consumes air at the mass rate m'_a and exhausts burned gas at the mass rate m'_g with a velocity u relative to the aircraft. Fuel is consumed at the constant rate m'_f. The total aerodynamic forces acting on the aircraft are the lift L, normal to the direction of flight, and the drag D, opposite to the direction of flight. Any force due to the static pressure across the inlet and exhaust surfaces is assumed to be included in D. Write the equation for the motion of the aircraft and identify the thrust T.

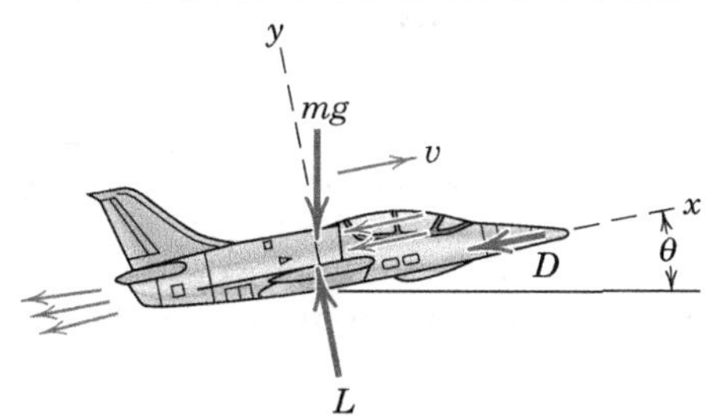

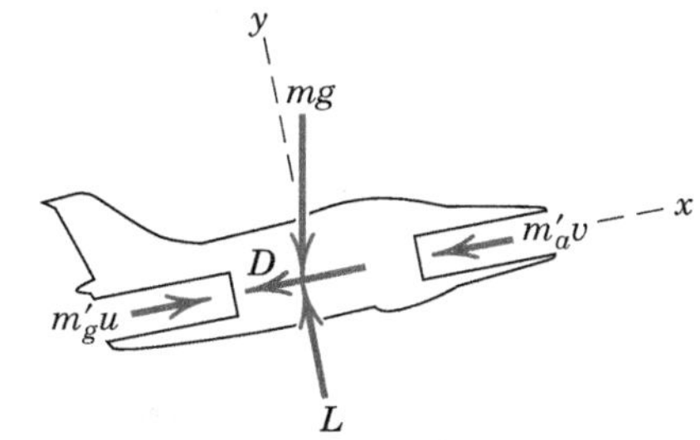

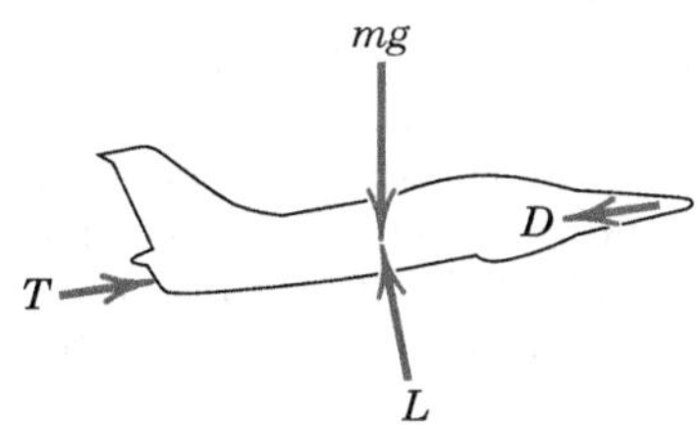

Solution. The free-body diagram of the aircraft together with the air, fuel,
① and exhaust gas within it is given and shows only the weight, lift, and drag forces as defined. We attach axes x-y to the aircraft and apply our momentum
② equation relative to the moving system.

The fuel will be treated as a steady stream entering the aircraft with no velocity relative to the system and leaving with a relative velocity u in the exhaust stream. We now apply Eq. 4/18 relative to the reference axes and treat the air and fuel flows separately. For the air flow, the change in velocity in the x-direction relative to the moving system is

③ $$\Delta v_a = -u - (-v) = -(u - v)$$

and for the fuel flow the x-change in velocity relative to x-y is

$$\Delta v_f = -u - (0) = -u$$

Thus, we have

$$[\Sigma F_x = m'\Delta v_x] \qquad -mg \sin\theta - D = -m'_a(u - v) - m'_f u$$
$$= -m'_g u + m'_a v$$

where the substitution $m'_g = m'_a + m'_f$ has been made. Changing signs gives

$$m'_g u - m'_a v = mg \sin\theta + D$$

which is the equation of motion of the system.

If we modify the boundaries of our system to expose the interior surfaces on which the air and gas act, we will have the simulated model shown, where the air exerts a force $m'_a v$ on the interior of the turbine and the exhaust gas reacts against the interior surfaces with the force $m'_g u$.

The commonly used model is shown in the final diagram, where the net effect of air and exhaust momentum changes is replaced by a simulated thrust

④ $$T = m'_g u - m'_a v \qquad \textit{Ans.}$$

applied to the aircraft from a presumed external source.

Inasmuch as m'_f is generally only 2 percent or less of m'_a, we can use the approximation $m'_g \cong m'_a$ and express the thrust as

$$T \cong m'_g(u - v) \qquad \textit{Ans.}$$

We have analyzed the case of constant velocity. Although our Newtonian principles do not generally hold relative to accelerating axes, it can be shown that we may use the $F = ma$ equation for the simulated model and write $T - mg \sin\theta - D = m\dot{v}$ with virtually no error.

Helpful Hints

① Note that the boundary of the system cuts across the air stream at the entrance to the air scoop and across the exhaust stream at the nozzle.

② We are permitted to use moving axes which translate with constant velocity. See Arts. 3/14 and 4/2.

③ Riding with the aircraft, we observe the air entering our system with a velocity $-v$ measured in the plus x-direction and leaving the system with an x-velocity of $-u$. The final value minus the initial one gives the expression cited, namely, $-u - (-v) = -(u - v)$.

④ We now see that the "thrust" is, in reality, not a force external to the entire airplane shown in the first figure but can be modeled as an external force.

PROBLEMS

Introductory Problems

4/35 The experimental race car is propelled by a rocket motor and is designed to reach a maximum speed $v = 300$ mi/hr under the thrust T of its motor. Prior wind-tunnel tests disclose that the wind resistance at this speed is 225 lb. If the rocket motor is burning fuel at the rate of 3.5 lb/sec, determine the velocity u of the exhaust gases relative to the car.

Problem 4/35

4/36 The jet aircraft has a mass of 4.6 Mg and a drag (air resistance) of 32 kN at a speed of 1000 km/h at a particular altitude. The aircraft consumes air at the rate of 106 kg/s through its intake scoop and uses fuel at the rate of 4 kg/s. If the exhaust has a rearward velocity of 680 m/s relative to the exhaust nozzle, determine the maximum angle of elevation α at which the jet can fly with a constant speed of 1000 km/h at the particular altitude in question.

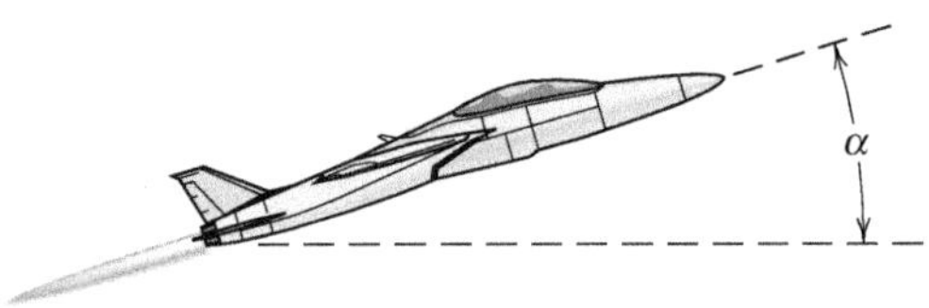

Problem 4/36

4/37 Fresh water issues from the nozzle with a velocity of 30 m/s at the rate of 0.05 m^3/s and is split into two equal streams by the fixed vane and deflected through 60° as shown. Calculate the force F required to hold the vane in place. The density of water is 1000 kg/m^3.

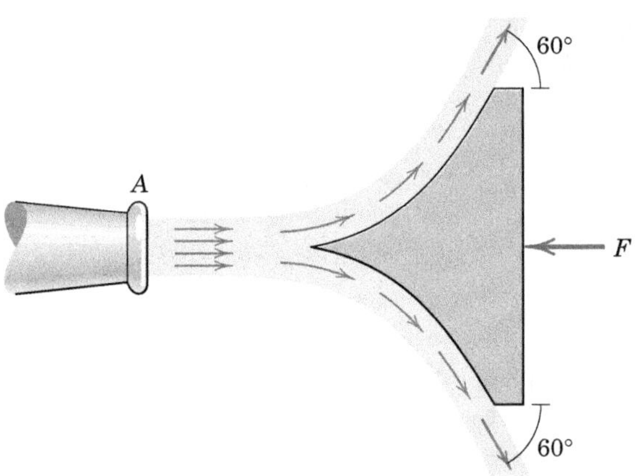

Problem 4/37

4/38 In an unwise effort to remove debris, a homeowner directs the nozzle of his backpack blower directly toward the garage door. The nozzle velocity is 130 mi/hr and the flow rate is 410 ft^3/min. Estimate the force F exerted by the airflow on the door. The specific weight of air is 0.0753 lb/ft^3.

Problem 4/38

4/39 The jet water ski has reached its maximum velocity of 70 km/h when operating in salt water. The water intake is in the horizontal tunnel in the bottom of the hull, so the water enters the intake at the velocity of 70 km/h relative to the ski. The motorized pump discharges water from the horizontal exhaust nozzle of 50-mm diameter at the rate of 0.082 m^3/s. Calculate the resistance R of the water to the hull at the operating speed.

Problem 4/39

4/40 The 25-mm steel slab 1.2 m wide enters the rolls at the speed of 0.4 m/s and is reduced in thickness to 19 mm. Calculate the small horizontal thrust T on the bearings of each of the two rolls.

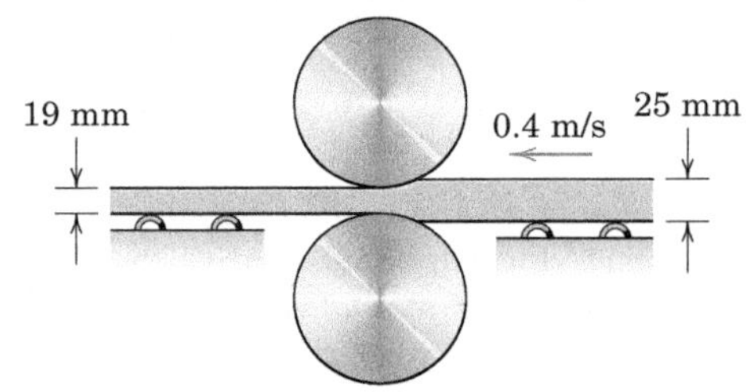

Problem 4/40

4/41 The fire tug discharges a stream of salt water (density 1030 kg/m^3) with a nozzle velocity of 40 m/s at the rate of 0.080 m^3/s. Calculate the propeller thrust T which must be developed by the tug to maintain a fixed position while pumping.

Problem 4/41

4/42 The pump shown draws air with a density ρ through the fixed duct A of diameter d with a velocity u and discharges it at high velocity v through the two outlets B. The pressure in the airstreams at A and B is atmospheric. Determine the expression for the tension T exerted on the pump unit through the flange at C.

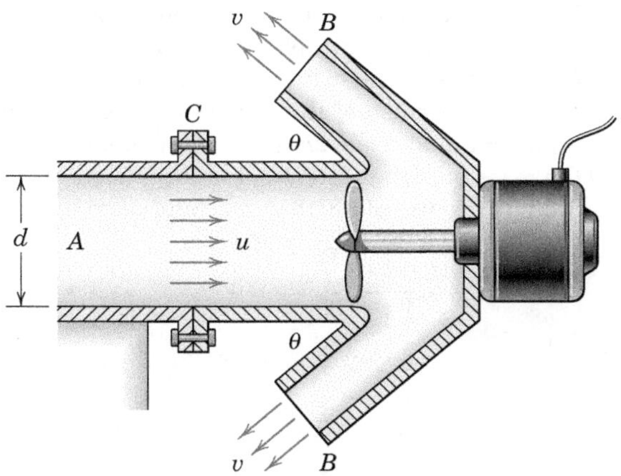

Problem 4/42

4/43 A jet-engine noise suppressor consists of a movable duct which is secured directly behind the jet exhaust by cable A and deflects the blast directly upward. During a ground test, the engine sucks in air at the rate of 43 kg/s and burns fuel at the rate of 0.8 kg/s. The exhaust velocity is 720 m/s. Determine the tension T in the cable.

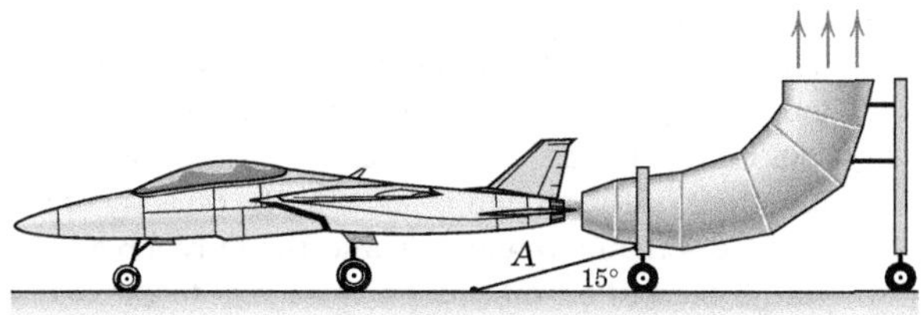

Problem 4/43

4/44 The 90° vane moves to the left with a constant velocity of 10 m/s against a stream of fresh water issuing with a velocity of 20 m/s from the 25-mm-diameter nozzle. Calculate the forces F_x and F_y on the vane required to support the motion.

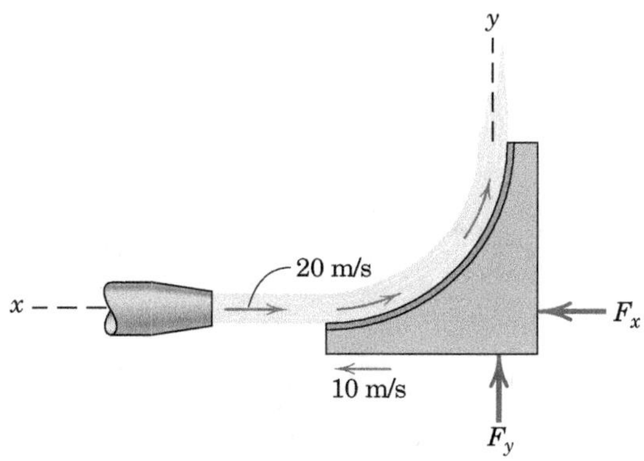

Problem 4/44

Representative Problems

4/45 The pipe bend shown has a cross-sectional area A and is supported in its plane by the tension T applied to its flanges by the adjacent connecting pipes (not shown). If the velocity of the liquid is v, its density ρ, and its static pressure p, determine T and show that it is independent of the angle θ.

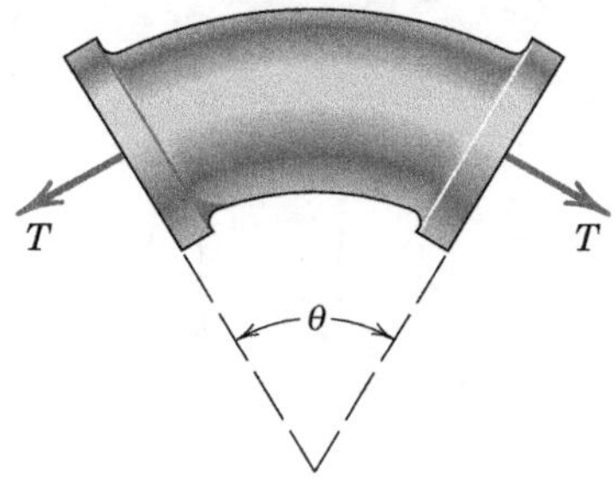

Problem 4/45

4/46 A jet of fluid with cross-sectional area A and mass density ρ issues from the nozzle with a velocity v and impinges on the inclined trough shown in section. Some of the fluid is diverted in each of the two directions. If the trough is smooth, the velocity of both diverted streams remains v, and the only force which can be exerted on the trough is normal to the bottom surface. Hence, the trough will be held in position by forces whose resultant is F normal to the trough. By writing impulse-momentum equations for the directions along and normal to the trough, determine the force F required to support the trough. Also find the volume rates of flow Q_1 and Q_2 for the two streams.

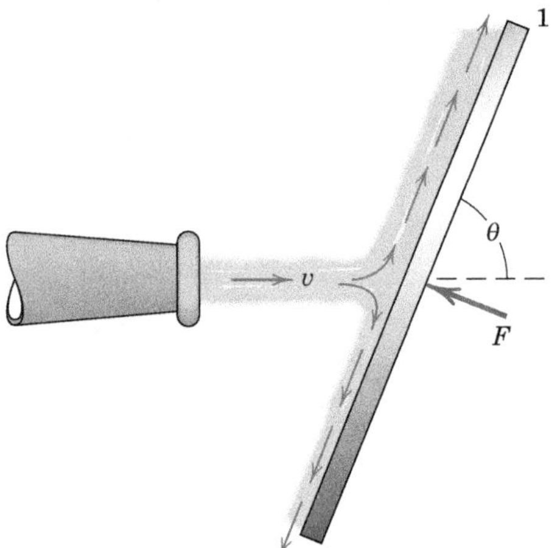

Problem 4/46

4/47 The 8-oz ball is supported by the vertical stream of fresh water which issues from the 1/2-in.-diameter nozzle with a velocity of 35 ft/sec. Calculate the height h of the ball above the nozzle. Assume that the stream remains intact and there is no energy lost in the jet stream.

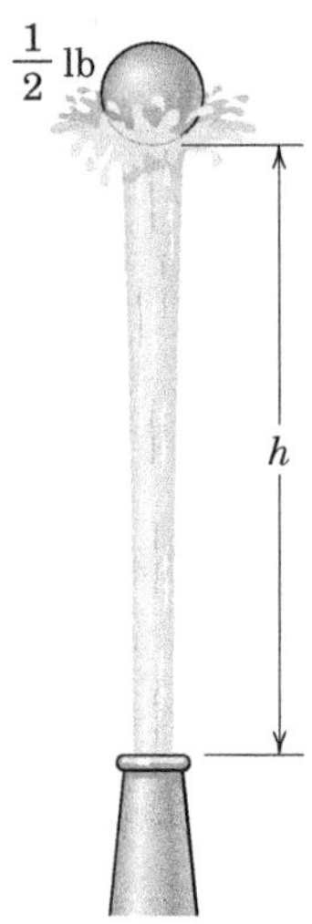

Problem 4/47

4/48 A jet-engine thrust reverser to reduce an aircraft speed of 200 km/h after landing employs folding vanes which deflect the exhaust gases in the direction indicated. If the engine is consuming 50 kg of air and 0.65 kg of fuel per second, calculate the braking thrust as a fraction n of the engine thrust without the deflector vanes. The exhaust gases have a velocity of 650 m/s relative to the nozzle.

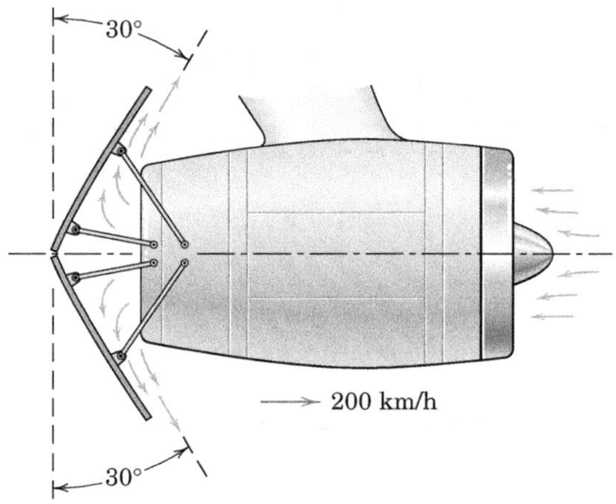

Problem 4/48

4/49 Water issues from a nozzle with an initial velocity v and supports a thin plate of mass m at a height h above the nozzle exit. A hole in the center of the plate allows some of the water to travel upward to a maximum altitude $2h$ above the plate. Determine the mass m of the plate. Neglect any effects of water falling onto the plate after reaching maximum altitude. Let ρ be the density of water.

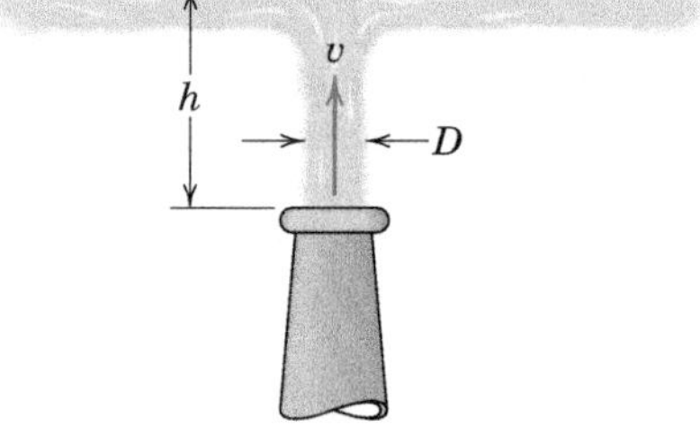

Problem 4/49

4/50 The axial-flow fan C pumps air through the duct of circular cross section and exhausts it with a velocity v at B. The air densities at A and B are ρ_A and ρ_B, respectively, and the corresponding pressures are p_A and p_B. The fixed deflecting blades at D restore axial flow to the air after it passes through the propeller blades C. Write an expression for the resultant horizontal force R exerted on the fan unit by the flange and bolts at A.

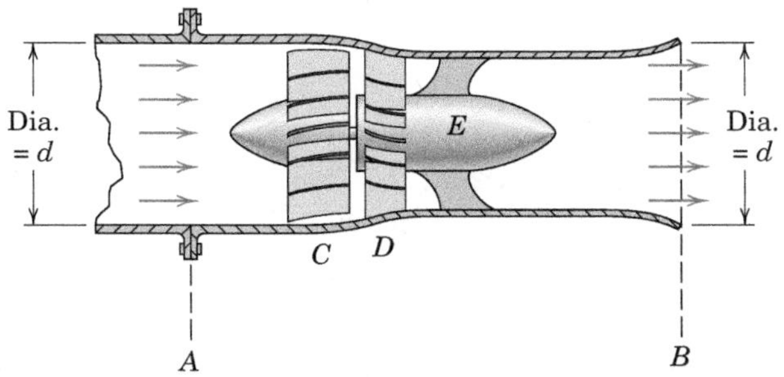

Problem 4/50

4/51 Air is pumped through the stationary duct A with a velocity of 50 ft/sec and exhausted through an experimental nozzle section BC. The average static pressure across section B is 150 lb/in.^{2} gage, and the specific weight of air at this pressure and at the temperature prevailing is 0.840 lb/ft^3. The average static pressure across the exit section C is measured to be 2 lb/in.^{2} gage, and the corresponding specific weight of air is 0.0760 lb/ft^3. Calculate the force T exerted on the nozzle flange at B by the bolts and the gasket to hold the nozzle in place.

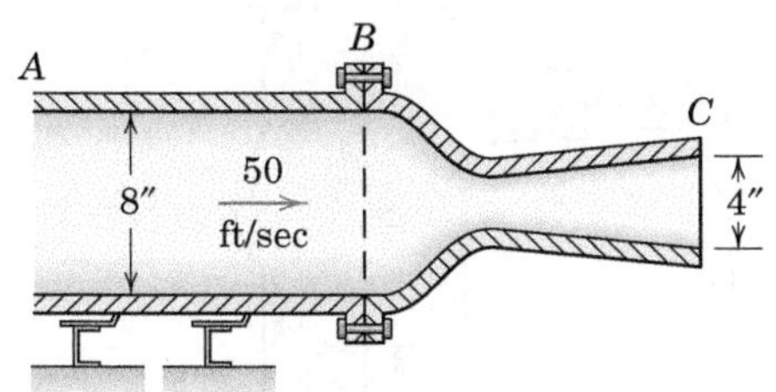

Problem 4/51

4/52 Air enters the pipe at A at the rate of 6 kg/s under a pressure of 1400 kPa gage and leaves the whistle at atmospheric pressure through the opening at B. The entering velocity of the air at A is 45 m/s, and the exhaust velocity at B is 360 m/s. Calculate the tension T, shear V, and bending moment M in the pipe at A. The net flow area at A is 7500 mm^2.

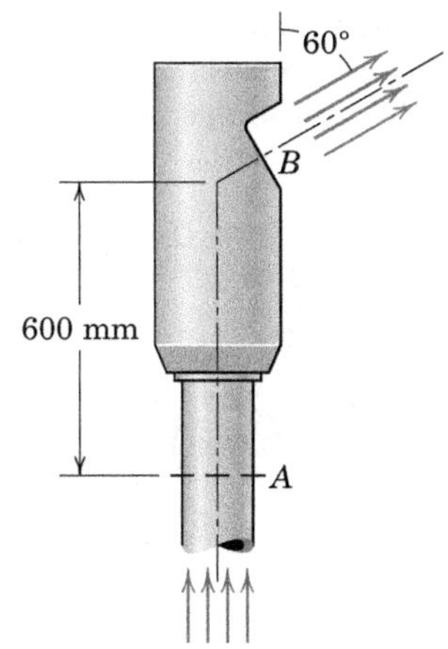

Problem 4/52

4/53 The sump pump has a net mass of 310 kg and pumps fresh water against a 6-m head at the rate of 0.125 m^3/s. Determine the vertical force R between the supporting base and the pump flange at A during operation. The mass of water in the pump may be taken as the equivalent of a 200-mm-diameter column 6 m in height.

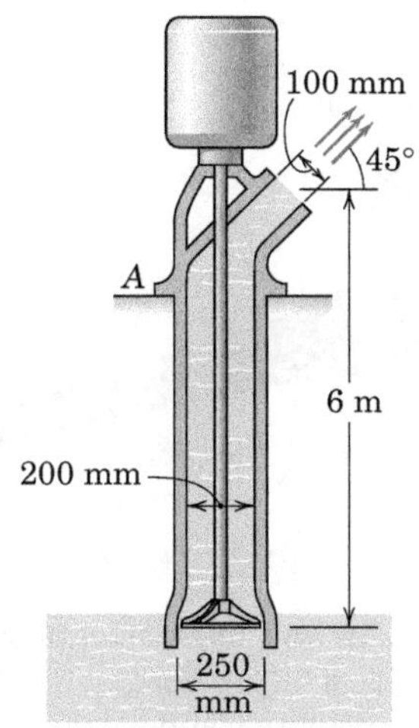

Problem 4/53

4/54 The experimental ground-effect machine has a total weight of 4200 lb. It hovers 1 or 2 ft off the ground by pumping air at atmospheric pressure through the circular intake duct at B and discharging it horizontally under the periphery of the skirt C. For an intake velocity v of 150 ft/sec, calculate the average air pressure p under the 18-ft-diameter machine at ground level. The specific weight of the air is 0.076 lb/ft^3.

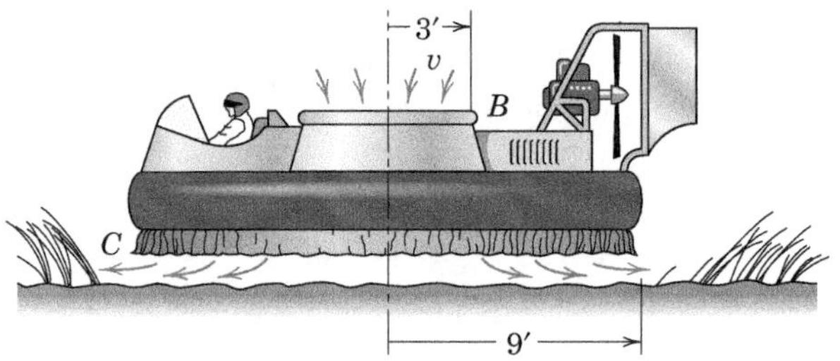

Problem 4/54

4/55 The leaf blower draws in air at a rate of 400 ft^3/min and discharges it at a speed v = 240 mi/hr. If the specific weight of the air being drawn into the blower is 7.53(10^{-2}) lb/ft^3, determine the added torque which the man must exert on the handle of the blower when it is running, compared with that when it is off, to maintain a steady orientation.

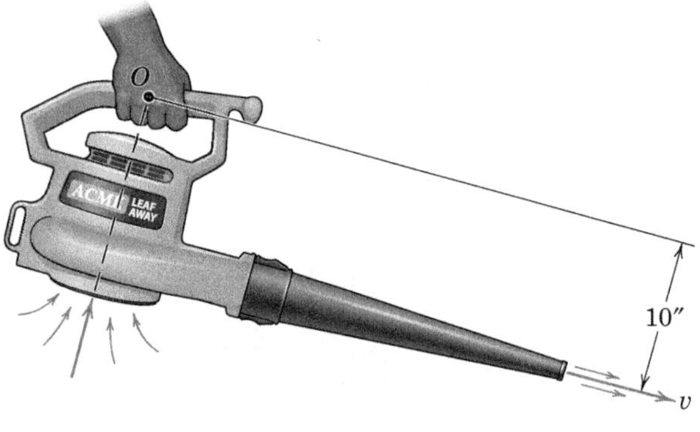

Problem 4/55

4/56 The ducted fan unit of mass m is supported in the vertical position on its flange at A. The unit draws in air with a density ρ and a velocity u through section A and discharges it through section B with a velocity v. Both inlet and outlet pressures are atmospheric. Write an expression for the force R applied to the flange of the fan unit by the supporting slab.

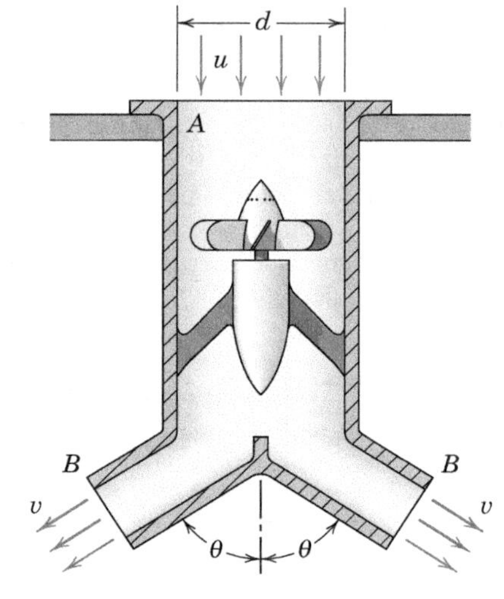

Problem 4/56

4/57 The fire hydrant is tested under a high standpipe pressure. The total flow of 10 ft^3/sec is divided equally between the two outlets, each of which has a cross-sectional area of 0.040 ft^2. The inlet cross-sectional area at the base is 0.75 ft^2. Neglect the weight of the hydrant and water within it and compute the tension T, the shear V, and the bending moment M in the base of the standpipe at B. The specific weight of water is 62.4 lb/ft^3. The static pressure of the water as it enters the base at B is 120 lb/in.^{2}

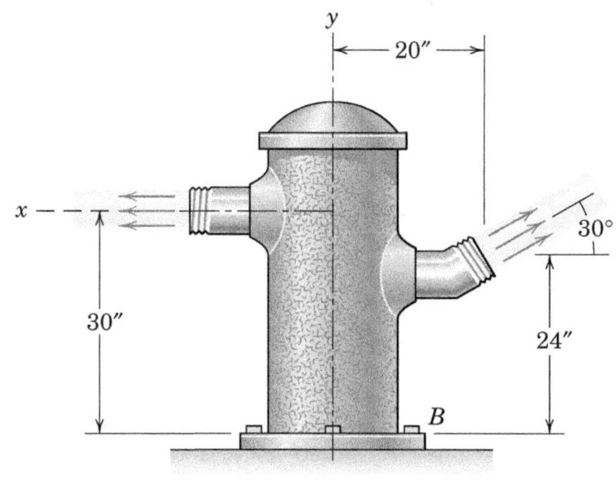

Problem 4/57

4/58 A rotary snow plow mounted on a large truck eats its way through a snow drift on a level road at a constant speed of 20 km/h. The plow discharges 60 Mg of snow per minute from its 45° chute with a velocity of 12 m/s relative to the plow. Calculate the tractive force P on the tires in the direction of motion necessary to move the plow and find the corresponding lateral force R between the tires and the road.

Problem 4/58

4/59 Salt water flows through the fixed 12-in.-inside-diameter pipe at a speed $v_0 = 4$ ft/sec and enters the 150° bend with inside radius of 24 in. The water exits to the atmosphere through the 6-in.-diameter nozzle C. Determine the shear force V, axial force P, and bending moment M at flanges A and B which result from the flow of the salt water. The gage pressure in the pipe at flange A is 250 lb/in.^{2} and the pressure drop between A and B due to head loss may be neglected.

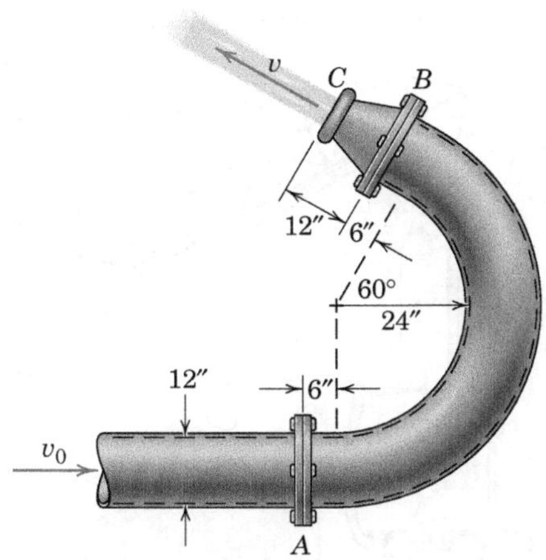

Problem 4/59

4/60 The industrial blower sucks in air through the axial opening A with a velocity v_1 and discharges it at atmospheric pressure and temperature through the 150-mm-diameter duct B with a velocity v_2. The blower handles 16 m^3 of air per minute with the motor and fan running at 3450 rev/min. If the motor requires 0.32 kW of power under no load (both ducts closed), calculate the power P consumed while air is being pumped.

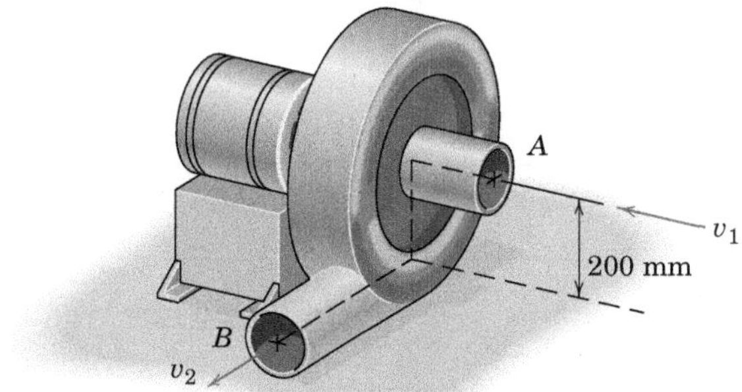

Problem 4/60

4/61 The feasibility of a one-passenger VTOL (vertical takeoff and landing) craft is under review. The preliminary design calls for a small engine with a high power-to-weight ratio driving an air pump that draws in air through the 70° ducts with an inlet velocity $v = 40$ m/s at a static gage pressure of -1.8 kPa across the inlet areas totaling 0.1320 m^2. The air is exhausted vertically down with a velocity $u = 420$ m/s. For a 90-kg passenger, calculate the maximum net mass m of the machine for which it can take off and hover. (See Table D/1 for air density.)

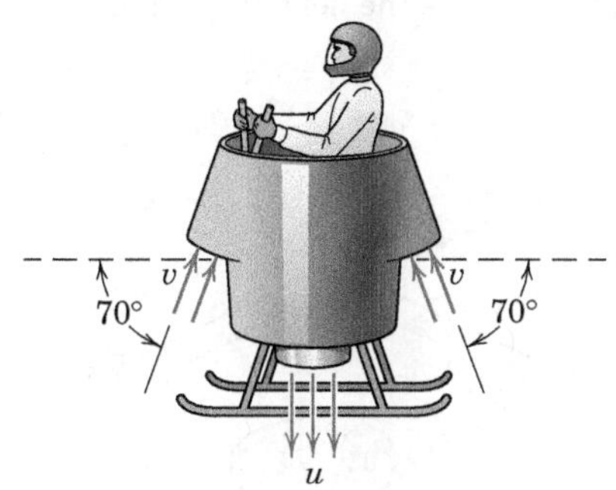

Problem 4/61

4/62 The helicopter shown has a mass m and hovers in position by imparting downward momentum to a column of air defined by the slipstream boundary shown. Find the downward velocity v given to the air by the rotor at a section in the stream below the rotor, where the pressure is atmospheric and the stream radius is r. Also find the power P required of the engine. Neglect the rotational energy of the air, any temperature rise due to air friction, and any change in air density ρ.

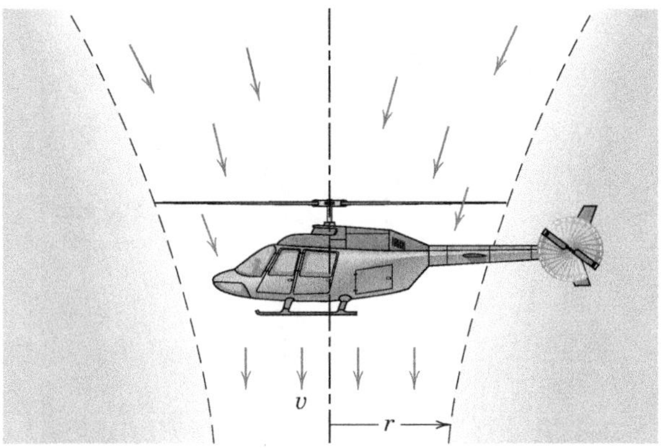

Problem 4/62

4/63 The sprinkler is made to rotate at the constant angular velocity ω and distributes water at the volume rate Q. Each of the four nozzles has an exit area A. Water is ejected from each nozzle at an angle ϕ that is measured in the horizontal plane as shown. Write an expression for the torque M on the shaft of the sprinkler necessary to maintain the given motion. For a given pressure and thus flow rate Q, at what speed ω_0 will the sprinkler operate with no applied torque? Let ρ be the density of water.

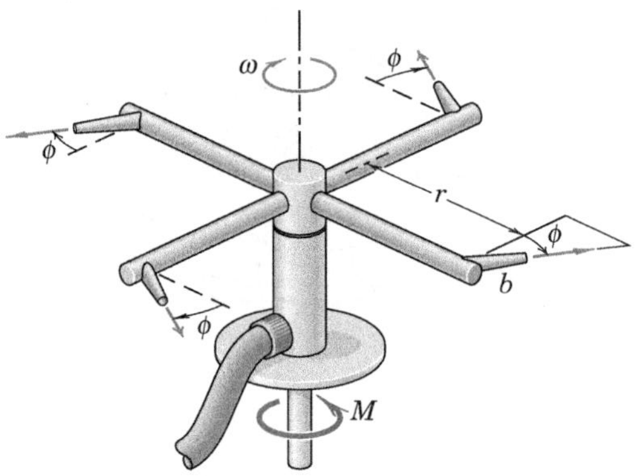

Problem 4/63

4/64 The VTOL (vertical takeoff and landing) military aircraft is capable of rising vertically under the action of its jet exhaust, which can be "vectored" from $\theta \cong 0$ for takeoff and hovering to $\theta = 90°$ for forward flight. The loaded aircraft has a mass of 8600 kg. At full takeoff power, its turbo-fan engine consumes air at the rate of 90 kg/s and has an air–fuel ratio of 18. Exhaust-gas velocity is 1020 m/s with essentially atmospheric pressure across the exhaust nozzles. Air with a density of 1.206 kg/m^3 is sucked into the intake scoops at a pressure of -2 kPa (gage) over the total inlet area of 1.10 m^2. Determine the angle θ for vertical takeoff and the corresponding vertical acceleration a_y of the aircraft.

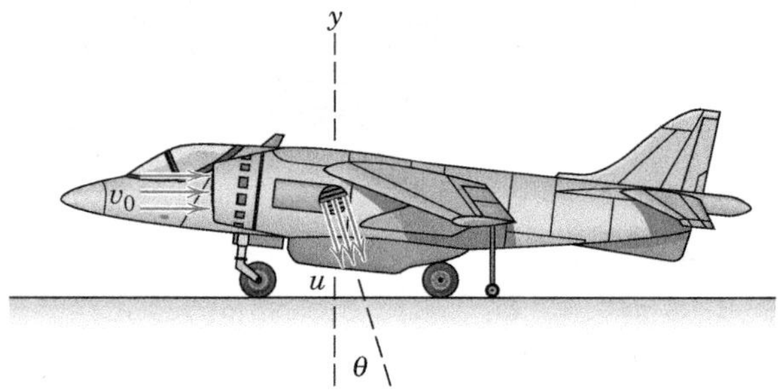

Problem 4/64

4/65 A marine terminal for unloading bulk wheat from a ship is equipped with a vertical pipe with a nozzle at A which sucks wheat up the pipe and transfers it to the storage building. Calculate the x- and y-components of the force **R** required to change the momentum of the flowing mass in rounding the bend. Identify all forces applied externally to the bend and mass within it. Air flows through the 14-in.-diameter pipe at the rate of 18 tons per hour under a vacuum of 9 in. of mercury ($p = -4.42$ lb/in.^{2}) and carries with it 150 tons of wheat per hour at a speed of 124 ft/sec.

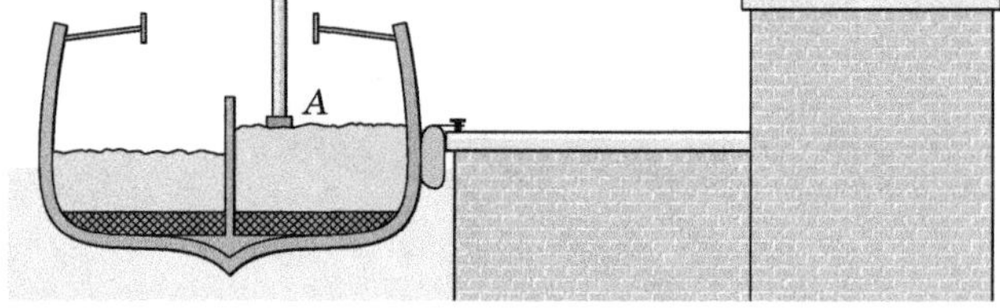

Problem 4/65

▶**4/66** An axial section of the suction nozzle A for a bulk wheat unloader is shown here. The outer pipe is secured to the inner pipe by several longitudinal webs which do not restrict the flow of air. A vacuum of 9 in. of mercury ($p = -4.42$ lb/in.^{2} gage) is maintained in the inner pipe, and the pressure across the bottom of the outer pipe is atmospheric ($p = 0$). Air at 0.075 lb/ft^3 is drawn in through the space between the pipes at a rate of 18 tons/hr at atmospheric pressure and draws with it 150 tons of wheat per hour up the pipe at a velocity of 124 ft/sec. If the nozzle unit below section *A-A* weighs 60 lb, calculate the compression *C* in the connection at *A-A*.

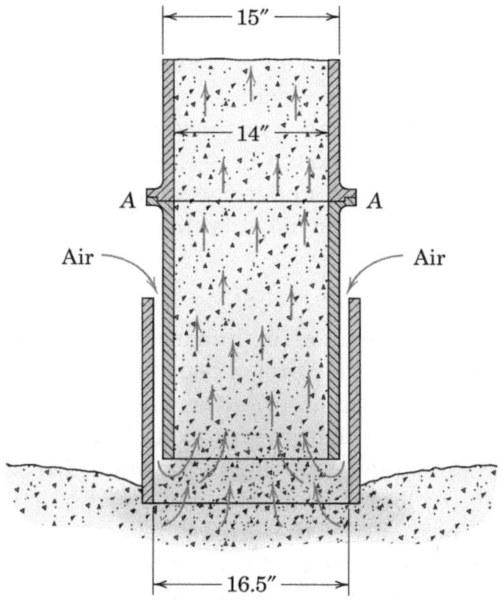

Problem 4/66

▶**4/67** The valve, which is screwed into the fixed pipe at section *A-A*, is designed to discharge fresh water at the rate of 340 gal/min into the atmosphere in the *x-y* plane as shown. Water pressure at *A-A* is 150 lb/in.^{2} gage. The flow area at *A-A* has a diameter of 2 in., and the diameter of the discharge area at *B* is 1 in. Neglect the weight of the valve and water within it and compute the shear *V*, tension *F*, torsion *T*, and bending moment *M* at section *A-A*. (1 gallon contains 231 in.^{3})

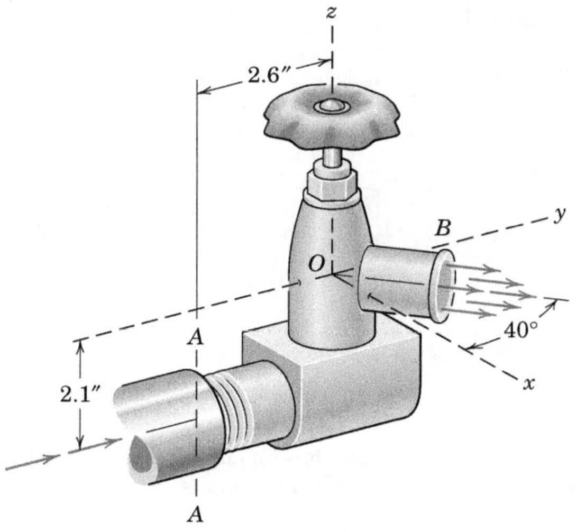

Problem 4/67

▶**4/68** In the figure is shown a detail of the stationary nozzle diaphragm *A* and the rotating blades *B* of a gas turbine. The products of combustion pass through the fixed diaphragm blades at the 27° angle and impinge on the moving rotor blades. The angles shown are selected so that the velocity of the gas relative to the moving blade at entrance is at the 20° angle for minimum turbulence, corresponding to a mean blade velocity of 315 m/s at a radius of 375 mm. If gas flows past the blades at the rate of 15 kg/s, determine the theoretical power output *P* of the turbine. Neglect fluid and mechanical friction with the resulting heat-energy loss and assume that all the gases are deflected along the surfaces of the blades with a velocity of constant magnitude relative to the blade.

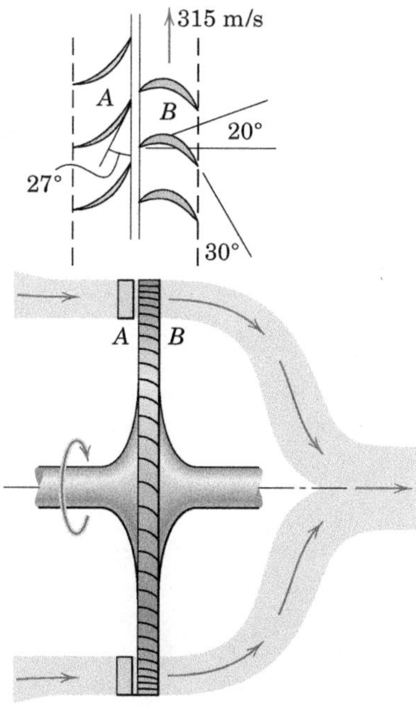

Problem 4/68

4/7 Variable Mass

In Art. 4/4 we extended the equations for the motion of a particle to include a system of particles. This extension led to the very general expressions $\Sigma\mathbf{F} = \dot{\mathbf{G}}$, $\Sigma\mathbf{M}_O = \dot{\mathbf{H}}_O$, and $\Sigma\mathbf{M}_G = \dot{\mathbf{H}}_G$, which are Eqs. 4/6, 4/7, and 4/9, respectively. In their derivation, the summations were taken over a fixed collection of particles, so that the mass of the system to be analyzed was constant.

In Art. 4/6 these momentum principles were extended in Eqs. 4/18 and 4/19*a* to describe the action of forces on a system defined by a geometric volume through which passes a steady flow of mass. Therefore, the amount of mass within this volume was constant with respect to time and thus we were able to use Eqs. 4/6, 4/7, and 4/9. When the mass within the boundary of a system under consideration is not constant, the foregoing relationships are no longer valid.*

Equation of Motion

We will now develop the equation for the linear motion of a system whose mass varies with time. Consider first a body which gains mass by overtaking and swallowing a stream of matter, Fig. 4/6*a*. The mass of the body and its velocity at any instant are m and v, respectively. The stream of matter is assumed to be moving in the same direction as m with a constant velocity v_0 less than v. By virtue of Eq. 4/18, the force exerted by m on the particles of the stream to accelerate them from a velocity v_0 to a greater velocity v is $R = m'(v - v_0) = \dot{m}u$, where the time rate of increase of m is $m' = \dot{m}$ and where u is the magnitude of the relative velocity with which the particles approach m. In addition to R, all other forces acting on m in the direction of its motion are

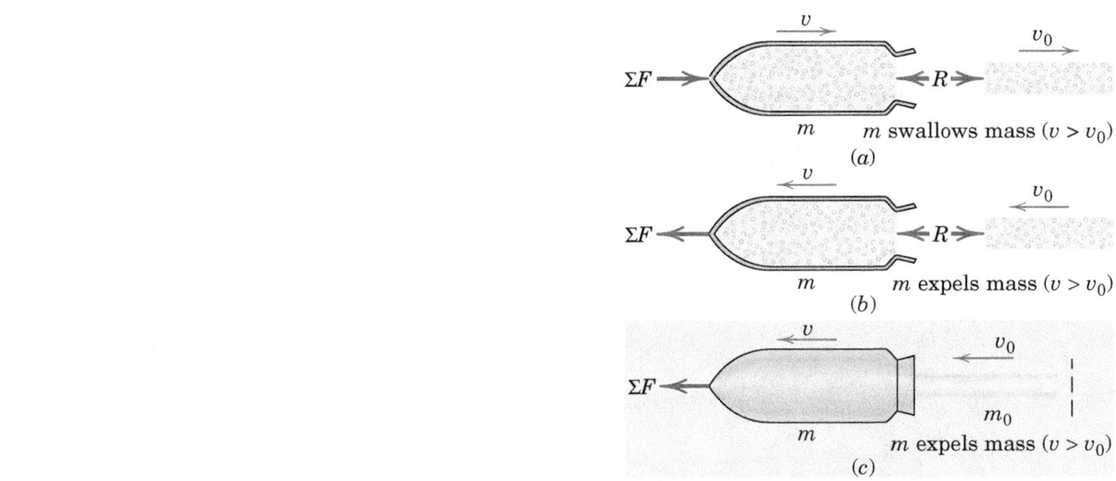

Figure 4/6

*In relativistic mechanics the mass is found to be a function of velocity, and its time derivative has a meaning different from that in Newtonian mechanics.

denoted by ΣF. The equation of motion of m from Newton's second law is, therefore, $\Sigma F - R = m\dot{v}$ or

$$\boxed{\Sigma F = m\dot{v} + \dot{m}u} \qquad \textbf{(4/20)}$$

Similarly, if the body loses mass by expelling it rearward so that its velocity v_0 is less than v, Fig. 4/6*b*, the force R required to decelerate the particles from a velocity v to a lesser velocity v_0 is $R = m'(-v_0 - [-v]) = m'(v - v_0)$. But $m' = -\dot{m}$ since m is decreasing. Also, the relative velocity with which the particles leave m is $u = v - v_0$. Thus, the force R becomes $R = -\dot{m}u$. If ΣF denotes the resultant of all other forces acting on m in the direction of its motion, Newton's second law requires that $\Sigma F + R = m\dot{v}$ or

$$\Sigma F = m\dot{v} + \dot{m}u$$

which is the same relationship as in the case where m is gaining mass. We may use Eq. 4/20, therefore, as the equation of motion of m, whether it is gaining or losing mass.

A frequent error in the use of the force-momentum equation is to express the partial force sum ΣF as

$$\Sigma F = \frac{d}{dt}(mv) = m\dot{v} + \dot{m}v$$

From this expansion we see that the direct differentiation of the linear momentum gives the correct force ΣF *only* when the body picks up mass initially at rest or when it expels mass which is left with zero absolute velocity. In both instances, $v_0 = 0$ and $u = v$.

The Super Scooper is a firefighting airplane which can quickly ingest water from a lake by skimming across the surface with just a bottom-mounted scoop entering the water. The mass within the aircraft boundary varies during the scooping operation as well as during the dumping operation shown.

Alternative Approach

We may also obtain Eq. 4/20 by a direct differentiation of the momentum from the basic relation $\Sigma F = \dot{G}$, provided a proper system of constant total mass is chosen. To illustrate this approach, we take the case where m is losing mass and use Fig. 4/6*c*, which shows the system of m and an arbitrary portion m_0 of the stream of ejected mass. The mass of this system is $m + m_0$ and is constant.

The ejected stream of mass is assumed to move undisturbed once separated from m, and the only force external to the entire system is ΣF which is applied directly to m as before. The reaction $R = -\dot{m}u$ is internal to the system and is not disclosed as an external force on the system. With constant total mass, the momentum principle $\Sigma F = \dot{G}$ is applicable and we have

$$\Sigma F = \frac{d}{dt}(mv + m_0v_0) = m\dot{v} + \dot{m}v + \dot{m}_0v_0 + m_0\dot{v}_0$$

Clearly, $\dot{m}_0 = -\dot{m}$, and the velocity of the ejected mass with respect to m is $u = v - v_0$. Also $\dot{v}_0 = 0$ since m_0 moves undisturbed with no acceleration once free of m. Thus, the relation becomes

$$\Sigma F = m\dot{v} + \dot{m}u$$

which is identical to the result of the previous formulation, Eq. 4/20.

Application to Rocket Propulsion

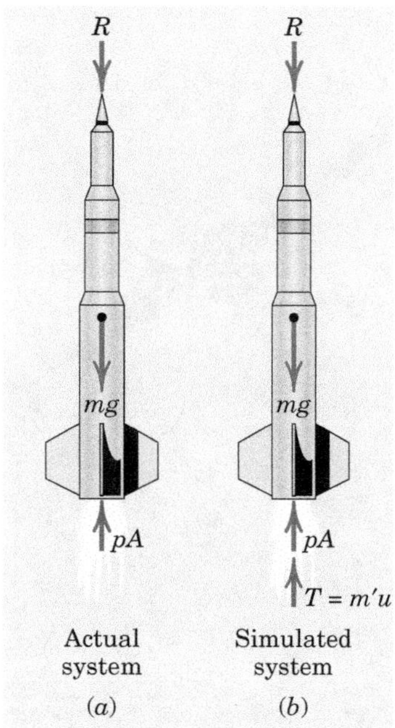

Figure 4/7

The case of m losing mass is clearly descriptive of rocket propulsion. Figure 4/7*a* shows a vertically ascending rocket, the system for which is the mass within the volume defined by the exterior surface of the rocket and the exit plane across the nozzle. External to this system, the free-body diagram discloses the instantaneous values of gravitational attraction mg, aerodynamic resistance R, and the force pA due to the average static pressure p across the nozzle exit plane of area A. The rate of mass flow is $m' = -\dot{m}$. Thus, we may write the equation of motion of the rocket, $\Sigma F = m\dot{v} + \dot{m}u$, as $pA - mg - R = m\dot{v} + \dot{m}u$, or

$$m'u + pA - mg - R = m\dot{v} \qquad \textbf{(4/21)}$$

Equation 4/21 is of the form "$\Sigma F = ma$" where the first term in "ΣF" is the thrust $T = m'u$. Thus, the rocket may be simulated as a body to which an external thrust T is applied, Fig. 4/7*b*, and the problem may then be analyzed like any other $F = ma$ problem, except that m is a function of time.

Observe that, during the initial stages of motion when the magnitude of the velocity v of the rocket is less than the relative exhaust velocity u, the absolute velocity v_0 of the exhaust gases will be directed rearward. On the other hand, when the rocket reaches a velocity v whose magnitude is greater than u, the absolute velocity v_0 of the exhaust gases will be directed forward. For a given mass rate of flow, the rocket thrust T depends only on the relative exhaust velocity u and not on the magnitude or on the direction of the absolute velocity v_0 of the exhaust gases.

In the foregoing treatment of bodies whose mass changes with time, we have assumed that all elements of the mass m of the body were moving with the same velocity v at any instant of time and that the particles of mass added to or expelled from the body underwent an abrupt transition of velocity upon entering or leaving the body. Thus, this velocity change has been modeled as a mathematical discontinuity. In reality, this change in velocity cannot be discontinuous even though the transition may be rapid. In the case of a rocket, for example, the velocity change occurs continuously in the space between the combustion zone and the exit plane of the exhaust nozzle. A more general analysis* of variable-mass dynamics removes this restriction of discontinuous velocity change and introduces a slight correction to Eq. 4/20.

*For a development of the equations which describe the general motion of a time-dependent system of mass, see Art. 53 of the first author's *Dynamics, 2nd Edition, SI Version*, 1975, John Wiley & Sons, Inc.

Sample Problem 4/10

The end of a chain of length L and mass ρ per unit length which is piled on a platform is lifted vertically with a constant velocity v by a variable force P. Find P as a function of the height x of the end above the platform. Also find the energy lost during the lifting of the chain.

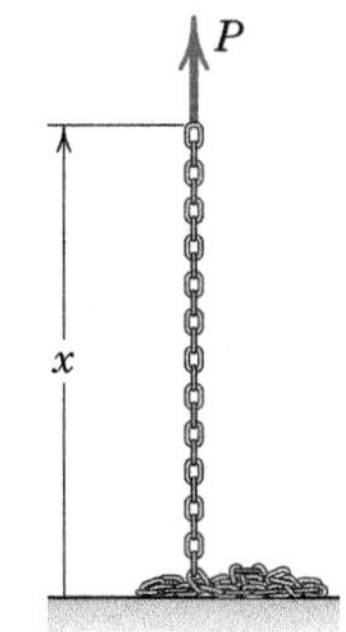

Solution I (Variable-Mass Approach). Equation 4/20 will be used and applied to the moving part of the chain of length x which is gaining mass. The force summation ΣF includes all forces acting on the moving part except the force exerted by the particles which are being attached. From the diagram we have

$$\Sigma F_x = P - \rho g x$$

The velocity is constant so that $\dot{v} = 0$. The rate of increase of mass is $\dot{m} = \rho v$, and the relative velocity with which the attaching particles approach the moving part is $u = v - 0 = v$. Thus, Eq. 4/20 becomes

① $[\Sigma F = m\dot{v} + \dot{m}u] \qquad P - \rho g x = 0 + \rho v(v) \qquad P = \rho(gx + v^2)$ *Ans.*

We now see that the force P consists of the two parts, $\rho g x$, which is the weight of the moving part of the chain, and ρv^2, which is the added force required to change the momentum of the links on the platform from a condition at rest to a velocity v.

Solution II (Constant-Mass Approach). The principle of impulse and momentum for a system of particles expressed by Eq. 4/6 will be applied to the entire chain considered as the system of constant mass. The free-body diagram of the system shows the unknown force P, the total weight of all links $\rho g L$, and the force $\rho g(L - x)$ exerted by the platform on those links which are at rest on it. The momentum of the system at any position is $G_x = \rho x v$ and the momentum equation gives

② $\left[\Sigma F_x = \dfrac{dG_x}{dt}\right] \qquad P + \rho g(L - x) - \rho g L = \dfrac{d}{dt}(\rho x v) \qquad P = \rho(gx + v^2)$ *Ans.*

Again the force P is seen to be equal to the weight of the portion of the chain which is off the platform plus the added term which accounts for the time rate of increase of momentum of the chain.

Energy Loss. Each link on the platform acquires its velocity abruptly through an impact with the link above it, which lifts it off the platform. The succession of impacts gives rise to an energy loss ΔE (negative work $-\Delta E$) so that the work-

③ energy equation becomes $U'_{1\text{-}2} = \int P\,dx - \Delta E = \Delta T + \Delta V_g$, where

$$\int P\,dx = \int_0^L (\rho g x + \rho v^2)dx = \tfrac{1}{2}\rho g L^2 + \rho v^2 L$$

$$\Delta T = \tfrac{1}{2}\rho L v^2 \qquad \Delta V_g = \rho g L \frac{L}{2} = \tfrac{1}{2}\rho g L^2$$

Substituting into the work-energy equation gives

$$\tfrac{1}{2}\rho g L^2 + \rho v^2 L - \Delta E = \tfrac{1}{2}\rho L v^2 + \tfrac{1}{2}\rho g L^2 \qquad \Delta E = \tfrac{1}{2}\rho L v^2 \qquad \textit{Ans.}$$

Helpful Hints

① The model of Fig. 4/6*a* shows the mass being added to the leading end of the moving part. With the chain the mass is added to the trailing end, but the effect is the same.

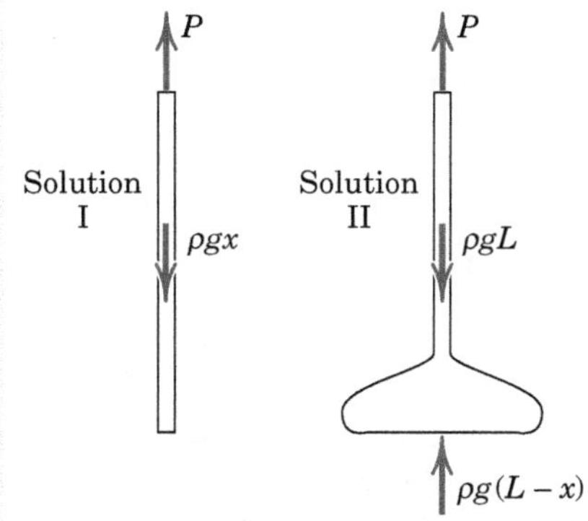

② We must be very careful not to use $\Sigma F = \dot{G}$ for a system whose mass is changing. Thus, we have taken the total chain as the system since its mass is constant.

③ Note that $U'_{1\text{-}2}$ includes work done by internal nonelastic forces, such as the link-to-link impact forces, where this work is converted into heat and acoustical energy loss ΔE.

Sample Problem 4/11

Replace the open-link chain of Sample Problem 4/10 by a flexible but inextensible rope or bicycle-type chain of length L and mass ρ per unit length. Determine the force P required to elevate the end of the rope with a constant velocity v and determine the corresponding reaction R between the coil and the platform.

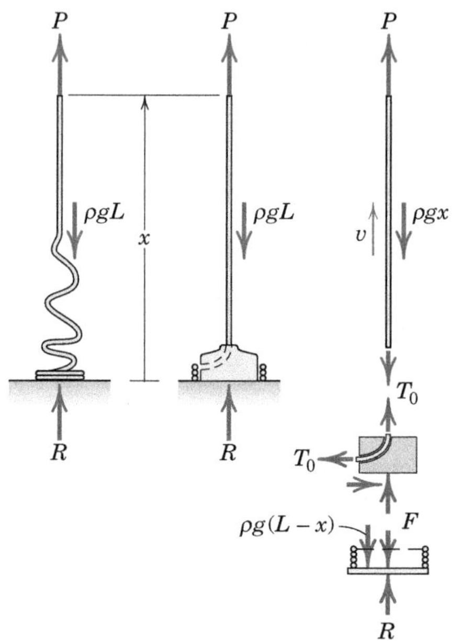

Solution. The free-body diagram of the coil and moving portion of the rope is shown in the left-hand figure. Because of some resistance to bending and some ① lateral motion, the transition from rest to vertical velocity v will occur over an appreciable segment of the rope. Nevertheless, assume first that all moving elements have the same velocity so that Eq. 4/6 for the system gives

② $$\left[\Sigma F_x = \frac{dG_x}{dt}\right] \qquad P + R - \rho g L = \frac{d}{dt}(\rho x v) \qquad P + R = \rho v^2 + \rho g L$$

We assume further that all elements of the coil of rope are at rest on the platform and transmit no force to the platform other than their weight, so that $R = \rho g(L - x)$. Substitution into the foregoing relation gives

$$P + \rho g(L - x) = \rho v^2 + \rho g L \qquad \text{or} \qquad P = \rho v^2 + \rho g x$$

which is the same result as that for the chain in Sample Problem 4/10.

The total work done on the rope by P becomes

$$U'_{1\text{-}2} = \int P\,dx = \int_0^x (\rho v^2 + \rho g x)\,dx = \rho v^2 x + \tfrac{1}{2}\rho g x^2$$

Substitution into the work-energy equation gives

$$[U'_{1\text{-}2} = \Delta T + \Delta V_g] \qquad \rho v^2 x + \tfrac{1}{2}\rho g x^2 = \Delta T + \rho g x \frac{x}{2} \qquad \Delta T = \rho x v^2$$

③ which is twice the kinetic energy $\frac{1}{2}\rho x v^2$ of vertical motion. Thus, an equal amount of kinetic energy is unaccounted for. This conclusion largely negates our assumption of one-dimensional x-motion.

In order to produce a one-dimensional model which retains the inextensibility property assigned to the rope, it is necessary to impose a physical constraint at the ④ base to guide the rope into vertical motion and at the same time preserve a smooth transition from rest to upward velocity v without energy loss. Such a guide is included in the free-body diagram of the entire rope in the middle figure and is represented schematically in the middle free-body diagram of the right-hand figure.

For a conservative system, the work-energy equation gives

⑤ $$[dU' = dT + dV_g] \qquad P\,dx = d(\tfrac{1}{2}\rho x v^2) + d\left(\rho g x \frac{x}{2}\right)$$

$$P = \tfrac{1}{2}\rho v^2 + \rho g x$$

Substitution into the impulse-momentum equation $\Sigma F_x = \dot{G}_x$ gives

$$\tfrac{1}{2}\rho v^2 + \rho g x + R - \rho g L = \rho v^2 \qquad R = \tfrac{1}{2}\rho v^2 + \rho g(L - x)$$

Although this force, which exceeds the weight by $\frac{1}{2}\rho v^2$, is unrealistic experimentally, it would be present in the idealized model.

Equilibrium of the vertical section requires

$$T_0 = P - \rho g x = \tfrac{1}{2}\rho v^2 + \rho g x - \rho g x = \tfrac{1}{2}\rho v^2$$

Because it requires a force of ρv^2 to change the momentum of the rope elements, the restraining guide must supply the balance $F = \frac{1}{2}\rho v^2$ which, in turn, is transmitted to the platform.

Helpful Hints

① Perfect flexibility would not permit any resistance to bending.

② Remember that v is constant and equals $\dot{x}$. Also note that this same relation applies to the chain of Sample Problem 4/10.

③ This added term of unaccounted-for kinetic energy exactly equals the energy lost by the chain during the impact of its links.

④ This restraining guide may be visualized as a canister of negligible mass rotating within the coil with an angular velocity v/r and connected to the platform through its shaft. As it turns, it feeds the rope from a rest position to an upward velocity v, as indicated in the accompanying figure.

⑤ Note that the mass center of the section of length x is a distance $x/2$ above the base.

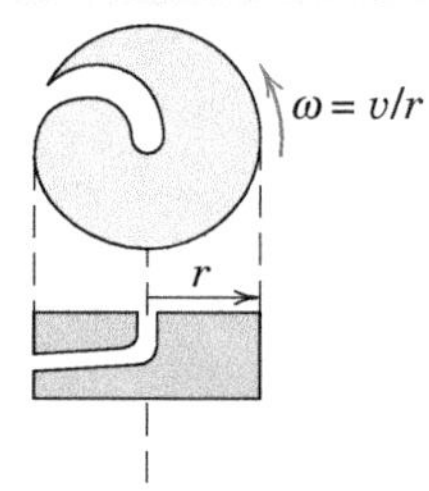

Sample Problem 4/12

A rocket of initial total mass m_0 is fired vertically up from the north pole and accelerates until the fuel, which burns at a constant rate, is exhausted. The relative nozzle velocity of the exhaust gas has a constant value u, and the nozzle exhausts at atmospheric pressure throughout the flight. If the residual mass of the rocket structure and machinery is m_b when burnout occurs, determine the expression for the maximum velocity reached by the rocket. Neglect atmospheric resistance and the variation of gravity with altitude.

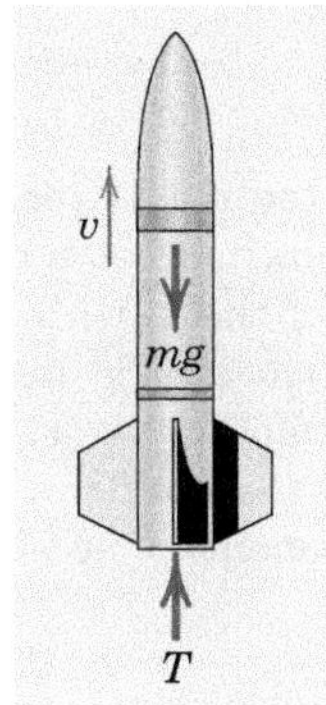

Solution I (F = ma Solution). We adopt the approach illustrated with Fig. 4/7*b* and treat the thrust as an external force on the rocket. With the neglect of the back pressure p across the nozzle and the atmospheric resistance R, Eq. 4/21 or Newton's second law gives ①

$$T - mg = m\dot{v}$$

But the thrust is $T = m'u = -\dot{m}u$ so that the equation of motion becomes

$$-\dot{m}u - mg = m\dot{v}$$

Multiplication by dt, division by m, and rearrangement give

$$dv = -u\frac{dm}{m} - g\,dt$$

which is now in a form which can be integrated. The velocity v corresponding to the time t is given by the integration

$$\int_0^v dv = -u\int_{m_0}^{m}\frac{dm}{m} - g\int_0^t dt$$

or

$$v = u\ln\frac{m_0}{m} - gt$$

Since the fuel is burned at the constant rate $m' = -\dot{m}$, the mass at any time t is $m = m_0 + \dot{m}t$. If we let m_b stand for the mass of the rocket when burnout occurs, then the time at burnout becomes $t_b = (m_b - m_0)/\dot{m} = (m_0 - m_b)/(-\dot{m})$. ② This time gives the condition for maximum velocity, which is

$$v_{max} = u\ln\frac{m_0}{m_b} + \frac{g}{\dot{m}}(m_0 - m_b) \qquad \textit{Ans.}$$

The quantity $\dot{m}$ is a negative number since the mass decreases with time.

Solution II (Variable-Mass Solution). If we use Eq. 4/20, then $\Sigma F = -mg$ and the equation becomes

$$[\Sigma F = m\dot{v} + \dot{m}u] \qquad -mg = m\dot{v} + \dot{m}u$$

But $\dot{m}u = -m'u = -T$ so that the equation of motion becomes

$$T - mg = m\dot{v}$$

which is the same as formulated with *Solution I*.

Helpful Hints

① The neglect of atmospheric resistance is not a bad assumption for a first approximation inasmuch as the velocity of the ascending rocket is smallest in the dense part of the atmosphere and greatest in the rarefied region. Also for an altitude of 320 km, the acceleration due to gravity is 91 percent of the value at the surface of the earth.

② Vertical launch from the north pole is taken only to eliminate any complication due to the earth's rotation in figuring the absolute trajectory of the rocket.

PROBLEMS

Introductory Problems

4/69 When the rocket reaches the position in its trajectory shown, it has a mass of 3 Mg and is beyond the effect of the earth's atmosphere. Gravitational acceleration is 9.60 m/s^2. Fuel is being consumed at the rate of 130 kg/s, and the exhaust velocity relative to the nozzle is 600 m/s. Compute the n- and t-components of acceleration of the rocket.

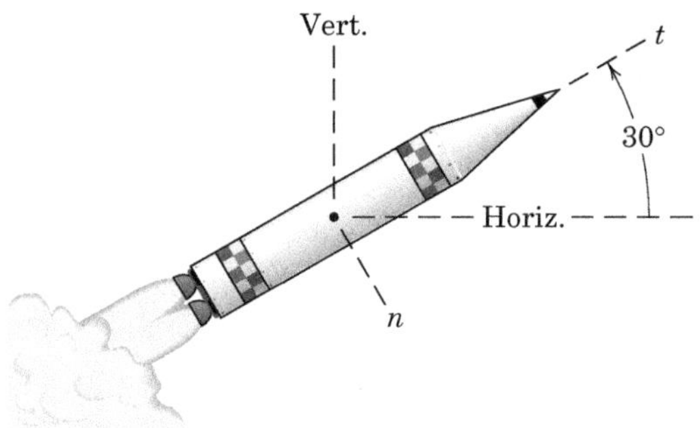

Problem 4/69

4/70 At the instant of vertical launch the rocket expels exhaust at the rate of 220 kg/s with an exhaust velocity of 820 m/s. If the initial vertical acceleration is 6.80 m/s^2, calculate the total mass of the rocket and fuel at launch.

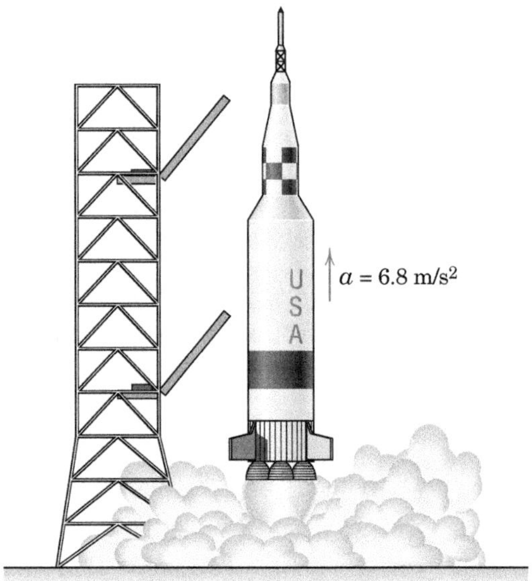

Problem 4/70

4/71 The space shuttle, together with its central fuel tank and two booster rockets, has a total mass of 2.04(10^6) kg at liftoff. Each of the two booster rockets produces a thrust of 11.80(10^6) N, and each of the three main engines of the shuttle produces a thrust of 2.00(10^6) N. The specific impulse (ratio of exhaust velocity to gravitational acceleration) for each of the three main engines of the shuttle is 455 s. Calculate the initial vertical acceleration a of the assembly with all five engines operating and find the rate at which fuel is being consumed by each of the shuttle's three engines.

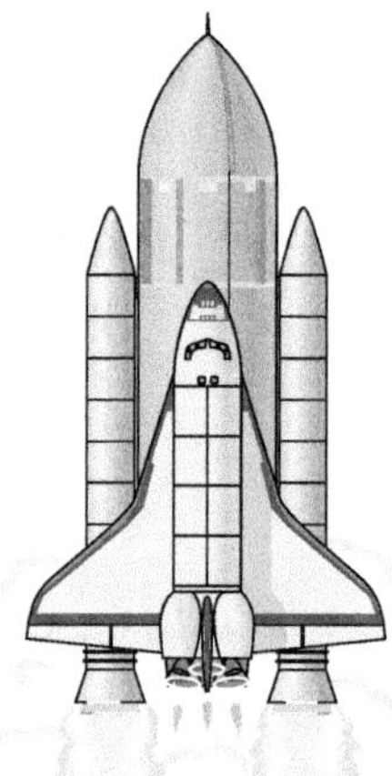

Problem 4/71

4/72 A small rocket of initial mass m_0 is fired vertically upward near the surface of the earth (g constant). If air resistance is neglected, determine the manner in which the mass m of the rocket must vary as a function of the time t after launching in order that the rocket may have a constant vertical acceleration a, with a constant relative velocity u of the escaping gases with respect to the nozzle.

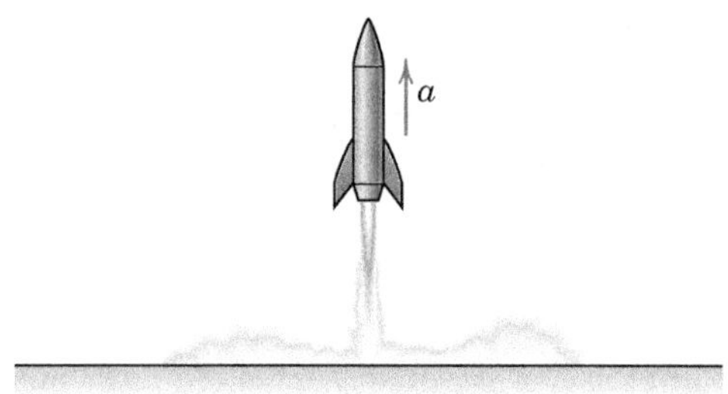

Problem 4/72

4/73 The mass m of a raindrop increases as it picks up moisture during its vertical descent through still air. If the air resistance to motion of the drop is R and its downward velocity is v, write the equation of motion for the drop and show that the relation $\Sigma F = d(mv)/dt$ is obeyed as a special case of the variable-mass equation.

4/74 A tank truck for washing down streets has a total weight of 20,000 lb when its tank is full. With the spray turned on, 80 lb of water per second issue from the nozzle with a velocity of 60 ft/sec relative to the truck at the 30° angle shown. If the truck is to accelerate at the rate of 2 ft/sec^2 when starting on a level road, determine the required tractive force P between the tires and the road when (*a*) the spray is turned on and (*b*) the spray is turned off.

Problem 4/74

4/75 A model rocket weighs 1.5 lb just before its vertical launch. Its experimental solid-fuel motor carries 0.1 lb of fuel, has an escape velocity of 3000 ft/sec, and burns the fuel for 0.9 sec. Determine the acceleration of the rocket at launch and its burnout velocity. Neglect aerodynamic drag and state any other assumptions.

Problem 4/75

4/76 The magnetometer boom for a spacecraft consists of a large number of triangular-shaped units which spring into their deployed configuration upon release from the canister in which they were folded and packed prior to release. Write an expression for the force F which the base of the canister must exert on the boom during its deployment in terms of the increasing length x and its time derivatives. The mass of the boom per unit of deployed length is ρ. Treat the supporting base on the spacecraft as a fixed platform and assume that the deployment takes place outside of any gravitational field. Neglect the dimension b compared with x.

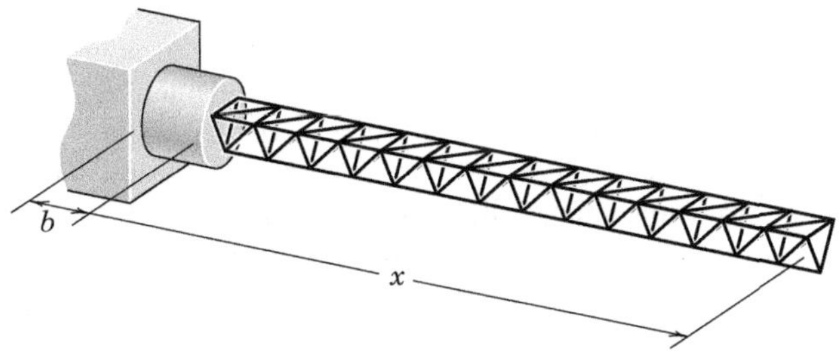

Problem 4/76

4/77 Fresh water issues from the two 30-mm-diameter holes in the bucket with a velocity of 2.5 m/s in the directions shown. Calculate the force P required to give the bucket an upward acceleration of 0.5 m/s^2 from rest if it contains 20 kg of water at that time. The empty bucket has a mass of 0.6 kg.

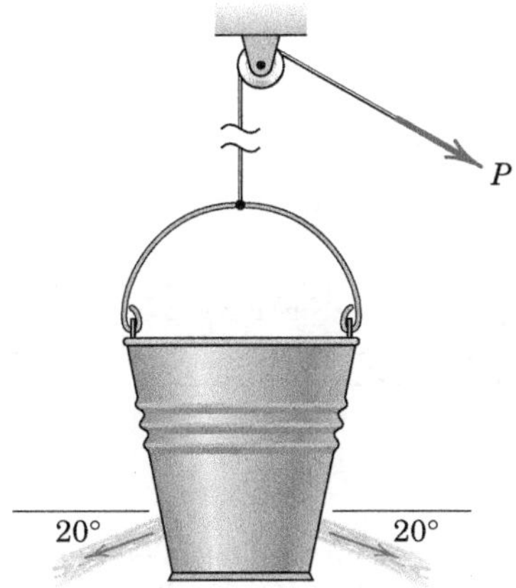

Problem 4/77

Representative Problems

4/78 The upper end of the open-link chain of length L and mass ρ per unit length is lowered at a constant speed v by the force P. Determine the reading R of the platform scale in terms of x.

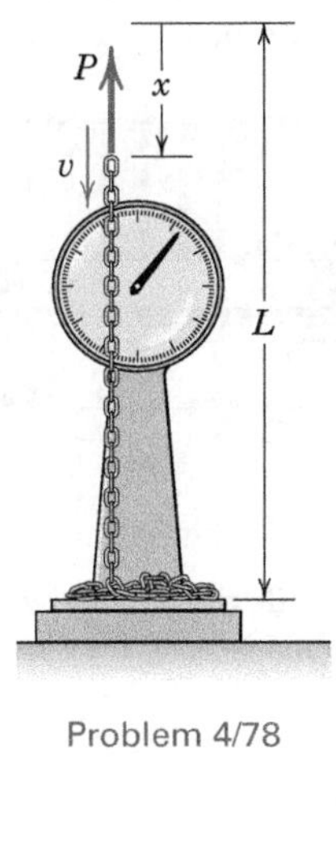

Problem 4/78

4/79 A rocket stage designed for deep-space missions consists of 200 kg of fuel and 300 kg of structure and payload combined. In terms of burnout velocity, what would be the advantage of reducing the structural/payload mass by 1 percent (3 kg) and using that mass for additional fuel? Express your answer in terms of a percent increase in burnout velocity. Repeat your calculation for a 5 percent reduction in the structural/payload mass.

4/80 At a bulk loading station, gravel leaves the hopper at the rate of 220 lb/sec with a velocity of 10 ft/sec in the direction shown and is deposited on the moving flatbed truck. The tractive force between the driving wheels and the road is 380 lb, which overcomes the 200 lb of frictional road resistance. Determine the acceleration a of the truck 4 seconds after the hopper is opened over the truck bed, at which instant the truck has a forward speed of 1.5 mi/hr. The empty weight of the truck is 12,000 lb.

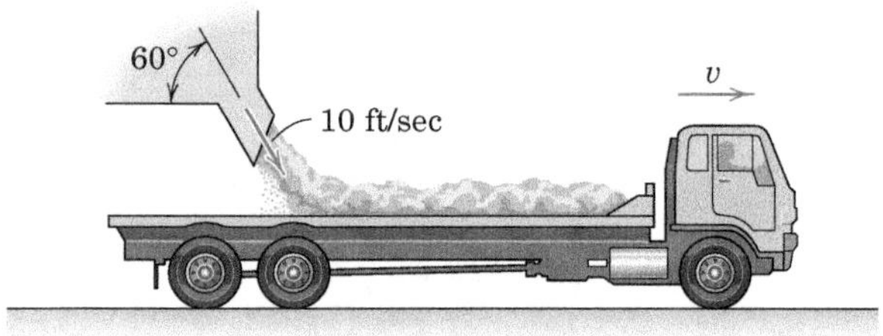

Problem 4/80

4/81 A railroad coal car weighs 54,600 lb empty and carries a total load of 180,000 lb of coal. The bins are equipped with bottom doors which permit discharging coal through an opening between the rails. If the car dumps coal at the rate of 20,000 lb/sec in a downward direction relative to the car, and if frictional resistance to motion is 4 lb per ton of total remaining weight, determine the coupler force P required to give the car an acceleration of 0.15 ft/sec^2 in the direction of P at the instant when half the coal has been dumped.

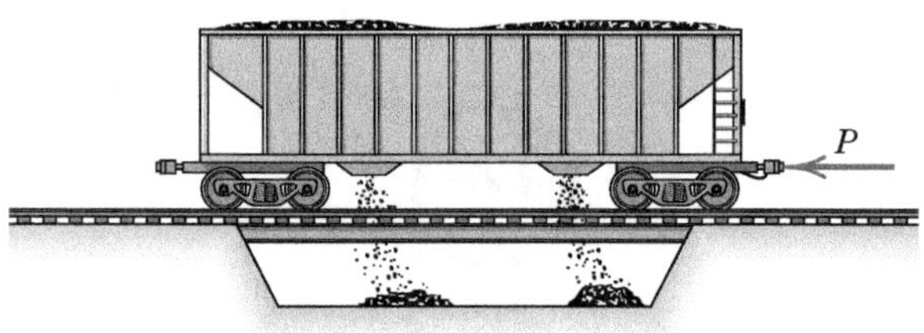

Problem 4/81

4/82 A coil of heavy flexible cable with a total length of 100 m and a mass of 1.2 kg/m is to be laid along a straight horizontal line. The end is secured to a post at A, and the cable peels off the coil and emerges through the horizontal opening in the cart as shown. The cart and drum together have a mass of 40 kg. If the cart is moving to the right with a velocity of 2 m/s when 30 m of cable remain in the drum and the tension in the rope at the post is 2.4 N, determine the force P required to give the cart and drum an acceleration of 0.3 m/s^2. Neglect all friction.

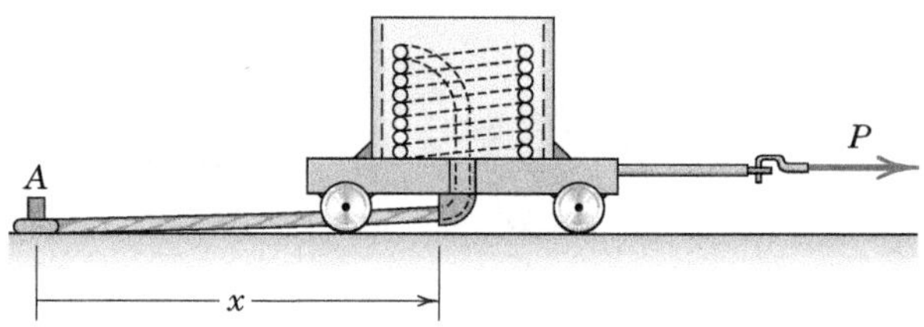

Problem 4/82

4/83 By lowering a scoop as it skims the surface of a body of water, the aircraft (nicknamed the "Super Scooper") is able to ingest 4.5 m^3 of fresh water during a 12-second run. The plane then flies to a fire area and makes a massive water drop with the ability to repeat the procedure as many times as necessary. The plane approaches its run with a velocity of 280 km/h and an initial mass of 16.4 Mg. As the scoop enters the water, the pilot advances the throttle to provide an additional 300 hp (223.8 kW) needed to prevent undue deceleration. Determine the initial deceleration when the scooping action starts. (Neglect the difference between the average and the initial rates of water intake.)

Problem 4/83

4/84 A small rocket-propelled vehicle weighs 125 lb, including 20 lb of fuel. Fuel is burned at the constant rate of 2 lb/sec with an exhaust velocity relative to the nozzle of 400 ft/sec. Upon ignition the vehicle is released from rest on the 10° incline. Calculate the maximum velocity v reached by the vehicle. Neglect all friction.

Problem 4/84

4/85 The end of a pile of loose-link chain of mass ρ per unit length is being pulled horizontally along the surface by a constant force P. If the coefficient of kinetic friction between the chain and the surface is μ_k, determine the acceleration a of the chain in terms of x and $\dot{x}$.

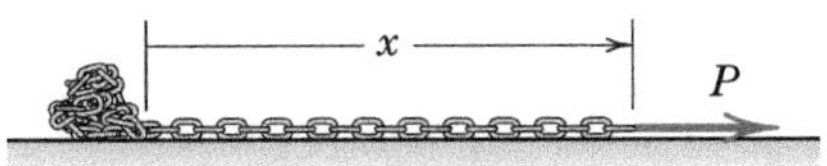

Problem 4/85

4/86 A coal car with an empty mass of 25 Mg is moving freely with a speed of 1.2 m/s under a hopper which opens and releases coal into the moving car at the constant rate of 4 Mg per second. Determine the distance x moved by the car during the time that 32 Mg of coal are deposited in the car. Neglect any frictional resistance to rolling along the horizontal track.

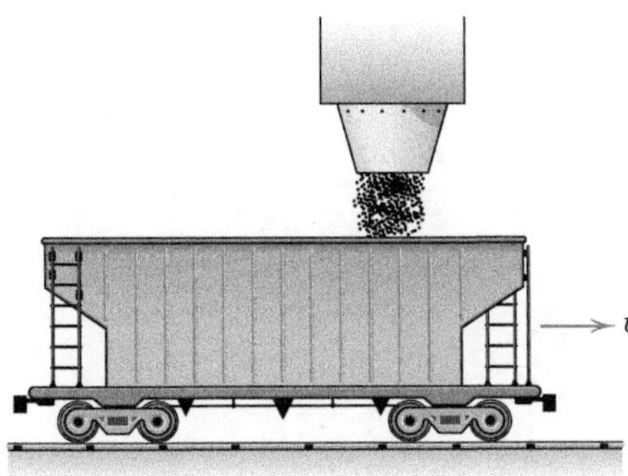

Problem 4/86

4/87 Sand is released from the hopper H with negligible velocity and then falls a distance h to the conveyor belt. The mass flow rate from the hopper is m'. Develop an expression for the steady-state belt speed v for the case $h = 0$. Assume that the sand quickly acquires the belt velocity with no rebound, and neglect friction at the pulleys A and B.

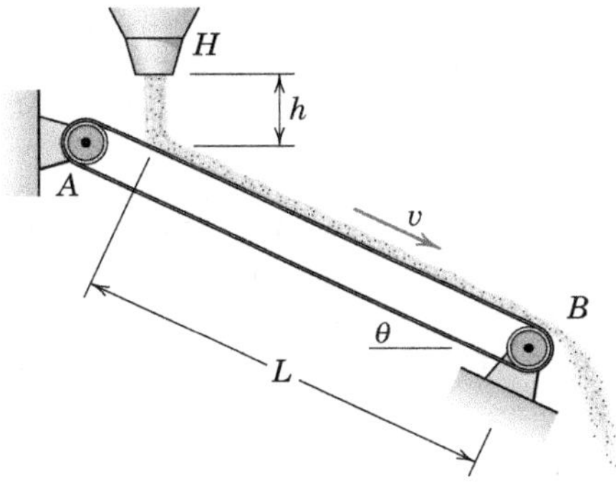

Problem 4/87

4/88 Repeat the previous problem, but now let $h \neq 0$. Then evaluate your expression for the conditions $h = 2$ m, $L = 10$ m, and $\theta = 25°$.

4/89 The open-link chain of length L and mass ρ per unit length is released from rest in the position shown, where the bottom link is almost touching the platform and the horizontal section is supported on a smooth surface. Friction at the corner guide is negligible. Determine (*a*) the velocity v_1 of end A as it reaches the corner and (*b*) its velocity v_2 as it strikes the platform. (*c*) Also specify the total loss Q of energy.

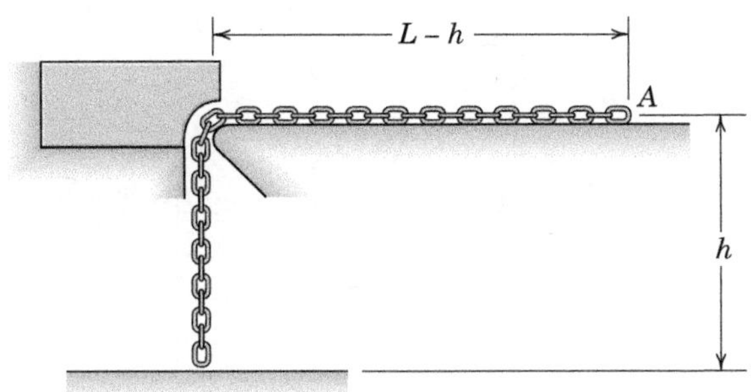

Problem 4/89

4/90 In the figure is shown a system used to arrest the motion of an airplane landing on a field of restricted length. The plane of mass m rolling freely with a velocity v_0 engages a hook which pulls the ends of two heavy chains, each of length L and mass ρ per unit length, in the manner shown. A conservative calculation of the effectiveness of the device neglects the retardation of chain friction on the ground and any other resistance to the motion of the airplane. With these assumptions, compute the velocity v of the airplane at the instant when the last link of each chain is put in motion. Also determine the relation between the displacement x and the time t after contact with the chain. Assume each link of the chain acquires its velocity v suddenly upon contact with the moving links.

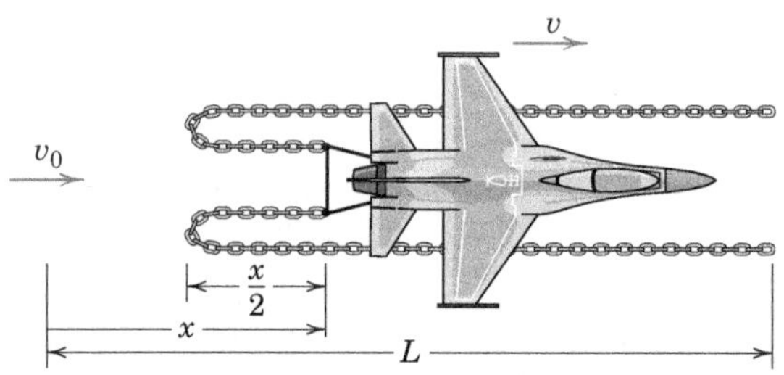

Problem 4/90

▶4/91 The free end of the flexible and inextensible rope of mass ρ per unit length and total length L is given a constant upward velocity v. Write expressions for P, the force R supporting the fixed end, and the tension T_1 in the rope at the loop in terms of x. (For the loop of negligible size, the tension is the same on both sides.)

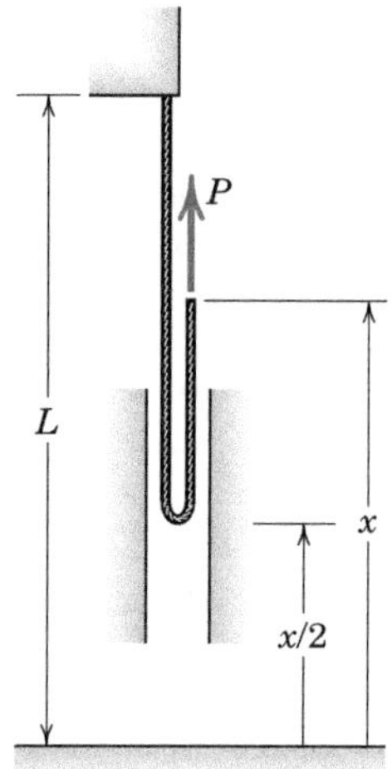

Problem 4/91

▶4/92 Replace the rope of Prob. 4/91 by an open-link chain with the same mass ρ per unit length. The free end is given a constant upward velocity v. Write expressions for P, the tension T_1 at the bottom of the moving part, and the force R supporting the fixed end in terms of x. Also find the energy loss Q in terms of x.

4/8 Chapter Review

In this chapter we have extended the principles of dynamics for the motion of a single mass particle to the motion of a general system of particles. Such a system can form a rigid body, a nonrigid (elastic) solid body, or a group of separate and unconnected particles, such as those in a defined mass of liquid or gaseous particles. The following summarizes the principal results of Chapter 4.

1. We derived the generalized form of Newton's second law, which is expressed as the *principle of motion of the mass center*, Eq. 4/1 in Art. 4/2. This principle states that the vector sum of the external forces acting on any system of mass particles equals the total system mass times the acceleration of the center of mass.
2. In Art. 4/3, we established a *work-energy principle* for a system of particles, Eq. 4/3a, and showed that the total kinetic energy of the system equals the energy of the mass-center translation plus the energy due to motion of the particles relative to the mass center.
3. The resultant of the external forces acting on any system equals the time rate of change of the linear momentum of the system, Eq. 4/6 in Art. 4/4.
4. For a fixed point O and the mass center G, the resultant vector moment of all external forces about the point equals the time rate of change of angular momentum about the point, Eq. 4/7 and Eq. 4/9 in Art. 4/4. The principle for an arbitrary point P, Eqs. 4/11 and 4/13, has an additional term and thus does not follow the form of the equations for O and G.
5. In Art. 4/5 we developed the *law of conservation of dynamical energy*, which applies to a system in which the internal kinetic friction is negligible.
6. *Conservation of linear momentum* applies to a system in the absence of an external linear impulse. Similarly, *conservation of angular momentum* applies when there is no external angular impulse.
7. For applications involving steady mass flow, we developed a relation, Eq. 4/18 in Art. 4/6, between the resultant force on a system, the corresponding mass flow rate, and the change in fluid velocity from entrance to exit.
8. Analysis of angular momentum in steady mass flow resulted in Eq. 4/19a in Art. 4/6, which is a relation between the resultant moment of all external forces about a fixed point O on or off the system, the mass flow rate, and the incoming and outgoing velocities.
9. Finally, in Art. 4/7 we developed the equation of linear motion for variable-mass systems, Eq. 4/20. Common examples of such systems are rockets and flexible chains and ropes.

The principles developed in this chapter enable us to treat the motion of both rigid and nonrigid bodies in a unified manner. In addition, the developments in Arts. 4/2–4/5 will serve to place on a rigorous basis the treatment of rigid-body kinetics in Chapters 6 and 7.

REVIEW PROBLEMS

4/93 The system of three particles has the indicated particle masses, velocities, and external forces. Determine $\bar{\mathbf{r}}$, $\dot{\bar{\mathbf{r}}}$, $\ddot{\bar{\mathbf{r}}}$, T, $\mathbf{H}_O$, and $\dot{\mathbf{H}}_O$ for this three-dimensional system.

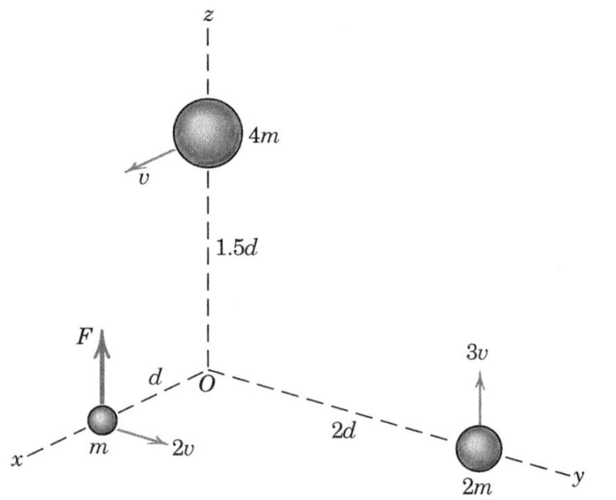

Problem 4/93

4/94 For the particle system of Prob. 4/93, determine $\mathbf{H}_G$ and $\dot{\mathbf{H}}_G$.

4/95 Each of the identical steel balls weighs 4 lb and is fastened to the other two by connecting bars of negligible weight and unequal length. In the absence of friction at the supporting horizontal surface, determine the initial acceleration $\bar{a}$ of the mass center of the assembly when it is subjected to the horizontal force $F = 20$ lb applied to the supporting ball. The assembly is initially at rest in the vertical plane. Can you show that $\bar{a}$ is initially horizontal?

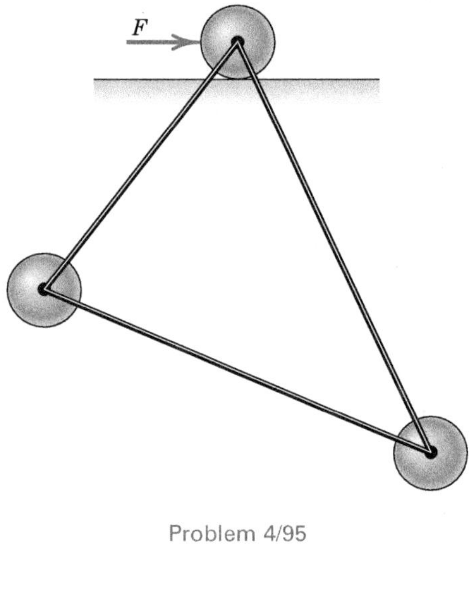

Problem 4/95

4/96 A 2-oz bullet is fired horizontally with a velocity $v = 1000$ ft/sec into the slender bar of a 3-lb pendulum initially at rest. If the bullet embeds itself in the bar, compute the resulting angular velocity of the pendulum immediately after the impact. Treat the sphere as a particle and neglect the mass of the bar. Why is the linear momentum of the system not conserved?

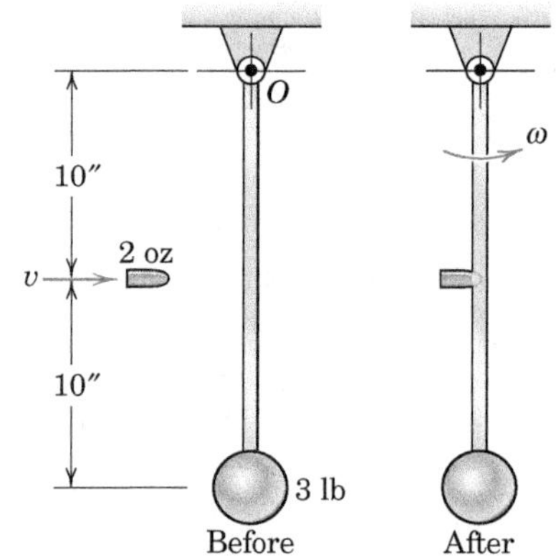

Problem 4/96

4/97 A large rocket ready for vertical launch has a total mass of $2.7(10^3)$ Mg. At launch, fuel is burned at the rate of 13 Mg/s with an exhaust velocity of 2400 m/s. Determine the initial acceleration a of the rocket. Assume atmospheric pressure across the exit plane of the exhaust nozzle.

4/98 In an operational design test of the equipment of the fire truck, the water cannon is delivering fresh water through its 2-in.-diameter nozzle at the rate of 1400 gal/min at the 20° angle. Calculate the total friction force F exerted by the pavement on the tires of the truck, which remains in a fixed position with its brakes locked. (There are 231 in.^{3} in 1 gal.)

Problem 4/98

4/99 The rocket shown is designed to test the operation of a new guidance system. When it has reached a certain altitude beyond the effective influence of the earth's atmosphere, its mass has decreased to 2.80 Mg, and its trajectory is 30° from the vertical. Rocket fuel is being consumed at the rate of 120 kg/s with an exhaust velocity of 640 m/s relative to the nozzle. Gravitational acceleration is 9.34 m/s^2 at its altitude. Calculate the n- and t-components of the acceleration of the rocket.

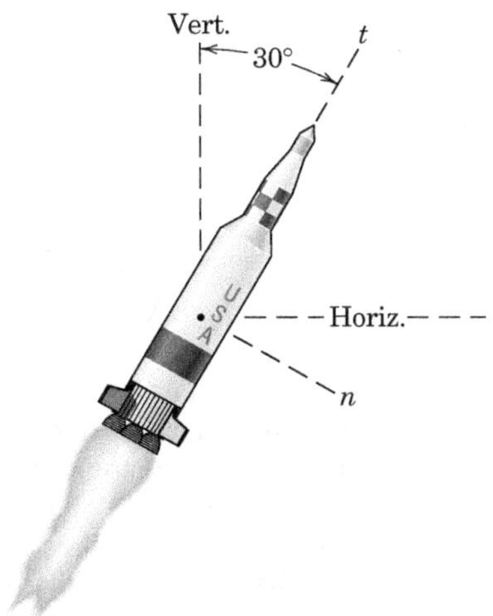

Problem 4/99

4/100 When only the air of a sand-blasting gun is turned on, the force of the air on a flat surface normal to the stream and close to the nozzle is 20 N. With the nozzle in the same position, the force increases to 30 N when sand is admitted to the stream. If sand is being consumed at the rate of 4.5 kg/min, calculate the velocity v of the sand particles as they strike the surface.

4/101 A two-stage rocket is fired vertically up and is above the atmosphere when the first stage burns out and the second stage separates and ignites. The second stage carries 1200 kg of fuel and has an empty mass of 200 kg. Upon ignition the second stage burns fuel at the rate of 5.2 kg/s and has a constant exhaust velocity of 3000 m/s relative to its nozzle. Determine the acceleration of the second stage 60 seconds after ignition and find the maximum acceleration and the time t after ignition at which it occurs. Neglect the variation of g and take it to be 8.70 m/s^2 for the range of altitude averaging about 400 km.

4/102 Water of density ρ issues from the nozzle of area A and impinges upon the block of mass m, which is at rest on the rough horizontal surface. If the block is on the verge of tipping rightward, determine the required minimum coefficient of friction μ between the block and the surface and the exit velocity v of the water. Water exits the nozzle at atmospheric pressure.

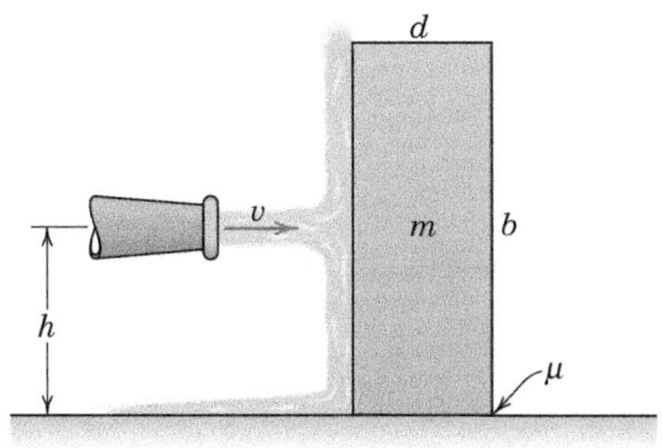

Problem 4/102

4/103 The block of Prob. 4/102 is now fitted with two small rollers. Determine the normal force under each roller and the initial rightward acceleration of the block if the exit velocity of the water is (*a*) 75 percent of the tipping velocity and (*b*) 50 percent of the tipping velocity. Refer to the printed answers for Prob. 4/102 as needed. (*Note:* Be sure to use Eq. 4/9 when summing moments.)

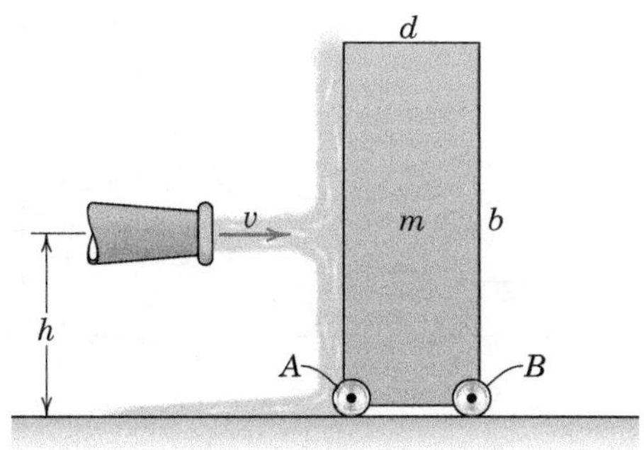

Problem 4/103

4/104 A jet of fresh water under pressure issues from the 3/4-in.-diameter fixed nozzle with a velocity v = 120 ft/sec and is diverted into the two equal streams. Neglect any energy loss in the streams and compute the force F required to hold the vane in place.

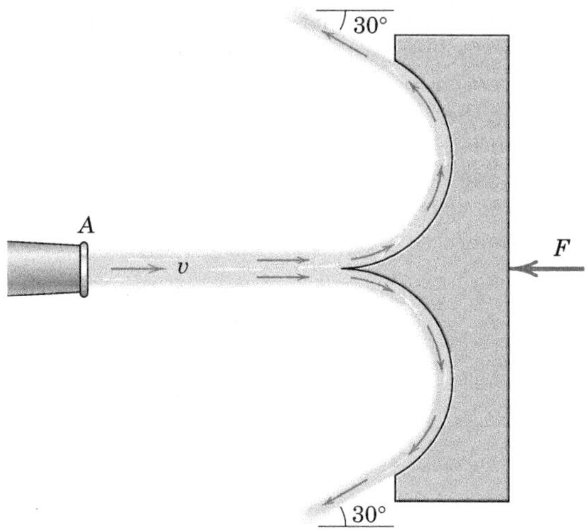

Problem 4/104

4/105 The flexible nonextensible rope of length $\pi r/2$ and mass ρ per unit length is attached at A to the fixed quarter-circular guide and allowed to fall from rest in the horizontal position. When the rope comes to rest in the dashed position, the system will have lost energy. Determine the loss ΔQ and explain what becomes of the lost energy.

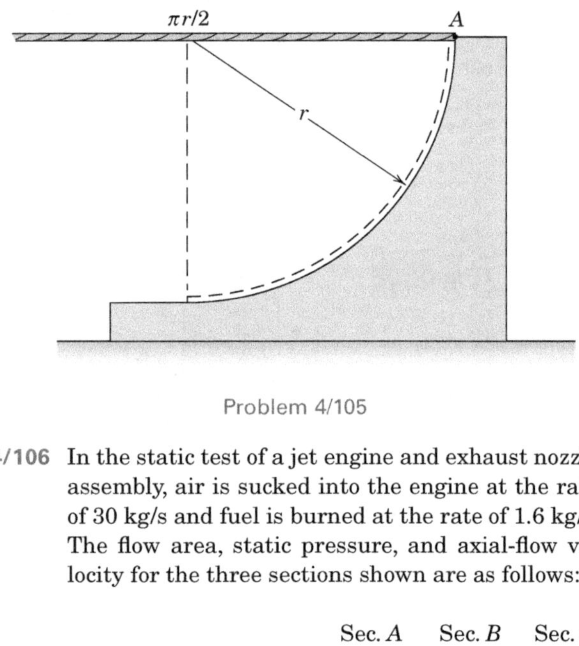

Problem 4/105

4/106 In the static test of a jet engine and exhaust nozzle assembly, air is sucked into the engine at the rate of 30 kg/s and fuel is burned at the rate of 1.6 kg/s. The flow area, static pressure, and axial-flow velocity for the three sections shown are as follows:

	Sec. A	Sec. B	Sec. C
Flow area, m^2	0.15	0.16	0.06
Static pressure, kPa	−14	140	14
Axial-flow velocity, m/s	120	315	600

Determine the tension T in the diagonal member of the supporting test stand and calculate the force F exerted on the nozzle flange at B by the bolts and gasket to hold the nozzle to the engine housing.

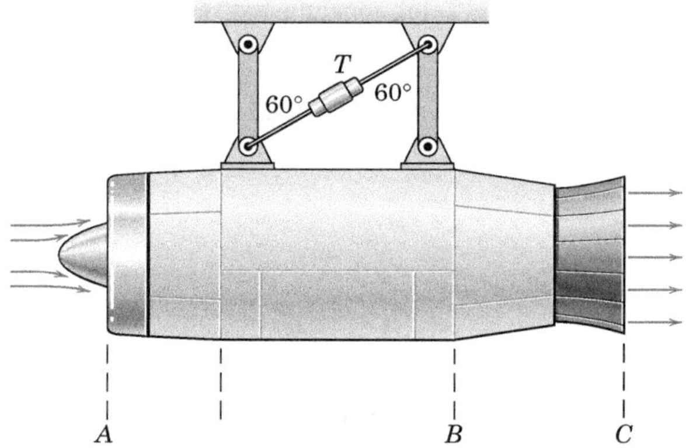

Problem 4/106

4/107 The upper end of the open-link chain of length L and mass ρ per unit length is released from rest with the lower end just touching the platform of the scale. Determine the expression for the force F read on the scale as a function of the distance x through which the upper end has fallen. (*Comment:* The chain acquires a free-fall velocity of $\sqrt{2gx}$ because the links on the scale exert no force on those above, which are still falling freely. Work the problem in two ways: first, by evaluating the time rate of change of momentum for the entire chain and second, by considering the force F to be composed of the weight of the links at rest on the scale plus the force necessary to divert an equivalent stream of fluid.)

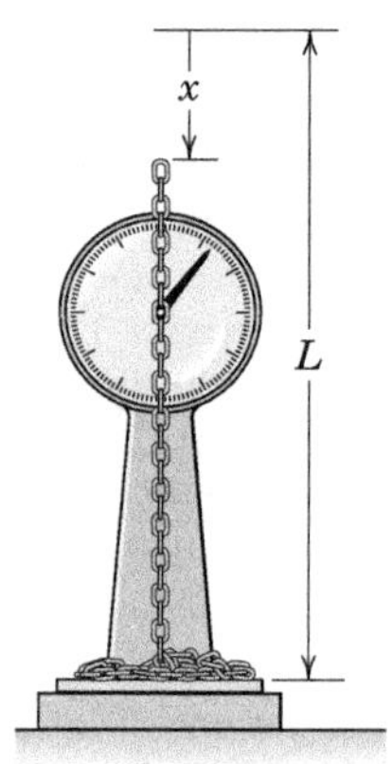

Problem 4/107

4/108 The chain of length L and mass ρ per unit length is released from rest on the smooth horizontal surface with a negligibly small overhang x to initiate motion. Determine (a) the acceleration a as a function of x, (b) the tension T in the chain at the smooth corner as a function of x, and (c) the velocity v of the last link A as it reaches the corner.

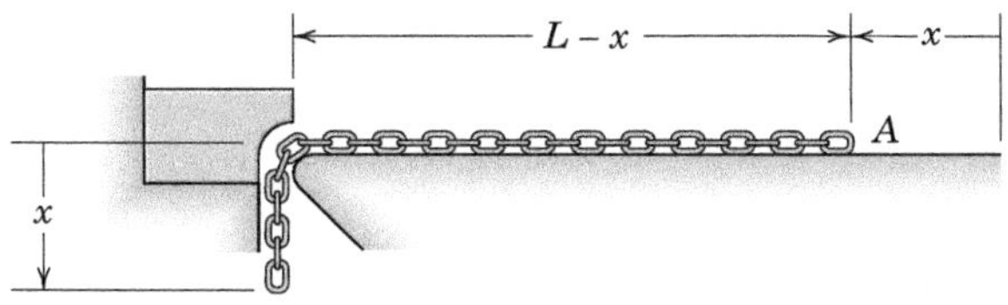

Problem 4/108

4/109 A vertical force P acting on the 2.5-lb cylindrical valve A, shown in section, serves to limit the flow of fresh water from the top of the vertical pipe B of 3-in. inside diameter. Water is fed through the bottom inlet of the pipe. Calculate the force P required to maintain the valve in the position shown under a flow rate of 600 gal/min and a static pressure of 12 lb/in.^{2} in the water at section C. Recall that 1 gal contains 231 in.^{3} (*Suggestion:* Choose the valve and the portion of water above section C as the combined free body.)

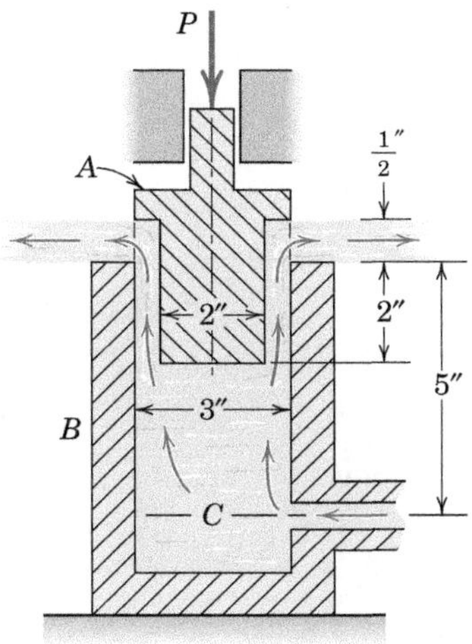

Problem 4/109

▶4/110 The chain of mass ρ per unit length passes over the small freely turning pulley and is released from rest with only a small imbalance h to initiate motion. Determine the acceleration a and velocity v of the chain and the force R supported by the hook at A, all in terms of h as it varies from essentially zero to H. Neglect the weight of the pulley and its supporting frame and the weight of the small amount of chain in contact with the pulley. (*Hint:* The force R does not equal two times the equal tensions T in the chain tangent to the pulley.)

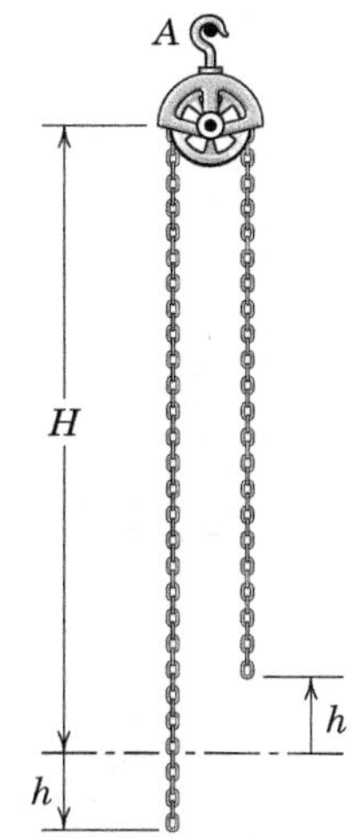

Problem 4/110

▶4/111 The centrifugal pump handles 20 m^3 of fresh water per minute with inlet and outlet velocities of 18 m/s. The impeller is turned clockwise through the shaft at O by a motor which delivers 40 kW at a pump speed of 900 rev/min. With the pump filled but not turning, the vertical reactions at C and D are each 250 N. Calculate the forces exerted by the foundation on the pump at C and D while the pump is running. The tensions in the connecting pipes at A and B are exactly balanced by the respective forces due to the static pressure in the water. (*Suggestion:* Isolate the entire pump and water within it between sections A and B and apply the momentum principle to the entire system.)

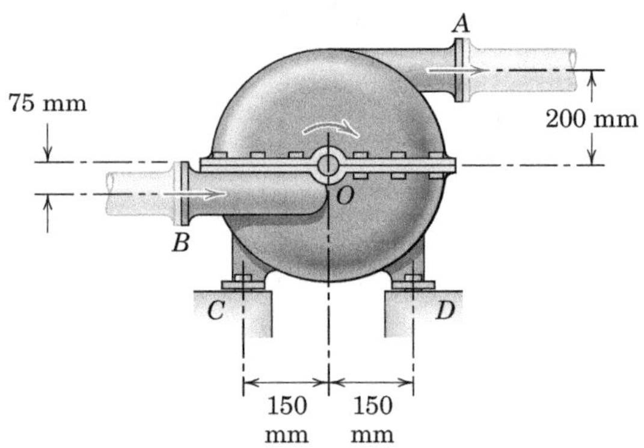

Problem 4/111

▶4/112 A rope or hinged-link bicycle-type chain of length L and mass ρ per unit length is released from rest with $x = 0$. Determine the expression for the total force R exerted on the fixed platform by the chain as a function of x. Note that the hinged-link chain is a conservative system during all but the last increment of motion.

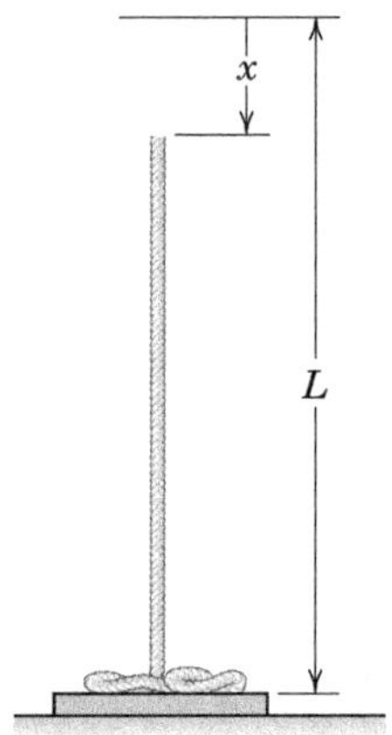

Problem 4/112

PART II

DYNAMICS OF RIGID BODIES

Rigid-body kinematics describes the relationships between the linear and angular motions of bodies without regard to the forces and moments associated with such motions. The designs of gears, cams, connecting links, and many other moving machine parts are largely kinematic problems.

5 Plane Kinematics of Rigid Bodies

CHAPTER OUTLINE

5/1 Introduction

In Chapter 2 on particle kinematics, we developed the relationships governing the displacement, velocity, and acceleration of points as they moved along straight or curved paths. In rigid-body kinematics we use these same relationships but must also account for the rotational motion of the body. Thus rigid-body kinematics involves both linear and angular displacements, velocities, and accelerations.

We need to describe the motion of rigid bodies for two important reasons. First, we frequently need to generate, transmit, or control certain motions by the use of cams, gears, and linkages of various types. Here we must analyze the displacement, velocity, and acceleration of the motion to determine the design geometry of the mechanical parts. Furthermore, as a result of the motion generated, forces may be developed which must be accounted for in the design of the parts.

Second, we must often determine the motion of a rigid body caused by the forces applied to it. Calculation of the motion of a rocket under the influence of its thrust and gravitational attraction is an example of such a problem.

We need to apply the principles of rigid-body kinematics in both situations. This chapter covers the kinematics of rigid-body motion which may be analyzed as occurring in a single plane. In Chapter 7 we will present an introduction to the kinematics of motion in three dimensions.

Rigid-Body Assumption

In the previous chapter we defined a *rigid body* as a system of particles for which the distances between the particles remain unchanged. Thus, if each particle of such a body is located by a position vector from reference axes attached to and rotating with the body, there will be no change in any position vector as measured from these axes. This is, of course, an ideal case since all solid materials change shape to some extent when forces are applied to them.

Nevertheless, if the movements associated with the changes in shape are very small compared with the movements of the body as a whole, then the assumption of rigidity is usually acceptable. The displacements due to the flutter of an aircraft wing, for instance, do not affect the description of the flight path of the aircraft as a whole, and thus the rigid-body assumption is clearly acceptable. On the other hand, if the problem is one of describing, as a function of time, the internal wing stress due to wing flutter, then the relative motions of portions of the wing cannot be neglected, and the wing may not be considered a rigid body. In this and the next two chapters, almost all of the material is based on the assumption of rigidity.

Plane Motion

A rigid body executes plane motion when all parts of the body move in parallel planes. For convenience, we generally consider the *plane of motion* to be the plane which contains the mass center, and we treat the body as a thin slab whose motion is confined to the plane of the slab. This idealization adequately describes a very large category of rigid-body motions encountered in engineering.

The plane motion of a rigid body may be divided into several categories, as represented in Fig. 5/1.

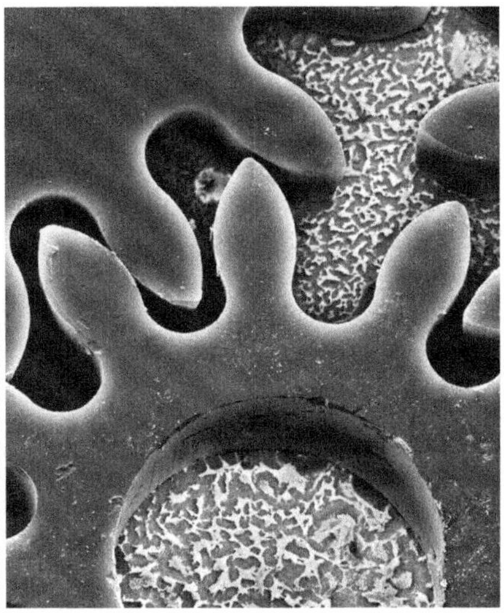

David Parker/Photo Researchers

These nickel microgears are only 150 microns ($150(10^{-6})$ m) thick and have potential application in microscopic robots.

Translation is defined as any motion in which every line in the body remains parallel to its original position at all times. In translation there is *no rotation of any line in the body*. In *rectilinear translation*, part a of Fig. 5/1, all points in the body move in parallel straight lines. In *curvilinear translation*, part b, all points move on congruent curves. We note that in each of the two cases of translation, the motion of the body is completely specified by the motion of any point in the body, since all points have the same motion. Thus, our earlier study of the motion of a point (particle) in Chapter 2 enables us to describe completely the translation of a rigid body.

Rotation about a fixed axis, part c of Fig. 5/1, is the angular motion about the axis. It follows that all particles in a rigid body move in circular paths about the axis of rotation, and all lines in the body which are perpendicular to the axis of rotation (including those which do not pass through the axis) rotate through the same angle in the same time. Again, our discussion in Chapter 2 on the circular motion of a point enables us to describe the motion of a rotating rigid body, which is treated in the next article.

General plane motion of a rigid body, part d of Fig. 5/1, is a combination of translation and rotation. We will utilize the principles of relative motion covered in Art. 2/8 to describe general plane motion.

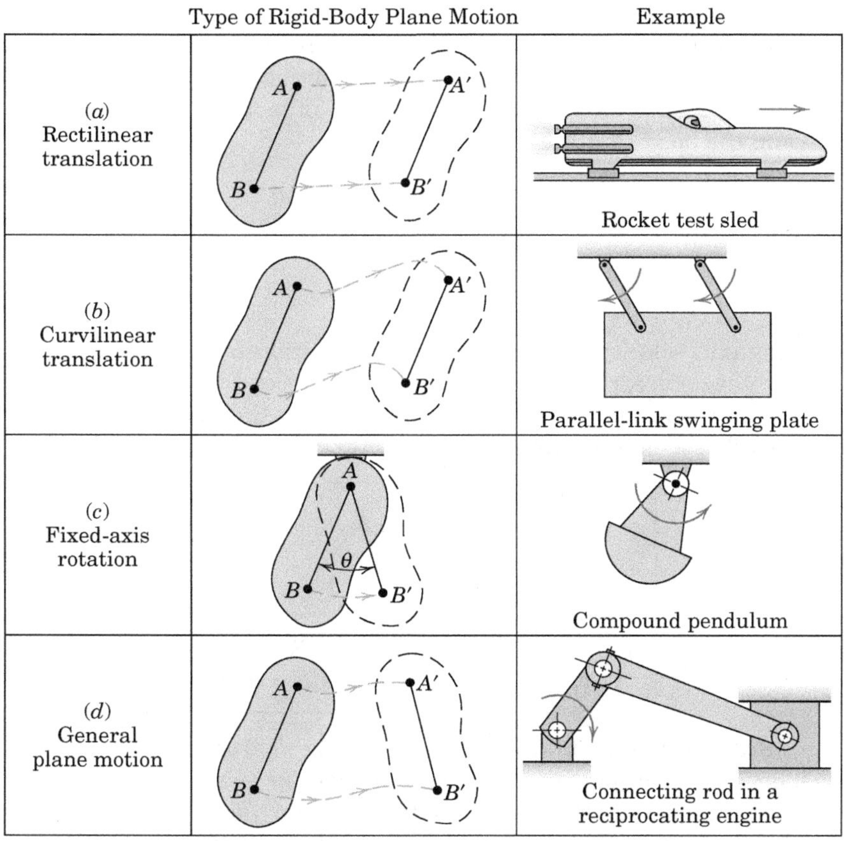

Figure 5/1

Note that in each of the examples cited, the actual paths of all particles in the body are projected onto the single plane of motion as represented in each figure.

Analysis of the plane motion of rigid bodies is accomplished either by directly calculating the absolute displacements and their time derivatives from the geometry involved or by utilizing the principles of relative motion. Each method is important and useful and will be covered in turn in the articles which follow.

5/2 Rotation

The rotation of a rigid body is described by its angular motion. Figure 5/2 shows a rigid body which is rotating as it undergoes plane motion in the plane of the figure. The angular positions of any two lines 1 and 2 attached to the body are specified by θ_1 and θ_2 measured from any convenient fixed reference direction. Because the angle β is invariant, the relation $\theta_2 = \theta_1 + \beta$ upon differentiation with respect to time gives $\dot{\theta}_2 = \dot{\theta}_1$ and $\ddot{\theta}_2 = \ddot{\theta}_1$ or, during a finite interval, $\Delta\theta_2 = \Delta\theta_1$. Thus, *all lines on a rigid body in its plane of motion have the same angular displacement, the same angular velocity, and the same angular acceleration.*

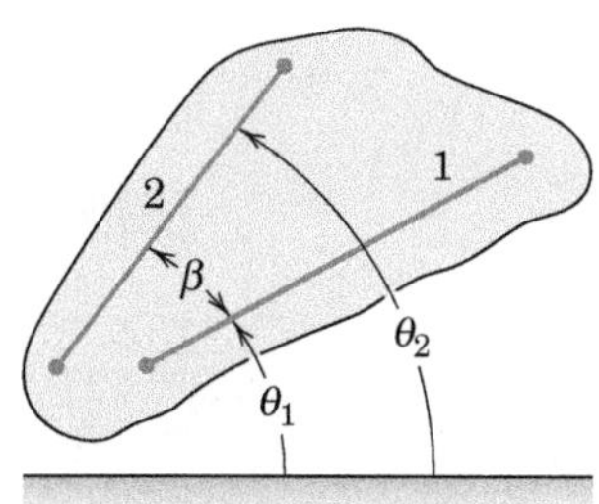

Figure 5/2

Note that the angular motion of a line depends only on its angular position with respect to any arbitrary fixed reference and on the time derivatives of the displacement. Angular motion does not require the presence of a fixed axis, normal to the plane of motion, about which the line and the body rotate.

Key Concepts

Angular-Motion Relations

The angular velocity ω and angular acceleration α of a rigid body in plane rotation are, respectively, the first and second time derivatives of the angular position coordinate θ of any line in the plane of motion of the body. These definitions give

$$\omega = \frac{d\theta}{dt} = \dot{\theta}$$

$$\alpha = \frac{d\omega}{dt} = \dot{\omega} \quad \text{or} \quad \alpha = \frac{d^2\theta}{dt^2} = \ddot{\theta} \qquad \textbf{(5/1)}$$

$$\omega\, d\omega = \alpha\, d\theta \quad \text{or} \quad \dot{\theta}\, d\dot{\theta} = \ddot{\theta}\, d\theta$$

The third relation is obtained by eliminating dt from the first two. In each of these relations, the positive direction for ω and α, clockwise or counterclockwise, is the same as that chosen for θ. Equations 5/1 should be recognized as analogous to the defining equations for the rectilinear motion of a particle, expressed by Eqs. 2/1, 2/2, and 2/3. In fact, all relations which were described for rectilinear motion in Art. 2/2 apply to the case of rotation in a plane if the linear quantities s, v, and a are replaced by their respective equivalent angular quantities θ, ω, and α. As we proceed further with rigid-body dynamics, we will find that the analogies between the relationships for linear and angular motion are almost complete throughout kinematics and kinetics. These relations are important to recognize, as they help to demonstrate the symmetry and unity found throughout mechanics.

For rotation with *constant* angular acceleration, the integrals of Eqs. 5/1 becomes

$$\omega = \omega_0 + \alpha t$$

$$\omega^2 = {\omega_0}^2 + 2\alpha(\theta - \theta_0)$$

$$\theta = \theta_0 + \omega_0 t + \tfrac{1}{2}\alpha t^2$$

Here θ_0 and ω_0 are the values of the angular position coordinate and angular velocity, respectively, at $t = 0$, and t is the duration of the motion considered. You should be able to carry out these integrations easily, as they are completely analogous to the corresponding equations for rectilinear motion with constant acceleration covered in Art. 2/2.

The graphical relationships described for s, v, a, and t in Figs. 2/3 and 2/4 may be used for θ, ω, and α merely by substituting the corresponding symbols. You should sketch these graphical relations for plane rotation. The mathematical procedures for obtaining rectilinear velocity and displacement from rectilinear acceleration may be applied to rotation by merely replacing the linear quantities by their corresponding angular quantities.

Rotation about a Fixed Axis

When a rigid body rotates about a fixed axis, all points other than those on the axis move in concentric circles about the fixed axis. Thus, for the rigid body in Fig. 5/3 rotating about a fixed axis normal to the plane of the figure through O, any point such as A moves in a circle of radius r. From the previous discussion in Art. 2/5, you should already be familiar with the relationships between the linear motion of A and the angular motion of the line normal to its path, which is also the angular motion of the rigid body. With the notation $\omega = \dot{\theta}$ and $\alpha = \dot{\omega} = \ddot{\theta}$ for the angular velocity and angular acceleration, respectively, of the body we have Eqs. 2/11, rewritten as

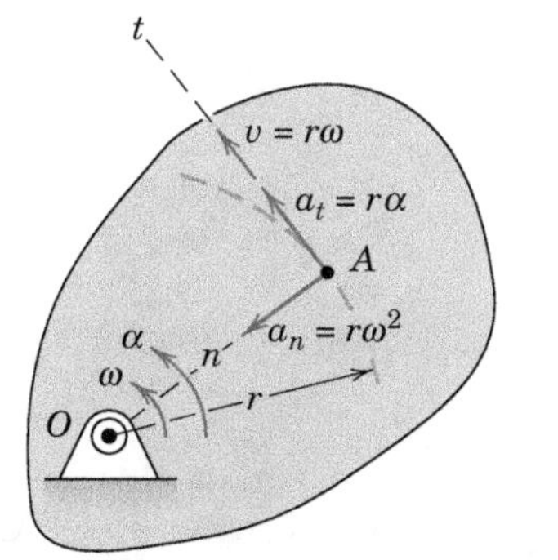

Figure 5/3

$$\begin{aligned} v &= r\omega \\ a_n &= r\omega^2 = v^2/r = v\omega \\ a_t &= r\alpha \end{aligned} \qquad \textbf{(5/2)}$$

These quantities may be expressed alternatively using the cross-product relationship of vector notation. The vector formulation is especially important in the analysis of three-dimensional motion. The angular velocity of the rotating body may be expressed by the vector $\boldsymbol{\omega}$ normal to the plane of rotation and having a sense governed by the right-hand rule, as shown in Fig. 5/4*a*. From the definition of the vector cross product, we see that the vector $\mathbf{v}$ is obtained by crossing $\boldsymbol{\omega}$ into $\mathbf{r}$. This cross product gives the correct magnitude and direction for $\mathbf{v}$ and we write

$$\mathbf{v} = \dot{\mathbf{r}} = \boldsymbol{\omega} \times \mathbf{r}$$

The order of the vectors to be crossed must be retained. The reverse order gives $\mathbf{r} \times \boldsymbol{\omega} = -\mathbf{v}$.

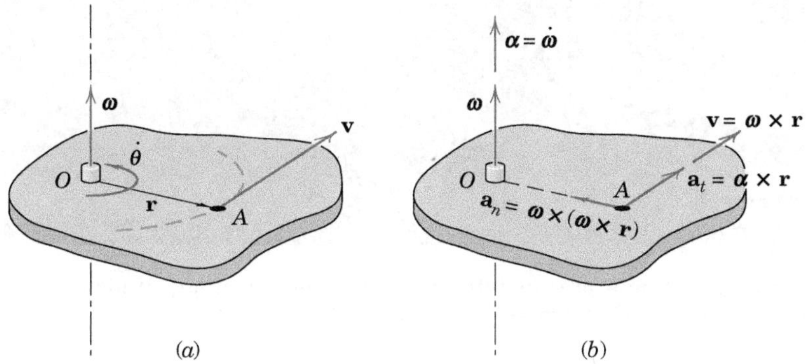

Figure 5/4

Michael Stokes/Shutterstock

A camshaft-drive system for an internal-combustion engine.

The acceleration of point A is obtained by differentiating the cross-product expression for $\mathbf{v}$, which gives

$$\begin{aligned}\mathbf{a} = \dot{\mathbf{v}} &= \boldsymbol{\omega} \times \dot{\mathbf{r}} + \dot{\boldsymbol{\omega}} \times \mathbf{r} \\ &= \boldsymbol{\omega} \times (\boldsymbol{\omega} \times \mathbf{r}) + \dot{\boldsymbol{\omega}} \times \mathbf{r} \\ &= \boldsymbol{\omega} \times \mathbf{v} + \boldsymbol{\alpha} \times \mathbf{r}\end{aligned}$$

Here $\boldsymbol{\alpha} = \dot{\boldsymbol{\omega}}$ stands for the angular acceleration of the body. Thus, the vector equivalents to Eqs. 5/2 are

$$\begin{aligned}\mathbf{v} &= \boldsymbol{\omega} \times \mathbf{r} \\ \mathbf{a}_n &= \boldsymbol{\omega} \times (\boldsymbol{\omega} \times \mathbf{r}) \\ \mathbf{a}_t &= \boldsymbol{\alpha} \times \mathbf{r}\end{aligned} \qquad \textbf{(5/3)}$$

and are shown in Fig. 5/4*b*.

For three-dimensional motion of a rigid body, the angular-velocity vector $\boldsymbol{\omega}$ may change direction as well as magnitude, and in this case, the angular acceleration, which is the time derivative of angular velocity, $\boldsymbol{\alpha} = \dot{\boldsymbol{\omega}}$, will no longer be in the same direction as $\boldsymbol{\omega}$.

Walter Bibikow/Photolibrary/Getty Images, Inc.

Cable-pulley systems are an integral component in many transportation systems including this tramway near Monte Solaro in Italy.

Sample Problem 5/1

A flywheel rotating freely at 1800 rev/min clockwise is subjected to a variable counterclockwise torque which is first applied at time $t = 0$. The torque produces a counterclockwise angular acceleration $\alpha = 4t$ rad/s^2, where t is the time in seconds during which the torque is applied. Determine (*a*) the time required for the flywheel to reduce its clockwise angular speed to 900 rev/min, (*b*) the time required for the flywheel to reverse its direction of rotation, and (*c*) the total number of revolutions, clockwise plus counterclockwise, turned by the flywheel during the first 14 seconds of torque application.

Solution. The counterclockwise direction will be taken arbitrarily as positive.

(a) Since α is a known function of the time, we may integrate it to obtain angular
① velocity. With the initial angular velocity of $-1800(2\pi)/60 = -60\pi$ rad/s, we have

$$[d\omega = \alpha\, dt] \qquad \int_{-60\pi}^{\omega} d\omega = \int_0^t 4t\, dt \qquad \omega = -60\pi + 2t^2$$

Substituting the clockwise angular speed of 900 rev/min or $\omega = -900(2\pi)/60 = -30\pi$ rad/s gives

$$-30\pi = -60\pi + 2t^2 \qquad t^2 = 15\pi \qquad t = 6.86 \text{ s} \qquad \textit{Ans.}$$

(b) The flywheel changes direction when its angular velocity is momentarily zero. Thus,

$$0 = -60\pi + 2t^2 \qquad t^2 = 30\pi \qquad t = 9.71 \text{ s} \qquad \textit{Ans.}$$

(c) The total number of revolutions through which the flywheel turns during 14 seconds is the number of clockwise turns N_1 during the first 9.71 seconds, plus the number of counterclockwise turns N_2 during the remainder of the interval. Integrating the expression for ω in terms of t gives us the angular displacement in radians. Thus, for the first interval

$$[d\theta = \omega\, dt] \qquad \int_0^{\theta_1} d\theta = \int_0^{9.71} (-60\pi + 2t^2)\, dt$$

② $$\theta_1 = [-60\pi t + \tfrac{2}{3}t^3]_0^{9.71} = -1220 \text{ rad}$$

or $N_1 = 1220/2\pi = 194.2$ revolutions clockwise.

For the second interval

$$\int_0^{\theta_2} d\theta = \int_{9.71}^{14} (-60\pi + 2t^2)\, dt$$

③ $$\theta_2 = [-60\pi t + \tfrac{2}{3}t^3]_{9.71}^{14} = 410 \text{ rad}$$

or $N_2 = 410/2\pi = 65.3$ revolutions counterclockwise. Thus, the total number of revolutions turned during the 14 seconds is

$$N = N_1 + N_2 = 194.2 + 65.3 = 259 \text{ rev} \qquad \textit{Ans.}$$

We have plotted ω versus t and we see that θ_1 is represented by the negative area and θ_2 by the positive area. If we had integrated over the entire interval in one step, we would have obtained $|\theta_2| - |\theta_1|$.

Helpful Hints

① We must be very careful to be consistent with our algebraic signs. The lower limit is the negative (clockwise) value of the initial angular velocity. Also we must convert revolutions to radians since α is in radian units.

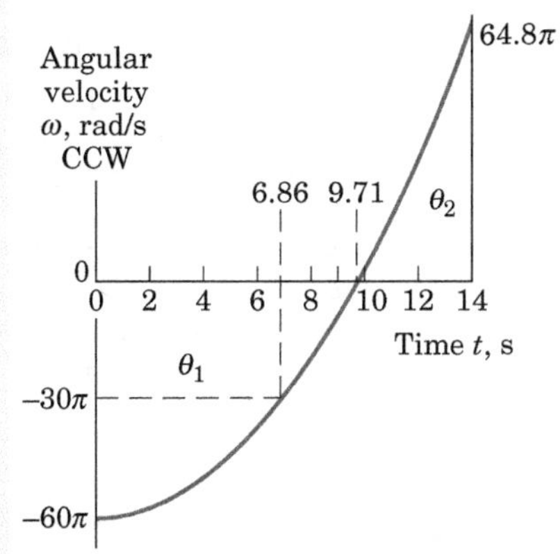

② Again note that the minus sign signifies clockwise in this problem.

③ We could have converted the original expression for α into the units of rev/s^2, in which case our integrals would have come out directly in revolutions.

Sample Problem 5/2

The pinion A of the hoist motor drives gear B, which is attached to the hoisting drum. The load L is lifted from its rest position and acquires an upward velocity of 3 ft/sec in a vertical rise of 4 ft with constant acceleration. As the load passes this position, compute (a) the acceleration of point C on the cable in contact with the drum and (b) the angular velocity and angular acceleration of the pinion A.

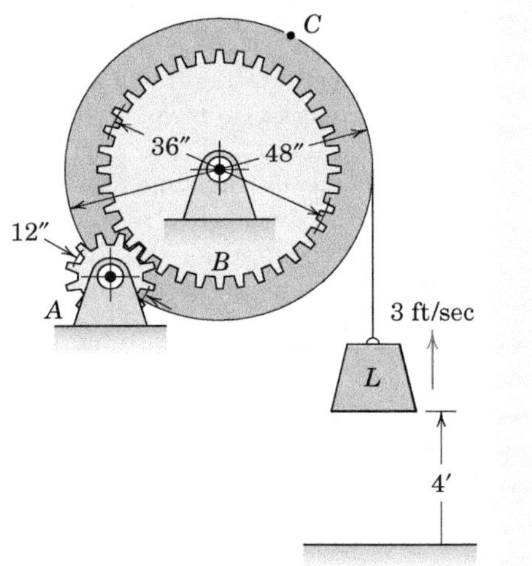

Solution. ***(a)*** If the cable does not slip on the drum, the vertical velocity and acceleration of the load L are, of necessity, the same as the tangential velocity v and tangential acceleration a_t of point C. For the rectilinear motion of L with constant acceleration, the n- and t-components of the acceleration of C become

$[v^2 = 2as]$ $\qquad a = a_t = v^2/2s = 3^2/[2(4)] = 1.125 \text{ ft/sec}^2$

① $[a_n = v^2/r]$ $\qquad a_n = 3^2/(24/12) = 4.5 \text{ ft/sec}^2$

$[a = \sqrt{a_n{}^2 + a_t{}^2}]$ $\qquad a_C = \sqrt{(4.5)^2 + (1.125)^2} = 4.64 \text{ ft/sec}^2$ *Ans.*

(b) The angular motion of gear A is determined from the angular motion of gear B by the velocity v_1 and tangential acceleration a_1 of their common point of contact. First, the angular motion of gear B is determined from the motion of point C on the attached drum. Thus,

$[v = r\omega]$ $\qquad \omega_B = v/r = 3/(24/12) = 1.5 \text{ rad/sec}$

$[a_t = r\alpha]$ $\qquad \alpha_B = a_t/r = 1.125/(24/12) = 0.562 \text{ rad/sec}^2$

Then from $v_1 = r_A\omega_A = r_B\omega_B$ and $a_1 = r_A\alpha_A = r_B\alpha_B$, we have

$$\omega_A = \frac{r_B}{r_A}\omega_B = \frac{18/12}{6/12}1.5 = 4.5 \text{ rad/sec CW}$$ *Ans.*

$$\alpha_A = \frac{r_B}{r_A}\alpha_B = \frac{18/12}{6/12}0.562 = 1.688 \text{ rad/sec}^2 \text{ CW}$$ *Ans.*

Helpful Hint

① Recognize that a point on the cable changes the direction of its velocity after it contacts the drum and acquires a normal component of acceleration.

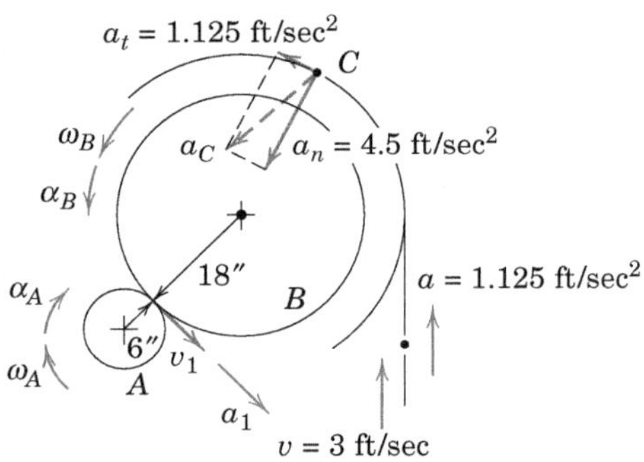

Sample Problem 5/3

The right-angle bar rotates clockwise with an angular velocity which is decreasing at the rate of 4 rad/s². Write the vector expressions for the velocity and acceleration of point A when $\omega = 2$ rad/s.

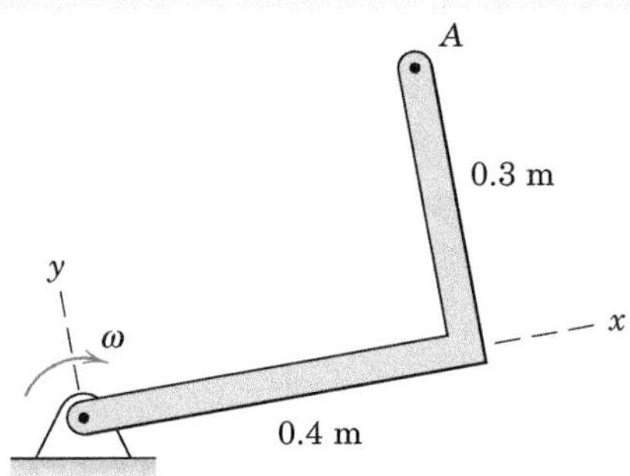

Solution. Using the right-hand rule gives

$$\boldsymbol{\omega} = -2\mathbf{k} \text{ rad/s} \qquad \text{and} \qquad \boldsymbol{\alpha} = +4\mathbf{k} \text{ rad/s}^2$$

The velocity and acceleration of A become

$[\mathbf{v} = \boldsymbol{\omega} \times \mathbf{r}]$ $\qquad \mathbf{v} = -2\mathbf{k} \times (0.4\mathbf{i} + 0.3\mathbf{j}) = 0.6\mathbf{i} - 0.8\mathbf{j} \text{ m/s}$ *Ans.*

$[\mathbf{a}_n = \boldsymbol{\omega} \times (\boldsymbol{\omega} \times \mathbf{r})]$ $\qquad \mathbf{a}_n = -2\mathbf{k} \times (0.6\mathbf{i} - 0.8\mathbf{j}) = -1.6\mathbf{i} - 1.2\mathbf{j} \text{ m/s}^2$

$[\mathbf{a}_t = \boldsymbol{\alpha} \times \mathbf{r}]$ $\qquad \mathbf{a}_t = 4\mathbf{k} \times (0.4\mathbf{i} + 0.3\mathbf{j}) = -1.2\mathbf{i} + 1.6\mathbf{j} \text{ m/s}^2$

$[\mathbf{a} = \mathbf{a}_n + \mathbf{a}_t]$ $\qquad \mathbf{a} = -2.8\mathbf{i} + 0.4\mathbf{j} \text{ m/s}^2$ *Ans.*

The magnitudes of $\mathbf{v}$ and $\mathbf{a}$ are

$$v = \sqrt{0.6^2 + 0.8^2} = 1 \text{ m/s} \qquad \text{and} \qquad a = \sqrt{2.8^2 + 0.4^2} = 2.83 \text{ m/s}^2$$

PROBLEMS

Introductory Problems

5/1 A torque applied to a flywheel causes it to accelerate uniformly from a speed of 300 rev/min to a speed of 900 rev/min in 6 seconds. Determine the number of revolutions N through which the wheel turns during this interval. (*Suggestion:* Use revolutions and minutes for units in your calculations.)

5/2 The circular sector rotates about a fixed axis through point O with angular velocity $\omega = 2$ rad/s and angular acceleration $\alpha = 4$ rad/s^2 with directions as indicated in the figure. Determine the instantaneous velocity and acceleration of point A.

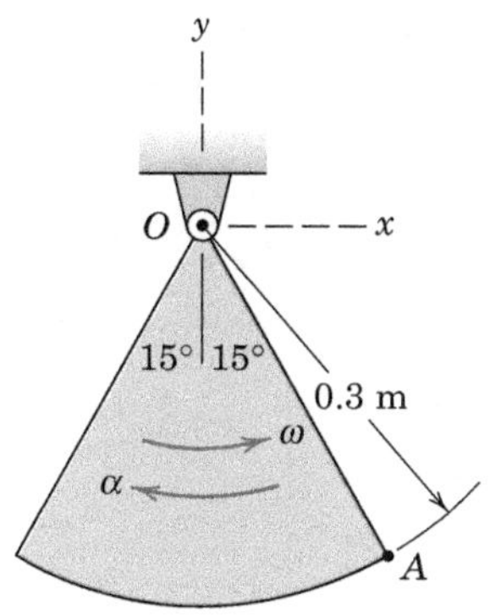

Problem 5/2

5/3 The angular velocity of a gear is controlled according to $\omega = 12 - 3t^2$ where ω, in radians per second, is positive in the clockwise sense and where t is the time in seconds. Find the net angular displacement $\Delta\theta$ from the time $t = 0$ to $t = 3$ s. Also find the total number of revolutions N through which the gear turns during the 3 seconds.

5/4 Magnetic tape is fed over and around the light pulleys mounted in a computer frame. If the speed v of the tape is constant and if the ratio of the magnitudes of the acceleration of points A and B is 2/3, determine the radius r of the larger pulley.

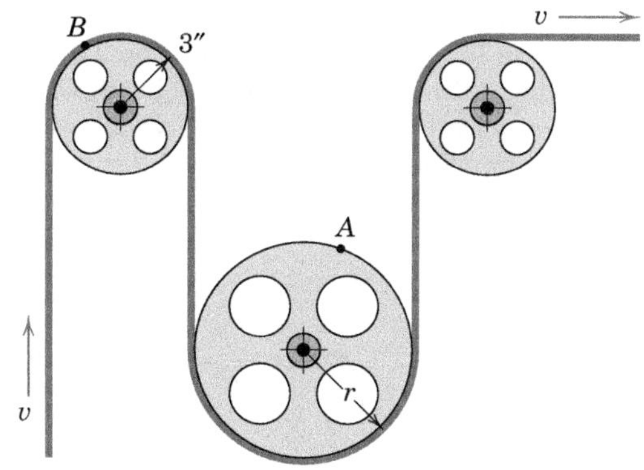

Problem 5/4

5/5 When switched on, the grinding machine accelerates from rest to its operating speed of 3450 rev/min in 6 seconds. When switched off, it coasts to rest in 32 seconds. Determine the number of revolutions turned during both the startup and shutdown periods. Assume uniform angular acceleration in both cases.

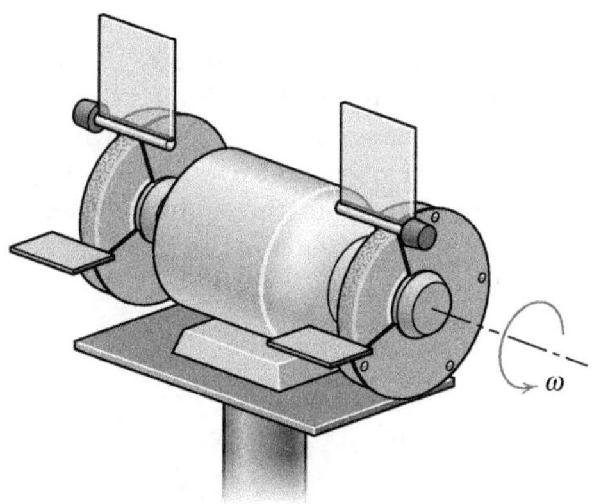

Problem 5/5

5/6 The small cart is released from rest in position 1 and requires 0.638 seconds to reach position 2 at the bottom of the path, where its center G has a velocity of 14.20 ft/sec. Determine the angular velocity ω of line AB in position 2 and the average angular velocity ω_{av} of AB during the interval.

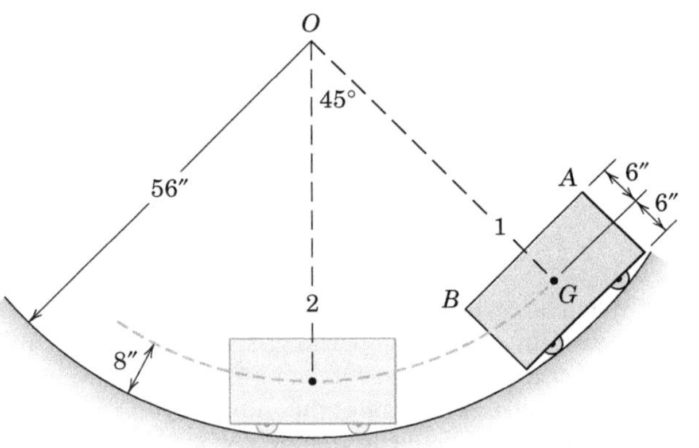

Problem 5/5

5/7 The flywheel has a diameter of 600 mm and rotates with increasing speed about its z-axis shaft. When point P on the rim crosses the y-axis with $\theta = 90°$, it has an acceleration given by $\mathbf{a} = -1.8\mathbf{i} - 4.8\mathbf{j}$ m/s^2. For this instant, determine the angular velocity ω and the angular acceleration α of the flywheel.

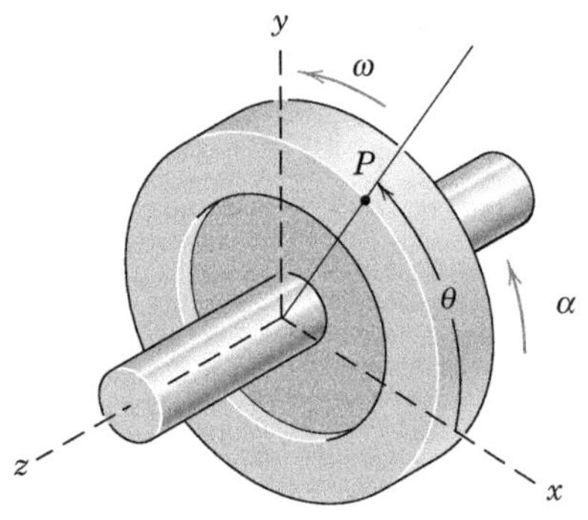

Problem 5/7

5/8 The drive mechanism imparts to the semicircular plate simple harmonic motion of the form $\theta = \theta_0 \sin \omega_0 t$, where θ_0 is the amplitude of the oscillation and ω_0 is its circular frequency. Determine the amplitudes of the angular velocity and angular acceleration and state where in the motion cycle these maxima occur. Note that this motion is *not* that of a freely pivoted and undriven body undergoing arbitrarily large-amplitude angular motion.

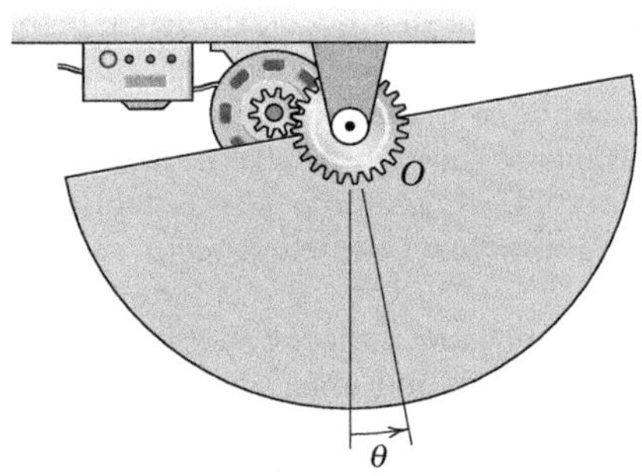

Problem 5/8

5/9 The cylinder rotates about the fixed z-axis in the direction indicated. If the speed of point A is $v_A = 2$ ft/sec and the magnitude of its acceleration is $a_A = 12$ ft/sec^2, determine the angular velocity and angular acceleration of the cylinder. Is knowledge of the angle θ necessary?

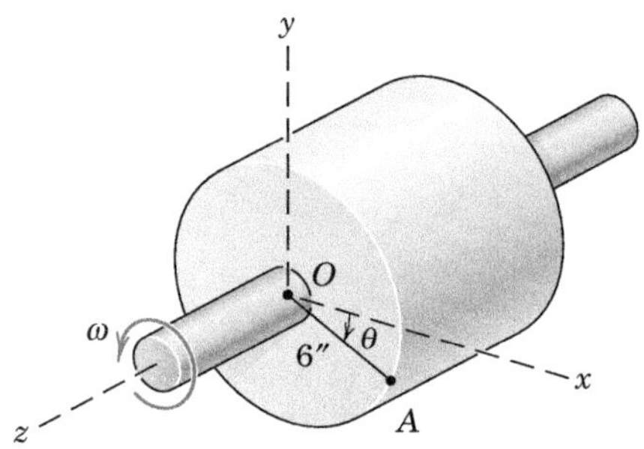

Problem 5/9

Representative Problems

5/10 The angular acceleration of a body which is rotating about a fixed axis is given by $\alpha = -k\omega^2$, where the constant $k = 0.1$ (no units). Determine the angular displacement and time elapsed when the angular velocity has been reduced to one-third its initial value $\omega_0 = 12$ rad/s.

5/11 The device shown rotates about the fixed z-axis with angular velocity $\omega = 20$ rad/s and angular acceleration $\alpha = 40$ rad/s^2 in the directions indicated. Determine the instantaneous velocity and acceleration of point B.

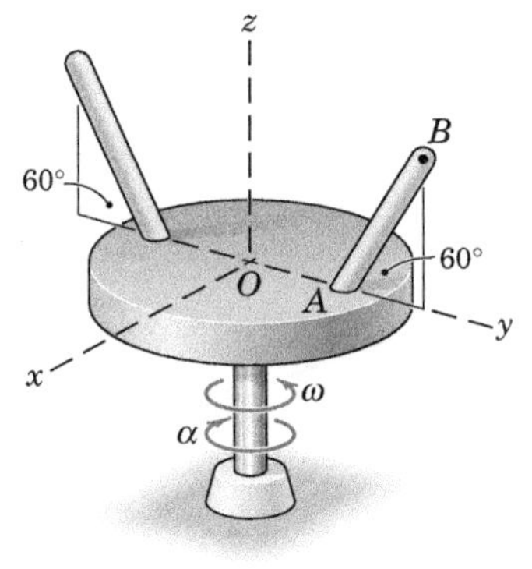

$\overline{OA} = 300$ mm, $\overline{AB} = 500$ mm

Problem 5/11

5/12 The circular disk rotates with a constant angular velocity $\omega = 40$ rad/sec about its axis, which is inclined in the y-z plane at the angle $\theta = \tan^{-1}\frac{3}{4}$. Determine the vector expressions for the velocity and acceleration of point P, whose position vector at the instant shown is $\mathbf{r} = 15\mathbf{i} + 16\mathbf{j} - 12\mathbf{k}$ in. (Check the magnitudes of your results from the scalar values $v = r\omega$ and $a_n = r\omega^2$.)

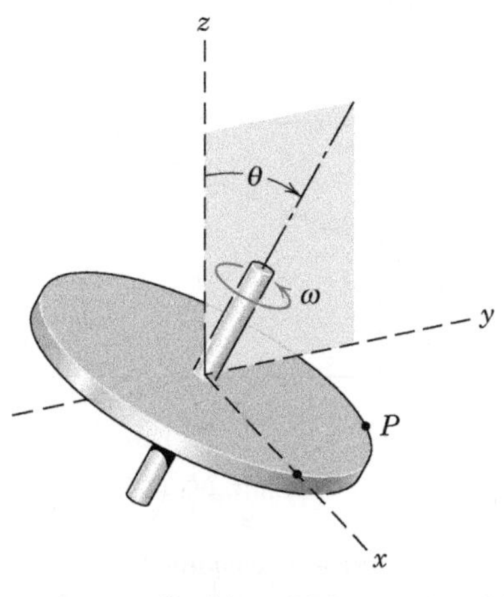

Problem 5/12

5/13 The T-shaped body rotates about a horizontal axis through point O. At the instant represented, its angular velocity is $\omega = 5$ rad/sec and its angular acceleration is $\alpha = 10$ rad/sec^2 in the directions indicated. Determine the velocity and acceleration of (a) point A and (b) point B. Express your results in terms of components along the x- and y-axes shown.

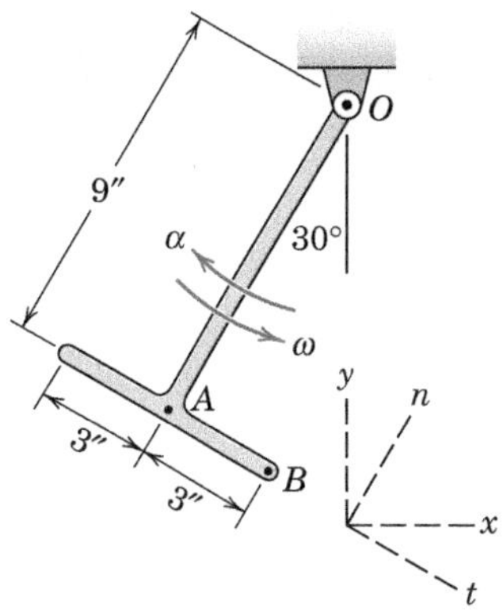

Problem 5/13

5/14 Repeat the previous problem, but now express your results in terms of components along the n- and t-axes.

5/15 In order to test an intentionally weak adhesive, the bottom of the small 0.3-kg block is coated with adhesive and then the block is pressed onto the turntable with a known force. The turntable starts from rest at time $t = 0$ and uniformly accelerates with $\alpha = 2$ rad/s^2. If the adhesive fails at exactly $t = 3$ s, determine the ultimate shear force which the adhesive supports. What is the angular displacement of the turntable at the time of failure?

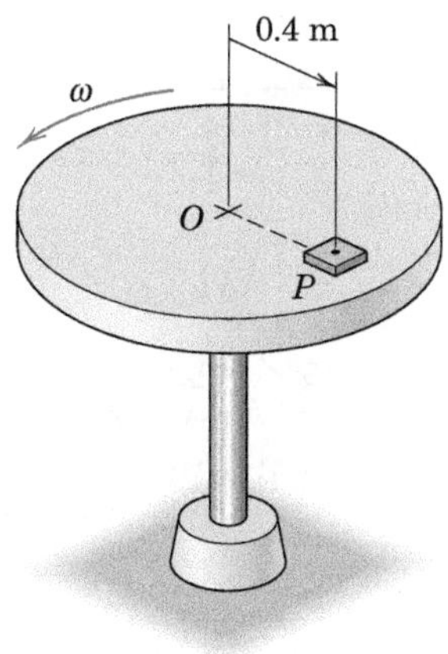

Problem 5/15

5/16 The two attached pulleys are driven by the belt with increasing speed. When the belt reaches a speed $v = 2$ ft/sec, the total acceleration of point P is 26 ft/sec^2. For this instant determine the angular acceleration α of the pulleys and the acceleration of point B on the belt.

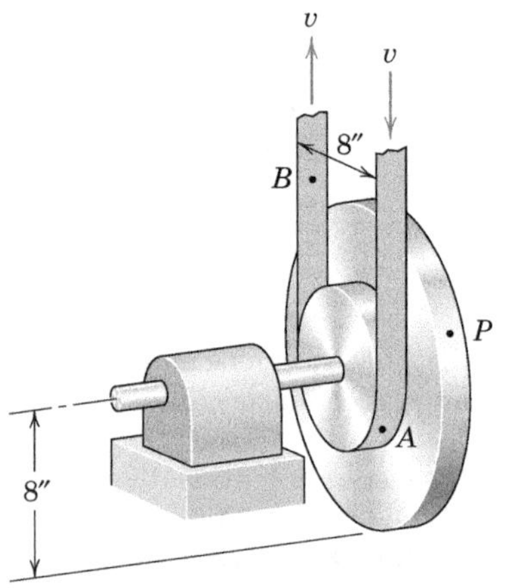

Problem 5/16

5/17 The bent flat bar rotates about a fixed axis through point O with the instantaneous angular properties indicated in the figure. Determine the velocity and acceleration of point A.

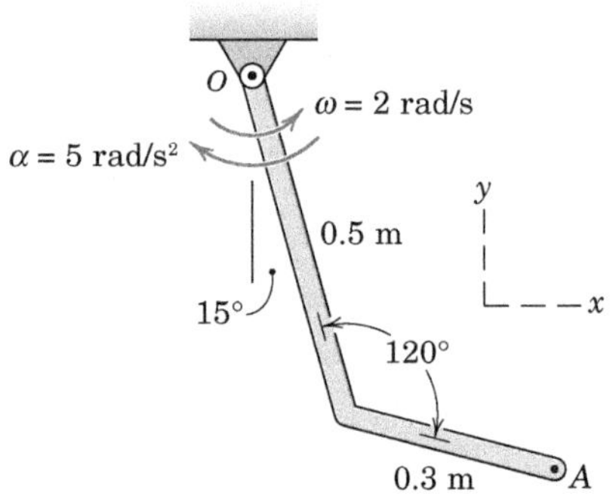

Problem 5/17

5/18 At time $t = 0$, the arm is rotating about the fixed z-axis with an angular velocity $\omega = 200$ rad/s in the direction shown. At that time, a constant angular deceleration begins and the arm comes to a stop in 10 seconds. At what time t does the acceleration of point P make a 15° angle with the arm AB?

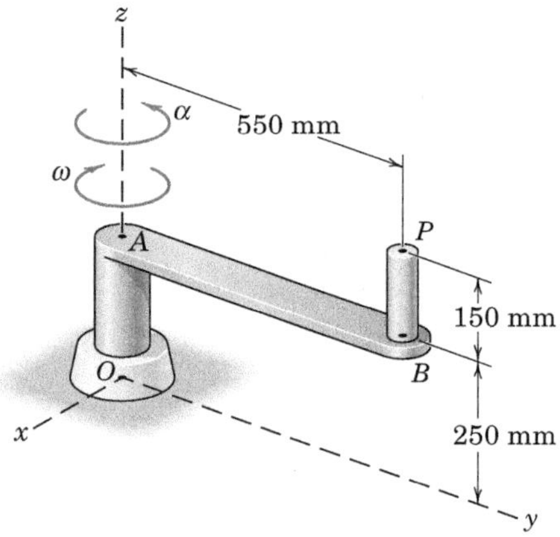

Problem 5/18

5/19 A variable torque is applied to a rotating wheel at time $t = 0$ and causes the clockwise angular acceleration to increase linearly with the clockwise angular displacement θ of the wheel during the next 30 revolutions. When the wheel has turned the additional 30 revolutions, its angular velocity is 90 rad/s. Determine its angular velocity ω_0 at the start of the interval at $t = 0$.

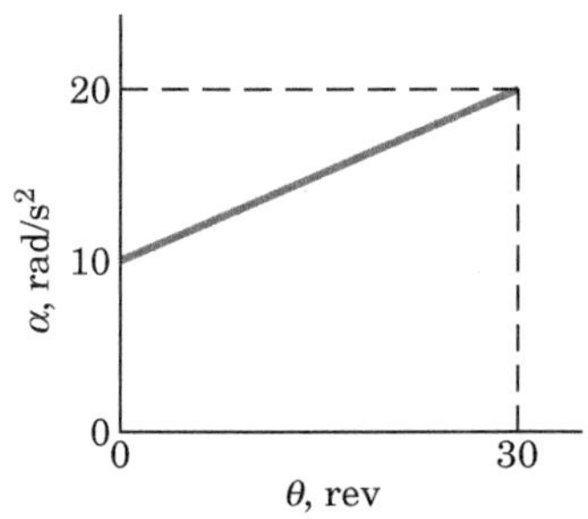

Problem 5/19

5/20 Develop general expressions for the instantaneous velocity and acceleration of point A of the square plate, which rotates about a fixed axis through point O. Take all variables to be positive. Then evaluate your expressions for $\theta = 30°$, $b = 0.2$ m, $\omega = 1.4$ rad/s, and $\alpha = 2.5$ rad/s^2.

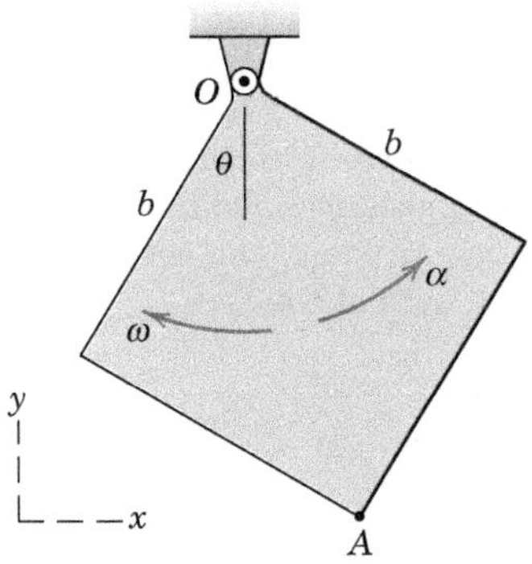

Problem 5/20

5/21 The motor A accelerates uniformly from zero to 3600 rev/min in 8 seconds after it is turned on at time $t = 0$. It drives a fan (not shown) which is attached to drum B. The effective pulley radii are shown in the figure. Determine (a) the number of revolutions turned by drum B during the 8-second startup period, (b) the angular velocity of drum B at time $t = 4$ s, and (c) the number of revolutions turned by drum B during the first 4 seconds of motion. Assume no belt slippage.

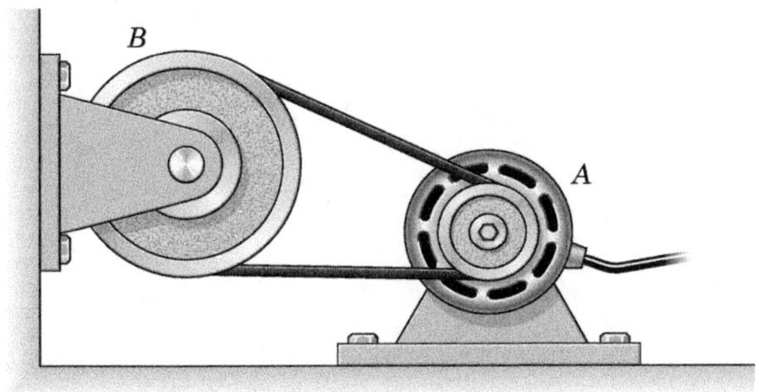

$r_A = 75$ mm, $r_B = 200$ mm

Problem 5/21

5/22 Point A of the circular disk is at the angular position $\theta = 0$ at time $t = 0$. The disk has angular velocity $\omega_0 = 0.1$ rad/s at $t = 0$ and subsequently experiences a constant angular acceleration $\alpha = 2$ rad/s^2. Determine the velocity and acceleration of point A in terms of fixed $\mathbf{i}$ and $\mathbf{j}$ unit vectors at time $t = 1$ s.

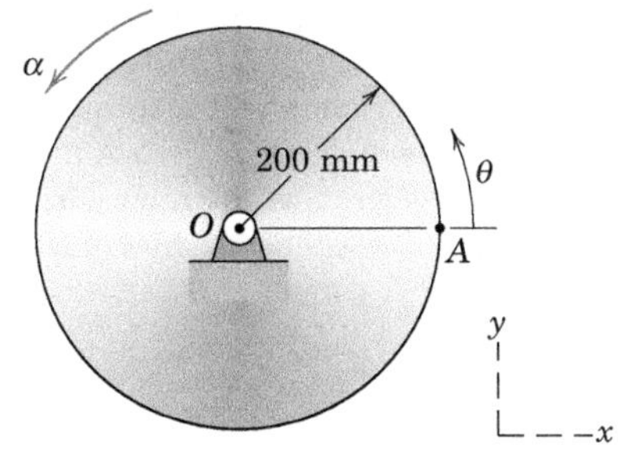

Problem 5/22

5/23 Repeat Prob. 5/22, except now the angular acceleration of the disk is given by $\alpha = 2t$, where t is in seconds and α is in radians per second squared. Determine the velocity and acceleration of point A in terms of fixed $\mathbf{i}$ and $\mathbf{j}$ unit vectors at time $t = 2$ s.

5/24 Repeat Prob. 5/22, except now the angular acceleration of the disk is given by $\alpha = 2\omega$, where ω is in radians per second and α is in radians per second squared. Determine the velocity and acceleration of point A in terms of fixed $\mathbf{i}$ and $\mathbf{j}$ unit vectors at time $t = 1$ s.

5/25 The disk of Prob. 5/22 is at the angular position $\theta = 0$ at time $t = 0$. Its angular velocity at $t = 0$ is $\omega_0 = 0.1$ rad/s, and then it experiences an angular acceleration given by $\alpha = 2\theta$, where θ is in radians and α is in radians per second squared. Determine the angular position of point A at time $t = 2$ s.

5/26 During its final spin cycle, a front-loading washing machine has a spin rate of 1200 rev/min. Once power is removed, the drum is observed to uniformly decelerate to rest in 25 s. Determine the number of revolutions made during this period as well as the number of revolutions made during the first half of it.

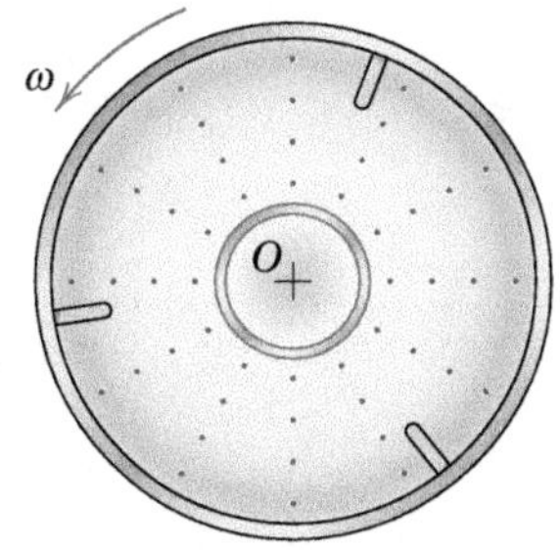

Problem 5/26

5/27 The design characteristics of a gear-reduction unit are under review. Gear B is rotating clockwise with a speed of 300 rev/min when a torque is applied to gear A at time $t = 2$ s to give gear A a counterclockwise angular acceleration α which varies with time for a duration of 4 seconds as shown. Determine the speed N_B of gear B when $t = 6$ s.

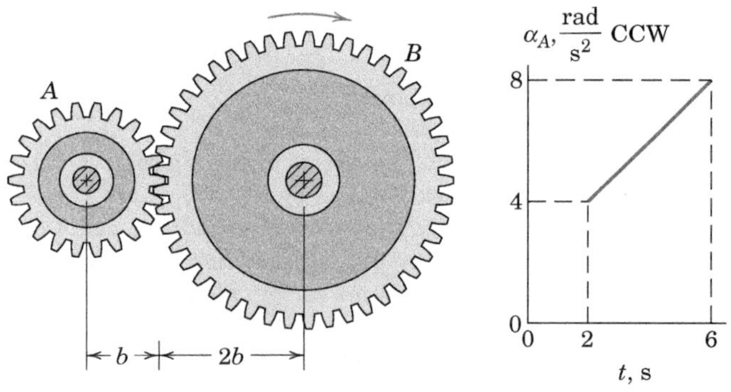

Problem 5/27

▶5/28 A V-belt speed-reduction drive is shown where pulley A drives the two integral pulleys B which in turn drive pulley C. If A starts from rest at time $t = 0$ and is given a constant angular acceleration α_1, derive expressions for the angular velocity of C and the magnitude of the acceleration of a point P on the belt, both at time t.

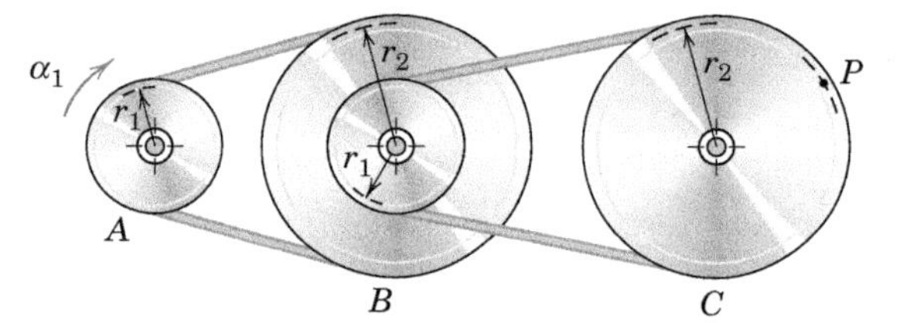

Problem 5/28

5/3 Absolute Motion

We now develop the approach of absolute-motion analysis to describe the plane kinematics of rigid bodies. In this approach, we make use of the geometric relations which define the configuration of the body involved and then proceed to take the time derivatives of the defining geometric relations to obtain velocities and accelerations.

In Art. 2/9 of Chapter 2 on particle kinematics, we introduced the application of absolute-motion analysis for the constrained motion of connected particles. For the pulley configurations treated, the relevant velocities and accelerations were determined by successive differentiation of the lengths of the connecting cables. In this earlier treatment, the geometric relations were quite simple, and no angular quantities had to be considered. Now that we will be dealing with rigid-body motion, however, we find that our defining geometric relations include both linear and angular variables and, therefore, the time derivatives of these quantities will involve both linear and angular velocities and linear and angular accelerations.

This piston and connecting-rod assembly is an example of a slider-crank mechanism.

In absolute-motion analysis, it is essential that we be consistent with the mathematics of the description. For example, if the angular position of a moving line in the plane of motion is specified by its counterclockwise angle θ measured from some convenient fixed reference axis, then the positive sense for both angular velocity $\dot{\theta}$ and angular acceleration $\ddot{\theta}$ will also be counterclockwise. A negative sign for either quantity will, of course, indicate a clockwise angular motion. The defining relations for linear motion, Eqs. 2/1, 2/2, and 2/3, and the relations involving angular motion, Eqs. 5/1 and 5/2 or 5/3, will find repeated use in the motion analysis and should be mastered.

The absolute-motion approach to rigid-body kinematics is quite straightforward, provided the configuration lends itself to a geometric description which is not overly complex. If the geometric configuration is awkward or complex, analysis by the principles of relative motion may be preferable. Relative-motion analysis is treated in this chapter beginning with Art. 5/4. The choice between absolute- and relative-motion analyses is best made after experience has been gained with both approaches.

The next three sample problems illustrate the application of absolute-motion analysis to three commonly encountered situations. The kinematics of a rolling wheel, treated in Sample Problem 5/4, is especially important and will be useful in much of the problem work because the rolling wheel in various forms is such a common element in mechanical systems.

Sample Problem 5/4

A wheel of radius r rolls on a flat surface without slipping. Determine the angular motion of the wheel in terms of the linear motion of its center O. Also determine the acceleration of a point on the rim of the wheel as the point comes into contact with the surface on which the wheel rolls.

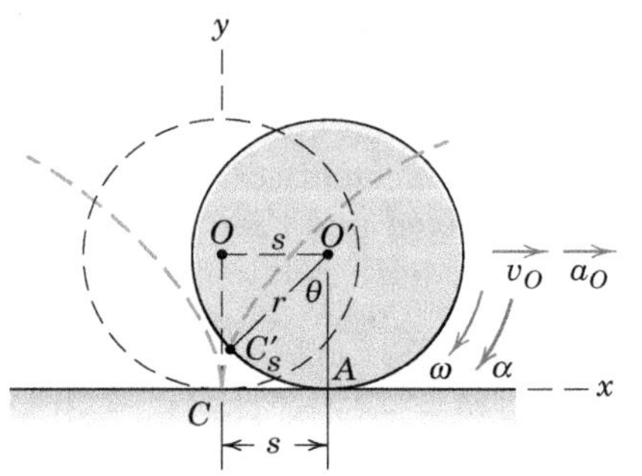

Solution. The figure shows the wheel rolling to the right from the dashed to the full position without slipping. The linear displacement of the center O is s, which is also the arc length $C'A$ along the rim on which the wheel rolls. The radial line CO rotates to the new position $C'O'$ through the angle θ, where θ is measured from the vertical direction. If the wheel does not slip, the arc $C'A$ must equal the distance s. Thus, the displacement relationship and its two time derivatives give

$$s = r\theta$$

① $$v_O = r\omega \qquad \textit{Ans.}$$

$$a_O = r\alpha$$

where $v_O = \dot{s}$, $a_O = \dot{v}_O = \ddot{s}$, $\omega = \dot{\theta}$, and $\alpha = \dot{\omega} = \ddot{\theta}$. The angle θ, of course, must be in radians. The acceleration a_O will be directed in the sense opposite to that of v_O if the wheel is slowing down. In this event, the angular acceleration α will have the sense opposite to that of ω.

The origin of fixed coordinates is taken arbitrarily but conveniently at the point of contact between C on the rim of the wheel and the ground. When point C has moved along its cycloidal path to C', its new coordinates and their time derivatives become

$$x = s - r\sin\theta = r(\theta - \sin\theta) \qquad y = r - r\cos\theta = r(1 - \cos\theta)$$

$$\dot{x} = r\dot{\theta}(1 - \cos\theta) = v_O(1 - \cos\theta) \qquad \dot{y} = r\dot{\theta}\sin\theta = v_O\sin\theta$$

$$\ddot{x} = \dot{v}_O(1 - \cos\theta) + v_O\dot{\theta}\sin\theta \qquad \ddot{y} = \dot{v}_O\sin\theta + v_O\dot{\theta}\cos\theta$$

$$= a_O(1 - \cos\theta) + r\omega^2\sin\theta \qquad = a_O\sin\theta + r\omega^2\cos\theta$$

For the desired instant of contact, $\theta = 0$ and

② $$\ddot{x} = 0 \quad \text{and} \quad \ddot{y} = r\omega^2 \qquad \textit{Ans.}$$

Thus, the acceleration of the point C on the rim at the instant of contact with the ground depends only on r and ω and is directed toward the center of the wheel. If desired, the velocity and acceleration of C at any position θ may be obtained by writing the expressions $\mathbf{v} = \dot{x}\mathbf{i} + \dot{y}\mathbf{j}$ and $\mathbf{a} = \ddot{x}\mathbf{i} + \ddot{y}\mathbf{j}$.

Application of the kinematic relationships for a wheel which rolls without slipping should be recognized for various configurations of rolling wheels such as those illustrated on the right. If a wheel slips as it rolls, the foregoing relations are no longer valid.

Helpful Hints

① These three relations are not entirely unfamiliar at this point, and their application to the rolling wheel should be mastered thoroughly.

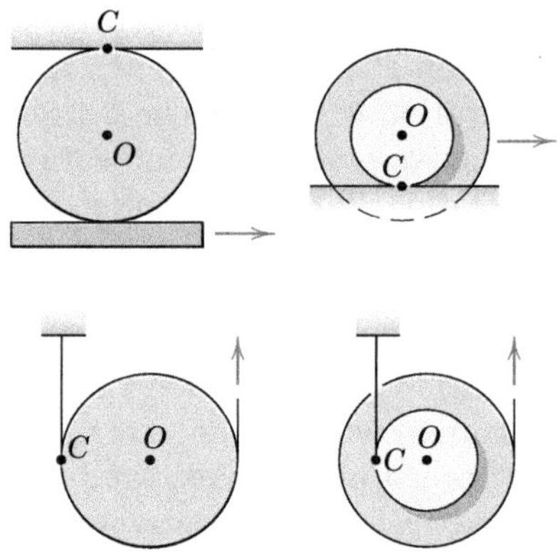

② Clearly, when $\theta = 0$, the point of contact has zero velocity so that $\dot{x} = \dot{y} = 0$. The acceleration of the contact point on the wheel will also be obtained by the principles of relative motion in Art. 5/6.

Sample Problem 5/5

The load L is being hoisted by the pulley and cable arrangement shown. Each cable is wrapped securely around its respective pulley so it does not slip. The two pulleys to which L is attached are fastened together to form a single rigid body. Calculate the velocity and acceleration of the load L and the corresponding angular velocity ω and angular acceleration α of the double pulley under the following conditions:

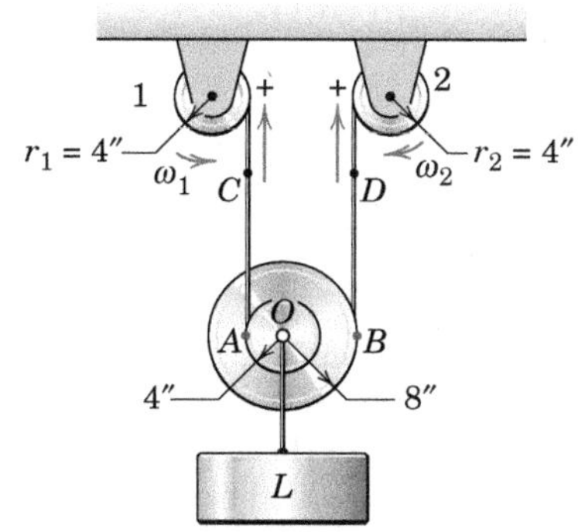

Case (a) Pulley 1: $\omega_1 = \dot{\omega}_1 = 0$ (pulley at rest)
Pulley 2: $\omega_2 = 2$ rad/sec, $\alpha_2 = \dot{\omega}_2 = -3$ rad/sec^2

Case (b) Pulley 1: $\omega_1 = 1$ rad/sec, $\alpha_1 = \dot{\omega}_1 = 4$ rad/sec^2
Pulley 2: $\omega_2 = 2$ rad/sec, $\alpha_2 = \dot{\omega}_2 = -2$ rad/sec^2

Solution. The tangential displacement, velocity, and acceleration of a point on the rim of pulley 1 or 2 equal the corresponding vertical motions of point A or B since the cables are assumed to be inextensible.

Case (a). With A momentarily at rest, line AB rotates to AB' through the angle $d\theta$ during time dt. From the diagram we see that the displacements and their time derivatives give

(1)
$$ds_B = \overline{AB}\, d\theta \qquad v_B = \overline{AB}\omega \qquad (a_B)_t = \overline{AB}\alpha$$
$$ds_O = \overline{AO}\, d\theta \qquad v_O = \overline{AO}\omega \qquad a_O = \overline{AO}\alpha$$

With $v_D = r_2\omega_2 = 4(2) = 8$ in./sec and $a_D = r_2\alpha_2 = 4(-3) = -12$ in./sec^2, we have for the angular motion of the double pulley

(2)
$$\omega = v_B/\overline{AB} = v_D/\overline{AB} = 8/12 = 2/3 \text{ rad/sec(CCW)} \qquad \textit{Ans.}$$
$$\alpha = (a_B)_t/\overline{AB} = a_D/\overline{AB} = -12/12 = -1 \text{ rad/sec}^2\text{(CW)} \qquad \textit{Ans.}$$

The corresponding motion of O and the load L is

(3)
$$v_O = \overline{AO}\omega = 4(2/3) = 8/3 \text{ in./sec} \qquad \textit{Ans.}$$
$$a_O = \overline{AO}\alpha = 4(-1) = -4 \text{ in./sec}^2 \qquad \textit{Ans.}$$

Case (b). With point C, and hence point A, in motion, line AB moves to $A'B'$ during time dt. From the diagram for this case, we see that the displacements and their time derivatives give

$$ds_B - ds_A = \overline{AB}\, d\theta \qquad v_B - v_A = \overline{AB}\omega \qquad (a_B)_t - (a_A)_t = \overline{AB}\alpha$$
$$ds_O - ds_A = \overline{AO}\, d\theta \qquad v_O - v_A = \overline{AO}\omega \qquad a_O - (a_A)_t = \overline{AO}\alpha$$

With $v_C = r_1\omega_1 = 4(1) = 4$ in./sec $\qquad v_D = r_2\omega_2 = 4(2) = 8$ in./sec

$a_C = r_1\alpha_1 = 4(4) = 16$ in./sec^2 $\qquad a_D = r_2\alpha_2 = 4(-2) = -8$ in./sec^2

we have for the angular motion of the double pulley

(4)
$$\omega = \frac{v_B - v_A}{\overline{AB}} = \frac{v_D - v_C}{\overline{AB}} = \frac{8-4}{12} = 1/3 \text{ rad/sec (CCW)} \qquad \textit{Ans.}$$
$$\alpha = \frac{(a_B)_t - (a_A)_t}{\overline{AB}} = \frac{a_D - a_C}{\overline{AB}} = \frac{-8-16}{12} = -2 \text{ rad/sec}^2\text{(CW)} \qquad \textit{Ans.}$$

The corresponding motion of O and the load L is

$$v_O = v_A + \overline{AO}\omega = v_C + \overline{AO}\omega = 4 + 4(1/3) = 16/3 \text{ in./sec} \qquad \textit{Ans.}$$
$$a_O = (a_A)_t + \overline{AO}\alpha = a_C + \overline{AO}\alpha = 16 + 4(-2) = 8 \text{ in./sec}^2 \qquad \textit{Ans.}$$

Helpful Hints

(1) Recognize that the inner pulley is a wheel rolling along the fixed line of the left-hand cable. Thus, the expressions of Sample Problem 5/4 hold.

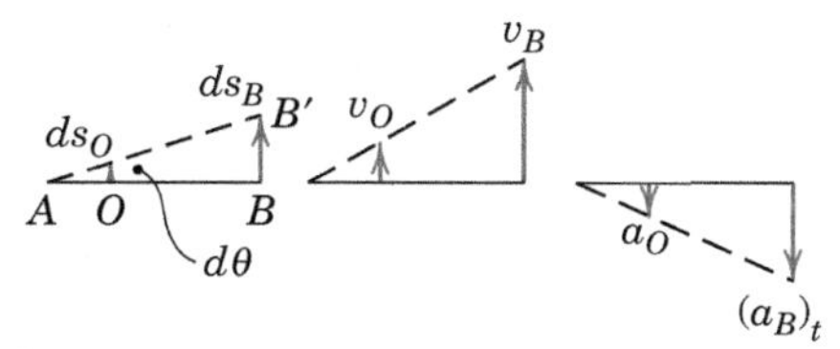

Case (a)

(2) Since B moves along a curved path, in addition to its tangential component of acceleration $(a_B)_t$, it will also have a normal component of acceleration toward O which does not affect the angular acceleration of the pulley.

(3) The diagrams show these quantities and the simplicity of their linear relationships. The visual picture of the motion of O and B as AB rotates through the angle $d\theta$ should clarify the analysis.

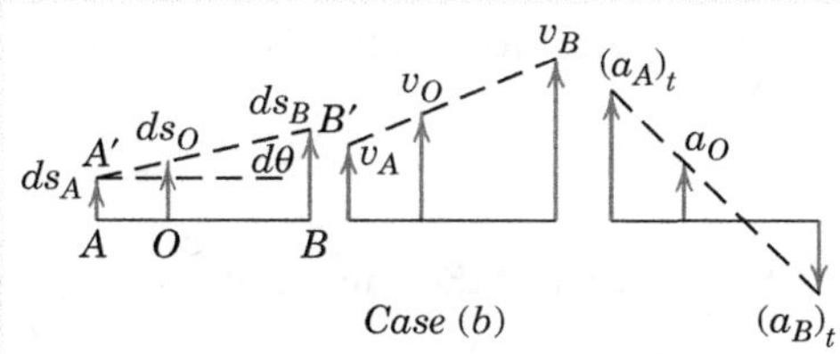

Case (b)

(4) Again, as in case (*a*), the differential rotation of line AB as seen from the figure establishes the relation between the angular velocity of the pulley and the linear velocities of points A, O, and B. The negative sign for $(a_B)_t = a_D$ produces the acceleration diagram shown but does not destroy the linearity of the relationships.

Sample Problem 5/6

Motion of the equilateral triangular plate ABC in its plane is controlled by the hydraulic cylinder D. If the piston rod in the cylinder is moving upward at the constant rate of 0.3 m/s during an interval of its motion, calculate for the instant when $\theta = 30°$ the velocity and acceleration of the center of the roller B in the horizontal guide and the angular velocity and angular acceleration of edge CB.

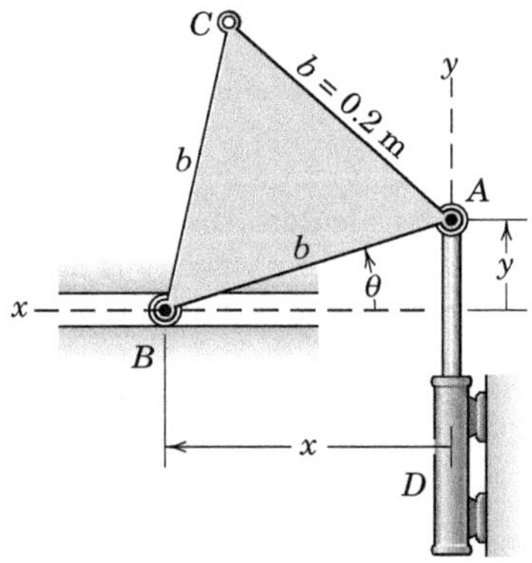

Solution. With the x-y coordinates chosen as shown, the given motion of A is $v_A = \dot{y} = 0.3$ m/s and $a_A = \ddot{y} = 0$. The accompanying motion of B is given by x and its time derivatives, which may be obtained from $x^2 + y^2 = b^2$. Differentiating gives

① $$x\dot{x} + y\dot{y} = 0 \qquad \dot{x} = -\frac{y}{x}\dot{y}$$

$$x\ddot{x} + \dot{x}^2 + y\ddot{y} + \dot{y}^2 = 0 \qquad \ddot{x} = -\frac{\dot{x}^2 + \dot{y}^2}{x} - \frac{y}{x}\ddot{y}$$

With $y = b \sin\theta$, $x = b\cos\theta$, and $\ddot{y} = 0$, the expressions become

$$v_B = \dot{x} = -v_A \tan\theta$$

$$a_B = \ddot{x} = -\frac{v_A{}^2}{b}\sec^3\theta$$

Substituting the numerical values $v_A = 0.3$ m/s and $\theta = 30°$ gives

$$v_B = -0.3\left(\frac{1}{\sqrt{3}}\right) = -0.1732 \text{ m/s} \qquad \textit{Ans.}$$

$$a_B = -\frac{(0.3)^2(2/\sqrt{3})^3}{0.2} = -0.693 \text{ m/s}^2 \qquad \textit{Ans.}$$

The negative signs indicate that the velocity and acceleration of B are both to the right since x and its derivatives are positive to the left.

The angular motion of CB is the same as that of every line on the plate, including AB. Differentiating $y = b\sin\theta$ gives

$$\dot{y} = b\dot{\theta}\cos\theta \qquad \omega = \dot{\theta} = \frac{v_A}{b}\sec\theta$$

The angular acceleration is

$$\alpha = \dot{\omega} = \frac{v_A}{b}\dot{\theta}\sec\theta\tan\theta = \frac{v_A{}^2}{b^2}\sec^2\theta\tan\theta$$

Substitution of the numerical values gives

$$\omega = \frac{0.3}{0.2}\frac{2}{\sqrt{3}} = 1.732 \text{ rad/s} \qquad \textit{Ans.}$$

$$\alpha = \frac{(0.3)^2}{(0.2)^2}\left(\frac{2}{\sqrt{3}}\right)^2\frac{1}{\sqrt{3}} = 1.732 \text{ rad/s}^2 \qquad \textit{Ans.}$$

Both ω and α are counterclockwise since their signs are positive in the sense of the positive measurement of θ.

Helpful Hint

① Observe that it is simpler to differentiate a product than a quotient. Thus, differentiate $x\dot{x} + y\dot{y} = 0$ rather than $\dot{x} = -y\dot{y}/x$.

PROBLEMS

Introductory Problems

5/29 Slider A moves in the horizontal slot with a constant speed v for a short interval of motion. Determine the angular velocity ω of bar AB in terms of the displacement x_A.

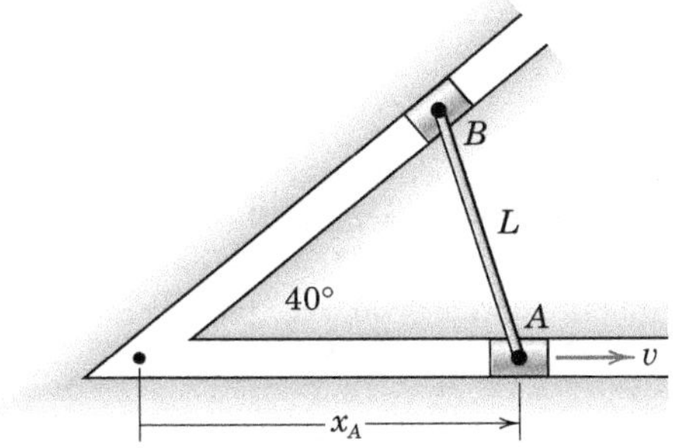

Problem 5/29

5/30 The fixed hydraulic cylinder C imparts a constant upward velocity v to the collar B, which slips freely on rod OA. Determine the resulting angular velocity ω_{OA} in terms of v, the displacement s of point B, and the fixed distance d.

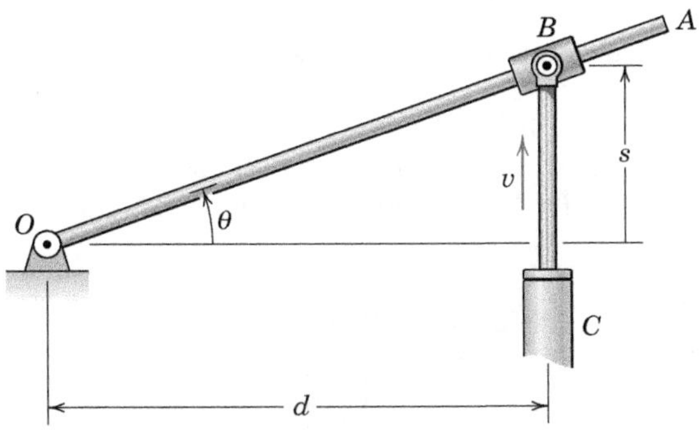

Problem 5/30

5/31 The concrete pier P is being lowered by the pulley and cable arrangement shown. If points A and B have velocities of 0.4 m/s and 0.2 m/s, respectively, compute the velocity of P, the velocity of point C for the instant represented, and the angular velocity of the pulley.

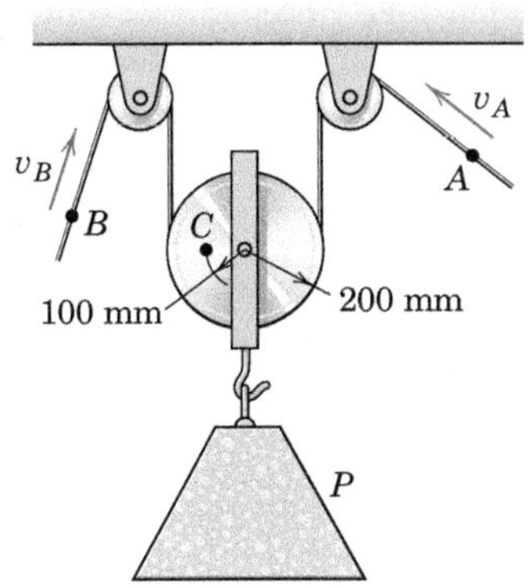

Problem 5/31

5/32 At the instant under consideration, the hydraulic cylinder AB has a length $L = 0.75$ m, and this length is momentarily increasing at a constant rate of 0.2 m/s. If $v_A = 0.6$ m/s and $\theta = 35°$, determine the velocity of slider B.

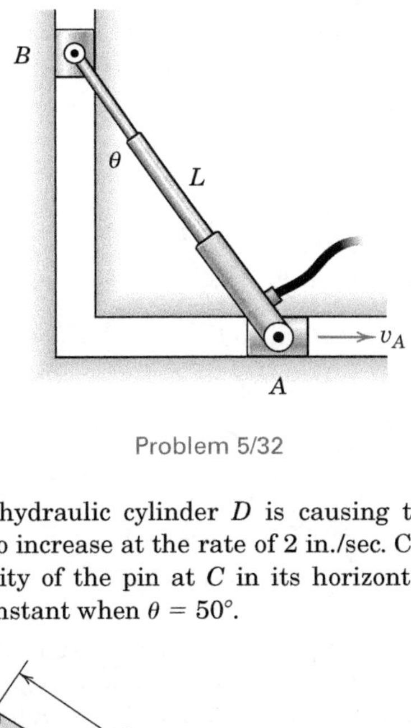

Problem 5/32

5/33 The hydraulic cylinder D is causing the distance OA to increase at the rate of 2 in./sec. Calculate the velocity of the pin at C in its horizontal guide for the instant when $\theta = 50°$.

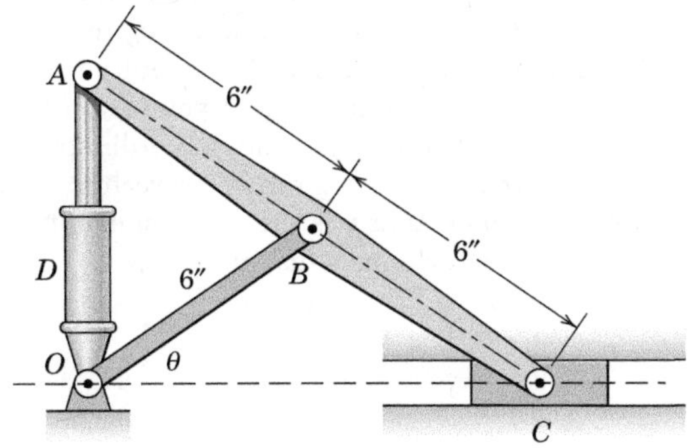

Problem 5/33

5/34 The Scotch-yoke mechanism converts rotational motion of the disk to oscillatory translation of the shaft. For given values of θ, ω, α, r, and d, determine the velocity and acceleration of point P of the shaft.

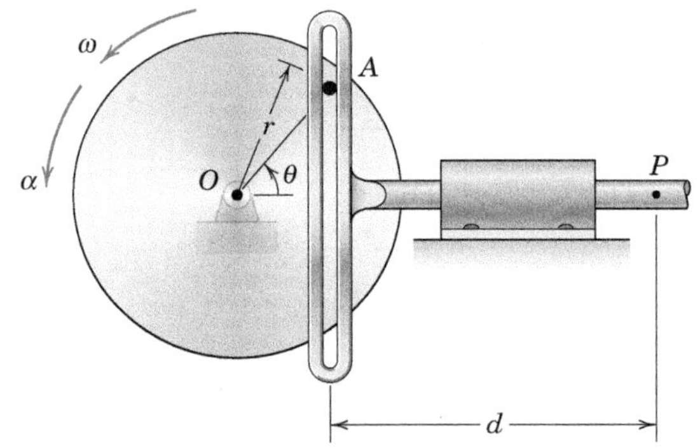

Problem 5/34

5/35 The Scotch-yoke mechanism of Prob. 5/34 is modified as shown in the figure. For given values of ω, α, r, θ, d, and β, determine the velocity and acceleration of point P of the shaft.

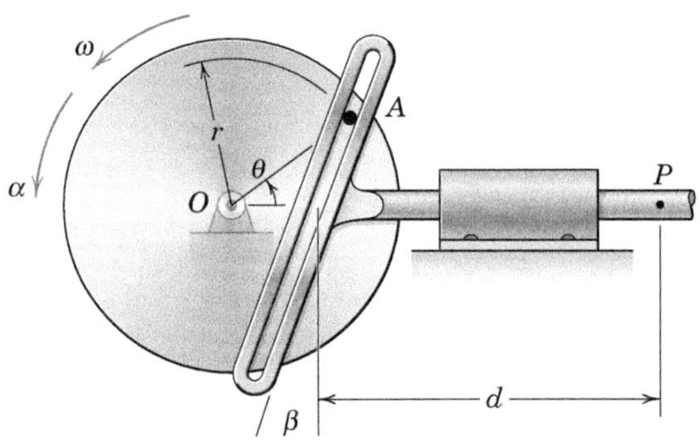

Problem 5/35

5/36 The wheel of radius r rolls without slipping, and its center O has a constant velocity v_O to the right. Determine expressions for the magnitudes of the velocity $\mathbf{v}$ and acceleration $\mathbf{a}$ of point A on the rim by differentiating its x- and y-coordinates. Represent your results graphically as vectors on your sketch and show that $\mathbf{v}$ is the vector sum of two vectors, each of which has a magnitude v_O.

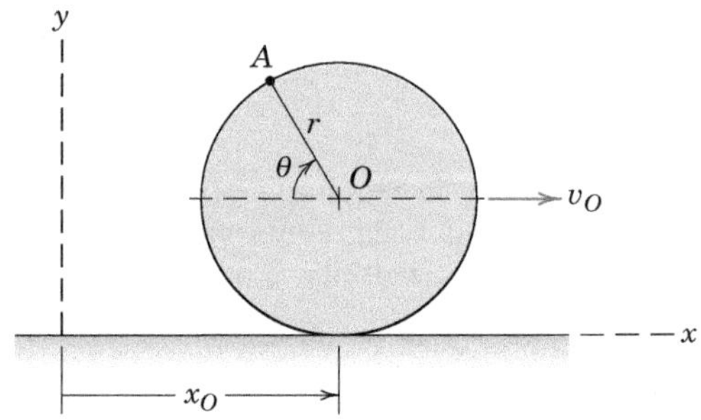

Problem 5/36

5/37 Link OA rotates with a clockwise angular velocity $\omega = 7$ rad/s. Determine the velocity of point B for the position $\theta = 30°$. Use the values $b = 80$ mm, $d = 100$ mm, and $h = 30$ mm.

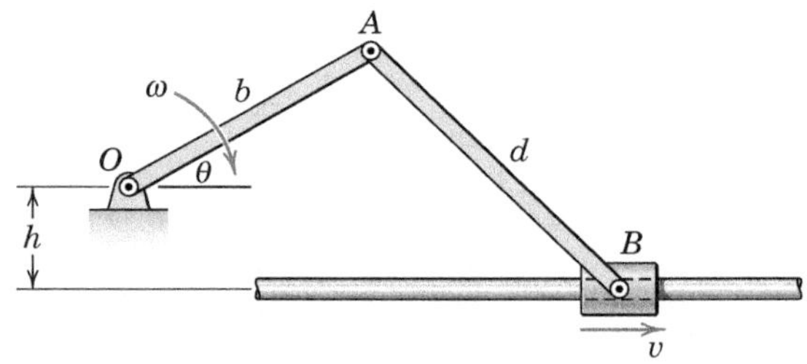

Problem 5/37

5/38 Determine the acceleration of the shaft B for $\theta = 60°$ if the crank OA has an angular acceleration $\ddot{\theta} = 8$ rad/s^2 and an angular velocity $\dot{\theta} = 4$ rad/s at this position. The spring maintains contact between the roller and the surface of the plunger.

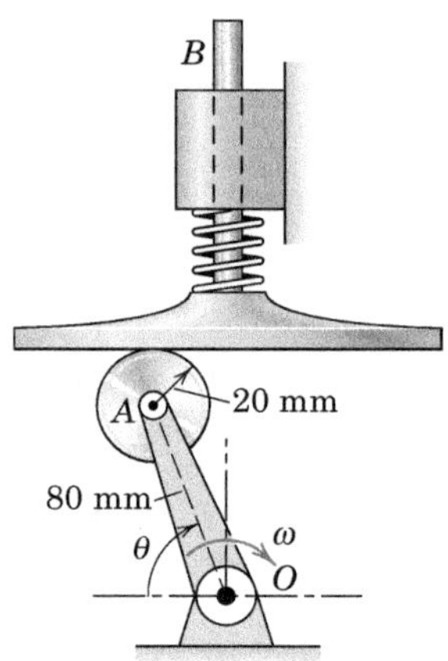

Problem 5/38

5/39 Link OA rotates with a counterclockwise angular velocity $\omega = 3$ rad/s. Determine the angular velocity of bar BC when $\theta = 20°$.

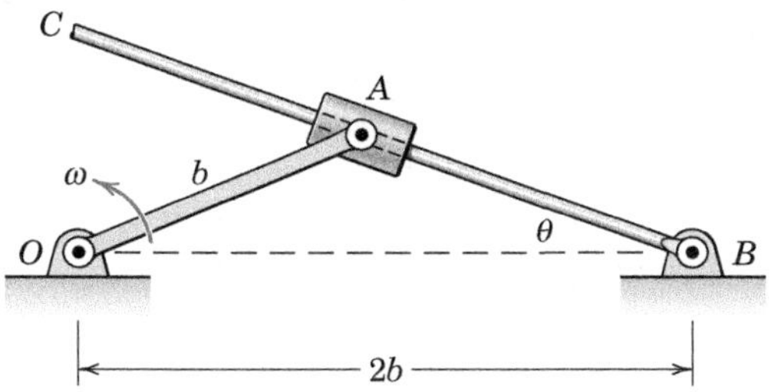

Problem 5/39

5/40 The collar C moves to the left on the fixed guide with speed v. Determine the angular velocity ω_{OA} as a function of v, the collar position s, and the height h.

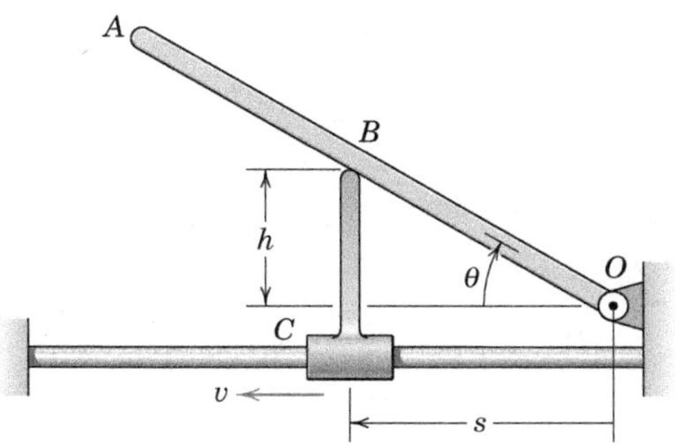

Problem 5/40

Representative Problems

5/41 Boom OA is being elevated by the rope-and-pulley arrangement shown. If point B on the rope is given a constant velocity $v_B = 9$ ft/sec, determine the angular velocity ω and angular acceleration α of the boom for $\theta = 30°$.

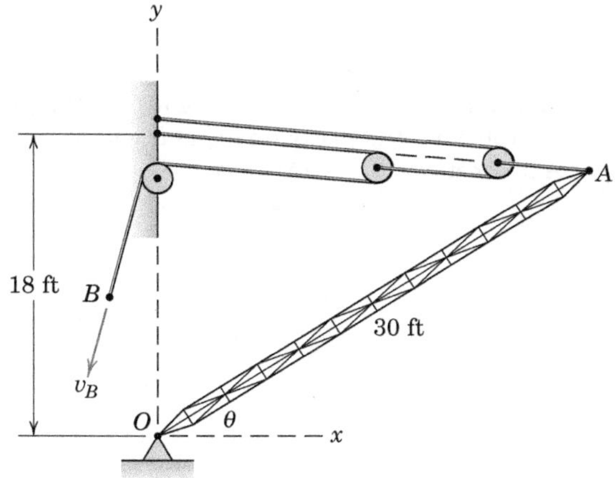

Problem 5/41

5/42 The hydraulic cylinder imparts a constant upward velocity $v_A = 0.2$ m/s to corner A of the rectangular container during an interval of its motion. For the instant when $\theta = 20°$, determine the velocity and acceleration of roller B. Also, determine the corresponding angular velocity of edge CD.

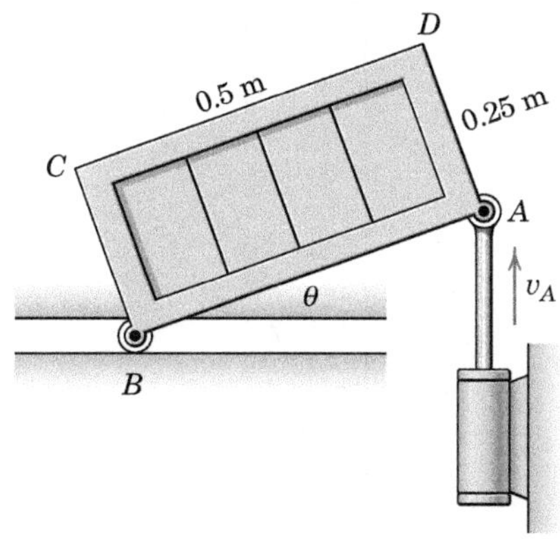

Problem 5/42

5/43 Vertical motion of the work platform is controlled by the horizontal motion of pin A. If A has a velocity v_0 to the left, determine the vertical velocity v of the platform for any value of θ.

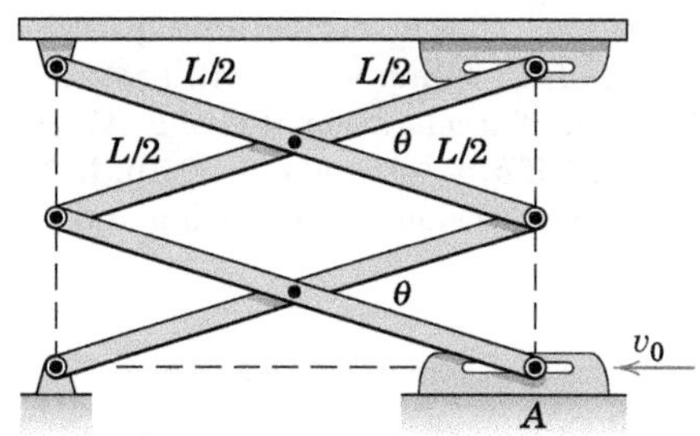

Problem 5/43

5/44 The rod OB slides through the collar pivoted to the rotating link at A. If CA has an angular velocity $\omega = 3$ rad/s for an interval of motion, calculate the angular velocity of OB when $\theta = 45°$.

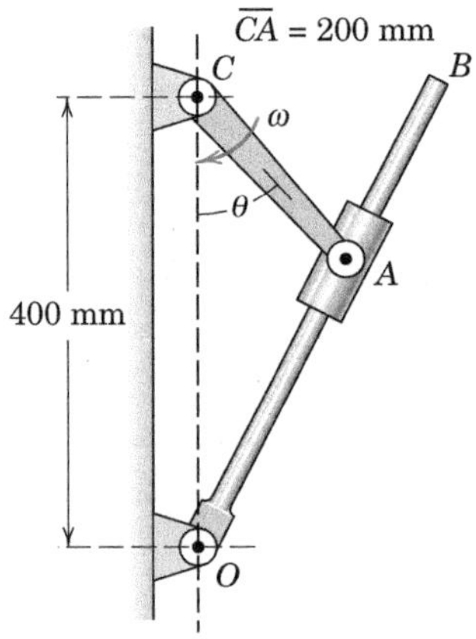

Problem 5/44

5/45 A roadway speed bump is being installed on a level road to remind motorists of the existing speed limit. If the driver of the car experiences at G a vertical acceleration of as much as g, up or down, he is expected to realize that his speed is bordering on being excessive. For the speed bump with the cosine contour shown, derive an expression for the height h of the bump which will produce a vertical component of acceleration at G of g at a car speed v. Compute h if $b = 1$ m and $v = 20$ km/h. Neglect the effects of suspension-spring flexing and finite wheel diameter.

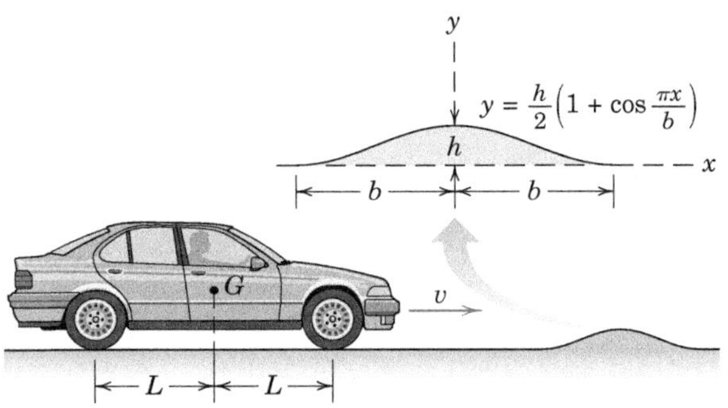

Problem 5/45

5/46 Motion of the wheel as it rolls up the fixed rack on its geared hub is controlled through the peripheral cable by the driving wheel D, which turns counterclockwise at the constant rate $\omega_0 = 4$ rad/s for a short interval of motion. By examining the geometry of a small (differential) rotation of line $AOCB$ as it pivots momentarily about the contact point C, determine the angular velocity ω of the wheel and the velocities of point A and the center O. Also find the acceleration of point C.

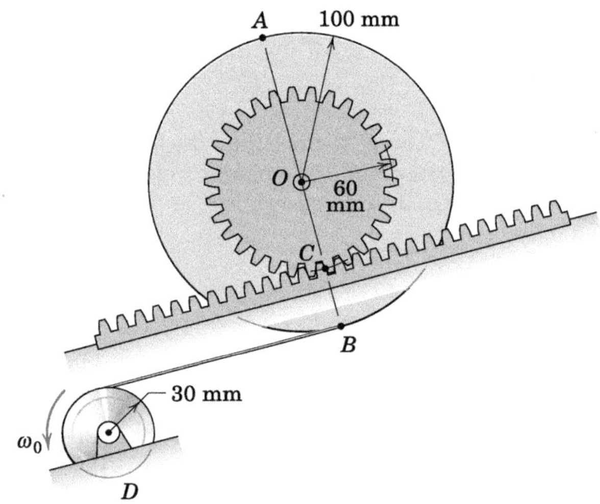

Problem 5/46

5/47 Link OA is given a clockwise angular velocity $\omega = 2$ rad/sec as indicated. Determine the velocity v of point C for the position $\theta = 30°$ if $b = 8$ in.

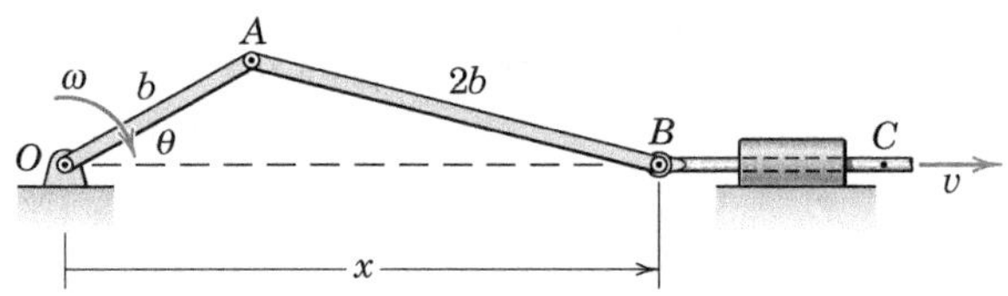

Problem 5/47

5/48 Determine the acceleration of point C of the previous problem if the clockwise angular velocity of link OA is constant at $\omega = 2$ rad/sec.

5/49 Derive an expression for the upward velocity v of the car hoist in terms of θ. The piston rod of the hydraulic cylinder is extending at the rate $\dot{s}$.

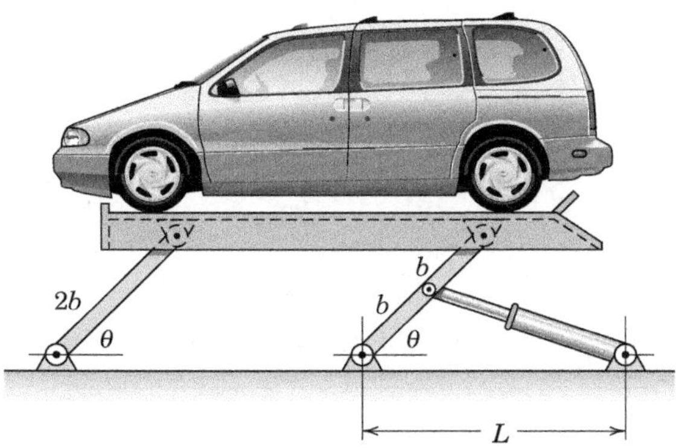

Problem 5/49

5/50 It is desired to design a system for controlling the rate of extension $\dot{x}$ of the fire-truck ladder during elevation of the ladder so that the bucket B will have vertical motion only. Determine $\dot{x}$ in terms of the elongation rate $\dot{c}$ of the hydraulic cylinder for given values of θ and x.

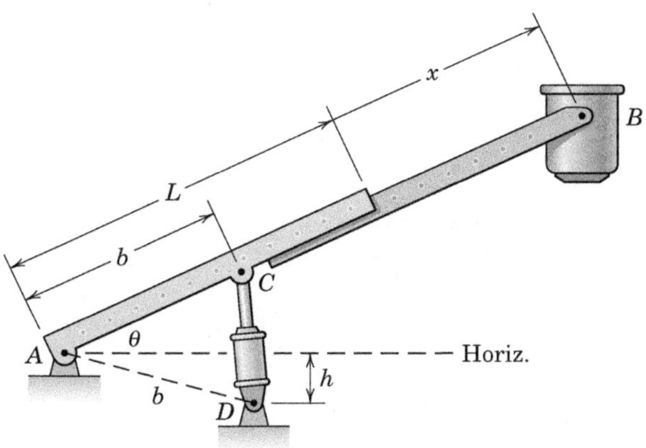

Problem 5/50

5/51 Show that the expressions $v = r\omega$ and $a_t = r\alpha$ hold for the motion of the center O of the wheel which rolls on the concave or convex circular arc, where ω and α are the absolute angular velocity and acceleration, respectively, of the wheel. (*Hint:* Follow the example of Sample Problem 5/4 and allow the wheel to roll a small distance. Be very careful to identify the correct *absolute* angle through which the wheel turns in each case in determining its angular velocity and angular acceleration.)

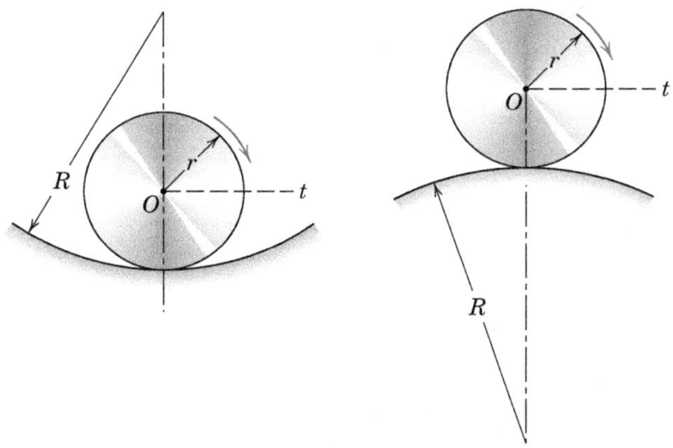

Problem 5/51

5/52 The rotation of link AO is controlled by the piston rod of hydraulic cylinder BC, which is elongating at the constant rate $\dot{s} = k$ for an interval of motion. Write the vector expression for the acceleration of end A for a given value of θ using unit vectors $\mathbf{e}_n$ and $\mathbf{e}_t$ with $n-t$ coordinates.

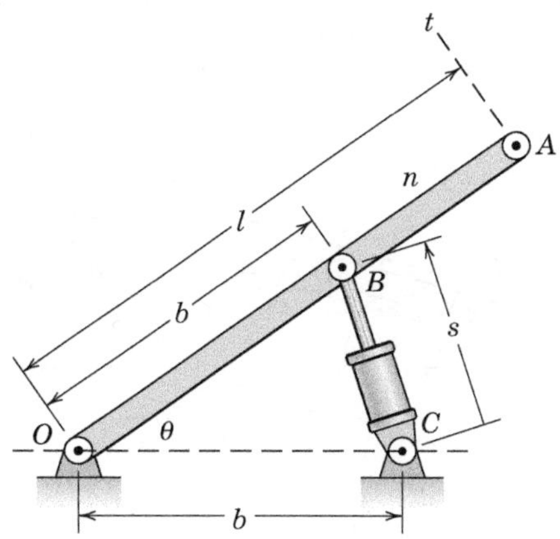

Problem 5/52

5/53 A variable-speed belt drive consists of the two pulleys, each of which is constructed of two cones which turn as a unit but are capable of being drawn together or separated so as to change the effective radius of the pulley. If the angular velocity ω_1 of pulley 1 is constant, determine the expression for the angular acceleration $\alpha_2 = \dot{\omega}_2$ of pulley 2 in terms of the rates of change $\dot{r}_1$ and $\dot{r}_2$ of the effective radii.

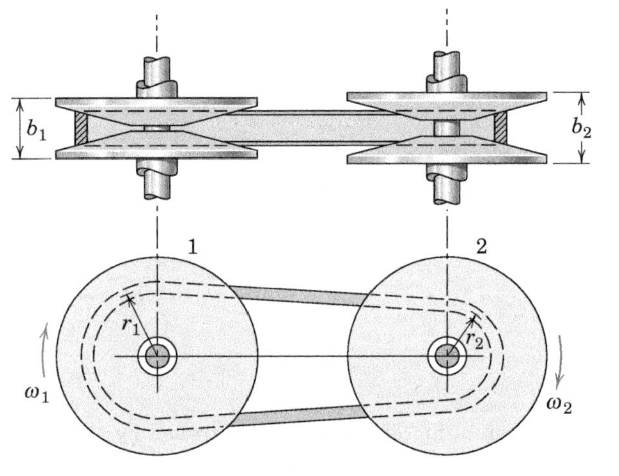

Problem 5/53

5/54 Angular oscillation of the slotted link is achieved by the crank *OA*, which rotates clockwise at the steady speed $N = 120$ rev/min. Determine an expression for the angular velocity $\dot{\beta}$ of the slotted link in terms of θ.

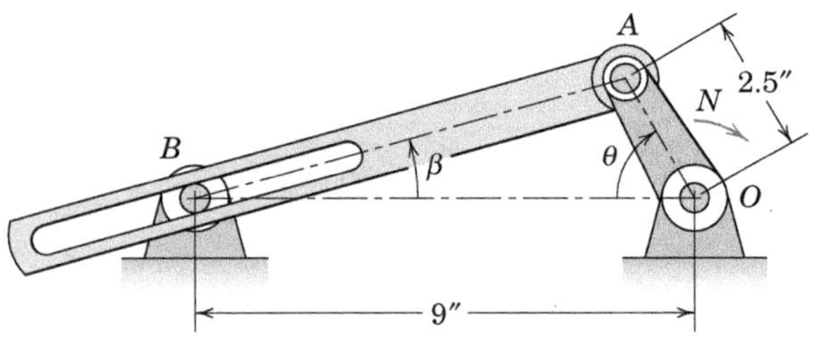

Problem 5/54

5/55 The two gears form an integral unit and roll on the fixed rack. The large gear has 48 teeth, and the worm turns with a speed of 120 rev/min. Find the velocity v_O of the center *O* of the gear.

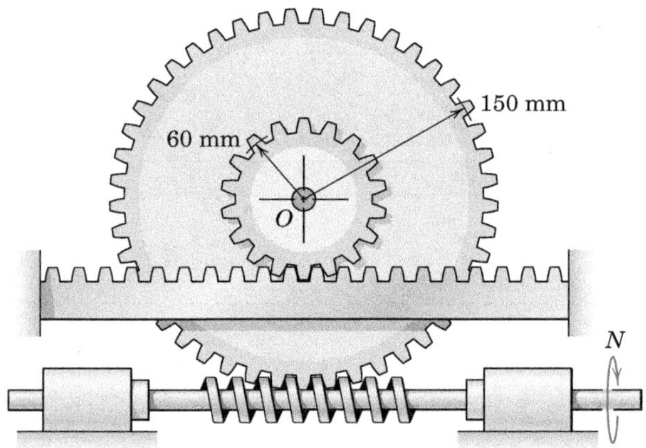

Problem 5/55

►5/56 One of the most common mechanisms is the slider-crank. Express the angular velocity ω_{AB} and angular acceleration α_{AB} of the connecting rod *AB* in terms of the crank angle θ for a given constant crank speed ω_0. Take ω_{AB} and α_{AB} to be positive counterclockwise.

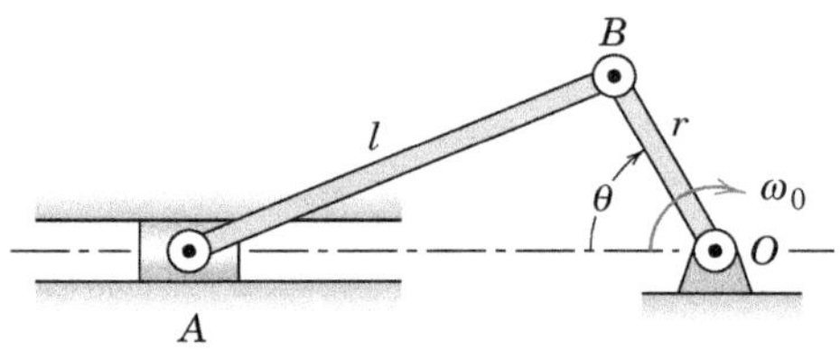

Problem 5/56

▶5/57 The Geneva wheel is a mechanism for producing intermittent rotation. Pin P in the integral unit of wheel A and locking plate B engages the radial slots in wheel C, thus turning wheel C one-fourth of a revolution for each revolution of the pin. At the engagement position shown, $\theta = 45°$. For a constant clockwise angular velocity $\omega_1 = 2$ rad/s of wheel A, determine the corresponding counterclockwise angular velocity ω_2 of wheel C for $\theta = 20°$. (Note that the motion during engagement is governed by the geometry of triangle O_1O_2P with changing θ.)

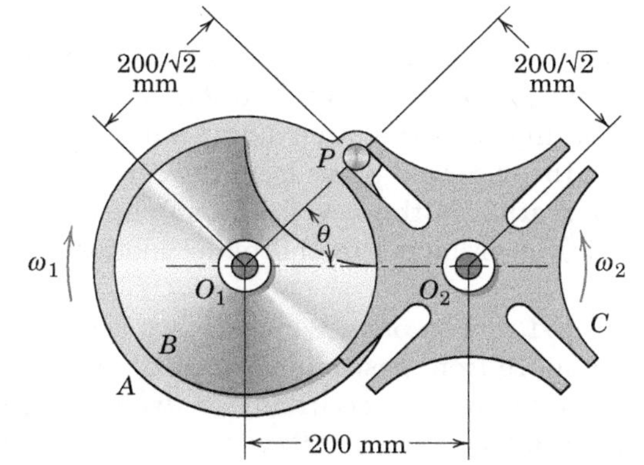

Problem 5/57

▶5/58 The punch is operated by a simple harmonic oscillation of the pivoted sector given by $\theta = \theta_0 \sin 2\pi t$ where the amplitude is $\theta_0 = \pi/12$ rad (15°) and the time for one complete oscillation is 1 second. Determine the acceleration of the punch when (*a*) $\theta = 0$ and (*b*) $\theta = \pi/12$.

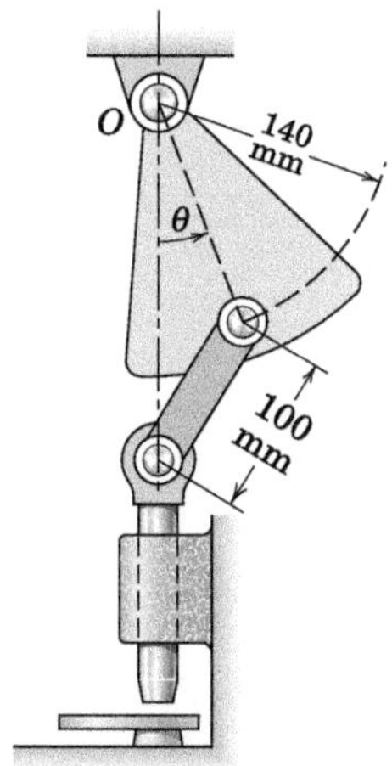

Problem 5/58

5/4 Relative Velocity

The second approach to rigid-body kinematics is to use the principles of relative motion. In Art. 2/8 we developed these principles for motion relative to translating axes and applied the relative-velocity equation

$$\mathbf{v}_A = \mathbf{v}_B + \mathbf{v}_{A/B} \qquad \textbf{[2/20]}$$

to the motions of two particles A and B.

Relative Velocity Due to Rotation

We now choose two points on the *same* rigid body for our two particles. The consequence of this choice is that the motion of one point as seen by an observer translating with the other point must be circular since the radial distance to the observed point from the reference point does not change. This observation is the *key* to the successful understanding of a large majority of problems in the plane motion of rigid bodies.

This concept is illustrated in Fig. 5/5*a*, which shows a rigid body moving in the plane of the figure from position AB to $A'B'$ during time Δt. This movement may be visualized as occurring in two parts. First, the body translates to the parallel position $A''B'$ with the displacement $\Delta\mathbf{r}_B$. Second, the body rotates about B' through the angle $\Delta\theta$. From the nonrotating reference axes x'-y' attached to the reference point B', you can see that this remaining motion of the body is one of simple rotation about B', giving rise to the displacement $\Delta\mathbf{r}_{A/B}$ of A with respect to B. To the nonrotating observer attached to B, the body appears to undergo fixed-axis rotation about B with A executing circular motion as emphasized in Fig. 5/5*b*. Therefore, the relationships developed for circular motion in Arts. 2/5 and 5/2 and cited as Eqs. 2/11 and 5/2 (or 5/3) describe the relative portion of the motion of point A.

Point B was arbitrarily chosen as the reference point for attachment of our nonrotating reference axes x-y. Point A could have been used just as well, in which case we would observe B to have circular motion about A considered fixed as shown in Fig. 5/5*c*. We see that the

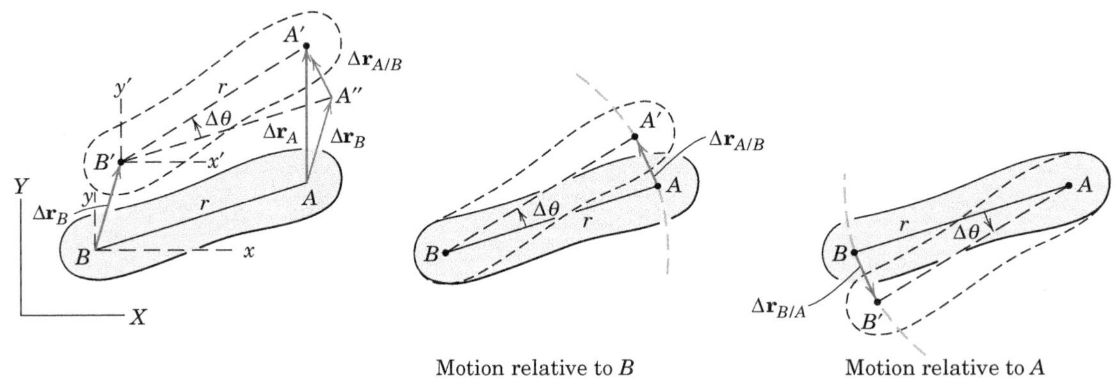

Motion relative to B

Motion relative to A

(*a*) (*b*) (*c*)

Figure 5/5

sense of the rotation, counterclockwise in this example, is the same whether we choose A or B as the reference, and we see that $\Delta\mathbf{r}_{B/A} = -\Delta\mathbf{r}_{A/B}$.

With B as the reference point, we see from Fig. 5/5*a* that the total displacement of A is

$$\Delta\mathbf{r}_A = \Delta\mathbf{r}_B + \Delta\mathbf{r}_{A/B}$$

where $\Delta\mathbf{r}_{A/B}$ has the magnitude $r\Delta\theta$ as $\Delta\theta$ approaches zero. We note that the *relative linear motion* $\Delta\mathbf{r}_{A/B}$ is accompanied by the *absolute angular motion* $\Delta\theta$, as seen from the translating axes x'-y'. Dividing the expression for $\Delta\mathbf{r}_A$ by the corresponding time interval Δt and passing to the limit, we obtain the relative-velocity equation

$$\boxed{\mathbf{v}_A = \mathbf{v}_B + \mathbf{v}_{A/B}} \qquad \textbf{(5/4)}$$

This expression is the same as Eq. 2/20, with the one restriction that the distance r between A and B remains constant. The magnitude of the relative velocity is thus seen to be $v_{A/B} = \lim_{\Delta t \to 0} (|\Delta\mathbf{r}_{A/B}|/\Delta t) = \lim_{\Delta t \to 0} (r\Delta\theta/\Delta t)$ which, with $\omega = \dot{\theta}$, becomes

$$\boxed{v_{A/B} = r\omega} \qquad \textbf{(5/5)}$$

Using $\mathbf{r}$ to represent the vector $\mathbf{r}_{A/B}$ from the first of Eqs. 5/3, we may write the relative velocity as the vector

$$\boxed{\mathbf{v}_{A/B} = \boldsymbol{\omega} \times \mathbf{r}} \qquad \textbf{(5/6)}$$

where $\boldsymbol{\omega}$ is the angular-velocity vector normal to the plane of the motion in the sense determined by the right-hand rule. A critical observation seen from Figs. 5/5*b* and *c* is that the relative linear velocity is always perpendicular to the line joining the two points in question.

Interpretation of the Relative-Velocity Equation

We can better understand the application of Eq. 5/4 by visualizing the separate translation and rotation components of the equation. These components are emphasized in Fig. 5/6, which shows a rigid body in

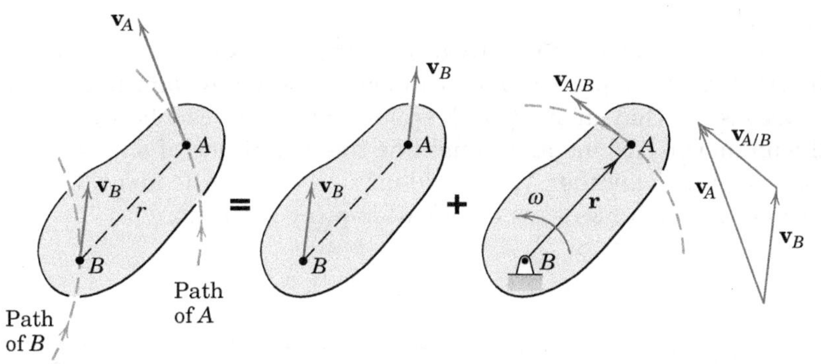

Figure 5/6

plane motion. With B chosen as the reference point, the velocity of A is the vector sum of the translational portion $\mathbf{v}_B$, plus the rotational portion $\mathbf{v}_{A/B} = \boldsymbol{\omega} \times \mathbf{r}$, which has the magnitude $v_{A/B} = r\omega$, where $|\boldsymbol{\omega}| = \dot{\theta}$, the *absolute* angular velocity of AB. The fact that the *relative linear velocity* is *always perpendicular* to the line joining the two points in question is an important key to the solution of many problems. To reinforce your understanding of this concept, you should draw the equivalent diagram where point A is used as the reference point rather than B.

Equation 5/4 may also be used to analyze constrained sliding contact between two links in a mechanism. In this case, we choose points A and B as coincident points, one on each link, for the instant under consideration. In contrast to the previous example, in this case, the two points are on different bodies so they are not a fixed distance apart. This second use of the relative-velocity equation is illustrated in Sample Problem 5/10.

Solution of the Relative-Velocity Equation

Solution of the relative-velocity equation may be carried out by scalar or vector algebra, or a graphical analysis may be employed. A sketch of the vector polygon which represents the vector equation should always be made to reveal the physical relationships involved. From this sketch, you can write scalar component equations by projecting the vectors along convenient directions. You can usually avoid solving simultaneous equations by a careful choice of the projections. Alternatively, each term in the relative-motion equation may be written in terms of its **i**- and **j**-components, from which you will obtain two scalar equations when the equality is applied, separately, to the coefficients of the **i**- and **j**-terms.

Many problems lend themselves to a graphical solution, particularly when the given geometry results in an awkward mathematical expression. In this case, we first construct the known vectors in their correct positions using a convenient scale. Then we construct the unknown vectors which complete the polygon and satisfy the vector equation. Finally, we measure the unknown vectors directly from the drawing.

The choice of method to be used depends on the particular problem at hand, the accuracy required, and individual preference and experience. All three approaches are illustrated in the sample problems which follow.

Regardless of which method of solution we employ, we note that the single vector equation in two dimensions is equivalent to two scalar equations, so that at most two scalar unknowns can be determined. The unknowns, for instance, might be the magnitude of one vector and the direction of another. We should make a systematic identification of the knowns and unknowns before attempting a solution.

Sample Problem 5/7

The wheel of radius $r = 300$ mm rolls to the right without slipping and has a velocity $v_O = 3$ m/s of its center O. Calculate the velocity of point A on the wheel for the instant represented.

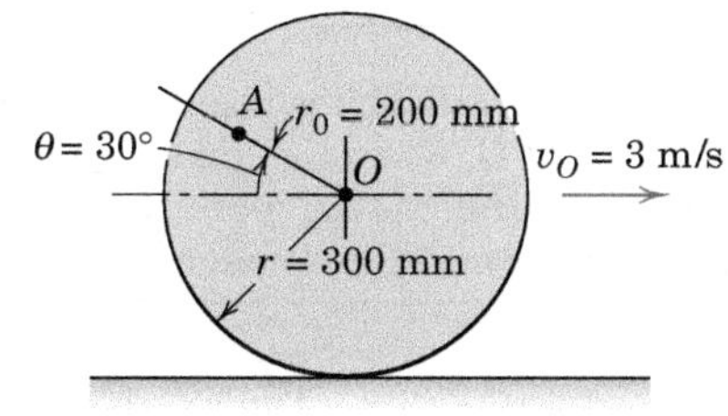

Solution I (Scalar-Geometric). The center O is chosen as the reference point for the relative-velocity equation since its motion is given. We therefore write

$$\mathbf{v}_A = \mathbf{v}_O + \mathbf{v}_{A/O}$$

where the relative-velocity term is observed from the translating axes x-y attached to O. The angular velocity of AO is the same as that of the wheel which, from Sample Problem 5/4, is $\omega = v_O/r = 3/0.3 = 10$ rad/s. Thus, from Eq. 5/5 we have

$$[v_{A/O} = r_0\dot{\theta}] \qquad v_{A/O} = 0.2(10) = 2 \text{ m/s}$$

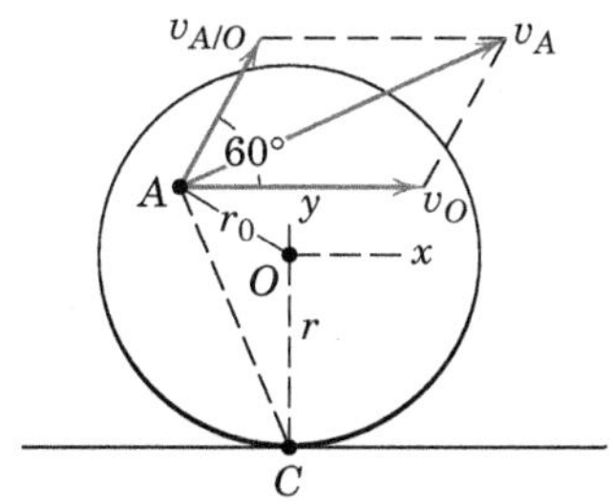

① which is normal to AO as shown. The vector sum $\mathbf{v}_A$ is shown on the diagram and may be calculated from the law of cosines. Thus,

② $$v_A{}^2 = 3^2 + 2^2 + 2(3)(2) \cos 60° = 19 \text{ (m/s)}^2 \qquad v_A = 4.36 \text{ m/s} \qquad \textit{Ans.}$$

The contact point C momentarily has zero velocity and can be used alternatively as the reference point, in which case, the relative-velocity equation becomes $\mathbf{v}_A = \mathbf{v}_C + \mathbf{v}_{A/C} = \mathbf{v}_{A/C}$ where

$$v_{A/C} = \overline{AC}\omega = \frac{\overline{AC}}{\overline{OC}} v_O = \frac{0.436}{0.300}(3) = 4.36 \text{ m/s} \qquad v_A = v_{A/C} = 4.36 \text{ m/s}$$

③ The distance $\overline{AC} = 436$ mm is calculated separately. We see that $\mathbf{v}_A$ is normal to AC since A is momentarily rotating about point C.

Solution II (Vector). We will now use Eq. 5/6 and write

$$\mathbf{v}_A = \mathbf{v}_O + \mathbf{v}_{A/O} = \mathbf{v}_O + \boldsymbol{\omega} \times \mathbf{r}_0$$

where

④ $$\boldsymbol{\omega} = -10\mathbf{k} \text{ rad/s}$$

$$\mathbf{r}_0 = 0.2(-\mathbf{i} \cos 30° + \mathbf{j} \sin 30°) = -0.1732\mathbf{i} + 0.1\mathbf{j} \text{ m}$$

$$\mathbf{v}_O = 3\mathbf{i} \text{ m/s}$$

We now solve the vector equation

$$\mathbf{v}_A = 3\mathbf{i} + \begin{vmatrix} \mathbf{i} & \mathbf{j} & \mathbf{k} \\ 0 & 0 & -10 \\ -0.1732 & 0.1 & 0 \end{vmatrix} = 3\mathbf{i} + 1.732\mathbf{j} + \mathbf{i}$$

$$= 4\mathbf{i} + 1.732\mathbf{j} \text{ m/s} \qquad \textit{Ans.}$$

The magnitude $v_A = \sqrt{4^2 + (1.732)^2} = \sqrt{19} = 4.36$ m/s and direction agree with the previous solution.

Helpful Hints

① Be sure to visualize $v_{A/O}$ as the velocity which A appears to have in its circular motion relative to O.

② The vectors may also be laid off to scale graphically and the magnitude and direction of v_A measured directly from the diagram.

③ The velocity of any point on the wheel is easily determined by using the contact point C as the reference point. You should construct the velocity vectors for a number of points on the wheel for practice.

④ The vector $\boldsymbol{\omega}$ is directed into the paper by the right-hand rule, whereas the positive z-direction is out from the paper; hence, the minus sign.

Sample Problem 5/8

Crank CB oscillates about C through a limited arc, causing crank OA to oscillate about O. When the linkage passes the position shown with CB horizontal and OA vertical, the angular velocity of CB is 2 rad/s counterclockwise. For this instant, determine the angular velocities of OA and AB.

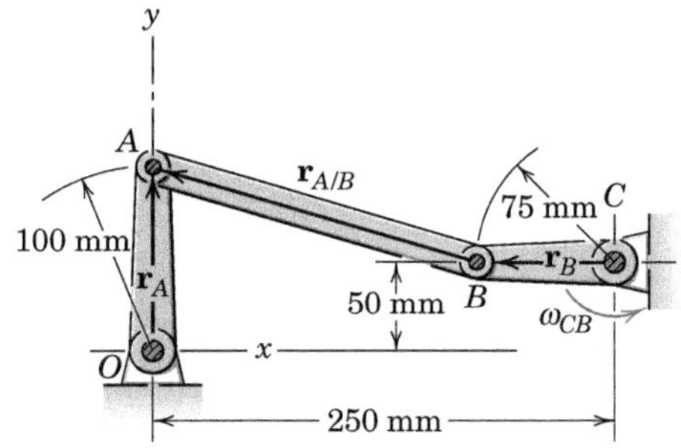

Solution I (Vector). The relative-velocity equation $\mathbf{v}_A = \mathbf{v}_B + \mathbf{v}_{A/B}$ is rewritten as

① $$\boldsymbol{\omega}_{OA} \times \mathbf{r}_A = \boldsymbol{\omega}_{CB} \times \mathbf{r}_B + \boldsymbol{\omega}_{AB} \times \mathbf{r}_{A/B}$$

where

$$\boldsymbol{\omega}_{OA} = \omega_{OA}\mathbf{k} \qquad \boldsymbol{\omega}_{CB} = 2\mathbf{k} \text{ rad/s} \qquad \boldsymbol{\omega}_{AB} = \omega_{AB}\mathbf{k}$$

$$\mathbf{r}_A = 100\mathbf{j} \text{ mm} \qquad \mathbf{r}_B = -75\mathbf{i} \text{ mm} \qquad \mathbf{r}_{A/B} = -175\mathbf{i} + 50\mathbf{j} \text{ mm}$$

Substitution gives

$$\omega_{OA}\mathbf{k} \times 100\mathbf{j} = 2\mathbf{k} \times (-75\mathbf{i}) + \omega_{AB}\mathbf{k} \times (-175\mathbf{i} + 50\mathbf{j})$$

$$-100\omega_{OA}\mathbf{i} = -150\mathbf{j} - 175\omega_{AB}\mathbf{j} - 50\omega_{AB}\mathbf{i}$$

Matching coefficients of the respective **i**- and **j**-terms gives

$$-100\omega_{OA} + 50\omega_{AB} = 0 \qquad 25(6 + 7\omega_{AB}) = 0$$

the solutions of which are

② $$\omega_{AB} = -6/7 \text{ rad/s} \quad \text{and} \quad \omega_{OA} = -3/7 \text{ rad/s} \qquad \textit{Ans.}$$

Solution II (Scalar-Geometric). Solution by the scalar geometry of the vector triangle is particularly simple here since $\mathbf{v}_A$ and $\mathbf{v}_B$ are at right angles for this special position of the linkages. First, we compute v_B, which is

$[v = r\omega]$ $$v_B = 0.075(2) = 0.150 \text{ m/s}$$

and represent it in its correct direction as shown. The vector $\mathbf{v}_{A/B}$ must be perpendicular to AB, and the angle θ between $\mathbf{v}_{A/B}$ and $\mathbf{v}_B$ is also the angle made by AB with the horizontal direction. This angle is given by

$$\tan\theta = \frac{100 - 50}{250 - 75} = \frac{2}{7}$$

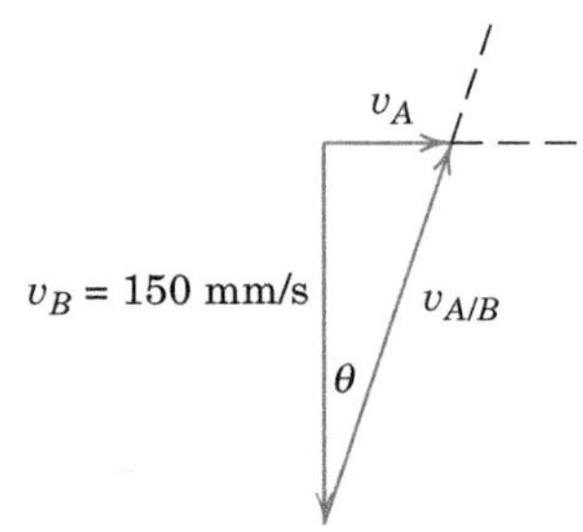

③ The horizontal vector $\mathbf{v}_A$ completes the triangle for which we have

$$v_{A/B} = v_B/\cos\theta = 0.150/\cos\theta$$

$$v_A = v_B \tan\theta = 0.150(2/7) = 0.30/7 \text{ m/s}$$

The angular velocities become

$[\omega = v/r]$ $$\omega_{AB} = \frac{v_{A/B}}{\overline{AB}} = \frac{0.150}{\cos\theta}\,\frac{\cos\theta}{0.250 - 0.075}$$

$$= 6/7 \text{ rad/s CW} \qquad \textit{Ans.}$$

$[\omega = v/r]$ $$\omega_{OA} = \frac{v_A}{\overline{OA}} = \frac{0.30}{7}\,\frac{1}{0.100} = 3/7 \text{ rad/s CW} \qquad \textit{Ans.}$$

Helpful Hints

① We are using here the first of Eqs. 5/3 and Eq. 5/6.

② The minus signs in the answers indicate that the vectors $\boldsymbol{\omega}_{AB}$ and $\boldsymbol{\omega}_{OA}$ are in the negative **k**-direction. Hence, the angular velocities are clockwise.

③ Always make certain that the sequence of vectors in the vector polygon agrees with the equality of vectors specified by the vector equation.

Sample Problem 5/9

The common configuration of a reciprocating engine is that of the slider-crank mechanism shown. If the crank OB has a clockwise rotational speed of 1500 rev/min, determine for the position where $\theta = 60°$ the velocity of the piston A, the velocity of point G on the connecting rod, and the angular velocity of the connecting rod.

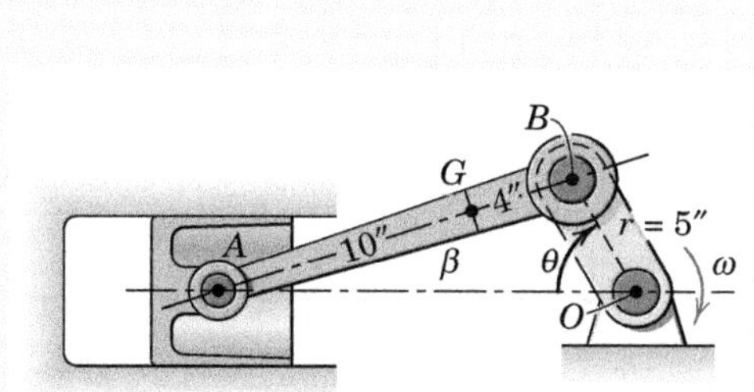

Solution. The velocity of the crank pin B as a point on AB is easily found, so that B will be used as the reference point for determining the velocity of A. The relative-velocity equation may now be written

$$\mathbf{v}_A = \mathbf{v}_B + \mathbf{v}_{A/B}$$

The crank-pin velocity is

① $[v = r\omega]$ $$v_B = \frac{5}{12}\frac{1500\,(2\pi)}{60} = 65.4 \text{ ft/sec}$$

and is normal to OB. The direction of $\mathbf{v}_A$ is, of course, along the horizontal cylinder axis. The direction of $\mathbf{v}_{A/B}$ must be perpendicular to the line AB as explained in the present article and as indicated in the lower diagram, where the reference point B is shown as fixed. We obtain this direction by computing angle β from the law of sines, which gives

$$\frac{5}{\sin\beta} = \frac{14}{\sin 60°} \qquad \beta = \sin^{-1} 0.309 = 18.02°$$

We now complete the sketch of the velocity triangle, where the angle between $\mathbf{v}_{A/B}$ and $\mathbf{v}_A$ is $90° - 18.02° = 72.0°$ and the third angle is $180° - 30° - 72.0° = 78.0°$. Vectors $\mathbf{v}_A$ and $\mathbf{v}_{A/B}$ are shown with their proper sense such that the head-to-tail sum of $\mathbf{v}_B$ and $\mathbf{v}_{A/B}$ equals $\mathbf{v}_A$. The magnitudes of the unknowns are now calculated from the trigonometry of the vector triangle or are scaled from the diagram if a graphical solution is used. Solving for v_A and $v_{A/B}$ by the law of sines gives

② $$\frac{v_A}{\sin 78.0°} = \frac{65.4}{\sin 72.0°} \qquad v_A = 67.3 \text{ ft/sec} \qquad \textit{Ans.}$$

$$\frac{v_{A/B}}{\sin 30°} = \frac{65.4}{\sin 72.0°} \qquad v_{A/B} = 34.4 \text{ ft/sec}$$

The angular velocity of AB is counterclockwise, as revealed by the sense of $\mathbf{v}_{A/B}$, and is

$[\omega = v/r]$ $$\omega_{AB} = \frac{v_{A/B}}{\overline{AB}} = \frac{34.4}{14/12} = 29.5 \text{ rad/sec} \qquad \textit{Ans.}$$

We now determine the velocity of G by writing

$$\mathbf{v}_G = \mathbf{v}_B + \mathbf{v}_{G/B}$$

where $$v_{G/B} = \overline{GB}\omega_{AB} = \frac{\overline{GB}}{\overline{AB}} v_{A/B} = \frac{4}{14}(34.4) = 9.83 \text{ ft/sec}$$

As seen from the diagram, $\mathbf{v}_{G/B}$ has the same direction as $\mathbf{v}_{A/B}$. The vector sum is shown on the last diagram. We can calculate v_G with some geometric labor or simply measure its magnitude and direction from the velocity diagram drawn to scale. For simplicity we adopt the latter procedure here and obtain

$$v_G = 64.1 \text{ ft/sec} \qquad \textit{Ans.}$$

As seen, the diagram may be superposed directly on the first velocity diagram.

Helpful Hints

① Remember always to convert ω to radians per unit time when using $v = r\omega$.

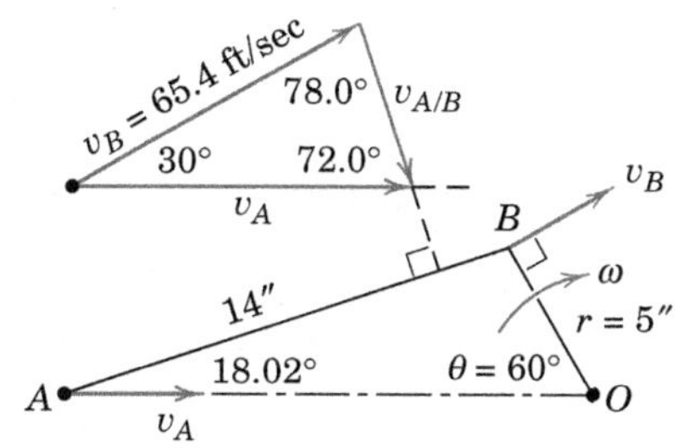

② A graphical solution to this problem is the quickest to achieve, although its accuracy is limited. Solution by vector algebra can, of course, be used but would involve somewhat more labor in this problem.

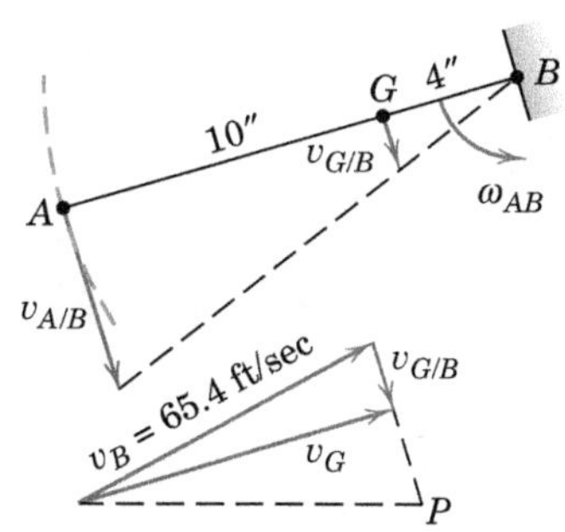

Sample Problem 5/10

The power screw turns at a speed which gives the threaded collar C a velocity of 0.8 ft/sec vertically down. Determine the angular velocity of the slotted arm when $\theta = 30°$.

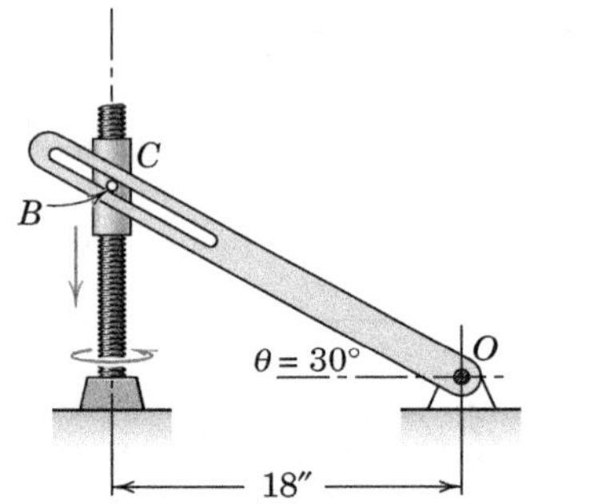

Solution. The angular velocity of the arm can be found if the velocity of a point on the arm is known. We choose a point A on the arm coincident with the
① pin B of the collar for this purpose. If we use B as our reference point and write $\mathbf{v}_A = \mathbf{v}_B + \mathbf{v}_{A/B}$, we see from the diagram, which shows the arm and points A and B an instant before and an instant after coincidence, that $\mathbf{v}_{A/B}$ has a direction along the slot away from O.

② The magnitudes of $\mathbf{v}_A$ and $\mathbf{v}_{A/B}$ are the only unknowns in the vector equation, so that it may now be solved. We draw the known vector $\mathbf{v}_B$ and then obtain the intersection P of the known directions of $\mathbf{v}_{A/B}$ and $\mathbf{v}_A$. The solution gives

$$v_A = v_B \cos\theta = 0.8 \cos 30° = 0.693 \text{ ft/sec}$$

$$[\omega = v/r] \qquad \omega = \frac{v_A}{\overline{OA}} = \frac{0.693}{\left(\frac{18}{12}\right)/\cos 30°}$$

$$= 0.400 \text{ rad/sec CCW} \qquad \textit{Ans.}$$

We note the difference between this problem of constrained sliding contact between two links and the three preceding sample problems of relative velocity, where no sliding contact occurred and where the points A and B were located on the same rigid body in each case.

Helpful Hints

① Physically, of course, this point does not exist, but we can imagine such a point in the middle of the slot and attached to the arm.

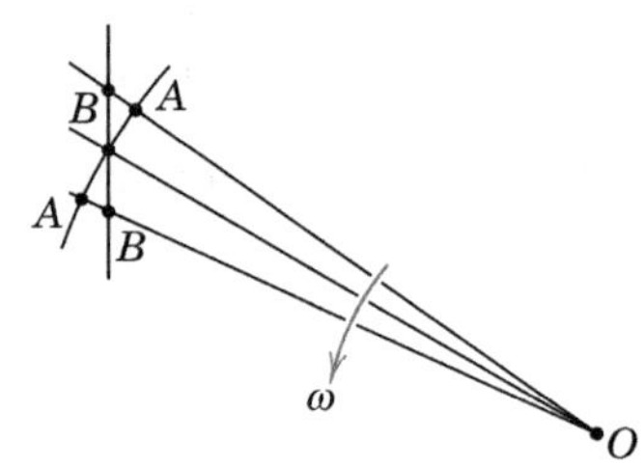

② Always identify the knowns and unknowns before attempting the solution of a vector equation.

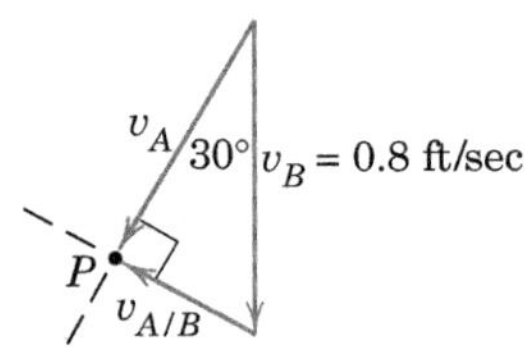

PROBLEMS

Introductory Problems

5/59 The right-angle link AB has a clockwise angular velocity $\omega = 2$ rad/sec when in the position shown. Determine the velocity of B with respect to A for this instant.

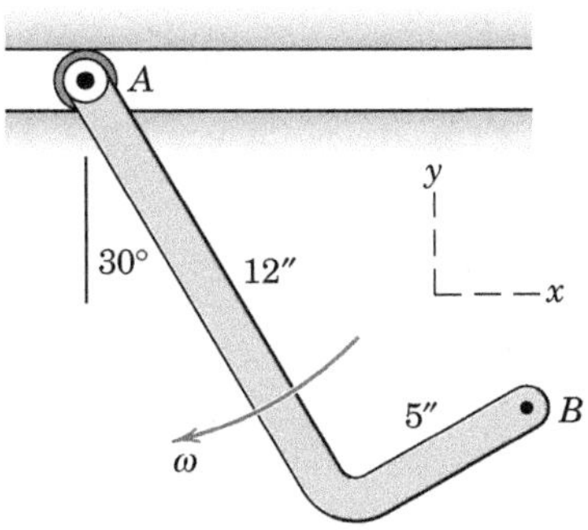

Problem 5/59

5/60 The uniform rectangular plate moves on the horizontal surface. Its mass center has a velocity $v_G = 3$ m/s directed parallel to the x-axis and the plate has a counterclockwise (as seen from above) angular velocity $\omega = 4$ rad/s. Determine the velocities of points A and B.

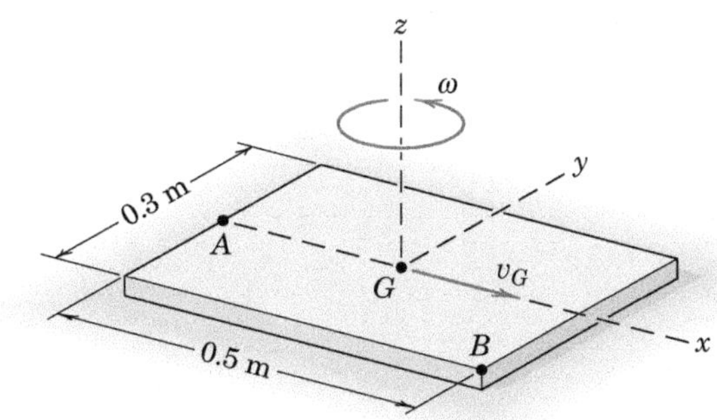

Problem 5/60

5/61 The cart has a velocity of 4 ft/sec to the right. Determine the angular speed N of the wheel so that point A on the top of the rim has a velocity (*a*) equal to 4 ft/sec to the left, (*b*) equal to zero, and (*c*) equal to 8 ft/sec to the right.

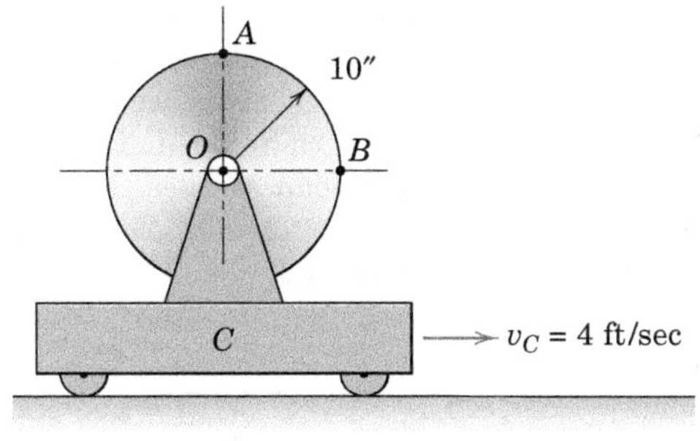

Problem 5/61

5/62 End A of the 2-ft link has a velocity of 5 ft/sec in the direction shown. At the same instant, end B has a velocity whose magnitude is 7 ft/sec as indicated. Find the angular velocity ω of the link in two ways.

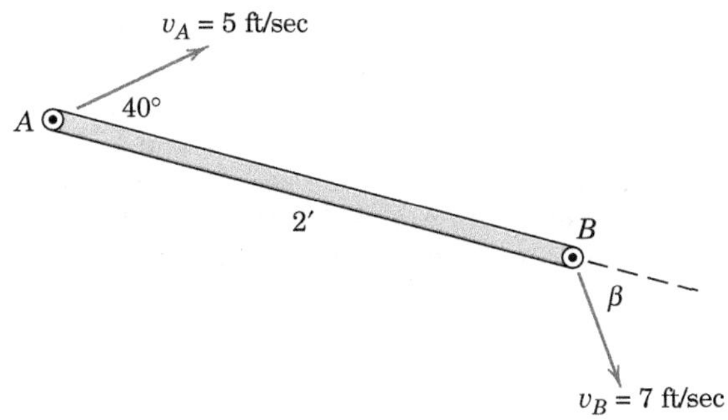

Problem 5/62

5/63 The speed of the center of the earth as it orbits the sun is $v = 107\,257$ km/h, and the absolute angular velocity of the earth about its north–south spin axis is $\omega = 7.292(10^{-5})$ rad/s. Use the value $R = 6371$ km for the radius of the earth and determine the velocities of points A, B, C, and D, all of which are on the equator. The inclination of the axis of the earth is neglected.

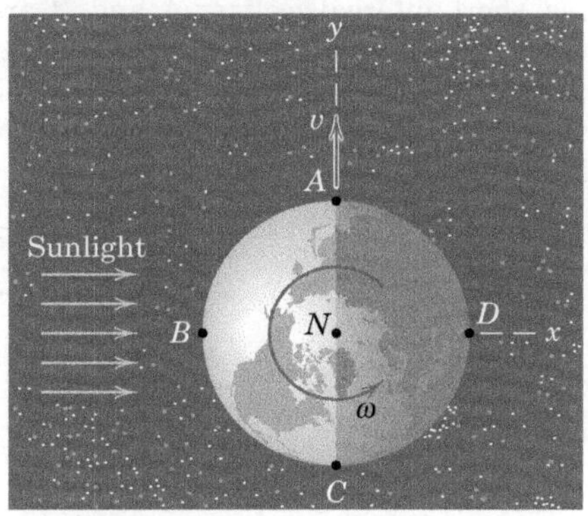

Problem 5/63

5/64 The center C of the smaller wheel has a velocity $v_C = 0.4$ m/s in the direction shown. The cord which connects the two wheels is securely wrapped around the respective peripheries and does not slip. Calculate the speed of point D when in the position shown. Also compute the change Δx which occurs per second if v_C is constant.

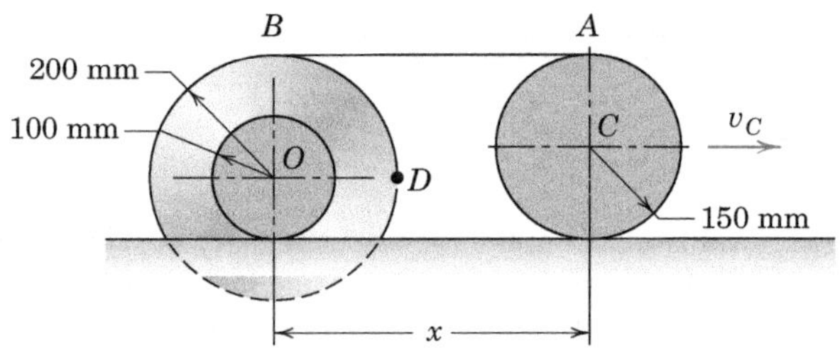

Problem 5/64

5/65 The circular disk of radius 8 in. is released very near the horizontal surface with a velocity of its center $v_O = 27$ in./sec to the right and a clockwise angular velocity $\omega = 2$ rad/sec. Determine the velocities of points A and P of the disk. Describe the motion upon contact with the ground.

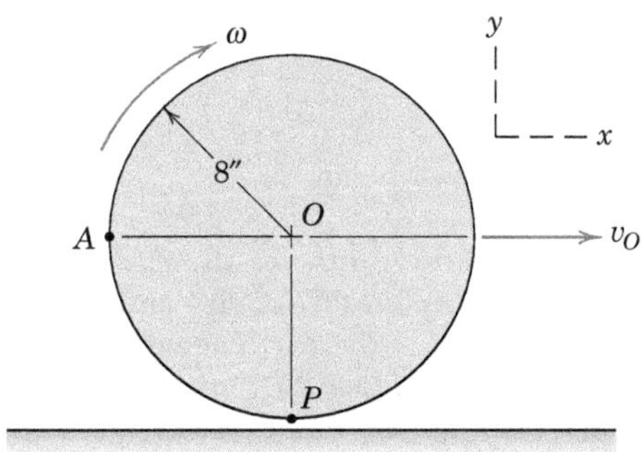

Problem 5/65

5/66 For a short interval, collars A and B are sliding along the fixed vertical shaft with velocities $v_A = 2$ m/s and $v_B = 3$ m/s in the directions shown. Determine the magnitude of the velocity of point C for the position $\theta = 60°$.

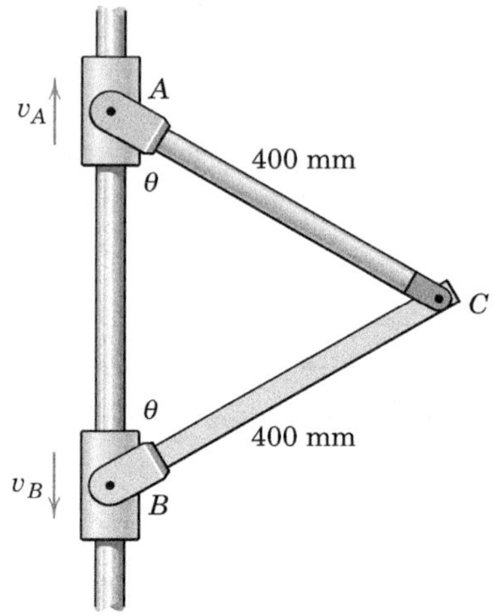

Problem 5/66

5/67 The right-angle link has a counterclockwise angular velocity of 3 rad/s at the instant represented, and point B has a velocity $\mathbf{v}_B = 2\mathbf{i} - 0.3\mathbf{j}$ m/s. Determine the velocity of A using vector notation. Sketch the vector polygon which corresponds to the terms in the relative-velocity equation and estimate or measure the magnitude of $\mathbf{v}_A$.

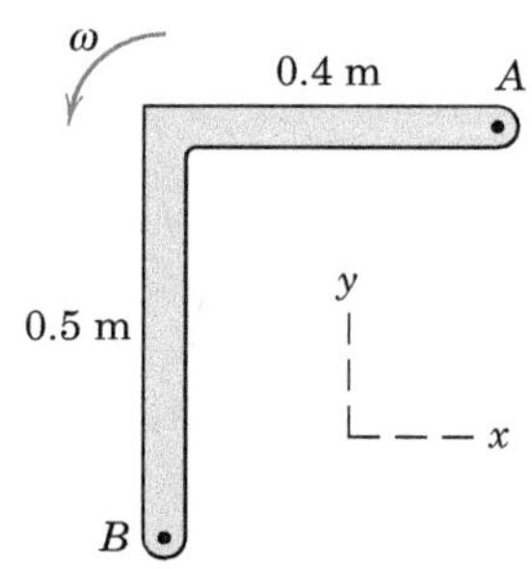

Problem 5/67

5/68 The magnitude of the absolute velocity of point A on the automobile tire is 12 m/s when A is in the position shown. What are the corresponding velocity v_O of the car and the angular velocity ω of the wheel? (The wheel rolls without slipping.)

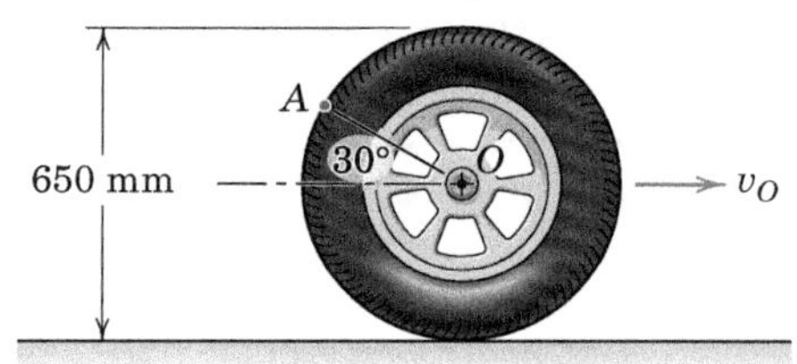

Problem 5/68

5/69 The two pulleys are riveted together to form a single rigid unit, and each of the two cables is securely wrapped around its respective pulley. If point A on the hoisting cable has a velocity $v = 3$ ft/sec, determine the magnitudes of the velocity of point O and the velocity of point B on the larger pulley for the position shown.

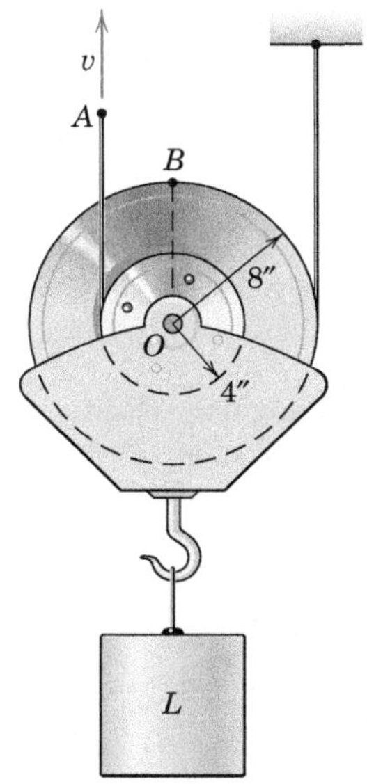

Problem 5/69

5/70 The rider of the bicycle shown pumps steadily to maintain a constant speed of 16 km/h against a slight head wind. Calculate the maximum and minimum magnitudes of the absolute velocity of the pedal A.

Problem 5/70

5/71 Determine the angular velocity of bar AB just after roller B has begun moving up the 15° incline. At the instant under consideration, the velocity of roller A is v_A.

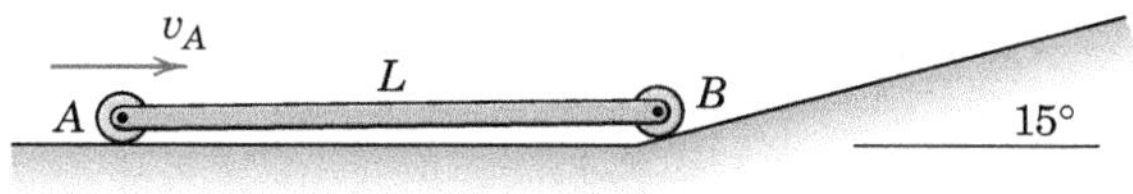

Problem 5/71

5/72 For the instant represented, point B crosses the horizontal axis through point O with a downward velocity $v = 0.6$ m/s. Determine the corresponding value of the angular velocity ω_{OA} of link OA.

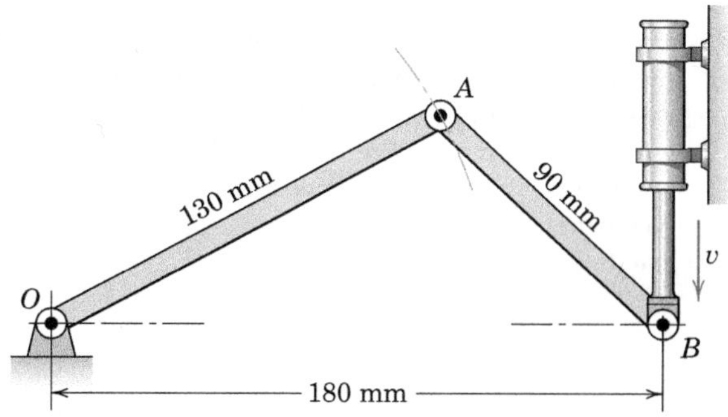

Problem 5/72

Representative Problems

5/73 The spoked wheel of radius r is made to roll up the incline by the cord wrapped securely around a shallow groove on its outer rim. For a given cord speed v at point P, determine the velocities of points A and B. No slipping occurs.

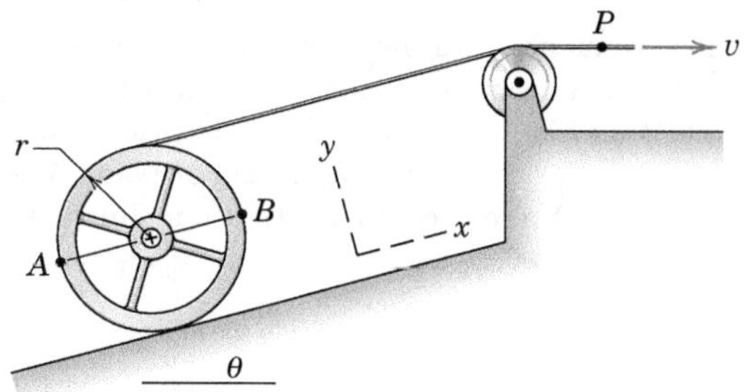

Problem 5/73

5/74 At the instant represented, the velocity of point A of the 1.2-m bar is 3 m/s to the right. Determine the speed v_B of point B and the angular velocity ω of the bar. The diameter of the small end wheels may be neglected.

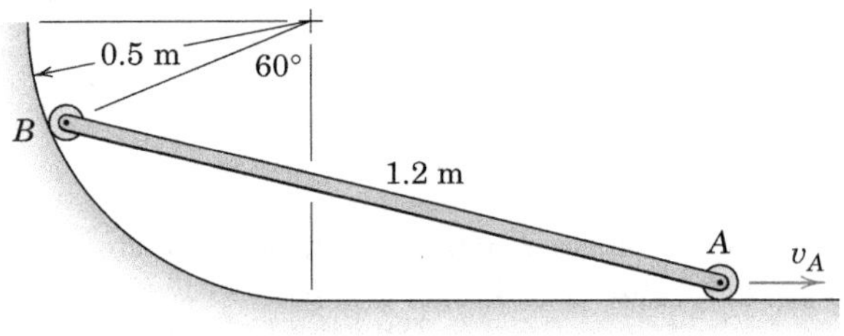

Problem 5/74

5/75 Determine the angular velocity of link BC for the instant indicated. In case (a), the center O of the disk is a fixed pivot, while in case (b), the disk rolls without slipping on the horizontal surface. In both cases, the disk has clockwise angular velocity ω. Neglect the small distance of pin A from the edge of the disk.

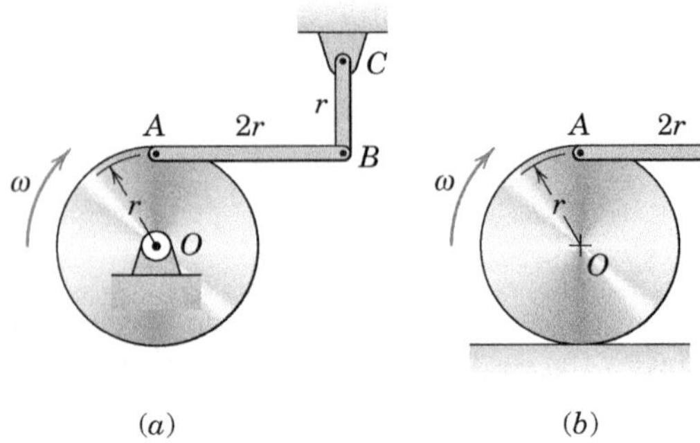

Problem 5/75

5/76 The elements of a switching device are shown. If the vertical control rod has a downward velocity $v =$ 2 ft/sec when the device is in the position shown, determine the corresponding speed of point A. Roller C is in continuous contact with the inclined surface.

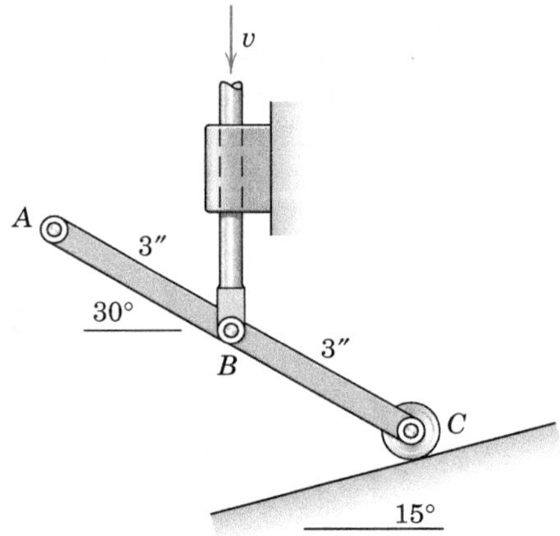

Problem 5/76

5/77 Determine the angular velocity ω_{AB} of link AB and the velocity v_B of collar B for the instant represented. Assume the quantities ω_0 and r to be known.

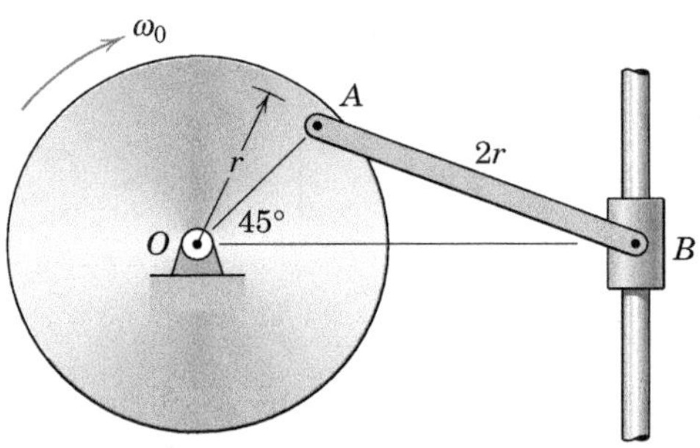

Problem 5/77

5/78 Determine the angular velocity ω_{AB} of link AB and the velocity v_B of collar B for the instant represented. Assume the quantities ω_0 and r to be known.

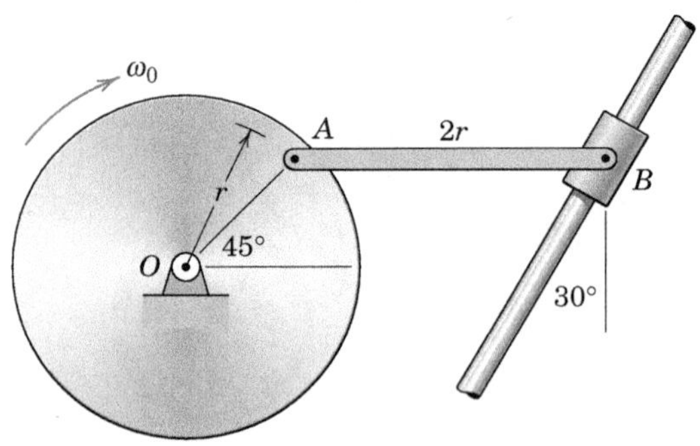

Problem 5/78

5/79 The rotation of the gear is controlled by the horizontal motion of end A of the rack AB. If the piston rod has a constant velocity $\dot{x} = 300$ mm/s during a short interval of motion, determine the angular velocity ω_0 of the gear and the angular velocity ω_{AB} of AB at the instant when $x = 800$ mm.

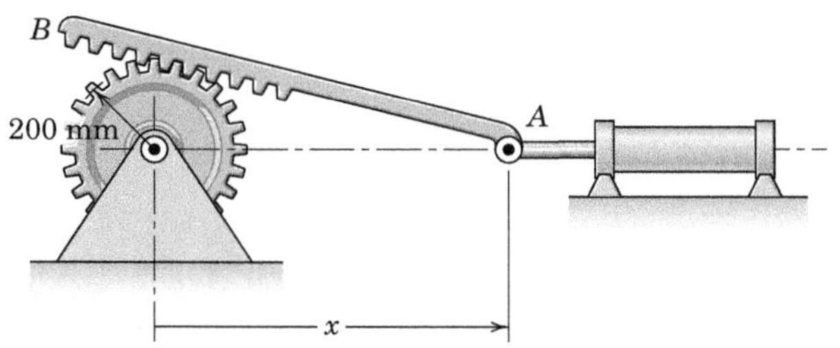

Problem 5/79

5/80 Motion of the rectangular plate P is controlled by the two links which cross without touching. For the instant represented where the links are perpendicular to each other, the plate has a counterclockwise angular velocity $\omega_P = 2$ rad/s. Determine the corresponding angular velocities of the two links.

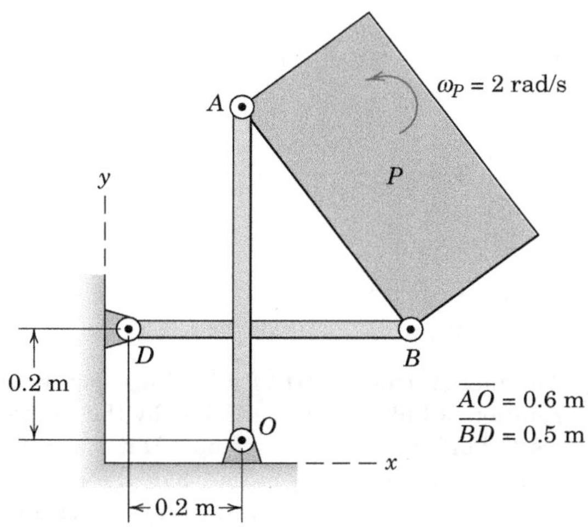

Problem 5/80

5/81 The elements of a simplified clam-shell bucket for a dredge are shown. The cable which opens and closes the bucket passes through the block at O. With O as a fixed point, determine the angular velocity ω of the bucket jaws when $\theta = 45°$ as they are closing. The upward velocity of the control cable is 0.5 m/s as it passes through the block.

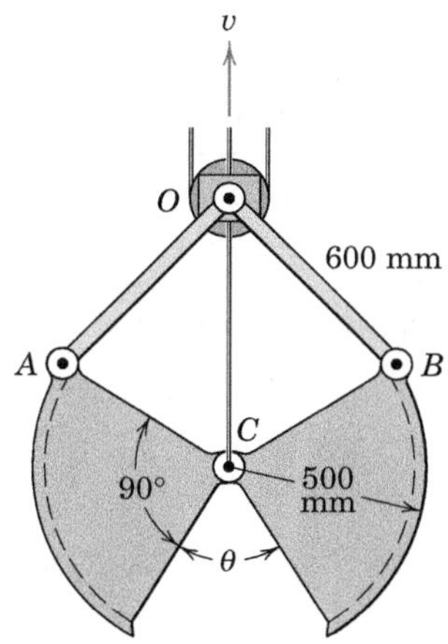

Problem 5/81

5/82 The ends of the 0.4-m slender bar remain in contact with their respective support surfaces. If end B has a velocity $v_B = 0.5$ m/s in the direction shown, determine the angular velocity of the bar and the velocity of end A.

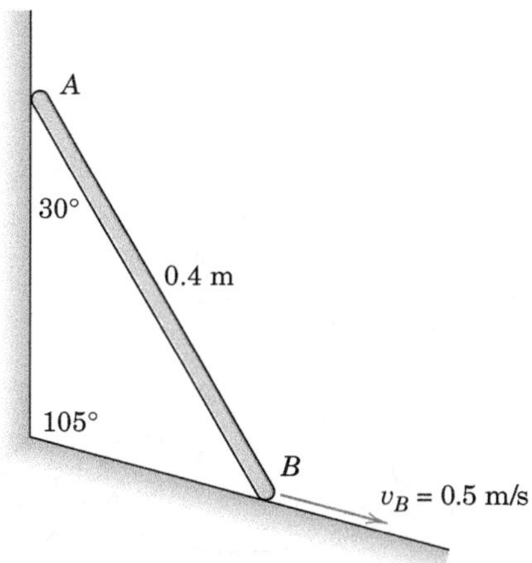

Problem 5/82

5/83 In the design of a produce-processing plant, roller trays of produce are to be oscillated under water spray by the action of the connecting link AB and crank OB. For the instant when $\theta = 15°$, the angular velocity of AB is 0.086 rad/s clockwise. Find the corresponding angular velocity $\dot{\theta}$ of the crank and the velocity v_A of the tray. Solve the relative-velocity equation by either vector algebra or vector geometry.

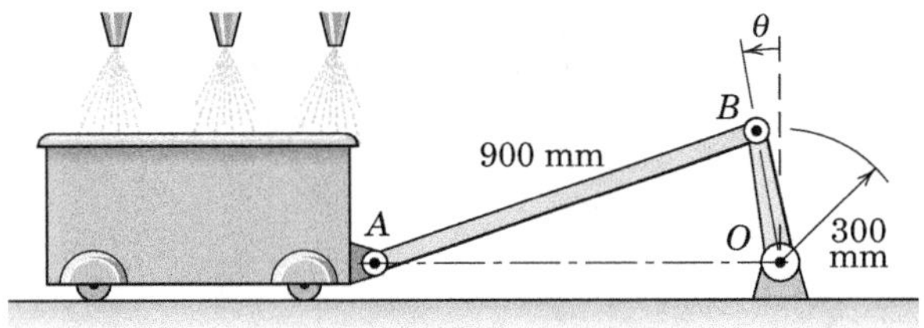

Problem 5/83

5/84 The vertical rod has a downward velocity $v = 0.8$ m/s when link AB is in the 30° position shown. Determine the corresponding angular velocity of AB and the speed of roller B if $R = 0.4$ m.

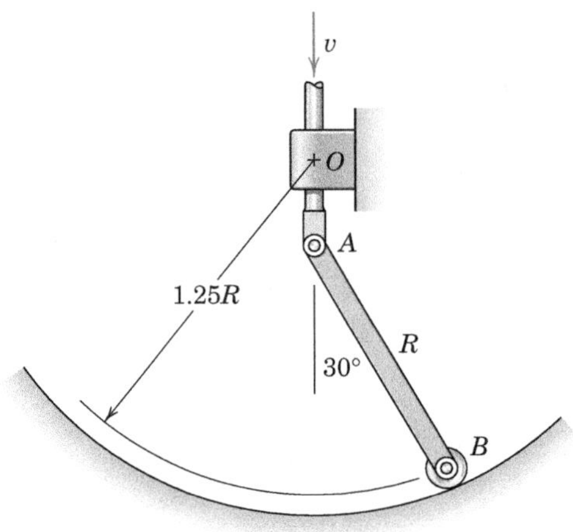

Problem 5/84

5/85 Pin P on the end of the horizontal rod slides freely in the slotted gear. The gear engages the moving rack A and the fixed rack B (teeth not shown) so it rolls without slipping. If A has a velocity of 0.4 ft/sec to the left for the instant shown, determine the velocity v_P of the rod for this position.

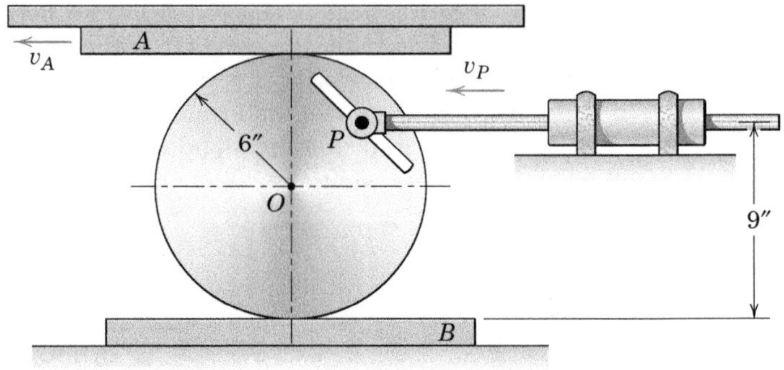

Problem 5/85

5/86 A four-bar linkage is shown in the figure (the ground "link" OC is considered the fourth bar). If the drive link OA has a counterclockwise angular velocity $\omega_0 = 10$ rad/s, determine the angular velocities of links AB and BC.

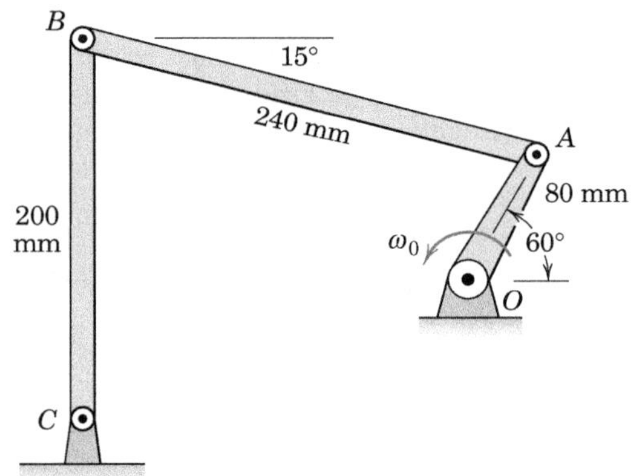

Problem 5/86

5/87 The mechanism is part of a latching device where rotation of link AOB is controlled by the rotation of slotted link D about C. If member D has a clockwise angular velocity of 1.5 rad/s when the slot is parallel to OC, determine the corresponding angular velocity of AOB. Solve graphically or geometrically.

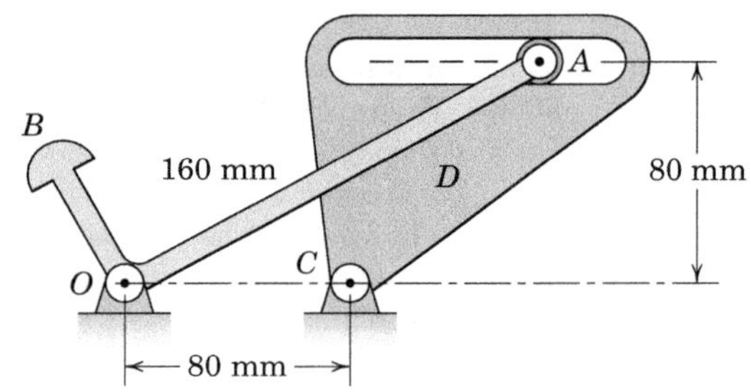

Problem 5/87

5/88 The elements of the mechanism for deployment of a spacecraft magnetometer boom are shown. Determine the angular velocity of the boom when the driving link OB crosses the y-axis with an angular velocity $\omega_{OB} = 0.5$ rad/sec if $\tan \theta = 4/3$ at this instant.

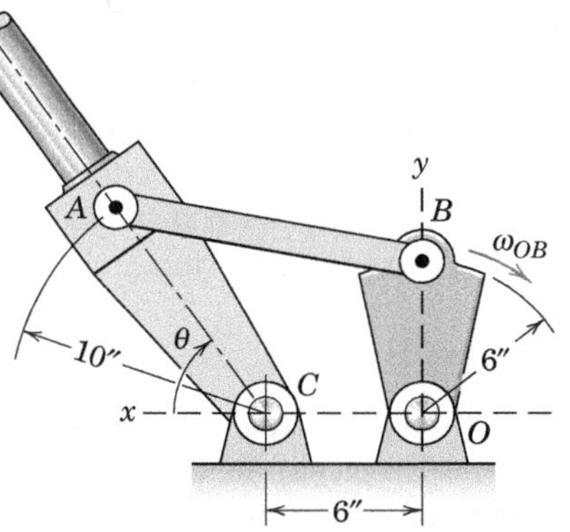

Problem 5/88

5/89 A mechanism for pushing small boxes from an assembly line onto a conveyor belt is shown with arm OD and crank CB in their vertical positions. The crank revolves clockwise at a constant rate of 1 revolution every 2 seconds. For the position shown, determine the speed at which the box is being shoved horizontally onto the conveyor belt.

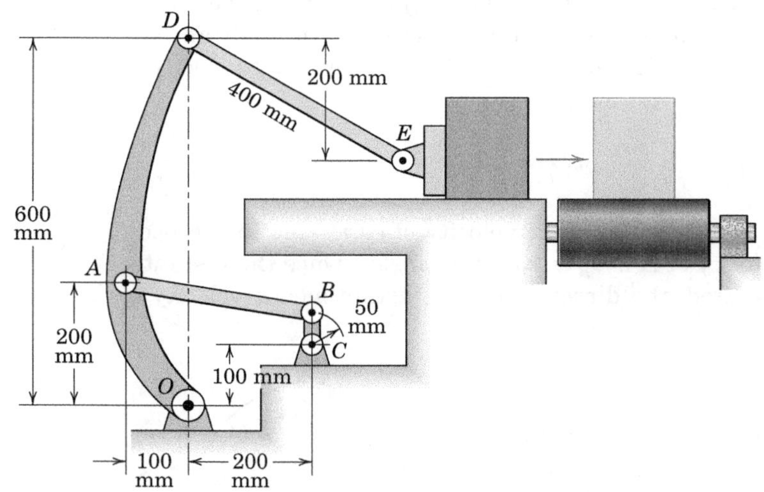

Problem 5/89

▶5/90 The wheel rolls without slipping. For the instant portrayed, when O is directly under point C, link OA has a velocity $v = 1.5$ m/s to the right and $\theta = 30°$. Determine the angular velocity ω of the slotted link.

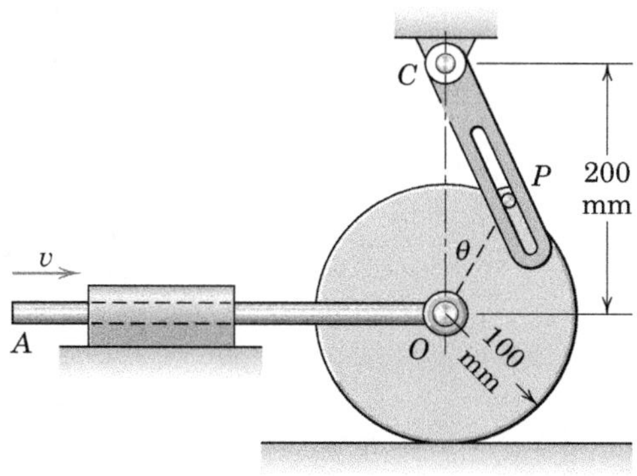

Problem 5/90

5/5 Instantaneous Center of Zero Velocity

In the previous article, we determined the velocity of a point on a rigid body in plane motion by adding the relative velocity due to rotation about a convenient reference point to the velocity of the reference point. We now solve the problem by choosing a unique reference point which momentarily has zero velocity. As far as velocities are concerned, the body may be considered to be in pure rotation about an axis, normal to the plane of motion, passing through this point. This axis is called the *instantaneous axis of zero velocity*, and the intersection of this axis with the plane of motion is known as the *instantaneous center of zero velocity*. This approach provides us with a valuable means for visualizing and analyzing velocities in plane motion.

Locating the Instantaneous Center

The existence of the instantaneous center is easily shown. For the body in Fig. 5/7, assume that the directions of the absolute velocities of any two points A and B on the body are known and are not parallel. If there is a point about which A has absolute circular motion at the instant considered, this point must lie on the normal to $\mathbf{v}_A$ through A. Similar reasoning applies to B, and the intersection of the two perpendiculars fulfills the requirement for an absolute center of rotation *at the instant considered*. Point C is the instantaneous center of zero velocity and may lie on or off the body. If it lies off the body, it may be visualized as lying on an imaginary extension of the body. The instantaneous center need not be a fixed point in the body or a fixed point in the plane.

If we also know the magnitude of the velocity of one of the points, say, v_A, we may easily obtain the angular velocity ω of the body and the linear velocity of every point in the body. Thus, the angular velocity of the body, Fig. 5/7*a*, is

$$\omega = \frac{v_A}{r_A}$$

which, of course, is also the angular velocity of *every* line in the body. Therefore, the velocity of B is $v_B = r_B\omega = (r_B/r_A)v_A$. Once the instantaneous center is located, the direction of the instantaneous velocity of

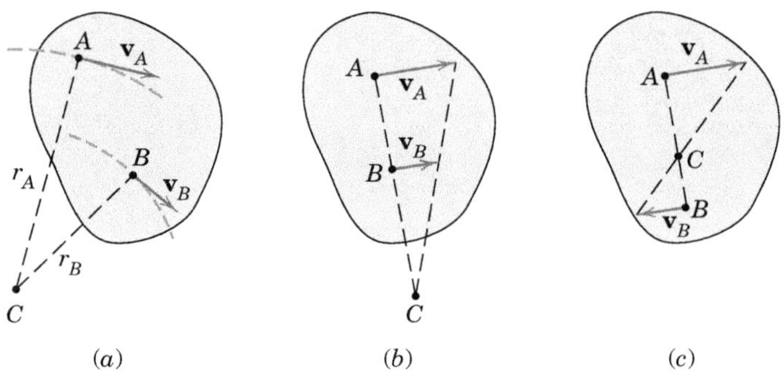

Figure 5/7

every point in the body is readily found since it must be perpendicular to the radial line joining the point in question with C.

If the velocities of two points in a body having plane motion are parallel, Fig. 5/7*b* or 5/7*c*, and the line joining the points is perpendicular to the direction of the velocities, the instantaneous center is located by direct proportion as shown. We can readily see from Fig. 5/7*b* that as the parallel velocities become equal in magnitude, the instantaneous center moves farther away from the body and approaches infinity in the limit as the body stops rotating and translates only.

Motion of the Instantaneous Center

As the body changes its position, the instantaneous center C also changes its position both in space and on the body. The locus of the instantaneous centers in space is known as the *space centrode*, and the locus of the positions of the instantaneous centers on the body is known as the *body centrode*. At the instant considered, the two curves are tangent at the position of point C. It can be shown that the body-centrode curve rolls on the space-centrode curve during the motion of the body, as indicated schematically in Fig. 5/8.

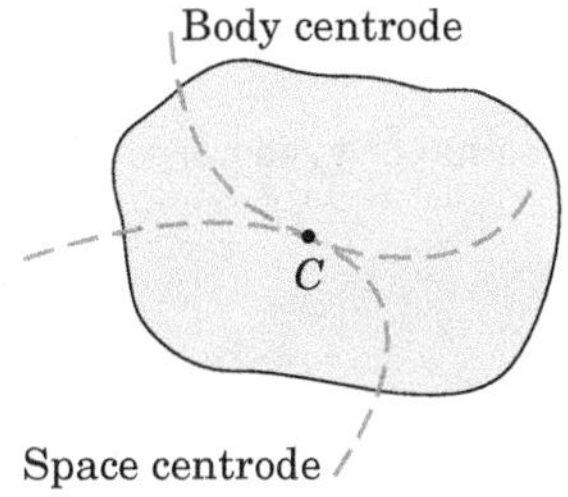

Figure 5/8

Although the instantaneous center of zero velocity is momentarily at rest, its acceleration generally is *not* zero. Thus, this point may *not* be used as an instantaneous center of zero acceleration in a manner analogous to its use for finding velocity. An instantaneous center of zero acceleration does exist for bodies in general plane motion, but its location and use represent a specialized topic in mechanism kinematics and will not be treated here.

IgorSinitsyn/age fotostock/SuperStock

This valve gear of a steam locomotive provides an interesting (albeit not cutting-edge) study in rigid-body kinematics.

Sample Problem 5/11

The wheel of Sample Problem 5/7, shown again here, rolls to the right without slipping, with its center O having a velocity $v_O = 3$ m/s. Locate the instantaneous center of zero velocity and use it to find the velocity of point A for the position indicated.

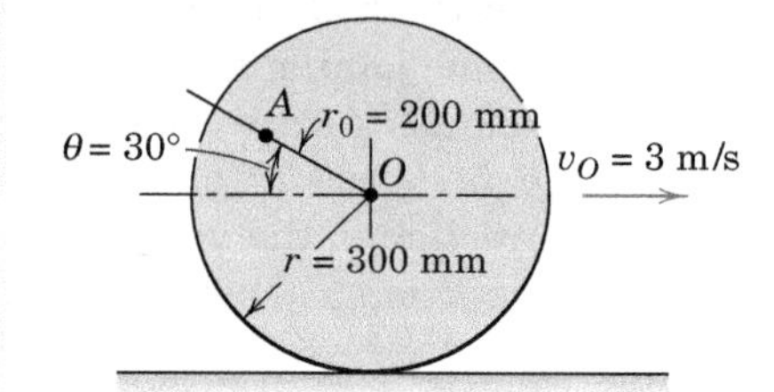

Solution. The point on the rim of the wheel in contact with the ground has no velocity if the wheel is not slipping; it is, therefore, the instantaneous center C of zero velocity. The angular velocity of the wheel becomes

$$[\omega = v/r] \qquad \omega = v_O / \overline{OC} = 3/0.300 = 10 \text{ rad/s}$$

The distance from A to C is

① $$\overline{AC} = \sqrt{(0.300)^2 + (0.200)^2 - 2(0.300)(0.200)\cos 120°} = 0.436 \text{ m}$$

The velocity of A becomes

② $$[v = r\omega] \qquad v_A = \overline{AC}\omega = 0.436(10) = 4.36 \text{ m/s} \qquad \textit{Ans.}$$

The direction of $\mathbf{v}_A$ is perpendicular to AC as shown.

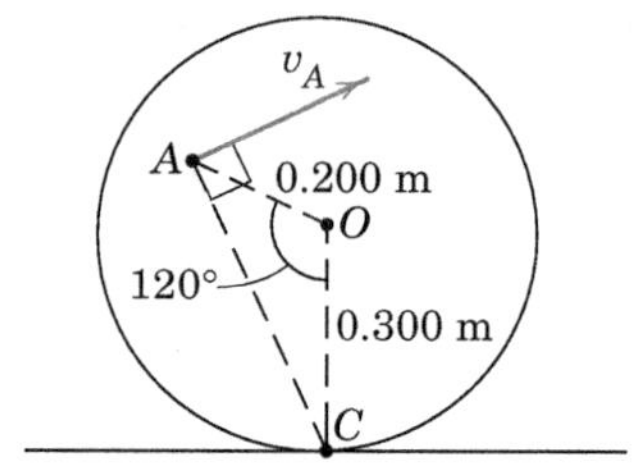

Helpful Hints

① Be sure to recognize that the cosine of 120° is itself negative.

② From the results of this problem, you should be able to visualize and sketch the velocities of all points on the wheel.

Sample Problem 5/12

Arm OB of the linkage has a clockwise angular velocity of 10 rad/sec in the position shown where $\theta = 45°$. Determine the velocity of A, the velocity of D, and the angular velocity of link AB for the position shown.

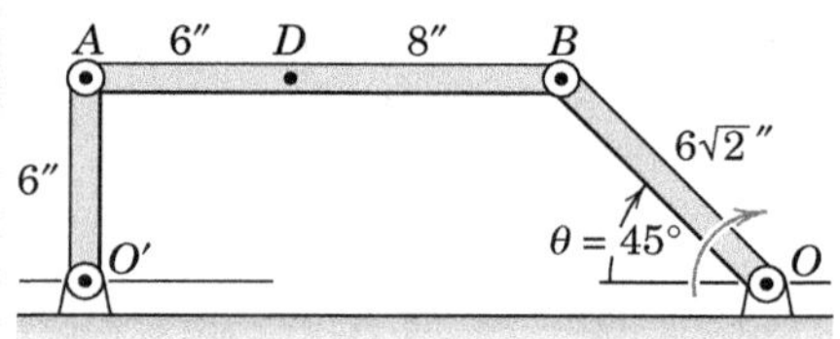

Solution. ① The directions of the velocities of A and B are tangent to their circular paths about the fixed centers O' and O as shown. The intersection of the two perpendiculars to the velocities from A and B locates the instantaneous center C for the link AB. The distances $\overline{AC}$, $\overline{BC}$, and $\overline{DC}$ shown on the diagram are computed or scaled from the drawing. The angular velocity of BC, considered a line on the body extended, is equal to the angular velocity of AC, DC, and AB and is

$$[\omega = v/r] \qquad \omega_{BC} = \frac{v_B}{\overline{BC}} = \frac{\overline{OB}\omega_{OB}}{\overline{BC}} = \frac{6\sqrt{2}(10)}{14\sqrt{2}}$$

$$= 4.29 \text{ rad/sec CCW} \qquad \textit{Ans.}$$

Thus, the velocities of A and D are

$$[v = r\omega] \qquad v_A = \frac{14}{12}(4.29) = 5.00 \text{ ft/sec} \qquad \textit{Ans.}$$

$$v_D = \frac{15.23}{12}(4.29) = 5.44 \text{ ft/sec} \qquad \textit{Ans.}$$

in the directions shown.

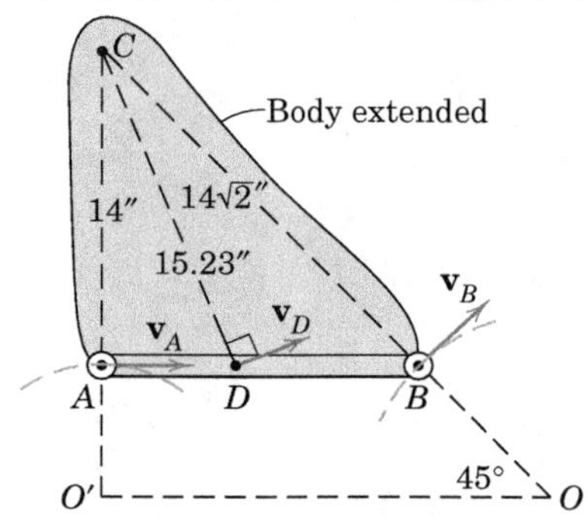

Helpful Hint

① For the instant depicted, we should visualize link AB and its body extended to be rotating as a single unit about point C.

PROBLEMS

Introductory Problems

5/91 The slender bar is moving in general plane motion with the indicated linear and angular properties. Locate the instantaneous center of zero velocity and determine the speeds of points A and B.

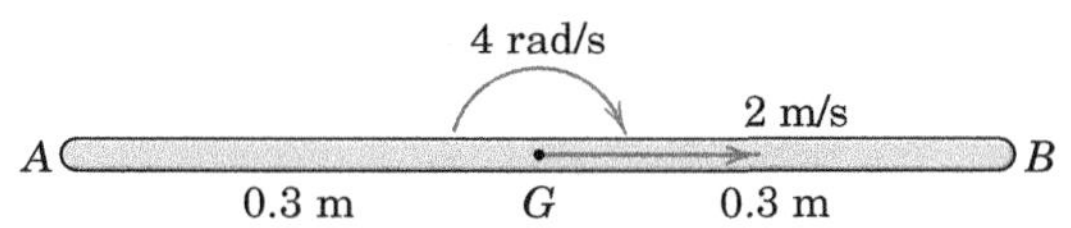

Problem 5/91

5/92 The slender bar is moving in general plane motion with the indicated linear and angular properties. Locate the instantaneous center of zero velocity and determine the velocities of points A and B.

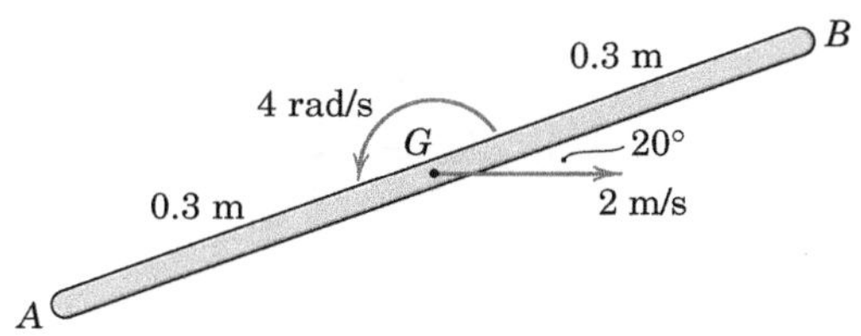

Problem 5/92

5/93 For the instant represented, corner A of the rectangular plate has a velocity $v_A = 2.8$ m/s and the plate has a clockwise angular velocity $\omega = 12$ rad/s. Determine the magnitude of the corresponding velocity of point B.

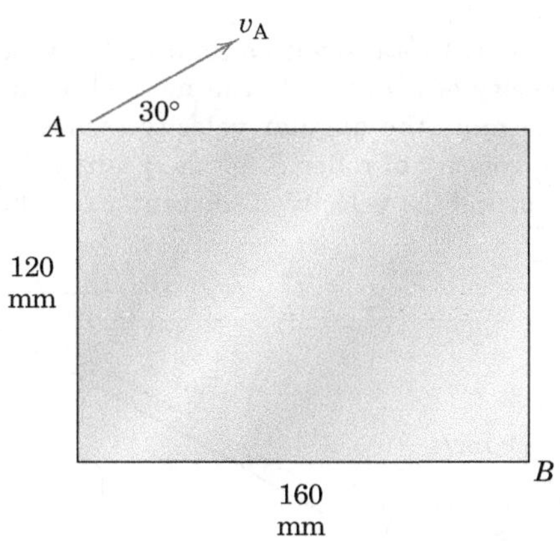

Problem 5/93

5/94 Roller B of the quarter-circular link has a velocity $v_B = 3$ ft/sec directed down the 15° incline. The link has a counterclockwise angular velocity $\omega = 2$ rad/sec. By the method of this article, determine the velocity of roller A.

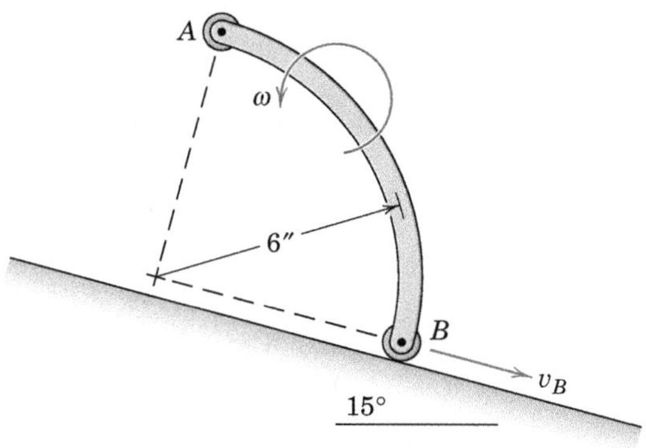

Problem 5/94

5/95 The bar of Prob. 5/82 is repeated here. By the method of this article, determine the velocity of end A. Both ends remain in contact with their respective support surfaces.

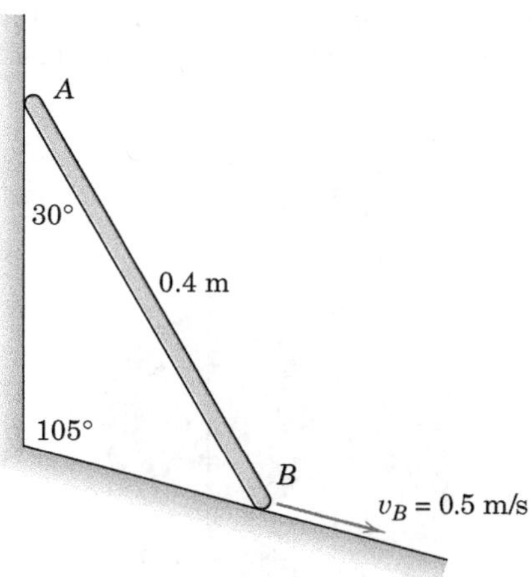

Problem 5/95

5/96 The bar AB has a counterclockwise angular velocity of 6 rad/sec. Construct the velocity vectors for points A and G of the bar and specify their magnitudes if the instantaneous center of zero velocity for the bar is (a) at C_1, (b) at C_2, and (c) at C_3.

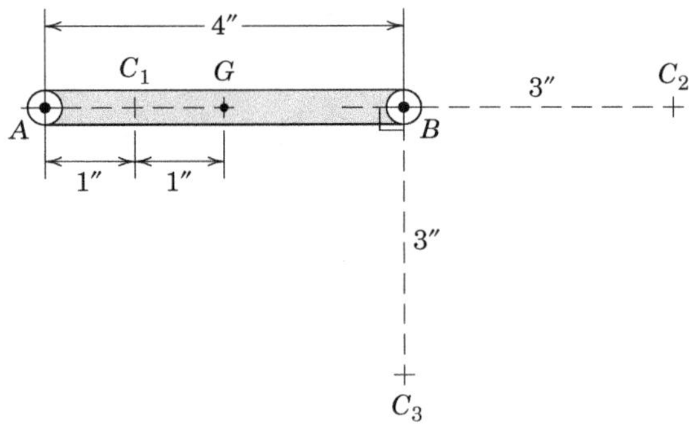

Problem 5/96

5/97 A car mechanic "walks" two wheel/tire units across a horizontal floor as shown. He walks with constant speed v and keeps the tires in the configuration shown with the same position relative to his body. If there is no slipping at any interface, determine (a) the angular velocity of the lower tire, (b) the angular velocity of the upper tire, and (c) the velocities of points A, B, C, and D. The radius of both tires is r.

Problem 5/97

5/98 At a certain instant vertex B of the right-triangular plate has a velocity of 200 mm/s in the direction shown. If the instantaneous center of zero velocity for the plate is 40 mm from point B and if the angular velocity of the plate is clockwise, determine the speed of point D.

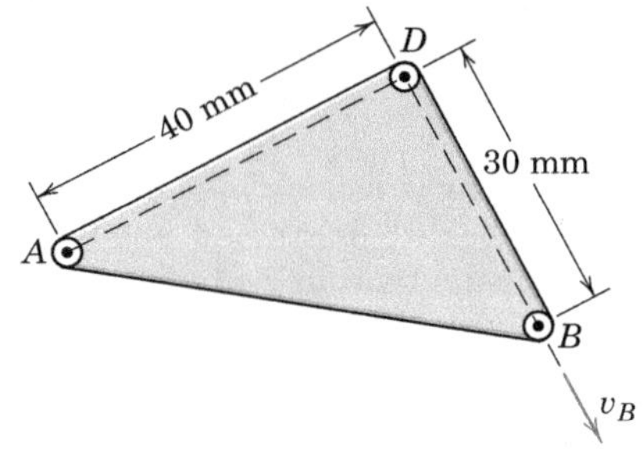

Problem 5/98

5/99 At the instant represented, crank OB has a clockwise angular velocity $\omega = 0.8$ rad/sec and is passing the horizontal position. By the method of this article, determine the corresponding speed of the guide roller A in the 20° slot and the speed of point C midway between A and B.

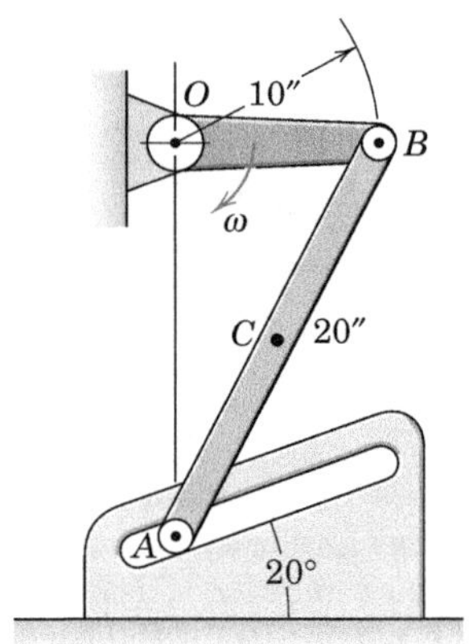

Problem 5/99

5/100 Crank OA rotates with a counterclockwise angular velocity of 9 rad/s. By the method of this article, determine the angular velocity ω of link AB and the velocity of roller B for the position illustrated. Also, find the velocity of the center G of link AB.

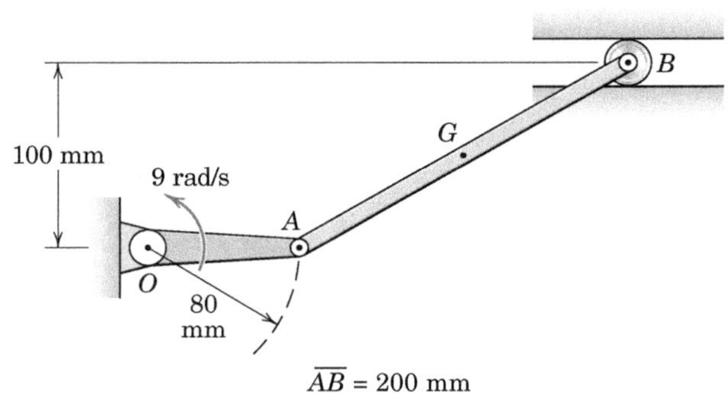

Problem 5/100

5/101 The mechanism of Prob. 5/100 is now shown in a different position, with the crank OA 30° below the horizontal as illustrated. Determine the angular velocity ω of link AB and the velocity of roller B.

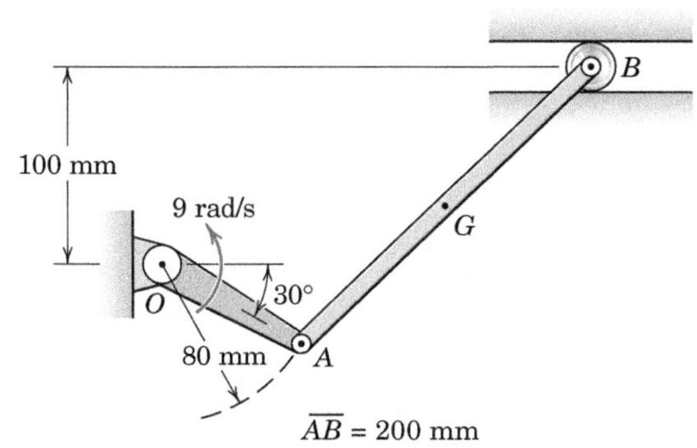

Problem 5/101

5/102 If link OA has a clockwise angular velocity of 2 rad/s in the position for which $x = 75$ mm, determine the velocity of the slider at B by the method of this article.

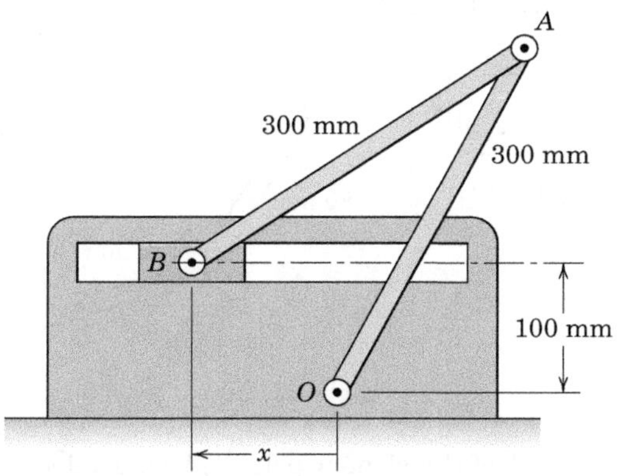

Problem 5/102

5/103 Motion of the bar is controlled by the constrained paths of A and B. If the angular velocity of the bar is 2 rad/s counterclockwise as the position $\theta = 45°$ is passed, determine the speeds of points A and P.

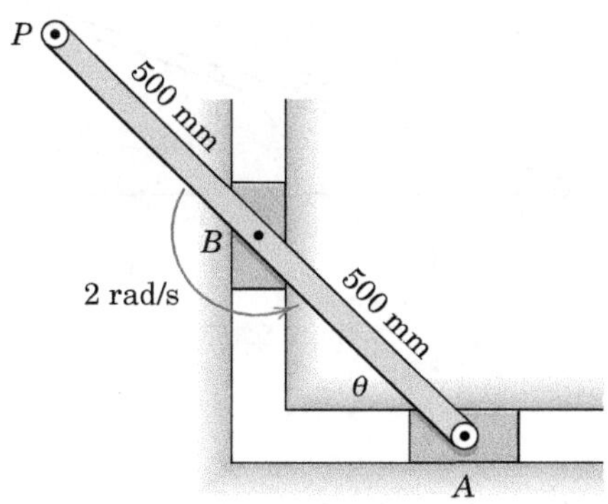

Problem 5/103

5/104 The switching device of Prob. 5/76 is repeated here. If the vertical control rod has a downward velocity $v = 2$ ft/sec when the device is in the position shown, determine the corresponding speed of point A by the method of this article. Roller C is in continuous contact with the inclined surface.

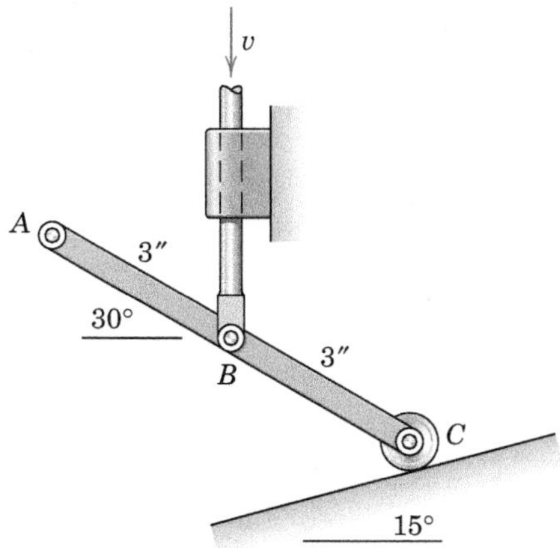

Problem 5/104

5/105 The shaft of the wheel unit rolls without slipping on the fixed horizontal surface, and point O has a velocity of 3 ft/sec to the right. By the method of this article, determine the velocities of points A, B, C, and D.

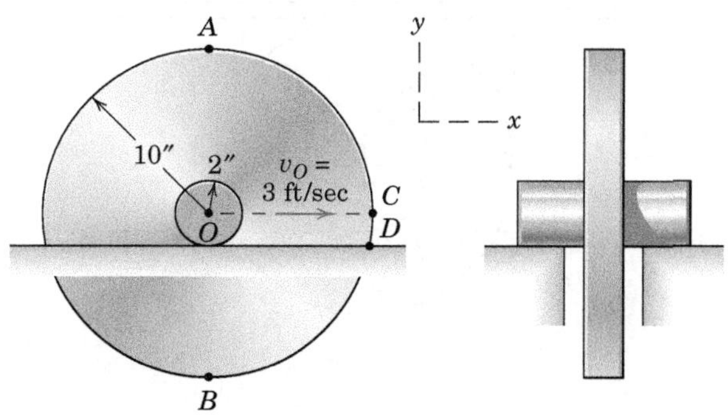

Problem 5/105

5/106 The center D of the car follows the centerline of the 100-ft skidpad. The speed of point D is $v = 45$ ft/sec. Determine the angular velocity of the car and the speeds of points A and B of the car.

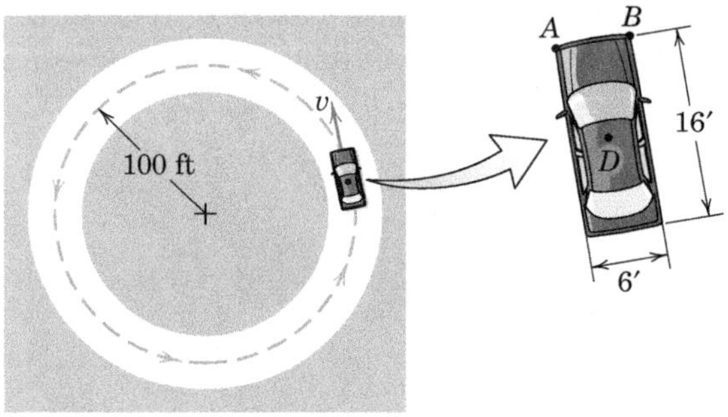

Problem 5/106

Representative Problems

5/107 The attached wheels roll without slipping on the plates A and B, which are moving in opposite directions as shown. If $v_A = 60$ mm/s to the right and $v_B = 200$ mm/s to the left, determine the speeds of the center O and the point P for the position shown.

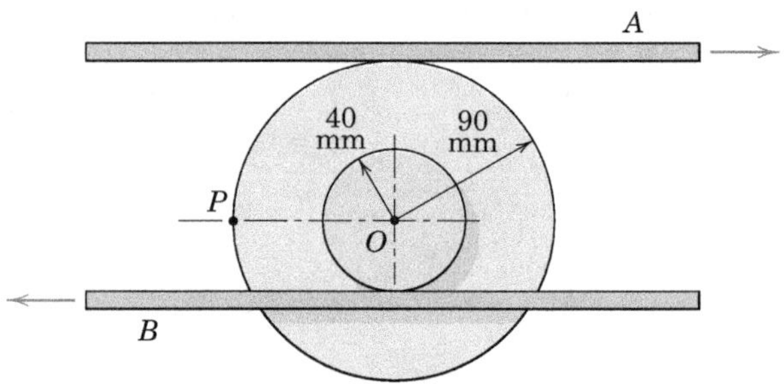

Problem 5/107

5/108 The mechanism of Prob. 5/77 is repeated here. By the method of this article, determine the angular velocity of link AB and the velocity of collar B for the position shown. Assume the quantities ω_0 and r to be known.

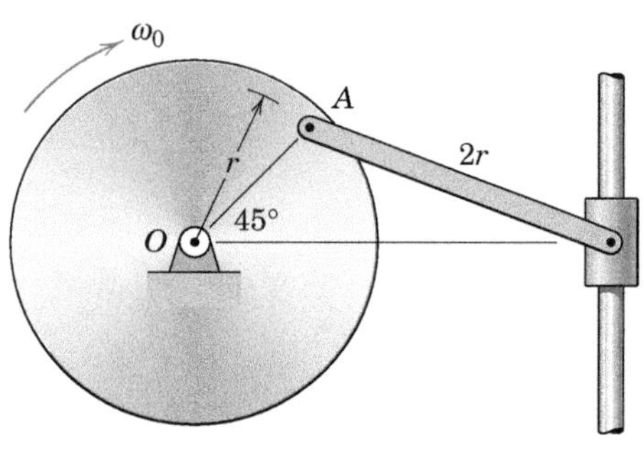

Problem 5/108

5/109 The mechanism of Prob. 5/78 is repeated here. By the method of this article, determine the angular velocity of link AB and the velocity of collar B for the instant depicted. Assume the quantities ω_0 and r to be known.

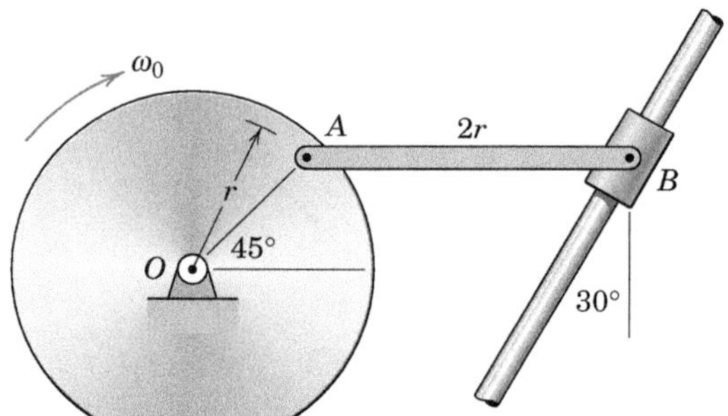

Problem 5/109

5/110 At the instant under consideration, the rod of the hydraulic cylinder is extending at the rate $v_A = 2$ m/s. Determine the corresponding angular velocity ω_{OB} of link OB.

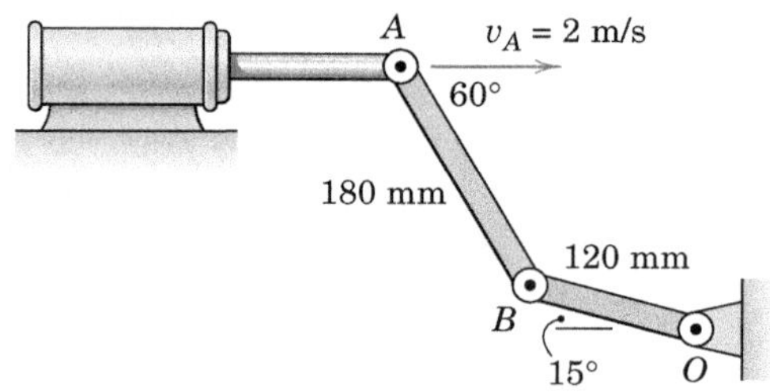

Problem 5/110

5/111 End A of the slender pole is given a velocity v_A to the right along the horizontal surface. Show that the magnitude of the velocity of end B equals v_A when the midpoint M of the pole comes in contact with the semicircular obstruction.

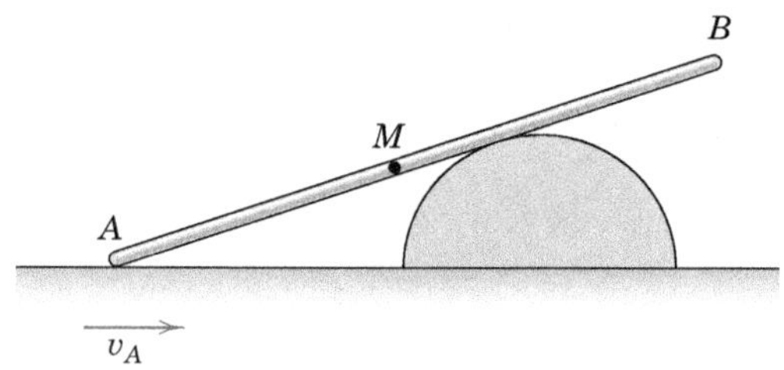

Problem 5/111

5/112 The flexible band F is attached at E to the rotating sector and leads over the guide pulley G. Determine the angular velocities of links AB and BD for the position shown if the band has a speed of 2 m/s.

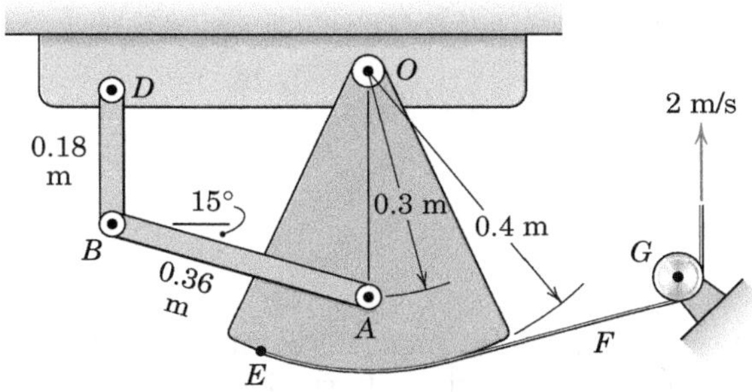

Problem 5/112

5/113 The rear driving wheel of a car has a diameter of 26 in. and has an angular speed N of 200 rev/min on an icy road. If the instantaneous center of zero velocity is 4 in. above the point of contact of the tire with the road, determine the velocity v of the car and the slipping velocity v_s of the tire on the ice.

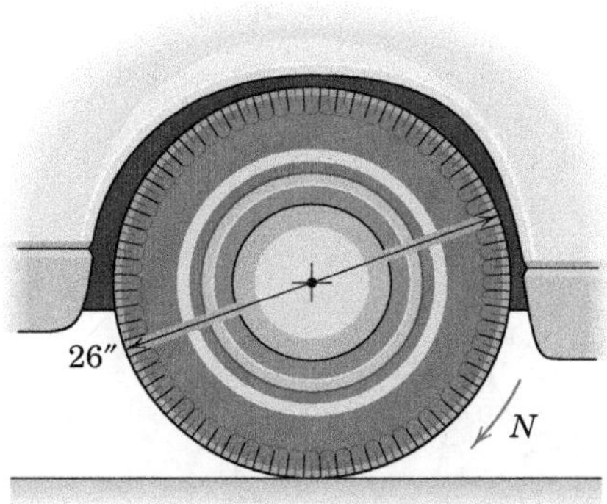

Problem 5/113

5/114 Solve for the speed of point D in Prob. 5/64 by the method of Art. 5/5.

5/115 Link OA has a counterclockwise angular velocity $\dot{\theta} = 4$ rad/sec during an interval of its motion. Determine the angular velocity of link AB and of sector BD for $\theta = 45°$ at which instant AB is horizontal and BD is vertical.

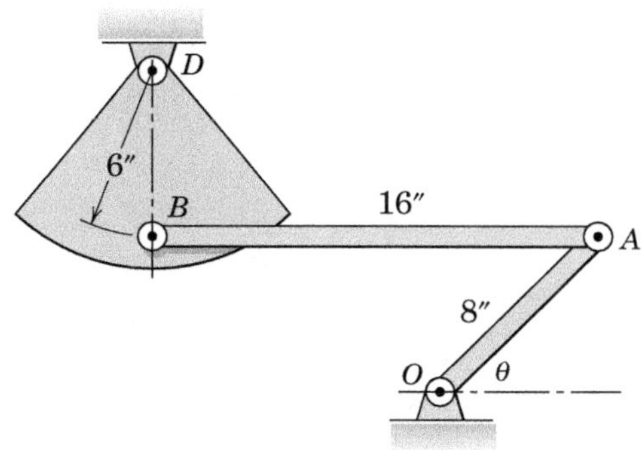

Problem 5/115

5/116 Vertical oscillation of the spring-loaded plunger F is controlled by a periodic change in pressure in the vertical hydraulic cylinder E. For the position $\theta = 60°$, determine the angular velocity of AD and the velocity of the roller A in its horizontal guide if the plunger F has a downward velocity of 2 m/s.

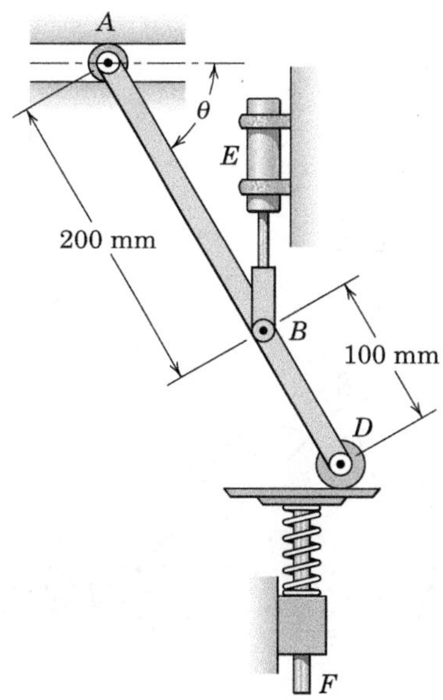

Problem 5/116

5/117 A device which tests the resistance to wear of two materials A and B is shown. If the link EO has a velocity of 4 ft/sec to the right when $\theta = 45°$, determine the rubbing velocity v_A.

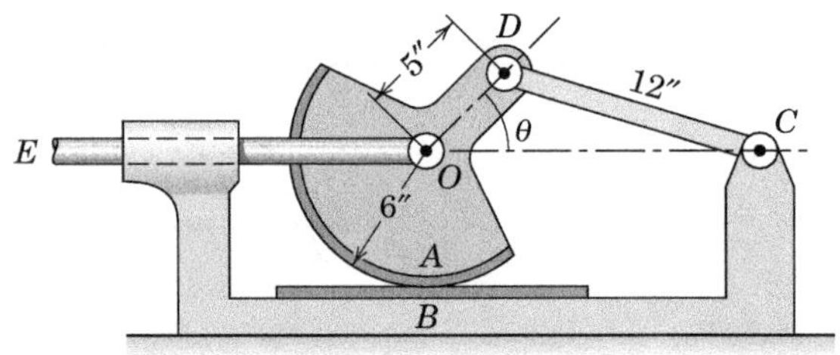

Problem 5/117

5/118 Motion of the roller A against its restraining spring is controlled by the downward motion of the plunger E. For an interval of motion the velocity of E is $v = 0.2$ m/s. Determine the velocity of A when θ becomes 90°.

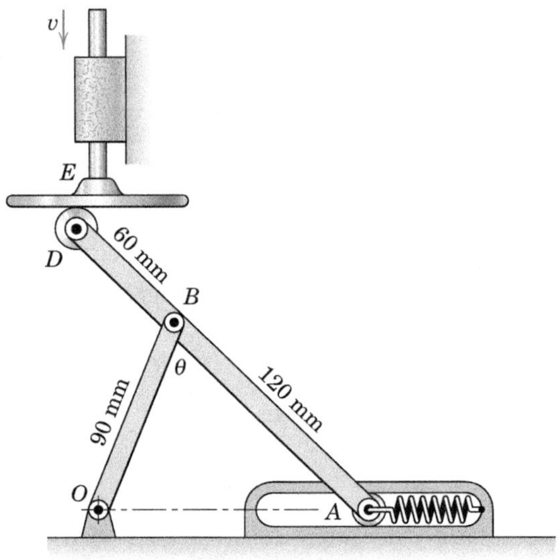

Problem 5/118

5/119 In the design of the mechanism shown, collar A is to slide along the fixed shaft as angle θ increases. When $\theta = 30°$, the control link at D is to have a downward component of velocity of 0.60 m/s. Determine the corresponding velocity of collar A by the method of this article.

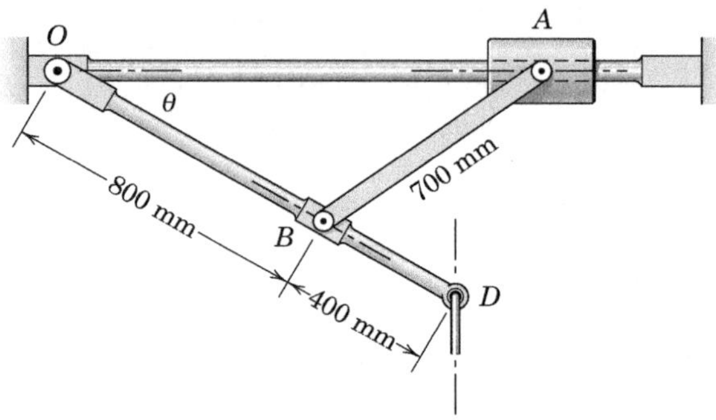

Problem 5/119

▶5/120 Determine the angular velocity ω of the ram head AE of the rock crusher in the position for which $\theta = 60°$. The crank OB has an angular speed of 60 rev/min. When B is at the bottom of its circle, D and E are on a horizontal line through F, and lines BD and AE are vertical. The dimensions are $\overline{OB} = 4$ in., $\overline{BD} = 30$ in., and $\overline{AE} = \overline{ED} = \overline{DF} = 15$ in. Carefully construct the configuration graphically, and use the method of this article.

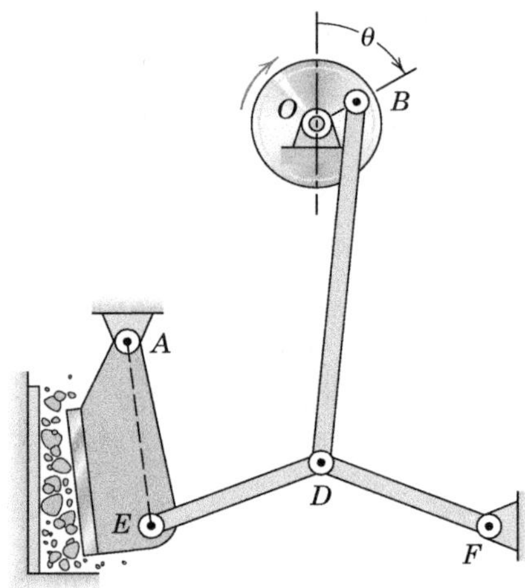

Problem 5/120

5/6 Relative Acceleration

Consider the equation $\mathbf{v}_A = \mathbf{v}_B + \mathbf{v}_{A/B}$, which describes the relative velocities of two points A and B in plane motion in terms of nonrotating reference axes. By differentiating the equation with respect to time, we may obtain the relative-acceleration equation, which is $\dot{\mathbf{v}}_A = \dot{\mathbf{v}}_B + \dot{\mathbf{v}}_{A/B}$ or

$$\boxed{\mathbf{a}_A = \mathbf{a}_B + \mathbf{a}_{A/B}} \qquad \textbf{(5/7)}$$

In words, Eq. 5/7 states that the acceleration of point A equals the vector sum of the acceleration of point B and the acceleration which A appears to have to a nonrotating observer moving with B.

Relative Acceleration Due to Rotation

If points A and B are located on the same rigid body and in the plane of motion, the distance r between them remains constant so that the observer moving with B perceives A to have circular motion about B, as we saw in Art. 5/4 with the relative-velocity relationship. Because the relative motion is circular, it follows that the relative-acceleration term will have both a normal component directed from A toward B due to the change of direction of $\mathbf{v}_{A/B}$ and a tangential component perpendicular to AB due to the change in magnitude of $\mathbf{v}_{A/B}$. These acceleration components for circular motion, cited in Eqs. 5/2, were covered earlier in Art. 2/5 and should be thoroughly familiar by now.

Thus we may write

$$\boxed{\mathbf{a}_A = \mathbf{a}_B + (\mathbf{a}_{A/B})_n + (\mathbf{a}_{A/B})_t} \qquad \textbf{(5/8)}$$

where the magnitudes of the relative-acceleration components are

$$\begin{aligned} (a_{A/B})_n &= v_{A/B}{}^2/r = r\omega^2 \\ (a_{A/B})_t &= \dot{v}_{A/B} = r\alpha \end{aligned} \qquad \textbf{(5/9)}$$

In vector notation the acceleration components are

$$\begin{aligned} (\mathbf{a}_{A/B})_n &= \boldsymbol{\omega} \times (\boldsymbol{\omega} \times \mathbf{r}) \\ (\mathbf{a}_{A/B})_t &= \boldsymbol{\alpha} \times \mathbf{r} \end{aligned} \qquad \textbf{(5/9}\boldsymbol{a}\textbf{)}$$

In these relationships, $\boldsymbol{\omega}$ is the angular velocity and $\boldsymbol{\alpha}$ is the angular acceleration of the body. The vector locating A from B is $\mathbf{r}$. It is important to observe that the *relative* acceleration terms depend on the respective *absolute* angular velocity and *absolute* angular acceleration.

Interpretation of the Relative-Acceleration Equation

The meaning of Eqs. 5/8 and 5/9 is illustrated in Fig. 5/9, which shows a rigid body in plane motion with points A and B moving along separate curved paths with absolute accelerations $\mathbf{a}_A$ and $\mathbf{a}_B$. Contrary to the case with velocities, the accelerations $\mathbf{a}_A$ and $\mathbf{a}_B$ are, in general,

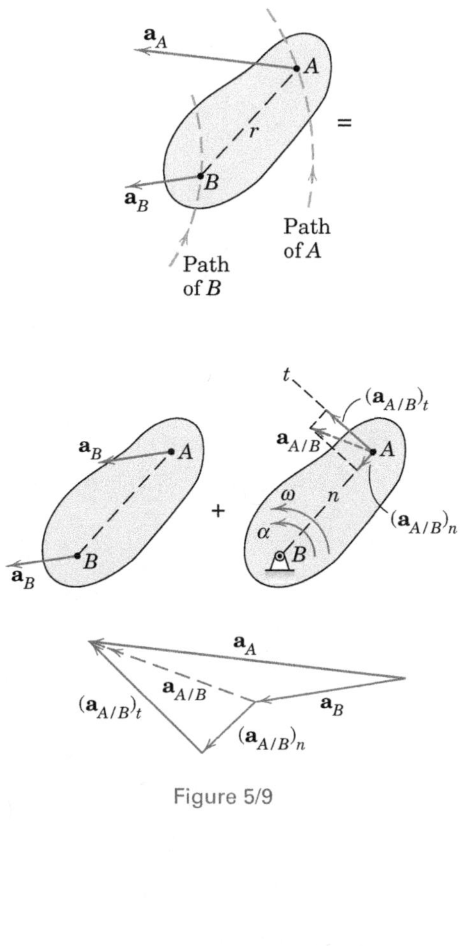

Figure 5/9

not tangent to the paths described by A and B when these paths are curvilinear. The figure shows the acceleration of A to be composed of two parts: the acceleration of B and the acceleration of A with respect to B. A sketch showing the reference point as fixed is useful in disclosing the correct sense of each of the two components of the relative-acceleration term.

Alternatively, we may express the acceleration of B in terms of the acceleration of A, which puts the nonrotating reference axes on A rather than B. This order gives

$$\mathbf{a}_B = \mathbf{a}_A + \mathbf{a}_{B/A}$$

Here $\mathbf{a}_{B/A}$ and its n- and t-components are the negatives of $\mathbf{a}_{A/B}$ and its n- and t-components. To understand this analysis better, you should make a sketch corresponding to Fig. 5/9 for this choice of terms.

Solution of the Relative-Acceleration Equation

As in the case of the relative-velocity equation, we can handle the solution to Eq. 5/8 in three different ways, namely, by scalar algebra and geometry, by vector algebra, or by graphical construction. It is helpful to be familiar with all three techniques. You should make a sketch of the vector polygon representing the vector equation and pay close attention to the head-to-tail combination of vectors so that it agrees with the equation. Known vectors should be added first, and the unknown vectors will become the closing legs of the vector polygon. It is vital that you visualize the vectors in their geometrical sense, as only then can you understand the full significance of the acceleration equation.

Before attempting a solution, identify the knowns and unknowns, keeping in mind that a solution to a vector equation in two dimensions can be carried out when the unknowns have been reduced to two scalar quantities. These quantities may be the magnitude or direction of any of the terms of the equation. When both points move on curved paths, there will, in general, be six scalar quantities to account for in Eq. 5/8.

Because the normal acceleration components depend on velocities, it is generally necessary to solve for the velocities before the acceleration calculations can be made. Choose the reference point in the relative-acceleration equation as some point on the body in question whose acceleration is either known or can be easily found. Be careful *not* to use the instantaneous center of zero velocity as the reference point unless its acceleration is known and accounted for.

An instantaneous center of zero acceleration exists for a rigid body in general plane motion, but will not be discussed here since its use is somewhat specialized.

Sample Problem 5/13

The wheel of radius r rolls to the left without slipping and, at the instant considered, the center O has a velocity $\mathbf{v}_O$ and an acceleration $\mathbf{a}_O$ to the left. Determine the acceleration of points A and C on the wheel for the instant considered.

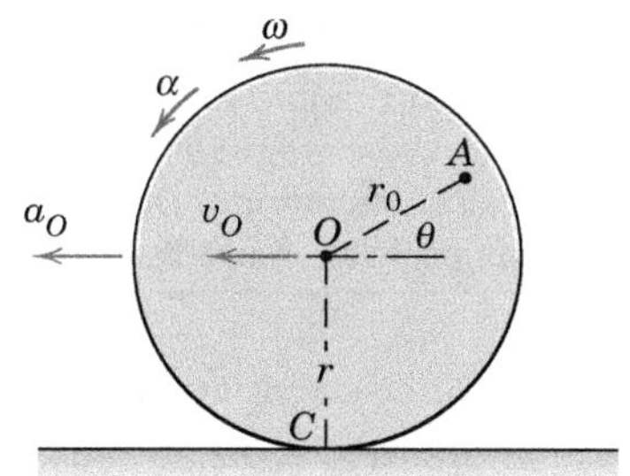

Solution. From our previous analysis of Sample Problem 5/4, we know that the angular velocity and angular acceleration of the wheel are

$$\omega = v_O/r \qquad \text{and} \qquad \alpha = a_O/r$$

The acceleration of A is written in terms of the given acceleration of O. Thus,

$$\mathbf{a}_A = \mathbf{a}_O + \mathbf{a}_{A/O} = \mathbf{a}_O + (\mathbf{a}_{A/O})_n + (\mathbf{a}_{A/O})_t$$

The relative-acceleration terms are viewed as though O were fixed, and for this relative circular motion they have the magnitudes

$$(a_{A/O})_n = r_0\omega^2 = r_0\left(\frac{v_O}{r}\right)^2$$

$$(a_{A/O})_t = r_0\alpha = r_0\left(\frac{a_O}{r}\right)$$

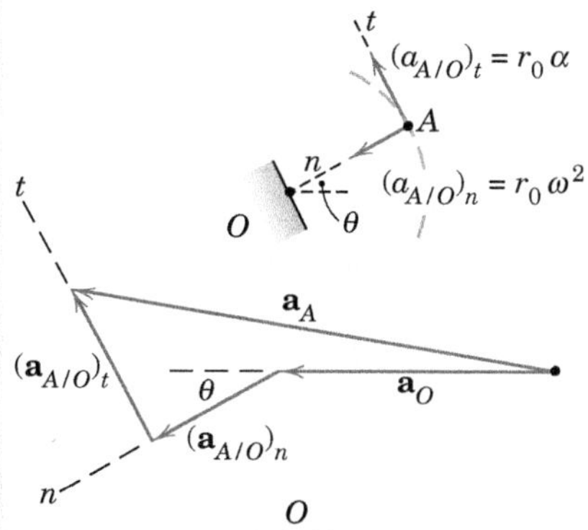

(1) and the directions shown.

Adding the vectors head-to-tail gives $\mathbf{a}_A$ as shown. In a numerical problem, we may obtain the combination algebraically or graphically. The algebraic expression for the magnitude of $\mathbf{a}_A$ is found from the square root of the sum of the squares of its components. If we use n- and t-directions, we have

(2)
$$\begin{aligned} a_A &= \sqrt{(a_A)_n{}^2 + (a_A)_t{}^2} \\ &= \sqrt{[a_O \cos\theta + (a_{A/O})_n]^2 + [a_O \sin\theta + (a_{A/O})_t]^2} \\ &= \sqrt{(r\alpha \cos\theta + r_0\omega^2)^2 + (r\alpha \sin\theta + r_0\alpha)^2} \end{aligned} \qquad \textit{Ans.}$$

The direction of $\mathbf{a}_A$ can be computed if desired.

The acceleration of the instantaneous center C of zero velocity, considered a point on the wheel, is obtained from the expression

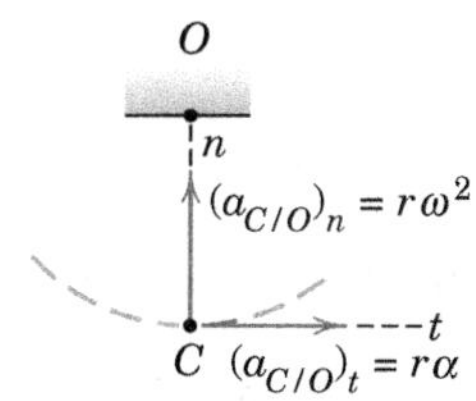

$$\mathbf{a}_C = \mathbf{a}_O + \mathbf{a}_{C/O}$$

where the components of the relative-acceleration term are $(a_{C/O})_n = r\omega^2$ directed from C to O and $(a_{C/O})_t = r\alpha$ directed to the right because of the counterclockwise angular acceleration of line CO about O. The terms are added together in the lower diagram and it is seen that

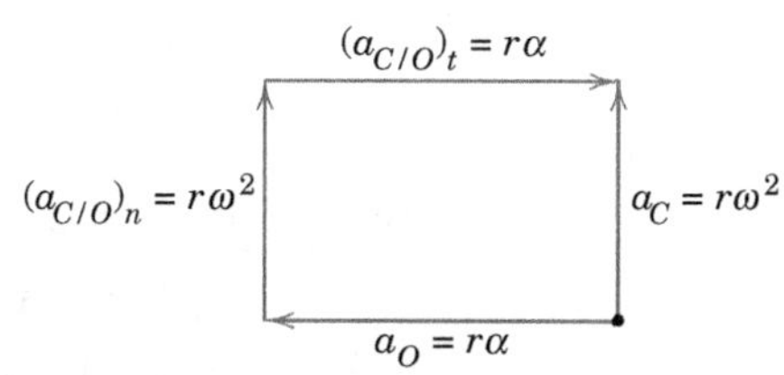

(3)
$$a_C = r\omega^2 \qquad \textit{Ans.}$$

Helpful Hints

(1) The counterclockwise angular acceleration α of OA determines the positive direction of $(a_{A/O})_t$. The normal component $(a_{A/O})_n$ is, of course, directed toward the reference center O.

(2) If the wheel were rolling to the right with the same velocity v_O but still had an acceleration a_O to the left, note that the solution for a_A would be unchanged.

(3) We note that the acceleration of the instantaneous center of zero velocity is independent of α and is directed toward the center of the wheel. This conclusion is a useful result to remember.

Sample Problem 5/14

The linkage of Sample Problem 5/8 is repeated here. Crank CB has a constant counterclockwise angular velocity of 2 rad/s in the position shown during a short interval of its motion. Determine the angular acceleration of links AB and OA for this position. Solve by using vector algebra.

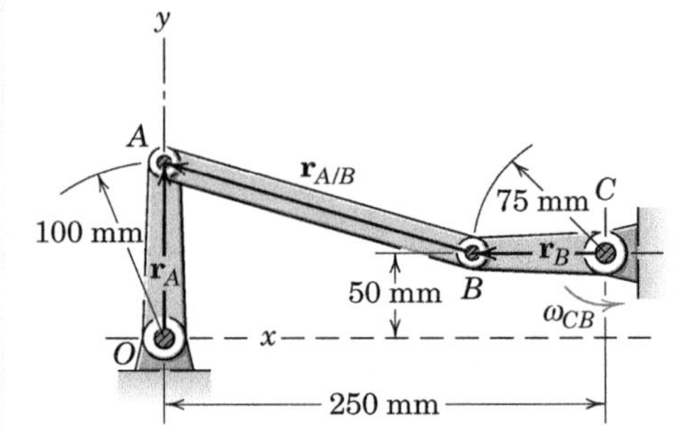

Solution. We first solve for the velocities which were obtained in Sample Problem 5/8. They are

$$\omega_{AB} = -6/7 \text{ rad/s} \qquad \text{and} \qquad \omega_{OA} = -3/7 \text{ rad/s}$$

where the counterclockwise direction (+**k**-direction) is taken as positive. The acceleration equation is

$$\mathbf{a}_A = \mathbf{a}_B + (\mathbf{a}_{A/B})_n + (\mathbf{a}_{A/B})_t$$

where, from Eqs. 5/3 and 5/9*a*, we may write

① $$\mathbf{a}_A = \boldsymbol{\alpha}_{OA} \times \mathbf{r}_A + \boldsymbol{\omega}_{OA} \times (\boldsymbol{\omega}_{OA} \times \mathbf{r}_A)$$
$$= \alpha_{OA}\mathbf{k} \times 100\mathbf{j} + (-\tfrac{3}{7}\mathbf{k}) \times (-\tfrac{3}{7}\mathbf{k} \times 100\mathbf{j})$$
$$= -100\alpha_{OA}\mathbf{i} - 100(\tfrac{3}{7})^2\mathbf{j} \text{ mm/s}^2$$

$$\mathbf{a}_B = \boldsymbol{\alpha}_{CB} \times \mathbf{r}_B + \boldsymbol{\omega}_{CB} \times (\boldsymbol{\omega}_{CB} \times \mathbf{r}_B)$$
$$= \mathbf{0} + 2\mathbf{k} \times (2\mathbf{k} \times [-75\mathbf{i}])$$
$$= 300\mathbf{i} \text{ mm/s}^2$$

② $$(\mathbf{a}_{A/B})_n = \boldsymbol{\omega}_{AB} \times (\boldsymbol{\omega}_{AB} \times \mathbf{r}_{A/B})$$
$$= -\tfrac{6}{7}\mathbf{k} \times [(-\tfrac{6}{7}\mathbf{k}) \times (-175\mathbf{i} + 50\mathbf{j})]$$
$$= (\tfrac{6}{7})^2(175\mathbf{i} - 50\mathbf{j}) \text{ mm/s}^2$$

$$(\mathbf{a}_{A/B})_t = \boldsymbol{\alpha}_{AB} \times \mathbf{r}_{A/B}$$
$$= \alpha_{AB}\mathbf{k} \times (-175\mathbf{i} + 50\mathbf{j})$$
$$= -50\alpha_{AB}\mathbf{i} - 175\alpha_{AB}\mathbf{j} \text{ mm/s}^2$$

We now substitute these results into the relative-acceleration equation and equate separately the coefficients of the **i**-terms and the coefficients of the **j**-terms to give

$$-100\alpha_{OA} = 429 - 50\alpha_{AB}$$
$$-18.37 = -36.7 - 175\alpha_{AB}$$

The solutions are

$$\alpha_{AB} = -0.1050 \text{ rad/s}^2 \qquad \text{and} \qquad \alpha_{OA} = -4.34 \text{ rad/s}^2 \qquad \textit{Ans.}$$

Since the unit vector **k** points out from the paper in the positive z-direction, we see that the angular accelerations of AB and OA are both clockwise (negative).

It is recommended that the student sketch each of the acceleration vectors in its proper geometric relationship according to the relative-acceleration equation to help clarify the meaning of the solution.

Helpful Hints

① Remember to preserve the order of the factors in the cross products.

② In expressing the term $\mathbf{a}_{A/B}$ be certain that $\mathbf{r}_{A/B}$ is written as the vector from B to A and not the reverse.

Sample Problem 5/15

The slider-crank mechanism of Sample Problem 5/9 is repeated here. The crank OB has a constant clockwise angular speed of 1500 rev/min. For the instant when the crank angle θ is 60°, determine the acceleration of the piston A and the angular acceleration of the connecting rod AB.

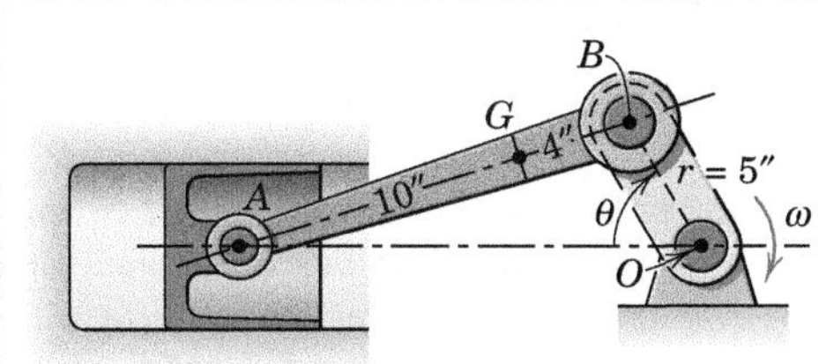

Solution. The acceleration of A may be expressed in terms of the acceleration of the crank pin B. Thus,

$$\mathbf{a}_A = \mathbf{a}_B + (\mathbf{a}_{A/B})_n + (\mathbf{a}_{A/B})_t$$

Point B moves in a circle of 5-in. radius with a constant speed so that it has
① only a normal component of acceleration directed from B to O.

$$[a_n = r\omega^2] \qquad a_B = \frac{5}{12}\left(\frac{1500[2\pi]}{60}\right)^2 = 10{,}280 \text{ ft/sec}^2$$

The relative-acceleration terms are visualized with A rotating in a circle relative to B, which is considered fixed, as shown. From Sample Problem 5/9, the angular velocity of AB for these same conditions is $\omega_{AB} = 29.5$ rad/sec so that

② $$[a_n = r\omega^2] \qquad (a_{A/B})_n = \frac{14}{12}(29.5)^2 = 1015 \text{ ft/sec}^2$$

directed from A to B. The tangential component $(\mathbf{a}_{A/B})_t$ is known in direction only since its magnitude depends on the unknown angular acceleration of AB. We also know the direction of $\mathbf{a}_A$ since the piston is confined to move along the horizontal axis of the cylinder. There are now only two scalar unknowns left in the equation, namely, the magnitudes of $\mathbf{a}_A$ and $(\mathbf{a}_{A/B})_t$, so the solution can be carried out.

If we adopt an algebraic solution using the geometry of the acceleration polygon, we first compute the angle between AB and the horizontal. With the law of sines, this angle becomes 18.02°. Equating separately the horizontal components and the vertical components of the terms in the acceleration equation, as seen from the acceleration polygon, gives

$$a_A = 10{,}280 \cos 60° + 1015 \cos 18.02° - (a_{A/B})_t \sin 18.02°$$

$$0 = 10{,}280 \sin 60° - 1015 \sin 18.02° - (a_{A/B})_t \cos 18.02°$$

The solution to these equations gives the magnitudes

$$(a_{A/B})_t = 9030 \text{ ft/sec}^2 \quad \text{and} \quad a_A = 3310 \text{ ft/sec}^2 \qquad \textit{Ans.}$$

With the sense of $(\mathbf{a}_{A/B})_t$ also determined from the diagram, the angular acceleration of AB is seen from the figure representing rotation relative to B to be

$$[\alpha = a_t/r] \qquad \alpha_{AB} = 9030/(14/12) = 7740 \text{ rad/sec}^2 \text{ clockwise} \qquad \textit{Ans.}$$

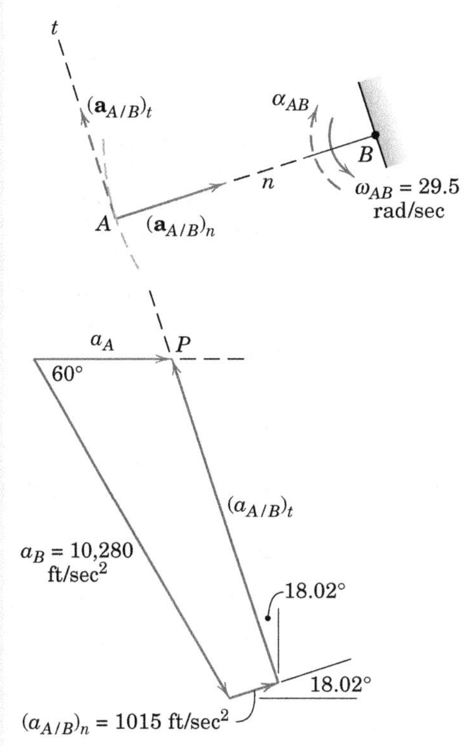

If we adopt a graphical solution, we begin with the known vectors $\mathbf{a}_B$ and
③ $(\mathbf{a}_{A/B})_n$ and add them head-to-tail using a convenient scale. Next we construct the direction of $(\mathbf{a}_{A/B})_t$ through the head of the last vector. The solution of the equation is obtained by the intersection P of this last line with a horizontal line through the starting point representing the known direction of the vector sum $\mathbf{a}_A$. Scaling the magnitudes from the diagram gives values which agree with the calculated results.

$$a_A = 3310 \text{ ft/sec}^2 \quad \text{and} \quad (a_{A/B})_t = 9030 \text{ ft/sec}^2 \qquad \textit{Ans.}$$

Helpful Hints

① If the crank OB had an angular acceleration, $\mathbf{a}_B$ would also have a tangential component of acceleration.

② Alternatively, the relation $a_n = v^2/r$ may be used for calculating $(a_{A/B})_n$, provided the relative velocity $v_{A/B}$ is used for v. The equivalence is easily seen when it is recalled that $v_{A/B} = r\omega$.

③ Except where extreme accuracy is required, do not hesitate to use a graphical solution, as it is quick and reveals the physical relationships among the vectors. The known vectors, of course, may be added in any order as long as the governing equation is satisfied.

PROBLEMS

Introductory Problems

5/121 For the instant represented, corner C of the rectangular plate has an acceleration of 5 m/s^2 in the negative y-direction, and the plate has a clockwise angular velocity of 4 rad/s which is decreasing by 12 rad/s each second. Determine the magnitude of the acceleration of A at this instant. Solve by scalar-geometric and by vector-algebraic methods.

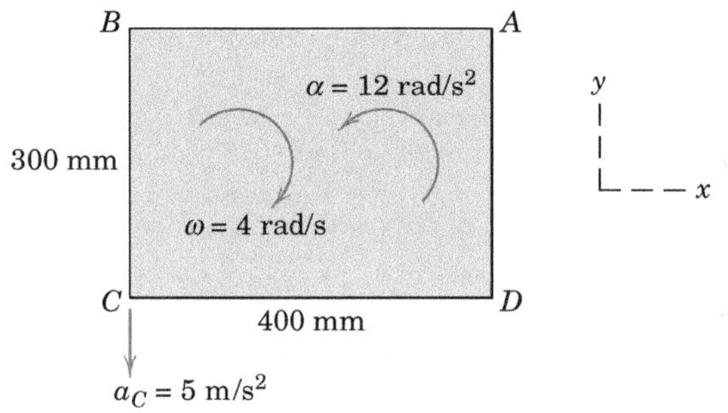

Problem 5/121

5/122 The two rotor blades of 800-mm radius rotate counterclockwise with a constant angular velocity $\omega = \dot{\theta} = 2$ rad/s about the shaft at O mounted in the sliding block. The acceleration of the block is $a_O = 3$ m/s^2. Determine the magnitude of the acceleration of the tip A of the blade when (*a*) $\theta = 0$, (*b*) $\theta = 90°$, and (*c*) $\theta = 180°$. Does the velocity of O or the sense of ω enter into the calculation?

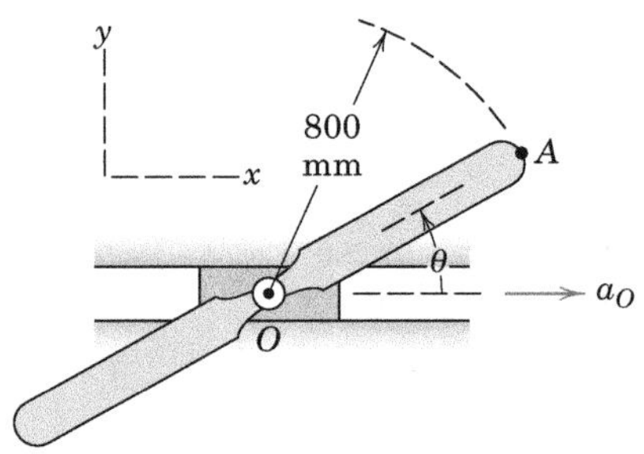

Problem 5/122

5/123 A container for waste materials is dumped by the hydraulically-activated linkage shown. If the piston rod starts from rest in the position indicated and has an acceleration of 1.5 ft/sec^2 in the direction shown, compute the initial angular acceleration of the container.

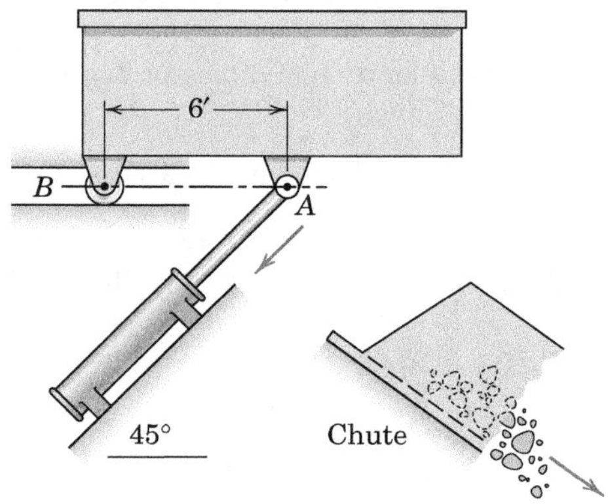

Problem 5/123

5/124 Determine the angular velocity and angular acceleration of the slender bar AB just after roller B passes point C and enters the circular portion of the support surface.

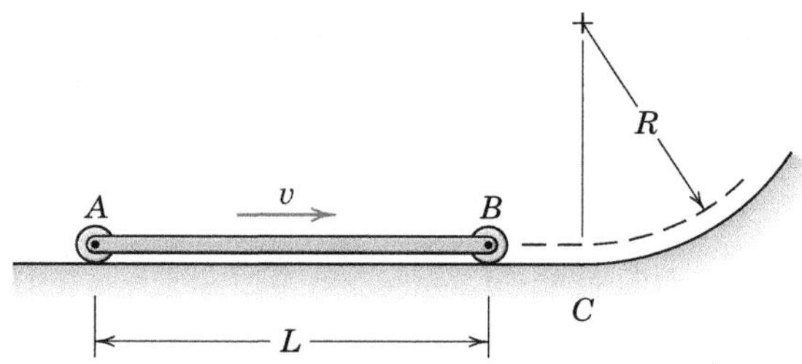

Problem 5/124

5/125 The wheel of radius R rolls without slipping, and its center O has an acceleration a_O. A point P on the wheel is a distance r from O. For given values of a_O, R, and r, determine the angle θ and the velocity v_O of the wheel for which P has no acceleration in this position.

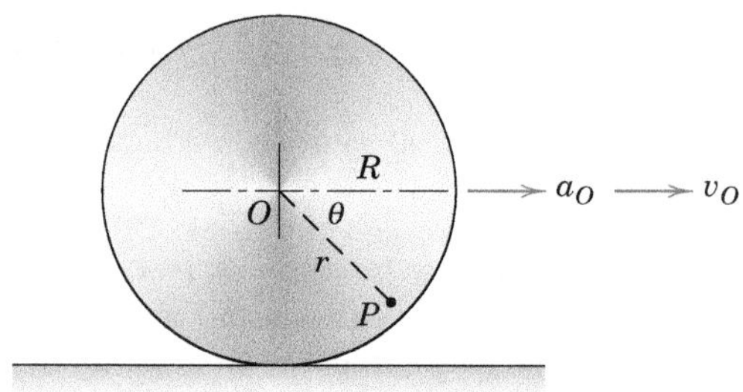

Problem 5/125

5/126 The 9-m steel beam is being hoisted from its horizontal position by the two cables attached at A and B. If the initial angular accelerations of the hoisting drums are $\alpha_1 = 0.5$ rad/s^2 and $\alpha_2 = 0.2$ rad/s^2 in the directions shown, determine the corresponding angular acceleration α of the beam, the acceleration of C, and the distance b from B to a point P on the beam centerline which has no acceleration.

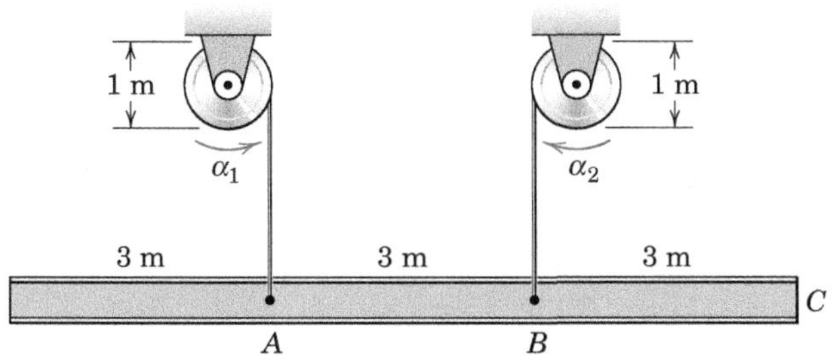

Problem 5/126

5/127 The bar of Prob. 5/82 is repeated here. The ends of the 0.4-m bar remain in contact with their respective support surfaces. End B has a velocity of 0.5 m/s and an acceleration of 0.3 m/s^2 in the directions shown. Determine the angular acceleration of the bar and the acceleration of end A.

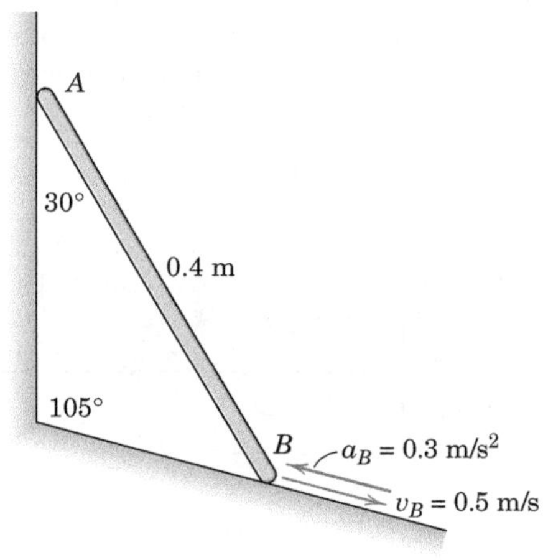

Problem 5/127

5/128 Determine the acceleration of point B on the equator of the earth, repeated here from Prob. 5/63. Use the data given with that problem and assume that the earth's orbital path is circular, consulting Table D/2 as necessary. Consider the center of the sun fixed and neglect the tilt of the axis of the earth.

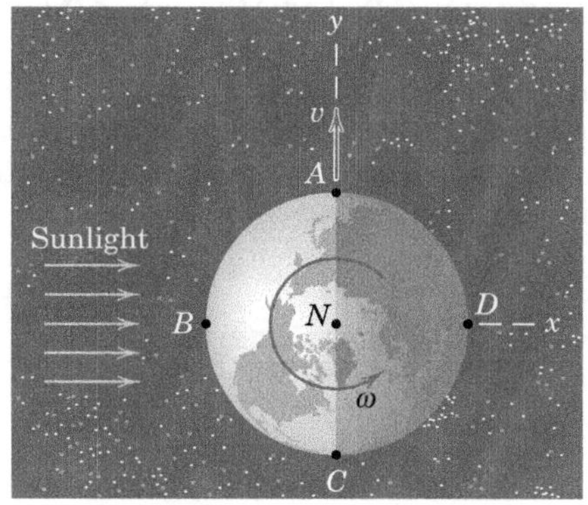

Problem 5/128

5/129 The spoked wheel of Prob. 5/73 is repeated here with additional information supplied. For a given cord speed v and acceleration a at point P and wheel radius r, determine the acceleration of point B with respect to point A.

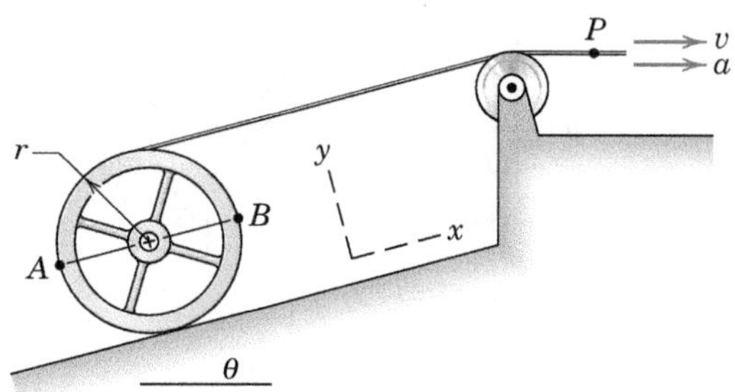

Problem 5/129

5/130 Calculate the angular acceleration of the plate in the position shown, where control link AO has a constant angular velocity $\omega_{OA} = 4$ rad/sec and $\theta = 60°$ for both links.

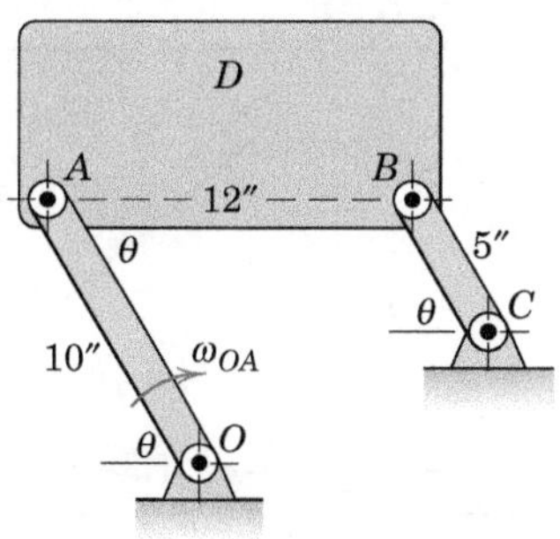

Problem 5/130

5/131 The bar AB of Prob. 5/71 is repeated here. At the instant under consideration, roller B has just begun moving on the 15° incline, and the velocity and acceleration of roller A are given. Determine the angular acceleration of bar AB and the acceleration of roller B.

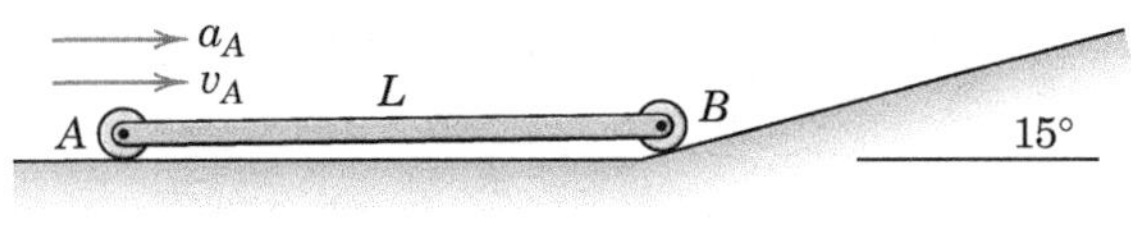

Problem 5/131

5/132 Determine the angular acceleration α_{AB} of AB for the position shown if link OB has a constant angular velocity ω.

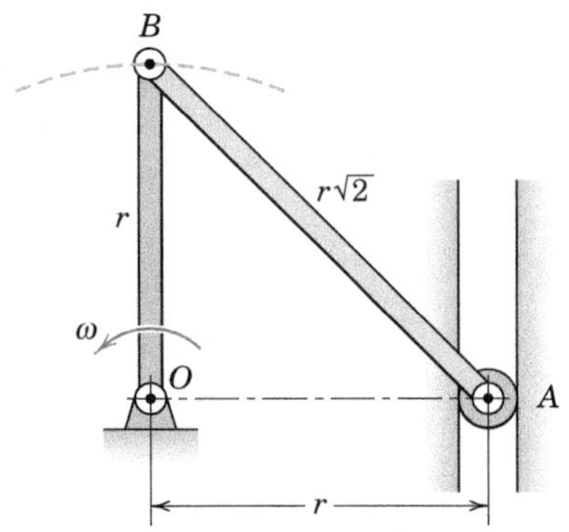

Problem 5/132

Representative Problems

5/133 Determine the angular acceleration of AB and the linear acceleration of A for the position $\theta = 90°$ if $\dot{\theta} = 4$ rad/s and $\ddot{\theta} = 0$ at that position.

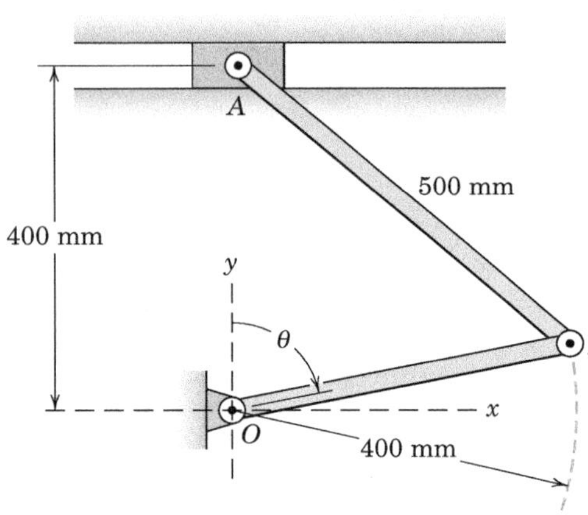

Problem 5/133

5/134 The switching device of Prob. 5/76 is repeated here. If the vertical control rod has a downward velocity $v = 2$ ft/sec and an upward acceleration $a = 1.2$ ft/sec^2 when the device is in the position shown, determine the magnitude of the acceleration of point A. Roller C is in continuous contact with the inclined surface.

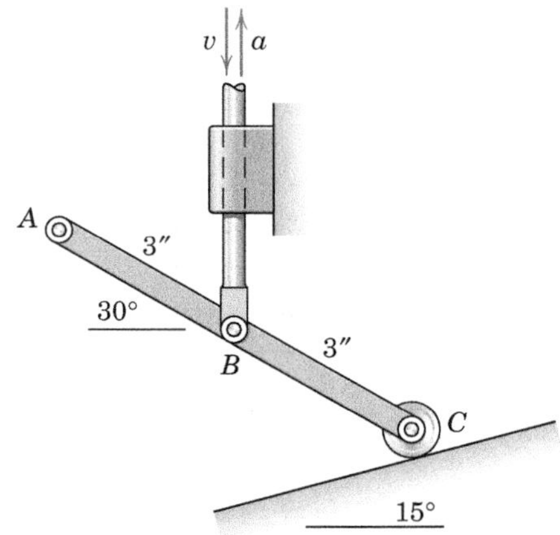

Problem 5/134

5/135 The two connected wheels of Prob. 5/64 are shown again here. Determine the magnitude of the acceleration of point D in the position shown if the center C of the smaller wheel has an acceleration to the right of 0.8 m/s^2 and has reached a velocity of 0.4 m/s at this instant.

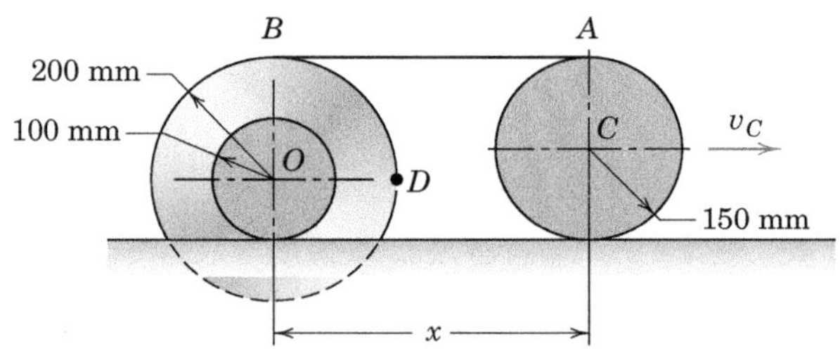

Problem 5/135

5/136 The end rollers of bar AB are constrained to the slot shown. If roller A has a downward velocity of 1.2 m/s and this speed is constant over a small motion interval, determine the tangential acceleration of roller B as it passes the topmost position. The value of R is 0.5 m.

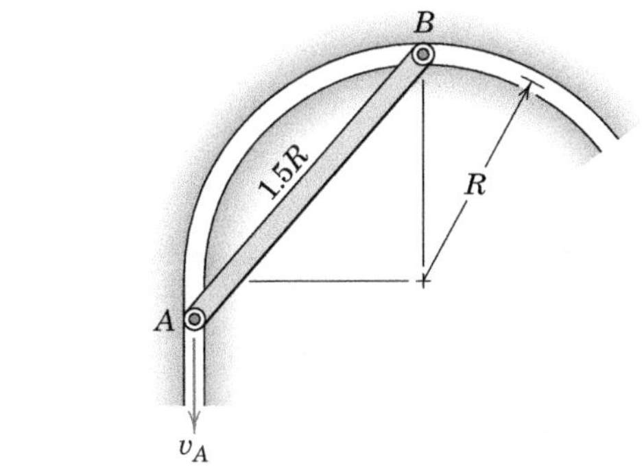

Problem 5/136

5/137 If the wheel in each case rolls on the circular surface without slipping, determine the acceleration of point C on the wheel momentarily in contact with the circular surface. The wheel has an angular velocity ω and an angular acceleration α.

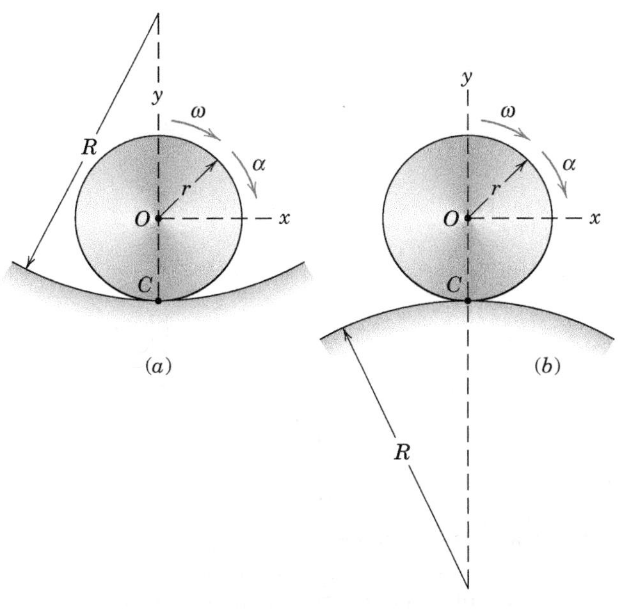

Problem 5/137

5/138 The system of Prob. 5/100 is repeated here. Crank OA rotates with a constant counterclockwise angular velocity of 9 rad/s. Determine the angular acceleration α_{AB} of link AB for the position shown.

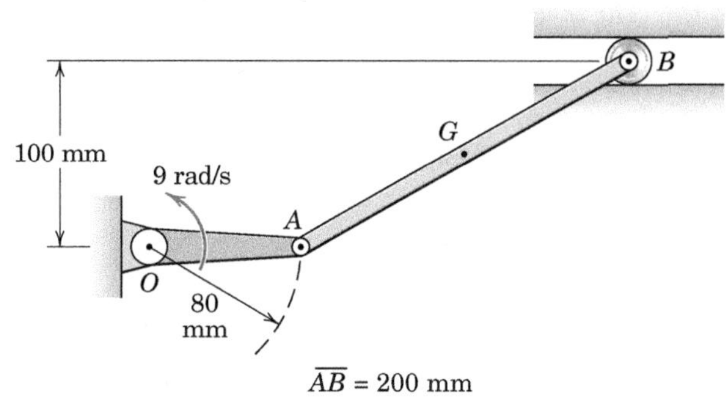

Problem 5/138

5/139 The system of Prob. 5/101 is repeated here. Crank OA is rotating at a counterclockwise angular rate of 9 rad/s, and this rate is decreasing at 5 rad/s^2. Determine the angular acceleration α_{AB} of link AB for the position shown.

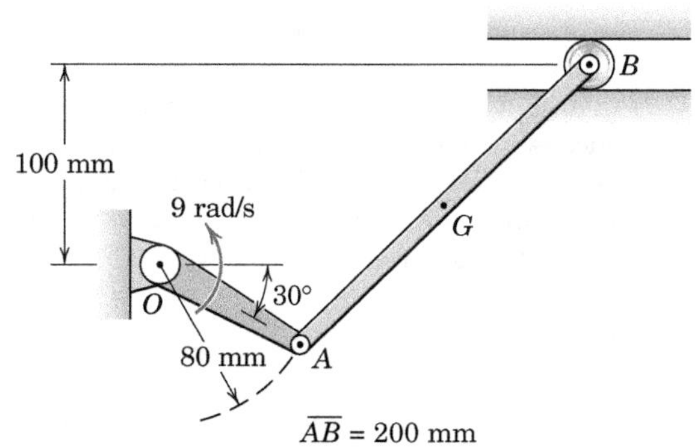

Problem 5/139

5/140 The triangular plate ABD has a clockwise angular velocity of 3 rad/sec and link OA has zero angular acceleration for the instant represented. Determine the angular accelerations of plate ABD and link BC for this instant.

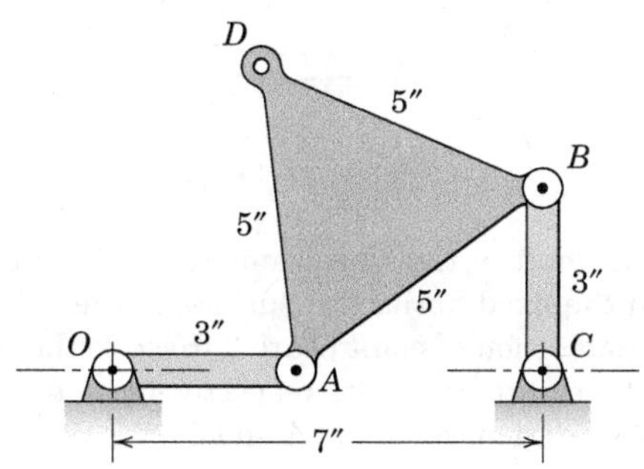

Problem 5/140

5/141 The mechanism of Prob. 5/77 is repeated here. The angular velocity ω_0 of the disk is constant. For the instant represented, determine the angular acceleration α_{AB} of link AB and the acceleration a_B of collar B. Assume the quantities ω_0 and r to be known.

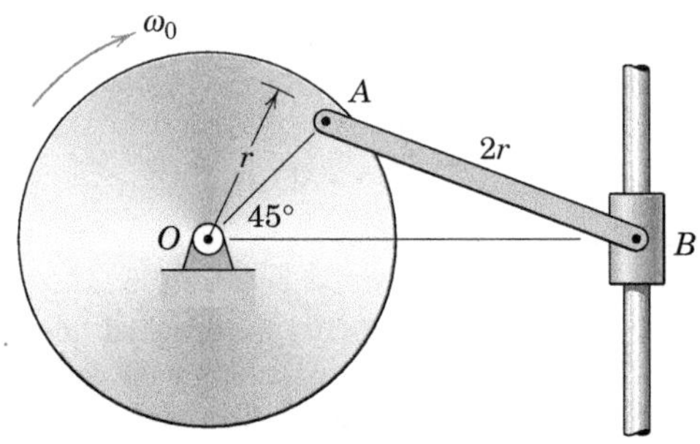

Problem 5/141

5/142 The system of Prob. 5/84 is repeated here. If the vertical rod has a downward velocity $v = 0.8$ m/s and an upward acceleration $a = 1.2$ m/s^2 when the device is in the position shown, determine the corresponding angular acceleration α of bar AB and the magnitude of the acceleration of roller B. The value of R is 0.4 m.

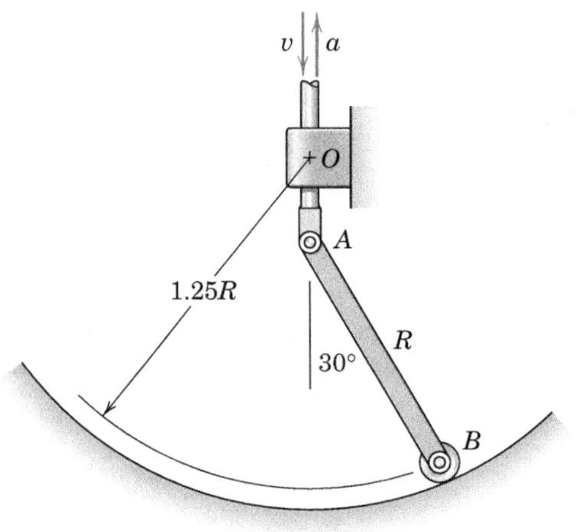

Problem 5/142

5/143 The shaft of the wheel unit rolls without slipping on the fixed horizontal surface. If the velocity and acceleration of point O are 3 ft/sec to the right and 4 ft/sec^2 to the left, respectively, determine the accelerations of points A and D.

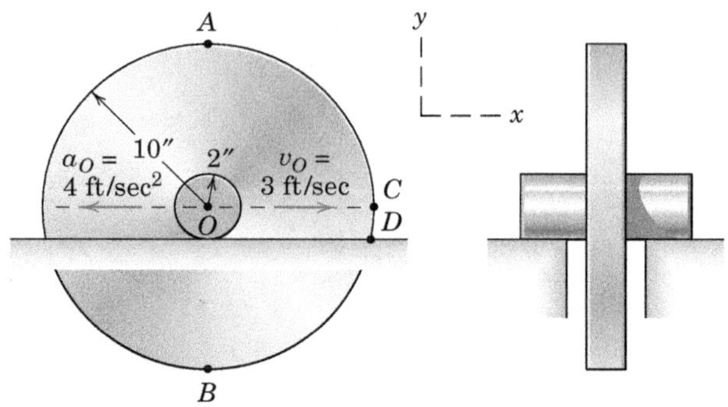

Problem 5/143

5/144 Plane motion of the triangular plate ABC is controlled by crank OA and link DB. For the instant represented, when OA and DB are vertical, OA has a clockwise angular velocity of 3 rad/s and a counterclockwise angular acceleration of 10 rad/s^2. Determine the angular acceleration of DB for this instant.

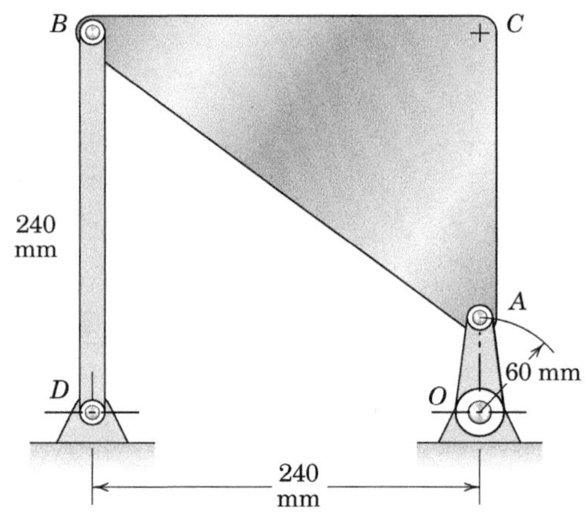

Problem 5/144

5/145 The system of Prob. 5/110 is repeated here. At the instant under consideration, the rod of the hydraulic cylinder is extending at the constant rate $v_A = 2$ m/s. Determine the angular acceleration α_{OB} of link OB.

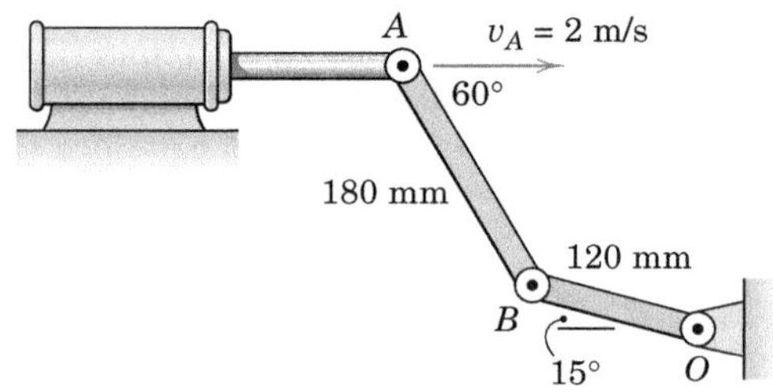

Problem 5/145

5/146 The velocity of roller A is $v_A = 0.5$ m/s to the right as shown, and this velocity is momentarily decreasing at a rate of 2 m/s^2. Determine the corresponding value of the angular acceleration α of bar AB as well as the tangential acceleration of roller B along the circular guide. The value of R is 0.6 m.

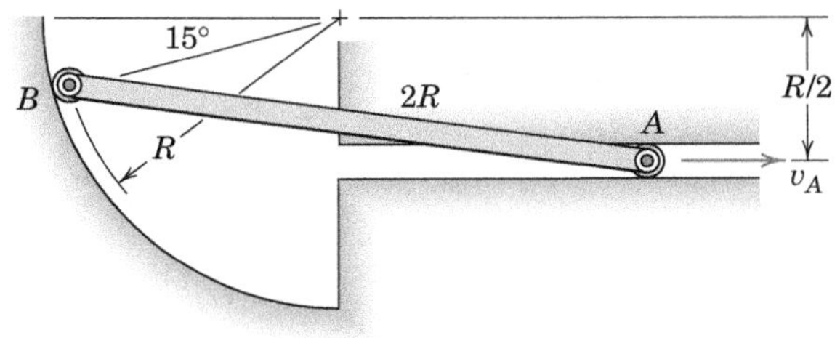

Problem 5/146

5/147 In the design of this linkage, motion of the square plate is controlled by the two pivoted links. Link OA has a constant angular velocity $\omega = 4$ rad/s during a short interval of motion. For the instant represented, $\theta = \tan^{-1} 4/3$ and AB is parallel to the x-axis. For this instant, determine the angular acceleration of both the plate and link CB.

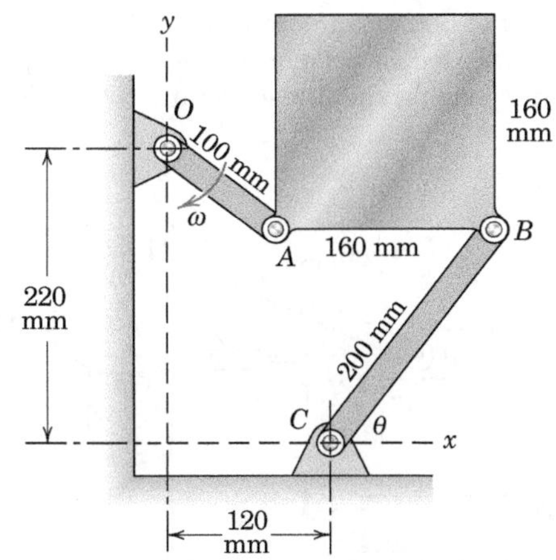

Problem 5/147

5/148 The mechanism of Prob. 5/112 is repeated here. If the band has a constant speed of 2 m/s as indicated in the figure, determine the angular acceleration α_{AB} of link AB.

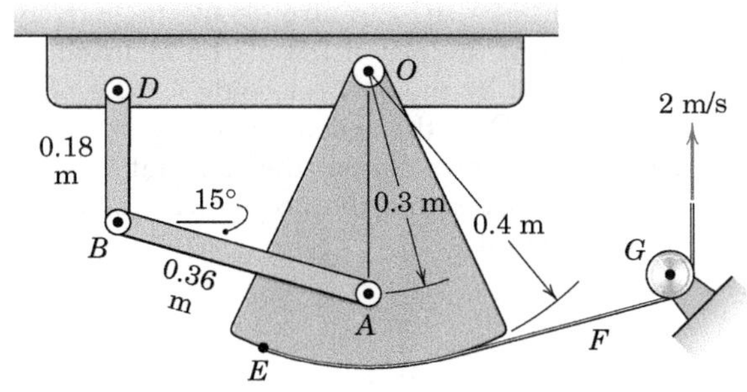

Problem 5/148

5/149 The bar AB from Prob. 5/74 is repeated here. If the velocity of point A is 3 m/s to the right and is constant for an interval including the position shown, determine the tangential acceleration of point B along its path and the angular acceleration of the bar.

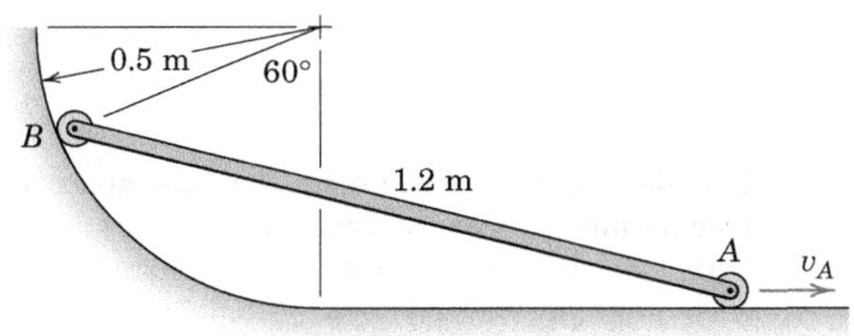

Problem 5/149

5/150 If the piston rod of the hydraulic cylinder C has a constant upward velocity of 0.5 m/s, calculate the acceleration of point D for the position where θ is 45°.

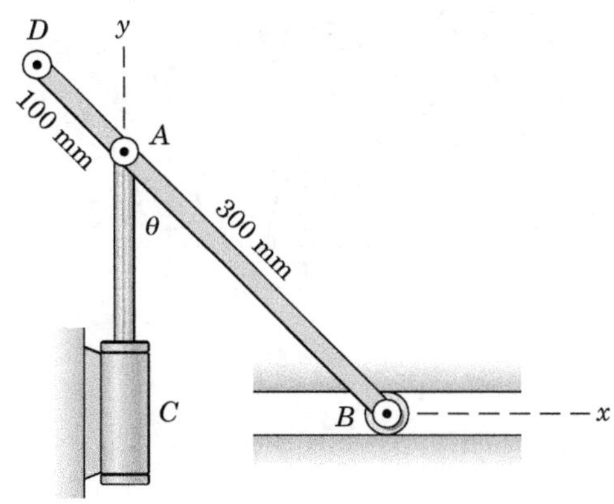

Problem 5/150

5/151 Motion of link *ABC* is controlled by the horizontal movement of the piston rod of the hydraulic cylinder *D* and by the vertical guide for the pinned slider at *B*. For the instant when $\theta = 45°$, the piston rod is retracting at the constant rate $v_C = 0.6$ ft/sec. Determine the acceleration of point *A* for this instant.

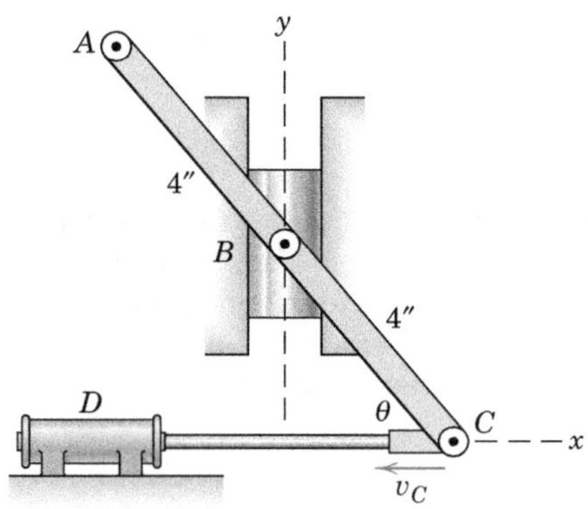

Problem 5/151

5/152 The deployment mechanism for the spacecraft magnetometer boom of Prob. 5/88 is shown again here. The driving link *OB* has a constant clockwise angular velocity ω_{OB} of 0.5 rad/sec as it crosses the vertical position. Determine the angular acceleration $\boldsymbol{\alpha}_{CA}$ of the boom for the position shown where $\tan\theta = 4/3$.

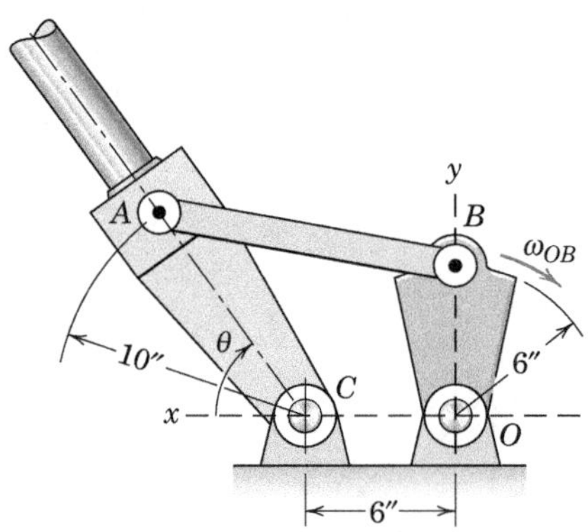

Problem 5/152

5/153 The four-bar linkage of Prob. 5/86 is repeated here. If the angular velocity and angular acceleration of drive link *OA* are 10 rad/s and 5 rad/s², respectively, both counterclockwise, determine the angular accelerations of bars *AB* and *BC* for the instant represented.

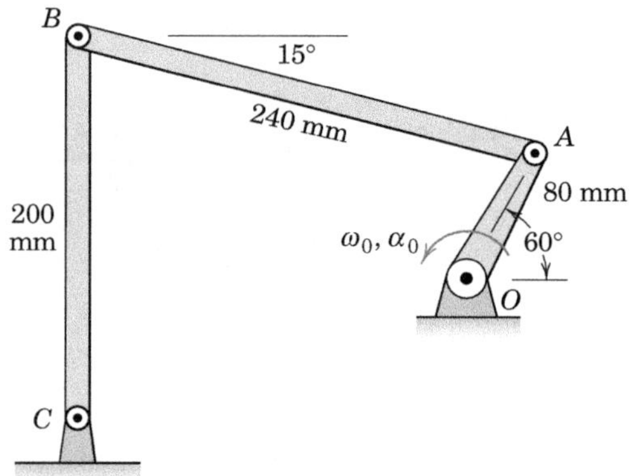

Problem 5/153

5/154 The elements of a power hacksaw are shown in the figure. The saw blade is mounted in a frame which slides along the horizontal guide. If the motor turns the flywheel at a constant counterclockwise speed of 60 rev/min, determine the acceleration of the blade for the position where $\theta = 90°$, and find the corresponding angular acceleration of the link *AB*.

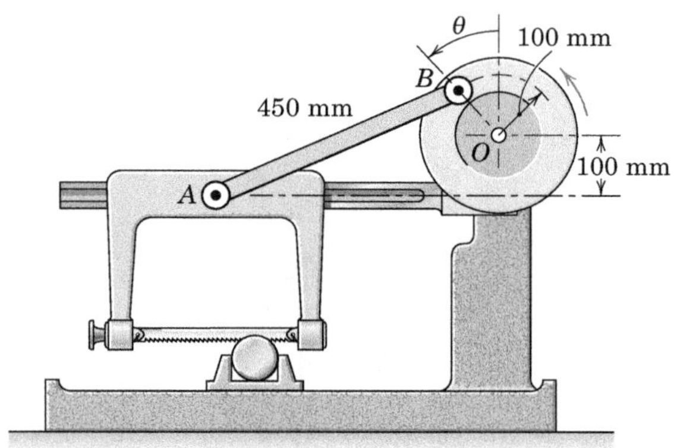

Problem 5/154

▶5/155 A mechanism for pushing small boxes from an assembly line onto a conveyor belt, repeated from Prob. 5/89, is shown with arm OD and crank CB in their vertical positions. For the configuration shown, crank CB has a constant clockwise angular velocity of π rad/s. Determine the acceleration of E.

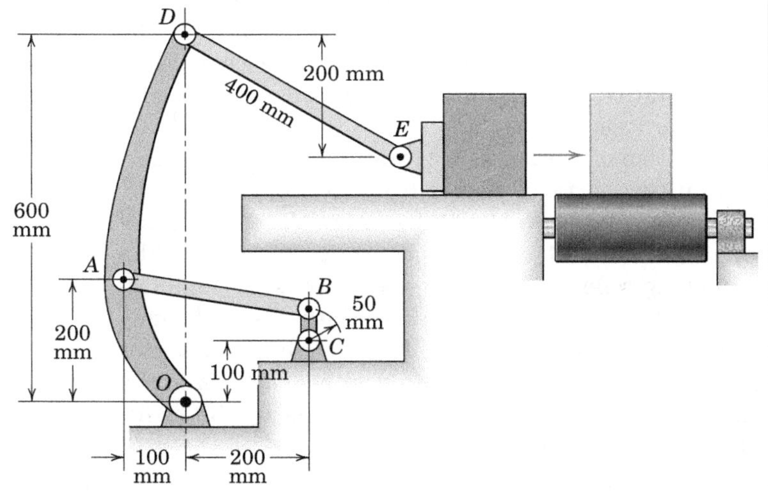

Problem 5/155

▶5/156 An intermittent-drive mechanism for perforated tape F consists of the link DAB driven by the crank OB. The trace of the motion of the finger at D is shown by the dashed line. Determine the magnitude of the acceleration of D at the instant represented when both OB and CA are horizontal if OB has a constant clockwise rotational velocity of 120 rev/min.

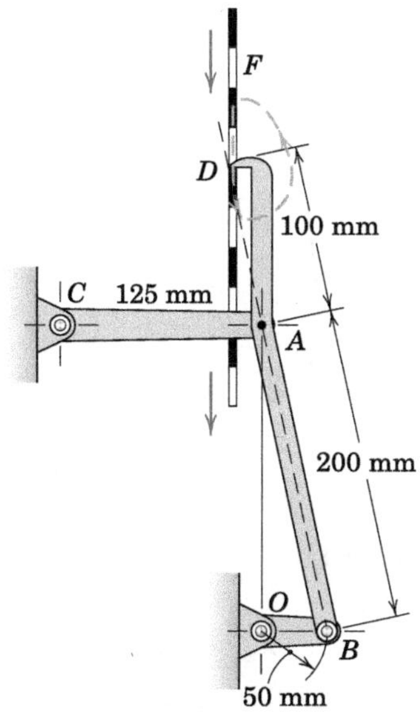

Problem 5/156

5/7 Motion Relative to Rotating Axes

In our discussion of the relative motion of particles in Art. 2/8 and in our use of the relative-motion equations for the plane motion of rigid bodies in this present chapter, we have used *nonrotating* reference axes to describe relative velocity and relative acceleration. Use of rotating reference axes greatly facilitates the solution of many problems in kinematics where motion is generated within a system or observed from a system which itself is rotating. An example of such a motion is the movement of a fluid particle along the curved vane of a centrifugal pump, where the path relative to the vanes of the impeller becomes an important design consideration.

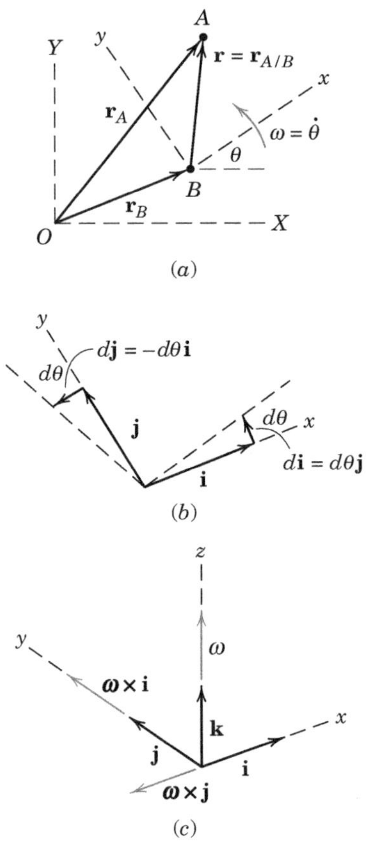

Figure 5/10

We begin the description of motion using rotating axes by considering the plane motion of two particles A and B in the fixed X-Y plane, Fig. 5/10a. For the time being, we will consider A and B to be moving independently of one another for the sake of generality. We observe the motion of A from a moving reference frame x-y which has its origin attached to B and which rotates with an angular velocity $\omega = \dot{\theta}$. We may write this angular velocity as the vector $\boldsymbol{\omega} = \omega\mathbf{k} = \dot{\theta}\mathbf{k}$, where the vector is normal to the plane of motion and where its positive sense is in the positive z-direction (out from the paper), as established by the right-hand rule. The absolute position vector of A is given by

$$\mathbf{r}_A = \mathbf{r}_B + \mathbf{r} = \mathbf{r}_B + (x\mathbf{i} + y\mathbf{j}) \tag{5/10}$$

where $\mathbf{i}$ and $\mathbf{j}$ are unit vectors attached to the x-y frame and $\mathbf{r} = x\mathbf{i} + y\mathbf{j}$ stands for $\mathbf{r}_{A/B}$, the position vector of A with respect to B.

Time Derivatives of Unit Vectors

To obtain the velocity and acceleration equations we must successively differentiate the position-vector equation with respect to time. In contrast to the case of translating axes treated in Art. 2/8, the unit vectors $\mathbf{i}$ and $\mathbf{j}$ are now rotating with the x-y axes and, therefore, have time derivatives which must be evaluated. These derivatives may be seen from Fig. 5/10b, which shows the infinitesimal change in each unit vector during time dt as the reference axes rotate through an angle $d\theta = \omega\,dt$. The differential change in $\mathbf{i}$ is $d\mathbf{i}$, and it has the direction of $\mathbf{j}$ and a magnitude equal to the angle $d\theta$ times the length of the vector $\mathbf{i}$, which is unity. Thus, $d\mathbf{i} = d\theta\,\mathbf{j}$.

Similarly, the unit vector $\mathbf{j}$ has an infinitesimal change $d\mathbf{j}$ which points in the negative x-direction, so that $d\mathbf{j} = -d\theta\,\mathbf{i}$. Dividing by dt and replacing $d\mathbf{i}/dt$ by $\dot{\mathbf{i}}$, $d\mathbf{j}/dt$ by $\dot{\mathbf{j}}$, and $d\theta/dt$ by $\dot{\theta} = \omega$ result in

$$\dot{\mathbf{i}} = \omega\mathbf{j} \qquad \text{and} \qquad \dot{\mathbf{j}} = -\omega\mathbf{i}$$

By using the cross product, we can see from Fig. 5/10c that $\boldsymbol{\omega} \times \mathbf{i} = \omega\mathbf{j}$ and $\boldsymbol{\omega} \times \mathbf{j} = -\omega\mathbf{i}$. Thus, the time derivatives of the unit vectors may be written as

$$\boxed{\dot{\mathbf{i}} = \boldsymbol{\omega} \times \mathbf{i} \qquad \text{and} \qquad \dot{\mathbf{j}} = \boldsymbol{\omega} \times \mathbf{j}} \tag{5/11}$$

Relative Velocity

We now use the expressions of Eqs. 5/11 when taking the time derivative of the position-vector equation for A and B to obtain the relative-velocity relation. Differentiation of Eq. 5/10 gives

$$\begin{aligned}\dot{\mathbf{r}}_A &= \dot{\mathbf{r}}_B + \frac{d}{dt}(x\mathbf{i} + y\mathbf{j})\\ &= \dot{\mathbf{r}}_B + (x\dot{\mathbf{i}} + y\dot{\mathbf{j}}) + (\dot{x}\mathbf{i} + \dot{y}\mathbf{j})\end{aligned}$$

But $x\dot{\mathbf{i}} + y\dot{\mathbf{j}} = \boldsymbol{\omega} \times x\mathbf{i} + \boldsymbol{\omega} \times y\mathbf{j} = \boldsymbol{\omega} \times (x\mathbf{i} + y\mathbf{j}) = \boldsymbol{\omega} \times \mathbf{r}$. Also, since the observer in x-y measures velocity components $\dot{x}$ and $\dot{y}$, we see that $\dot{x}\mathbf{i} + \dot{y}\mathbf{j} = \mathbf{v}_{\text{rel}}$, which is the velocity relative to the x-y frame of reference. Thus, the relative-velocity equation becomes

$$\boxed{\mathbf{v}_A = \mathbf{v}_B + \boldsymbol{\omega} \times \mathbf{r} + \mathbf{v}_{\text{rel}}} \qquad \textbf{(5/12)}$$

Comparison of Eq. 5/12 with Eq. 2/20 for nonrotating reference axes shows that $\mathbf{v}_{A/B} = \boldsymbol{\omega} \times \mathbf{r} + \mathbf{v}_{\text{rel}}$, from which we conclude that the term $\boldsymbol{\omega} \times \mathbf{r}$ is the difference between the relative velocities as measured from nonrotating and rotating axes.

To illustrate further the meaning of the last two terms in Eq. 5/12, the motion of particle A relative to the rotating x-y plane is shown in Fig. 5/11 as taking place in a curved slot in a plate which represents the rotating x-y reference system. The velocity of A as measured relative to the plate, $\mathbf{v}_{\text{rel}}$, would be tangent to the path fixed in the x-y plate and would have a magnitude $\dot{s}$, where s is measured along the path. This relative velocity may also be viewed as the velocity $\mathbf{v}_{A/P}$ relative to a point P attached to the plate and coincident with A at the instant under consideration. The term $\boldsymbol{\omega} \times \mathbf{r}$ has a magnitude $r\dot{\theta}$ and a direction normal to $\mathbf{r}$ and is the velocity relative to B of point P as seen from nonrotating axes attached to B.

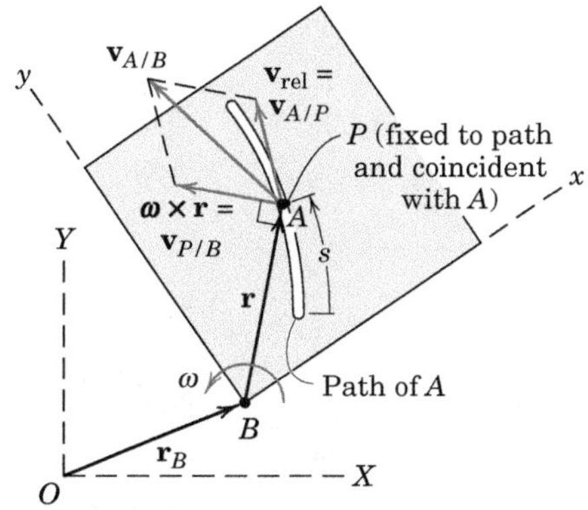

Figure 5/11

The following comparison will help establish the equivalence of, and clarify the differences between, the relative-velocity equations written for rotating and nonrotating reference axes:

$$\begin{aligned}\mathbf{v}_A &= \mathbf{v}_B + \boldsymbol{\omega} \times \mathbf{r} + \mathbf{v}_{\text{rel}}\\ \mathbf{v}_A &= \underbrace{\mathbf{v}_B + \mathbf{v}_{P/B}} + \mathbf{v}_{A/P}\\ \mathbf{v}_A &= \quad \mathbf{v}_P \quad \underbrace{\quad + \mathbf{v}_{A/P}}\\ \mathbf{v}_A &= \mathbf{v}_B + \quad \mathbf{v}_{A/B}\end{aligned} \qquad \textbf{(5/12}\boldsymbol{a}\textbf{)}$$

In the second equation, the term $\mathbf{v}_{P/B}$ is measured from a nonrotating position—otherwise, it would be zero. The term $\mathbf{v}_{A/P}$ is the same as $\mathbf{v}_{\text{rel}}$ and is the velocity of A as measured in the x-y frame. In the third equation, $\mathbf{v}_P$ is the absolute velocity of P and represents the effect of the moving coordinate system, both translational and rotational. The fourth equation is the same as that developed for nonrotating axes, Eq. 2/20, and it is seen that $\mathbf{v}_{A/B} = \mathbf{v}_{P/B} + \mathbf{v}_{A/P} = \boldsymbol{\omega} \times \mathbf{r} + \mathbf{v}_{\text{rel}}$.

Transformation of a Time Derivative

Equation 5/12 represents a transformation of the time derivative of the position vector between rotating and nonrotating axes. We may easily generalize this result to apply to the time derivative of any vector quantity $\mathbf{V} = V_x\mathbf{i} + V_y\mathbf{j}$. Accordingly, the total time derivative with respect to the X-Y system is

$$\left(\frac{d\mathbf{V}}{dt}\right)_{XY} = (\dot{V}_x\mathbf{i} + \dot{V}_y\mathbf{j}) + (V_x\dot{\mathbf{i}} + V_y\dot{\mathbf{j}})$$

The first two terms in the expression represent that part of the total derivative of $\mathbf{V}$ which is measured relative to the x-y reference system, and the second two terms represent that part of the derivative due to the rotation of the reference system.

With the expressions for $\dot{\mathbf{i}}$ and $\dot{\mathbf{j}}$ from Eqs. 5/11, we may now write

$$\boxed{\left(\frac{d\mathbf{V}}{dt}\right)_{XY} = \left(\frac{d\mathbf{V}}{dt}\right)_{xy} + \boldsymbol{\omega} \times \mathbf{V}} \tag{5/13}$$

Here $\boldsymbol{\omega} \times \mathbf{V}$ represents the difference between the time derivative of the vector as measured in a fixed reference system and its time derivative as measured in the rotating reference system. As we will see in Art. 7/2, where three-dimensional motion is introduced, Eq. 5/13 is valid in three dimensions, as well as in two dimensions.

The physical significance of Eq. 5/13 is illustrated in Fig. 5/12, which shows the vector $\mathbf{V}$ at time t as observed both in the fixed axes X-Y and in the rotating axes x-y. Because we are dealing with the effects of rotation only, we may draw the vector through the coordinate origin without loss of generality. During time dt, the vector swings to position $\mathbf{V}'$, and the observer in x-y measures the two components (a) dV due to its change in magnitude and (b) $V\,d\beta$ due to its rotation $d\beta$ relative to x-y. To the rotating observer, then, the derivative $(d\mathbf{V}/dt)_{xy}$ which the observer measures has the components dV/dt and $V\,d\beta/dt = V\dot{\beta}$. The remaining part of the total time derivative not measured by the rotating observer has the magnitude $V\,d\theta/dt$ and, expressed as a vector, is $\boldsymbol{\omega} \times \mathbf{V}$. Thus, we see from the diagram that

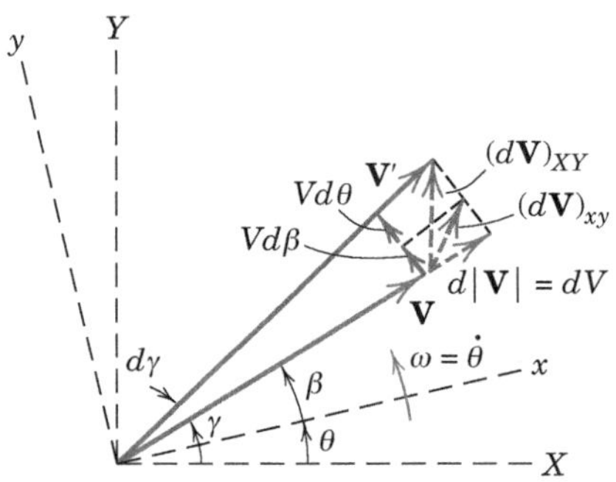

Figure 5/12

$$(\dot{\mathbf{V}})_{XY} = (\dot{\mathbf{V}})_{xy} + \boldsymbol{\omega} \times \mathbf{V}$$

which is Eq. 5/13.

Relative Acceleration

The relative-acceleration equation may be obtained by differentiating the relative-velocity relation, Eq. 5/12. Thus,

$$\mathbf{a}_A = \mathbf{a}_B + \dot{\boldsymbol{\omega}} \times \mathbf{r} + \boldsymbol{\omega} \times \dot{\mathbf{r}} + \dot{\mathbf{v}}_{\text{rel}}$$

In the derivation of Eq. 5/12 we saw that

$$\dot{\mathbf{r}} = \frac{d}{dt}(x\mathbf{i} + y\mathbf{j}) = (x\dot{\mathbf{i}} + y\dot{\mathbf{j}}) + (\dot{x}\mathbf{i} + \dot{y}\mathbf{j})$$
$$= \boldsymbol{\omega} \times \mathbf{r} + \mathbf{v}_{\text{rel}}$$

Therefore, the third term on the right side of the acceleration equation becomes

$$\boldsymbol{\omega} \times \dot{\mathbf{r}} = \boldsymbol{\omega} \times (\boldsymbol{\omega} \times \mathbf{r} + \mathbf{v}_{\text{rel}}) = \boldsymbol{\omega} \times (\boldsymbol{\omega} \times \mathbf{r}) + \boldsymbol{\omega} \times \mathbf{v}_{\text{rel}}$$

With the aid of Eqs. 5/11, the last term on the right side of the equation for $\mathbf{a}_A$ becomes

$$\dot{\mathbf{v}}_{\text{rel}} = \frac{d}{dt}(\dot{x}\mathbf{i} + \dot{y}\mathbf{j}) = (\dot{x}\dot{\mathbf{i}} + \dot{y}\dot{\mathbf{j}}) + (\ddot{x}\mathbf{i} + \ddot{y}\mathbf{j})$$
$$= \boldsymbol{\omega} \times (\dot{x}\mathbf{i} + \dot{y}\mathbf{j}) + (\ddot{x}\mathbf{i} + \ddot{y}\mathbf{j})$$
$$= \boldsymbol{\omega} \times \mathbf{v}_{\text{rel}} + \mathbf{a}_{\text{rel}}$$

Substituting this into the expression for $\mathbf{a}_A$ and collecting terms, we obtain

$$\mathbf{a}_A = \mathbf{a}_B + \dot{\boldsymbol{\omega}} \times \mathbf{r} + \boldsymbol{\omega} \times (\boldsymbol{\omega} \times \mathbf{r}) + 2\boldsymbol{\omega} \times \mathbf{v}_{\text{rel}} + \mathbf{a}_{\text{rel}} \qquad \textbf{(5/14)}$$

Equation 5/14 is the general vector expression for the absolute acceleration of a particle A in terms of its acceleration $\mathbf{a}_{\text{rel}}$ measured relative to a moving coordinate system which rotates with an angular velocity $\boldsymbol{\omega}$ and an angular acceleration $\dot{\boldsymbol{\omega}}$. The terms $\dot{\boldsymbol{\omega}} \times \mathbf{r}$ and $\boldsymbol{\omega} \times (\boldsymbol{\omega} \times \mathbf{r})$ are shown in Fig. 5/13. They represent, respectively, the tangential and normal components of the acceleration $\mathbf{a}_{P/B}$ of the coincident point P in its circular motion with respect to B. This motion would be observed from a set of nonrotating axes moving with B. The magnitude of $\dot{\boldsymbol{\omega}} \times \mathbf{r}$ is $r\ddot{\theta}$ and its direction is tangent to the circle. The magnitude of $\boldsymbol{\omega} \times (\boldsymbol{\omega} \times \mathbf{r})$ is $r\omega^2$ and its direction is from P to B along the normal to the circle.

The acceleration of A relative to the plate along the path, $\mathbf{a}_{\text{rel}}$, may be expressed in rectangular, normal and tangential, or polar coordinates in the rotating system. Frequently, n- and t-components are used, and these components are depicted in Fig. 5/13. The tangential component has the magnitude $(a_{\text{rel}})_t = \ddot{s}$, where s is the distance measured along the path to A. The normal component has the magnitude $(a_{\text{rel}})_n = {v_{\text{rel}}}^2/\rho$, where ρ is the radius of curvature of the path as measured in x-y. The sense of this vector is always toward the center of curvature.

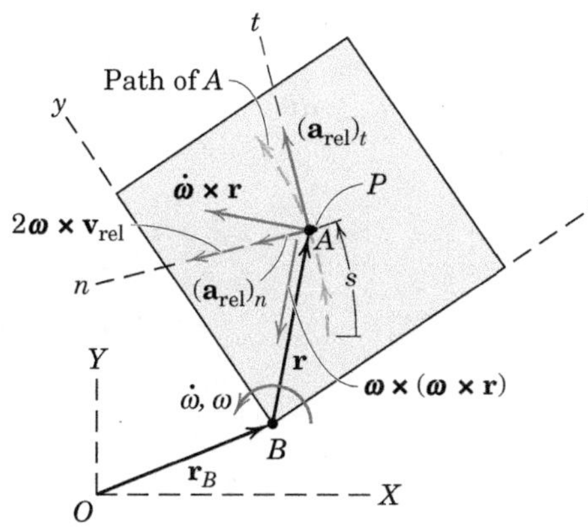

Figure 5/13

Coriolis Acceleration

The term $2\boldsymbol{\omega} \times \mathbf{v}_{\text{rel}}$, shown in Fig. 5/13, is called the *Coriolis* acceleration.* It represents the difference between the acceleration of A relative to P as measured from nonrotating axes and from rotating axes.

*Named after the French military engineer G. Coriolis (1792–1843), who was the first to call attention to this term.

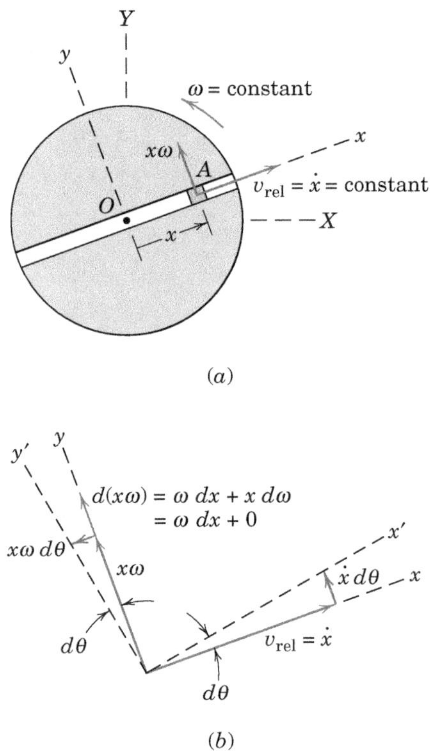

Figure 5/14

The direction is always normal to the vector $\mathbf{v}_{rel}$, and the sense is established by the right-hand rule for the cross product.

The Coriolis acceleration $\mathbf{a}_{Cor} = 2\boldsymbol{\omega} \times \mathbf{v}_{rel}$ is difficult to visualize because it is composed of two separate physical effects. To help with this visualization, we will consider the simplest possible motion in which this term appears. In Fig. 5/14*a* we have a rotating disk with a radial slot in which a small particle *A* is confined to slide. Let the disk turn with a constant angular velocity $\omega = \dot{\theta}$ and let the particle move along the slot with a constant speed $v_{rel} = \dot{x}$ relative to the slot. The velocity of *A* has the two components (*a*) $\dot{x}$ due to motion along the slot and (*b*) $x\omega$ due to the rotation of the slot. The changes in these two velocity components due to the rotation of the disk are shown in part *b* of the figure for the interval dt, during which the x-y axes rotate with the disk through the angle $d\theta$ to x'-y'.

The velocity increment due to the change in direction of $\mathbf{v}_{rel}$ is $\dot{x}\, d\theta$ and that due to the change in magnitude of $x\omega$ is $\omega\, dx$, both being in the y-direction normal to the slot. Dividing each increment by dt and adding give the sum $\omega\dot{x} + \dot{x}\omega = 2\dot{x}\omega$, which is the magnitude of the Coriolis acceleration $2\boldsymbol{\omega} \times \mathbf{v}_{rel}$.

Dividing the remaining velocity increment $x\omega\, d\theta$ due to the change in direction of $x\omega$ by dt gives $x\omega\dot{\theta}$ or $x\omega^2$, which is the acceleration of a point *P* fixed to the slot and momentarily coincident with the particle *A*.

We now see how Eq. 5/14 fits these results. With the origin *B* in that equation taken at the fixed center *O*, $\mathbf{a}_B = \mathbf{0}$. With constant angular velocity, $\dot{\boldsymbol{\omega}} \times \mathbf{r} = \mathbf{0}$. With $\mathbf{v}_{rel}$ constant in magnitude and no curvature to the slot, $\mathbf{a}_{rel} = \mathbf{0}$. We are left with

$$\mathbf{a}_A = \boldsymbol{\omega} \times (\boldsymbol{\omega} \times \mathbf{r}) + 2\boldsymbol{\omega} \times \mathbf{v}_{rel}$$

Replacing $\mathbf{r}$ by $x\mathbf{i}$, $\boldsymbol{\omega}$ by $\omega\mathbf{k}$, and $\mathbf{v}_{rel}$ by $\dot{x}\mathbf{i}$ gives

$$\mathbf{a}_A = -x\omega^2\mathbf{i} + 2\dot{x}\omega\mathbf{j}$$

which checks our analysis from Fig. 5/14.

We also note that this same result is contained in our polar-coordinate analysis of plane curvilinear motion in Eq. 2/14 when we let $\ddot{r} = 0$ and $\ddot{\theta} = 0$ and replace r by x and $\dot{\theta}$ by ω. If the slot in the disk of Fig. 5/14 had been curved, we would have had a normal component of acceleration relative to the slot so that $\mathbf{a}_{rel}$ would not be zero.

Rotating versus Nonrotating Systems

The following comparison will help to establish the equivalence of, and clarify the differences between, the relative-acceleration equations written for rotating and nonrotating reference axes:

$$\begin{aligned}
\mathbf{a}_A &= \mathbf{a}_B + \underbrace{\dot{\boldsymbol{\omega}} \times \mathbf{r} + \boldsymbol{\omega} \times (\boldsymbol{\omega} \times \mathbf{r})} + \underbrace{2\boldsymbol{\omega} \times \mathbf{v}_{rel} + \mathbf{a}_{rel}} \\
\mathbf{a}_A &= \underbrace{\mathbf{a}_B + \qquad \mathbf{a}_{P/B}} \qquad + \qquad \mathbf{a}_{A/P} \\
\mathbf{a}_A &= \qquad \mathbf{a}_P \qquad + \underbrace{\qquad \mathbf{a}_{A/P}} \\
\mathbf{a}_A &= \mathbf{a}_B + \qquad \mathbf{a}_{A/B}
\end{aligned} \qquad \textbf{(5/14a)}$$

The equivalence of $\mathbf{a}_{P/B}$ and $\dot{\boldsymbol{\omega}} \times \mathbf{r} + \boldsymbol{\omega} \times (\boldsymbol{\omega} \times \mathbf{r})$, as shown in the second equation, has already been described. From the third equation where $\mathbf{a}_B + \mathbf{a}_{P/B}$ has been combined to give $\mathbf{a}_P$, it is seen that the relative-acceleration term $\mathbf{a}_{A/P}$, unlike the corresponding relative-velocity term, is not equal to the relative acceleration $\mathbf{a}_{\text{rel}}$ measured from the rotating x-y frame of reference.

The Coriolis term is, therefore, the difference between the acceleration $\mathbf{a}_{A/P}$ of A relative to P as measured in a nonrotating system and the acceleration $\mathbf{a}_{\text{rel}}$ of A relative to P as measured in a rotating system. From the fourth equation, it is seen that the acceleration $\mathbf{a}_{A/B}$ of A with respect to B as measured in a nonrotating system, Eq. 2/21, is a combination of the last four terms in the first equation for the rotating system.

The results expressed by Eq. 5/14 may be visualized somewhat more simply by writing the acceleration of A in terms of the acceleration of the coincident point P. Because the acceleration of P is $\mathbf{a}_P = \mathbf{a}_B + \dot{\boldsymbol{\omega}} \times \mathbf{r} + \boldsymbol{\omega} \times (\boldsymbol{\omega} \times \mathbf{r})$, we may rewrite Eq. 5/14 as

$$\mathbf{a}_A = \mathbf{a}_P + 2\boldsymbol{\omega} \times \mathbf{v}_{\text{rel}} + \mathbf{a}_{\text{rel}} \qquad \textbf{(5/14b)}$$

When the equation is written in this form, point P may not be picked at random because it is the one point attached to the rotating reference frame coincident with A at the instant of analysis. Again, reference to Fig. 5/13 should be made to clarify the meaning of each of the terms in Eq. 5/14 and its equivalent, Eq. 5/14*b*.

Jeff Schultes/Shutterstock

The relative motion of one airplane with respect to an observer fixed in the other translating and rotating airplane is a subject treated in this article.

Key Concepts

In summary, once we have chosen our rotating reference system, we must recognize the following quantities in Eqs. 5/12 and 5/14:

$\mathbf{v}_B$ = absolute velocity of the origin B of the rotating axes

$\mathbf{a}_B$ = absolute acceleration of the origin B of the rotating axes

$\mathbf{r}$ = position vector of the coincident point P measured from B

$\boldsymbol{\omega}$ = angular velocity of the rotating axes

$\dot{\boldsymbol{\omega}}$ = angular acceleration of the rotating axes

$\mathbf{v}_{\text{rel}}$ = velocity of A measured relative to the rotating axes

$\mathbf{a}_{\text{rel}}$ = acceleration of A measured relative to the rotating axes

Also, keep in mind that our vector analysis depends on the consistent use of a right-handed set of coordinate axes. Finally, note that Eqs. 5/12 and 5/14, developed here for plane motion, hold equally well for space motion. The extension to space motion will be covered in Art. 7/6.

Sample Problem 5/16

At the instant represented, the disk with the radial slot is rotating about O with a counterclockwise angular velocity of 4 rad/sec which is decreasing at the rate of 10 rad/sec^2. The motion of slider A is separately controlled, and at this instant, $r = 6$ in., $\dot{r} = 5$ in./sec, and $\ddot{r} = 81$ in./sec^2. Determine the absolute velocity and acceleration of A for this position.

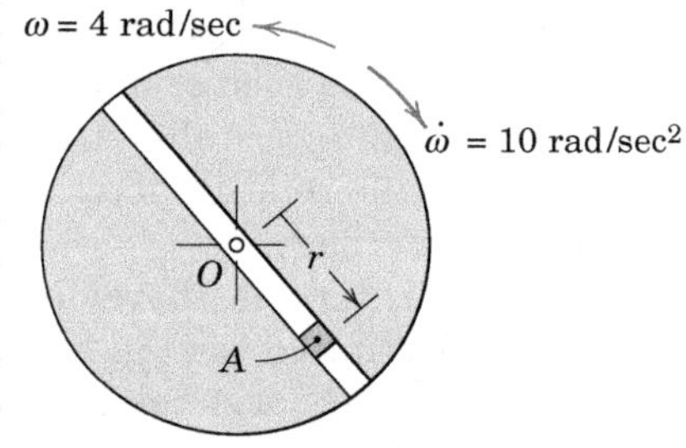

Solution. We have motion relative to a rotating path, so that a rotating coordinate system with origin at O is indicated. We attach x-y axes to the disk and use the unit vectors **i** and **j**.

Velocity. With the origin at O, the term $\mathbf{v}_B$ of Eq. 5/12 disappears and we have

① $$\mathbf{v}_A = \boldsymbol{\omega} \times \mathbf{r} + \mathbf{v}_{\text{rel}}$$

② The angular velocity as a vector is $\boldsymbol{\omega} = 4\mathbf{k}$ rad/sec, where $\mathbf{k}$ is the unit vector normal to the x-y plane in the $+z$-direction. Our relative-velocity equation becomes

$$\mathbf{v}_A = 4\mathbf{k} \times 6\mathbf{i} + 5\mathbf{i} = 24\mathbf{j} + 5\mathbf{i} \text{ in./sec} \qquad \textit{Ans.}$$

in the direction indicated and has the magnitude

$$v_A = \sqrt{(24)^2 + (5)^2} = 24.5 \text{ in./sec} \qquad \textit{Ans.}$$

Acceleration. Equation 5/14 written for zero acceleration of the origin of the rotating coordinate system is

$$\mathbf{a}_A = \boldsymbol{\omega} \times (\boldsymbol{\omega} \times \mathbf{r}) + \dot{\boldsymbol{\omega}} \times \mathbf{r} + 2\boldsymbol{\omega} \times \mathbf{v}_{\text{rel}} + \mathbf{a}_{\text{rel}}$$

The terms become

③ $$\boldsymbol{\omega} \times (\boldsymbol{\omega} \times \mathbf{r}) = 4\mathbf{k} \times (4\mathbf{k} \times 6\mathbf{i}) = 4\mathbf{k} \times 24\mathbf{j} = -96\mathbf{i} \text{ in./sec}^2$$

$$\dot{\boldsymbol{\omega}} \times \mathbf{r} = -10\mathbf{k} \times 6\mathbf{i} = -60\mathbf{j} \text{ in./sec}^2$$

$$2\boldsymbol{\omega} \times \mathbf{v}_{\text{rel}} = 2(4\mathbf{k}) \times 5\mathbf{i} = 40\mathbf{j} \text{ in./sec}^2$$

$$\mathbf{a}_{\text{rel}} = 81\mathbf{i} \text{ in./sec}^2$$

The total acceleration is, therefore,

$$\mathbf{a}_A = (81 - 96)\mathbf{i} + (40 - 60)\mathbf{j} = -15\mathbf{i} - 20\mathbf{j} \text{ in./sec}^2 \qquad \textit{Ans.}$$

in the direction indicated and has the magnitude

$$a_A = \sqrt{(15)^2 + (20)^2} = 25 \text{ in./sec}^2 \qquad \textit{Ans.}$$

Vector notation is certainly not essential to the solution of this problem. The student should be able to work out the steps with scalar notation just as easily. The correct direction of the Coriolis-acceleration term can always be found by the direction in which the head of the $\mathbf{v}_{\text{rel}}$ vector would move if rotated about its tail in the sense of $\boldsymbol{\omega}$ as shown.

Helpful Hints

① This equation is the same as $\mathbf{v}_A = \mathbf{v}_P + \mathbf{v}_{A/P}$, where P is a point attached to the disk coincident with A at this instant.

② Note that the x-y-z axes chosen constitute a right-handed system.

③ Be sure to recognize that $\boldsymbol{\omega} \times (\boldsymbol{\omega} \times \mathbf{r})$ and $\dot{\boldsymbol{\omega}} \times \mathbf{r}$ represent the normal and tangential components of acceleration of a point P on the disk coincident with A. This description becomes that of Eq. 5/14b.

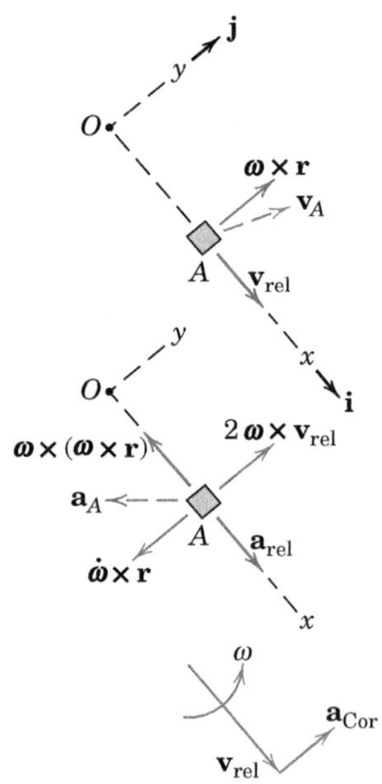

Sample Problem 5/17

The pin A of the hinged link AC is confined to move in the rotating slot of link OD. The angular velocity of OD is $\omega = 2$ rad/s clockwise and is constant for the interval of motion concerned. For the position where $\theta = 45°$ with AC horizontal, determine the velocity of pin A and the velocity of A relative to the rotating slot in OD.

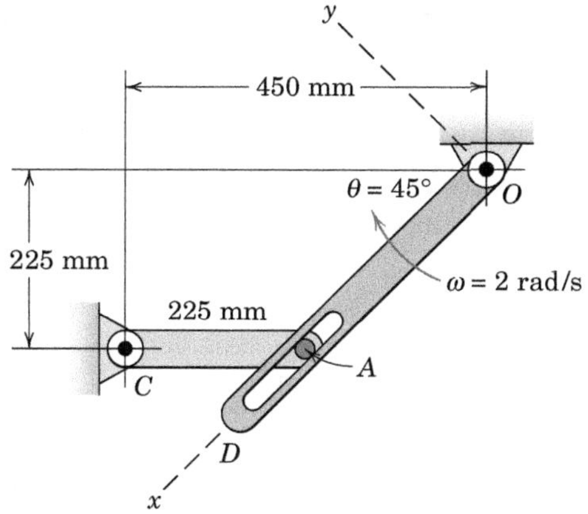

Solution. Motion of a point (pin A) along a rotating path (the slot) suggests the use of rotating coordinate axes x-y attached to arm OD. With the origin at the fixed point O, the term $\mathbf{v}_B$ of Eq. 5/12 vanishes, and we have $\mathbf{v}_A = \boldsymbol{\omega} \times \mathbf{r} + \mathbf{v}_{\text{rel}}$.

The velocity of A in its circular motion about C is

$$\mathbf{v}_A = \boldsymbol{\omega}_{CA} \times \mathbf{r}_{CA} = \omega_{CA}\mathbf{k} \times (225/\sqrt{2})(-\mathbf{i} - \mathbf{j}) = (225/\sqrt{2})\omega_{CA}\,(\mathbf{i} - \mathbf{j})$$

(1) where the angular velocity $\boldsymbol{\omega}_{CA}$ is arbitrarily assigned in a clockwise sense in the positive z-direction ($+\mathbf{k}$).

The angular velocity $\boldsymbol{\omega}$ of the rotating axes is that of the arm OD and, by the right-hand rule, is $\boldsymbol{\omega} = \omega\mathbf{k} = 2\mathbf{k}$ rad/s. The vector from the origin to the point P on OD coincident with A is $\mathbf{r} = \overline{OP}\mathbf{i} = \sqrt{(450 - 225)^2 + (225)^2}\,\mathbf{i} = 225\sqrt{2}\mathbf{i}$ mm. Thus,

$$\boldsymbol{\omega} \times \mathbf{r} = 2\mathbf{k} \times 225\sqrt{2}\mathbf{i} = 450\sqrt{2}\mathbf{j} \text{ mm/s}$$

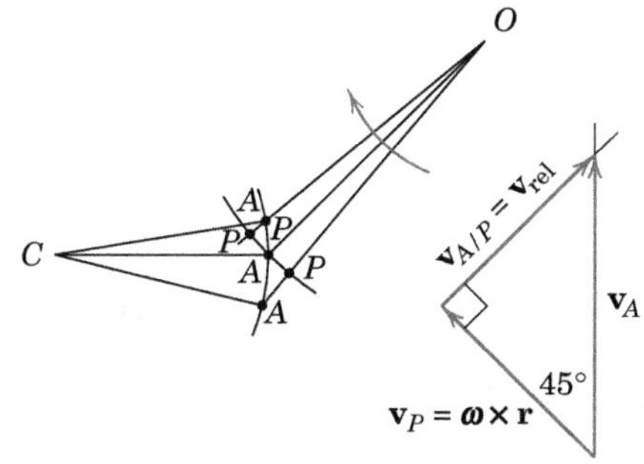

Finally, the relative-velocity term $\mathbf{v}_{\text{rel}}$ is the velocity measured by an observer attached to the rotating reference frame and is $\mathbf{v}_{\text{rel}} = \dot{x}\mathbf{i}$. Substitution into the relative-velocity equation gives

$$(225/\sqrt{2})\omega_{CA}\,(\mathbf{i} - \mathbf{j}) = 450\sqrt{2}\mathbf{j} + \dot{x}\mathbf{i}$$

Equating separately the coefficients of the $\mathbf{i}$ and $\mathbf{j}$ terms yields

$$(225/\sqrt{2})\omega_{CA} = \dot{x} \qquad \text{and} \qquad -(225/\sqrt{2})\omega_{CA} = 450\sqrt{2}$$

giving

$$\omega_{CA} = -4 \text{ rad/s} \qquad \text{and} \qquad \dot{x} = v_{\text{rel}} = -450\sqrt{2} \text{ mm/s} \qquad \textit{Ans.}$$

With a negative value for ω_{CA}, the actual angular velocity of CA is counterclockwise, so the velocity of A is up with a magnitude of

(2) $$v_A = 225(4) = 900 \text{ mm/s} \qquad \textit{Ans.}$$

Geometric clarification of the terms is helpful and is easily shown. Using the equivalence between the third and the first of Eqs. 5/12*a* with $\mathbf{v}_B = \mathbf{0}$ enables us to write $\mathbf{v}_A = \mathbf{v}_P + \mathbf{v}_{A/P}$, where P is the point on the rotating arm OD coincident with A. Clearly, $v_P = \overline{OP}\omega = 225\sqrt{2}(2) = 450\sqrt{2}$ mm/s and its direction is normal to OD. The relative velocity $\mathbf{v}_{A/P}$, which is the same as $\mathbf{v}_{\text{rel}}$, is seen from the figure to be along the slot toward O. This conclusion becomes clear when it is observed that A is approaching P along the slot from below before coincidence and is receding from P upward along the slot following coincidence. The velocity of A is tangent to its circular arc about C. The vector equation can now be satisfied since there are only two remaining scalar unknowns, namely, the magnitude of $\mathbf{v}_{A/P}$ and the magnitude of $\mathbf{v}_A$. For the 45° position, the figure requires $v_{A/P} = 450\sqrt{2}$ mm/s and $v_A = 900$ mm/s, each in its direction shown. The angular velocity of AC is

$$[\omega = v/r] \qquad \omega_{AC} = v_A/\overline{AC} = 900/225 = 4 \text{ rad/s counterclockwise}$$

Helpful Hints

(1) It is clear enough physically that CA will have a counterclockwise angular velocity for the conditions specified, so we anticipate a negative value for ω_{CA}.

(2) Solution of the problem is not restricted to the reference axes used. Alternatively, the origin of the x-y axes, still attached to OD, could be chosen at the coincident point P on OD. This choice would merely replace the $\boldsymbol{\omega} \times \mathbf{r}$ term by its equal, $\mathbf{v}_P$. As a further selection, all vector quantities could be expressed in terms of X-Y components using unit vectors $\mathbf{I}$ and $\mathbf{J}$.

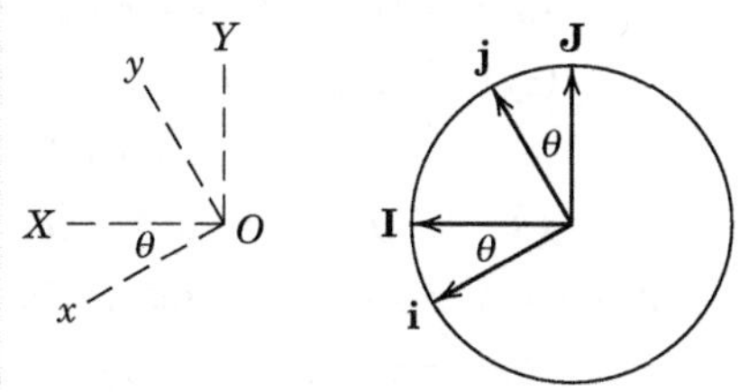

A direct conversion between the two reference systems is obtained from the geometry of the unit circle and gives

$$\mathbf{i} = \mathbf{I}\cos\theta - \mathbf{J}\sin\theta$$

and $$\mathbf{j} = \mathbf{I}\sin\theta + \mathbf{J}\cos\theta$$

Sample Problem 5/18

For the conditions of Sample Problem 5/17, determine the angular acceleration of AC and the acceleration of A relative to the rotating slot in arm OD.

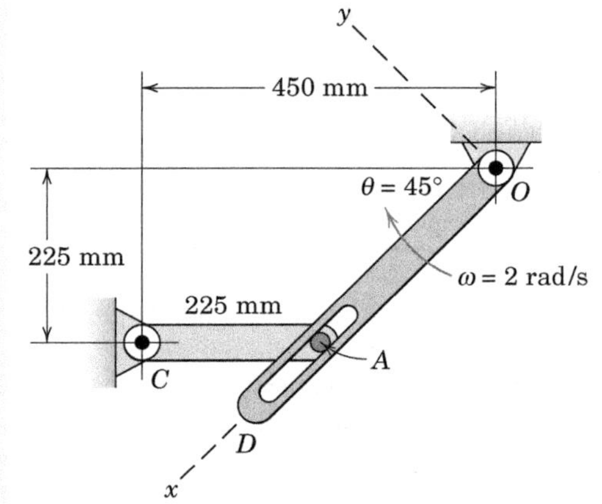

Solution. We attach the rotating coordinate system x-y to arm OD and use Eq. 5/14. With the origin at the fixed point O, the term $\mathbf{a}_B$ becomes zero so that

$$\mathbf{a}_A = \dot{\boldsymbol{\omega}} \times \mathbf{r} + \boldsymbol{\omega} \times (\boldsymbol{\omega} \times \mathbf{r}) + 2\boldsymbol{\omega} \times \mathbf{v}_{\text{rel}} + \mathbf{a}_{\text{rel}}$$

From the solution to Sample Problem 5/17, we make use of the values $\boldsymbol{\omega} = 2\mathbf{k}$ rad/s, $\boldsymbol{\omega}_{CA} = -4\mathbf{k}$ rad/s, and $\mathbf{v}_{\text{rel}} = -450\sqrt{2}\mathbf{i}$ mm/s and write

$$\mathbf{a}_A = \dot{\boldsymbol{\omega}}_{CA} \times \mathbf{r}_{CA} + \boldsymbol{\omega}_{CA} \times (\boldsymbol{\omega}_{CA} \times \mathbf{r}_{CA})$$

$$= \dot{\omega}_{CA}\mathbf{k} \times \frac{225}{\sqrt{2}}(-\mathbf{i} - \mathbf{j}) - 4\mathbf{k} \times \left(-4\mathbf{k} \times \frac{225}{\sqrt{2}}[-\mathbf{i} - \mathbf{j}]\right)$$

$$\dot{\boldsymbol{\omega}} \times \mathbf{r} = \mathbf{0} \text{ since } \boldsymbol{\omega} = \text{constant}$$

(1) $$\boldsymbol{\omega} \times (\boldsymbol{\omega} \times \mathbf{r}) = 2\mathbf{k} \times (2\mathbf{k} \times 225\sqrt{2}\mathbf{i}) = -900\sqrt{2}\mathbf{i} \text{ mm/s}^2$$

$$2\boldsymbol{\omega} \times \mathbf{v}_{\text{rel}} = 2(2\mathbf{k}) \times (-450\sqrt{2}\mathbf{i}) = -1800\sqrt{2}\mathbf{j} \text{ mm/s}^2$$

$$\mathbf{a}_{\text{rel}} = \ddot{x}\mathbf{i}$$

Substitution into the relative-acceleration equation yields

$$\frac{1}{\sqrt{2}}(225\dot{\omega}_{CA} + 3600)\mathbf{i} + \frac{1}{\sqrt{2}}(-225\dot{\omega}_{CA} + 3600)\mathbf{j} = -900\sqrt{2}\mathbf{i} - 1800\sqrt{2}\mathbf{j} + \ddot{x}\mathbf{i}$$

Equating separately the **i** and **j** terms gives

$$(225\dot{\omega}_{CA} + 3600)/\sqrt{2} = -900\sqrt{2} + \ddot{x}$$

and

$$(-225\dot{\omega}_{CA} + 3600)/\sqrt{2} = -1800\sqrt{2}$$

Solving for the two unknowns gives

$$\dot{\omega}_{CA} = 32 \text{ rad/s}^2 \qquad \text{and} \qquad \ddot{x} = a_{\text{rel}} = 8910 \text{ mm/s}^2 \qquad \textit{Ans.}$$

If desired, the acceleration of A may also be written as

$$\mathbf{a}_A = (225/\sqrt{2})(32)(\mathbf{i} - \mathbf{j}) + (3600/\sqrt{2})(\mathbf{i} + \mathbf{j}) = 7640\mathbf{i} - 2550\mathbf{j} \text{ mm/s}^2$$

We make use here of the geometric representation of the relative-acceleration equation to further clarify the problem. The geometric approach may be used as an alternative solution. Again, we introduce point P on OD coincident with A. The equivalent scalar terms are

$$(a_A)_t = |\dot{\boldsymbol{\omega}}_{CA} \times \mathbf{r}_{CA}| = r\dot{\omega}_{CA} = r\alpha_{CA} \text{ normal to } CA \text{, sense unknown}$$

$$(a_A)_n = |\boldsymbol{\omega}_{CA} \times (\boldsymbol{\omega}_{CA} \times \mathbf{r}_{CA})| = r\omega_{CA}^2 \text{ from } A \text{ to } C$$

$$(a_P)_n = |\boldsymbol{\omega} \times (\boldsymbol{\omega} \times \mathbf{r})| = \overline{OP}\omega^2 \text{ from } P \text{ to } O$$

$$(a_P)_t = |\dot{\boldsymbol{\omega}} \times \mathbf{r}| = r\dot{\omega} = 0 \text{ since } \omega = \text{constant}$$

$$|2\boldsymbol{\omega} \times \mathbf{v}_{\text{rel}}| = 2\omega v_{\text{rel}} \text{ directed as shown}$$

$$a_{\text{rel}} = \ddot{x} \text{ along } OD \text{, sense unknown}$$

We start with the known vectors and add them head-to-tail for each side of the equation beginning at R and ending at S, where the intersection of the known directions of $(\mathbf{a}_A)_t$ and $\mathbf{a}_{\text{rel}}$ establishes the solution. Closure of the polygon determines the sense of each of the two unknown vectors, and their magnitudes are
(2) easily calculated from the figure geometry.

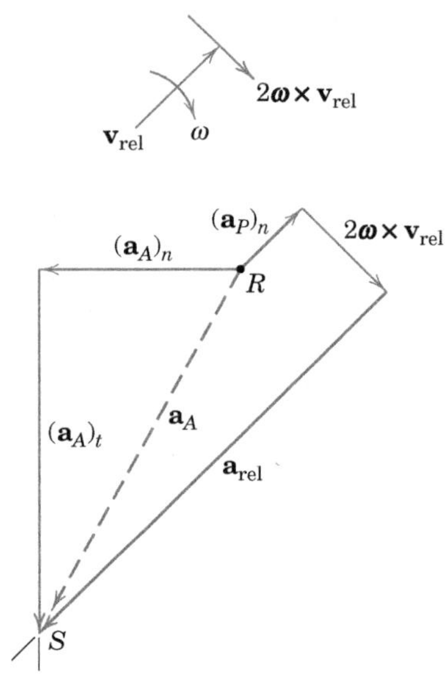

Helpful Hints

(1) If the slot had been curved with a radius of curvature ρ, the term $\mathbf{a}_{\text{rel}}$ would have had a component v_{rel}^2/ρ normal to the slot and directed toward the center of curvature in addition to its component along the slot.

(2) It is always possible to avoid a simultaneous solution by projecting the vectors onto the perpendicular to one of the unknowns.

Sample Problem 5/19

Aircraft B has a constant speed of 150 m/s as it passes the bottom of a circular loop of 400-m radius. Aircraft A flying horizontally in the plane of the loop passes 100 m directly below B at a constant speed of 100 m/s. (a) Determine the instantaneous velocity and acceleration which A appears to have to the pilot of B, who is fixed to his rotating aircraft. (b) Compare your results for part (a) with the case of erroneously treating the pilot of aircraft B as nonrotating.

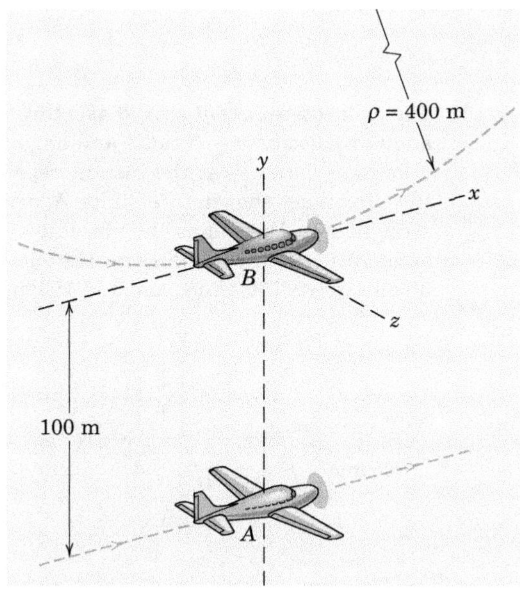

Solution (a). We begin by clearly defining the rotating coordinate system x-y-z which best helps us to answer the questions. With x-y-z attached to aircraft B as shown, the terms $\mathbf{v}_{rel}$ and $\mathbf{a}_{rel}$ in Eqs. 5/12 and 5/14 will be the desired results. The terms in Eq. 5/12 are

$$\mathbf{v}_A = 100\mathbf{i} \text{ m/s}$$

$$\mathbf{v}_B = 150\mathbf{i} \text{ m/s}$$

① $$\boldsymbol{\omega} = \frac{v_B}{\rho}\mathbf{k} = \frac{150}{400}\mathbf{k} = 0.375\mathbf{k} \text{ rad/s}$$

$$\mathbf{r} = \mathbf{r}_{A/B} = -100\mathbf{j} \text{ m}$$

Eq. 5/12: $$\mathbf{v}_A = \mathbf{v}_B + \boldsymbol{\omega} \times \mathbf{r} + \mathbf{v}_{rel}$$

$$100\mathbf{i} = 150\mathbf{i} + 0.375\mathbf{k} \times (-100\mathbf{j}) + \mathbf{v}_{rel}$$

Solving for $\mathbf{v}_{rel}$ gives $$\mathbf{v}_{rel} = -87.5\mathbf{i} \text{ m/s}$$ *Ans.*

The terms in Eq. 5/14, in addition to those listed above, are

$$\mathbf{a}_A = \mathbf{0}$$

$$\mathbf{a}_B = \frac{v_B^2}{\rho}\mathbf{j} = \frac{150^2}{400}\mathbf{j} = 56.2\mathbf{j} \text{ m/s}^2$$

$$\dot{\boldsymbol{\omega}} = \mathbf{0}$$

Eq. 5/14: $$\mathbf{a}_A = \mathbf{a}_B + \dot{\boldsymbol{\omega}} \times \mathbf{r} + \boldsymbol{\omega} \times (\boldsymbol{\omega} \times \mathbf{r}) + 2\boldsymbol{\omega} \times \mathbf{v}_{rel} + \mathbf{a}_{rel}$$

$$\mathbf{0} = 56.2\mathbf{j} + \mathbf{0} \times (-100\mathbf{j}) + 0.375\mathbf{k} \times [0.375\mathbf{k} \times (-100\mathbf{j})] + 2[0.375\mathbf{k} \times (-87.5\mathbf{i})] + \mathbf{a}_{rel}$$

Solving for $\mathbf{a}_{rel}$ gives $$\mathbf{a}_{rel} = -4.69\mathbf{k} \text{ m/s}^2$$ *Ans.*

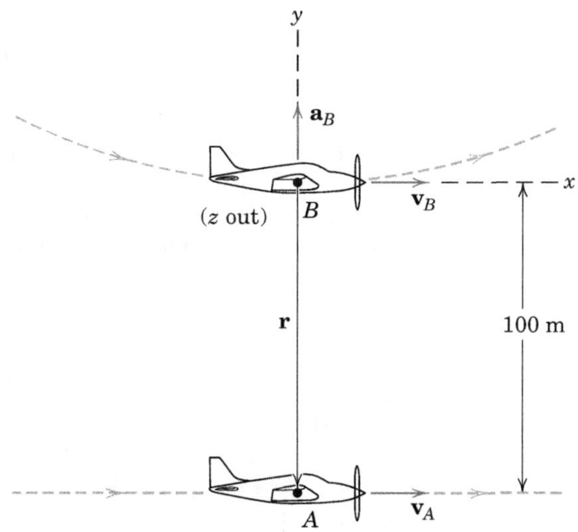

(b) For motion relative to translating frames, we use Eqs. 2/20 and 2/21 of Chapter 2:

$$\mathbf{v}_{A/B} = \mathbf{v}_A - \mathbf{v}_B = 100\mathbf{i} - 150\mathbf{i} = -50\mathbf{i} \text{ m/s}$$

$$\mathbf{a}_{A/B} = \mathbf{a}_A - \mathbf{a}_B = \mathbf{0} - 56.2\mathbf{j} = -56.2\mathbf{j} \text{ m/s}^2$$

Again, we see that $\mathbf{v}_{rel} \neq \mathbf{v}_{A/B}$ and $\mathbf{a}_{rel} \neq \mathbf{a}_{A/B}$. The rotation of pilot B makes a difference in what he observes!

The scalar result $\omega = \dfrac{v_B}{\rho}$ can be obtained by considering a complete circular motion of aircraft B, during which it rotates 2π radians in a time $t = \dfrac{2\pi\rho}{v_B}$:

$$\omega = \frac{2\pi}{2\pi\rho/v_B} = \frac{v_B}{\rho}$$

Because the speed of aircraft B is constant, there is no tangential acceleration and thus the angular acceleration $\boldsymbol{\alpha} = \dot{\boldsymbol{\omega}}$ of this aircraft is zero.

Helpful Hint

① Because we choose the rotating frame x-y-z to be fixed to aircraft B, the angular velocity of the aircraft and the term $\boldsymbol{\omega}$ in Eqs. 5/12 and 5/14 are identical.

PROBLEMS

Introductory Problems

5/157 The disk rotates about a fixed axis through O with angular velocity $\omega = 5$ rad/s and angular acceleration $\alpha = 3$ rad/s^2 at the instant represented, in the directions shown. The slider A moves in the straight slot. Determine the absolute velocity and acceleration of A for the same instant, when $x = 36$ mm, $\dot{x} = -100$ mm/s, and $\ddot{x} = 150$ mm/s^2.

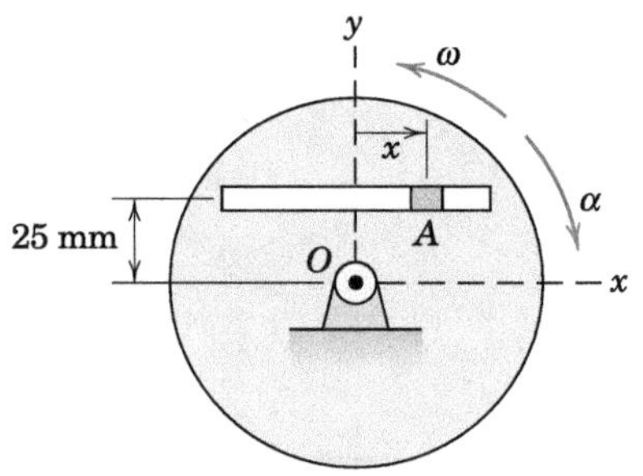

Problem 5/157

5/158 The sector rotates with the indicated angular quantities about a fixed axis through point B. Simultaneously, the particle A moves in the curved slot with constant speed u relative to the sector. Determine the absolute velocity and acceleration of particle A, and identify the Coriolis acceleration.

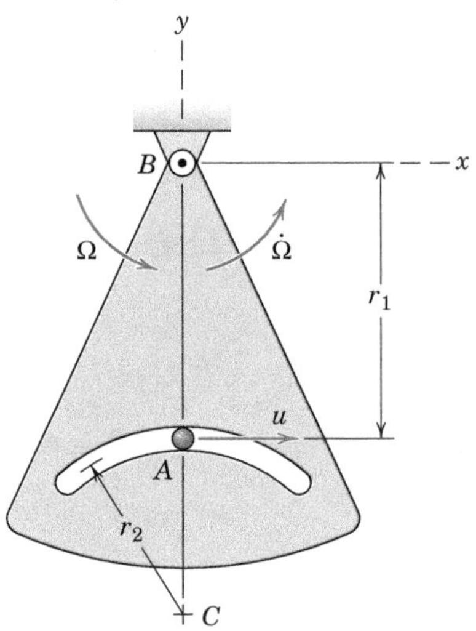

Problem 5/158

5/159 The slotted wheel rolls to the right without slipping, with a constant speed $v = 2$ ft/sec of its center O. Simultaneously, motion of the sliding block A is controlled by a mechanism not shown so that $\dot{x} = 1.5$ ft/sec with $\ddot{x} = 0$. Determine the magnitude of the acceleration of A for the instant when $x = 6$ in. and $\theta = 30°$.

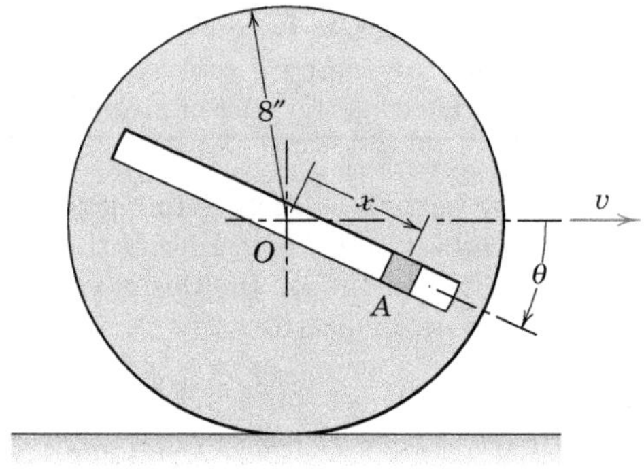

Problem 5/159

5/160 The disk rolls without slipping on the horizontal surface, and at the instant represented, the center O has the velocity and acceleration shown in the figure. For this instant, the particle A has the indicated speed u and time rate of change of speed $\dot{u}$, both relative to the disk. Determine the absolute velocity and acceleration of particle A.

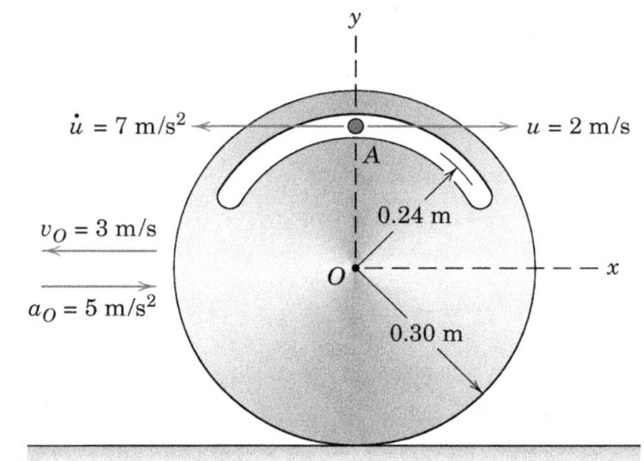

Problem 5/160

5/161 The cars of the roller coaster have a speed $v = 25$ ft/sec at the instant under consideration. As rider B passes the topmost point, she observes a stationary friend A. What velocity of A does she observe? At the position under consideration, the center of curvature of the path of rider B is point C.

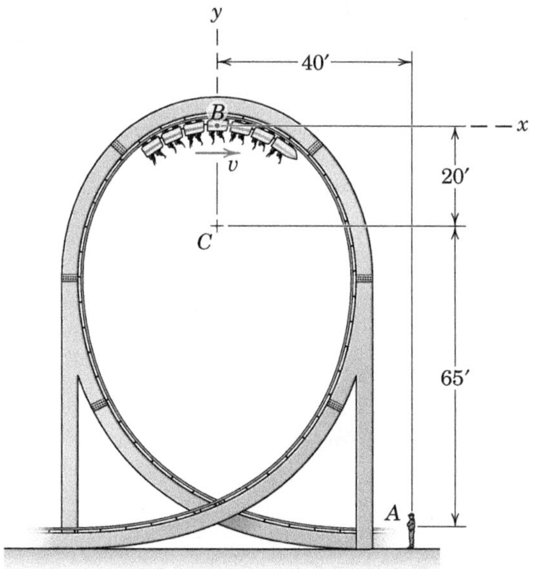

Problem 5/161

5/162 An experimental vehicle A travels with constant speed v relative to the earth along a north–south track. Determine the Coriolis acceleration $\mathbf{a}_{\text{Cor}}$ as a function of the latitude θ. Assume an earth-fixed rotating frame $Bxyz$ and a spherical earth. If the vehicle speed is $v = 500$ km/h, determine the magnitude of the Coriolis acceleration at (a) the equator and (b) the north pole.

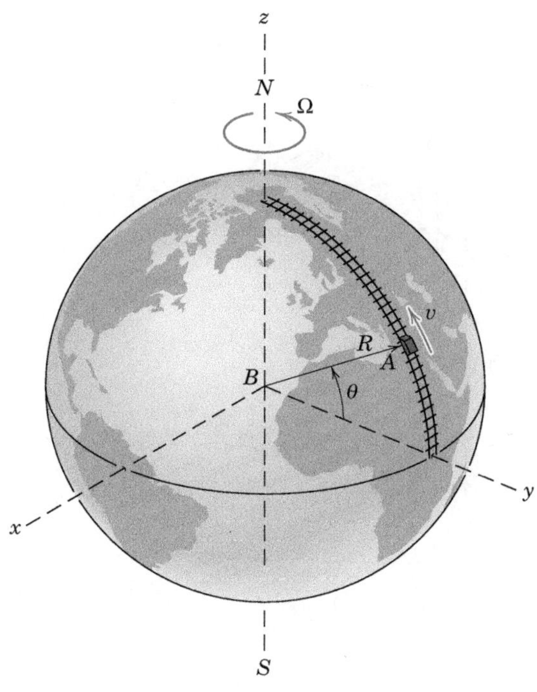

Problem 5/162

5/163 Car B is rounding the curve with a constant speed of 54 km/h, and car A is approaching car B in the intersection with a constant speed of 72 km/h. Determine the velocity which car A appears to have to an observer riding in and turning with car B. The x-y axes are attached to car B. Is this apparent velocity the negative of the velocity which B appears to have to a nonrotating observer in car A? The distance separating the two cars at the instant depicted is 40 m.

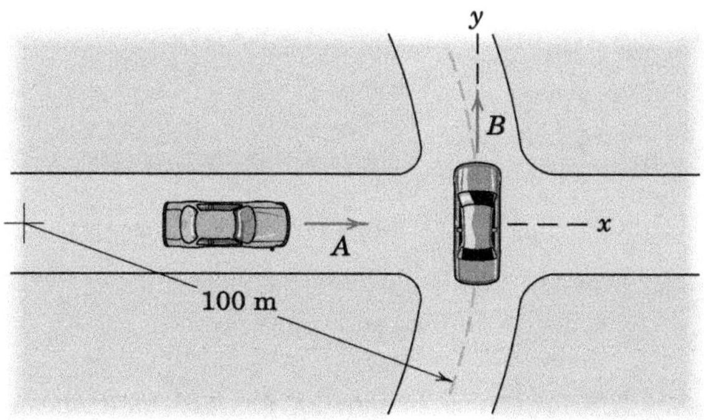

Problem 5/163

5/164 For the cars of Prob. 5/163 traveling with constant speed, determine the acceleration which car A appears to have to an observer riding in and turning with car B.

5/165 The small collar A is sliding on the bent bar with speed u relative to the bar as shown. Simultaneously, the bar is rotating with angular velocity ω about the fixed pivot B. Take the x-y axes to be fixed to the bar and determine the Coriolis acceleration of the slider for the instant represented. Interpret your result.

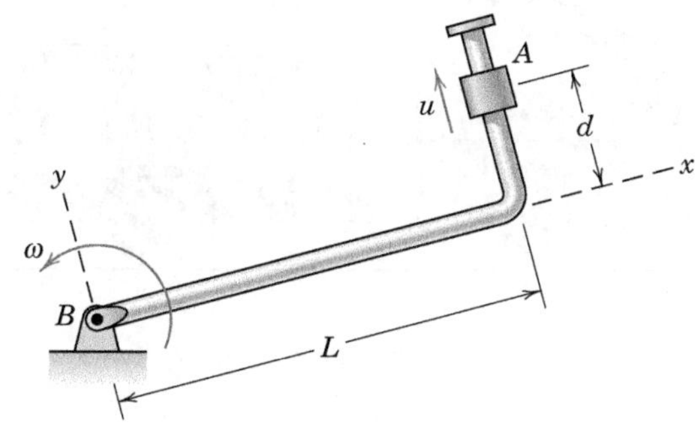

Problem 5/165

5/166 A train traveling at a constant speed $v = 25$ mi/hr has entered a circular portion of track with a radius $R = 200$ ft. Determine the velocity and acceleration of point A of the train as observed by the engineer B, who is fixed to the locomotive. Use the axes given in the figure.

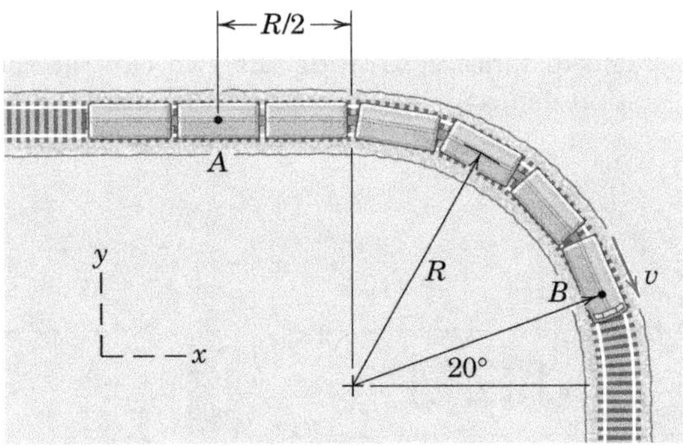

Problem 5/166

5/167 The fire truck is moving forward at a speed of 35 mi/hr and is decelerating at the rate of 10 ft/sec². Simultaneously, the ladder is being raised and extended. At the instant considered the angle θ is 30° and is increasing at the constant rate of 10 deg/sec. Also at this instant the extension b of the ladder is 5 ft, with $\dot{b} = 2$ ft/sec and $\ddot{b} = -1$ ft/sec². For this instant determine the acceleration of the end A of the ladder (*a*) with respect to the truck and (*b*) with respect to the ground.

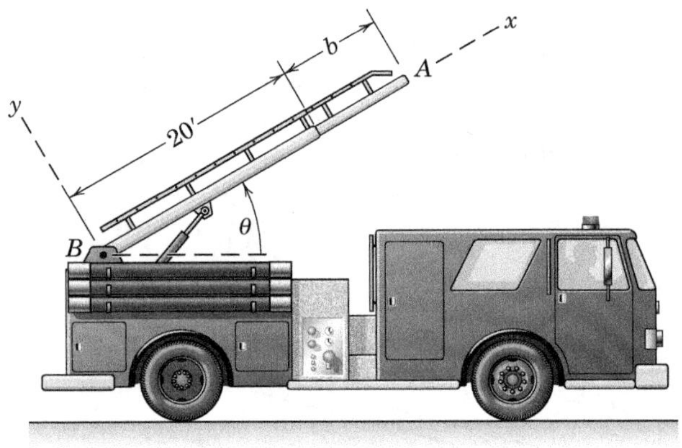

Problem 5/167

Representative Problems

5/168 Vehicle A travels west at high speed on a perfectly straight road B which is tangent to the surface of the earth at the equator. The road has no curvature whatsoever in the vertical plane. Determine the necessary speed v_{rel} of the vehicle relative to the road which will give rise to zero acceleration of the vehicle in the vertical direction. Assume that the center of the earth has no acceleration.

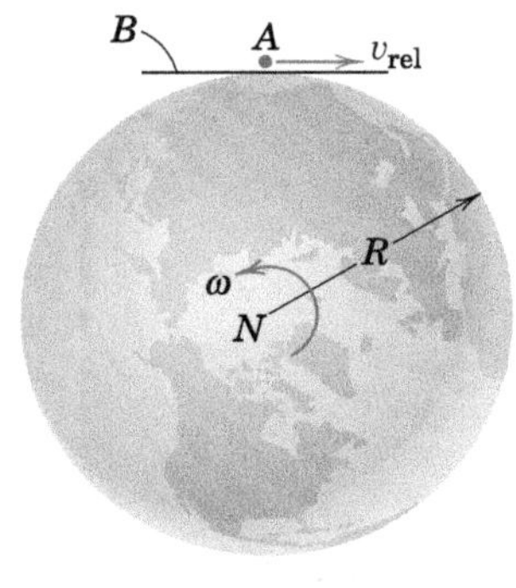

Problem 5/168

5/169 Aircraft B has a constant speed of 540 km/h at the bottom of a circular loop of 400-m radius. Aircraft A flying horizontally in the plane of the loop passes 100 m directly under B at a constant speed of 360 km/h. With coordinate axes attached to B as shown, determine the acceleration which A appears to have to the pilot of B for this instant.

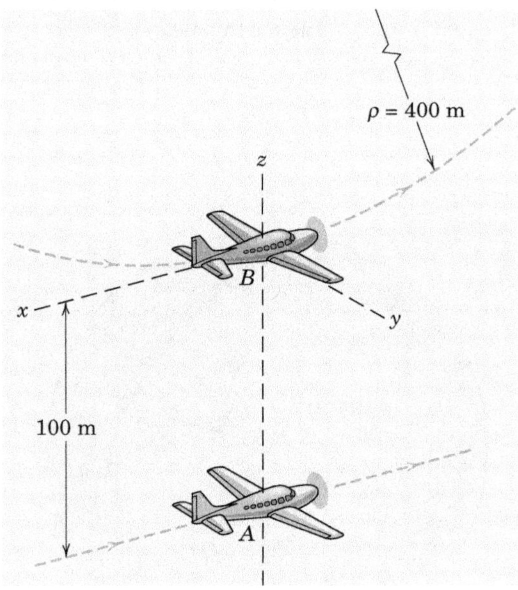

Problem 5/169

5/170 Bar OC rotates with a clockwise angular velocity $\omega_{OC} = 2$ rad/s. The pin A attached to bar OC engages the straight slot of the sector. Determine the angular velocity ω of the sector and the velocity of pin A relative to the sector for the instant represented.

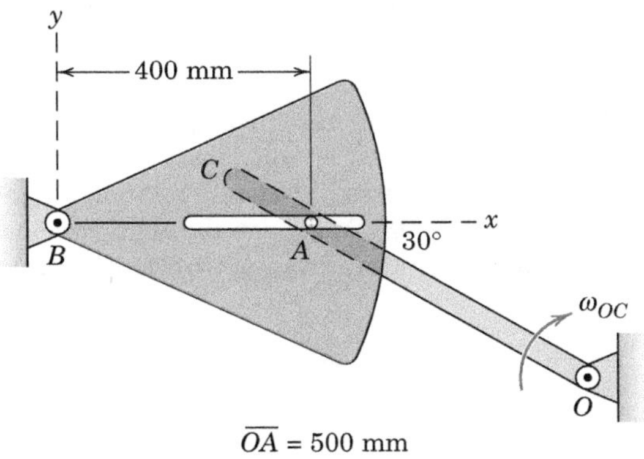

Problem 5/170

5/171 If the bar OC of the previous problem rotates with a clockwise angular velocity $\omega_{OC} = 2$ rad/s and a counterclockwise angular acceleration $\alpha_{OC} = 4$ rad/s^2, determine the angular acceleration α of the sector and the acceleration of point A relative to the sector.

5/172 The system of Prob. 5/170 is modified in that OC is now a slotted member which accommodates the pin A attached to the sector. If bar OC rotates with a clockwise angular velocity $\omega_{OC} = 2$ rad/s and a counterclockwise angular acceleration $\alpha_{OC} = 4$ rad/s^2, determine the angular velocity ω and the angular acceleration α of the sector.

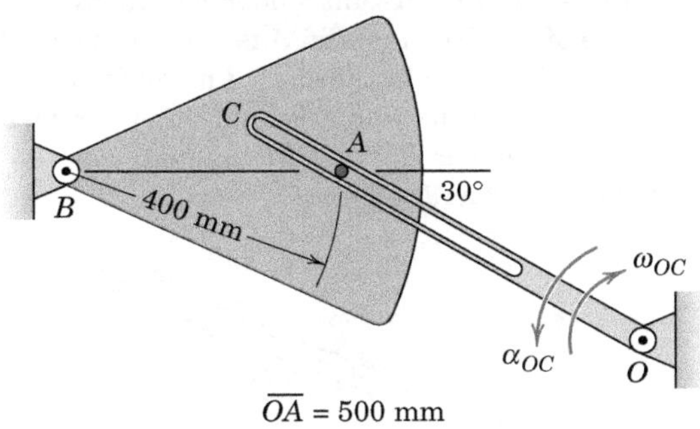

Problem 5/172

5/173 A smooth bowling alley is oriented north–south as shown. A ball A is released with speed v along the lane as shown. Because of the Coriolis effect, it will deflect a distance δ as shown. Develop a general expression for δ. The bowling alley is located at a latitude θ in the northern hemisphere. Evaluate your expression for the conditions $L = 60$ ft, $v = 15$ ft/sec, and $\theta = 40°$. Should bowlers prefer east–west alleys? State any assumptions.

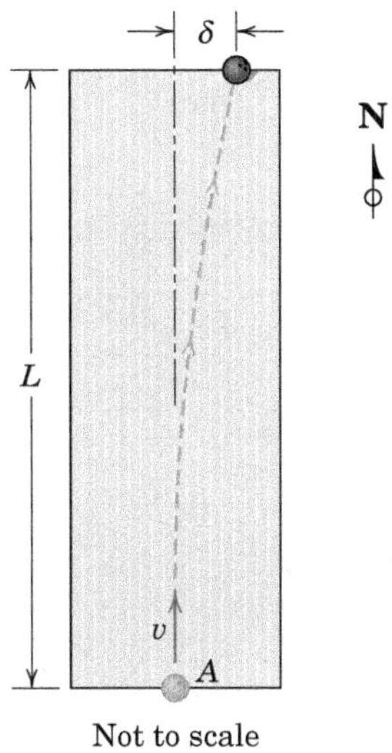

Problem 5/173

5/174 Under the action of its stern and starboard bow thrusters, the cruise ship has the velocity $v_B = 1$ m/s of its mass center B and angular velocity $\omega = 1$ deg/s about a vertical axis. The velocity of B is constant, but the angular rate ω is decreasing at 0.5 deg/s^2. Person A is stationary on the dock. What velocity and acceleration of A are observed by a passenger fixed to and rotating with the ship? Treat the problem as two-dimensional.

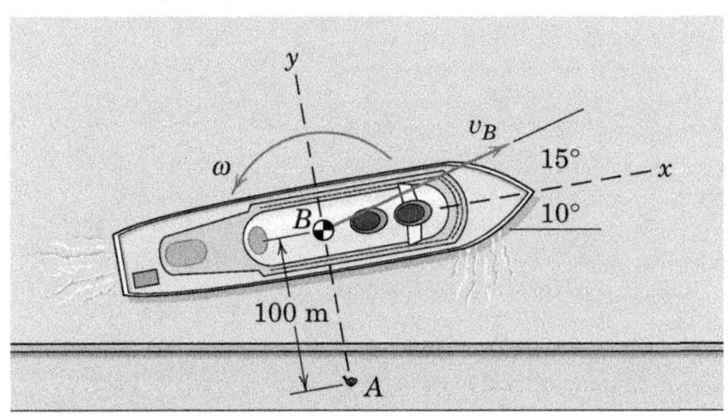

Problem 5/174

5/175 The air transport B is flying with a constant speed of 480 mi/hr in a horizontal arc of 9-mi radius. When B reaches the position shown, aircraft A, flying southwest at a constant speed of 360 mi/hr, crosses the radial line from B to the center of curvature C of its path. Write the vector expression, using the x-y axes attached to B, for the velocity of A as measured by an observer in and turning with B.

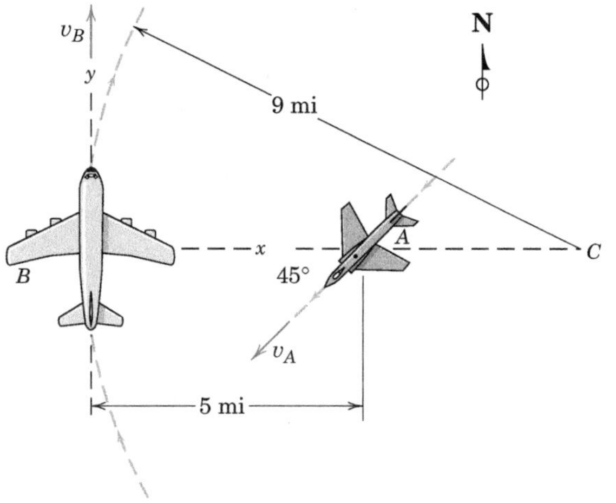

Problem 5/175

5/176 For the conditions of Prob. 5/175, obtain the vector expression for the acceleration which aircraft A appears to have to an observer in and turning with aircraft B, to which axes x-y are attached.

5/177 Car A is traveling along the straightaway with constant speed v. Car B is moving along the circular on-ramp with constant speed $v/2$. Determine the velocity and acceleration of car A as seen by an observer fixed to car B. Use the values $v = 60$ mi/hr and $R = 200$ ft, and utilize the x-y coordinates shown in the figure.

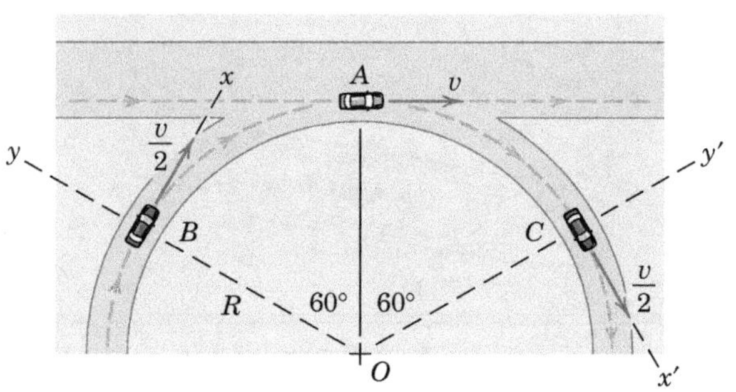

Problem 5/177

5/178 Refer to the figure for Prob. 5/177. Car A is traveling along the straightaway with speed v, and this speed is decreasing at a rate a. Car C is moving along the circular off-ramp with speed $v/2$, and this speed is decreasing at a rate $a/2$. Determine the velocity and acceleration which car A appears to have to an observer fixed to car C. Use the values $v = 60$ mi/hr, $a = 10$ ft/sec^2, and $R = 200$ ft, and utilize the x'-y' coordinates shown in the figure.

5/179 For the instant represented, link CB is rotating counterclockwise at a constant rate $N = 4$ rad/s, and its pin A causes a clockwise rotation of the slotted member ODE. Determine the angular velocity ω and angular acceleration α of ODE for this instant.

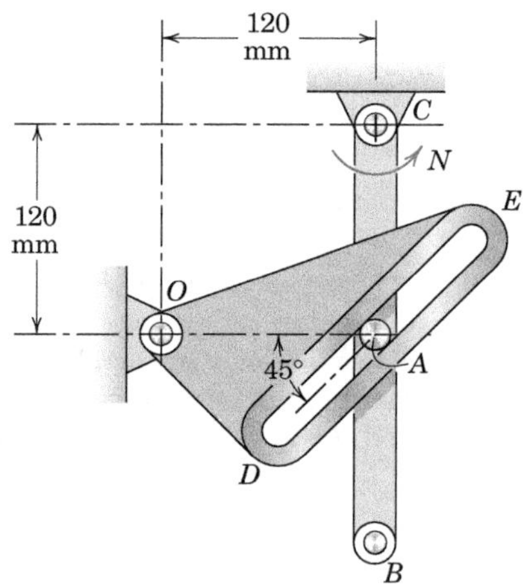

Problem 5/179

5/180 The disk rotates about a fixed axis through point O with a clockwise angular velocity $\omega_0 = 20$ rad/s and a counterclockwise angular acceleration $\alpha_0 = 5$ rad/s^2 at the instant under consideration. The value of r is 200 mm. Pin A is fixed to the disk but slides freely within the slotted member BC. Determine the velocity and acceleration of A relative to slotted member BC and the angular velocity and angular acceleration of BC.

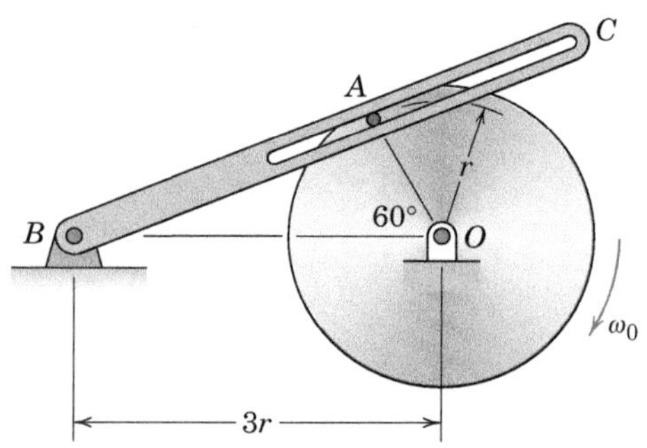

Problem 5/180

5/181 All conditions of the previous problem remain the same, except now, rather than rotating about a fixed center, the disk rolls without slipping on the horizontal surface. If the disk has a clockwise angular velocity of 20 rad/s and a counterclockwise angular acceleration of 5 rad/s^2, determine the velocity and acceleration of pin A relative to the slotted member BC and the angular velocity and angular acceleration of BC. The value of r is 200 mm. Neglect the distance from the center of pin A to the edge of the disk.

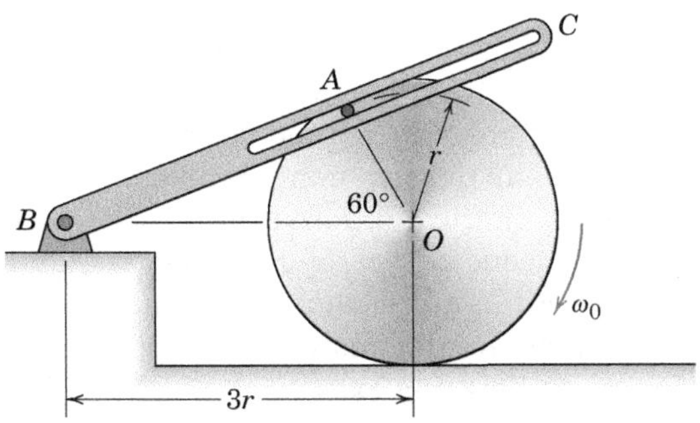

Problem 5/181

▶5/182 The space shuttle A is in an equatorial circular orbit of 240-km altitude and is moving from west to east. Determine the velocity and acceleration which it appears to have to an observer B fixed to and rotating with the earth at the equator as the shuttle passes overhead. Use $R = 6378$ km for the radius of the earth. Also use Fig. 1/1 for the appropriate value of g and carry out your calculations to four-figure accuracy.

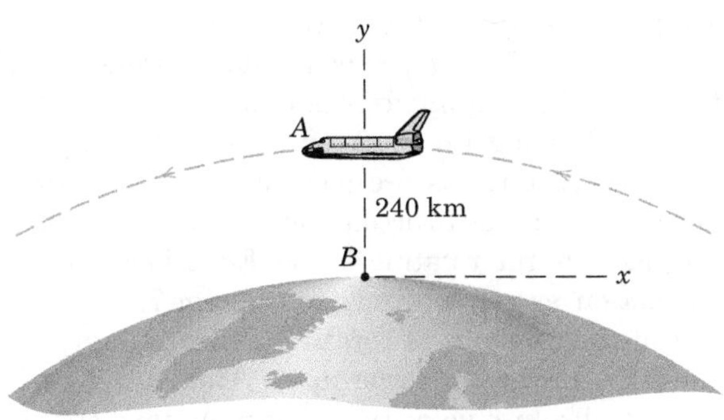

Problem 5/182

▶5/183 Determine the angular acceleration of link EC in the position shown, where $\omega = \dot{\beta} = 2$ rad/sec and $\ddot{\beta} = 6$ rad/sec^2 when $\theta = \beta = 60°$. Pin A is fixed to link EC. The circular slot in link DO has a radius of curvature of 6 in. In the position shown, the tangent to the slot at the point of contact is parallel to AO.

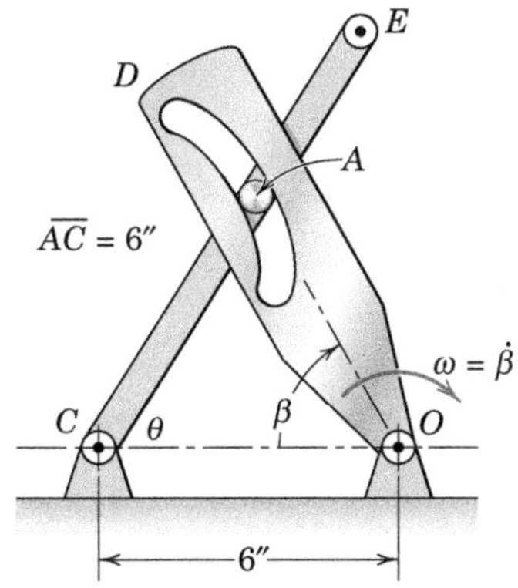

Problem 5/183

▶5/184 One wheel of an experimental vehicle F, which has a constant velocity $v = 36$ km/h, is shown. The wheel rolls without slipping and causes an oscillation of the slotted arm through the action of its pin A. Control rod DB, in turn, moves back and forth relative to the vehicle by virtue of the motion imparted to pin B. For the position shown, determine the acceleration a_B of the control rod DB. (*Suggestion:* Consider the justification and advantage of using a reference frame attached to the vehicle.)

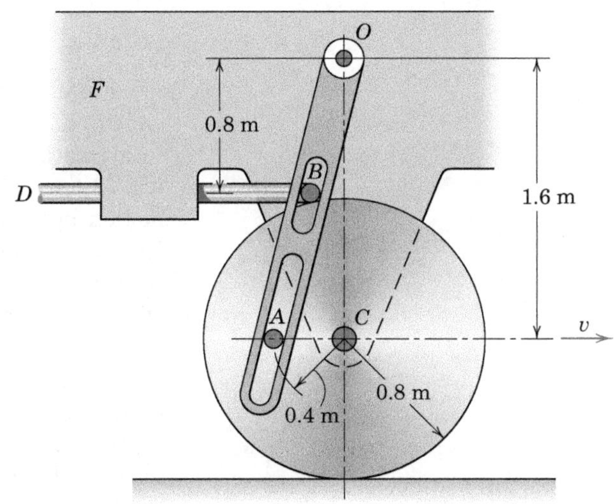

Problem 5/184

5/8 Chapter Review

In Chapter 5 we have applied our knowledge of basic kinematics from Chapter 2 to the plane motion of rigid bodies. We approached the problem in two ways.

1. Absolute-Motion Analysis

First, we wrote an equation which describes the general geometric configuration of a given problem in terms of knowns and unknowns. Then we differentiated this equation with respect to time to obtain velocities and accelerations, both linear and angular.

2. Relative-Motion Analysis

We applied the principles of relative motion to rigid bodies and found that this approach enables us to solve many problems which are too awkward to handle by mathematical differentiation. The relative-velocity equation, the instantaneous center of zero velocity, and the relative-acceleration equation all require that we visualize clearly and analyze correctly the case of circular motion of one point around another point, as viewed from nonrotating axes.

Solution of the Velocity and Acceleration Equations

The relative-velocity and relative-acceleration relationships are vector equations which we may solve in any one of three ways:

1. by a scalar-geometric analysis of the vector polygon,
2. by vector algebra, or
3. by a graphical construction of the vector polygon.

Rotating Coordinate Systems

Finally, in Chapter 5 we introduced rotating coordinate systems which enable us to solve problems where the motion is observed relative to a rotating frame of reference. Whenever a point moves along a path which itself is turning, analysis by rotating axes is indicated if a relative-motion approach is used. In deriving Eq. 5/12 for velocity and Eq. 5/14 for acceleration, where the relative terms are measured from a rotating reference system, it was necessary for us to account for the time derivatives of the unit vectors $\mathbf{i}$ and $\mathbf{j}$ fixed to the rotating frame. Equations 5/12 and 5/14 also apply to spatial motion, as will be shown in Chapter 7.

An important result of the analysis of rotating coordinate systems is the identification of the *Coriolis acceleration*. This acceleration represents the fact that the absolute velocity vector may have changes in both direction and magnitude due to rotation of the relative-velocity vector and change in position of the particle along the rotating path.

In Chapter 6 we will study the kinetics of rigid bodies in plane motion. There we will find that the ability to analyze the linear and angular accelerations of rigid bodies is necessary in order to apply the force and moment equations which relate the applied forces to the associated motions. Thus, the material of Chapter 5 is essential to that in Chapter 6.

REVIEW PROBLEMS

5/185 The circular disk of radius r rotates about a fixed axis through point O with the angular properties indicated in the figure. Determine the instantaneous velocity and acceleration of point A. Take all given quantities to be positive.

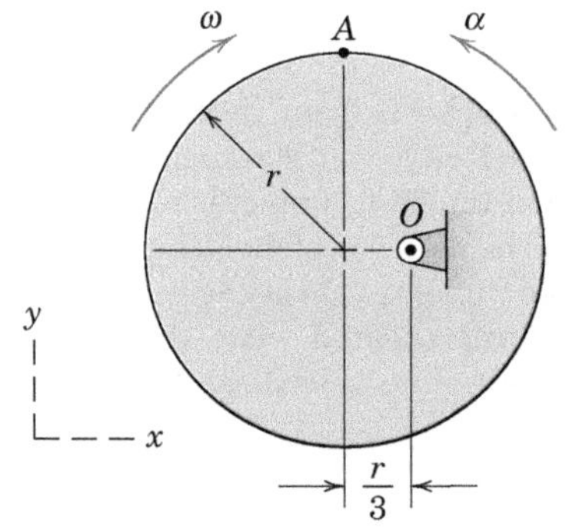

Problem 5/185

5/186 The circular disk rotates about its z-axis with an angular velocity $\omega = 2$ rad/s. A point P located on the rim has a velocity given by $\mathbf{v} = -0.8\mathbf{i} - 0.6\mathbf{j}$ m/s. Determine the coordinates of P and the radius r of the disk.

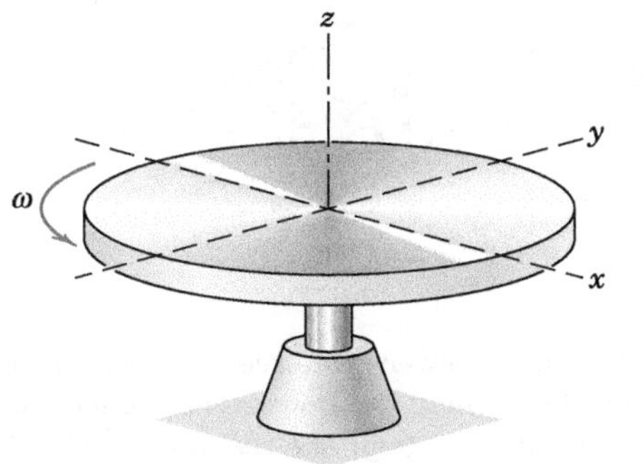

Problem 5/186

5/187 The frictional resistance to the rotation of a flywheel consists of a retardation due to air friction which varies as the square of the angular velocity and a constant frictional retardation in the bearing. As a result the angular acceleration of the flywheel while it is allowed to coast is given by $\alpha = -K - k\omega^2$, where K and k are constants. Determine an expression for the time required for the flywheel to come to rest from an initial angular velocity ω_0.

5/188 What angular velocity ω of bar AC will result in point B having zero velocity? What would be the corresponding velocity of point C? Take the length L of the bar and the velocity v of the collar as given quantities.

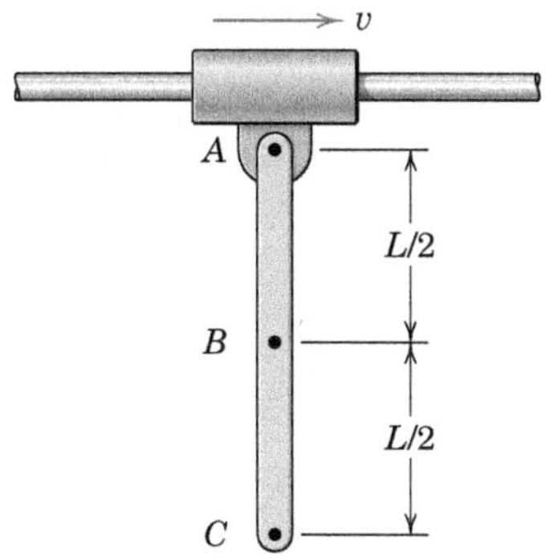

Problem 5/188

5/189 The rectangular plate rotates about its fixed z-axis. At the instant considered its angular velocity is $\omega = 3$ rad/s and is decreasing at the rate of 6 rad/s per second. For this instant write the vector expressions for the velocity of P and its normal and tangential components of acceleration.

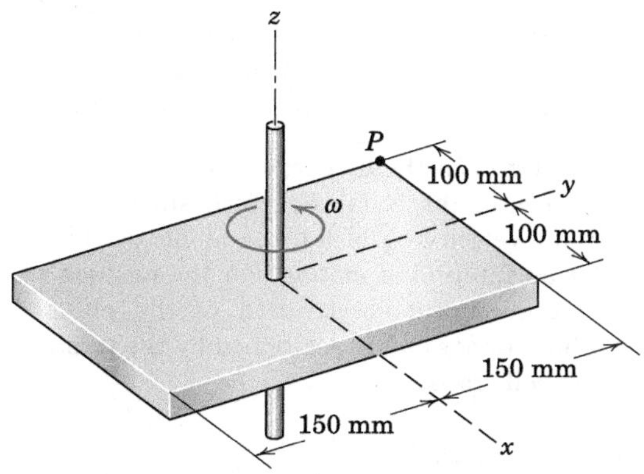

Problem 5/189

5/190 Roller B of the linkage has a velocity of 0.75 m/s to the right as the angle θ passes 60° and bar AB also makes an angle of 60° with the horizontal. Locate the instantaneous center of zero velocity for bar AB and determine its angular velocity ω_{AB}.

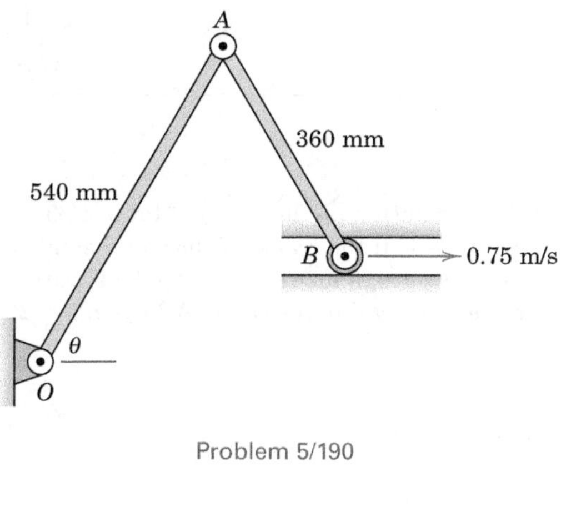

Problem 5/190

5/191 Rotation of the slotted bar OA is controlled by the lead screw that imparts a horizontal velocity v to collar C. Pin P is attached to the collar. Determine the angular velocity ω of bar OA in terms of v and the displacement x.

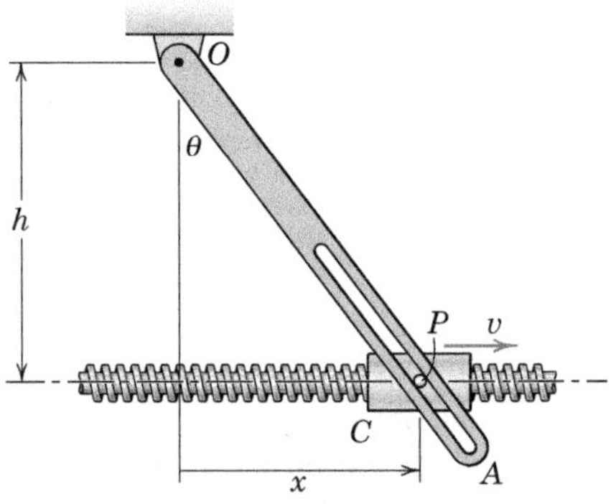

Problem 5/191

5/192 Consider the system of the previous problem. If the speed of collar C is decreasing at the rate a for the instant represented, determine the angular acceleration α of bar OA in terms of a, v, and x.

5/193 The pin A in the bell crank AOD is guided by the flanges of the collar B, which slides with a constant velocity v_B of 3 ft/sec along the fixed shaft for an interval of motion. For the position $\theta = 30°$ determine the acceleration of the plunger CE, whose upper end is positioned by the radial slot in the bell crank.

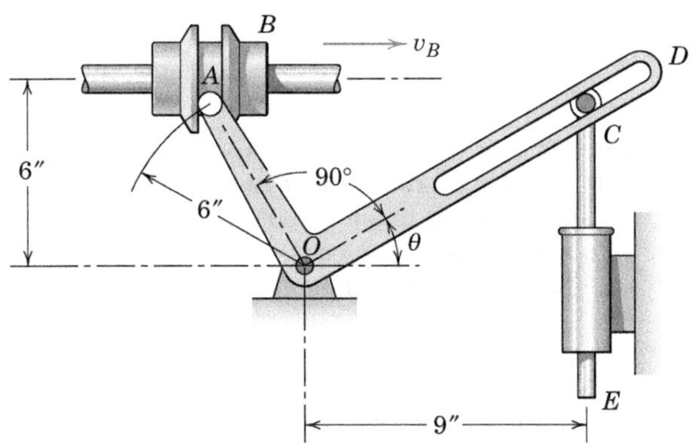

Problem 5/193

5/194 The oscillating produce tray of Prob. 5/83 is shown again here. If the crank OB has a constant counterclockwise angular velocity of 0.944 rad/s, determine the angular velocity of AB when $\theta = 20°$.

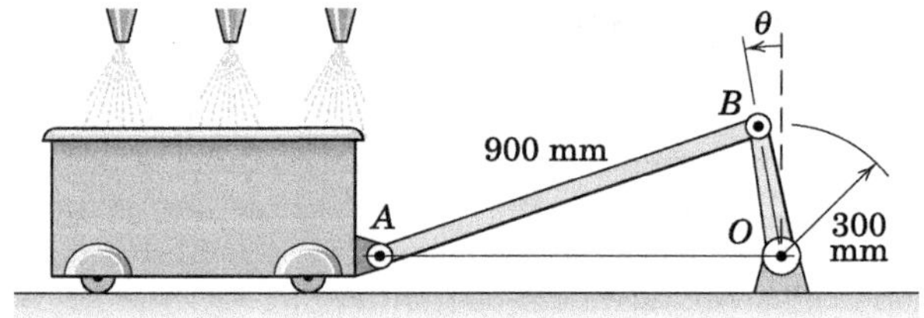

Problem 5/194

5/195 The helicopter is flying in the horizontal x-direction with a velocity $v = 120$ mi/hr, and the plane of rotation of the 26-ft-diameter rotor is tilted 10° from the horizontal x-y plane. The rotor blades rotate with an angular velocity $\Omega = 800$ rev/min. For the instant represented write the vector expressions for the absolute velocities of rotor tip A and rotor tip B.

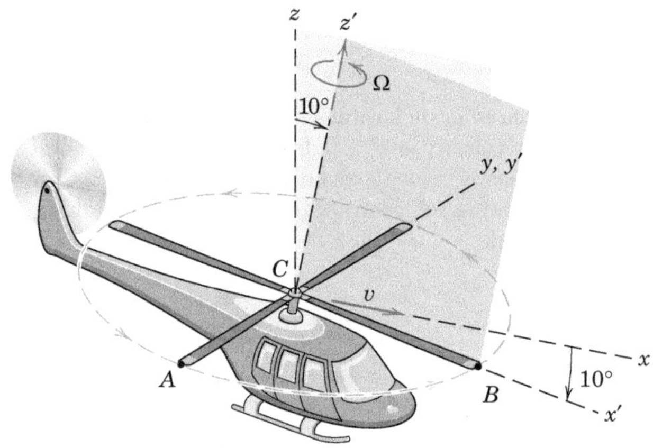

Problem 5/195

5/196 The large power-cable reel is rolled up the incline by the vehicle as shown. The vehicle starts from rest with $x = 0$ for the reel and accelerates at the constant rate of 2 ft/sec². For the instant when $x = 6$ ft, calculate the magnitude of the acceleration of point P on the reel in the position shown.

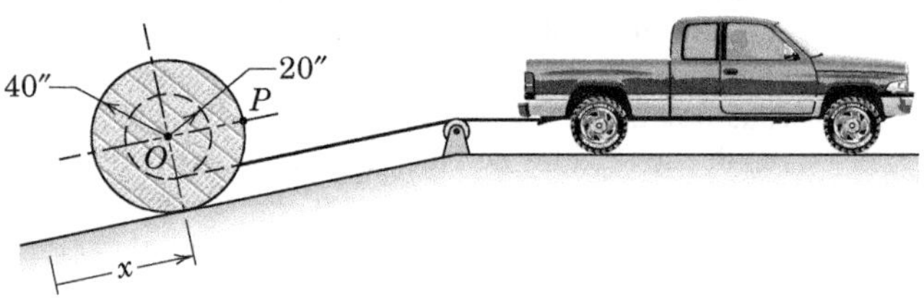

Problem 5/196

5/197 The isosceles triangular plate is guided by the two vertex rollers A and B which are confined to move in the perpendicular slots. The control rod gives A a constant velocity v_A to the left for an interval of motion. Determine the value of θ for which the horizontal component of the velocity of C is zero.

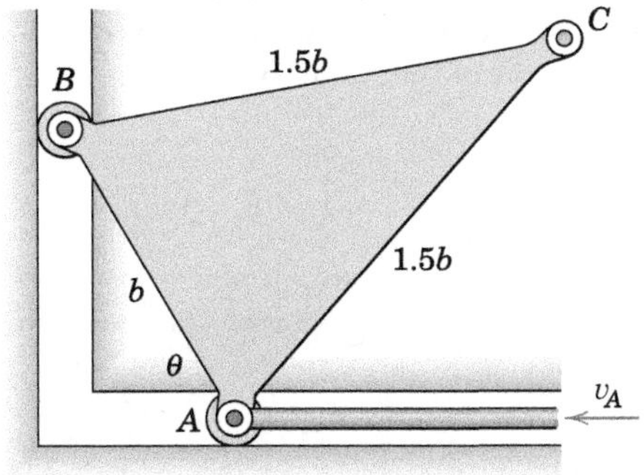

Problem 5/197

5/198 The load L is being elevated by the downward velocities of ends A and B of the cable. Determine the magnitude of the acceleration of point P on the top of the sheave for the instant when $v_A = 2$ ft/sec, $\dot{v}_A = 0.5$ ft/sec^2, $v_B = 3$ ft/sec, and $\dot{v}_B = -0.5$ ft/sec^2.

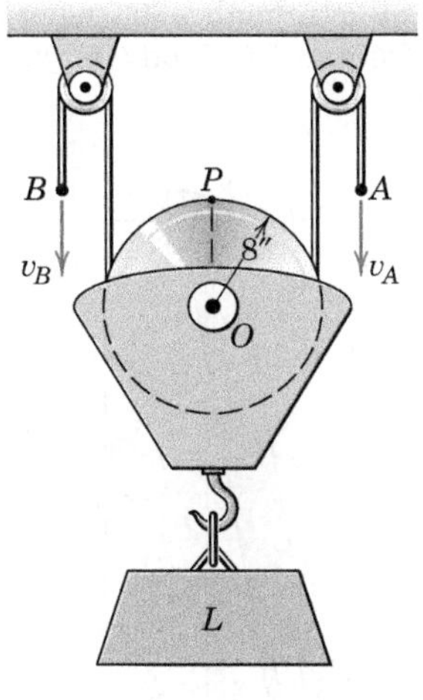

Problem 5/198

▶5/199 The hydraulic cylinder C imparts a velocity v to pin B in the direction shown. The collar slips freely on rod OA. Determine the resulting angular velocity of rod OA in terms of v, the displacement s of pin B, and the fixed distance d, for the angle $\beta = 15°$.

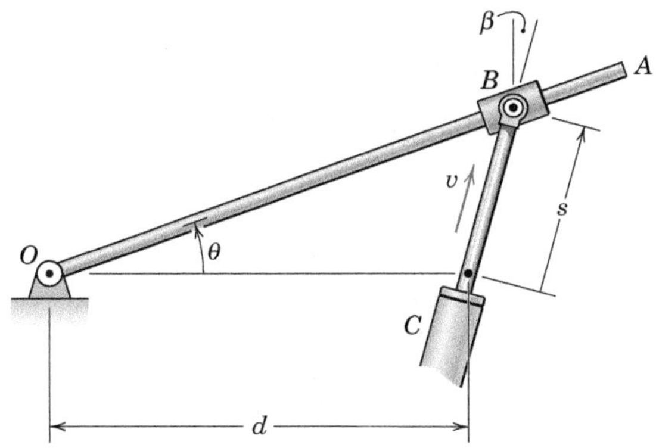

Problem 5/199

5/200 The end rollers of the bent bar ADB are confined to the slots shown. If $v_B = 0.3$ m/s, determine the velocity of roller A and the angular velocity of the bar.

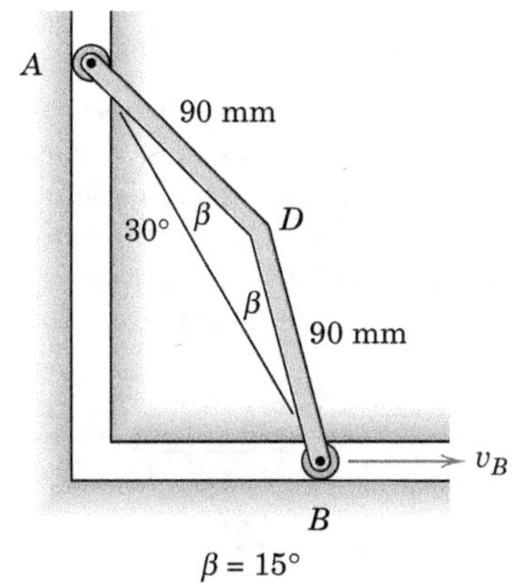

Problem 5/200

5/201 The figure illustrates a commonly used quick-return mechanism which produces a slow cutting stroke of the tool (attached to D) and a rapid return stroke. If the driving crank OA is turning at the constant rate $\dot{\theta} = 3$ rad/s, determine the magnitude of the velocity of point B for the instant when $\theta = 30°$.

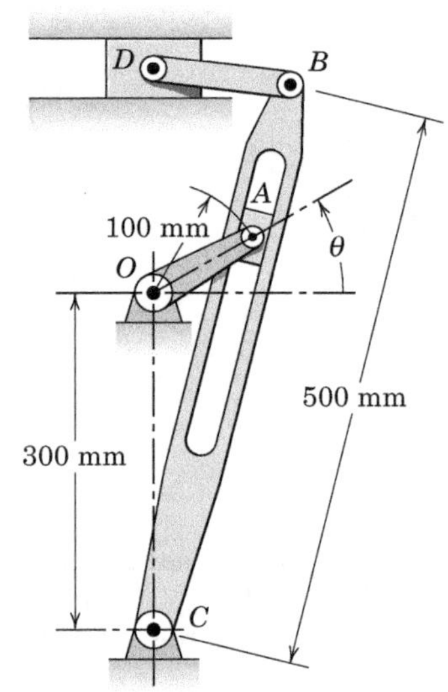

Problem 5/201

5/202 For the position shown where $\theta = 30°$, point A on the sliding collar has a constant velocity $v = 0.3$ m/s with corresponding lengthening of the hydraulic cylinder AC. For this same position BD is horizontal and DE is vertical. Determine the angular acceleration α_{DE} of DE at this instant.

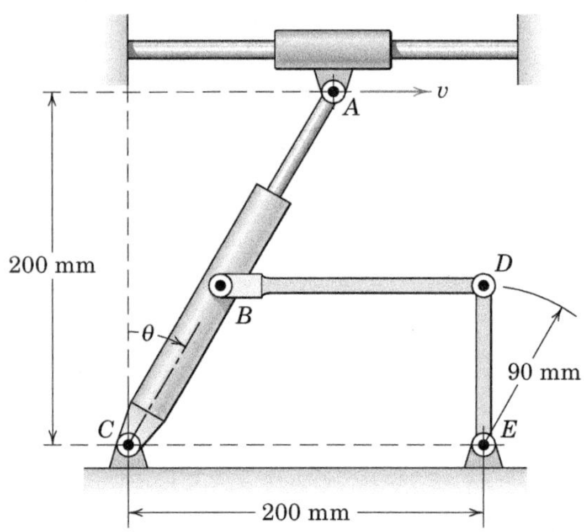

Problem 5/202

5/203 A radar station B situated at the equator observes a satellite A in a circular equatorial orbit of 200-km altitude and moving from west to east. For the instant when the satellite is 30° above the horizon, determine the difference between the velocity of the satellite relative to the radar station, as measured from a nonrotating frame of reference, and the velocity as measured relative to the reference frame of the radar system.

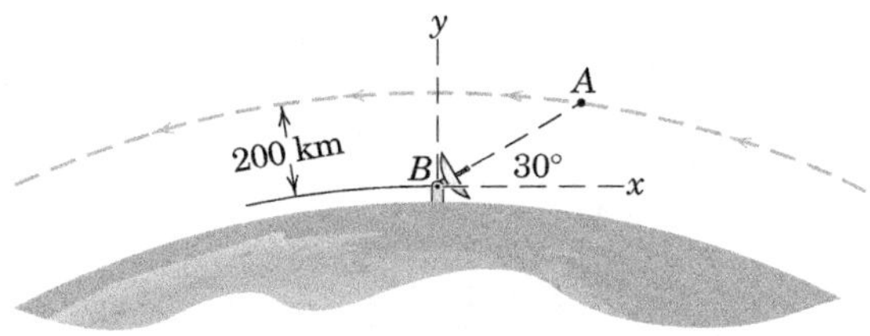

Problem 5/203

Computer-Oriented Problems

***5/204** Slotted arm OB oscillates about the vertical by the action of the rotating crank CA of 5-in. length, where the pin A engages the slot. For a constant speed $N = 120$ rev/min of crank CA, determine and plot the angular velocity $\dot{\beta}$ of arm OB as a function of θ through 360°, where β is the angle between OC and OB. Find θ for zero angular velocity of OB.

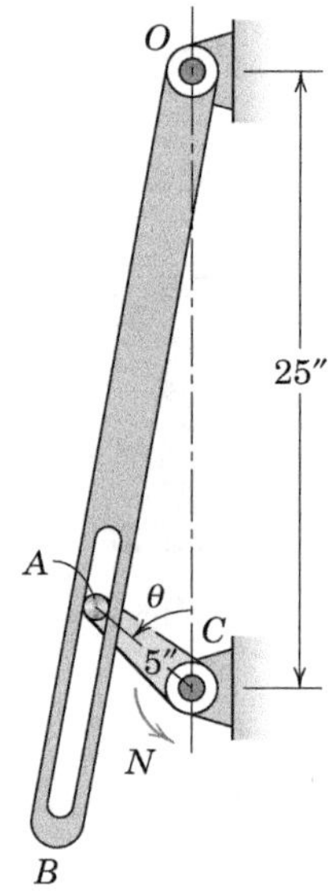

Problem 5/204

***5/205** The disk rotates about a fixed axis with a constant angular velocity $\omega_0 = 10$ rad/s. Pin A is fixed to the disk. Determine and plot the magnitudes of the velocity and acceleration of pin A relative to the slotted member BC as functions of the disk angle θ over the range $0 \leq \theta \leq 360°$. State the maximum and minimum values and also the values of θ at which they occur. The value of r is 200 mm.

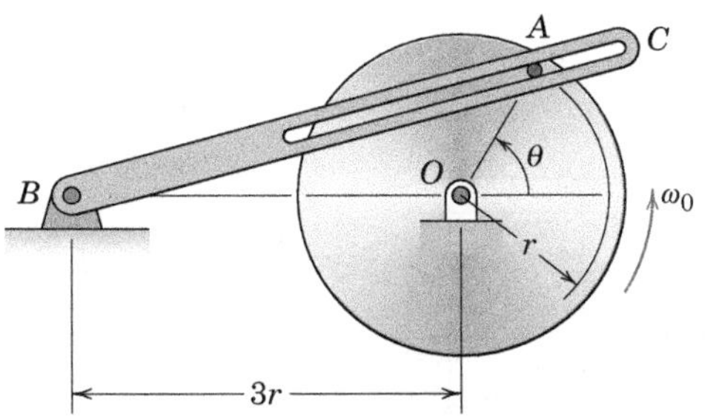

Problem 5/205

***5/206** A constant torque M exceeds the moment about O due to the force F on the plunger, and an angular acceleration $\ddot{\theta} = 100(1 - \cos\theta)$ rad/s^2 results. If the crank OA is released from rest at B, where $\theta = 30°$, and strikes the stop at C, where $\theta = 150°$, plot the angular velocity $\dot{\theta}$ as a function of θ and find the time t for the crank to rotate from $\theta = 90°$ to $\theta = 150°$.

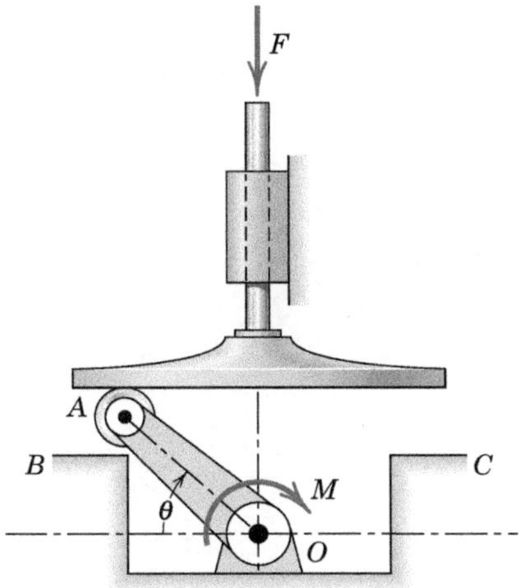

Problem 5/206

***5/207** The crank OA of the four-bar linkage is driven at a constant counterclockwise angular velocity $\omega_0 = 10$ rad/s. Determine and plot as functions of the crank angle θ the angular velocities of bars AB and BC over the range $0 \leq \theta \leq 360°$. State the maximum absolute value of each angular velocity and the value of θ at which it occurs.

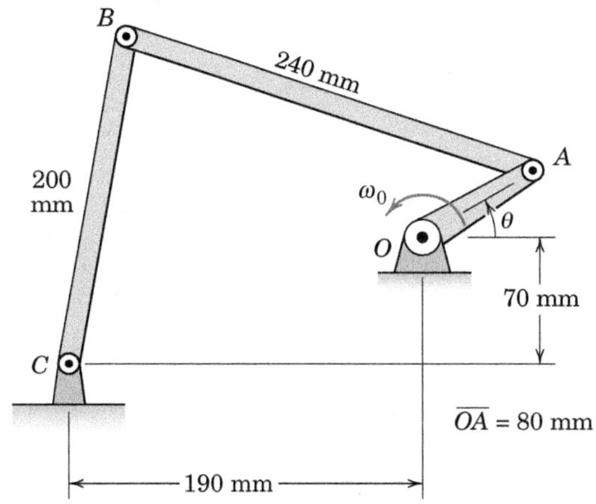

Problem 5/207

***5/208** If all conditions in the previous problem remain the same, determine and plot as functions of the crank angle θ the angular accelerations of bars AB and BC over the range $0 \leq \theta \leq 360°$. State the maximum absolute value of each angular acceleration and the value of θ at which it occurs.

***5/209** All conditions of Prob. 5/207 remain the same, except the counterclockwise angular velocity of crank OA is 10 rad/s when $\theta = 0$ and the constant counterclockwise angular acceleration of the crank is 20 rad/s^2. Determine and plot as functions of the crank angle θ the angular velocities of bars AB and BC over the range $0 \leq \theta \leq 360°$. State the maximum absolute value of each angular velocity and the value of θ at which it occurs.

*5/210 Bar OA rotates about the fixed pivot O with constant angular velocity $\dot{\beta} = 0.8$ rad/s. Pin A is fixed to bar OA and is engaged in the slot of member BD, which rotates about a fixed axis through point B. Determine and plot over the range $0 \leq \beta \leq 360°$ the angular velocity and angular acceleration of BD and the velocity and acceleration of pin A relative to member BD. State the magnitude and direction of the acceleration of pin A relative to member BD for $\beta = 180°$.

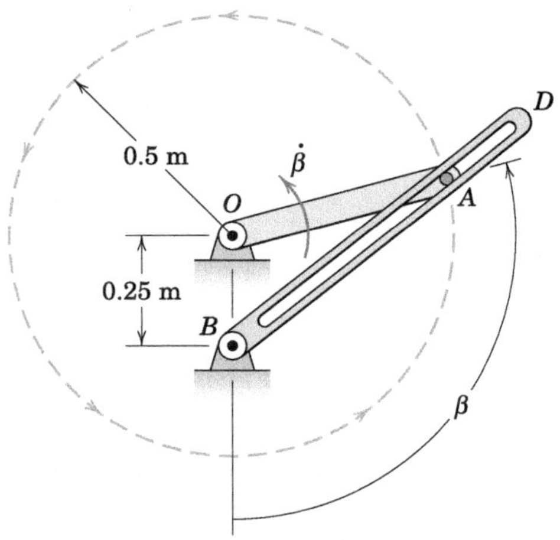

Problem 5/210

*5/211 For the slider-crank configuration shown, derive the expression for the velocity v_A of the piston (taken positive to the right) as a function of θ. Substitute the numerical data of Sample Problem 5/15 and calculate v_A as a function of θ for $0 \leq \theta \leq 180°$. Plot v_A versus θ and find its maximum magnitude and the corresponding value of θ. (By symmetry anticipate the results for $180° \leq \theta \leq 360°$.)

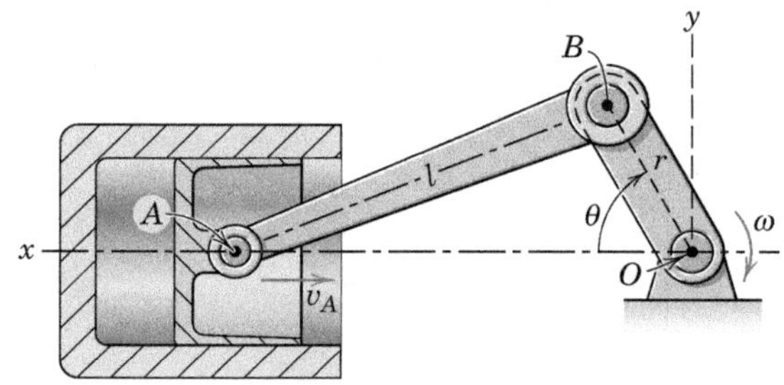

Problem 5/211

*5/212 For the slider-crank of Prob. 5/211, derive the expression for the acceleration a_A of the piston (taken positive to the right) as a function of θ for $\omega = \dot{\theta} =$ constant. Substitute the numerical data of Sample Problem 5/15 and calculate a_A as a function of θ for $0 \leq \theta \leq 180°$. Plot a_A versus θ and find the value of θ for which $a_A = 0$. (By symmetry anticipate the results for $180° \leq \theta \leq 360°$.)

The principles of this chapter must be applied during the design of the massive blades of large wind turbines.

6 Plane Kinetics of Rigid Bodies

Chapter Outline

6/1 Introduction

The *kinetics* of rigid bodies treats the relationships between the external forces acting on a body and the corresponding translational and rotational motions of the body. In Chapter 5 we developed the kinematic relationships for the plane motion of rigid bodies, and we will use these relationships extensively in this present chapter, where the effects of forces on the two-dimensional motion of rigid bodies are examined.

For our purpose in this chapter, a body which can be approximated as a thin slab with its motion confined to the plane of the slab will be considered to be in plane motion. The plane of motion will contain the mass center, and all forces which act on the body will be projected onto the plane of motion. A body which has appreciable dimensions normal to the plane of motion but is symmetrical about that plane of motion through the mass center may be treated as having plane motion. These idealizations clearly fit a very large category of rigid-body motions.

Background for the Study of Kinetics

In Chapter 3 we found that two force equations of motion were required to define the motion of a particle whose motion is confined to a plane. For the plane motion of a rigid body, an additional equation is needed to specify the state of rotation of the body. Thus, two force equations and one moment equation or their equivalent are required to determine the state of rigid-body plane motion.

The kinetic relationships which form the basis for most of the analysis of rigid-body motion were developed in Chapter 4 for a general system of particles. Frequent reference will be made to these equations as they are further developed in Chapter 6 and applied specifically to the plane motion of rigid bodies. You should refer to Chapter 4 frequently as you study Chapter 6. Also, before proceeding make sure that you have a firm grasp of the calculation of velocities and accelerations as developed in Chapter 5 for rigid-body plane motion. Unless you can determine accelerations correctly from the principles of kinematics, you frequently will be unable to apply the force and moment principles of kinetics. Consequently, you should master the necessary kinematics, including the calculation of relative accelerations, before proceeding.

Successful application of kinetics requires that you isolate the body or system to be analyzed. The isolation technique was illustrated and used in Chapter 3 for particle kinetics and will be employed consistently in the present chapter. For problems involving the instantaneous relationships among force, mass, and acceleration, the body or system should be explicitly defined by isolating it with its *free-body diagram*. When the principles of work and energy are employed, an *active-force diagram* which shows only those external forces which do work on the system may be used in lieu of the free-body diagram. The impulse-momentum diagram should be constructed when impulse-momentum methods are used. *No solution of a problem should be attempted without first defining the complete external boundary of the body or system and identifying all external forces which act on it.*

In the kinetics of rigid bodies which have angular motion, we must introduce a property of the body which accounts for the radial distribution of its mass with respect to a particular axis of rotation normal to the plane of motion. This property is known as the *mass moment of inertia* of the body, and it is essential that we be able to calculate this property in order to solve rotational problems. We assume that you are familiar with the calculation of mass moments of inertia. Appendix B treats this topic for those who need instruction or review.

Organization of the Chapter

Chapter 6 is organized in the same three sections in which we treated the kinetics of particles in Chapter 3. Section A relates the forces and moments to the instantaneous linear and angular accelerations. Section B treats the solution of problems by the method of work and energy. Section C covers the methods of impulse and momentum.

Virtually all of the basic concepts and approaches covered in these three sections were treated in Chapter 3 on particle kinetics. This repetition will help you with the topics of Chapter 6, provided you understand the kinematics of rigid-body plane motion. In each of the three sections, we will treat three types of motion: *translation, fixed-axis rotation*, and *general plane motion*.

SECTION A FORCE, MASS, AND ACCELERATION

6/2 General Equations of Motion

In Arts. 4/2 and 4/4 we derived the force and moment vector equations of motion for a general system of mass. We now apply these results by starting, first, with a general rigid body in three dimensions. The force equation, Eq. 4/1,

$$\Sigma \mathbf{F} = m\overline{\mathbf{a}} \qquad [4/1]$$

tells us that the resultant $\Sigma\mathbf{F}$ of the external forces acting on the body equals the mass m of the body times the acceleration $\overline{\mathbf{a}}$ of its mass center G. The moment equation taken about the mass center, Eq. 4/9,

$$\Sigma \mathbf{M}_G = \dot{\mathbf{H}}_G \qquad [4/9]$$

shows that the resultant moment about the mass center of the external forces on the body equals the time rate of change of the angular momentum of the body about the mass center.

Recall from our study of statics that a general system of forces acting on a rigid body may be replaced by a resultant force applied at a chosen point and a corresponding couple. By replacing the external forces by their equivalent force-couple system in which the resultant force acts through the mass center, we may visualize the action of the forces and the corresponding dynamic response of the body with the aid of Fig. 6/1.

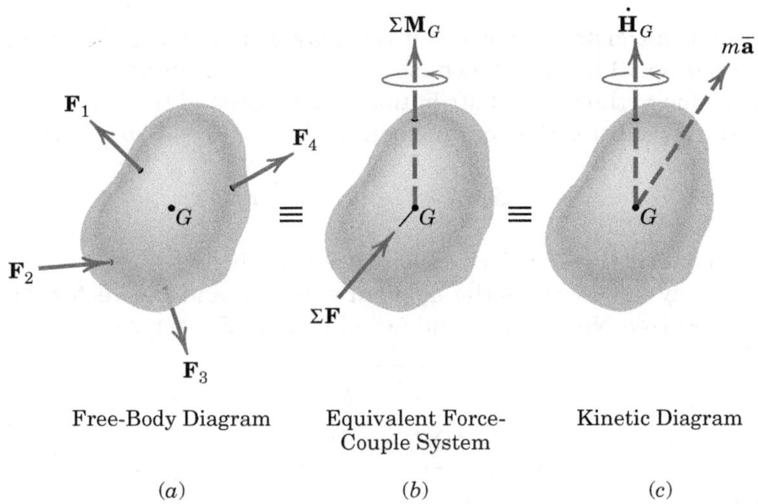

Figure 6/1

Part a of the figure shows the relevant free-body diagram. Part b of the figure shows the equivalent force-couple system with the resultant force applied through G. Part c of the figure is a *kinetic diagram*, which represents the resulting dynamic effects as specified by Eqs. 4/1 and 4/9. The equivalence between the free-body diagram and the kinetic diagram enables us to clearly visualize and easily remember the separate translational and rotational effects of the forces applied to a rigid body. We will express this equivalence mathematically as we apply these results to the treatment of rigid-body plane motion.

Plane-Motion Equations

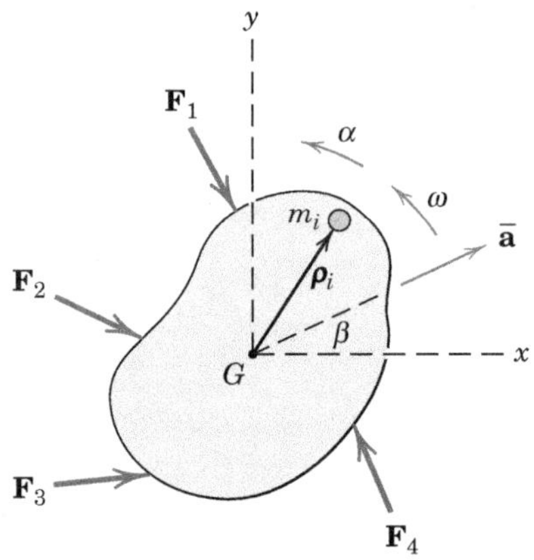

Figure 6/2

We now apply the foregoing relationships to the case of plane motion. Figure 6/2 represents a rigid body moving with plane motion in the x-y plane. The mass center G has an acceleration $\bar{\mathbf{a}}$, and the body has an angular velocity $\boldsymbol{\omega} = \omega\mathbf{k}$ and an angular acceleration $\boldsymbol{\alpha} = \alpha\mathbf{k}$, both taken positive in the z-direction. Because the z-direction of both $\boldsymbol{\omega}$ and $\boldsymbol{\alpha}$ remains perpendicular to the plane of motion, we may use scalar notation ω and $\alpha = \dot{\omega}$ to represent the angular velocity and angular acceleration.

The angular momentum about the mass center for the general system was expressed in Eq. 4/8a as $\mathbf{H}_G = \Sigma\boldsymbol{\rho}_i \times m_i\dot{\boldsymbol{\rho}}_i$ where $\boldsymbol{\rho}_i$ is the position vector relative to G of the representative particle of mass m_i. For our rigid body, the velocity of m_i relative to G is $\dot{\boldsymbol{\rho}}_i = \boldsymbol{\omega} \times \boldsymbol{\rho}_i$, which has a magnitude $\rho_i\omega$ and lies in the plane of motion normal to $\boldsymbol{\rho}_i$. The product $\boldsymbol{\rho}_i \times \dot{\boldsymbol{\rho}}_i$ is then a vector normal to the x-y plane in the sense of $\boldsymbol{\omega}$, and its magnitude is $\rho_i^2\omega$. Thus, the magnitude of $\mathbf{H}_G$ becomes $H_G = \Sigma\rho_i^2 m_i\omega = \omega\Sigma\rho_i^2 m_i$. The summation, which may also be written as $\int\rho^2\,dm$, is defined as the *mass moment of inertia* $\bar{I}$ of the body about the z-axis through G. (See Appendix B for a discussion of the calculation of mass moments of inertia.)

We may now write

$$H_G = \bar{I}\omega$$

where $\bar{I}$ is a constant property of the body. This property is a measure of the rotational inertia, which is the resistance to change in rotational velocity due to the radial distribution of mass around the z-axis through G. With this substitution, our moment equation, Eq. 4/9, becomes

$$\Sigma M_G = \dot{H}_G = \bar{I}\dot{\omega} = \bar{I}\alpha$$

where $\alpha = \dot{\omega}$ is the angular acceleration of the body.

We may now express the moment equation and the vector form of the generalized Newton's second law of motion, Eq. 4/1, as

$$\boxed{\begin{aligned} \Sigma\mathbf{F} &= m\bar{\mathbf{a}} \\ \Sigma M_G &= \bar{I}\alpha \end{aligned}} \qquad \textbf{(6/1)}$$

Equations 6/1 are the general equations of motion for a rigid body in plane motion. In applying Eqs. 6/1, we express the vector force equation in terms of its two scalar components using x-y, n-t, or r-θ coordinates, whichever is most convenient for the problem at hand.

Alternative Derivation

It is instructive to use an alternative approach to derive the moment equation by referring directly to the forces which act on the representative particle of mass m_i, as shown in Fig. 6/3. The acceleration of m_i equals the vector sum of $\bar{a}$ and the relative terms $\rho_i\omega^2$ and $\rho_i\alpha$, where the mass center G is used as the reference point. It follows that the resultant of all forces on m_i has the components $m_i\bar{a}$, $m_i\rho_i\omega^2$, and $m_i\rho_i\alpha$ in the directions shown. The sum of the moments of these force components about G in the sense of α becomes

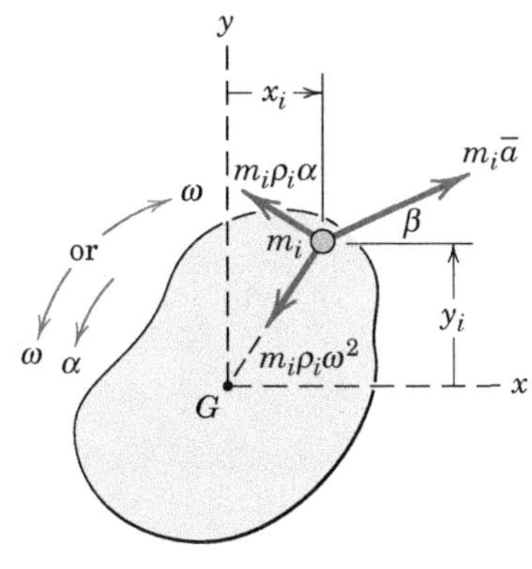

Figure 6/3

$$M_{G_i} = m_i\rho_i^{\,2}\alpha + (m_i\bar{a}\sin\beta)x_i - (m_i\bar{a}\cos\beta)y_i$$

Similar moment expressions exist for all particles in the body, and the sum of these moments about G for the resultant forces acting on all particles may be written as

$$\Sigma M_G = \Sigma m_i\rho_i^{\,2}\alpha + \bar{a}\sin\beta\;\Sigma m_i x_i - \bar{a}\cos\beta\;\Sigma m_i y_i$$

But the origin of coordinates is taken at the mass center, so that $\Sigma m_i x_i = m\bar{x} = 0$ and $\Sigma m_i y_i = m\bar{y} = 0$. Thus, the moment sum becomes

$$\Sigma M_G = \Sigma m_i\rho_i^{\,2}\alpha = \bar{I}\alpha$$

as before. The contribution to ΣM_G of the forces internal to the body is, of course, zero since they occur in pairs of equal and opposite forces of action and reaction between interacting particles. Thus, ΣM_G, as before, represents the sum of moments about the mass center G of only the external forces acting on the body, as disclosed by the free-body diagram.

We note that the force component $m_i\rho_i\omega^2$ has no moment about G and conclude, therefore, that the angular velocity ω has no influence on the moment equation about the mass center.

The results embodied in our basic equations of motion for a rigid body in plane motion, Eqs. 6/1, are represented diagrammatically in Fig. 6/4,

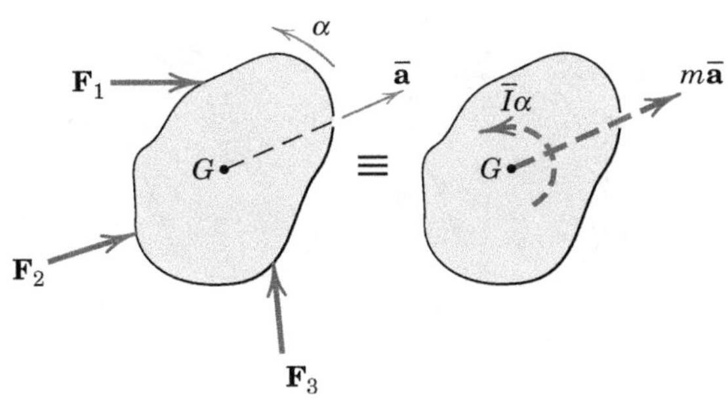

Figure 6/4

which is the two-dimensional counterpart of parts *a* and *c* of Fig. 6/1 for a general three-dimensional body. The free-body diagram discloses the forces and moments appearing on the left-hand side of our equations of motion. The kinetic diagram discloses the resulting dynamic response in terms of the translational term $m\bar{\mathbf{a}}$ and the rotational term $\bar{I}\alpha$ which appear on the right-hand side of Eqs. 6/1.

As previously mentioned, the translational term $m\bar{\mathbf{a}}$ will be expressed by its *x-y*, *n-t*, or *r-θ* components once the appropriate inertial reference system is designated. The equivalence depicted in Fig. 6/4 is basic to our understanding of the kinetics of plane motion and will be employed frequently in the solution of problems.

Representation of the resultants $m\bar{\mathbf{a}}$ and $\bar{I}\alpha$ will help ensure that the force and moment sums determined from the free-body diagram are equated to their proper resultants.

Alternative Moment Equations

In Art. 4/4 of Chapter 4 on systems of particles, we developed a general equation for moments about an arbitrary point *P*, Eq. 4/11, which is

$$\Sigma\mathbf{M}_P = \dot{\mathbf{H}}_G + \bar{\boldsymbol{\rho}} \times m\bar{\mathbf{a}} \qquad \textbf{[4/11]}$$

where $\bar{\boldsymbol{\rho}}$ is the vector from *P* to the mass center *G* and $\bar{\mathbf{a}}$ is the mass-center acceleration. As we have shown earlier in this article, for a rigid body in plane motion $\dot{\mathbf{H}}_G$ becomes $\bar{I}\alpha$. Also, the cross product $\bar{\boldsymbol{\rho}} \times m\bar{\mathbf{a}}$ is simply the moment of magnitude $m\bar{a}d$ of $m\bar{\mathbf{a}}$ about *P*. Therefore, for the two-dimensional body illustrated in Fig. 6/5 with its free-body diagram and kinetic diagram, we may rewrite Eq. 4/11 simply as

$$\boxed{\Sigma M_P = \bar{I}\alpha + m\bar{a}d} \qquad \textbf{(6/2)}$$

Clearly, all three terms are positive in the counterclockwise sense for the example shown, and the choice of *P* eliminates reference to $\mathbf{F}_1$ and $\mathbf{F}_3$.

If we had wished to eliminate reference to $\mathbf{F}_2$ and $\mathbf{F}_3$, for example, by choosing their intersection as the reference point, then *P* would lie

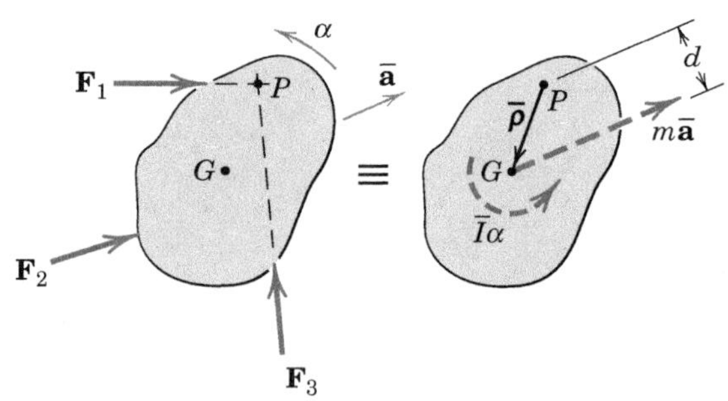

Figure 6/5

on the opposite side of the $m\bar{\mathbf{a}}$ vector, and the clockwise moment of $m\bar{\mathbf{a}}$ about P would be a negative term in the equation. Equation 6/2 is easily remembered as it is merely an expression of the familiar principle of moments, where the sum of the moments about P equals the combined moment about P of their sum, expressed by the resultant couple $\Sigma M_G = \bar{I}\alpha$ and the resultant force $\Sigma \mathbf{F} = m\bar{\mathbf{a}}$.

In Art. 4/4 we also developed an alternative moment equation about P, Eq. 4/13, which is

$$\Sigma \mathbf{M}_P = (\dot{\mathbf{H}}_P)_{\text{rel}} + \bar{\boldsymbol{\rho}} \times m\mathbf{a}_P \qquad \textbf{[4/13]}$$

For rigid-body plane motion, if P is chosen as a point *fixed* to the body, then in scalar form $(\dot{\mathbf{H}}_P)_{\text{rel}}$ becomes $I_P\alpha$, where I_P is the mass moment of inertia about an axis through P and α is the angular acceleration of the body. So we may write the equation as

$$\boxed{\Sigma \mathbf{M}_P = I_P\boldsymbol{\alpha} + \bar{\boldsymbol{\rho}} \times m\mathbf{a}_P} \qquad \textbf{(6/3)}$$

where the acceleration of P is $\mathbf{a}_P$ and the position vector from P to G is $\bar{\boldsymbol{\rho}}$.

When $\bar{\boldsymbol{\rho}} = \mathbf{0}$, point P becomes the mass center G, and Eq. 6/3 reduces to the scalar form $\Sigma M_G = \bar{I}\alpha$, previously derived. When point P becomes a point O fixed in an inertial reference system and attached to the body (or body extended), then $\mathbf{a}_P = \mathbf{0}$, and Eq. 6/3 in scalar form reduces to

$$\boxed{\Sigma M_O = I_O\alpha} \qquad \textbf{(6/4)}$$

Equation 6/4 then applies to the rotation of a rigid body about a nonaccelerating point O fixed to the body and is the two-dimensional simplification of Eq. 4/7.

Unconstrained and Constrained Motion

The motion of a rigid body may be unconstrained or constrained. The rocket moving in a vertical plane, Fig. 6/6a, is an example of unconstrained

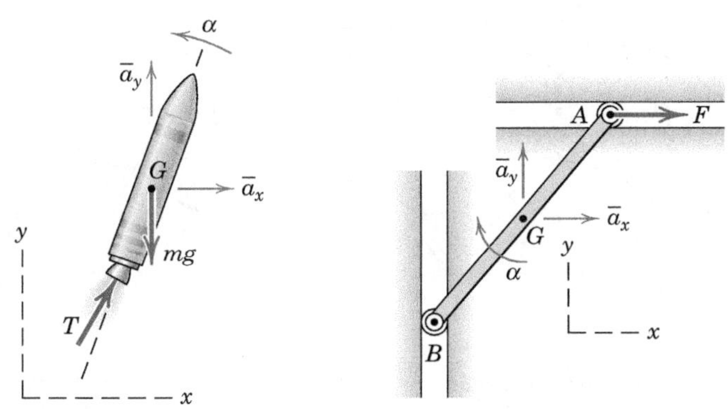

(a) Unconstrained Motion (b) Constrained Motion

Figure 6/6

motion as there are no physical confinements to its motion. The two components $\bar{a}_x$ and $\bar{a}_y$ of the mass-center acceleration and the angular acceleration α may be determined independently of one another by direct application of Eqs. 6/1.

The bar in Fig. 6/6*b*, on the other hand, undergoes a constrained motion, where the vertical and horizontal guides for the ends of the bar impose a kinematic relationship between the acceleration components of the mass center and the angular acceleration of the bar. Thus, it is necessary to determine this kinematic relationship from the principles established in Chapter 5 and to combine it with the force and moment equations of motion before a solution can be carried out.

In general, dynamics problems which involve physical constraints to motion require a kinematic analysis relating linear to angular acceleration before the force and moment equations of motion can be solved. It is for this reason that an understanding of the principles and methods of Chapter 5 is so vital to the work of Chapter 6.

Systems of Interconnected Bodies

Upon occasion, in problems dealing with two or more connected rigid bodies whose motions are related kinematically, it is convenient to analyze the bodies as an entire system.

Figure 6/7 illustrates two rigid bodies hinged at A and subjected to the external forces shown. The forces in the connection at A are internal to the system and are not disclosed. The resultant of all external forces must equal the vector sum of the two resultants $m_1\bar{\mathbf{a}}_1$ and $m_2\bar{\mathbf{a}}_2$, and the sum of the moments about some arbitrary point such as P of all external forces must equal the moment of the resultants, $\bar{I}_1\alpha_1 + \bar{I}_2\alpha_2 + m_1\bar{a}_1d_1 + m_2\bar{a}_2d_2$. Thus, we may state

$$\begin{aligned} \Sigma\mathbf{F} &= \Sigma m\bar{\mathbf{a}} \\ \Sigma M_P &= \Sigma\bar{I}\alpha + \Sigma m\bar{a}d \end{aligned} \tag{6/5}$$

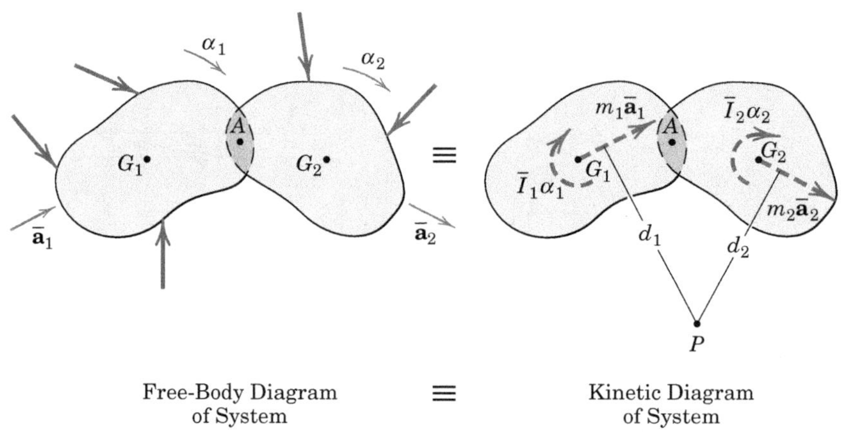

Figure 6/7

where the summations on the right-hand side of the equations represent as many terms as there are separate bodies.

If there are more than three remaining unknowns in a system, however, the three independent scalar equations of motion, when applied to the system, are not sufficient to solve the problem. In this case, more advanced methods such as virtual work (Art. 6/7) or Lagrange's equations (not discussed in this book*) could be employed, or else the system could be dismembered and each part analyzed separately with the resulting equations solved simultaneously.

Key Concepts

ANALYSIS PROCEDURE

In the solution of force-mass-acceleration problems for the plane motion of rigid bodies, the following steps should be taken once you understand the conditions and requirements of the problem:

1. Kinematics. First, identify the class of motion and then solve for any needed linear and angular accelerations which can be determined solely from given kinematic information. In the case of constrained plane motion, it is usually necessary to establish the relation between the linear acceleration of the mass center and the angular acceleration of the body by first solving the appropriate relative-velocity and relative-acceleration equations. Again, we emphasize that success in working force-mass-acceleration problems in this chapter is contingent on the ability to describe the necessary kinematics, so that frequent review of Chapter 5 is recommended.

2. Diagrams. Always draw the complete free-body diagram of the body to be analyzed. Assign a convenient inertial coordinate system and label all known and unknown quantities. The kinetic diagram should also be constructed so as to clarify the equivalence between the applied forces and the resulting dynamic response.

3. Equations of Motion. Apply the three equations of motion from Eqs. 6/1, being consistent with the algebraic signs in relation to the choice of reference axes. Equation 6/2 or 6/3 may be employed as an alternative to the second of Eqs. 6/1. Combine these relations with the results from any needed kinematic analysis. Count the number of unknowns and be certain that there are an equal number of independent equations available. For a solvable rigid-body problem in plane motion, there can be no more than the five scalar unknowns which can be determined from the three scalar equations of motion, obtained from Eqs. 6/1, and the two scalar component relations which come from the relative-acceleration equation.

*When an interconnected system has more than one degree of freedom, that is, requires more than one coordinate to specify completely the configuration of the system, the more advanced equations of Lagrange are generally used. See the first author's *Dynamics, 2nd Edition, SI Version*, 1975, John Wiley & Sons, for a treatment of Lagrange's equations.

In the following three articles the foregoing developments will be applied to three cases of motion in a plane: *translation, fixed-axis rotation,* and *general plane motion.*

6/3 Translation

Rigid-body translation in plane motion was described in Art. 5/1 and illustrated in Figs. 5/1*a* and 5/1*b*, where we saw that every line in a translating body remains parallel to its original position at all times. In rectilinear translation all points move in straight lines, whereas in curvilinear translation all points move on congruent curved paths. In either case, there is no angular motion of the translating body, so that both ω and α are zero. Therefore, from the moment relation of Eqs. 6/1, we see that all reference to the moment of inertia is eliminated for a translating body.

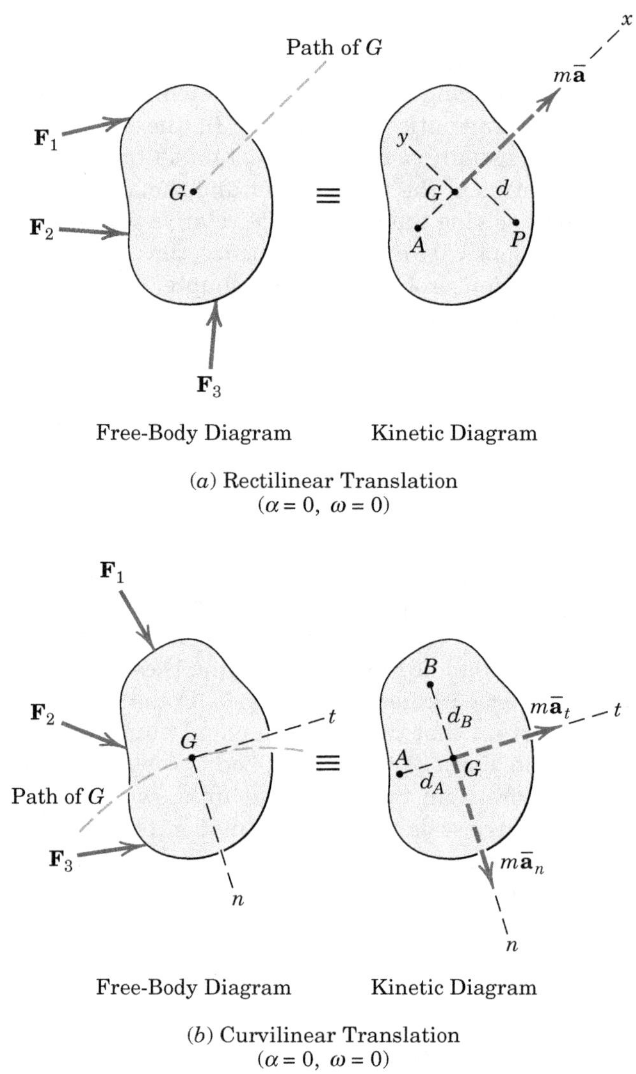

Figure 6/8

For a translating body, then, our general equations for plane motion, Eqs. 6/1, may be written

$$\Sigma \mathbf{F} = m\bar{\mathbf{a}}$$
$$\Sigma M_G = \bar{I}\alpha = 0 \qquad \textbf{(6/6)}$$

For rectilinear translation, illustrated in Fig. 6/8*a*, if the *x*-axis is chosen in the direction of the acceleration, then the two scalar force equations become $\Sigma F_x = m\bar{a}_x$ and $\Sigma F_y = m\bar{a}_y = 0$. For curvilinear translation, Fig. 6/8*b*, if we use *n-t* coordinates, the two scalar force equations become $\Sigma F_n = m\bar{a}_n$ and $\Sigma F_t = m\bar{a}_t$. In both cases, $\Sigma M_G = 0$.

We may also employ the alternative moment equation, Eq. 6/2, with the aid of the kinetic diagram. For rectilinear translation we see that $\Sigma M_P = m\bar{a}d$ and $\Sigma M_A = 0$. For curvilinear translation the kinetic diagram permits us to write $\Sigma M_A = m\bar{a}_n d_A$ in the clockwise sense and $\Sigma M_B = m\bar{a}_t d_B$ in the counterclockwise sense. Thus, we have complete freedom to choose a convenient moment center.

Stuart Westmorland/The Image Bank/Getty Images, Inc.

The methods of this article apply to these motorcycles if their roll (lean) angles are constant for an interval of time.

Sample Problem 6/1

The pickup truck weighs 3220 lb and reaches a speed of 30 mi/hr from rest in a distance of 200 ft up the 10-percent incline with constant acceleration. Calculate the normal force under each pair of wheels and the friction force under the rear driving wheels. The effective coefficient of friction between the tires and the road is known to be at least 0.80.

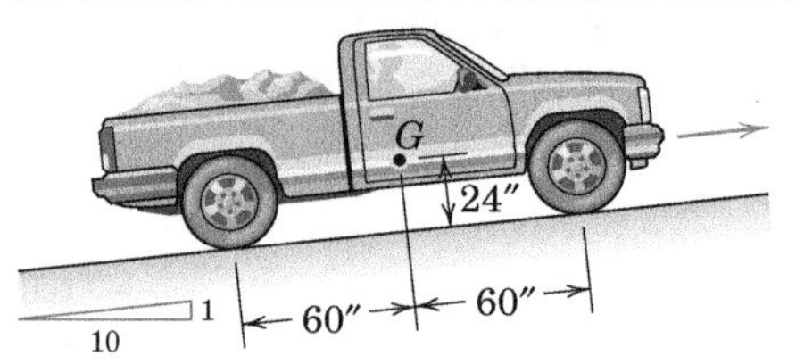

Solution. We will assume that the mass of the wheels is negligible compared
① with the total mass of the truck. The truck may now be simulated by a single rigid body in rectilinear translation with an acceleration of

② $[v^2 = 2as]$ $$\bar{a} = \frac{(44)^2}{2(200)} = 4.84 \text{ ft/sec}^2$$

The free-body diagram of the complete truck shows the normal forces N_1 and N_2, the friction force F in the direction to oppose the slipping of the driving wheels, and the weight W represented by its two components. With $\theta = \tan^{-1} 1/10 = 5.71°$, these components are $W \cos\theta = 3220 \cos 5.71° = 3200$ lb and $W \sin\theta = 3220 \sin 5.71° = 320$ lb. The kinetic diagram shows the resultant, which passes through the mass center and is in the direction of its acceleration. Its magnitude is

$$m\bar{a} = \frac{3220}{32.2}(4.84) = 484 \text{ lb}$$

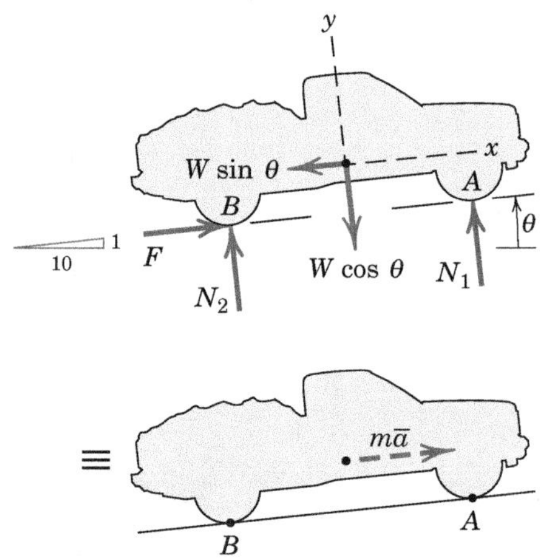

Applying the three equations of motion, Eqs. 6/1, for the three unknowns gives

③ $[\Sigma F_x = m\bar{a}_x]$ $\quad F - 320 = 484 \qquad F = 804$ lb *Ans.*

$[\Sigma F_y = m\bar{a}_y = 0]$ $\quad N_1 + N_2 - 3200 = 0$ *(a)*

$[\Sigma M_G = \bar{I}\alpha = 0]$ $\quad 60N_1 + 804(24) - N_2(60) = 0$ *(b)*

Solving *(a)* and *(b)* simultaneously gives

$$N_1 = 1441 \text{ lb} \qquad N_2 = 1763 \text{ lb}$$ *Ans.*

In order to support a friction force of 804 lb, a coefficient of friction of at least $F/N_2 = 804/1763 = 0.46$ is required. Since our coefficient of friction is at least 0.80, the surfaces are rough enough to support the calculated value of F so that our result is correct.

Alternative Solution. From the kinetic diagram we see that N_1 and N_2 can be obtained independently of one another by writing separate moment equations about A and B.

$[\Sigma M_A = m\bar{a}d]$ $\quad 120N_2 - 60(3200) - 24(320) = 484(24)$

④ $$N_2 = 1763 \text{ lb}$$ *Ans.*

$[\Sigma M_B = m\bar{a}d]$ $\quad 3200(60) - 320(24) - 120N_1 = 484(24)$

$$N_1 = 1441 \text{ lb}$$ *Ans.*

Helpful Hints

① Without this assumption, we would be obliged to account for the relatively small additional forces which produce moments to give the wheels their angular acceleration.

② Recall that 30 mi/hr is 44 ft/sec.

③ We must be careful not to use the friction equation $F = \mu N$ here since we do not have a case of slipping or impending slipping. If the given coefficient of friction were less than 0.46, the friction force would be μN_2, and the car would be unable to attain the acceleration of 4.84 ft/sec^2. In this case, the unknowns would be N_1, N_2, and a.

④ The left-hand side of the equation is evaluated from the free-body diagram, and the right-hand side from the kinetic diagram. The positive sense for the moment sum is arbitrary but must be the same for both sides of the equation. In this problem, we have taken the clockwise sense as positive for the moment of the resultant force about B.

Sample Problem 6/2

The vertical bar AB has a mass of 150 kg with center of mass G midway between the ends. The bar is elevated from rest at $\theta = 0$ by means of the parallel links of negligible mass, with a constant couple $M = 5$ kN·m applied to the lower link at C. Determine the angular acceleration α of the links as a function of θ and find the force B in the link DB at the instant when $\theta = 30°$.

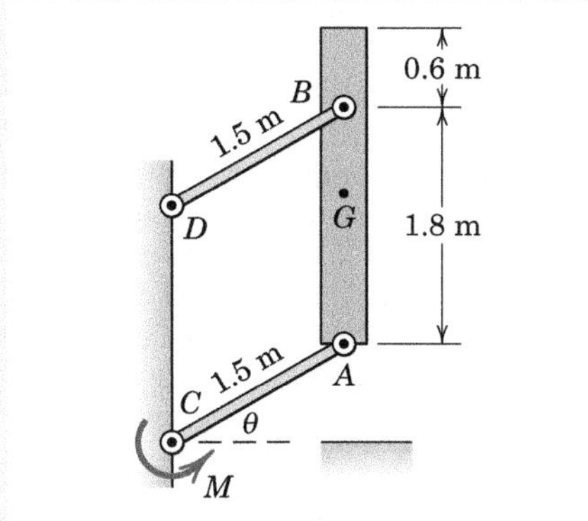

Solution. The motion of the bar is seen to be curvilinear translation since the bar itself does not rotate during the motion. With the circular motion of the mass center G, we choose n- and t-coordinates as the most convenient de-
① scription. With negligible mass of the links, the tangential component A_t of the force at A is obtained from the free-body diagram of AC, where $\Sigma M_C \cong 0$ and
② $A_t = M/\overline{AC} = 5/1.5 = 3.33$ kN. The force at B is along the link. All applied forces are shown on the free-body diagram of the bar, and the kinetic diagram is also indicated, where the $m\bar{\mathbf{a}}$ resultant is shown in terms of its two components.

The sequence of solution is established by noting that A_n and B depend on the n-summation of forces and, hence, on $m\bar{r}\omega^2$ at $\theta = 30°$. The value of ω depends on the variation of $\alpha = \ddot{\theta}$ with θ. This dependency is established from a force summation in the t-direction for a general value of θ, where $\bar{a}_t = (\bar{a}_t)_A = \overline{AC}\alpha$. Thus, we begin with

$$[\Sigma F_t = m\bar{a}_t] \qquad 3.33 - 0.15(9.81)\cos\theta = 0.15(1.5\alpha)$$

$$\alpha = 14.81 - 6.54\cos\theta \text{ rad/s}^2 \qquad \textit{Ans.}$$

With α a known function of θ, the angular velocity ω of the links is obtained from

$$[\omega\, d\omega = \alpha\, d\theta] \qquad \int_0^\omega \omega\, d\omega = \int_0^\theta (14.81 - 6.54\cos\theta)\, d\theta$$

$$\omega^2 = 29.6\theta - 13.08\sin\theta$$

Substitution of $\theta = 30°$ gives

$$(\omega^2)_{30°} = 8.97 \text{ (rad/s)}^2 \qquad \alpha_{30°} = 9.15 \text{ rad/s}^2$$

and

$$m\bar{r}\omega^2 = 0.15(1.5)(8.97) = 2.02 \text{ kN}$$

$$m\bar{r}\alpha = 0.15(1.5)(9.15) = 2.06 \text{ kN}$$

The force B may be obtained by a moment summation about A, which eliminates A_n and A_t and the weight. Or a moment summation may be taken about the intersection of A_n and the line of action of $m\bar{r}\alpha$, which eliminates A_n and $m\bar{r}\alpha$. Using A as a moment center gives

$$[\Sigma M_A = m\bar{a}d] \qquad 1.8\cos 30°\, B = 2.02(1.2)\cos 30° + 2.06(0.6)$$

$$B = 2.14 \text{ kN} \qquad \textit{Ans.}$$

The component A_n could be obtained from a force summation in the n-direction or from a moment summation about G or about the intersection of B and the line of action of $m\bar{r}\alpha$.

Helpful Hints

① Generally speaking, the best choice of reference axes is to make them coincide with the directions in which the components of the mass-center acceleration are expressed. Examine the consequences of choosing horizontal and vertical axes.

② The force and moment equations for a body of negligible mass become the same as the equations of equilibrium. Link BD, therefore, acts as a two-force member in equilibrium.

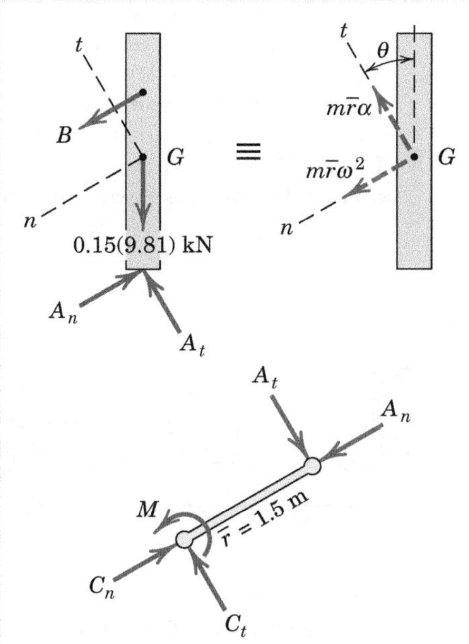

PROBLEMS

Introductory Problems

6/1 The right-angle bar with equal legs weighs 6 lb and is freely hinged to the vertical plate at C. The bar is prevented from rotating by the two pegs A and B fixed to the plate. Determine the acceleration a of the plate for which no force is exerted on the bar by either peg A or B.

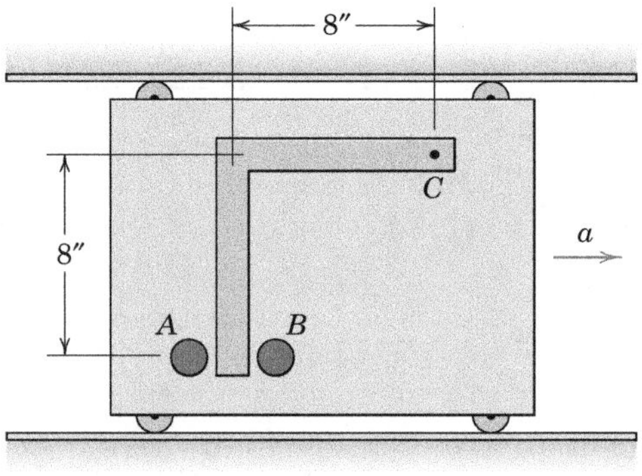

Problem 6/1

6/2 In Prob. 6/1, if the plate is given a horizontal acceleration $a = 2g$, calculate the force exerted on the bar by either peg A or B.

6/3 The driver of a pickup truck accelerates from rest to a speed of 45 mi/hr over a horizontal distance of 225 ft with constant acceleration. The truck is hauling an empty 500-lb trailer with a uniform 60-lb gate hinged at O and held in the slightly tilted position by two pegs, one on each side of the trailer frame at A. Determine the maximum shearing force developed in each of the two pegs during the acceleration.

Problem 6/3

6/4 A passenger car of an overhead monorail system is driven by one of its two small wheels A or B. Select the one for which the car can be given the greater acceleration without slipping the driving wheel and compute the maximum acceleration if the effective coefficient of friction is limited to 0.25 between the wheels and the rail. Neglect the small mass of the wheels.

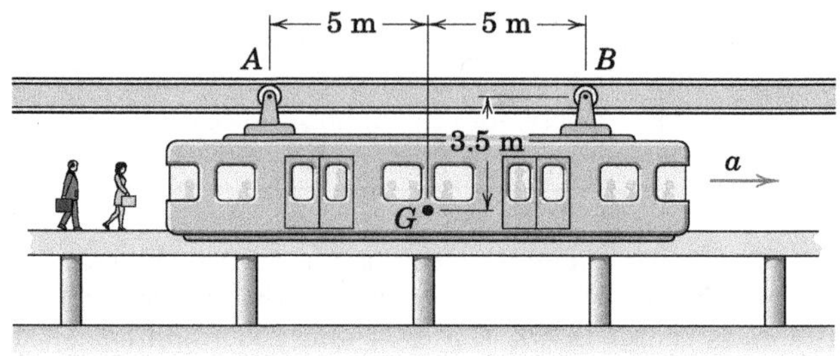

Problem 6/4

6/5 The uniform box of mass m slides down the rough incline. Determine the location d of the effective normal force N. The effective normal force is located at the centroid of the nonuniform pressure distribution which the incline exerts on the bottom surface of the block.

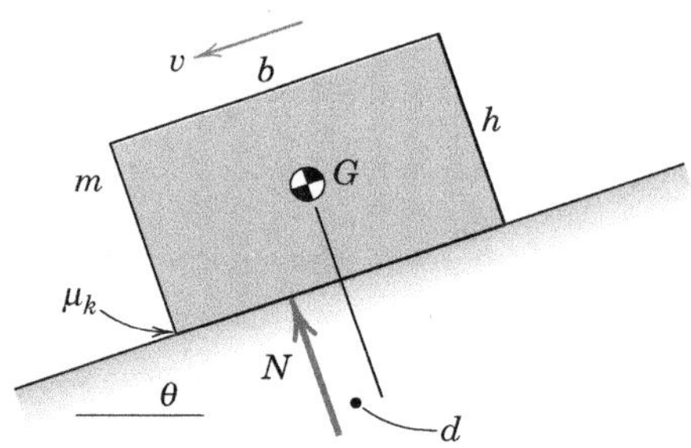

Problem 6/5

6/6 The uniform slender bar of mass m and length L is held in the position shown by the stop at A. What acceleration a will cause the normal force acting on the roller at B to become (a) one-half of the static value, (b) one-fourth of the static value, and (c) zero?

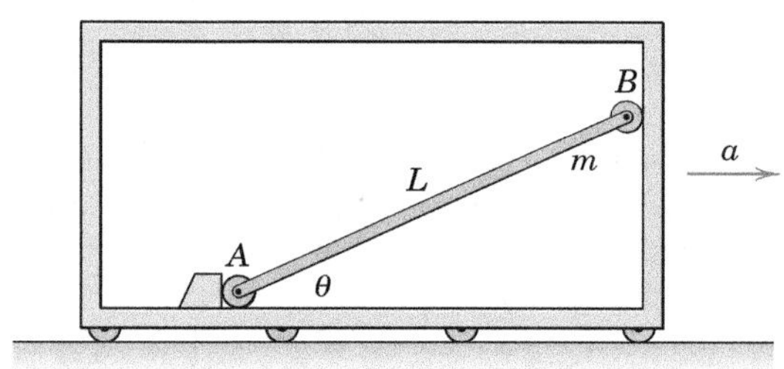

Problem 6/6

6/7 The homogeneous crate of mass m is mounted on small wheels as shown. Determine the maximum force P which can be applied without overturning the crate about (a) its lower front edge with $h = b$ and (b) its lower back edge with $h = 0$.

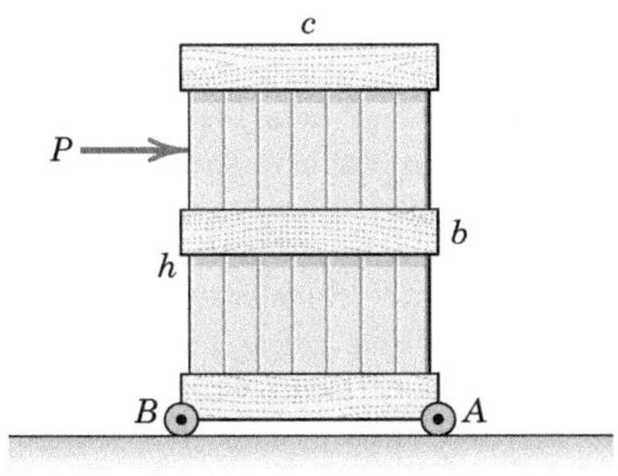

Problem 6/7

6/8 The frame is made from uniform rod which has a mass ρ per unit length. A smooth recessed slot constrains the small rollers at A and B to travel horizontally. Force P is applied to the frame through a cable attached to an adjustable collar C. Determine the magnitudes and directions of the normal forces which act on the rollers if (*a*) $h = 0.3L$, (*b*) $h = 0.5L$, and (*c*) $h = 0.9L$. Evaluate your results for $\rho = 2$ kg/m, $L = 500$ mm, and $P = 60$ N. What is the acceleration of the frame in each case?

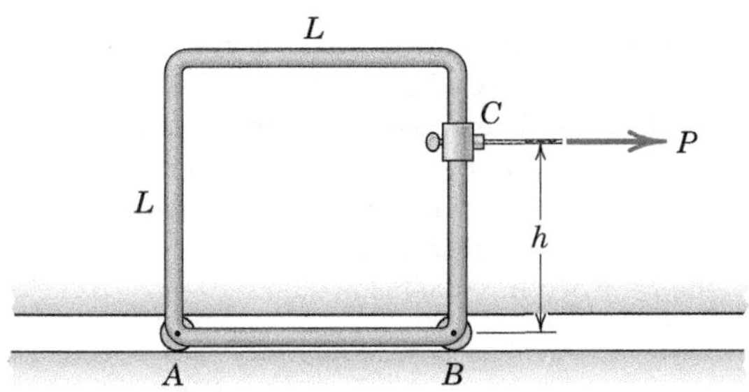

Problem 6/8

6/9 A uniform slender rod rests on a car seat as shown. Determine the deceleration a for which the rod will begin to tip forward. Assume that friction at B is sufficient to prevent slipping.

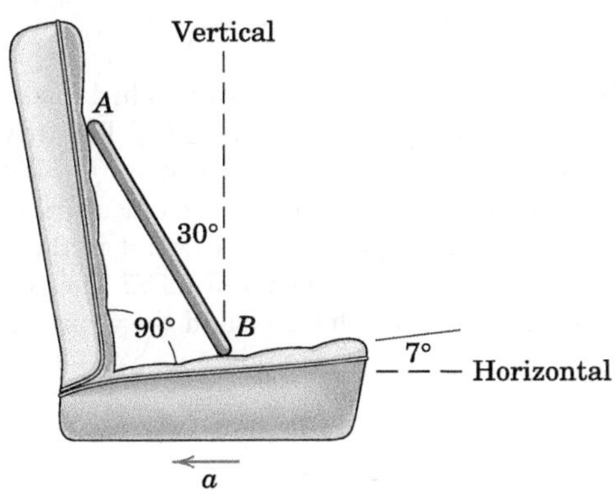

Problem 6/9

6/10 Determine the value of P which will cause the homogeneous cylinder to begin to roll up out of its rectangular recess. The mass of the cylinder is m and that of the cart is M. The cart wheels have negligible mass and friction.

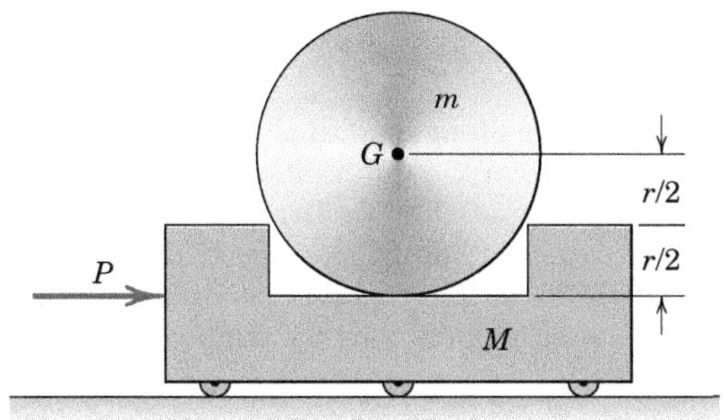

Problem 6/10

6/11 The uniform 5-kg bar AB is suspended in a vertical position from an accelerating vehicle and restrained by the wire BC. If the acceleration is $a = 0.6g$, determine the tension T in the wire and the magnitude of the total force supported by the pin at A.

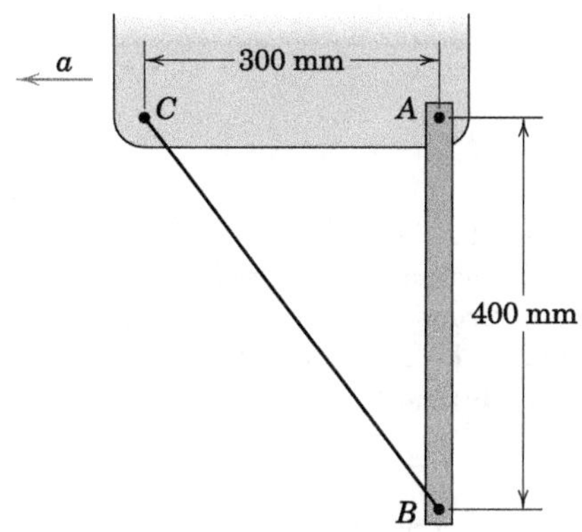

Problem 6/11

6/12 If the collar P is given a constant acceleration $a = 3g$ to the right, the pendulum will assume a steady-state deflection $\theta = 30°$. Determine the stiffness k_T of the torsional spring which will allow this to happen. The torsional spring is undeformed when the pendulum is in the vertical position.

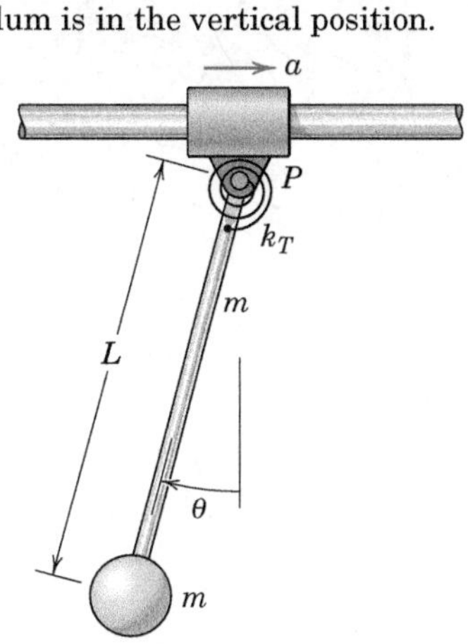

Problem 6/12

*6/13 If the collar P of the pendulum of Prob. 6/12 is given a constant acceleration $a = 5g$, what will be the steady-state deflection of the pendulum from the vertical? Use the value $k_T = 7mgL$.

6/14 The uniform 30-kg bar OB is secured to the accelerating frame in the 30° position from the horizontal by the hinge at O and roller at A. If the horizontal acceleration of the frame is $a = 20$ m/s^2, compute the force F_A on the roller and the x- and y-components of the force supported by the pin at O.

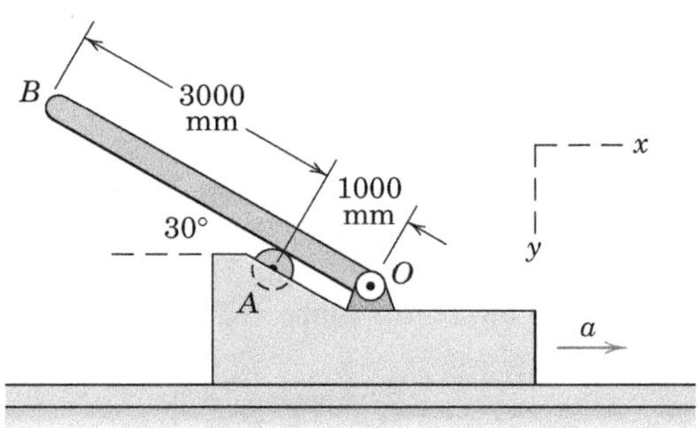

Problem 6/14

6/15 The bicyclist applies the brakes as he descends the 10° incline. What deceleration a would cause the dangerous condition of tipping about the front wheel A? The combined center of mass of the rider and bicycle is at G.

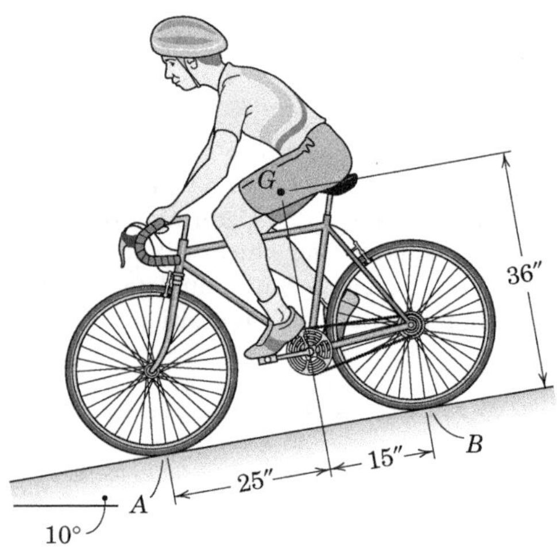

Problem 6/15

Representative Problems

6/16 The right-angle bar acts to control the maximum acceleration of an experimental vehicle by depressing the spring-loaded limit switch A with a vertical force. If a 60-percent reduction in the force applied to the button at A (relative to the static value) results in the switch being tripped, determine the maximum acceleration a which the vehicle is permitted to experience.

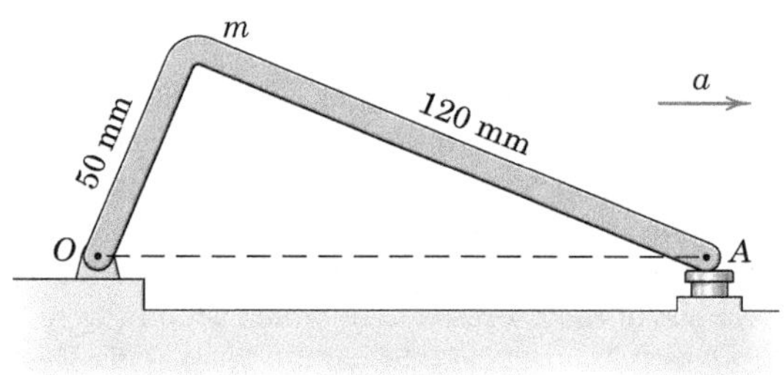

Problem 6/16

6/17 The 1650-kg car has its mass center at G. Calculate the normal forces N_A and N_B between the road and the front and rear pairs of wheels under conditions of maximum acceleration. The mass of the wheels is small compared with the total mass of the car. The coefficient of static friction between the road and the rear driving wheels is 0.80.

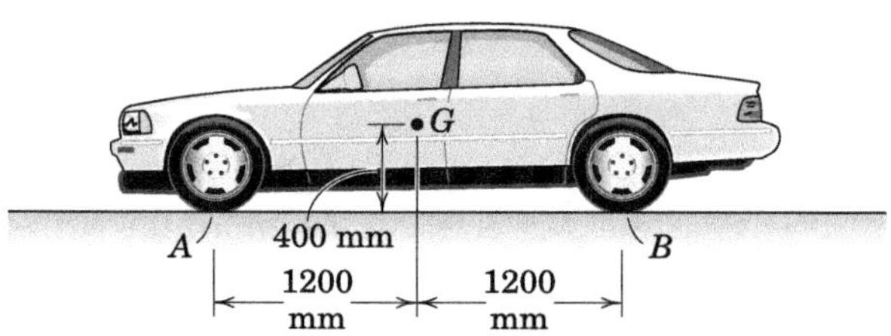

Problem 6/17

6/18 The four-wheel-drive all-terrain vehicle has a mass of 300 kg with center of mass G_2. The driver has a mass of 85 kg with center of mass G_1. If all four wheels are observed to spin momentarily as the driver attempts to go forward, what is the forward acceleration of the driver and ATV? The coefficient of friction between the tires and the ground is 0.40.

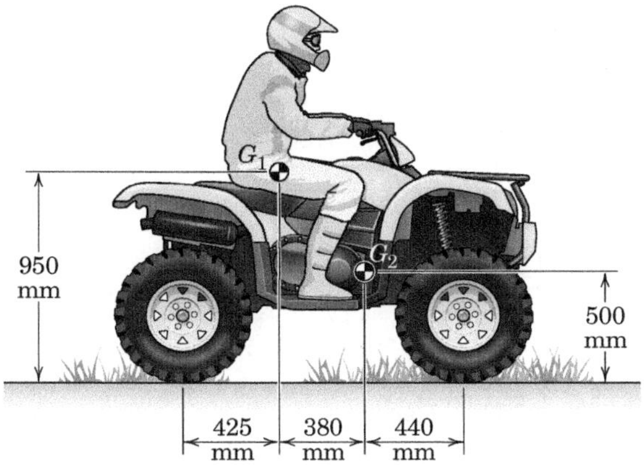

Problem 6/18

6/19 A cleated conveyor belt transports solid homogeneous cylinders up a 15° incline. The diameter of each cylinder is half its height. Determine the maximum acceleration which the belt may have without tipping the cylinders as it starts.

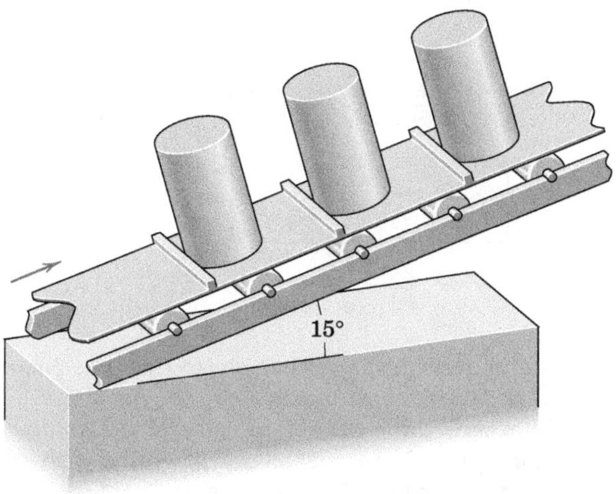

Problem 6/19

6/20 The thin hoop of negligible mass and radius r contains a homogeneous semicylinder of mass m which is rigidly attached to the hoop and positioned such that its diametral face is vertical. The assembly is centered on the top of a cart of mass M which rolls freely on the horizontal surface. If the system is released from rest, what x-directed force P must be applied to the cart to keep the hoop and semicylinder stationary with respect to the cart, and what is the resulting acceleration a of the cart? Motion takes place in the x-y plane. Neglect the mass of the cart wheels and any friction in the wheel bearings. What is the requirement on the coefficient of static friction between the hoop and cart?

Problem 6/20

6/21 Determine the magnitude P and direction θ of the force required to impart a rearward acceleration $a = 5$ ft/sec^2 to the loaded wheelbarrow with no rotation from the position shown. The combined weight of the wheelbarrow and its load is 500 lb with center of gravity at G. Compare the normal force at B under acceleration with that for static equilibrium in the position shown. Neglect the friction and mass of the wheel.

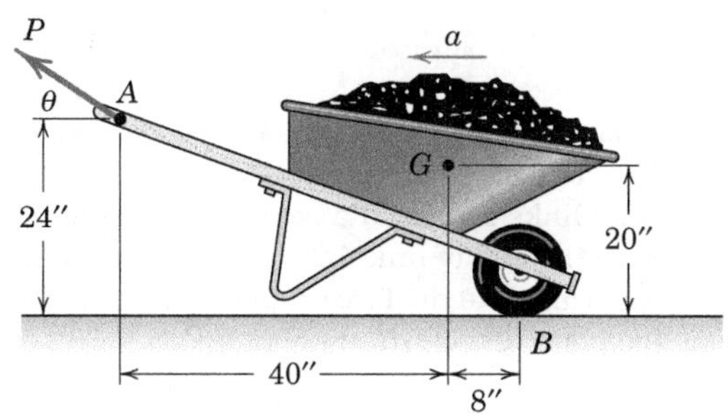

Problem 6/21

6/22 The mine skip has a loaded mass of 2000 kg and is attached to the towing vehicle by the light hinged link CD. If the towing vehicle has an acceleration of 3 m/s^2, calculate the corresponding reactions under the small wheels at A and B.

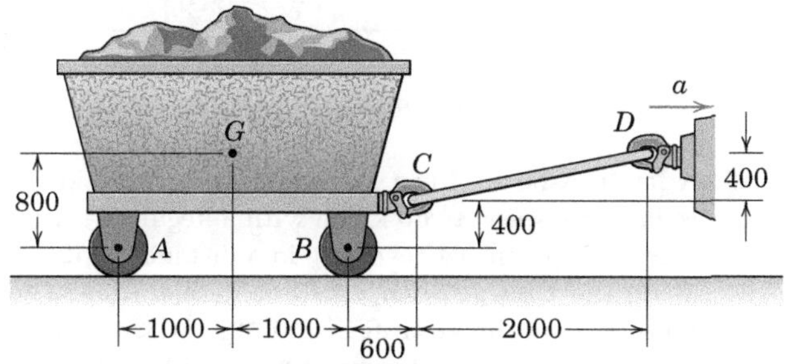

Problem 6/22

6/23 The block A and attached rod have a combined mass of 60 kg and are confined to move along the 60° guide under the action of the 800-N applied force. The uniform horizontal rod has a mass of 20 kg and is welded to the block at B. Friction in the guide is negligible. Compute the bending moment M exerted by the weld on the rod at B.

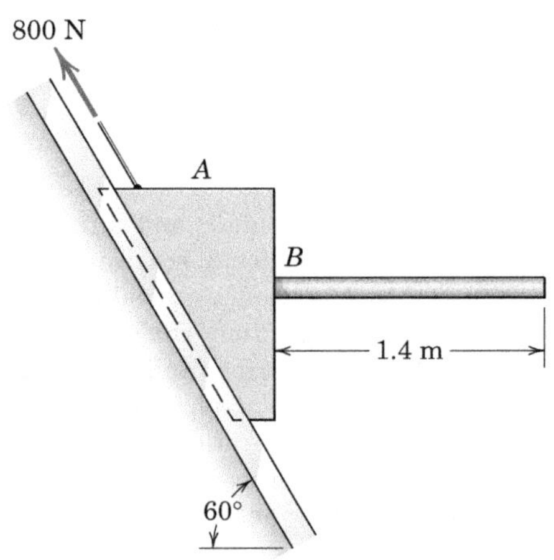

Problem 6/23

6/24 The homogeneous rectangular plate weighs 40 lb and is supported in the vertical plane by the light parallel links shown. If a couple $M = 80$ lb-ft is applied to the end of link AB with the system initially at rest, calculate the force supported by the pin at C as the plate lifts off its support with $\theta = 30°$.

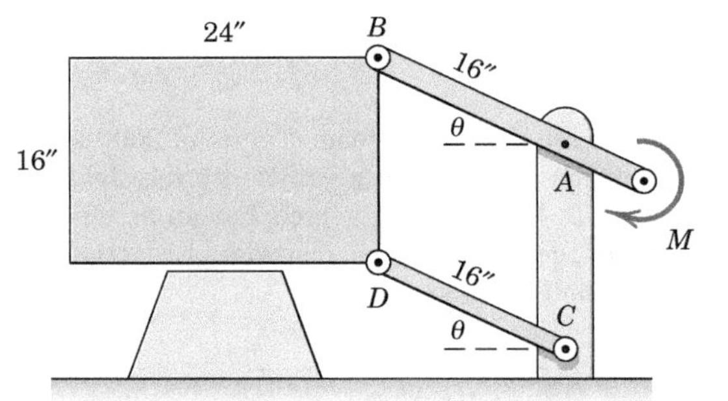

Problem 6/24

6/25 A jet transport with a landing speed of 200 km/h reduces its speed to 60 km/h with a negative thrust R from its jet thrust reversers in a distance of 425 m along the runway with constant deceleration. The total mass of the aircraft is 140 Mg with mass center at G. Compute the reaction N under the nose wheel B toward the end of the braking interval and prior to the application of mechanical braking. At the lower speed, aerodynamic forces on the aircraft are small and may be neglected.

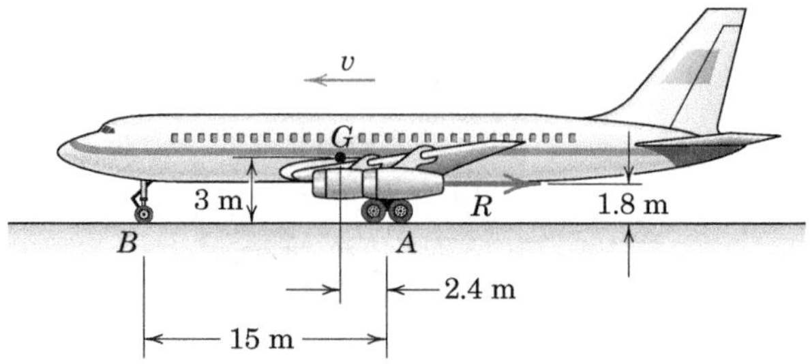

Problem 6/25

6/26 The 1300-lb homogeneous plate is suspended from the overhead carriage by the two parallel steel cables. What acceleration a will cause the tensions in the two cables to be equal? What is the resulting steady-state deflection θ of the cables from the vertical? Evaluate your results for the case where $b = 2h$.

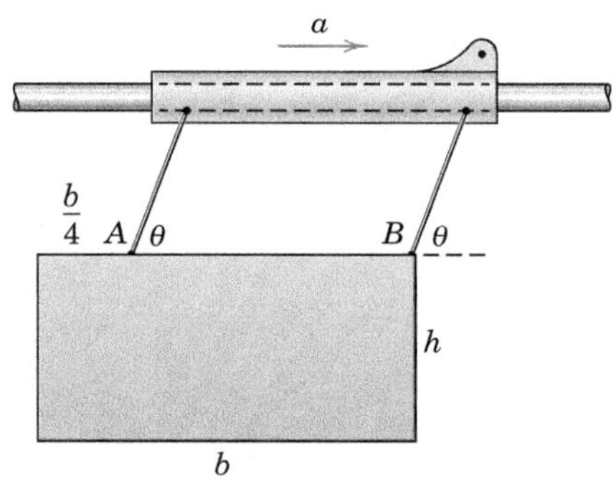

Problem 6/26

6/27 The uniform L-shaped bar pivots freely at point P of the slider, which moves along the horizontal rod. Determine the steady-state value of the angle θ if (*a*) $a = 0$ and (*b*) $a = g/2$. For what value of a would the steady-state value of θ be zero?

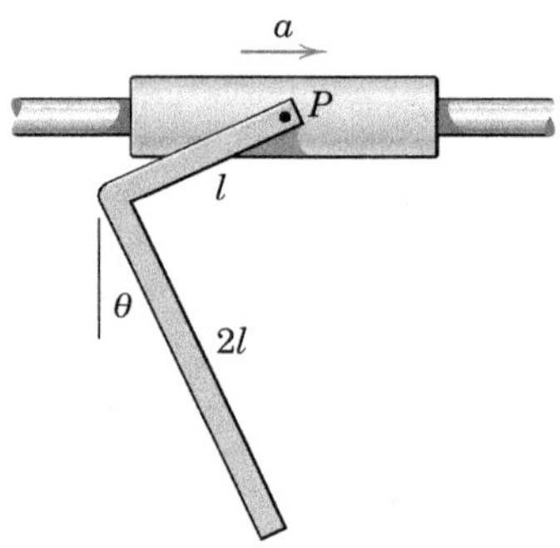

Problem 6/27

▶6/28 The 30,000-lb concrete pipe section is being transported on a flatbed truck. Five inextensible cables are passed across the top of the pipe and tightened securely to the flatbed with an initial tension of 2000 lb. What is the maximum deceleration a which the truck can experience if the pipe is to remain stationary relative to the truck? The coefficient of static friction between the concrete and the flatbed is 0.80, and that between the cables and the concrete is 0.75.

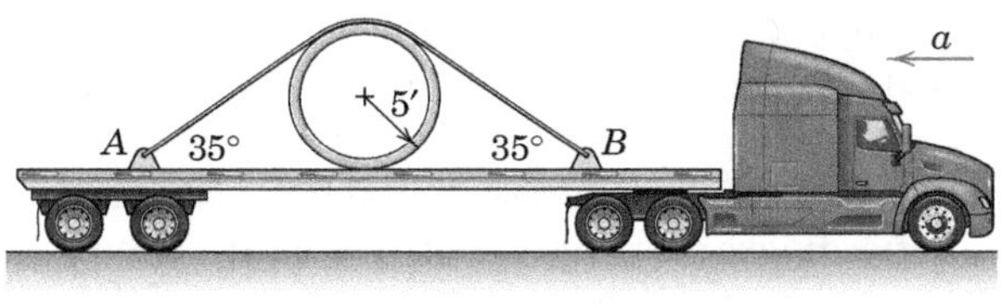

Problem 6/28

6/29 Determine the maximum counterweight W for which the loaded 4000-lb coal car will not overturn about the rear wheels B. Neglect the mass of all pulleys and wheels. (Note that the tension in the cable at C is not $2W$.)

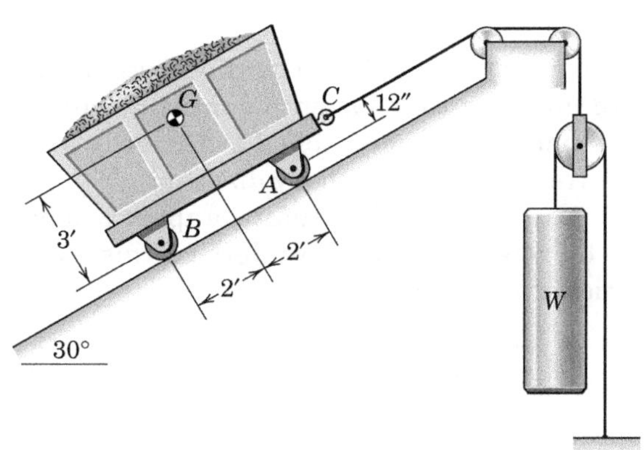

Problem 6/29

6/30 The 1800-kg rear-wheel-drive car accelerates forward at a rate of $g/2$. If the modulus of each of the rear and front springs is 35 kN/m, estimate the resulting momentary nose-up pitch angle θ. (This upward pitch angle during acceleration is called *squat*, while the downward pitch angle during braking is called *dive*!) Neglect the unsprung mass of the wheels and tires. (*Hint:* Begin by assuming a rigid vehicle.)

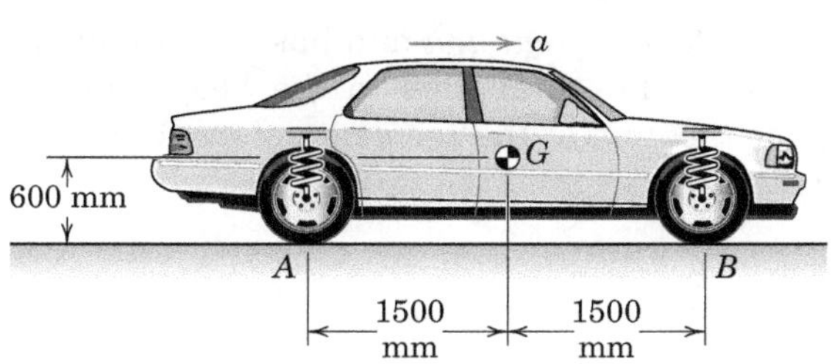

Problem 6/30

6/31 The experimental Formula One race car is traveling at 300 km/h when the driver begins braking to investigate the behavior of the extreme-grip tires. An accelerometer in the car records a maximum deceleration of $4g$ when both the front and rear tires are on the verge of slipping. The car and driver have a combined mass of 690 kg with mass center G. The horizontal drag acting on the car at this speed is 4 kN and may be assumed to pass through the mass center G. The downforce acting over the body of the car at this speed is 13 kN. For simplicity, assume that 35% of this force acts directly over the front wheels, 40% acts directly over the rear wheels, and the remaining portion acts at the mass center. What is the necessary coefficient of friction μ between the tires and the road for this condition? Compare your results with those for passenger car tires.

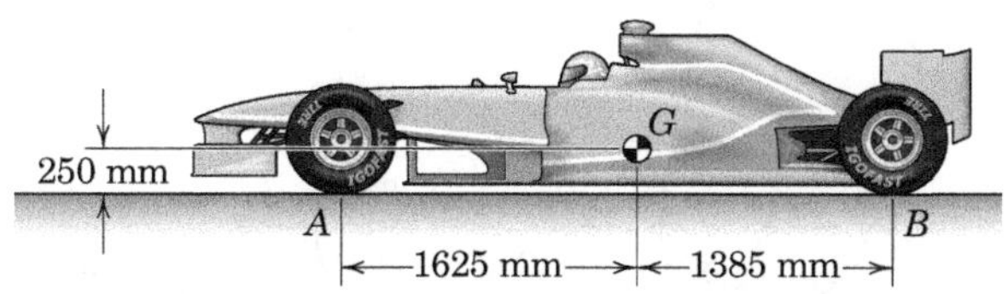

Problem 6/31

▶6/32 The uniform 225-lb crate is supported by the thin homogeneous 40-lb platform BF and light support links whose motion is controlled by the hydraulic cylinder CD. If the cylinder is extending at a constant rate of 6 in./sec when $\theta = 75°$, determine the magnitudes of the forces supported by the pins at B and F. Additionally, determine the total friction force acting on the crate. The crate is centered on the 6-ft platform, and friction is sufficient to keep the crate motionless relative to the platform. (*Hint:* Be careful with the location of the resultant normal force beneath the crate.)

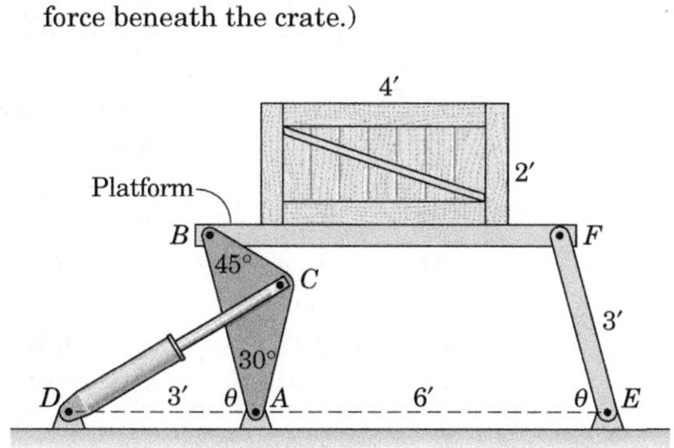

Problem 6/32

6/4 Fixed-Axis Rotation

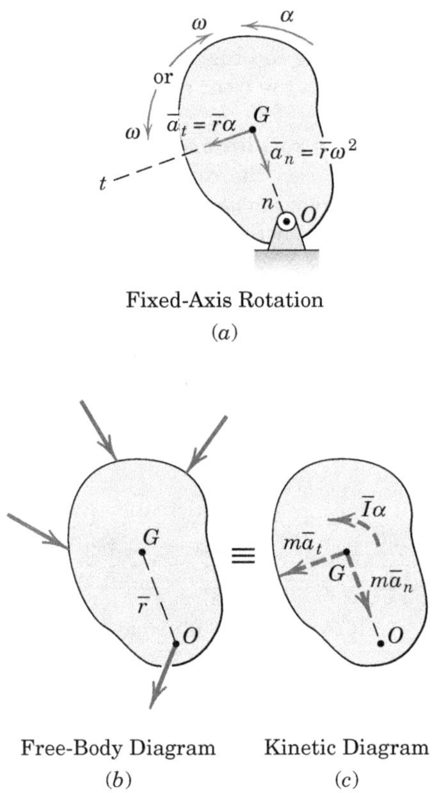

Fixed-Axis Rotation
(*a*)

Free-Body Diagram
(*b*)

Kinetic Diagram
(*c*)

Figure 6/9

Rotation of a rigid body about a fixed axis O was described in Art. 5/2 and illustrated in Fig. 5/1*c*. For this motion, we saw that all points in the body describe circles about the rotation axis, and all lines of the body in the plane of motion have the same angular velocity ω and angular acceleration α.

The acceleration components of the mass center for circular motion are most easily expressed in *n*-*t* coordinates, so we have $\bar{a}_n = \bar{r}\omega^2$ and $\bar{a}_t = \bar{r}\alpha$, as shown in Fig. 6/9*a* for rotation of the rigid body about the fixed axis through O. Part *b* of the figure represents the free-body diagram, and the equivalent kinetic diagram in part *c* of the figure shows the force resultant $m\bar{\mathbf{a}}$ in terms of its *n*- and *t*-components and the resultant couple $\bar{I}\alpha$.

Our general equations for plane motion, Eqs. 6/1, are directly applicable and are repeated here.

$$\Sigma\mathbf{F} = m\bar{\mathbf{a}} \qquad \Sigma M_G = \bar{I}\alpha \tag{6/1}$$

Thus, the two scalar components of the force equation become $\Sigma F_n = m\bar{r}\omega^2$ and $\Sigma F_t = m\bar{r}\alpha$. In applying the moment equation about G, we must account for the moment of the force applied to the body at O, so this force must not be omitted from the free-body diagram.

For fixed-axis rotation, it is generally useful to apply a moment equation directly about the rotation axis O. We derived this equation previously as Eq. 6/4, which is repeated here.

$$\Sigma M_O = I_O\alpha \tag{6/4}$$

From the kinetic diagram in Fig. 6/9*c*, we may obtain Eq. 6/4 very easily by evaluating the moment of the resultants about O, which becomes $\Sigma M_O = \bar{I}\alpha + m\bar{a}_t\bar{r}$. Application of the parallel-axis theorem for mass moments of inertia, $I_O = \bar{I} + m\bar{r}^2$, gives $\Sigma M_O = (I_O - m\bar{r}^2)\alpha + m\bar{r}^2\alpha = I_O\alpha$.

For the common case of rotation of a rigid body about a fixed axis through its mass center G, clearly, $\bar{\mathbf{a}} = \mathbf{0}$, and therefore $\Sigma\mathbf{F} = \mathbf{0}$. The resultant of the applied forces then is the couple $\bar{I}\alpha$.

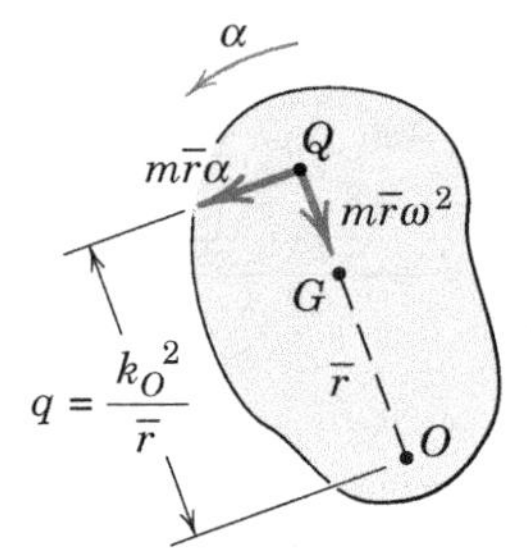

Figure 6/10

We may combine the resultant-force component $m\bar{a}_t$ and resultant couple $\bar{I}\alpha$ by moving $m\bar{a}_t$ to a parallel position through point Q on line OG, Fig. 6/10, located by $m\bar{r}\alpha q = \bar{I}\alpha + m\bar{r}\alpha(\bar{r})$. Using the parallel-axis theorem and $I_O = k_O{}^2 m$ gives $q = k_O{}^2/\bar{r}$.

Point Q is called the *center of percussion* and has the unique property that the resultant of all forces applied to the body must pass through it. It follows that the sum of the moments of all forces about the center of percussion is always zero, $\Sigma M_Q = 0$.

Sample Problem 6/3

The concrete block weighing 644 lb is elevated by the hoisting mechanism shown, where the cables are securely wrapped around the respective drums. The drums, which are fastened together and turn as a single unit about their mass center at O, have a combined weight of 322 lb and a radius of gyration about O of 18 in. If a constant tension $P = 400$ lb is maintained by the power unit at A, determine the vertical acceleration of the block and the resultant force on the bearing at O.

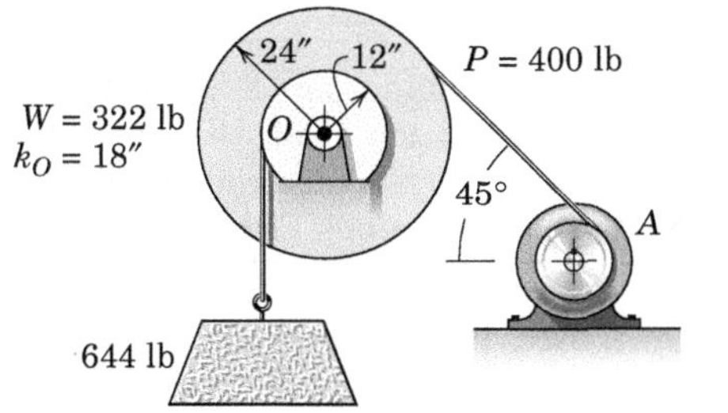

Solution I. The free-body and kinetic diagrams of the drums and concrete block are drawn showing all forces which act, including the components
① O_x and O_y of the bearing reaction. The resultant of the force system on the drums for centroidal rotation is the couple $\bar{I}\alpha = I_O\alpha$, where

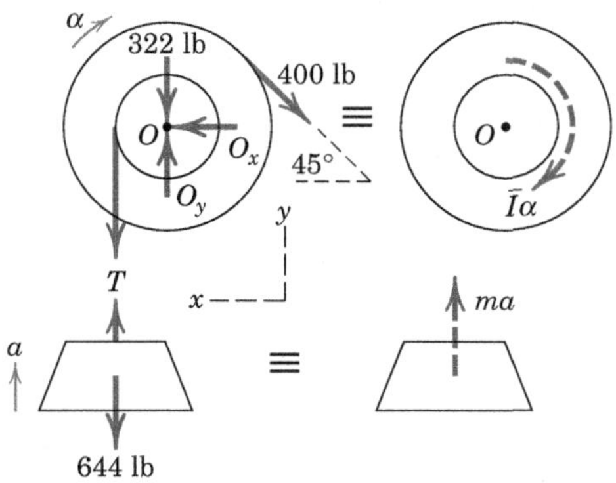

② $[I = k^2m]$ $\qquad \bar{I} = I_O = \left(\frac{18}{12}\right)^2 \frac{322}{32.2} = 22.5 \text{ lb-ft-sec}^2$

Taking moments about the mass center O for the pulley in the sense of the angular acceleration α gives

$$[\Sigma M_G = \bar{I}\alpha] \qquad 400\left(\frac{24}{12}\right) - T\left(\frac{12}{12}\right) = 22.5\alpha \qquad (a)$$

The acceleration of the block is described by

$$[\Sigma F_y = ma_y] \qquad T - 644 = \frac{644}{32.2}a \qquad (b)$$

From $a_t = r\alpha$, we have $a = (12/12)\alpha$. With this substitution, Eqs. (a) and (b) are combined to give

$$T = 717 \text{ lb} \qquad \alpha = 3.67 \text{ rad/sec}^2 \qquad a = 3.67 \text{ ft/sec}^2 \qquad \textit{Ans.}$$

The bearing reaction is computed from its components. Since $\bar{a} = 0$, we use the equilibrium equations

$$[\Sigma F_x = 0] \qquad O_x - 400 \cos 45° = 0 \qquad O_x = 283 \text{ lb}$$

$$[\Sigma F_y = 0] \qquad O_y - 322 - 717 - 400 \sin 45° = 0 \qquad O_y = 1322 \text{ lb}$$

$$O = \sqrt{(283)^2 + (1322)^2} = 1352 \text{ lb} \qquad \textit{Ans.}$$

Solution II. We may use a more condensed approach by drawing the free-body diagram of the entire system, thus eliminating reference to T, which becomes internal to the new system. From the kinetic diagram for this system, we see that the moment sum about O must equal the resultant couple $\bar{I}\alpha$ for the drums, plus the moment of the resultant ma for the block. Thus, from the principle of Eq. 6/5 we have

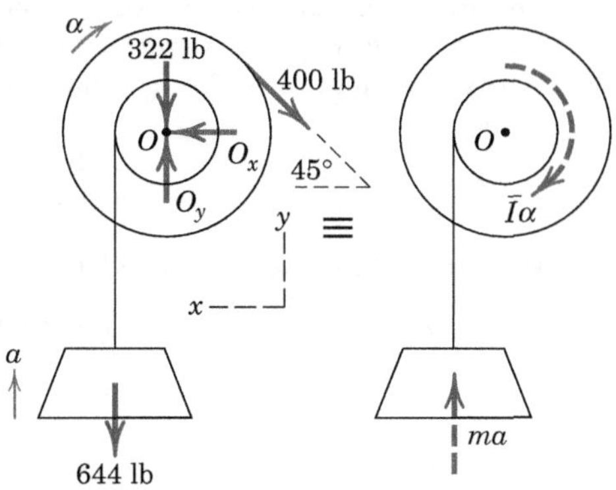

$$[\Sigma M_O = \Sigma\bar{I}\alpha + \Sigma m\bar{a}d] \qquad 400\left(\frac{24}{12}\right) - 644\left(\frac{12}{12}\right) = 22.5\alpha + \frac{644}{32.2}\alpha\left(\frac{12}{12}\right)$$

With $a = (12/12)\alpha$, the solution gives, as before, $a = 3.67$ ft/sec^2.

We may equate the force sums on the entire system to the sums of the resultants. Thus,

$$[\Sigma F_y = \Sigma m\bar{a}_y] \qquad O_y - 322 - 644 - 400 \sin 45° = \frac{322}{32.2}(0) + \frac{644}{32.2}(3.67)$$

$$O_y = 1322 \text{ lb}$$

$$[\Sigma F_x = \Sigma m\bar{a}_x] \qquad O_x - 400 \cos 45° = 0 \qquad O_x = 283 \text{ lb}$$

Helpful Hints

① Be alert to the fact that the tension T is not 644 lb. If it were, the block would not accelerate.

② Do not overlook the need to express k_O in feet when using g in ft/sec^2.

Sample Problem 6/4

The pendulum has a mass of 7.5 kg with center of mass at G and has a radius of gyration about the pivot O of 295 mm. If the pendulum is released from rest at $\theta = 0$, determine the total force supported by the bearing at the instant when $\theta = 60°$. Friction in the bearing is negligible.

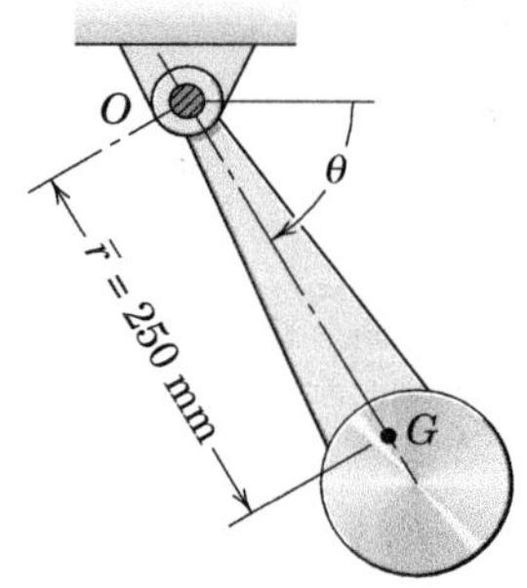

Solution. The free-body diagram of the pendulum in a general position is shown along with the corresponding kinetic diagram, where the components of
① the resultant force have been drawn through G.

The normal component O_n is found from a force equation in the n-direction, which involves the normal acceleration $\bar{r}\omega^2$. Since the angular velocity ω of the pendulum is found from the integral of the angular acceleration and since O_t depends on the tangential acceleration $\bar{r}\alpha$, it follows that α should be obtained first. To this end with $I_O = k_O{}^2 m$, the moment equation about O gives

② $[\Sigma M_O = I_O\alpha]$ $\quad 7.5(9.81)(0.25)\cos\theta = (0.295)^2(7.5)\alpha$

$$\alpha = 28.2 \cos\theta \text{ rad/s}^2$$

and for $\theta = 60°$

$[\omega\, d\omega = \alpha\, d\theta]$ $\quad \int_0^{\omega} \omega\, d\omega = \int_0^{\pi/3} 28.2 \cos\theta\, d\theta$

$$\omega^2 = 48.8 \text{ (rad/s)}^2$$

The remaining two equations of motion applied to the 60° position yield

$[\Sigma F_n = m\bar{r}\omega^2]$ $\quad O_n - 7.5(9.81)\sin 60° = 7.5(0.25)(48.8)$

③ $$O_n = 155.2 \text{ N}$$

$[\Sigma F_t = m\bar{r}\alpha]$ $\quad -O_t + 7.5(9.81)\cos 60° = 7.5(0.25)(28.2)\cos 60°$

$$O_t = 10.37 \text{ N}$$

$$O = \sqrt{(155.2)^2 + (10.37)^2} = 155.6 \text{ N}$$ *Ans.*

The proper sense for O_t may be observed at the outset by applying the moment equation $\Sigma M_G = \bar{I}\alpha$, where the moment about G due to O_t must be clockwise to agree with α. The force O_t may also be obtained initially by a moment equation about the center of percussion Q, shown in the lower figure, which avoids the necessity of computing α. First, we must obtain the distance q, which is

$[q = k_O{}^2/\bar{r}]$ $\quad q = \dfrac{(0.295)^2}{0.250} = 0.348 \text{ m}$

$[\Sigma M_Q = 0]$ $\quad O_t(0.348) - 7.5(9.81)(\cos 60°)(0.348 - 0.250) = 0$

$$O_t = 10.37 \text{ N}$$ *Ans.*

Helpful Hints

① The acceleration components of G are, of course, $\bar{a}_n = \bar{r}\omega^2$ and $\bar{a}_t = \bar{r}\alpha$.

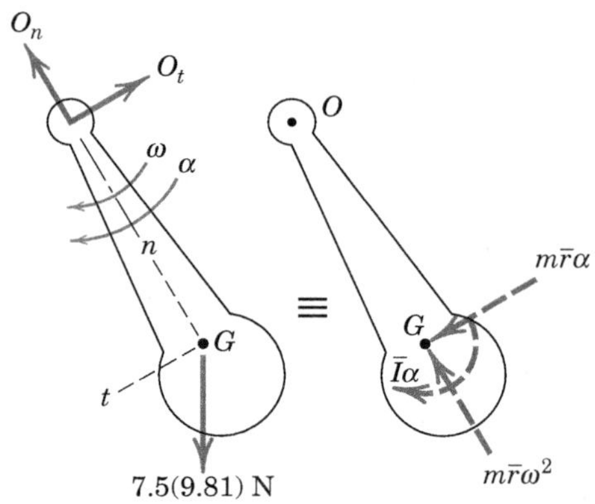

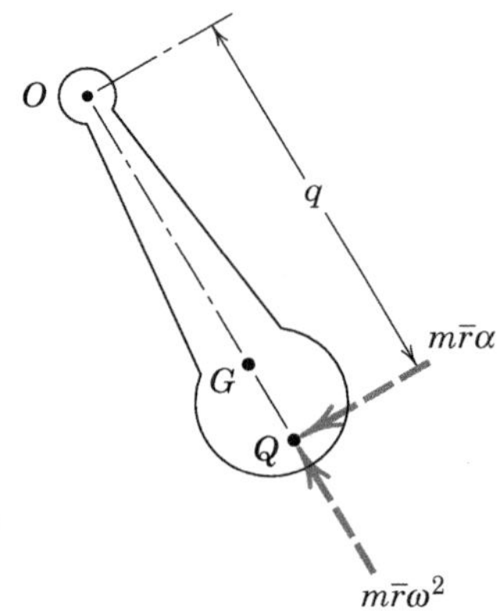

② Review the theory again and satisfy yourself that $\Sigma M_O = I_O\alpha = \bar{I}\alpha + m\bar{r}^2\alpha = m\bar{r}\alpha q$.

③ Note especially here that the force summations are taken in the positive direction of the acceleration components of the mass center G.

PROBLEMS

Introductory Problems

6/33 Two pulleys are fastened together to form an integral unit. At a certain instant, the indicated belt tensions act on the unit and the unit is turning counterclockwise. Determine the angular acceleration of the unit for this instant if the moment due to friction in the bearing at O is 2.5 N·m.

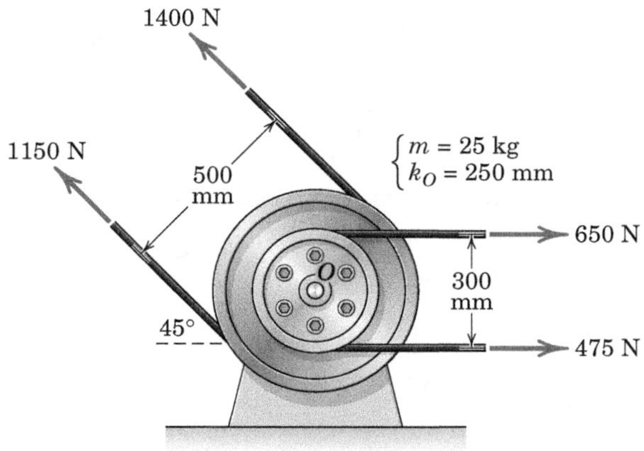

Problem 6/33

6/34 The uniform 20-kg slender bar is pivoted at O and swings freely in the vertical plane. If the bar is released from rest in the horizontal position, calculate the initial value of the force R exerted by the bearing on the bar an instant after release.

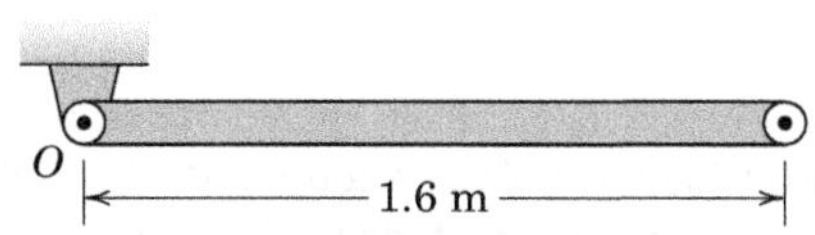

Problem 6/34

6/35 The figure shows an overhead view of a hydraulically-operated gate. As fluid enters the piston side of the cylinder near A, the rod at B extends causing the gate to rotate about a vertical axis through O. For a 2-in.-diameter piston, what fluid pressure p will give the gate an initial counterclockwise angular acceleration of 4 rad/sec? The radius of gyration about O for the 500-lb gate is k_O = 38 in.

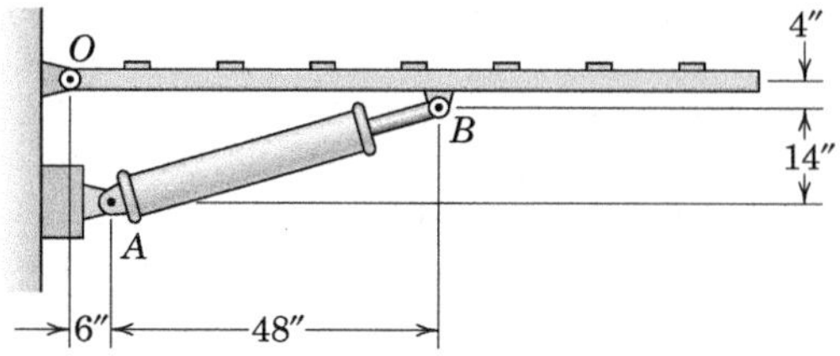

Problem 6/35

6/36 The uniform 100-kg beam is freely hinged about its upper end A and is initially at rest in the vertical position with $\theta = 0$. Determine the initial angular acceleration α of the beam and the magnitude F_A of the force supported by the pin at A due to the application of the force P = 300 N on the attached cable.

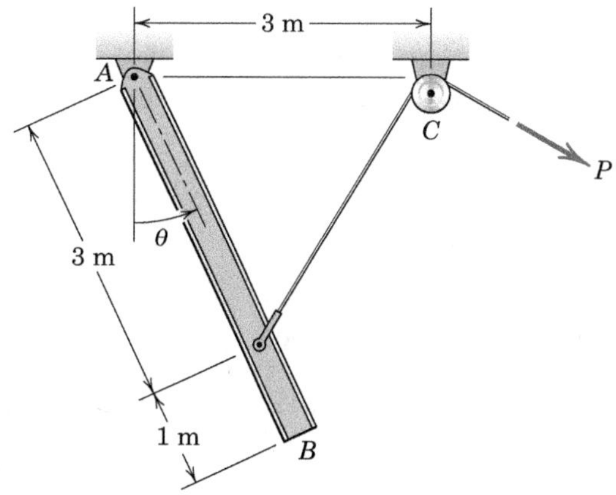

Problem 6/36

6/37 The motor M is used to hoist the 12,000-lb stadium panel (centroidal radius of gyration $\bar{k}$ = 6.5 ft) into position by pivoting the panel about its corner A. If the motor is capable of producing 5000 lb-ft of torque, what pulley diameter d will give the panel an initial counterclockwise angular acceleration of 1.5 deg/sec^2? Neglect all friction.

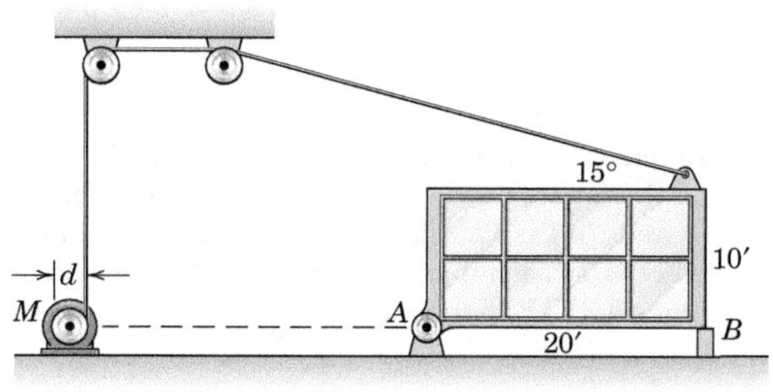

Problem 6/37

6/38 A momentum wheel for dynamics-class demonstrations is shown. It is basically a bicycle wheel modified with rim band-weighting, handles, and a pulley for cord startup. The heavy rim band causes the radius of gyration of the 7-lb wheel to be 11 in. If a steady 10-lb pull T is applied to the cord, determine the angular acceleration of the wheel. Neglect bearing friction.

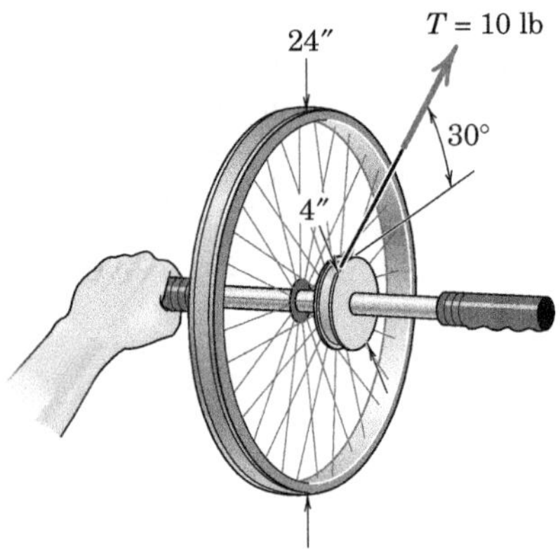

Problem 6/38

6/39 Each of the two drums and connected hubs of 8-in. radius weighs 200 lb and has a radius of gyration about its center of 15 in. Calculate the angular acceleration of each drum. Friction in each bearing is negligible.

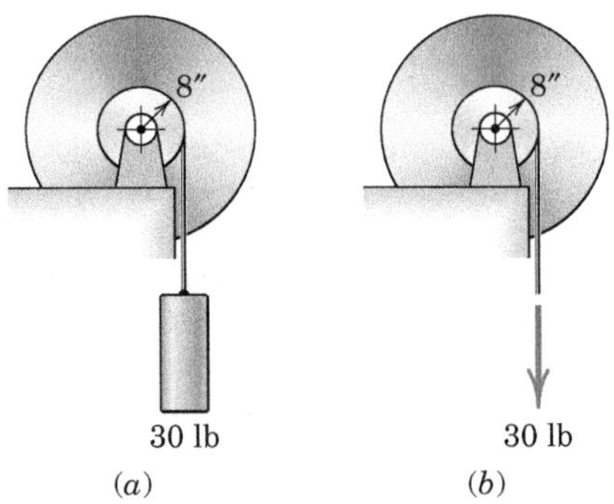

Problem 6/39

6/40 Determine the angular acceleration and the force on the bearing at O for (a) the narrow ring of mass m and (b) the flat circular disk of mass m immediately after each is released from rest in the vertical plane with OC horizontal.

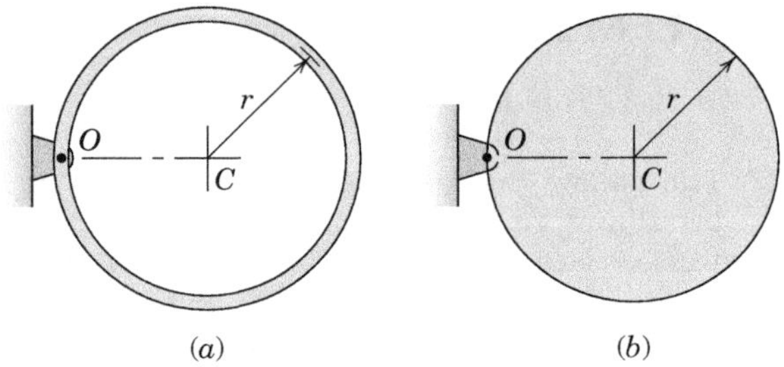

Problem 6/40

6/41 The uniform 5-kg portion of a circular hoop is released from rest while in the position shown where the torsional spring of stiffness $k_T = 15$ N·m/rad has been twisted 90° clockwise from its undeformed position. Determine the magnitude of the pin force at O at the instant of release. Motion takes place in a vertical plane and the hoop radius is $r = 150$ mm.

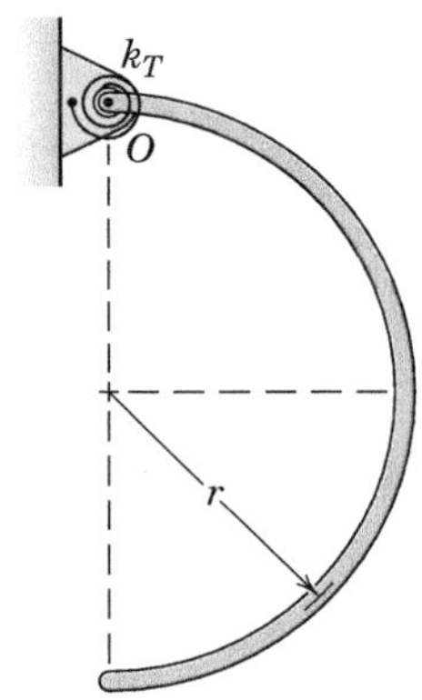

Problem 6/41

6/42 The 30-in. slender bar weighs 20 lb and is mounted on a vertical shaft at O. If a torque $M = 100$ lb-in. is applied to the bar through its shaft, calculate the horizontal force R on the bearing as the bar starts to rotate.

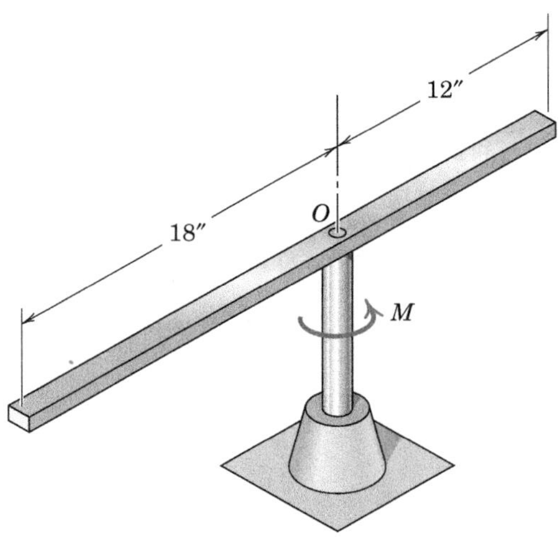

Problem 6/42

6/43 The half ring of mass m and radius r is welded to a small horizontal shaft mounted in a bearing as shown. Neglect the mass of the shaft and determine the angular acceleration of the ring when a torque M is applied to the shaft.

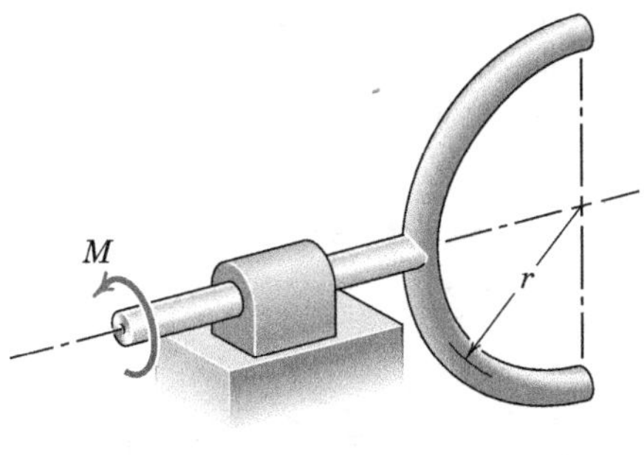

Problem 6/43

6/44 The uniform plate of mass m is released from rest while in the position shown. Determine the initial angular acceleration α of the plate and the magnitude of the force supported by the pin at O. The axis of rotation is horizontal.

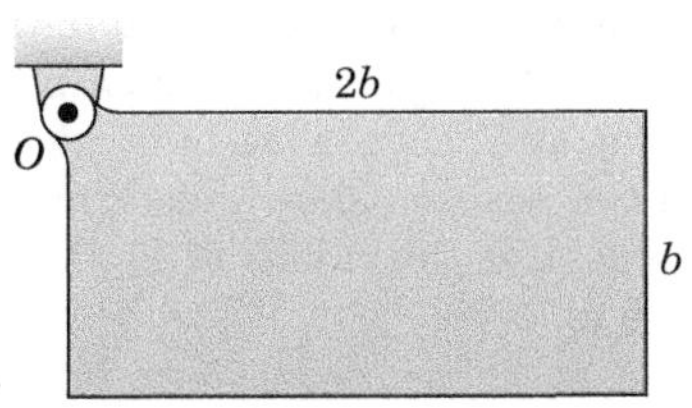

Problem 6/44

6/45 The uniform slender bar AB has a mass of 8 kg and swings in a vertical plane about the pivot at A. If $\dot{\theta} = 2$ rad/s when $\theta = 30°$, compute the force supported by the pin at A at that instant.

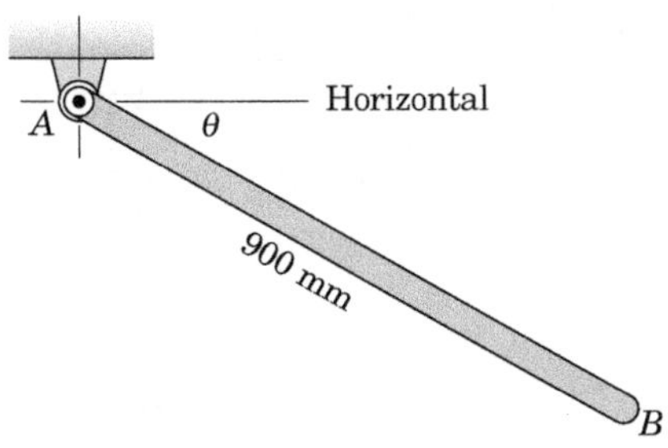

Problem 6/45

6/46 The uniform 16.1-lb slender bar is hinged about a horizontal axis through O and released from rest in the horizontal position. Determine the distance b from the mass center to O which will result in an initial angular acceleration of 16.1 rad/sec^2, and find the force R on the bar at O just after release.

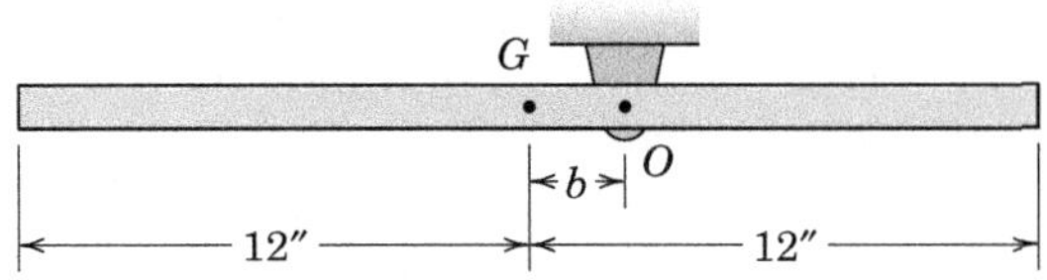

Problem 6/46

6/47 The uniform quarter-circular sector of mass m is released from rest with one straight edge vertical as shown. Determine the initial angular acceleration and the horizontal and vertical components of the reaction at the ideal pivot at O.

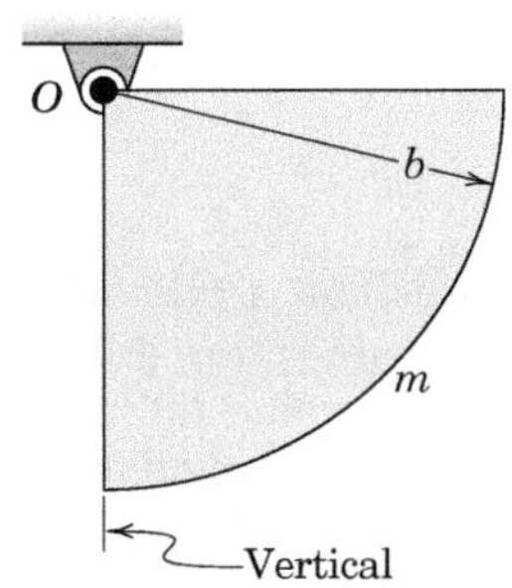

Problem 6/47

Representative Problems

6/48 The 15-kg uniform steel plate is freely hinged about the horizontal z-axis. Calculate the force supported by each of the bearings at A and B an instant after the plate is released from rest while in the horizontal position shown.

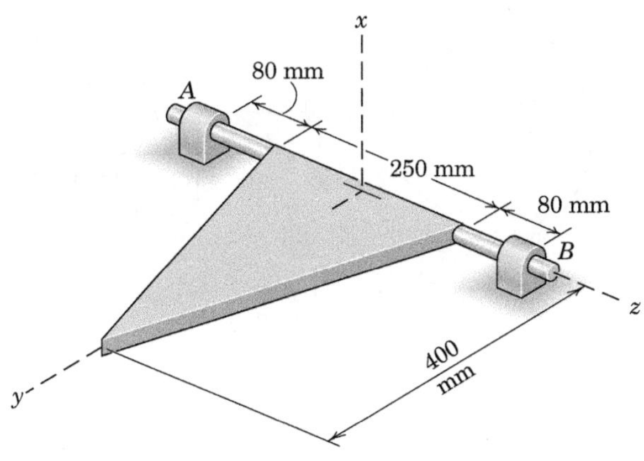

Problem 6/48

6/49 The square frame is composed of four equal lengths of uniform slender rod, and the ball attachment at O is suspended in a socket (not shown). Beginning from the position shown, the assembly is rotated 45° about axis A-A and released. Determine the initial angular acceleration of the frame. Repeat for a 45° rotation about axis B-B. Neglect the small mass, offset, and friction of the ball.

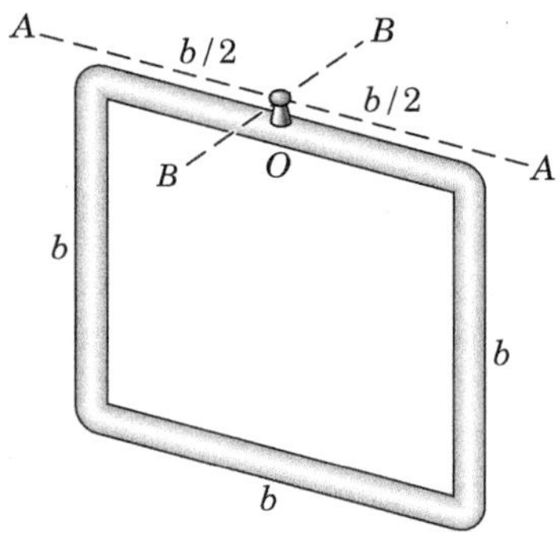

Problem 6/49

6/50 A uniform torus and a cylindrical ring, each solid and of mass m, are released from rest in the positions shown. Determine the magnitude of the pin reaction at O and the angular acceleration of each body an instant after release. Neglect friction in the pivot at O for each case.

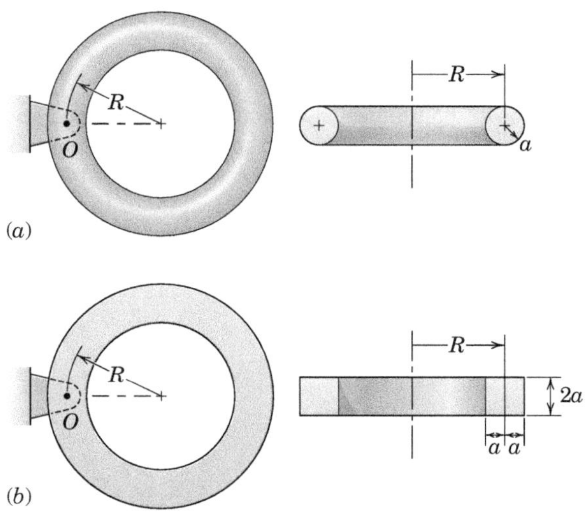

Problem 6/50

6/51 A reel of flexible power cable is mounted on the dolly, which is fixed in position. There are 200 ft of cable weighing 0.436 lb per foot of length wound on the reel at a radius of 15 in. The empty spool weighs 62 lb and has a radius of gyration about its axis of 12 in. A tension T of 20 lb is required to overcome frictional resistance to turning. Calculate the angular acceleration α of the reel if a tension of 40 lb is applied to the free end of the cable.

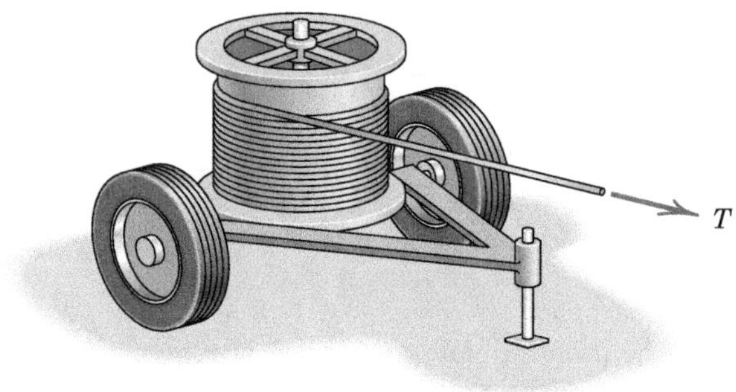

Problem 6/51

6/52 The uniform bar of mass m is supported by the smooth pin at O and is connected to the cylinder of mass m_1 by the light cable which passes over the light pulley at C. If the system is released from rest while in the position shown, determine the tension in the cable. Use the values $m = 30$ kg, $m_1 = 20$ kg, and $L = 6$ m.

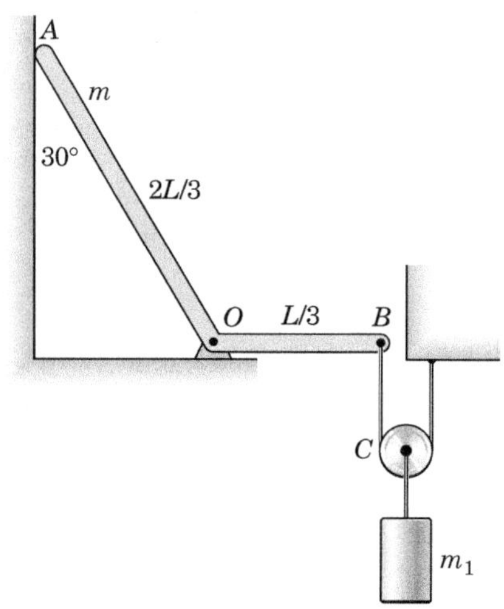

Problem 6/52

6/53 An *air table* is used to study the elastic motion of flexible spacecraft models. Pressurized air escaping from numerous small holes in the horizontal surface provides a supporting air cushion which largely eliminates friction. The model shown consists of a cylindrical hub of radius r and four appendages of length l and small thickness t. The hub and the four appendages all have the same depth d and are constructed of the same material of density ρ. Assume that the spacecraft is rigid and determine the moment M which must be applied to the hub to spin the model from rest to an angular velocity ω in a time period of τ seconds. (Note that for a spacecraft with highly flexible appendages, the moment must be judiciously applied to the rigid hub to avoid undesirable large elastic deflections of the appendages.)

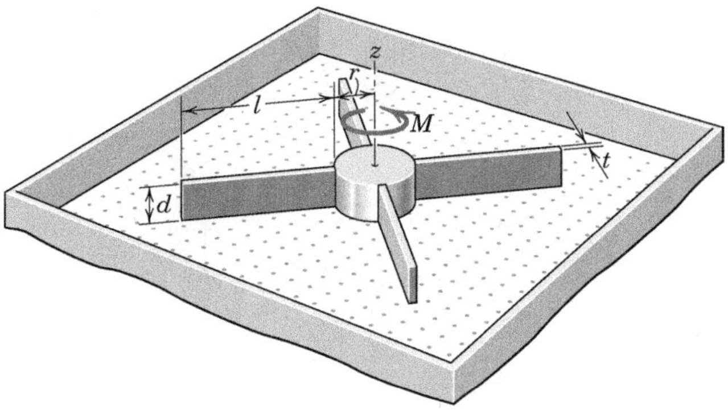

Problem 6/53

6/54 A vibration test is run to check the design adequacy of bearings A and B. The unbalanced rotor and attached shaft have a combined mass of 2.8 kg. To locate the mass center, a torque of 0.660 N·m is applied to the shaft to hold it in equilibrium in a position rotated 90° from that shown. A constant torque M = 1.5 N·m is then applied to the shaft, which reaches a speed of 1200 rev/min in 18 revolutions starting from rest. (During each revolution the angular acceleration varies, but its average value is the same as for constant acceleration.) Determine (*a*) the radius of gyration k of the rotor and shaft about the rotation axis, (*b*) the force F which each bearing exerts on the shaft immediately after M is applied, and (*c*) the force R exerted by each bearing when the speed of 1200 rev/min is reached and M is removed. Neglect any frictional resistance and the bearing forces due to static equilibrium.

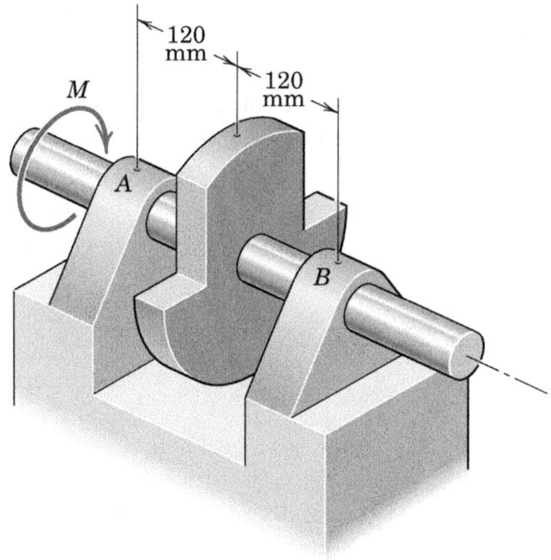

Problem 6/54

6/55 The solid cylindrical rotor B has a mass of 43 kg and is mounted on its central axis C-C. The frame A rotates about the fixed vertical axis O-O under the applied torque M = 30 N·m. The rotor may be unlocked from the frame by withdrawing the locking pin P. Calculate the angular acceleration α of the frame A if the locking pin is (*a*) in place and (*b*) withdrawn. Neglect all friction and the mass of the frame.

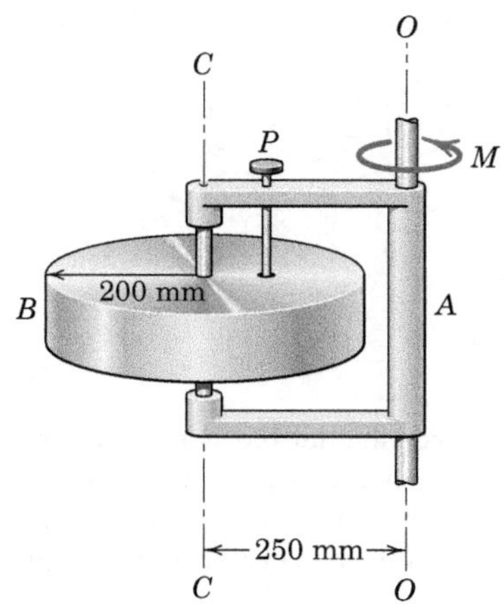

Problem 6/55

6/56 The right-angle body is made of uniform slender bar of mass m and length L. It is released from rest while in the position shown. Determine the initial angular acceleration α of the body and the magnitude of the force supported by the pivot at O.

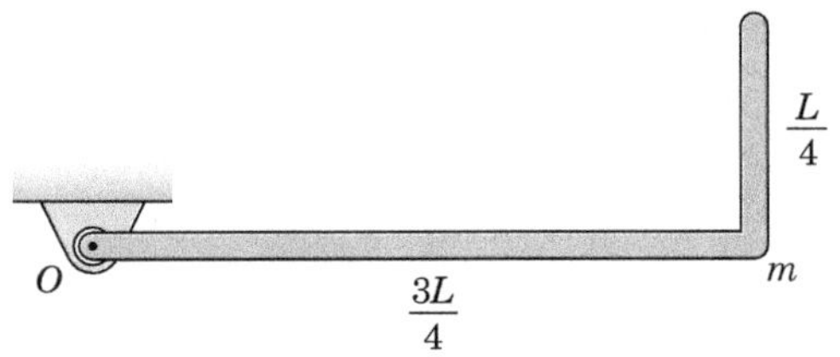

Problem 6/56

6/57 Each of the two grinding wheels has a diameter of 6 in., a thickness of 3/4 in., and a specific weight of 425 lb/ft^3. When switched on, the machine accelerates from rest to its operating speed of 3450 rev/min in 5 sec. When switched off, it comes to rest in 35 sec. Determine the motor torque and frictional moment, assuming that each is constant. Neglect the effects of the inertia of the rotating motor armature.

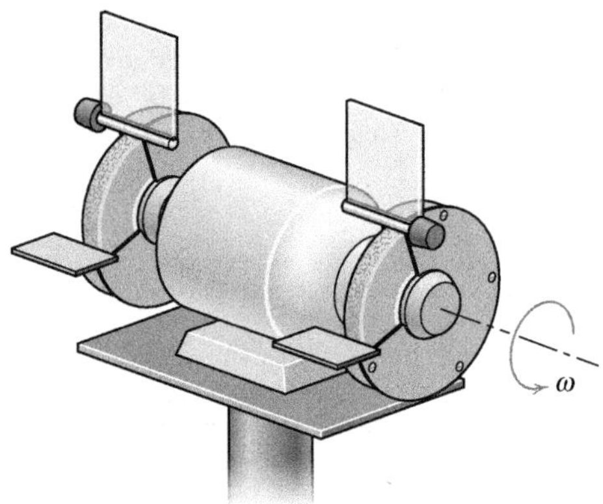

Problem 6/57

6/58 The uniform slender bar is released from rest in the horizontal position shown. Determine the value of x for which the angular acceleration is a maximum, and determine the corresponding angular acceleration α.

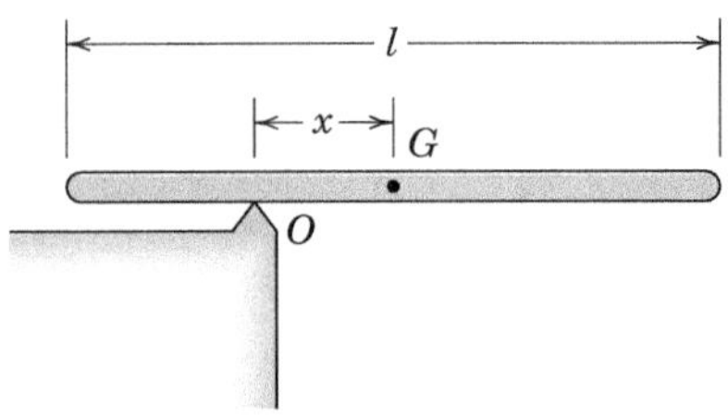

Problem 6/58

6/59 The assembly from Prob. 3/219 is repeated here with the following additional information. The 2-kg collar at C has an outer diameter of 80 mm and is press fitted to the light 50-mm-diameter shaft. Each spoke has a mass of 1.5 kg and carries a 3-kg sphere with a radius of 40 mm attached to its end. The pulley at D has a mass of 5 kg with a centroidal radius of gyration of 60 mm. If a tension $T = 20$ N is applied to the end of the securely wrapped cable with the assembly initially at rest, determine the initial angular acceleration of the assembly. Neglect friction in the bearings at A and B and state any assumptions.

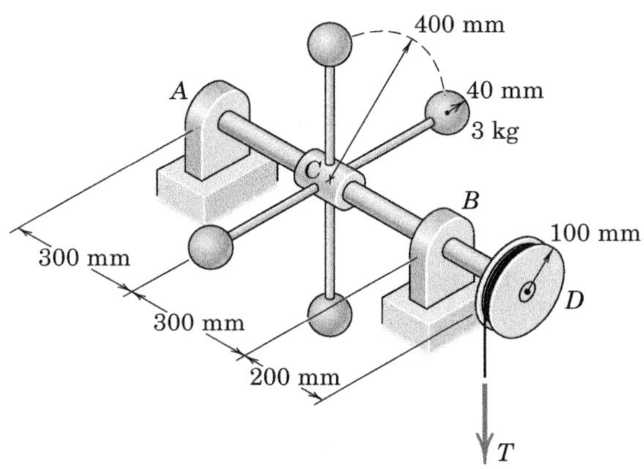

Problem 6/59

6/60 The right-angle plate is formed from a flat plate having a mass ρ per unit area and is welded to the horizontal shaft mounted in the bearing at O. If the shaft is free to rotate, determine the initial angular acceleration α of the plate when it is released from rest with the upper surface in the horizontal plane. Also determine the y- and z-components of the resultant force on the shaft at O.

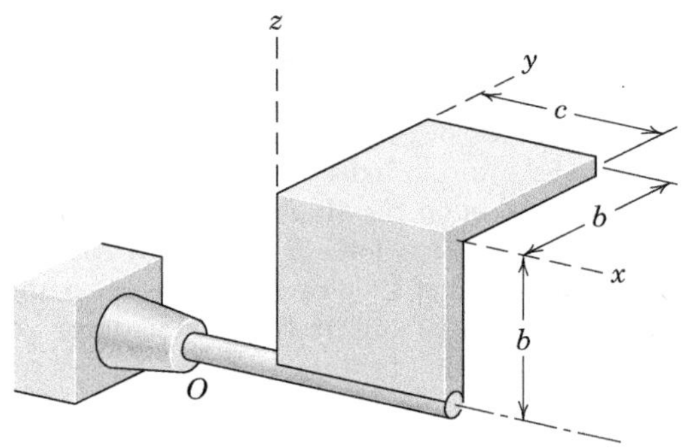

Problem 6/60

6/61 The semicircular disk of mass m and radius r is released from rest at $\theta = 0$ and rotates freely in the vertical plane about its fixed bearing at O. Derive expressions for the n- and t-components of the force F on the bearing as functions of θ.

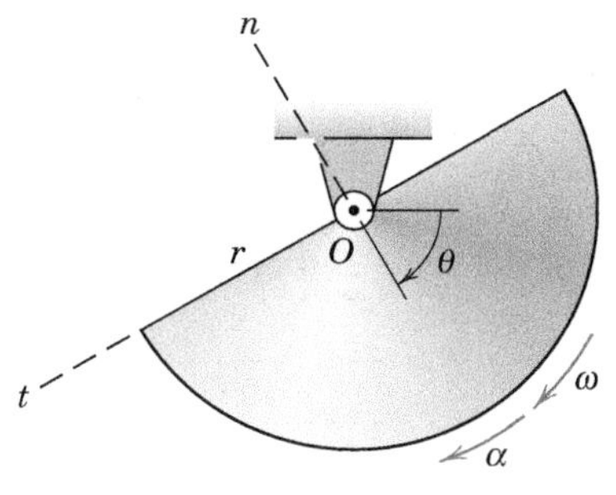

Problem 6/61

6/62 The uniform steel I-beam has a mass of 300 kg and is supported in the vertical plane as shown. Calculate the force R supported by the pin at O for the condition immediately after the support at B is suddenly removed. The mass of the bracket on the left end is small and may be neglected. Also treat the beam as a uniform slender bar.

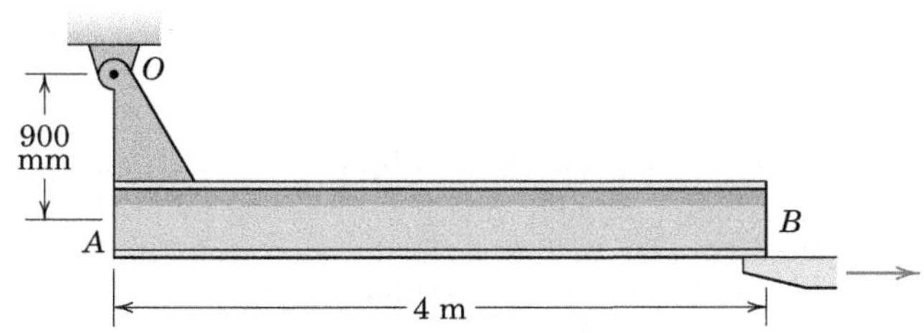

Problem 6/62

6/63 The gear train shown operates in a horizontal plane and is used to transmit motion to the rack D of mass m_D. If an input torque M is applied to gear A, what will be the resulting acceleration a of the unloaded rack? (The mechanism which it normally drives has been disengaged.) Gear C is keyed to the same shaft as gear B. Gears A, B, and C have pitch diameters d_A, d_B, and d_C, and centroidal mass moments of inertia I_A, I_B, and I_C, respectively. All friction is negligible.

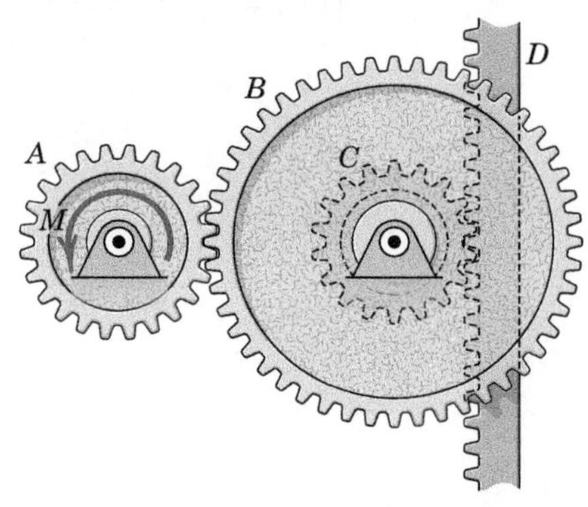

Problem 6/63

6/64 The uniform semicircular ring of mass $m = 2.5$ kg and mean radius $r = 200$ mm is mounted on spokes of negligible mass and pivoted about a horizontal axis through O. If the ring is released from rest in the position $\theta = 30°$, determine the force R supported by the bearing O just after release.

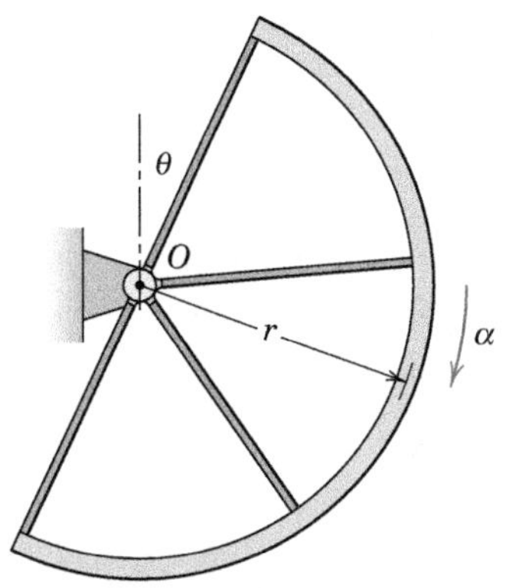

Problem 6/64

6/65 The link B weighs 0.80 lb with center of mass 2.20 in. from O-O and has a radius of gyration about O-O of 2.76 in. The link is welded to the steel tube and is free to rotate about the fixed horizontal shaft at O-O. The tube weighs 1.84 lb. If the tube is released from rest with the link in the horizontal position, calculate the initial angular acceleration α of the assembly and the corresponding reaction O exerted by the shaft on the link.

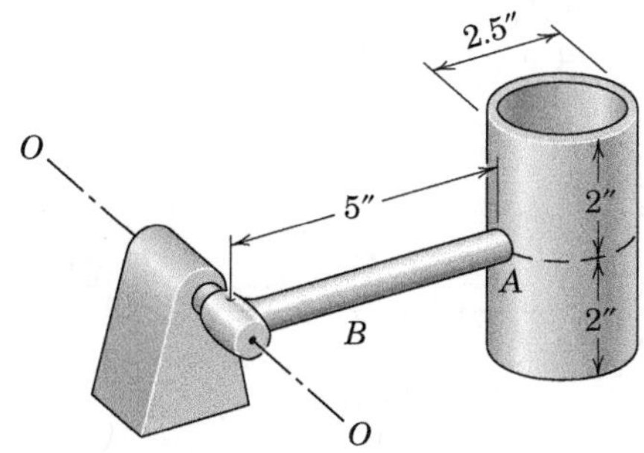

Problem 6/65

6/66 A flexible cable 60 meters long with a mass of 0.160 kg per meter of length is wound around the reel. With $y = 0$, the weight of the 4-kg cylinder is required to start turning the reel to overcome friction in its bearings. Determine the downward acceleration a in meters per second squared of the cylinder as a function of y in meters. The empty reel has a mass of 16 kg with a radius of gyration about its bearing of 200 mm.

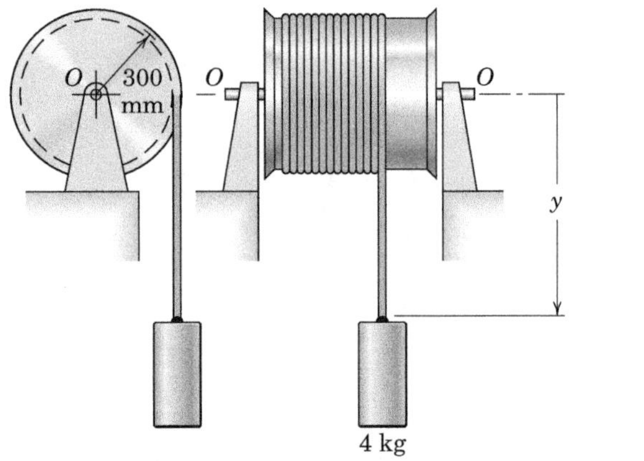

Problem 6/66

6/67 The uniform 72-ft mast weighs 600 lb and is hinged at its lower end to a fixed support at O. If the winch C develops a starting torque of 900 lb-ft, calculate the total force supported by the pin at O as the mast begins to lift off its support at B. Also find the corresponding angular acceleration α of the mast. The cable at A is horizontal, and the mass of the pulleys and winch is negligible.

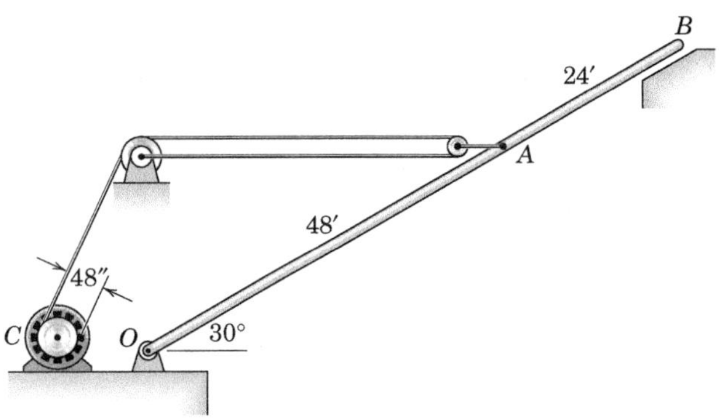

Problem 6/67

6/68 The robotic device consists of the stationary pedestal OA, arm AB pivoted at A, and arm BC pivoted at B. The rotation axes are normal to the plane of the figure. Estimate (a) the moment M_A applied to arm AB required to rotate it about joint A at 4 rad/s^2 counterclockwise from the position shown with joint B locked and (b) the moment M_B applied to arm BC required to rotate it about joint B at the same rate with joint A locked. The mass of arm AB is 25 kg and that of BC is 4 kg, with the stationary portion of joint A excluded entirely and the mass of joint B divided equally between the two arms. Assume that the centers of mass G_1 and G_2 are in the geometric centers of the arms and model the arms as slender rods.

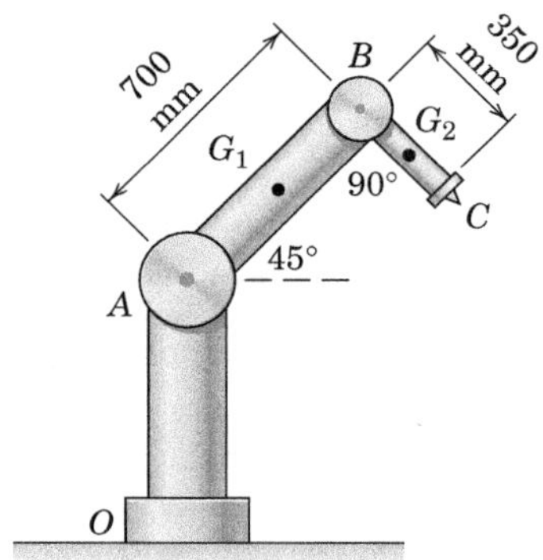

Problem 6/68

▶6/69 For the beam described in Prob. 6/62, determine the maximum angular velocity ω reached by the beam as it rotates in the vertical plane about the bearing at O. Also determine the corresponding force R supported by the pin at O for this condition. Again, treat the beam as a slender 300-kg bar and neglect the mass of the supporting bracket.

▶6/70 The uniform parabolic plate has a mass per unit area of 225 kg/m^2. If the plate is released from rest while in the horizontal position shown, determine the initial angular acceleration $\boldsymbol{\alpha}$ about the bearing axis AB.

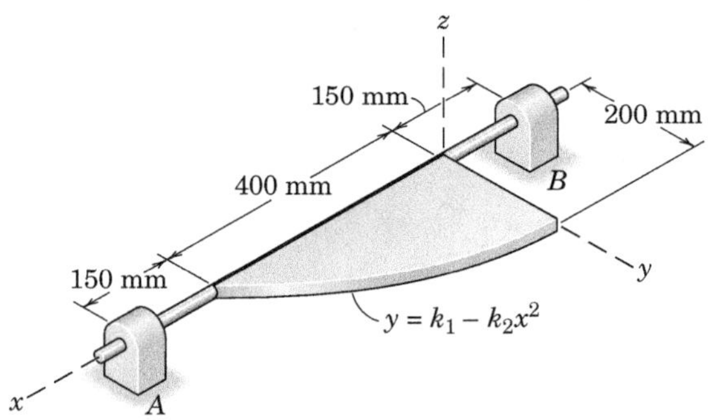

Problem 6/70

6/5 General Plane Motion

The dynamics of a rigid body in general plane motion combines translation and rotation. In Art. 6/2 we represented such a body in Fig. 6/4 with its free-body diagram and its kinetic diagram, which discloses the dynamic resultants of the applied forces. Figure 6/4 and Eqs. 6/1, which apply to general plane motion, are repeated here for convenient reference.

$$\Sigma \mathbf{F} = m\bar{\mathbf{a}}$$
$$\Sigma M_G = \bar{I}\alpha \qquad \textbf{[6/1]}$$

Direct application of these equations expresses the equivalence between the externally applied forces, as disclosed by the free-body diagram, and their force and moment resultants, as represented by the kinetic diagram.

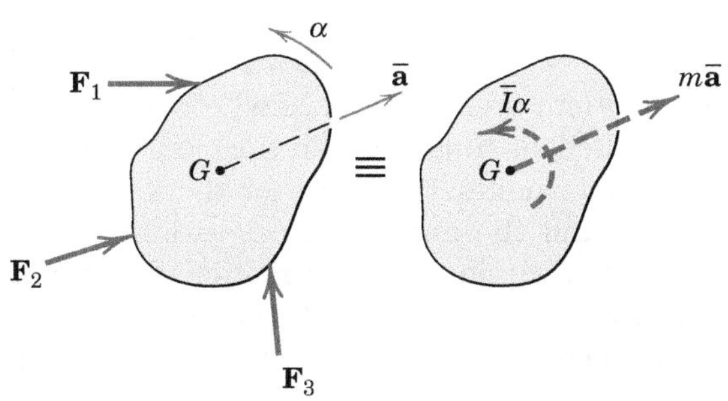

Free-Body Diagram Kinetic Diagram

Figure 6/4 repeated

Solving Plane-Motion Problems

Keep in mind the following considerations when solving plane-motion problems.

Choice of Coordinate System. The force equation of Eq. 6/1 should be expressed in whatever coordinate system most readily describes the acceleration of the mass center. You should consider rectangular, normal-tangential, and polar coordinates.

Choice of Moment Equation. In Art. 6/2 we also showed, with the aid of Fig. 6/5, the application of the alternative relation for moments about any point P, Eq. 6/2. This figure and this equation are also repeated here for easy reference.

$$\Sigma M_P = \bar{I}\alpha + m\bar{a}d \qquad \textbf{[6/2]}$$

In some instances, it may be more convenient to use the alternative moment relation of Eq. 6/3 when moments are taken about a point P whose acceleration is known. Note also that the equation for moments about a nonaccelerating point O on the body, Eq. 6/4, constitutes still

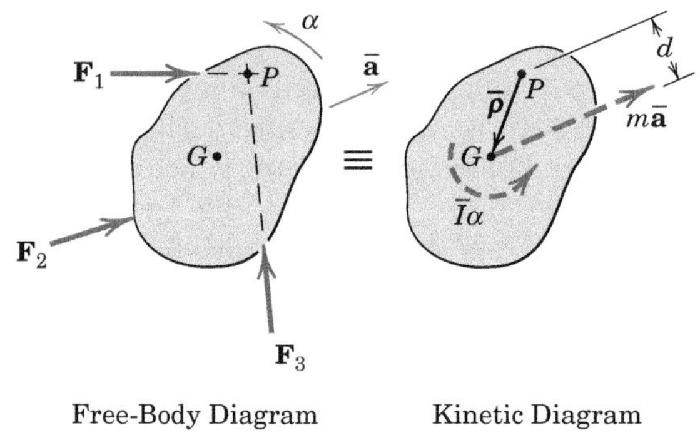

Figure 6/5 repeated

another alternative moment relation and at times may be used to advantage.

Constrained versus Unconstrained Motion. In working a problem in general plane motion, we first observe whether the motion is unconstrained or constrained, as illustrated in the examples of Fig. 6/6. If the motion is constrained, we must account for the kinematic relationship between the linear and the angular accelerations and incorporate it into our force and moment equations of motion. If the motion is unconstrained, the accelerations can be determined independently of one another by direct application of the three motion equations, Eqs. 6/1.

Number of Unknowns. In order for a rigid-body problem to be solvable, the number of unknowns cannot exceed the number of independent equations available to describe them, and a check on the sufficiency of the relationships should always be made. At the most, for plane motion we have three scalar equations of motion and two scalar components of the vector relative-acceleration equation for constrained motion. Thus, we can handle as many as five unknowns for each rigid body.

Identification of the Body or System. We emphasize the importance of clearly choosing the body to be isolated and representing this isolation by a correct free-body diagram. Only after this vital step has been completed can we properly evaluate the equivalence between the external forces and their resultants.

Kinematics. Of equal importance in the analysis of plane motion is a clear understanding of the kinematics involved. Very often, the difficulties experienced at this point have to do with kinematics, and a thorough review of the relative-acceleration relations for plane motion will be most helpful.

Consistency of Assumptions. In formulating the solution to a problem, we recognize that the directions of certain forces or accelerations may not be known at the outset, so that it may be necessary to make initial assumptions whose validity will be proved or disproved when the solution is carried out. It is essential, however, that all assumptions made be consistent with the principle of action and reaction and with any kinematic requirements, which are also called *conditions of constraint*.

Thus, for example, if a wheel is rolling on a horizontal surface, its center is constrained to move on a horizontal line. Furthermore, if the unknown linear acceleration a of the center of the wheel is assumed positive to the right, then the unknown angular acceleration α will be positive in a clockwise sense in order that $a = +r\alpha$, if we assume the wheel does not slip. Also, we note that, for a wheel which rolls without slipping, the static friction force between the wheel and its supporting surface is generally *less* than its maximum value, so that $F \neq \mu_s N$. But if the wheel slips as it rolls, $a \neq r\alpha$, and a kinetic friction force is generated which is given by $F = \mu_k N$. It may be necessary to test the validity of either assumption, slipping or no slipping, in a given problem. The difference between the coefficients of static and kinetic friction, μ_s and μ_k, is sometimes ignored, in which case, μ is used for either or both coefficients.

Look ahead to Prob. 6/104 to see a special-case problem involving a dummy such as the one in this car undergoing an impact test.

Sample Problem 6/5

A metal hoop with a radius $r = 6$ in. is released from rest on the 20° incline. If the coefficients of static and kinetic friction are $\mu_s = 0.15$ and $\mu_k = 0.12$, determine the angular acceleration α of the hoop and the time t for the hoop to move a distance of 10 ft down the incline.

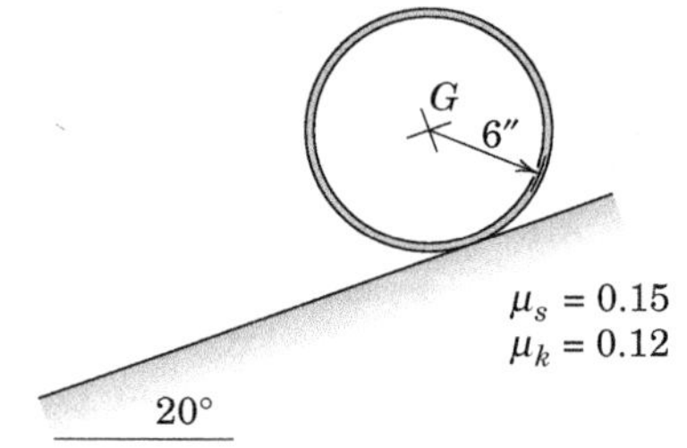

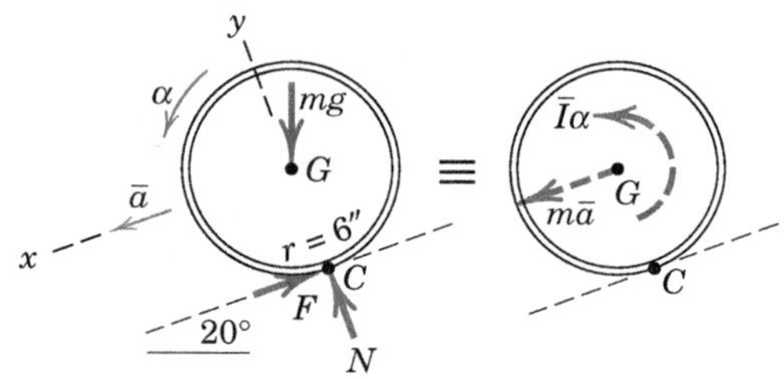

Solution. The free-body diagram shows the unspecified weight mg, the normal force N, and the friction force F acting on the hoop at the contact point C with the incline. The kinetic diagram shows the resultant force $m\bar{a}$ through G in the direction of its acceleration and the couple $\bar{I}\alpha$. The counterclockwise angular acceleration requires a counterclockwise moment about G, so F must be up the incline.

Assume that the hoop rolls without slipping, so that $\bar{a} = r\alpha$. Application of the components of Eqs. 6/1 with x- and y-axes assigned gives

$[\Sigma F_x = m\bar{a}_x]$ $\qquad mg \sin 20° - F = m\bar{a}$

$[\Sigma F_y = m\bar{a}_y = 0]$ $\qquad N - mg \cos 20° = 0$

(1) $[\Sigma M_G = \bar{I}\alpha]$ $\qquad Fr = mr^2\alpha$

Elimination of F between the first and third equations and substitution of the kinematic assumption $\bar{a} = r\alpha$ give

(2) $$\bar{a} = \frac{g}{2}\sin 20° = \frac{32.2}{2}(0.342) = 5.51 \text{ ft/sec}^2$$

Alternatively, with our assumption of $\bar{a} = r\alpha$ for pure rolling, a moment sum about C by Eq. 6/2 gives $\bar{a}$ directly. Thus,

$[\Sigma M_C = \bar{I}\alpha + m\bar{a}d]$ $\qquad mgr \sin 20° = mr^2\frac{\bar{a}}{r} + m\bar{a}r \qquad \bar{a} = \frac{g}{2}\sin 20°$

To check our assumption of no slipping, we calculate F and N and compare F with its limiting value. From the above equations,

$$F = mg \sin 20° - m\frac{g}{2}\sin 20° = 0.1710mg$$

$$N = mg \cos 20° = 0.940mg$$

But the maximum possible friction force is

$[F_{max} = \mu_s N]$ $\qquad F_{max} = 0.15(0.940mg) = 0.1410mg$

Because our calculated value of $0.1710mg$ exceeds the limiting value of $0.1410mg$, we conclude that our assumption of pure rolling was wrong. Therefore, the hoop slips as it rolls and $\bar{a} \neq r\alpha$. The friction force then becomes the kinetic value

$[F = \mu_k N]$ $\qquad F = 0.12(0.940mg) = 0.1128mg$

The motion equations now give

$[\Sigma F_x = m\bar{a}_x]$ $\qquad mg \sin 20° - 0.1128mg = m\bar{a}$

$$\bar{a} = 0.229(32.2) = 7.38 \text{ ft/sec}^2$$

(3) $[\Sigma M_G = \bar{I}\alpha]$ $\qquad 0.1128mg(r) = mr^2\alpha$

$$\alpha = \frac{0.1128(32.2)}{6/12} = 7.26 \text{ rad/sec}^2 \qquad \textit{Ans.}$$

The time required for the center G of the hoop to move 10 ft from rest with constant acceleration is

$[x = \frac{1}{2}at^2]$ $\qquad t = \sqrt{\frac{2x}{a}} = \sqrt{\frac{2(10)}{7.38}} = 1.646 \text{ sec} \qquad \textit{Ans.}$

Helpful Hints

(1) Because all of the mass of a hoop is a distance r from its center G, its moment of inertia about G must be mr^2.

(2) Note that $\bar{a}$ is independent of both m and r.

(3) Note that α is independent of m but dependent on r.

Sample Problem 6/6

The drum A is given a constant angular acceleration α_0 of 3 rad/s^2 and causes the 70-kg spool B to roll on the horizontal surface by means of the connecting cable, which wraps around the inner hub of the spool. The radius of gyration $\bar{k}$ of the spool about its mass center G is 250 mm, and the coefficient of static friction between the spool and the horizontal surface is 0.25. Determine the tension T in the cable and the friction force F exerted by the horizontal surface on the spool.

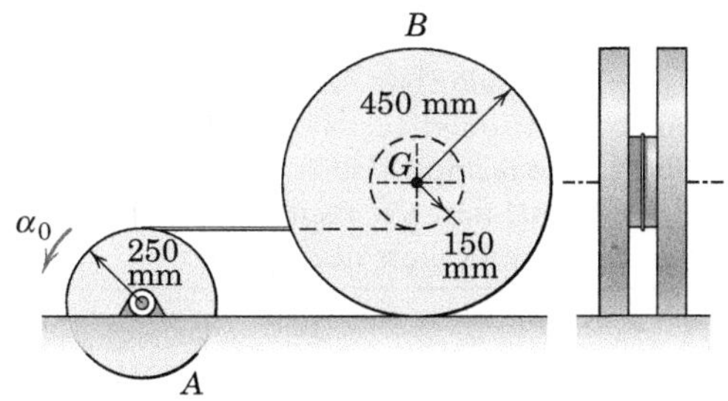

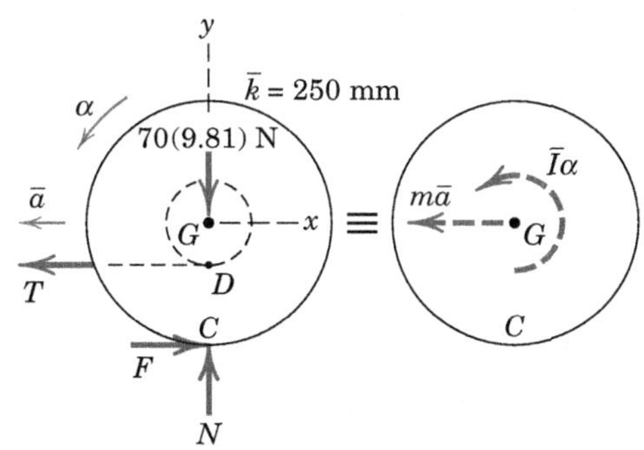

Solution. The free-body diagram and the kinetic diagram of the spool are drawn as shown. The correct direction of the friction force may be assigned in this problem by observing from both diagrams that with counterclockwise angular acceleration, a moment sum about point G (and also about point D) must be counterclockwise. A point on the connecting cable has an acceleration $a_t = r\alpha = 0.25(3) = 0.75$ m/s^2, which is also the horizontal component of the acceleration of point D on the spool. It will be assumed initially that the spool rolls without slipping, in which case it
(1) has a counterclockwise angular acceleration $\alpha = (a_D)_x/\overline{DC} = 0.75/0.30 = 2.5$ rad/s^2. The acceleration of the mass center G is, therefore, $\bar{a} = r\alpha = 0.45(2.5) = 1.125$ m/s^2.

With the kinematics determined, we now apply the three equations of motion, Eqs. 6/1,

$$[\Sigma F_x = m\bar{a}_x] \qquad F - T = 70(-1.125) \qquad (a)$$

$$[\Sigma F_y = m\bar{a}_y] \qquad N - 70(9.81) = 0 \qquad N = 687 \text{ N}$$

(2) $$[\Sigma M_G = \bar{I}\alpha] \qquad F(0.450) - T(0.150) = 70(0.250)^2(2.5) \qquad (b)$$

Solving (a) and (b) simultaneously gives

$$F = 75.8 \text{ N} \quad \text{and} \quad T = 154.6 \text{ N} \qquad \textit{Ans.}$$

To establish the validity of our assumption of no slipping, we see that the surfaces are capable of supporting a maximum friction force $F_{max} = \mu_s N = 0.25(687) = 171.7$ N. Since only 75.8 N of friction force is required, we conclude that our assumption of rolling without slipping is valid.

If the coefficient of static friction had been 0.10, for example, then the friction force would have been limited to $0.10(687) = 68.7$ N, which is less than 75.8 N, and the spool would slip. In this event, the kinematic relation $\bar{a} = r\alpha$ would
(3) no longer hold. With $(a_D)_x$ known, the angular acceleration would be $\alpha = [\bar{a} - (a_D)_x]/\overline{GD}$. Using this relation along with $F = \mu_k N = 68.7$ N, we would then re-solve the three equations of motion for the unknowns T, $\bar{a}$, and α.

Alternatively, with point C as a moment center in the case of pure rolling, we may use Eq. 6/2 and obtain T directly. Thus,

$$[\Sigma M_C = \bar{I}\alpha + m\bar{a}r] \quad 0.3T = 70(0.25)^2(2.5) + 70(1.125)(0.45)$$

$$T = 154.6 \text{ N} \qquad \textit{Ans.}$$

(4) where the previous kinematic results for no slipping have been incorporated. We could also write a moment equation about point D to obtain F directly.

Helpful Hints

(1) The relation between $\bar{a}$ and α is the kinematic constraint which accompanies the assumption that the spool rolls without slipping.

(2) Be careful not to make the mistake of using $\frac{1}{2}mr^2$ for $\bar{I}$ of the spool, which is not a uniform circular disk.

(3) Our principles of relative acceleration are a necessity here. Hence, the relation $(a_{G/D})_t = \overline{GD}\alpha$ should be recognized.

(4) The flexibility in the choice of moment centers provided by the kinetic diagram can usually be employed to simplify the analysis.

Sample Problem 6/7

The slender bar AB weighs 60 lb and moves in the vertical plane, with its ends constrained to follow the smooth horizontal and vertical guides. If the 30-lb force is applied at A with the bar initially at rest in the position for which $\theta = 30°$, calculate the resulting angular acceleration of the bar and the forces on the small end rollers at A and B.

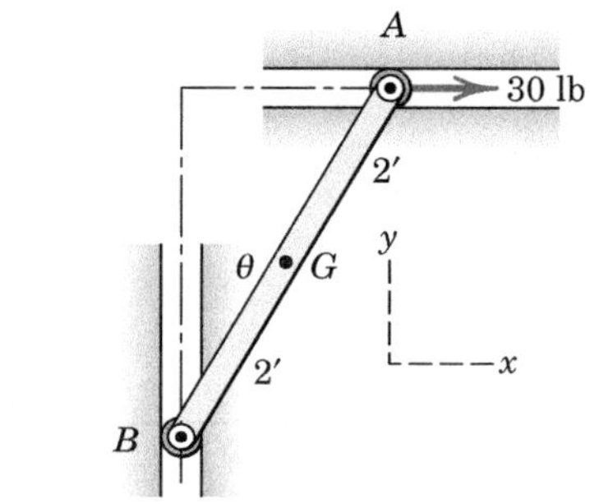

Solution. The bar undergoes constrained motion, so that we must establish the relationship between the mass-center acceleration and the angular accel-
① eration. The relative-acceleration equation $\mathbf{a}_A = \mathbf{a}_B + \mathbf{a}_{A/B}$ must be solved first, and then the equation $\bar{\mathbf{a}} = \mathbf{a}_G = \mathbf{a}_B + \mathbf{a}_{G/B}$ is next solved to obtain expressions relating $\bar{a}$ and α. With α assigned in its clockwise physical sense, the acceleration polygons which represent these equations are shown, and their solution gives

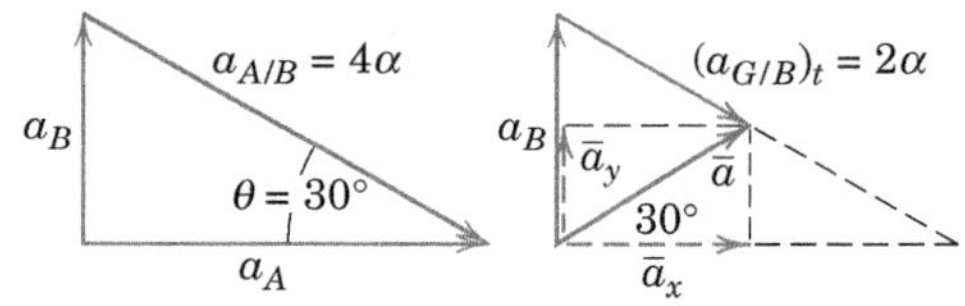

$$\bar{a}_x = \bar{a}\cos 30° = 2\alpha \cos 30° = 1.732\alpha \text{ ft/sec}^2$$

$$\bar{a}_y = \bar{a}\sin 30° = 2\alpha \sin 30° = 1.0\alpha \text{ ft/sec}^2$$

Next we construct the free-body diagram and the kinetic diagram as shown. With $\bar{a}_x$ and $\bar{a}_y$ now known in terms of α, the remaining unknowns are α and the forces A and B. We now apply Eqs. 6/1, which give

② $[\Sigma M_G = \bar{I}\alpha]$

$$30(2\cos 30°) - A(2\sin 30°) + B(2\cos 30°) = \frac{1}{12}\frac{60}{32.2}(4^2)\alpha$$

$[\Sigma F_x = m\bar{a}_x]$ $$30 - B = \frac{60}{32.2}(1.732\alpha)$$

$[\Sigma F_y = m\bar{a}_y]$ $$A - 60 = \frac{60}{32.2}(1.0\alpha)$$

Solving the three equations simultaneously gives us the results

$$A = 68.2 \text{ lb} \qquad B = 15.74 \text{ lb} \qquad \alpha = 4.42 \text{ rad/sec}^2 \qquad \textit{Ans.}$$

As an alternative solution, we can use Eq. 6/2 with point C as the moment center and avoid the necessity of solving three equations simultaneously. This choice eliminates reference to forces A and B and gives α directly. Thus,

③ $[\Sigma M_C = \bar{I}\alpha + \Sigma m\bar{a}d]$

$$30(4\cos 30°) - 60(2\sin 30°) = \frac{1}{12}\frac{60}{32.2}(4^2)\alpha$$
$$+ \frac{60}{32.2}(1.732\alpha)(2\cos 30°) + \frac{60}{32.2}(1.0\alpha)(2\sin 30°)$$

$$43.9 = 9.94\alpha \qquad \alpha = 4.42 \text{ rad/sec}^2 \qquad \textit{Ans.}$$

With α determined, we can now apply the force equations independently and get

$[\Sigma F_y = m\bar{a}_y]$ $$A - 60 = \frac{60}{32.2}(1.0)(4.42) \qquad A = 68.2 \text{ lb} \qquad \textit{Ans.}$$

$[\Sigma F_x = m\bar{a}_x]$ $$30 - B = \frac{60}{32.2}(1.732)(4.42) \qquad B = 15.74 \text{ lb} \qquad \textit{Ans.}$$

Helpful Hints

① If the application of the relative-acceleration equations is not perfectly clear at this point, then review Art. 5/6. Note that the relative normal acceleration term is absent since there is no angular velocity of the bar.

② Recall that the moment of inertia of a slender rod about its center is $\frac{1}{12}ml^2$.

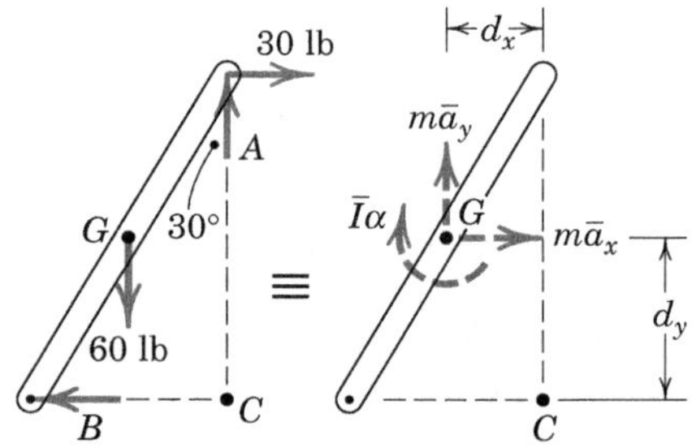

③ From the kinetic diagram, $\Sigma m\bar{a}d = m\bar{a}_x d_y + m\bar{a}_y d_x$. Since both terms of the sum are clockwise, in the same sense as $\bar{I}\alpha$, they are positive.

Sample Problem 6/8

A car door is inadvertently left slightly open when the brakes are applied to give the car a constant rearward acceleration a. Derive expressions for the angular velocity of the door as it swings past the 90° position and the components of the hinge reactions for any value of θ. The mass of the door is m, its mass center is a distance $\bar{r}$ from the hinge axis O, and the radius of gyration about O is k_O.

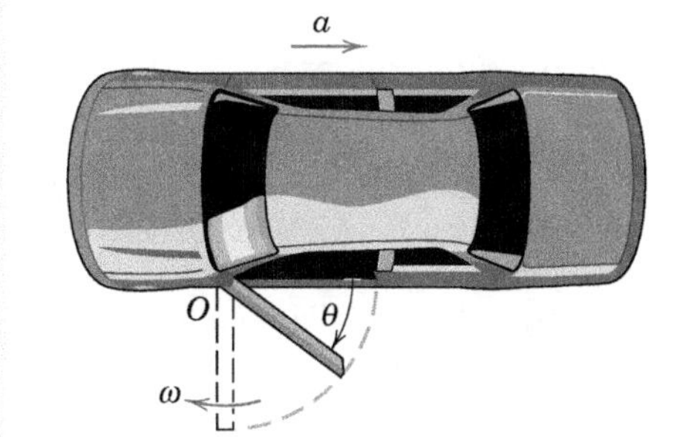

Solution. Because the angular velocity ω increases with θ, we need to find how the angular acceleration α varies with θ so that we may integrate it over the interval to obtain ω. We obtain α from a moment equation about O. First, we draw the free-body diagram of the door in the horizontal plane for a general position θ. The only forces in this plane are the components of the hinge reaction shown here in the x- and y-directions. On the kinetic diagram, in addition to the resultant couple $\bar{I}\alpha$ shown in the sense of α, we represent the resultant force $m\bar{\mathbf{a}}$ in terms of its components by using an equation of relative acceleration with
① respect to O. This equation becomes the kinematic equation of constraint and is

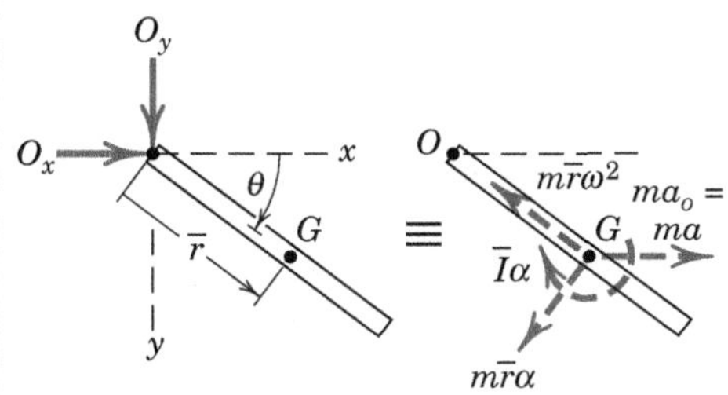

$$\bar{\mathbf{a}} = \mathbf{a}_G = \mathbf{a}_O + (\mathbf{a}_{G/O})_n + (\mathbf{a}_{G/O})_t$$

The magnitudes of the $m\bar{\mathbf{a}}$ components are then

② $$ma_O = ma \qquad m(a_{G/O})_n = m\bar{r}\omega^2 \qquad m(a_{G/O})_t = m\bar{r}\alpha$$

where $\omega = \dot{\theta}$ and $\alpha = \ddot{\theta}$.

For a given angle θ, the three unknowns are α, O_x, and O_y. We can eliminate O_x and O_y by a moment equation about O, which gives

③ $[\Sigma M_O = \bar{I}\alpha + \Sigma m\bar{a}d]$ $\qquad 0 = m({k_O}^2 - \bar{r}^2)\alpha + m\bar{r}\alpha(\bar{r}) - ma(\bar{r}\sin\theta)$

④ Solving for α gives $\qquad \alpha = \dfrac{a\bar{r}}{{k_O}^2}\sin\theta$

Now we integrate α first to a general position and get

$[\omega\, d\omega = \alpha\, d\theta]$
$$\int_0^\omega \omega\, d\omega = \int_0^\theta \frac{a\bar{r}}{{k_O}^2}\sin\theta\, d\theta$$

$$\omega^2 = \frac{2a\bar{r}}{{k_O}^2}(1 - \cos\theta)$$

For $\theta = \pi/2$, $\qquad \omega = \dfrac{1}{k_O}\sqrt{2a\bar{r}}$ *Ans.*

⑤ To find O_x and O_y for any given value of θ, the force equations give

$[\Sigma F_x = m\bar{a}_x]$ $\qquad O_x = ma - m\bar{r}\omega^2\cos\theta - m\bar{r}\alpha\sin\theta$

$$= m\left[a - \frac{2a\bar{r}^2}{{k_O}^2}(1 - \cos\theta)\cos\theta - \frac{a\bar{r}^2}{{k_O}^2}\sin^2\theta\right]$$

$$= ma\left[1 - \frac{\bar{r}^2}{{k_O}^2}(1 + 2\cos\theta - 3\cos^2\theta)\right]$$ *Ans.*

$[\Sigma F_y = m\bar{a}_y]$ $\qquad O_y = m\bar{r}\alpha\cos\theta - m\bar{r}\omega^2\sin\theta$

$$= m\bar{r}\frac{a\bar{r}}{{k_O}^2}\sin\theta\cos\theta - m\bar{r}\frac{2a\bar{r}}{{k_O}^2}(1 - \cos\theta)\sin\theta$$

$$= \frac{ma\bar{r}^2}{{k_O}^2}(3\cos\theta - 2)\sin\theta$$ *Ans.*

Helpful Hints

① Point O is chosen because it is the only point on the door whose acceleration is known.

② Be careful to place $m\bar{r}\alpha$ in the sense of positive α with respect to rotation about O.

③ The free-body diagram shows that there is zero moment about O. We use the transfer-of-axis theorem here and substitute ${k_O}^2 = \bar{k}^2 + \bar{r}^2$. If this relation is not totally familiar, review Art. B/1 in Appendix B.

④ We may also use Eq. 6/3 with O as a moment center

$$\Sigma\mathbf{M}_O = I_O\boldsymbol{\alpha} + \bar{\boldsymbol{\rho}} \times m\mathbf{a}_O$$

where the scalar values of the terms are $I_O\alpha = m{k_O}^2\alpha$ and $\bar{\boldsymbol{\rho}} \times m\mathbf{a}_O$ becomes $-\bar{r}ma\sin\theta$.

⑤ The kinetic diagram shows clearly the terms which make up $m\bar{a}_x$ and $m\bar{a}_y$.

PROBLEMS

Introductory Problems

6/71 The uniform slender bar rests on a smooth horizontal surface when a force F is applied normal to the bar at point A. Point A is observed to have an initial acceleration a_A of 20 m/s^2, and the bar has a corresponding angular acceleration α of 18 rad/s^2. Determine the distance b.

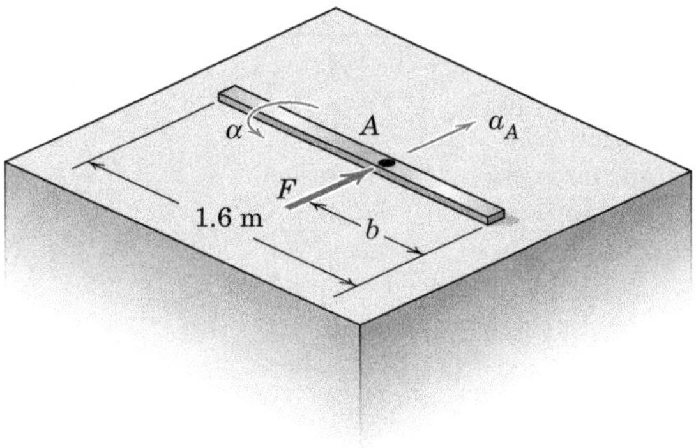

Problem 6/71

6/72 The 64.4-lb solid circular disk is initially at rest on the horizontal surface when a 3-lb force P, constant in magnitude and direction, is applied to the cord wrapped securely around its periphery. Friction between the disk and the surface is negligible. Calculate the angular velocity ω of the disk after the 3-lb force has been applied for 2 seconds and find the linear velocity v of the center of the disk after it has moved 3 feet from rest.

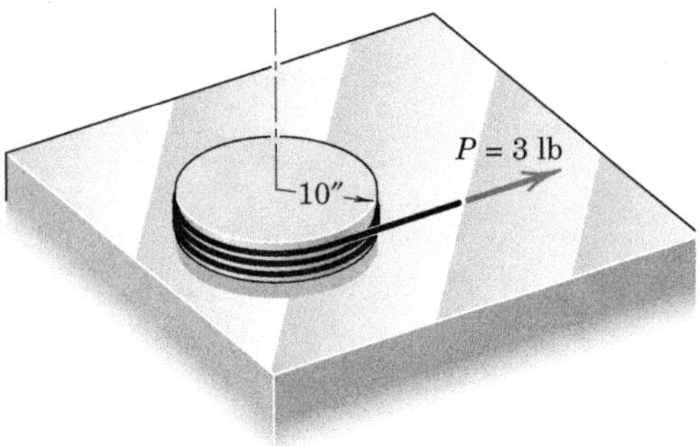

Problem 6/72

6/73 A long cable of length L and mass ρ per unit length is wrapped around the periphery of a spool of negligible mass. One end of the cable is fixed, and the spool is released from rest in the position shown. Find the initial acceleration a of the center of the spool.

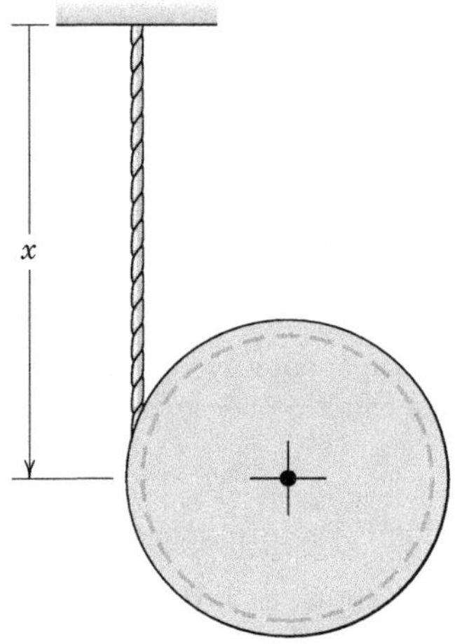

Problem 6/73

6/74 The uniform semicircular rod of mass m is lying motionless on the smooth horizontal surface when the force F is applied at B as shown. Determine the resulting initial acceleration of point A.

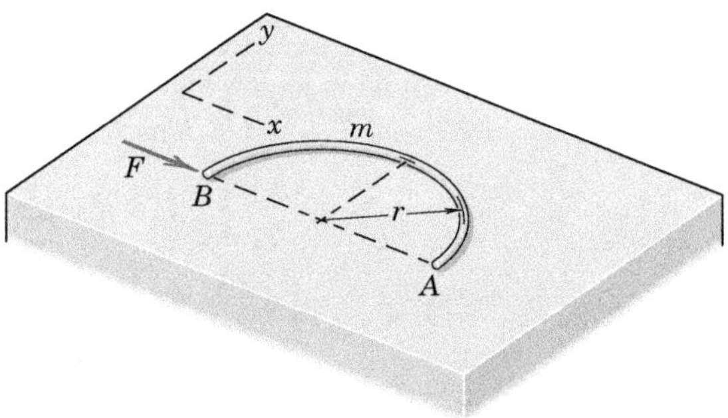

Problem 6/74

6/75 Repeat Prob. 6/74, except that the direction of the applied force has been changed as shown in the figure.

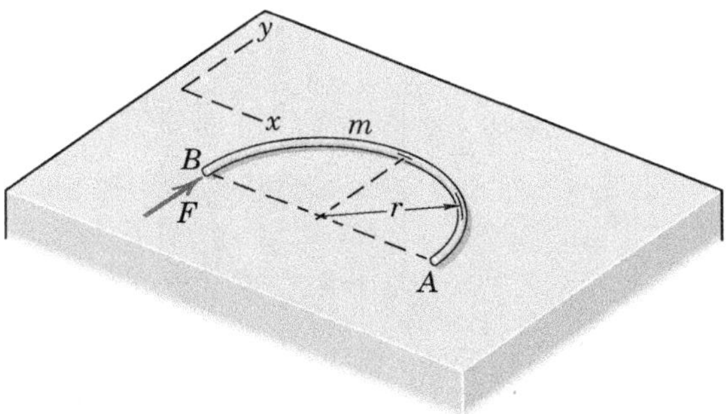

Problem 6/75

6/76 The spacecraft is spinning with a constant angular velocity ω about the z-axis at the same time that its mass center O is traveling with a velocity v_O in the y-direction. If a tangential hydrogen-peroxide jet is fired when the craft is in the position shown, determine the expression for the absolute acceleration of point A on the spacecraft rim at the instant the jet force is F. The radius of gyration of the craft about the z-axis is k, and its mass is m.

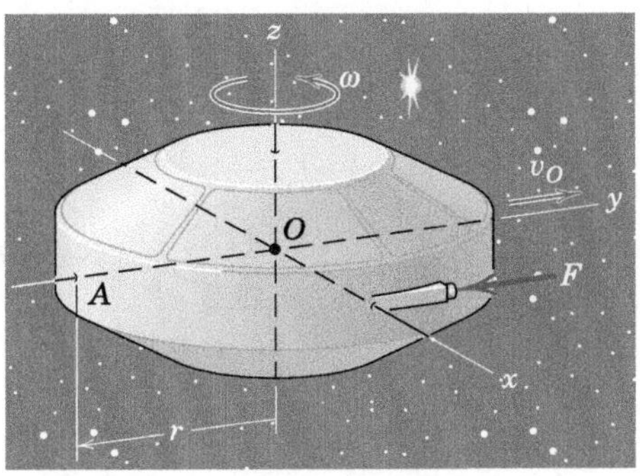

Problem 6/76

6/77 The body consists of a uniform slender bar and a uniform disk, each of mass $m/2$. It rests on a smooth surface. Determine the angular acceleration α and the acceleration of the mass center of the body when the force $P = 6$ N is applied as shown. The value of the mass m of the entire body is 1.2 kg.

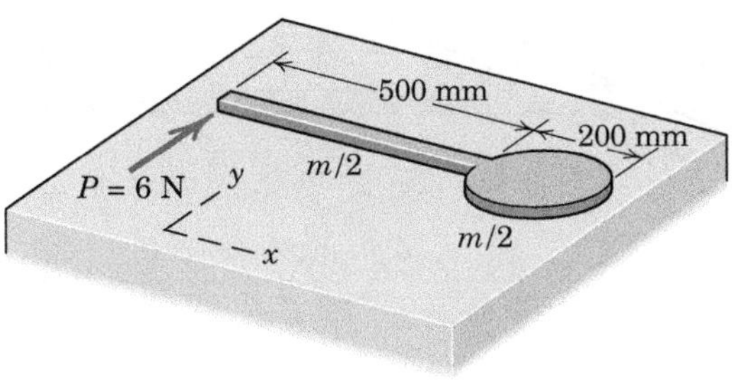

Problem 6/77

6/78 Determine the angular acceleration of each of the two wheels as they roll without slipping down the inclines. For wheel A investigate the case where the mass of the rim and spokes is negligible and the mass of the bar is concentrated along its centerline. For wheel B assume that the thickness of the rim is negligible compared with its radius so that all of the mass is concentrated in the rim. Also specify the minimum coefficient of static friction μ_s required to prevent each wheel from slipping.

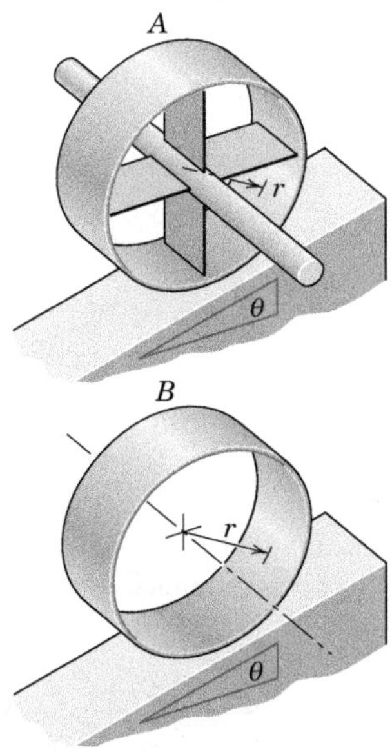

Problem 6/78

6/79 The solid homogeneous cylinder is released from rest on the ramp. If $\theta = 40°$, $\mu_s = 0.30$, and $\mu_k = 0.20$, determine the acceleration of the mass center G and the friction force exerted by the ramp on the cylinder.

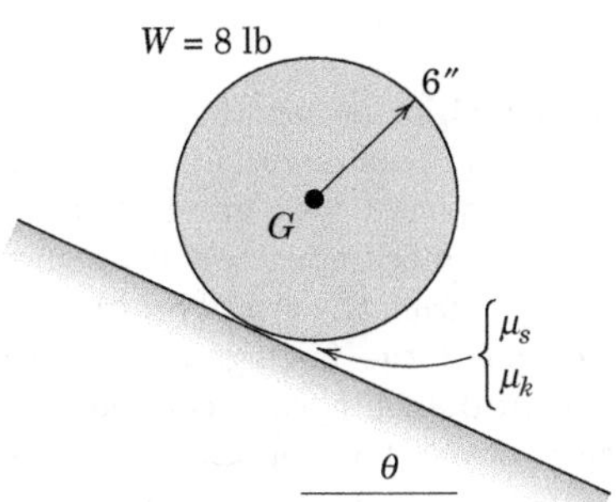

Problem 6/79

6/80 The 30-kg spool of outer radius $r_o = 450$ mm has a centroidal radius of gyration $\bar{k} = 275$ mm and a central shaft of radius $r_i = 200$ mm. The spool is at rest on the incline when a tension $T = 300$ N is applied to the end of a cable which is wrapped securely around the central shaft as shown. Determine the acceleration of the spool center G and the magnitude and direction of the friction force acting at the interface of the spool and incline. The friction coefficients there are $\mu_s = 0.45$ and $\mu_k = 0.30$. The tension T is applied parallel to the incline and the angle $\theta = 20°$.

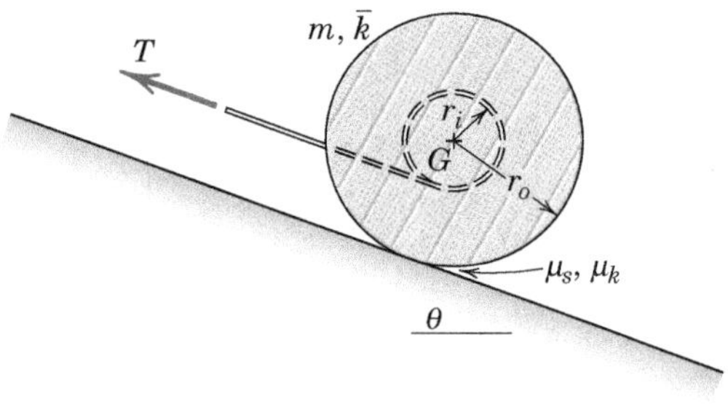

Problem 6/80

6/81 Repeat Prob. 6/80 for the case where the cable configuration has been changed as shown in the figure.

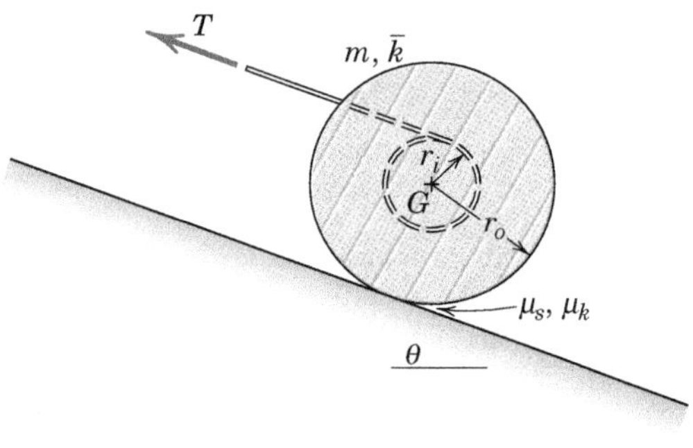

Problem 6/81

6/82 The fairing which covers the spacecraft package in the nose of the booster rocket is jettisoned when the rocket is in space where gravitational attraction is negligible. A mechanical actuator moves the two halves slowly from the closed position I to position II at which point the fairings are released to rotate freely about their hinges at O under the influence of the constant acceleration a of the rocket. When position III is reached, the hinge at O is released and the fairings drift away from the rocket. Determine the angular velocity ω of the fairing at the 90° position. The mass of each fairing is m with center of mass at G and radius of gyration k_O about O.

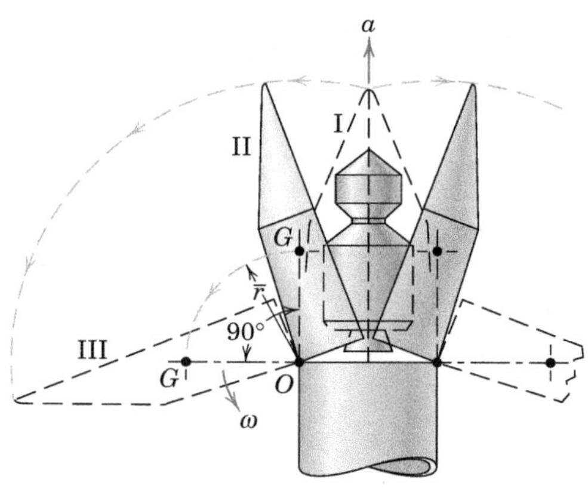

Problem 6/82

6/83 The uniform heavy bar AB of mass m is moving on its light end rollers along the horizontal with a velocity v when end A passes point C and begins to move on the curved portion of the path with radius r. Determine the force exerted by the path on roller A immediately after it passes C.

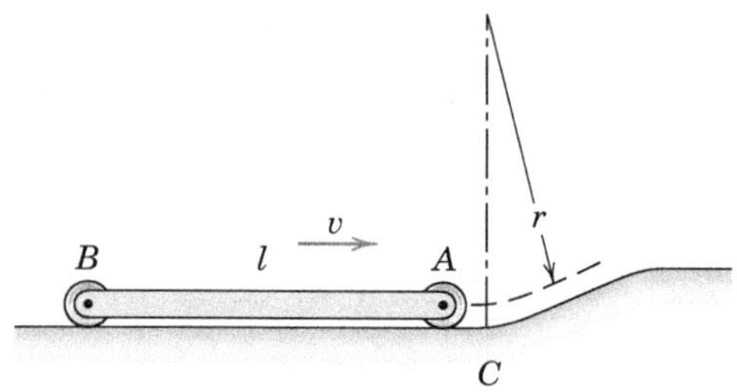

Problem 6/83

6/84 The uniform slender bar of mass m and total length L is released from rest in the position shown. Determine the force supported by the small roller at A and the acceleration of roller A along the smooth guide. Evaluate your results for $\theta = 15°$.

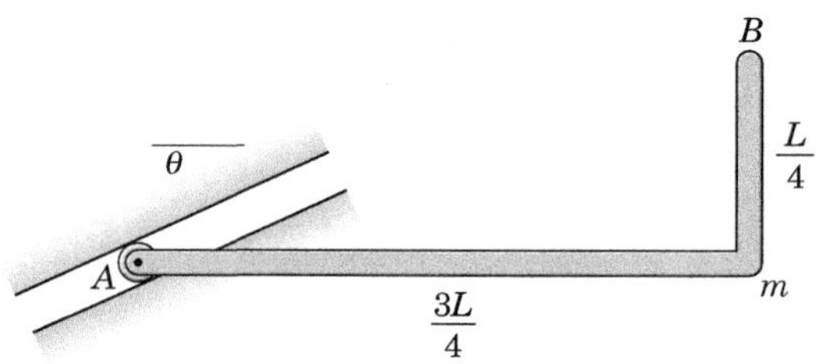

Problem 6/84

6/85 During a test, the car travels in a horizontal circle of radius R and has a forward tangential acceleration a. Determine the lateral reactions at the front and rear wheel pairs if (a) the car speed $v = 0$ and (b) the speed $v \neq 0$. The car mass is m and its polar moment of inertia (about a vertical axis through G) is $\bar{I}$. Assume that $R \gg d$.

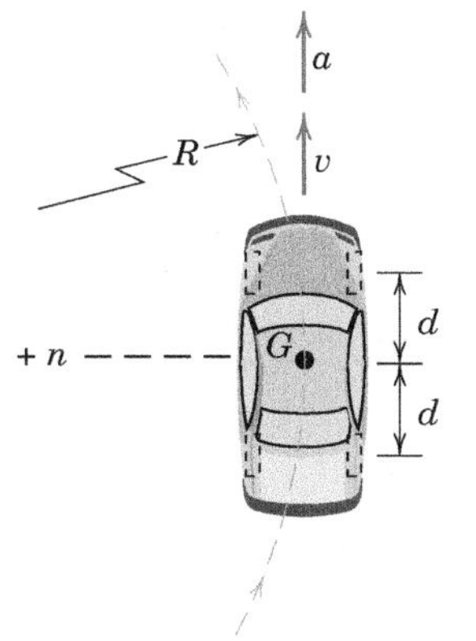

Problem 6/85

6/86 The system of Prob. 6/20 is repeated here. If the hoop- and semicylinder-assembly is centered on the top of the stationary cart and the system is released from rest, determine the initial acceleration a of the cart and the angular acceleration α of the hoop and semicylinder. Friction between the hoop and cart is sufficient to prevent slip. Motion takes place in the x-y plane. Neglect the mass of the cart wheels and any friction in the wheel bearings.

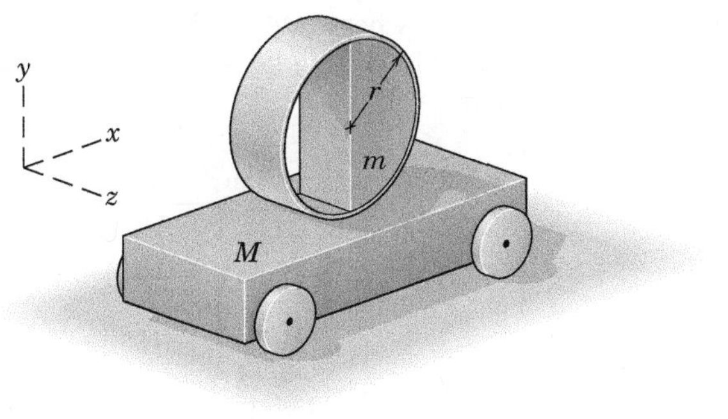

Problem 6/86

6/87 The 9-ft steel beam weighs 280 lb and is hoisted from rest where the tension in each of the cables is 140 lb. If the hoisting drums are given initial angular accelerations $\alpha_1 = 4$ rad/sec^2 and $\alpha_2 = 6$ rad/sec^2, calculate the corresponding tensions T_A and T_B in the cables. The beam may be treated as a slender bar.

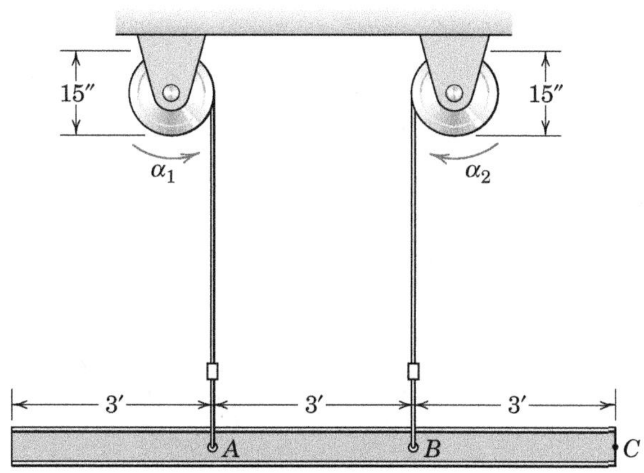

Problem 6/87

Representative Problems

6/88 The system is released from rest with the cable taut, and the homogeneous cylinder does not slip on the rough incline. Determine the angular acceleration of the cylinder and the minimum coefficient μ_s of friction for which the cylinder will not slip.

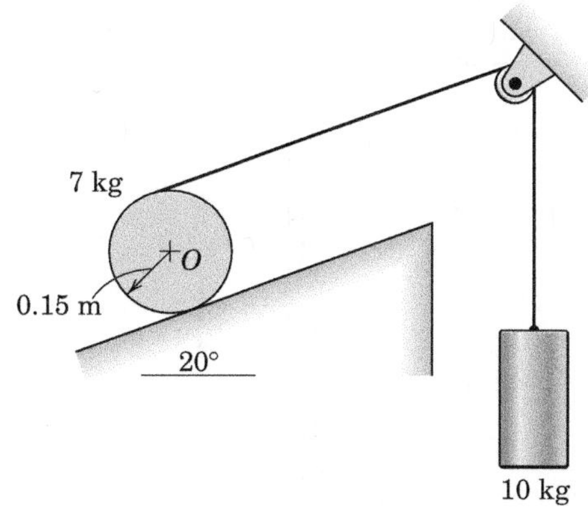

Problem 6/88

6/89 The mass center G of the 20-lb wheel is off center by 0.50 in. If G is in the position shown as the wheel rolls without slipping through the bottom of the circular path of 6-ft radius with an angular velocity ω of 10 rad/sec, compute the force P exerted by the path on the wheel. (Be careful to use the correct mass-center acceleration.)

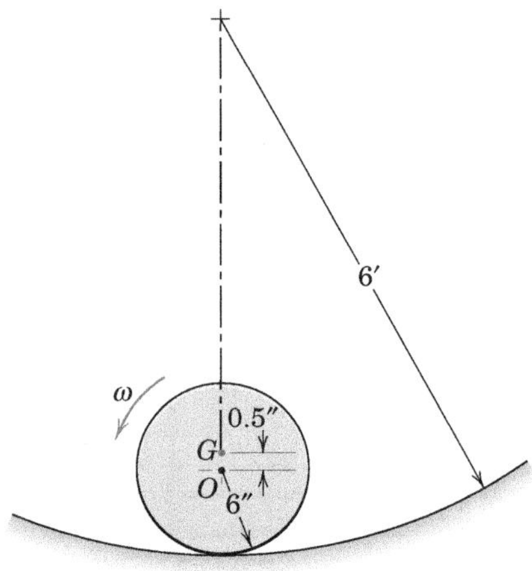

Problem 6/89

6/90 End A of the uniform 5-kg bar is pinned freely to the collar, which has an acceleration $a = 4$ m/s^2 along the fixed horizontal shaft. If the bar has a clockwise angular velocity $\omega = 2$ rad/s as it swings past the vertical, determine the components of the force on the bar at A for this instant.

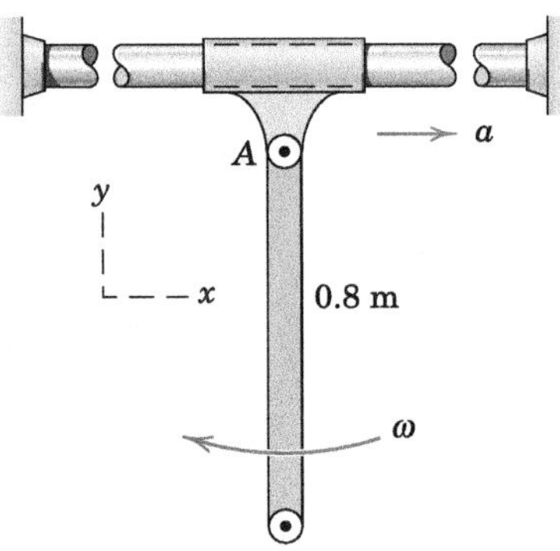

Problem 6/90

6/91 The uniform rectangular panel of mass m is moving to the right when wheel B drops off the horizontal support rail. Determine the resulting angular acceleration and the force T_A in the strap at A immediately after wheel B rolls off the rail. Neglect friction and the mass of the small straps and wheels.

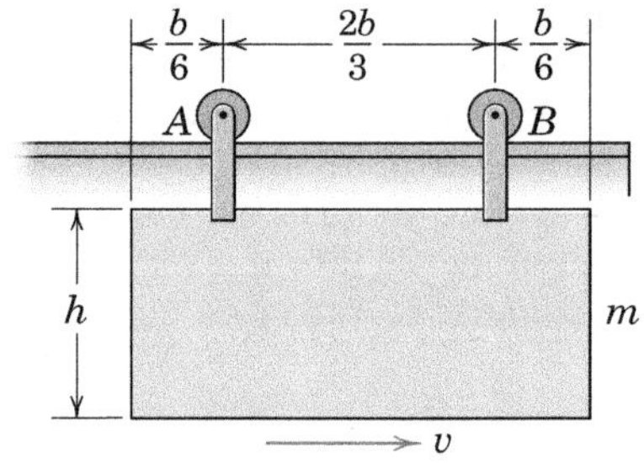

Problem 6/91

6/92 The truck, initially at rest with a solid cylindrical roll of paper in the position shown, moves forward with a constant acceleration a. Find the distance s which the truck goes before the paper rolls off the edge of its horizontal bed. Friction is sufficient to prevent slipping.

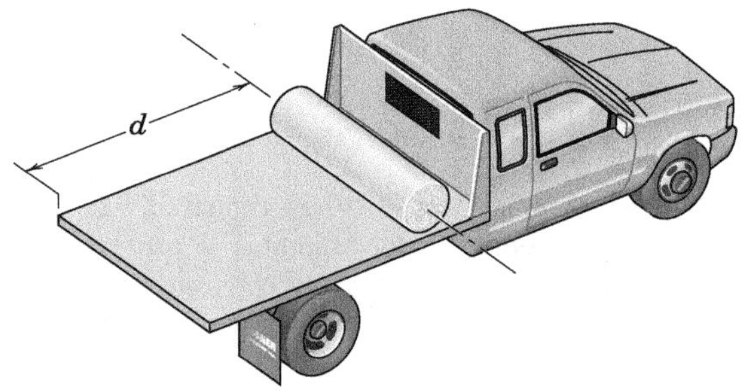

Problem 6/92

6/93 The crank OA rotates in the vertical plane with a constant clockwise angular velocity ω_0 of 4.5 rad/s. For the position where OA is horizontal, calculate the force under the light roller B of the 10-kg slender bar AB.

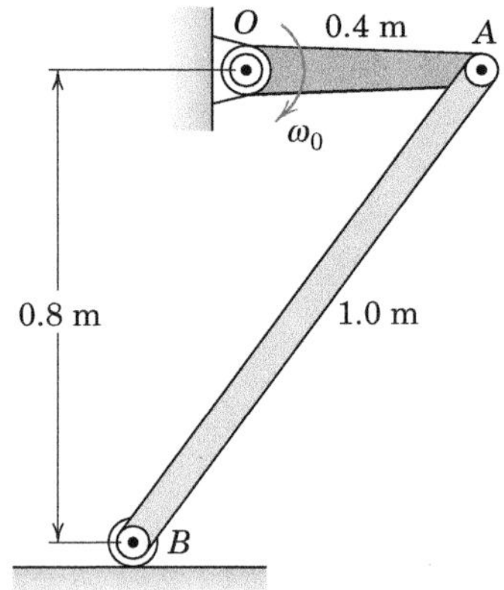

Problem 6/93

6/94 The uniform rectangular 300-lb plate is held in the horizontal position by two cables each of length $L = 3$ ft. If the cable at A suddenly breaks, calculate the tension T_B in the cable at B an instant after the break occurs.

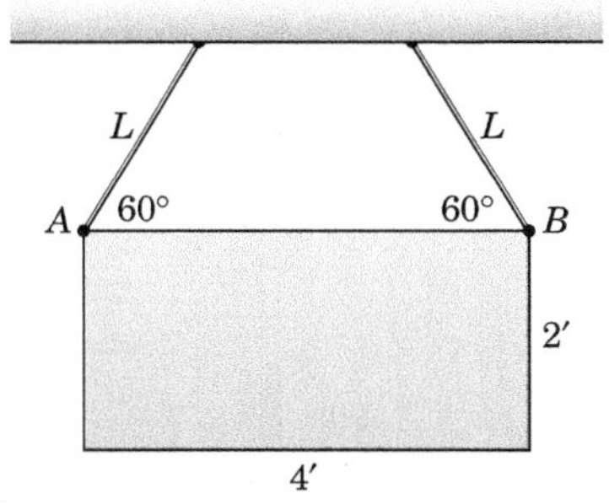

Problem 6/94

6/95 The rectangular plate of Prob. 6/94 is repeated here. The cables at A and B are now attached to a 50-lb trolley which is constrained to move in the horizontal guide. If the cable at A suddenly breaks, calculate the tension T_B in the cable at B an instant after the break occurs and the acceleration a_T of the trolley in the guide. Neglect friction and the mass of the trolley wheels.

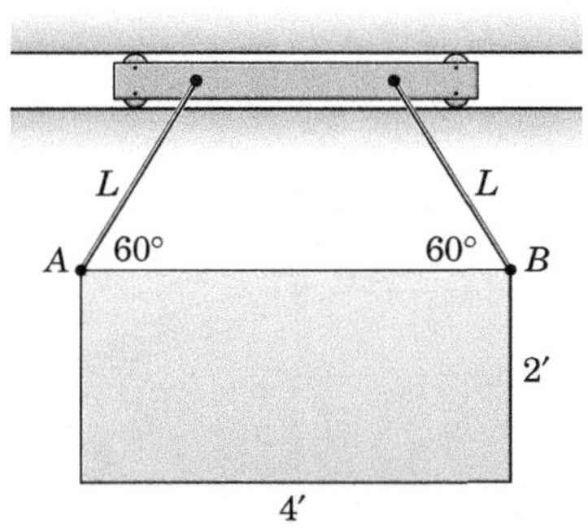

Problem 6/95

6/96 The robotic device of Prob. 6/68 is repeated here. Member AB is rotating about joint A with a counterclockwise angular velocity of 2 rad/s, and this rate is increasing at 4 rad/s^2. Determine the moment M_B exerted by arm AB on arm BC if joint B is held in a locked condition. The mass of arm BC is 4 kg, and the arm may be treated as a uniform slender rod.

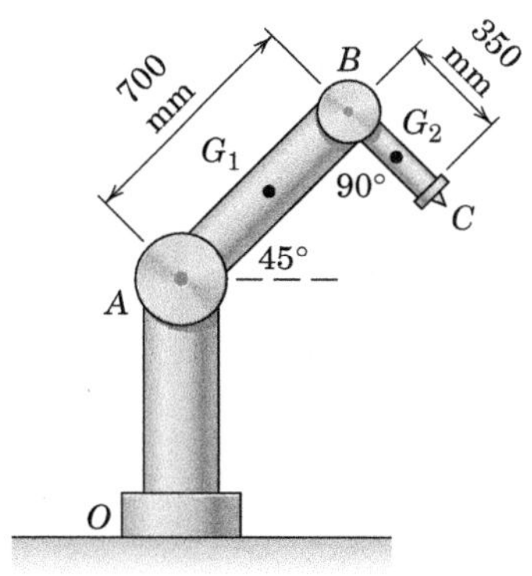

Problem 6/96

6/97 Small ball-bearing rollers mounted on the ends of the slender bar of mass m and length l constrain the motion of the bar in the horizontal x-y slots. If a couple M is applied to the bar initially at rest at $\theta = 45°$, determine the forces exerted on the rollers at A and B as the bar starts to move.

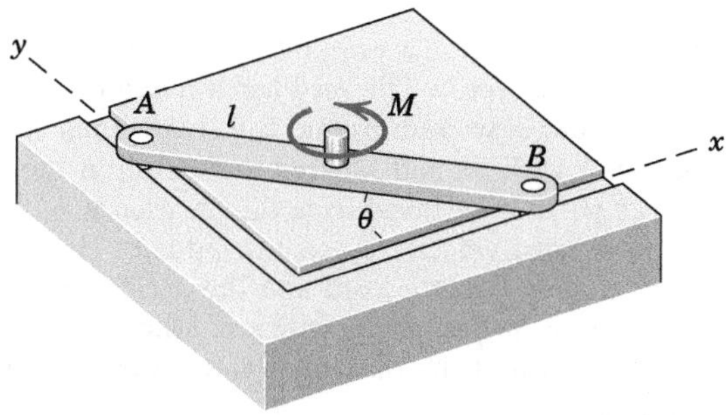

Problem 6/97

6/98 The assembly consisting of a uniform slender bar (mass $m/5$) and a rigidly attached uniform disk (mass $4m/5$) is freely pinned to point O on the collar that in turn slides on the fixed horizontal guide. The assembly is at rest when the collar is given a sudden acceleration a to the left as shown. Determine the initial angular acceleration of the assembly.

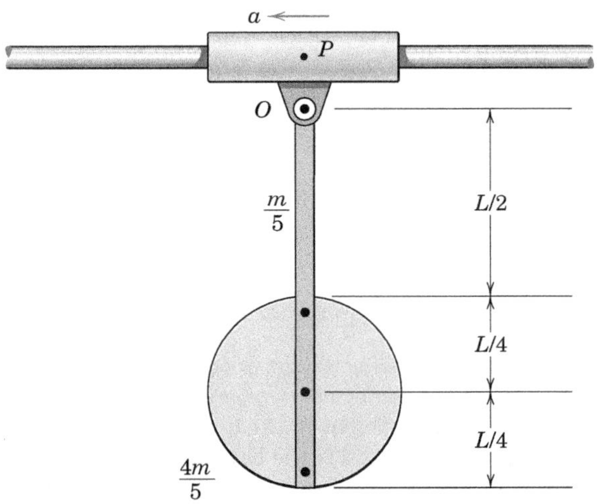

Problem 6/98

6/99 The yo-yo has a mass m and a radius of gyration k about its center O. The cord has a maximum length $y = L$ and is wound around the small inner hub of radius r with its end secured to a point on the hub. If the yo-yo is released from the position $y = 0$ with a downward velocity v_O of its center O, determine the tension T in the cord and the acceleration a of its center during its downward and upward motions. Also find the maximum downward velocity v of its center.

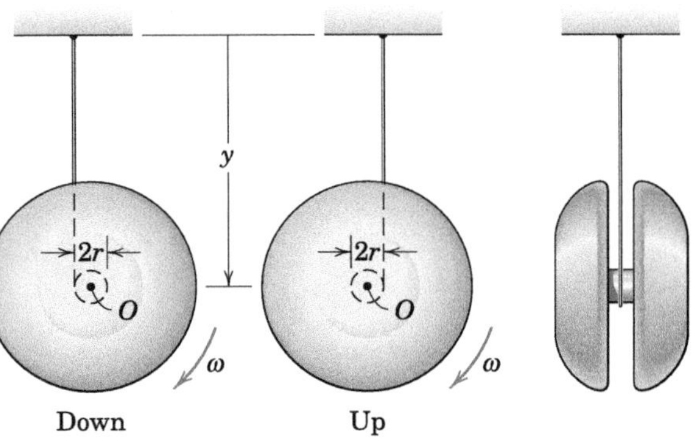

Problem 6/99

6/100 The uniform 12-ft pole is hinged to the truck bed and released from the vertical position as the truck starts from rest with an acceleration of 3 ft/sec^2. If the acceleration remains constant during the motion of the pole, calculate the angular velocity ω of the pole as it reaches the horizontal position.

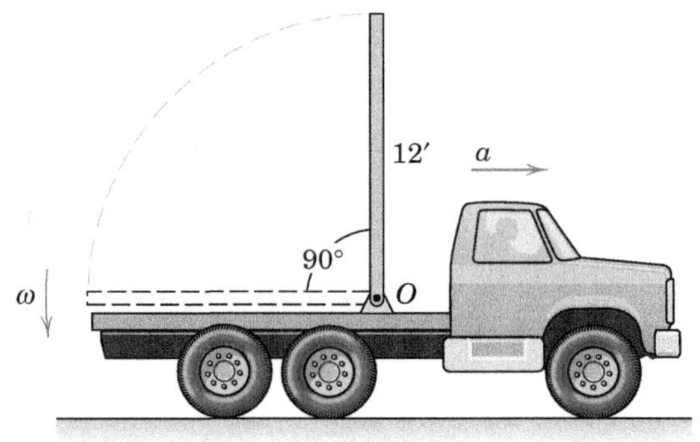

Problem 6/100

6/101 The uniform bar of mass m is constrained by the light rollers which move in the smooth guide, which lies in a vertical plane. If the bar is released from rest, what is the force at each roller an instant after release? Use the values $m = 18$ kg and $r = 150$ mm.

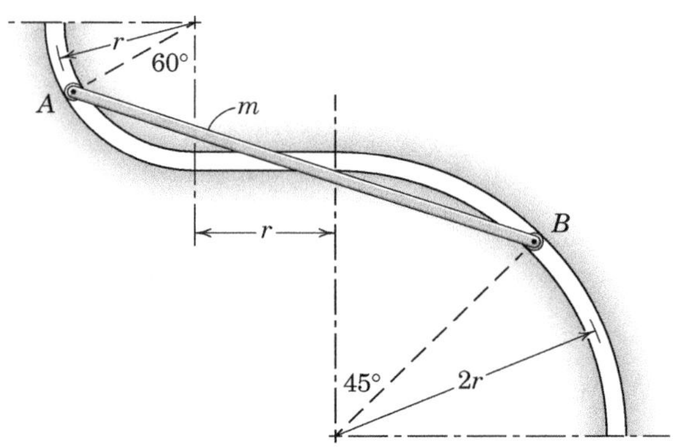

Problem 6/101

6/102 A bowling ball with a circumference of 27 in. weighs 14 lb and has a centroidal radius of gyration of 3.28 in. If the ball is released with a velocity of 20 ft/sec but with no angular velocity as it touches the alley floor, compute the distance traveled by the ball before it begins to roll without slipping. The coefficient of friction between the ball and the floor is 0.20.

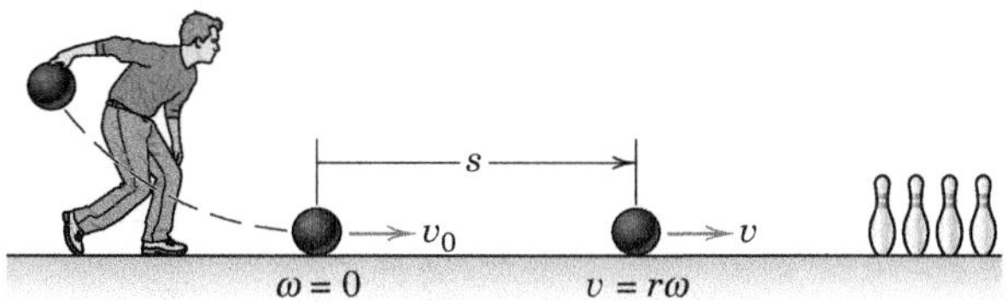

Problem 6/102

6/103 The figure shows the edge view of a uniform concrete slab with a mass of 12 Mg. The slab is being hoisted slowly by the winch D with cable attached to the dolly. At the position $\theta = 60°$, the distance x from the fixed ground position to the dolly is equal to the length $L = 4$ m of the slab. If the hoisting cable should break at this position, determine the initial acceleration a_A of the small dolly, whose mass is negligible, and the initial tension T in the fixed cable. End A of the slab will not slip on the dolly.

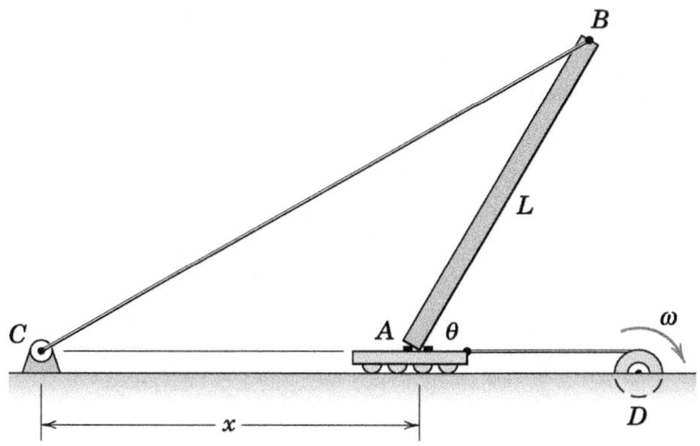

Problem 6/103

6/104 In a study of head injury against the instrument panel of a car during sudden or crash stops where lap belts without shoulder straps or airbags are used, the segmented human model shown in the figure is analyzed. The hip joint O is assumed to remain fixed relative to the car, and the torso above the hip is treated as a rigid body of mass m freely pivoted at O. The center of mass of the torso is at G with the initial position of OG taken as vertical. The radius of gyration of the torso about O is k_O. If the car is brought to a sudden stop with a constant deceleration a, determine the speed v relative to the car with which the model's head strikes the instrument panel. Substitute the values $m = 50$ kg, $\bar{r} = 450$ mm, $r = 800$ mm, $k_O = 550$ mm, $\theta = 45°$, and $a = 10g$ and compute v.

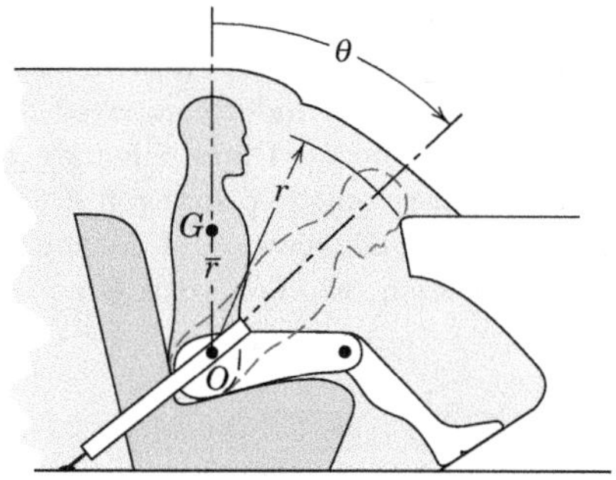

Problem 6/104

6/105 The 25-mm-thick uniform steel plate is being pulled very slowly into position by the cable connected to a winch. If the cable breaks at the instant represented, what are the force under each roller and the acceleration $\mathbf{a}_G$ of the center of mass? Neglect friction and the mass of the rollers, and consider the rollers small. Evaluate your results for $a = 3$ m, $b = 6$ m, and $r = 6$ m. (*Note:* Arc BC is of radius r, the same as arc BD.)

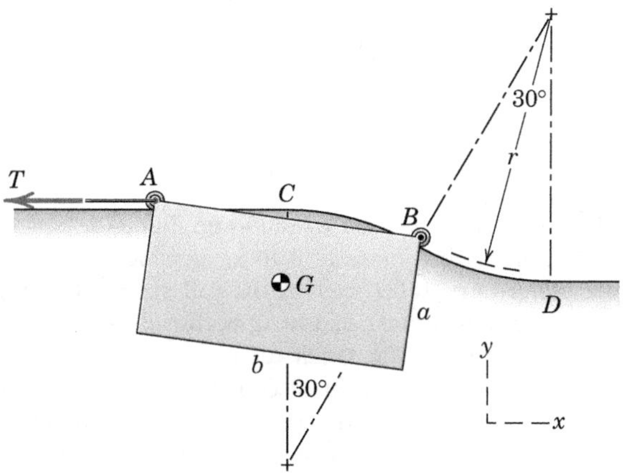

Problem 6/105

6/106 The connecting rod AB of a certain internal-combustion engine weighs 1.2 lb with mass center at G and has a radius of gyration about G of 1.12 in. The piston and piston pin A together weigh 1.80 lb. The engine is running at a constant speed of 3000 rev/min, so that the angular velocity of the crank is $3000(2\pi)/60 = 100\pi$ rad/sec. Neglect the weights of the components and the force exerted by the gas in the cylinder compared with the dynamic forces generated and calculate the magnitude of the force on the piston pin A for the crank angle $\theta = 90°$. (*Suggestion:* Use the alternative moment relation, Eq. 6/3, with B as the moment center.)

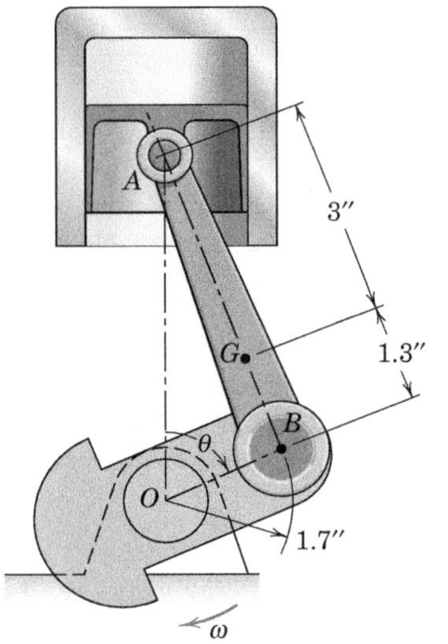

Problem 6/106

6/107 The truck carries a 1500-mm-diameter spool of cable with a mass of 0.75 kg per meter of length. There are 150 turns on the full spool. The empty spool has a mass of 140 kg with radius of gyration of 530 mm. The truck alone has a mass of 2030 kg with mass center at G. If the truck starts from rest with an initial acceleration of $0.2g$, determine (a) the tension T in the cable where it attaches to the wall and (b) the normal reaction under each pair of wheels. Neglect the rotational inertia of the truck wheels.

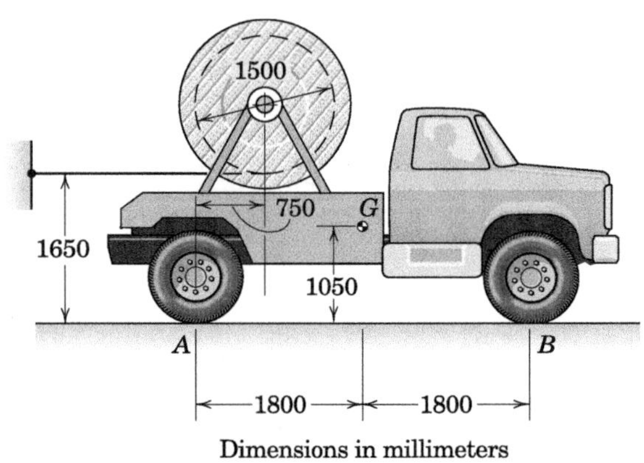

Problem 6/107

▶6/108 The four-bar mechanism lies in a vertical plane and is controlled by crank OA which rotates counterclockwise at a steady rate of 60 rev/min. Determine the torque M which must be applied to the crank at O when the crank angle $\theta = 45°$. The uniform coupler AB has a mass of 7 kg, and the masses of crank OA and the output arm BC may be neglected.

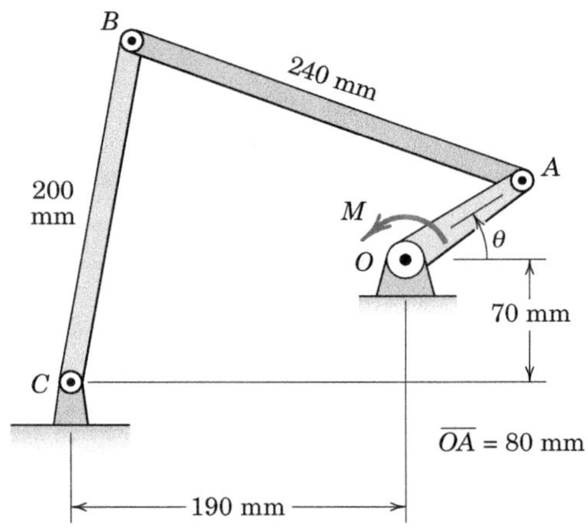

Problem 6/108

▶6/109 Repeat the analysis of Prob. 6/108 with the added information that the mass of crank OA is 1.2 kg and the mass of the output arm BC is 1.8 kg. Each of these bars may be considered uniform for this analysis.

▶**6/110** The Ferris wheel at an amusement park has an even number n of gondolas, each freely pivoted at its point of support on the wheel periphery. Each gondola has a loaded mass m, a radius of gyration k about its point of support A, and a mass center a distance h from A. The wheel structure has a moment of inertia I_O about its bearing at O. Determine an expression for the tangential force F which must be transmitted to the wheel periphery at C in order to give the wheel an initial angular acceleration α starting from rest. *Suggestion:* Analyze the gondolas in pairs A and B. Be careful not to assume that the initial angular acceleration of the gondolas is the same as that of the wheel. (*Note:* An American engineer named George Washington Gale Ferris, Jr., created a giant amusement-wheel ride for the World's Columbian Exposition in Chicago in 1893. The wheel was 250 ft in diameter with 36 gondolas, each of which carried up to 60 passengers. Fully loaded, the wheel and gondolas had a mass of 1200 tons. The ride was powered by a 1000-hp steam engine.)

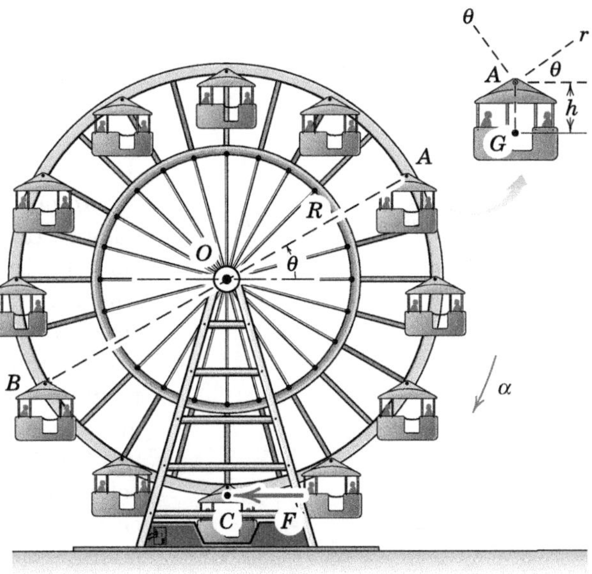

Problem 6/110

SECTION B WORK AND ENERGY

6/6 Work-Energy Relations

In our study of the kinetics of particles in Arts. 3/6 and 3/7, we developed the principles of work and energy and applied them to the motion of a particle and to selected cases of connected particles. We found that these principles were especially useful in describing motion which resulted from the cumulative effect of forces acting through distances. Furthermore, when the forces were conservative, we were able to determine velocity changes by analyzing the energy conditions at the beginning and end of the motion interval. For finite displacements, the work-energy method eliminates the necessity for determining the acceleration and integrating it over the interval to obtain the velocity change. These same advantages are realized when we extend the work-energy principles to describe rigid-body motion.

Before carrying out this extension, you should review the definitions and concepts of work, kinetic energy, gravitational and elastic potential energy, conservative forces, and power treated in Arts. 3/6 and 3/7 because we will apply them to rigid-body problems. You should also review Arts. 4/3 and 4/4 on the kinetics of systems of particles, in which we extended the principles of Arts. 3/6 and 3/7 to encompass any general system of mass particles, which includes rigid bodies.

Work of Forces and Couples

The work done by a force $\mathbf{F}$ has been treated in detail in Art. 3/6 and is given by

$$U = \int \mathbf{F} \cdot d\mathbf{r} \qquad \text{or} \qquad U = \int (F \cos \alpha)\, ds$$

where $d\mathbf{r}$ is the infinitesimal vector displacement of the point of application of $\mathbf{F}$, as shown in Fig. 3/2a. In the equivalent scalar form of the integral, α is the angle between $\mathbf{F}$ and the direction of the displacement, and ds is the magnitude of the vector displacement $d\mathbf{r}$.

We frequently need to evaluate the work done by a couple M which acts on a rigid body during its motion. Figure 6/11 shows a couple $M = Fb$ acting on a rigid body which moves in the plane of the couple. During time dt the body rotates through an angle $d\theta$, and line AB moves to $A'B'$. We may consider this motion in two parts, first a translation to $A'B''$ and then a rotation $d\theta$ about A'. We see immediately that during the translation the work done by one of the forces cancels that done by the other force, so that the net work done is $dU = F(b\, d\theta) = M\, d\theta$ due to the rotational part of the motion. If the couple acts in the sense opposite to the rotation, the work done is negative. During a finite rotation, the work done by a couple M whose plane is parallel to the plane of motion is, therefore,

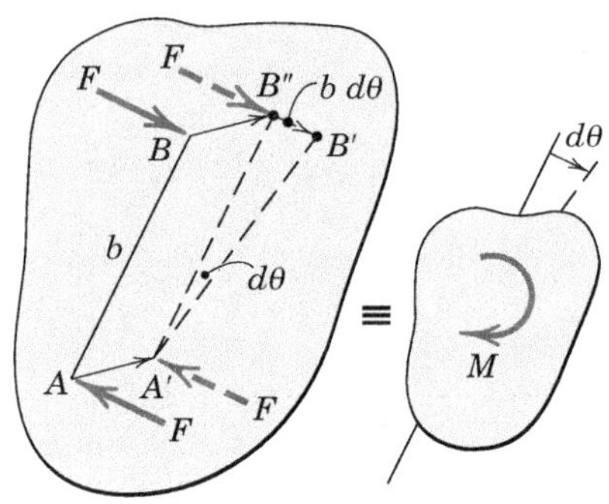

Figure 6/11

$$U = \int M\, d\theta$$

Kinetic Energy

We now use the familiar expression for the kinetic energy of a particle to develop expressions for the kinetic energy of a rigid body for each of the three classes of rigid-body plane motion illustrated in Fig. 6/12.

(a) Translation. The translating rigid body of Fig. 6/12*a* has a mass m and all of its particles have a common velocity v. The kinetic energy of any particle of mass m_i of the body is $T_i = \frac{1}{2}m_i v^2$, so for the entire body $T = \Sigma\frac{1}{2}m_i v^2 = \frac{1}{2}v^2\Sigma m_i$ or

$$T = \frac{1}{2}mv^2 \tag{6/7}$$

This expression holds for both rectilinear and curvilinear translation.

(b) Fixed-axis rotation. The rigid body in Fig. 6/12*b* rotates with an angular velocity ω about the fixed axis through O. The kinetic energy of a representative particle of mass m_i is $T_i = \frac{1}{2}m_i(r_i\omega)^2$. Thus, for the entire body $T = \frac{1}{2}\omega^2\Sigma m_i r_i^{\,2}$. But the moment of inertia of the body about O is $I_O = \Sigma m_i r_i^{\,2}$, so

$$T = \frac{1}{2}I_O\omega^2 \tag{6/8}$$

Note the similarity in the forms of the kinetic energy expressions for translation and rotation. You should verify that the dimensions of the two expressions are identical.

(c) General plane motion. The rigid body in Fig. 6/12*c* executes plane motion where, at the instant considered, the velocity of its mass center G is $\bar{v}$ and its angular velocity is ω. The velocity v_i of a representative particle of mass m_i may be expressed in terms of the mass-center velocity $\bar{v}$ and the velocity $\rho_i\omega$ relative to the mass center as shown. With the aid of the law of cosines, we write the kinetic energy of the body as the sum ΣT_i of the kinetic energies of all its particles. Thus,

$$T = \Sigma\frac{1}{2}m_i v_i^{\,2} = \Sigma\frac{1}{2}m_i(\bar{v}^2 + \rho_i^{\,2}\omega^2 + 2\bar{v}\rho_i\omega\cos\theta)$$

Because ω and $\bar{v}$ are common to all terms in the third summation, we may factor them out. Thus, the third term in the expression for T becomes

$$\omega\bar{v}\,\Sigma m_i\rho_i\cos\theta = \omega\bar{v}\,\Sigma m_i y_i = 0$$

since $\Sigma m_i y_i = m\bar{y} = 0$. The kinetic energy of the body is then $T = \frac{1}{2}\bar{v}^2\Sigma m_i + \frac{1}{2}\omega^2\Sigma m_i\rho_i^{\,2}$ or

$$T = \frac{1}{2}m\bar{v}^2 + \frac{1}{2}\bar{I}\omega^2 \tag{6/9}$$

where $\bar{I}$ is the moment of inertia of the body about its mass center. This expression for kinetic energy clearly shows the separate contributions to the total kinetic energy resulting from the translational velocity $\bar{v}$ of the mass center and the rotational velocity ω about the mass center.

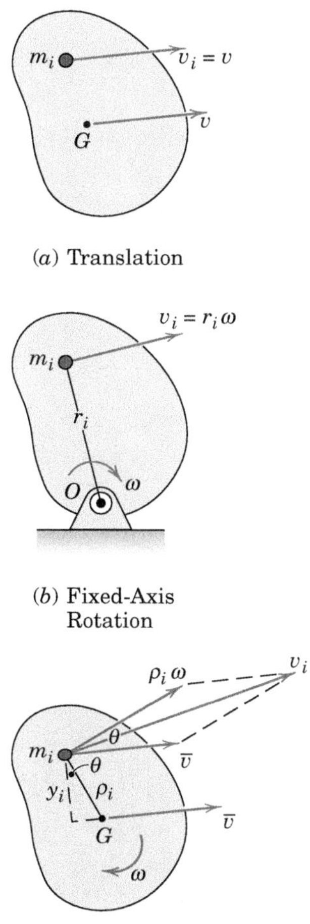

(*a*) Translation

(*b*) Fixed-Axis Rotation

(*c*) General Plane Motion

Figure 6/12

The kinetic energy of plane motion may also be expressed in terms of the rotational velocity about the instantaneous center C of zero velocity. Because C momentarily has zero velocity, the proof leading to Eq. 6/8 for the fixed point O holds equally well for point C, so that, alternatively, we may write the kinetic energy of a rigid body in plane motion as

$$T = \frac{1}{2}I_C\omega^2 \tag{6/10}$$

In Art. 4/3 we derived Eq. 4/4 for the kinetic energy of any system of mass. We now see that this expression is equivalent to Eq. 6/9 when the mass system is rigid. For a rigid body, the quantity $\dot{\boldsymbol{\rho}}_i$ in Eq. 4/4 is the velocity of the representative particle relative to the mass center and is the vector $\boldsymbol{\omega} \times \boldsymbol{\rho}_i$, which has the magnitude $\rho_i\omega$. The summation term in Eq. 4/4 becomes $\Sigma\frac{1}{2}m_i(\rho_i\omega)^2 = \frac{1}{2}\omega^2\Sigma m_i{\rho_i}^2 = \frac{1}{2}\bar{I}\omega^2$, which brings Eq. 4/4 into agreement with Eq. 6/9.

Potential Energy and the Work-Energy Equation

Gravitational potential energy V_g and elastic potential energy V_e were covered in detail in Art. 3/7. Recall that the symbol U' (rather than U) is used to denote the work done by all forces except the weight and elastic forces, which are accounted for in the potential-energy terms.

The work-energy relation, Eq. 3/15*a*, was introduced in Art. 3/6 for particle motion and was generalized in Art. 4/3 to include the motion of a general system of particles. This equation

$$T_1 + U_{1\text{-}2} = T_2 \qquad [4/2]$$

applies to any mechanical system. For application to the motion of a single rigid body, the terms T_1 and T_2 must include the effects of translaton and rotation as given by Eqs. 6/7, 6/8, 6/9, or 6/10, and $U_{1\text{-}2}$ is the work done by all external forces. On the other hand, if we choose to express the effects of weight and springs by means of potential energy rather than work, we may rewrite the above equation as

$$T_1 + V_1 + U'_{1\text{-}2} = T_2 + V_2 \qquad [4/3a]$$

where the prime denotes the work done by all forces other than weight and spring forces.

When applied to an interconnected system of rigid bodies, Eq. 4/3*a* includes the effect of stored elastic energy in the connections, as well as that of gravitational potential energy for the various members. The term $U'_{1\text{-}2}$ includes the work of all forces external to the system (other than gravitational forces), as well as the negative work of internal friction forces, if any. The terms T_1 and T_2 are the initial and final kinetic energies of all moving parts over the interval of motion in question.

When the work-energy principle is applied to a single rigid body, either a *free-body diagram* or an *active-force diagram* should be used. In the case of an interconnected system of rigid bodies, an active-force diagram of the entire system should be drawn in order to isolate the system and disclose all forces which do work on the system. Diagrams should also be drawn to disclose the initial and final positions of the system for the given interval of motion.

The work-energy equation provides a direct relationship between the forces which do work and the corresponding changes in the motion of a mechanical system. However, if there is appreciable internal mechanical friction, then the system must be dismembered in order to disclose the kinetic-friction forces and account for the negative work which they do. When the system is dismembered, however, one of the primary advantages of the work-energy approach is automatically lost. The work-energy method is most useful for analyzing conservative systems of interconnected bodies, where energy loss due to the negative work of friction forces is negligible.

Power

The concept of power was discussed in Art. 3/6, which treated work-energy for particle motion. Recall that power is the time rate at which work is performed. For a force $\mathbf{F}$ acting on a rigid body in plane motion, the power developed by that force at a given instant is given by Eq. 3/16 and is the rate at which the force is doing work. The power is given by

$$P = \frac{dU}{dt} = \frac{\mathbf{F}\cdot d\mathbf{r}}{dt} = \mathbf{F}\cdot\mathbf{v}$$

where $d\mathbf{r}$ and $\mathbf{v}$ are, respectively, the differential displacement and the velocity of the point of application of the force.

Similarly, for a couple M acting on the body, the power developed by the couple at a given instant is the rate at which it is doing work, and is given by

$$P = \frac{dU}{dt} = \frac{M\,d\theta}{dt} = M\omega$$

where $d\theta$ and ω are, respectively, the differential angular displacement and the angular velocity of the body. If the senses of M and ω are the same, the power is positive and energy is supplied to the body. Conversely, if M and ω have opposite senses, the power is negative and energy is removed from the body. If the force $\mathbf{F}$ and the couple M act simultaneously, the total instantaneous power is

$$P = \mathbf{F}\cdot\mathbf{v} + M\omega$$

We may also express power by evaluating the rate at which the total mechanical energy of a rigid body or a system of rigid bodies is changing. The work-energy relation, Eq. 4/3, for an infinitesimal displacement is

$$dU' = dT + dV$$

where dU' is the work of the active forces and couples applied to the body or to the system of bodies. Excluded from dU' are the work of gravitational forces and that of spring forces, which are accounted for in the dV term. Dividing by dt gives the total power of the active forces and couples as

$$P = \frac{dU'}{dt} = \dot{T} + \dot{V} = \frac{d}{dt}(T + V)$$

Thus, we see that the power developed by the active forces and couples equals the rate of change of the total mechanical energy of the body or system of bodies.

We note from Eq. 6/9 that, for a given body, the first term may be written

$$\begin{aligned}\dot{T} = \frac{dT}{dt} &= \frac{d}{dt}\left(\frac{1}{2}m\bar{\mathbf{v}}\cdot\bar{\mathbf{v}} + \frac{1}{2}\bar{I}\omega^2\right) \\ &= \frac{1}{2}m(\bar{\mathbf{a}}\cdot\bar{\mathbf{v}} + \bar{\mathbf{v}}\cdot\bar{\mathbf{a}}) + \bar{I}\omega\dot{\omega} \\ &= m\bar{\mathbf{a}}\cdot\bar{\mathbf{v}} + \bar{I}\alpha(\omega) = \mathbf{R}\cdot\bar{\mathbf{v}} + \overline{M}\omega\end{aligned}$$

where $\mathbf{R}$ is the resultant of *all* forces acting on the body and $\overline{M}$ is the resultant moment about the mass center G of *all* forces. The dot product accounts for the case of curvilinear motion of the mass center, where $\bar{\mathbf{a}}$ and $\bar{\mathbf{v}}$ are not in the same direction.

Transtock/SuperStock

Many cars are now using regenerative braking, which means that the overall kinetic energy of the car is captured via generators attached to the wheels rather than being wasted through heat energy as is the case with conventional friction brakes.

Sample Problem 6/9

The wheel rolls up the incline on its hubs without slipping and is pulled by the 100-N force applied to the cord wrapped around its outer rim. If the wheel starts from rest, compute its angular velocity ω after its center has moved a distance of 3 m up the incline. The wheel has a mass of 40 kg with center of mass at O and has a centroidal radius of gyration of 150 mm. Determine the power input from the 100-N force at the end of the 3-m motion interval.

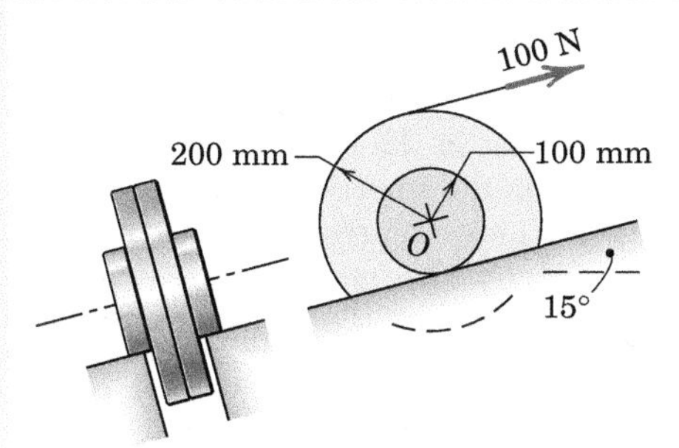

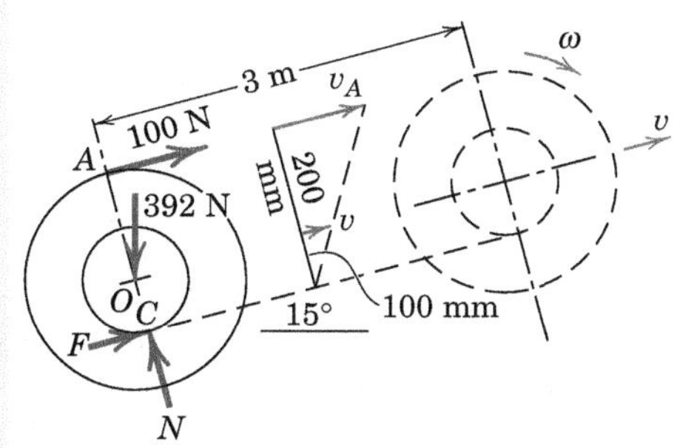

Solution. Of the four forces shown on the free-body diagram of the wheel, ① only the 100-N pull and the weight of 40(9.81) = 392 N do work. The friction force does no work as long as the wheel does not slip. By use of the concept of the instantaneous center C of zero velocity, we see that a point A on the cord to which the 100-N force is applied has a velocity $v_A = [(200 + 100)/100]v$. Hence, point A on the cord moves a distance of (200 + 100)/100 = 3 times as far as the center O. Thus, with the effect of the weight included in the U-term, the work done on the wheel becomes

② $$U_{1\text{-}2} = 100\frac{200 + 100}{100}(3) - (392 \sin 15°)(3) = 595 \text{ J}$$

The wheel has general plane motion, so that the initial and final kinetic energies are

③ $$[T = \tfrac{1}{2}m\bar{v}^2 + \tfrac{1}{2}\bar{I}\omega^2] \qquad T_1 = 0 \qquad T_2 = \tfrac{1}{2}40(0.10\omega)^2 + \tfrac{1}{2}40(0.15)^2\omega^2 = 0.650\omega^2$$

The work-energy equation gives

$$[T_1 + U_{1\text{-}2} = T_2] \qquad 0 + 595 = 0.650\omega^2 \qquad \omega = 30.3 \text{ rad/s}$$

Alternatively, the kinetic energy of the wheel may be written

④ $$[T = \tfrac{1}{2}I_C\omega^2] \qquad T = \tfrac{1}{2}40[(0.15)^2 + (0.10)^2]\omega^2 = 0.650\omega^2$$

The power input from the 100-N force when $\omega = 30.3$ rad/s is

⑤ $$[P = \mathbf{F}\cdot\mathbf{v}] \qquad P_{100} = 100(0.3)(30.3) = 908 \text{ W}$$ *Ans.*

Helpful Hints

① Since the velocity of the instantaneous center C on the wheel is zero, it follows that the rate at which the friction force does work is continuously zero. Hence, F does no work as long as the wheel does not slip. If the wheel were rolling on a moving platform, however, the friction force would do work, even if the wheel were not slipping.

② Note that the component of the weight down the plane does negative work.

③ Be careful to use the correct radius in the expression $v = r\omega$ for the velocity of the center of the wheel.

④ Recall that $I_C = \bar{I} + m\overline{OC}^2$, where $\bar{I} = I_O = mk_O^{\,2}$.

⑤ The velocity here is that of the application point of the 100-N force.

Sample Problem 6/10

The 4-ft slender bar weighs 40 lb with mass center at B and is released from rest in the position for which θ is essentially zero. Point B is confined to move in the smooth vertical guide, while end A moves in the smooth horizontal guide and compresses the spring as the bar falls. Determine (*a*) the angular velocity of the bar as the position $\theta = 30°$ is passed and (*b*) the velocity with which B strikes the horizontal surface if the stiffness of the spring is 30 lb/in.

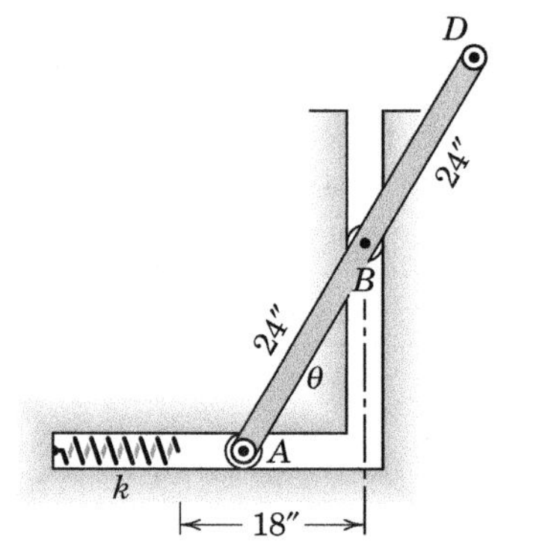

Solution. With the friction and mass of the small rollers at A and B neglected, the system may be treated as being conservative.

Part (a). For the first interval of motion from $\theta = 0$ (state 1) to $\theta = 30°$ (state 2), the spring is not engaged, so that there is no V_e term in the energy equation. If ① we adopt the alternative of treating the work of the weight in the V_g term, then there are no other forces which do work, and $U'_{1\text{-}2} = 0$.

Since we have a constrained plane motion, there is a kinematic relation between the velocity v_B of the center of mass and the angular velocity ω of the bar. This relation is easily obtained by using the instantaneous center C of zero velocity and noting that $v_B = \overline{CB}\omega$. Thus, the kinetic energy of the bar in the 30° position becomes

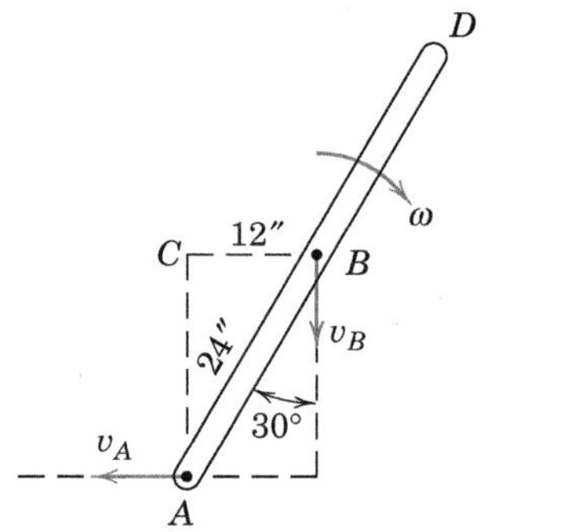

$$[T = \tfrac{1}{2}m\bar{v}^2 + \tfrac{1}{2}\bar{I}\omega^2] \qquad T = \frac{1}{2}\frac{40}{32.2}\left(\frac{12}{12}\omega\right)^2 + \frac{1}{2}\left(\frac{1}{12}\frac{40}{32.2}4^2\right)\omega^2 = 1.449\omega^2$$

With a datum established at the initial position of the mass center B, our initial and final gravitational potential energies are

$$V_1 = 0 \qquad V_2 = 40(2\cos 30° - 2) = -10.72 \text{ ft-lb}$$

We now substitute into the energy equation and obtain

$$[T_1 + V_1 + U'_{1\text{-}2} = T_2 + V_2] \qquad 0 + 0 + 0 = 1.449\omega^2 - 10.72$$

$$\omega = 2.72 \text{ rad/sec} \qquad \textit{Ans.}$$

Part (b). We define state 3 as that for which $\theta = 90°$. The initial and final spring potential energies are

② $$[V_e = \tfrac{1}{2}kx^2] \qquad V_1 = 0 \qquad V_3 = \frac{1}{2}(30)(24 - 18)^2\frac{1}{12} = 45 \text{ ft-lb}$$

In the final horizontal position, point A has no velocity, so that the bar is, in effect, rotating about A. Hence, its final kinetic energy is

$$[T = \tfrac{1}{2}I_A\omega^2] \qquad T_3 = \frac{1}{2}\left(\frac{1}{3}\frac{40}{32.2}4^2\right)\left(\frac{v_B}{24/12}\right)^2 = 0.828{v_B}^2$$

The final gravitational potential energy is

$$[V_g = Wh] \qquad V_3 = 40(-2) = -80 \text{ ft-lb}$$

Substituting into the energy equation gives

$$[T_1 + V_1 + U'_{1\text{-}3} = T_3 + V_3] \qquad 0 + 0 + 0 = 0.828{v_B}^2 + 45 - 80$$

$$v_B = 6.50 \text{ ft/sec} \qquad \textit{Ans.}$$

Alternatively, if the bar alone constitutes the system, the active-force diagram shows the weight, which does positive work, and the spring force kx, which does negative work. We would then write

$$[T_1 + U_{1\text{-}3} = T_3] \qquad 80 - 45 = 0.828{v_B}^2$$

which is identical with the previous result.

Helpful Hints

① We recognize that the forces acting on the bar at A and B are normal to the respective directions of motion and, hence, do no work.

② If we convert k to lb/ft, we have

$$V_e = \frac{1}{2}\left(30\frac{\text{lb}}{\text{in.}}\right)\left(12\frac{\text{in.}}{\text{ft}}\right)\left(\frac{24-18}{12}\text{ft}\right)^2 = 45 \text{ ft-lb}$$

Always check the consistency of your units.

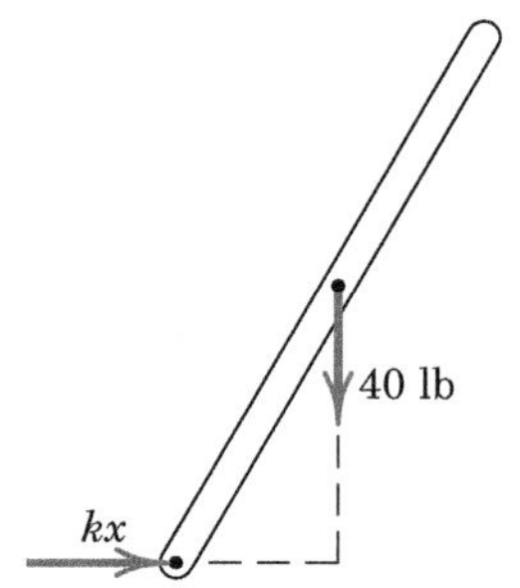

(Alternative Active-Force Diagram)

Sample Problem 6/11

In the mechanism shown, each of the two wheels has a mass of 30 kg and a centroidal radius of gyration of 100 mm. Each link OB has a mass of 10 kg and may be treated as a slender bar. The 7-kg collar at B slides on the fixed vertical shaft with negligible friction. The spring has a stiffness $k = 30$ kN/m and is contacted by the bottom of the collar when the links reach the horizontal position. If the collar is released from rest at the position $\theta = 45°$ and if friction is sufficient to prevent the wheels from slipping, determine (*a*) the velocity v_B of the collar as it first strikes the spring and (*b*) the maximum deformation x of the spring.

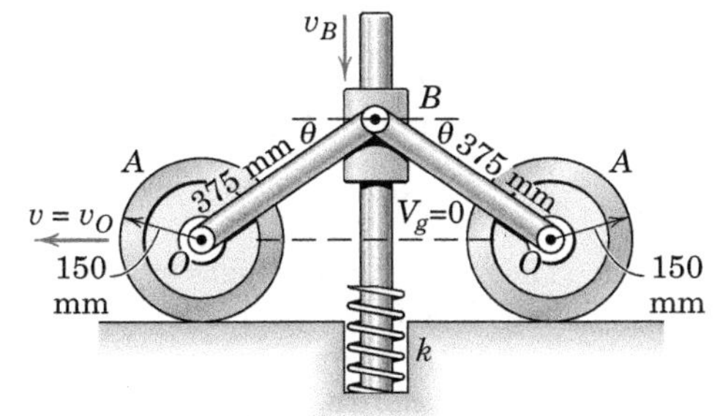

Solution. The mechanism executes plane motion and is conservative with the neglect of kinetic friction losses. We define states 1, 2, and 3 to be at $\theta = 45°$, $\theta = 0$, and maximum spring deflection, respectively. The datum for zero gravitational potential energy V_g is conveniently taken through O as shown.

(a) For the interval from $\theta = 45°$ to $\theta = 0$, we note that the initial and final kinetic energies of the wheels are zero since each wheel starts from rest and momentarily comes to rest at $\theta = 0$. Also, at position 2, each link is merely rotating about its point O so that

$$T_2 = [2(\tfrac{1}{2}I_O\omega^2)]_{\text{links}} + [\tfrac{1}{2}mv^2]_{\text{collar}}$$

$$= \frac{1}{3}10(0.375)^2\left(\frac{v_B}{0.375}\right)^2 + \frac{1}{2}7{v_B}^2 = 6.83{v_B}^2$$

The collar at B drops a distance $0.375/\sqrt{2} = 0.265$ m so that

$$V_1 = 2(10)(9.81)\frac{0.265}{2} + 7(9.81)(0.265) = 44.2 \text{ J} \qquad V_2 = 0$$

① Also, $U'_{1\text{-}2} = 0$. Hence,

$$[T_1 + V_1 + U'_{1\text{-}2} = T_2 + V_2] \qquad 0 + 44.2 + 0 = 6.83{v_B}^2 + 0$$

$$v_B = 2.54 \text{ m/s} \qquad \textit{Ans.}$$

(b) At the condition of maximum deformation x of the spring, all parts are momentarily at rest, which makes $T_3 = 0$. Thus,

$$[T_1 + V_1 + U'_{1\text{-}3} = T_3 + V_3] \qquad 0 + 2(10)(9.81)\frac{0.265}{2} + 7(9.81)(0.265) + 0$$

$$= 0 - 2(10)(9.81)\left(\frac{x}{2}\right) - 7(9.81)x + \tfrac{1}{2}(30)(10^3)x^2$$

Solution for the positive value of x gives

$$x = 60.1 \text{ mm} \qquad \textit{Ans.}$$

It should be noted that the results of parts (*a*) and (*b*) involve a very simple net energy change despite the fact that the mechanism has undergone a fairly complex sequence of motions. Solution of this and similar problems by other than a work-energy approach is not an inviting prospect.

Helpful Hint

① With the work of the weight of the collar B included in the potential-energy terms, there are no other forces external to the system which do work. The friction force acting under each wheel does no work since the wheel does not slip, and, of course, the normal force does no work here. Hence, $U'_{1\text{-}2} = 0$.

PROBLEMS

(In the following problems neglect any energy loss due to kinetic friction unless otherwise instructed.)

Introductory Problems

6/111 The slender rod of mass m and length l has a particle (negligible radius, mass $2m$) attached to its end. If the body is released from rest when in the position shown, determine its angular velocity as it passes the vertical position.

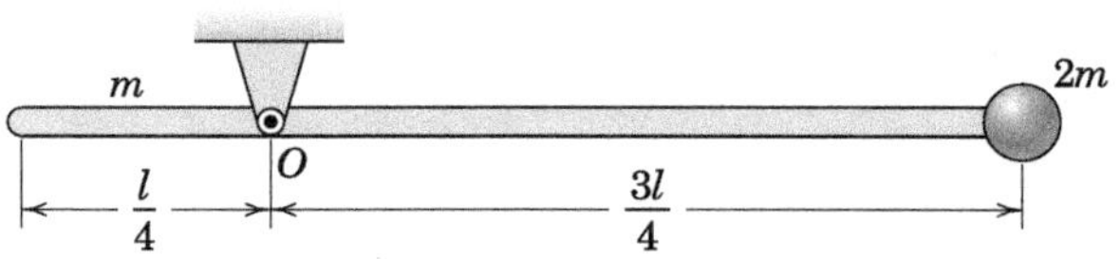

Problem 6/111

6/112 The log is suspended by the two parallel 5-m cables and used as a battering ram. At what angle θ should the log be released from rest in order to strike the object to be smashed with a velocity of 4 m/s?

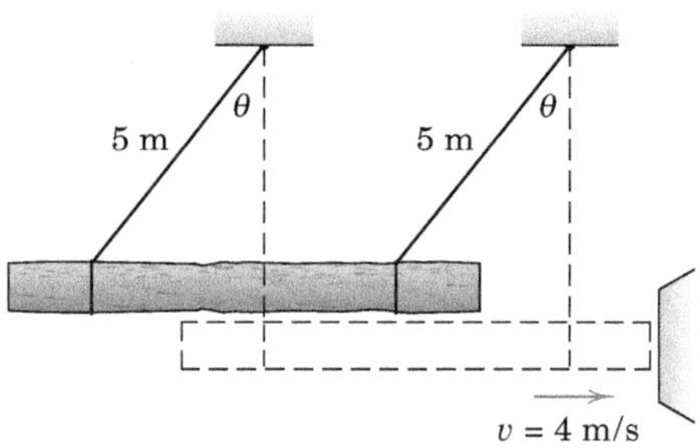

Problem 6/112

6/113 The assembly is constructed of homogeneous slender rod which has a mass ρ per unit length, and it rotates freely about a horizontal axis through the pivot at O. If the assembly is nudged from the starting position shown, determine its angular velocity after it has rotated (*a*) 45°, (*b*) 90°, and (*c*) 180°.

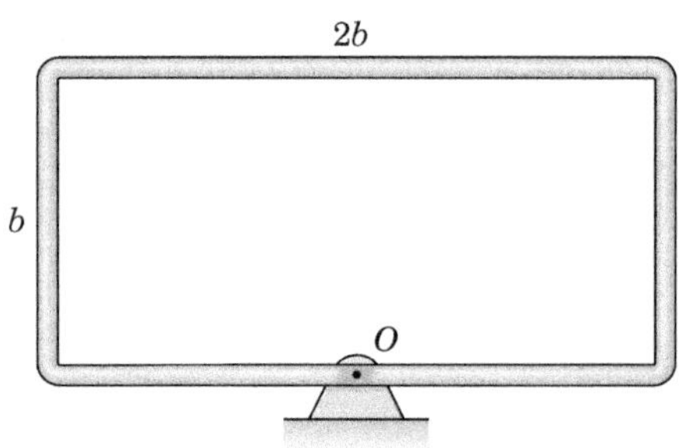

Problem 6/113

6/114 The velocity of the 8-kg cylinder is 0.3 m/s at a certain instant. What is its speed v after dropping an additional 1.5 m? The mass of the grooved drum is 12 kg, its centroidal radius of gyration is $\bar{k}$ = 210 mm, and the radius of its groove is r_i = 200 mm. The frictional moment at O is a constant 3 N·m.

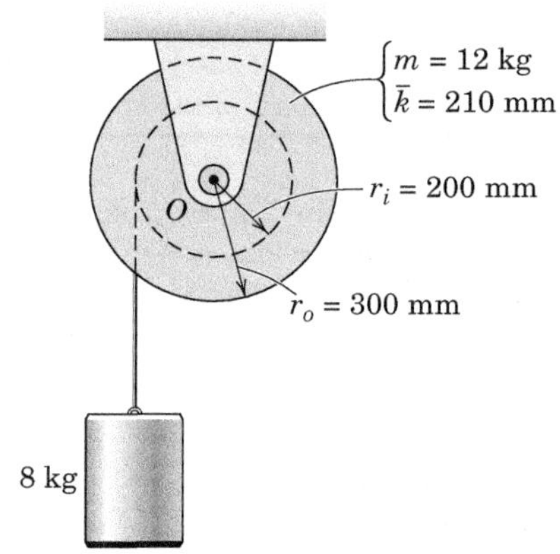

Problem 6/114

6/115 The 32.2-lb wheel is released from rest and rolls on its hubs without slipping. Calculate the speed v of the center O of the wheel after it has moved a distance x = 10 ft down the incline. The radius of gyration of the wheel about O is 5 in.

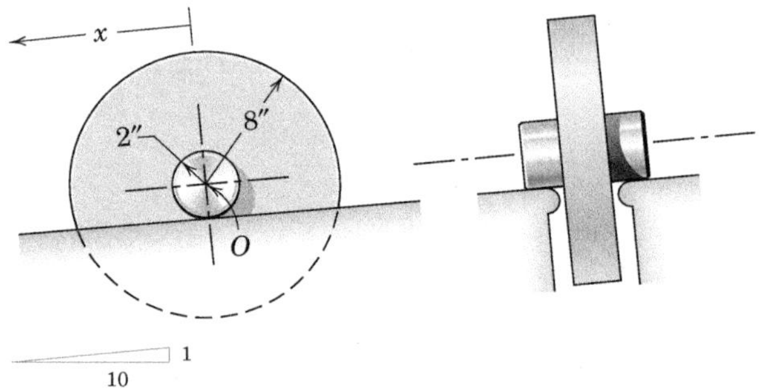

Problem 6/115

6/116 The uniform semicircular bar of radius r = 75 mm and mass m = 3 kg rotates freely about a horizontal axis through the pivot O. The bar is initially held in position 1 against the action of the torsional spring and then suddenly released. Determine the spring stiffness k_T which will give the bar a counterclockwise angular velocity ω = 4 rad/s when it reaches position 2, at which the spring is undeformed.

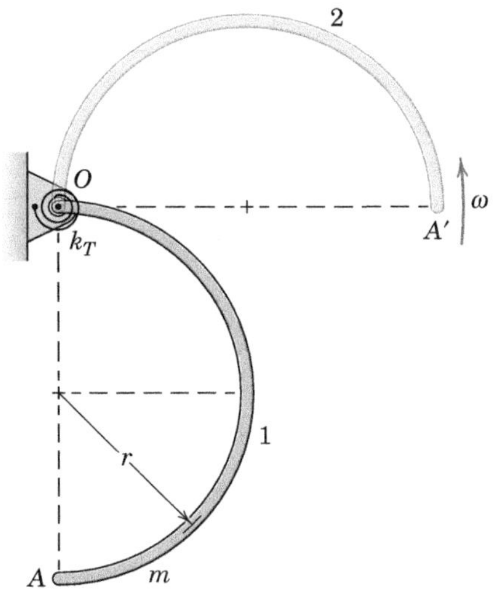

Problem 6/116

6/117 The homogeneous rectangular crate weighs 250 lb and is supported in the horizontal position by the cable at A and the corner hinge at O. If the cable at A is suddenly released, calculate the angular velocity ω of the crate just before it strikes the 30° incline. Does the weight of the crate influence the results, other quantities unchanged?

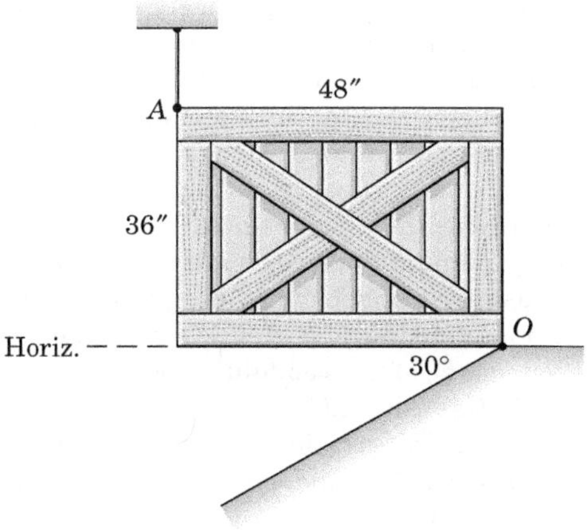

Problem 6/117

6/118 The 24-lb disk is rigidly attached to the 7-lb bar OA, which is pivoted freely about a horizontal axis through point O. If the system is released from rest in the position shown, determine the angular velocity of the bar and the magnitude of the pin reaction at O after the bar has rotated 90°.

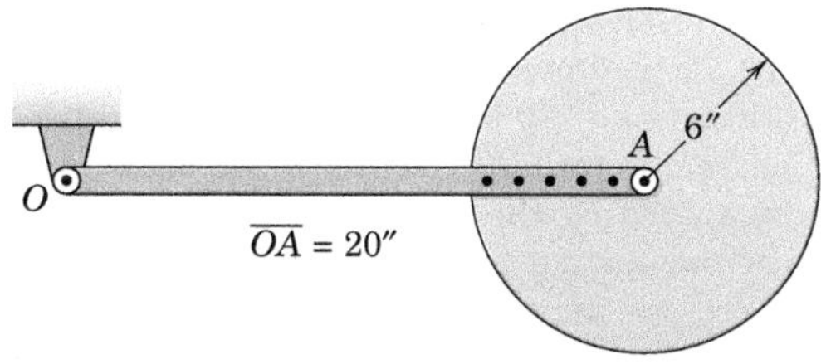

Problem 6/118

6/119 The two wheels of Prob. 6/78, shown again here, represent two extreme conditions of distribution of mass. For case A all of the mass m is assumed to be concentrated in the center of the hoop in the axial bar of negligible diameter. For case B all of the mass m is assumed to be concentrated in the rim. Determine the speed of the center of each hoop after it has traveled a distance x down the incline from rest. The hoops roll without slipping.

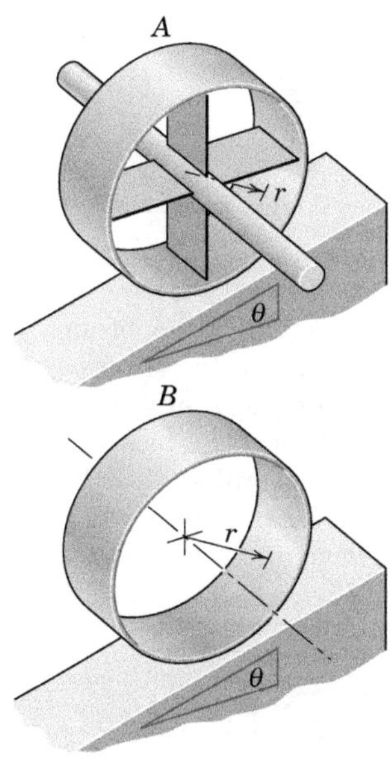

Problem 6/119

6/120 The 15-kg slender bar OA is released from rest in the vertical position and compresses the spring of stiffness $k = 20$ kN/m as the horizontal position is passed. Determine the proper setting of the spring, by specifying the distance h, which will result in the bar having an angular velocity $\omega = 4$ rad/s as it crosses the horizontal position. What is the effect of x on the dynamics of the problem?

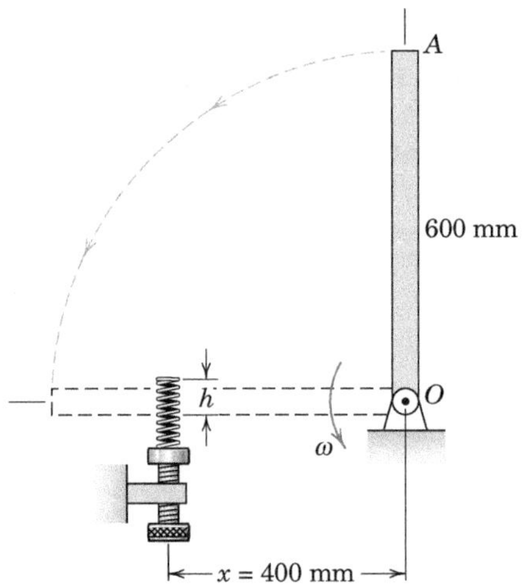

Problem 6/120

6/121 The light circular hoop of radius r contains a quarter-circular sector of mass m and is initially at rest on the horizontal surface. A couple M is applied to the hoop, which rolls without slipping along the horizontal surface. Determine the velocity v of the hoop after it has rolled through one-half of a revolution. Evaluate for $mg = 6$ lb, $r = 9$ in., and $M = 4$ lb-ft.

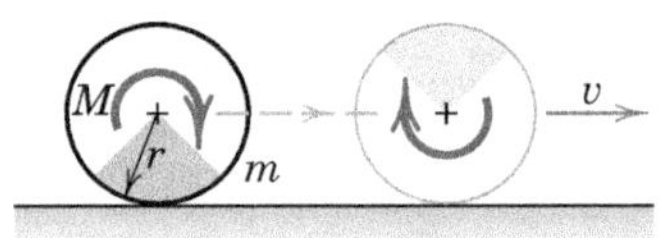

Problem 6/121

6/122 A steady 5-lb force is applied normal to the handle of the hand-operated grinder. The gear inside the housing with its shaft and attached handle together weigh 3.94 lb and have a radius of gyration about their axis of 2.85 in. The grinding wheel with its attached shaft and pinion (inside housing) together weigh 1.22 lb and have a radius of gyration of 2.14 in. If the gear ratio between gear and pinion is 4:1, calculate the speed N of the grinding wheel after 6 complete revolutions of the handle starting from rest.

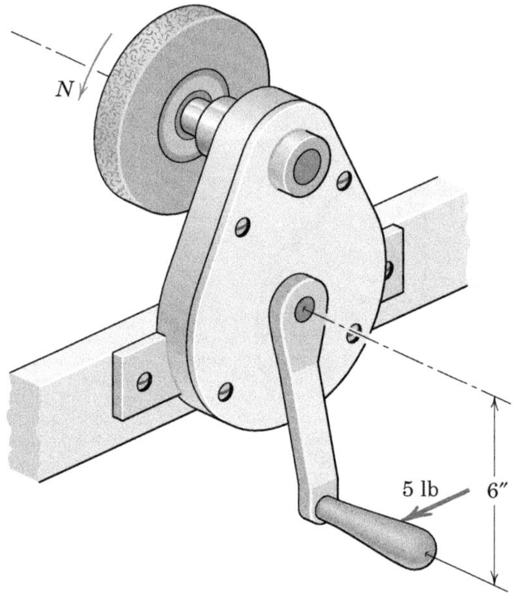

Problem 6/122

Representative Problems

6/123 The figure shows an impact tester used in studying material response to shock loads. The 60-lb pendulum is released from rest and swings through the vertical with negligible resistance. At the bottommost point of the ensuing motion, the pendulum strikes a notched material specimen A. After impact with the specimen, the pendulum swings upward to a height $\bar{h}' = 3.17$ ft. If the impact-energy capacity of the pendulum is 300 ft-lb, determine the change in the angular velocity of the pendulum during the interval from just before to just after impact with the specimen. The center of mass distance $\bar{r}$ from O and the radius of gyration k_O of the pendulum about O are both 35.5 in. (*Note:* Positioning the center of mass directly at the radius of gyration eliminates shock loads on the bearing at O and extends the life of the tester significantly.)

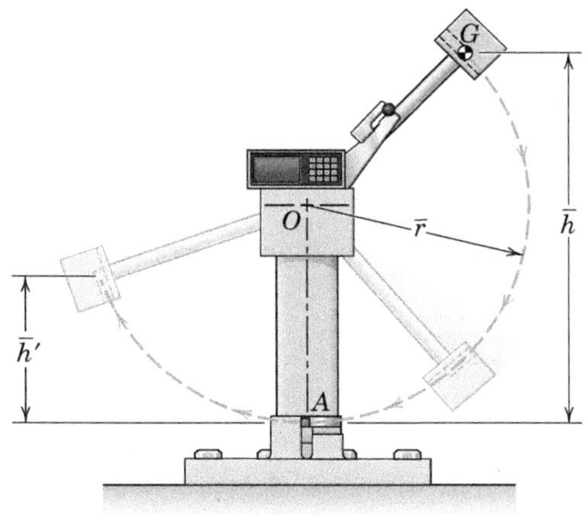

Problem 6/123

6/124 The uniform rectangular plate is released from rest in the position shown. Determine the maximum angular velocity ω during the ensuing motion. Friction at the pivot is negligible.

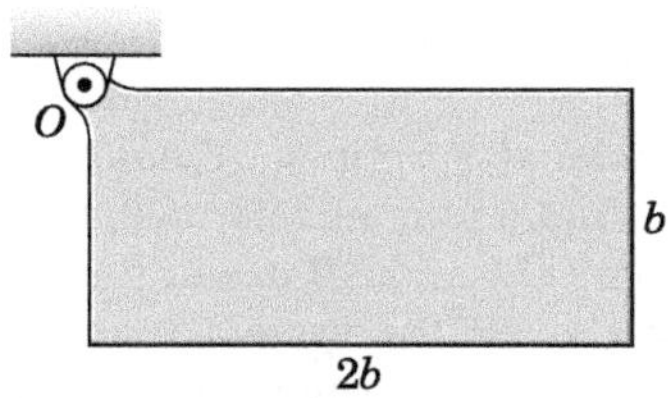

Problem 6/124

6/125 The 50-kg flywheel has a radius of gyration $\bar{k} = 0.4$ m about its shaft axis and is subjected to the torque $M = 2(1 - e^{-0.1\theta})$ N·m, where θ is in radians. If the flywheel is at rest when $\theta = 0$, determine its angular velocity after 5 revolutions.

Problem 6/125

6/126 The 20-kg wheel has an eccentric mass which places the center of mass G a distance $\bar{r} = 70$ mm away from the geometric center O. A constant couple $M = 6$ N·m is applied to the initially stationary wheel, which rolls without slipping along the horizontal surface and enters the curve of radius $R = 600$ mm. Determine the normal force under the wheel just before it exits the curve at C. The wheel has a rolling radius $r = 100$ mm and a radius of gyration $k_O = 65$ mm.

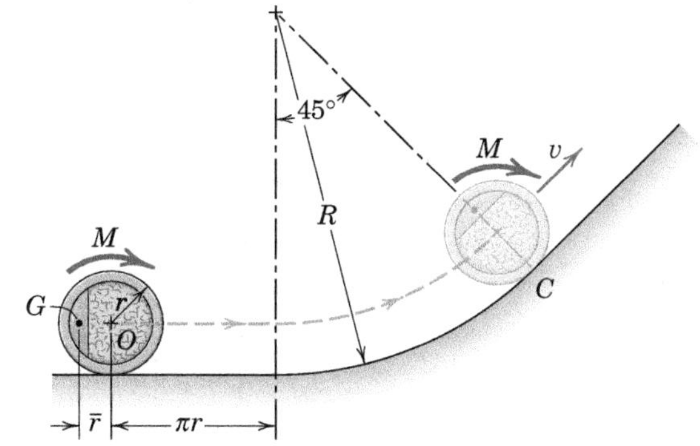

Problem 6/126

6/127 The figure shows the cross section AB of a garage door which is a rectangular 2.5-m by 5-m panel of uniform thickness with a mass of 200 kg. The door is supported by the struts of negligible mass and hinged at O. Two spring-and-cable assemblies, one on each side of the door, control the movement. When the door is in the horizontal open position, each spring is unextended. If the door is given a slight imbalance from the open position and allowed to fall, determine the value of the spring constant k for each spring which will limit the angular velocity of the door to 1.5 rad/s when edge B strikes the floor.

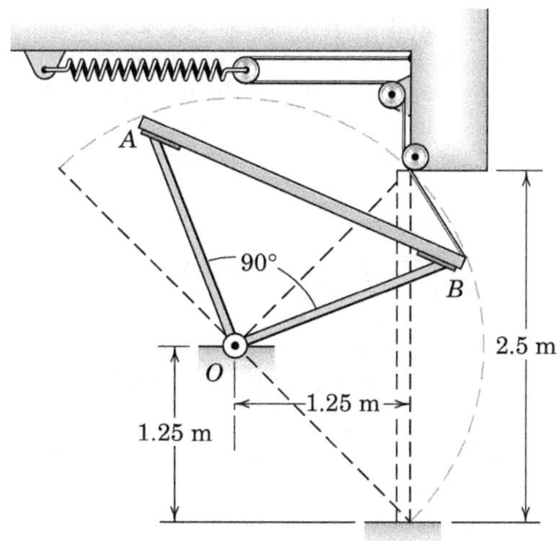

Problem 6/127

6/128 The uniform 40-lb bar with attached 12-lb wheels is released from rest when $\theta = 60°$. If the wheels roll without slipping on the horizontal and vertical surfaces, determine the angular velocity of the bar when $\theta = 45°$. Each wheel has a centroidal radius of gyration of 4.5 inches.

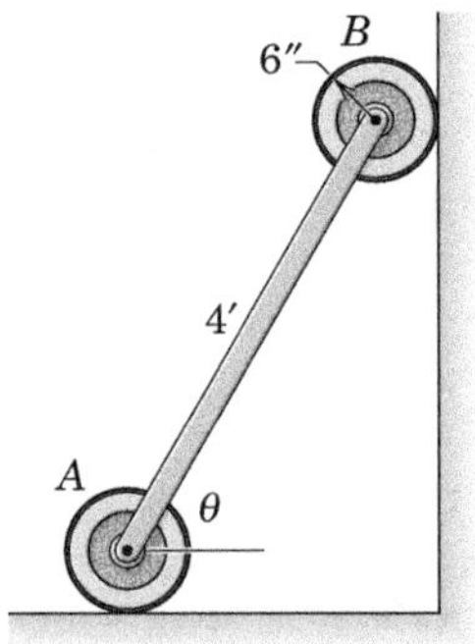

Problem 6/128

6/129 A 1200-kg flywheel with a radius of gyration of 400 mm has its speed reduced from 5000 to 3000 rev/min during a 2-min interval. Calculate the average power supplied by the flywheel. Express your answer both in kilowatts and in horsepower.

6/130 The wheel consists of a 4-kg rim of 250-mm radius with hub and spokes of negligible mass. The wheel is mounted on the 3-kg yoke OA with mass center at G and with a radius of gyration about O of 350 mm. If the assembly is released from rest in the horizontal position shown and if the wheel rolls on the circular surface without slipping, compute the velocity of point A when it reaches A'.

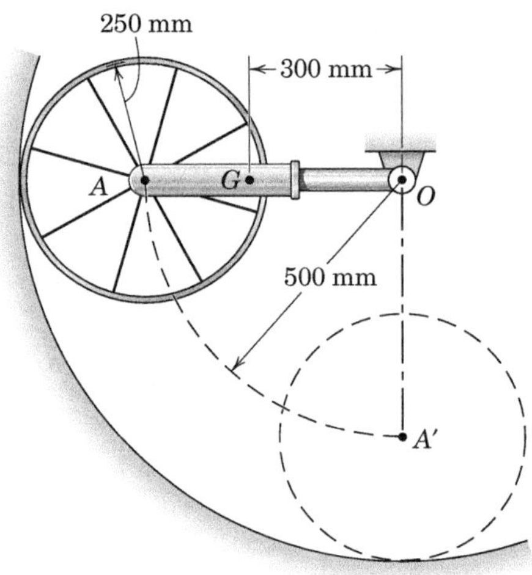

Problem 6/130

6/131 The uniform slender bar ABC weighs 6 lb and is initially at rest with end A bearing against the stop in the horizontal guide. When a constant couple $M = 72$ lb-in. is applied to end C, the bar rotates causing end A to strike the side of the vertical guide with a velocity of 10 ft/sec. Calculate the loss of energy ΔE due to friction in the guides and rollers. The mass of the rollers may be neglected.

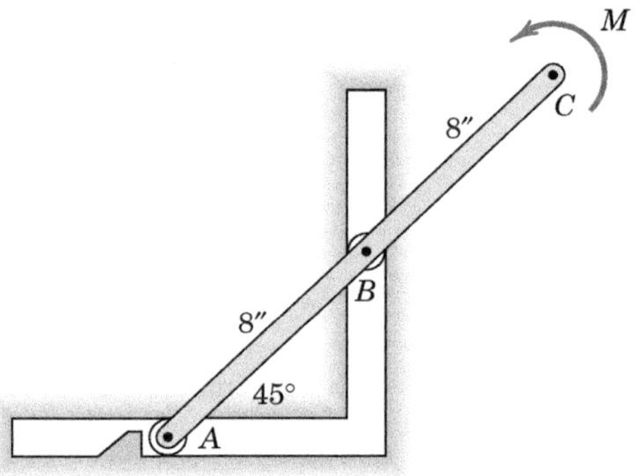

Problem 6/131

6/132 The torsional spring at A has a stiffness $k_T = 10$ N·m/rad and is undeformed when the uniform 10-kg bars OA and AB are in the vertical position and overlap. If the system is released from rest with $\theta = 60°$, determine the angular velocity of wheel B when $\theta = 30°$. The 6-kg wheel at B has a centroidal radius of gyration of 50 mm and is observed to roll without slipping on the horizontal surface.

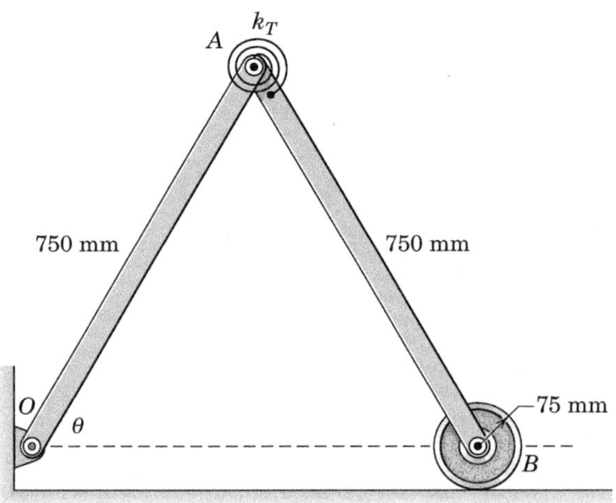

Problem 6/132

6/133 The uniform slender bar of mass m pivots freely about a horizontal axis through O. If the bar is released from rest in the horizontal position shown where the spring is unstretched, it is observed to rotate a maximum of 30° clockwise. The spring constant $k = 200$ N/m and the distance $b = 200$ mm. Determine (a) the mass m of the bar and (b) the angular velocity ω of the bar when the angular displacement is 15° clockwise from the release position.

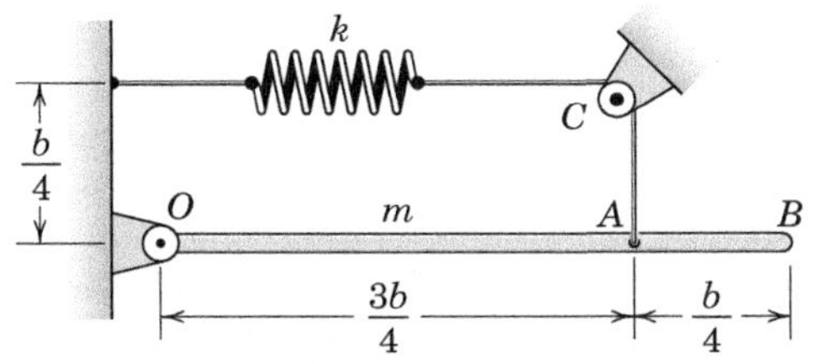

Problem 6/133

6/134 The system is released from rest when the angle $\theta = 90°$. Determine the angular velocity of the uniform slender bar when θ equals 60°. Use the values $m_1 = 1$ kg, $m_2 = 1.25$ kg, and $b = 0.4$ m.

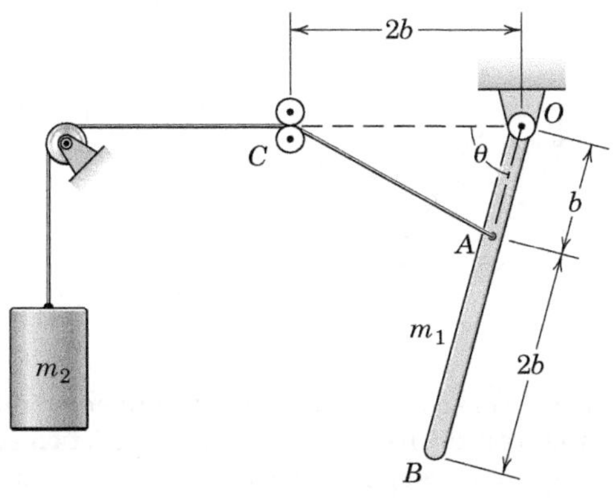

Problem 6/134

6/135 The homogeneous torus and cylindrical ring are released from rest and roll without slipping down the incline. Determine an expression for the velocity difference v_{diff} which develops between the two objects during the ensuing motion as a function of the distance x they have traveled down the incline. Assume that the masses roll straight down the incline and evaluate your expression for the case where $a = 0.2R$. Which object is in the lead, and does the relative size of a ever alter the finishing order?

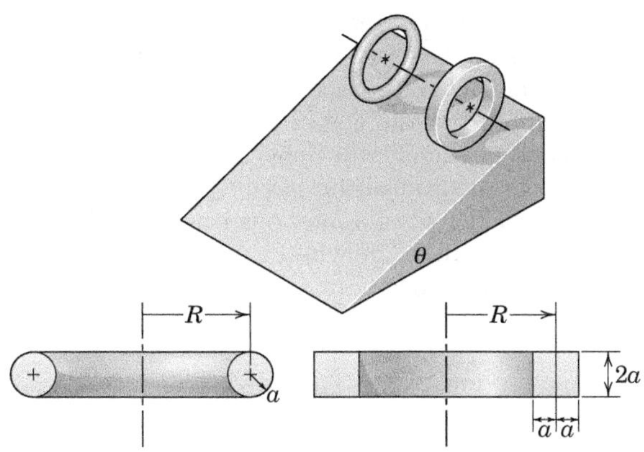

Problem 6/135

6/136 The uniform 12-lb disk pivots freely about a horizontal axis through O. A 4-lb slender bar is fastened to the disk as shown. If the system is nudged from rest while in the position shown, determine its angular velocity ω after it has rotated 180°.

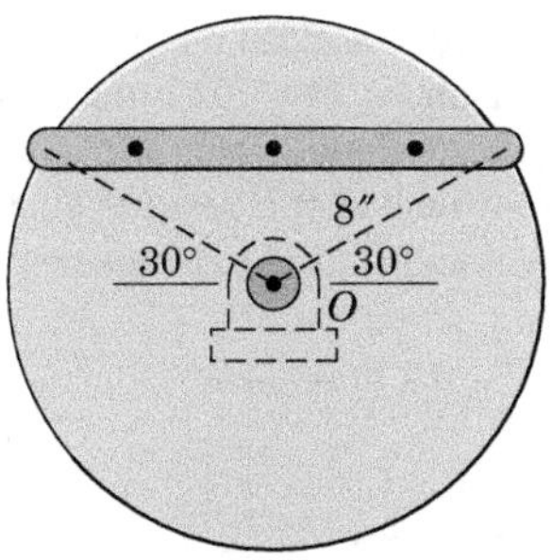

Problem 6/136

6/137 Under active development is the storage of energy in high-speed rotating disks where friction is effectively eliminated by encasing the rotor in an evacuated enclosure and by using magnetic bearings. For a 10-kg rotor with a radius of gyration of 90 mm rotating initially at 80 000 rev/min, calculate the power P which can be extracted from the rotor by applying a constant 2.10-N·m retarding torque (a) when the torque is first applied and (b) at the instant when the torque has been applied for 120 seconds.

6/138 For the pivoted slender rod of length l, determine the distance x for which the angular velocity will be a maximum as the bar passes the vertical position after being released in the horizontal position shown. State the corresponding angular velocity.

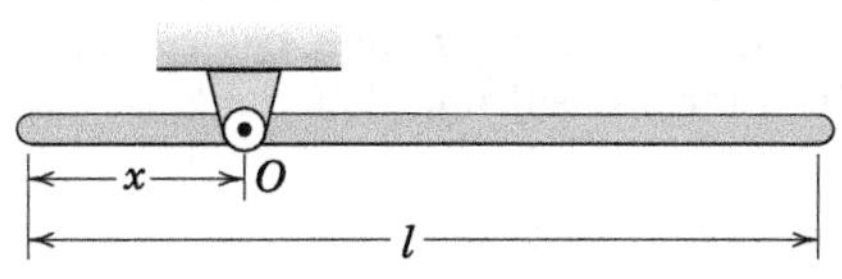

Problem 6/138

6/139 The wheel has mass m and a centroidal radius of gyration $\bar{k}$ and rolls without slipping up the incline under the action of a force P. The force is applied to the end of a cord which is wrapped securely around the inner hub of the wheel as shown. Determine the speed v_O of the wheel center O after the wheel center has traveled a distance d up the incline. The wheel is at rest when the force P is first applied.

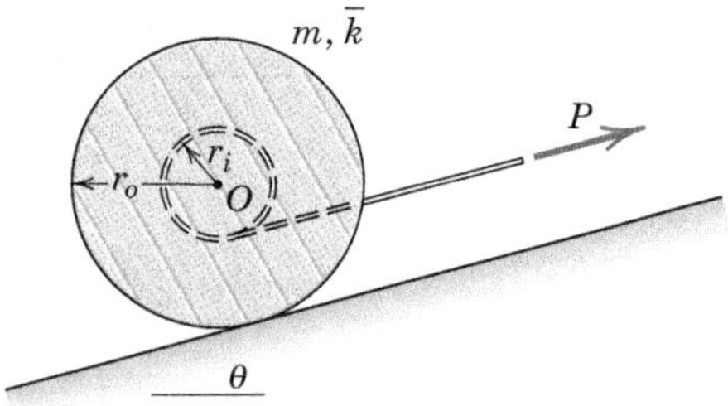

Problem 6/139

6/140 The 8-kg crank OA, with mass center at G and radius of gyration about O of 0.22 m, is connected to the 12-kg uniform slender bar AB. If the linkage is released from rest in the position shown, compute the velocity v of end B as OA swings through the vertical.

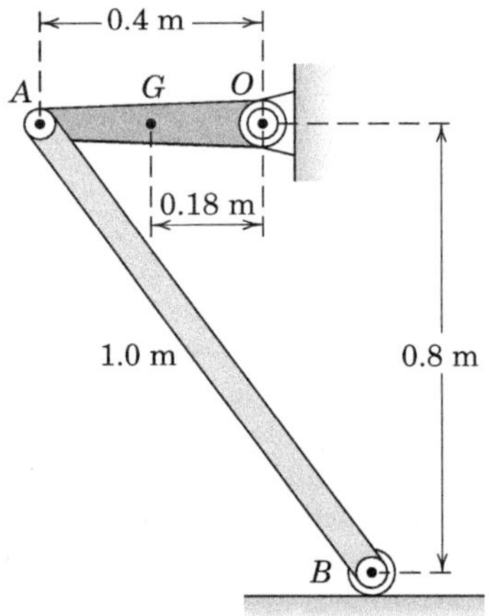

Problem 6/140

6/141 The sheave of 400-mm radius has a mass of 50 kg and a radius of gyration of 300 mm. The sheave and its 100-kg load are suspended by the cable and the spring, which has a stiffness of 1.5 kN/m. If the system is released from rest with the spring initially stretched 100 mm, determine the velocity of O after it has dropped 50 mm.

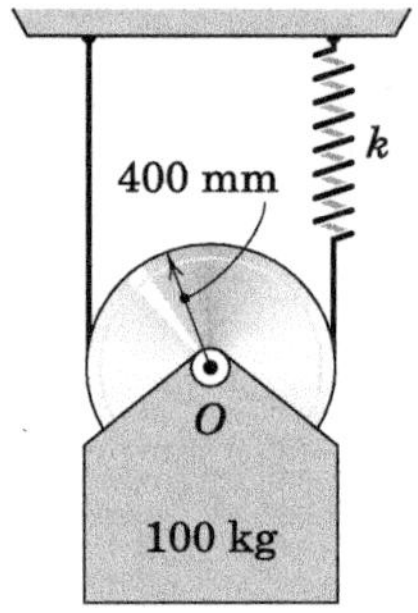

Problem 6/141

6/142 Motive power for the experimental 10-Mg bus comes from the energy stored in a rotating flywheel which it carries. The flywheel has a mass of 1500 kg and a radius of gyration of 500 mm and is brought up to a maximum speed of 4000 rev/min. If the bus starts from rest and acquires a speed of 72 km/h at the top of a hill 20 m above the starting position, compute the reduced speed N of the flywheel. Assume that 10 percent of the energy taken from the flywheel is lost. Neglect the rotational energy of the wheels of the bus. The 10-Mg mass includes the flywheel.

Problem 6/142

6/143 The homogeneous solid semicylinder is released from rest in the position shown. If friction is sufficient to prevent slipping, determine the maximum angular velocity ω reached by the cylinder as it rolls on the horizontal surface.

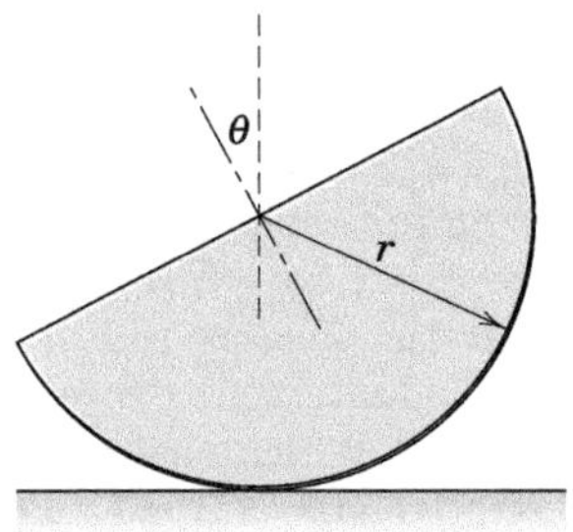

Problem 6/143

6/144 The figure shows the side view of a door to a storage compartment. As the 40-kg uniform door is opened, the light rod slides through the collar at C and compresses the spring of stiffness k. With the door closed ($\theta = 0$), a constant force $P = 225$ N is applied to the end of the door via a cable. If the door has a clockwise angular velocity of 1 rad/s as the position $\theta = 60°$ is passed, determine (*a*) the stiffness of the spring and (*b*) the angular velocity of the door as it passes the position $\theta = 45°$. Neglect all friction and the mass of the pulleys at C and D. The spring is uncompressed when the door is vertical, and $b = 1.25$ m.

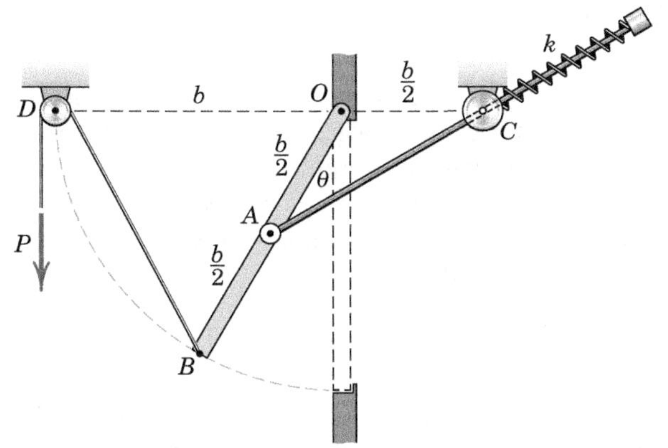

Problem 6/144

***6/145** Reconsider the door of Prob. 6/144. If the door is in the closed vertical position when a constant input force $P = 225$ N is applied through the end of the cable, determine the maximum angle θ_{max} reached by the door before it comes to a stop. Plot the angular velocity of the door over this period and determine the maximum angular velocity of the door along with the corresponding angle θ at which it occurs. The uniform door has a mass $m = 40$ kg, dimension $b = 1.25$ m, and the spring constant is $k = 2280$ N/m.

6/146 A small experimental vehicle has a total mass m of 500 kg including wheels and driver. Each of the four wheels has a mass of 40 kg and a centroidal radius of gyration of 400 mm. Total frictional resistance R to motion is 400 N and is measured by towing the vehicle at a constant speed on a level road with engine disengaged. Determine the power output of the engine for a speed of 72 km/h up the 10-percent grade (*a*) with zero acceleration and (*b*) with an acceleration of 3 m/s^2. (*Hint:* Power equals the time rate of increase of the total energy of the vehicle plus the rate at which frictional work is overcome.)

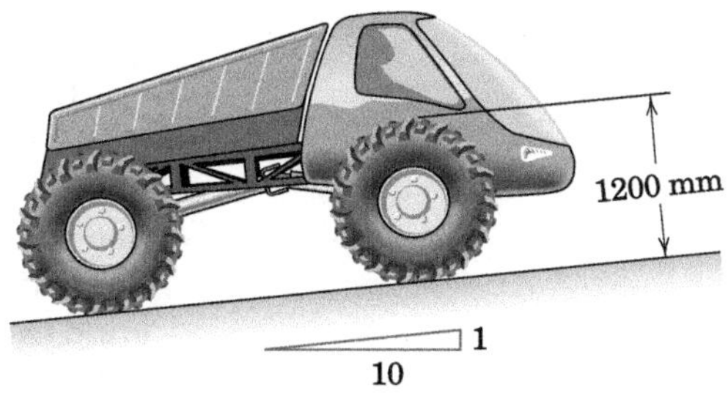

Problem 6/146

6/147 The two slender bars each of mass m and length b are pinned together and move in the vertical plane. If the bars are released from rest in the position shown and move together under the action of a couple M of constant magnitude applied to AB, determine the velocity of A as it strikes O.

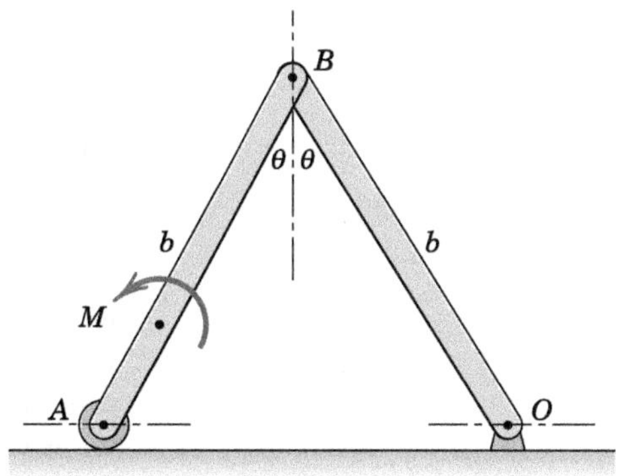

Problem 6/147

6/148 The open square frame is constructed of four identical slender rods, each of length b. If the frame is released from rest in the position shown, determine the speed of corner A (*a*) after A has dropped a distance b and (*b*) after A has dropped a distance $2b$. The small wheels roll without friction in the slots of the vertical surface.

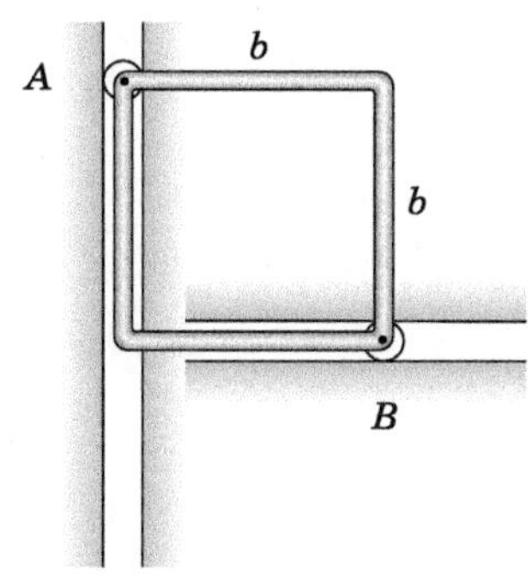

Problem 6/148

6/7 Acceleration from Work-Energy; Virtual Work

In addition to using the work-energy equation to determine the velocities due to the action of forces acting over finite displacements, we may also use the equation to establish the instantaneous accelerations of the members of a system of interconnected bodies as a result of the active forces applied. We may also modify the equation to determine the configuration of such a system when it undergoes a constant acceleration.

Work-Energy Equation for Differential Motions

For an infinitesimal interval of motion, Eq. 4/3 becomes

$$dU' = dT + dV$$

The term dU' represents the total work done by all active nonpotential forces acting on the system under consideration during the infinitesimal displacement of the system. The work of potential forces is included in the dV-term. If we use the subscript i to denote a representative body of the interconnected system, the differential change in kinetic energy T for the entire system becomes

$$dT = d(\Sigma\tfrac{1}{2}m_i\bar{v}_i{}^2 + \Sigma\tfrac{1}{2}\bar{I}_i\omega_i{}^2) = \Sigma m_i\bar{v}_i\,d\bar{v}_i + \Sigma\bar{I}_i\omega_i\,d\omega_i$$

where $d\bar{v}_i$ and $d\omega_i$ are the respective changes in the magnitudes of the velocities and where the summation is taken over all bodies of the system. But for each body, $m_i\bar{v}_i\,d\bar{v}_i = m_i\bar{\mathbf{a}}_i\cdot d\bar{\mathbf{s}}_i$ and $\bar{I}_i\omega_i\,d\omega_i = \bar{I}_i\alpha_i\,d\theta_i$, where $d\bar{\mathbf{s}}_i$ represents the infinitesimal linear displacement of the center of mass and where $d\theta_i$ represents the infinitesimal angular displacement of the body in the plane of motion. We note that $\bar{\mathbf{a}}_i\cdot d\bar{\mathbf{s}}_i$ is identical to $(\bar{a}_i)_t\,d\bar{s}_i$, where $(\bar{a}_i)_t$ is the component of $\bar{\mathbf{a}}_i$ along the tangent to the curve described by the mass center of the body in question. Also α_i represents $\ddot{\theta}_i$, the angular acceleration of the representative body. Consequently, for the entire system

$$dT = \Sigma m_i\bar{\mathbf{a}}_i\cdot d\bar{\mathbf{s}}_i + \Sigma\bar{I}_i\alpha_i\,d\theta_i$$

This change may also be written as

$$dT = \Sigma\mathbf{R}_i\cdot d\bar{\mathbf{s}}_i + \Sigma\mathbf{M}_{G_i}\cdot d\boldsymbol{\theta}_i$$

where $\mathbf{R}_i$ and $\mathbf{M}_{G_i}$ are the resultant force and resultant couple acting on body i and where $d\boldsymbol{\theta}_i = d\theta_i\mathbf{k}$. These last two equations merely show us that the differential change in kinetic energy equals the differential work done on the system by the resultant forces and resultant couples acting on all the bodies of the system.

The term dV represents the differential change in the total gravitational potential energy V_g and the total elastic potential energy V_e and has the form

$$dV = d(\Sigma m_igh_i + \Sigma\tfrac{1}{2}k_jx_j{}^2) = \Sigma m_ig\,dh_i + \Sigma k_jx_j\,dx_j$$

where h_i represents the vertical distance of the center of mass of the representative body of mass m_i above any convenient datum plane and where x_j stands for the deformation, tensile or compressive, of a representative elastic member of the system (spring) whose stiffness is k_j.

The complete expression for dU' may now be written as

$$dU' = \Sigma m_i \bar{\mathbf{a}}_i \cdot d\bar{\mathbf{s}}_i + \Sigma \bar{I}_i \alpha_i \, d\theta_i + \Sigma m_i g \, dh_i + \Sigma k_j x_j \, dx_j \qquad \textbf{(6/11)}$$

When Eq. 6/11 is applied to a system of one degree of freedom, the terms $m_i \bar{\mathbf{a}}_i \cdot d\bar{\mathbf{s}}_i$ and $\bar{I}_i \alpha_i \, d\theta_i$ will be positive if the accelerations are in the same direction as the respective displacements and negative if they are in the opposite direction. Equation 6/11 has the advantage of relating the accelerations to the active forces directly, which eliminates the need for dismembering the system and then eliminating the internal forces and reactive forces by simultaneous solution of the force-mass-acceleration equations for each member.

Virtual Work

In Eq. 6/11 the differential motions are differential changes in the real or actual displacements which occur. For a mechanical system which assumes a steady-state configuration during constant acceleration, we often find it convenient to introduce the concept of *virtual work*. The concepts of virtual work and virtual displacement were introduced and used to establish equilibrium configurations for static systems of interconnected bodies (see Chapter 7 of *Vol. 1 Statics*).

A *virtual displacement* is any assumed and arbitrary displacement, linear or angular, away from the natural or actual position. For a system of connected bodies, the virtual displacements must be consistent with the constraints of the system. For example, when one end of a link is hinged about a fixed pivot, the virtual displacement of the other end must be normal to the line joining the two ends. Such requirements for displacements consistent with the constraints are purely kinematic and provide what are known as the *equations of constraint*.

If a set of virtual displacements satisfying the equations of constraint and therefore consistent with the constraints is assumed for a mechanical system, the proper relationship between the coordinates which specify the configuration of the system will be determined by applying the work-energy relationship of Eq. 6/11, expressed in terms of virtual changes. Thus,

$$\delta U' = \Sigma m_i \bar{\mathbf{a}}_i \cdot \delta\bar{\mathbf{s}}_i + \Sigma \bar{I}_i \alpha_i \, \delta\theta_i + \Sigma m_i g \, \delta h_i + \Sigma k_j x_j \, \delta x_j \qquad \textbf{(6/11a)}$$

It is customary to use the differential symbol d to refer to differential changes in the *real* displacements, whereas the symbol δ is used to signify virtual changes, that is, differential changes which are *assumed* rather than real.

Rhonda Roth/Shutterstock

The kinetics of this elevated work platform are best analyzed using the principles of virtual work.

Sample Problem 6/12

The movable rack A has a mass of 3 kg, and rack B is fixed. The gear has a mass of 2 kg and a radius of gyration of 60 mm. In the position shown, the spring, which has a stiffness of 1.2 kN/m, is stretched a distance of 40 mm. For the instant represented, determine the acceleration a of rack A under the action of the 80-N force. The plane of the figure is vertical.

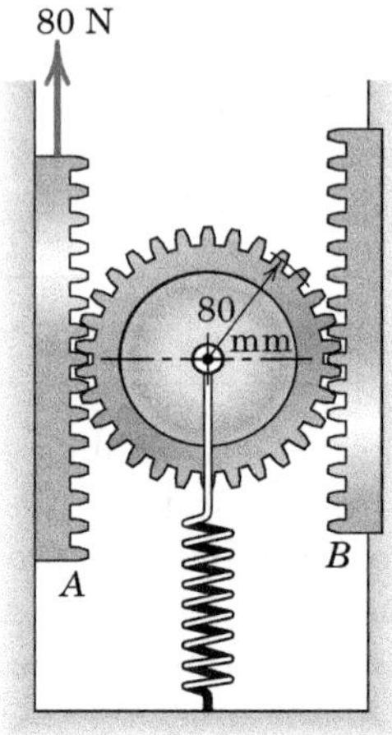

Solution. The given figure represents the active-force diagram for the entire system, which is conservative. ①

During an infinitesimal upward displacement dx of rack A, the work dU' done on the system is $80\,dx$, where x is in meters, and this work equals the sum of the corresponding changes in the total energy of the system. These changes, which appear in Eq. 6/11, are as follows:

$$[dT = \Sigma m_i \bar{\mathbf{a}}_i \cdot d\bar{\mathbf{s}}_i + \Sigma \bar{I}_i \alpha_i \, d\theta_i]$$

$$dT_{\text{rack}} = 3a\,dx$$

② $$dT_{\text{gear}} = 2\frac{a}{2}\frac{dx}{2} + 2(0.06)^2 \frac{a/2}{0.08}\frac{dx/2}{0.08} = 0.781a\,dx$$

The change in potential energies of the system, from Eq. 6/11, becomes

$$[dV = \Sigma m_i g\,dh_i + \Sigma k_j x_j\,dx_j]$$

$$dV_{\text{rack}} = 3g\,dx = 3(9.81)\,dx = 29.4\,dx$$

$$dV_{\text{gear}} = 2g(dx/2) = g\,dx = 9.81\,dx$$

③ $$dV_{\text{spring}} = k_j x_j\,dx_j = 1200(0.04)\,dx/2 = 24\,dx$$

Substitution into Eq. 6/11 gives us

$$80\,dx = 3a\,dx + 0.781a\,dx + 29.4\,dx + 9.81\,dx + 24\,dx$$

Canceling dx and solving for a give

$$a = 16.76/3.78 = 4.43 \text{ m/s}^2 \qquad \textit{Ans.}$$

We see that using the work-energy method for an infinitesimal displacement has given us the direct relation between the applied force and the resulting acceleration. It was unnecessary to dismember the system, draw two free-body diagrams, apply $\Sigma F = m\bar{a}$ twice, apply $\Sigma M_G = \bar{I}\alpha$ and $F = kx$, eliminate unwanted terms, and finally solve for a.

Helpful Hints

① Note that none of the remaining forces external to the system do any work. The work done by the weight and by the spring is accounted for in the potential-energy terms.

② Note that $\bar{a}_i$ for the gear is its mass-center acceleration, which is half that for the rack A. Also, its displacement is $dx/2$. For the rolling gear, the angular acceleration from $a = r\alpha$ becomes $\alpha_i = (a/2)/0.08$, and the angular displacement from $ds = r\,d\theta$ becomes $d\theta_i = (dx/2)/0.08$.

③ Note here that the displacement of the spring is one-half that of the rack. Hence, $x_i = x/2$.

Sample Problem 6/13

A constant force P is applied to end A of the two identical and uniform links and causes them to move to the right in their vertical plane with a horizontal acceleration a. Determine the steady-state angle θ made by the bars with one another.

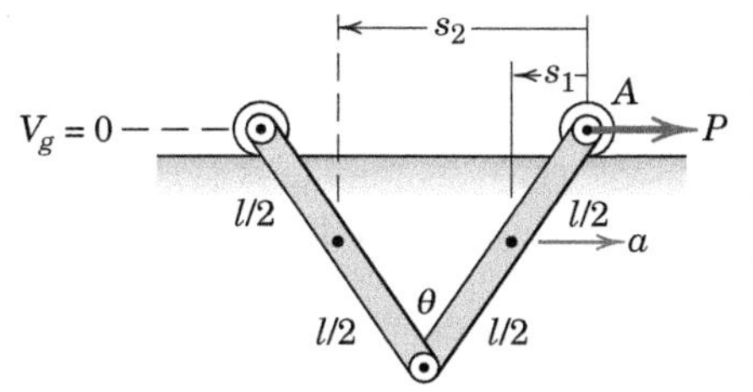

Solution. The figure constitutes the active-force diagram for the system. To find the steady-state configuration, consider a virtual displacement of each bar from the natural position assumed during the acceleration. Measurement of the displacement with respect to end A eliminates any work done by force P during the virtual displacement. Thus,

① $$\delta U' = 0$$

② The terms involving acceleration in Eq. 6/11a reduce to

$$m\bar{\mathbf{a}}\cdot\delta\bar{\mathbf{s}} = ma(-\delta s_1) + ma(-\delta s_2)$$

$$= -ma\left[\delta\left(\frac{l}{2}\sin\frac{\theta}{2}\right) + \delta\left(\frac{3l}{2}\sin\frac{\theta}{2}\right)\right]$$

③ $$= -ma\left(l\cos\frac{\theta}{2}\,\delta\theta\right)$$

④ We choose the horizontal line through A as the datum for zero potential energy. Thus, the potential energy of the links is

$$V_g = 2mg\left(-\frac{l}{2}\cos\frac{\theta}{2}\right)$$

and the virtual change in potential energy becomes

$$\delta V_g = \delta\left(-2mg\frac{l}{2}\cos\frac{\theta}{2} = \frac{mgl}{2}\sin\frac{\theta}{2}\,\delta\theta\right)$$

Substitution into the work-energy equation for virtual changes, Eq. 6/11a, gives

$$0 = -mal\cos\frac{\theta}{2}\,\delta\theta + \frac{mgl}{2}\sin\frac{\theta}{2}\,\delta\theta$$

from which

$$\theta = 2\tan^{-1}\frac{2a}{g} \qquad \textit{Ans.}$$

Again, in this problem we see that the work-energy approach obviated the necessity for dismembering the system, drawing separate free-body diagrams, applying motion equations, eliminating unwanted terms, and solving for θ.

Helpful Hints

① Note that we use the symbol δ to refer to an assumed or virtual differential change rather than the symbol d, which refers to an infinitesimal change in the real displacement.

② Here we are evaluating the work done by the resultant forces and couples in the virtual displacement. Note that $\alpha = 0$ for both bars.

③ We have chosen to use the angle θ to describe the configuration of the links, although we could have used the distance between the two ends of the links just as well.

④ The last two terms in Eq. 6/11a express the virtual changes in gravitational and elastic potential energy.

PROBLEMS

Introductory Problems

6/149 The load of mass m is supported by the light parallel links and the fixed stop A. Determine the initial angular acceleration α of the links due to the application of the couple M to one end as shown.

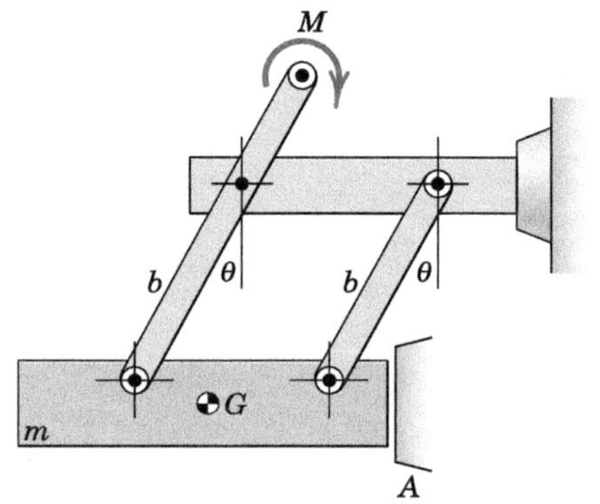

Problem 6/149

6/150 The uniform slender bar of mass m is shown in its equilibrium configuration before the force P is applied. Compute the initial angular acceleration of the bar upon application of P.

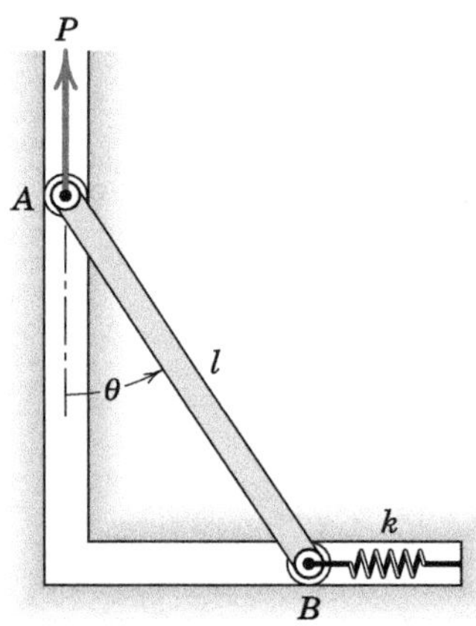

Problem 6/150

6/151 The two uniform slender bars are hinged at O and supported on the horizontal surface by their end rollers of negligible mass. If the bars are released from rest in the position shown, determine their initial angular acceleration α as they collapse in the vertical plane. (*Suggestion:* Make use of the instantaneous center of zero velocity in writing the expression for dT.)

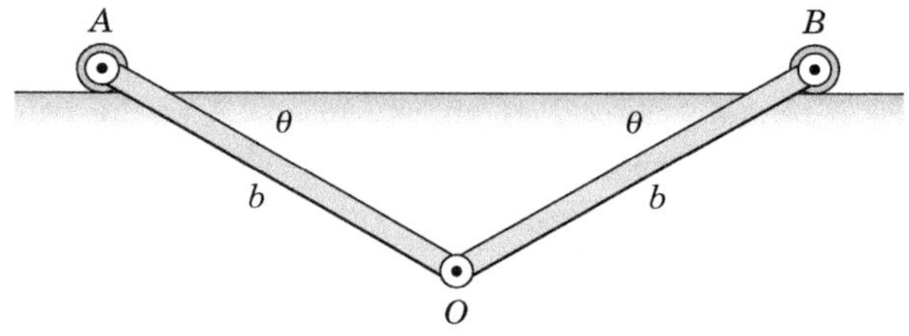

Problem 6/151

6/152 Links A and B each weigh 8 lb, and bar C weighs 12 lb. Calculate the angle θ assumed by the links if the body to which they are pinned is given a steady horizontal acceleration a of 4 ft/sec^2.

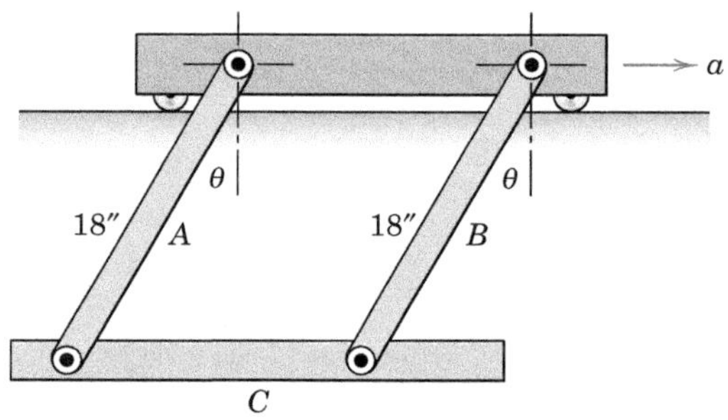

Problem 6/152

6/153 The mechanism shown moves in the vertical plane. The vertical bar AB weighs 10 lb, and each of the two links weighs 6 lb with mass center at G and with a radius of gyration of 10 in. about its bearing (O or C). The spring has a stiffness of 15 lb/ft and an unstretched length of 18 in. If the support at D is suddenly withdrawn, determine the initial angular acceleration α of the links.

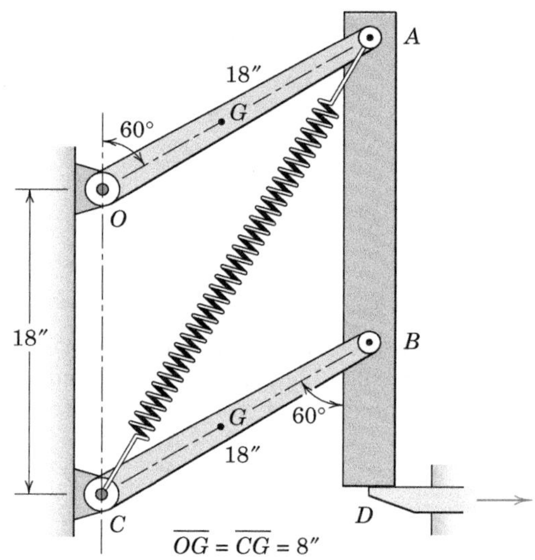

Problem 6/153

Representative Problems

6/154 The load of mass m is given an upward acceleration a from its supported rest position by the application of the forces P. Neglect the mass of the links compared with m and determine the initial acceleration a.

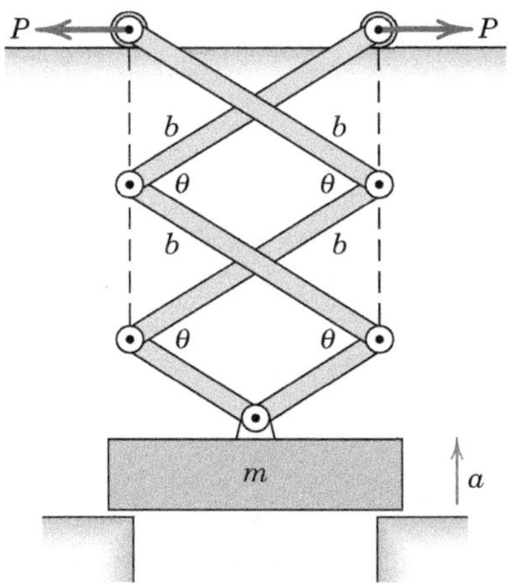

Problem 6/154

6/155 The cargo box of the food-delivery truck for aircraft servicing has a loaded mass m and is elevated by the application of a couple M on the lower end of the link which is hinged to the truck frame. The horizontal slots allow the linkage to unfold as the cargo box is elevated. Determine the upward acceleration of the box in terms of h for a given value of M. Neglect the mass of the links.

Problem 6/155

6/156 The sliding block is given a horizontal acceleration to the right that is slowly increased to a steady value a. The attached pendulum of mass m and mass center G assumes a steady angular deflection θ. The torsion spring at O exerts a moment $M = k_T\theta$ on the pendulum to oppose the angular deflection. Determine the torsional stiffness k_T that will allow a steady deflection θ.

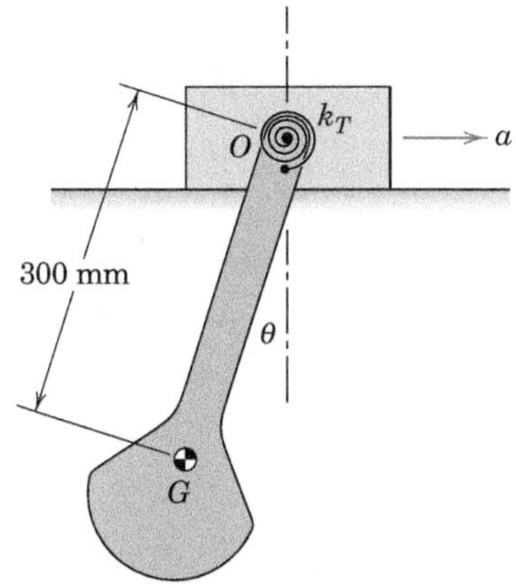

Problem 6/156

6/157 Each of the uniform bars OA and OB has a mass of 2 kg and is freely hinged at O to the vertical shaft, which is given an upward acceleration $a = g/2$. The links which connect the light collar C to the bars have negligible mass, and the collar slides freely on the shaft. The spring has a stiffness k = 130 N/m and is uncompressed for the position equivalent to $\theta = 0$. Calculate the angle θ assumed by the bars under conditions of steady acceleration.

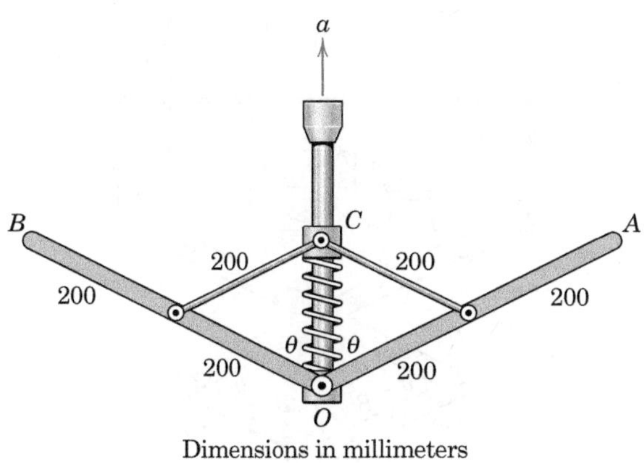

Dimensions in millimeters

Problem 6/157

6/158 The linkage consists of the two slender bars and moves in the horizontal plane under the influence of force P. Link OC has a mass m and link AC has a mass $2m$. The sliding block at B has negligible mass. Without dismembering the system, determine the initial angular acceleration α of the links as P is applied at A with the links initially at rest. (*Suggestion:* Replace P by its equivalent force-couple system.)

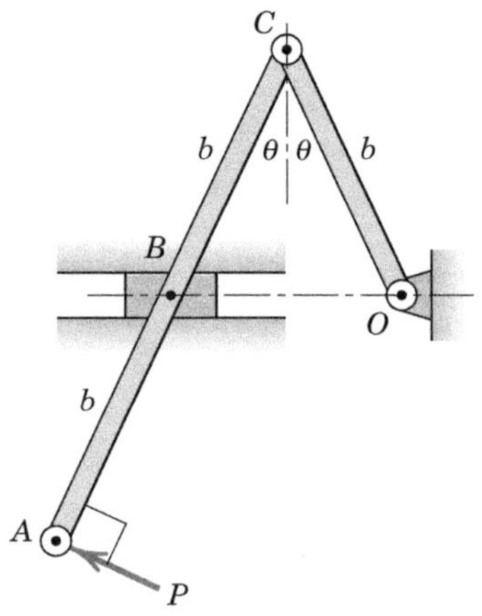

Problem 6/158

6/159 The portable work platform is elevated by means of the two hydraulic cylinders articulated at points C. The pressure in each cylinder produces a force F. The platform, man, and load have a combined mass m, and the mass of the linkage is small and may be neglected. Determine the upward acceleration a of the platform and show that it is independent of both b and θ.

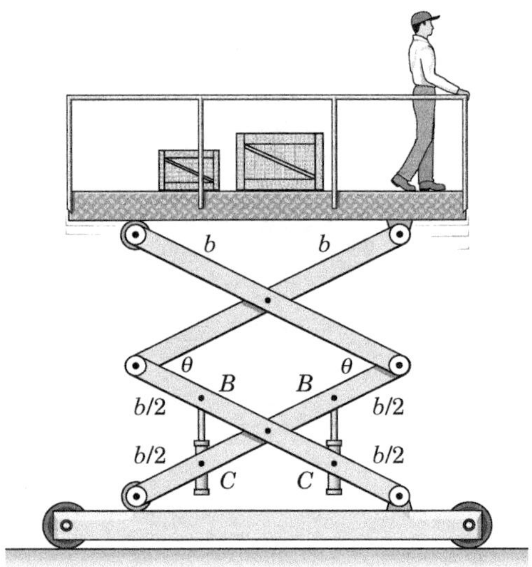

Problem 6/159

6/160 Each of the three identical uniform panels of a segmented industrial door has mass m and is guided in the tracks (one shown dashed). Determine the horizontal acceleration a of the upper panel under the action of the force P. Neglect any friction in the guide rollers.

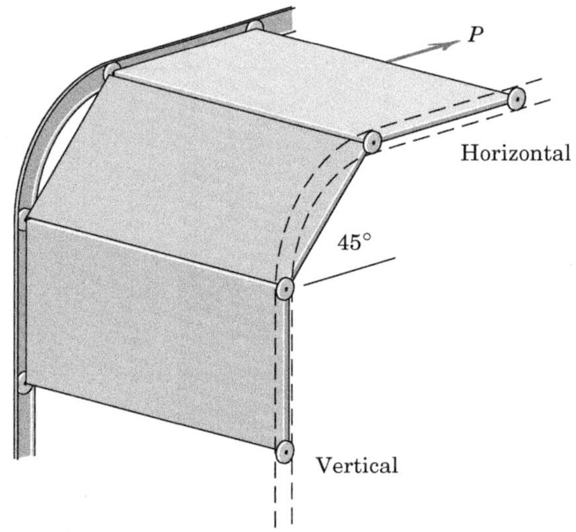

Problem 6/160

6/161 The mechanical tachometer measures the rotational speed N of the shaft by the horizontal motion of the collar B along the rotating shaft. This movement is caused by the centrifugal action of the two 12-oz weights A, which rotate with the shaft. Collar C is fixed to the shaft. Determine the rotational speed N of the shaft for a reading $\beta = 15°$. The stiffness of the spring is 5 lb/in., and it is uncompressed when $\theta = 0$ and $\beta = 0$. Neglect the weights of the links.

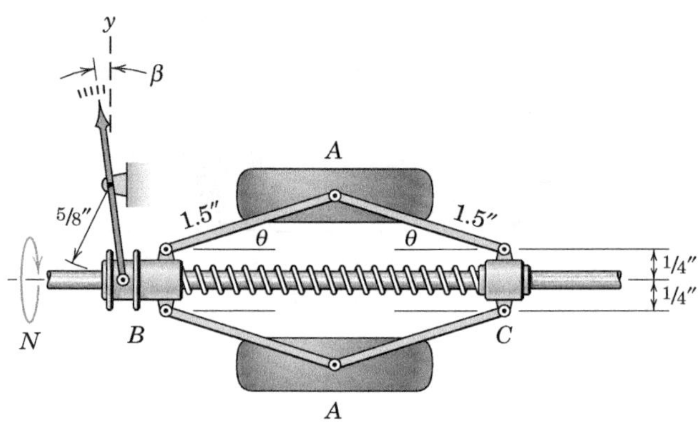

Problem 6/161

6/162 A planetary gear system is shown, where the gear teeth are omitted from the figure. Each of the three identical planet gears A, B, and C has a mass of 0.8 kg, a radius $r = 50$ mm, and a radius of gyration of 30 mm about its center. The spider E has a mass of 1.2 kg and a radius of gyration about O of 60 mm. The ring gear D has a radius $R = 150$ mm and is fixed. If a torque $M = 5$ N·m is applied to the shaft of the spider at O, determine the initial angular acceleration α of the spider.

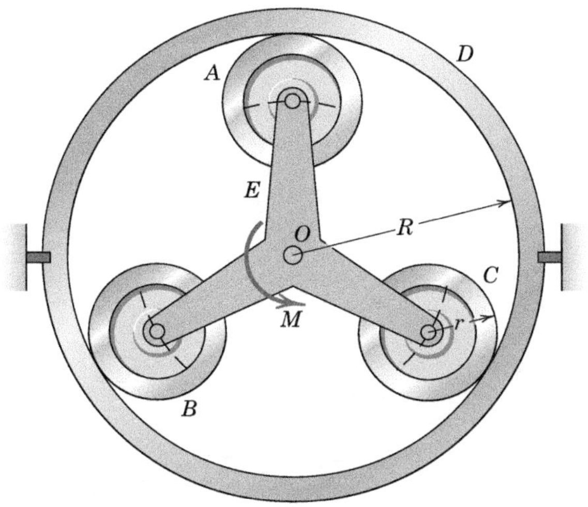

Problem 6/162

6/163 The sector and attached wheels are released from rest in the position shown in the vertical plane. Each wheel is a solid circular disk weighing 12 lb and rolls on the fixed circular path without slipping. The sector weighs 18 lb and is closely approximated by one-fourth of a solid circular disk of 16-in. radius. Determine the initial angular acceleration α of the sector.

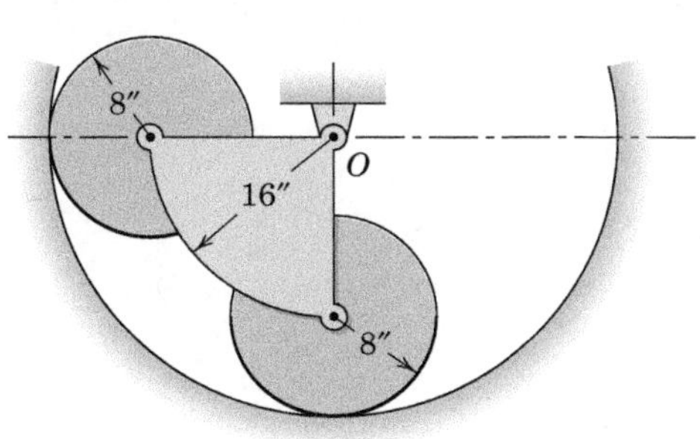

Problem 6/163

6/164 The aerial tower shown is designed to elevate a workman in a vertical direction. An internal mechanism at B maintains the angle between AB and BC at twice the angle θ between BC and the ground. If the combined mass of the man and the cab is 200 kg and if all other masses are neglected, determine the torque M applied to BC at C and the torque M_B in the joint at B required to give the cab an initial vertical acceleration of 1.2 m/s^2 when it is started from rest in the position $\theta = 30°$.

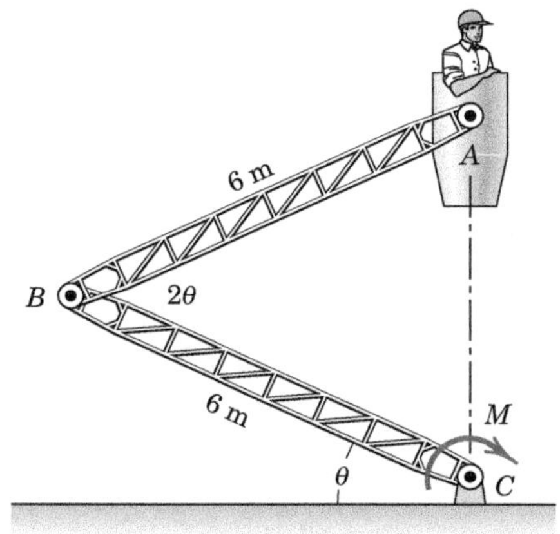

Problem 6/164

6/165 The uniform arm OA has a mass of 4 kg, and the gear D has a mass of 5 kg with a radius of gyration about its center of 64 mm. The large gear B is fixed and cannot rotate. If the arm and small gear are released from rest in the position shown in the vertical plane, calculate the initial angular acceleration α of OA.

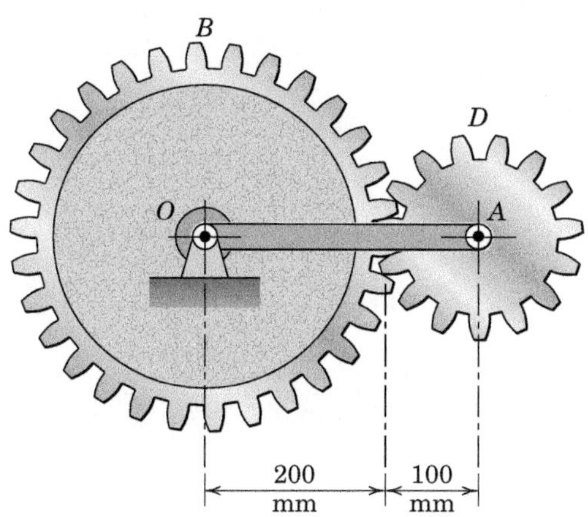

Problem 6/165

6/166 The vehicle is used to transport supplies to and from the bottom of the 25-percent grade. Each pair of wheels, one at A and the other at B, has a mass of 140 kg with a radius of gyration of 150 mm. The drum C has a mass of 40 kg and a radius of gyration of 100 mm. The total mass of the vehicle is 520 kg. The vehicle is released from rest with a restraining force T of 500 N in the control cable which passes around the drum and is secured at D. Determine the initial acceleration a of the vehicle. The wheels roll without slipping.

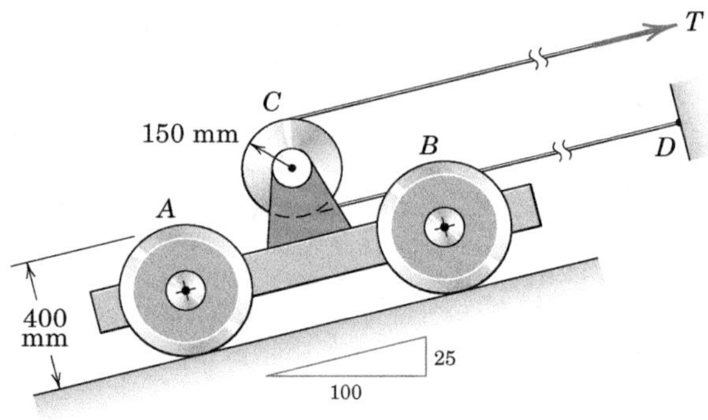

Problem 6/166

SECTION C IMPULSE AND MOMENTUM

6/8 Impulse-Momentum Equations

The principles of impulse and momentum were developed and used in Articles 3/9 and 3/10 for the description of particle motion. In that treatment, we observed that those principles were of particular importance when the applied forces were expressible as functions of the time and when interactions between particles occurred during short periods of time, such as with impact. Similar advantages result when the impulse-momentum principles are applied to the motion of rigid bodies.

In Art. 4/2 the impulse-momentum principles were extended to cover any defined system of mass particles without restriction as to the connections between the particles of the system. These extended relations all apply to the motion of a rigid body, which is merely a special case of a general system of mass. We will now apply these equations directly to rigid-body motion in two dimensions.

Linear Momentum

In Art. 4/4 we defined the linear momentum of a mass system as the vector sum of the linear momenta of all its particles and wrote $\mathbf{G} = \Sigma m_i \mathbf{v}_i$. With $\mathbf{r}_i$ representing the position vector to m_i, we have $\mathbf{v}_i = \dot{\mathbf{r}}_i$ and $\mathbf{G} = \Sigma m_i \dot{\mathbf{r}}_i$ which, for a system whose total mass is constant, may be written as $\mathbf{G} = d(\Sigma m_i \mathbf{r}_i)/dt$. When we substitute the principle of moments $m\bar{\mathbf{r}} = \Sigma m_i \mathbf{r}_i$ to locate the mass center, the momentum becomes $\mathbf{G} = d(m\bar{\mathbf{r}})/dt = m\dot{\bar{\mathbf{r}}}$, where $\dot{\bar{\mathbf{r}}}$ is the velocity $\bar{\mathbf{v}}$ of the mass center. Therefore, as before, we find that the linear momentum of any mass system, rigid or nonrigid, is

$$\boxed{\mathbf{G} = m\bar{\mathbf{v}}} \qquad \textbf{[4/5]}$$

In the derivation of Eq. 4/5, we note that it was unnecessary to employ the kinematic condition for a rigid body, Fig. 6/13, which is $\mathbf{v}_i = \bar{\mathbf{v}} + \boldsymbol{\omega} \times \boldsymbol{\rho}_i$. In that case, we obtain the same result by writing $\mathbf{G} = \Sigma m_i(\bar{\mathbf{v}} + \boldsymbol{\omega} \times \boldsymbol{\rho}_i)$. The first sum is $\bar{\mathbf{v}}\Sigma m_i = m\bar{\mathbf{v}}$, and the second sum becomes $\boldsymbol{\omega} \times \Sigma m_i \boldsymbol{\rho}_i = \boldsymbol{\omega} \times m\bar{\boldsymbol{\rho}} = \mathbf{0}$ since $\boldsymbol{\rho}_i$ is measured from the mass center, making $\bar{\boldsymbol{\rho}}$ zero.

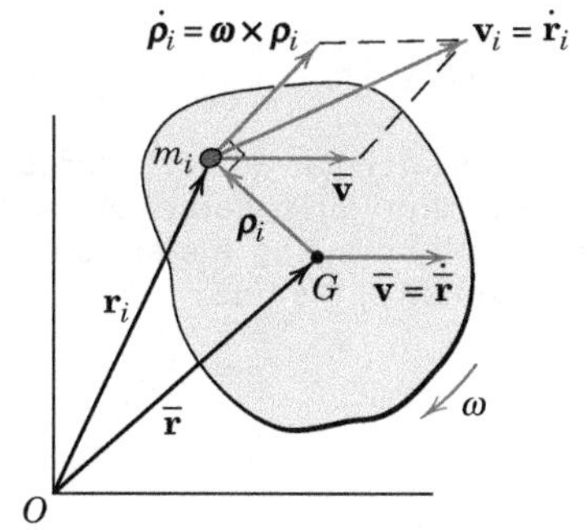

Figure 6/13

Next in Art. 4/4 we rewrote Newton's generalized second law as Eq. 4/6. This equation and its integrated form are

$$\boxed{\Sigma\mathbf{F} = \dot{\mathbf{G}}} \quad \text{and} \quad \boxed{\mathbf{G}_1 + \int_{t_1}^{t_2} \Sigma\mathbf{F}\, dt = \mathbf{G}_2} \qquad \textbf{(6/12)}$$

Equation 6/12 may be written in its scalar-component form, which, for plane motion in the *x-y* plane, gives

$$\boxed{\begin{aligned} \Sigma F_x &= \dot{G}_x \\ \Sigma F_y &= \dot{G}_y \end{aligned}} \quad \text{and} \quad \boxed{\begin{aligned} (G_x)_1 + \int_{t_1}^{t_2} \Sigma F_x\, dt &= (G_x)_2 \\ (G_y)_1 + \int_{t_1}^{t_2} \Sigma F_y\, dt &= (G_y)_2 \end{aligned}} \qquad \textbf{(6/12}\boldsymbol{a}\textbf{)}$$

In words, the first of Eqs. 6/12 and 6/12*a* states that the resultant force equals the time rate of change of momentum. The integrated form of Eqs. 6/12 and 6/12*a* states that the initial linear momentum plus the linear impulse acting on the body equals the final linear momentum.

As in the force-mass-acceleration formulation, the force summations in Eqs. 6/12 and 6/12*a* must include *all* forces acting externally on the body considered. We emphasize again, therefore, that in the use of the impulse-momentum equations, it is essential to construct the complete impulse-momentum diagrams so as to disclose all external impulses. In contrast to the method of work and energy, all forces exert impulses, whether they do work or not.

Angular Momentum

July Store/Shutterstock

This ice skater can effect a large increase in angular speed about a vertical axis by drawing his arms closer to the center of his body.

Angular momentum is defined as the moment of linear momentum. In Art. 4/4 we expressed the angular momentum about the mass center of any prescribed system of mass as $\mathbf{H}_G = \Sigma \boldsymbol{\rho}_i \times m_i \mathbf{v}_i$, which is merely the vector sum of the moments about G of the linear momenta of all particles. We showed in Art. 4/4 that this vector sum could also be written as $\mathbf{H}_G = \Sigma \boldsymbol{\rho}_i \times m_i \dot{\boldsymbol{\rho}}_i$, where $\dot{\boldsymbol{\rho}}_i$ is the velocity of m_i with respect to G.

Although we have simplified this expression in Art. 6/2 in the course of deriving the moment equation of motion, we will pursue this same expression again for sake of emphasis by using the rigid body in plane motion represented in Fig. 6/13. The relative velocity becomes $\dot{\boldsymbol{\rho}}_i = \boldsymbol{\omega} \times \boldsymbol{\rho}_i$, where the angular velocity of the body is $\boldsymbol{\omega} = \omega \mathbf{k}$. The unit vector $\mathbf{k}$ is directed into the paper for the sense of $\boldsymbol{\omega}$ shown. Because $\boldsymbol{\rho}_i$, $\dot{\boldsymbol{\rho}}_i$, and $\boldsymbol{\omega}$ are at right angles to one another, the magnitude of $\dot{\boldsymbol{\rho}}_i$ is $\rho_i \omega$, and the magnitude of $\boldsymbol{\rho}_i \times m_i \dot{\boldsymbol{\rho}}_i$ is $\rho_i^2 \omega m_i$. Thus, we may write $\mathbf{H}_G = \Sigma \rho_i^2 m_i \omega \mathbf{k} = \bar{I} \omega \mathbf{k}$, where $\bar{I} = \Sigma m_i \rho_i^2$ is the mass moment of inertia of the body about its mass center.

Because the angular-momentum vector is always normal to the plane of motion, vector notation is generally unnecessary, and we may write the angular momentum about the mass center as the scalar

$$\boxed{H_G = \bar{I}\omega} \qquad \textbf{(6/13)}$$

This angular momentum appears in the moment-angular-momentum relation, Eq. 4/9, which in scalar notation for plane motion, along with its integrated form, is

$$\boxed{\Sigma M_G = \dot{H}_G} \quad \text{and} \quad \boxed{(H_G)_1 + \int_{t_1}^{t_2} \Sigma M_G \, dt = (H_G)_2} \qquad \textbf{(6/14)}$$

In words, the first of Eqs. 6/14 states that the sum of the moments about the mass center of *all* forces acting on the body equals the time rate of change of angular momentum about the mass center. The integrated form of Eq. 6/14 states that the initial angular momentum about the mass center G plus the external angular impulse about G equals the final angular momentum about G.

The sense for positive rotation must be clearly established, and the algebraic signs of ΣM_G, $(H_G)_1$, and $(H_G)_2$ must be consistent with this choice. The impulse-momentum diagram (see Art. 3/9) is again essential. See the Sample Problems which accompany this article for examples of these diagrams.

With the moments about G of the linear momenta of all particles accounted for by $H_G = \bar{I}\omega$, it follows that we may represent the linear momentum $\mathbf{G} = m\bar{\mathbf{v}}$ as a vector through the mass center G, as shown in Fig. 6/14a. Thus, $\mathbf{G}$ and $\mathbf{H}_G$ have vector properties analogous to those of the resultant force and couple.

With the establishment of the linear- and angular-momentum resultants in Fig. 6/14a, which represents the momentum diagram, the angular momentum H_O about any point O is easily written as

$$H_O = \bar{I}\omega + m\bar{v}d \tag{6/15}$$

This expression holds at any particular instant of time about O, which may be a fixed or moving point on or off the body.

When a body rotates about a fixed point O on the body or body extended, as shown in Fig. 6/14b, the relations $\bar{v} = \bar{r}\omega$ and $d = \bar{r}$ may be substituted into the expression for H_O, giving $H_O = (\bar{I}\omega + m\bar{r}^2\omega)$. But $\bar{I} + m\bar{r}^2 = I_O$ so that

$$H_O = I_O\omega \tag{6/16}$$

In Art. 4/2 we derived Eq. 4/7, which is the moment-angular-momentum equation about a fixed point O. This equation, written in scalar notation for plane motion along with its integrated form, is

$$\Sigma M_O = \dot{H}_O \quad \text{and} \quad (H_O)_1 + \int_{t_1}^{t_2} \Sigma M_O\, dt = (H_O)_2 \tag{6/17}$$

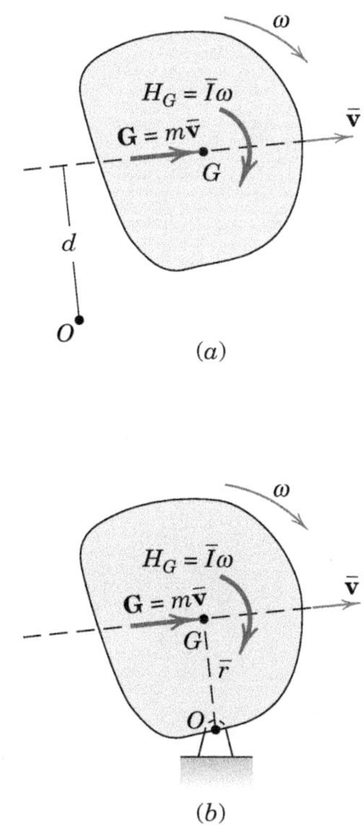

Figure 6/14

Note that you should not add linear momentum and angular momentum for the same reason that force and moment cannot be added directly.

Interconnected Rigid Bodies

The equations of impulse and momentum may also be used for a system of interconnected rigid bodies since the momentum principles are applicable to any general system of constant mass. Figure 6/15 shows the combined free-body diagram and momentum diagram for two interconnected bodies a and b. Equations 4/6 and 4/7, which are $\Sigma\mathbf{F} = \dot{\mathbf{G}}$ and $\Sigma\mathbf{M}_O = \dot{\mathbf{H}}_O$ where O is a fixed reference point, may be written for each member of the system and added. The sums are

$$\begin{aligned} \Sigma\mathbf{F} &= \dot{\mathbf{G}}_a + \dot{\mathbf{G}}_b + \cdots \\ \Sigma\mathbf{M}_O &= (\dot{\mathbf{H}}_O)_a + (\dot{\mathbf{H}}_O)_b + \cdots \end{aligned} \tag{6/18}$$

In integrated form for a finite time interval, these expressions become

$$\int_{t_1}^{t_2} \Sigma\mathbf{F}\, dt = (\Delta\mathbf{G})_{\text{system}} \qquad \int_{t_1}^{t_2} \Sigma\mathbf{M}_O\, dt = (\Delta\mathbf{H}_O)_{\text{system}} \tag{6/19}$$

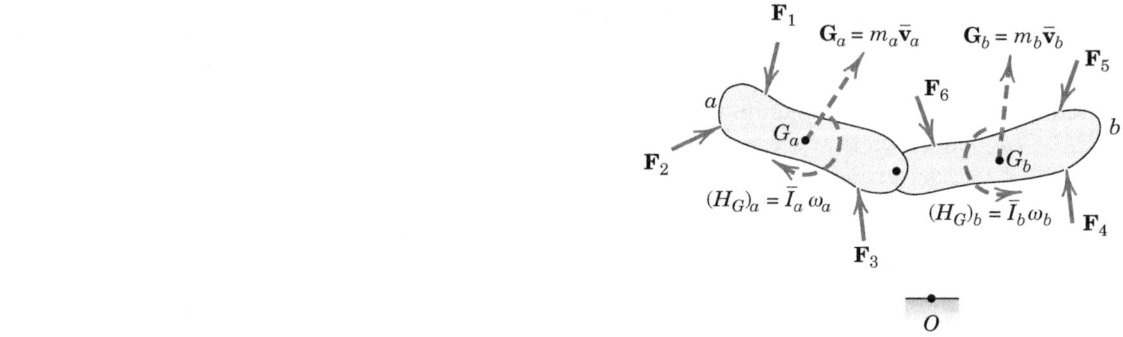

Figure 6/15

We note that the equal and opposite actions and reactions in the connections are internal to the system and cancel one another so they are not involved in the force and moment summations. Also, point O is one fixed reference point for the entire system.

Conservation of Momentum

In Art. 4/5, we expressed the principles of conservation of momentum for a general mass system by Eqs. 4/15 and 4/16. These principles are applicable to either a single rigid body or a system of interconnected rigid bodies. Thus, if $\Sigma\mathbf{F} = \mathbf{0}$ for a given interval of time, then

$$\boxed{\mathbf{G}_1 = \mathbf{G}_2} \qquad \textbf{[4/15]}$$

which says that the linear-momentum vector undergoes no change in the absence of a resultant linear impulse. For the system of interconnected rigid bodies, there may be linear-momentum changes of individual parts of the system during the interval, but there will be no resultant momentum change for the system as a whole if there is no resultant linear impulse.

Similarly, if the resultant moment about a given fixed point O or about the mass center is zero during a particular interval of time for a single rigid body or for a system of interconnected rigid bodies, then

$$\boxed{(\mathbf{H}_O)_1 = (\mathbf{H}_O)_2} \quad \text{or} \quad \boxed{(\mathbf{H}_G)_1 = (\mathbf{H}_G)_2} \qquad \textbf{[4/16]}$$

which says that the angular momentum either about the fixed point or about the mass center undergoes no change in the absence of a corresponding resultant angular impulse. Again, in the case of the interconnected system, there may be angular-momentum changes of individual components during the interval, but there will be no resultant angular-momentum change for the system as a whole if there is no resultant angular impulse about the fixed point or the mass center. Either of Eqs. 4/16 may hold without the other.

In the case of an interconnected system, the system center of mass is generally inconvenient to use.

As was illustrated previously in Articles 3/9 and 3/10 in the chapter on particle motion, the use of momentum principles greatly facilitates the analysis of situations where forces and couples act for very short periods of time.

Impact of Rigid Bodies

Impact phenomena involve a fairly complex interrelationship of energy and momentum transfer, energy dissipation, elastic and plastic deformation, relative impact velocity, and body geometry. In Art. 3/12 we treated the impact of bodies modeled as particles and considered only the case of central impact, where the contact forces of impact passed through the mass centers of the bodies, as would always happen with colliding smooth spheres, for example. To relate the conditions after impact to those before impact required the introduction of the so-called coefficient of restitution e or impact coefficient, which compares the relative separation velocity with the relative approach velocity measured along the direction of the contact forces. Although in the classical theory of impact, e was considered a constant for given materials, more modern investigations show that e is highly dependent on geometry and impact velocity as well as on materials. At best, even for spheres and rods under direct central and longitudinal impact, the coefficient of restitution is a complex and variable factor of limited use.

Any attempt to extend this simplified theory of impact utilizing a coefficient of restitution for the noncentral impact of rigid bodies of varying shape is a gross oversimplification which has little practical value. For this reason, we do not include such an exercise in this book, even though such a theory is easily developed and appears in certain references. We can and do, however, make full use of the principles of conservation of linear and angular momentum when they are applicable in discussing impact and other interactions of rigid bodies.

There are small reaction wheels inside the Hubble Space Telescope that make precision attitude control possible. The principles of angular momentum are fundamental to the design and operation of such a control system.

Sample Problem 6/14

The force P, which is applied to the cable wrapped around the central hub of the symmetrical wheel, is increased slowly according to $P = 1.5t$, where P is in pounds and t is the time in seconds after P is first applied. Determine the angular velocity ω_2 of the wheel 10 seconds after P is applied if the wheel is rolling to the left with a velocity of its center of 3 ft/sec at time $t = 0$. The wheel weighs 120 lb with a radius of gyration about its center of 10 in. and rolls without slipping.

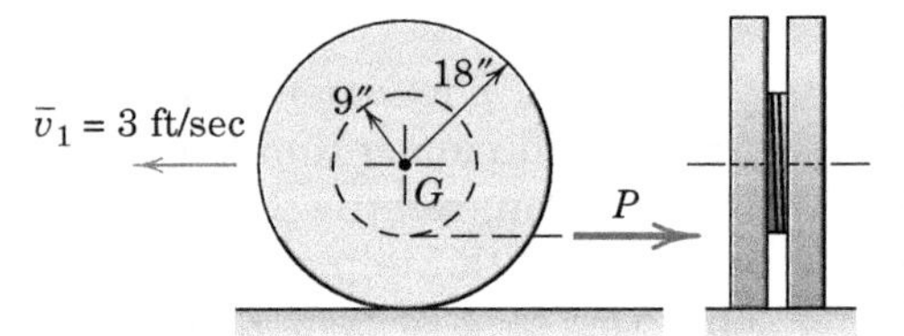

Solution. The impulse-momentum diagram of the wheel discloses the initial linear and angular momenta at time $t_1 = 0$, all external impulses, and the final linear and angular momenta at time $t_2 = 10$ sec. The correct direction of the
① friction force F is that to oppose the slipping which would occur without friction.

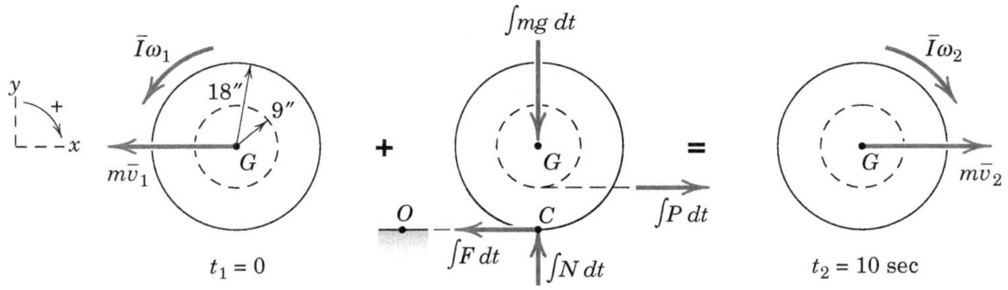

Application of the linear impulse-momentum equation and the angular impulse-momentum equation over the *entire* interval gives

② $$\left[(G_x)_1 + \int_{t_1}^{t_2} \Sigma F_x\,dt = (G_x)_2\right] \qquad \frac{120}{32.2}(-3) + \int_0^{10} (1.5t - F)\,dt = \frac{120}{32.2}\left[\frac{18}{12}\omega_2\right]$$

③ $$\left[(H_G)_1 + \int_{t_1}^{t_2} \Sigma M_G\,dt = (H_G)_2\right]$$

$$\frac{120}{32.2}\left(\frac{10}{12}\right)^2\left(-\frac{3}{18/12}\right) + \int_0^{10}\left[\frac{18}{12}F - \frac{9}{12}(1.5t)\right]dt = \frac{120}{32.2}\left(\frac{10}{12}\right)^2\left[\omega_2\right]$$

Since the force F is variable, it must remain under the integral sign. We eliminate F between the two equations by multiplying the second one by $\frac{12}{18}$ and adding to the first one. Integrating and solving for ω_2 give

$$\omega_2 = 3.13 \text{ rad/sec clockwise} \qquad \textit{Ans.}$$

Alternative Solution. We could avoid the necessity of a simultaneous solution by applying the second of Eqs. 6/17 about a fixed point O on the horizontal surface. The moments of the 120-lb weight and the equal and opposite force N cancel one another, and F is eliminated since its moment about O is zero. Thus, the angular momentum about O becomes $H_O = \bar{I}\omega + m\bar{v}r = m\bar{k}^2\omega + mr^2\omega = m(\bar{k}^2 + r^2)\omega$, where $\bar{k}$ is the centroidal radius of gyration and r is the 18-in. rolling radius. Thus, we see that $H_O = H_C$ since $\bar{k}^2 + r^2 = k_C{}^2$ and $H_C = I_C\omega = mk_C{}^2\omega$. Equation 6/17 now gives

$$\left[(H_O)_1 + \int_{t_1}^{t_2} \Sigma M_O dt = (H_O)_2\right]$$

$$\frac{120}{32.2}\left[\left(\frac{10}{12}\right)^2 + \left(\frac{18}{12}\right)^2\right]\left[-\frac{3}{18/12}\right] + \int_0^{10} 1.5t\left(\frac{18-9}{12}\right)dt$$

$$= \frac{120}{32.2}\left[\left(\frac{10}{12}\right)^2 + \left(\frac{18}{12}\right)^2\right]\left[\omega_2\right]$$

Solution of this one equation is equivalent to the simultaneous solution of the two previous equations.

Helpful Hints

① Also, we note the clockwise imbalance of moments about C, which causes a clockwise angular acceleration as the wheel rolls without slipping. Since the moment sum about G must also be in the clockwise sense of α, the friction force must act to the left to provide it.

② Note carefully the signs of the momentum terms. The final linear velocity is assumed in the positive x-direction, so $(G_x)_2$ is positive. The initial linear velocity is negative, so $(G_x)_1$ is negative.

③ Since the wheel rolls without slipping, a positive x-velocity requires a clockwise angular velocity, and vice versa.

Sample Problem 6/15

The sheave E of the hoisting rig shown has a mass of 30 kg and a centroidal radius of gyration of 250 mm. The 40-kg load D which is carried by the sheave has an initial downward velocity $v_1 = 1.2$ m/s at the instant when a clockwise torque is applied to the hoisting drum A to maintain essentially a constant force $F = 380$ N in the cable at B. Compute the angular velocity ω_2 of the sheave 5 seconds after the torque is applied to the drum and find the tension T in the cable at O during the interval. Neglect all friction.

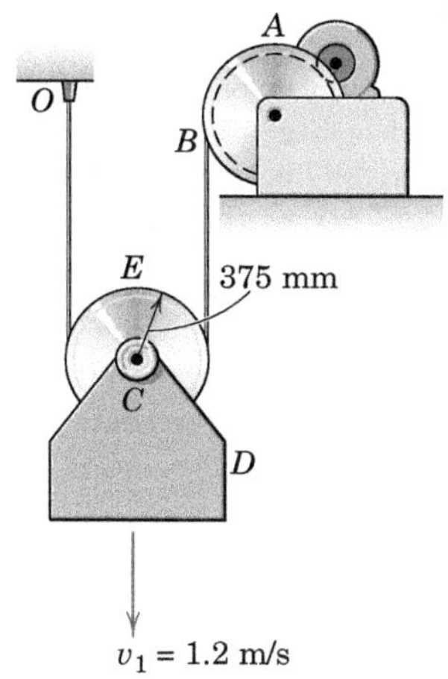

Solution. The load and the sheave taken together constitute the system, and its impulse-momentum diagram is shown. The tension T in the cable at O and the final angular velocity ω_2 of the sheave are the two unknowns. We eliminate T initially by applying the moment-angular-momentum equation about the fixed point O, taking counterclockwise as positive.

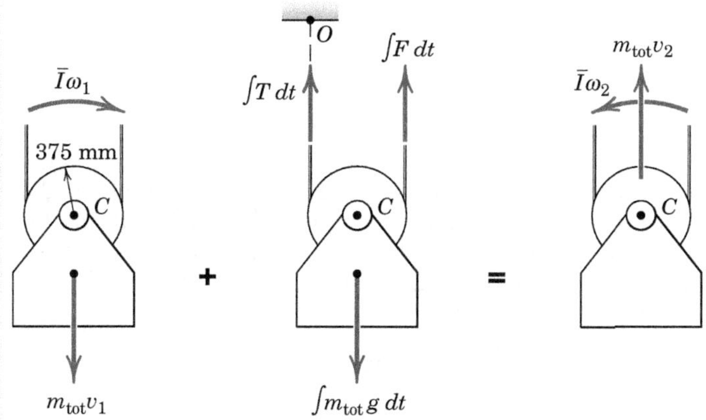

$$\left[(H_O)_1 + \int_{t_1}^{t_2} \Sigma M_O\,dt = (H_O)_2\right]$$

$$\int_{t_1}^{t_2} \Sigma M_O\,dt = \int_0^5 [380(0.750) - (30 + 40)(9.81)(0.375)]\,dt$$

$$= 137.4 \text{ N·m·s}$$

$$(H_O)_1 = -(m_E + m_D)v_1 d - \bar{I}\omega_1$$

$$= -(30 + 40)(1.2)(0.375) - 30(0.250)^2\left(\frac{1.2}{0.375}\right)$$

① $$= -37.5 \text{ N·m·s}$$

$$(H_O)_2 = (m_E + m_D)v_2 d + \bar{I}\omega_2$$

$$= +(30 + 40)(0.375\omega_2)(0.375) + 30(0.250)^2\omega_2$$

$$= 11.72\omega_2$$

Substituting into the momentum equation gives

$$-37.5 + 137.4 = 11.72\omega_2$$

$$\omega_2 = 8.53 \text{ rad/s counterclockwise} \qquad \textit{Ans.}$$

The linear-impulse-momentum equation is now applied to the system to determine T. With the positive direction up, we have

$$\left[G_1 + \int_{t_1}^{t_2} \Sigma F\,dt = G_2\right]$$

$$70(-1.2) + \int_0^5 [T + 380 - 70(9.81)]\,dt = 70[0.375(8.53)]$$

$$5T = 1841 \qquad T = 368 \text{ N} \qquad \textit{Ans.}$$

If we had taken our moment equation around the center C of the sheave instead of point O, it would contain both unknowns T and ω, and we would be obliged to solve it simultaneously with the foregoing force equation, which would also contain the same two unknowns.

Helpful Hint

① The units of angular momentum, which are those of angular impulse, may also be written as kg·m^2/s.

Sample Problem 6/16

The uniform rectangular block of dimensions shown is sliding to the left on the horizontal surface with a velocity v_1 when it strikes the small step at O. Assume negligible rebound at the step and compute the minimum value of v_1 which will permit the block to pivot freely about O and just reach the standing position A with no velocity. Compute the percentage energy loss n for $b = c$.

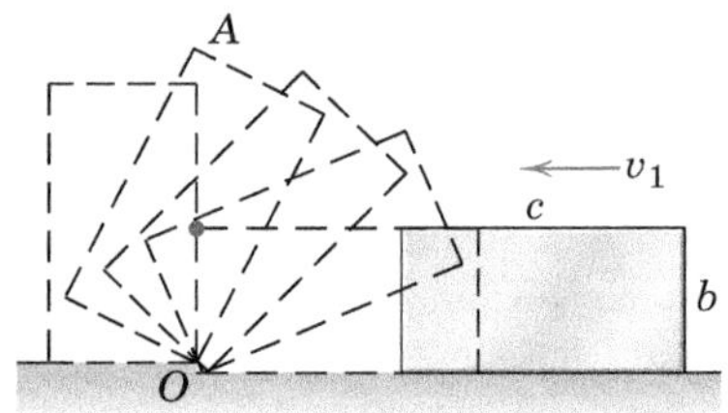

Solution. We break the overall process into two subevents: the collision (I) and the subsequent rotation (II).

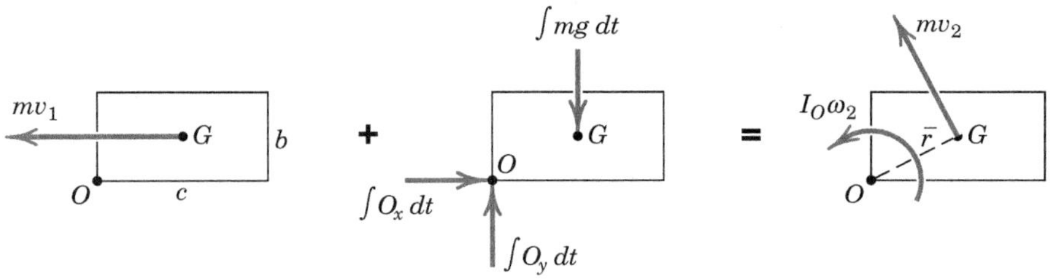

I. Collision. With the assumption that the weight mg is nonimpulsive, an-
(1) gular momentum about O is conserved. The initial angular momentum of the block about O just before impact is the moment about O of its linear momen-
(2) tum and is $(H_O)_1 = mv_1(b/2)$. The angular momentum about O just after impact when the block is starting its rotation about O is

(3) $[H_O = I_O\omega]$ $\quad (H_O)_2 = \left\{\frac{1}{12}m(b^2 + c^2) + m\left[\left(\frac{c}{2}\right)^2 + \left(\frac{b}{2}\right)^2\right]\right\}\omega_2$

$$= \frac{m}{3}(b^2 + c^2)\omega_2$$

Conservation of angular momentum gives

$$[(H_O)_1 = (H_O)_2] \qquad mv_1\frac{b}{2} = \frac{m}{3}(b^2 + c^2)\omega_2 \qquad \omega_2 = \frac{3v_1 b}{2(b^2 + c^2)}$$

II. Rotation about O. With the assumptions that the rotation is like that about a fixed frictionless pivot and that the location of the effective pivot O is at ground level, mechanical energy is conserved during the rotation according to

(4) $[T_2 + V_2 = T_3 + V_3]$ $\quad \frac{1}{2}I_O\omega_2^2 + 0 = 0 + mg\left[\sqrt{\left(\frac{b}{2}\right)^2 + \left(\frac{c}{2}\right)^2} - \frac{b}{2}\right]$

$$\frac{1}{2}\frac{m}{3}(b^2 + c^2)\left[\frac{3v_1 b}{2(b^2 + c^2)}\right]^2 = \frac{mg}{2}(\sqrt{b^2 + c^2} - b)$$

$$v_1 = 2\sqrt{\frac{g}{3}\left(1 + \frac{c^2}{b^2}\right)(\sqrt{b^2 + c^2} - b)} \qquad \textit{Ans.}$$

The percentage loss of energy during the impact is

$$n = \frac{|\Delta E|}{E} = \frac{\frac{1}{2}mv_1^2 - \frac{1}{2}I_O\omega_2^2}{\frac{1}{2}mv_1^2} = 1 - \frac{k_O^2\omega_2^2}{v_1^2} = 1 - \left(\frac{b^2 + c^2}{3}\right)\left[\frac{3b}{2(b + c^2)}\right]^2$$

$$= 1 - \frac{3}{4\left(1 + \frac{c^2}{b^2}\right)} \qquad n = 62.5\% \text{ for } b = c \qquad \textit{Ans.}$$

Helpful Hints

(1) If the corner of the block struck a spring instead of the rigid step, then the time of the interaction during compression of the spring could become appreciable, and the angular impulse about the fixed point at the end of the spring due to the moment of the weight would have to be accounted for.

(2) Notice the abrupt change in direction and magnitude of the velocity of G during the impact.

(3) Be sure to use the transfer theorem $I_O = \bar{I} + m\bar{r}^2$ correctly here.

(4) The datum is taken at the initial altitude of the mass center G. State 3 is taken to be the standing position A, at which the diagonal of the block is vertical.

PROBLEMS

Introductory Problems

6/167 A person who walks through the revolving door exerts a 90-N horizontal force on one of the four door panels and keeps the 15° angle constant relative to a line which is normal to the panel. If each panel is modeled by a 60-kg uniform rectangular plate which is 1.2 m in length as viewed from above, determine the final angular velocity ω of the door if the person exerts the force for 3 seconds. The door is initially at rest and friction may be neglected.

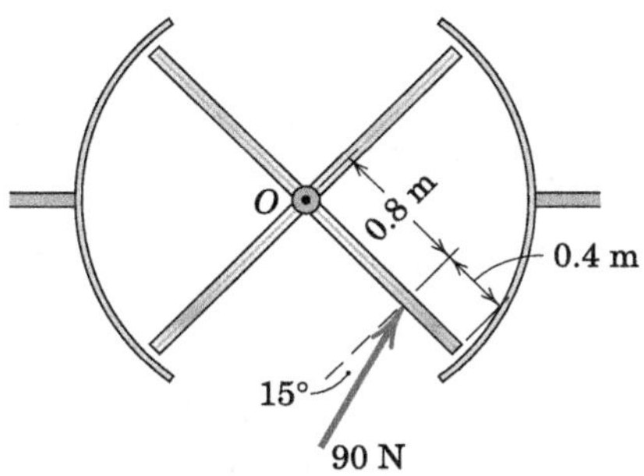

Problem 6/167

6/168 The 75-kg flywheel has a radius of gyration about its shaft axis of $\bar{k} = 0.50$ m and is subjected to the torque $M = 10(1 - e^{-t})$ N·m, where t is in seconds. If the flywheel is at rest at time $t = 0$, determine its angular velocity ω at $t = 3$ s.

Problem 6/168

6/169 Determine the angular momentum of the earth about the center of the sun. Assume a homogeneous earth and a circular earth orbit of radius $149.6(10^6)$ km. Consult Table D/2 of Appendix D for other needed information. Comment on the relative contributions of the terms $\bar{I}\omega$ and $m\bar{v}d$.

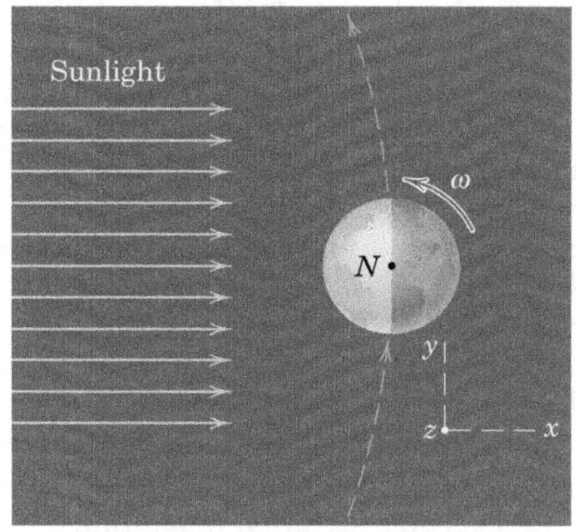

Problem 6/169

6/170 The frame of mass m is welded together from uniform slender rods. The frame is released from rest in the upper position shown and constrained to fall vertically by two light rollers which travel along the smooth slots. The roller at A catches in the support at O without rebounding and serves as a hinge for the frame as it rotates thereafter. Determine the angular velocity of the frame an instant after the roller at A engages the support at O.

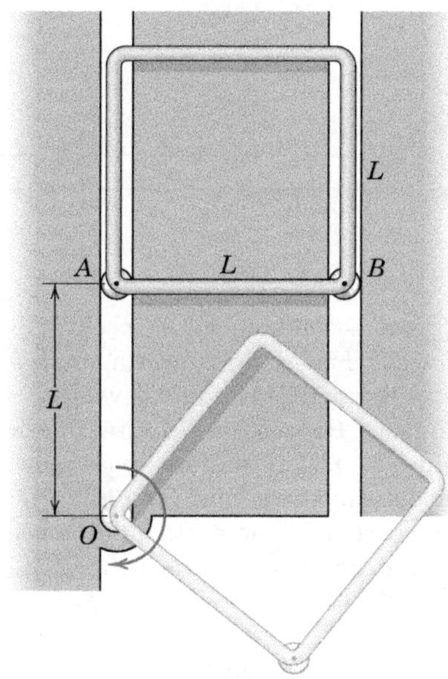

Problem 6/170

6/171 The frictional moment M_f acting on a rotating turbine disk and its shaft is given by $M_f = k\omega^2$ where ω is the angular velocity of the turbine. If the source of power is cut off while the turbine is running with an angular velocity ω_0, determine the time t for the speed of the turbine to drop to one-half of its initial value. The moment of inertia of the turbine disk and shaft is I.

6/172 The cable drum has a mass of 800 kg with radius of gyration of 480 mm about its center O and is mounted in bearings on the 1200-kg carriage. The carriage is initially moving to the left with a speed of 1.5 m/s, and the drum is rotating counterclockwise with an angular velocity of 3 rad/s when a constant horizontal tension $T = 400$ N is applied to the cable at time $t = 0$. Determine the velocity v of the carriage and the angular velocity ω of the drum when $t = 10$ s. Neglect the mass of the carriage wheels.

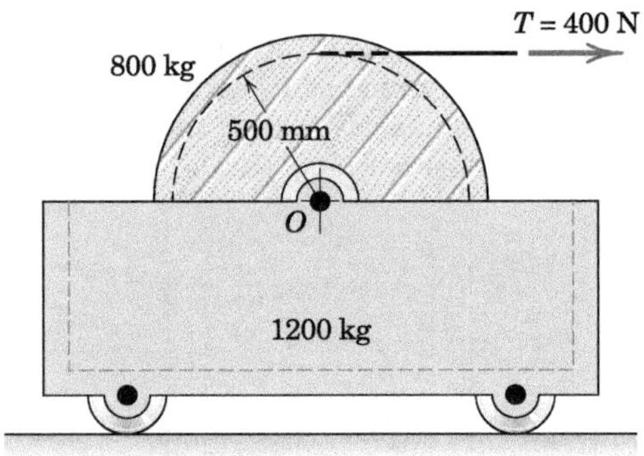

Problem 6/172

6/173 The man is walking with speed $v_1 = 1.2$ m/s to the right when he trips over a small floor discontinuity. Estimate his angular velocity ω just after the impact. His mass is 76 kg with center-of-mass height $h = 0.87$ m, and his mass moment of inertia about the ankle joint O is 66 kg·m^2, where all are properties of the portion of his body above O; i.e., both the mass and moment of inertia do not include the foot.

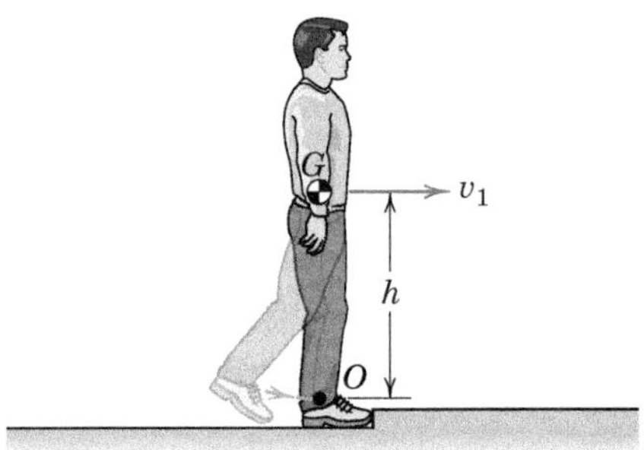

Problem 6/173

6/174 The 15-kg wheel with 150-mm outer radius and 115-mm centroidial radius of gyration is rolling without slipping down the 15° incline at a speed of 2 m/s when a tension $T = 30$ N is applied to a cable wrapped securely around an inner hub with a radius of 100 mm. Determine the time t required for the wheel to come to a stop if the tension is applied first in configuration (*a*) and then in configuration (*b*). The wheel is observed to roll without slipping throughout the entire motion for both cases.

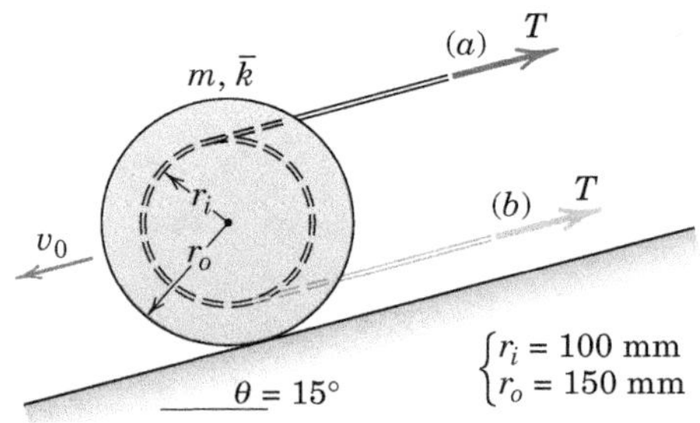

Problem 6/174

6/175 A uniform slender bar of mass M and length L is translating on the smooth horizontal x-y plane with a velocity v_M when a particle of mass m traveling with a velocity v_m as shown strikes and becomes embedded in the bar. Determine the final linear and angular velocities of the bar with its embedded particle.

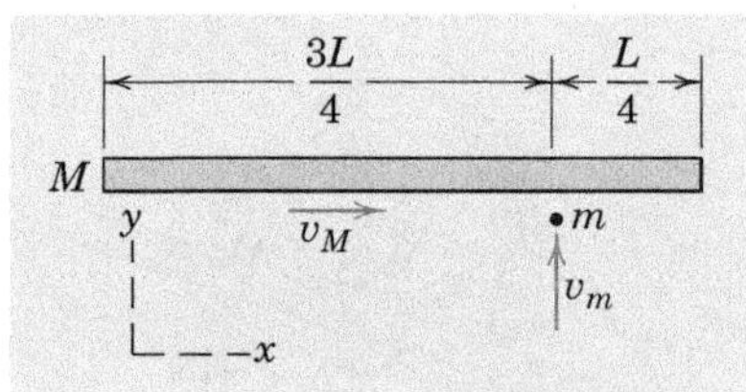

Problem 6/175

6/176 The homogeneous circular cylinder of mass m and radius R carries a slender rod of mass $m/2$ attached to it as shown. If the cylinder rolls on the surface without slipping with a velocity v_O of its center O, determine the angular momenta H_G and H_O of the system about its center of mass G and about O for the instant shown.

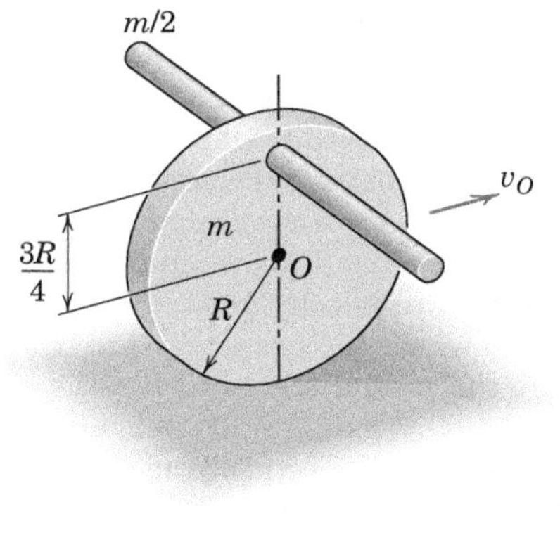

Problem 6/176

6/177 The grooved pulley of mass m is acted on by a constant force F through a cable which is wrapped securely around the exterior of the pulley. The pulley supports a cylinder of mass M which is attached to the end of a cable which is wrapped securely around an inner hub. If the system is stationary when the force F is first applied, determine the upward velocity of the supported mass after 3 seconds. Use the values $m = 40$ kg, $M = 10$ kg, $r_o = 225$ mm, $r_i = 150$ mm, $k_O = 160$ mm, and $F = 75$ N. Assume no mechanical interference for the indicated time frame and neglect friction in the bearing at O. What is the time-averaged value of the force in the cable which supports the 10-kg mass?

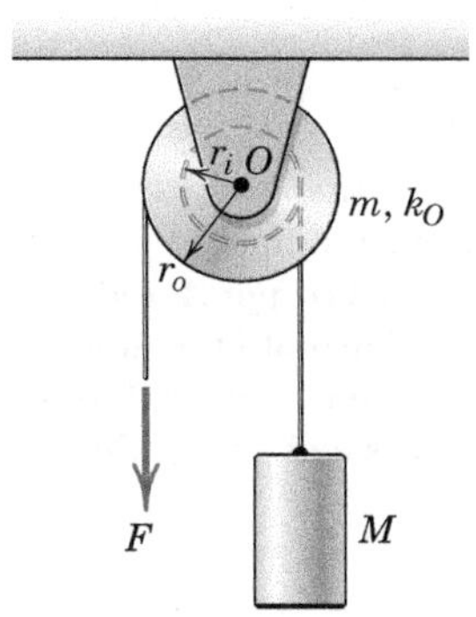

Problem 6/177

6/178 The wad of clay of mass m is initially moving with a horizontal velocity v_1 when it strikes and sticks to the initially stationary uniform slender bar of mass M and length L. Determine the final angular velocity of the combined body and the x-component of the linear impulse applied to the body by the pivot O during the impact.

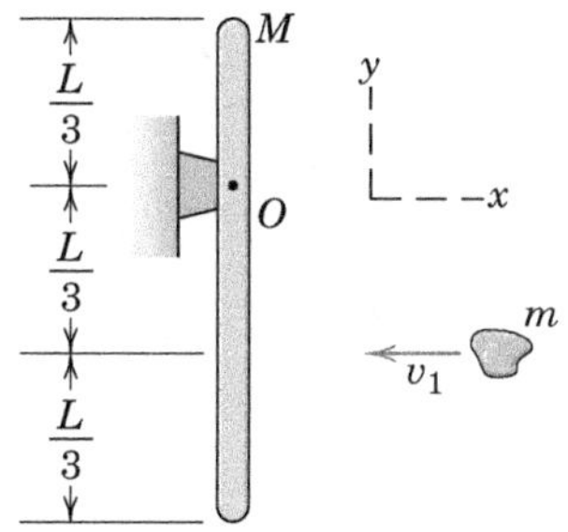

Problem 6/178

Representative Problems

6/179 The 30-lb uniform plate has the indicated velocities at corners A and B. What is the angular momentum of the plate about the mass center G?

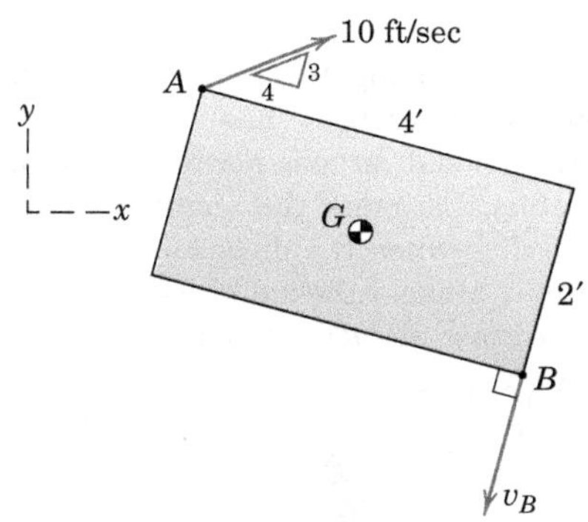

Problem 6/179

6/180 The plate of Prob. 6/179 is repeated here where the coordinates of corner C are established. The plate is falling freely in the x-y vertical plane. What is the linear momentum of the plate and its angular momentum about point O? Additionally, determine the time rate of change of the angular momentum of the plate about point O.

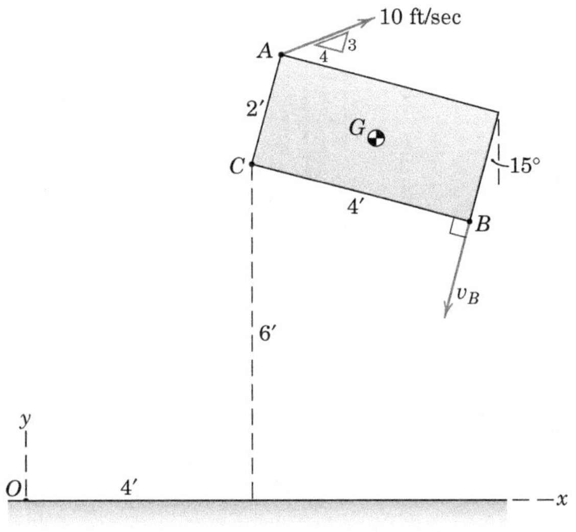

Problem 6/180

6/181 Just after leaving the platform, the diver's fully extended 80-kg body has a rotational speed of 0.3 rev/s about an axis normal to the plane of the trajectory. Estimate the angular velocity N later in the dive when the diver has assumed the tuck position. Make reasonable assumptions concerning the mass moment of inertia of the body in each configuration.

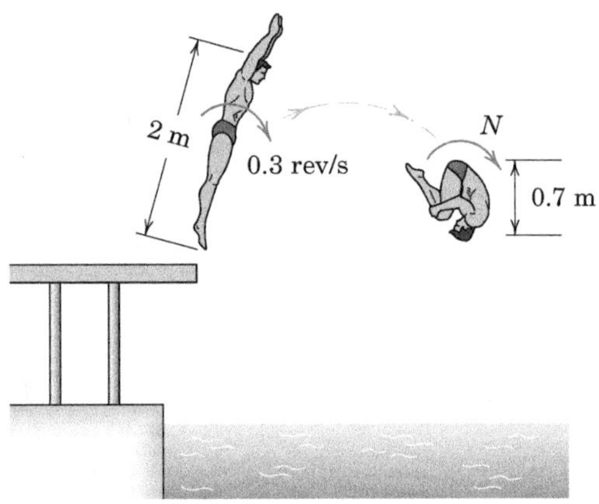

Problem 6/181

6/182 The device shown is a simplified model of an amusement-park ride in which passengers are rotated about the vertical axis of the central post at an angular speed Ω while sitting in a pod which is capable of rotating the occupants 360° about the longitudinal axis of the connecting arm attached to the central collar. Determine the percent increase n in angular velocity between configurations (a) and (b), where the passenger pod has rotated 90° about the connecting arm. For the model, $m = 1.2$ kg, $r = 75$ mm, $l = 300$ mm, and $L = 650$ mm. The post and connecting arms rotate freely about the z-axis at an initial angular speed $\Omega = 120$ rev/min and have a combined mass moment of inertia about the z-axis of $30(10^{-3})$ kg·m^2.

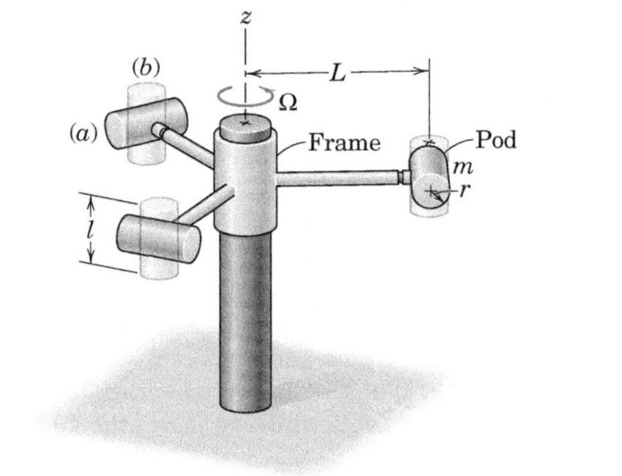

Problem 6/182

6/183 The uniform concrete block, which weighs 171 lb and falls from rest in the horizontal position shown, strikes the fixed corner A and pivots around it with no rebound. Calculate the angular velocity ω of the block immediately after it hits the corner and the percentage loss n of energy due to the impact.

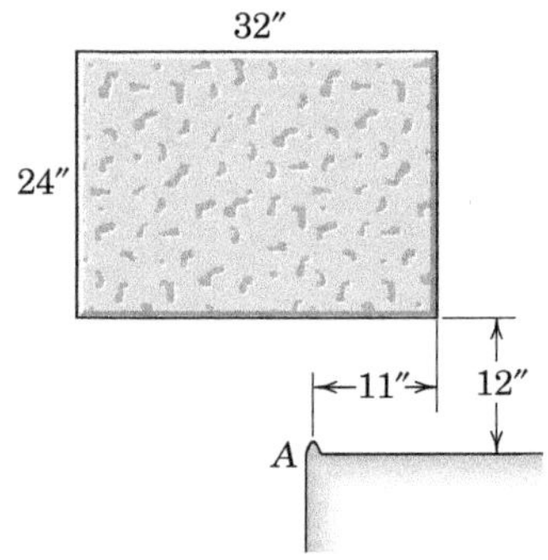

Problem 6/183

6/184 Two small variable-thrust jets are actuated to keep the spacecraft angular velocity about the z-axis constant at $\omega_0 = 1.25$ rad/s as the two telescoping booms are extended from $r_1 = 1.2$ m to $r_2 = 4.5$ m at a constant rate over a 2-min period. Determine the necessary thrust T for each jet as a function of time where $t = 0$ is the time when the telescoping action is begun. The small 10-kg experiment modules at the ends of the booms may be treated as particles, and the mass of the rigid booms is negligible.

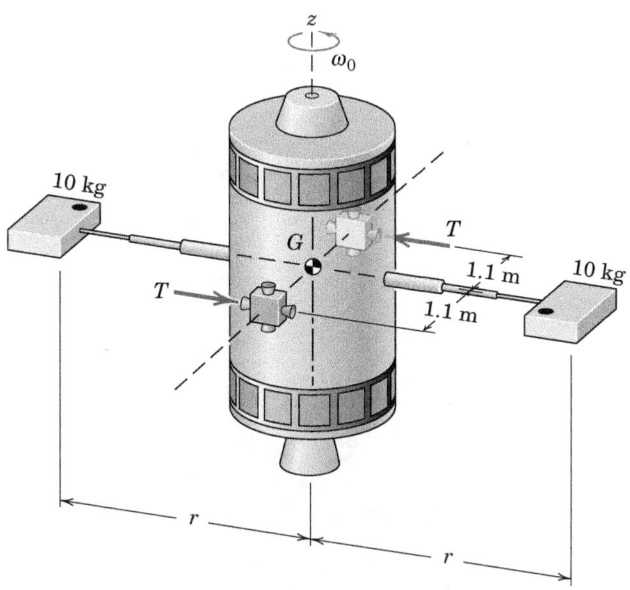

Problem 6/184

6/185 The body composed of slender rods of mass ρ per unit length is lying motionless on the smooth horizontal surface when a linear impluse $\int P\,dt$ is applied as shown. Determine the velocity $\mathbf{v}_B$ of corner B immediately following the application of the impulse if $l = 500$ mm, $\rho = 3$ kg/m, and $\int P\,dt = 8$ N·s.

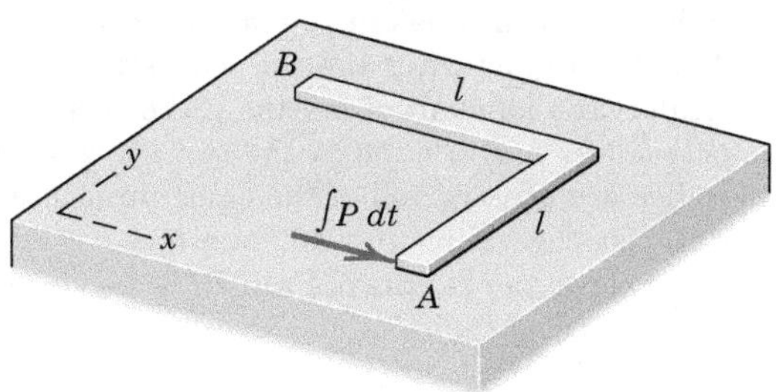

Problem 6/185

6/186 Each of the two 300-mm uniform rods A has a mass of 1.5 kg and is hinged at its end to the rotating base B. The 4-kg base has a radius of gyration of 40 mm and is initially rotating freely about its vertical axis with a speed of 300 rev/min and with the rods latched in the vertical positions. If the latches are released and the rods assume the horizontal positions, calculate the new rotational speed N of the assembly.

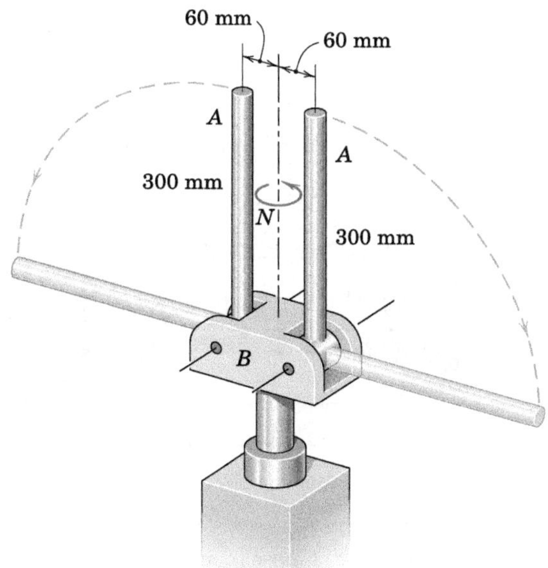

Problem 6/186

6/187 The phenomenon of vehicle "tripping" is investigated here. The sport-utility vehicle is sliding sideways with speed v_1 and no angular velocity when it strikes a small curb. Assume no rebound of the right-side tires and estimate the minimum speed v_1 which will cause the vehicle to roll completely over to its right side. The mass of the SUV is 2300 kg and its mass moment of inertia about a longitudinal axis through the mass center G is 900 kg·m^2.

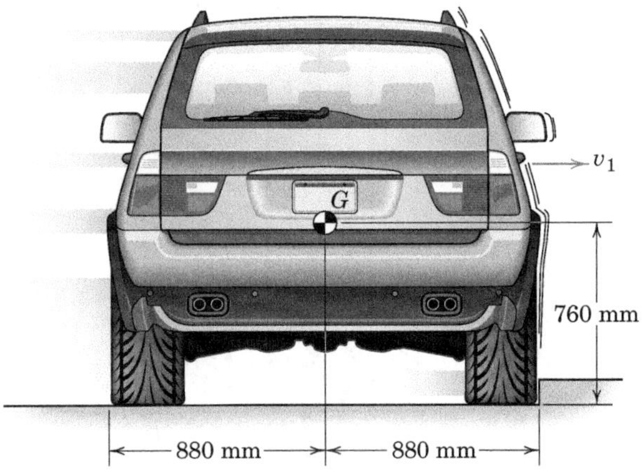

Problem 6/187

6/188 The slender bar of mass m and length l is released from rest in the horizontal position shown. If point A of the bar becomes attached to the pivot at B upon impact, determine the angular velocity ω of the bar immediately after impact in terms of the distance x. Evaluate your expression for $x = 0$, $l/2$, and l.

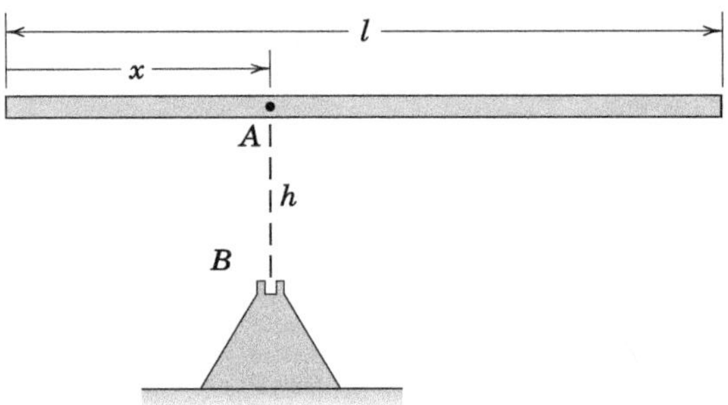

Problem 6/188

6/189 Child A weighs 75 lb and is sitting at rest relative to the merry-go-round which is rotating counterclockwise with an angular velocity $\Omega = 15$ rev/min. Child B, with a weight of 65 lb, runs toward the merry-go-round with a speed $v = 12$ ft/sec and jumps onto the edge. Compute the angular velocity Ω' of the merry-go-round when child B has just jumped aboard the merry-go-round and is standing in position (a). If child B moves to position (b), what is the new angular velocity of the merry-go-round? For simplicity, model the inertia of child A as a 20-in.-diameter uniform disk (horizontal) and model child B as a 14-in.-diameter uniform and upright cylinder. The merry-go-round itself has a centroidal mass moment of inertia $I_O = 110$ lb-ft-sec^2.

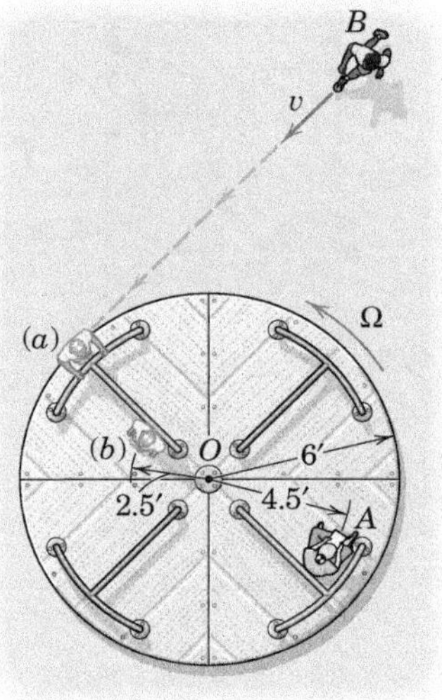

Problem 6/189

6/190 The system is initially rotating freely with angular velocity $\omega_1 = 10$ rad/s when the inner rod A is centered lengthwise within the hollow cylinder B as shown in the figure. Determine the angular velocity of the system (a) if the inner rod A has moved so that a length $b/2$ is protruding from the cylinder, (b) just before the rod leaves the cylinder, and (c) just after the rod leaves the cylinder. Neglect the moment of inertia of the vertical support shafts and friction in the two bearings. Both bodies are constructed of the same uniform material. Use the values $b = 400$ mm and $r = 20$ mm, and refer to the results of Prob. B/34 as needed.

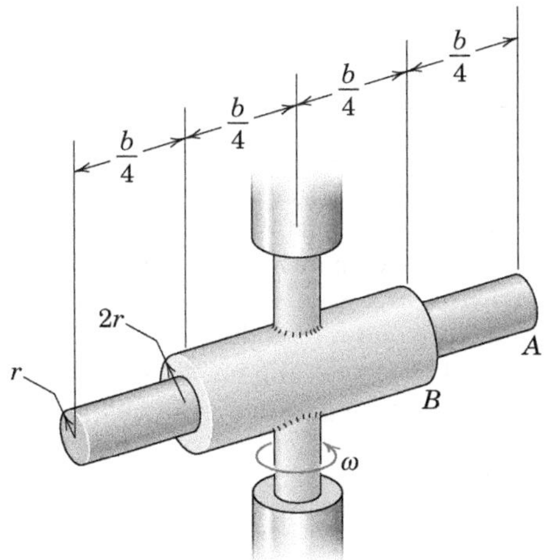

Problem 6/190

6/191 The homogeneous sphere of mass m and radius r is projected along the incline of angle θ with an initial speed v_0 and no angular velocity ($\omega_0 = 0$). If the coefficient of kinetic friction is μ_k, determine the time duration t of the period of slipping. In addition, state the velocity v of the mass center G and the angular velocity ω at the end of the period of slipping.

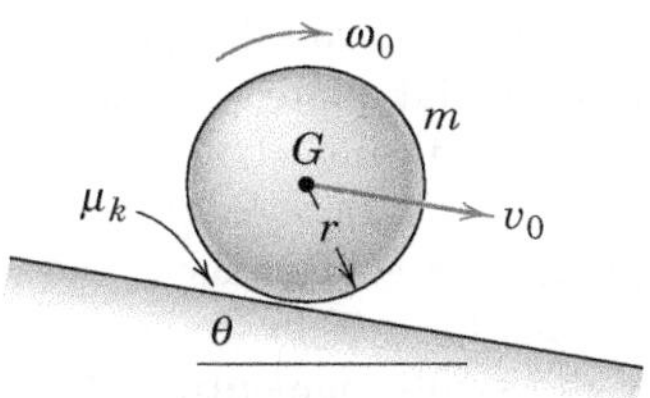

Problem 6/191

6/192 The homogeneous sphere of Prob. 6/191 is placed on the incline with a clockwise angular velocity ω_0 but no linear velocity of its center ($v_0 = 0$). Determine the time duration t of the period of slipping. In addition, state the velocity v of the mass center G and angular velocity ω at the end of the period of slipping.

6/193 The 100-lb platform rolls without slipping along the 10° incline on two pairs of 16-in.-diameter wheels. Each pair of wheels with attached axle weighs 25 lb and has a centroidal radius of gyration of 5.5 in. The platform has an initial speed of 3 ft/sec down the incline when a tension T is applied through a cable attached to the platform. If the platform acquires a speed of 3 ft/sec up the incline after the tension has been applied for 8 seconds, what is the average value of the tension in the cable?

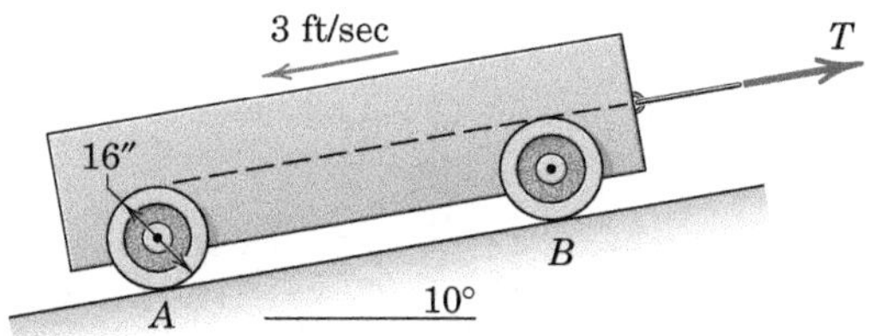

Problem 6/193

6/194 The uniform slender bar of mass m and length l has no angular velocity as end A strikes the ground against the stop with no rebound. If $\alpha = 15°$, what is the minimum magnitude of the initial velocity $\mathbf{v}_1$ for which the bar will rotate about A to the vertical position?

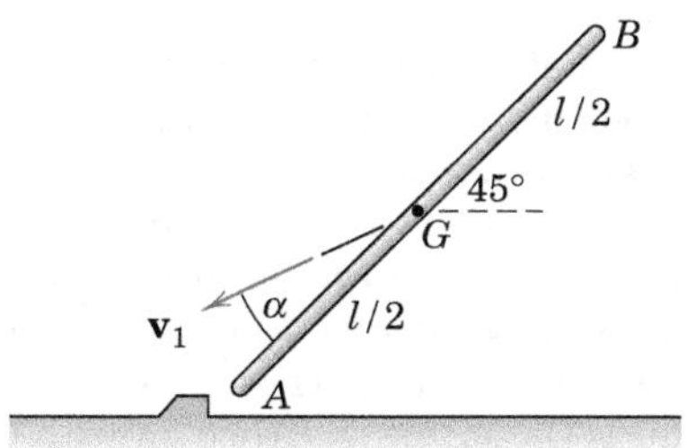

Problem 6/194

6/195 The 165-lb ice skater with arms extended horizontally spins about a vertical axis with a rotational speed of 1 rev/sec. Estimate his rotational speed N if he fully retracts his arms, bringing his hands very close to the centerline of his body. As a reasonable approximation, model the extended arms as uniform slender rods, each of which is 27 in. long and weighs 15 lb. Model the torso as a solid 135-lb cylinder 13 in. in diameter. Treat the man with arms retracted as a solid 165-lb cylinder of 13-in. diameter. Neglect friction at the skate–ice interface.

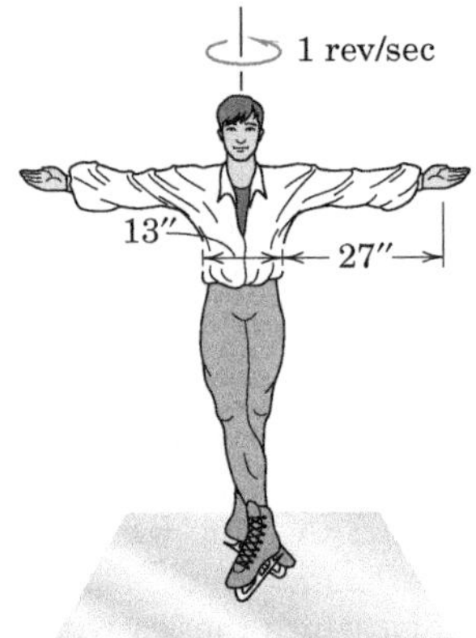

Problem 6/195

6/196 In the rotating assembly shown, arm OA and the attached motor housing B together weigh 10 lb and have a radius of gyration about the z-axis of 7 in. The motor armature and attached 5-in.-radius disk have a combined weight of 15 lb and a radius of gyration of 4 in. about their own axis. The entire assembly is free to rotate about the z-axis. If the motor is turned on with OA initially at rest, determine the angular speed N of OA when the motor has reached a speed of 300 rev/min *relative* to arm OA.

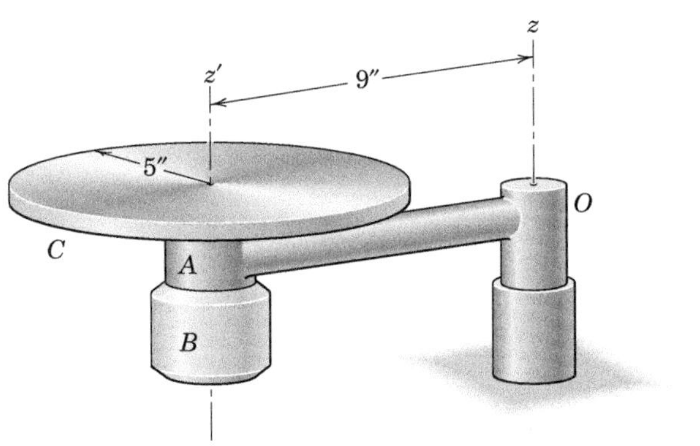

Problem 6/196

6/197 The motor at B supplies a constant torque M which is applied to a 375-mm-diameter internal drum around which is wound the cable shown. This cable then wraps around an 80-kg pulley attached to a 125-kg cart carrying 600 kg of rock. The motor is able to bring the loaded cart to a cruising speed of 1.5 m/s in 3 seconds. What torque M is the motor able to supply, and what is the average value of the tension in each side of the cable which is wrapped around the pulley at O during the speed-up period? The cable does not slip on the pulley and the centroidal radius of gyration of the pulley is 450 mm. What is the power output of the motor when the cart reaches its cruising speed?

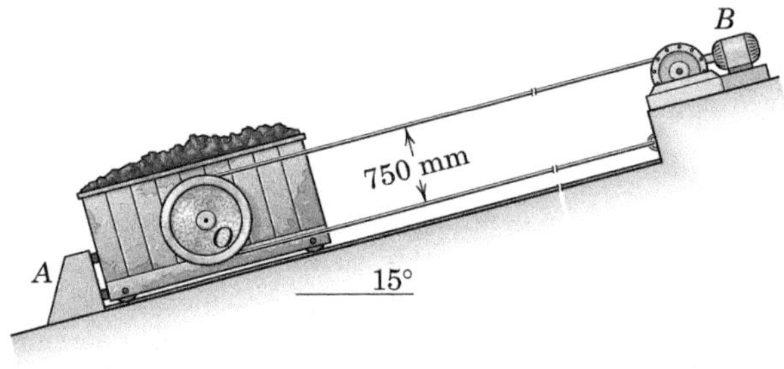

Problem 6/197

6/198 The body of the spacecraft weighs 322 lb on earth and has a radius of gyration about its z-axis of 1.5 ft. Each of the two solar panels may be treated as a uniform flat plate weighing 16.1 lb. If the spacecraft is rotating about its z-axis at the angular rate of 1.0 rad/sec with $\theta = 0$, determine the angular rate ω after the panels are rotated to the position $\theta = \pi/2$ by an internal mechanism. Neglect the small momentum change of the body about the y-axis.

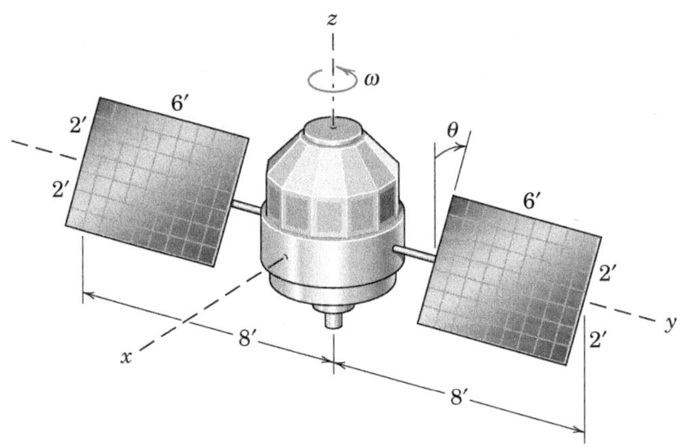

Problem 6/198

6/199 A 55-kg dynamics instructor is demonstrating the principles of angular momentum to her class. She stands on a freely rotating platform with her body aligned with the vertical platform axis. With the platform not rotating, she holds a modified bicycle wheel so that its axis is vertical. She then turns the wheel axis to a horizontal orientation without changing the 600-mm distance from the centerline of her body to the wheel center, and her students observe a platform rotation rate of 30 rev/min. If the rim-weighted wheel has a mass of 10 kg and a centroidal radius of gyration $\bar{k}$ = 300 mm, and is spinning at a fairly constant rate of 250 rev/min, estimate the mass moment of inertia I of the instructor (in the posture shown) about the vertical platform axis.

Problem 6/199

6/200 The 8-lb slotted circular disk has a radius of gyration about its center O of 6 in. and initially is rotating freely about a fixed vertical axis through O with a speed N_1 = 600 rev/min. The 2-lb uniform slender bar A is initially at rest relative to the disk in the centered slot position as shown. A slight disturbance causes the bar to slide to the end of the slot where it comes to rest relative to the disk. Calculate the new angular speed N_2 of the disk, assuming the absence of friction in the shaft bearing at O. Does the presence of any friction in the slot affect the final result?

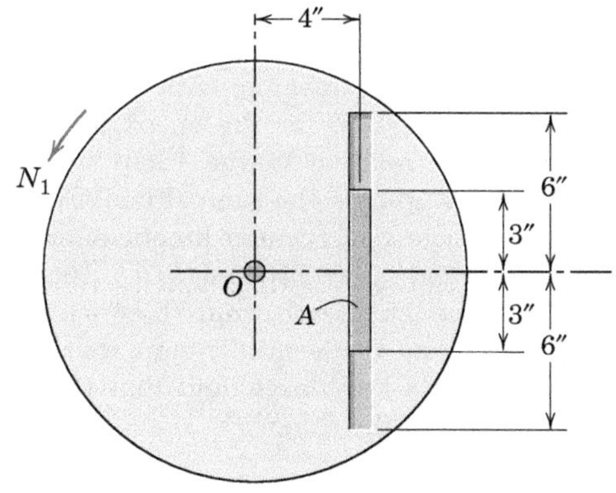

Problem 6/200

6/201 The gear train shown starts from rest and reaches an output speed of ω_C = 240 rev/min in 2.25 s. Rotation of the train is resisted by a constant 150 N·m moment at the output gear C. Determine the required input power to the 86% efficient motor at A just before the final speed is reached. The gears have masses m_A = 6 kg, m_B = 10 kg, and m_C = 24 kg, pitch diameters d_A = 120 mm, d_B = 160 mm, and d_C = 240 mm, and centroidal radii of gyration k_A = 48 mm, k_B = 64 mm, and k_C = 96 mm.

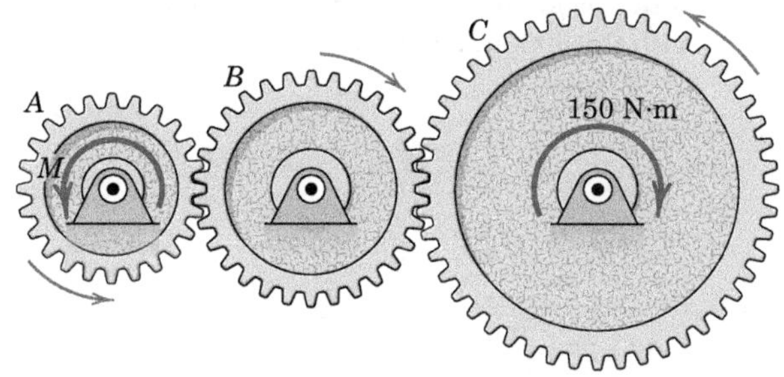

Problem 6/201

6/202 The uniform cylinder is rolling without slip with a velocity v along the horizontal surface when it overtakes a ramp traveling with speed v_0. Determine an expression for the speed v' which the cylinder has relative to the ramp immediately after it rolls up onto the ramp. Finally, determine the percentage n of cylinder kinetic energy lost if (*a*) $\theta = 10°$ and $v_0 = 0.25v$ and (*b*) $\theta = 10°$ and $v_0 = 0.5v$. Assume that the clearance between the ramp and the ground is essentially zero, that the mass of the ramp is very large, and that the cylinder does not slip on the ramp.

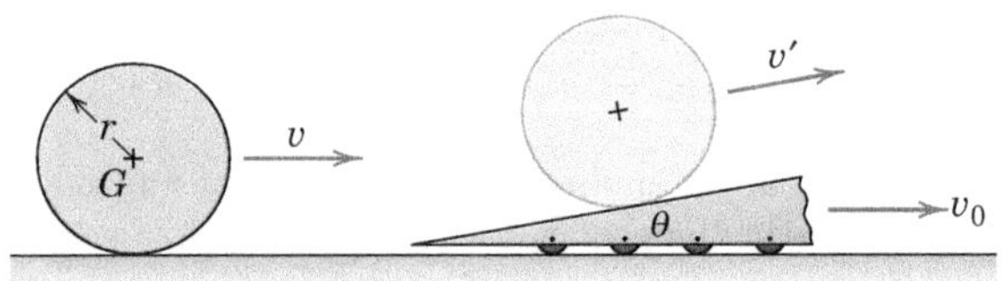

Problem 6/202

6/203 A frozen-juice can rests on the horizontal rack of a freezer door as shown. With what maximum angular velocity Ω can the door be "slammed" shut against its seal and not dislodge the can? Assume that the can rolls without slipping on the corner of the rack, and neglect the dimension d compared with the 500-mm distance.

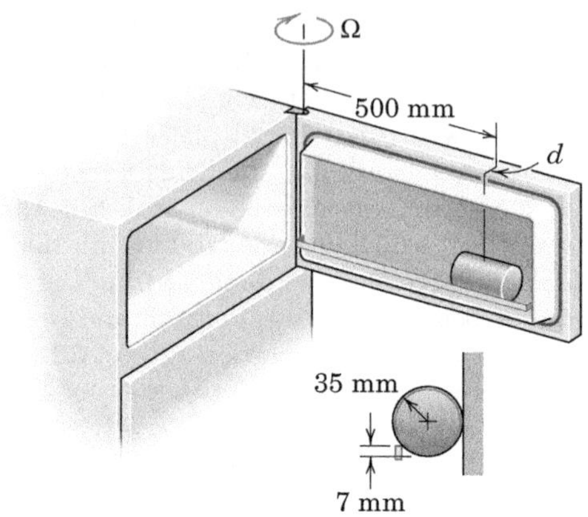

Problem 6/203

►6/204 The 30-kg wheel has a radius of gyration about its center of 75 mm and is rotating clockwise at the rate of 300 rev/min when it is released onto the incline with no velocity of its center O. While the wheel is slipping, it is observed that the center O remains in a fixed position. Determine the coefficient of kinetic friction μ_k and the time t during which slipping occurs. Also determine the velocity v of the center 4 seconds after the wheel has stopped slipping.

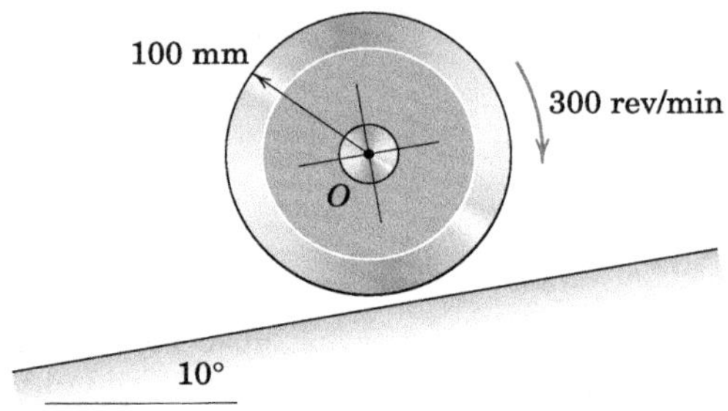

Problem 6/204

6/9 Chapter Review

In Chapter 6 we have made use of essentially all the elements of dynamics studied so far. We noted that a knowledge of kinematics, using both absolute- and relative-motion analysis, is an essential part of the solution to problems in rigid-body kinetics. Our approach in Chapter 6 paralleled Chapter 3, where we developed the kinetics of particles using force-mass-acceleration, work-energy, and impulse-momentum methods.

The following is a summary of the important considerations in the solution of rigid-body kinetics problems in plane motion:

1. ***Identification of the body or system.*** It is essential to make an unambiguous decision as to which body or system of bodies is to be analyzed and then isolate the selected body or system by drawing the free-body and kinetic diagrams, the active-force diagram, or the impulse-momentum diagram, whichever is appropriate.
2. ***Type of motion.*** Next identify the category of motion as rectilinear translation, curvilinear translation, fixed-axis rotation, or general plane motion. Always make sure that the kinematics of the problem is properly described before attempting to solve the kinetic equations.
3. ***Coordinate system.*** Choose an appropriate coordinate system. The geometry of the particular motion involved is usually the deciding factor. Designate the positive sense for moment and force summations and be consistent with the choice.
4. ***Principle and method.*** If the instantaneous relationship between the applied forces and the acceleration is desired, then the equivalence between the forces and their $m\bar{\mathbf{a}}$ and $\bar{I}\alpha$ resultants, as disclosed by the free-body and kinetic diagrams, will indicate the most direct approach to a solution.

   When motion occurs over an interval of displacement, the work-energy approach is indicated, and we relate initial to final velocities without calculating the acceleration. We have seen the advantage of this approach for interconnected mechanical systems with negligible internal friction.

   If the interval of motion is specified in terms of time rather than displacement, the impulse-momentum approach is indicated. When the angular motion of a rigid body is suddenly changed, the principle of conservation of angular momentum may apply.
5. ***Assumptions and approximations.*** By now you should have acquired a feel for the practical significance of certain assumptions and approximations, such as treating a rod as an ideal slender bar and neglecting friction when it is minimal. These and other idealizations are important to the process of obtaining solutions to real problems.

REVIEW PROBLEMS

6/205 The mass m is traveling with speed v when it strikes the corner of the plate of mass M. If the mass sticks to the side of the plate, determine the maximum angle θ reached by the plate. Use the values $m = 500$ g, $M = 20$ kg, $v = 30$ m/s, $b = 400$ mm, and $h = 800$ mm.

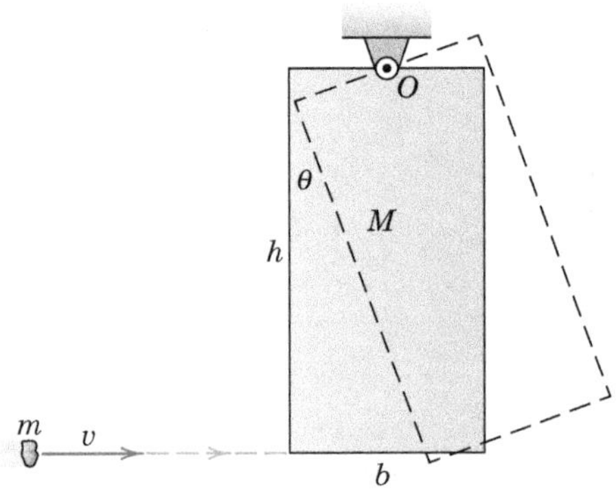

Problem 6/205

6/206 The 5-kg bar is released from rest while in the position shown and its end rollers travel in the vertical-plane slot shown. If the speed of roller A is 3.25 m/s as it passes point C, determine the work done by friction on the system over this portion of the motion. The bar has a length $l = 700$ mm.

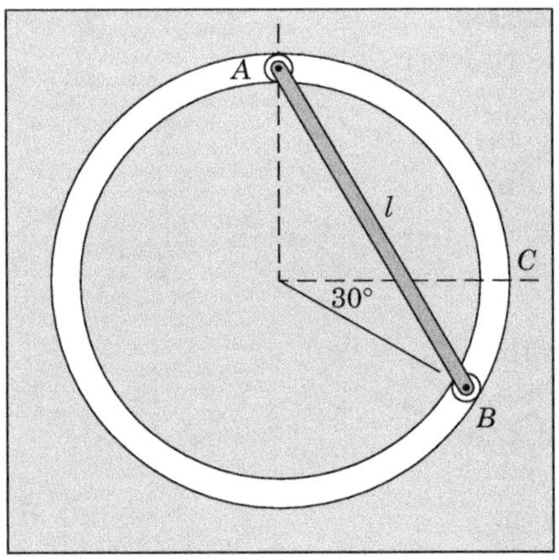

Problem 6/206

6/207 The body of mass $m = 1.2$ kg is lying motionless on the smooth horizontal surface when an impulse $\int P dt = 6$ N·s is applied as shown. Determine the linear momentum $\mathbf{G}$ and the angular momentum $\mathbf{H}_G$ of the body about the mass center G immediately following the application of the impulse.

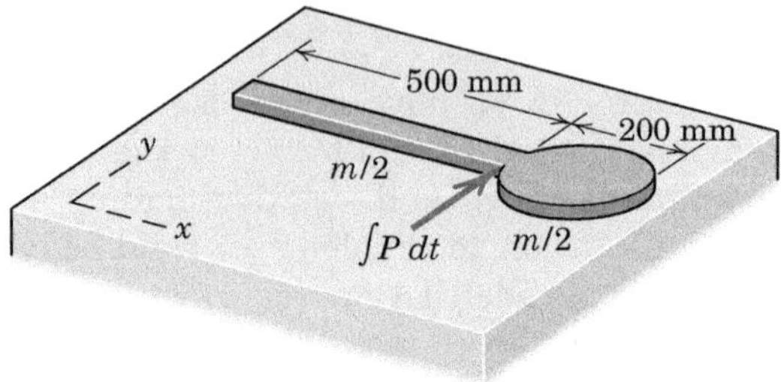

Problem 6/207

6/208 The preliminary design of a unit for automatically reducing the speed of a freely rotating assembly is shown. Initially the unit is rotating freely about a vertical axis through O at a speed of 600 rev/min with the arms secured in the positions shown by AB. When the arms are released, they swing outward and become latched in the dashed positions shown. The disk has a mass of 30 kg with a radius of gyration of 90 mm about O. Each arm has a length of 160 mm and a mass of 0.84 kg and may be treated as a uniform slender rod. Determine the new speed N of rotation and calculate the loss $|\Delta E|$ of energy of the system. Would the results be affected by either the direction of rotation or the sequence of release of the rods?

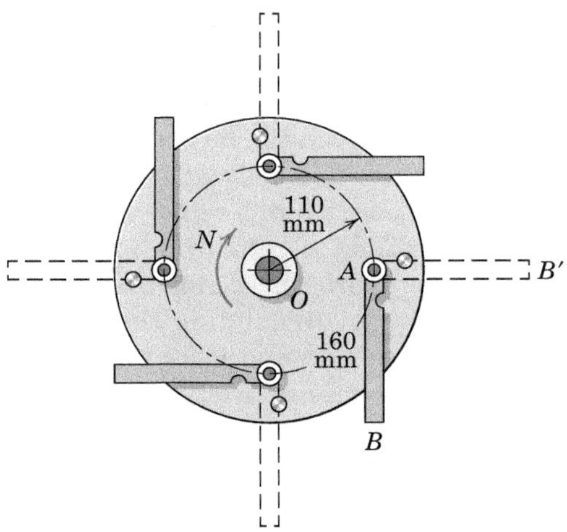

Problem 6/208

6/209 The nose-wheel assembly is raised by the application of a torque M to link BC through the shaft at B. The arm and wheel AO have a combined weight of 100 lb with center of mass at G, and a centroidal radius of gyration of 14 in. If the angle $\theta = 30°$, determine the torque M necessary to rotate link AO with a counterclockwise angular velocity of 10 deg/sec that is increasing at the rate of 5 deg/sec every second. Additionally, determine the total force supported by the pin at A. The mass of links BC and CD may be neglected for this analysis.

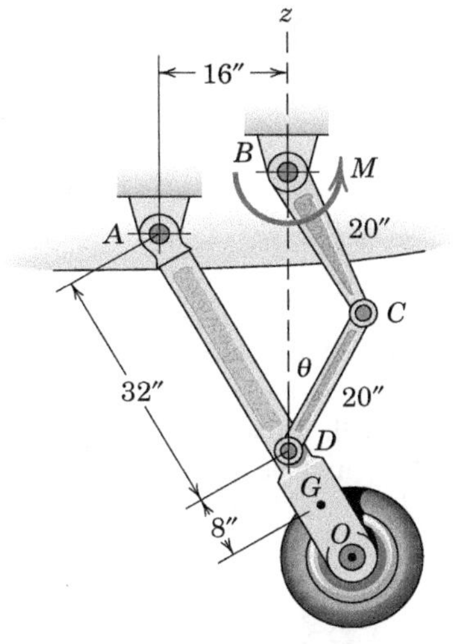

Problem 6/209

6/210 Each of the solid circular disk wheels has a mass of 2 kg, and the inner solid cylinder has a mass of 3 kg. The disks and cylinder are mounted on the small central shaft so that each can rotate independently of the other with negligible friction in the bearings. Calculate the acceleration of the center of the wheels when the 20-N force is applied as shown. The coefficients of friction between the wheels and the horizontal surface are $\mu_k = 0.30$ and $\mu_s = 0.40$.

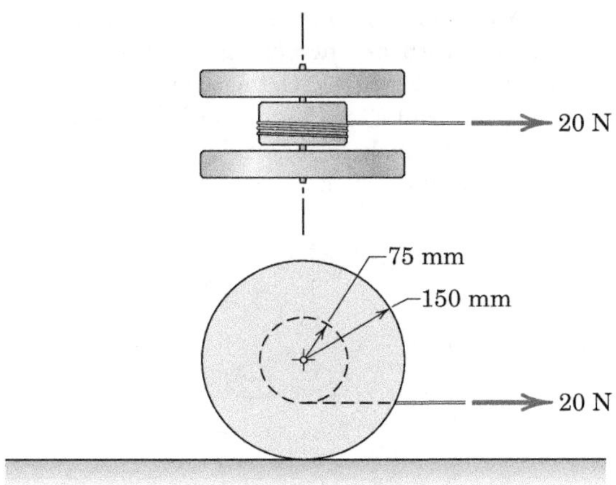

Problem 6/210

6/211 The dump truck carries 5 m^3 of dirt with a density of 1600 kg/m^3, and the elevating mechanism rotates the dump about the pivot A at a constant angular rate of 4 deg/s. The mass center of the dump and load is at G. Determine the maximum power P required during the tilting of the load.

Problem 6/211

6/212 Four identical slender rods each of mass m are welded at their ends to form a square, and the corners are then welded to a light metal hoop of radius r. If the rigid assembly of rods and hoop is allowed to roll down the incline, determine the minimum value of the coefficient of static friction which will prevent slipping.

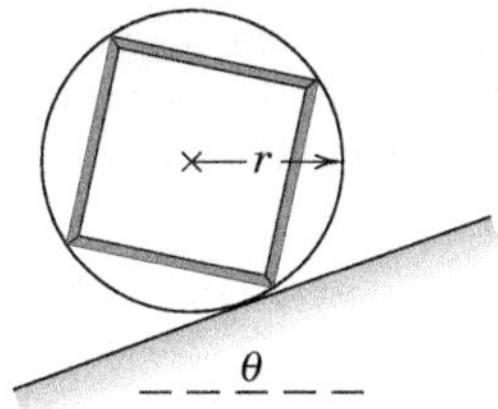

Problem 6/212

6/213 The gear train shown operates in a horizontal plane at a steady speed and receives 6 hp from a motor at A to move rack D against a 4500-pound load L. At what speed will the rack move if gears A, B, and C have pitch diameters $d_A = 12$ in., $d_B = 24$ in., and $d_C = 12$ in.? Gear C is keyed to the same shaft as gear B and friction in the bearings is negligible.

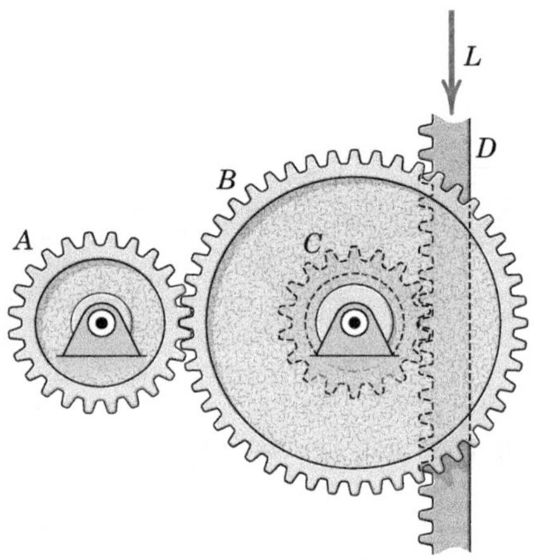

Problem 6/213

6/214 The uniform slender bar weighs 60 lb and is released from rest in the near-vertical position shown, where the spring of stiffness 10 lb/ft is unstretched. Calculate the speed with which end A strikes the horizontal surface.

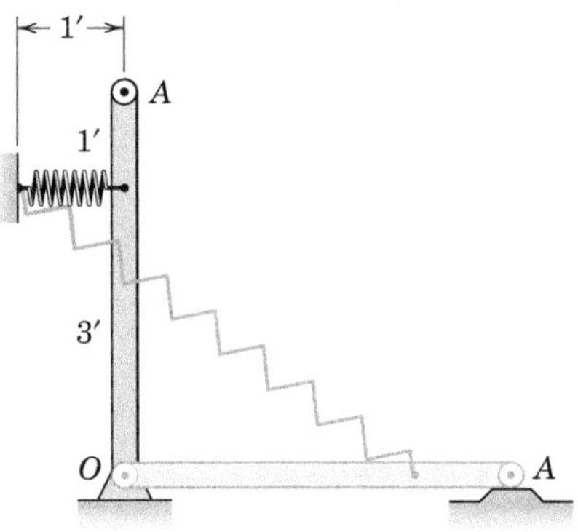

Problem 6/214

6/215 The body of mass m is supported by feet of negligible size and is at rest relative to the tilted surface, which is contained within an experimental vehicle. If the vehicle is brought to rest with a slowly increasing deceleration a, determine the ratio h/b which will just allow the body to tip forward about the front feet before it slides forward relative to the vehicle. For this ratio, what is the deceleration a at which tipping occurs?

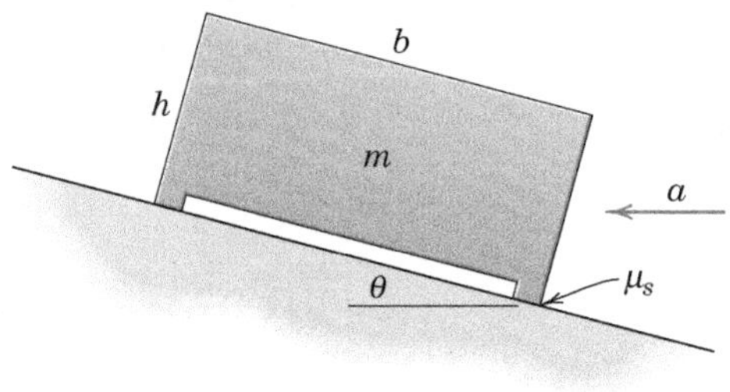

Problem 6/215

6/216 Repeat Prob. 6/215 for the case of a slowly increasing forward acceleration a. Determine the ratio h/b which will allow the package to tip backward about the rear feet before it slides backward relative to the vehicle. In terms of this ratio, what is the acceleration a?

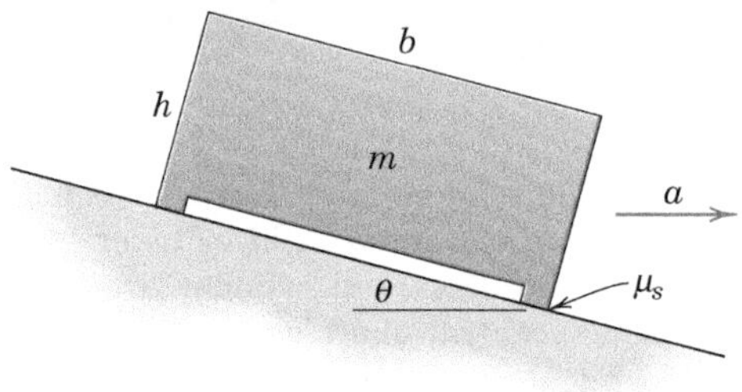

Problem 6/216

6/217 The small block of mass m slides along the radial slot of the disk while the disk rotates in the horizontal plane about its center O. The block is released from rest relative to the disk and moves outward with an increasing velocity $\dot{r}$ along the slot as the disk turns. Determine the expression in terms of r and $\dot{r}$ for the torque M that must be applied to the disk to maintain a constant angular velocity ω of the disk.

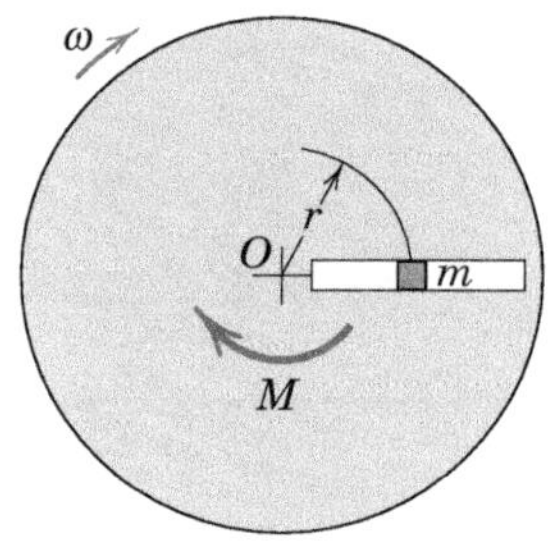

Problem 6/217

6/218 The forklift truck with center of mass at G_1 has a weight of 3200 lb including the vertical mast. The fork and load have a combined weight of 1800 lb with center of mass at G_2. The roller guide at B is capable of supporting horizontal force only, whereas the connection at C, in addition to supporting horizontal force, also transmits the vertical elevating force. If the fork is given an upward acceleration which is sufficient to reduce the force under the rear wheels at A to zero, calculate the corresponding reaction at B.

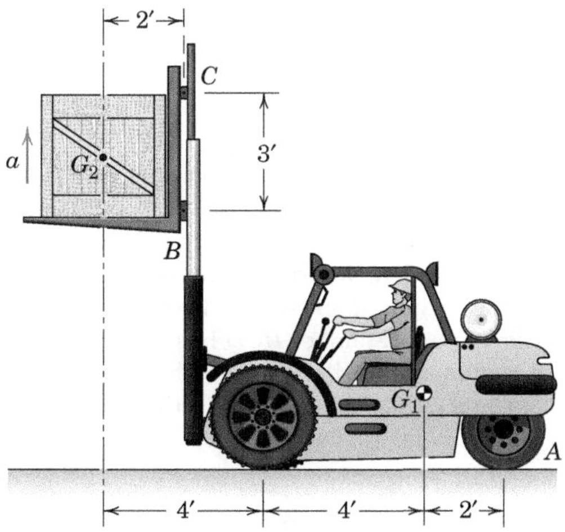

Problem 6/218

6/219 A space telescope is shown in the figure. One of the reaction wheels of its attitude-control system is spinning as shown at 10 rad/s, and at this speed the friction in the wheel bearing causes an internal moment of 10^{-6} N·m. Both the wheel speed and the friction moment may be considered constant over a time span of several hours. If the mass moment of inertia of the entire spacecraft about the x-axis is $150(10^3)$ kg·m^2, determine how much time passes before the line of sight of the initially stationary spacecraft drifts by 1 arc-second, which is 1/3600 degree. All other elements are fixed relative to the spacecraft, and no torquing of the reaction wheel shown is performed to correct the attitude drift. Neglect external torques.

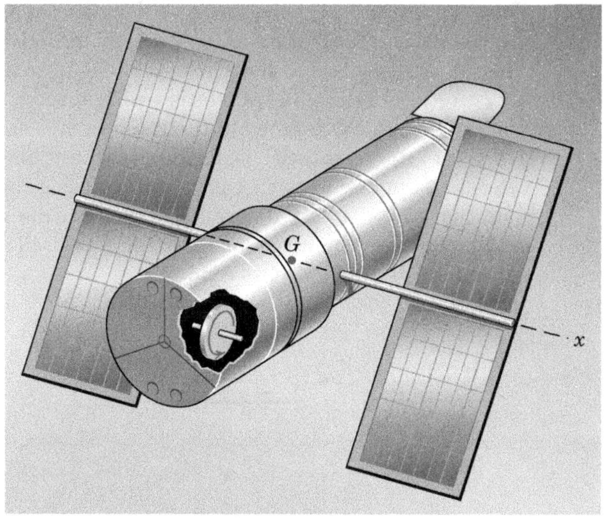

Problem 6/219

6/220 The uniform semicircular plate is at rest on the smooth horizontal surface when the force F is applied at B. Determine the coordinates of the point P in the plate which has zero initial acceleration.

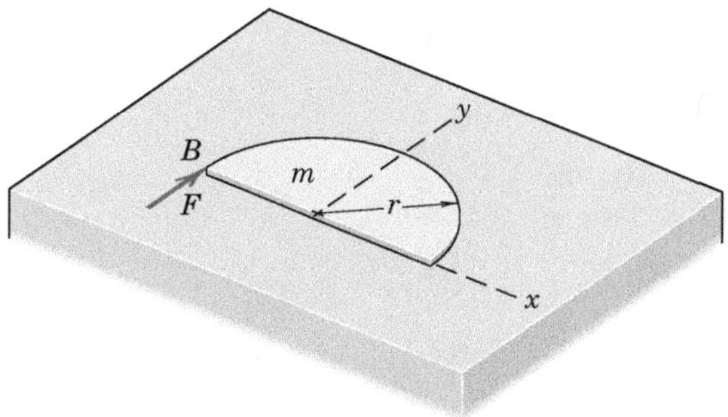

Problem 6/220

6/221 The slender rod of mass m_1 and length L has a movable slider of mass m_2 which can be tightened at any location x along the rod. The assembly is initially falling in translation with speed v_1. A small peg on the left end of the rod becomes engaged in the receptacle. Determine the angular velocity ω_2 of the body just after impact. For the condition $m_2 = m_1/2$, determine the maximum value of ω_2 and the corresponding value of x. Plot ω_2 versus x/L for this mass condition.

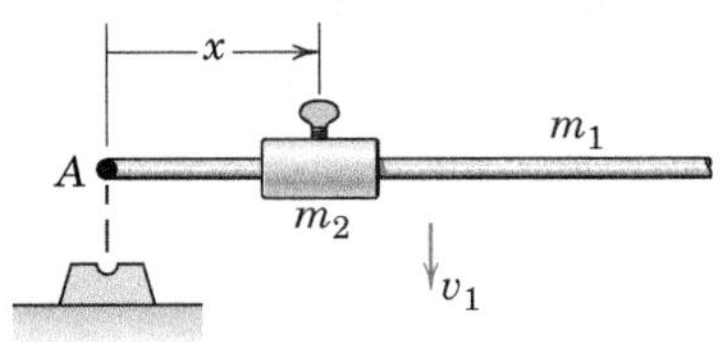

Problem 6/221

6/222 The 6-lb pendulum with mass center at G is pivoted at A to the fixed support CA. It has a radius of gyration of 17 in. about O-O and swings through an amplitude $\theta = 60°$. For the instant when the pendulum is in the extreme position, calculate the moments M_x, M_y, and M_z applied by the base support to the column at C.

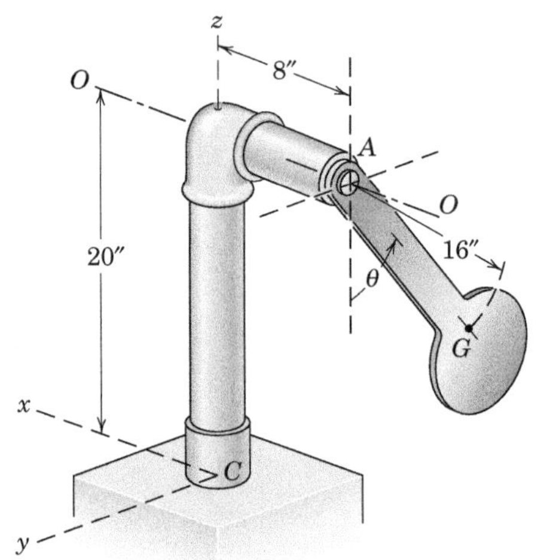

Problem 6/222

▶6/223 The uniform 40-lb bar with attached 12-lb wheels is released from rest in the orientation shown. The wheels have a centroidal radius of gyration of 4.5 in., and the coefficients of static and kinetic friction between the wheels and the horizontal and vertical surfaces are $\mu_s = 0.65$ and $\mu_k = 0.50$. Friction may be neglected in the pins connecting the wheels to the bar. Determine the acceleration components of the mass center of the bar at the instant of release.

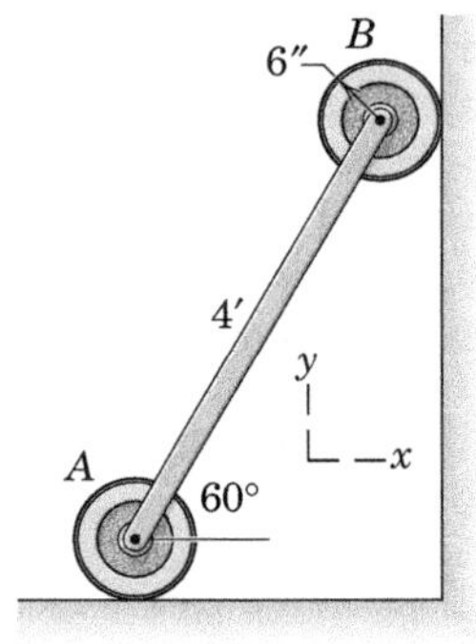

Problem 6/223

▶6/224 The four-bar mechanism operates in a horizontal plane. At the instant illustrated, $\theta = 30°$ and crank OA has a constant counterclockwise angular velocity of 3 rad/s. Determine the required magnitude of the couple M necessary to drive the system at this instant. Member BCD has a mass of 8 kg with a radius of gyration of 450 mm about point C. The mass of crank OA and connecting link AB may be neglected for this analysis.

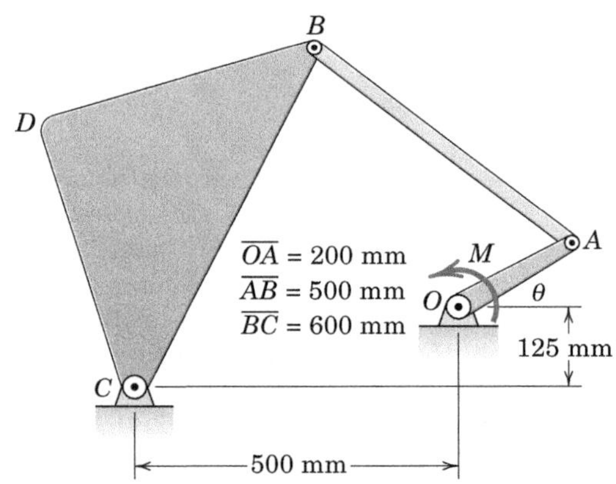

Problem 6/224

*Computer-Oriented Problems

***6/225** If the 1.2-kg uniform slender bar is released from rest in the position $\theta = 0$ where the spring is unstretched, determine and plot its angular velocity as a function of θ over the range $0 \leq \theta \leq \theta_{max}$, where θ_{max} is the value of θ at which the bar momentarily comes to rest. The value of the spring constant k is 100 N/m, and friction can be neglected. State the maximum angular speed and the value of θ at which it occurs.

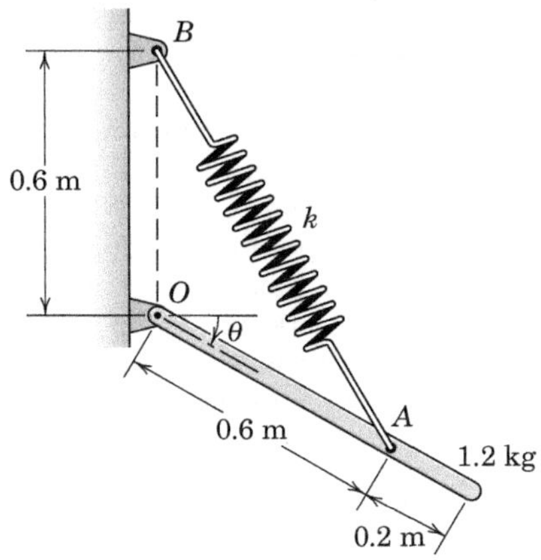

Problem 6/225

***6/226** The steel I-beam is to be transported by the overhead trolley to which it is hinged at O. If the trolley starts from rest with $\theta = \dot{\theta} = 0$ and is given a constant horizontal acceleration $a = 2\ \text{m/s}^2$, find the maximum values of $\dot{\theta}$ and θ. The magnitude of the initial swing would constitute a shop safety consideration.

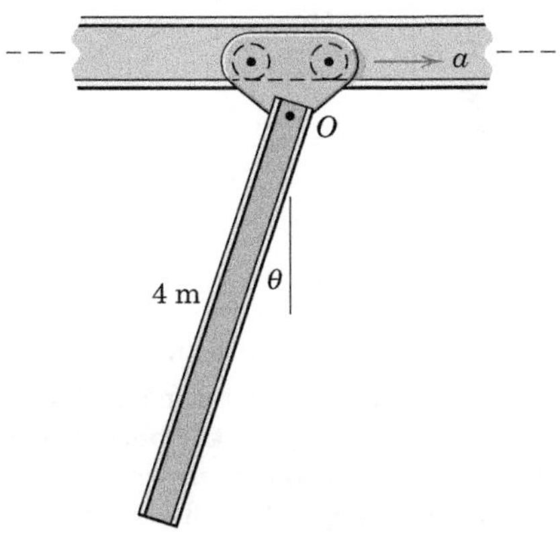

Problem 6/226

***6/227** The uniform power pole of mass m and length L is hoisted into a vertical position with its lower end supported by a fixed pivot at O. The guy wires supporting the pole are accidentally released, and the pole falls to the ground. Plot the x- and y-components of the force exerted on the pole at O in terms of θ from 0 to 90°. Can you explain why O_y increases again after going to zero?

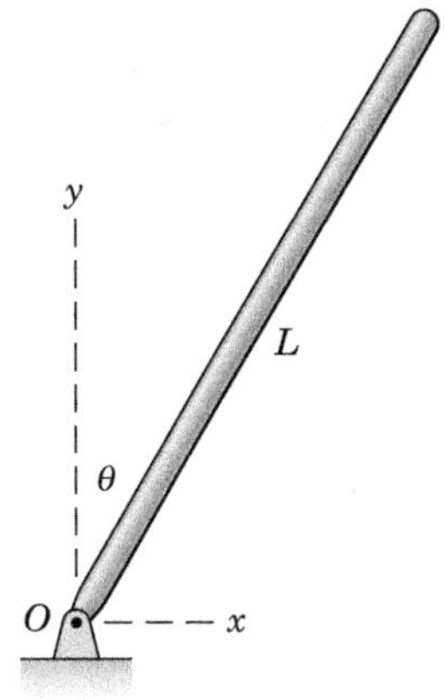

Problem 6/227

***6/228** The compound pendulum is composed of a uniform slender rod of length l and mass $2m$ to which is fastened a uniform disk of diameter $l/2$ and mass m. The body pivots freely about a horizontal axis through O. If the pendulum has a clockwise angular velocity of 3 rad/s when $\theta = 0$ at time $t = 0$, determine the time t at which the pendulum passes the position $\theta = 90°$. The pendulum length $l = 0.8$ m.

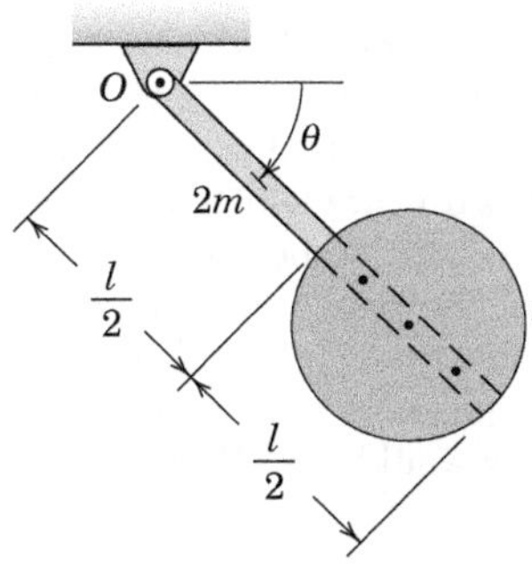

Problem 6/228

*6/229 The uniform 100-kg beam AB is hanging initially at rest with $\theta = 0$ when the constant force $P = 300$ N is applied to the cable. Determine (*a*) the maximum angular velocity reached by the beam with the corresponding angle θ and (*b*) the maximum angle θ_{max} reached by the beam.

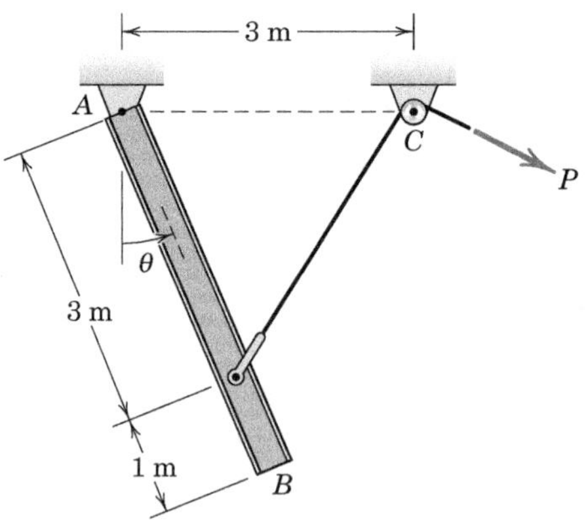

Problem 6/229

*6/230 For a train traveling at 160 km/h around a horizontal curve of radius 1.9 km, calculate the elevation angle β of the track so that passengers will feel only a force normal to their seats and the rails will exert no side thrust against the wheels, as indicated in part (*a*) of the figure. An experimental train rounds this same curve at a speed of 260 km/h with cars which are automatically tilted an angle θ with respect to the rails, as shown in part (*b*) of the figure. This angle reduces the side thrust F felt by the passengers. Determine the tilt angle θ required to limit F to 30 percent of the side thrust they would feel if both θ and β were zero.

Problem 6/230

*6/231 The 60-ft telephone pole of essentially uniform diameter is being hoisted into the vertical position by two cables attached at B as shown. The end O rests on a fixed support and cannot slip. When the pole is nearly vertical, the fitting at B suddenly breaks, releasing both cables. When the angle θ reaches 10°, the speed of the upper end A of the pole is 4.5 ft/sec. From this point, calculate the time t which the workman would have to get out of the way before the pole hits the ground. With what speed v_A does end A hit the ground?

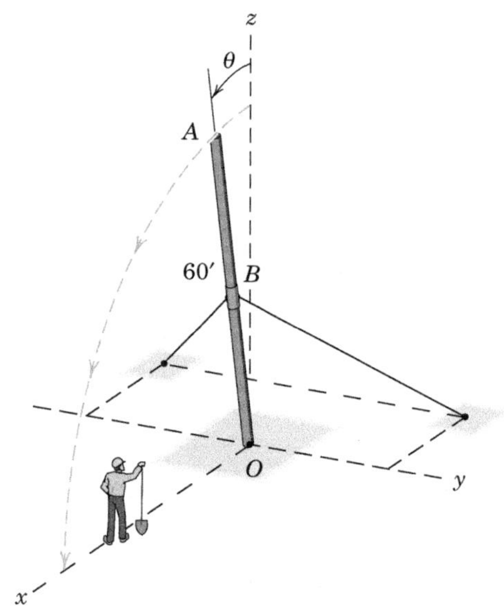

Problem 6/231

*6/232 The four-bar mechanism of Prob. 6/108 is repeated here. The coupler AB has a mass of 7 kg, and the masses of crank OA and the output arm BC may be neglected. Determine and plot the torque M which must be applied to the crank at O in order to keep the speed of the crank steady at 60 rev/min over the range $0 \leq \theta \leq 2\pi$. What is the largest magnitude M which occurs during this motion, and at what angle θ does it occur? Additionally, plot the magnitude of the force on each pin over this range and state the maximum magnitude of each pin force along with the corresponding crank angle θ at which it occurs.

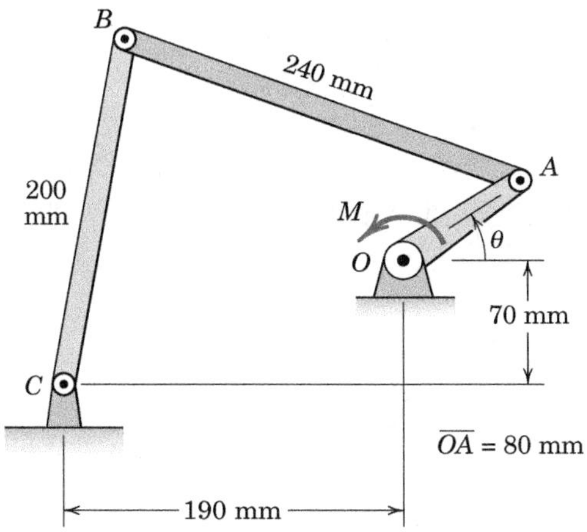

Problem 6/232

***6/233** Reconsider the mechanism of Prob. 6/232. If crank OA now starts from rest and acquires a speed of 60 rev/min in one complete revolution with constant angular acceleration, determine and plot the torque M which must be applied to the crank over the range $0 \leq \theta \leq 2\pi$. What is the largest magnitude M which occurs during this motion, and at what angle θ does it occur? Additionally, plot the magnitude of the force in each pin over this range and state the maximum magnitude of each pin force along with the corresponding crank angle θ at which it occurs.

***6/234** Reconsider the basic mechanism of Prob. 6/232, only now the mass of the crank OA is 1.2 kg and that of the uniform output arm BC is 1.8 kg. For simplicity, treat the crank OA as uniform. Determine and plot the torque M which must be applied to the crank at O in order to keep the speed of the crank steady at 60 rev/min over the range $0 \leq \theta \leq 2\pi$. What is the largest magnitude M which occurs during this motion, and at what angle θ does it occur? Additionally, plot the magnitude of the force on each pin over this range and state the maximum magnitude of each pin force along with the corresponding crank angle θ at which it occurs.

These robots execute a variety of three-dimensional motions as they perform a range of tasks during the manufacture of automobiles.

7 Introduction to Three-Dimensional Dynamics of Rigid Bodies

Chapter Outline

7/1 Introduction

Although a large percentage of dynamics problems in engineering can be solved by the principles of plane motion, modern developments have focused increasing attention on problems which call for the analysis of motion in three dimensions. Inclusion of the third dimension adds considerable complexity to the kinematic and kinetic relationships. Not only does the added dimension introduce a third component to vectors which represent force, linear velocity, linear acceleration, and linear momentum, but the introduction of the third dimension also adds the possibility of two additional components for vectors representing angular quantities including moments of forces, angular velocity, angular acceleration, and angular momentum. It is in three-dimensional motion that the full power of vector analysis is utilized.

A good background in the dynamics of plane motion is extremely useful in the study of three-dimensional dynamics, where the approach to problems and many of the terms are the same as or analogous to those in two dimensions. If the study of three-dimensional dynamics is undertaken without the benefit of prior study of plane-motion dynamics, more time will be required to master the principles and to become familiar with the approach to problems.

The treatment presented in Chapter 7 is not intended as a complete development of the three-dimensional motion of rigid bodies but merely as a basic introduction to the subject. This introduction should, however, be sufficient to solve many of the more common problems in three-dimensional motion and also to lay the foundation for more advanced study. We will proceed as we did for particle motion and for rigid-body plane motion by first examining the necessary kinematics and then proceeding to the kinetics.

SECTION A **KINEMATICS**

7/2 Translation

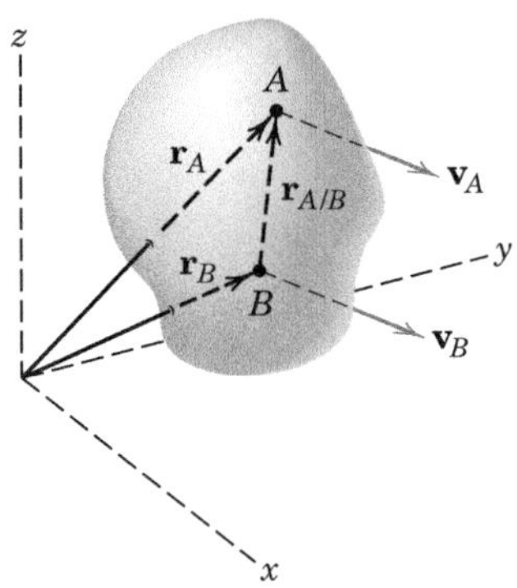

Figure 7/1

Figure 7/1 shows a rigid body translating in three-dimensional space. Any two points in the body, such as A and B, will move along parallel straight lines if the motion is one of *rectilinear translation* or will move along congruent curves if the motion is one of *curvilinear translation*. In either case, every line in the body, such as AB, remains parallel to its original position.

The position vectors and their first and second time derivatives are

$$\mathbf{r}_A = \mathbf{r}_B + \mathbf{r}_{A/B} \qquad \mathbf{v}_A = \mathbf{v}_B \qquad \mathbf{a}_A = \mathbf{a}_B$$

where $\mathbf{r}_{A/B}$ remains constant, and therefore its time derivative is zero. Thus, all points in the body have the same velocity and the same acceleration. The kinematics of translation presents no special difficulty, and further elaboration is unnecessary.

7/3 Fixed-Axis Rotation

Consider now the *rotation* of a rigid body about a fixed axis n-n in space with an angular velocity $\boldsymbol{\omega}$, as shown in Fig. 7/2. The angular velocity is a vector in the direction of the rotation axis with a sense established by the familiar right-hand rule. For fixed-axis rotation, $\boldsymbol{\omega}$ does not change its direction since it lies along the axis. We choose the origin O of the fixed coordinate system on the rotation axis for convenience. Any point such as A which is not on the axis moves in a circular arc in a plane normal to the axis and has a velocity

$$\boxed{\mathbf{v} = \boldsymbol{\omega} \times \mathbf{r}} \qquad \textbf{(7/1)}$$

which may be seen by replacing $\mathbf{r}$ by $\mathbf{h} + \mathbf{b}$ and noting that $\boldsymbol{\omega} \times \mathbf{h} = \mathbf{0}$.

The acceleration of A is given by the time derivative of Eq. 7/1. Thus,

$$\boxed{\mathbf{a} = \dot{\boldsymbol{\omega}} \times \mathbf{r} + \boldsymbol{\omega} \times (\boldsymbol{\omega} \times \mathbf{r})} \qquad \textbf{(7/2)}$$

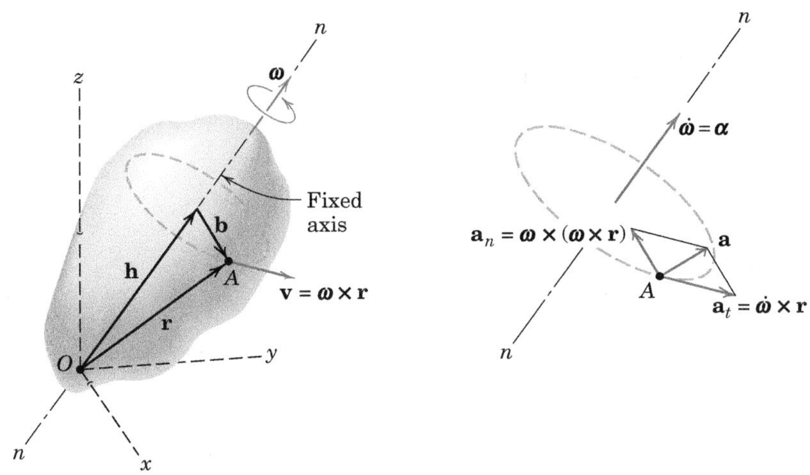

Figure 7/2

where $\dot{\mathbf{r}}$ has been replaced by its equal, $\mathbf{v} = \boldsymbol{\omega} \times \mathbf{r}$. The normal and tangential components of $\mathbf{a}$ for the circular motion have the familiar magnitudes $a_n = |\boldsymbol{\omega} \times (\boldsymbol{\omega} \times \mathbf{r})| = b\omega^2$ and $a_t = |\dot{\boldsymbol{\omega}} \times \mathbf{r}| = b\alpha$, where $\alpha = \dot{\omega}$. Inasmuch as both $\mathbf{v}$ and $\mathbf{a}$ are perpendicular to $\boldsymbol{\omega}$ and $\dot{\boldsymbol{\omega}}$, it follows that $\mathbf{v}\cdot\boldsymbol{\omega} = 0$, $\mathbf{v}\cdot\dot{\boldsymbol{\omega}} = 0$, $\mathbf{a}\cdot\boldsymbol{\omega} = 0$, and $\mathbf{a}\cdot\dot{\boldsymbol{\omega}} = 0$ for fixed-axis rotation.

7/4 Parallel-Plane Motion

When all points in a rigid body move in planes which are parallel to a fixed plane P, Fig. 7/3, we have a general form of plane motion. The reference plane is customarily taken through the mass center G and is called the *plane of motion*. Because each point in the body, such as A', has a motion identical with the motion of the corresponding point (A) in plane P, it follows that the kinematics of plane motion covered in Chapter 5 provides a complete description of the motion when applied to the reference plane.

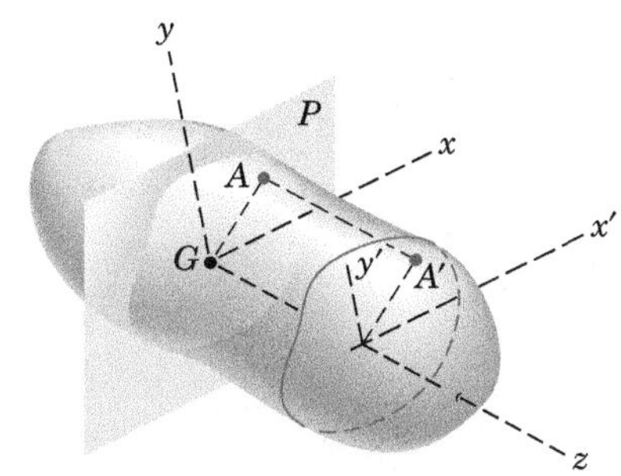

Figure 7/3

7/5 Rotation about a Fixed Point

When a body rotates about a fixed point, the angular-velocity vector no longer remains fixed in direction, and this change calls for a more general concept of rotation.

Rotation and Proper Vectors

We must first examine the conditions under which rotation vectors obey the parallelogram law of addition and may, therefore, be treated as proper vectors. Consider a solid sphere, Fig. 7/4, which is cut from a rigid body confined to rotate about the fixed point O.

The x-y-z axes here are taken as fixed in space and do not rotate with the body. In part a of the figure, two successive 90° rotations of the sphere about, first, the x-axis and, second, the y-axis result in the motion of a point which is initially on the y-axis in position 1, to positions

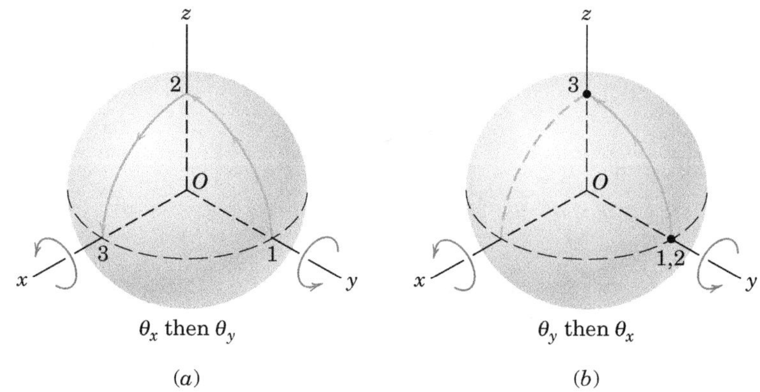

Figure 7/4

2 and 3, successively. On the other hand, if the order of the rotations is reversed, the point undergoes no motion during the y-rotation but moves to point 3 during the 90° rotation about the x-axis. Thus, the two cases do not produce the same final position, and it is evident from this one special example that finite rotations do not generally obey the parallelogram law of vector addition and are not commutative. Thus, finite rotations may *not* be treated as proper vectors.

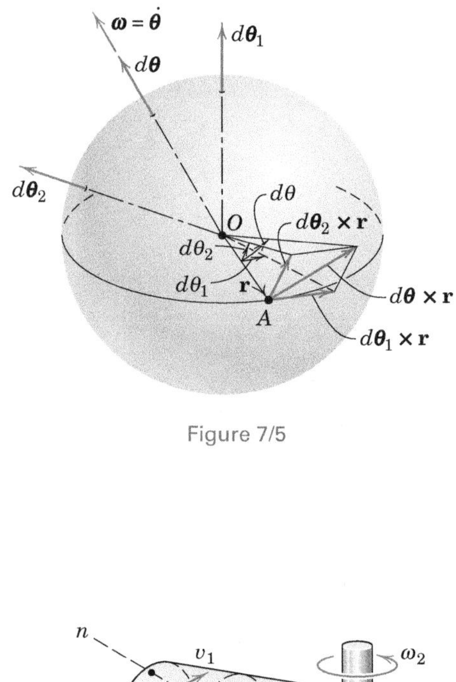

Figure 7/5

Infinitesimal rotations, however, do obey the parallelogram law of vector addition. This fact is shown in Fig. 7/5, which represents the combined effect of two infinitesimal rotations $d\boldsymbol{\theta}_1$ and $d\boldsymbol{\theta}_2$ of a rigid body about the respective axes through the fixed point O. As a result of $d\boldsymbol{\theta}_1$, point A has a displacement $d\boldsymbol{\theta}_1 \times \mathbf{r}$, and likewise $d\boldsymbol{\theta}_2$ causes a displacement $d\boldsymbol{\theta}_2 \times \mathbf{r}$ of point A. Either order of addition of these infinitesimal displacements clearly produces the same resultant displacement, which is $d\boldsymbol{\theta}_1 \times \mathbf{r} + d\boldsymbol{\theta}_2 \times \mathbf{r} = (d\boldsymbol{\theta}_1 + d\boldsymbol{\theta}_2) \times \mathbf{r}$. Thus, the two rotations are equivalent to the single rotation $d\boldsymbol{\theta} = d\boldsymbol{\theta}_1 + d\boldsymbol{\theta}_2$. It follows that the angular velocities $\boldsymbol{\omega}_1 = \dot{\boldsymbol{\theta}}_1$ and $\boldsymbol{\omega}_2 = \dot{\boldsymbol{\theta}}_2$ may be added vectorially to give $\boldsymbol{\omega} = \dot{\boldsymbol{\theta}} = \boldsymbol{\omega}_1 + \boldsymbol{\omega}_2$. We conclude, therefore, that at any instant of time a body with one fixed point is rotating instantaneously about a particular axis passing through the fixed point.

Instantaneous Axis of Rotation

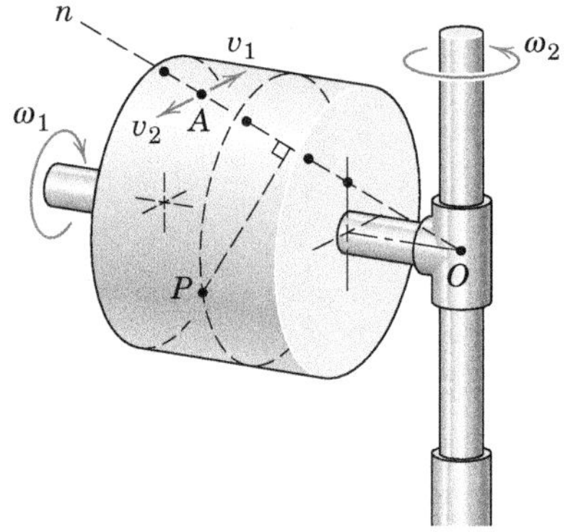

Figure 7/6

To aid in visualizing the concept of the instantaneous axis of rotation, we will cite a specific example. Figure 7/6 represents a solid cylindrical rotor made of clear plastic containing many black particles embedded in the plastic. The rotor is spinning about its shaft axis at the steady rate ω_1, and its shaft, in turn, is rotating about the fixed vertical axis at the steady rate ω_2, with rotations in the directions indicated. If the rotor is photographed at a certain instant during its motion, the resulting picture would show one line of black dots sharply defined, indicating that, momentarily, their velocity was zero. This line of points with no velocity establishes the instantaneous position of the axis of rotation O-n. Any dot on this line, such as A, would have equal and opposite velocity components, v_1 due to ω_1 and v_2 due to ω_2. All other dots, such as the one at P, would appear blurred, and their movements would show as short streaks in the form of small circular arcs in planes normal to the axis O-n. Thus, all particles of the body, except those on line O-n, are momentarily rotating in circular arcs about the instantaneous axis of rotation.

If a succession of photographs were taken, we would observe in each photograph that the rotation axis would be defined by a new series of sharply-defined dots and that the axis would change position both in space and relative to the body. For rotation of a rigid body about a fixed point, then, it is seen that the rotation axis is, in general, not a line fixed in the body.

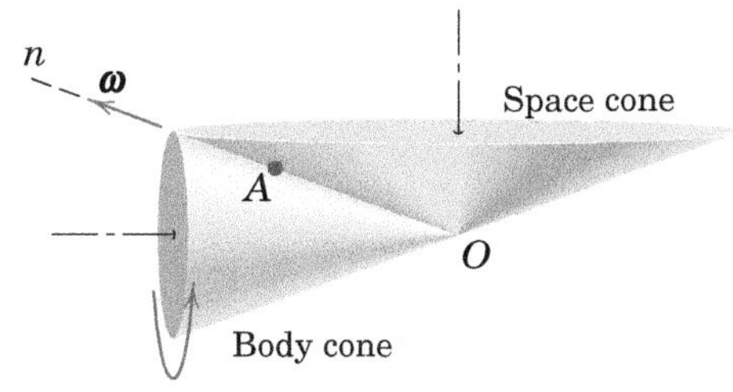

Figure 7/7

Body and Space Cones

Relative to the plastic cylinder of Fig. 7/6, the instantaneous axis of rotation O-A-n generates a right-circular cone about the cylinder axis called the *body cone*. As the two rotations continue and the cylinder swings around the vertical axis, the instantaneous axis of rotation also generates a right-circular cone about the vertical axis called the *space cone*. These cones are shown in Fig. 7/7 for this particular example.

We see that the body cone rolls on the space cone and that the angular velocity $\boldsymbol{\omega}$ of the body is a vector which lies along the common element of the two cones. For a more general case where the rotations are not steady, the space and body cones are not right-circular cones, Fig. 7/8, but the body cone still rolls on the space cone.

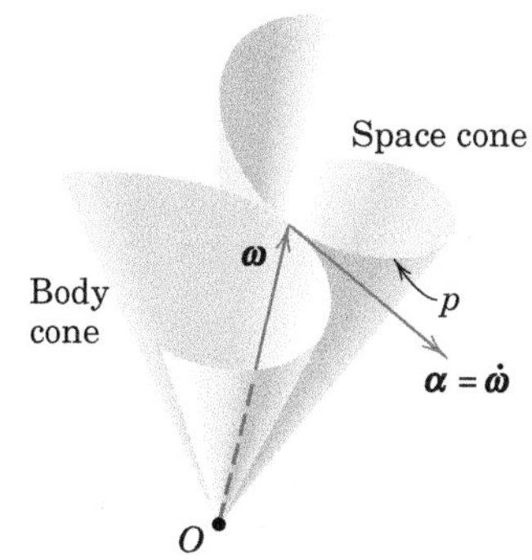

Figure 7/8

Angular Acceleration

The angular acceleration $\boldsymbol{\alpha}$ of a rigid body in three-dimensional motion is the time derivative of its angular velocity, $\boldsymbol{\alpha} = \dot{\boldsymbol{\omega}}$. In contrast to the case of rotation in a single plane where the scalar α measures only the change in magnitude of the angular velocity, in three-dimensional motion the vector $\boldsymbol{\alpha}$ reflects the change in direction of $\boldsymbol{\omega}$ as well as its change in magnitude. Thus in Fig. 7/8 where the tip of the angular velocity vector $\boldsymbol{\omega}$ follows the space curve p and changes in both magnitude and direction, the angular acceleration $\boldsymbol{\alpha}$ becomes a vector tangent to this curve in the direction of the change in $\boldsymbol{\omega}$.

When the magnitude of $\boldsymbol{\omega}$ remains constant, the angular acceleration $\boldsymbol{\alpha}$ is normal to $\boldsymbol{\omega}$. For this case, if we let $\boldsymbol{\Omega}$ stand for the angular velocity with which the vector $\boldsymbol{\omega}$ itself rotates (*precesses*) as it forms the space cone, the angular acceleration may be written

$$\boldsymbol{\alpha} = \boldsymbol{\Omega} \times \boldsymbol{\omega} \tag{7/3}$$

This relation is easily seen from Fig. 7/9. The upper part of the figure relates the velocity of a point A on a rigid body to its position vector from O and the angular velocity of the body. The vectors $\boldsymbol{\alpha}$, $\boldsymbol{\omega}$, and $\boldsymbol{\Omega}$ in the lower figure bear exactly the same relationship to each other as do the vectors $\mathbf{v}$, $\mathbf{r}$, and $\boldsymbol{\omega}$ in the upper figure.

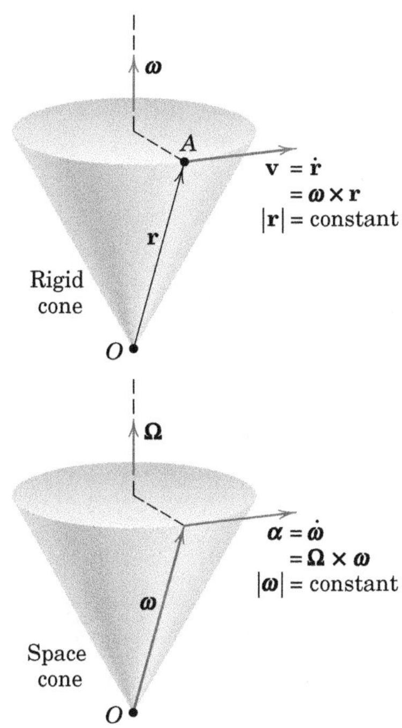

Figure 7/9

If we use Fig. 7/2 to represent a rigid body rotating about a fixed point O with the instantaneous axis of rotation n-n, we see that the velocity $\mathbf{v}$ and acceleration $\mathbf{a} = \dot{\mathbf{v}}$ of any point A in the body are given by the same expressions as apply to the case in which the axis is fixed, namely,

$$\mathbf{v} = \boldsymbol{\omega} \times \mathbf{r} \tag{7/1}$$

$$\mathbf{a} = \dot{\boldsymbol{\omega}} \times \mathbf{r} + \boldsymbol{\omega} \times (\boldsymbol{\omega} \times \mathbf{r}) \tag{7/2}$$

The one difference between the case of rotation about a fixed axis and rotation about a fixed point lies in the fact that for rotation about a fixed point, the angular acceleration $\boldsymbol{\alpha} = \dot{\boldsymbol{\omega}}$ will have a component normal to $\boldsymbol{\omega}$ due to the change in direction of $\boldsymbol{\omega}$, as well as a component in the direction of $\boldsymbol{\omega}$ to reflect any change in the magnitude of $\boldsymbol{\omega}$. Although any point on the rotation axis n-n momentarily will have zero velocity, it will *not* have zero acceleration as long as $\boldsymbol{\omega}$ is changing its direction. On the other hand, for rotation about a fixed axis, $\boldsymbol{\alpha} = \dot{\boldsymbol{\omega}}$ has only the one component along the fixed axis to reflect the change in the magnitude of $\boldsymbol{\omega}$. Furthermore, points which lie on the fixed rotation axis clearly have no velocity or acceleration.

Although the development in this article is for the case of rotation about a fixed point, we observe that rotation is a function *solely* of angular change, so that the expressions for $\boldsymbol{\omega}$ and $\boldsymbol{\alpha}$ do not depend on the fixity of the point around which rotation occurs. Thus, rotation may take place independently of the linear motion of the rotation point. This conclusion is the three-dimensional counterpart of the concept of rotation of a rigid body in plane motion described in Art. 5/2 and used throughout Chapters 5 and 6.

Ed Darack/SuperStock

The engine/propeller units at the wingtips of this V-22 Osprey can tilt from a vertical takeoff position to a horizontal position for forward flight.

Sample Problem 7/1

The 0.8-m arm OA for a remote-control mechanism is pivoted about the horizontal x-axis of the clevis, and the entire assembly rotates about the z-axis with a constant speed $N = 60$ rev/min. Simultaneously, the arm is being raised at the constant rate $\dot{\beta} = 4$ rad/s. For the position where $\beta = 30°$, determine (a) the angular velocity of OA, (b) the angular acceleration of OA, (c) the velocity of point A, and (d) the acceleration of point A. If, in addition to the motion described, the vertical shaft and point O had a linear motion, say, in the z-direction, would that motion change the angular velocity or angular acceleration of OA?

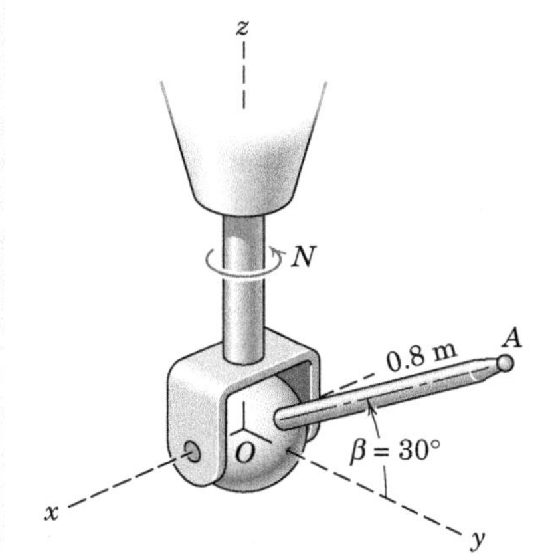

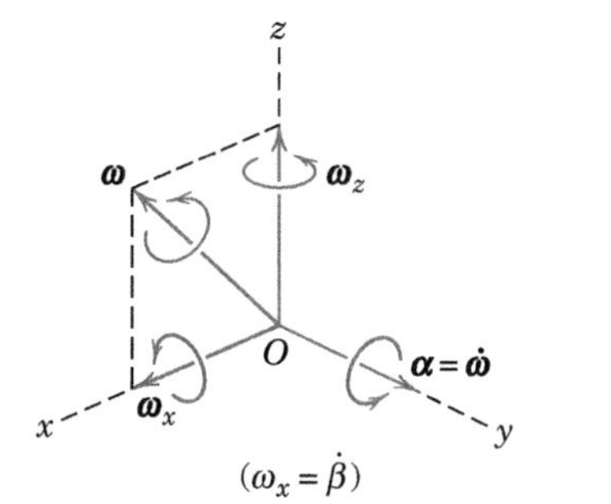

***Solution.* (a)** Since the arm OA is rotating about both the x- and the z-axes, it has the components $\omega_x = \dot{\beta} = 4$ rad/s and $\omega_z = 2\pi N/60 = 2\pi(60)/60 = 6.28$ rad/s. The angular velocity is

$$\boldsymbol{\omega} = \boldsymbol{\omega}_x + \boldsymbol{\omega}_z = 4\mathbf{i} + 6.28\mathbf{k} \text{ rad/s} \qquad \textit{Ans.}$$

(b) The angular acceleration of OA is

$$\boldsymbol{\alpha} = \dot{\boldsymbol{\omega}} = \dot{\boldsymbol{\omega}}_x + \dot{\boldsymbol{\omega}}_z$$

Since $\boldsymbol{\omega}_z$ is not changing in magnitude or direction, $\dot{\boldsymbol{\omega}}_z = \mathbf{0}$. But $\boldsymbol{\omega}_x$ is changing direction and thus has a derivative which, from Eq. 7/3, is

$$\dot{\boldsymbol{\omega}}_x = \boldsymbol{\omega}_z \times \boldsymbol{\omega}_x = 6.28\mathbf{k} \times 4\mathbf{i} = 25.1\mathbf{j} \text{ rad/s}^2$$

① Therefore,

$$\boldsymbol{\alpha} = 25.1\mathbf{j} + \mathbf{0} = 25.1\mathbf{j} \text{ rad/s}^2 \qquad \textit{Ans.}$$

(c) With the position vector of A given by $\mathbf{r} = 0.693\mathbf{j} + 0.4\mathbf{k}$ m, the velocity of A from Eq. 7/1 becomes

$$\mathbf{v} = \boldsymbol{\omega} \times \mathbf{r} = \begin{vmatrix} \mathbf{i} & \mathbf{j} & \mathbf{k} \\ 4 & 0 & 6.28 \\ 0 & 0.693 & 0.4 \end{vmatrix} = -4.35\mathbf{i} - 1.60\mathbf{j} + 2.77\mathbf{k} \text{ m/s} \qquad \textit{Ans.}$$

(d) The acceleration of A from Eq. 7/2 is

$$\mathbf{a} = \dot{\boldsymbol{\omega}} \times \mathbf{r} + \boldsymbol{\omega} \times (\boldsymbol{\omega} \times \mathbf{r})$$

$$= \boldsymbol{\alpha} \times \mathbf{r} + \boldsymbol{\omega} \times \mathbf{v}$$

$$= \begin{vmatrix} \mathbf{i} & \mathbf{j} & \mathbf{k} \\ 0 & 25.1 & 0 \\ 0 & 0.693 & 0.4 \end{vmatrix} + \begin{vmatrix} \mathbf{i} & \mathbf{j} & \mathbf{k} \\ 4 & 0 & 6.28 \\ -4.35 & -1.60 & 2.77 \end{vmatrix}$$

$$= (10.05\mathbf{i}) + (10.05\mathbf{i} - 38.4\mathbf{j} - 6.40\mathbf{k})$$

② $$= 20.1\mathbf{i} - 38.4\mathbf{j} - 6.40\mathbf{k} \text{ m/s}^2 \qquad \textit{Ans.}$$

The angular motion of OA depends only on the angular changes N and $\dot{\beta}$, so any linear motion of O does not affect $\boldsymbol{\omega}$ and $\boldsymbol{\alpha}$.

Helpful Hints

① Alternatively, consider axes x-y-z to be attached to the vertical shaft and clevis so that they rotate. The derivative of ω_x becomes $\dot{\boldsymbol{\omega}}_x = 4\dot{\mathbf{i}}$. But from Eq. 5/11, we have $\dot{\mathbf{i}} = \boldsymbol{\omega}_z \times \mathbf{i} = 6.28\mathbf{k} \times \mathbf{i} = 6.28\mathbf{j}$. Thus, $\boldsymbol{\alpha} = \dot{\boldsymbol{\omega}}_x = 4(6.28)\mathbf{j} = 25.1\mathbf{j}$ rad/s^2 as before.

② To compare methods, it is suggested that these results for $\mathbf{v}$ and $\mathbf{a}$ be obtained by applying Eqs. 2/18 and 2/19 for particle motion in spherical coordinates, changing symbols as necessary.

Sample Problem 7/2

The electric motor with an attached disk is running at a constant low speed of 120 rev/min in the direction shown. Its housing and mounting base are initially at rest. The entire assembly is next set in rotation about the vertical Z-axis at the constant rate $N = 60$ rev/min with a fixed angle γ of 30°. Determine (a) the angular velocity and angular acceleration of the disk, (b) the space and body cones, and (c) the velocity and acceleration of point A at the top of the disk for the instant shown.

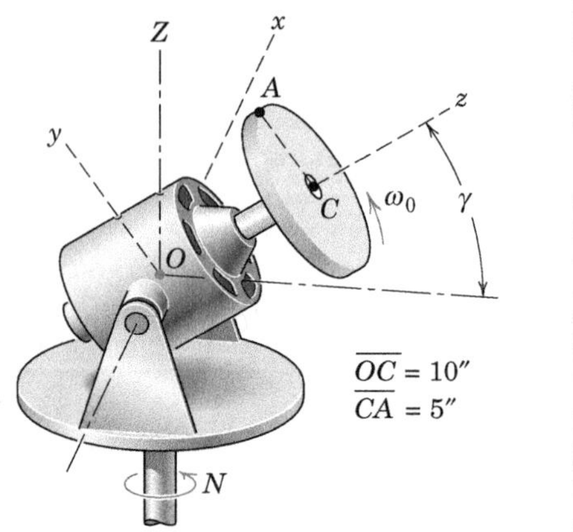

Solution. The axes x-y-z with unit vectors $\mathbf{i}$, $\mathbf{j}$, $\mathbf{k}$ are attached to the motor frame, with the z-axis coinciding with the rotor axis and the x-axis coinciding with the horizontal axis through O about which the motor tilts. The Z-axis is vertical and carries the unit vector $\mathbf{K} = \mathbf{j}\cos\gamma + \mathbf{k}\sin\gamma$.

(a) The rotor and disk have two components of angular velocity: $\omega_0 = 120(2\pi)/60 = 4\pi$ rad/sec about the z-axis and $\Omega = 60(2\pi)/60 = 2\pi$ rad/sec about the Z-axis. Thus, the angular velocity becomes

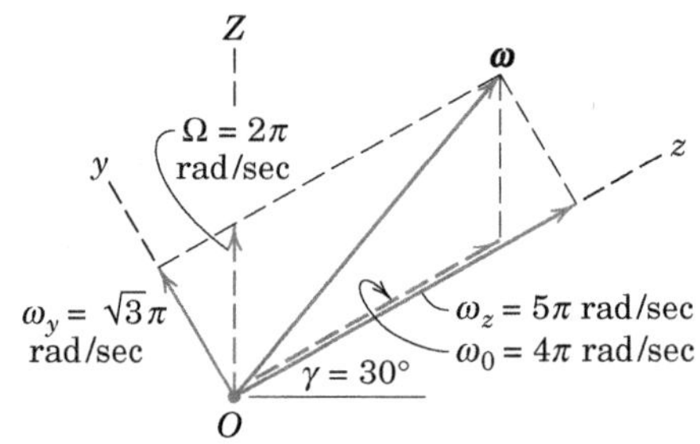

(1)
$$\boldsymbol{\omega} = \boldsymbol{\omega}_0 + \boldsymbol{\Omega} = \omega_0\mathbf{k} + \Omega\mathbf{K}$$
$$= \omega_0\mathbf{k} + \Omega(\mathbf{j}\cos\gamma + \mathbf{k}\sin\gamma) = (\Omega\cos\gamma)\mathbf{j} + (\omega_0 + \Omega\sin\theta)\mathbf{k}$$
$$= (2\pi\cos 30°)\mathbf{j} + (4\pi + 2\pi\sin 30°)\mathbf{k} = \pi(\sqrt{3}\mathbf{j} + 5.0\mathbf{k}) \text{ rad/sec} \qquad \textit{Ans.}$$

The angular acceleration of the disk from Eq. 7/3 is

(2)
$$\boldsymbol{\alpha} = \dot{\boldsymbol{\omega}} = \boldsymbol{\Omega} \times \boldsymbol{\omega}$$
$$= \Omega(\mathbf{j}\cos\gamma + \mathbf{k}\sin\gamma) \times [(\Omega\cos\gamma)\mathbf{j} + (\omega_0 + \Omega\sin\gamma)\mathbf{k}]$$
$$= \Omega(\omega_0\cos\gamma + \Omega\sin\gamma\cos\gamma)\mathbf{i} - (\Omega^2\sin\gamma\cos\gamma)\mathbf{i}$$

(3)
$$= (\Omega\,\omega_0\cos\gamma)\mathbf{i} = \mathbf{i}(2\pi)(4\pi)\cos 30° = 68.4\mathbf{i} \text{ rad/sec}^2 \qquad \textit{Ans.}$$

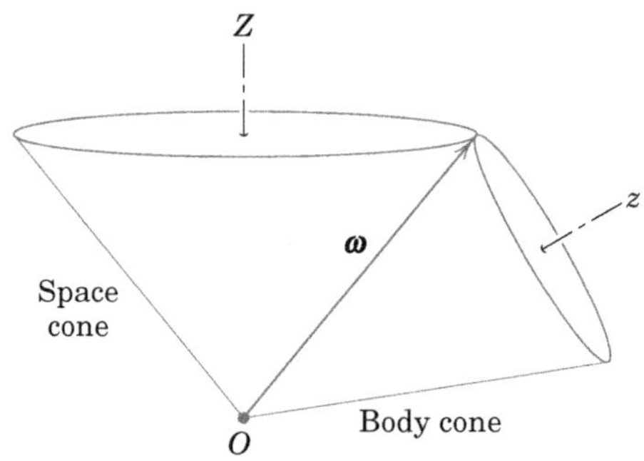

(b) The angular velocity vector $\boldsymbol{\omega}$ is the common element of the space and body cones which may now be constructed as shown.

(c) The position vector of point A for the instant considered is

$$\mathbf{r} = 5\mathbf{j} + 10\mathbf{k} \text{ in.}$$

From Eq. 7/1 the velocity of A is

$$\mathbf{v} = \boldsymbol{\omega} \times \mathbf{r} = \begin{vmatrix} \mathbf{i} & \mathbf{j} & \mathbf{k} \\ 0 & \sqrt{3}\pi & 5\pi \\ 0 & 5 & 10 \end{vmatrix} = -7.68\pi\mathbf{i} \text{ in./sec} \qquad \textit{Ans.}$$

From Eq. 7/2 the acceleration of point A is

$$\mathbf{a} = \dot{\boldsymbol{\omega}} \times \mathbf{r} + \boldsymbol{\omega} \times (\boldsymbol{\omega} \times \mathbf{r}) = \boldsymbol{\alpha} \times \mathbf{r} + \boldsymbol{\omega} \times \mathbf{v}$$
$$= 68.4\mathbf{i} \times (5\mathbf{j} + 10\mathbf{k}) + \pi(\sqrt{3}\mathbf{j} + 5\mathbf{k}) \times (-7.68\pi\mathbf{i})$$
$$= -1063\mathbf{j} + 473\mathbf{k} \text{ in./sec}^2 \qquad \textit{Ans.}$$

Helpful Hints

(1) Note that $\boldsymbol{\omega}_0 + \boldsymbol{\Omega} = \boldsymbol{\omega} = \boldsymbol{\omega}_y + \boldsymbol{\omega}_z$ as shown on the vector diagram.

(2) Remember that Eq. 7/3 gives the complete expression for $\boldsymbol{\alpha}$ only for steady precession where $|\boldsymbol{\omega}|$ is constant, which applies to this problem.

(3) Since the magnitude of $\boldsymbol{\omega}$ is constant, $\boldsymbol{\alpha}$ must be tangent to the base circle of the space cone, which puts it in the plus x-direction in agreement with our calculated conclusion.

PROBLEMS

Introductory Problems

7/1 Place your textbook on your desk, with fixed axes oriented as shown. Rotate the book about the x-axis through a 90° angle and then from this new position rotate it 90° about the y-axis. Sketch the final position of the book. Repeat the process but reverse the order of rotation. From your results, state your conclusion concerning the vector addition of finite rotations. Reconcile your observations with Fig. 7/4.

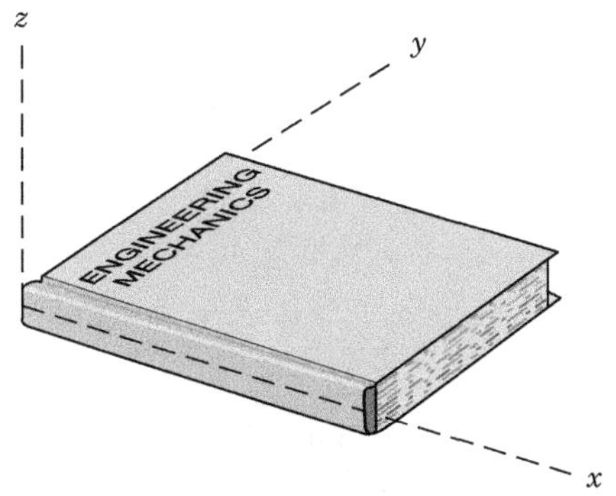

Problem 7/1

7/2 Repeat the experiment of Prob. 7/1 but use a small angle of rotation, say, 5°. Note the near-equal final positions for the two different rotation sequences. What does this observation lead you to conclude for the combination of infinitesimal rotations and for the time derivatives of angular quantities? Reconcile your observations with Fig. 7/5.

7/3 The solid cylinder is rotating about the fixed axis OA with a constant speed $N = 600$ rev/min in the direction shown. If the x- and y-components of the velocity of point P are 12 ft/sec and -6 ft/sec, determine its z-component of velocity and the radial distance R from P to the rotation axis. Also find the magnitude of the acceleration of P.

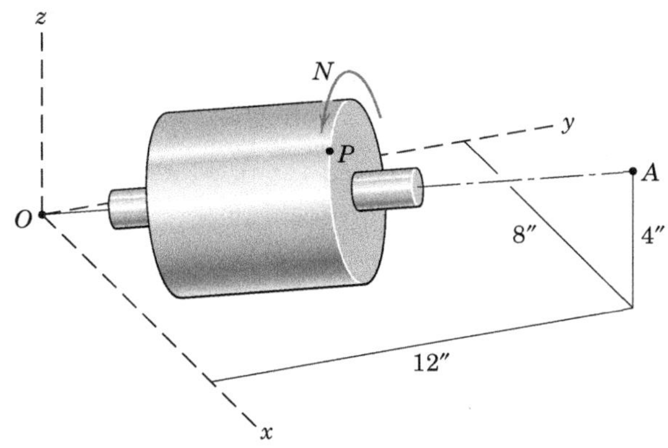

Problem 7/3

7/4 A timing mechanism consists of the rotating distributor arm AB and the fixed contact C. If the arm rotates about the fixed axis OA with a constant angular velocity $\boldsymbol{\omega} = 30(3\mathbf{i} + 2\mathbf{j} + 6\mathbf{k})$ rad/s, and if the coordinates of the contact C expressed in millimeters are (20, 30, 80), determine the magnitude of the acceleration of the tip B of the distributor arm as it passes point C.

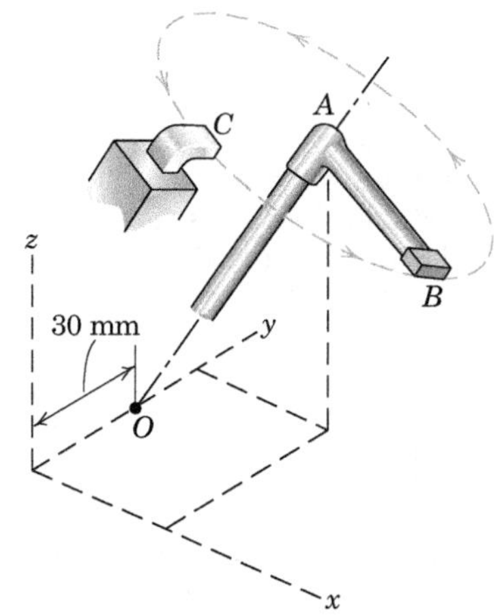

Problem 7/4

7/5 The rotor and shaft are mounted in a clevis which can rotate about the z-axis with an angular velocity Ω. With $\Omega = 0$ and θ constant, the rotor has an angular velocity $\boldsymbol{\omega}_0 = -4\mathbf{j} - 3\mathbf{k}$ rad/s. Find the velocity $\mathbf{v}_A$ of point A on the rim if its position vector at this instant is $\mathbf{r} = 0.5\mathbf{i} + 1.2\mathbf{j} + 1.1\mathbf{k}$ m. What is the rim speed v_B of any point B?

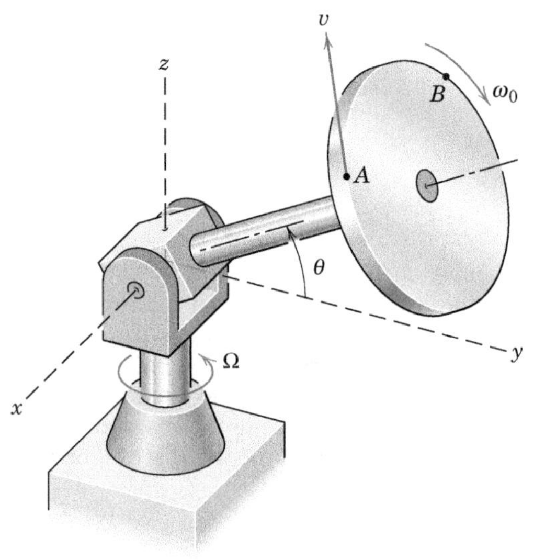

Problem 7/5

7/6 The disk rotates with a spin velocity of 15 rad/s about its horizontal z-axis first in the direction (a) and second in the direction (b). The assembly rotates with the angular velocity $N = 10$ rad/s about the vertical axis. Construct the space and body cones for each case.

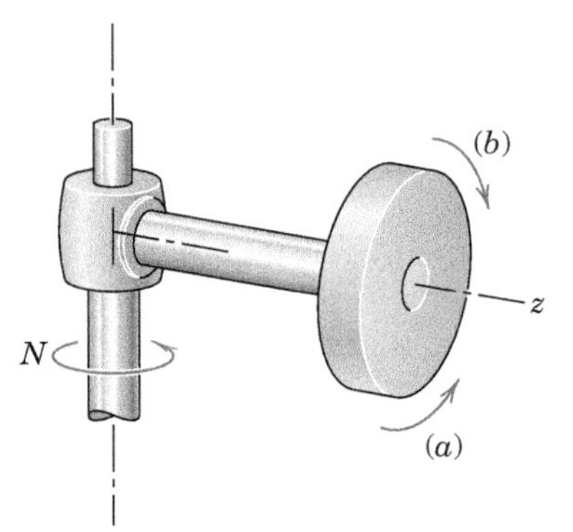

Problem 7/6

7/7 The rotor B spins about its inclined axis OA at the angular speed $N_1 = 200$ rev/min, where $\beta = 30°$. Simultaneously, the assembly rotates about the vertical z-axis at the rate N_2. If the total angular velocity of the rotor has a magnitude of 40 rad/s, determine N_2.

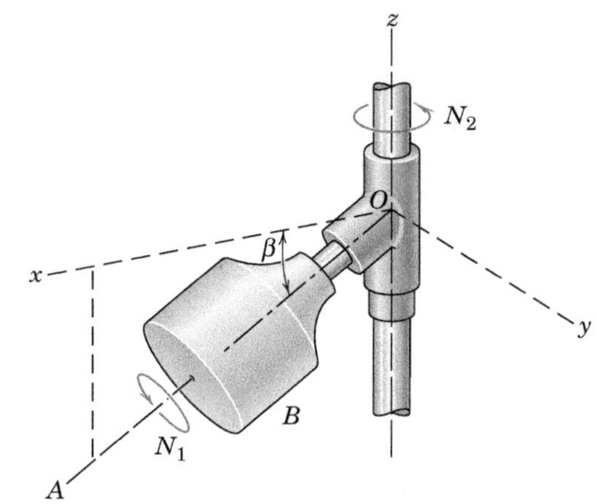

Problem 7/7

7/8 A slender rod bent into the shape shown rotates about the fixed line CD at a constant angular rate ω. Determine the velocity and acceleration of point A.

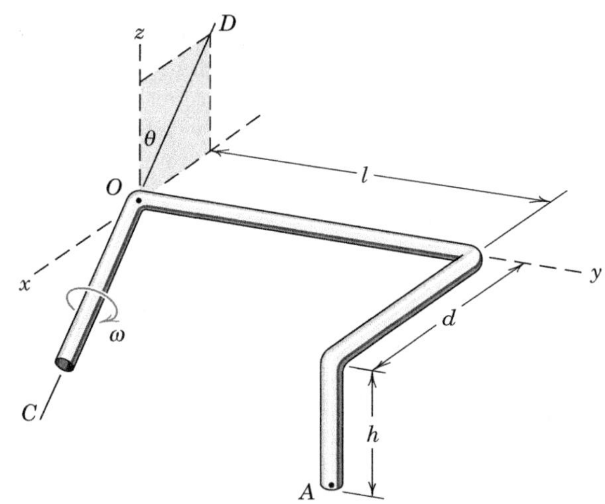

Problem 7/8

7/9 The rod is hinged about the axis *O-O* of the clevis, which is attached to the end of the vertical shaft. The shaft rotates with a constant angular velocity ω_0 as shown. If θ is decreasing at the constant rate $-\dot{\theta} = p$, write expressions for the angular velocity $\boldsymbol{\omega}$ and angular acceleration $\boldsymbol{\alpha}$ of the rod.

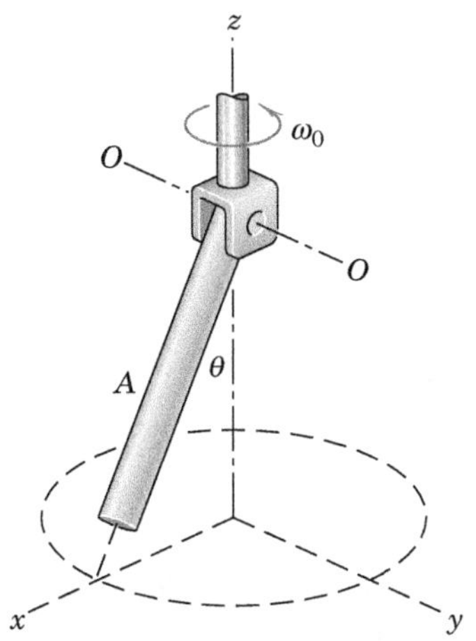

Problem 7/9

7/10 The panel assembly and attached *x-y-z* axes rotate with a constant angular velocity $\Omega = 0.6$ rad/sec about the vertical *z*-axis. Simultaneously, the panels rotate about the *y*-axis as shown with a constant rate $\omega_0 = 2$ rad/sec. Determine the angular acceleration $\boldsymbol{\alpha}$ of panel *A* and find the acceleration of point *P* for the instant when $\beta = 90°$.

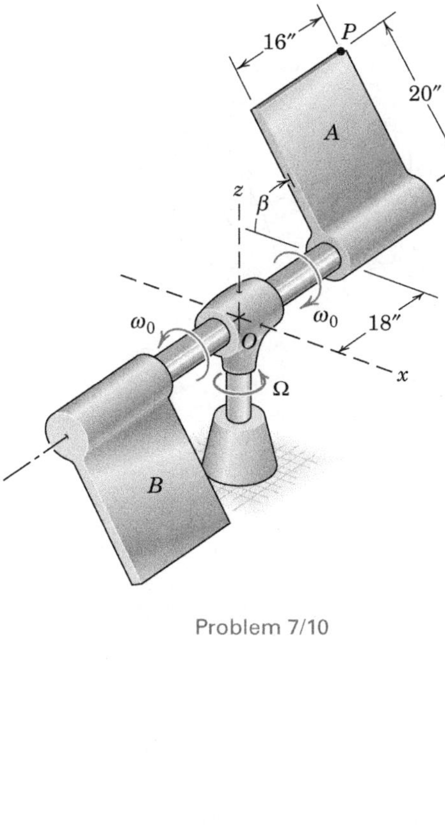

Problem 7/10

Representative Problems

7/11 The motor of Sample Problem 7/2 is shown again here. If the motor pivots about the *x*-axis at the constant rate $\dot{\gamma} = 3\pi$ rad/sec with no rotation about the *Z*-axis ($N = 0$), determine the angular acceleration $\boldsymbol{\alpha}$ of the rotor and disk as the position $\gamma = 30°$ is passed. The constant speed of the motor is 120 rev/min. Also find the velocity and acceleration of point *A*, which is on the top of the disk at this instant.

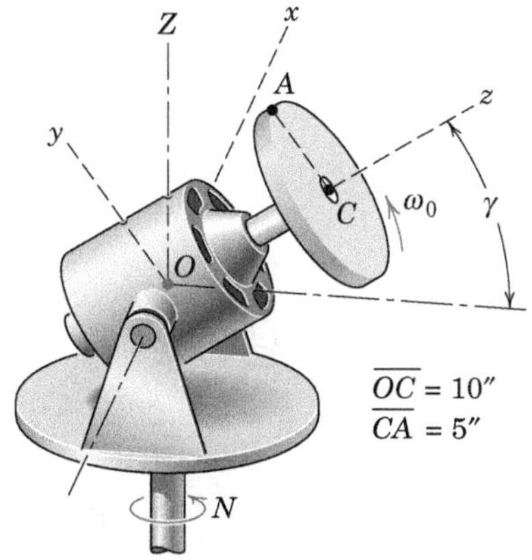

Problem 7/11

7/12 If the motor of Sample Problem 7/2, repeated in Prob. 7/11, reaches a speed of 3000 rev/min in 2 seconds from rest with constant acceleration, determine the total angular acceleration of the rotor and disk $\frac{1}{3}$ second after it is turned on if the turntable is rotating at a constant rate $N = 30$ rev/min. The angle $\gamma = 30°$ is constant.

7/13 The spool *A* rotates about its axis with an angular velocity of 20 rad/s, first in the sense of ω_a and second in the sense of ω_b. Simultaneously, the assembly rotates about the vertical axis with an angular velocity $\omega_1 = 10$ rad/s. Determine the magnitude ω of the total angular velocity of the spool and construct the body and space cones for the spool for each case.

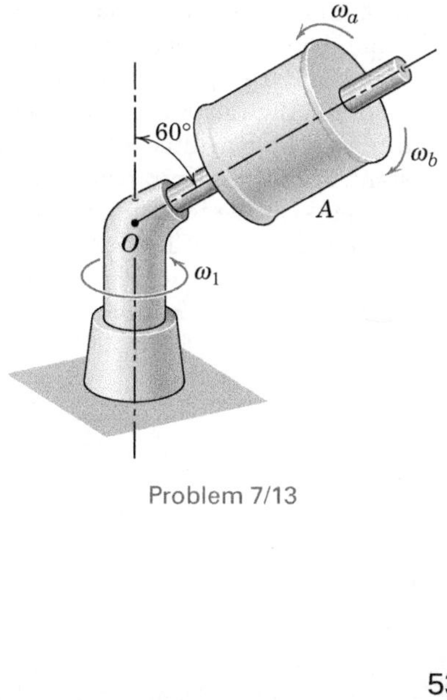

Problem 7/13

7/14 In manipulating the dumbbell, the jaws of the robotic device have an angular velocity $\omega_p = 2$ rad/s about the axis OG with γ fixed at 60°. The entire assembly rotates about the vertical Z-axis at the constant rate $\Omega = 0.8$ rad/s. Determine the angular velocity $\boldsymbol{\omega}$ and angular acceleration $\boldsymbol{\alpha}$ of the dumbbell. Express the results in terms of the given orientation of axes x-y-z, where the y-axis is parallel to the Y-axis.

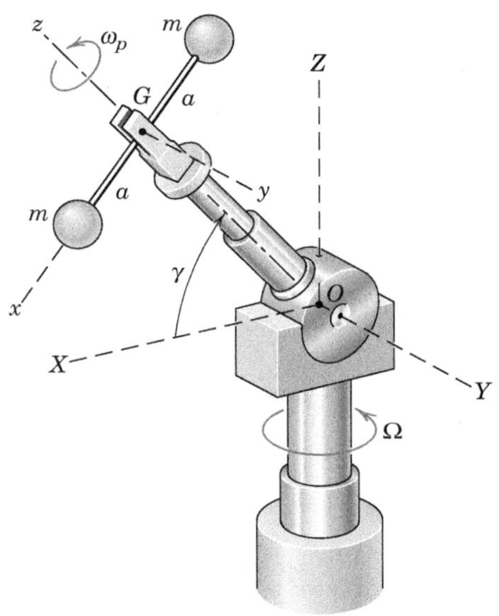

Problem 7/14

7/15 Determine the angular acceleration $\boldsymbol{\alpha}$ of the dumbbell of Prob. 7/14 for the conditions stated, except that Ω is increasing at the rate of 3 rad/s^2 for the instant under consideration.

7/16 The robot shown has five degrees of rotational freedom. The x-y-z axes are attached to the base ring, which rotates about the z-axis at the rate ω_1. The arm O_1O_2 rotates about the x-axis at the rate $\omega_2 = \dot{\theta}$. The control arm O_2A rotates about axis O_1-O_2 at the rate ω_3 and about a perpendicular axis through O_2 which is momentarily parallel to the x-axis at the rate $\omega_4 = \dot{\beta}$. Finally, the jaws rotate about axis O_2-A at the rate ω_5. The magnitudes of all angular rates are constant. For the configuration shown, determine the magnitude ω of the total angular velocity of the jaws for $\theta = 60°$ and $\beta = 45°$ if $\omega_1 = 2$ rad/s, $\dot{\theta} = 1.5$ rad/s, and $\omega_3 = \omega_4 = \omega_5 = 0$. Also express the angular acceleration $\boldsymbol{\alpha}$ of arm O_1O_2 as a vector.

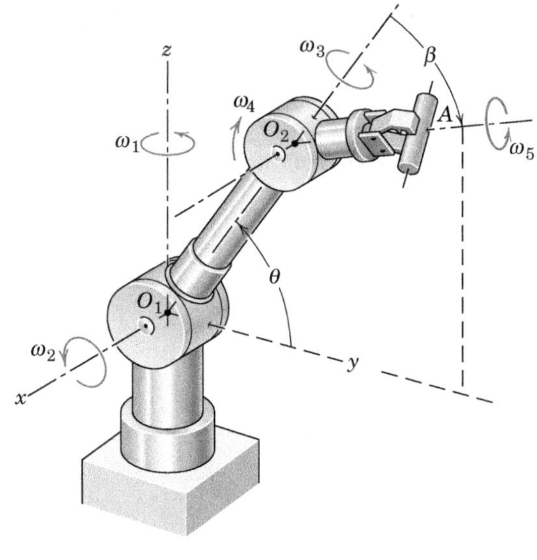

Problem 7/16

7/17 For the robot of Prob. 7/16, determine the angular velocity $\boldsymbol{\omega}$ and angular acceleration $\boldsymbol{\alpha}$ of the jaws A if $\theta = 60°$ and $\beta = 30°$, both constant, and if $\omega_1 = 2$ rad/s, $\omega_2 = \omega_3 = \omega_4 = 0$, and $\omega_5 = 0.8$ rad/s, all constant.

7/18 The wheel rolls without slipping in a circular arc of radius R and makes one complete turn about the vertical y-axis with constant speed in time τ. Determine the vector expression for the angular acceleration $\boldsymbol{\alpha}$ of the wheel and construct the space and body cones.

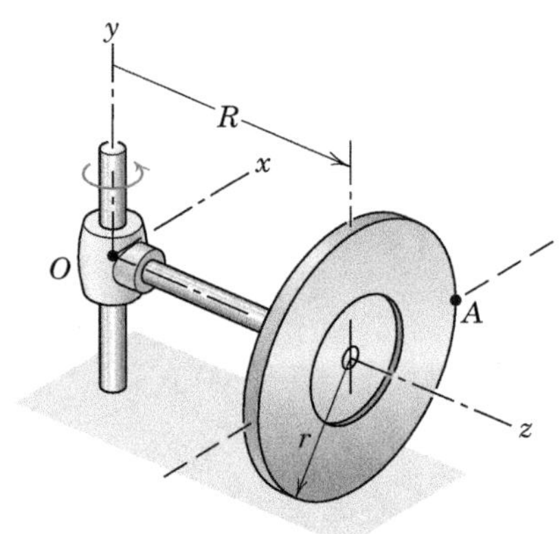

Problem 7/18

7/19 Determine expressions for the velocity $\mathbf{v}$ and acceleration $\mathbf{a}$ of point A on the wheel of Prob. 7/18 for the position shown, where A crosses the horizontal line through the center of the wheel.

7/20 The circular disk of 120-mm radius rotates about the z-axis at the constant rate $\omega_z = 20$ rad/s, and the entire assembly rotates about the fixed x-axis at the constant rate $\omega_x = 10$ rad/s. Calculate the magnitudes of the velocity $\mathbf{v}$ and acceleration $\mathbf{a}$ of point B for the instant when $\theta = 30°$.

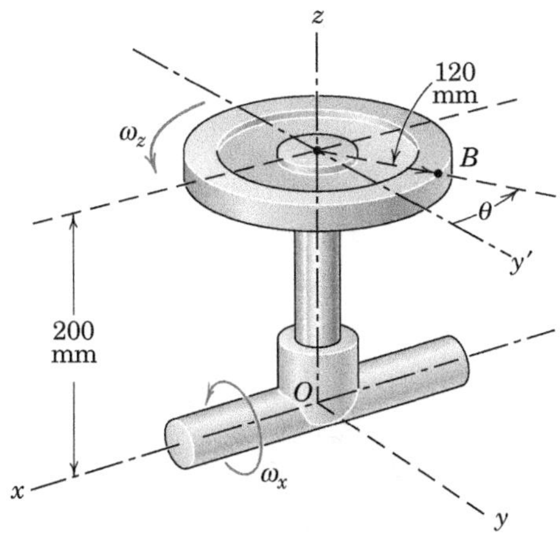

Problem 7/20

7/21 The crane has a boom of length $\overline{OP} = 24$ m and is revolving about the vertical axis at the constant rate of 2 rev/min in the direction shown. Simultaneously, the boom is being lowered at the constant rate $\dot{\beta} = 0.10$ rad/s. Calculate the magnitudes of the velocity and acceleration of the end P of the boom for the instant when it passes the position $\beta = 30°$.

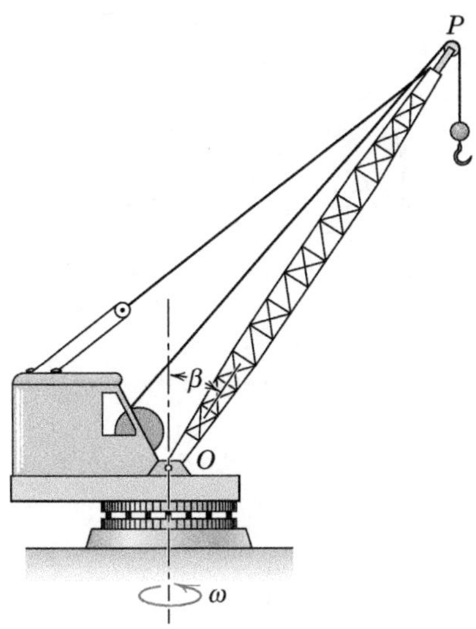

Problem 7/21

7/22 The design of the rotating arm OA of a control mechanism requires that it rotate about the vertical Z-axis at the constant rate $\Omega = \dot{\beta} = \pi$ rad/s. Simultaneously, OA oscillates according to $\theta = \theta_0 \sin 4\Omega t$, where $\theta_0 = \pi/6$ radians and t is in seconds measured from the time when $\beta = 0$. Determine the angular velocity $\boldsymbol{\omega}$ and the angular acceleration $\boldsymbol{\alpha}$ of OA for the instant (a) when $t = 1/2$ s and (b) when $t = 1/8$ s. The x-y reference axes rotate in the X-Y plane with the angular velocity Ω.

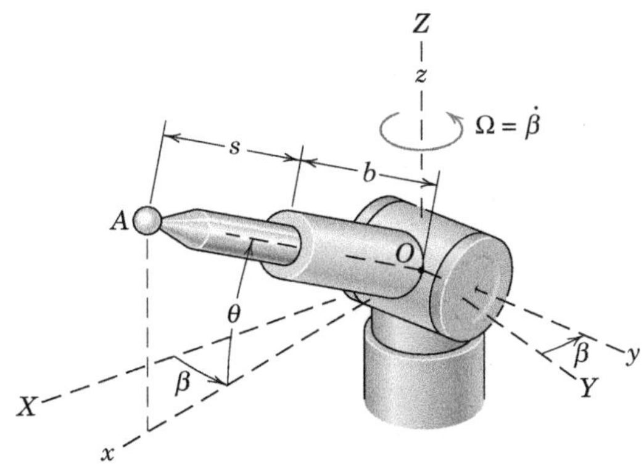

Problem 7/22

7/23 For the rotating and oscillating control arm OA of Prob. 7/22, determine the velocity $\mathbf{v}$ and acceleration $\mathbf{a}$ of the ball tip A for the condition when $t = 1/2$ s. Distance $b = 120$ mm, $s = 100$ mm, and $\theta = \theta_0 \sin 4\Omega t$ as defined in Prob. 7/22 with $\Omega = \pi$ rad/s and $\theta_0 = \pi/6$ rad.

7/24 If the angular velocity $\boldsymbol{\omega}_0 = -4\mathbf{j} - 3\mathbf{k}$ rad/s of the rotor in Prob. 7/5 is constant in magnitude, determine the angular acceleration $\boldsymbol{\alpha}$ of the rotor for (a) $\Omega = 0$ and $\dot{\theta} = 2$ rad/s (both constant) and (b) $\theta = \tan^{-1}(\frac{3}{4})$ and $\Omega = 2$ rad/s (both constant). Find the magnitude of the acceleration of point A in each case, where A has the position vector $\mathbf{r} = 0.5\mathbf{i} + 1.2\mathbf{j} + 1.1\mathbf{k}$ m at the instant represented.

7/25 The vertical shaft and attached clevis rotate about the z-axis at the constant rate $\Omega = 4$ rad/s. Simultaneously, the shaft B revolves about its axis OA at the constant rate $\omega_0 = 3$ rad/s, and the angle γ is decreasing at the constant rate of $\pi/4$ rad/s. Determine the angular velocity $\boldsymbol{\omega}$ and the magnitude of the angular acceleration $\boldsymbol{\alpha}$ of shaft B when $\gamma = 30°$. The x-y-z axes are attached to the clevis and rotate with it.

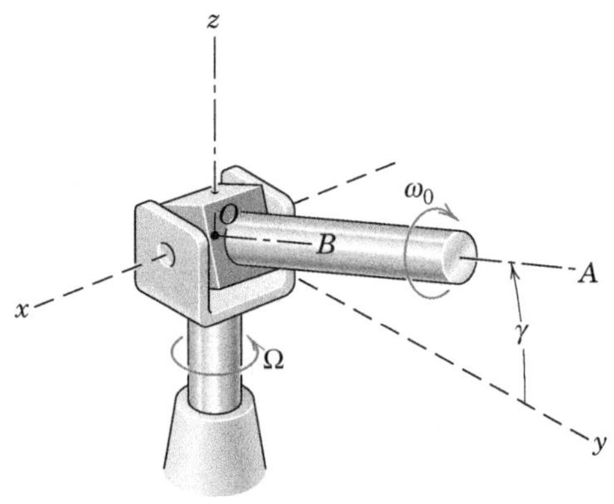

Problem 7/25

▶7/26 The right-circular cone A rolls on the fixed right-circular cone B at a constant rate and makes one complete trip around B every 4 seconds. Compute the magnitude of the angular acceleration $\boldsymbol{\alpha}$ of cone A during its motion.

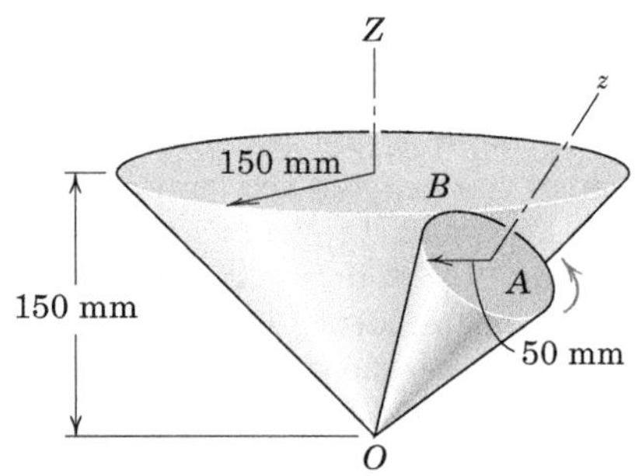

Problem 7/26

▶7/27 The pendulum oscillates about the x-axis according to $\theta = \frac{\pi}{6} \sin 3\pi t$ radians, where t is the time in seconds. Simultaneously, the shaft OA revolves about the vertical z-axis at the constant rate $\omega_z = 2\pi$ rad/sec. Determine the velocity $\mathbf{v}$ and acceleration $\mathbf{a}$ of the center B of the pendulum as well as its angular acceleration $\boldsymbol{\alpha}$ for the instant when $t = 0$.

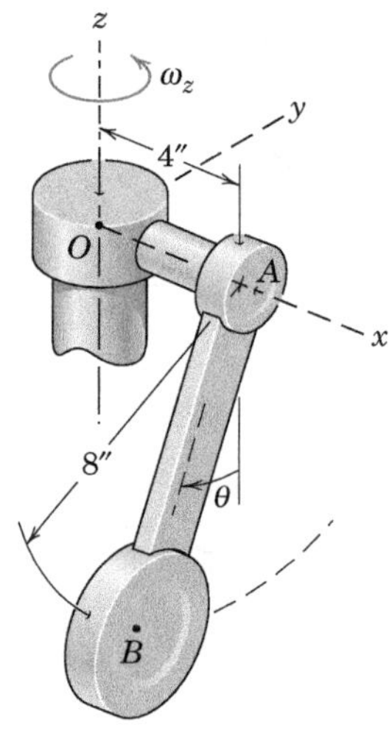

Problem 7/27

▶7/28 The solid right-circular cone of base radius r and height h rolls on a flat surface without slipping. The center B of the circular base moves in a circular path around the z-axis with a constant speed v. Determine the angular velocity $\boldsymbol{\omega}$ and the angular acceleration $\boldsymbol{\alpha}$ of the solid cone.

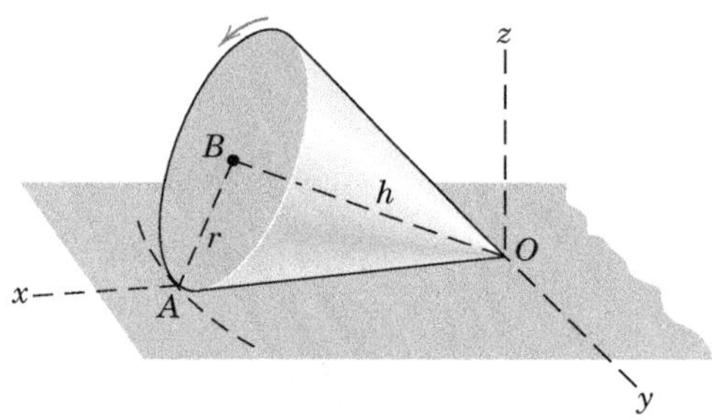

Problem 7/28

7/6 General Motion

The kinematic analysis of a rigid body which has general three-dimensional motion is best accomplished with the aid of our principles of relative motion. We have applied these principles to problems in plane motion and now extend them to space motion. We will make use of both translating axes and rotating reference axes.

Translating Reference Axes

Figure 7/10 shows a rigid body which has an angular velocity $\boldsymbol{\omega}$. We may choose any convenient point B as the origin of a translating reference system x-y-z. The velocity $\mathbf{v}$ and acceleration $\mathbf{a}$ of any other point A in the body are given by the relative-velocity and relative-acceleration expressions

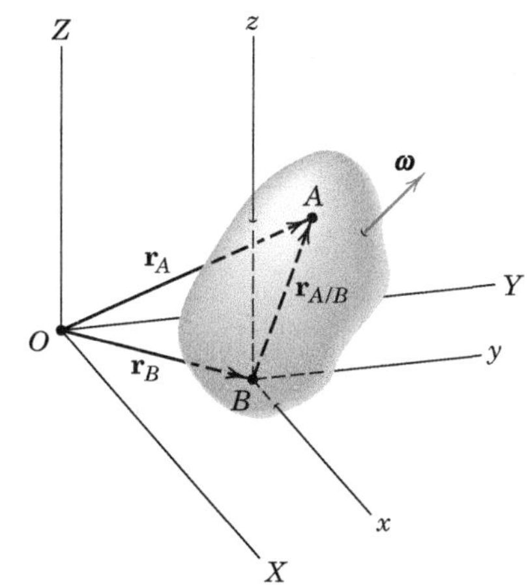

Figure 7/10

$$\mathbf{v}_A = \mathbf{v}_B + \mathbf{v}_{A/B} \qquad [5/4]$$

$$\mathbf{a}_A = \mathbf{a}_B + \mathbf{a}_{A/B} \qquad [5/7]$$

which were developed in Arts. 5/4 and 5/6 for the plane motion of rigid bodies. These expressions also hold in three dimensions, where the three vectors for each of the equations are also coplanar.

In applying these relations to rigid-body motion in space, we note from Fig. 7/10 that the distance $\overline{AB}$ remains constant. Thus, from an observer's position on x-y-z, the body appears to rotate about the point B and point A appears to lie on a spherical surface with B as the center. Consequently, we may view the general motion as a translation of the body with the motion of B plus a rotation of the body about B.

The relative-motion terms represent the effect of the rotation about B and are identical to the velocity and acceleration expressions discussed in the previous article for rotation of a rigid body about a fixed point. Therefore, the relative-velocity and relative-acceleration equations may be written

$$\boxed{\begin{aligned} \mathbf{v}_A &= \mathbf{v}_B + \boldsymbol{\omega} \times \mathbf{r}_{A/B} \\ \mathbf{a}_A &= \mathbf{a}_B + \dot{\boldsymbol{\omega}} \times \mathbf{r}_{A/B} + \boldsymbol{\omega} \times (\boldsymbol{\omega} \times \mathbf{r}_{A/B}) \end{aligned}} \qquad (7/4)$$

where $\boldsymbol{\omega}$ and $\dot{\boldsymbol{\omega}}$ are the instantaneous angular velocity and angular acceleration of the body, respectively.

The selection of the reference point B is quite arbitrary in theory. In practice, point B is chosen for convenience as some point in the body whose motion is known in whole or in part. If point A is chosen as the reference point, the relative-motion equations become

$$\mathbf{v}_B = \mathbf{v}_A + \boldsymbol{\omega} \times \mathbf{r}_{B/A}$$

$$\mathbf{a}_B = \mathbf{a}_A + \dot{\boldsymbol{\omega}} \times \mathbf{r}_{B/A} + \boldsymbol{\omega} \times (\boldsymbol{\omega} \times \mathbf{r}_{B/A})$$

where $\mathbf{r}_{B/A} = -\mathbf{r}_{A/B}$. It should be clear that $\boldsymbol{\omega}$ and, thus, $\dot{\boldsymbol{\omega}}$ are the same vectors for either formulation since the absolute angular motion of the body is independent of the choice of reference point. When we come to

By proper management of the hydraulic cylinders which support and move this aircraft flight simulator, a variety of three-dimensional translational and rotational accelerations can be produced.

the kinetic equations for general motion, we will see that the mass center of a body is frequently the most convenient reference point to choose.

If points A and B in Fig. 7/10 represent the ends of a rigid control link in a spatial mechanism where the end connections act as ball-and-socket joints (as in Sample Problem 7/3), it is necessary to impose certain kinematic requirements. Clearly, any rotation of the link about its own axis AB does not affect the action of the link. Thus, the angular velocity $\boldsymbol{\omega}_n$ whose vector is normal to the link describes its action. It is necessary, therefore, that $\boldsymbol{\omega}_n$ and $\mathbf{r}_{A/B}$ be at right angles, and this condition is satisfied if $\boldsymbol{\omega}_n \cdot \mathbf{r}_{A/B} = 0$.

Similarly, it is only the component $\boldsymbol{\alpha}_n$* of the angular acceleration of the link normal to AB which affects its action, so that $\boldsymbol{\alpha}_n \cdot \mathbf{r}_{A/B} = 0$ must also hold.

Rotating Reference Axes

A more general formulation of the motion of a rigid body in space calls for the use of reference axes which rotate as well as translate. The description of Fig. 7/10 is modified in Fig. 7/11 to show reference axes whose origin is attached to the reference point B as before, but which rotate with an absolute angular velocity $\boldsymbol{\Omega}$ which may be different from the absolute angular velocity $\boldsymbol{\omega}$ of the body.

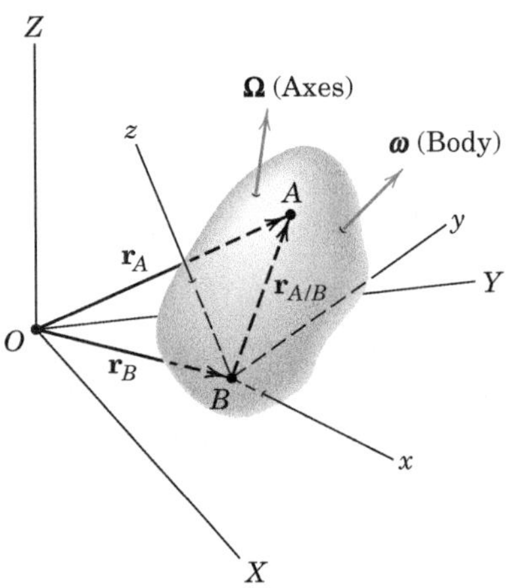

Figure 7/11

We now make use of Eqs. 5/11, 5/12, 5/13, and 5/14 developed in Art. 5/7 for describing the plane motion of a rigid body with the use of rotating axes. The extension of these relations from two to three dimensions is easily accomplished by merely including the z-component of the vectors, and this step is left to the student to carry out. Replacing $\boldsymbol{\omega}$ in these equations by the angular velocity $\boldsymbol{\Omega}$ of our rotating x-y-z axes gives us

$$\dot{\mathbf{i}} = \boldsymbol{\Omega} \times \mathbf{i} \qquad \dot{\mathbf{j}} = \boldsymbol{\Omega} \times \mathbf{j} \qquad \dot{\mathbf{k}} = \boldsymbol{\Omega} \times \mathbf{k} \tag{7/5}$$

for the time derivatives of the rotating unit vectors attached to x-y-z. The expression for the velocity and acceleration of point A become

$$\begin{aligned} \mathbf{v}_A &= \mathbf{v}_B + \boldsymbol{\Omega} \times \mathbf{r}_{A/B} + \mathbf{v}_{\text{rel}} \\ \mathbf{a}_A &= \mathbf{a}_B + \dot{\boldsymbol{\Omega}} \times \mathbf{r}_{A/B} + \boldsymbol{\Omega} \times (\boldsymbol{\Omega} \times \mathbf{r}_{A/B}) + 2\boldsymbol{\Omega} \times \mathbf{v}_{\text{rel}} + \mathbf{a}_{\text{rel}} \end{aligned} \tag{7/6}$$

where $\mathbf{v}_{\text{rel}} = \dot{x}\mathbf{i} + \dot{y}\mathbf{j} + \dot{z}\mathbf{k}$ and $\mathbf{a}_{\text{rel}} = \ddot{x}\mathbf{i} + \ddot{y}\mathbf{j} + \ddot{z}\mathbf{k}$ are, respectively, the velocity and acceleration of point A measured relative to x-y-z by an observer attached to x-y-z.

We again note that $\boldsymbol{\Omega}$ is the angular velocity of the axes and may be different from the angular velocity $\boldsymbol{\omega}$ of the body. Also we note that $\mathbf{r}_{A/B}$ remains constant in magnitude for points A and B fixed to a rigid body, but it will change direction with respect to x-y-z when the angular velocity $\boldsymbol{\Omega}$ of the axes is different from the angular velocity $\boldsymbol{\omega}$ of the

*It may be shown that $\boldsymbol{\alpha}_n = \dot{\boldsymbol{\omega}}_n$ if the angular velocity of the link about its own axis is not changing. See the first author's *Dynamics, 2nd Edition, SI Version*, 1975, John Wiley & Sons, Art. 37.

body. We observe further that, if x-y-z are rigidly attached to the body, $\mathbf{\Omega} = \boldsymbol{\omega}$ and $\mathbf{v}_{\text{rel}}$ and $\mathbf{a}_{\text{rel}}$ are both zero, which makes the equations identical to Eqs. 7/4.

In Art. 5/7 we also developed the relationship (Eq. 5/13) between the time derivative of a vector $\mathbf{V}$ as measured in the fixed X-Y system and the time derivative of $\mathbf{V}$ as measured relative to the rotating x-y system. For our three-dimensional case, this relation becomes

$$\left(\frac{d\mathbf{V}}{dt}\right)_{XYZ} = \left(\frac{d\mathbf{V}}{dt}\right)_{xyz} + \mathbf{\Omega} \times \mathbf{V} \tag{7/7}$$

When we apply this transformation to the relative-position vector $\mathbf{r}_{A/B} = \mathbf{r}_A - \mathbf{r}_B$ for our rigid body of Fig. 7/11, we obtain

$$\left(\frac{d\mathbf{r}_A}{dt}\right)_{XYZ} = \left(\frac{d\mathbf{r}_B}{dt}\right)_{XYZ} + \left(\frac{d\mathbf{r}_{A/B}}{dt}\right)_{xyz} + \mathbf{\Omega} \times \mathbf{r}_{A/B}$$

or

$$\mathbf{v}_A = \mathbf{v}_B + \mathbf{v}_{\text{rel}} + \mathbf{\Omega} \times \mathbf{r}_{A/B}$$

which gives us the first of Eqs. 7/6.

Equations 7/6 are particularly useful when the reference axes are attached to a moving body within which relative motion occurs.

Equation 7/7 may be recast as the vector operator

$$\left(\frac{d[\ \ ]}{dt}\right)_{XYZ} = \left(\frac{d[\ \ ]}{dt}\right)_{xyz} + \mathbf{\Omega} \times [\ \ ] \tag{7/7a}$$

where [ ] stands for any vector $\mathbf{V}$ expressible both in X-Y-Z and in x-y-z. If we apply the operator to itself, we obtain the second time derivative, which becomes

$$\left(\frac{d^2[\ \ ]}{dt^2}\right)_{XYZ} = \left(\frac{d^2[\ \ ]}{dt^2}\right)_{xyz} + \dot{\mathbf{\Omega}} \times [\ \ ] + \mathbf{\Omega} \times (\mathbf{\Omega} \times [\ \ ]) + 2\mathbf{\Omega} \times \left(\frac{d[\ \ ]}{dt}\right)_{xyz} \tag{7/7b}$$

This exercise is left to the student. Note that the form of Eq. 7/7*b* is the same as that of the second of Eqs. 7/6 expressed for $\mathbf{a}_{A/B} = \mathbf{a}_A - \mathbf{a}_B$.

Flirt/SuperStock

Robot performing a precision weld.

Sample Problem 7/3

Crank CB rotates about the horizontal axis with an angular velocity $\omega_1 = 6$ rad/s which is constant for a short interval of motion which includes the position shown. The link AB has a ball-and-socket fitting on each end and connects crank DA with CB. For the instant shown, determine the angular velocity ω_2 of crank DA and the angular velocity $\boldsymbol{\omega}_n$ of link AB.

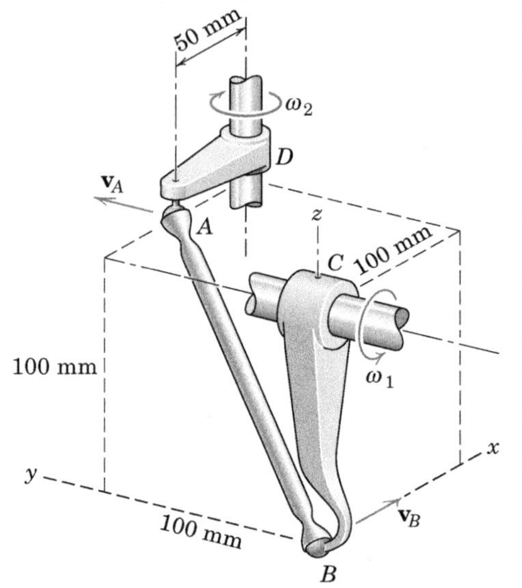

Solution. The relative-velocity relation, Eq. 7/4, will be solved first using
① translating reference axes attached to B. The equation is

$$\mathbf{v}_A = \mathbf{v}_B + \boldsymbol{\omega}_n \times \mathbf{r}_{A/B}$$

② where $\boldsymbol{\omega}_n$ is the angular velocity of link AB taken normal to AB. The velocities of A and B are

$[v = r\omega]$ $\qquad \mathbf{v}_A = 50\omega_2\mathbf{j} \qquad \mathbf{v}_B = 100(6)\mathbf{i} = 600\mathbf{i}$ mm/s

Also $\mathbf{r}_{A/B} = 50\mathbf{i} + 100\mathbf{j} + 100\mathbf{k}$ mm. Substitution into the velocity relation gives

$$50\omega_2\mathbf{j} = 600\mathbf{i} + \begin{vmatrix} \mathbf{i} & \mathbf{j} & \mathbf{k} \\ \omega_{n_x} & \omega_{n_y} & \omega_{n_z} \\ 50 & 100 & 100 \end{vmatrix}$$

Expanding the determinant and equating the coefficients of the **i**, **j**, **k** terms give

$$\begin{aligned} -6 &= \qquad\quad + \omega_{n_y} - \omega_{n_z} \\ \omega_2 &= -2\omega_{n_x} \qquad\quad + \omega_{n_z} \\ 0 &= \;\; 2\omega_{n_x} - \omega_{n_y} \end{aligned}$$

These equations may be solved for ω_2, which becomes

$$\omega_2 = 6 \text{ rad/s} \qquad \textit{Ans.}$$

As they stand, the three equations incorporate the fact that $\boldsymbol{\omega}_n$ is normal to $\mathbf{v}_{A/B}$, but they cannot be solved until the requirement that $\boldsymbol{\omega}_n$ be normal to $\mathbf{r}_{A/B}$ is
③ included. Thus,

$[\boldsymbol{\omega}_n \cdot \mathbf{r}_{A/B} = 0]$ $\qquad 50\omega_{n_x} + 100\omega_{n_y} + 100\omega_{n_z} = 0$

Combination with two of the three previous equations yields the solutions

$$\omega_{n_x} = -\tfrac{4}{3} \text{ rad/s} \qquad \omega_{n_y} = -\tfrac{8}{3} \text{ rad/s} \qquad \omega_{n_z} = \tfrac{10}{3} \text{ rad/s}$$

Thus,

$$\boldsymbol{\omega}_n = \tfrac{2}{3}(-2\mathbf{i} - 4\mathbf{j} + 5\mathbf{k}) \text{ rad/s} \qquad \textit{Ans.}$$

Helpful Hints

① We select B as the reference point since its motion can easily be determined from the given angular velocity ω_1 of CB.

② The angular velocity $\boldsymbol{\omega}$ of AB is taken as a vector $\boldsymbol{\omega}_n$ normal to AB since any rotation of the link about its own axis AB has no influence on the behavior of the linkage.

③ The relative-velocity equation may be written as $\mathbf{v}_A - \mathbf{v}_B = \mathbf{v}_{A/B} = \boldsymbol{\omega}_n \times \mathbf{r}_{A/B}$, which requires that $\mathbf{v}_{A/B}$ be perpendicular to both $\boldsymbol{\omega}_n$ and $\mathbf{r}_{A/B}$. This equation alone does not incorporate the additional requirement that $\boldsymbol{\omega}_n$ be perpendicular to $\mathbf{r}_{A/B}$. Thus, we must also satisfy $\boldsymbol{\omega}_n \cdot \mathbf{r}_{A/B} = 0$.

Sample Problem 7/4

Determine the angular acceleration $\dot{\omega}_2$ of crank AD in Sample Problem 7/3 for the conditions cited. Also find the angular acceleration $\dot{\boldsymbol{\omega}}_n$ of link AB.

Solution. The accelerations of the links may be found from the second of Eqs. 7/4, which may be written

$$\mathbf{a}_A = \mathbf{a}_B + \dot{\boldsymbol{\omega}}_n \times \mathbf{r}_{A/B} + \boldsymbol{\omega}_n \times (\boldsymbol{\omega}_n \times \mathbf{r}_{A/B})$$

where $\boldsymbol{\omega}_n$, as in Sample Problem 7/3, is the angular velocity of AB taken normal ① to AB. The angular acceleration of AB is written as $\dot{\boldsymbol{\omega}}_n$.

In terms of their normal and tangential components, the accelerations of A and B are

$$\mathbf{a}_A = 50\omega_2{}^2\mathbf{i} + 50\dot{\omega}_2\mathbf{j} = 1800\mathbf{i} + 50\dot{\omega}_2\mathbf{j}\ \text{mm/s}^2$$

$$\mathbf{a}_B = 100\omega_1{}^2\mathbf{k} + (0)\mathbf{i} = 3600\mathbf{k}\ \text{mm/s}^2$$

Also

$$\boldsymbol{\omega}_n \times (\boldsymbol{\omega}_n \times \mathbf{r}_{A/B}) = -\omega_n{}^2\mathbf{r}_{A/B} = -20(50\mathbf{i} + 100\mathbf{j} + 100\mathbf{k})\ \text{mm/s}^2$$

$$\dot{\boldsymbol{\omega}}_n \times \mathbf{r}_{A/B} = (100\dot{\omega}_{n_y} - 100\dot{\omega}_{n_z})\mathbf{i} + (50\dot{\omega}_{n_z} - 100\dot{\omega}_{n_x})\mathbf{j} + (100\dot{\omega}_{n_x} - 50\dot{\omega}_{n_y})\mathbf{k}$$

Substitution into the relative-acceleration equation and equating respective coefficients of $\mathbf{i}, \mathbf{j}, \mathbf{k}$ give

$$28 = \dot{\omega}_{n_y} - \dot{\omega}_{n_z}$$

$$\dot{\omega}_2 + 40 = -2\dot{\omega}_{n_x} + \dot{\omega}_{n_z}$$

$$-32 = 2\dot{\omega}_{n_x} - \dot{\omega}_{n_y}$$

Solution of these equations for $\dot{\omega}_2$ gives

$$\dot{\omega}_2 = -36\ \text{rad/s}^2 \qquad \textit{Ans.}$$

② The vector $\dot{\boldsymbol{\omega}}_n$ is normal to $\mathbf{r}_{A/B}$ but is not normal to $\mathbf{v}_{A/B}$, as was the case with $\boldsymbol{\omega}_n$.

$$[\dot{\boldsymbol{\omega}}_n \cdot \mathbf{r}_{A/B} = 0] \qquad 2\dot{\omega}_{n_x} + 4\dot{\omega}_{n_y} + 4\dot{\omega}_{n_z} = 0$$

which, when combined with the preceding relations for these same quantities, gives

$$\dot{\omega}_{n_x} = -8\ \text{rad/s}^2 \qquad \dot{\omega}_{n_y} = 16\ \text{rad/s}^2 \qquad \dot{\omega}_{n_z} = -12\ \text{rad/s}^2$$

Thus,

$$\dot{\boldsymbol{\omega}}_n = 4(-2\mathbf{i} + 4\mathbf{j} - 3\mathbf{k})\ \text{rad/s}^2 \qquad \textit{Ans.}$$

and

$$|\dot{\boldsymbol{\omega}}_n| = 4\sqrt{2^2 + 4^2 + 3^2} = 4\sqrt{29}\ \text{rad/s}^2$$

Helpful Hints

① If the link AB had an angular velocity component along AB, then a change in both magnitude and direction of this component could occur which would contribute to the actual angular acceleration of the link as a rigid body. However, since any rotation about its own axis AB has no influence on the motion of the cranks at C and D, we will concern ourselves only with $\dot{\omega}_n$.

② The component of $\dot{\omega}_n$ which is not normal to $\mathbf{v}_{A/B}$ gives rise to the change in direction of $\mathbf{v}_{A/B}$.

Sample Problem 7/5

The motor housing and its bracket rotate about the Z-axis at the constant rate $\Omega = 3$ rad/s. The motor shaft and disk have a constant angular velocity of spin $p = 8$ rad/s with respect to the motor housing in the direction shown. If γ is constant at 30°, determine the velocity and acceleration of point A at the top of the disk and the angular acceleration $\boldsymbol{\alpha}$ of the disk.

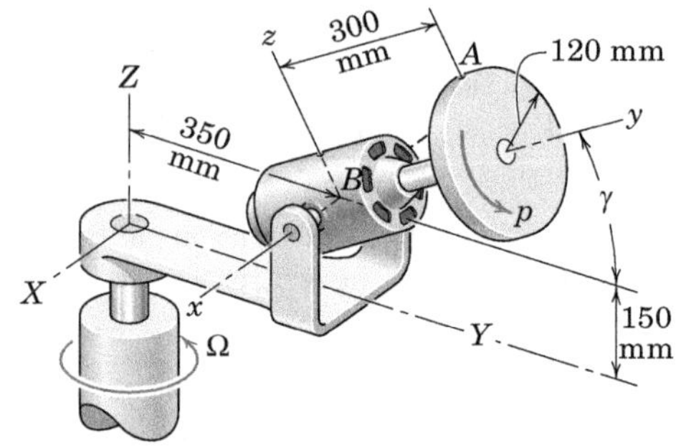

Solution. The rotating reference axes x-y-z are attached to the motor housing, and the rotating base for the motor has the momentary orientation shown with ① respect to the fixed axes X-Y-Z. We will use both X-Y-Z components with unit vectors $\mathbf{I}, \mathbf{J}, \mathbf{K}$ and x-y-z components with unit vectors $\mathbf{i}, \mathbf{j}, \mathbf{k}$. The angular velocity of the x-y-z axes becomes $\boldsymbol{\Omega} = \Omega\mathbf{K} = 3\mathbf{K}$ rad/s.

Velocity. The velocity of A is given by the first of Eqs. 7/6

$$\mathbf{v}_A = \mathbf{v}_B + \boldsymbol{\Omega} \times \mathbf{r}_{A/B} + \mathbf{v}_{\text{rel}}$$

where

$$\mathbf{v}_B = \boldsymbol{\Omega} \times \mathbf{r}_B = 3\mathbf{K} \times 0.350\mathbf{J} = -1.05\mathbf{I} = -1.05\mathbf{i} \text{ m/s}$$

② $$\boldsymbol{\Omega} \times \mathbf{r}_{A/B} = 3\mathbf{K} \times (0.300\mathbf{j} + 0.120\mathbf{k})$$
$$= (-0.9 \cos 30°)\mathbf{i} + (0.36 \sin 30°)\mathbf{i} = -0.599\mathbf{i} \text{ m/s}$$

$$\mathbf{v}_{\text{rel}} = \mathbf{p} \times \mathbf{r}_{A/B} = 8\mathbf{j} \times (0.300\mathbf{j} + 0.120\mathbf{k}) = 0.960\mathbf{i} \text{ m/s}$$

Thus,

$$\mathbf{v}_A = -1.05\mathbf{i} - 0.599\mathbf{i} + 0.960\mathbf{i} = -0.689\mathbf{i} \text{ m/s} \qquad \textit{Ans.}$$

Acceleration. The acceleration of A is given by the second of Eqs. 7/6

$$\mathbf{a}_A = \mathbf{a}_B + \dot{\boldsymbol{\Omega}} \times \mathbf{r}_{A/B} + \boldsymbol{\Omega} \times (\boldsymbol{\Omega} \times \mathbf{r}_{A/B}) + 2\boldsymbol{\Omega} \times \mathbf{v}_{\text{rel}} + \mathbf{a}_{\text{rel}}$$

where

$$\mathbf{a}_B = \boldsymbol{\Omega} \times (\boldsymbol{\Omega} \times \mathbf{r}_B) = 3\mathbf{K} \times (3\mathbf{K} \times 0.350\mathbf{J}) = -3.15\mathbf{J}$$
$$= 3.15(-\mathbf{j} \cos 30° + \mathbf{k} \sin 30°) = -2.73\mathbf{j} + 1.575\mathbf{k} \text{ m/s}^2$$

$$\dot{\boldsymbol{\Omega}} = \mathbf{0}$$

② $$\boldsymbol{\Omega} \times (\boldsymbol{\Omega} \times \mathbf{r}_{A/B}) = 3\mathbf{K} \times [3\mathbf{K} \times (0.300\mathbf{j} + 0.120\mathbf{k})]$$
$$= 3\mathbf{K} \times (-0.599\mathbf{i}) = -1.557\mathbf{j} + 0.899\mathbf{k} \text{ m/s}^2$$

$$2\boldsymbol{\Omega} \times \mathbf{v}_{\text{rel}} = 2(3\mathbf{K}) \times 0.960\mathbf{i} = 5.76\mathbf{J}$$
$$= 5.76(\mathbf{j} \cos 30° - \mathbf{k} \sin 30°) = 4.99\mathbf{j} - 2.88\mathbf{k} \text{ m/s}^2$$

$$\mathbf{a}_{\text{rel}} = \mathbf{p} \times (\mathbf{p} \times \mathbf{r}_{A/B}) = 8\mathbf{j} \times [8\mathbf{j} \times (0.300\mathbf{j} + 0.120\mathbf{k})]$$
$$= -7.68\mathbf{k} \text{ m/s}^2$$

Substituting into the expression for $\mathbf{a}_A$ and collecting terms give us

$$\mathbf{a}_A = 0.703\mathbf{j} - 8.09\mathbf{k} \text{ m/s}^2$$

and

$$a_A = \sqrt{(0.703)^2 + (8.09)^2} = 8.12 \text{ m/s}^2 \qquad \textit{Ans.}$$

Angular Acceleration. Since the precession is steady, we may use Eq. 7/3 to give us

$$\boldsymbol{\alpha} = \dot{\boldsymbol{\omega}} = \boldsymbol{\Omega} \times \boldsymbol{\omega} = 3\mathbf{K} \times (3\mathbf{K} + 8\mathbf{j})$$
$$= \mathbf{0} + (-24 \cos 30°)\mathbf{i} = -20.8\mathbf{i} \text{ rad/s}^2 \qquad \textit{Ans.}$$

Helpful Hints

① This choice for the reference axes provides a simple description for the motion of the disk relative to these axes.

② Note that $\mathbf{K} \times \mathbf{i} = \mathbf{J} = \mathbf{j} \cos \gamma - \mathbf{k} \sin \gamma$, $\mathbf{K} \times \mathbf{j} = -\mathbf{i} \cos \gamma$, and $\mathbf{K} \times \mathbf{k} = \mathbf{i} \sin \gamma$.

PROBLEMS

Introductory Problems

7/29 The solid cylinder has a body cone with a semi-vertex angle of 20°. Momentarily the angular velocity $\boldsymbol{\omega}$ has a magnitude of 30 rad/s and lies in the y-z plane. Determine the rate p at which the cylinder is spinning about its z-axis and write the vector expression for the velocity of B with respect to A.

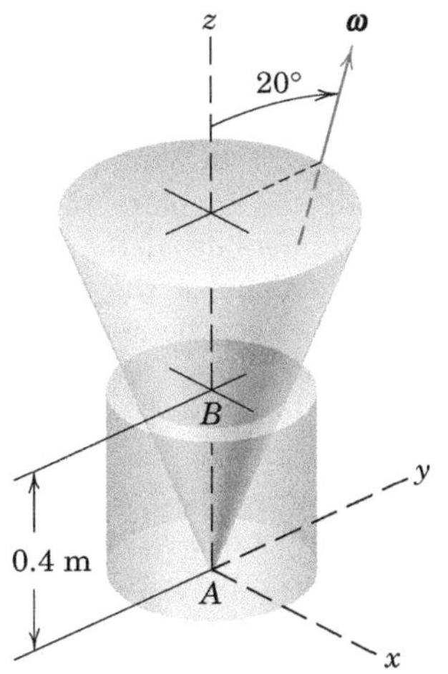

Problem 7/29

7/30 The helicopter is nosing over at the constant rate q rad/s. If the rotor blades revolve at the constant speed p rad/s, write the expression for the angular acceleration $\boldsymbol{\alpha}$ of the rotor. Take the y-axis to be attached to the fuselage and pointing forward perpendicular to the rotor axis.

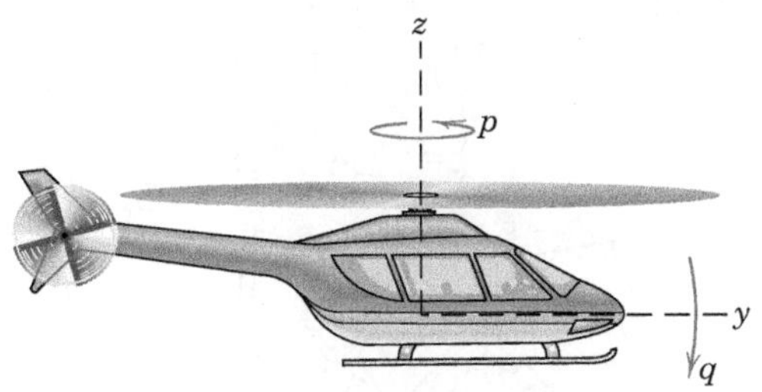

Problem 7/30

7/31 The collar at O and attached shaft OC rotate about the fixed x_0-axis at the constant rate $\Omega = 4$ rad/s. Simultaneously, the circular disk rotates about OC at the constant rate $p = 10$ rad/s. Determine the magnitude of the total angular velocity $\boldsymbol{\omega}$ of the disk and find its angular acceleration $\boldsymbol{\alpha}$.

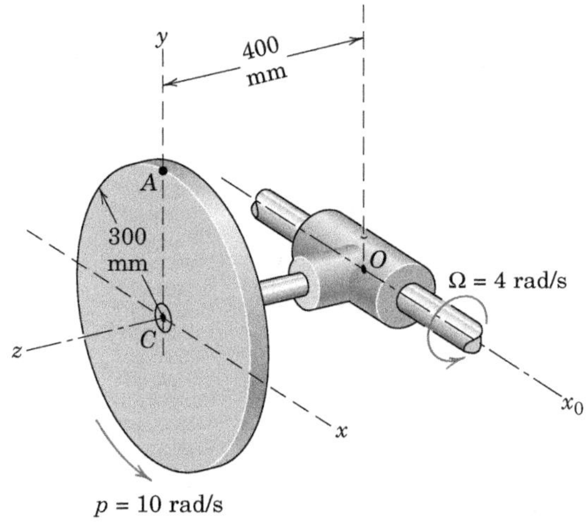

Problem 7/31

7/32 If the angular rate p of the disk in Prob. 7/31 is increasing at the rate of 6 rad/s per second and if Ω remains constant at 4 rad/s, determine the angular acceleration $\boldsymbol{\alpha}$ of the disk at the instant when p reaches 10 rad/s.

7/33 For the conditions of Prob. 7/31, determine the velocity $\mathbf{v}_A$ and acceleration $\mathbf{a}_A$ of point A on the disk as it passes the position shown. Reference axes x-y-z are attached to the collar at O and its shaft OC.

7/34 An unmanned radar-radio controlled aircraft with tilt-rotor propulsion is being designed for reconnaissance purposes. Vertical rise begins with $\theta = 0$ and is followed by horizontal flight as θ approaches 90°. If the rotors turn at a constant speed N of 360 rev/min, determine the angular acceleration $\boldsymbol{\alpha}$ of rotor A for $\theta = 30°$ if $\dot{\theta}$ is constant at 0.2 rad/s.

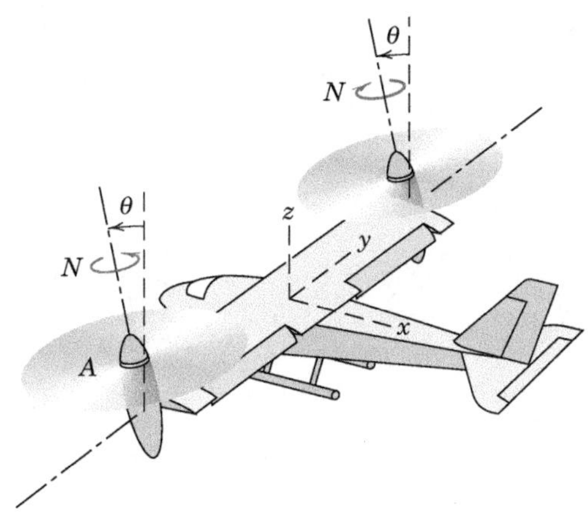

Problem 7/34

7/35 End A of the rigid link is confined to move in the $-x$-direction while end B is confined to move along the z-axis. Determine the component $\boldsymbol{\omega}_n$ normal to AB of the angular velocity of the link as it passes the position shown with $v_A = 3$ ft/sec.

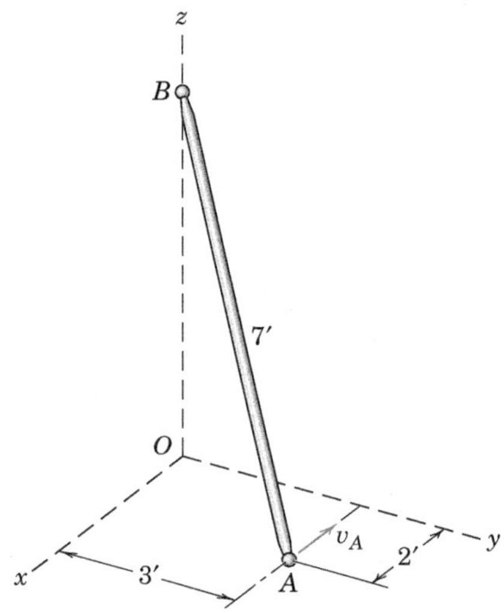

Problem 7/35

Representative Problems

7/36 The small motor M is pivoted about the x-axis through O and gives its shaft OA a constant speed p rad/s in the direction shown relative to its housing. The entire unit is then set into rotation about the vertical Z-axis at the constant angular velocity Ω rad/s. Simultaneously, the motor pivots about the x-axis at the constant rate $\dot{\beta}$ for an interval of motion. Determine the angular acceleration $\boldsymbol{\alpha}$ of the shaft OA in terms of β. Express your result in terms of the unit vectors for the rotating x-y-z axes.

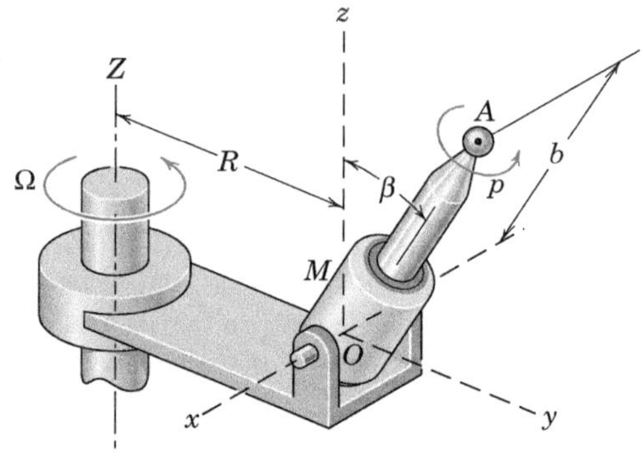

Problem 7/36

7/37 The flight simulator is mounted on six hydraulic actuators connected in pairs to their attachment points on the underside of the simulator. By programming the actions of the actuators, a variety of flight conditions can be simulated with translational and rotational displacements through a limited range of motion. Axes x-y-z are attached to the simulator with origin B at the center of the volume. For the instant represented, B has a velocity and an acceleration in the horizontal y-direction of 3.2 ft/sec and 4 ft/sec^2, respectively. Simultaneously, the angular velocities and their time rates of change are $\omega_x = 1.4$ rad/sec, $\dot{\omega}_x = 2$ rad/sec^2, $\omega_y = 1.2$ rad/sec, $\dot{\omega}_y = 3$ rad/sec^2, $\omega_z = \dot{\omega}_z = 0$. For this instant determine the magnitudes of the velocity and acceleration of point A.

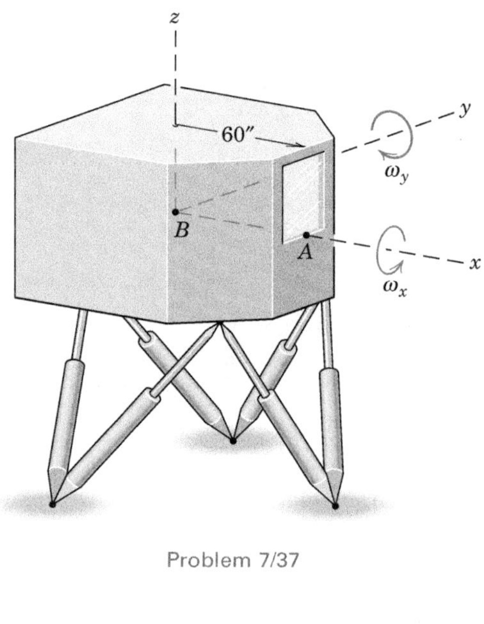

Problem 7/37

7/38 The robot of Prob. 7/16 is shown again here, where the coordinate system x-y-z with origin at O_2 rotates about the X-axis at the rate $\dot{\theta}$. Nonrotating axes X-Y-Z oriented as shown have their origin at O_1. If $\omega_2 = \dot{\theta} = 3$ rad/s constant, $\omega_3 = 1.5$ rad/s constant, $\omega_1 = \omega_5 = 0$, $\overline{O_1O_2} = 1.2$ m, and $\overline{O_2A} = 0.6$ m, determine the velocity of the center A of the jaws for the instant when $\theta = 60°$. The angle β lies in the y-z plane and is constant at 45°.

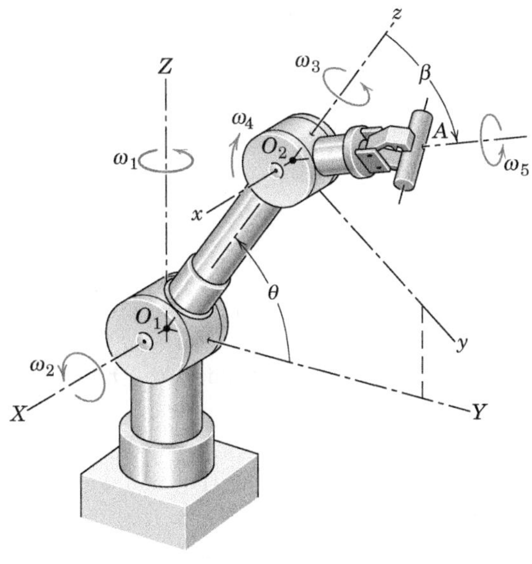

Problem 7/38

7/39 For the instant represented collar B is moving along the fixed shaft in the X-direction with a constant velocity $v_B = 4$ m/s. Also at this instant $X = 0.3$ m and $Y = 0.2$ m. Calculate the velocity of collar A, which moves along the fixed shaft parallel to the Y-axis. Solve, first, by differentiating the relation $X^2 + Y^2 + Z^2 = L^2$ with respect to time and, second, by using the first of Eqs. 7/4 with translating axes attached to B. Each clevis is free to rotate about the axis of the rod.

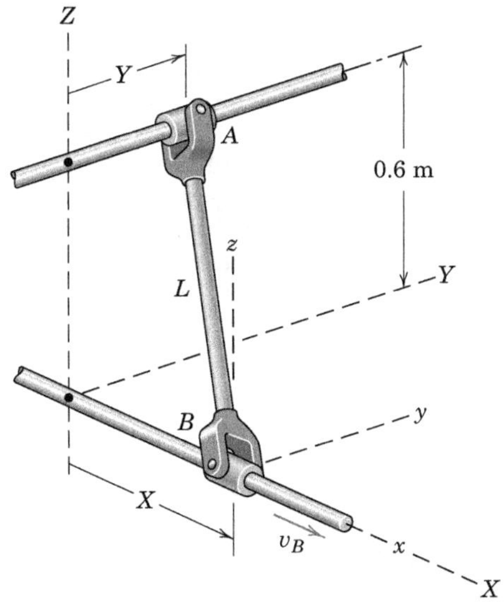

Problem 7/39

7/40 The spacecraft is revolving about its z-axis, which has a fixed space orientation, at the constant rate $p = \frac{1}{10}$ rad/s. Simultaneously, its solar panels are unfolding at the rate $\dot{\beta}$ which is programmed to vary with β as shown in the graph. Determine the angular acceleration $\boldsymbol{\alpha}$ of panel A an instant (a) before and an instant (b) after it reaches the position $\beta = 18°$.

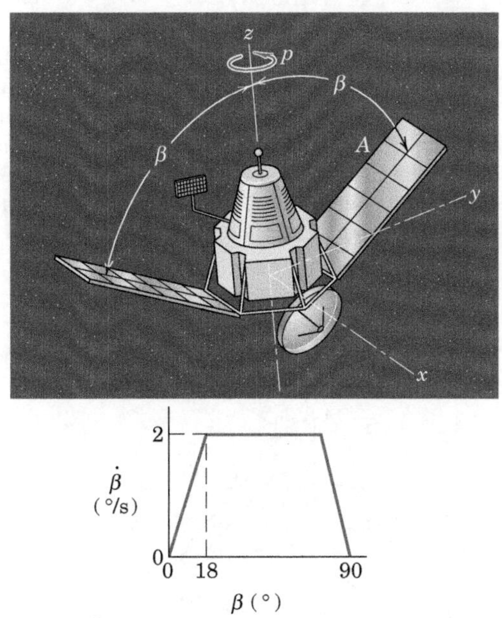

Problem 7/40

7/41 The disk has a constant angular velocity p about its z-axis, and the yoke A has a constant angular velocity ω_2 about its shaft as shown. Simultaneously, the entire assembly revolves about the fixed X-axis with a constant angular velocity ω_1. Determine the expression for the angular acceleration of the disk as the yoke brings it into the vertical plane in the position shown. Solve by picturing the vector changes in the angular-velocity components.

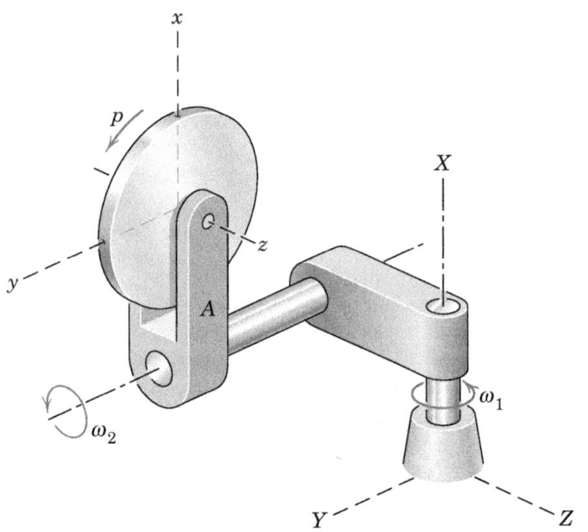

Problem 7/41

7/42 The collar and clevis A are given a constant upward velocity of 8 in./sec for an interval of motion and cause the ball end of the bar to slide in the radial slot in the rotating disk. Determine the angular acceleration of the bar when the bar passes the position for which z = 3 in. The disk turns at the constant rate of 2 rad/sec.

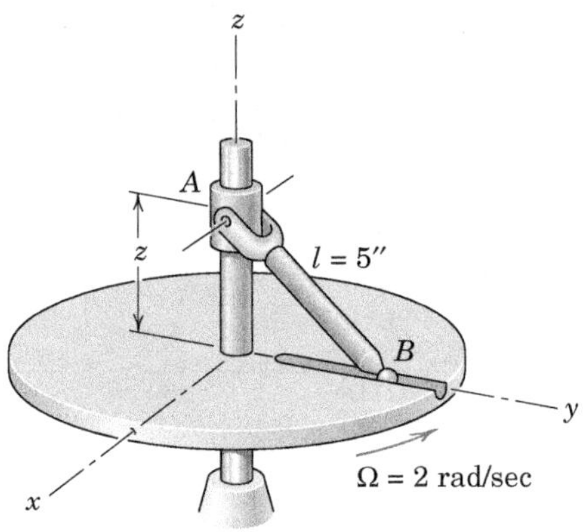

Problem 7/42

7/43 The circular disk of 100-mm radius rotates about its z-axis at the constant speed p = 240 rev/min, and arm OCB rotates about the Y-axis at the constant speed N = 30 rev/min. Determine the velocity **v** and acceleration **a** of point A on the disk as it passes the position shown. Use reference axes x-y-z attached to the arm OCB.

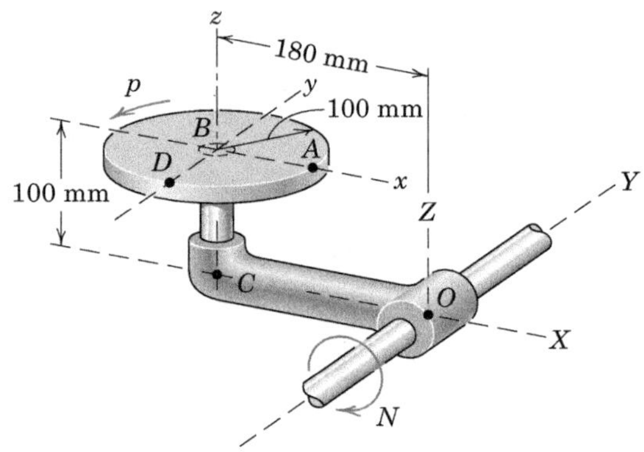

Problem 7/43

7/44 Solve Prob. 7/43 by attaching the reference axes x-y-z to the rotating disk.

7/45 For the conditions described in Prob. 7/36, determine the velocity **v** and acceleration **a** of the center A of the ball tool in terms of β.

7/46 The circular disk is spinning about its own axis (y-axis) at the constant rate p = 10π rad/s. Simultaneously, the frame is rotating about the Z-axis at the constant rate Ω = 4π rad/s. Calculate the angular acceleration $\boldsymbol{\alpha}$ of the disk and the acceleration of point A at the top of the disk. Axes x-y-z are attached to the frame, which has the momentary orientation shown with respect to the fixed axes X-Y-Z.

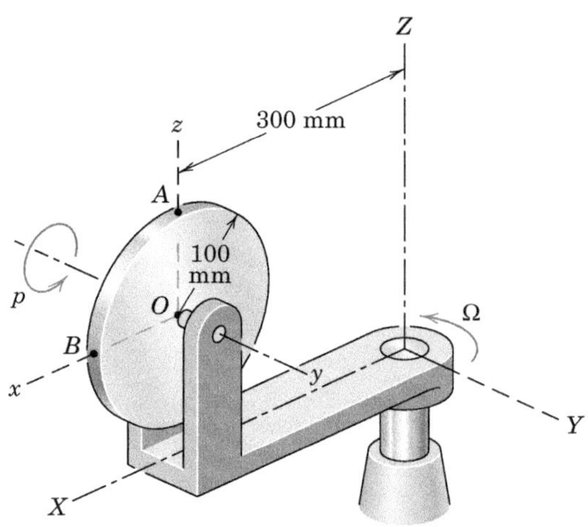

Problem 7/46

7/47 The center O of the spacecraft is moving through space with a constant velocity. During the period of motion prior to stabilization, the spacecraft has a constant rotational rate $\Omega = 0.5$ rad/sec about its z-axis. The x-y-z axes are attached to the body of the craft, and the solar panels rotate about the y-axis at the constant rate $\dot{\theta} = 0.25$ rad/sec with respect to the spacecraft. If $\boldsymbol{\omega}$ is the absolute angular velocity of the solar panels, determine $\dot{\boldsymbol{\omega}}$. Also find the acceleration of point A when $\theta = 30°$.

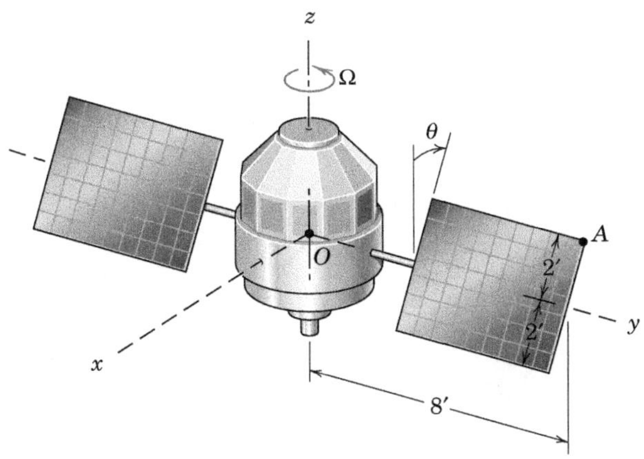

Problem 7/47

7/48 The thin circular disk of mass m and radius r is rotating about its z-axis with a constant angular velocity p, and the yoke in which it is mounted rotates about the x-axis through OB with a constant angular velocity ω_1. Simultaneously, the entire assembly rotates about the fixed Y-axis through O with a constant angular velocity ω_2. Determine the velocity $\mathbf{v}$ and acceleration $\mathbf{a}$ of point A on the rim of the disk as it passes the position shown where the x-y plane of the disk coincides with the X-Y plane. The x-y-z axes are attached to the yoke.

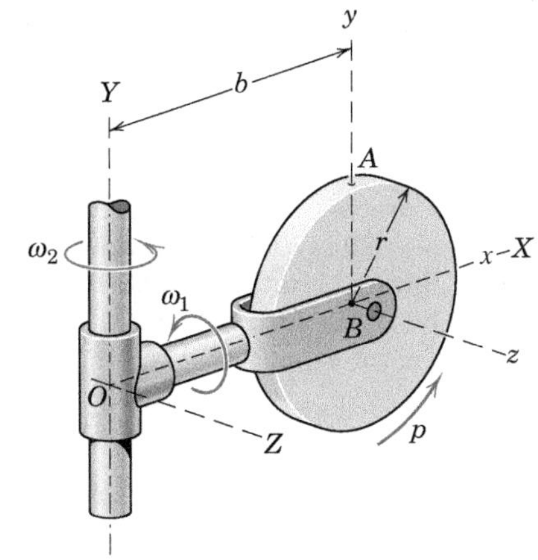

Problem 7/48

▶7/49 For the conditions specified with Sample Problem 7/2, except that γ is increasing at the steady rate of 3π rad/sec, determine the angular velocity $\boldsymbol{\omega}$ and the angular acceleration $\boldsymbol{\alpha}$ of the rotor when the position $\gamma = 30°$ is passed. (*Suggestion:* Apply Eq. 7/7 to the vector $\boldsymbol{\omega}$ to find $\boldsymbol{\alpha}$. Note that $\boldsymbol{\Omega}$ in Sample Problem 7/2 is no longer the complete angular velocity of the axes.)

▶7/50 The wheel of radius r is free to rotate about the bent axle CO which turns about the vertical axis at the constant rate p rad/s. If the wheel rolls without slipping on the horizontal circle of radius R, determine the expressions for the angular velocity $\boldsymbol{\omega}$ and angular acceleration $\boldsymbol{\alpha}$ of the wheel. The x-axis is always horizontal.

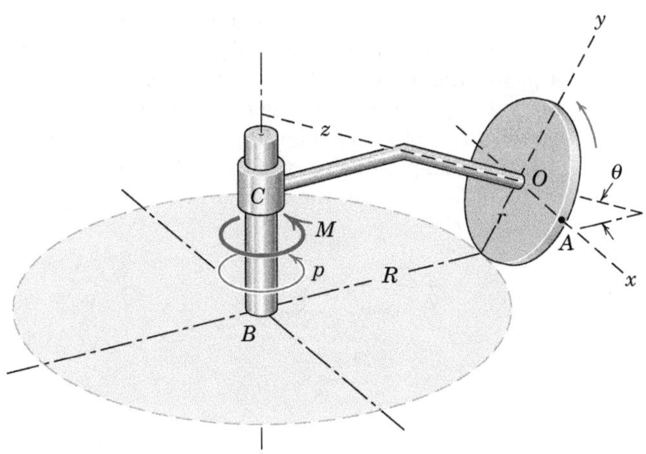

Problem 7/50

▶7/51 The gyro rotor shown is spinning at the constant rate of 100 rev/min relative to the x-y-z axes in the direction indicated. If the angle γ between the gimbal ring and the horizontal X-Y plane is made to increase at the constant rate of 4 rad/s and if the unit is forced to precess about the vertical at the constant rate $N = 20$ rev/min, calculate the magnitude of the angular acceleration $\boldsymbol{\alpha}$ of the rotor when $\gamma = 30°$. Solve by using Eq. 7/7 applied to the angular velocity of the rotor.

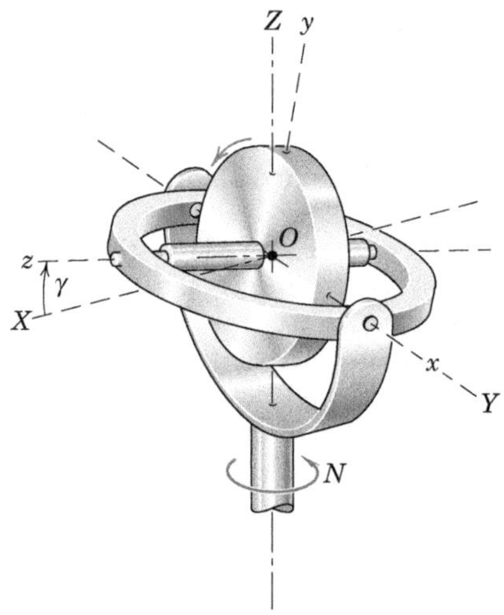

Problem 7/51

▶7/52 For a short interval of motion, collar A moves along its fixed shaft with a velocity $v_A = 2$ m/s in the Y-direction. Collar B, in turn, slides along its fixed vertical shaft. Link AB is 700 mm in length and can turn within the clevis at A to allow for the angular change between the clevises. For the instant when A passes the position where $y = 200$ mm, determine the velocity of collar B using nonrotating axes attached to B and find the component $\boldsymbol{\omega}_n$, normal to AB, of the angular velocity of the link. Also solve for $\mathbf{v}_B$ by differentiating the appropriate relation $x^2 + y^2 + z^2 = l^2$.

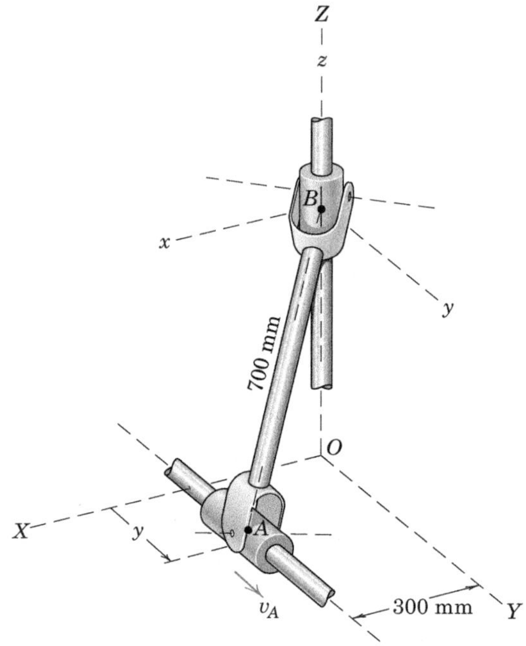

Problem 7/52

SECTION B KINETICS

7/7 Angular Momentum

The force equation for a mass system, rigid or nonrigid, Eq. 4/1 or 4/6, is the generalization of Newton's second law for the motion of a particle and should require no further explanation. The moment equation for three-dimensional motion, however, is not nearly as simple as the third of Eqs. 6/1 for plane motion since the change of angular momentum has a number of additional components which are absent in plane motion.

We now consider a rigid body moving with any general motion in space, Fig. 7/12*a*. Axes *x-y-z* are *attached to the body* with origin at the mass center *G*. Thus, the angular velocity $\boldsymbol{\omega}$ of the body becomes the angular velocity of the *x-y-z* axes as observed from the fixed reference axes *X-Y-Z*. The absolute angular momentum $\mathbf{H}_G$ of the body about its mass center *G* is the sum of the moments about *G* of the linear momenta of all elements of the body and was expressed in Art. 4/4 as $\mathbf{H}_G = \Sigma(\boldsymbol{\rho}_i \times m_i\mathbf{v}_i)$, where $\mathbf{v}_i$ is the absolute velocity of the mass element m_i.

But for the rigid body, $\mathbf{v}_i = \overline{\mathbf{v}} + \boldsymbol{\omega} \times \boldsymbol{\rho}_i$, where $\boldsymbol{\omega} \times \boldsymbol{\rho}_i$ is the relative velocity of m_i with respect to *G* as seen from nonrotating axes. Thus, we may write

$$\mathbf{H}_G = -\overline{\mathbf{v}} \times \Sigma m_i\boldsymbol{\rho}_i + \Sigma[\boldsymbol{\rho}_i \times m_i(\boldsymbol{\omega} \times \boldsymbol{\rho}_i)]$$

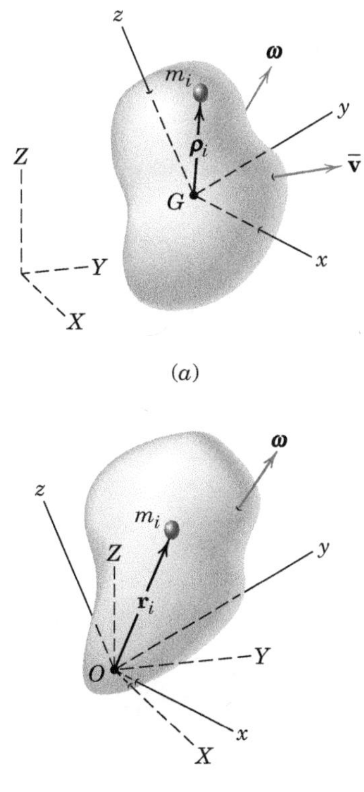

Figure 7/12

where we have factored out $\overline{\mathbf{v}}$ from the first summation terms by reversing the order of the cross product and changing the sign. With the origin at the mass center *G*, the first term in $\mathbf{H}_G$ is zero since $\Sigma m_i\boldsymbol{\rho}_i = m\overline{\boldsymbol{\rho}} = \mathbf{0}$. The second term with the substitution of dm for m_i and $\boldsymbol{\rho}$ for $\boldsymbol{\rho}_i$ gives

$$\mathbf{H}_G = \int [\boldsymbol{\rho} \times (\boldsymbol{\omega} \times \boldsymbol{\rho})]\, dm \qquad \textbf{(7/8)}$$

Before expanding the integrand of Eq. 7/8, we consider also the case of a rigid body rotating about a fixed point *O*, Fig. 7/12*b*. The *x-y-z* axes are attached to the body, and both body and axes have an angular velocity $\boldsymbol{\omega}$. The angular momentum about *O* was expressed in Art. 4/4 and is $\mathbf{H}_O = \Sigma(\mathbf{r}_i \times m_i\mathbf{v}_i)$, where, for the rigid body, $\mathbf{v}_i = \boldsymbol{\omega} \times \mathbf{r}_i$. Thus, with the substitution of dm for m_i and $\mathbf{r}$ for $\mathbf{r}_i$, the angular momentum is

$$\mathbf{H}_O = \int [\mathbf{r} \times (\boldsymbol{\omega} \times \mathbf{r})]\, dm \qquad \textbf{(7/9)}$$

Moments and Products of Inertia

We observe now that for the two cases of Figs. 7/12*a* and 7/12*b*, the position vectors $\boldsymbol{\rho}_i$ and $\mathbf{r}_i$ are given by the same expression $x\mathbf{i} + y\mathbf{j} + z\mathbf{k}$. Thus, Eqs. 7/8 and 7/9 are identical in form, and the symbol $\mathbf{H}$ will be used here for either case. We now carry out the expansion of the integrand in the two expressions for angular momentum, recognizing that the components of $\boldsymbol{\omega}$ are invariant with respect to the integrals over the body and thus become constant multipliers of the integrals. The cross-product expansion applied to the triple vector product gives, upon collection of terms,

$$
\begin{aligned}
d\mathbf{H} = \mathbf{i}[(y^2 + z^2)\omega_x \quad & -xy\omega_y \quad -xz\omega_z]\, dm \\
+\mathbf{j}[\quad -yx\omega_x + (z^2 + x^2)\omega_y \quad & -yz\omega_z]\, dm \\
+\mathbf{k}[\quad -zx\omega_x \quad -zy\omega_y + & (x^2 + y^2)\omega_z]\, dm
\end{aligned}
$$

Now let

$$
\begin{aligned}
I_{xx} &= \int (y^2 + z^2)\, dm \qquad & I_{xy} &= \int xy\, dm \\
I_{yy} &= \int (z^2 + x^2)\, dm \qquad & I_{xz} &= \int xz\, dm \\
I_{zz} &= \int (x^2 + y^2)\, dm \qquad & I_{yz} &= \int yz\, dm
\end{aligned}
\tag{7/10}
$$

The quantities I_{xx}, I_{yy}, I_{zz} are called the *moments of inertia* of the body about the respective axes, and I_{xy}, I_{xz}, I_{yz} are the *products of inertia* with respect to the coordinate axes. These quantities describe the manner in which the mass of a rigid body is distributed with respect to the chosen axes. The calculation of moments and products of inertia is explained fully in Appendix B. The double subscripts for the moments and products of inertia preserve a symmetry of notation which has special meaning in their description by tensor notation.*

Observe that $I_{xy} = I_{yx}$, $I_{xz} = I_{zx}$, and $I_{yz} = I_{zy}$. With the substitutions of Eqs. 7/10, the expression for **H** becomes

$$
\begin{aligned}
\mathbf{H} = &\;(\;\; I_{xx}\omega_x - I_{xy}\omega_y - I_{xz}\omega_z)\mathbf{i} \\
&+(-I_{yx}\omega_x + I_{yy}\omega_y - I_{yz}\omega_z)\mathbf{j} \\
&+(-I_{zx}\omega_x - I_{zy}\omega_y + I_{zz}\omega_z)\mathbf{k}
\end{aligned}
\tag{7/11}
$$

and the components of **H** are clearly

$$
\begin{aligned}
H_x &= \;\; I_{xx}\omega_x - I_{xy}\omega_y - I_{xz}\omega_z \\
H_y &= -I_{yx}\omega_x + I_{yy}\omega_y - I_{yz}\omega_z \\
H_z &= -I_{zx}\omega_x - I_{zy}\omega_y + I_{zz}\omega_z
\end{aligned}
\tag{7/12}
$$

Equation 7/11 is the general expression for the angular momentum either about the mass center G or about a fixed point O for a rigid body rotating with an instantaneous angular velocity $\boldsymbol{\omega}$.

Remember that in each of the two cases represented, the reference axes x-y-z are *attached* to the rigid body. This attachment makes the

*See, for example, the first author's *Dynamics, 2nd Edition, SI Version*, 1975, John Wiley & Sons, Art. 41.

moment-of-inertia integrals and the product-of-inertia integrals of Eqs. 7/10 invariant with time. If the x-y-z axes were to rotate with respect to an irregular body, then these inertia integrals would be functions of the time, which would introduce an undesirable complexity into the angular-momentum relations. An important exception occurs when a rigid body is spinning about an axis of symmetry, in which case, the inertia integrals are not affected by the angular position of the body about its spin axis. Thus, for a body rotating about an axis of symmetry, it is frequently convenient to choose one axis of the reference system to coincide with the axis of rotation and allow the other two axes not to turn with the body. In addition to the momentum components due to the angular velocity $\mathbf{\Omega}$ of the reference axes, then, an added angular-momentum component along the spin axis due to the relative spin about the axis would have to be accounted for.

Principal Axes

The array of moments and products of inertia

$$\begin{bmatrix} I_{xx} & -I_{xy} & -I_{xz} \\ -I_{yx} & I_{yy} & -I_{yz} \\ -I_{zx} & -I_{zy} & I_{zz} \end{bmatrix}$$

which appear in Eq. 7/12 is called the *inertia matrix* or *inertia tensor*. As we change the orientation of the axes relative to the body, the moments and products of inertia will also change in value. It can be shown* that there is one unique orientation of axes x-y-z for a given origin for which the products of inertia vanish and the moments of inertia I_{xx}, I_{yy}, I_{zz} take on stationary values. For this orientation, the inertia matrix takes the form

$$\begin{bmatrix} I_{xx} & 0 & 0 \\ 0 & I_{yy} & 0 \\ 0 & 0 & I_{zz} \end{bmatrix}$$

and is said to be diagonalized. The axes x-y-z for which the products of inertia vanish are called the *principal axes of inertia*, and I_{xx}, I_{yy}, and I_{zz} are called the *principal moments of inertia*. The principal moments of inertia for a given origin represent the maximum, the minimum, and an intermediate value of the moments of inertia.

If the coordinate axes coincide with the principal axes of inertia, Eq. 7/11 for the angular momentum about the mass center or about a fixed point becomes

$$\boxed{\mathbf{H} = I_{xx}\omega_x\mathbf{i} + I_{yy}\omega_y\mathbf{j} + I_{zz}\omega_z\mathbf{k}} \qquad \textbf{(7/13)}$$

It is always possible to locate the principal axes of inertia for a general three-dimensional rigid body. Thus, we can express its angular momentum by Eq. 7/13, although it may not always be convenient to do so

*See, for example, the first author's *Dynamics, 2nd Edition, SI Version*, 1975, John Wiley & Sons, Art. 41.

for geometric reasons. Except when the body rotates about one of the principal axes of inertia or when $I_{xx} = I_{yy} = I_{zz}$, the vectors $\mathbf{H}$ and $\boldsymbol{\omega}$ have different directions.

Transfer Principle for Angular Momentum

The momentum properties of a rigid body may be represented by the resultant linear-momentum vector $\mathbf{G} = m\overline{\mathbf{v}}$ through the mass center and the resultant angular-momentum vector $\mathbf{H}_G$ about the mass center, as shown in Fig. 7/13. Although $\mathbf{H}_G$ has the properties of a free vector, we represent it through G for convenience.

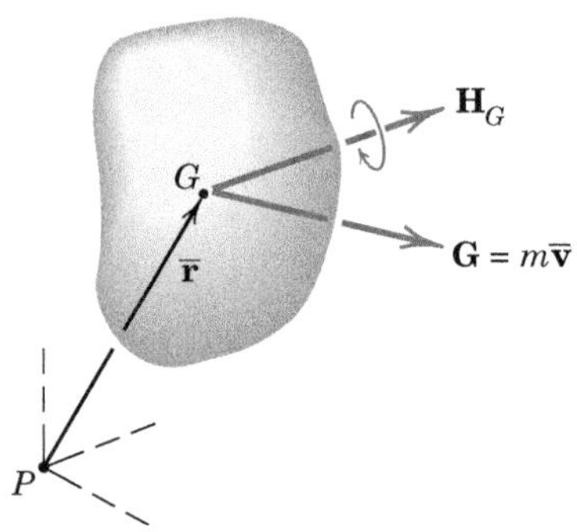

Figure 7/13

These vectors have properties analogous to those of a force and a couple. Thus, the angular momentum about any point P equals the free vector $\mathbf{H}_G$ plus the moment of the linear-momentum vector $\mathbf{G}$ about P. Therefore, we may write

$$\boxed{\mathbf{H}_P = \mathbf{H}_G + \overline{\mathbf{r}} \times \mathbf{G}} \qquad \textbf{(7/14)}$$

This relation, which was derived previously in Chapter 4 as Eq. 4/10, also applies to a fixed point O on the body or body extended, where O merely replaces P. Equation 7/14 constitutes a transfer theorem for angular momentum.

7/8 Kinetic Energy

In Art. 4/3 on the dynamics of systems of particles, we developed the expression for the kinetic energy T of any general system of mass, rigid or nonrigid, and obtained the result

$$T = \frac{1}{2}m\bar{v}^2 + \Sigma\frac{1}{2}m_i|\dot{\boldsymbol{\rho}}_i|^2 \qquad [4/4]$$

where $\bar{v}$ is the velocity of the mass center and $\boldsymbol{\rho}_i$ is the position vector of a representative element of mass m_i with respect to the mass center. We identified the first term as the kinetic energy due to the translation of the system and the second term as the kinetic energy associated with the motion relative to the mass center. The translational term may be written alternatively as

$$\frac{1}{2}m\bar{v}^2 = \frac{1}{2}m\dot{\overline{\mathbf{r}}}\cdot\dot{\overline{\mathbf{r}}} = \frac{1}{2}\overline{\mathbf{v}}\cdot\mathbf{G}$$

where $\dot{\overline{\mathbf{r}}}$ is the velocity $\overline{\mathbf{v}}$ of the mass center and $\mathbf{G}$ is the linear momentum of the body.

For a rigid body, the relative term becomes the kinetic energy due to rotation about the mass center. Because $\dot{\boldsymbol{\rho}}_i$ is the velocity of the representative particle with respect to the mass center, then for the rigid body we may write it as $\dot{\boldsymbol{\rho}}_i = \boldsymbol{\omega} \times \boldsymbol{\rho}_i$, where $\boldsymbol{\omega}$ is the angular velocity of the body. With this substitution, the relative term in the kinetic energy expression becomes

$$\Sigma\frac{1}{2}m_i|\dot{\boldsymbol{\rho}}_i|^2 = \Sigma\frac{1}{2}m_i(\boldsymbol{\omega} \times \boldsymbol{\rho}_i)\cdot(\boldsymbol{\omega} \times \boldsymbol{\rho}_i)$$

If we use the fact that the dot and the cross may be interchanged in the triple scalar product, that is, $\mathbf{P} \times \mathbf{Q} \cdot \mathbf{R} = \mathbf{P} \cdot \mathbf{Q} \times \mathbf{R}$, we may write

$$(\boldsymbol{\omega} \times \boldsymbol{\rho}_i) \cdot (\boldsymbol{\omega} \times \boldsymbol{\rho}_i) = \boldsymbol{\omega} \cdot \boldsymbol{\rho}_i \times (\boldsymbol{\omega} \times \boldsymbol{\rho}_i)$$

Because $\boldsymbol{\omega}$ is the same factor in all terms of the summation, it may be factored out to give

$$\Sigma \frac{1}{2} m_i |\dot{\boldsymbol{\rho}}_i|^2 = \frac{1}{2} \boldsymbol{\omega} \cdot \Sigma \boldsymbol{\rho}_i \times m_i (\boldsymbol{\omega} \times \boldsymbol{\rho}_i) = \frac{1}{2} \boldsymbol{\omega} \cdot \mathbf{H}_G$$

where $\mathbf{H}_G$ is the same as the integral expressed by Eq. 7/8. Thus, the general expression for the kinetic energy of a rigid body moving with mass-center velocity $\overline{\mathbf{v}}$ and angular velocity $\boldsymbol{\omega}$ is

$$T = \frac{1}{2} \overline{\mathbf{v}} \cdot \mathbf{G} + \frac{1}{2} \boldsymbol{\omega} \cdot \mathbf{H}_G \qquad \textbf{(7/15)}$$

Expansion of this vector equation by substitution of the expression for $\mathbf{H}_G$ written from Eq. 7/11 yields

$$T = \frac{1}{2} m\bar{v}^2 + \frac{1}{2} (\bar{I}_{xx}\omega_x{}^2 + \bar{I}_{yy}\omega_y{}^2 + \bar{I}_{zz}\omega_z{}^2)$$
$$- (\bar{I}_{xy}\omega_x\omega_y + \bar{I}_{xz}\omega_x\omega_z + \bar{I}_{yz}\omega_y\omega_z) \qquad \textbf{(7/16)}$$

If the axes coincide with the principal axes of inertia, the kinetic energy is merely

$$T = \frac{1}{2} m\bar{v}^2 + \frac{1}{2} (\bar{I}_{xx}\omega_x{}^2 + \bar{I}_{yy}\omega_y{}^2 + \bar{I}_{zz}\omega_z{}^2) \qquad \textbf{(7/17)}$$

When a rigid body is pivoted about a fixed point O or when there is a point O in the body which momentarily has zero velocity, the kinetic energy is $T = \Sigma \frac{1}{2} m_i \dot{\mathbf{r}}_i \cdot \dot{\mathbf{r}}_i$. This expression reduces to

$$T = \frac{1}{2} \boldsymbol{\omega} \cdot \mathbf{H}_O \qquad \textbf{(7/18)}$$

where $\mathbf{H}_O$ is the angular momentum about O, as may be seen by replacing $\boldsymbol{\rho}_i$ in the previous derivation by $\mathbf{r}_i$, the position vector from O. Equations 7/15 and 7/18 are the three-dimensional counterparts of Eqs. 6/9 and 6/8 for plane motion.

Portions of the landing gear for a large aircraft undergo three-dimensional motion during retraction and deployment.

Sample Problem 7/6

The bent plate has a mass of 70 kg per square meter of surface area and revolves about the z-axis at the rate $\omega = 30$ rad/s. Determine (a) the angular momentum $\mathbf{H}$ of the plate about point O and (b) the kinetic energy T of the plate. Neglect the mass of the hub and the thickness of the plate compared with its surface dimensions.

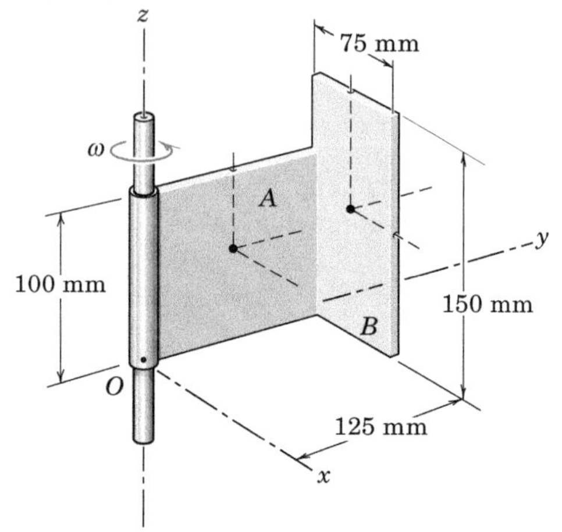

Solution. The moments and products of inertia are written with the aid of ① Eqs. B/3 and B/9 in Appendix B by transfer from the parallel centroidal axes for each part. First, the masses of the parts are $m_A = (0.100)(0.125)(70) = 0.875$ kg and $m_B = (0.075)(0.150)(70) = 0.788$ kg.

Part A

$$[I_{xx} = \bar{I}_{xx} + md^2] \qquad I_{xx} = \frac{0.875}{12}[(0.100)^2 + (0.125)^2] + 0.875[(0.050)^2 + (0.0625)^2] = 0.007\,47 \text{ kg}\cdot\text{m}^2$$

$$[I_{yy} = \tfrac{1}{3}ml^2] \qquad I_{yy} = \frac{0.875}{3}(0.100)^2 = 0.002\,92 \text{ kg}\cdot\text{m}^2$$

$$[I_{zz} = \tfrac{1}{3}ml^2] \qquad I_{zz} = \frac{0.875}{3}(0.125)^2 = 0.004\,56 \text{ kg}\cdot\text{m}^2$$

$$\left[I_{xy} = \int xy\,dm, \quad I_{xz} = \int xz\,dm\right] \qquad I_{xy} = 0 \qquad I_{xz} = 0$$

$$[I_{yz} = \bar{I}_{yz} + md_yd_z] \qquad I_{yz} = 0 + 0.875(0.0625)(0.050) = 0.002\,73 \text{ kg}\cdot\text{m}^2$$

Part B

$$[I_{xx} = \bar{I}_{xx} + md^2] \qquad I_{xx} = \frac{0.788}{12}(0.150)^2 + 0.788[(0.125)^2 + (0.075)^2] = 0.018\,21 \text{ kg}\cdot\text{m}^2$$

$$[I_{yy} = \bar{I}_{yy} + md^2] \qquad I_{yy} = \frac{0.788}{12}[(0.075)^2 + (0.150)^2] + 0.788[(0.0375)^2 + (0.075)^2] = 0.007\,38 \text{ kg}\cdot\text{m}^2$$

$$[I_{zz} = \bar{I}_{zz} + md^2] \qquad I_{zz} = \frac{0.788}{12}(0.075)^2 + 0.788[(0.125)^2 + (0.0375)^2] = 0.013\,78 \text{ kg}\cdot\text{m}^2$$

$$[I_{xy} = \bar{I}_{xy} + md_xd_y] \qquad I_{xy} = 0 + 0.788(0.0375)(0.125) = 0.003\,69 \text{ kg}\cdot\text{m}^2$$

$$[I_{xz} = \bar{I}_{xz} + md_xd_z] \qquad I_{xz} = 0 + 0.788(0.0375)(0.075) = 0.002\,21 \text{ kg}\cdot\text{m}^2$$

$$[I_{yz} = \bar{I}_{yz} + md_yd_z] \qquad I_{yz} = 0 + 0.788(0.125)(0.075) = 0.007\,38 \text{ kg}\cdot\text{m}^2$$

The sum of the respective inertia terms gives for the two plates together

$$I_{xx} = 0.0257 \text{ kg}\cdot\text{m}^2 \qquad I_{xy} = 0.003\,69 \text{ kg}\cdot\text{m}^2$$
$$I_{yy} = 0.010\,30 \text{ kg}\cdot\text{m}^2 \qquad I_{xz} = 0.002\,21 \text{ kg}\cdot\text{m}^2$$
$$I_{zz} = 0.018\,34 \text{ kg}\cdot\text{m}^2 \qquad I_{yz} = 0.010\,12 \text{ kg}\cdot\text{m}^2$$

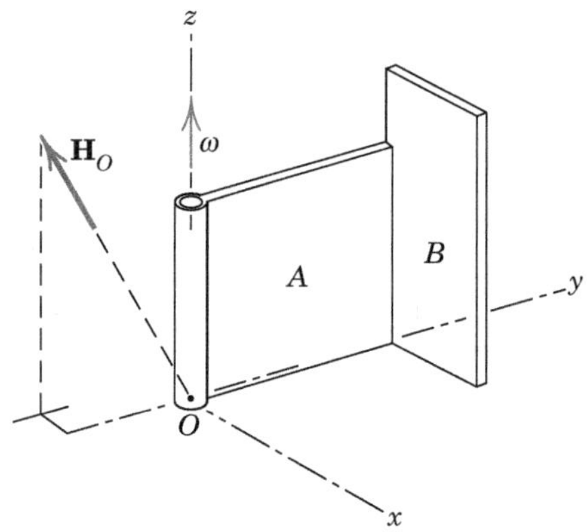

(a) The angular momentum of the body is given by Eq. 7/11, where $\omega_z = 30$ rad/s and ω_x and ω_y are zero. Thus,

② $$\mathbf{H}_O = 30(-0.002\,21\mathbf{i} - 0.010\,12\mathbf{j} + 0.018\,34\mathbf{k}) \text{ N}\cdot\text{m}\cdot\text{s} \qquad \textit{Ans.}$$

(b) The kinetic energy from Eq. 7/18 becomes

$$T = \tfrac{1}{2}\boldsymbol{\omega}\cdot\mathbf{H}_O = \tfrac{1}{2}(30\mathbf{k})\cdot 30(-0.002\,21\mathbf{i} - 0.010\,12\mathbf{j} + 0.018\,34\mathbf{k})$$
$$= 8.25 \text{ J} \qquad \textit{Ans.}$$

Helpful Hints

① The parallel-axis theorems for transferring moments and products of inertia from centroidal axes to parallel axes are explained in Appendix B and are most useful relations.

② Recall that the units of angular momentum may also be written in the base units as kg·m²/s.

PROBLEMS

Introductory Problems

7/53 The three small spheres, each of mass m, are rigidly mounted to the horizontal shaft which rotates with the angular velocity ω as shown. Neglect the radius of each sphere compared with the other dimensions and write expressions for the magnitudes of their linear momentum $\mathbf{G}$ and their angular momentum $\mathbf{H}_O$ about the origin O of the coordinates.

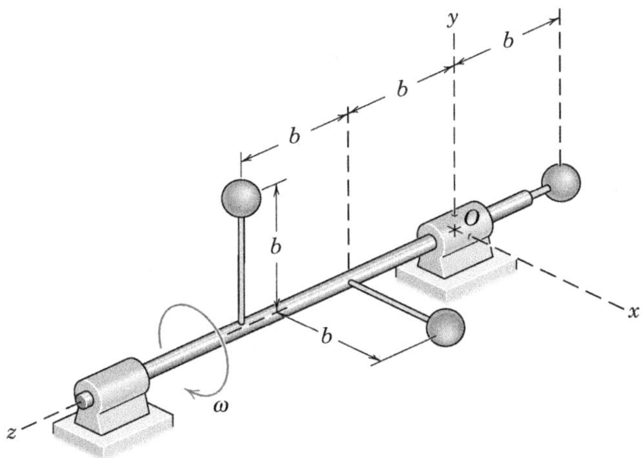

Problem 7/53

7/54 The spheres of Prob. 7/53 are replaced by three rods, each of mass m and length l, mounted at their centers to the shaft, which rotates with the angular velocity ω as shown. The axes of the rods are, respectively, in the x-, y-, and z-directions, and their diameters are negligible compared with the other dimensions. Determine the angular momentum $\mathbf{H}_O$ of the three rods with respect to the coordinate origin O.

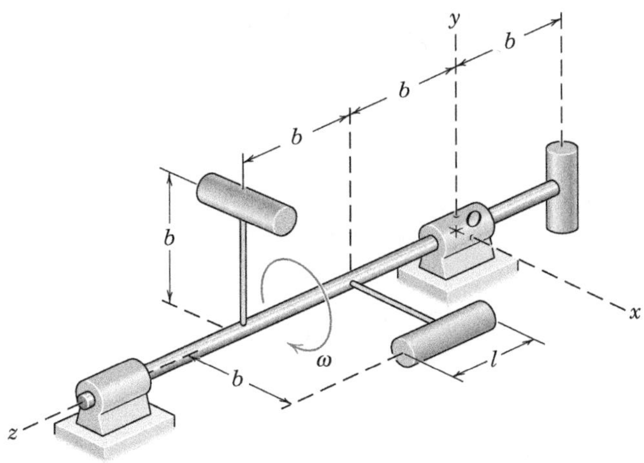

Problem 7/54

7/55 The aircraft landing gear viewed from the front is being retracted immediately after takeoff, and the wheel is spinning at the rate corresponding to the takeoff speed of 200 km/h. The 45-kg wheel has a radius of gyration about its z-axis of 370 mm. Neglect the thickness of the wheel and calculate the angular momentum of the wheel about G and about A for the position where θ is increasing at the rate of 30° per second.

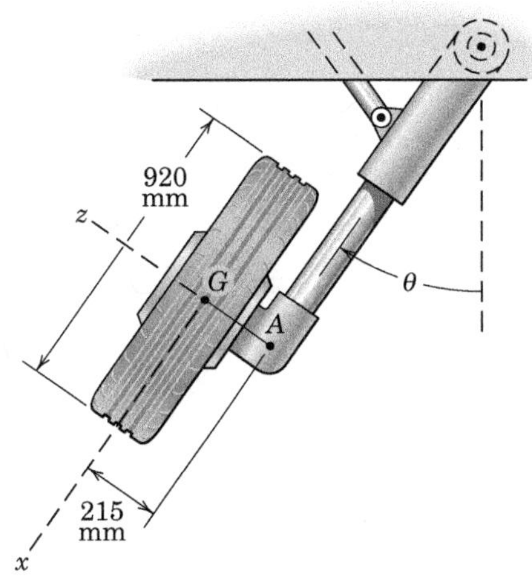

Problem 7/55

7/56 The bent rod has a mass ρ per unit length and rotates about the z-axis with an angular velocity ω. Determine the angular momentum $\mathbf{H}_O$ of the rod about the fixed origin O of the axes, which are attached to the rod. Also find the kinetic energy T of the rod.

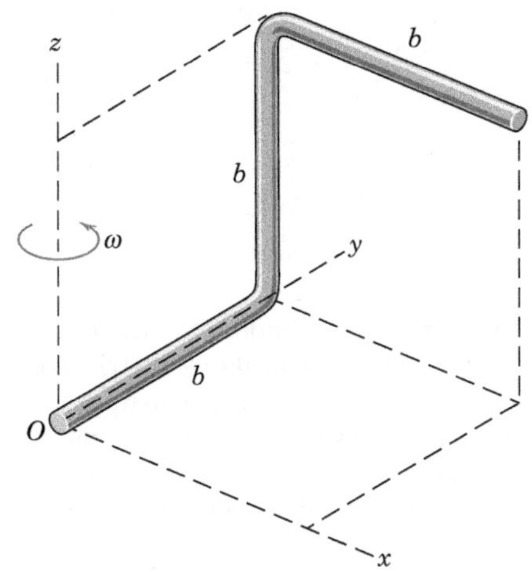

Problem 7/56

7/57 Use the results of Prob. 7/56 and determine the angular momentum $\mathbf{H}_G$ of the bent rod of that problem about its mass center G using the given reference axes.

7/58 The slender rod of mass m and length l rotates about the y-axis as the element of a right-circular cone. If the angular velocity about the y-axis is ω, determine the expression for the angular momentum of the rod with respect to the x-y-z axes for the particular position shown.

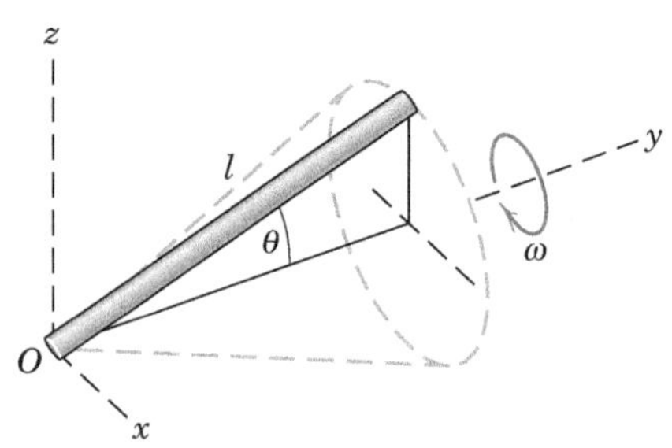

Problem 7/58

Representative Problems

7/59 The solid half-circular cylinder of mass m revolves about the z-axis with an angular velocity ω as shown. Determine its angular momentum $\mathbf{H}$ with respect to the x-y-z axes.

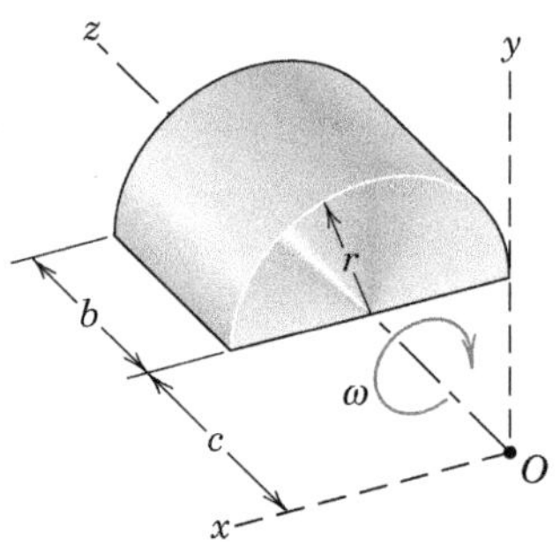

Problem 7/59

7/60 The solid circular cylinder of mass m, radius r, and length b revolves about its geometric axis at an angular rate p rad/s. Simultaneously, the bracket and attached shaft revolve about the x-axis at the rate ω rad/s. Write the expression for the angular momentun $\mathbf{H}_O$ of the cylinder about O with reference axes as shown.

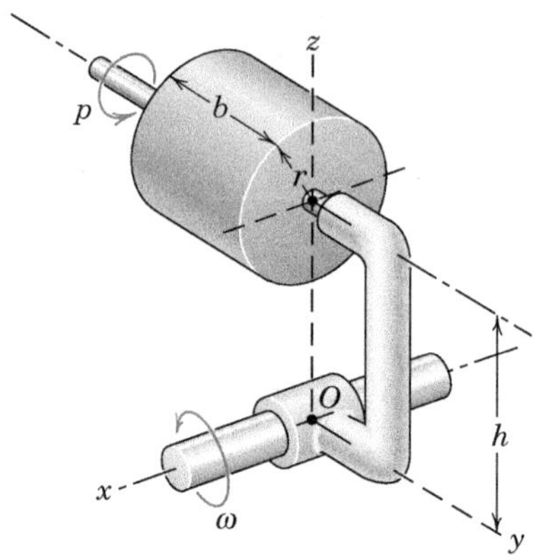

Problem 7/60

7/61 The elements of a reaction-wheel attitude-control system for a spacecraft are shown in the figure. Point G is the center of mass for the system of the spacecraft and wheels, and x, y, z are principal axes for the system. Each wheel has a mass m and a moment of inertia I about its own axis and spins with a relative angular velocity p in the direction indicated. The center of each wheel, which may be treated as a thin disk, is a distance b from G. If the spacecraft has angular velocity components Ω_x, Ω_y, and Ω_z, determine the angular momentum $\mathbf{H}_G$ of the three wheels as a unit.

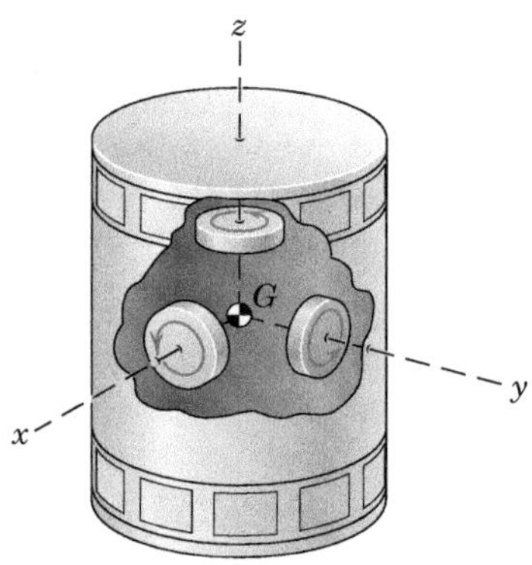

Problem 7/61

7/62 The gyro rotor is spinning at the constant rate p = 100 rev/min relative to the x-y-z axes in the direction indicated. If the angle γ between the gimbal ring and horizontal X-Y plane is made to increase at the rate of 4 rad/sec and if the unit is forced to precess about the vertical at the constant rate N = 20 rev/min, calculate the angular momentum $\mathbf{H}_O$ of the rotor when $\gamma = 30°$. The axial and transverse moments of inertia are $I_{zz} = 5(10^{-3})$ lb-ft-sec^2 and $I_{xx} = I_{yy} = 2.5(10^{-3})$ lb-ft-sec^2.

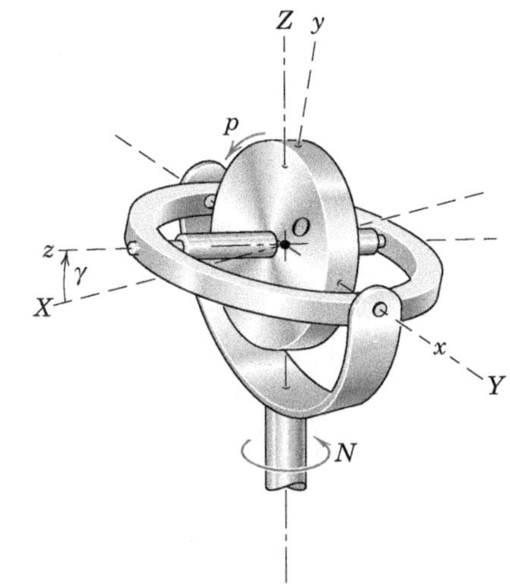

Problem 7/62

7/63 The slender steel rod AB weighs 6.20 lb and is secured to the rotating shaft by the rod OG and its fittings at O and G. The angle β remains constant at 30°, and the entire rigid assembly rotates about the z-axis at the steady rate N = 600 rev/min. Calculate the angular momentum $\mathbf{H}_O$ of AB and its kinetic energy T.

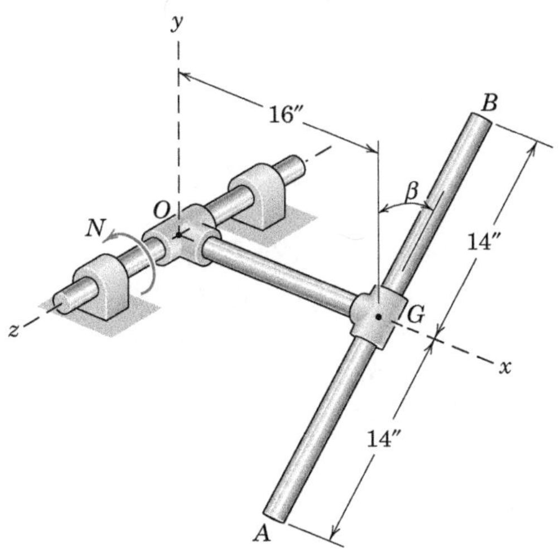

Problem 7/63

7/64 The rectangular plate, with a mass of 3 kg and a uniform small thickness, is welded at the 45° angle to the vertical shaft, which rotates with the angular velocity of 20π rad/s. Determine the angular momentum $\mathbf{H}$ of the plate about O and find the kinetic energy of the plate.

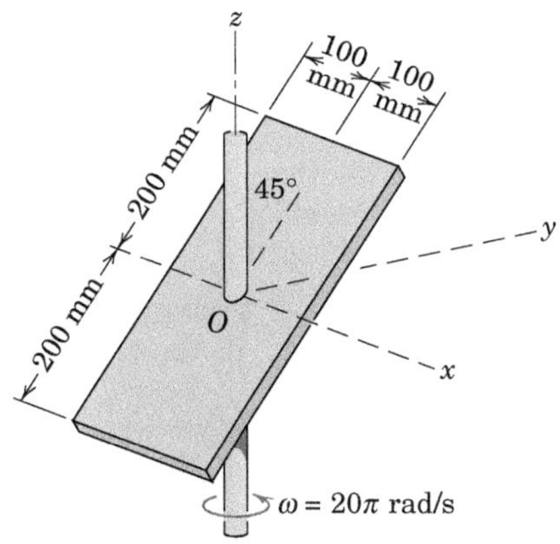

Problem 7/64

7/65 The circular disk of mass m and radius r is mounted on the vertical shaft with an angle α between its plane and the plane of rotation of the shaft. Determine an expression for the angular momentum $\mathbf{H}$ of the disk about O. Find the angle β which the angular momentum $\mathbf{H}$ makes with the shaft if $\alpha = 10°$.

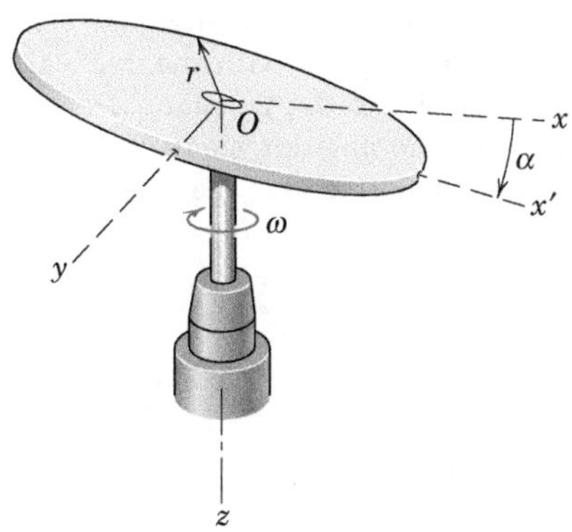

Problem 7/65

7/66 The right-circular cone of height h and base radius r spins about its axis of symmetry with an angular rate p. Simultaneously, the entire cone revolves about the x-axis with angular rate Ω. Determine the angular momentum $\mathbf{H}_O$ of the cone about the origin O of the x-y-z axes and the kinetic energy T for the position shown. The mass of the cone is m.

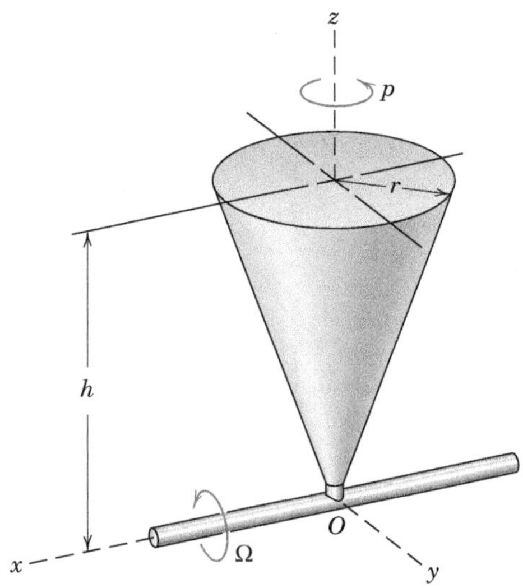

Problem 7/66

7/67 Each of the slender rods of length l and mass m is welded to the circular disk which rotates about the vertical z-axis with an angular velocity ω. Each rod makes an angle β with the vertical and lies in a plane parallel to the y-z plane. Determine an expression for the angular momentum $\mathbf{H}_O$ of the two rods about the origin O of the axes.

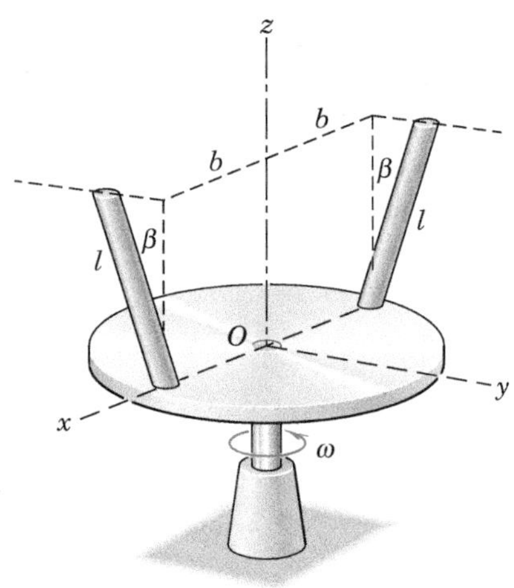

Problem 7/67

7/68 The spacecraft shown has a mass m with mass center G. Its radius of gyration about its z-axis of rotational symmetry is k and that about either the x- or y-axis is k'. In space, the spacecraft spins within its x-y-z reference frame at the rate $p = \dot{\phi}$. Simultaneously, a point C on the z-axis moves in a circle about the z_0-axis with a frequency f (rotations per unit time). The z_0-axis has a constant direction in space. Determine the angular momentum $\mathbf{H}_G$ of the spacecraft relative to the axes designated. Note that the x-axis always lies in the z-z_0 plane and that the y-axis is therefore normal to z_0.

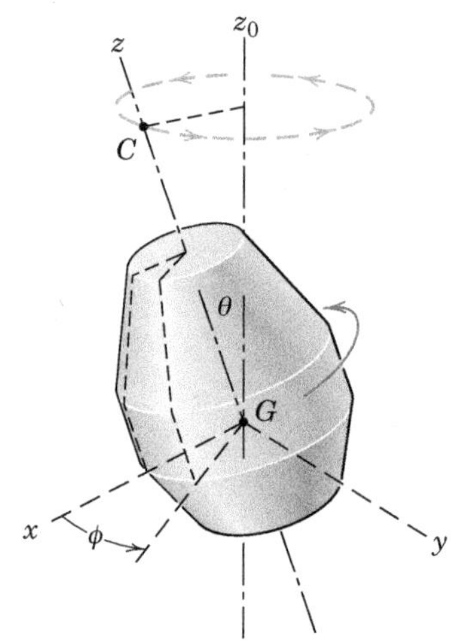

Problem 7/68

7/69 The uniform circular disk of Prob. 7/48 with the three components of angular velocity is shown again here. Determine the kinetic energy T and the angular momentum $\mathbf{H}_O$ with respect to O of the disk for the instant represented, when the x-y plane coincides with the X-Y plane. The mass of the disk is m.

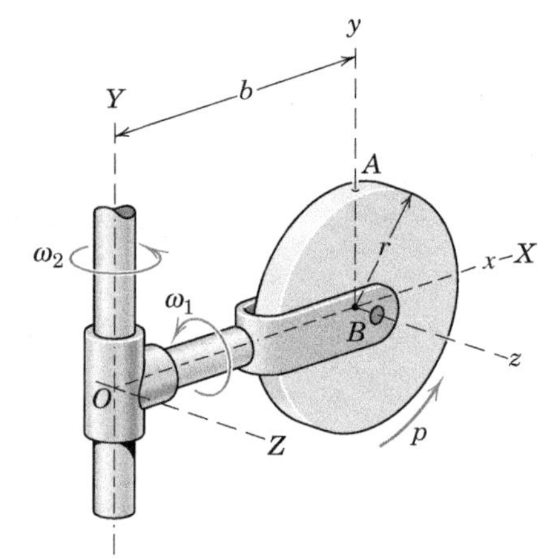

Problem 7/69

7/70 The 4-in.-radius wheel weighs 6 lb and turns about its y'-axis with an angular velocity $p = 40\pi$ rad/sec in the direction shown. Simultaneously, the fork rotates about its x-axis shaft with an angular velocity $\omega = 10\pi$ rad/sec as indicated. Calculate the angular momentum of the wheel about its center O'. Also compute the kinetic energy of the wheel.

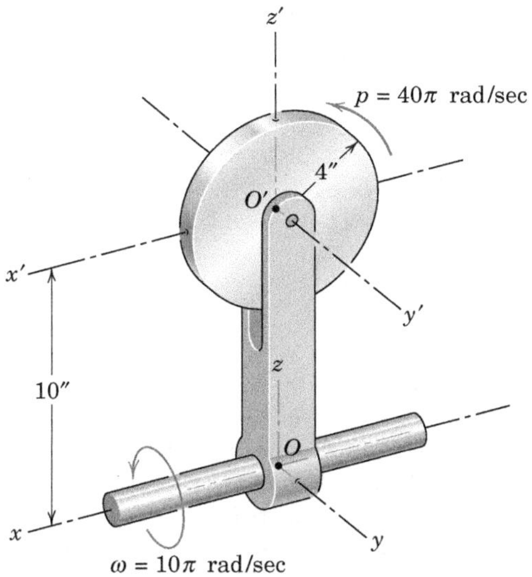

Problem 7/70

7/71 The assembly, consisting of the solid sphere of mass m and the uniform rod of length $2c$ and equal mass m, revolves about the vertical z-axis with an angular velocity ω. The rod of length $2c$ has a diameter which is small compared with its length and is perpendicular to the horizontal rod to which it is welded with the inclination β shown. Determine the combined angular momentum $\mathbf{H}_O$ of the sphere and inclined rod.

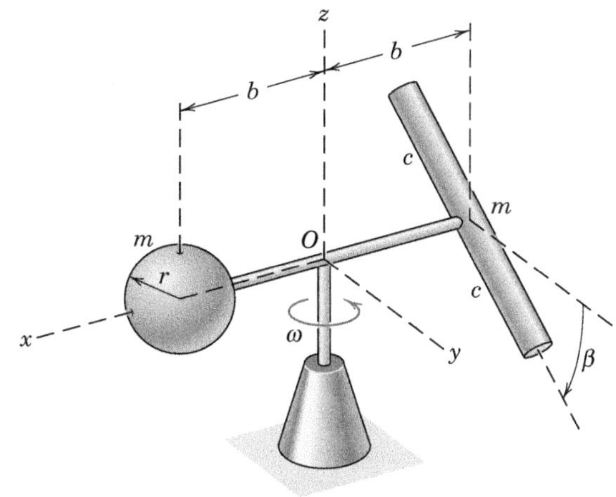

Problem 7/71

7/72 In a test of the solar panels for a spacecraft, the model shown is rotated about the vertical axis at the angular rate ω. If the mass per unit area of panel is ρ, write the expression for the angular momentum $\mathbf{H}_O$ of the assembly about the axes shown in terms of θ. Also determine the maximum, minimum, and intermediate values of the moment of inertia about the axes through O. The combined mass of both panels is m.

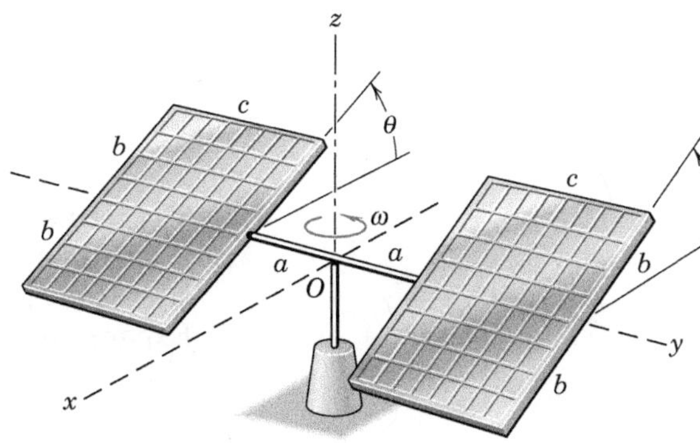

Problem 7/72

7/9 Momentum and Energy Equations of Motion

With the description of angular momentum, inertial properties, and kinetic energy of a rigid body established in the previous two articles, we are ready to apply the general momentum and energy equations of motion.

Momentum Equations

In Art. 4/4 of Chapter 4, we established the general linear- and angular-momentum equations for a system of constant mass. These equations are

$$\Sigma \mathbf{F} = \dot{\mathbf{G}} \qquad [4/6]$$

$$\Sigma \mathbf{M} = \dot{\mathbf{H}} \qquad [4/7] \text{ or } [4/9]$$

The general moment relation, Eq. 4/7 or 4/9, is expressed here by the single equation $\Sigma \mathbf{M} = \dot{\mathbf{H}}$, where the terms are taken either about a fixed point O or about the mass center G. In the derivation of the moment principle, the derivative of $\mathbf{H}$ was taken with respect to an absolute coordinate system. When $\mathbf{H}$ is expressed in terms of components measured relative to a moving coordinate system x-y-z which has an angular velocity $\mathbf{\Omega}$, then by Eq. 7/7 the moment relation becomes

$$\begin{aligned}\Sigma \mathbf{M} &= \left(\frac{d\mathbf{H}}{dt}\right)_{xyz} + \mathbf{\Omega} \times \mathbf{H} \\ &= (\dot{H}_x \mathbf{i} + \dot{H}_y \mathbf{j} + \dot{H}_z \mathbf{k}) + \mathbf{\Omega} \times \mathbf{H}\end{aligned}$$

The terms in parentheses represent that part of $\dot{\mathbf{H}}$ due to the change in magnitude of the components of $\mathbf{H}$, and the cross-product term represents that part due to the changes in direction of the components of $\mathbf{H}$. Expansion of the cross product and rearrangement of terms give

$$\begin{aligned}\Sigma \mathbf{M} = &(\dot{H}_x - H_y \Omega_z + H_z \Omega_y)\mathbf{i} \\ &+ (\dot{H}_y - H_z \Omega_x + H_x \Omega_z)\mathbf{j} \\ &+ (\dot{H}_z - H_x \Omega_y + H_y \Omega_x)\mathbf{k}\end{aligned} \qquad (7/19)$$

Equation 7/19 is the most general form of the moment equation about a fixed point O or about the mass center G. The Ω's are the angular velocity components of rotation of the reference axes, and the H-components in the case of a rigid body are as defined in Eq. 7/12, where the ω's are the components of the angular velocity of the body.

We now apply Eq. 7/19 to a rigid body where the coordinate axes are *attached to the body*. Under these conditions, when expressed in the x-y-z coordinates, the *moments and products of inertia are invariant with time*, and $\boldsymbol{\Omega} = \boldsymbol{\omega}$. Thus, for axes attached to the body, the three scalar components of Eq. 7/19 become

$$\begin{aligned}
\Sigma M_x &= \dot{H}_x - H_y\omega_z + H_z\omega_y \\
\Sigma M_y &= \dot{H}_y - H_z\omega_x + H_x\omega_z \\
\Sigma M_z &= \dot{H}_z - H_x\omega_y + H_y\omega_x
\end{aligned} \tag{7/20}$$

Equations 7/20 are the general moment equations for rigid-body motion with axes *attached to the body*. They hold with respect to axes through a fixed point O or through the mass center G.

In Art. 7/7 it was mentioned that, in general, for any origin fixed to a rigid body, there are three principal axes of inertia with respect to which the products of inertia vanish. If the reference axes coincide with the principal axes of inertia with origin at the mass center G or at a point O fixed to the body and fixed in space, the factors I_{xy}, I_{yz}, I_{xz} will be zero, and Eqs. 7/20 become

$$\begin{aligned}
\Sigma M_x &= I_{xx}\dot{\omega}_x - (I_{yy} - I_{zz})\omega_y\omega_z \\
\Sigma M_y &= I_{yy}\dot{\omega}_y - (I_{zz} - I_{xx})\omega_z\omega_x \\
\Sigma M_z &= I_{zz}\dot{\omega}_z - (I_{xx} - I_{yy})\omega_x\omega_y
\end{aligned} \tag{7/21}$$

These relations, known as *Euler's equations*,* are extremely useful in the study of rigid-body motion.

Energy Equations

The resultant of all external forces acting on a rigid body may be replaced by the resultant force $\Sigma\mathbf{F}$ acting through the mass center and a resultant couple $\Sigma\mathbf{M}_G$ acting about the mass center. Work is done by the resultant force and the resultant couple at the respective rates $\Sigma\mathbf{F}\cdot\overline{\mathbf{v}}$ and $\Sigma\mathbf{M}_G\cdot\boldsymbol{\omega}$, where $\overline{\mathbf{v}}$ is the linear velocity of the mass center and $\boldsymbol{\omega}$ is the angular velocity of the body. Integration over the time from condition 1 to condition 2 gives the total work done during the time interval. Equating the works done to the respective changes in kinetic energy as expressed in Eq. 7/15 gives

$$\int_{t_1}^{t_2} \Sigma\mathbf{F}\cdot\overline{\mathbf{v}}\,dt = \frac{1}{2}\overline{\mathbf{v}}\cdot\mathbf{G}\Big|_1^2 \qquad \int_{t_1}^{t_2} \Sigma\mathbf{M}_G\cdot\boldsymbol{\omega}\,dt = \frac{1}{2}\boldsymbol{\omega}\cdot\mathbf{H}_G\Big|_1^2 \tag{7/22}$$

These equations express the change in translational kinetic energy and the change in rotational kinetic energy, respectively, for the interval during which $\Sigma\mathbf{F}$ or $\Sigma\mathbf{M}_G$ acts, and the sum of the two expressions equals ΔT.

*Named after Leonhard Euler (1707–1783), a Swiss mathematician.

The work-energy relationship, developed in Chapter 4 for a general system of particles and given by

$$U'_{1\text{-}2} = \Delta T + \Delta V \qquad \textbf{[4/3]}$$

was used in Chapter 6 for rigid bodies in plane motion. The equation is equally applicable to rigid-body motion in three dimensions. As we have seen previously, the work-energy approach is of great advantage when we analyze the initial and final end-point conditions of motion. Here the work $U'_{1\text{-}2}$ done during the interval by all active forces external to the body or system is equated to the sum of the corresponding changes in kinetic energy ΔT and potential energy ΔV. The potential-energy change is determined in the usual way, as described previously in Art. 3/7.

We will limit our application of the equations developed in this article to two problems of special interest, parallel-plane motion and gyroscopic motion, discussed in the next two articles.

7/10 Parallel-Plane Motion

When all particles of a rigid body move in planes which are parallel to a fixed plane, the body has a general form of plane motion, as described in Art. 7/4 and pictured in Fig. 7/3. Every line in such a body which is normal to the fixed plane remains parallel to itself at all times. We take the mass center G as the origin of coordinates x-y-z which are attached to the body, with the x-y plane coinciding with the plane of motion P. The components of the angular velocity of both the body and the attached axes become $\omega_x = \omega_y = 0$, $\omega_z \neq 0$. For this case, the angular-momentum components from Eq. 7/12 become

$$H_x = -I_{xz}\omega_z \qquad H_y = -I_{yz}\omega_z \qquad H_z = I_{zz}\omega_z$$

and the moment relations of Eqs. 7/20 reduce to

$$\boxed{\begin{aligned} \Sigma M_x &= -I_{xz}\dot{\omega}_z + I_{yz}\omega_z{}^2 \\ \Sigma M_y &= -I_{yz}\dot{\omega}_z - I_{xz}\omega_z{}^2 \\ \Sigma M_z &= I_{zz}\dot{\omega}_z \end{aligned}} \qquad \textbf{(7/23)}$$

We see that the third moment equation is equivalent to the second of Eqs. 6/1, where the z-axis passes through the mass center, or to Eq. 6/4 if the z-axis passes through a fixed point O.

Equations 7/23 hold for an origin of coordinates at the mass center, as shown in Fig. 7/3, or for any origin on a fixed axis of rotation. The three independent force equations of motion which also apply to parallel-plane motion are clearly

$$\Sigma F_x = m\bar{a}_x \qquad \Sigma F_y = m\bar{a}_y \qquad \Sigma F_z = 0$$

Equations 7/23 find special use in describing the effect of dynamic imbalance in rotating machinery and in rolling bodies.

Sample Problem 7/7

The two circular disks, each of mass m_1, are connected by the curved bar bent into quarter-circular arcs and welded to the disks. The bar has a mass m_2. The total mass of the assembly is $m = 2m_1 + m_2$. If the disks roll without slipping on a horizontal plane with a constant velocity v of the disk centers, determine the value of the friction force under each disk at the instant represented when the plane of the curved bar is horizontal.

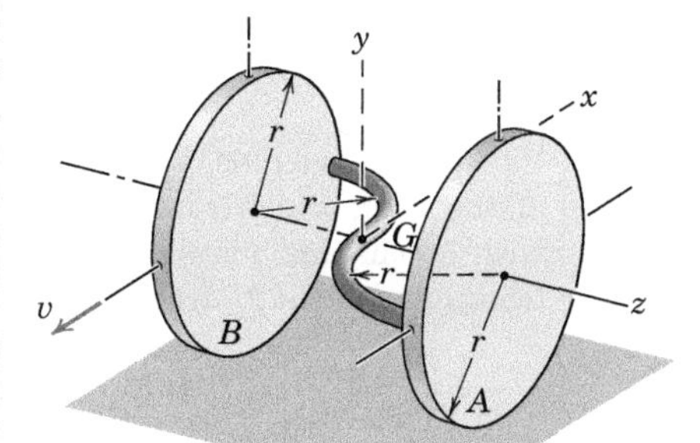

Solution. The motion is identified as parallel-plane motion since the planes of motion of all parts of the system are parallel. The free-body diagram shows the normal forces and friction forces at A and B and the total weight mg acting through the mass center G, which we take as the origin of coordinates which rotate with the body.

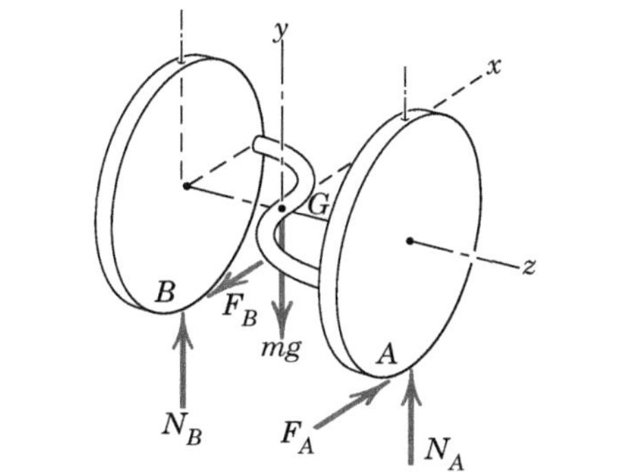

We now apply Eqs. 7/23, where $I_{yz} = 0$ and $\dot{\omega}_z = 0$. The moment equation about the y-axis requires determination of I_{xz}. From the diagram showing the geometry of the curved rod and with ρ standing for the mass of the rod per unit length, we have

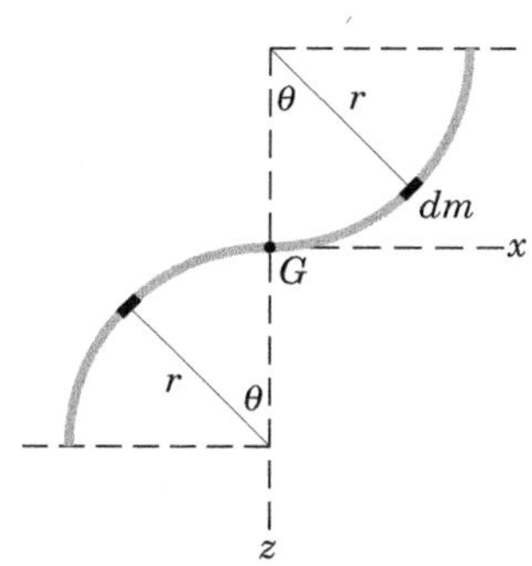

① $\left[I_{xz} = \int xz\, dm\right]$ $$I_{xz} = \int_0^{\pi/2} (r \sin \theta)(-r + r \cos \theta)\rho r\, d\theta + \int_0^{\pi/2} (-r \sin \theta)(r - r \cos \theta)\rho r\, d\theta$$

Evaluating the integrals gives

$$I_{xz} = -\rho r^3/2 - \rho r^3/2 = -\rho r^3 = -\frac{m_2 r^2}{\pi}$$

The second of Eqs. 7/23 with $\omega_z = v/r$ and $\dot{\omega}_z = 0$ gives

$[\Sigma M_y = -I_{xz}\omega_z^{\,2}]$ $$F_A r + F_B r = -\left(-\frac{m_2 r^2}{\pi}\right)\frac{v^2}{r^2}$$

$$F_A + F_B = \frac{m_2 v^2}{\pi r}$$

But with $\bar{v} = v$ constant, $\bar{a}_x = 0$ so that

$[\Sigma F_x = 0]$ $$F_A - F_B = 0 \qquad F_A = F_B$$

Thus,

$$F_A = F_B = \frac{m_2 v^2}{2\pi r}$$ *Ans.*

We also note for the given position that with $I_{yz} = 0$ and $\dot{\omega}_z = 0$, the moment equation about the x-axis gives

② $[\Sigma M_x = 0]$ $$-N_A r + N_B r = 0 \qquad N_A = N_B = mg/2$$

Helpful Hints

① We must be very careful to observe the correct sign for each of the coordinates of the mass element dm which make up the product xz.

② When the plane of the curved bar is not horizontal, the normal forces under the disks are no longer equal.

PROBLEMS

Introductory Problems

7/73 Each of the two rods of mass m is welded to the face of the disk, which rotates about the vertical axis with a constant angular velocity ω. Determine the bending moment M acting on each rod at its base.

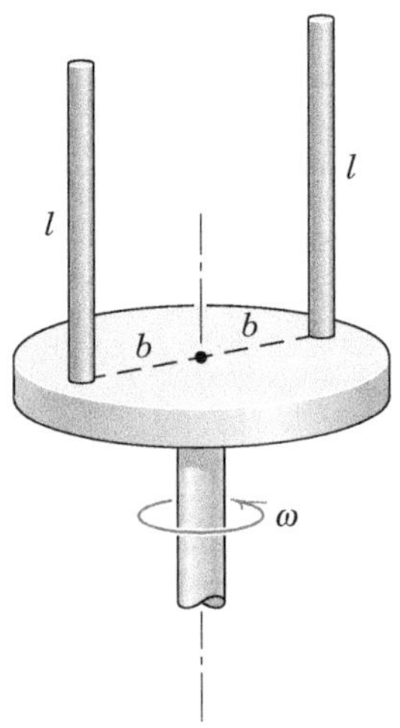

Problem 7/73

7/74 The slender shaft carries two offset particles, each of mass m, and rotates about the z-axis with the constant angular rate ω as indicated. Determine the x- and y-components of the bearing reactions at A and B due to the dynamic imbalance of the shaft for the position shown.

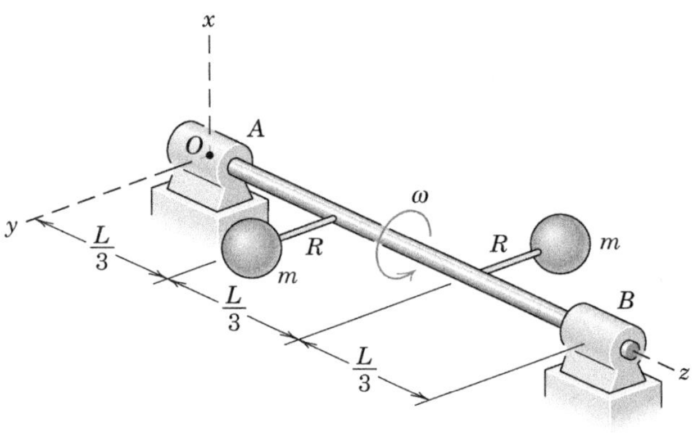

Problem 7/74

7/75 The uniform slender bar of length l and mass m is welded to the shaft, which rotates in bearings A and B with a constant angular velocity ω. Determine the expression for the force supported by the bearing at B as a function of θ. Consider only the force due to the dynamic imbalance and assume that the bearings can support radial forces only.

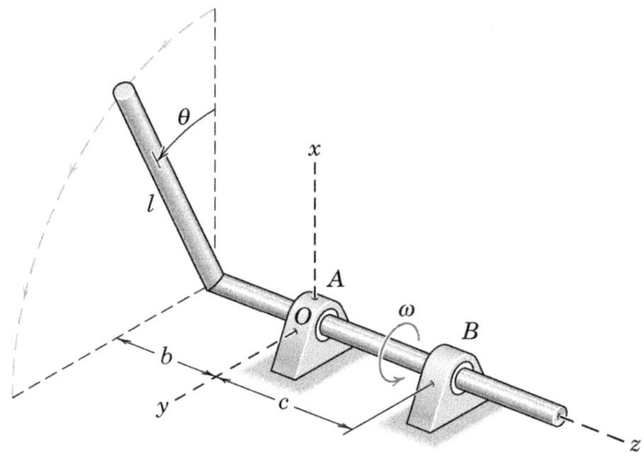

Problem 7/75

7/76 If a torque $\mathbf{M} = M\mathbf{k}$ is applied to the shaft in Prob. 7/75, determine the x- and y-components of the force supported by the bearing B as the bar and shaft start from rest in the position shown. Neglect the mass of the shaft and consider dynamic forces only.

7/77 The paint stirrer shown in the figure is made from a rod of length $7b$ and mass ρ per unit length. Before immersion in the paint, the stirrer is rotating freely at a constant high angular velocity ω about its z-axis. Determine the bending moment $\mathbf{M}$ in the rod at the base O of the chuck.

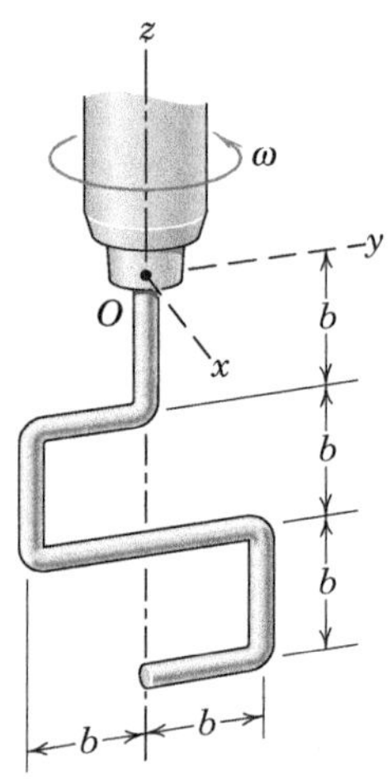

Problem 7/77

7/78 The 6-kg circular disk and attached shaft rotate at a constant speed $\omega = 10\ 000$ rev/min. If the center of mass of the disk is 0.05 mm off center, determine the magnitudes of the horizontal forces A and B supported by the bearings because of the rotational imbalance.

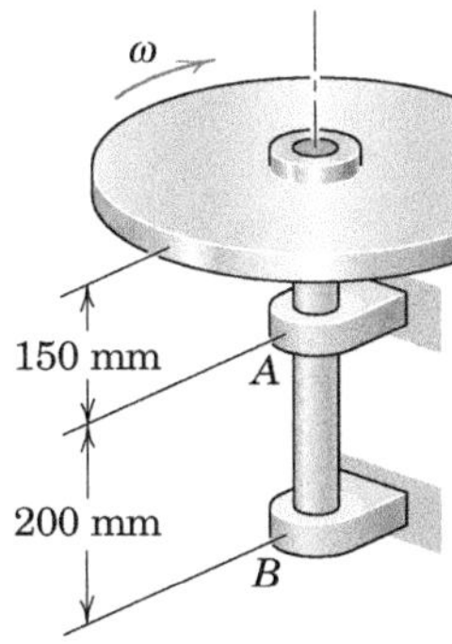

Problem 7/78

Representative Problems

7/79 Determine the bending moment $\mathbf{M}$ at the tangency point A in the semicircular rod of radius r and mass m as it rotates about the tangent axis with a constant and large angular velocity ω. Neglect the moment mgr produced by the weight of the rod.

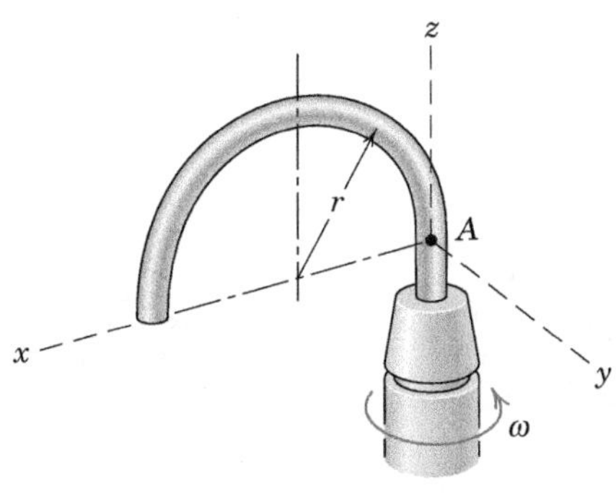

Problem 7/79

7/80 If the semicircular rod of Prob. 7/79 starts from rest under the action of a torque M_O applied through the collar about its z-axis of rotation, determine the initial bending moment $\mathbf{M}$ in the rod at A.

7/81 The large satellite-tracking antenna has a moment of inertia I about its z-axis of symmetry and a moment of inertia I_O about each of the x- and y-axes. Determine the angular acceleration α of the antenna about the vertical Z-axis caused by a torque M applied about Z by the drive mechanism for a given orientation θ.

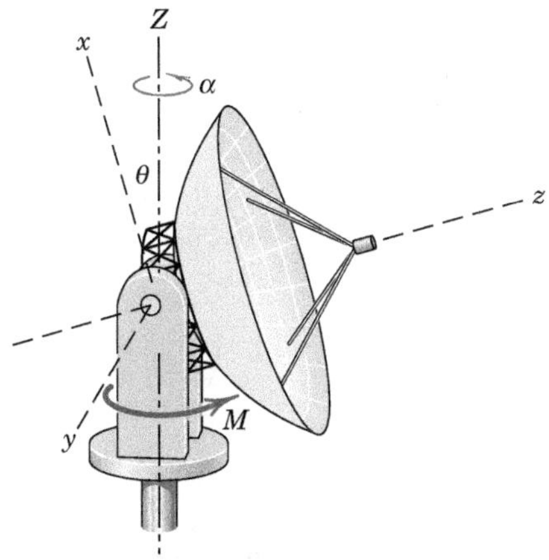

Problem 7/81

7/82 The plate has a mass of 3 kg and is welded to the fixed vertical shaft, which rotates at the constant speed of 20π rad/s. Compute the moment $\mathbf{M}$ applied *to* the shaft *by* the plate due to dynamic imbalance.

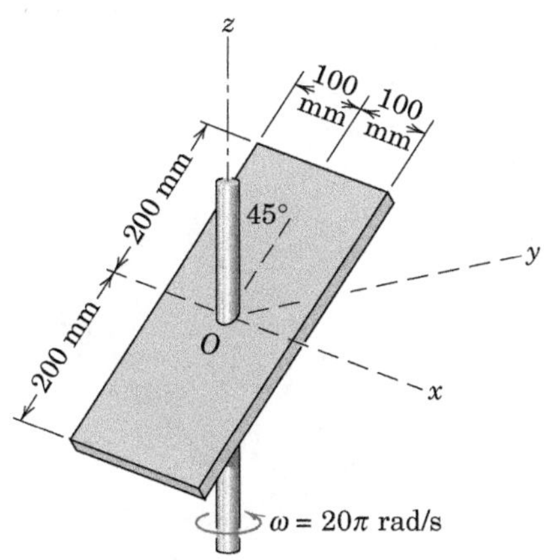

Problem 7/82

7/83 Each of the two semicircular disks has a mass of 1.20 kg and is welded to the shaft supported in bearings A and B as shown. Calculate the forces applied to the shaft by the bearings for a constant angular speed $N = 1200$ rev/min. Neglect the forces of static equilibrium.

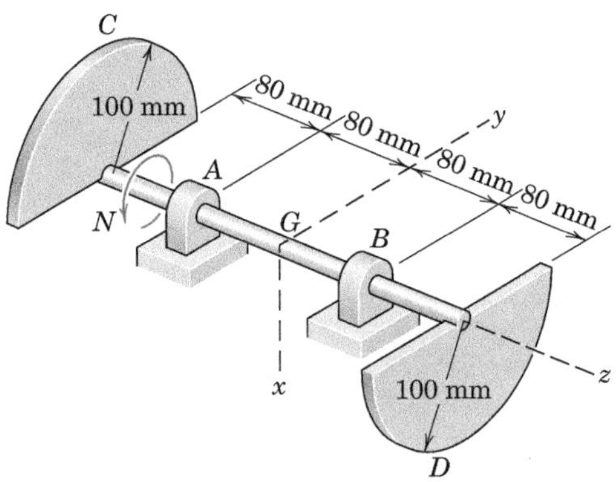

Problem 7/83

7/84 Solve Prob. 7/83 for the case where the assembly starts from rest with an initial angular acceleration $\alpha = 900$ rad/s^2 as a result of a starting torque (couple) M applied to the shaft in the same sense as N. Neglect the moment of inertia of the shaft about its z-axis and calculate M.

7/85 The uniform slender bar of mass ρ per unit length is freely pivoted about the y-axis at the clevis, which rotates about the fixed vertical z-axis with a constant angular velocity ω. Determine the steady-state angle θ assumed by the bar. Length b is greater than length c.

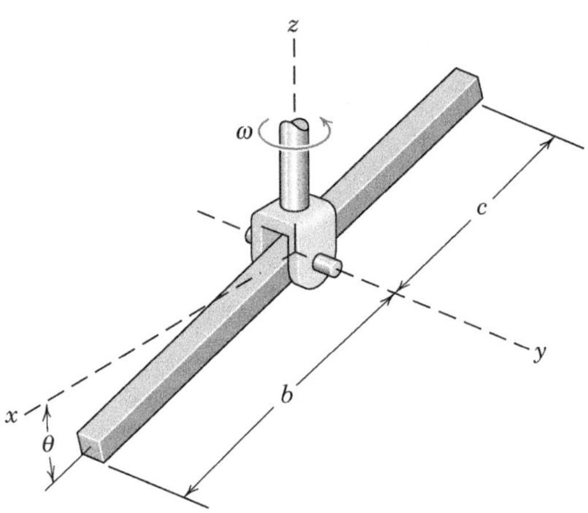

Problem 7/85

7/86 The circular disk of mass m and radius r is mounted on the vertical shaft with a small angle α between its plane and the plane of rotation of the shaft. Determine the expression for the bending moment $\mathbf{M}$ acting *on* the shaft due to the wobble of the disk at a shaft speed of ω rad/s.

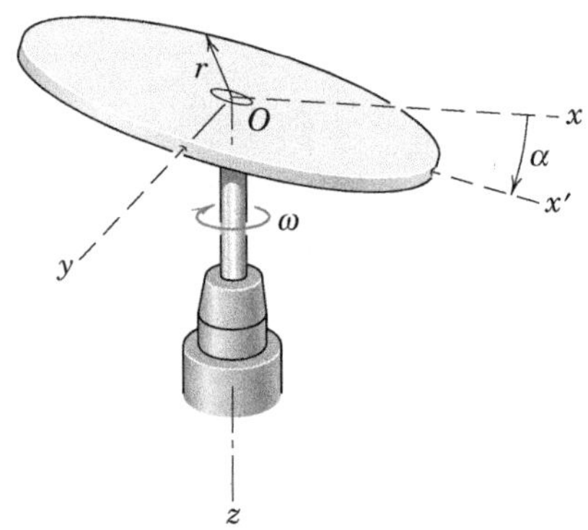

Problem 7/86

7/87 The thin circular disk of mass m and radius R is hinged about its horizontal tangent axis to the end of a shaft rotating about its vertical axis with an angular velocity ω. Determine the steady-state angle β assumed by the plane of the disk with the vertical axis. Observe any limitation on ω to ensure that $\beta > 0$.

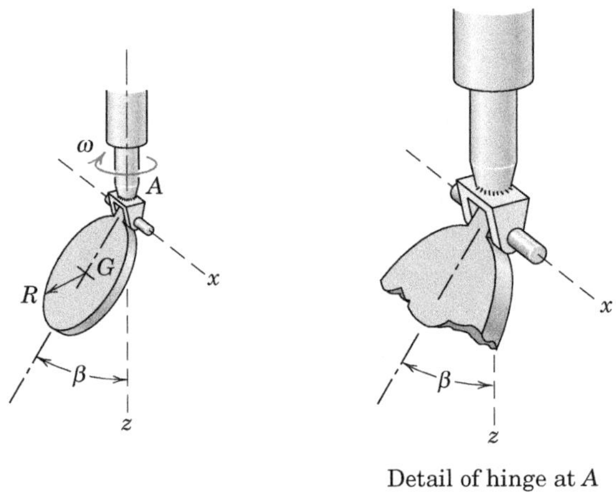

Problem 7/87

7/88 Determine the normal forces under the two disks of Sample Problem 7/7 for the position where the plane of the curved bar is vertical. Take the curved bar to be at the top of disk A and at the bottom of disk B.

7/89 The uniform square plate of mass m is welded at O to the end of the shaft, which rotates about the vertical z-axis with a constant angular velocity ω. Determine the moment applied to the plate by the weld due only to the rotation.

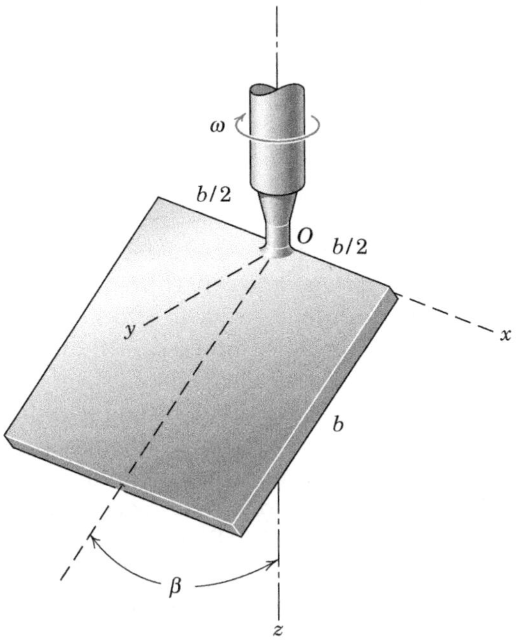

Problem 7/89

7/90 For the plate of mass m in Prob. 7/89, determine the y- and z-components of the moment applied to the plate by the weld at O necessary to give the plate an angular acceleration $\alpha = \dot{\omega}$ starting from rest. Neglect the moment due to the weight.

7/91 The uniform slender rod of length l is welded to the bracket at A on the underside of the disk B. The disk rotates about a vertical axis with a constant angular velocity ω. Determine the value of ω which will result in a zero moment supported by the weld at A for the position $\theta = 60°$ with $b = l/4$.

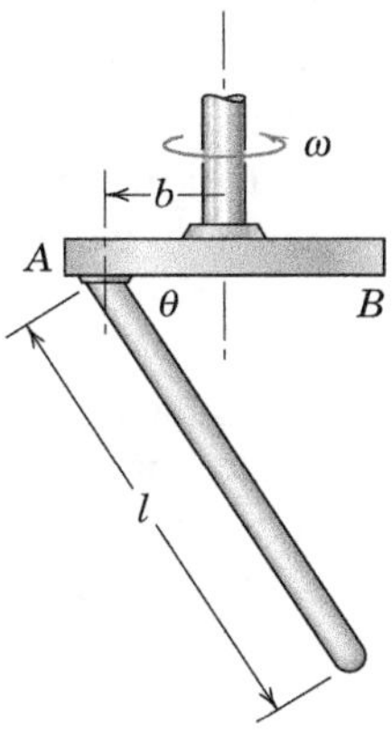

Problem 7/91

7/92 The half-cylindrical shell of radius r, length $2b$, and mass m revolves about the vertical z-axis with a constant angular velocity ω as indicated. Determine the magnitude M of the bending moment in the shaft at A due to both the weight and the rotational motion of the shell.

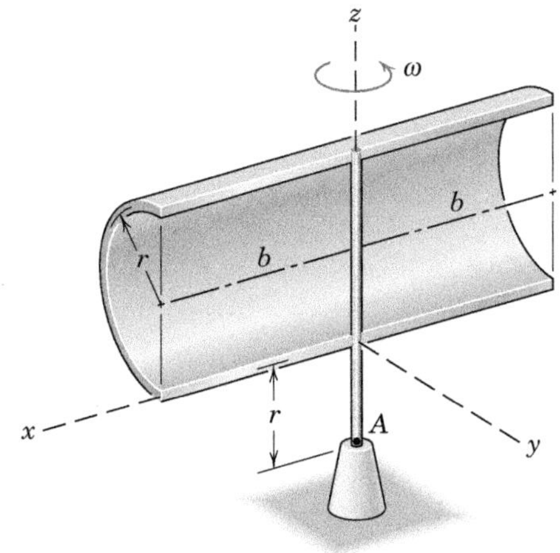

Problem 7/92

▶7/93 The homogeneous thin triangular plate of mass m is welded to the horizontal shaft, which rotates freely in the bearings at A and B. If the plate is released from rest in the horizontal position shown, determine the magnitude of the bearing reaction at A for the instant just after release.

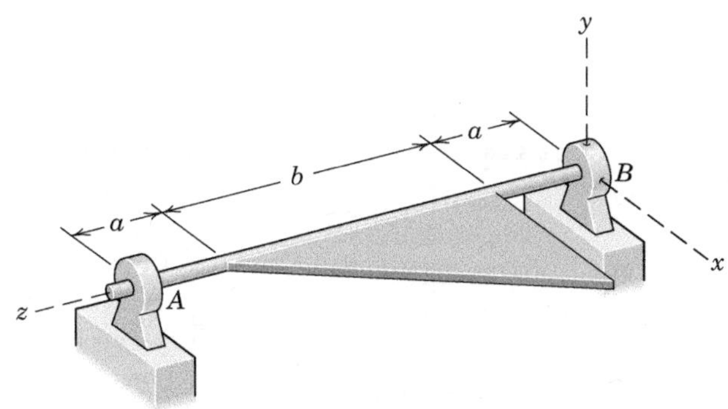

Problem 7/93

▶7/94 If the homogeneous triangular plate of Prob. 7/93 is released from rest in the position shown, determine the magnitude of the bearing reaction at A after the plate has rotated 90°.

7/11 Gyroscopic Motion: Steady Precession

One of the most interesting of all problems in dynamics is that of gyroscopic motion. This motion occurs whenever the axis about which a body is spinning is itself rotating about another axis. Although the complete description of this motion involves considerable complexity, the most common and useful examples of gyroscopic motion occur when the axis of a rotor spinning at constant speed turns (precesses) about another axis at a steady rate. Our discussion in this article will focus on this special case.

The gyroscope has important engineering applications. With a mounting in gimbal rings (see Fig. 7/19*b*), the gyro is free from external moments, and its axis will retain a fixed direction in space regardless of the rotation of the structure to which it is attached. In this way, the gyro is used for inertial guidance systems and other directional control devices. With the addition of a pendulous mass to the inner gimbal ring, the earth's rotation causes the gyro to precess so that the spin axis will always point north, and this action forms the basis of the gyro compass. The gyroscope has also found important use as a stabilizing device. The controlled precession of a large gyro mounted in a ship is used to produce a gyroscopic moment to counteract the rolling of a ship at sea. The gyroscopic effect is also an extremely important consideration in the design of bearings for the shafts of rotors which are subjected to forced precessions.

We will first describe gyroscopic action with a simple physical approach which relies on our previous experience with the vector changes encountered in particle kinetics. This approach will help us gain a direct physical insight into gyroscopic action. Next, we will make use of the general momentum relation, Eq. 7/19, for a more complete description.

Simplified Approach

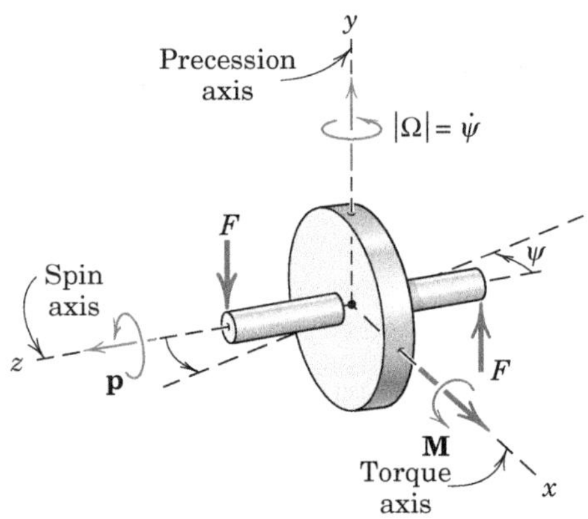

Figure 7/14

Figure 7/14 shows a symmetrical rotor spinning about the z-axis with a large angular velocity $\mathbf{p}$, known as the *spin velocity*. If we apply two forces F to the rotor axle to form a couple $\mathbf{M}$ whose vector is directed along the x-axis, we will find that the rotor shaft rotates in the x-z plane about the y-axis in the sense indicated, with a relatively slow angular velocity $\Omega = \dot{\psi}$ known as the *precession velocity*. Thus, we identify the spin axis ($\mathbf{p}$), the torque axis ($\mathbf{M}$), and the precession axis ($\mathbf{\Omega}$), where the usual right-hand rule identifies the sense of the rotation vectors. The rotor shaft does *not* turn about the x-axis in the sense of $\mathbf{M}$, as it would if the rotor were not spinning. To aid understanding of this phenomenon, a direct analogy may be made between the rotation vectors and the familiar vectors which describe the curvilinear motion of a particle.

Figure 7/15*a* shows a particle of mass m moving in the x-z plane with constant speed $|\mathbf{v}| = v$. The application of a force $\mathbf{F}$ normal to its linear momentum $\mathbf{G} = m\mathbf{v}$ causes a change $d\mathbf{G} = d(m\mathbf{v})$ in its momentum. We see that $d\mathbf{G}$, and thus $d\mathbf{v}$, is a vector in the direction of the normal force $\mathbf{F}$ according to Newton's second law $\mathbf{F} = \dot{\mathbf{G}}$, which may be written as $\mathbf{F}\,dt = d\mathbf{G}$.

From Fig. 7/15*b* we see that, in the limit, $\tan d\theta = d\theta = F\,dt/mv$ or $F = mv\dot{\theta}$. In vector notation with $\boldsymbol{\omega} = \dot{\theta}\mathbf{j}$, the force becomes

$$\mathbf{F} = m\boldsymbol{\omega} \times \mathbf{v}$$

which is the vector equivalent of our familiar scalar relation $F_n = ma_n$ for the normal force on the particle, as treated extensively in Chapter 3.

With these relations in mind, we now turn to our problem of rotation. Recall now the analogous equation $\mathbf{M} = \dot{\mathbf{H}}$ which we developed for any prescribed mass system, rigid or nonrigid, referred to its mass center (Eq. 4/9) or to a fixed point O (Eq. 4/7). We now apply this relation to our symmetrical rotor, as shown in Fig. 7/15*c*. For a high rate of spin $\mathbf{p}$ and a low precession rate $\boldsymbol{\Omega}$ about the y-axis, the angular momentum is represented by the vector $\mathbf{H} = I\mathbf{p}$, where $I = I_{zz}$ is the moment of inertia of the rotor about the spin axis.

Initially, we neglect the small component of angular momentum about the y-axis which accompanies the slow precession. The application of the couple $\mathbf{M}$ normal to $\mathbf{H}$ causes a change $d\mathbf{H} = d(I\mathbf{p})$ in the angular momentum. We see that $d\mathbf{H}$, and thus $d\mathbf{p}$, is a vector in the direction of the couple $\mathbf{M}$ since $\mathbf{M} = \dot{\mathbf{H}}$, which may also be written $\mathbf{M}\,dt = d\mathbf{H}$. Just as the change in the linear-momentum vector of the particle is in the direction of the applied force, so is the change in the angular-momentum vector of the gyro in the direction of the couple. Thus, we see that the vectors $\mathbf{M}$, $\mathbf{H}$, and $d\mathbf{H}$ are analogous to the vectors $\mathbf{F}$, $\mathbf{G}$, and $d\mathbf{G}$. With this insight, it is no longer strange to see the rotation vector undergo a change which is in the direction of $\mathbf{M}$, thereby causing the axis of the rotor to precess about the y-axis.

In Fig. 7/15*d* we see that during time dt the angular-momentum vector $I\mathbf{p}$ has swung through the angle $d\psi$, so that in the limit with $\tan d\psi = d\psi$, we have

$$d\psi = \frac{M\,dt}{Ip} \qquad \text{or} \qquad M = I\frac{d\psi}{dt}p$$

Substituting $\Omega = d\psi/dt$ for the magnitude of the precession velocity gives us

$$\boxed{M = I\Omega p} \qquad \textbf{(7/24)}$$

We note that $\mathbf{M}$, $\boldsymbol{\Omega}$, and $\mathbf{p}$ as vectors are mutually perpendicular, and that their vector relationship may be represented by writing the equation in the cross-product form

$$\boxed{\mathbf{M} = I\boldsymbol{\Omega} \times \mathbf{p}} \qquad \textbf{(7/24a)}$$

which is completely analogous to the foregoing relation $\mathbf{F} = m\boldsymbol{\omega} \times \mathbf{v}$ for the curvilinear motion of a particle as developed from Figs. 7/15*a* and *b*.

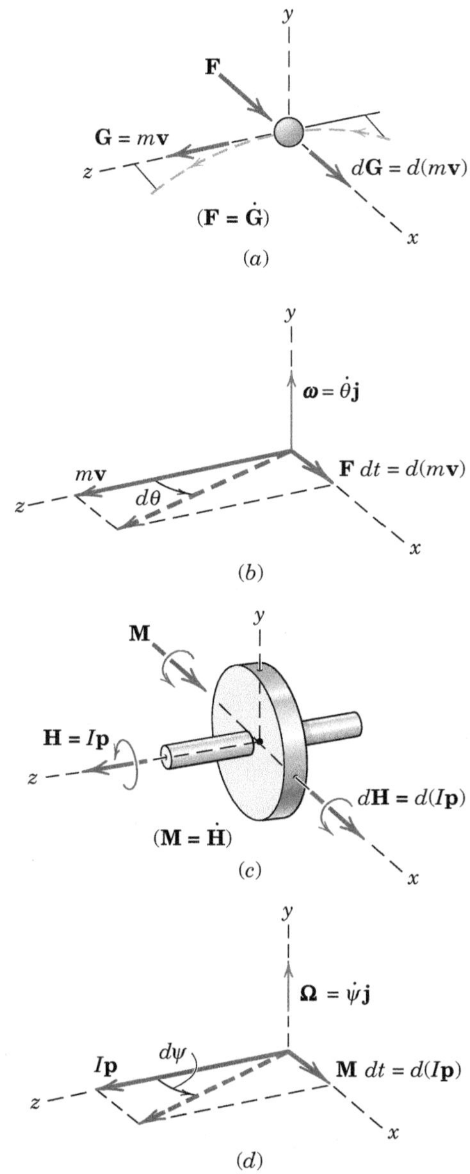

Figure 7/15

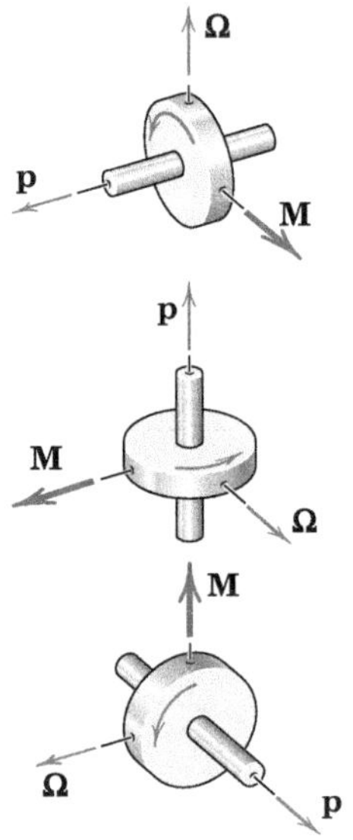

Figure 7/16

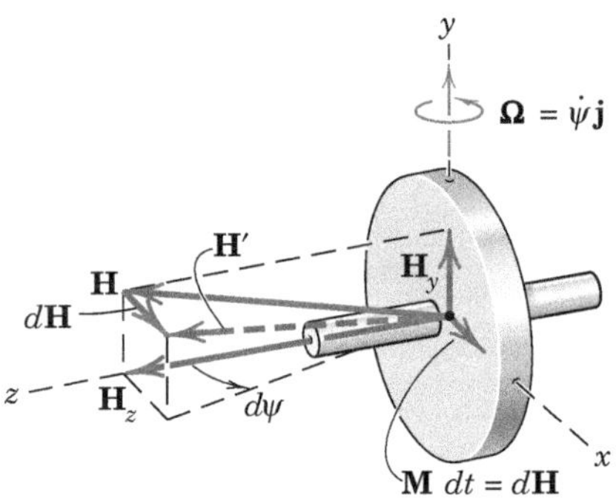

Figure 7/17

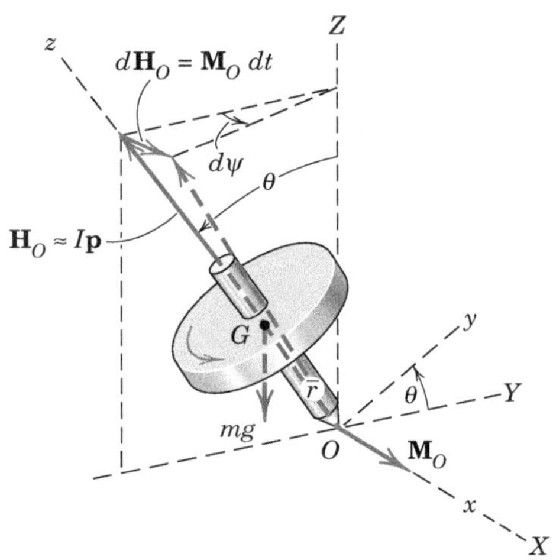

Figure 7/18

Equations 7/24 and 7/24*a* apply to moments taken about the mass center or about a fixed point on the axis of rotation.

The correct spatial relationship among the three vectors may be remembered from the fact that $d\mathbf{H}$, and thus $d\mathbf{p}$, is in the direction of $\mathbf{M}$, which establishes the correct sense for the precession $\boldsymbol{\Omega}$. Therefore, the spin vector $\mathbf{p}$ always tends to rotate toward the torque vector $\mathbf{M}$. Figure 7/16 represents three orientations of the three vectors which are consistent with their correct order. Unless we establish this order correctly in a given problem, we are likely to arrive at a conclusion directly opposite to the correct one. Remember that Eq. 7/24, like $\mathbf{F} = m\mathbf{a}$ and $M = I\alpha$, is an equation of motion, so that the couple $\mathbf{M}$ represents the couple due to *all* forces acting *on* the rotor, as disclosed by a correct *free-body diagram of the rotor*. Also note that, when a rotor is forced to precess, as occurs with the turbine in a ship which is executing a turn, the motion will generate a *gyroscopic couple* $\mathbf{M}$ which obeys Eq. 7/24*a* in both magnitude and sense.

In the foregoing discussion of gyroscopic motion, it was assumed that the spin was large and the precession was small. Although we can see from Eq. 7/24 that for given values of I and M, the precession Ω must be small if p is large, let us now examine the influence of Ω on the momentum relations. Again, we restrict our attention to steady precession, where $\boldsymbol{\Omega}$ has a constant magnitude.

Figure 7/17 shows our same rotor again. Because it has a moment of inertia about the y-axis and an angular velocity of precession about this axis, there will be an additional component of angular momentum about the y-axis. Thus, we have the two components $H_z = Ip$ and $H_y = I_0\Omega$, where I_0 stands for I_{yy} and, again, I stands for I_{zz}. The total angular momentum is $\mathbf{H}$ as shown. The change in $\mathbf{H}$ remains $d\mathbf{H} = \mathbf{M}\,dt$ as previously, and the precession during time dt is the angle $d\psi = M\,dt/H_z = M\,dt/(Ip)$ as before. Thus, Eq. 7/24 is still valid and for steady precession is an exact description of the motion as long as the spin axis is perpendicular to the axis around which precession occurs.

Consider now the steady precession of a symmetrical top, Fig. 7/18, spinning about its axis with a high angular velocity p and supported at its point O. Here the spin axis makes an angle θ with the vertical Z-axis around which precession occurs. Again, we will neglect the small angular-momentum component due to the precession and consider $\mathbf{H}$ equal to $I\mathbf{p}$, the angular momentum about the axis of the top associated with the spin only. The moment about O is due to the weight and is $mg\bar{r}\sin\theta$, where $\bar{r}$ is the distance from O to the mass center G. From the diagram, we see that the angular-momentum vector $\mathbf{H}_O$ has a change $d\mathbf{H}_O = \mathbf{M}_O\,dt$ in the direction of $\mathbf{M}_O$ during time dt and that θ is unchanged. The increment in precessional angle around the Z-axis is

$$d\psi = \frac{M_O\,dt}{Ip\sin\theta}$$

Substituting the values $M_O = mg\bar{r}\sin\theta$ and $\Omega = d\psi/dt$ gives

$$mg\bar{r}\sin\theta = I\Omega p\sin\theta \qquad \text{or} \qquad mg\bar{r} = I\Omega p$$

which is independent of θ. Introducing the radius of gyration so that $I = mk^2$ and solving for the precessional velocity give

$$\Omega = \frac{g\bar{r}}{k^2 p} \tag{7/25}$$

Unlike Eq. 7/24, which is an exact description for the rotor of Fig. 7/17 with precession confined to the x-z plane, Eq. 7/25 is an approximation based on the assumption that the angular momentum associated with Ω is negligible compared with that associated with p. We will see the amount of the error associated with this approximation when we reconsider steady-state precession later in this article. On the basis of our analysis, the top will have a steady precession at the constant angle θ only if it is set in motion with a value of Ω which satisfies Eq. 7/25. When these conditions are not met, the precession becomes unsteady, and θ may oscillate with an amplitude which increases as the spin velocity decreases. The corresponding rise and fall of the rotation axis is called *nutation*.

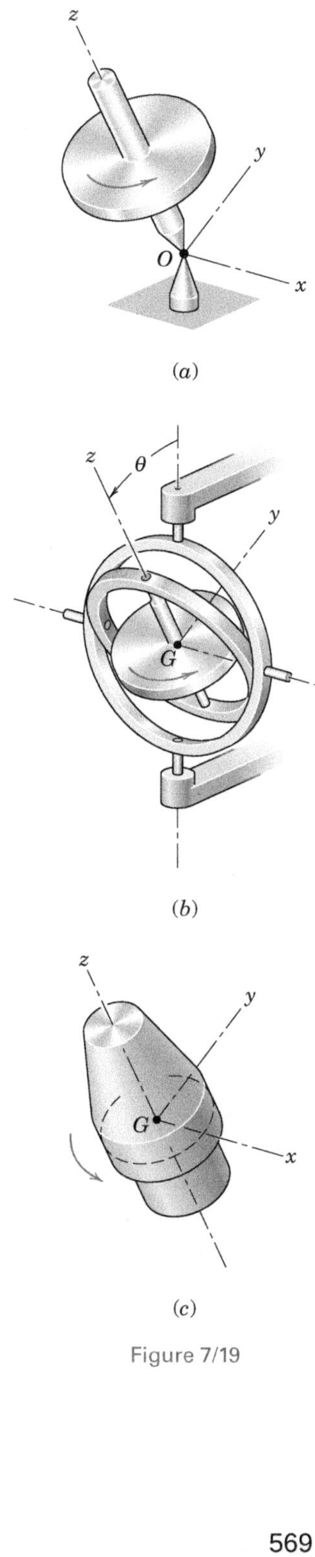

Figure 7/19

More Detailed Analysis

We now make direct use of Eq. 7/19, which is the general angular-momentum equation for a rigid body, by applying it to a body spinning about its axis of rotational symmetry. This equation is valid for rotation about a fixed point or for rotation about the mass center. A spinning top, the rotor of a gyroscope, and a spacecraft are examples of bodies whose motions can be described by the equations for rotation about a point. The general moment equations for this class of problems are fairly complex, and their complete solutions involve the use of elliptic integrals and somewhat lengthy computations. However, a large fraction of engineering problems where the motion is one of rotation about a point involves the steady precession of bodies of revolution which are spinning about their axes of symmetry. These conditions greatly simplify the equations and thus facilitate their solution.

Consider a body with axial symmetry, Fig. 7/19*a*, rotating about a fixed point O on its axis, which is taken to be the z-direction. With O as origin, the x- and y-axes automatically become principal axes of inertia along with the z-axis. This same description may be used for the rotation of a similar symmetrical body about its center of mass G, which is taken as the origin of coordinates as shown with the gimbaled gyroscope rotor of Fig. 7/19*b*. Again, the x- and y-axes are principal axes of inertia for point G. The same description may also be used to represent the rotation about the mass center of an axially symmetric body in space, such as the spacecraft in Fig. 7/19*c*. In each case, we note that, regardless of the rotation of the axes or of the body relative to the axes (spin about the z-axis), the moments of inertia about the x- and y-axes remain constant with time. The principal moments of inertia are again designated $I_{zz} = I$ and $I_{xx} = I_{yy} = I_0$. The products of inertia are, of course, zero.

Before applying Eq. 7/19, we introduce a set of coordinates which provide a natural description for our problem. These coordinates are shown in Fig. 7/20 for the example of rotation about a fixed point O. The axes X-Y-Z are fixed in space, and plane A contains the X-Y axes and the fixed point O on the rotor axis. Plane B contains point O and is always normal to the rotor axis. Angle θ measures the inclination of the rotor axis from the vertical Z-axis and is also a measure of the angle between planes A and B. The intersection of the two planes is the x-axis, which is located by the angle ψ from the X-axis. The y-axis lies in plane B, and the z-axis coincides with the rotor axis. The angles θ and ψ completely specify the position of the rotor axis. The angular displacement of the rotor with respect to axes x-y-z is specified by the angle ϕ measured from the x-axis to the x'-axis, which is attached to the rotor. The spin velocity becomes $p = \dot{\phi}$.

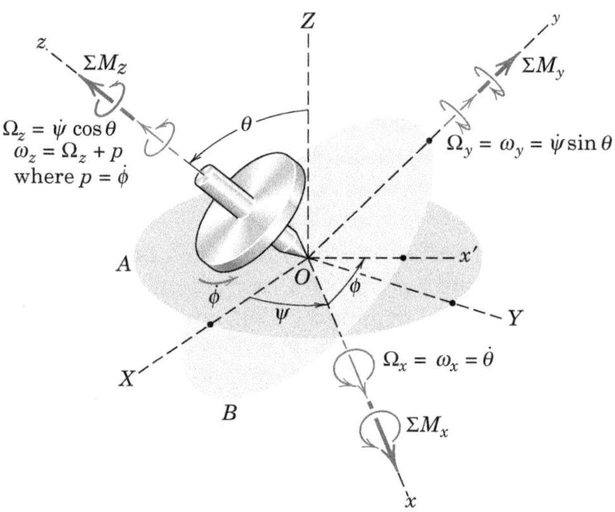

Figure 7/20

The components of the angular velocity $\boldsymbol{\omega}$ of the rotor and the angular velocity $\boldsymbol{\Omega}$ of the axes x-y-z from Fig. 7/20 become

$$\Omega_x = \dot{\theta} \qquad \omega_x = \dot{\theta}$$

$$\Omega_y = \dot{\psi} \sin \theta \qquad \omega_y = \dot{\psi} \sin \theta$$

$$\Omega_z = \dot{\psi} \cos \theta \qquad \omega_z = \dot{\psi} \cos \theta + p$$

It is important to note that the axes and the body have identical x- and y-components of angular velocity, but that the z-components differ by the relative angular velocity p.

The angular-momentum components from Eq. 7/12 become

$$H_x = I_{xx}\omega_x = I_0\dot{\theta}$$

$$H_y = I_{yy}\omega_y = I_0\dot{\psi} \sin \theta$$

$$H_z = I_{zz}\omega_z = I(\dot{\psi} \cos \theta + p)$$

Substitution of the angular-velocity and angular-momentum components into Eq. 7/19 yields

$$\Sigma M_x = I_0(\ddot{\theta} - \dot{\psi}^2 \sin\theta\cos\theta) + I\dot{\psi}(\dot{\psi}\cos\theta + p)\sin\theta$$
$$\Sigma M_y = I_0(\ddot{\psi}\sin\theta + 2\dot{\psi}\,\dot{\theta}\cos\theta) - I\dot{\theta}(\dot{\psi}\cos\theta + p) \qquad \textbf{(7/26)}$$
$$\Sigma M_z = I\frac{d}{dt}(\dot{\psi}\cos\theta + p)$$

Equations 7/26 are the general equations of rotation of a symmetrical body about either a fixed point O or the mass center G. In a given problem, the solution to the equations will depend on the moment sums applied to the body about the three coordinate axes. We will confine our use of these equations to two particular cases of rotation about a point which are described in the following sections.

Steady-State Precession

We now examine the conditions under which the rotor precesses at a steady rate $\dot{\psi}$ at a constant angle θ and with constant spin velocity p. Thus,

$$\dot{\psi} = \text{constant}, \qquad \ddot{\psi} = 0$$
$$\theta = \text{constant}, \qquad \dot{\theta} = \ddot{\theta} = 0$$
$$p = \text{constant}, \qquad \dot{p} = 0$$

and Eqs. 7/26 become

$$\Sigma M_x = \dot{\psi}\sin\theta[I(\dot{\psi}\cos\theta + p) - I_0\,\dot{\psi}\cos\theta]$$
$$\Sigma M_y = 0 \qquad \textbf{(7/27)}$$
$$\Sigma M_z = 0$$

From these results, we see that the required moment acting on the rotor about O (or about G) must be in the x-direction since the y- and z-components are zero. Furthermore, with the constant values of θ, $\dot{\psi}$, and p, the moment is constant in magnitude. It is also important to note that the moment axis is perpendicular to the plane defined by the precession axis (Z-axis) and the spin axis (z-axis).

We may also obtain Eqs. 7/27 by recognizing that the components of $\mathbf{H}$ remain constant as observed in x-y-z so that $(\dot{\mathbf{H}})_{xyz} = \mathbf{0}$. Because in general $\Sigma\mathbf{M} = (\dot{\mathbf{H}})_{xyz} + \boldsymbol{\Omega} \times \mathbf{H}$, we have for the case of steady precession

$$\Sigma\mathbf{M} = \boldsymbol{\Omega} \times \mathbf{H} \qquad \textbf{(7/28)}$$

which reduces to Eqs. 7/27 upon substitution of the values of $\boldsymbol{\Omega}$ and $\mathbf{H}$.

By far the most common engineering examples of gyroscopic motion occur when precession takes place about an axis which is normal to the rotor axis, as in Fig. 7/14. Thus with the substitution $\theta = \pi/2$, $\omega_z = p$, $\dot{\psi} = \Omega$, and $\Sigma M_x = M$, we have from Eqs. 7/27

$$M = I\Omega p \qquad \textbf{[7/24]}$$

which we derived initially in this article from a direct analysis of this special case.

Now let us examine the steady precession of the rotor (symmetrical top) of Fig. 7/20 for any constant value of θ other than $\pi/2$. The moment ΣM_x about the x-axis is due to the weight of the rotor and is $mg\bar{r}\sin\theta$. Substitution into Eqs. 7/27 and rearrangement of terms give us

$$mg\bar{r} = I\dot{\psi}p - (I_0 - I)\dot{\psi}^2\cos\theta$$

We see that $\dot{\psi}$ is small when p is large, so that the second term on the right-hand side of the equation becomes very small compared with $I\dot{\psi}p$. If we neglect this smaller term, we have $\dot{\psi} = mg\bar{r}/(Ip)$ which, upon use of the previous substitution $\Omega = \dot{\psi}$ and $mk^2 = I$, becomes

$$\Omega = \frac{g\bar{r}}{k^2p} \qquad [7/25]$$

We derived this same relation earlier by assuming that the angular momentum was entirely along the spin axis.

Steady Precession with Zero Moment

Consider now the motion of a symmetrical rotor with no external moment about its mass center. Such motion is encountered with spacecraft and projectiles which both spin and precess during flight.

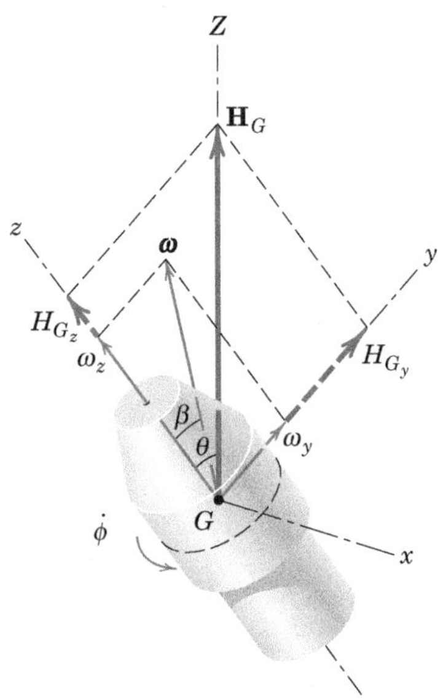

Figure 7/21

Figure 7/21 represents such a body. Here the Z-axis, which has a fixed direction in space, is chosen to coincide with the direction of the angular momentum $\mathbf{H}_G$, which is constant since $\Sigma\mathbf{M}_G = \mathbf{0}$. The x-y-z axes are attached in the manner described in Fig. 7/20. From Fig. 7/21 the three components of momentum are $H_{G_x} = 0$, $H_{G_y} = H_G\sin\theta$, $H_{G_z} = H_G\cos\theta$. From the defining relations, Eqs. 7/12, with the notation of this article, these components are also given by $H_{G_x} = I_0\omega_x$, $H_{G_y} = I_0\omega_y$, $H_{G_z} = I\omega_z$. Thus, $\omega_x = \Omega_x = 0$ so that θ is constant. This result means that the motion is one of steady precession about the constant $\mathbf{H}_G$ vector.

With no x-component, the angular velocity $\boldsymbol{\omega}$ of the rotor lies in the y-z plane along with the Z-axis and makes an angle β with the z-axis. The relationship between β and θ is obtained from $\tan\theta = H_{G_y}/H_{G_z} = I_0\omega_y/(I\omega_z)$, which is

$$\tan\theta = \frac{I_0}{I}\tan\beta \qquad (7/29)$$

Thus, the angular velocity $\boldsymbol{\omega}$ makes a constant angle β with the spin axis.

The rate of precession is easily obtained from Eq. 7/27 with $M = 0$, which gives

$$\dot{\psi} = \frac{Ip}{(I_0 - I)\cos\theta} \qquad (7/30)$$

It is clear from this relation that the direction of the precession depends on the relative magnitudes of the two moments of inertia.

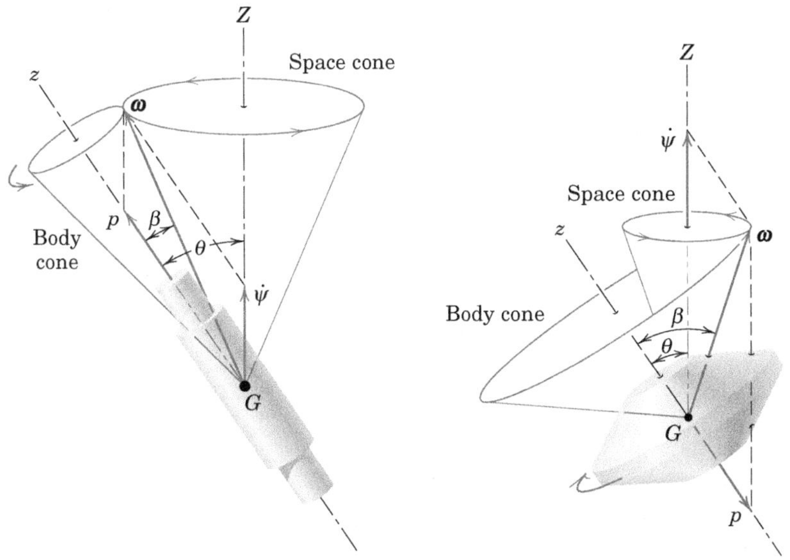

Direct precession $I_0 > I$

(*a*)

Retrograde precession $I_0 < I$

(*b*)

Figure 7/22

If $I_0 > I$, then $\beta < \theta$, as indicated in Fig. 7/22*a*, and the precession is said to be *direct*. Here the body cone rolls on the outside of the space cone.

If $I > I_0$, then $\theta < \beta$, as indicated in Fig. 7/22*b*, and the precession is said to be *retrograde*. In this instance, the space cone is internal to the body cone, and $\dot{\psi}$ and p have opposite signs.

If $I = I_0$, then $\theta = \beta$ from Eq. 7/29, and Fig. 7/22 shows that both angles must be zero to be equal. For this case, the body has no precession and merely rotates with an angular velocity $\mathbf{p}$. This condition occurs for a body with point symmetry, such as with a homogeneous sphere.

This toy top is useful in demonstrating the principles of this article.

Sample Problem 7/8

The turbine rotor in a ship's power plant has a mass of 1000 kg, with center of mass at G and a radius of gyration of 200 mm. The rotor shaft is mounted in bearings A and B with its axis in the horizontal fore-and-aft direction and turns counterclockwise at a speed of 5000 rev/min when viewed from the stern. Determine the vertical components of the bearing reactions at A and B if the ship is making a turn to port (left) of 400-m radius at a speed of 25 knots (1 knot = 0.514 m/s). Does the bow of the ship tend to rise or fall because of the gyroscopic action?

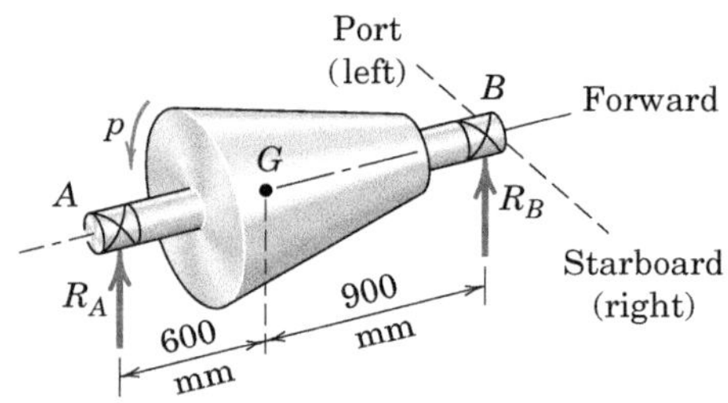

Solution. The vertical component of the bearing reactions will equal the static reactions R_1 and R_2 due to the weight of the rotor, plus or minus the increment ΔR due to the gyroscopic effect. The moment principle from statics easily gives
① $R_1 = 5890$ N and $R_2 = 3920$ N. The given directions of the spin velocity $\mathbf{p}$ and the precession velocity $\mathbf{\Omega}$ are shown with the free-body diagram of the rotor. Because the spin axis always tends to rotate toward the torque axis, we see that the torque axis $\mathbf{M}$ points in the starboard direction as shown. The sense of the ΔR's is, therefore, up at B and down at A to produce the couple $\mathbf{M}$. Thus, the bearing reactions at A and B are

$$R_A = R_1 - \Delta R \qquad \text{and} \qquad R_B = R_2 + \Delta R$$

The precession velocity Ω is the speed of the ship divided by the radius of its turn.

$$[v = \rho\Omega] \qquad \Omega = \frac{25(0.514)}{400} = 0.0321 \text{ rad/s}$$

Equation 7/24 is now applied around the mass center G of the rotor to give

$$[M = I\Omega p] \qquad 1.500(\Delta R) = 1000(0.200)^2(0.0321)\left[\frac{5000(2\pi)}{60}\right]$$

$$\Delta R = 449 \text{ N}$$

The required bearing reactions become

$$R_A = 5890 - 449 = 5440 \text{ N} \qquad \text{and} \qquad R_B = 3920 + 449 = 4370 \text{ N} \quad \textit{Ans.}$$

We now observe that the forces just computed are those exerted *on* the
② rotor shaft *by* the structure of the ship. Consequently, from the principle of action and reaction, the equal and opposite forces are applied to the ship *by* the rotor shaft, as shown in the bottom sketch. Therefore, the effect of the gyroscopic couple is to generate the increments ΔR shown, and the bow will tend to fall and the stern to rise (but only slightly).

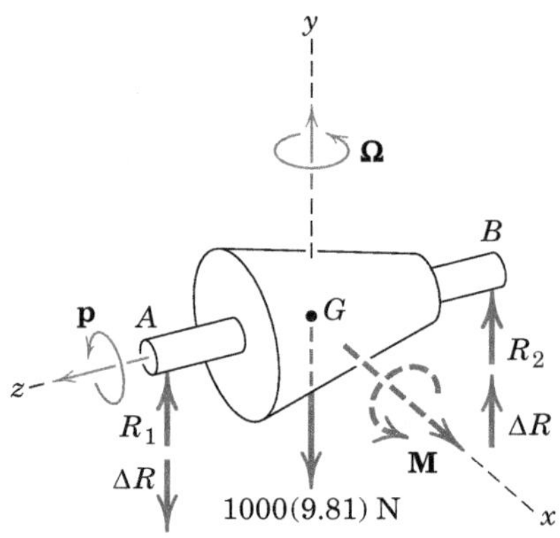

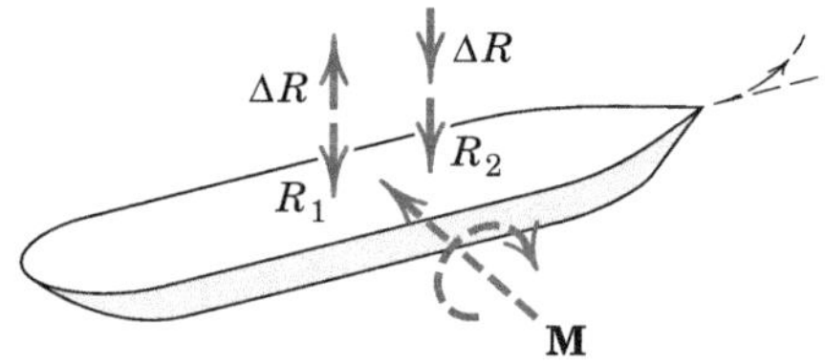

Helpful Hints

① If the ship is making a left turn, the rotation is counterclockwise as viewed from above, and the precession vector $\mathbf{\Omega}$ is up by the right-hand rule.

② After figuring the correct sense of $\mathbf{M}$ *on* the rotor, the common mistake is to apply it to the ship in the same sense, forgetting the action- and-reaction principle. Clearly, the results are then reversed. (Be certain not to make this mistake when operating a vertical gyro stabilizer in your yacht to counteract its roll!)

Sample Problem 7/9

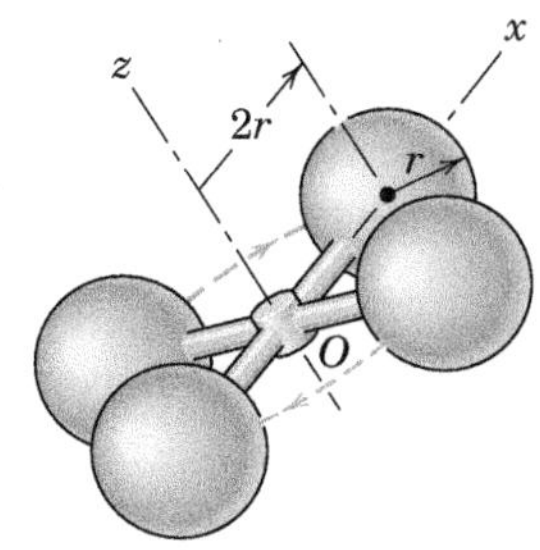

A proposed space station is closely approximated by four uniform spherical shells, each of mass m and radius r. The mass of the connecting structure and internal equipment may be neglected as a first approximation. If the station is designed to rotate about its z-axis at the rate of one revolution every 4 seconds, determine (*a*) the number n of complete cycles of precession for each revolution about the z-axis if the plane of rotation deviates only slightly from a fixed orientation, and (*b*) find the period τ of precession if the spin axis z makes an angle of 20° with respect to the axis of fixed orientation about which precession occurs. Draw the space and body cones for this latter condition.

Solution. ***(a)*** The number of precession cycles or wobbles for each revolution of the station about the z-axis would be the ratio of the precessional velocity $\dot{\psi}$ to the spin velocity p, which, from Eq. 7/30, is

$$\frac{\dot{\psi}}{p} = \frac{I}{(I_0 - I)\cos\theta}$$

The moments of inertia are

$$I_{zz} = I = 4[\tfrac{2}{3}mr^2 + m(2r)^2] = \tfrac{56}{3}mr^2$$

① $$I_{xx} = I_0 = 2(\tfrac{2}{3})mr^2 + 2[\tfrac{2}{3}mr^2 + m(2r)^2] = \tfrac{32}{3}mr^2$$

With θ very small, $\cos\theta \cong 1$, and the ratio of angular rates becomes

$$n = \frac{\dot{\psi}}{p} = \frac{\frac{56}{3}}{\frac{32}{3} - \frac{56}{3}} = -\frac{7}{3} \qquad \textit{Ans.}$$

The minus sign indicates retrograde precession where, in the present case, $\dot{\psi}$ and p are essentially of opposite sense. Thus, the station will make seven wobbles for every three revolutions.

(b) For $\theta = 20°$ and $p = 2\pi/4$ rad/s, the period of precession or wobble is $\tau = 2\pi/|\dot{\psi}|$, so that from Eq. 7/30

$$\tau = \frac{2\pi}{2\pi/4}\left|\frac{I_0 - I}{I}\cos\theta\right| = 4(\tfrac{3}{7})\cos 20° = 1.611 \text{ s} \qquad \textit{Ans.}$$

The precession is retrograde, and the body cone is external to the space cone as shown in the illustration where the body-cone angle, from Eq. 7/29, is

$$\tan\beta = \frac{I}{I_0}\tan\theta = \frac{56/3}{32/3}(0.364) = 0.637 \qquad \beta = 32.5°$$

Helpful Hint

① Our theory is based on the assumption that $I_{xx} = I_{yy}$ = the moment of inertia about any axis through G perpendicular to the z-axis. Such is the case here, and you should prove it to your own satisfaction.

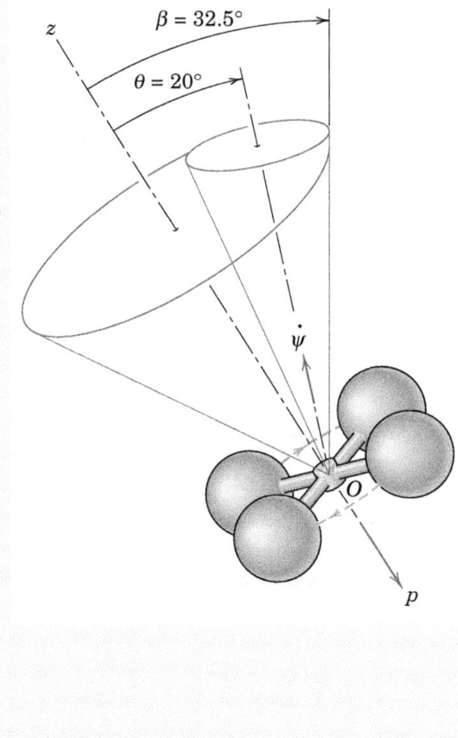

PROBLEMS

Introductory Problems

7/95 A dynamics instructor demonstrates gyroscopic principles to his students. He suspends a rapidly spinning wheel with a string attached to one end of its horizontal axle. Describe the precession motion of the wheel.

Problem 7/95

7/96 The student has volunteered to assist in a classroom demonstration involving a momentum wheel which is rapidly spinning with angular speed p as shown. The instructor has asked her to hold the axle of the wheel in the horizontal position shown and then attempt to tilt the axis upward in a vertical plane. What motion tendency of the wheel assembly will the student sense?

Problem 7/96

7/97 A car makes a turn to the right on a level road. Determine whether the normal reaction under the right rear wheel is increased or decreased as a result of the gyroscopic effect of the precessing wheels.

7/98 The 50-kg wheel is a solid circular disk which rolls on the horizontal plane in a circle of 600-mm radius. The wheel shaft is pivoted about the axis O-O and is driven by the vertical shaft at the constant rate $N = 48$ rev/min about the Z-axis. Determine the normal force R between the wheel and the horizontal surface. Neglect the weight of the horizontal shaft.

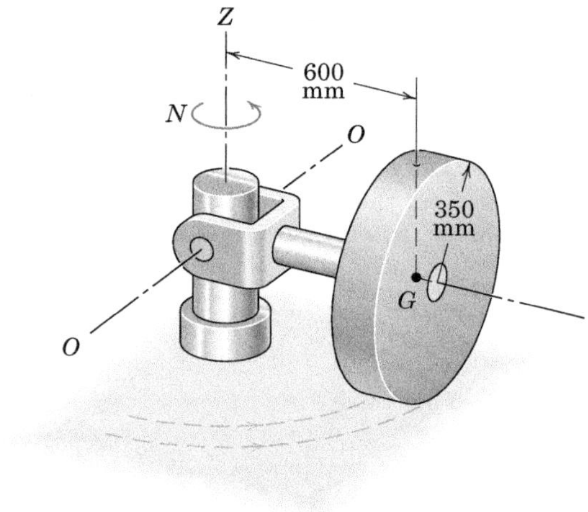

Problem 7/98

7/99 The special-purpose fan is mounted as shown. The motor armature, shaft, and blades have a combined mass of 2.2 kg with radius of gyration of 60 mm. The axial position b of the 0.8-kg block A can be adjusted. With the fan turned off, the unit is balanced about the x-axis when $b = 180$ mm. The motor and fan operate at 1725 rev/min in the direction shown. Determine the value of b which will produce a steady precession of 0.2 rad/s about the positive y-axis.

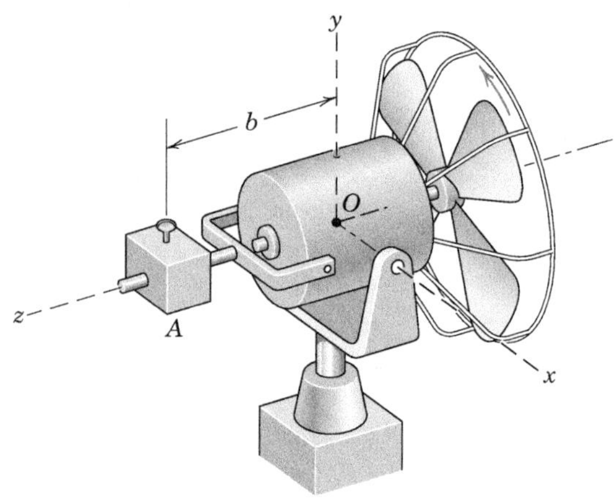

Problem 7/99

7/100 An airplane has just cleared the runway with a takeoff speed v. Each of its freely spinning wheels has a mass m, with a radius of gyration k about its axle. As seen from the front of the airplane, the wheel precesses at the angular rate Ω as the landing strut is folded into the wing about its pivot O. As a result of the gyroscopic action, the supporting member A exerts a torsional moment M on B to prevent the tubular member from rotating in the sleeve at B. Determine M and identify whether it is in the sense of M_1 or M_2.

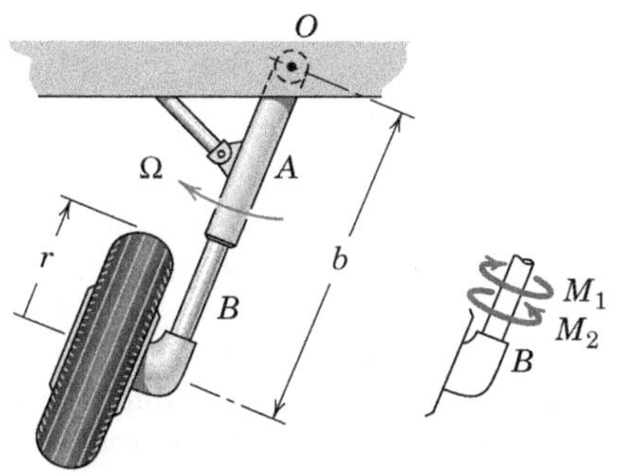

Problem 7/100

7/101 An experimental antipollution bus is powered by the kinetic energy stored in a large flywheel which spins at a high speed p in the direction indicated. As the bus encounters a short upward ramp, the front wheels rise, thus causing the flywheel to precess. What changes occur to the forces between the tires and the road during this sudden change?

Problem 7/101

7/102 The 210-kg rotor of a turbojet aircraft engine has a radius of gyration of 220 mm and rotates counterclockwise at 18 000 rev/min as viewed from the front. If the aircraft is traveling at 1200 km/h and starts to execute an inside vertical loop of 3800-m radius, compute the gyroscopic moment M transmitted to the airframe. What correction to the controls does the pilot have to make in order to remain in the vertical plane?

Representative Problems

7/103 A small air compressor for an aircraft cabin consists of the 3.50-kg turbine A which drives the 2.40-kg blower B at a speed of 20 000 rev/min. The shaft of the assembly is mounted transversely to the direction of flight and is viewed from the rear of the aircraft in the figure. The radii of gyration of A and B are 79.0 and 71.0 mm, respectively. Calculate the radial forces exerted on the shaft by the bearings at C and D if the aircraft executes a clockwise roll (rotation about the longitudinal flight axis) of 2 rad/s viewed from the rear of the aircraft. Neglect the small moments caused by the weights of the rotors. Draw a free-body diagram of the shaft as viewed from above and indicate the shape of its deflected centerline.

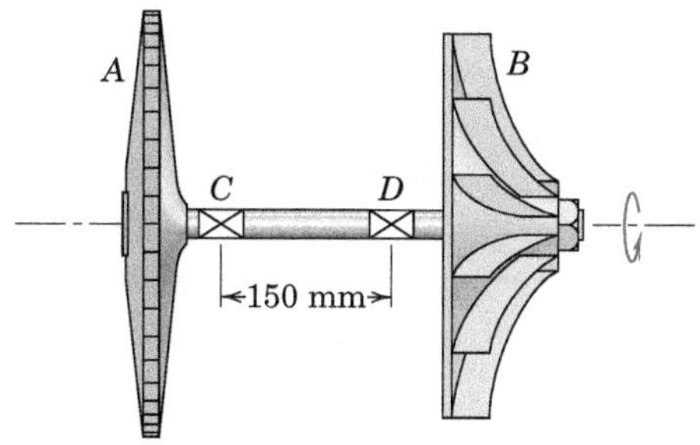

Problem 7/103

7/104 The two solid cones with the same base and equal altitudes are spinning in space about their common axis at the rate p. For what ratio h/r will precession of their spin axis be impossible?

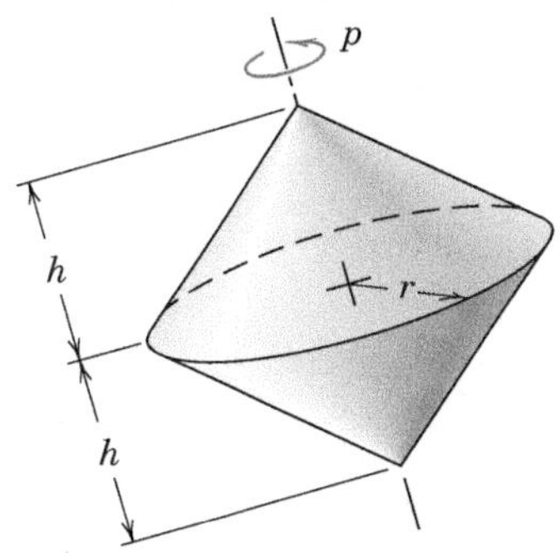

Problem 7/104

7/105 The blades and hub of the helicopter rotor weigh 140 lb and have a radius of gyration of 10 ft about the z-axis of rotation. With the rotor turning at 500 rev/min during a short interval following vertical liftoff, the helicopter tilts forward at the rate $\dot{\theta} = 10$ deg/sec in order to acquire forward velocity. Determine the gyroscopic moment M transmitted to the body of the helicopter by its rotor and indicate whether the helicopter tends to deflect clockwise or counterclockwise, as viewed by a passenger facing forward.

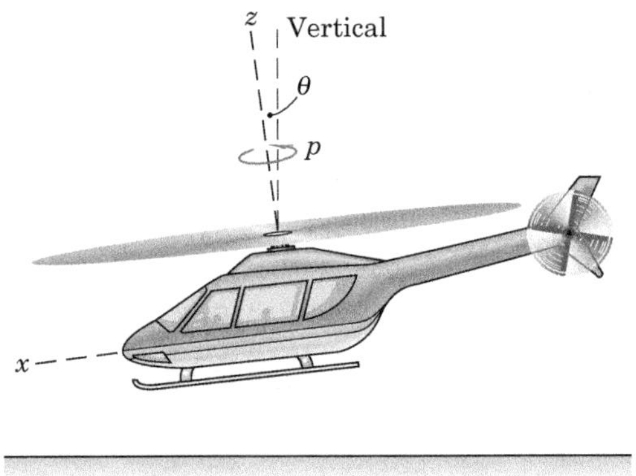

Problem 7/105

7/106 The 4-oz top with radius of gyration about its spin axis of 0.62 in. is spinning at the rate $p = 3600$ rev/min in the sense shown, with its spin axis making an angle $\theta = 20°$ with the vertical. The distance from its tip O to its mass center G is $\bar{r} = 2.5$ in. Determine the precession $\boldsymbol{\Omega}$ of the top and explain why θ gradually decreases as long as the spin rate remains large. An enlarged view of the contact of the tip is shown.

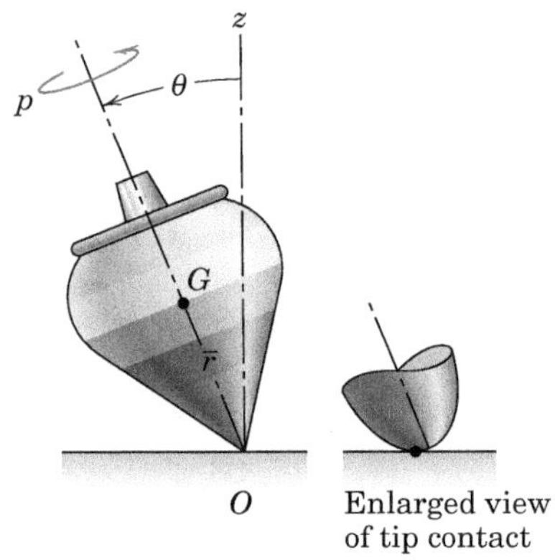

Problem 7/106

7/107 The figure shows a gyro mounted with a vertical axis and used to stabilize a hospital ship against rolling. The motor A turns the pinion which precesses the gyro by rotating the large precession gear B and attached rotor assembly about a horizontal transverse axis in the ship. The rotor turns inside the housing at a clockwise speed of 960 rev/min as viewed from the top and has a mass of 80 Mg with radius of gyration of 1.45 m. Calculate the moment exerted on the hull structure by the gyro if the motor turns the precession gear B at the rate of 0.320 rad/s. In which of the two directions, (a) or (b), should the motor turn in order to counteract a roll of the ship to port?

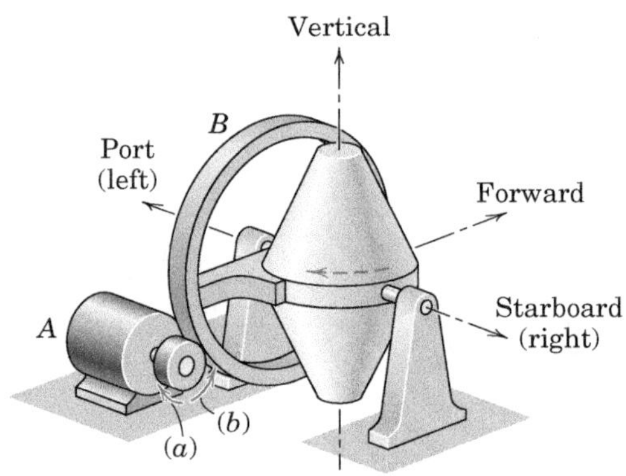

Problem 7/107

7/108 Each of the identical wheels has a mass of 4 kg and a radius of gyration $k_z = 120$ mm and is mounted on a horizontal shaft AB secured to the vertical shaft at O. In case (a), the horizontal shaft is fixed to a collar at O which is free to rotate about the vertical y-axis. In case (b), the shaft is secured by a yoke hinged about the x-axis to the collar. If the wheel has a large angular velocity $p = 3600$ rev/min about its z-axis in the position shown, determine any precession which occurs and the bending moment M_A in the shaft at A for each case. Neglect the small mass of the shaft and fitting at O.

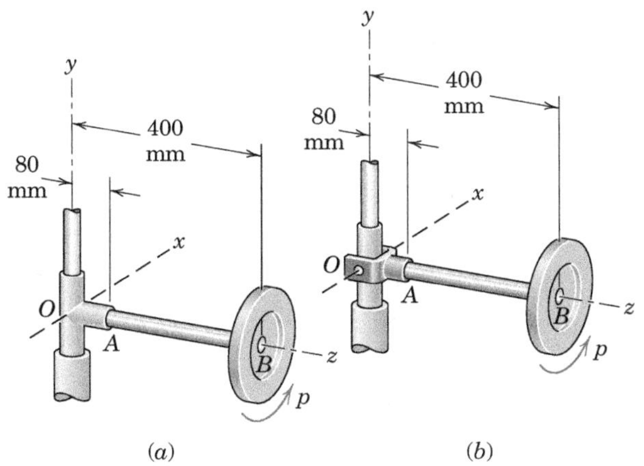

Problem 7/108

7/109 If the wheel in case (a) of Prob. 7/108 is forced to precess about the vertical by a mechanical drive at the steady rate $\mathbf{\Omega} = 2\mathbf{j}$ rad/s, determine the bending moment in the horizontal shaft at A. In the absence of friction, what torque M_O is applied to the collar at O to sustain this motion?

7/110 The figure shows the side view of the wheel carriage (truck) of a railway passenger car where the vertical load is transmitted to the frame in which the journal wheel bearings are located. The lower view shows only one pair of wheels and their axle which rotates with the wheels. Each of the 33-in.-diameter wheels weighs 560 lb, and the axle weighs 300 lb with a diameter of 5 in. Both wheels and axle are made of steel with a specific weight of 489 lb/ft^3. If the train is traveling at 80 mi/hr while rounding an 8° curve to the right (radius of curvature 717 ft), calculate the change ΔR in the vertical force supported by each wheel due only to the gyroscopic action. As a close approximation, treat each wheel as a uniform circular disk and the axle as a uniform solid cylinder. Also assume that both rails are in the same horizontal plane.

Side view of carriage

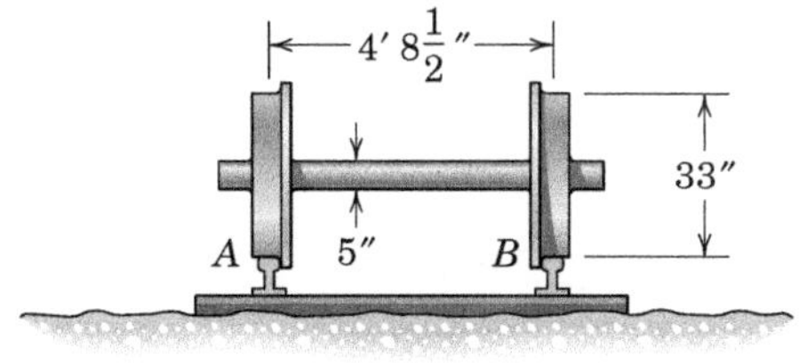

View of wheels and axle

Problem 7/110

7/111 The primary structure of a proposed space station consists of five spherical shells connected by tubular spokes. The moment of inertia of the structure about its geometric axis A-A is twice as much as that about any axis through O normal to A-A. The station is designed to rotate about its geometric axis at the constant rate of 3 rev/min. If the spin axis A-A precesses about the Z-axis of fixed orientation and makes a very small angle with it, calculate the rate $\dot{\psi}$ at which the station wobbles. The mass center O has negligible acceleration.

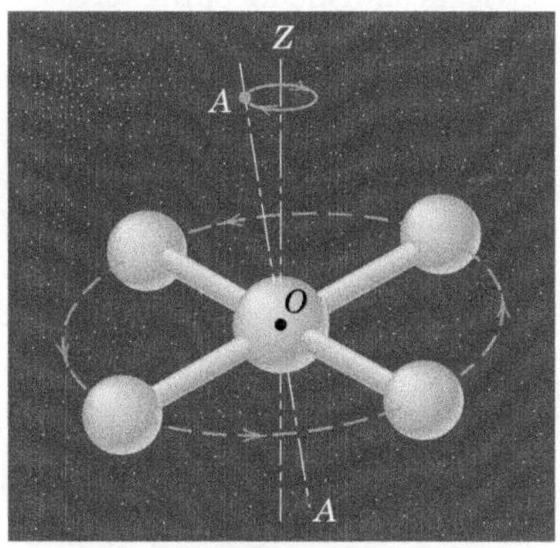

Problem 7/111

7/112 The uniform 640-mm rod has a mass of 3 kg and is welded centrally to the uniform 160-mm-radius circular disk which has a mass of 8 kg. The unit is given a spin velocity $p = 60$ rad/s in the direction shown. The axis of the rod is seen to wobble through a total angle of 30°. Calculate the angular velocity $\dot{\psi}$ of precession and determine whether it is $\dot{\psi}_1$ or $\dot{\psi}_2$.

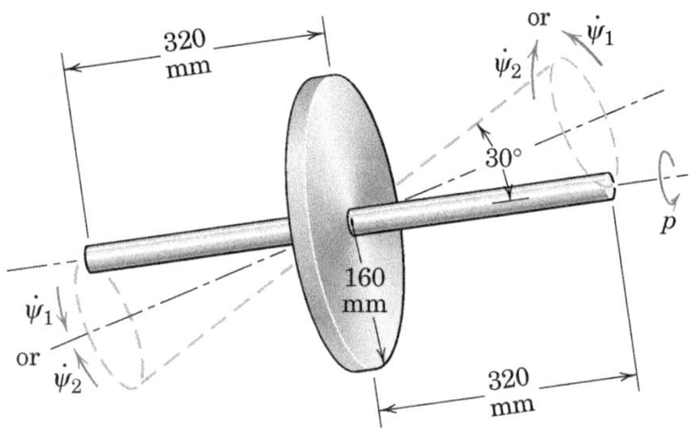

Problem 7/112

7/113 The electric motor has a total weight of 20 lb and is supported by the mounting brackets A and B attached to the rotating disk. The armature of the motor has a weight of 5 lb and a radius of gyration of 1.5 in. and turns counterclockwise at a speed of 1725 rev/min as viewed from A to B. The turntable revolves about its vertical axis at the constant rate of 48 rev/min in the direction shown. Determine the vertical components of the forces supported by the mounting brackets at A and B.

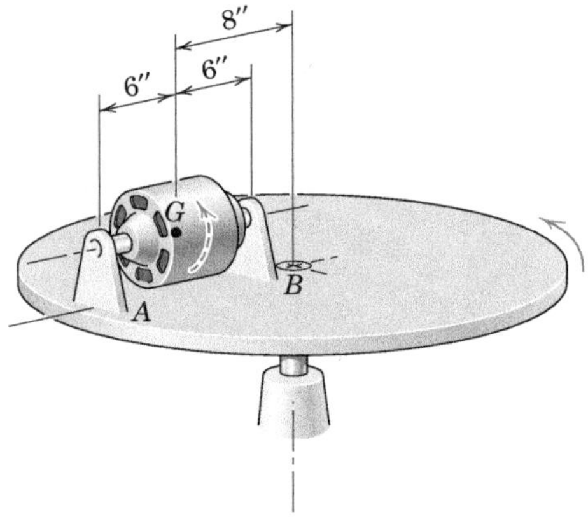

Problem 7/113

7/114 The spacecraft shown is symmetrical about its z-axis and has a radius of gyration of 720 mm about this axis. The radii of gyration about the x- and y-axes through the mass center are both equal to 540 mm. When moving in space, the z-axis is observed to generate a cone with a total vertex angle of 4° as it precesses about the axis of total angular momentum. If the spacecraft has a spin velocity $\dot{\phi}$ about its z-axis of 1.5 rad/s, compute the period τ of each full precession. Is the spin vector in the positive or negative z-direction?

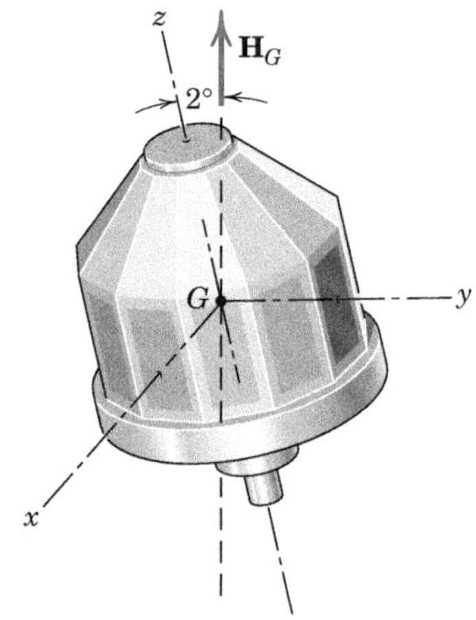

Problem 7/114

7/115 The 8-lb rotor with radius of gyration of 3 in. rotates on ball bearings at a speed of 3000 rev/min about its shaft OG. The shaft is free to pivot about the X-axis, as well as to rotate about the Z-axis. Calculate the vector $\mathbf{\Omega}$ for precession about the Z-axis. Neglect the mass of shaft OG and compute the gyroscopic couple $\mathbf{M}$ exerted by the shaft on the rotor at G.

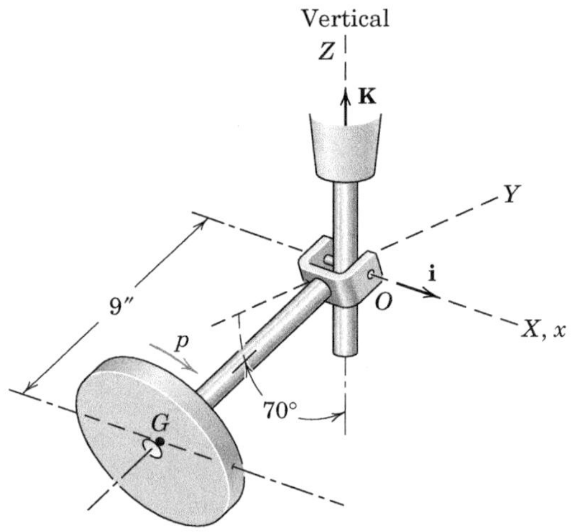

Problem 7/115

7/116 The housing of the electric motor is freely pivoted about the horizontal x-axis, which passes through the mass center G of the rotor. If the motor is turning at the constant rate $\dot{\phi} = p$, determine the angular acceleration $\ddot{\psi}$ which will result from the application of the moment M about the vertical shaft if $\dot{\gamma} = \dot{\psi} = 0$. The mass of the frame and housing is considered negligible compared with the mass m of the rotor. The radius of gyration of the rotor about the z-axis is k_z and that about the x-axis is k_x.

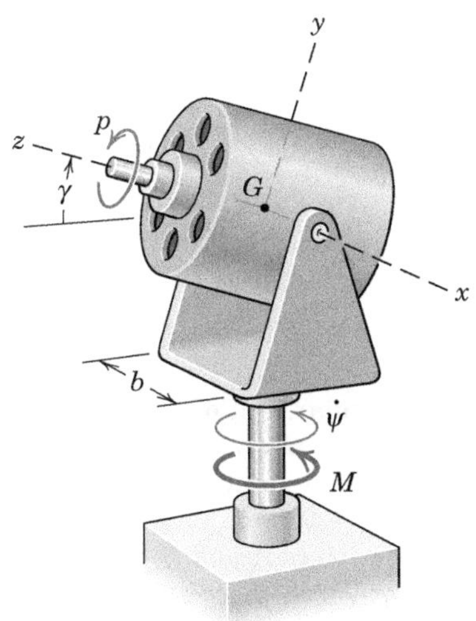

Problem 7/116

7/117 The thin ring is projected into the air with a spin velocity of 300 rev/min. If its geometric axis is observed to have a very slight precessional wobble, determine the frequency f of the wobble.

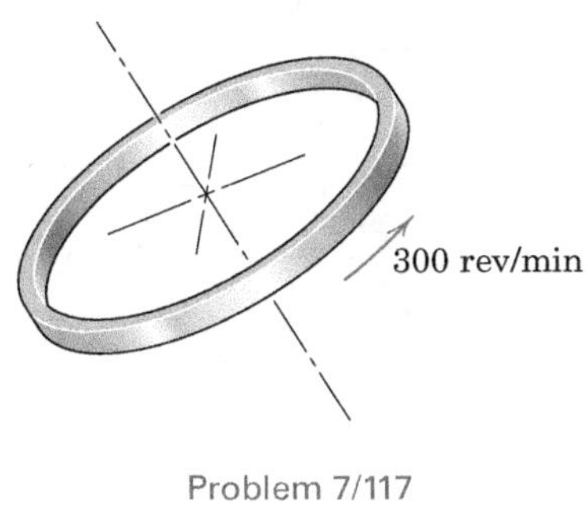

Problem 7/117

7/118 A boy throws a thin circular disk (like a Frisbee) with a spin rate of 300 rev/min. The plane of the disk is seen to wobble through a total angle of 10°. Calculate the period τ of the wobble and indicate whether the precession is direct or retrograde.

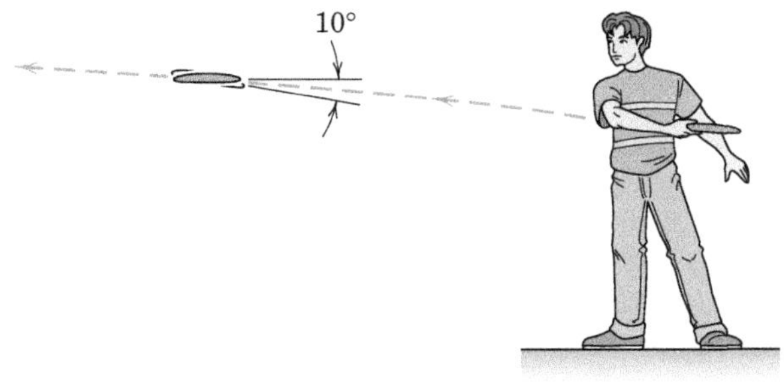

Problem 7/118

7/119 The figure shows a football in three common in-flight configurations. Case (a) is a perfectly thrown spiral pass with a spin rate of 120 rev/min. Case (b) is a wobbly spiral pass again with a spin rate of 120 rev/min about its own axis, but with the axis wobbling through a total angle of 20°. Case (c) is an end-over-end place kick with a rotational rate of 120 rev/min. For each case, specify the values of p, θ, β, and $\dot{\psi}$ as defined in this article. The moment of inertia about the long axis of the ball is 0.3 of that about the transverse axis of symmetry.

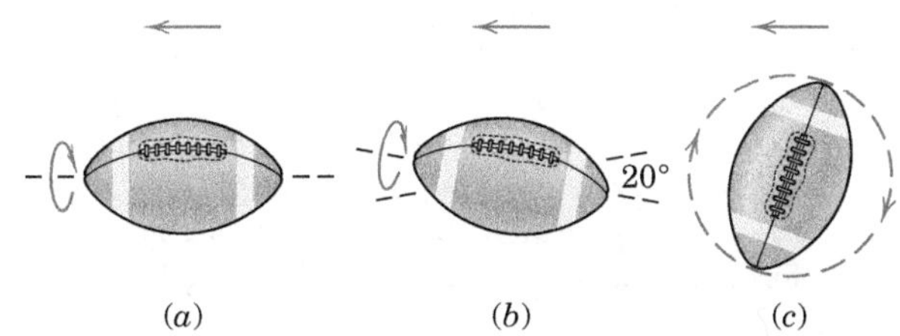

Problem 7/119

7/120 The rectangular bar is spinning in space about its longitudinal axis at the rate $p = 200$ rev/min. If its axis wobbles through a total angle of 20° as shown, calculate the period τ of the wobble.

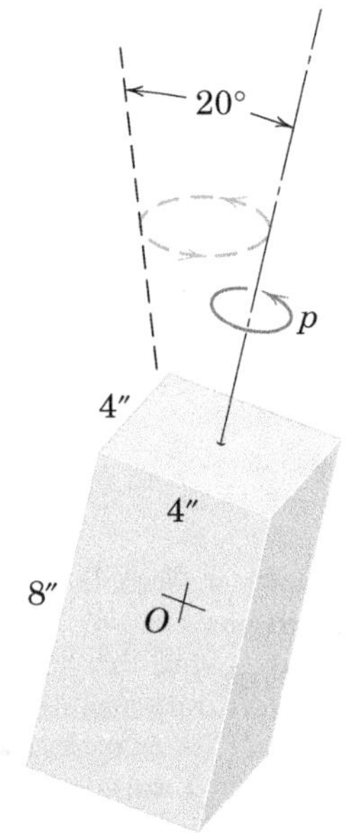

Problem 7/120

7/121 The 5-kg disk and hub A have a radius of gyration of 85 mm about the z_0-axis and spin at the rate $p = 1250$ rev/min. Simultaneously, the assembly rotates about the vertical z-axis at the rate $\Omega = 400$ rev/min. Calculate the gyroscopic moment **M** exerted *on* the shaft at C *by* the disk and the bending moment M_O in the shaft at O. Neglect the mass of the shaft but otherwise account for all forces acting on it.

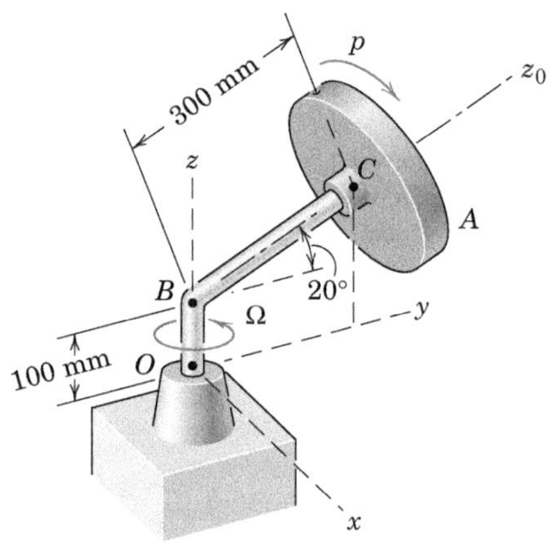

Problem 7/121

7/122 The uniform slender bar of mass m and length l is centrally mounted on the shaft A-A, about which it rotates with a constant speed $\dot{\phi} = p$. Simultaneously, the yoke is forced to rotate about the x-axis with a constant speed ω_0. As a function of ϕ, determine the magnitude of the torque M required to maintain the constant speed ω_0. (*Hint:* Apply Eq. 7/19 to obtain the x-component of M.)

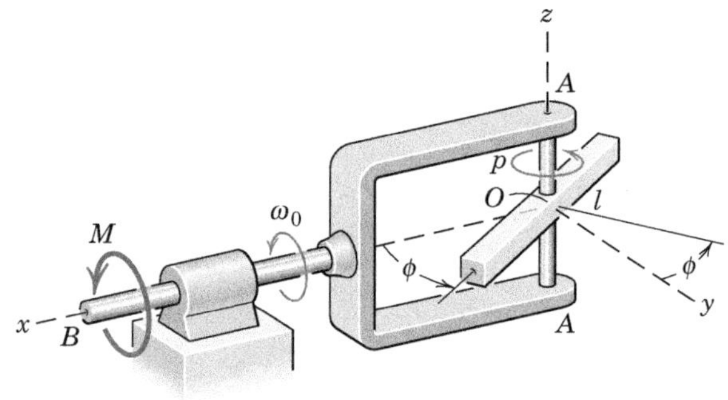

Problem 7/122

▶7/123 The solid circular disk of mass m and small thickness is spinning freely on its shaft at the rate p. If the assembly is released in the vertical position at $\theta = 0$ with $\dot{\theta} = 0$, determine the horizontal components of the forces A and B exerted by the respective bearings on the horizontal shaft as the position $\theta = \pi/2$ is passed. Neglect the mass of the two shafts compared with m and neglect all friction. Solve by using the appropriate moment equations.

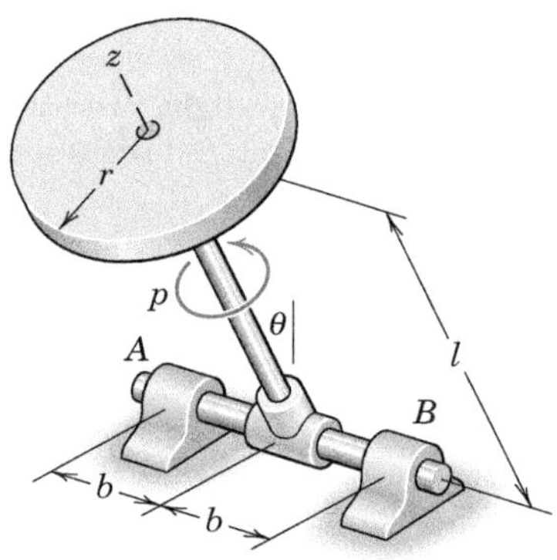

Problem 7/123

▶**7/124** The earth-scanning satellite is in a circular orbit of period τ. The angular velocity of the satellite about its y- or pitch-axis is $\omega = 2\pi/\tau$, and the angular rates about the x- and z-axes are zero. Thus, the x-axis of the satellite always points to the center of the earth. The satellite has a reaction-wheel attitude-control system consisting of the three wheels shown, each of which may be variably torqued by its individual motor. The angular rate Ω_z of the z-wheel relative to the satellite is Ω_0 at time $t = 0$, and the x- and y-wheels are at rest relative to the satellite at $t = 0$. Determine the axial torques M_x, M_y, and M_z which must be exerted by the motors on the shafts of their respective wheels in order that the angular velocity $\boldsymbol{\omega}$ of the satellite will remain constant. The moment of inertia of each reaction wheel about its axis is I. The x and z reaction-wheel speeds are harmonic functions of the time with a period equal to that of the orbit. Plot the variations of the torques and the relative wheel speeds Ω_x, Ω_y, and Ω_z as functions of the time during one orbit period. (*Hint:* The torque to accelerate the x-wheel equals the reaction of the gyroscopic moment on the z-wheel, and vice versa.)

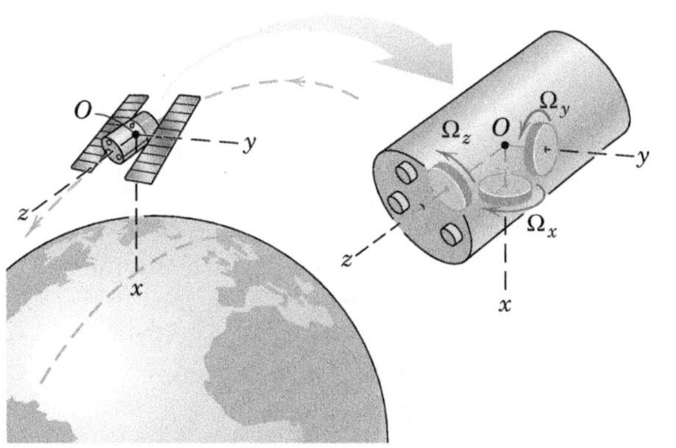

Problem 7/124

▶**7/125** The two solid homogeneous right-circular cones, each of mass m, are fastened together at their vertices to form a rigid unit and are spinning about their axis of radial symmetry at the rate $p = 200$ rev/min. (*a*) Determine the ratio h/r for which the rotation axis will not precess. (*b*) Sketch the space and body cones for the case where h/r is less than the critical ratio. (*c*) Sketch the space and body cones when $h = r$ and the precessional velocity is $\dot{\psi} = 18$ rad/s.

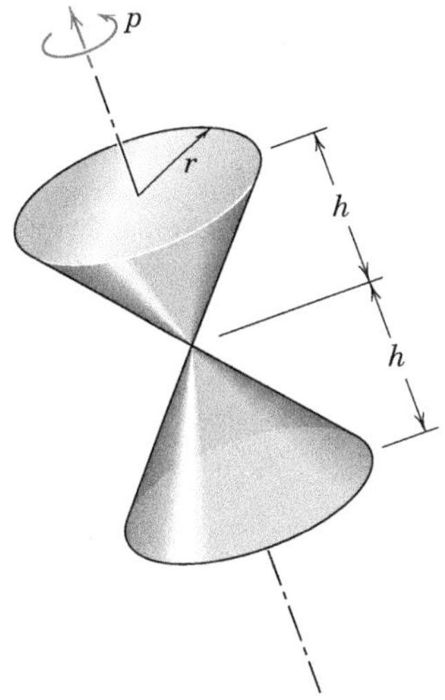

Problem 7/125

▶**7/126** The solid cylindrical rotor weighs 64.4 lb and is mounted in bearings A and B of the frame which rotates about the vertical Z-axis. If the rotor spins at the constant rate $p = 50$ rad/sec relative to the frame and if the frame itself rotates at the constant rate $\Omega = 30$ rad/sec, compute the bending moment $\mathbf{M}$ in the shaft at C which the lower portion of the shaft exerts on the upper portion. Also compute the kinetic energy T of the rotor. Neglect the mass of the frame.

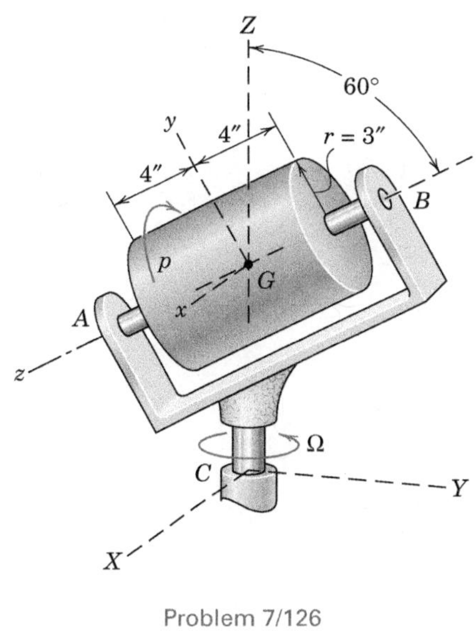

Problem 7/126

7/12 Chapter Review

In Chapter 7 we have studied the three-dimensional dynamics of rigid bodies. Motion in three dimensions adds considerable complexity to the kinematic and kinetic relationships. Compared with plane motion, there is now the possibility of two additional components of the vectors describing angular quantities such as moment, angular velocity, angular momentum, and angular acceleration. For this reason, the full power of vector analysis becomes apparent in the study of three-dimensional dynamics.

We divided our study of three-dimensional dynamics into kinematics, which is covered in Section A of the chapter, and kinetics, which is treated in Section B.

Kinematics

We arranged our coverage of three-dimensional kinematics in order of increasing complexity of the type of motion. These types are:

1. ***Translation.*** As in plane motion, covered in Chapter 5 (Plane Kinematics of Rigid Bodies), any two points on a rigid body have the same velocity and acceleration.
2. ***Fixed-Axis Rotation.*** In this case the angular-velocity vector does not change orientation, and the expressions for the velocity and acceleration of a point are easily obtained as Eqs. 7/1 and 7/2, which are identical in form to the corresponding plane-motion equations in Chapter 5.
3. ***Parallel-Plane Motion.*** This case occurs when all points in a rigid body move in planes which are parallel to a fixed plane. Thus, in each plane, the results of Chapter 5 hold.
4. ***Rotation about a Fixed Point.*** In this case, both the magnitude and the direction of the angular-velocity vector may vary. Once the angular acceleration is established by careful differentiation of the angular-velocity vector, Eqs. 7/1 and 7/2 may be used to determine the velocity and acceleration of a point.
5. ***General Motion.*** The principles of relative motion are useful in analyzing this type of motion. Relative velocity and relative acceleration are expressed in terms of translating reference axes by Eqs. 7/4. When rotating reference axes are used, the unit vectors of the reference system have nonzero time derivatives. Equations 7/6 express the velocity and acceleration in terms of quantities referred to rotating axes; these equations are identical in form to the corresponding results for plane motion, Eqs. 5/12 and 5/14. Equations 7/7*a* and 7/7*b* are the expressions relating the time derivatives of a vector as measured in a fixed system and as measured relative to a rotating system. These expressions are useful in the analysis of general motion.

Kinetics

We applied momentum and energy principles to analyze three-dimensional kinetics, as follows.

1. ***Angular Momentum.*** In three dimensions the vector expression for angular momentum has numerous additional components which are absent in plane motion. The components of angular momentum are expressed by Eqs. 7/12 and depend on both moments and products of inertia. There is a unique set of axes, called *principal axes*, for which the products of inertia are zero and the moments of inertia have stationary values. These values are called the *principal moments of inertia*.
2. ***Kinetic Energy.*** The kinetic energy of three-dimensional motion can be expressed either in terms of the motion of and about the mass center (Eq. 7/15) or in terms of the motion about a fixed point (Eq. 7/18).
3. ***Momentum Equations of Motion.*** By using the principal axes we may simplify the momentum equations of motion to obtain *Euler's equations*, Eqs. 7/21.
4. ***Energy Equations.*** The work-energy principle for three-dimensional motion is identical to that for plane motion.

Applications

In Chapter 7 we studied two applications of special interest, namely, parallel-plane motion and gyroscopic motion.

1. ***Parallel-Plane Motion.*** In such motion all points in a rigid body move in planes which are parallel to a fixed plane. The equations of motion are Eqs. 7/23. These equations are useful for analyzing the effects of dynamic imbalance in rotating machinery and in bodies which roll along straight paths.
2. ***Gyroscopic Motion.*** This type of motion occurs whenever the axis about which the body is spinning is itself rotating about another axis. Common applications include inertial guidance systems, stabilizing devices, spacecraft attitude motion, and any situation in which a rapidly spinning rotor (such as that of an aircraft engine) is being reoriented. In the case where an external torque is present, a basic analysis can be based upon the equation $\mathbf{M} = \dot{\mathbf{H}}$. For the case of torque-free motion of a body spinning about its axis of symmetry, the axis of symmetry is found to execute a coning motion about the fixed angular-momentum vector.

REVIEW PROBLEMS

7/127 The cylindrical shell is rotating in space about its geometric axis. If the axis has a slight wobble, for what ratios of l/r will the motion be direct or retrograde precession?

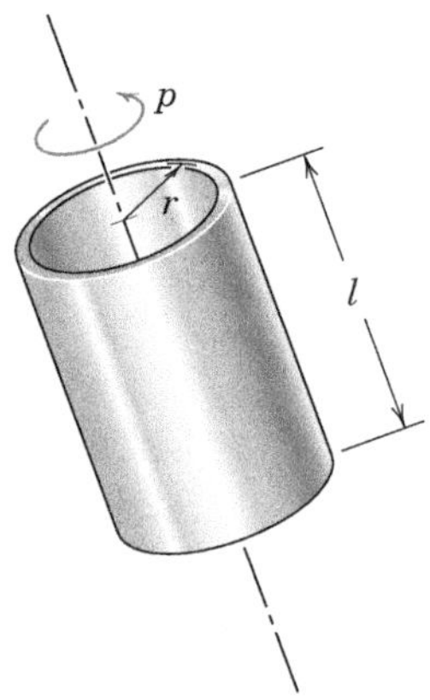

Problem 7/127

7/128 The solid cube of mass m and side a revolves about an axis M-M through a diagonal with an angular velocity ω. Write the expression for the angular momentum $\mathbf{H}$ of the cube with respect to the axes indicated.

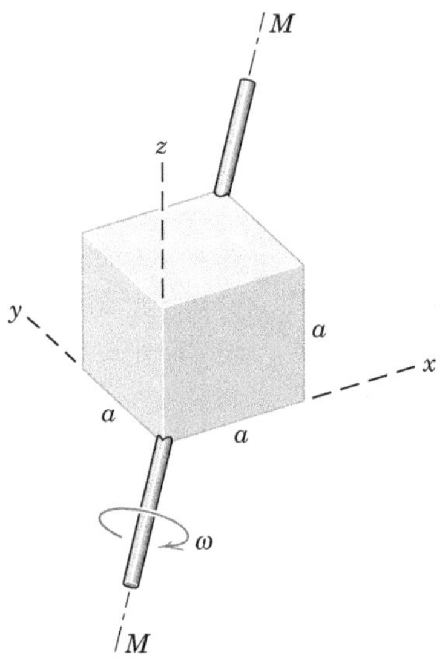

Problem 7/128

7/129 An experimental car is equipped with a gyro stabilizer to counteract completely the tendency of the car to tip when rounding a curve (no change in normal force between tires and road). The rotor of the gyro has a mass m_0 and a radius of gyration k, and is mounted in fixed bearings on a shaft which is parallel to the rear axle of the car. The center of mass of the car is a distance h above the road, and the car is rounding an unbanked level turn at a speed v. At what speed p should the rotor turn and in what direction to counteract completely the tendency of the car to overturn for either a right or a left turn? The combined mass of car and rotor is m.

7/130 The wheels of the jet plane are spinning at their angular rate corresponding to a takeoff speed of 150 km/h. The retracting mechanism operates with θ increasing at the rate of 30° per second. Calculate the angular acceleration $\boldsymbol{\alpha}$ of the wheels for these conditions.

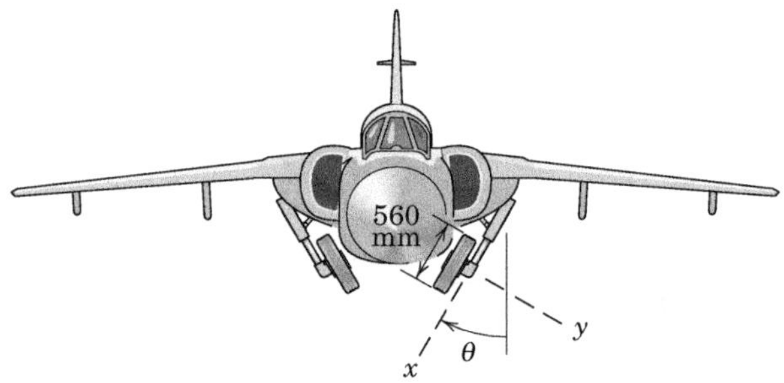

Problem 7/130

7/131 The motor turns the disk at the constant speed $p = 30$ rad/sec. The motor is also swiveling about the horizontal axis B-O (y-axis) at the constant speed $\dot{\theta} = 2$ rad/sec. Simultaneously, the entire assembly is rotating about the vertical axis C-C at the constant rate $q = 8$ rad/sec. For the instant when $\theta = 30°$, determine the angular acceleration $\boldsymbol{\alpha}$ of the disk and the acceleration $\mathbf{a}$ of point A at the bottom of the disk. Axes x-y-z are attached to the motor housing, and plane O-x_0-y is horizontal.

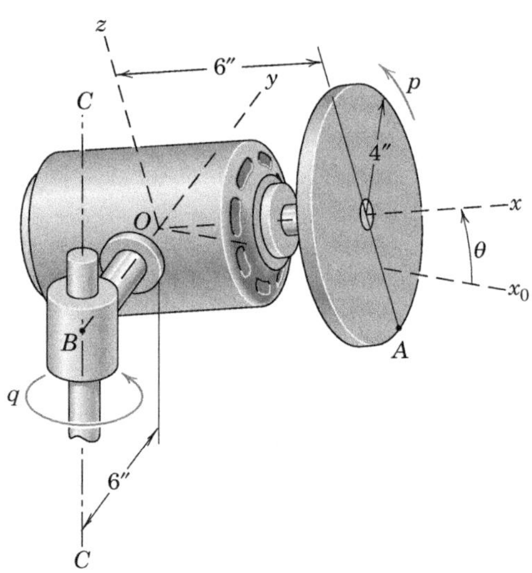

Problem 7/131

7/132 The collars at the ends of the telescoping link AB slide along the fixed shafts shown. During an interval of motion, $v_A = 5$ in./sec and $v_B = 2$ in./sec. Determine the vector expression for the angular velocity $\boldsymbol{\omega}_n$ of the centerline of the link for the position where $y_A = 4$ in. and $y_B = 2$ in.

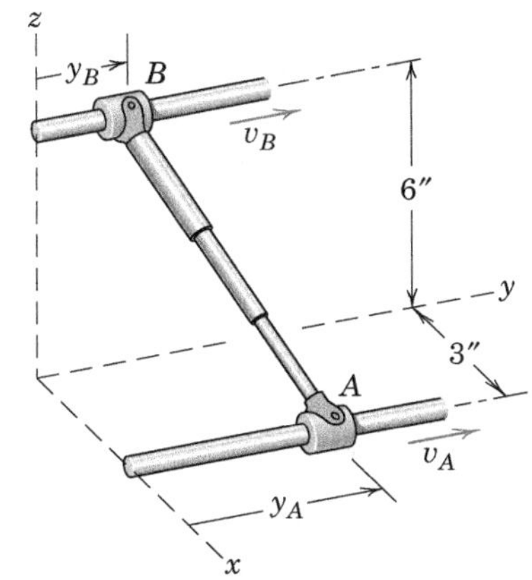

Problem 7/132

7/133 The solid cone of mass m, base radius r, and altitude h is spinning at a high rate p about its own axis and is released with its vertex O supported by a horizontal surface. Friction is sufficient to prevent the vertex from slipping in the x-y plane. Determine the direction of the precession Ω and the period τ of one complete rotation about the vertical z-axis.

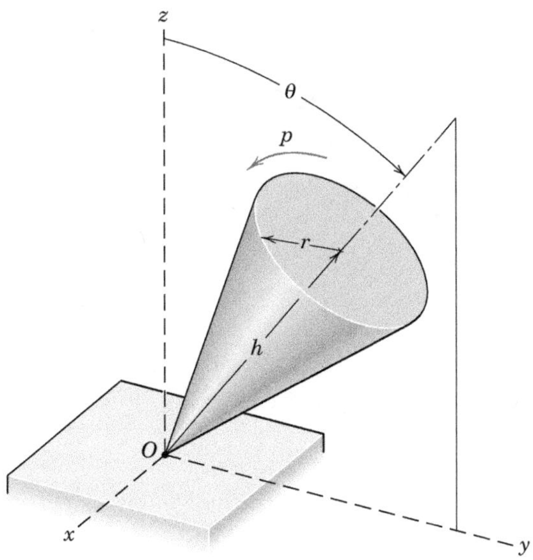

Problem 7/133

7/134 The rectangular steel plate of mass 12 kg is welded to the shaft with its plane tilted 15° from the plane (x-y) normal to the shaft axis. The shaft and plate are rotating about the fixed z-axis at the rate $N = 300$ rev/min. Determine the angular momentum $\mathbf{H}_O$ of the plate about the given axes and find its kinetic energy T.

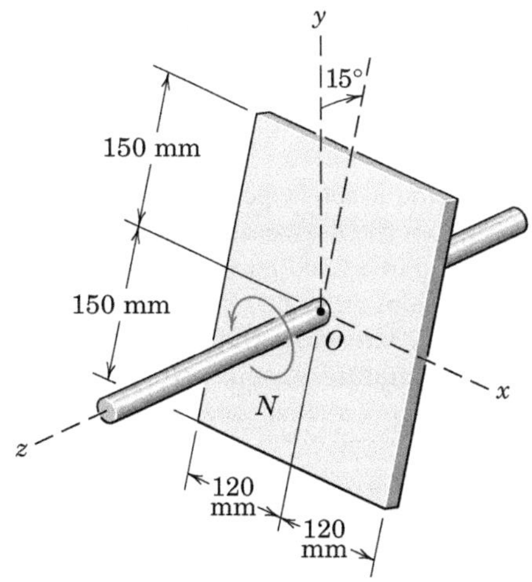

Problem 7/134

7/135 The circular disk of radius r is mounted on its shaft which is pivoted at O so that it may rotate about the vertical z_0-axis. If the disk rolls at constant speed without slipping and makes one complete turn around the circle of radius R in time τ, determine the expression for the absolute angular velocity $\boldsymbol{\omega}$ of the disk. Use axes x-y-z which rotate around the z_0-axis. (*Hint:* The absolute angular velocity of the disk equals the angular velocity of the axes plus (vectorially) the angular velocity relative to the axes as seen by holding x-y-z fixed and rotating the circular disk of radius R at the rate of $2\pi/\tau$.)

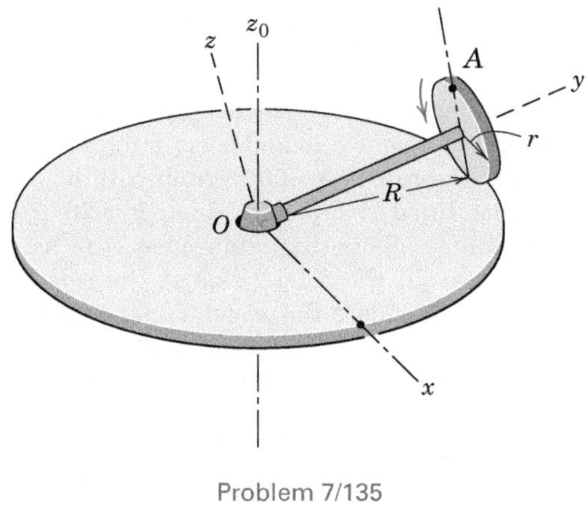

Problem 7/135

7/136 Determine the angular acceleration $\boldsymbol{\alpha}$ for the rolling circular disk of Prob. 7/135. Use the results cited in the answer for that problem.

7/137 Determine the velocity $\mathbf{v}$ of point A on the disk of Prob. 7/135 for the position shown.

7/138 Determine the acceleration $\mathbf{a}$ of point A on the disk of Prob. 7/135 for the position shown.

7/139 A top consists of a ring of mass $m = 0.52$ kg and mean radius $r = 60$ mm mounted on its central pointed shaft with spokes of negligible mass. The top is given a spin velocity of 10 000 rev/min and released on the horizontal surface with the point O remaining in a fixed position. The axis of the top is seen to make an angle of 15° with the vertical as it precesses. Determine the number N of precession cycles per minute. Also identify the direction of the precession and sketch the body and space cones.

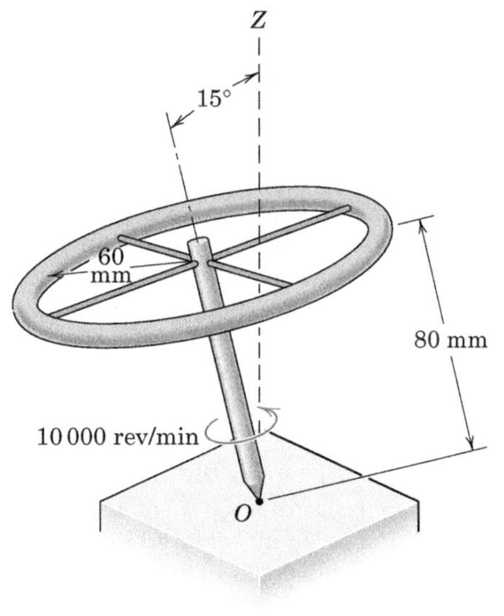

Problem 7/139

7/140 The uniform circular disk of 4-in. radius and small thickness weighs 8 lb and is spinning about its y'-axis at the rate $N = 300$ rev/min with its plane of rotation tilted at a constant angle $\beta = 20°$ from the vertical x-z plane. Simultaneously, the assembly rotates about the fixed z-axis at the rate $p = 60$ rev/min. Calculate the angular momentum $\mathbf{H}_O$ of the disk alone about the origin O of the x-y-z coordinates. Also calculate the kinetic energy T of the disk.

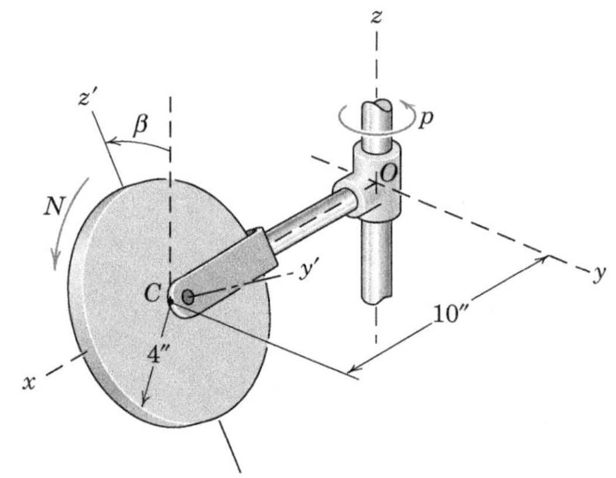

Problem 7/140

7/141 Rework Prob. 7/140 if β, instead of being constant at 20°, is increasing at the steady rate of 120 rev/min. Find the angular momentum $\mathbf{H}_O$ of the disk for the instant when $\beta = 20°$. Also compute the kinetic energy T of the disk. Is T dependent on β?

7/142 The dynamic imbalance of a certain crankshaft is approximated by the physical model shown, where the shaft carries three small 1.5-lb spheres attached by rods of negligible mass. If the shaft rotates at the constant speed of 1200 rev/min, calculate the forces R_A and R_B acting on the bearings. Neglect the gravitational forces.

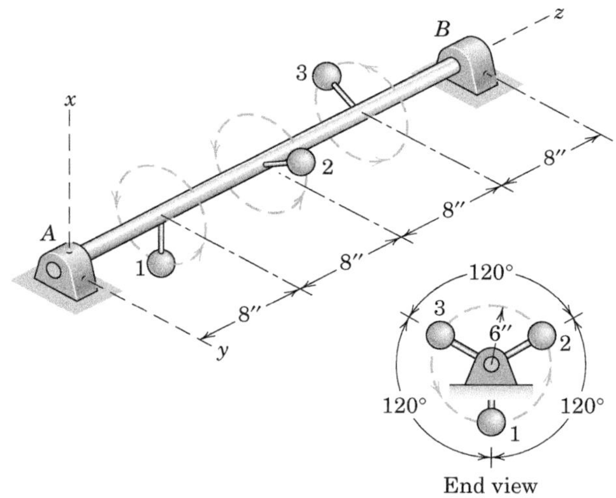

Problem 7/142

7/143 Each of the two right-angle bent rods weighs 2.80 lb and is parallel to the horizontal *x*-*y* plane. The rods are welded to the vertical shaft, which rotates about the *z*-axis with a constant angular speed $N = 1200$ rev/min. Calculate the bending moment M in the shaft at its base O.

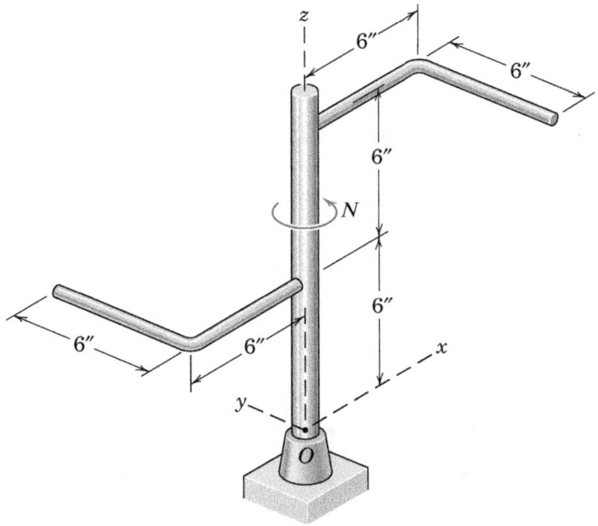

Problem 7/143

7/144 Each of the quarter-circular plates has a mass of 2 kg and is secured to the vertical shaft mounted in the fixed bearing at O. Calculate the magnitude M of the bending moment in the shaft at O for a constant rotational speed $N = 300$ rev/min. Treat the plates as exact quarter-circular shapes.

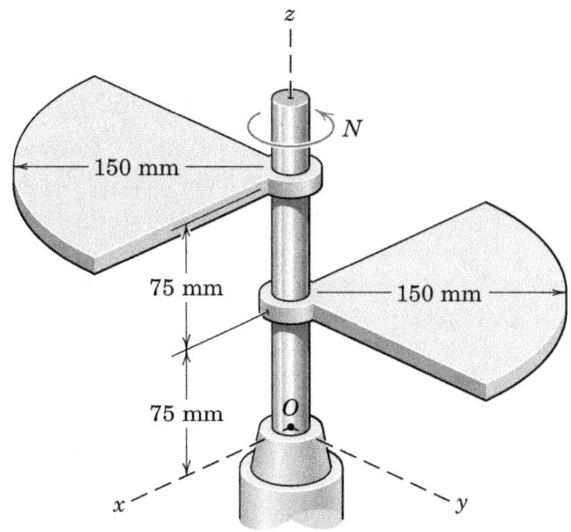

Problem 7/144

7/145 Calculate the bending moment M in the shaft at O for the rotating assembly of Prob. 7/144 as it starts from rest with an initial angular acceleration of 200 rad/s^2.

▶7/146 The half-cylindrical shell of mass m, radius r, and length b revolves about one edge along the *z*-axis with a constant rate ω as shown. Determine the angular momentum $\mathbf{H}$ of the shell with respect to the *x*-*y*-*z* axes.

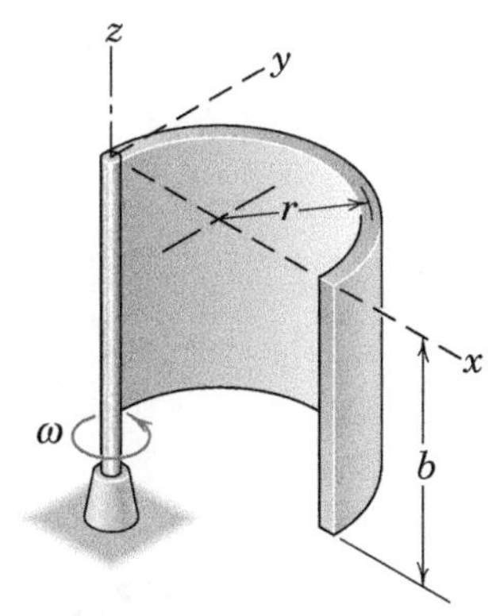

Problem 7/146

The damper and coil spring of this race-car suspension must be carefully selected to provide optimal vehicle handling.

8 VIBRATION AND TIME RESPONSE

CHAPTER OUTLINE

8/1 INTRODUCTION

An important and special class of problems in dynamics concerns the linear and angular motions of bodies which oscillate or otherwise respond to applied disturbances in the presence of restoring forces. A few examples of this class of dynamics problems are the response of an engineering structure to earthquakes, the vibration of an unbalanced rotating machine, the time response of the plucked string of a musical instrument, the wind-induced vibration of power lines, and the flutter of aircraft wings. In many cases, excessive vibration levels must be reduced to accommodate material limitations or human factors.

In the analysis of every engineering problem, we must represent the system under scrutiny by a physical model. We may often represent a *continuous* or *distributed-parameter system* (one in which the mass and spring elements are continuously spread over space) by a *discrete* or *lumped-parameter model* (one in which the mass and spring elements are separate and concentrated). The resulting simplified model is especially accurate when some portions of a continuous system are relatively massive in comparison with other portions. For example, the physical model of a ship propeller shaft is often assumed to be a massless but twistable rod with a disk rigidly attached to each end—one disk representing the turbine and the other representing the propeller. As a second example, we observe that the mass of springs may often be neglected in comparison with that of attached bodies.

Not every system is reducible to a discrete model. For example, the transverse vibration of a diving board after the departure of the diver is a somewhat difficult problem of distributed-parameter vibration. In this chapter, we will begin the study of discrete systems, limiting our discussion to those whose configurations may be described with one displacement variable. Such systems are said to possess *one degree of freedom*. For a more detailed study which includes the treatment of two or more degrees of freedom and continuous systems, you should consult one of the many textbooks devoted solely to the subject of vibrations.

The remainder of Chapter 8 is divided into four sections: Article 8/2 treats the free vibration of particles and Art. 8/3 introduces the forced vibration of particles. Each of these two articles is subdivided into undamped- and damped-motion categories. In Art. 8/4 we discuss the vibration of rigid bodies. Finally, an energy approach to the solution of vibration problems is presented in Art. 8/5.

The topic of vibrations is a direct application of the principles of kinetics as developed in Chapters 3 and 6. In particular, construction of a complete free-body diagram *drawn for an arbitrary positive value of the displacement variable*, followed by application of the appropriate governing equations of dynamics, will yield the equation of motion. From this equation of motion, which is a second-order ordinary differential equation, you can obtain all information of interest, such as the motion frequency, period, or the motion itself as a function of time.

8/2 Free Vibration of Particles

When a spring-mounted body is disturbed from its equilibrium position, its ensuing motion in the absence of any imposed external forces is termed *free vibration*. In every actual case of free vibration, there exists some retarding or damping force which tends to diminish the motion. Common damping forces are those due to mechanical and fluid friction. In this article we first consider the ideal case where the damping forces are small enough to be neglected. Then we treat the case where the damping is appreciable and must be accounted for.

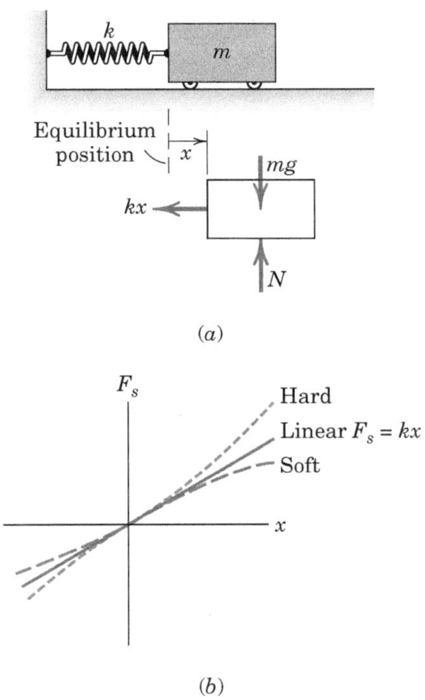

Figure 8/1

Equation of Motion for Undamped Free Vibration

We begin by considering the horizontal vibration of the simple frictionless spring-mass system of Fig. 8/1*a*. Note that the variable x denotes the displacement of the mass from the equilibrium position, which, for this system, is also the position of zero spring deflection. Figure 8/1*b* shows a plot of the force F_s necessary to deflect the spring versus the corresponding spring deflection for three types of springs. Although nonlinear hard and soft springs are useful in some applications, we will restrict our attention to the linear spring. Such a spring exerts a restoring force $-kx$ on the mass—that is, when the mass is displaced to the right, the spring force is to the left, and vice versa. We must be careful to distinguish between the forces of magnitude F_s which must be applied to both ends of the massless spring to cause tension or compression and the force $F = -kx$ of equal magnitude which the spring exerts on the mass. The constant of proportionality k is called the *spring constant*, *modulus*, or *stiffness* and has the units N/m or lb/ft.

The equation of motion for the body of Fig. 8/1*a* is obtained by first drawing its free-body diagram. Applying Newton's second law in the form $\Sigma F_x = m\ddot{x}$ gives

$$-kx = m\ddot{x} \qquad \text{or} \qquad m\ddot{x} + kx = 0 \tag{8/1}$$

The oscillation of a mass subjected to a linear restoring force as described by this equation is called *simple harmonic motion* and is characterized by acceleration which is proportional to the displacement but of opposite sign. Equation 8/1 is normally written as

$$\boxed{\ddot{x} + \omega_n{}^2 x = 0} \tag{8/2}$$

where

$$\boxed{\omega_n = \sqrt{k/m}} \tag{8/3}$$

is a convenient substitution whose physical significance will be clarified shortly.

Solution for Undamped Free Vibration

Because we anticipate an oscillatory motion, we look for a solution which gives x as a periodic function of time. Thus, a logical choice is

$$x = A \cos \omega_n t + B \sin \omega_n t \tag{8/4}$$

or, alternatively,

$$x = C \sin (\omega_n t + \psi) \tag{8/5}$$

Direct substitution of these expressions into Eq. 8/2 verifies that each expression is a valid solution to the equation of motion. We determine the constants A and B, or C and ψ, from knowledge of the initial displacement x_0 and initial velocity $\dot{x}_0$ of the mass. For example, if we work with the solution form of Eq. 8/4 and evaluate x and $\dot{x}$ at time $t = 0$, we obtain

$$x_0 = A \qquad \text{and} \qquad \dot{x}_0 = B\omega_n$$

Substitution of these values of A and B into Eq. 8/4 yields

$$x = x_0 \cos \omega_n t + \frac{\dot{x}_0}{\omega_n} \sin \omega_n t \tag{8/6}$$

The constants C and ψ of Eq. 8/5 can be determined in terms of given initial conditions in a similar manner. Evaluation of Eq. 8/5 and its first time derivative at $t = 0$ gives

$$x_0 = C \sin \psi \qquad \text{and} \qquad \dot{x}_0 = C\omega_n \cos \psi$$

Solving for C and ψ yields

$$C = \sqrt{x_0^2 + (\dot{x}_0/\omega_n)^2} \qquad \psi = \tan^{-1}(x_0\omega_n/\dot{x}_0)$$

Substitution of these values into Eq. 8/5 gives

$$x = \sqrt{x_0^2 + (\dot{x}_0/\omega_n)^2} \sin [\omega_n t + \tan^{-1}(x_0\omega_n/\dot{x}_0)] \qquad \textbf{(8/7)}$$

Equations 8/6 and 8/7 represent two different mathematical expressions for the same time-dependent motion. We observe that $C = \sqrt{A^2 + B^2}$ and $\psi = \tan^{-1}(A/B)$.

Graphical Representation of Motion

The motion may be represented graphically, Fig. 8/2, where x is seen to be the projection onto a vertical axis of the rotating vector of length C. The vector rotates at the constant angular velocity $\omega_n = \sqrt{k/m}$, which is called the *natural circular frequency* and has the units radians per second. The number of complete cycles per unit time is the *natural frequency* $f_n = \omega_n/2\pi$ and is expressed in hertz (1 hertz (Hz) = 1 cycle per second). The time required for one complete motion cycle (one rotation of the reference vector) is the *period* of the motion and is given by $\tau = 1/f_n = 2\pi/\omega_n$.

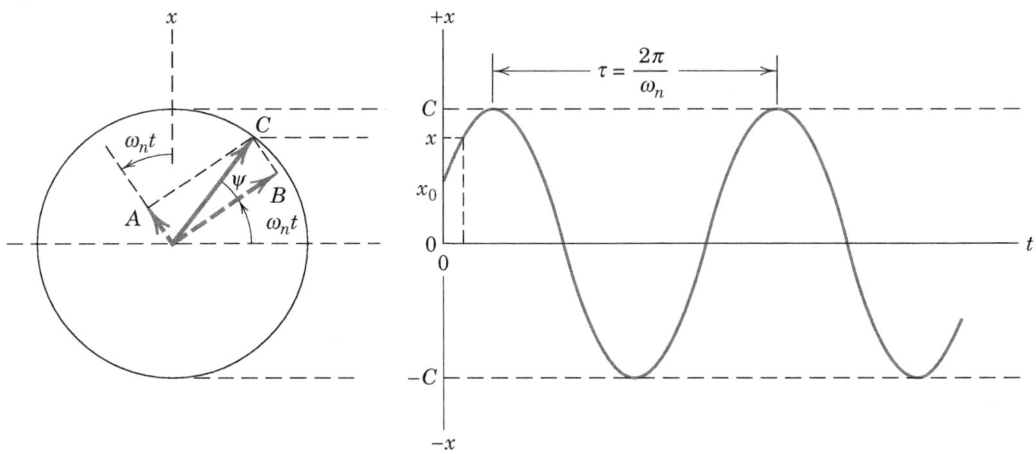

Figure 8/2

We also see from the figure that x is the sum of the projections onto the vertical axis of two perpendicular vectors whose magnitudes are A and B and whose vector sum C is the *amplitude*. Vectors A, B, and C rotate together with the constant angular velocity ω_n. Thus, as we have already seen, $C = \sqrt{A^2 + B^2}$ and $\psi = \tan^{-1}(A/B)$.

Equilibrium Position as Reference

As a further note on the free undamped vibration of particles, we see that, if the system of Fig. 8/1a is rotated 90° clockwise to obtain the system of Fig. 8/3 where the motion is vertical rather than horizontal, the equation of motion (and therefore all system properties) is unchanged if we continue to define x as the displacement from the equilibrium position. The equilibrium position now involves a nonzero spring deflection δ_{st}. From the free-body diagram of Fig. 8/3, Newton's second law gives

$$-k(\delta_{st} + x) + mg = m\ddot{x}$$

At the equilibrium position $x = 0$, the force sum must be zero, so that

$$-k\delta_{st} + mg = 0$$

Thus, we see that the pair of forces $-k\delta_{st}$ and mg on the left side of the motion equation cancel, giving

$$m\ddot{x} + kx = 0$$

which is identical to Eq. 8/1.

The lesson here is that by defining the displacement variable to be zero at equilibrium rather than at the position of zero spring deflection, we may ignore the equal and opposite forces associated with equilibrium.*

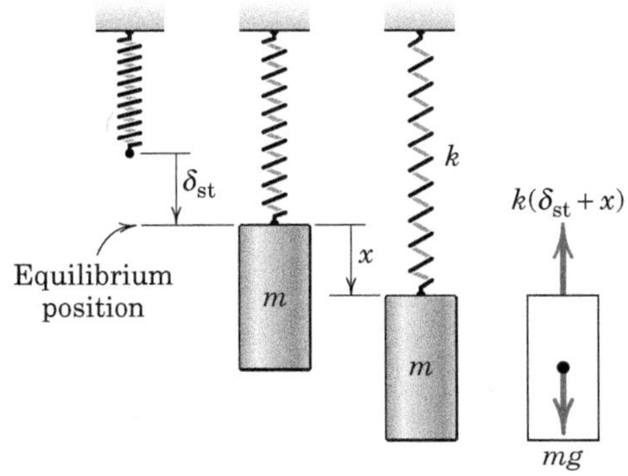

Figure 8/3

Equation of Motion for Damped Free Vibration

Every mechanical system possesses some inherent degree of friction, which dissipates mechanical energy. Precise mathematical models of the dissipative friction forces are usually complex. The dashpot or viscous damper is a device intentionally added to systems for the purpose of limiting or retarding vibration. It consists of a cylinder filled with a viscous fluid and a piston with holes or other passages by which the fluid can flow from one side of the piston to the other. Simple dashpots arranged as shown schematically in Fig. 8/4*a* exert a force F_d whose magnitude is proportional to the velocity of the mass, as depicted in Fig. 8/4*b*. The constant of proportionality c is called the *viscous damping coefficient* and has units of N·s/m or lb-sec/ft. The direction of the damping force as applied to the mass is opposite that of the velocity $\dot{x}$. Thus, the force on the mass is $-c\dot{x}$.

Complex dashpots with internal flow-rate-dependent one-way valves can produce different damping coefficients in extension and in compression; nonlinear characteristics are also possible. We will restrict our attention to the simple linear dashpot.

The equation of motion for the body with damping is determined from the free-body diagram as shown in Fig. 8/4*a*. Newton's second law gives

$$-kx - c\dot{x} = m\ddot{x} \qquad \text{or} \qquad m\ddot{x} + c\dot{x} + kx = 0 \qquad \textbf{(8/8)}$$

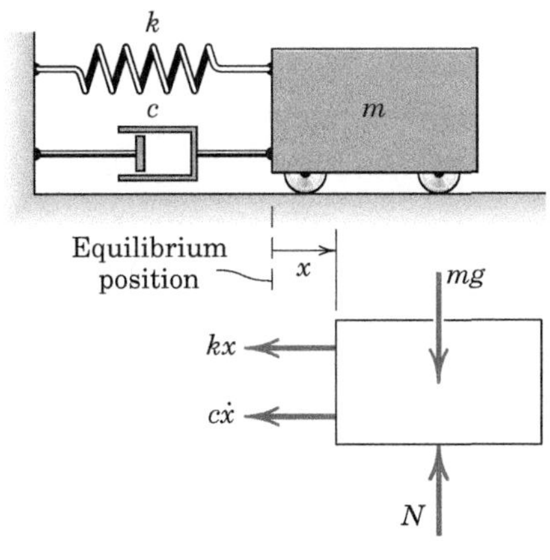

(*a*)

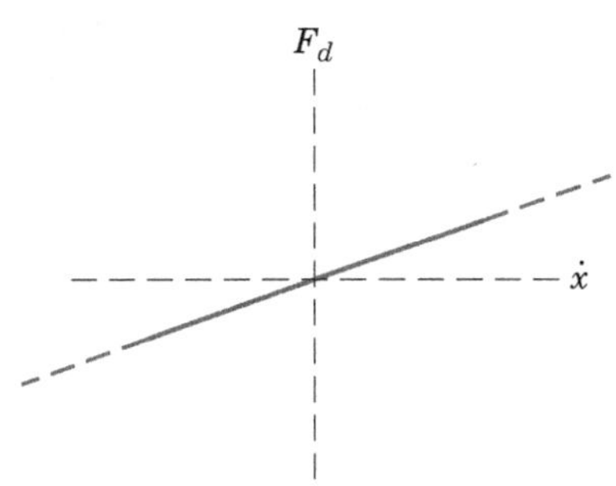

(*b*)

Figure 8/4

*For nonlinear systems, all forces, including the static forces associated with equilibrium, should be included in the analysis.

In addition to the substitution $\omega_n = \sqrt{k/m}$, it is convenient, for reasons which will shortly become evident, to introduce the combination of constants

$$\zeta = c/(2m\omega_n)$$

The quantity ζ (zeta) is called the *viscous damping factor* or *damping ratio* and is a measure of the severity of the damping. You should verify that ζ is nondimensional. Equation 8/8 may now be written as

$$\ddot{x} + 2\zeta\omega_n\dot{x} + \omega_n^2 x = 0 \qquad \textbf{(8/9)}$$

Solution for Damped Free Vibration

In order to solve the equation of motion, Eq. 8/9, we assume solutions of the form

$$x = Ae^{\lambda t}$$

Substitution into Eq. 8/9 yields

$$\lambda^2 + 2\zeta\omega_n\lambda + \omega_n^2 = 0$$

which is called the *characteristic equation*. Its roots are

$$\lambda_1 = \omega_n(-\zeta + \sqrt{\zeta^2 - 1}) \qquad \lambda_2 = \omega_n(-\zeta - \sqrt{\zeta^2 - 1})$$

Linear systems have the property of *superposition*, which means that the general solution is the sum of the individual solutions each of which corresponds to one root of the characteristic equation. Thus, the general solution is

$$\begin{aligned} x &= A_1 e^{\lambda_1 t} + A_2 e^{\lambda_2 t} \\ &= A_1 e^{(-\zeta + \sqrt{\zeta^2 - 1})\omega_n t} + A_2 e^{(-\zeta - \sqrt{\zeta^2 - 1})\omega_n t} \end{aligned} \qquad \textbf{(8/10)}$$

Categories of Damped Motion

Because $0 \le \zeta \le \infty$, the radicand $(\zeta^2 - 1)$ may be positive, negative, or even zero, giving rise to the following three categories of damped motion:

I. $\zeta > 1$ (*overdamped*). The roots λ_1 and λ_2 are distinct, real, and negative numbers. The motion as given by Eq. 8/10 decays so that x approaches zero for large values of time t. There is no oscillation and therefore no period associated with the motion.

II. $\zeta = 1$ (*critically damped*). The roots λ_1 and λ_2 are equal, real, and negative numbers ($\lambda_1 = \lambda_2 = -\omega_n$). The solution to the differential equation for the special case of equal roots is given by

$$x = (A_1 + A_2 t)e^{-\omega_n t}$$

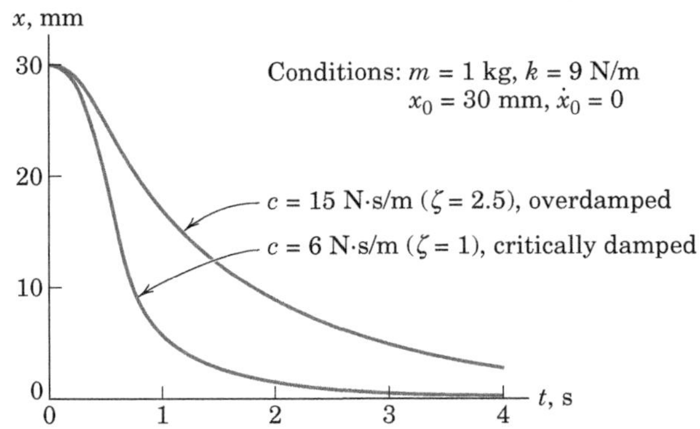

Figure 8/5

Again, the motion decays with x approaching zero for large time, and the motion is nonperiodic. A critically damped system, when excited with an initial velocity or displacement (or both), will approach equilibrium faster than will an overdamped system. Figure 8/5 depicts actual responses for both an overdamped and a critically damped system to an initial displacement x_0 and no initial velocity ($\dot{x}_0 = 0$).

III. $\zeta < 1$ (*underdamped*). Noting that the radicand $(\zeta^2 - 1)$ is negative and recalling that $e^{(a+b)} = e^a e^b$, we may rewrite Eq. 8/10 as

$$x = \{A_1 e^{i\sqrt{1-\zeta^2}\omega_n t} + A_2 e^{-i\sqrt{1-\zeta^2}\omega_n t}\} e^{-\zeta\omega_n t}$$

where $i = \sqrt{-1}$. It is convenient to let a new variable ω_d represent the combination $\omega_n\sqrt{1-\zeta^2}$. Thus,

$$x = \{A_1 e^{i\omega_d t} + A_2 e^{-i\omega_d t}\} e^{-\zeta\omega_n t}$$

Use of the Euler formula $e^{\pm ix} = \cos x \pm i \sin x$ allows the previous equation to be written as

$$\begin{aligned} x &= \{A_1(\cos \omega_d t + i \sin \omega_d t) + A_2(\cos \omega_d t - i \sin \omega_d t)\} e^{-\zeta\omega_n t} \\ &= \{(A_1 + A_2) \cos \omega_d t + i(A_1 - A_2) \sin \omega_d t\} e^{-\zeta\omega_n t} \\ &= \{A_3 \cos \omega_d t + A_4 \sin \omega_d t\} e^{-\zeta\omega_n t} \qquad \textbf{(8/11)} \end{aligned}$$

where $A_3 = (A_1 + A_2)$ and $A_4 = i(A_1 - A_2)$. We have shown with Eqs. 8/4 and 8/5 that the sum of two equal-frequency harmonics, such as those in the braces of Eq. 8/11, can be replaced by a single trigonometric function which involves a phase angle. Thus, Eq. 8/11 can be written as

$$x = \{C \sin (\omega_d t + \psi)\} e^{-\zeta\omega_n t}$$

or

$$x = Ce^{-\zeta\omega_n t} \sin (\omega_d t + \psi) \qquad \textbf{(8/12)}$$

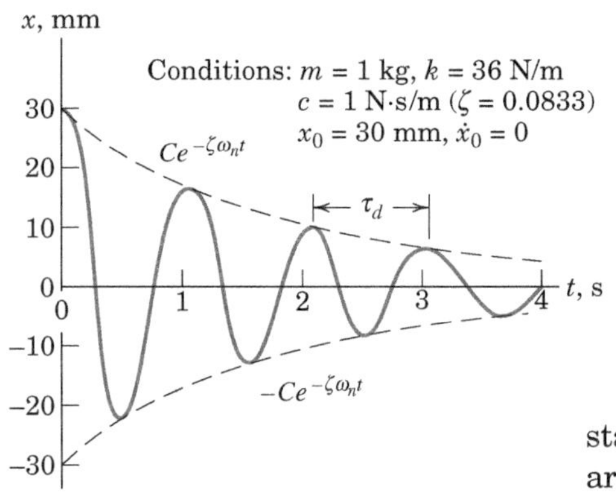

Figure 8/6

Equation 8/12 represents an exponentially decreasing harmonic function, as shown in Fig. 8/6 for specific numerical values. The frequency

$$\omega_d = \omega_n\sqrt{1 - \zeta^2}$$

is called the *damped natural frequency*. The *damped period* is given by $\tau_d = 2\pi/\omega_d = 2\pi/(\omega_n\sqrt{1 - \zeta^2})$.

It is important to note that the expressions developed for the constants C and ψ in terms of initial conditions for the case of no damping are not valid for the case of damping. To find C and ψ if damping is present, you must begin anew, setting the general displacement expression of Eq. 8/12 and its first time derivative, both evaluated at time $t = 0$, equal to the initial displacement x_0 and initial velocity $\dot{x}_0$, respectively.

Determination of Damping by Experiment

We often need to experimentally determine the value of the damping ratio ζ for an underdamped system. The usual reason is that the value of the viscous damping coefficient c is not otherwise well known. To determine the damping, we may excite the system by initial conditions and obtain a plot of the displacement x versus time t, such as that shown schematically in Fig. 8/7. We then measure two successive amplitudes x_1 and x_2 a full cycle apart and compute their ratio

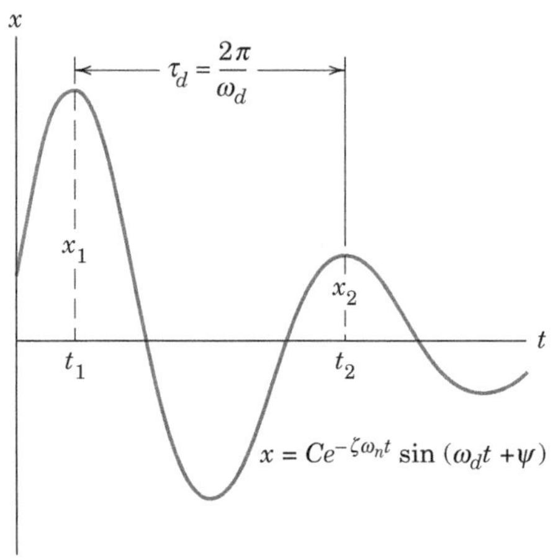

Figure 8/7

$$\frac{x_1}{x_2} = \frac{Ce^{-\zeta\omega_n t_1}}{Ce^{-\zeta\omega_n(t_1+\tau_d)}} = e^{\zeta\omega_n\tau_d}$$

The *logarithmic decrement* δ is defined as

$$\delta = \ln\left(\frac{x_1}{x_2}\right) = \zeta\omega_n\tau_d = \zeta\omega_n\frac{2\pi}{\omega_n\sqrt{1-\zeta^2}} = \frac{2\pi\zeta}{\sqrt{1-\zeta^2}}$$

From this equation, we may solve for ζ and obtain

$$\zeta = \frac{\delta}{\sqrt{(2\pi)^2 + \delta^2}}$$

For a small damping ratio, $x_1 \cong x_2$ and $\delta << 1$, so that $\zeta \cong \delta/2\pi$. If x_1 and x_2 are so close in value that experimental distinction between them is impractical, the above analysis may be modified by using two observed amplitudes which are n cycles apart.

Sample Problem 8/1

A body weighing 25 lb is suspended from a spring of constant $k = 160$ lb/ft. At time $t = 0$, it has a downward velocity of 2 ft/sec as it passes through the position of static equilibrium. Determine

(*a*) the static spring deflection δ_{st}

(*b*) the natural frequency of the system in both rad/sec (ω_n) and cycles/sec (f_n)

(*c*) the system period τ

(*d*) the displacement x as a function of time, where x is measured from the position of static equilibrium

(*e*) the maximum velocity v_{max} attained by the mass

(*f*) the maximum acceleration a_{max} attained by the mass.

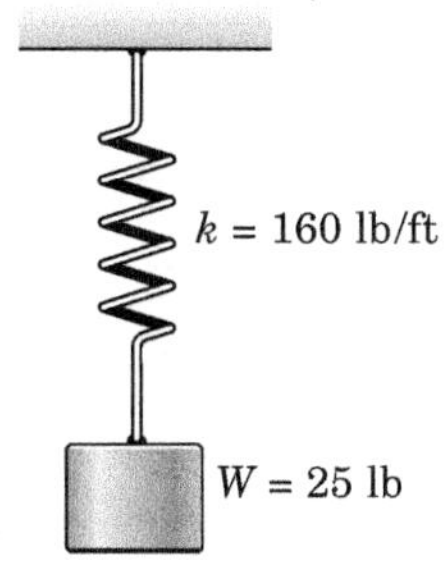

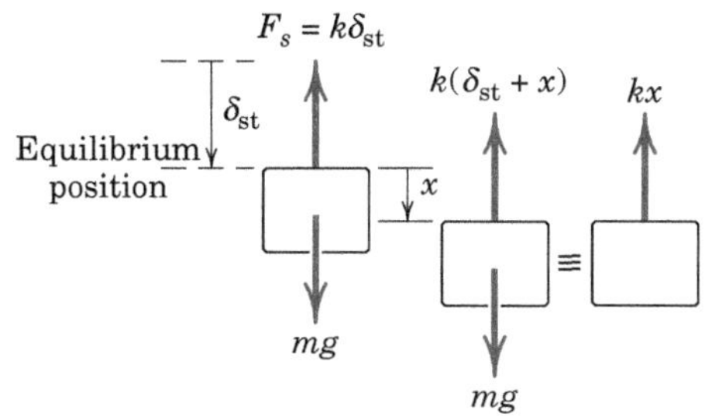

Solution. (*a*) From the spring relationship $F_s = kx$, we see that at equilibrium

① $$mg = k\delta_{st} \qquad \delta_{st} = \frac{mg}{k} = \frac{25}{160} = 0.1562 \text{ ft or } 1.875 \text{ in.} \qquad \textit{Ans.}$$

(*b*) $$\omega_n = \sqrt{\frac{k}{m}} = \sqrt{\frac{160}{25/32.2}} = 14.36 \text{ rad/sec} \qquad \textit{Ans.}$$

$$f_n = (14.36)\left(\frac{1}{2\pi}\right) = 2.28 \text{ cycles/sec} \qquad \textit{Ans.}$$

(*c*) $$\tau = \frac{1}{f_n} = \frac{1}{2.28} = 0.438 \text{ sec} \qquad \textit{Ans.}$$

② (*d*) From Eq. 8/6:

$$x = x_0 \cos \omega_n t + \frac{\dot{x}_0}{\omega_n} \sin \omega_n t$$

$$= (0) \cos 14.36t + \frac{2}{14.36} \sin 14.36t$$

$$= 0.1393 \sin 14.36t \text{ ft} \qquad \textit{Ans.}$$

As an exercise, let us determine x from the alternative Eq. 8/7:

$$x = \sqrt{{x_0}^2 + (\dot{x}_0/\omega_n)^2} \sin\left[\omega_n t + \tan^{-1}\left(\frac{x_0\omega_n}{\dot{x}_0}\right)\right]$$

$$= \sqrt{0^2 + \left(\frac{2}{14.36}\right)^2} \sin\left[14.36t + \tan^{-1}\left(\frac{(0)(14.36)}{2}\right)\right]$$

$$= 0.1393 \sin 14.36t \text{ ft}$$

(*e*) The velocity is $\dot{x} = 14.36(0.1393) \cos 14.36t = 2 \cos 14.36t$ ft/sec. Because the cosine function cannot be greater than 1 or less than -1, the maximum velocity v_{max} is 2 ft/sec, which, in this case, is the initial velocity. *Ans.*

(*f*) The acceleration is

$$\ddot{x} = -14.36(2) \sin 14.36t = -28.7 \sin 14.36t \text{ ft/sec}^2$$

The maximum acceleration a_{max} is 28.7 ft/sec^2. *Ans.*

Helpful Hints

① You should always exercise extreme caution in the matter of units. In the subject of vibrations, it is quite easy to commit errors due to mixing of feet and inches, cycles and radians, and other pairs which frequently enter the calculations.

② Recall that when we refer the motion to the position of static equilibrium, the equation of motion, and therefore its solution, for the present system is identical to that for the horizontally vibrating system.

Sample Problem 8/2

The 8-kg body is moved 0.2 m to the right of the equilibrium position and released from rest at time $t = 0$. Determine its displacement at time $t = 2$ s. The viscous damping coefficient c is 20 N·s/m, and the spring stiffness k is 32 N/m.

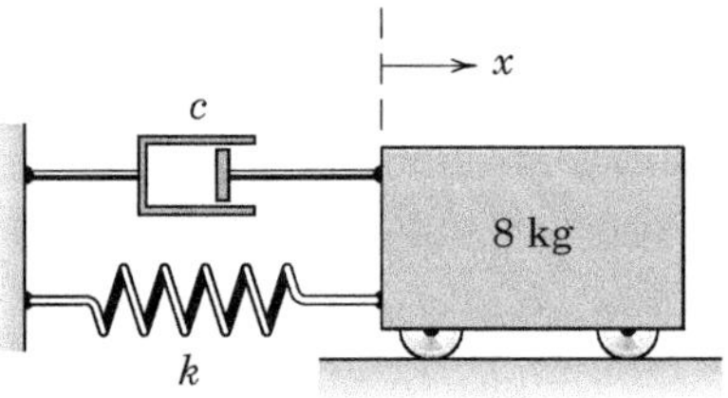

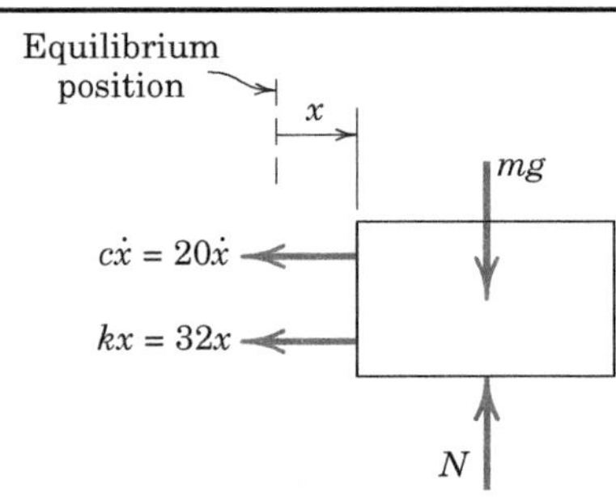

Solution. We must first determine whether the system is underdamped, critically damped, or overdamped. For that purpose, we compute the damping ratio ζ.

$$\omega_n = \sqrt{k/m} = \sqrt{32/8} = 2 \text{ rad/s} \qquad \zeta = \frac{c}{2m\omega_n} = \frac{20}{2(8)(2)} = 0.625$$

Since $\zeta < 1$, the system is underdamped. The damped natural frequency is $\omega_d = \omega_n\sqrt{1 - \zeta^2} = 2\sqrt{1 - (0.625)^2} = 1.561$ rad/s. The motion is given by Eq. 8/12 and is

$$x = Ce^{-\zeta\omega_n t}\sin(\omega_d t + \psi) = Ce^{-1.25t}\sin(1.561t + \psi)$$

The velocity is then

$$\dot{x} = -1.25Ce^{-1.25t}\sin(1.561t + \psi) + 1.561Ce^{-1.25t}\cos(1.561t + \psi)$$

Evaluating the displacement and velocity at time $t = 0$ gives

$$x_0 = C\sin\psi = 0.2 \qquad \dot{x}_0 = -1.25C\sin\psi + 1.561C\cos\psi = 0$$

Solving the two equations for C and ψ yields $C = 0.256$ m and $\psi = 0.896$ rad. Therefore, the displacement in meters is

$$x = 0.256e^{-1.25t}\sin(1.561t + 0.896) \text{ m}$$

① Evaluation for time $t = 2$ s gives $x_2 = -0.01616$ m. *Ans.*

Helpful Hint

① We note that the exponential factor $e^{-1.25t}$ is 0.0821 at $t = 2$ s. Thus, $\zeta = 0.625$ represents severe damping, although the motion is still oscillatory.

Sample Problem 8/3

The two fixed counterrotating pulleys are driven at the same angular speed ω_0. A round bar is placed off center on the pulleys as shown. Determine the natural frequency of the resulting bar motion. The coefficient of kinetic friction between the bar and pulleys is μ_k.

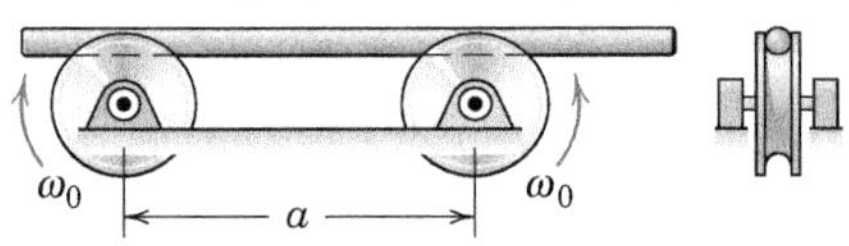

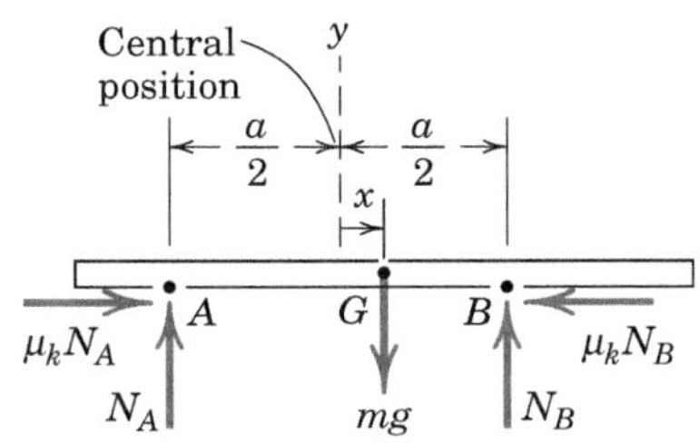

Solution. The free-body diagram of the bar is constructed for an arbitrary displacement x from the central position as shown. The governing equations are

$$[\Sigma F_x = m\ddot{x}] \qquad \mu_k N_A - \mu_k N_B = m\ddot{x}$$

$$[\Sigma F_y = 0] \qquad N_A + N_B - mg = 0$$

① $$[\Sigma M_A = 0] \qquad aN_B - \left(\frac{a}{2} + x\right)mg = 0$$

Eliminating N_A and N_B from the first equation yields

② $$\ddot{x} + \frac{2\mu_k g}{a}x = 0$$

We recognize the form of this equation as that of Eq. 8/2, so that the natural frequency in radians per second is $\omega_n = \sqrt{2\mu_k g/a}$ and the natural frequency in cycles per second is

$$f_n = \frac{1}{2\pi}\sqrt{2\mu_k g/a}$$ *Ans.*

Helpful Hints

① Because the bar is slender and does not rotate, the use of a moment equilibrium equation is justified.

② We note that the angular speed ω_0 does not enter the equation of motion. The reason for this is our assumption that the kinetic friction force does not depend on the relative velocity at the contacting surface.

PROBLEMS

(Unless otherwise indicated, all motion variables are referred to the equilibrium position.)

Undamped, Free Vibrations

8/1 When a 3-kg collar is placed upon the pan which is attached to the spring of unknown constant, the additional static deflection of the pan is observed to be 42 mm. Determine the spring constant k in N/m, lb/in., and lb/ft.

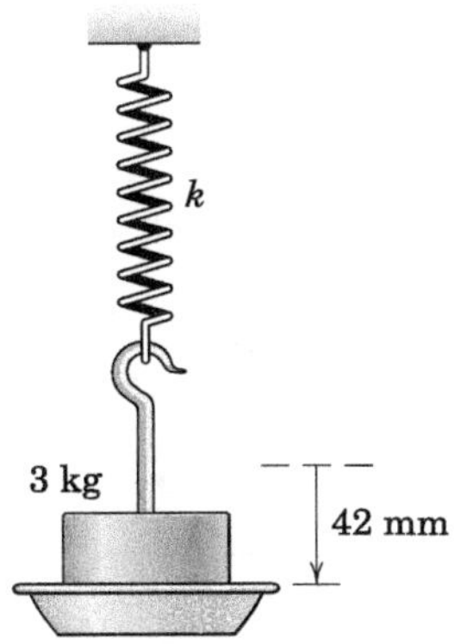

Problem 8/1

8/2 Determine the natural frequency of the spring-mass system in both radians per second and cycles per second (Hz).

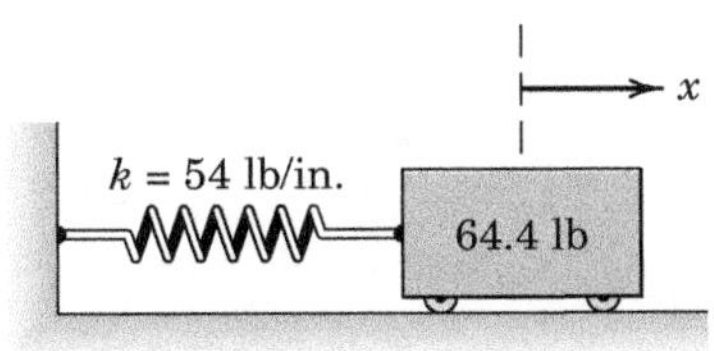

Problem 8/2

8/3 For the system of Prob. 8/2, determine the position x of the mass as a function of time if the mass is released from rest at time $t = 0$ from a position 2 inches to the left of the equilibrium position. Determine the maximum velocity and maximum acceleration of the mass over one cycle of motion.

8/4 For the system of Prob. 8/2, determine the position x as a function of time if the mass is released at time $t = 0$ from a position 2 inches to the right of the equilibrium position with an initial velocity of 9 in./sec to the left. Determine the amplitude C and period τ of the motion.

8/5 For the spring-mass system shown, determine the static deflection δ_{st}, the system period τ, and the maximum velocity v_{max} which result if the cylinder is displaced 100 mm downward from its equilibrium position and released from rest.

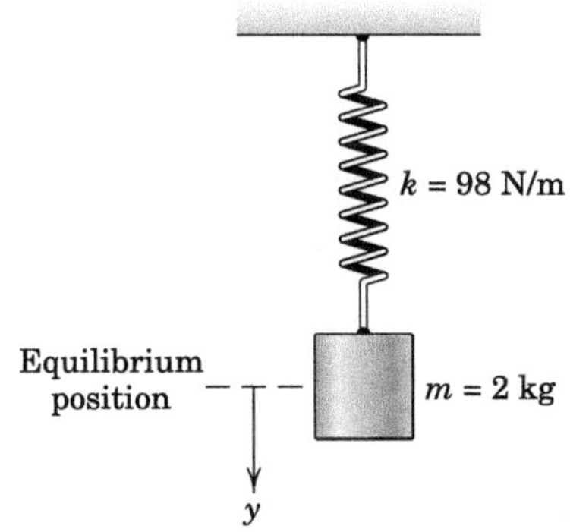

Problem 8/5

8/6 The cylinder of the system of Prob. 8/5 is displaced 100 mm downward from its equilibrium position and released at time $t = 0$. Determine the position y, velocity v, and acceleration a when $t = 3$ s. What is the maximum acceleration?

8/7 Determine the natural frequency in cycles per second for the system shown. Neglect the mass and friction of the pulleys. Assume that the block of mass m remains horizontal.

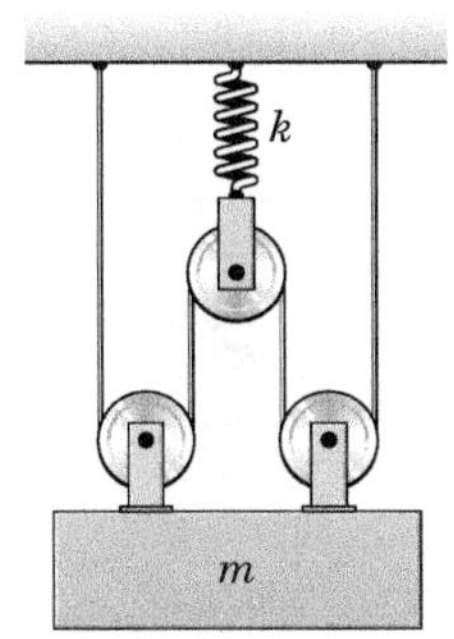

Problem 8/7

8/8 The vertical plunger has a mass of 2.5 kg and is supported by the two springs, which are always in compression. Calculate the natural frequency f_n of vibration of the plunger if it is deflected from the equilibrium position and released from rest. Friction in the guide is negligible.

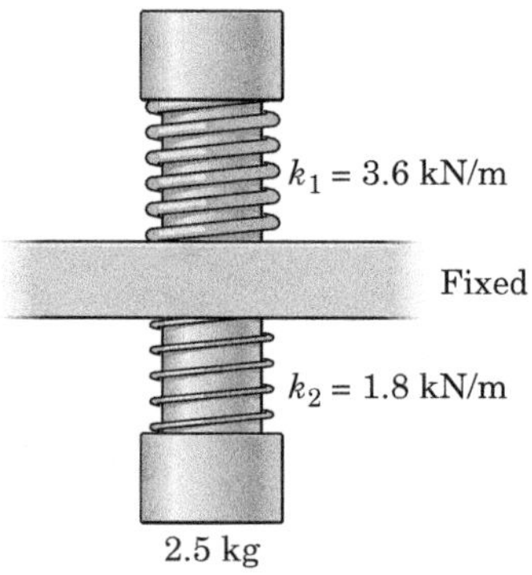

Problem 8/8

8/9 Determine the period τ for the system shown. The cable is always taut, and the mass and friction of the pulley are to be neglected.

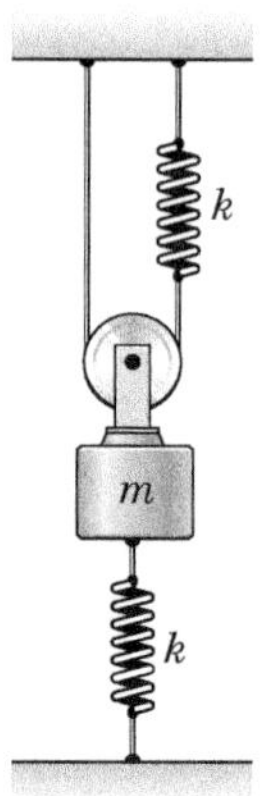

Problem 8/9

8/10 In the equilibrium position, the 30-kg cylinder causes a static deflection of 50 mm in the coiled spring. If the cylinder is depressed an additional 25 mm and released from rest, calculate the resulting natural frequency f_n of vertical vibration of the cylinder in cycles per second (Hz).

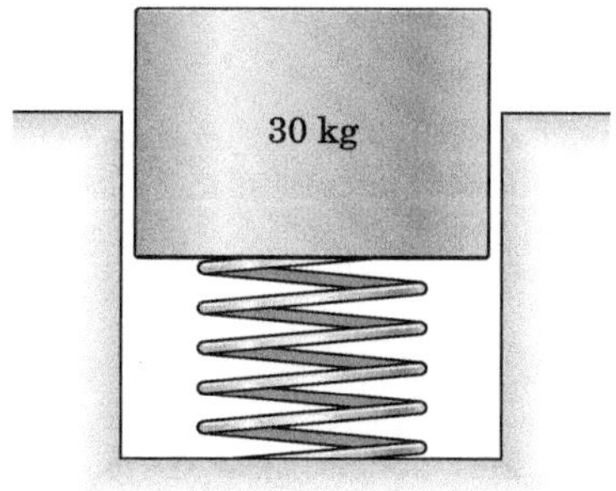

Problem 8/10

8/11 For the cylinder of Prob. 8/10, determine the vertical displacement x, measured positive down in millimeters from the equilibrium position, in terms of the time t in seconds measured from the instant of release from the position of 25 mm added deflection.

8/12 Determine the natural frequency in radians per second for the system shown. Neglect the mass and friction of the pulleys.

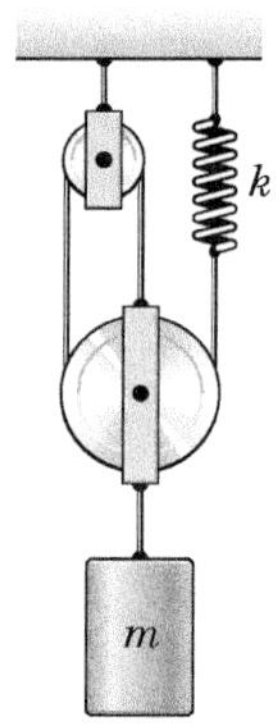

Problem 8/12

8/13 An old car being moved by a magnetic crane pickup is dropped from a short distance above the ground. Neglect any damping effects of its worn-out shock absorbers and calculate the natural frequency f_n in cycles per second (Hz) of the vertical vibration which occurs after impact with the ground. Each of the four springs on the 1000-kg car has a constant of 17.5 kN/m. Because the center of mass is located midway between the axles and the car is level when dropped, there is no rotational motion. State any assumptions.

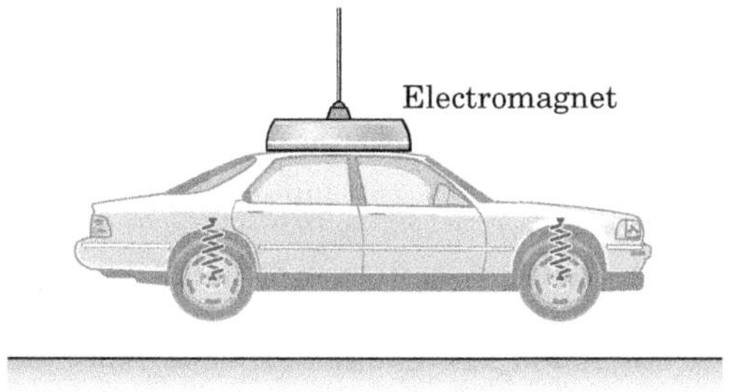

Problem 8/13

8/14 The 4-oz slider oscillates in the fixed slot under the action of the three springs, each of stiffness $k = 0.5$ lb/in. If the initial conditions at time $t = 0$ are $x_0 = 0.1$ in. and $\dot{x}_0 = 0.5$ in./sec, determine the position and velocity of the slider at time $t = 2$ sec. What is the system period?

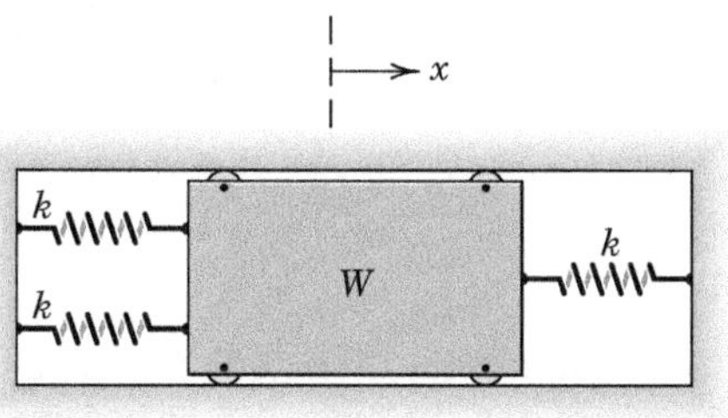

Problem 8/14

8/15 During the design of the spring-support system for the 4000-kg weighing platform, it is decided that the frequency of free vertical vibration in the unloaded condition shall not exceed 3 cycles per second. (*a*) Determine the maximum acceptable spring constant k for each of the three identical springs. (*b*) For this spring constant, what would be the natural frequency f_n of vertical vibration of the platform loaded by the 40-Mg truck?

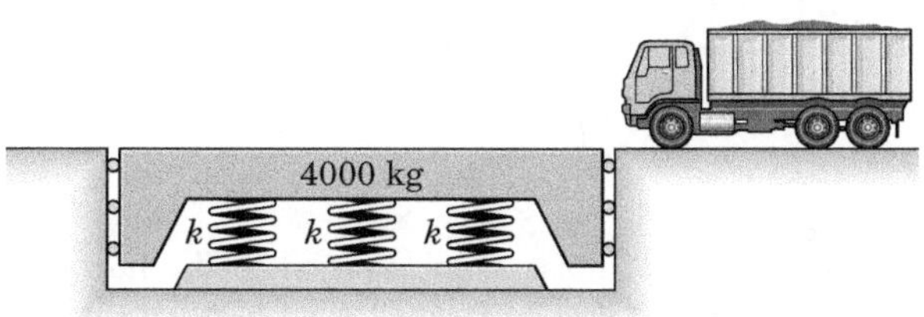

Problem 8/15

8/16 Calculate the natural frequency f_n of vibration if the mass is deflected from its equilibrium position and released from rest. Each pair of springs is connected by an inextensible cable. Evaluate your results for $m = 15$ kg, $k_1 = 225$ N/m, and $k_2 = 150$ N/m.

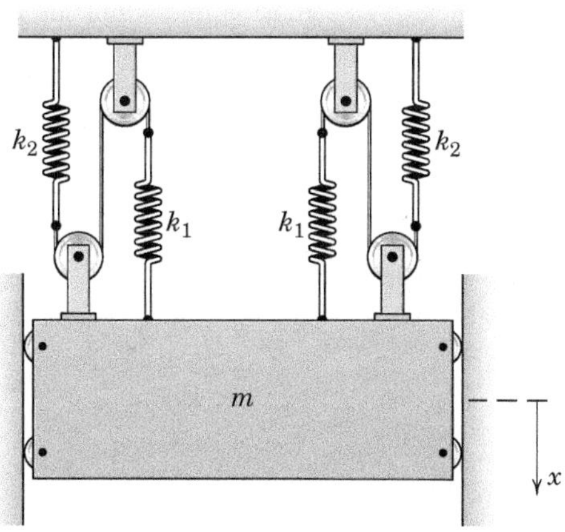

Problem 8/16

8/17 Replace the springs in each of the two cases shown by a single spring of stiffness k (equivalent spring stiffness) which will cause each mass to vibrate with its original frequency.

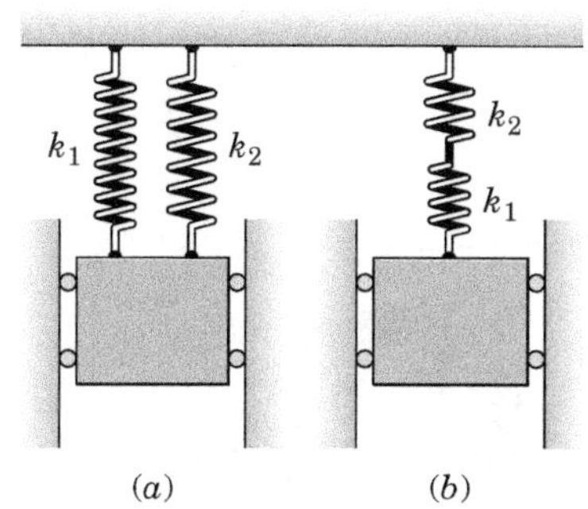

Problem 8/17

8/18 With the assumption of no slipping, determine the mass m of the block which must be placed on the top of the 6-kg cart in order that the system period be 0.75 s. What is the minimum coefficient μ_s of static friction for which the block will not slip relative to the cart if the cart is displaced 50 mm from the equilibrium position and released?

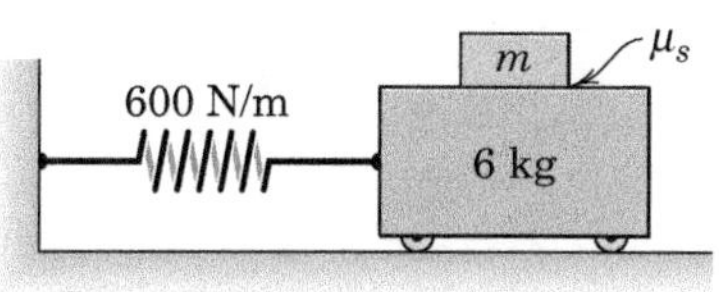

Problem 8/18

8/19 If both springs are unstretched when the mass is in the central position shown, determine the static deflection δ_{st} of the mass. What is the period of oscillatory motion about the position of static equilibrium?

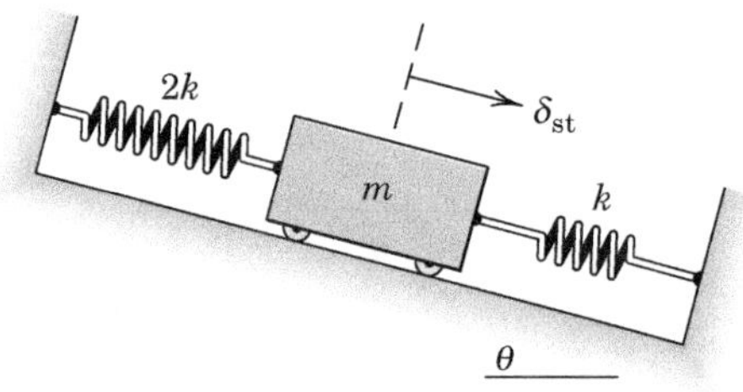

Problem 8/19

8/20 An energy-absorbing car bumper with its springs initially undeformed has an equivalent spring constant of 3000 lb/in. If the 2500-lb car approaches a massive wall with a speed of 5 mi/hr, determine (*a*) the velocity v of the car as a function of time during contact with the wall, where $t = 0$ is the beginning of the impact, and (*b*) the maximum deflection x_{max} of the bumper.

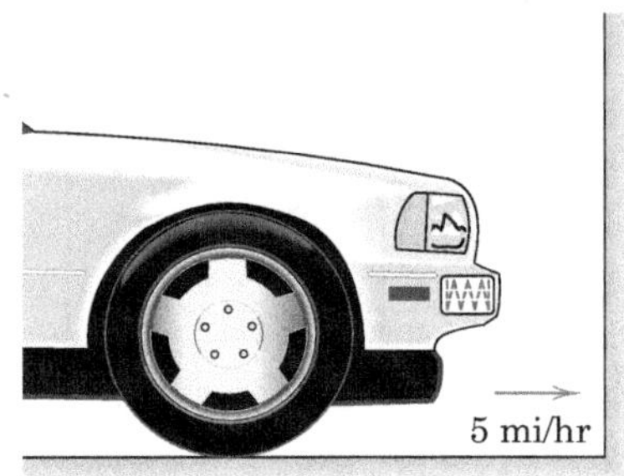

Problem 8/20

8/21 A 90-kg man stands at the end of a diving board and causes a vertical oscillation which is observed to have a period of 0.6 s. What is the static deflection δ_{st} at the end of the board? Neglect the mass of the board.

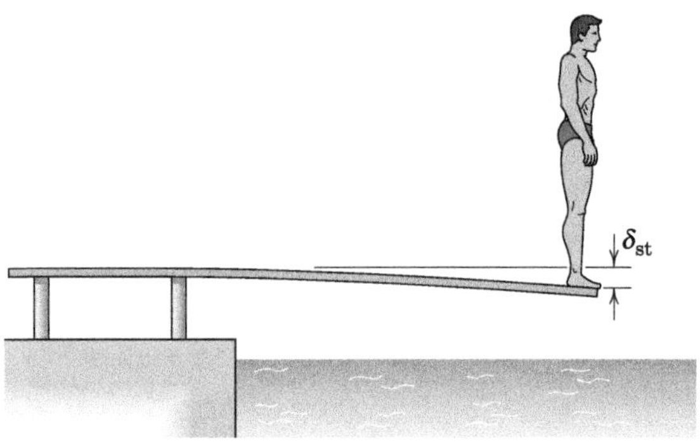

Problem 8/21

8/22 Shown in the figure is a model of a one-story building. The bar of mass m is supported by two light elastic upright columns whose upper and lower ends are fixed against rotation. For each column, if a force P and corresponding moment M were applied as shown in the right-hand part of the figure, the deflection δ would be given by $\delta = PL^3/12EI$, where L is the effective column length, E is Young's modulus, and I is the area moment of inertia of the column cross section with respect to its neutral axis. Determine the natural frequency of horizontal oscillation of the bar when the columns bend as shown in the figure.

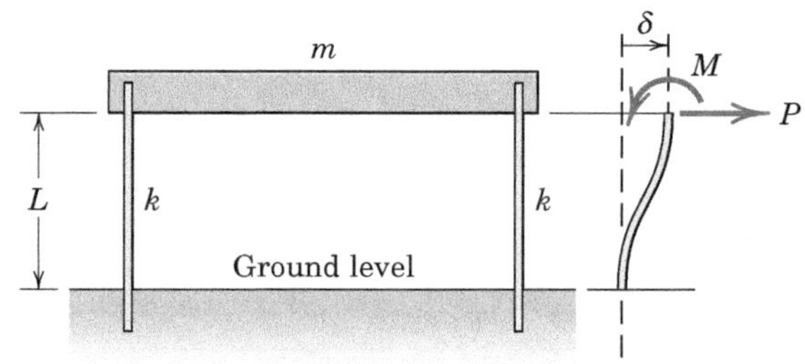

Problem 8/22

8/23 Calculate the natural circular frequency ω_n of the system shown in the figure. The mass and friction of the pulleys are negligible.

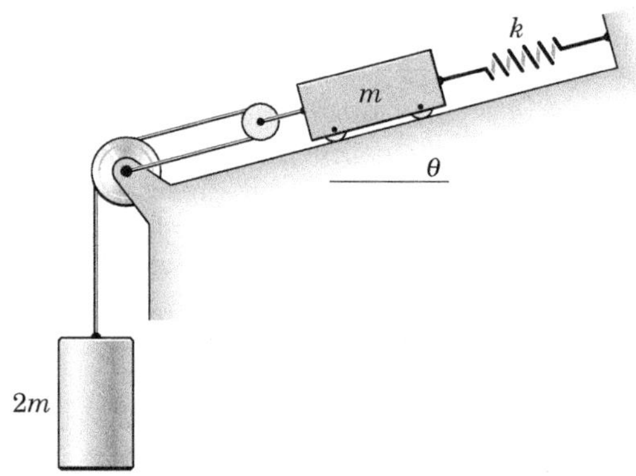

Problem 8/23

►**8/24** The slider of mass m is confined to the horizontal slot shown. The two springs each of constant k are linear. Derive the nonlinear equation of motion for small values of y, retaining terms of order y^3 and larger. Both springs are unstretched when $y = 0$. Neglect friction.

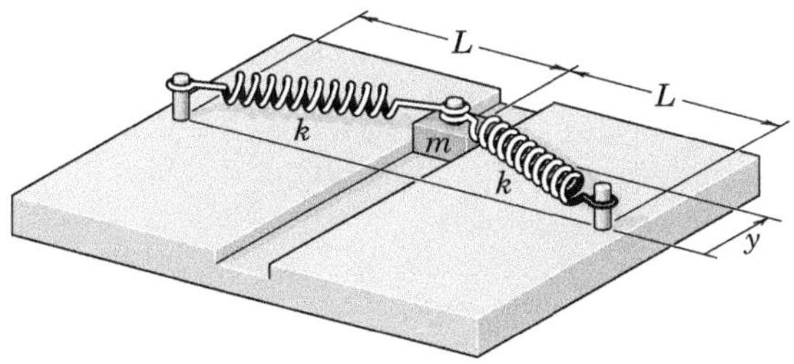

Problem 8/24

Damped, Free Vibrations

8/25 Determine the value of the damping ratio ζ for the simple spring-mass-dashpot system shown.

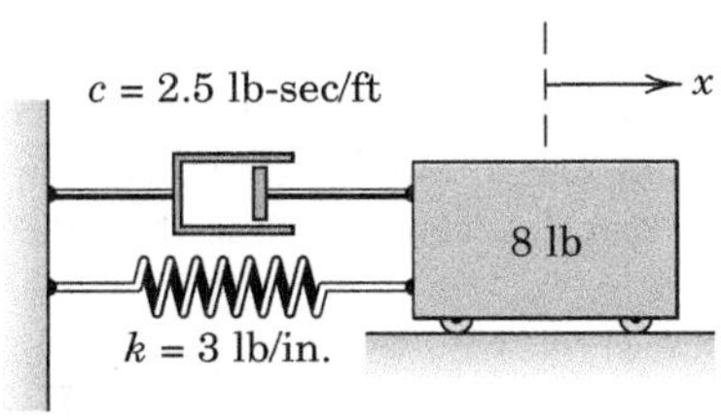

Problem 8/25

8/26 The period τ_d of damped linear oscillation for a certain 1-kg mass is 0.3 s. If the stiffness of the supporting linear spring is 800 N/m, calculate the damping coefficient c.

8/27 Viscous damping is added to an initially undamped spring-mass system. For what value of the damping ratio ζ will the damped natural frequency ω_d be equal to 90 percent of the natural frequency of the original undamped system?

8/28 The addition of damping to an undamped spring-mass system causes its period to increase by 25 percent. Determine the damping ratio ζ.

8/29 Determine the value of the viscous damping coefficient c for which the system shown is critically damped.

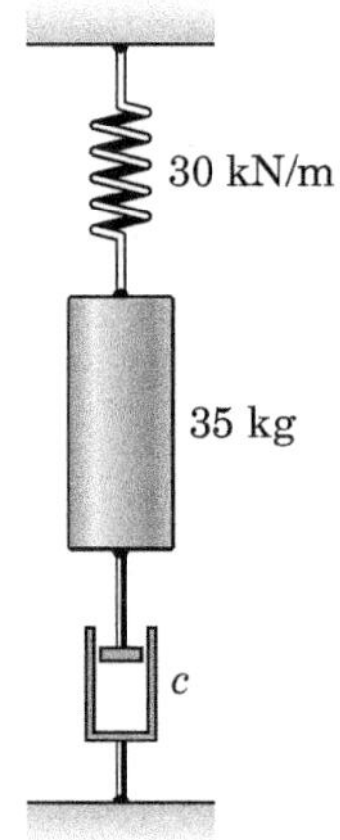

Problem 8/29

8/30 The 8-lb body of Prob. 8/25 is released from rest a distance x_0 to the right of the equilibrium position. Determine the displacement x as a function of time t, where $t = 0$ is the time of release.

8/31 The figure represents the measured displacement-time relationship for a vibration with small damping where it is impractical to achieve accurate results by measuring the nearly equal amplitudes of two successive cycles. Modify the expression for the viscous damping factor ζ based on the measured amplitudes x_0 and x_N which are N cycles apart.

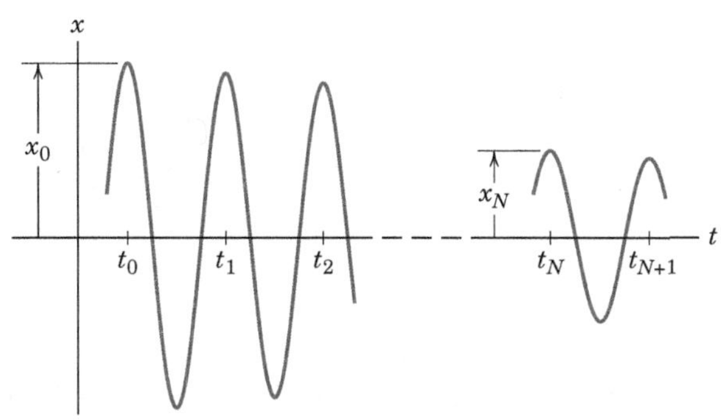

Problem 8/31

8/32 A linear harmonic oscillator having a mass of 1.10 kg is set into motion with viscous damping. If the frequency is 10 Hz and if two successive amplitudes a full cycle apart are measured to be 4.65 mm and 4.30 mm as shown, compute the viscous damping coefficient c.

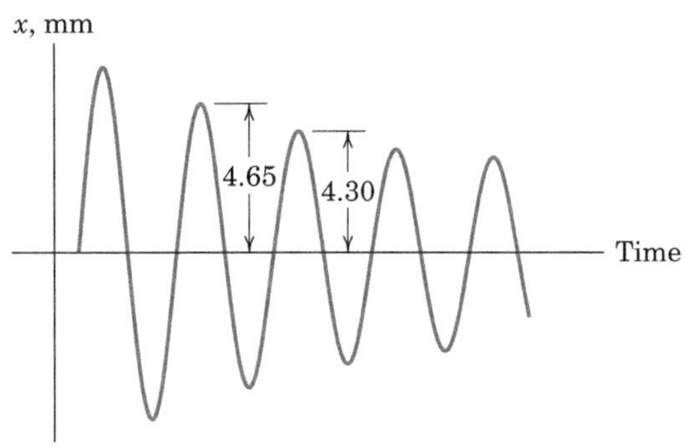

Problem 8/32

8/33 Determine the damping ratio ζ for the system shown. The system parameters are $m = 4$ kg, $k = 500$ N/m, and $c = 100$ N·s/m. Neglect the mass and friction of all pulleys, and assume that the cord remains taut throughout a motion cycle.

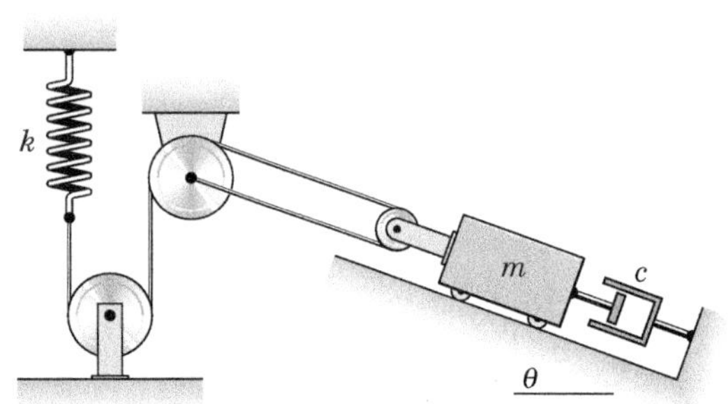

Problem 8/33

8/34 Further design refinement for the weighing platform of Prob. 8/15 is shown here where two viscous dampers are to be added to limit the ratio of successive positive amplitudes of vertical vibration in the unloaded condition to 4. Determine the necessary viscous damping coefficient c for each of the dampers.

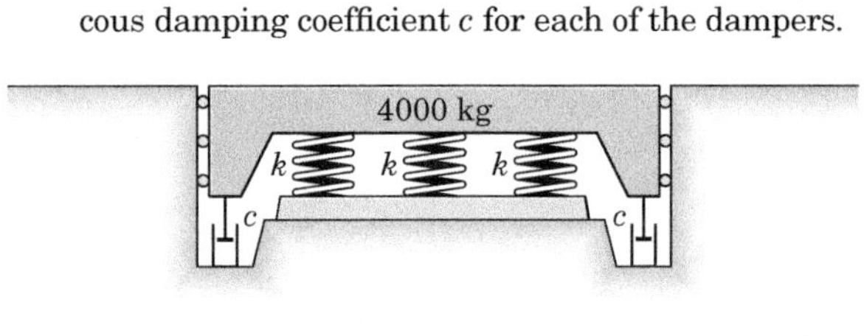

Problem 8/34

8/35 Derive the differential equation of motion for the system shown in terms of the variable x_1. Neglect friction and the mass of the linkage.

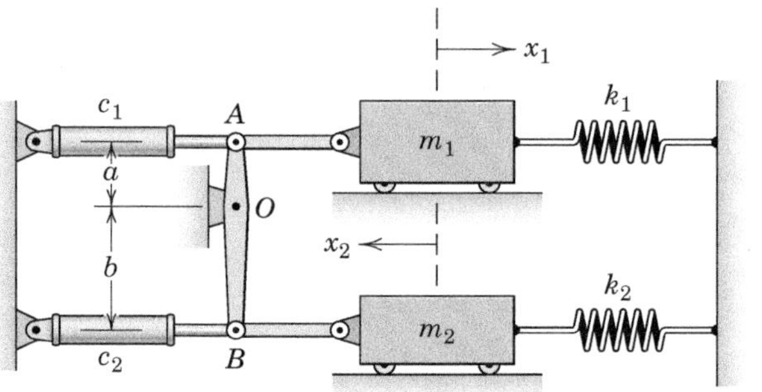

Problem 8/35

8/36 The system shown is released from rest from an initial position x_0. Determine the overshoot displacement x_1. Assume translational motion in the x-direction.

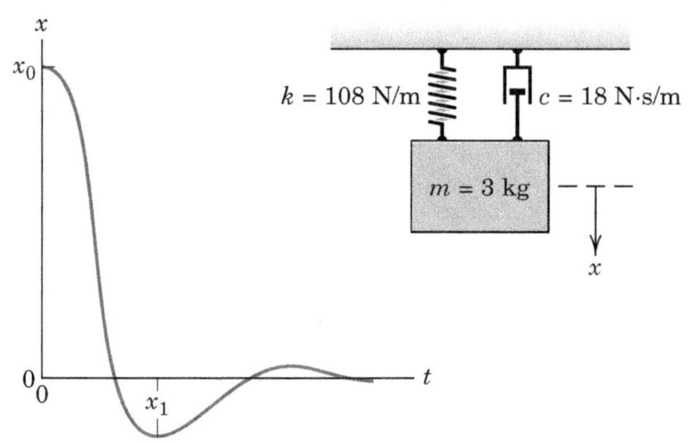

Problem 8/36

8/37 Determine the equation of motion for the system in terms of the variable x. The cables remain taut at all times, and the pulleys turn independently. Neglect friction and the mass of the pulleys. Additionally, determine expressions for the natural circular frequency ω_n and the damping ratio ζ.

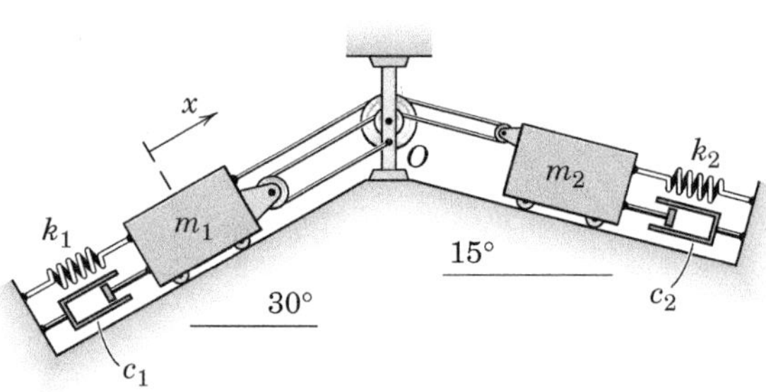

Problem 8/37

8/38 The mass of a given critically damped system is released at time $t = 0$ from the position $x_0 > 0$ with a negative initial velocity. Determine the critical value $(\dot{x}_0)_c$ of the initial velocity below which the mass will pass through the equilibrium position.

8/39 The mass of the system shown is released from rest at $x_0 = 6$ in. when $t = 0$. Determine the displacement x at $t = 0.5$ sec if (*a*) $c = 12$ lb-sec/ft and (*b*) $c = 18$ lb-sec/ft.

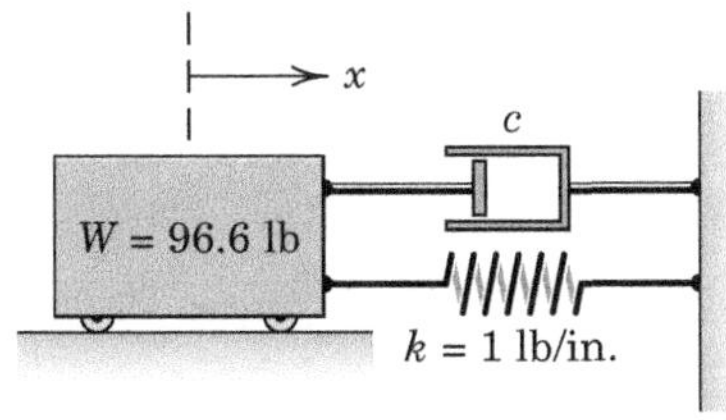

Problem 8/39

8/40 Derive the equation of motion for the system shown in terms of the displacement x. The masses are coupled through the light connecting rod *ABC* which pivots about the smooth bearing at point *O*. Neglect all friction, consider the rollers on m_2 and m_3 to be light, and assume small oscillations about the equilibrium position. State the system natural circular frequency ω_n and the viscous damping ratio ζ for $m_1 = 15$ kg, $m_2 = 12$ kg, $m_3 = 8$ kg, $k_1 = 400$ N/m, $k_2 = 650$ N/m, $k_3 = 225$ N/m, $c_1 = 44$ N·s/m, $c_2 = 36$ N·s/m, $c_3 = 52$ N·s/m, $a = 1.2$ m, $b = 1.8$ m, and $c = 0.9$ m.

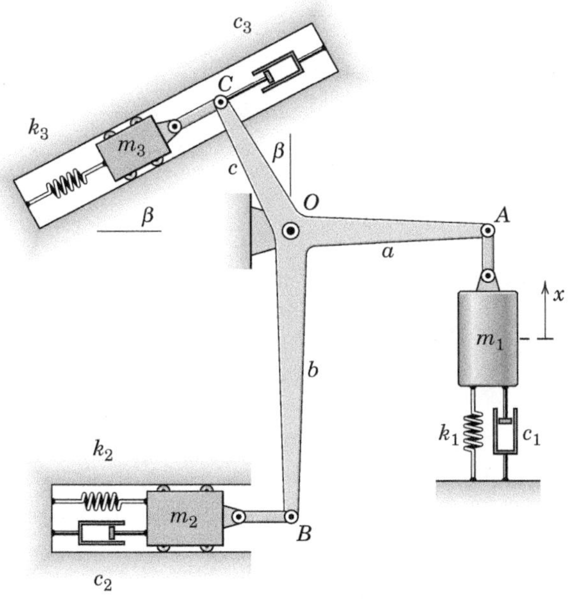

Problem 8/40

8/41 The owner of a 3400-lb pickup truck tests the action of his rear-wheel shock absorbers by applying a steady 100-lb force to the rear bumper and measuring a static deflection of 3 in. Upon sudden release of the force, the bumper rises and then falls to a maximum of $\frac{1}{2}$ in. below the unloaded equilibrium position of the bumper on the first rebound. Treat the action as a one-dimensional problem with an equivalent mass of half the truck mass. Find the viscous damping factor ζ for the rear end and the viscous damping coefficient c for each shock absorber assuming its action to be vertical.

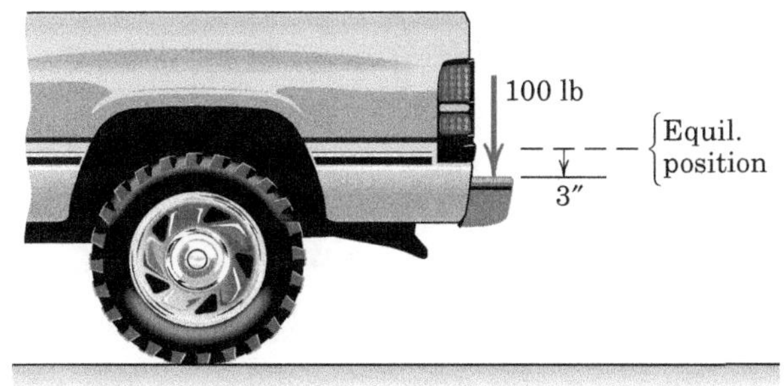

Problem 8/41

8/42 The 2-kg mass is released from rest at a distance x_0 to the right of the equilibrium position. Determine the displacement x as a function of time.

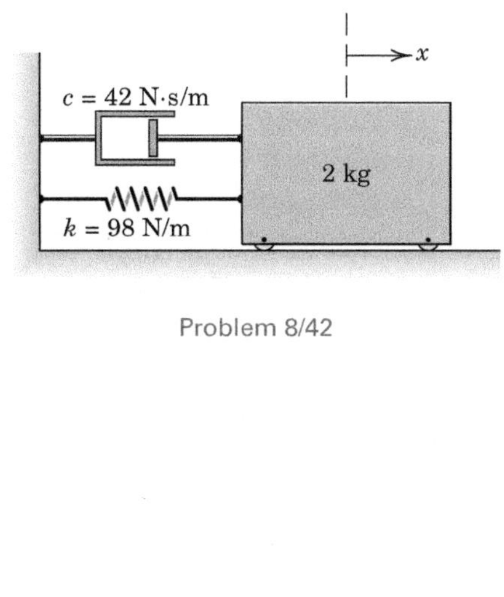

Problem 8/42

8/43 Develop the equation of motion in terms of the variable x for the system shown. Determine an expression for the damping ratio ζ in terms of the given system properties. Neglect the mass of the crank AB and assume small oscillations about the equilibrium position shown.

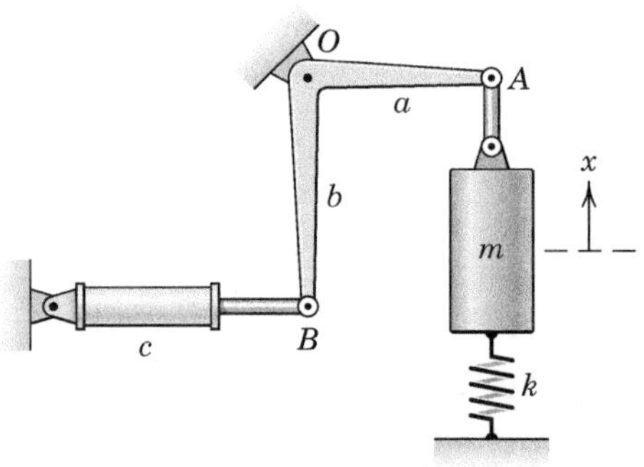

Problem 8/43

▶8/44 Investigate the case of Coulomb damping for the block shown, where the coefficient of kinetic friction is μ_k and each spring has a stiffness $k/2$. The block is displaced a distance x_0 from the neutral position and released. Determine and solve the differential equation of motion. Plot the resulting vibration and indicate the rate r of decay of the amplitude with time.

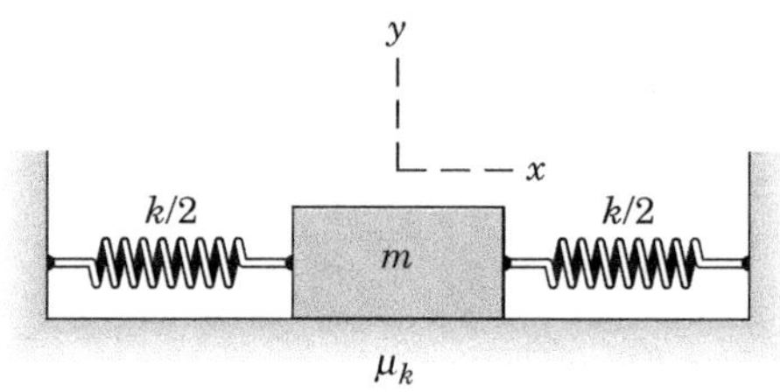

Problem 8/44

8/3 Forced Vibration of Particles

Although there are many significant applications of free vibrations, the most important class of vibration problems is that where the motion is continuously excited by a disturbing force. The force may be externally applied or may be generated within the system by such means as unbalanced rotating parts. Forced vibrations may also be excited by the motion of the system foundation.

Courtesy of MTS Systems Corporation

An automobile undergoing vibration testing of its suspension system.

Harmonic Excitation

Various forms of forcing functions $F = F(t)$ and foundation displacements $x_B = x_B(t)$ are depicted in Fig. 8/8. The harmonic force shown in part a of the figure occurs frequently in engineering practice, and the understanding of the analysis associated with harmonic forces is a necessary first step in the study of more complex forms. For this reason, we will focus our attention on harmonic excitation.

We first consider the system of Fig. 8/9a, where the body is subjected to the external harmonic force $F = F_0 \sin \omega t$, in which F_0 is the force amplitude and ω is the driving frequency (in radians per second). Be sure to distinguish between $\omega_n = \sqrt{k/m}$, which is a property of the system, and ω, which is a property of the force applied to the system. We also note that for a force $F = F_0 \cos \omega t$, one merely substitutes $\cos \omega t$ for $\sin \omega t$ in the results about to be developed.

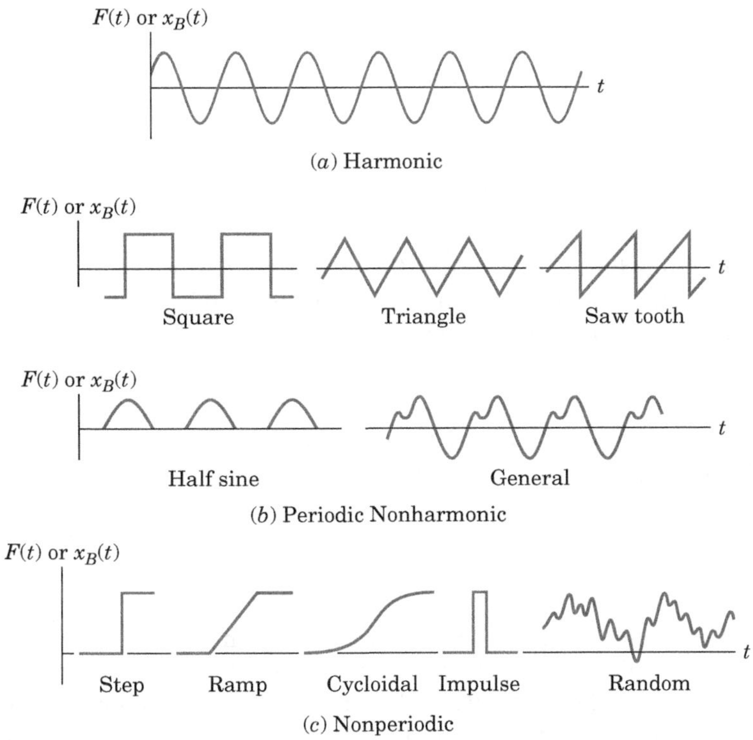

Figure 8/8

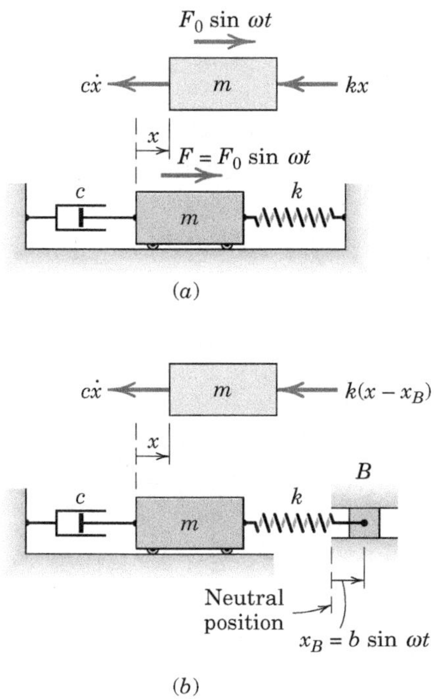

Figure 8/9

From the free-body diagram of Fig. 8/9*a*, we may apply Newton's second law to obtain

$$-kx - c\dot{x} + F_0 \sin \omega t = m\ddot{x}$$

In standard form, with the same variable substitutions made in Art. 8/2, the equation of motion becomes

$$\ddot{x} + 2\zeta\omega_n\dot{x} + \omega_n{}^2 x = \frac{F_0 \sin \omega t}{m} \tag{8/13}$$

Base Excitation

In many cases, the excitation of the mass is due not to a directly applied force but to the movement of the base or foundation to which the mass is connected by springs or other compliant mountings. Examples of such applications are seismographs, vehicle suspensions, and structures shaken by earthquakes.

Harmonic movement of the base is equivalent to the direct application of a harmonic force. To show this, consider the system of Fig. 8/9*b* where the spring is attached to the movable base. The free-body diagram shows the mass displaced a distance x from the neutral or equilibrium position it would have if the base were in its neutral position. The base, in turn, is assumed to have a harmonic movement $x_B = b \sin \omega t$. Note that the spring deflection is the difference between the inertial displacements of the mass and the base. From the free-body diagram, Newton's second law gives

$$-k(x - x_B) - c\dot{x} = m\ddot{x}$$

or

$$\ddot{x} + 2\zeta\omega_n\dot{x} + \omega_n{}^2 x = \frac{kb \sin \omega t}{m} \tag{8/14}$$

We see immediately that Eq. 8/14 is exactly the same as our basic equation of motion, Eq. 8/13, in that F_0 is replaced by kb. Consequently, all the results about to be developed apply to either Eq. 8/13 or 8/14.

Undamped Forced Vibration

First, we treat the case where damping is negligible ($c = 0$). Our basic equation of motion, Eq. 8/13, becomes

$$\ddot{x} + \omega_n{}^2 x = \frac{F_0}{m} \sin \omega t \tag{8/15}$$

The complete solution to Eq. 8/15 is the sum of the complementary solution x_c, which is the general solution of Eq. 8/15 with the right side set to zero, and the particular solution x_p, which is *any* solution to the complete equation. Thus, $x = x_c + x_p$. We developed the complementary solution in Art. 8/2. A particular solution is investigated by assuming that the form of the response to the force should resemble that of the force term. To that end, we assume

$$x_p = X \sin \omega t \qquad \textbf{(8/16)}$$

where X is the amplitude (in units of length) of the particular solution. Substituting this expression into Eq. 8/15 and solving for X yield

$$X = \frac{F_0/k}{1 - (\omega/\omega_n)^2} \qquad \textbf{(8/17)}$$

Thus, the particular solution becomes

$$x_p = \frac{F_0/k}{1 - (\omega/\omega_n)^2} \sin \omega t \qquad \textbf{(8/18)}$$

The complementary solution, known as the *transient solution*, is of no special interest here since, with time, it dies out with the small amount of damping which is always unavoidably present. The particular solution x_p describes the continuing motion and is called the *steady-state solution*. Its period is $\tau = 2\pi/\omega$, the same as that of the forcing function.

Of primary interest is the amplitude X of the motion. If we let δ_{st} stand for the magnitude of the static deflection of the mass under a static load F_0, then $\delta_{st} = F_0/k$, and we may form the ratio

$$M = \frac{X}{\delta_{st}} = \frac{1}{1 - (\omega/\omega_n)^2} \qquad \textbf{(8/19)}$$

The ratio M is called the *amplitude ratio* or *magnification factor* and is a measure of the severity of the vibration. We especially note that *M approaches infinity* as ω approaches ω_n. Consequently, if the system possesses no damping and is excited by a harmonic force whose frequency ω approaches the natural frequency ω_n of the system, then M, and thus X, increase without limit. Physically, this means that the motion amplitude would reach the limits of the attached spring, which is a condition to be avoided.

The value ω_n is called the *resonant* or *critical frequency* of the system, and the condition of ω being close in value to ω_n with the resulting large displacement amplitude X is called *resonance*. For $\omega < \omega_n$, the magnification factor M is positive, and the vibration is in phase with the force F. For $\omega > \omega_n$, the magnification factor is negative, and the vibration is 180° out of phase with F. Figure 8/10 shows a plot of the absolute value of M as a function of the driving-frequency ratio ω/ω_n.

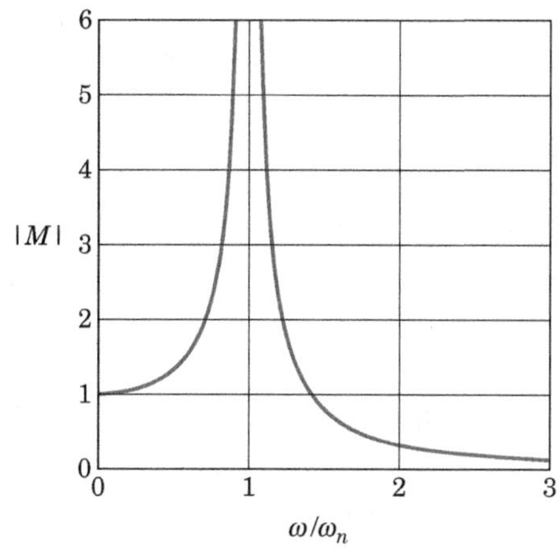

Figure 8/10

Damped Forced Vibration

We now reintroduce damping in our expressions for forced vibration. Our basic differential equation of motion is

$$\ddot{x} + 2\zeta\omega_n\dot{x} + \omega_n^2 x = \frac{F_0 \sin \omega t}{m} \qquad \textbf{[8/13]}$$

Again, the complete solution is the sum of the complementary solution x_c, which is the general solution of Eq. 8/13 with the right side equal to zero, and the particular solution x_p, which is *any* solution to the complete equation. We have already developed the complementary solution x_c in Art. 8/2. When damping is present, we find that a single sine or cosine term, such as we were able to use for the undamped case, is not sufficiently general for the particular solution. So we try

$$x_p = X_1 \cos \omega t + X_2 \sin \omega t \qquad \text{or} \qquad x_p = X \sin (\omega t - \phi)$$

Substitute the latter expression into Eq. 8/13, match coefficients of sin ωt and cos ωt, and solve the resulting two equations to obtain

$$X = \frac{F_0/k}{\{[1 - (\omega/\omega_n)^2]^2 + [2\zeta\omega/\omega_n]^2\}^{1/2}} \tag{8/20}$$

$$\phi = \tan^{-1}\left[\frac{2\zeta\omega/\omega_n}{1 - (\omega/\omega_n)^2}\right] \tag{8/21}$$

The complete solution is now known, and for underdamped systems it can be written as

$$x = Ce^{-\zeta\omega_n t} \sin (\omega_d t + \psi) + X \sin (\omega t - \phi) \tag{8/22}$$

Because the first term on the right side diminishes with time, it is known as the *transient solution*. The particular solution x_p is the *steady-state solution* and is the part of the solution in which we are primarily interested. All quantities on the right side of Eq. 8/22 are properties of the system and the applied force, except for C and ψ (which are determinable from initial conditions) and the running time variable t.

Magnification Factor and Phase Angle

Near resonance the magnitude X of the steady-state solution is a strong function of the damping ratio ζ and the nondimensional frequency ratio ω/ω_n. It is again convenient to form the nondimensional ratio $M = X/(F_0/k)$, which is called the *amplitude ratio* or *magnification factor*

$$M = \frac{1}{\{[1 - (\omega/\omega_n)^2]^2 + [2\zeta\omega/\omega_n]^2\}^{1/2}} \tag{8/23}$$

An accurate plot of the magnification factor M versus the frequency ratio ω/ω_n for various values of the damping ratio ζ is shown in Fig. 8/11. This figure reveals the most essential information pertinent to the forced vibration of a single-degree-of-freedom system under harmonic excitation. It is clear from the graph that, if a motion amplitude is excessive, two possible remedies would be to (a) increase the damping (to obtain a larger value of ζ) or (b) alter the driving frequency so that ω is farther from the resonant frequency ω_n. The addition of damping is most effective near resonance. Figure 8/11 also shows that, except for $\zeta = 0$, the magnification-factor curves do not actually peak at $\omega/\omega_n = 1$. The peak for any given value of ζ can be calculated by finding the maximum value of M from Eq. 8/23.

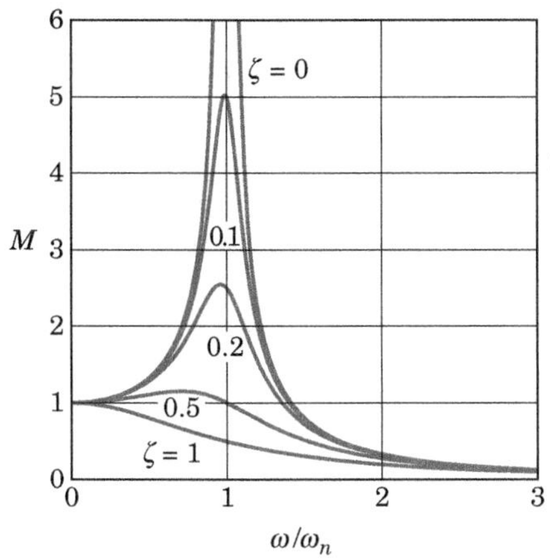

Figure 8/11

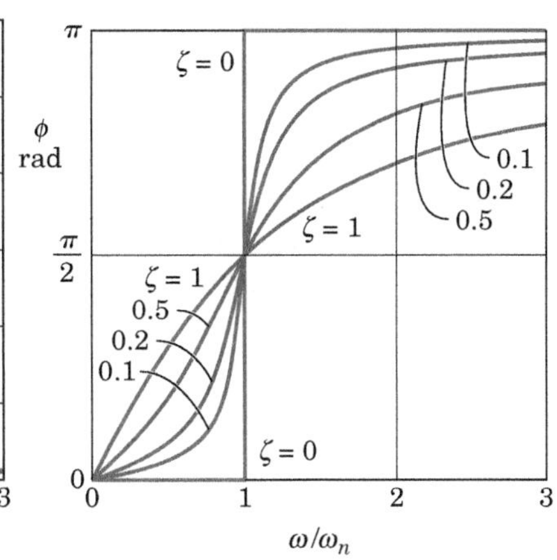

Figure 8/12

The phase angle ϕ, given by Eq. 8/21, can vary from 0 to π and represents the part of a cycle (and thus the time) by which the response x_p lags the forcing function F. Figure 8/12 shows how the phase angle ϕ varies with the frequency ratio for various values of the damping ratio ζ. Note that the value of ϕ when $\omega/\omega_n = 1$ is 90° for all values of ζ. To further illustrate the phase difference between the response and the forcing function, we show in Fig. 8/13 two examples of the variation of F and x_p with ωt. In the first example, $\omega < \omega_n$ and ϕ is taken to be $\pi/4$. In the second example, $\omega > \omega_n$ and ϕ is taken to be $3\pi/4$.

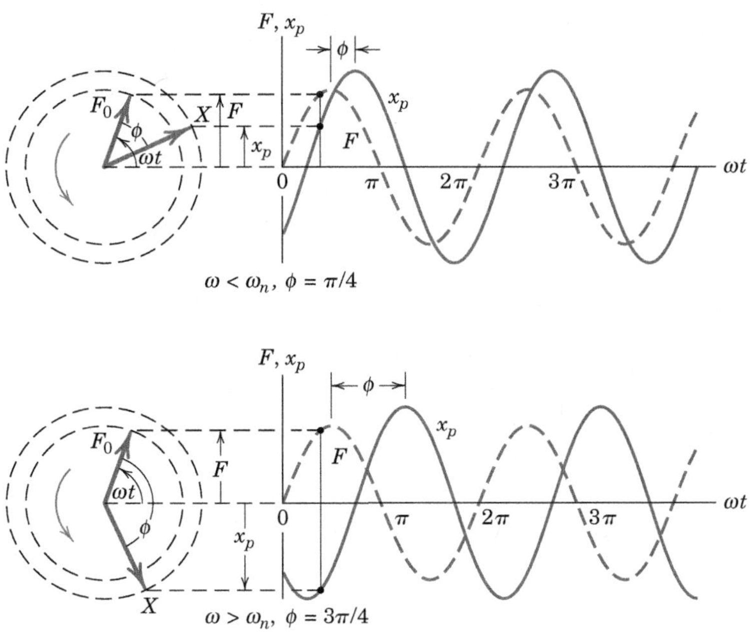

Figure 8/13

The seismograph is a useful application of the principles of this article.

Applications

Vibration-measuring instruments such as seismometers and accelerometers are frequently encountered applications of harmonic excitation. The elements of this class of instruments are shown in Fig. 8/14*a*. We note that the entire system is subjected to the motion x_B of the frame. Letting x denote the position of the mass *relative* to the frame, we may apply Newton's second law and obtain

$$-c\dot{x} - kx = m\frac{d^2}{dt^2}(x + x_B) \qquad \text{or} \qquad \ddot{x} + \frac{c}{m}\dot{x} + \frac{k}{m}x = -\ddot{x}_B$$

where $(x + x_B)$ is the inertial displacement of the mass. If $x_B = b \sin \omega t$, then our equation of motion with the usual notation is

$$\ddot{x} + 2\zeta\omega_n\dot{x} + \omega_n{}^2x = b\omega^2 \sin \omega t$$

which is the same as Eq. 8/13 if $b\omega^2$ is substituted for F_0/m.

Again, we are interested only in the steady-state solution x_p. Thus, from Eq. 8/20, we have

$$x_p = \frac{b(\omega/\omega_n)^2}{\{[1 - (\omega/\omega_n)^2]^2 + [2\zeta\omega/\omega_n]^2\}^{1/2}} \sin(\omega t - \phi)$$

If X represents the amplitude of the relative response x_p, then the nondimensional ratio X/b is

$$X/b = (\omega/\omega_n)^2 M$$

where M is the magnification ratio of Eq. 8/23. A plot of X/b as a function of the driving-frequency ratio ω/ω_n is shown in Fig. 8/14*b*. The similarities and differences between the magnification ratios of Figs. 8/14*b* and 8/11 should be noted.

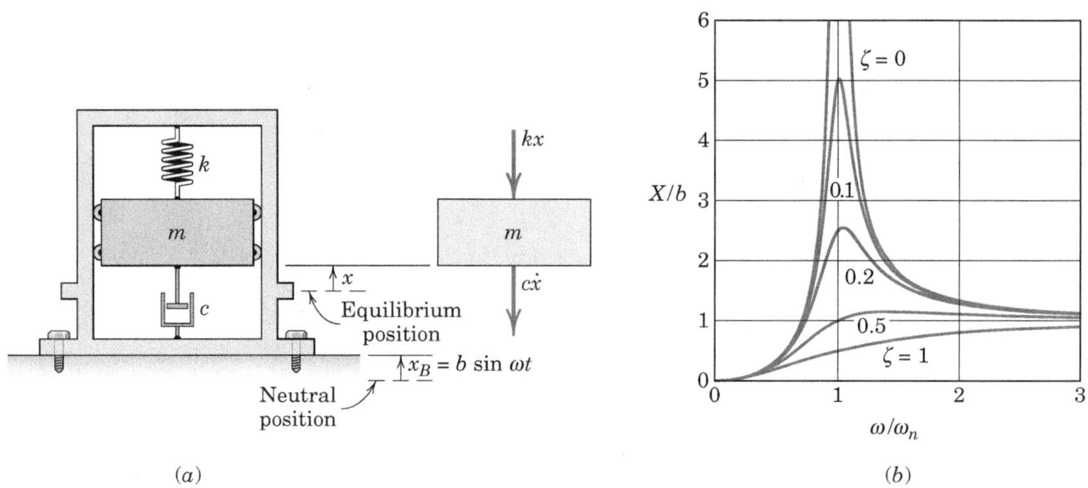

Figure 8/14

If the frequency ratio ω/ω_n is large, then $X/b \cong 1$ for all values of the damping ratio ζ. Under these conditions, the displacement of the mass relative to the frame is approximately the same as the absolute displacement of the frame, and the instrument acts as a *displacement meter*. To obtain a high value of ω/ω_n, we need a small value of $\omega_n = \sqrt{k/m}$, which means a soft spring and a large mass. With such a combination, the mass will tend to stay inertially fixed. Displacement meters generally have very light damping.

On the other hand, if the frequency ratio ω/ω_n is small, then M approaches unity (see Fig. 8/11) and $X/b \cong (\omega/\omega_n)^2$ or $X \cong b(\omega/\omega_n)^2$. But $b\omega^2$ is the maximum acceleration of the frame. Thus, X is proportional to the maximum acceleration of the frame, and the instrument may be used as an *accelerometer*. The damping ratio is generally selected so that M approximates unity over the widest possible range of ω/ω_n. From Fig. 8/11, we see that a damping factor somewhere between $\zeta = 0.5$ and $\zeta = 1$ would meet this criterion.

Electric Circuit Analogy

An important analogy exists between electric circuits and mechanical spring-mass systems. Figure 8/15 shows a series circuit consisting of a voltage E which is a function of time, an inductance L, a capacitance C, and a resistance R. If we denote the charge by the symbol q, the equation which governs the charge is

$$L\ddot{q} + R\dot{q} + \frac{1}{C}q = E \tag{8/24}$$

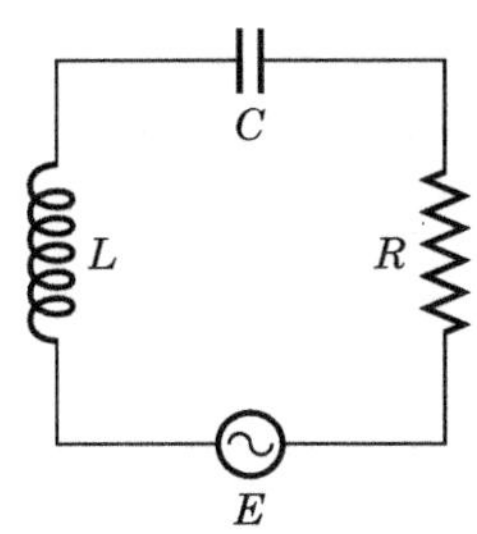

Figure 8/15

This equation has the same form as the equation for the mechanical system. Thus, by a simple interchange of symbols, the behavior of the electrical circuit may be used to predict the behavior of the mechanical system, or vice versa. The mechanical and electrical equivalents in the following table are worth noting:

MECHANICAL-ELECTRICAL EQUIVALENTS

MECHANICAL			ELECTRICAL			
QUANTITY	SYMBOL	SI UNIT	QUANTITY	SYMBOL		SI UNIT
Mass	m	kg	Inductance	L	H	henry
Spring stiffness	k	N/m	1/Capacitance	$1/C$	$1/F$	1/farad
Force	F	N	Voltage	E	V	volt
Velocity	$\dot{x}$	m/s	Current	I	A	ampere
Displacement	x	m	Charge	q	C	coulomb
Viscous damping constant	c	N·s/m	Resistance	R	Ω	ohm

Sample Problem 8/4

A 50-kg instrument is supported by four springs, each of stiffness 7500 N/m. If the instrument foundation undergoes harmonic motion given in meters by $x_B = 0.002 \cos 50t$, determine the amplitude of the steady-state motion of the instrument. Damping is negligible.

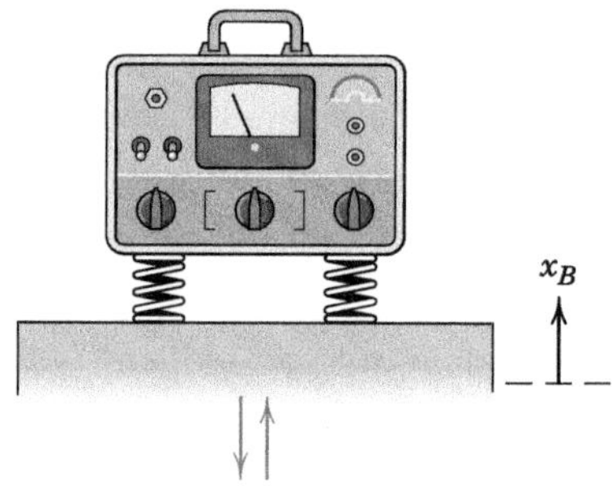

Solution. For harmonic oscillation of the base, we substitute kb for F_0 in our particular-solution results, so that, from Eq. 8/17, the steady-state amplitude becomes

① $$X = \frac{b}{1 - (\omega/\omega_n)^2}$$

The resonant frequency is $\omega_n = \sqrt{k/m} = \sqrt{4(7500)/50} = 24.5$ rad/s, and the impressed frequency $\omega = 50$ rad/s is given. Thus,

② $$X = \frac{0.002}{1 - (50/24.5)^2} = -6.32(10^{-4})\text{ m} \quad \text{or} \quad -0.632\text{ mm} \qquad \textit{Ans.}$$

Note that the frequency ratio ω/ω_n is approximately 2, so that the condition of resonance is avoided.

Helpful Hints

① Note that either $\sin 50t$ or $\cos 50t$ can be used for the forcing function with this same result.

② The minus sign indicates that the motion is 180° out of phase with the applied excitation.

Sample Problem 8/5

The spring attachment point B is given a horizontal motion $x_B = b \cos \omega t$. Determine the critical driving frequency ω_c for which the oscillations of the mass m tend to become excessively large. Neglect the friction and mass associated with the pulleys. The two springs have the same stiffness k.

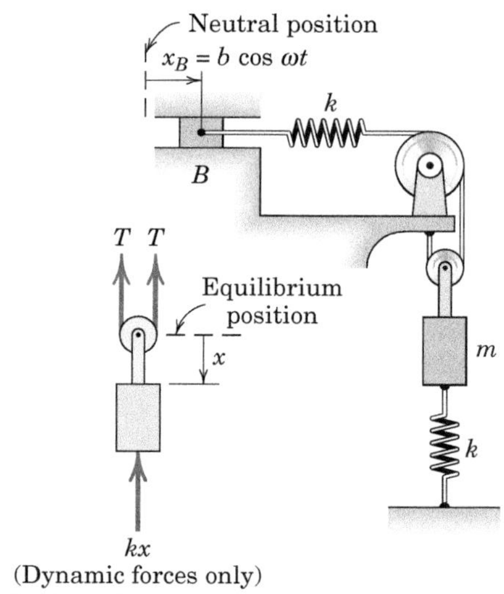

Solution. The free-body diagram is drawn for arbitrary positive displacements x and x_B. The motion variable x is measured downward from the position of static equilibrium defined as that which exists when $x_B = 0$. The additional stretch
① in the upper spring, beyond that which exists at static equilibrium, is $2x - x_B$. Therefore, the *dynamic* spring force in the upper spring, and hence the *dynamic*
② tension T in the cable, is $k(2x - x_B)$. Summing forces in the x-direction gives

$$[\Sigma F_x = m\ddot{x}] \qquad -2k(2x - x_B) - kx = m\ddot{x}$$

which becomes

$$\ddot{x} + \frac{5k}{m}x = \frac{2kb \cos \omega t}{m}$$

The natural frequency of the system is $\omega_n = \sqrt{5k/m}$. Thus,

$$\omega_c = \omega_n = \sqrt{5k/m} \qquad \textit{Ans.}$$

Helpful Hints

① If a review of the kinematics of constrained motion is necessary, see Art. 2/9.

② We learned from the discussion in Art. 8/2 that the equal and opposite forces associated with the position of static equilibrium may be omitted from the analysis. Our use of the terms *dynamic* spring force and *dynamic* tension stresses that only the force increments in addition to the static values need be considered.

Sample Problem 8/6

The 100-lb piston is supported by a spring of modulus $k = 200$ lb/in. A dashpot of damping coefficient $c = 85$ lb-sec/ft acts in parallel with the spring. A fluctuating pressure $p = 0.625 \sin 30t$ in lb/in.^{2} acts on the piston, whose top surface area is 80 in.^{2} Determine the steady-state displacement as a function of time and the maximum force transmitted to the base.

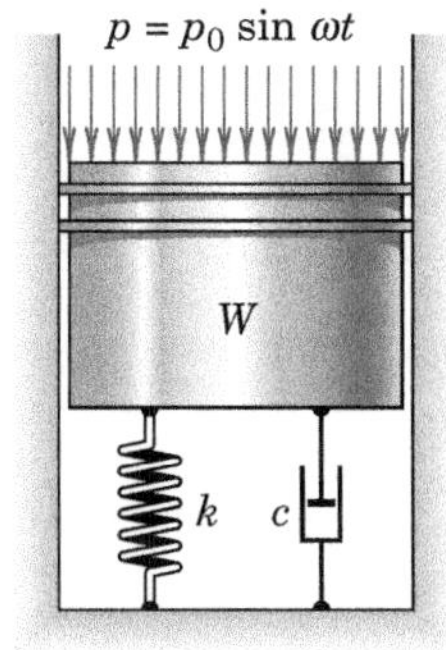

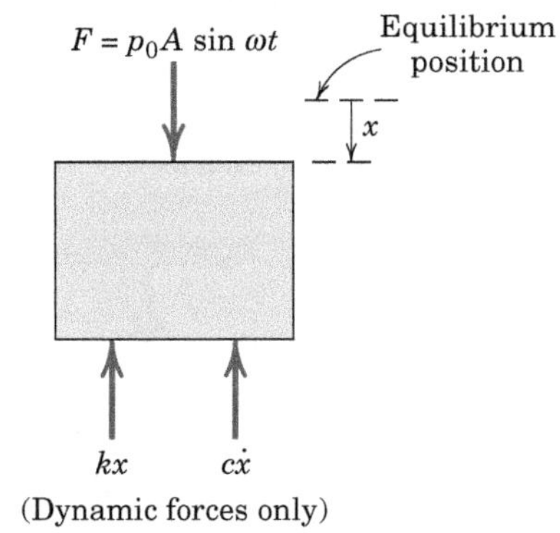

Solution. We begin by computing the system natural frequency and damping ratio:

$$\omega_n = \sqrt{\frac{k}{m}} = \sqrt{\frac{(200)(12)}{100/32.2}} = 27.8 \text{ rad/sec}$$

$$\zeta = \frac{c}{2m\omega_n} = \frac{85}{2\left(\frac{100}{32.2}\right)(27.8)} = 0.492 \text{ (underdamped)}$$

The steady-state amplitude, from Eq. 8/20, is

$$X = \frac{F_0/k}{\{[1-(\omega/\omega_n)^2]^2 + [2\zeta\omega/\omega_n]^2\}^{1/2}}$$

$$= \frac{(0.625)(80)/[(200)(12)]}{\{[1-(30/27.8)^2]^2 + [2(0.492)(30/27.8)]^2\}^{1/2}}$$

① $$= 0.01938 \text{ ft}$$

The phase angle, from Eq. 8/21, is

$$\phi = \tan^{-1}\left[\frac{2\zeta\omega/\omega_n}{1-(\omega/\omega_n)^2}\right]$$

② $$= \tan^{-1}\left[\frac{2(0.492)(30/27.8)}{1-(30/27.8)^2}\right]$$

$$= 1.724 \text{ rad}$$

The steady-state motion is then given by the second term on the right side of Eq. 8/22:

$$x_p = X \sin(\omega t - \phi) = 0.01938 \sin(30t - 1.724) \text{ ft} \qquad \textit{Ans.}$$

The force F_{tr} transmitted to the base is the sum of the spring and damper forces, or

$$F_{tr} = kx_p + c\dot{x}_p = kX\sin(\omega t - \phi) + c\omega X \cos(\omega t - \phi)$$

The maximum value of F_{tr} is

$$(F_{tr})_{max} = \sqrt{(kX)^2 + (c\omega X)^2} = X\sqrt{k^2 + c^2\omega^2}$$

$$= 0.01938\sqrt{[(200)(12)]^2 + (85)^2(30)^2}$$

① $$= 67.9 \text{ lb} \qquad \textit{Ans.}$$

Helpful Hints

① You are encouraged to repeat these calculations with the damping coefficient c set to zero so as to observe the influence of the relatively large amount of damping present.

② Note that the argument of the inverse tangent expression for ϕ has a positive numerator and a negative denominator for the case at hand, thus placing ϕ in the second quadrant. Recall that the defined range of ϕ is $0 \le \phi \le \pi$.

PROBLEMS

(Unless otherwise instructed, assume that the damping is light to moderate so that the amplitude of the forced response is a maximum at $\omega/\omega_n \cong 1$.)

Introductory Problems

8/45 A viscously damped spring-mass system is excited by a harmonic force of constant amplitude F_0 but varying frequency ω. If the amplitude of the steady-state motion is observed to decrease by a factor of 8 as the frequency ratio ω/ω_n is varied from 1 to 2, determine the damping ratio ζ of the system.

8/46 Determine the amplitude X of the steady-state motion of the 10-kg mass if (*a*) $c = 500$ N·s/m and (*b*) $c = 0$.

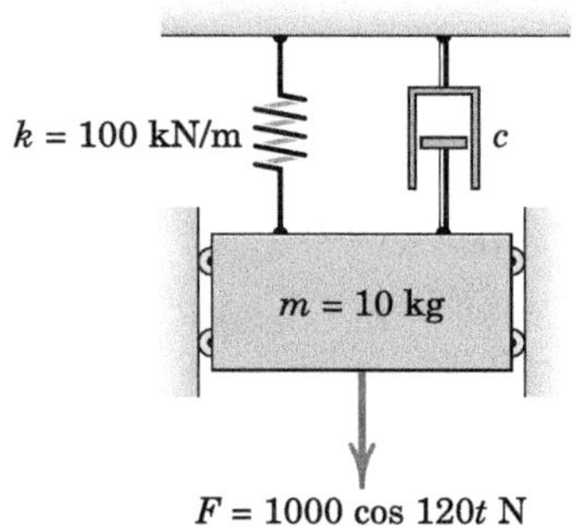

Problem 8/46

8/47 The 30-kg cart is acted upon by the harmonic force shown in the figure. If $c = 0$, determine the range of the driving frequency ω for which the magnitude of the steady-state response is less than 75 mm.

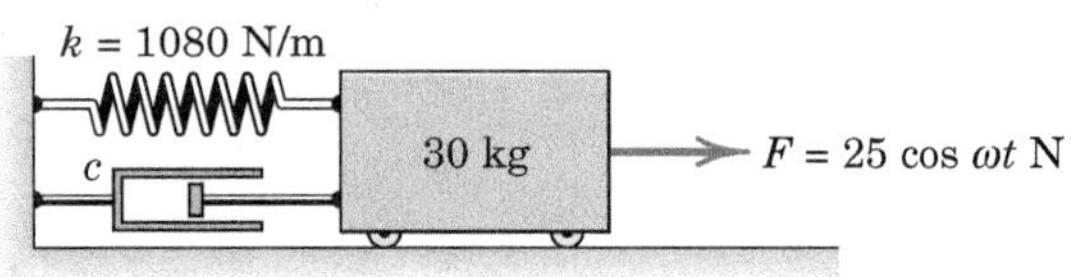

Problem 8/47

8/48 If the viscous damping coefficient of the damper in the system of Prob. 8/47 is $c = 36$ N·s/m, determine the range of the driving frequency ω for which the magnitude of the steady-state response is less than 75 mm.

8/49 If the driving frequency for the system of Prob. 8/47 is $\omega = 6$ rad/s, determine the required value of the damping coefficient c if the steady-state amplitude is not to exceed 75 mm.

8/50 A spring-mounted machine with a mass of 24 kg is observed to vibrate harmonically in the vertical direction with an amplitude of 0.30 mm under the action of a vertical force which varies harmonically between F_0 and $-F_0$ with a frequency of 4 Hz. Damping is negligible. If a static force of magnitude F_0 causes a deflection of 0.60 mm, calculate the equivalent spring constant k for the springs which support the machine.

8/51 The block of weight $W = 100$ lb is suspended by two springs each of stiffness $k = 200$ lb/ft and is acted upon by the force $F = 75 \cos 15t$ lb where t is the time in seconds. Determine the amplitude X of the steady-state motion if the viscous damping coefficient c is (*a*) 0 and (*b*) 60 lb-sec/ft. Compare these amplitudes to the static spring deflection δ_{st}.

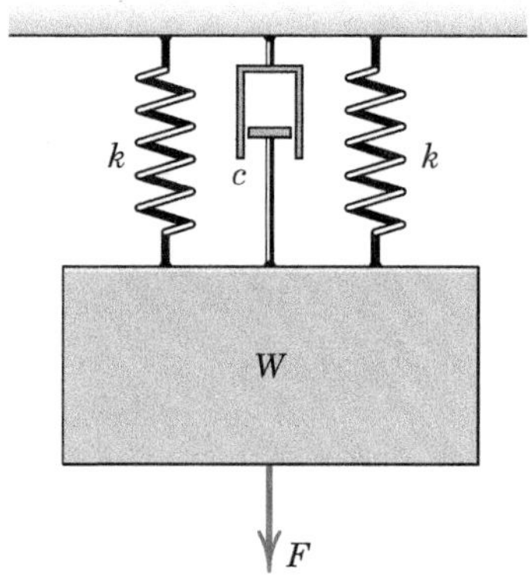

Problem 8/51

8/52 An external force $F = F_0 \sin \omega t$ is applied to the cylinder as shown. What value ω_c of the driving frequency would cause excessively large oscillations of the system?

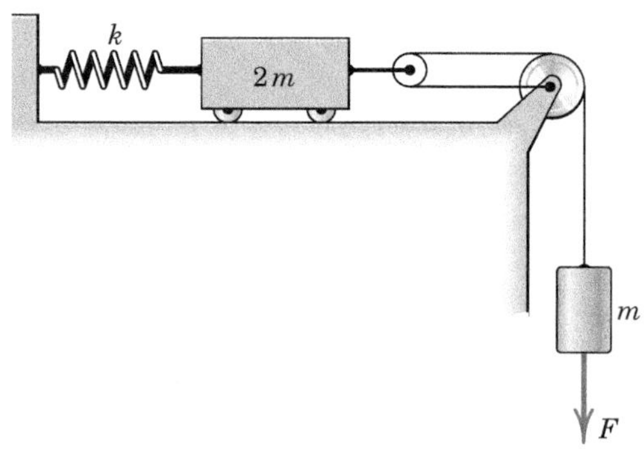

Problem 8/52

8/53 A viscously damped spring-mass system is forced harmonically at the undamped natural frequency ($\omega/\omega_n = 1$). If the damping ratio ζ is doubled from 0.1 to 0.2, compute the percentage reduction R_1 in the steady-state amplitude. Compare with the result R_2 of a similar calculation for the condition $\omega/\omega_n = 2$. Verify your results by inspecting Fig. 8/11.

8/54 The 4-lb body is attached to two springs, each of which has a stiffness of 6 lb/in. The body is mounted on a shake table which vibrates harmonically in the horizontal direction with an amplitude of 0.5 in. and a frequency f which can be varied. Power to the shake table is turned off when electrical contact is made at A or B. Determine the maximum value of the frequency f at which the shake table may be operated without turning the power off as it starts from rest and increases its frequency gradually. Damping may be neglected. The equilibrium position is centered between the fixed contacts.

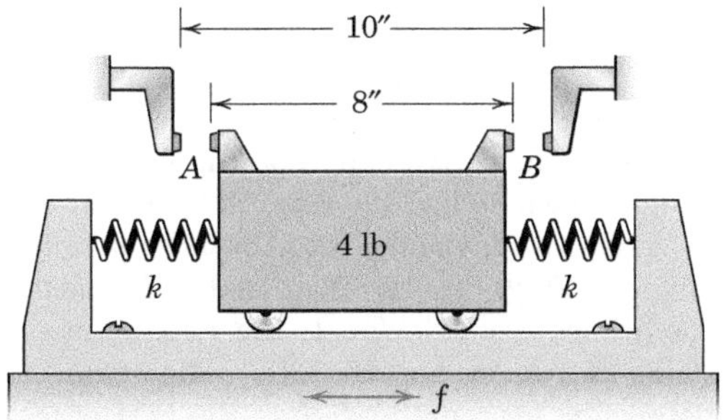

Problem 8/54

Representative Problems

8/55 It was noted in the text that the maxima of the curves for the magnification factor M are not located at $\omega/\omega_n = 1$. Determine an expression in terms of the damping ratio ζ for the frequency ratio at which the maxima occur.

8/56 The motion of the outer frame B is given by $x_B = b \sin \omega t$. For what range of the driving frequency ω is the amplitude of the motion of the mass m relative to the frame less than $2b$?

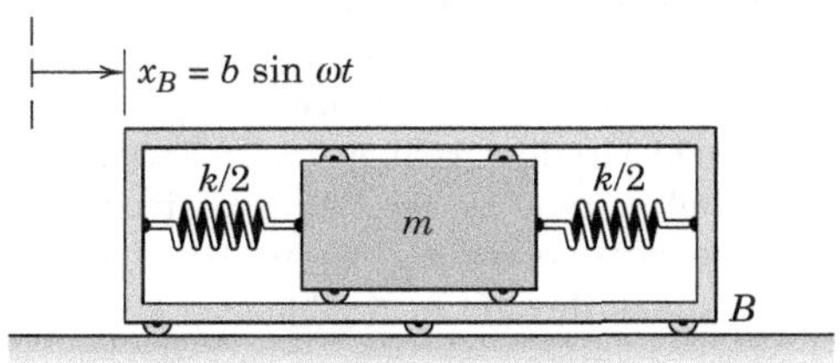

Problem 8/56

8/57 The 20-kg variable-speed motorized unit is restrained in the horizontal direction by two springs, each of which has a stiffness of 2.1 kN/m. Each of the two dashpots has a viscous damping coefficient $c = 58$ N·s/m. In what ranges of speeds N can the motor be run for which the magnification factor M will not exceed 2?

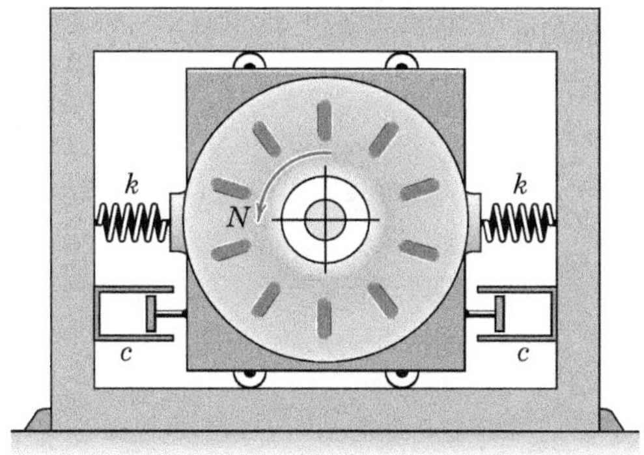

Problem 8/57

8/58 A single-cylinder four-stroke gasoline engine with a mass of 90 kg is mounted on four stiff spring pads, each with a stiffness of $30(10^3)$ kN/m, and is designed to run at 3600 rev/min. The mounting system is equipped with viscous dampers which have a large enough combined viscous damping coefficient c so that the system is critically damped when it is given a vertical displacement and then released while not running. When the engine is running, it fires every other revolution, causing a periodic vertical displacement modeled by $1.2 \cos \omega t$ mm with t in seconds. Determine the magnification factor M and the overall damping coefficient c.

8/59 When the person stands in the center of the floor system shown, he causes a static deflection δ_{st} of the floor under his feet. If he walks (or runs quickly!) in the same area, how many steps per second would cause the floor to vibrate with the greatest vertical amplitude?

Problem 8/59

8/60 The instrument shown has a mass of 43 kg and is spring-mounted to the horizontal base. If the amplitude of vertical vibration of the base is 0.10 mm, calculate the range of frequencies f_n of the base vibration which must be prohibited if the amplitude of vertical vibration of the instrument is not to exceed 0.15 mm. Each of the four identical springs has a stiffness of 7.2 kN/m.

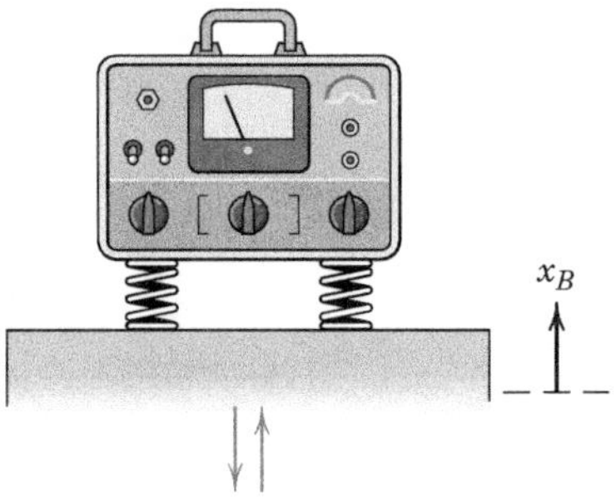

Problem 8/60

8/61 Derive the equation of motion for the *inertial* displacement x_i of the mass of Fig. 8/14. Comment on, but do not carry out, the solution to the equation of motion.

8/62 Attachment B is given a horizontal motion $x_B = b \cos \omega t$. Derive the equation of motion for the mass m and state the critical frequency ω_c for which the oscillations of the mass become excessively large.

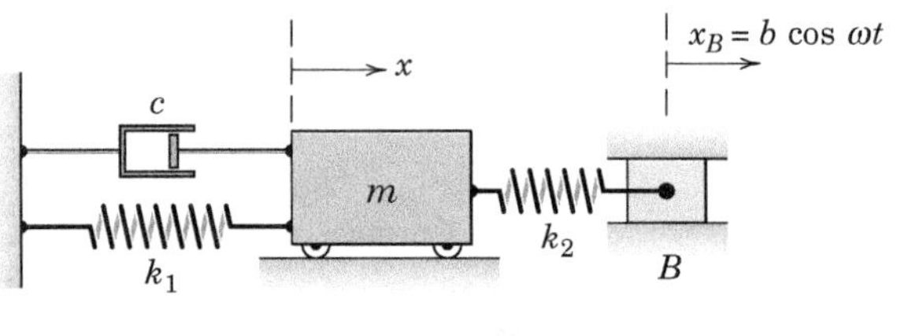

Problem 8/62

8/63 Attachment B is given a horizontal motion $x_B = b \cos \omega t$. Derive the equation of motion for the mass m and state the critical frequency ω_c for which the oscillations of the mass become excessively large. What is the damping ratio ζ for the system?

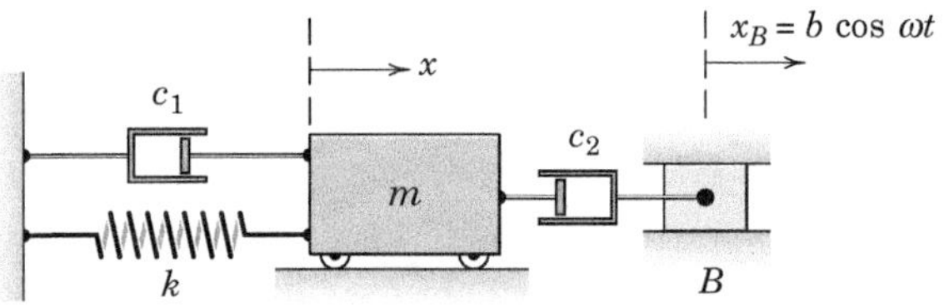

Problem 8/63

8/64 A device to produce vibrations consists of the two counter-rotating wheels, each carrying an eccentric mass $m_0 = 1$ kg with a center of mass at a distance $e = 12$ mm from its axis of rotation. The wheels are synchronized so that the vertical positions of the unbalanced masses are always identical. The total mass of the device is 10 kg. Determine the two possible values of the equivalent spring constant k for the mounting which will permit the amplitude of the periodic force transmitted to the fixed mounting to be 1500 N due to the imbalance of the rotors at a speed of 1800 rev/min. Neglect damping.

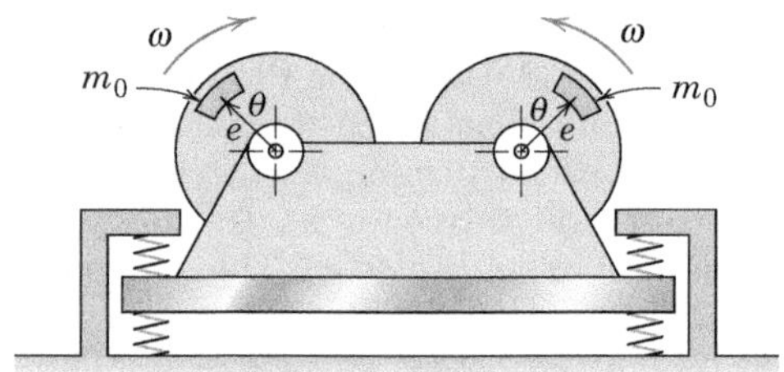

Problem 8/64

8/65 The seismic instrument shown is attached to a structure which has a horizontal harmonic vibration at 3 Hz. The instrument has a mass $m = 0.5$ kg, a spring stiffness $k = 20$ N/m, and a viscous damping coefficient $c = 3$ N·s/m. If the maximum recorded value of x in its steady-state motion is $X = 2$ mm, determine the amplitude b of the horizontal movement x_B of the structure.

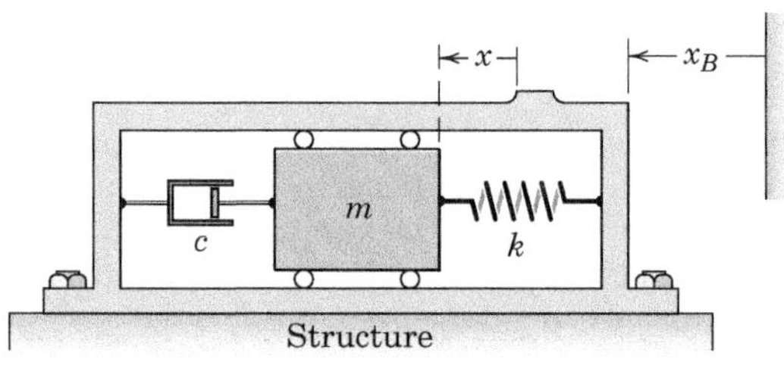

Problem 8/65

8/66 The equilibrium position of the mass m occurs where $y = 0$ and $y_B = 0$. When the attachment B is given a steady vertical motion $y_B = b \sin \omega t$, the mass m will acquire a steady vertical oscillation. Derive the differential equation of motion for m and specify the circular frequency ω_c for which the oscillations of m tend to become excessively large. The stiffness of the spring is k, and the mass and friction of the pulley are negligible.

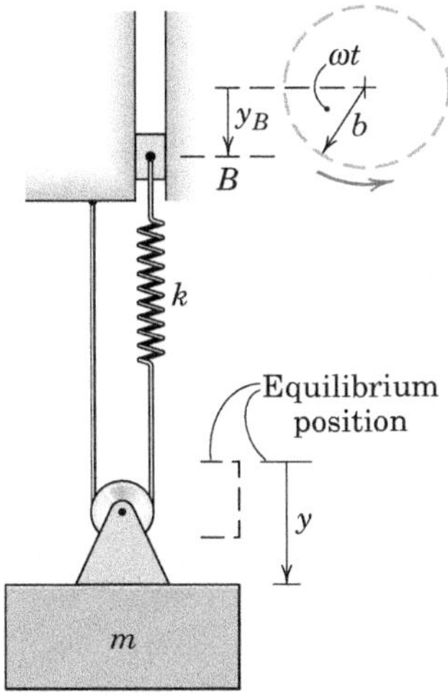

Problem 8/66

8/67 Derive and solve the equation of motion for the mass m in terms of the variable x for the system shown. Neglect the mass of the lever AOC and assume small oscillations.

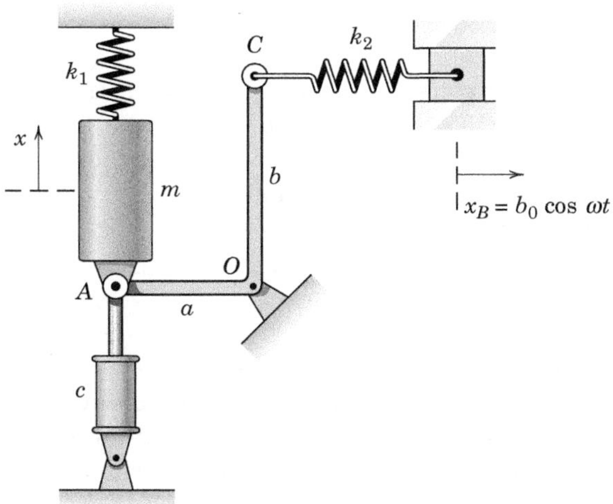

Problem 8/67

8/68 The seismic instrument is mounted on a structure which has a vertical vibration with a frequency of 5 Hz and a double amplitude of 18 mm. The sensing element has a mass $m = 2$ kg, and the spring stiffness is $k = 1.5$ kN/m. The motion of the mass relative to the instrument base is recorded on a revolving drum and shows a double amplitude of 24 mm during the steady-state condition. Calculate the viscous damping constant c.

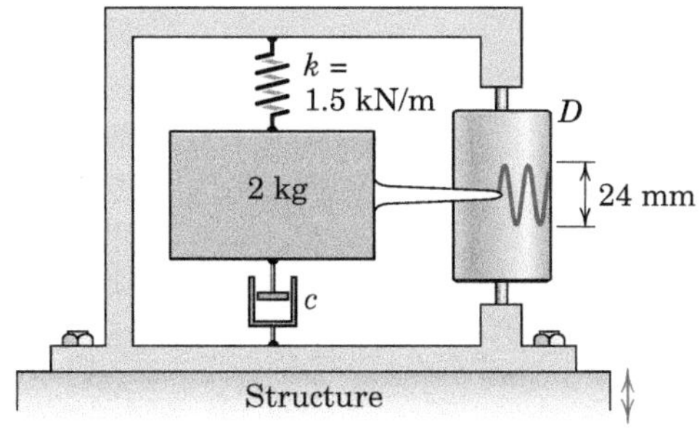

Problem 8/68

►8/69 Derive the expression for the energy loss E over a complete steady-state cycle due to the frictional dissipation of energy in a viscously damped linear oscillator. The forcing function is $F_0 \sin \omega t$, and the displacement-time relation for steady-state motion is $x_P = X \sin (\omega t - \phi)$ where the amplitude X is given by Eq. 8/20. (*Hint:* The frictional energy loss during a displacement dx is $c\dot{x}\,dx$, where c is the viscous damping coefficient. Integrate this expression over a complete cycle.)

►8/70 Determine the amplitude of vertical vibration of the car as it travels at a velocity $v = 40$ km/h over the wavy road whose contour may be expressed as a sine or cosine function with a double amplitude $2b = 50$ mm. The mass of the car is 1800 kg and the stiffness of each of the four car springs is 35 kN/m. Assume that all four wheels are in continuous contact with the road, and neglect damping. Note that the wheelbase of the car and the spatial period of the road are the same at $L = 3$ m, so that it may be assumed that the car translates but does not rotate. At what critical speed v_c is the vertical vibration of the car at its maximum?

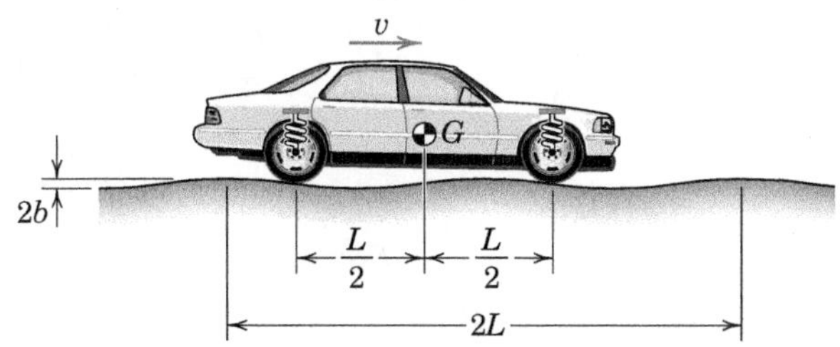

Problem 8/70

8/4 Vibration of Rigid Bodies

The subject of planar rigid-body vibrations is entirely analogous to that of particle vibrations. In particle vibrations, the variable of interest is one of translation (x), while in rigid-body vibrations, the variable of primary concern may be one of rotation (θ). Thus, the principles of rotational dynamics play a central role in the development of the equation of motion.

We will see that the equation of motion for rotational vibration of rigid bodies has a mathematical form identical to that developed in Arts. 8/2 and 8/3 for translational vibration of particles. As was the case with particles, it is convenient to draw the free-body diagram for an arbitrary positive value of the displacement variable, because a negative displacement value easily leads to sign errors in the equation of motion. The practice of measuring the displacement from the position of static equilibrium rather than from the position of zero spring deflection continues to simplify the formulation for linear systems because the equal and opposite forces and moments associated with the static equilibrium position cancel from the analysis.

Rather than individually treating the cases of (*a*) free vibration, undamped and damped, and (*b*) forced vibrations, undamped and damped, as was done with particles in Arts. 8/2 and 8/3, we will go directly to the damped, forced problem.

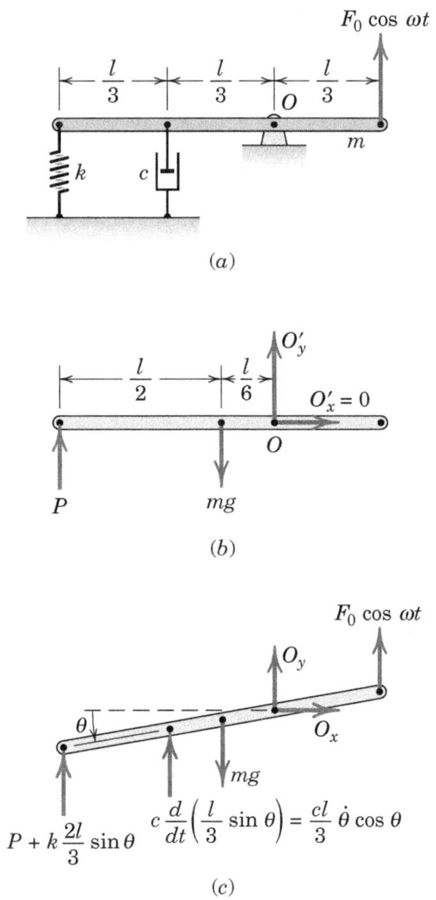

Figure 8/16

Rotational Vibration of a Bar

As an illustrative example, consider the rotational vibration of the uniform slender bar of Fig. 8/16*a*. Figure 8/16*b* depicts the free-body diagram associated with the horizontal position of static equilibrium. Equating to zero the moment sum about O yields

$$-P\left(\frac{l}{2} + \frac{l}{6}\right) + mg\left(\frac{l}{6}\right) = 0 \qquad P = \frac{mg}{4}$$

where P is the magnitude of the static spring force.

Figure 8/16*c* depicts the free-body diagram associated with an arbitrary positive angular displacement θ. Using the equation of rotational motion $\Sigma M_O = I_O\ddot{\theta}$ as developed in Chapter 6, we write

$$(mg)\left(\frac{l}{6}\cos\theta\right) - \left(\frac{cl}{3}\dot{\theta}\cos\theta\right)\left(\frac{l}{3}\cos\theta\right) - \left(P + k\frac{2l}{3}\sin\theta\right)\left(\frac{2l}{3}\cos\theta\right)$$
$$+ (F_0\cos\omega t)\left(\frac{l}{3}\cos\theta\right) = \frac{1}{9}ml^2\ddot{\theta}$$

where $I_O = \bar{I} + md^2 = ml^2/12 + m(l/6)^2 = ml^2/9$ is obtained from the parallel-axis theorem for mass moments of inertia.

For small angular deflections, the approximations $\sin\theta \cong \theta$ and $\cos\theta \cong 1$ may be used. With $P = mg/4$, the equation of motion, upon rearrangement and simplification, becomes

$$\ddot{\theta} + \frac{c}{m}\dot{\theta} + 4\frac{k}{m}\theta = \frac{(F_0 l/3)\cos\omega t}{ml^2/9} \tag{8/25}$$

The right side has been left unsimplified in the form $M_0(\cos \omega t)/I_O$, where $M_0 = F_0 l/3$ is the magnitude of the moment about point O of the externally applied force. Note that the two equal and opposite moments associated with static equilibrium forces canceled on the left side of the equation of motion. Thus, it is not necessary to include the static-equilibrium forces and moments in the analysis.

Rotational Counterparts of Translational Vibration

At this point, we observe that Eq. 8/25 is identical in form to Eq. 8/13 for the translational case, so we may write

$$\ddot{\theta} + 2\zeta\omega_n\dot{\theta} + \omega_n{}^2\theta = \frac{M_0 \cos \omega t}{I_O} \tag{8/26}$$

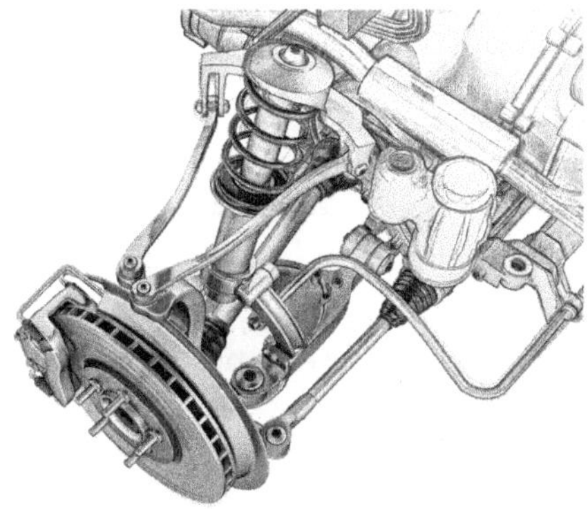

As in the case of the chapter-opening photograph, the spring and damper are coaxial in this multi-link automobile suspension.

Thus, we may use all of the relations developed in Arts. 8/2 and 8/3 merely by replacing the translational quantities with their rotational counterparts. The following table shows the results of this procedure as applied to the rotating bar of Fig. 8/16:

TRANSLATIONAL	ANGULAR (for current problem)
$\ddot{x} + \frac{c}{m}\dot{x} + \frac{k}{m}x = \frac{F_0 \cos \omega t}{m}$	$\ddot{\theta} + \frac{c}{m}\dot{\theta} + \frac{4k}{m}\theta = \frac{M_0 \cos \omega t}{I_O}$
$\omega_n = \sqrt{k/m}$	$\omega_n = \sqrt{4k/m} = 2\sqrt{k/m}$
$\zeta = \frac{c}{2m\omega_n} = \frac{c}{2\sqrt{km}}$	$\zeta = \frac{c}{2m\omega_n} = \frac{c}{4\sqrt{km}}$
$\omega_d = \omega_n\sqrt{1-\zeta^2} = \frac{1}{2m}\sqrt{4km - c^2}$	$\omega_d = \omega_n\sqrt{1-\zeta^2} = \frac{1}{2m}\sqrt{16km - c^2}$
$x_c = Ce^{-\zeta\omega_n t}\sin(\omega_d t + \psi)$	$\theta_c = Ce^{-\zeta\omega_n t}\sin(\omega_d t + \psi)$
$x_p = X\cos(\omega t - \phi)$	$\theta_p = \Theta\cos(\omega t - \phi)$
$X = M\left(\frac{F_0}{k}\right)$	$\Theta = M\left(\frac{M_0}{k_\theta}\right) = M\frac{F_0(l/3)}{\frac{4}{9}kl^2} = M\frac{3F_0}{4kl}$

In the preceding table, the variable k_θ in the expression for Θ represents the equivalent torsional spring constant of the system of Fig. 8/16 and is determined by writing the restoring moment of the spring. For a small angle θ, this moment about O is

$$M_k = -[k(2l/3)\sin\theta][(2l/3)\cos\theta] \cong -(\tfrac{4}{9}kl^2)\theta$$

Thus, $k_\theta = \frac{4}{9}kl^2$. Note that M_0/k_θ is the static angular deflection which would be produced by a constant external moment M_0.

We conclude that an exact analogy exists between particle vibration and the small angular vibration of rigid bodies. Furthermore, the utilization of this analogy can save the labor of complete rederivation of the governing relationships for a given problem of general rigid-body vibration.

Sample Problem 8/7

A simplified version of a pendulum used in impact tests is shown in the figure. Derive the equation of motion and determine the period for small oscillations about the pivot. The mass center G is located a distance $\bar{r} = 0.9$ m from O, and the radius of gyration about O is $k_O = 0.95$ m. The friction of the bearing is negligible.

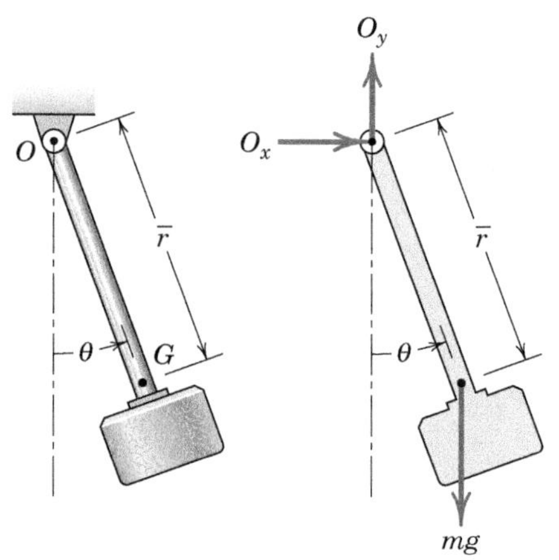

Solution. We draw the free-body diagram for an arbitrary, positive value of the angular-displacement variable θ, which is measured counterclockwise for the coordinate system chosen. Next we apply the governing equation of motion to obtain

① $[\Sigma M_O = I_O\ddot{\theta}]$ $\qquad -mg\bar{r}\sin\theta = mk_O{}^2\ddot{\theta}$

or $$\ddot{\theta} + \frac{g\bar{r}}{k_O{}^2}\sin\theta = 0 \qquad \textit{Ans.}$$

Note that the governing equation is independent of the mass. When θ is small, $\sin\theta \cong \theta$, and our equation of motion may be written as

$$\ddot{\theta} + \frac{g\bar{r}}{k_O{}^2}\theta = 0$$

② The frequency in cycles per second and the period in seconds are

$$f_n = \frac{1}{2\pi}\sqrt{\frac{g\bar{r}}{k_O{}^2}} \qquad \tau = \frac{1}{f_n} = 2\pi\sqrt{\frac{k_O{}^2}{g\bar{r}}} \qquad \textit{Ans.}$$

For the given properties: $$\tau = 2\pi\sqrt{\frac{(0.95)^2}{(9.81)(0.9)}} = 2.01 \text{ s} \qquad \textit{Ans.}$$

Helpful Hints

① With our choice of point O as the moment center, the bearing reactions O_x and O_y never enter the equation of motion.

② For large angles of oscillation, determining the period for the pendulum requires the evaluation of an elliptic integral.

Sample Problem 8/8

The uniform bar of mass m and length l is pivoted at its center. The spring of constant k at the left end is attached to a stationary surface, but the right-end spring, also of constant k, is attached to a support which undergoes a harmonic motion given by $y_B = b \sin \omega t$. Determine the driving frequency ω_c which causes resonance.

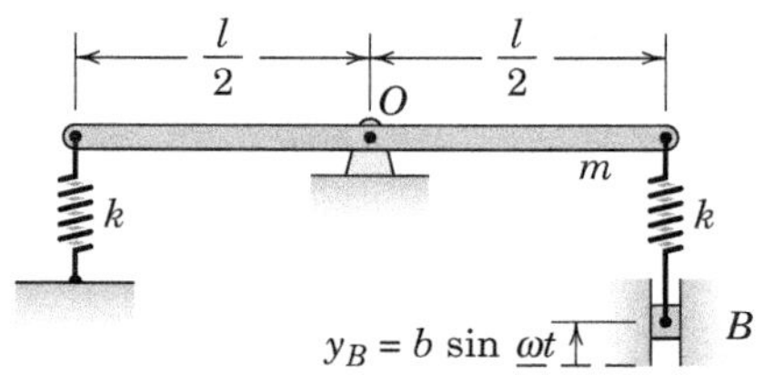

Solution. We use the moment equation of motion about the fixed point O to obtain

① $$-\left(k\frac{l}{2}\sin\theta\right)\frac{l}{2}\cos\theta - k\left(\frac{l}{2}\sin\theta - y_B\right)\frac{l}{2}\cos\theta = \frac{1}{12}ml^2\ddot{\theta}$$

Assuming small deflections and simplifying give us

$$\ddot{\theta} + \frac{6k}{m}\theta = \frac{6kb}{ml}\sin\omega t$$

② The natural frequency should be recognized from the now-familiar form of the equation to be

$$\omega_n = \sqrt{6k/m}$$

Thus, $\omega_c = \omega_n = \sqrt{6k/m}$ will result in resonance (as well as violation of our small-angle assumption!). *Ans.*

Helpful Hints

① As previously, we consider only the changes in the forces due to a movement away from the equilibrium position.

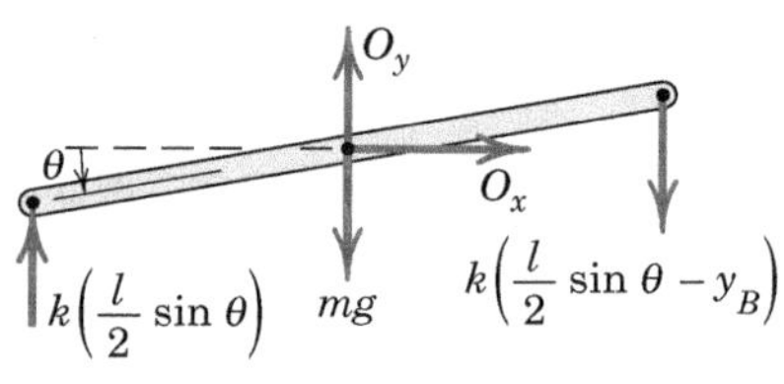

② The standard form here is $\ddot{\theta} + \omega_n{}^2\theta = \frac{M_0 \sin\omega t}{I_O}$, where $M_0 = \frac{klb}{2}$ and $I_O = \frac{1}{12}ml^2$. The natural frequency ω_n of a system does not depend on the external disturbance.

Sample Problem 8/9

Derive the equation of motion for the homogeneous circular cylinder, which rolls without slipping. If the cylinder mass is 50 kg, the cylinder radius 0.5 m, the spring constant 75 N/m, and the damping coefficient 10 N·s/m, determine

(*a*) the undamped natural frequency

(*b*) the damping ratio

(*c*) the damped natural frequency

(*d*) the period of the damped system.

In addition, determine x as a function of time if the cylinder is released from rest at the position $x = -0.2$ m when $t = 0$.

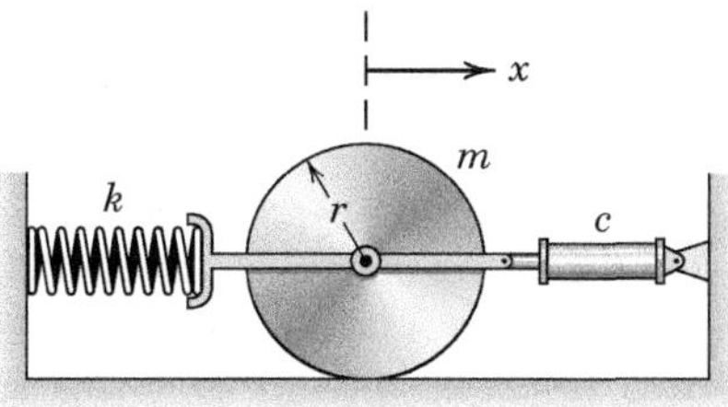

Solution. We have a choice of motion variables in that either x or the angular displacement θ of the cylinder may be used. Since the problem statement involves x, we draw the free-body diagram for an arbitrary, positive value of x and write the two motion equations for the cylinder as ①

② $[\Sigma F_x = m\ddot{x}]$ $\qquad -c\dot{x} - kx + F = m\ddot{x}$

$[\Sigma M_G = \bar{I}\ddot{\theta}]$ $\qquad -Fr = \frac{1}{2}mr^2\ddot{\theta}$

The condition of rolling with no slip is $\ddot{x} = r\ddot{\theta}$. Substitution of this condition into the moment equation gives $F = -\frac{1}{2}m\ddot{x}$. Inserting this expression for the friction force into the force equation for the x-direction yields

$$-c\dot{x} - kx - \frac{1}{2}m\ddot{x} = m\ddot{x} \qquad \text{or} \qquad \ddot{x} + \frac{2}{3}\frac{c}{m}\dot{x} + \frac{2}{3}\frac{k}{m}x = 0$$

Comparing the above equation with that for the standard damped oscillator, Eq. 8/9, allows us to state directly

(*a*) $$\omega_n^2 = \frac{2}{3}\frac{k}{m} \qquad \omega_n = \sqrt{\frac{2}{3}\frac{k}{m}} = \sqrt{\frac{2}{3}\frac{75}{50}} = 1 \text{ rad/s} \qquad \textit{Ans.}$$

(*b*) $$2\zeta\omega_n = \frac{2}{3}\frac{c}{m} \qquad \zeta = \frac{1}{3}\frac{c}{m\omega_n} = \frac{10}{3(50)(1)} = 0.0667 \qquad \textit{Ans.}$$

Hence, the damped natural frequency and the damped period are

(*c*) $$\omega_d = \omega_n\sqrt{1-\zeta^2} = (1)\sqrt{1-(0.0667)^2} = 0.998 \text{ rad/s} \qquad \textit{Ans.}$$

(*d*) $$\tau_d = 2\pi/\omega_d = 2\pi/0.998 = 6.30 \text{ s} \qquad \textit{Ans.}$$

From Eq. 8/12, the underdamped solution to the equation of motion is

$$x = Ce^{-\zeta\omega_n t}\sin(\omega_d t + \psi) = Ce^{-(0.0667)(1)t}\sin(0.998t + \psi)$$

The velocity is $\qquad \dot{x} = -0.0667Ce^{-0.0667t}\sin(0.998t+\psi)$
$\qquad\qquad +0.998Ce^{-0.0667t}\cos(0.998t+\psi)$

At time $t = 0$, x and $\dot{x}$ become

$$x_0 = C\sin\psi = -0.2$$

$$\dot{x}_0 = -0.0667C\sin\psi + 0.998C\cos\psi = 0$$

The solution to the two equations in C and ψ gives

$$C = -0.200 \text{ m} \qquad \psi = 1.504 \text{ rad}$$

Thus, the motion is given by

$$x = -0.200e^{-0.0667t}\sin(0.998t + 1.504) \text{ m} \qquad \textit{Ans.}$$

Helpful Hints

① The angle θ is taken positive clockwise to be kinematically consistent with x.

② The friction force F may be assumed in either direction. We will find that the *actual* direction is to the right for $x > 0$ and to the left for $x < 0$; $F = 0$ when $x = 0$.

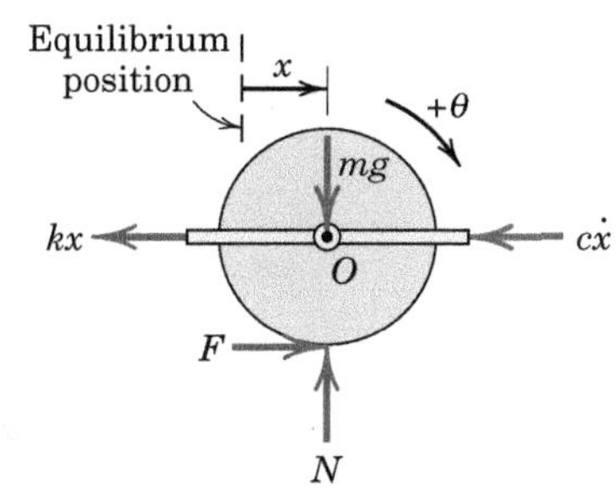

PROBLEMS

Introductory Problems

8/71 The light rod and attached small spheres of mass m each are shown in the equilibrium position, where all four springs are equally precompressed. Determine the natural frequency ω_n and period τ for small oscillations about the frictionless pivot O.

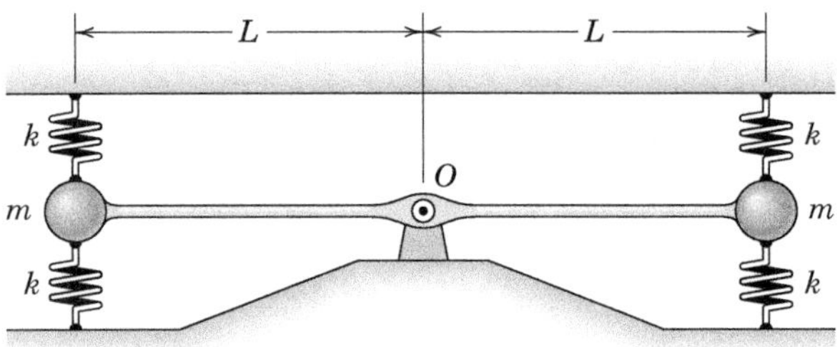

Problem 8/71

8/72 A uniform rectangular plate pivots about a horizontal axis through one of its corners as shown. Determine the natural frequency ω_n of small oscillations.

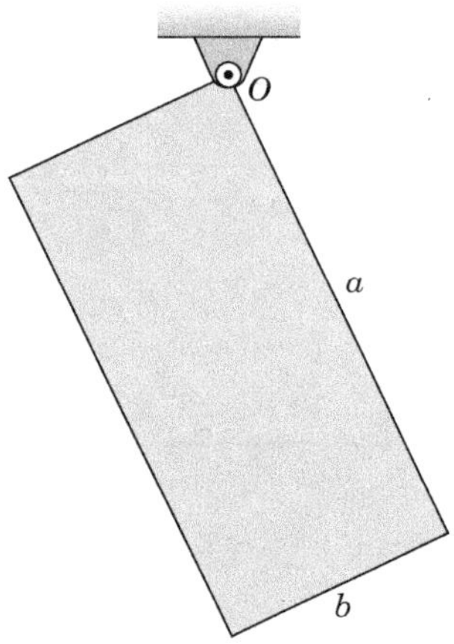

Problem 8/72

8/73 The thin square plate is suspended from a socket (not shown) which fits the small ball attachment at O. If the plate is made to swing about axis A-A, determine the period for small oscillations. Neglect the small offset, mass, and friction of the ball.

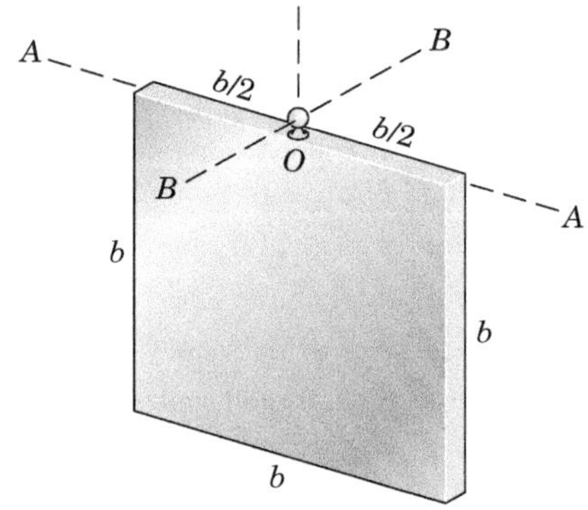

Problem 8/73

8/74 If the square plate of Prob. 8/73 is made to oscillate about axis B-B, determine the period of small oscillations.

8/75 The 20-lb spoked wheel has a centroidal radius of gyration k = 6 in. A torsional spring of constant k_T = 160 lb-ft/rad resists rotation about the smooth bearing. If an external torque of form $M = M_0 \cos \omega t$ is applied to the wheel, what is the magnitude of its steady-state angular displacement? The moment magnitude is M_0 = 8 lb-ft and the driving frequency is ω = 25 rad/sec.

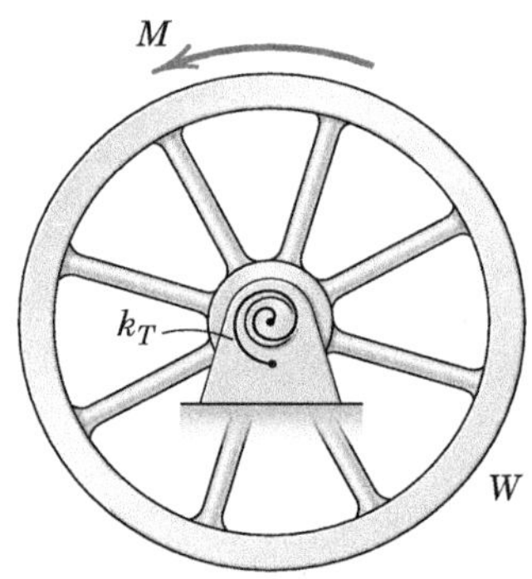

Problem 8/75

8/76 The uniform rod of length l and mass m is suspended at its midpoint by a wire of length L. The resistance of the wire to torsion is proportional to its angle of twist θ and equals $(JG/L)\theta$ where J is the polar moment of inertia of the wire cross section and G is the shear modulus of elasticity. Derive the expression for the period τ of oscillation of the bar when it is set into rotation about the axis of the wire.

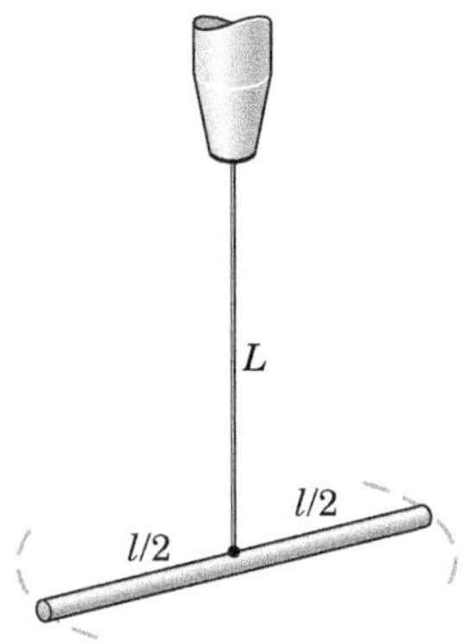

Problem 8/76

8/77 The uniform sector has mass m and is freely hinged about a horizontal axis through point O. Determine the equation of motion of the sector for large-amplitude vibrations about the equilibrium position. State the period τ for small oscillations about the equilibrium position if $r = 325$ mm and $\beta = 45°$.

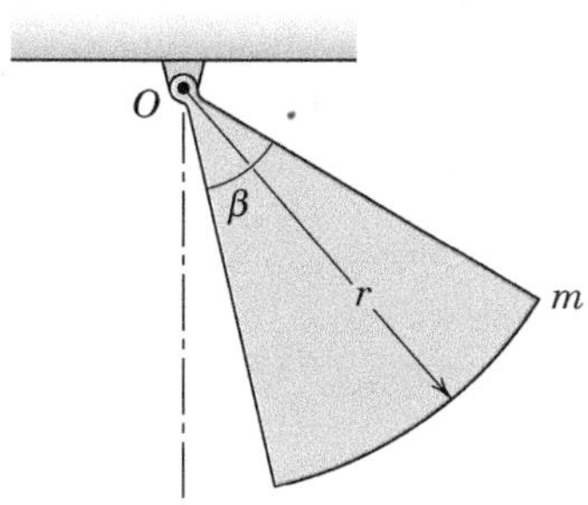

Problem 8/77

8/78 The assembly of mass m is formed from uniform and slender welded rods and is freely hinged about a horizontal axis through O. Determine the equation of motion of the assembly for large-amplitude vibrations about the equilibrium position. State the period τ for small oscillations about the equilibrium position if $r = 325$ mm and $\beta = 45°$.

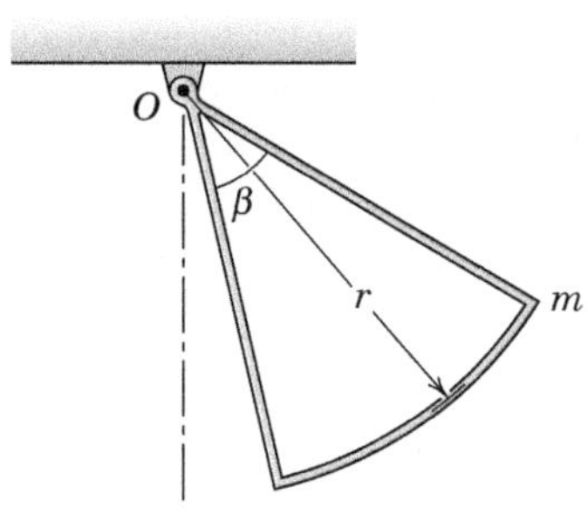

Problem 8/78

8/79 The thin-walled cylindrical shell of radius r and height h is welded to the small shaft at its upper end as shown. Determine the natural circular frequency ω_n for small oscillations of the shell about the y-axis.

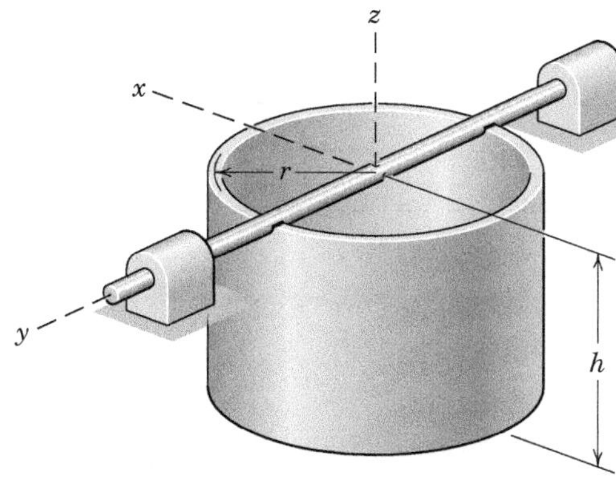

Problem 8/79

8/80 Determine the system equation of motion in terms of the variable θ shown in the figure. Assume small angular motion of bar OA, and neglect the mass of link CD.

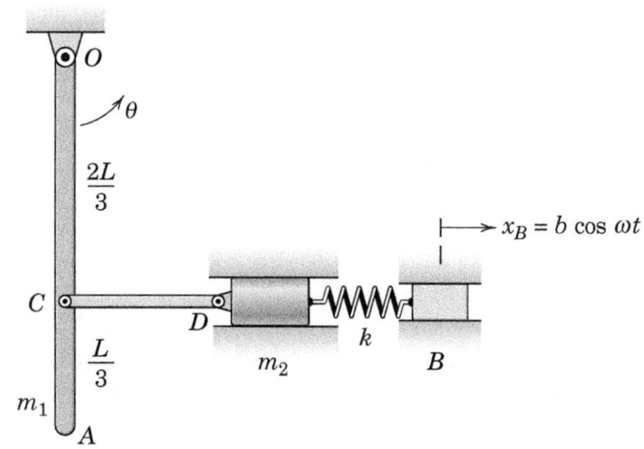

Problem 8/80

8/81 The uniform rod of mass m is freely pivoted about a horizontal axis through point O. Assume small oscillations and determine an expression for the damping ratio ζ. For what value c_{cr} of the damping coefficient c will the system be critically damped?

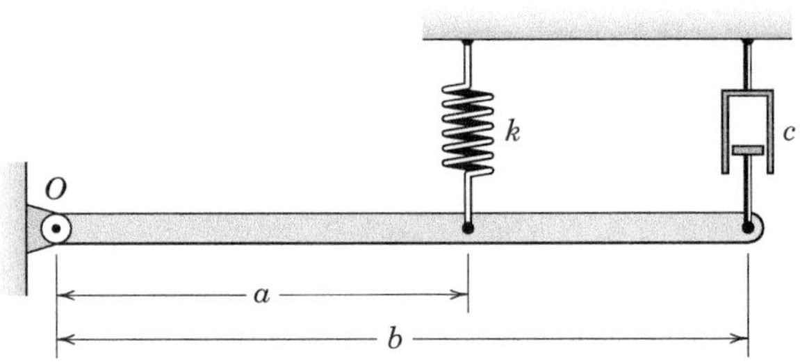

Problem 8/81

8/82 The mass of the uniform slender rod is 3 kg. Determine the position x for the 1.2-kg slider such that the system period is 1 s. Assume small oscillations about the horizontal equilibrium position shown.

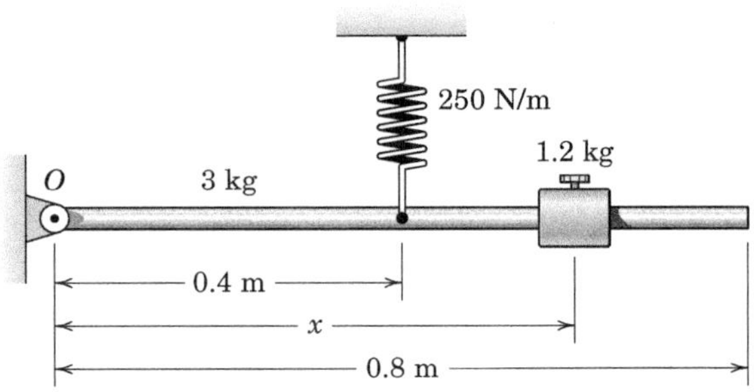

Problem 8/82

Representative Problems

8/83 The triangular frame of mass m is formed from uniform slender rod and is suspended from a socket (not shown) which fits the small ball attachment at O. If the frame is made to swing about axis A-A, determine the natural circular frequency ω_n for small oscillations. Neglect the small offset, mass, and friction of the ball. Evaluate for $l = 200$ mm.

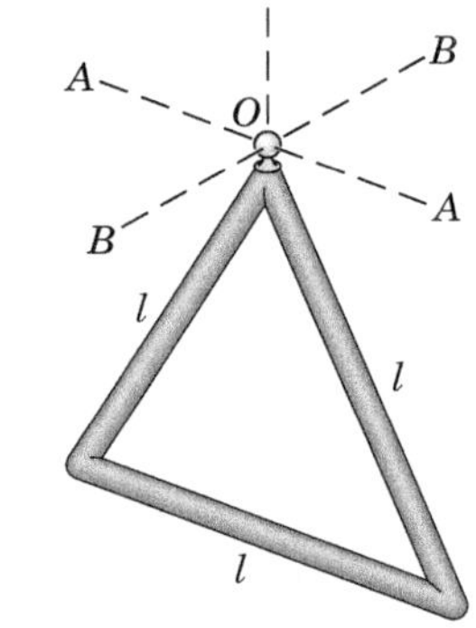

Problem 8/83

8/84 If the triangular frame of Prob. 8/83 is made to oscillate about axis B-B, determine the natural circular frequency ω_n for small oscillations. Evaluate for $l = 200$ mm and compare your answer with that of Prob. 8/83.

8/85 The uniform rod of mass m is freely pivoted about point O. Assume small oscillations and determine an expression for the damping ratio ζ. For what value c_{cr} of the damping coefficient c will the system be critically damped?

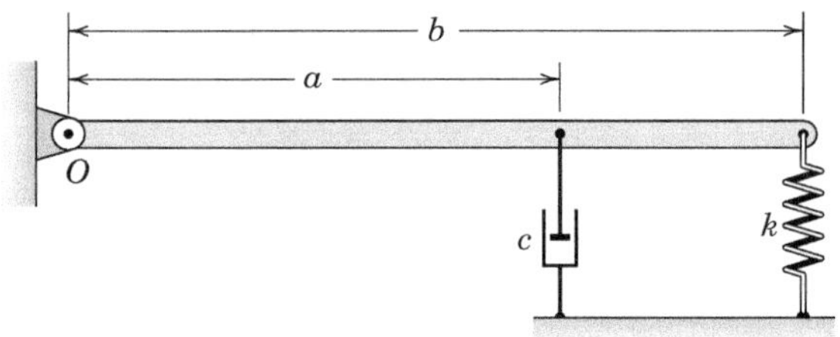

Problem 8/85

8/86 The mechanism shown oscillates in the vertical plane about the pivot O. The springs of equal stiffness k are both compressed in the equilibrium position $\theta = 0$. Determine an expression for the period τ of small oscillations about O. The mechanism has a mass m with mass center G, and the radius of gyration of the assembly about O is k_O.

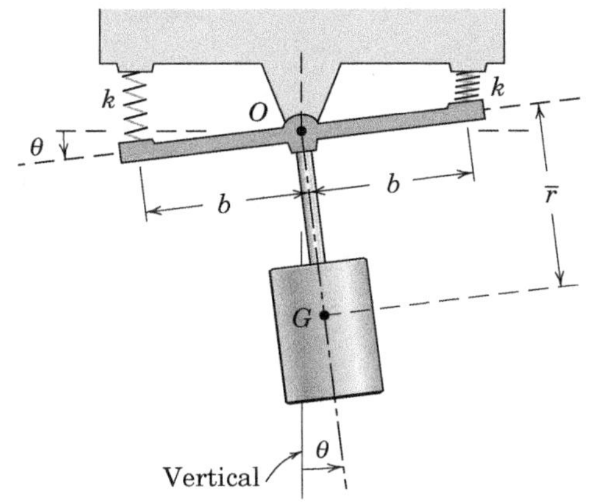

Problem 8/86

8/87 When the motor is slowly brought up to speed, a rather large vibratory oscillation of the entire motor about O-O occurs at a speed of 360 rev/min, which shows that this speed corresponds to the natural frequency of free oscillation of the motor. If the motor has a mass of 43 kg and radius of gyration of 100 mm about O-O, determine the stiffness k of each of the four identical spring mounts.

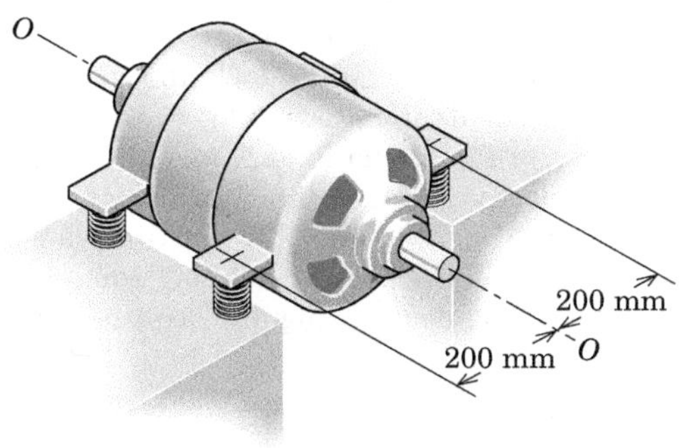

Problem 8/87

8/88 The system of Prob. 8/35 is repeated here with the added information that link AOB now has mass m_3 and radius of gyration k_O about point O. Ignore friction and derive the differential equation of motion for the system shown in terms of the variable x_1.

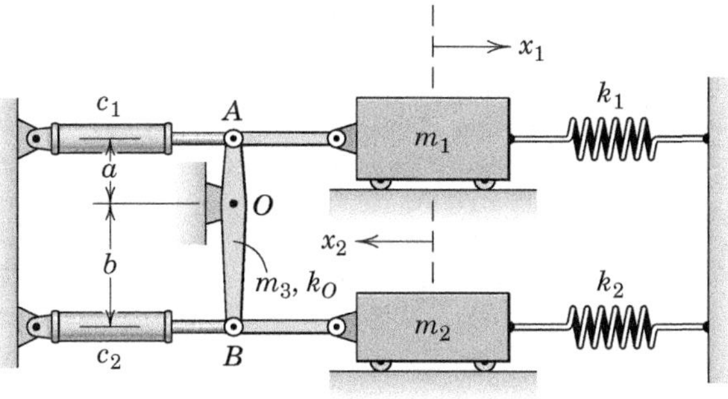

Problem 8/88

8/89 Determine the value m_{eff} of the mass of system (b) so that the frequency of system (b) is equal to that of system (a). Note that the two springs are identical and that the wheel of system (a) is a solid homogeneous cylinder of mass m_2. The cord does not slip on the cylinder.

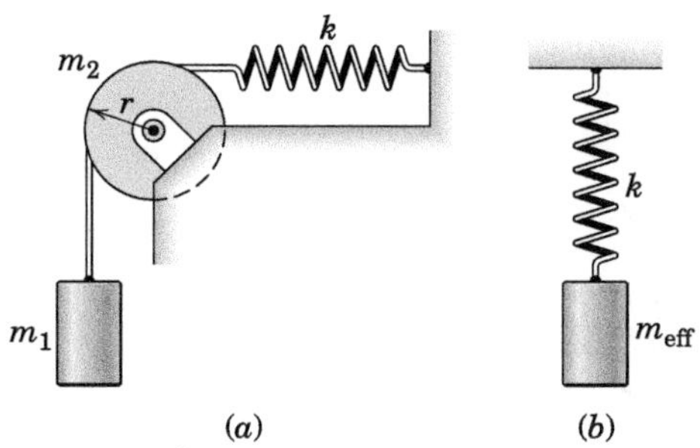

Problem 8/89

8/90 The system of Prob. 8/43 is repeated here. If the crank AB now has mass m_2 and a radius of gyration k_O about point O, determine expressions for the undamped natural frequency ω_n and the damping ratio ζ in terms of the given system properties. Assume small oscillations. The damping coefficient for the damper is c.

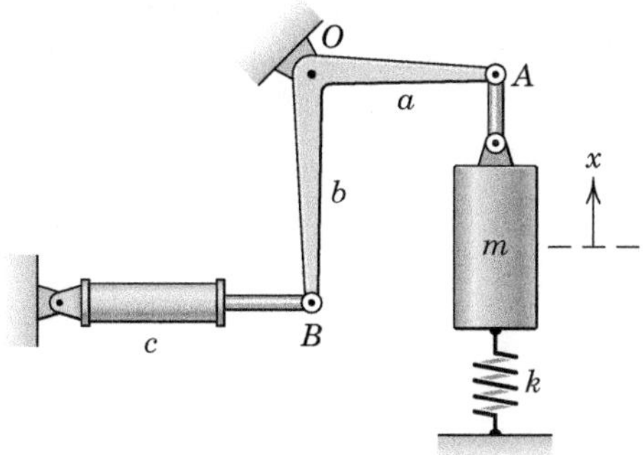

Problem 8/90

8/91 The two masses are connected by an inextensible cable which passes securely over the periphery of the cylindrical pulley of mass m_3, radius r, and radius of gyration k_O. Determine the equation of motion for the system in terms of the variable x. State the critical driving frequency ω_c of the block B which will result in excessively large oscillations of the assembly. Evaluate for $m_1 = 25$ kg, $m_2 = 10$ kg, $m_3 = 15$ kg, $k_1 = 450$ N/m, $k_2 = 300$ N/m, $r = 300$ mm, and $k_O = 200$ mm.

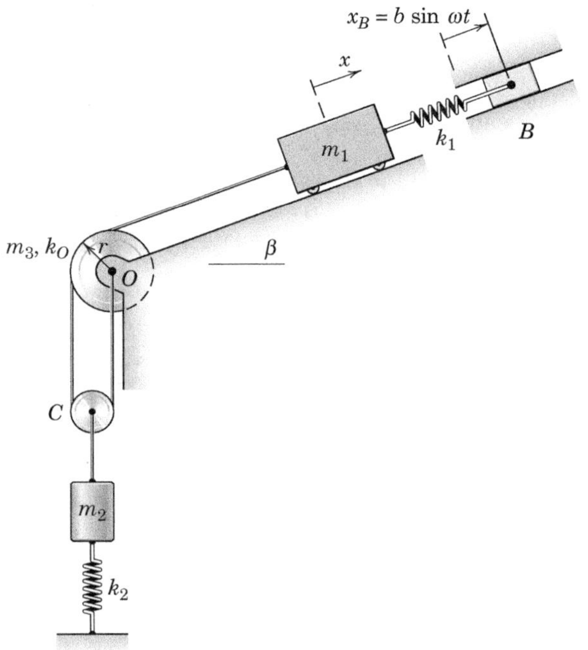

Problem 8/91

8/92 The lower spring of Prob. 8/91 is replaced by a damper. Determine the equation of motion for the system in terms of the variable x. State the value of the viscous damping coefficient c which will give a damping ratio $\zeta = 0.2$. Evaluate for $m_1 = 25$ kg, $m_2 = 10$ kg, $m_3 = 15$ kg, $k = 450$ N/m, $r = 300$ mm, and $k_O = 200$ mm.

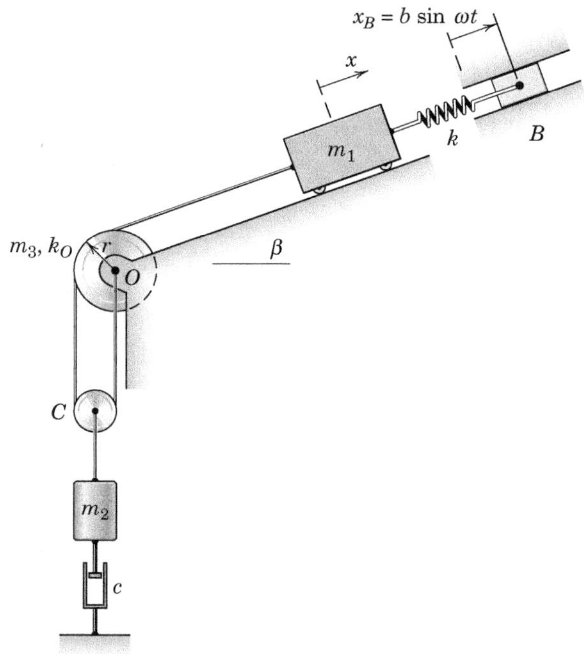

Problem 8/92

8/93 The uniform solid cylinder of mass m and radius r rolls without slipping during its oscillation on the circular surface of radius R. If the motion is confined to small amplitudes $\theta = \theta_0$, determine the period τ of the oscillations. Also determine the angular velocity ω of the cylinder as it crosses the vertical centerline. (*Caution:* Do not confuse ω with $\dot{\theta}$ or with ω_n as used in the defining equations. Note also that θ is not the angular displacement of the cylinder.)

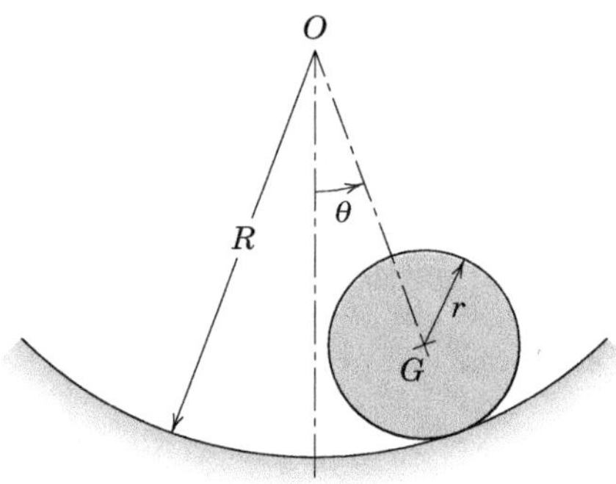

Problem 8/93

8/94 The cart B is given the harmonic displacement $x_B = b \sin \omega t$. Determine the steady-state amplitude Θ of the periodic oscillation of the uniform slender bar which is pinned to the cart at P. Assume small angles and neglect friction at the pivot. The torsional spring is undeformed when $\theta = 0$.

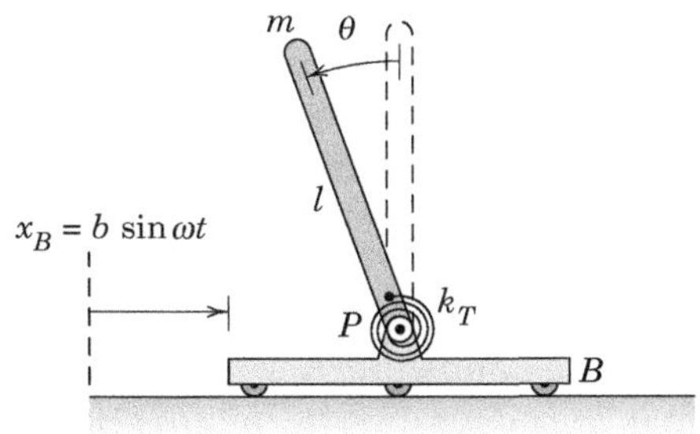

Problem 8/94

▶**8/95** The assembly of Prob. 8/40 is repeated here with the additional information that body ABC now has mass m_4 and a radius of gyration k_O about its pivot at O, about which it is balanced. If a harmonic torque $M = M_0 \cos \omega t$ is applied to body ABC, determine the equation of motion for the system in terms of the variable x. State the critical frequency ω_c of the harmonic torque which will result in an excessively large system response. Evaluate ω_c for $m_1 = 15$ kg, $m_2 = 12$ kg, $m_3 = 8$ kg, $m_4 = 6$ kg, $k_1 = 400$ N/m, $k_2 = 650$ N/m, $k_3 = 225$ N/m, $c_1 = 44$ N·s/m, $c_2 = 36$ N·s/m, $c_3 = 52$ N·s/m, $a = 1.2$ m, $b = 1.8$ m, $c = 0.9$ m, and $k_O = 0.75$ m. What is the damping ratio ζ for these conditions?

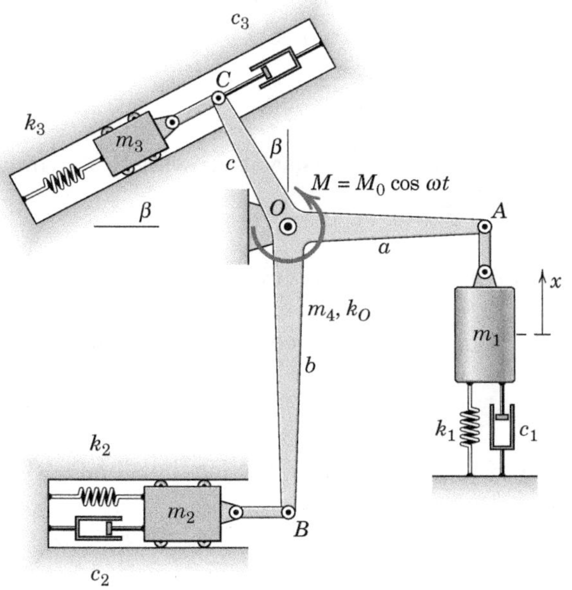

Problem 8/90

▶**8/96** The elements of the "swing-axle" type of independent rear suspension for automobiles are depicted in the figure. The differential D is rigidly attached to the car frame. The half-axles are pivoted at their inboard ends (point O for the half-axle shown) and are rigidly attached to the wheels. Suspension elements not shown constrain the wheel motion to the plane of the figure. The weight of the wheel–tire assembly is $W = 100$ lb, and its mass moment of inertia about a diametral axis passing through its mass center G is 1 lb-ft-sec^2. The weight of the half-axle is negligible. The spring rate and shock-absorber damping coefficient are $k = 50$ lb/in. and $c = 200$ lb-sec/ft, respectively. If a static tire imbalance is present, as represented by the additional concentrated weight $w = 0.5$ lb as shown, determine the angular velocity ω which results in the suspension system being driven at its undamped natural frequency. What would be the corresponding vehicle speed v? Determine the damping ratio ζ. Assume small angular deflections and neglect gyroscopic effects and any car frame vibration. In order to avoid the complications associated with the varying normal force exerted by the road on the tire, treat the vehicle as being on a lift with the wheels hanging free.

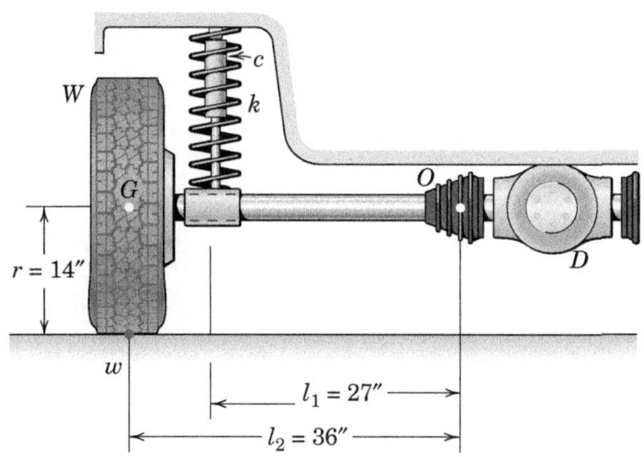

Problem 8/96

8/5 Energy Methods

In Arts. 8/2 through 8/4 we derived and solved the equations of motion for vibrating bodies by isolating the body with a free-body diagram and applying Newton's second law of motion. With this approach, we were able to account for the actions of all forces acting on the body, including frictional damping forces. There are many problems where the effect of damping is small and may be neglected, so that the total energy of the system is essentially conserved. For such systems, we find that the principle of conservation of energy may frequently be applied with considerable advantage in establishing the equation of motion and, when the motion is simple harmonic, in determining the frequency of vibration.

Determining the Equation of Motion

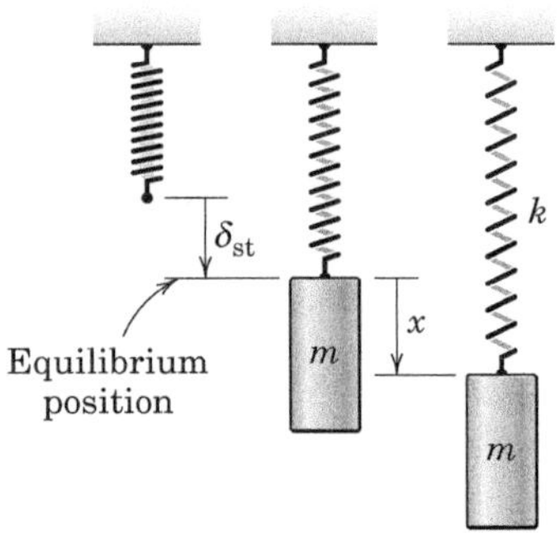

Figure 8/17

To illustrate this alternative approach, consider first the simple case of the body of mass m attached to the spring of stiffness k and vibrating in the vertical direction without damping, Fig. 8/17. As previously, we find it convenient to measure the motion variable x from the equilibrium position. With this datum, the total potential energy of the system, elastic plus gravitational, becomes

$$V = V_e + V_g = \frac{1}{2}k(x + \delta_{st})^2 - \frac{1}{2}k\delta_{st}^{\ 2} - mgx$$

where $\delta_{st} = mg/k$ is the initial static displacement. Substituting $k\delta_{st} = mg$ and simplifying give

$$V = \frac{1}{2}kx^2$$

Thus, the total energy of the system becomes

$$T + V = \frac{1}{2}m\dot{x}^2 + \frac{1}{2}kx^2$$

Because $T + V$ is constant for a conservative system, its time derivative is zero. Consequently,

$$\frac{d}{dt}(T + V) = m\dot{x}\ddot{x} + kx\dot{x} = 0$$

Canceling $\dot{x}$ gives us our basic differential equation of motion

$$m\ddot{x} + kx = 0$$

which is identical to Eq. 8/1 derived in Art. 8/2 for the same system of Fig. 8/3.

Determining the Frequency of Vibration

Conservation of energy may also be used to determine the period or frequency of vibration for a linear conservative system, without having to derive and solve the equation of motion. For a system which oscillates with simple harmonic motion about the equilibrium position, from which the displacement x is measured, the energy changes from maximum kinetic and zero potential at the equilibrium position $x = 0$ to zero kinetic and maximum potential at the position of maximum displacement $x = x_{max}$. Thus, we may write

$$T_{max} = V_{max}$$

The maximum kinetic energy is $\frac{1}{2}m(\dot{x}_{max})^2$, and the maximum potential energy is $\frac{1}{2}k(x_{max})^2$.

For the harmonic oscillator of Fig. 8/17, we know that the displacement may be written as $x = x_{max} \sin(\omega_n t + \psi)$, so that the maximum velocity is $\dot{x}_{max} = \omega_n x_{max}$. Thus, we may write

$$\frac{1}{2}m(\omega_n x_{max})^2 = \frac{1}{2}k(x_{max})^2$$

where x_{max} is the maximum displacement, at which the potential energy is a maximum. From this energy balance, we easily obtain

$$\omega_n = \sqrt{k/m}$$

This method of directly determining the frequency may be used for any linear undamped vibration.

The main advantage of the energy approach for the free vibration of conservative systems is that it becomes unnecessary to dismember the system and account for all of the forces which act on each member. In Art. 3/7 of Chapter 3 and in Arts. 6/6 and 6/7 of Chapter 6, we learned for a system of interconnected bodies that an active-force diagram of the complete system enabled us to evaluate the work U' of the external active forces and to equate it to the change in the total mechanical energy $T + V$ of the system.

Thus, for a conservative mechanical system of interconnected parts with a single degree of freedom where $U' = 0$, we may obtain its equation of motion simply by setting the time derivative of its constant total mechanical energy to zero, giving

$$\frac{d}{dt}(T + V) = 0$$

Here $V = V_e + V_g$ is the sum of the elastic and gravitational potential energies of the system.

Also, for an interconnected mechanical system, as for a single body, the natural frequency of vibration is obtained by equating the expression for its maximum total kinetic energy to the expression for its maximum potential energy, where the potential energy is taken to be zero at the equilibrium position. This approach to the determination of natural frequency is valid only if it can be determined that the system vibrates with simple harmonic motion.

Sample Problem 8/10

The small sphere of mass m is mounted on the light rod pivoted at O and supported at end A by the vertical spring of stiffness k. End A is displaced a small distance y_0 below the horizontal equilibrium position and released. By the energy method, derive the differential equation of motion for small oscillations of the rod and determine the expression for its natural frequency ω_n of vibration. Damping is negligible.

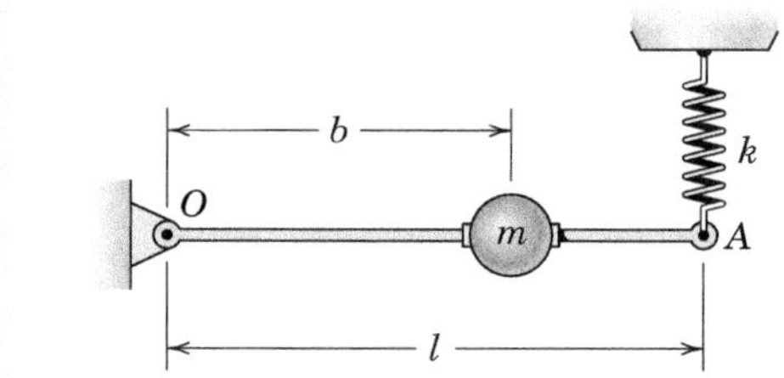

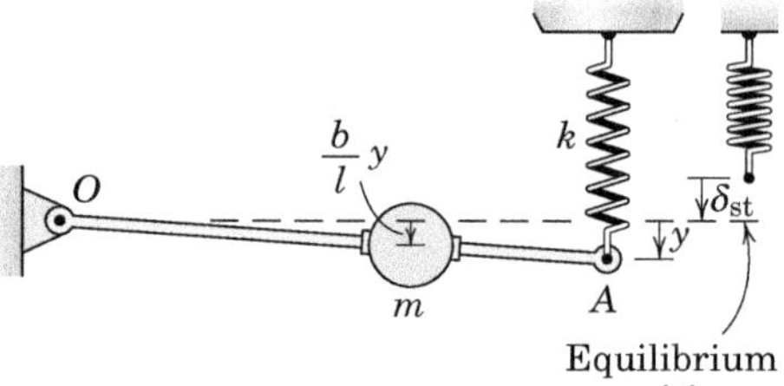

Solution. With the displacement y of the end of the bar measured from the equilibrium position, the potential energy in the displaced position for small values of y becomes

① $$V = V_e + V_g = \frac{1}{2}k(y + \delta_{st})^2 - \frac{1}{2}k\delta_{st}^{\,2} - mg\left(\frac{b}{l}y\right)$$

where δ_{st} is the static deflection of the spring at equilibrium. But the force in the spring in the equilibrium position, from a zero moment sum about O, is $(b/l)mg = k\delta_{st}$. Substituting this value in the expression for V and simplifying yield

② $$V = \frac{1}{2}ky^2$$

The kinetic energy in the displaced position is

$$T = \frac{1}{2}m\left(\frac{b}{l}\dot{y}\right)^2$$

where we see that the vertical displacement of m is $(b/l)y$. Thus, with the energy sum constant, its time derivative is zero, and we have

$$\frac{d}{dt}(T + V) = \frac{d}{dt}\left[\frac{1}{2}m\left(\frac{b}{l}\dot{y}\right)^2 + \frac{1}{2}ky^2\right] = 0$$

which yields

$$\ddot{y} + \frac{l^2}{b^2}\frac{k}{m}y = 0 \qquad \textit{Ans.}$$

when $\dot{y}$ is canceled. By analogy with Eq. 8/2, we may write the motion frequency directly as

$$\omega_n = \frac{l}{b}\sqrt{k/m} \qquad \textit{Ans.}$$

Alternatively, we can obtain the frequency by equating the maximum kinetic energy, which occurs at $y = 0$, to the maximum potential energy, which occurs at $y = y_0 = y_{max}$, where the deflection is a maximum. Thus,

$$T_{max} = V_{max} \qquad \text{gives} \qquad \frac{1}{2}m\left(\frac{b}{l}\dot{y}_{max}\right)^2 = \frac{1}{2}ky_{max}^{\;2}$$

Knowing that we have a harmonic oscillation, which can be expressed as $y = y_{max}\sin\omega_n t$, we have $\dot{y}_{max} = y_{max}\omega_n$. Substituting this relation into our energy balance gives us

$$\frac{1}{2}m\left(\frac{b}{l}y_{max}\omega_n\right)^2 = \frac{1}{2}ky_{max}^{\;2} \qquad \text{so that} \qquad \omega_n = \frac{l}{b}\sqrt{k/m} \qquad \textit{Ans.}$$

as before.

Helpful Hints

① For large values of y, the circular motion of the end of the bar would cause our expression for the deflection of the spring to be in error.

② Here again, we note the simplicity of the expression for potential energy when the displacement is measured from the equilibrium position.

Sample Problem 8/11

Determine the natural frequency ω_n of vertical vibration of the 3-kg collar to which are attached the two uniform 1.2-kg links, which may be treated as slender bars. The stiffness of the spring, which is attached to both the collar and the foundation, is $k = 1.5$ kN/m, and the bars are both horizontal in the equilibrium position. A small roller on end B of each link permits end A to move with the collar. Frictional retardation is negligible.

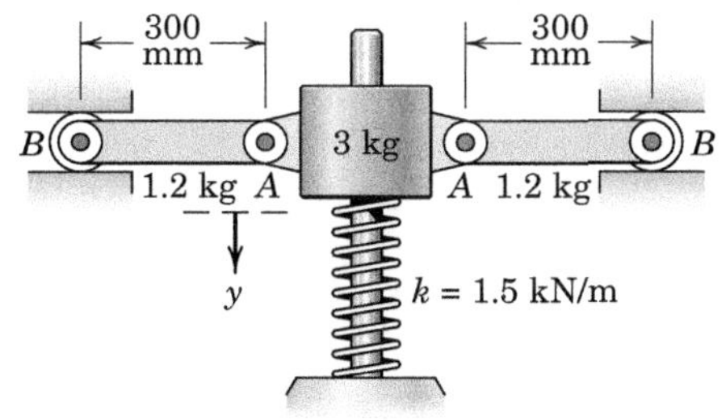

Solution. In the equilibrium position, the compression P in the spring equals the weight of the 3-kg collar, plus half the weight of each link or $P = 3(9.81) + 2(\frac{1}{2})(1.2)(9.81) = 41.2$ N. The corresponding static deflection of the spring is $\delta_{st} = P/k = 41.2/1.5(10^3) = 27.5(10^{-3})$ m. With the displacement variable y measured downward from the equilibrium position, which becomes the position of zero potential energy, the potential energy for each member in the displaced position is

$$\text{(Spring)}\quad V_e = \frac{1}{2}k(y + \delta_{st})^2 - \frac{1}{2}k\delta_{st}^{\,2} = \frac{1}{2}ky^2 + k\delta_{st}y$$

$$= \frac{1}{2}(1.5)(10^3)y^2 + 1.5(10^3)(27.5)(10^{-3})y$$

$$= 750y^2 + 41.2y \text{ J}$$

$$\text{(Collar)}\quad V_g = -m_c gy = -3(9.81)y = -29.4y \text{ J}$$

① $$\text{(Each link)}\quad V_g = -m_l g\frac{y}{2} = -1.2(9.81)\frac{y}{2} = -5.89y \text{ J}$$

The total potential energy of the system then becomes

② $$V = 750y^2 + 41.2y - 29.4y - 2(5.89)y = 750y^2 \text{ J}$$

The maximum kinetic energy occurs at the equilibrium position, where the velocity $\dot{y}$ of the collar has its maximum value. In that position, in which links
③ AB are horizontal, end B is the instantaneous center of zero velocity for each link, and each link rotates with an angular velocity $\dot{y}/0.3$. Thus, the kinetic energy of each part is

$$\text{(Collar)}\quad T = \frac{1}{2}m_c\dot{y}^2 = \frac{3}{2}\dot{y}^2 \text{ J}$$

$$\text{(Each link)}\quad T = \frac{1}{2}I_B\omega^2 = \frac{1}{2}\left(\frac{1}{3}m_l l^2\right)(\dot{y}/l)^2 = \frac{1}{6}m_l\dot{y}^2$$

$$= \frac{1}{6}(1.2)\dot{y}^2 = 0.2\dot{y}^2$$

Thus, the kinetic energy of the collar and both links is

$$T = \frac{3}{2}\dot{y}^2 + 2(0.2\dot{y}^2) = 1.9\dot{y}^2$$

With the harmonic motion expressed by $y = y_{max}\sin\omega_n t$, we have $\dot{y}_{max} = y_{max}\omega_n$,
④ so that the energy balance $T_{max} = V_{max}$ with $\dot{y} = \dot{y}_{max}$ becomes

⑤ $$1.9(y_{max}\omega_n)^2 = 750y_{max}^{\,2} \quad \text{or} \quad \omega_n = \sqrt{750/1.9} = 19.87 \text{ Hz} \qquad \textit{Ans.}$$

Helpful Hints

① Note that the mass center of each link moves down only half as far as the collar.

② We note again that measurement of the motion variable y from the equilibrium position results in the total potential energy being simply $V = \frac{1}{2}ky^2$.

③ Our knowledge of rigid-body kinematics is essential at this point.

④ To appreciate the advantage of the work-energy method for this and similar problems of interconnected systems, you are encouraged to explore the steps required for solution by the force and moment equations of motion of the separate parts.

⑤ If the oscillations were large, we would find that the angular velocity of each link in its general position would equal $\dot{y}/\sqrt{0.09 - y^2}$, which would cause a nonlinear response no longer described by $y = y_{max}\sin\omega t$.

PROBLEMS

(Solve the following problems by the energy method of Art. 8/5.)

Introductory Problems

8/97 The 1.5-kg bar OA is suspended vertically from the bearing O and is constrained by the two springs each of stiffness $k = 120$ N/m and both equally precompressed with the bar in the vertical equilibrium position. Treat the bar as a uniform slender rod and compute the natural frequency f_n of small oscillations about O.

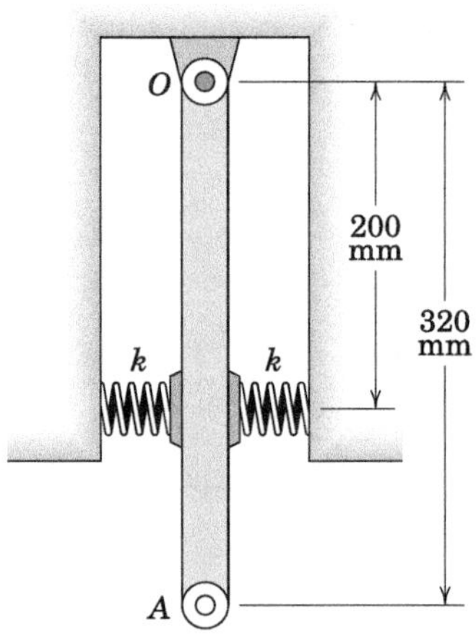

Problem 8/97

8/98 The light rod and attached sphere of mass m are at rest in the horizontal position shown. Determine the period τ for small oscillations in the vertical plane about the pivot O.

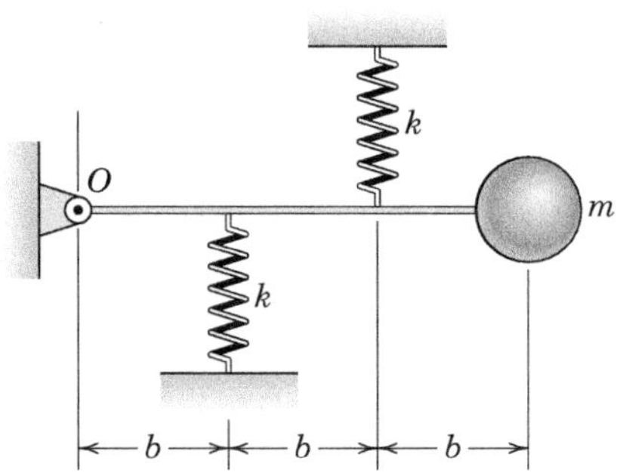

Problem 8/98

8/99 A uniform rod of mass m and length l is welded at one end to the rim of a light circular hoop of radius l. The other end lies at the center of the hoop. Determine the period τ for small oscillations about the vertical position of the bar if the hoop rolls on the horizontal surface without slipping.

Problem 8/99

8/100 The spoked wheel of radius r, mass m, and centroidal radius of gyration $\bar{k}$ rolls without slipping on the incline. Determine the natural frequency of oscillation and explore the limiting cases of $\bar{k} = 0$ and $\bar{k} = r$.

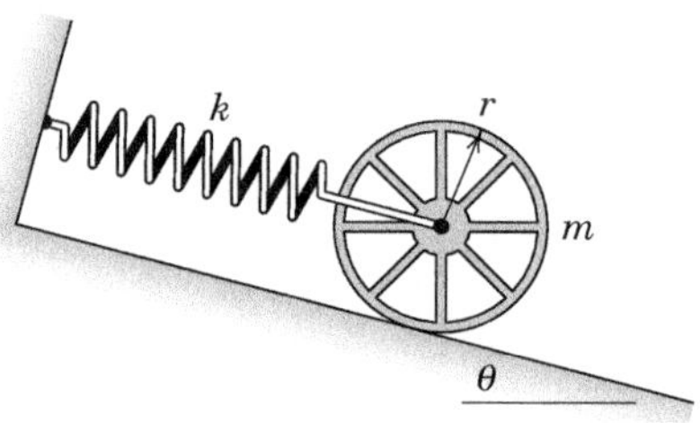

Problem 8/100

8/101 Determine the period τ for the uniform circular hoop of radius r as it oscillates with small amplitude about the horizontal knife edge.

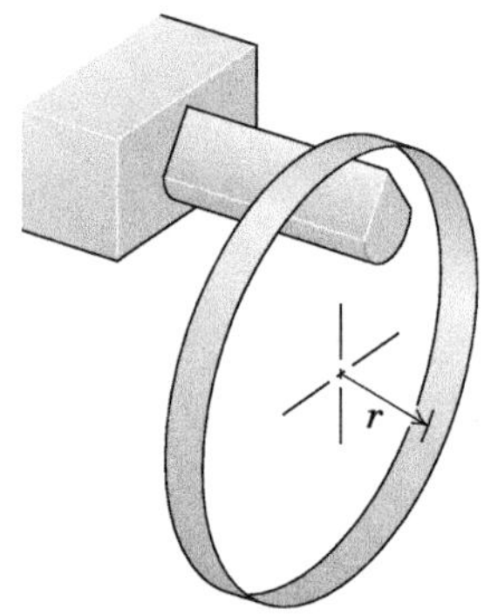

Problem 8/101

8/102 The length of the spring is adjusted so that the equilibrium position of the arm is horizontal as shown. Neglect the mass of the spring and the arm and calculate the natural frequency f_n for small oscillations.

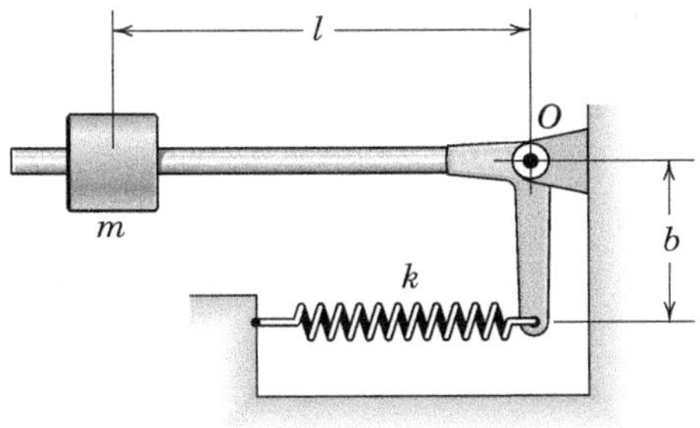

Problem 8/102

8/103 The body consists of two slender uniform rods which have a mass ρ per unit length. The rods are welded together and pivot about a horizontal axis through O against the action of a torsional spring of stiffness k_T. By the method of this article, determine the natural circular frequency ω_n for small oscillations about the equilibrium position. The spring is undeformed when $\theta = 0$, and friction in the pivot at O is negligible.

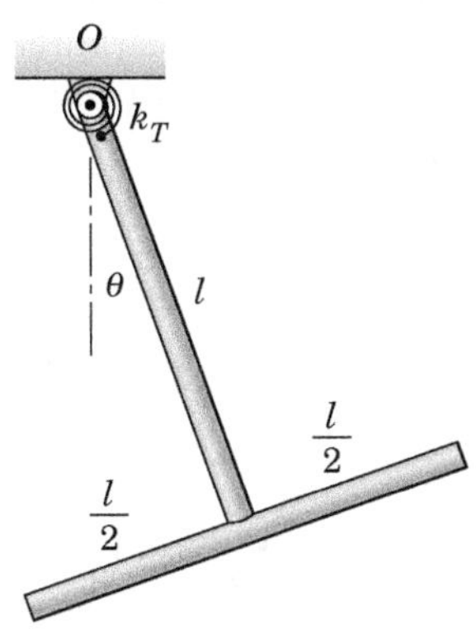

Problem 8/103

8/104 By the method of this article, determine the period of vertical oscillation. Each spring has a stiffness of 6 lb/in., and the mass of the pulleys may be neglected.

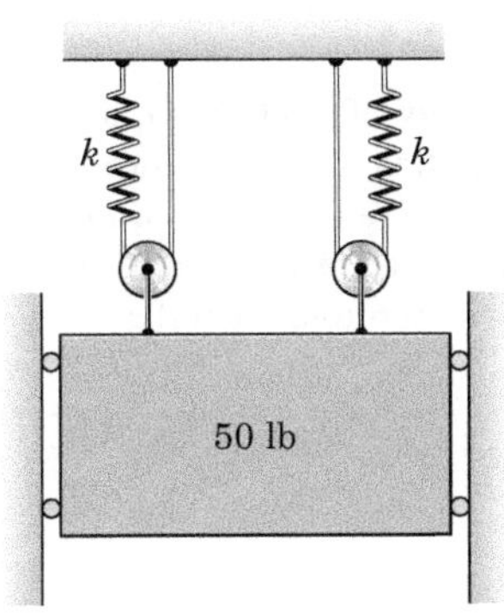

Problem 8/104

Representative Problems

8/105 The homogeneous circular cylinder of Prob. 8/93, repeated here, rolls without slipping on the track of radius R. Determine the period τ for small oscillations.

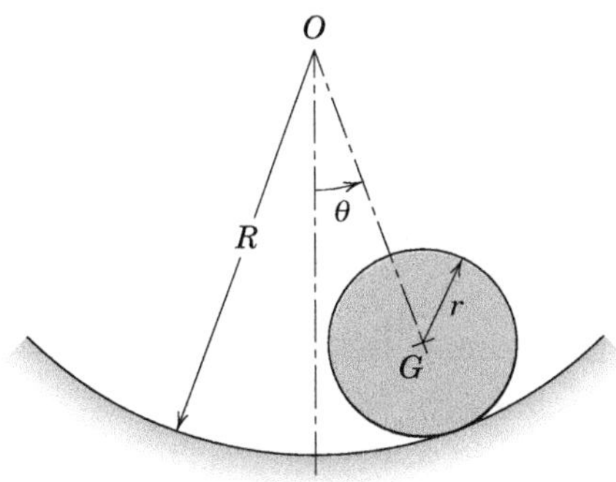

Problem 8/105

8/106 The disk has mass moment of inertia I_O about O and is acted upon by a torsional spring of constant k_T. The position of the small sliders, each of which has mass m, is adjustable. Determine the value of x for which the system has a given period τ.

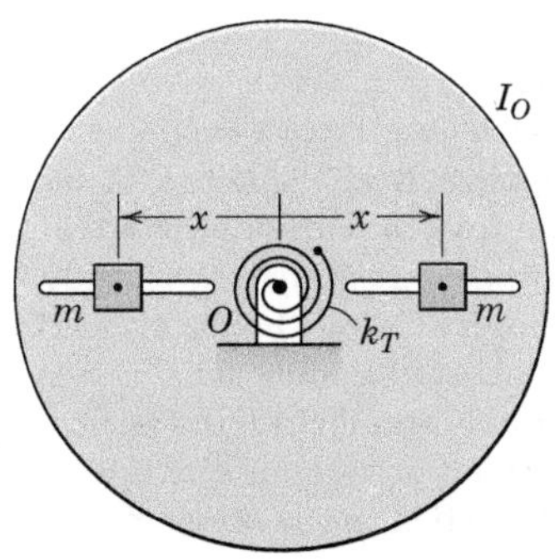

Problem 8/106

8/107 The uniform slender rod of length l and mass m_2 is secured to the uniform disk of radius $l/5$ and mass m_1. If the system is shown in its equilibrium position, determine the natural frequency ω_n and the maximum angular velocity ω for small oscillations of amplitude θ_0 about the pivot O.

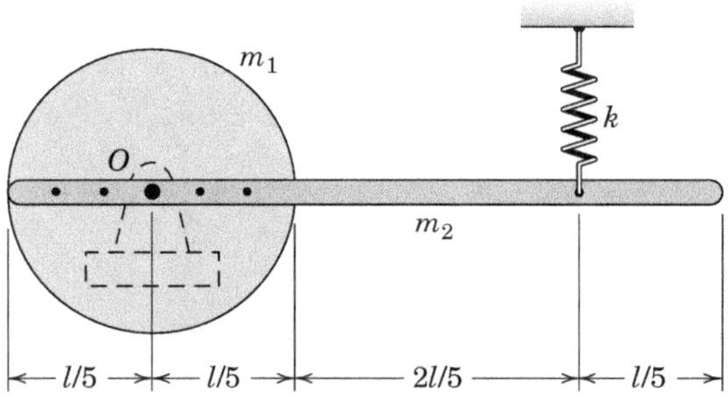

Problem 8/107

8/108 The assembly shown consists of two sheaves of mass $m_1 = 35$ kg and $m_2 = 15$ kg, outer groove radii $r_1 = 525$ mm and $r_2 = 250$ mm, and centroidal radii of gyration $(k_O)_1 = 350$ mm and $(k_O)_2 = 150$ mm. The sheaves are fitted to a central shaft at O with bearings which allow them to rotate independently of each other. Attached to the central shaft is a carriage of mass $m_3 = 25$ kg. Each sheave has an inextensible cable wrapped securely within its outer groove. Each cable is attached to a spring at one end and to a fixed support at the other end. The springs have stiffnesses $k_1 = 800$ N/m and $k_2 = 650$ N/m. By the method of this article, determine the equation of motion for the system in terms of the variable x and state the period τ for small vertical oscillations about the equilibrium position. Neglect friction in the bearings at O.

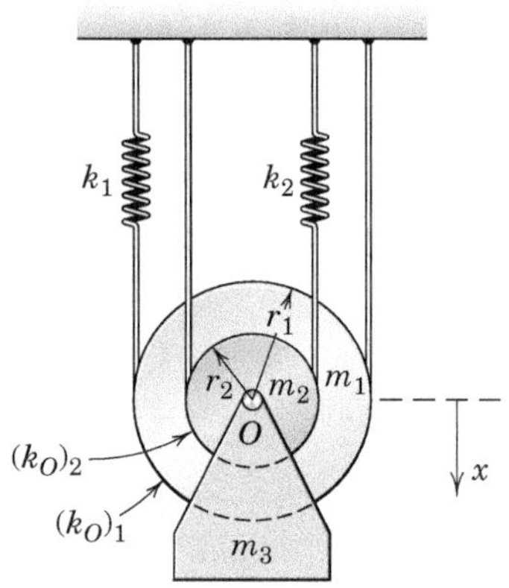

Problem 8/108

8/109 Derive the natural frequency f_n of the system composed of two homogeneous circular cylinders, each of mass M, and the connecting link AB of mass m. Assume small oscillations.

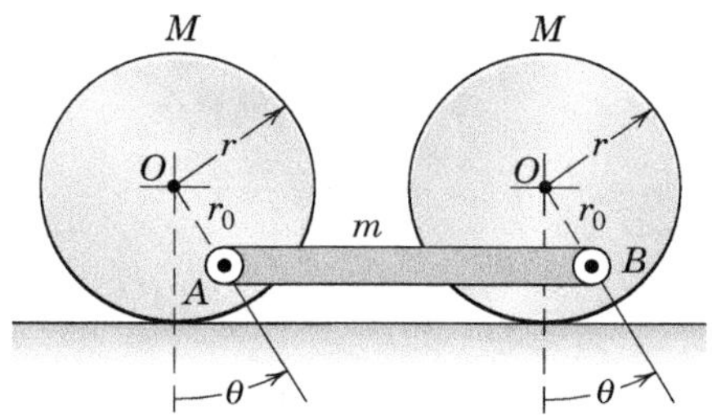

Problem 8/109

8/110 The rotational axis of the turntable is inclined at an angle α from the vertical. The turntable shaft pivots freely in bearings which are not shown. If a small block of mass m is placed a distance r from point O, determine the natural frequency ω_n for small rotational oscillations through the angle θ. The mass moment of inertia of the turntable about the axis of its shaft is I.

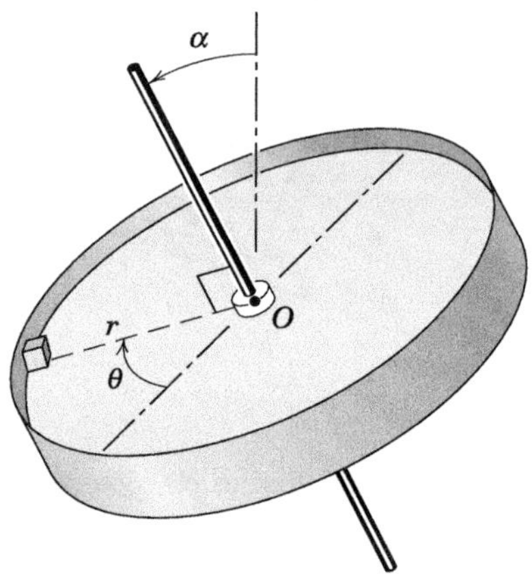

Problem 8/110

8/111 The assembly of Prob. 8/95 is repeated here without the applied harmonic torque. By the method of this article, determine the equation of motion of the system in terms of x for small oscillations about the equilibrium configuration if $c_1 = c_2 = c_3 = 0$.

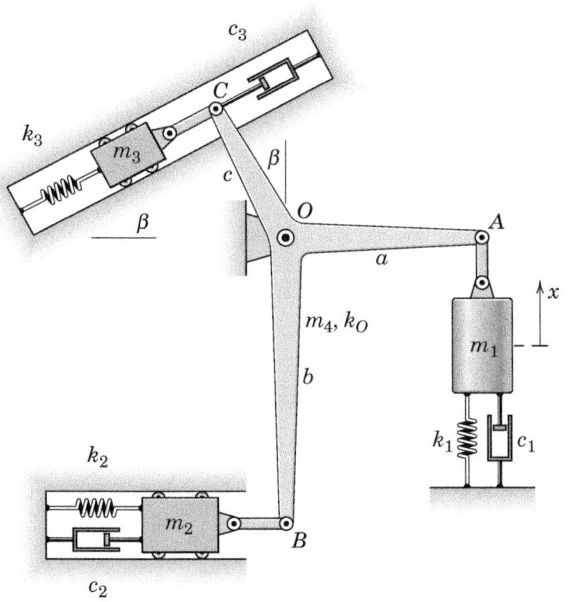

Problem 8/111

8/112 The ends of the uniform bar of mass m slide freely in the vertical and horizontal slots as shown. If the bar is in static equilibrium when $\theta = 0$, determine the natural frequency ω_n of small oscillations. What condition must be imposed on the spring constant k in order that oscillations take place?

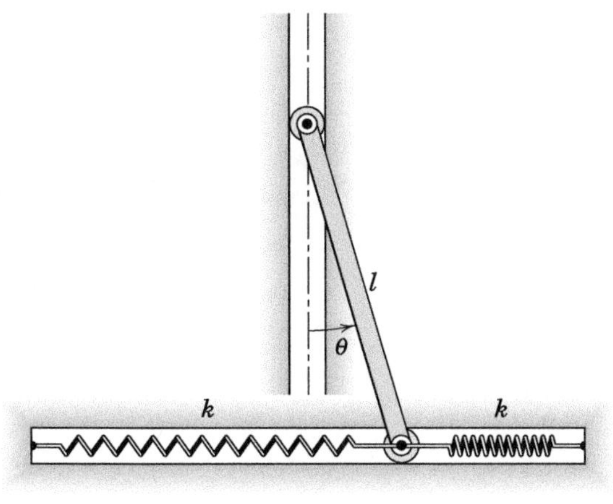

Problem 8/112

8/113 The 12-kg block is supported by the two 5-kg links with two torsion springs, each of constant $k_T = 500$ N·m/rad, arranged as shown. The springs are sufficiently stiff so that stable equilibrium is established in the position shown. Determine the natural frequency f_n for small oscillations about this equilibrium position.

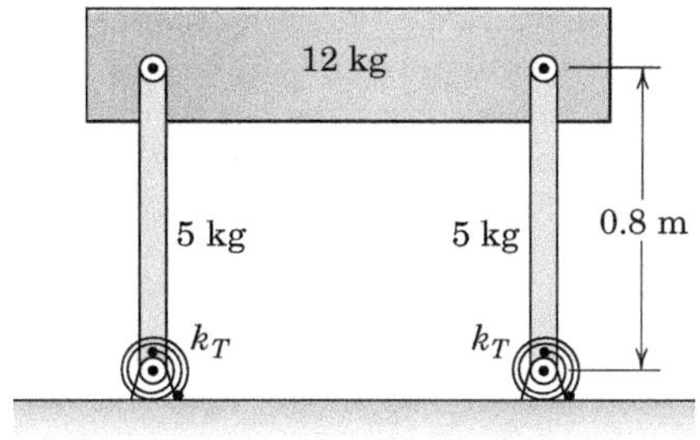

Problem 8/113

8/114 The front-end suspension of an automobile is shown. Each of the coil springs has a stiffness of 270 lb/in. If the weight of the front-end frame and equivalent portion of the body attached to the front end is 1800 lb, determine the natural frequency f_n of vertical oscillation of the frame and body in the absence of shock absorbers. (*Hint:* To relate the spring deflection to the deflection of the frame and body, consider the frame fixed and let the ground and wheels move vertically.)

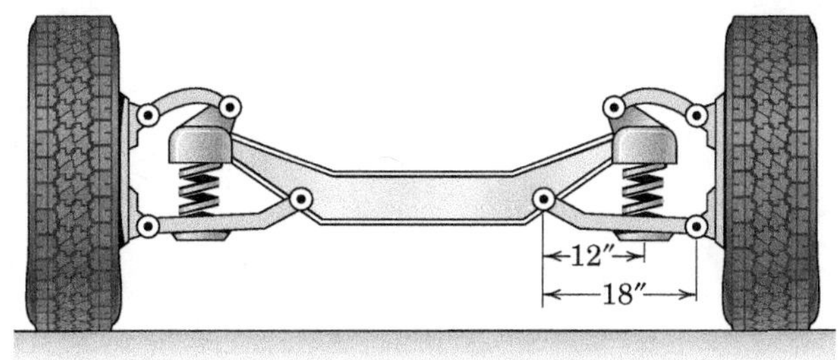

Problem 8/114

8/115 The system shown features a nonlinear spring whose resisting force F increases with deflection from the neutral position according to the graph shown. Determine the equation of motion for the system by the method of this article.

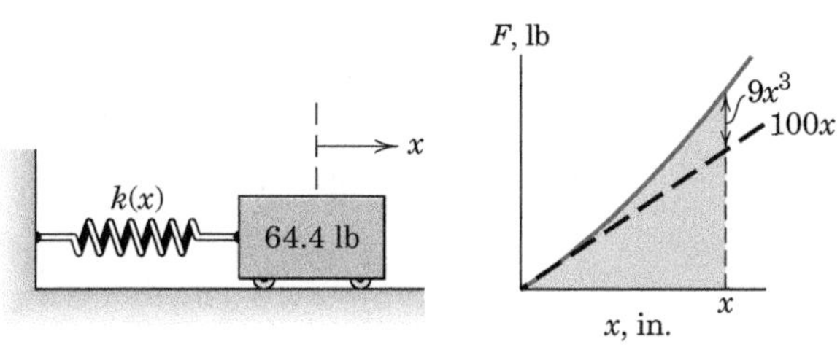

Problem 8/115

8/116 The semicircular cylindrical shell of radius r with small but uniform wall thickness is set into small rocking oscillation on the horizontal surface. If no slipping occurs, determine the expression for the period τ of each complete oscillation.

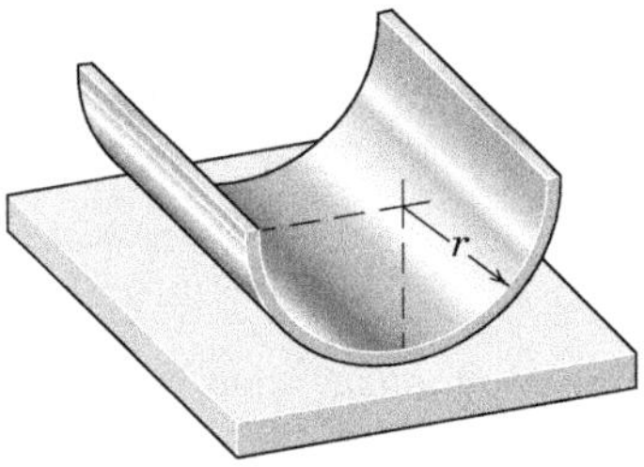

Problem 8/116

▶8/117 A hole of radius $R/4$ is drilled through a cylinder of radius R to form a body of mass m as shown. If the body rolls on the horizontal surface without slipping, determine the period τ for small oscillations.

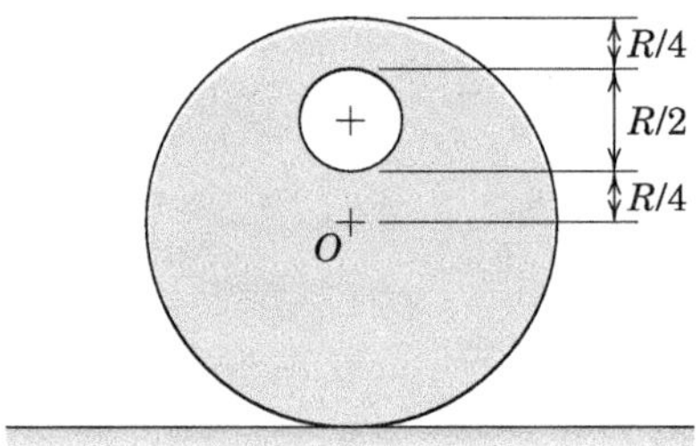

Problem 8/117

▶8/118 The quarter-circular sector of mass m and radius r is set into small rocking oscillation on the horizontal surface. If no slipping occurs, determine the expression for the period τ of each complete oscillation.

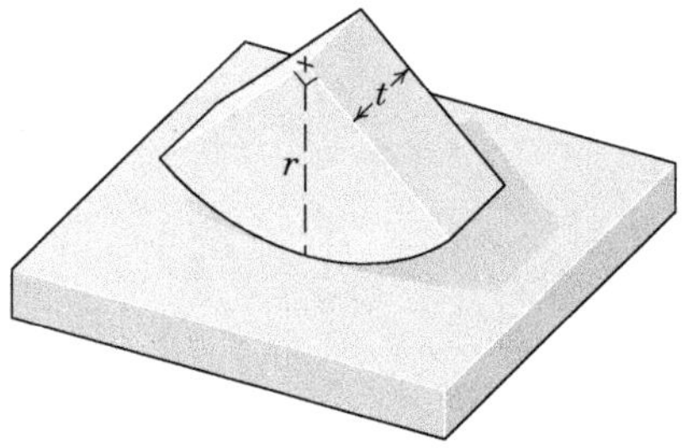

Problem 8/118

8/6 Chapter Review

In studying the vibrations of particles and rigid bodies in Chapter 8, we have observed that the subject is simply a direct application of the fundamental principles of dynamics as presented in Chapters 3 and 6. However, in these previous chapters, we determined the dynamic behavior of a body only at a particular instant of time or found the changes in motion resulting from only finite intervals of displacement or time. Chapter 8, on the other hand, has treated the solution of the defining differential equations of motion, so that the linear or angular displacement can be fully expressed as a function of time.

Particle Vibration

We divided our study of the time response of particles into the two categories of free and forced motion, with the further subdivisions of negligible and significant damping. We saw that the damping ratio ζ is a convenient parameter for determining the nature of unforced but viscously damped vibrations.

The prime lesson associated with harmonic forcing is that driving a lightly damped system with a force whose frequency is near the natural frequency can cause motion of excessively large amplitude—a condition called resonance, which usually must be carefully avoided.

Rigid-Body Vibration

In our study of rigid-body vibrations, we observed that the equation of small angular motion has a form identical to that for particle vibrations. Whereas particle vibrations may be described completely by the equations governing translational motion, rigid-body vibrations usually require the equations of rotational dynamics.

Energy Methods

In the final article of Chapter 8, we saw how the energy method can facilitate the determination of the natural frequency ω_n in free vibration problems where damping may be neglected. Here the total mechanical energy of the system is assumed to be constant. Setting its first time derivative to zero leads directly to the differential equation of motion for the system. The energy approach permits the analysis of a conservative system of interconnected parts without dismembering the system.

Degrees of Freedom

Throughout the chapter, we have restricted our attention to systems having one degree of freedom, where the position of the system can be specified by a single variable. If a system possesses n degrees of freedom, it has n natural frequencies. Thus, if a harmonic force is applied to such a system which is lightly damped, there are n driving frequencies which can cause motion of large amplitude. By a process called modal analysis, a complex system with n degrees of freedom can be reduced to n single-degree-of-freedom systems. For this reason, the thorough understanding of the material of this chapter is vital for the further study of vibrations.

REVIEW PROBLEMS

8/119 Determine the natural frequency f_n of the inverted pendulum. Assume small oscillations, and note any restrictions on your solution.

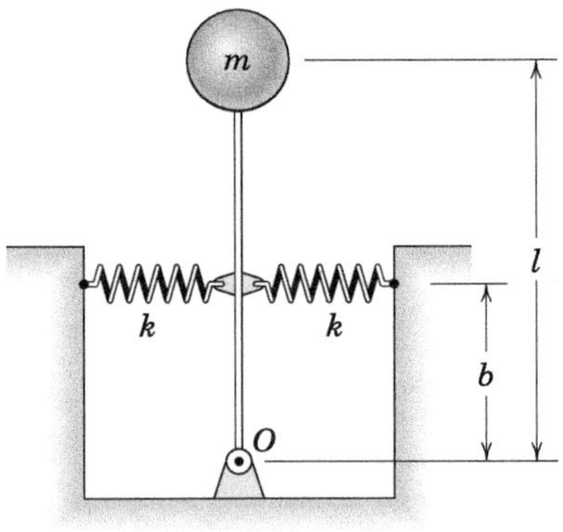

Problem 8/119

8/120 Determine the period τ of small oscillations for the uniform sector of mass m. The torsional spring of modulus k_T is undeflected when the sector is in the position shown.

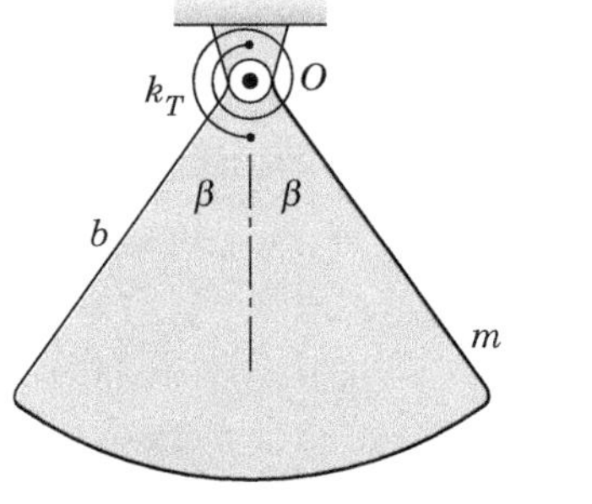

Problem 8/120

8/121 The 0.1-kg projectile is fired into the 10-kg block which is initially at rest with no force in the spring. The spring is attached at both ends. Calculate the maximum horizontal displacement X of the spring and the ensuing period of oscillation of the block and embedded projectile.

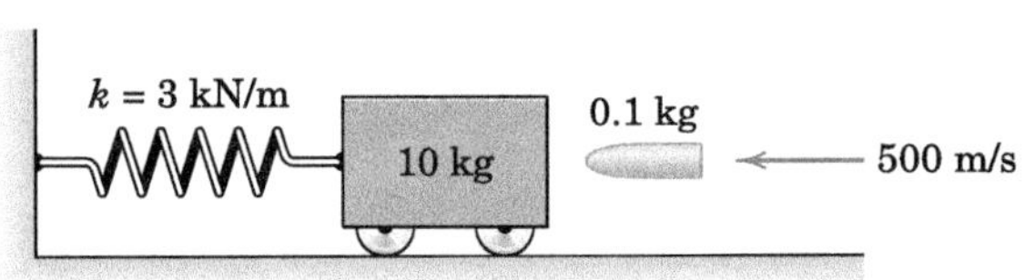

Problem 8/121

8/122 The uniform circular disk is suspended by a socket (not shown) which fits over the small ball attachment at O. Determine the frequency of small motion if the disk swings freely about (*a*) axis *A*-*A* and (*b*) axis *B*-*B*. Neglect the small offset, mass, and friction of the ball.

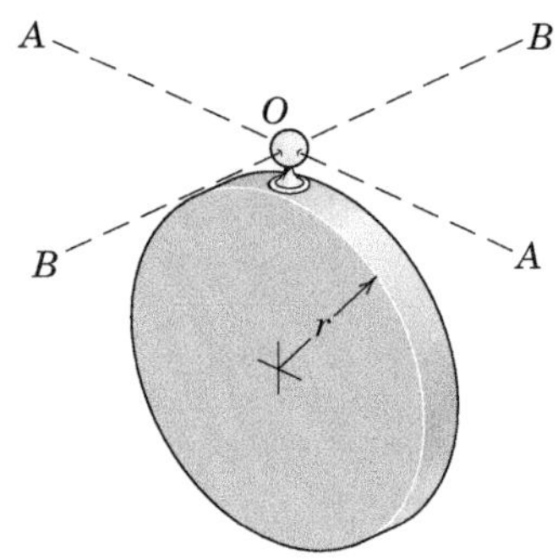

Problem 8/122

8/123 Determine the natural frequency f_n for small oscillations of the semicircular slender rod about the equilibrium position if (*a*) $r = 150$ mm, (*b*) $r = 300$ mm, and (*c*) $r = 600$ mm. Motion takes place in a vertical plane and friction in the pivot at O is negligible.

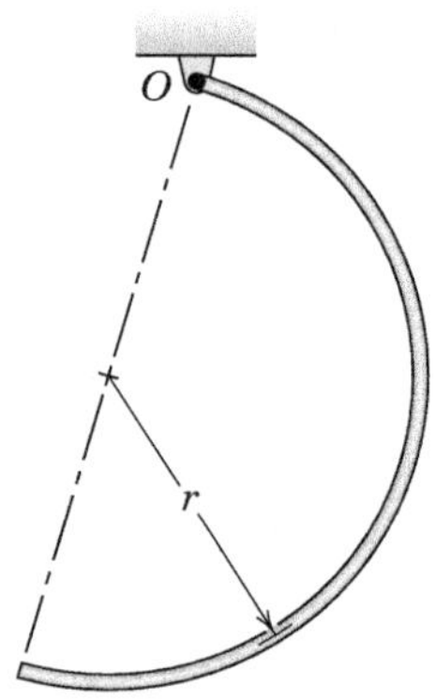

Problem 8/123

8/124 The triangular frame is constructed of uniform slender rod and pivots about a horizontal axis through point O. Determine the critical driving frequency ω_c of the block B which will result in excessively large oscillations of the assembly. The total mass of the frame is m.

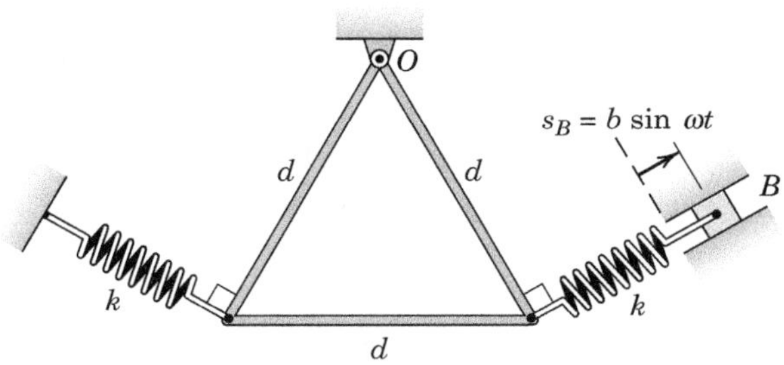

Problem 8/124

8/125 A linear oscillator with mass m, spring constant k, and viscous damping coefficient c is set into motion when released from a displaced position. Derive an expression for the energy loss Q during one complete cycle in terms of the amplitude x_1 at the start of the cycle. (See Fig. 8/7.)

8/126 Calculate the damping ratio ζ of the system shown if the weight and radius of gyration of the stepped cylinder are $W = 20$ lb and $\bar{k} = 5.5$ in., the spring constant is $k = 15$ lb/in., and the damping coefficient of the hydraulic cylinder is $c = 2$ lb-sec/ft. The cylinder rolls without slipping on the radius $r = 6$ in. and the spring can support tension as well as compression.

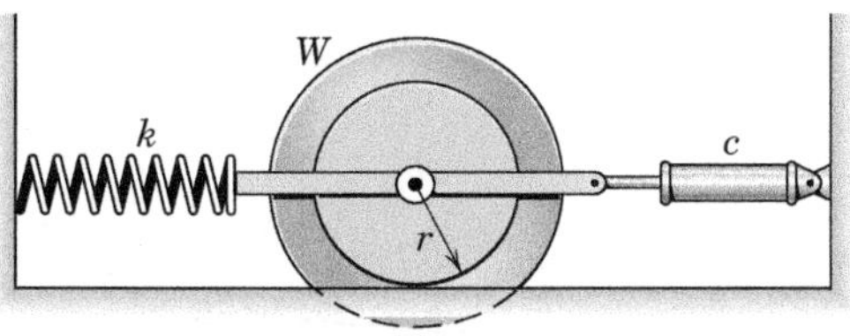

Problem 8/126

8/127 Determine the value of the viscous damping coefficient c for which the system is critically damped. The cylinder mass is $m = 2$ kg and the spring constant is $k = 150$ N/m. Neglect the mass and friction of the pulley.

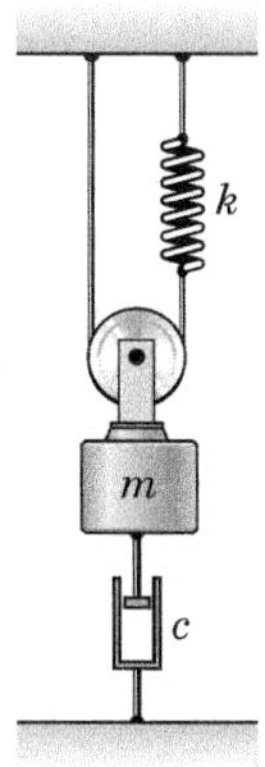

Problem 8/127

8/128 The seismic instrument shown is secured to a ship's deck near the stern where propeller-induced vibration is most pronounced. The ship has a single three-bladed propeller which turns at 180 rev/min and operates partly out of water, thus causing a shock as each blade breaks the surface. The damping ratio of the instrument is $\zeta = 0.5$, and its undamped natural frequency is 3 Hz. If the measured amplitude of A relative to its frame is 0.75 mm, compute the amplitude δ_0 of the vertical vibration of the deck.

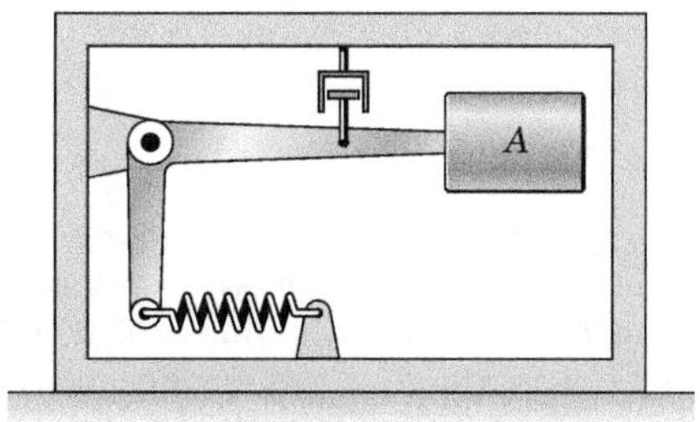

Problem 8/128

8/129 The square plate of side length $2b$ pivots about its center point O against the action of four springs, each of stiffness k. Each spring is attached to the corner of the plate at one end and to the center of a slotted collar at the other end. As the plate rotates, a rod slides through the smooth slot in the collar and deflects the spring. Determine the frequency f_n of small oscillations about the equilibrium configuration.

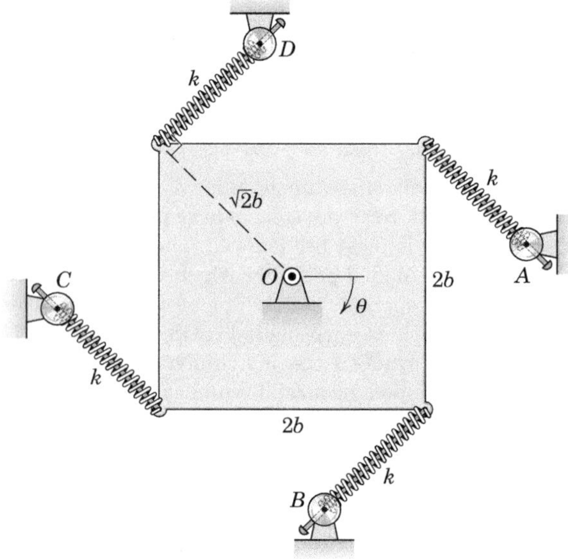

Problem 8/129

8/130 An experimental engine weighing 480 lb is mounted on a test stand with spring mounts at A and B, each with a stiffness of 600 lb/in. The radius of gyration of the engine about its mass center G is 4.60 in. With the motor not running, calculate the natural frequency $(f_n)_y$ of vertical vibration and $(f_n)_\theta$ of rotation about G. If vertical motion is suppressed and a light rotational imbalance occurs, at what speed N should the engine not be run?

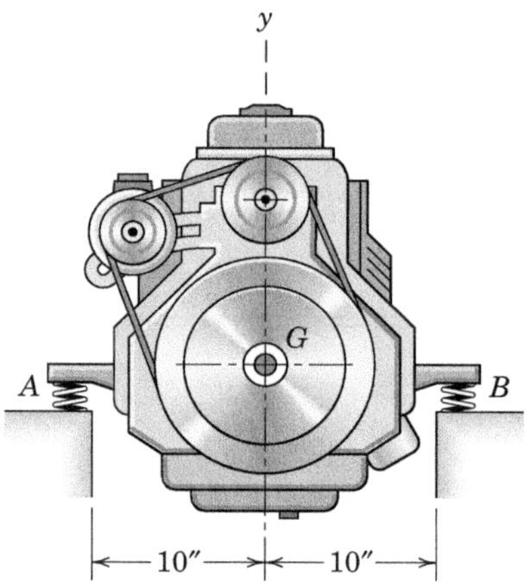

Problem 8/130

8/131 The uniform bar of mass M and length l has a small roller of mass m with negligible bearing friction at each end. Determine the period τ of the system for small oscillations on the curved track. The radius of gyration of the rollers is negligible.

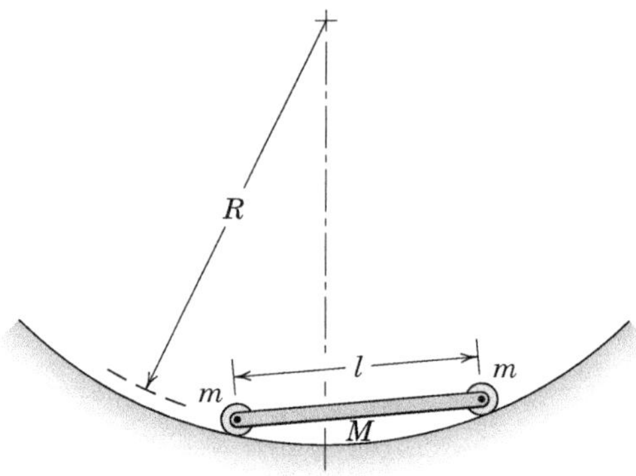

Problem 8/131

▶8/132 A 200-kg machine rests on four floor mounts, each of which has an effective spring constant $k = 250$ kN/m and an effective viscous damping coefficient $c = 1000$ N·s/m. The floor is known to vibrate vertically with a frequency of 24 Hz. What would be the effect on the amplitude of the absolute machine oscillation if the mounts were replaced with new ones which have the same effective spring constant but twice the effective damping coefficient?

*Computer-Oriented Problems

***8/133** The mass of a critically damped system having a natural frequency $\omega_n = 4$ rad/s is released from rest at an initial displacement x_0. Determine the time t required for the mass to reach the position $x = 0.1x_0$.

***8/134** The uniform sector of Prob. 8/77 is repeated here with $m = 4$ kg, $r = 325$ mm, and $\beta = 45°$. If the sector is released from rest with $\theta_0 = 90°$, plot the value of θ for the time period $0 \leq t \leq 6$ s. Friction in the pivot at O results in a resistive torque of magnitude $M = c\dot{\theta}$, where the constant $c = 0.35$ N·m·s/rad. Compare your large-angle results with those for the small-angle approximation of $\sin\theta \cong \theta$ and state the value of θ when $t = 1$ s for both large-angle and small-angle cases. (*Note:* The solution to this problem is of considerable difficulty and involves elliptic integrals. A numerical solution utilizing appropriate mathematics software is recommended.)

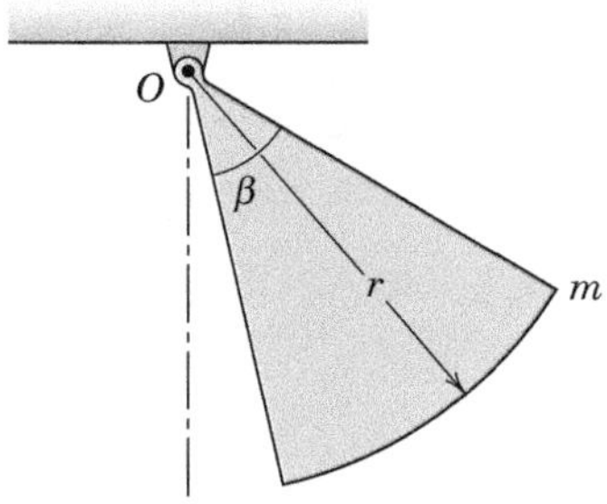

Problem 8/134

***8/135** The mass of the system shown is released with the initial conditions $x_0 = 0.1$ m and $\dot{x}_0 = -5$ m/s at $t = 0$. Plot the response of the system and determine the time(s) (if any) at which the displacement $x = -0.05$ m.

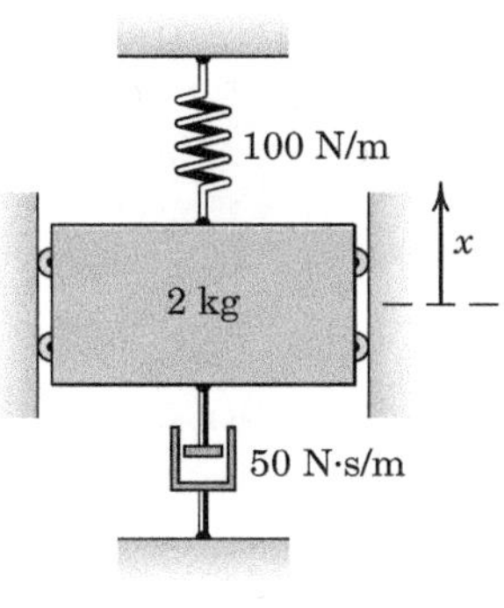

Problem 8/135

***8/136** Shown in the figure are the elements of a displacement meter used to study the motion $y_B = b \sin \omega t$ of the base. The motion of the mass relative to the frame is recorded on the rotating drum. If $l_1 = 1.2$ ft, $l_2 = 1.6$ ft, $l_3 = 2$ ft, $W = 2$ lb, $c = 0.1$ lb-sec/ft, and $\omega = 10$ rad/sec, determine the range of the spring constant k over which the magnitude of the recorded relative displacement is less than $1.5b$. It is assumed that the ratio ω/ω_n must remain greater than unity.

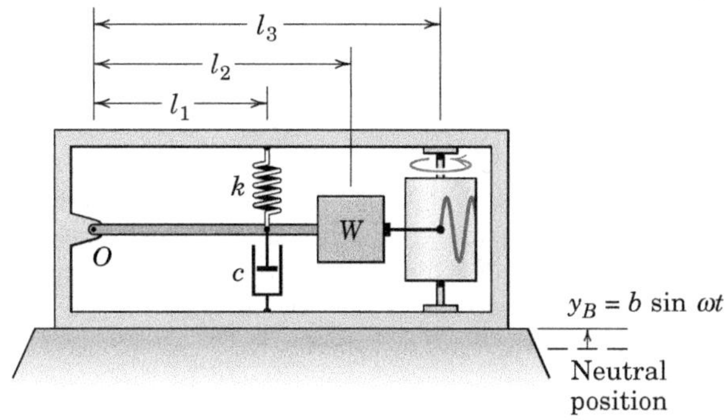

Problem 8/136

***8/137** The 4-kg mass is suspended by the spring of stiffness $k = 350$ N/m and is initially at rest in the equilibrium position. If a downward force $F = Ct$ is applied to the body and reaches a value of 40 N when $t = 1$ s, derive the differential equation of motion, obtain its solution, and plot the displacement y in millimeters as a function of time during the first second. Damping is negligible.

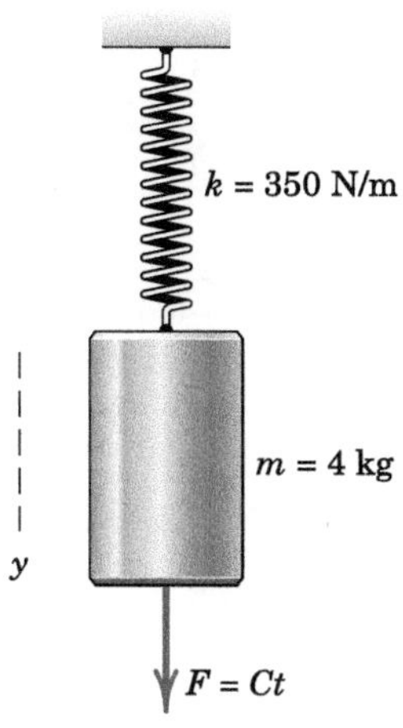

Problem 8/137

*8/138 Plot the response x of the 50-lb body over the time interval $0 \leq t \leq 1$ second. Determine the maximum and minimum values of x and their respective times. The initial conditions are $x_0 = 0$ and $\dot{x}_0 = 6$ ft/sec.

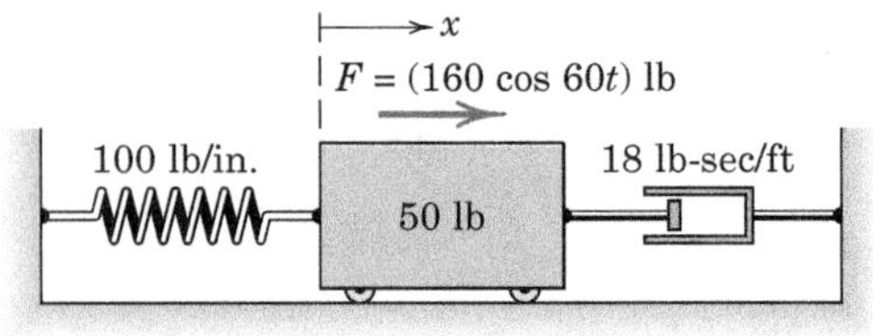

Problem 8/138

*8/139 Determine and plot the response x as a function of time for the undamped linear oscillator subjected to the force F which varies linearly with time for the first $\frac{3}{4}$ second as shown. The mass is initially at rest with $x = 0$ at time $t = 0$.

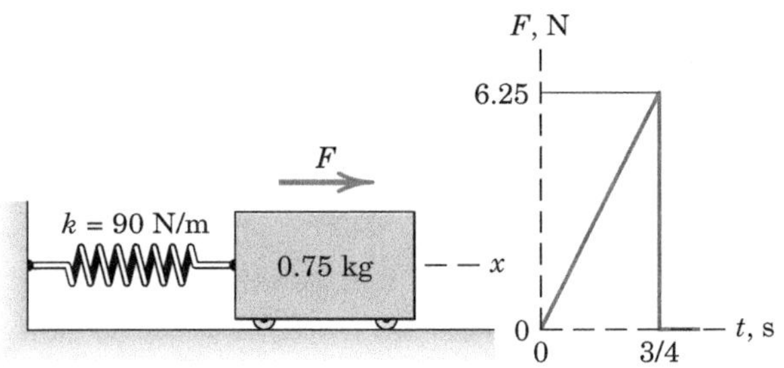

Problem 8/139

*8/140 The 4-kg cylinder is attached to a viscous damper and to the spring of stiffness $k = 800$ N/m. If the cylinder is released from rest at time $t = 0$ from the position where it is displaced a distance $y = 100$ mm from its equilibrium position, plot the displacement y as a function of time for the first second for the two cases where the viscous damping coefficient is (*a*) $c = 124$ N·s/m and (*b*) $c = 80$ N·s/m.

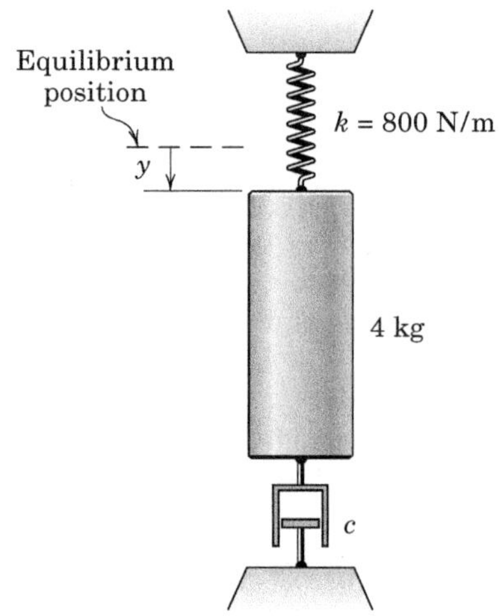

Problem 8/140

A

AREA MOMENTS OF INERTIA

See Appendix A of *Vol. 1 Statics* for a treatment of the theory and calculation of area moments of inertia. Because this quantity plays an important role in the design of structures, especially those dealt with in statics, we present only a brief definition in this *Dynamics* volume so that the student can appreciate the basic differences between area and mass moments of inertia.

The moments of inertia of a plane area A about the x- and y-axes in its plane and about the z-axis normal to its plane, Fig. A/1, are defined by

$$I_x = \int y^2\,dA \qquad I_y = \int x^2\,dA \qquad I_z = \int r^2\,dA$$

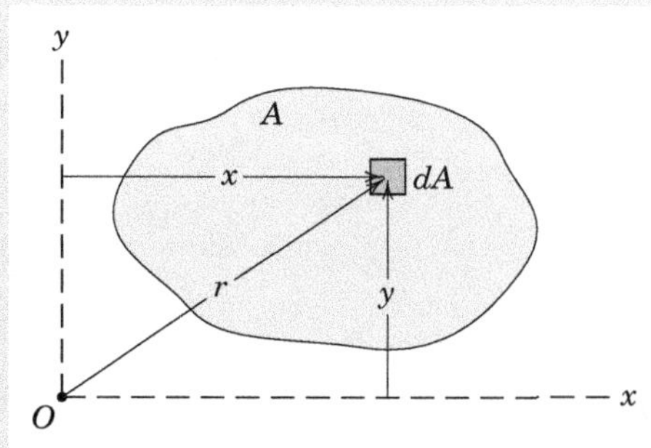

Figure A/1

where dA is the differential element of area and $r^2 = x^2 + y^2$. Clearly, the polar moment of inertia I_z equals the sum $I_x + I_y$ of the rectangular moments of inertia. For thin flat plates, the area moment of inertia is useful in the calculation of the mass moment of inertia, as explained in Appendix B.

The area moment of inertia is a measure of the distribution of area about the axis in question and, for that axis, is a constant property of the area. The dimensions of area moment of inertia are (distance)^4 expressed in m^4 or mm^4 in SI units and ft^4 or in.^{4} in U.S. customary units. In contrast, mass moment of inertia is a measure of the distribution of mass about the axis in question, and its dimensions are (mass) (distance)^2, which are expressed in kg·m^2 in SI units and in lb-ft-sec^2 or lb-in.-sec^2 in U.S. customary units.

B Mass Moments of Inertia

APPENDIX OUTLINE

B/1 Mass Moments of Inertia about an Axis

The equation of rotational motion about an axis normal to the plane of motion for a rigid body in plane motion contains an integral which depends on the distribution of mass with respect to the moment axis. This integral occurs whenever a rigid body has an angular acceleration about its axis of rotation. Thus, to study the dynamics of rotation, you should be thoroughly familiar with the calculation of mass moments of inertia for rigid bodies.

Consider a body of mass m, Fig. B/1, rotating about an axis O-O with an angular acceleration α. All particles of the body move in parallel planes which are normal to the rotation axis O-O. We may choose any one of the planes as the plane of motion, although the one containing the center of mass is usually the one so designated. An element of mass dm has a component of acceleration tangent to its circular path equal to $r\alpha$, and by Newton's second law of motion the resultant tangential force on this element equals $r\alpha\, dm$. The moment of this force about the axis O-O is $r^2\alpha\, dm$, and the sum of the moments of these forces for all elements is $\int r^2\alpha\, dm$.

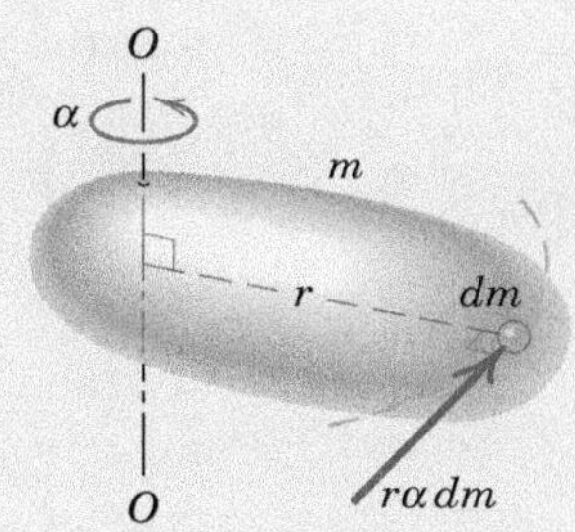

Figure B/1

For a rigid body, α is the same for all radial lines in the body and we may take it outside the integral sign. The remaining integral is called the mass moment of inertia I of the body about the axis O-O and is

$$I = \int r^2\, dm \qquad \textbf{(B/1)}$$

This integral represents an important property of a body and is involved in the analysis of any body which has rotational acceleration about a given axis. Just as the mass m of a body is a measure of the resistance to translational acceleration, the moment of inertia I is a measure of resistance to rotational acceleration of the body.

The moment-of-inertia integral may be expressed alternatively as

$$I = \Sigma r_i^2 m_i \qquad \textbf{(B/1}\boldsymbol{a}\textbf{)}$$

where r_i is the radial distance from the inertia axis to the representative particle of mass m_i and where the summation is taken over all particles of the body.

If the density ρ is constant throughout the body, the moment of inertia becomes

$$I = \rho \int r^2 \, dV$$

where dV is the element of volume. In this case, the integral by itself defines a purely geometrical property of the body. When the density is not constant but is expressed as a function of the coordinates of the body, it must be left within the integral sign and its effect accounted for in the integration process.

In general, the coordinates which best fit the boundaries of the body should be used in the integration. It is particularly important that we make a good choice of the element of volume dV. To simplify the integration, an element of lowest possible order should be chosen, and the correct expression for the moment of inertia of the element about the axis involved should be used. For example, in finding the moment of inertia of a solid right-circular cone about its central axis, we may choose an element in the form of a circular slice of infinitesimal thickness, Fig. B/2*a*. The differential moment of inertia for this element is the expression for the moment of inertia of a circular cylinder of infinitesimal altitude about its central axis. (This expression will be obtained in Sample Problem B/1.)

Alternatively, we could choose an element in the form of a cylindrical shell of infinitesimal thickness as shown in Fig. B/2*b*. Because all of the mass of the element is at the same distance r from the inertia axis, the differential moment of inertia for this element is merely $r^2 \, dm$ where dm is the differential mass of the elemental shell.

From the definition of mass moment of inertia, its dimensions are (mass)(distance)^2 and are expressed in the units kg·m^2 in SI units and lb-ft-sec^2 in U.S. customary units.

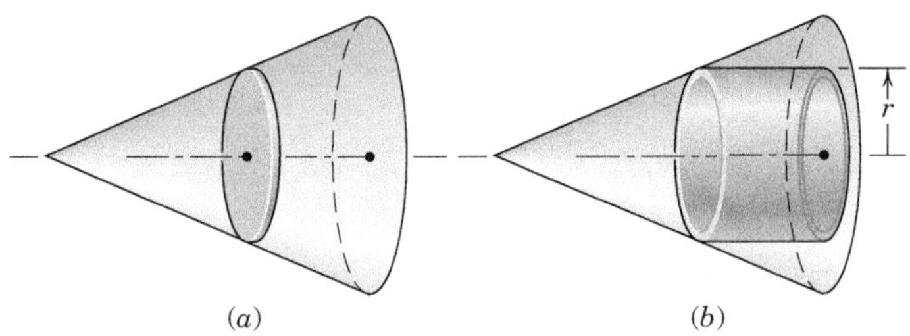

Figure B/2

Radius of Gyration

The radius of gyration k of a mass m about an axis for which the moment of inertia is I is defined as

$$k = \sqrt{\frac{1}{m}} \qquad \text{or} \qquad I = k^2 m \tag{B/2}$$

Thus, k is a measure of the distribution of mass of a given body about the axis in question, and its definition is analogous to the definition of the radius of gyration for area moments of inertia. If all the mass m of a body could be concentrated at a distance k from the axis, the moment of inertia would be unchanged.

The moment of inertia of a body about a particular axis is frequently indicated by specifying the mass of the body and the radius of gyration of the body about the axis. The moment of inertia is then calculated from Eq. B/2.

Transfer of Axes

If the moment of inertia of a body is known about an axis passing through the mass center, it may be determined easily about any parallel axis. To prove this statement, consider the two parallel axes in Fig. B/3, one being an axis through the mass center G and the other a parallel axis through some other point C. The radial distances from the two axes to any element of mass dm are r_0 and r, and the separation of the axes is d. Substituting the law of cosines $r^2 = r_0{}^2 + d^2 + 2r_0 d \cos\theta$ into the definition for the moment of inertia about the axis through C gives

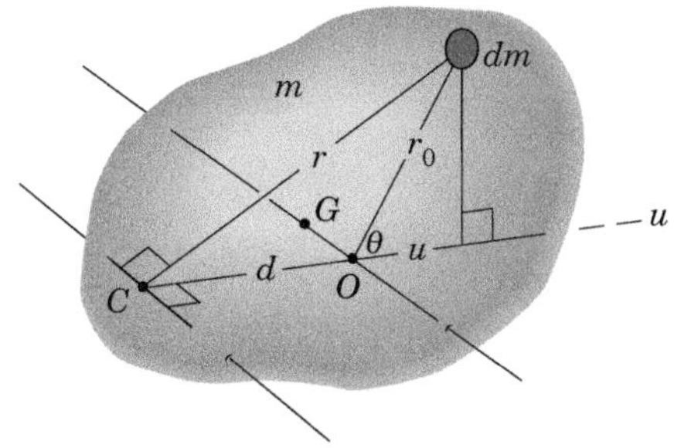

Figure B/3

$$\begin{aligned} I &= \int r^2\,dm = \int (r_0{}^2 + d^2 + 2r_0 d\cos\theta)\,dm \\ &= \int r_0{}^2\,dm + d^2 \int dm + 2d \int u\,dm \end{aligned}$$

The first integral is the moment of inertia $\bar{I}$ about the mass-center axis, the second term is md^2, and the third integral equals zero, since the u-coordinate of the mass center with respect to the axis through G is zero. Thus, the parallel-axis theorem is

$$I = \bar{I} + md^2 \tag{B/3}$$

Remember that the transfer cannot be made unless one axis passes through the center of mass and unless the axes are parallel.

When the expressions for the radii of gyration are substituted in Eq. B/3, there results

$$k^2 = \bar{k}^2 + d^2 \tag{B/3a}$$

Equation B/3a is the parallel-axis theorem for obtaining the radius of gyration k about an axis which is a distance d from a parallel axis through the mass center, for which the radius of gyration is $\bar{k}$.

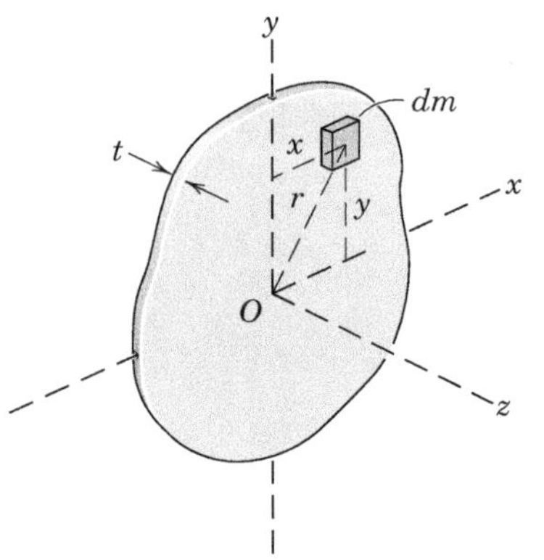

Figure B/4

For plane-motion problems where rotation occurs about an axis normal to the plane of motion, a single subscript for I is sufficient to designate the inertia axis. Thus, if the plate of Fig. B/4 has plane motion in the x-y plane, the moment of inertia of the plate about the z-axis through O is designated I_O. For three-dimensional motion, however, where components of rotation may occur about more than one axis, we use a double subscript to preserve notational symmetry with product-of-inertia terms, which are described in Art. B/2. Thus, the moments of inertia about the x-, y-, and z-axes are labeled I_{xx}, I_{yy}, and I_{zz}, respectively, and from Fig. B/5 we see that they become

$$\begin{aligned} I_{xx} &= \int r_x^{\,2}\,dm = \int (y^2 + z^2)\,dm \\ I_{yy} &= \int r_y^{\,2}\,dm = \int (z^2 + x^2)\,dm \\ I_{zz} &= \int r_z^{\,2}\,dm = \int (x^2 + y^2)\,dm \end{aligned} \qquad \textbf{(B/4)}$$

These integrals are cited in Eqs. 7/10 of Art. 7/7 on angular momentum in three-dimensional rotation.

The defining expressions for mass moments of inertia and area moments of inertia are similar. An exact relationship between the two moment-of-inertia expressions exists in the case of flat plates. Consider the flat plate of uniform thickness in Fig. B/4. If the constant thickness is t and the density is ρ, the mass moment of inertia I_{zz} of the plate about the z-axis normal to the plate is

$$I_{zz} = \int r^2\,dm = \rho t \int r^2\,dA = \rho t I_z \qquad \textbf{(B/5)}$$

Thus, the mass moment of inertia about the z-axis equals the mass per unit area ρt times the polar moment of inertia I_z of the plate area about the z-axis. If t is small compared with the dimensions of the plate in its

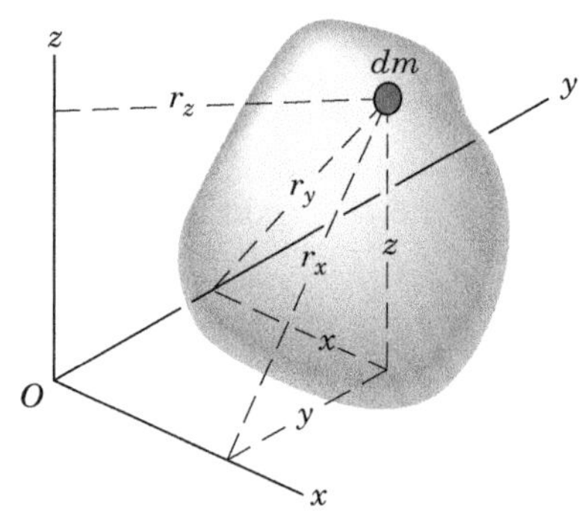

Figure B/5

plane, the mass moments of inertia I_{xx} and I_{yy} of the plate about the x- and y-axes are closely approximated by

$$I_{xx} = \int y^2\,dm = \rho t \int y^2\,dA = \rho t I_x$$
$$I_{yy} = \int x^2\,dm = \rho t \int x^2\,dA = \rho t I_y \qquad \textbf{(B/6)}$$

Thus, the mass moments of inertia equal the mass per unit area ρt times the corresponding area moments of inertia. The double subscripts for mass moments of inertia distinguish these quantities from area moments of inertia.

Inasmuch as $I_z = I_x + I_y$ for area moments of inertia, we have

$$I_{zz} = I_{xx} + I_{yy} \qquad \textbf{(B/7)}$$

which holds *only* for a thin flat plate. This restriction is observed from Eqs. B/6, which do not hold true unless the thickness t or the z-coordinate of the element is negligible compared with the distance of the element from the corresponding x- or y-axis. Equation B/7 is very useful when dealing with a differential mass element taken as a flat slice of differential thickness, say, dz. In this case, Eq. B/7 holds exactly and becomes

$$dI_{zz} = dI_{xx} + dI_{yy} \qquad \textbf{(B/7}\boldsymbol{a}\textbf{)}$$

for axes x and y in the plane of the plate.

Composite Bodies

As in the case of area moments of inertia, the mass moment of inertia of a composite body is the sum of the moments of inertia of the individual parts about the same axis. It is often convenient to treat a composite body as defined by positive volumes and negative volumes. The moment of inertia of a negative element, such as the material removed to form a hole, must be considered a negative quantity.

A summary of some of the more useful formulas for mass moments of inertia of various masses of common shape is given in Table D/4, Appendix D.

The problems which follow the sample problems are divided into the categories *Integration Exercises* and *Composite and Parallel-Axis Exercises*. The parallel-axis theorem will also be useful in some of the problems in the first category.

Sample Problem B/1

Determine the moment of inertia and radius of gyration of a homogeneous right-circular cylinder of mass m and radius r about its central axis O-O.

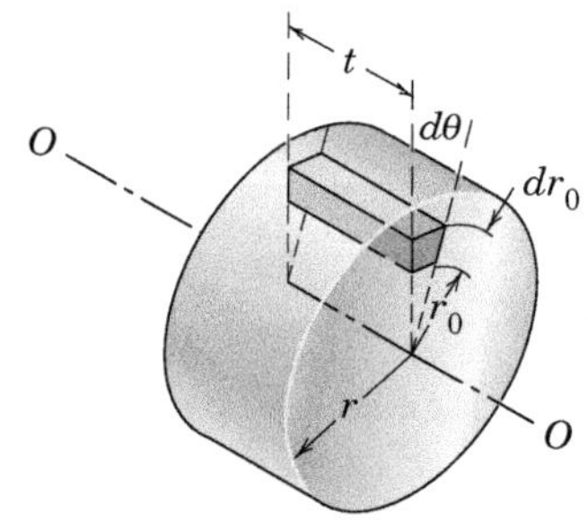

Solution. An element of mass in cylindrical coordinates is $dm = \rho\, dV =$
① $\rho t r_0\, dr_0\, d\theta$, where ρ is the density of the cylinder. The moment of inertia about the axis of the cylinder is

② $$I = \int r_0^2\, dm = \rho t \int_0^{2\pi} \int_0^r r_0^3\, dr_0\, d\theta = \rho t \frac{\pi r^4}{2} = \tfrac{1}{2} m r^2 \qquad \textit{Ans.}$$

The radius of gyration is

$$k = \sqrt{\frac{I}{m}} = \frac{r}{\sqrt{2}} \qquad \textit{Ans.}$$

Helpful Hints

① If we had started with a cylindrical shell of radius r_0 and axial length t as our mass element dm, then $dI = r_0^2\, dm$ directly. You should evaluate the integral.

② The result $I = \frac{1}{2}mr^2$ applies *only* to a solid homogeneous circular cylinder and cannot be used for any other wheel of circular periphery.

Sample Problem B/2

Determine the moment of inertia and radius of gyration of a homogeneous solid sphere of mass m and radius r about a diameter.

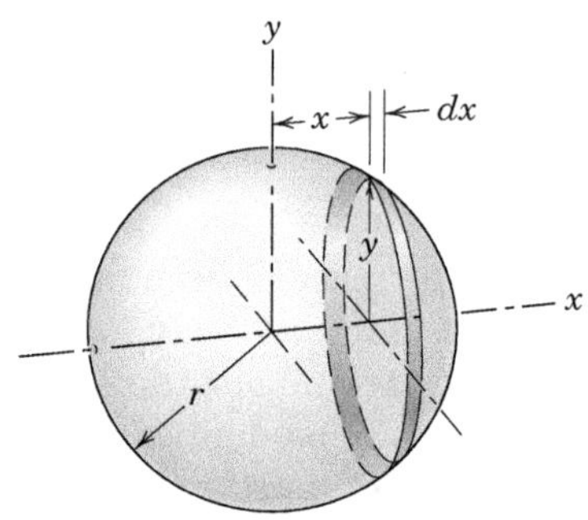

Solution. A circular slice of radius y and thickness dx is chosen as the volume element. From the results of Sample Problem B/1, the moment of inertia about the x-axis of the elemental cylinder is

① $$dI_{xx} = \tfrac{1}{2}(dm)y^2 = \tfrac{1}{2}(\pi\rho y^2\, dx)y^2 = \frac{\pi\rho}{2}(r^2 - x^2)^2\, dx$$

where ρ is the constant density of the sphere. The total moment of inertia about the x-axis is

$$I_{xx} = \frac{\pi\rho}{2}\int_{-r}^{r} (r^2 - x^2)^2\, dx = \tfrac{8}{15}\pi\rho r^5 = \tfrac{2}{5}mr^2 \qquad \textit{Ans.}$$

The radius of gyration about the x-axis is

$$k_x = \sqrt{\frac{I_{xx}}{m}} = \sqrt{\frac{2}{5}}\,r \qquad \textit{Ans.}$$

Helpful Hint

① Here is an example where we utilize a previous result to express the moment of inertia of the chosen element, which in this case is a right-circular cylinder of differential axial length dx. It would be foolish to start with a third-order element, such as $\rho\, dx\, dy\, dz$, when we can easily solve the problem with a first-order element.

Sample Problem B/3

Determine the moments of inertia of the homogeneous rectangular parallelepiped of mass m about the centroidal x_0- and z-axes and about the x-axis through one end.

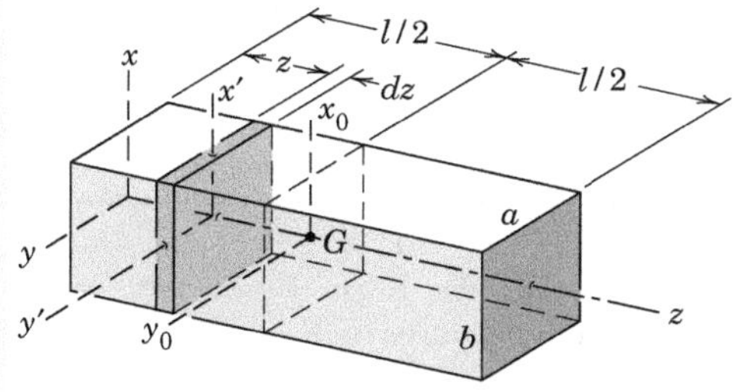

Solution. A transverse slice of thickness dz is selected as the element of volume. The moment of inertia of this slice of infinitesimal thickness equals the moment of inertia of the area of the section times the mass per unit area $\rho\, dz$. Thus, the moment of inertia of the transverse slice about the y'-axis is

$$dI_{y'y'} = (\rho\, dz)(\tfrac{1}{12}ab^3)$$

and that about the x'-axis is

① $$dI_{x'x'} = (\rho\, dz)(\tfrac{1}{12}a^3b)$$

As long as the element is a plate of differential thickness, the principle given by Eq. B/7*a* may be applied to give

$$dI_{zz} = dI_{x'x'} + dI_{y'y'} = (\rho\, dz)\frac{ab}{12}(a^2 + b^2)$$

These expressions may now be integrated to obtain the desired results.

The moment of inertia about the z-axis is

$$I_{zz} = \int dI_{zz} = \frac{\rho ab}{12}(a^2 + b^2)\int_0^l dz = \tfrac{1}{12}m(a^2 + b^2) \qquad \textit{Ans.}$$

where m is the mass of the block. By interchange of symbols, the moment of inertia about the x_0-axis is

$$I_{x_0x_0} = \tfrac{1}{12}m(a^2 + l^2) \qquad \textit{Ans.}$$

The moment of inertia about the x-axis may be found by the parallel-axis theorem, Eq. B/3. Thus,

$$I_{xx} = I_{x_0x_0} + m\left(\frac{l}{2}\right)^2 = \tfrac{1}{12}m(a^2 + 4l^2) \qquad \textit{Ans.}$$

This last result may be obtained by expressing the moment of inertia of the elemental slice about the x-axis and integrating the expression over the length of the bar. Again, by the parallel-axis theorem

$$dI_{xx} = dI_{x'x'} + z^2\, dm = (\rho\, dz)(\tfrac{1}{12}a^3b) + z^2\rho ab\, dz = \rho ab\left(\frac{a^2}{12} + z^2\right)dz$$

Integrating gives the result obtained previously:

$$I_{xx} = \rho ab\int_0^l\left(\frac{a^2}{12} + z^2\right)dz = \frac{\rho abl}{3}\left(l^2 + \frac{a^2}{4}\right) = \tfrac{1}{12}m(a^2 + 4l^2)$$

The expression for I_{xx} may be simplified for a long prismatic bar or slender rod whose transverse dimensions are small compared with the length. In this case, a^2 may be neglected compared with $4l^2$, and the moment of inertia of such a slender bar about an axis through one end normal to the bar becomes $I = \frac{1}{3}ml^2$. By the same approximation, the moment of inertia about a centroidal axis normal to the bar is $I = \frac{1}{12}ml^2$.

Helpful Hint

① Refer to Eqs. B/6 and recall the expression for the area moment of inertia of a rectangle about an axis through its center parallel to its base.

Sample Problem B/4

The upper edge of the thin homogeneous plate of mass m is parabolic with a vertical slope at the origin O. Determine its mass moments of inertia about the x-, y-, and z-axes.

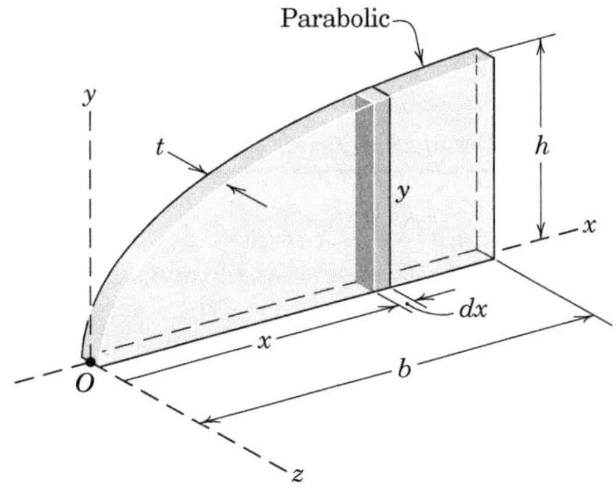

Solution. We begin by clearly establishing the function associated with ① the upper boundary. From $y = k\sqrt{x}$ evaluated at $(x, y) = (b, h)$, we find that $k = h/\sqrt{b}$ so that $y = \frac{h}{\sqrt{b}}\sqrt{x}$. We choose a transverse slice of thickness dx for the integrations leading to I_{xx} and I_{yy}. The mass of this slice is

$$dm = \rho ty\,dx$$

and the total mass of the plate is

② $$m = \int dm = \int \rho ty\,dx = \int_0^b \rho t \frac{h}{\sqrt{b}}\sqrt{x}\,dx = \tfrac{2}{3}\rho thb$$

The moment of inertia of the slice about the x-axis is

③ $$dI_{xx} = \tfrac{1}{3}\,dm\,y^2 = \tfrac{1}{3}(\rho ty\,dx)y^2 = \tfrac{1}{3}\rho ty^3\,dx$$

For the entire plate, we have

$$I_{xx} = \int dI_{xx} = \int_0^b \tfrac{1}{3}\rho t\left(\frac{h}{\sqrt{b}}\sqrt{x}\right)^3 dx = \tfrac{2}{15}\rho th^3 b$$

In terms of the mass m:

④ $$I_{xx} = \tfrac{2}{15}\rho th^3 b\left(\frac{m}{\tfrac{2}{3}\rho thb}\right) = \tfrac{1}{5}mh^2 \qquad \textit{Ans.}$$

The moment of inertia of the element about the y-axis is

$$dI_{yy} = dm\,y^2 = (\rho ty\,dx)y^2 = \rho ty^3\,dx$$

For the entire plate,

⑤ $$I_{yy} = \int dI_{yy} = \int_0^b \rho t\left(\frac{h}{\sqrt{b}}\sqrt{x}\right)^3 dx = \tfrac{2}{7}\rho thb^3\left(\frac{m}{\tfrac{2}{3}\rho thb}\right) = \tfrac{3}{7}mb^2 \qquad \textit{Ans.}$$

For thin plates which lie in the x-y plane,

$$I_{zz} = I_{xx} + I_{yy} = \tfrac{1}{5}mh^2 + \tfrac{3}{7}mb^2$$

$$I_{zz} = m\left(\frac{h^2}{5} + \frac{3b^2}{7}\right) \qquad \textit{Ans.}$$

Helpful Hints

① If we have $y = kx^2$, saying that "y gets large faster than x" helps establish that the parabola opens upward. Here, we have $y^2 = k^2x$, which says that "x gets large faster than y", helping establish that the parabola opens rightward.

② For a full b by h rectangular plate of thickness t, the mass would be ρthb (density times volume). So the factor of $\frac{2}{3}$ for the parabolic plate makes sense.

③ Recall that for a slender rod of mass m and length l, the moment of inertia about an axis perpendicular to the rod and passing through one end is $\frac{1}{3}ml^2$.

④ Note that I_{xx} is independent of the width b.

⑤ Note that I_{yy} is independent of the height h.

Sample Problem B/5

The radius of the homogeneous solid of revolution is proportional to the square of its x-coordinate. If the mass of the body is m, determine its mass moments of inertia about the x- and y-axes.

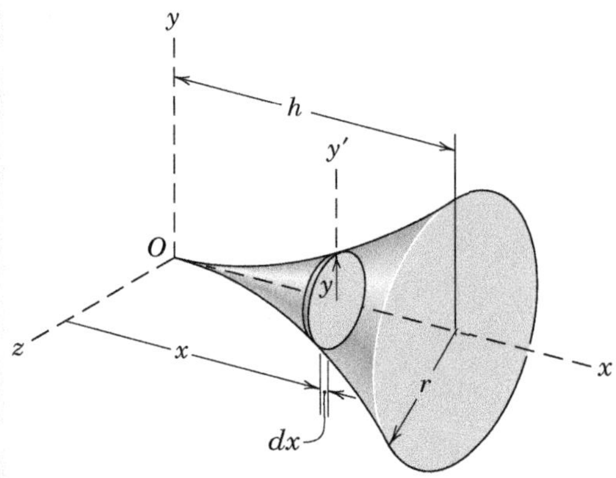

Solution. We begin by writing the boundary in the x-y plane as $y = kx^2$. The constant k is determined by evaluating this equation at the point $(x, y) = (h, r)$: $r = kh^2$, which gives $k = r/h^2$, so that $y = \frac{r}{h^2}x^2$.

As is usually convenient for bodies with axial symmetry, we choose a disk-shaped slice as our differential element, as shown in the given figure. The mass of this element is

① $$dm = \rho\pi y^2\, dx$$

where ρ represents the density of the body. The moment of inertia of the element about the x-axis is

② $$dI_{xx} = \tfrac{1}{2}\, dm\, y^2 = \tfrac{1}{2}(\rho\pi y^2\, dx)\, y^2 = \tfrac{1}{2}\rho\pi y^4\, dx$$

The mass of the entire body is

③ $$m = \int dm = \int_0^h \rho\pi y^2\, dx = \int_0^h \rho\pi\left(\frac{r}{h^2}x^2\right)^2 dx = \rho\pi\frac{r^2}{h^4}\frac{x^5}{5}\Bigg|_0^h = \tfrac{1}{5}\rho\pi r^2 h$$

and the moment of inertia of the entire body is

$$I_{xx} = \int dI_{xx} = \int_0^h \tfrac{1}{2}\rho\pi y^4\, dx = \int_0^h \tfrac{1}{2}\rho\pi\left(\frac{r}{h^2}x^2\right)^4 dx = \tfrac{1}{18}\rho\pi r^4 h$$

All that remains is to express I_{xx} more conventionally in terms of its mass. We do so by writing

④
⑤ $$I_{xx} = \tfrac{1}{18}\rho\pi r^4 h\left(\frac{m}{\frac{1}{5}\rho\pi r^2 h}\right) = \frac{5}{18}mr^2 \qquad \textit{Ans.}$$

By the parallel-axis theorem, the moment of inertia of the disk-shaped element about the y-axis is

$$\begin{aligned} dI_{yy} &= dI_{y'y'} + x^2\, dm = \tfrac{1}{4}\, dm\, y^2 + x^2\, dm \\ &= dm\left(\frac{1}{4}\left(\frac{r}{h^2}x^2\right)^2 + x^2\right) = \rho\pi y^2\, dx\left(\frac{1}{4}\frac{r^2}{h^4}x^4 + x^2\right) \\ &= \rho\pi\left(\frac{r}{h^2}x^2\right)^2\left(\frac{1}{4}\frac{r^2}{h^4}x^4 + x^2\right) dx = \rho\pi\frac{r^2}{h^4}\left(\frac{1}{4}\frac{r^2}{h^4}x^8 + x^6\right) dx \end{aligned}$$

For the entire body, we have

$$\begin{aligned} I_{yy} &= \int dI_{yy} = \int_0^h \rho\pi\frac{r^2}{h^4}\left(\frac{1}{4}\frac{r^2}{h^4}x^8 + x^6\right) dx = \rho\pi\frac{r^2}{h^4}\left(\frac{1}{4}\frac{r^2}{h^4}\frac{x^9}{9} + \frac{x^7}{7}\right)\Bigg|_0^h \\ &= \rho\pi r^2 h\left(\frac{r^2}{36} + \frac{h^2}{7}\right) \end{aligned}$$

Finally, we multiply by the same unit expression as above to obtain a result in terms of the body mass m.

$$I_{yy} = \rho\pi r^2 h\left(\frac{r^2}{36} + \frac{h^2}{7}\right)\left(\frac{m}{\frac{1}{5}\rho\pi r^2 h}\right) = 5m\left(\frac{r^2}{36} + \frac{h^2}{7}\right) \qquad \textit{Ans.}$$

Helpful Hints

① The volume of a disk is the area of its face times its thickness. Then density times volume gives mass.

② From Sample Problem B/1, the mass moment of inertia of a uniform cylinder (or disk) about its longitudinal axis is $\frac{1}{2}mr^2$.

③ Remember to regard an integral operation as an infinite summation.

④ The parenthetical expression here is unity, because its numerator and denominator are equal.

⑤ We note that I_{xx} is independent of h. So the body could be compressed to $h \cong 0$ or elongated to a large value of h with no resulting change in I_{xx}. This is true because no particle of the body would be changing its distance from the x-axis.

PROBLEMS

Integration Exercises

B/1 Determine the mass moment of inertia of the bent uniform slender rod about the x- and y-axes shown, and about the z-axis. The rod has a mass ρ per unit length.

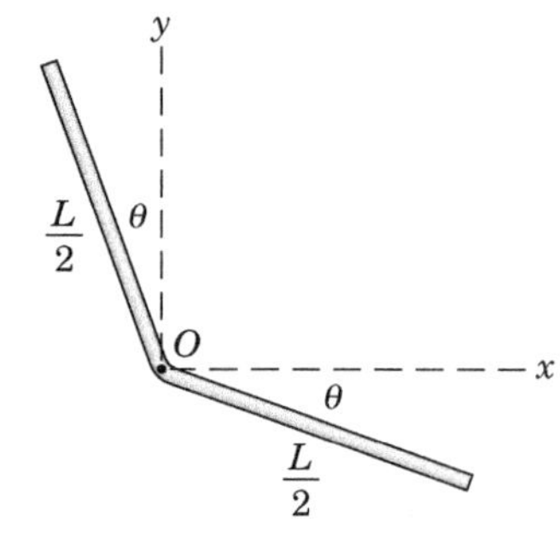

Problem B/1

B/2 Determine the mass moment of inertia of the uniform thin triangular plate of mass m about the x-axis. Also determine the radius of gyration about the x-axis. By analogy state I_{yy} and k_y. Then determine I_{zz} and k_z.

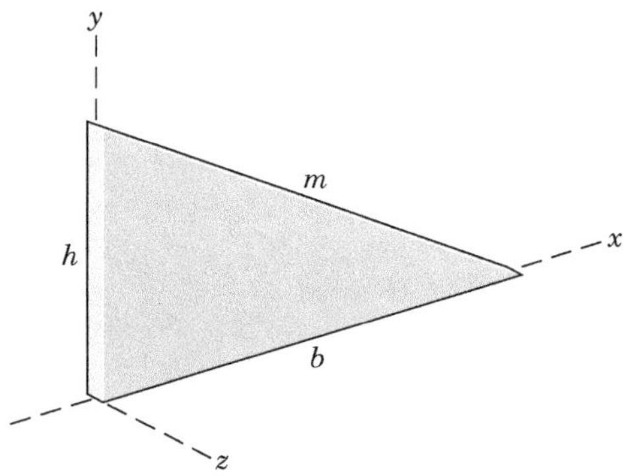

Problem B/2

B/3 Determine the mass moment of inertia about the y-axis for the uniform thin equilateral triangular plate of mass m. Also determine its radius of gyration about the y-axis.

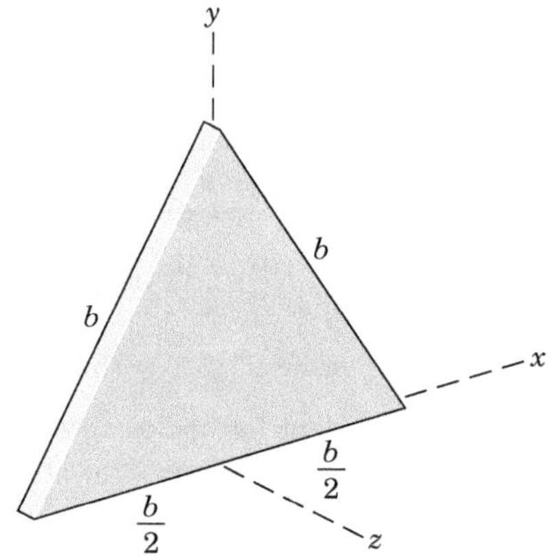

Problem B/3

B/4 Calculate the mass moment of inertia of the homogeneous right-circular cone of mass m, base radius r, and altitude h about the cone axis x and about the y-axis through its vertex.

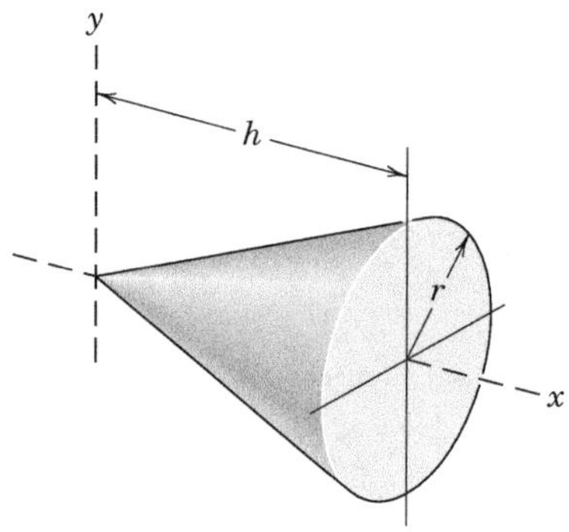

Problem B/4

B/5 Determine the mass moment of inertia of the uniform thin parabolic plate of mass m about the x-axis. State the corresponding radius of gyration.

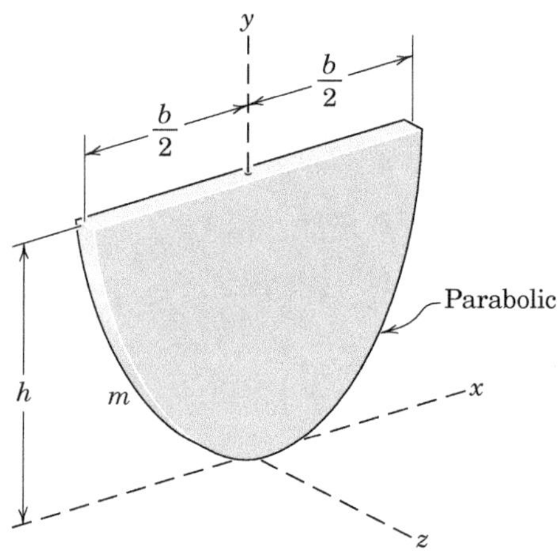

Problem B/5

B/6 Determine the mass moment of inertia about the y-axis for the parabolic plate of the previous problem. State the radius of gyration about the y-axis.

B/7 For the thin homogeneous plate of uniform thickness t and mass m, determine the mass moments of inertia about the axes x'-, y'-, and z'-through the end of the pate at A. Refer to the results of Sample Problem B/4 and Table D/3 in Appendix D as needed.

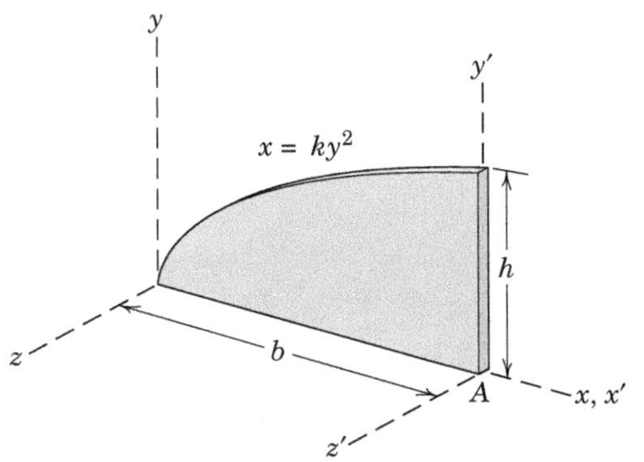

Problem B/7

B/8 Determine the mass moment of inertia about the x-axis of the thin elliptical plate of mass m.

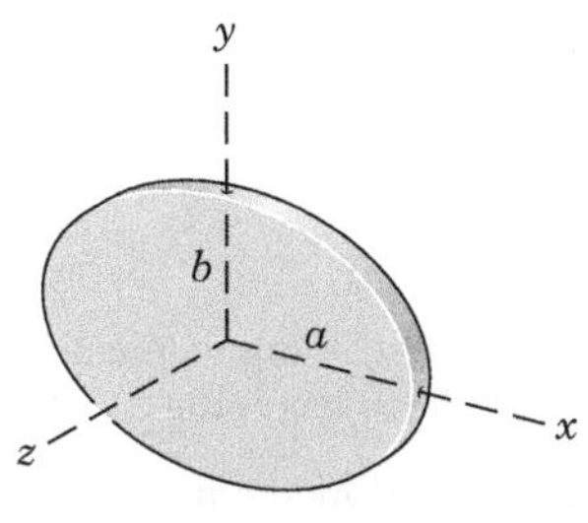

Problem B/8

B/9 Determine the mass moment of inertia of the homogeneous solid of revolution of mass m about the x-axis.

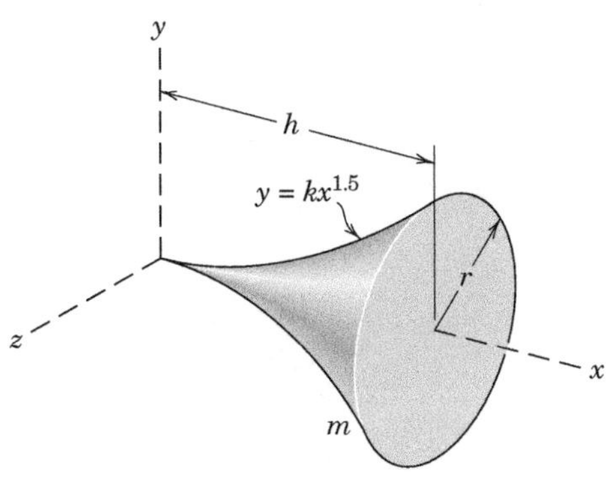

Problem B/9

B/10 Determine the mass moment of inertia of the homogeneous solid of revolution of the previous problem about the y- and z-axes.

B/11 Determine the radius of gyration about the y-axis for the steel part shown in section. The part is formed by revolving one of the trapezoidal areas 360° around the y-axis.

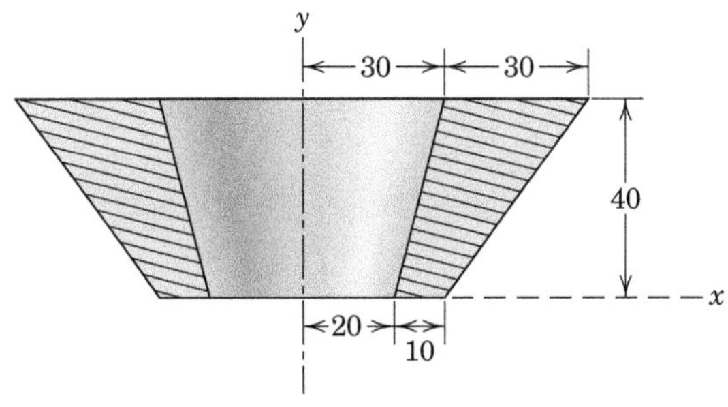

Dimensions in millimeters

Problem B/11

B/12 For the steel part of Prob. B/11, determine the radius of gyration about the x-axis.

B/13 Determine the radius of gyration about the z-axis of the paraboloid of revolution shown. The mass of the homogenous body is m.

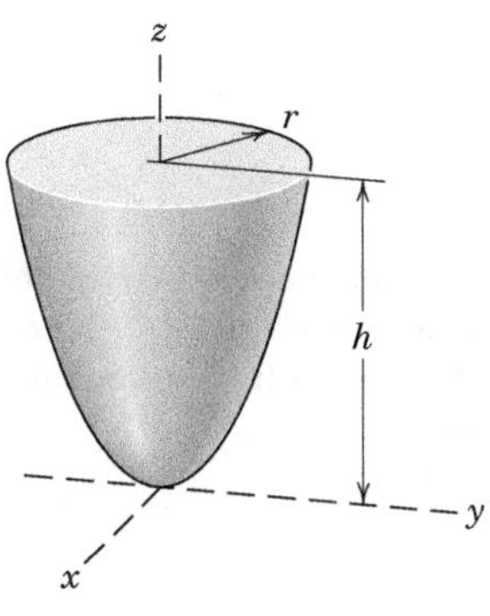

Problem B/13

B/14 Determine the moment of inertia about the y-axis for the paraboloid of revolution of Prob. B/13.

B/15 Develop an expression for the mass moment of inertia of the homogeneous solid of revolution of mass m about the y-axis.

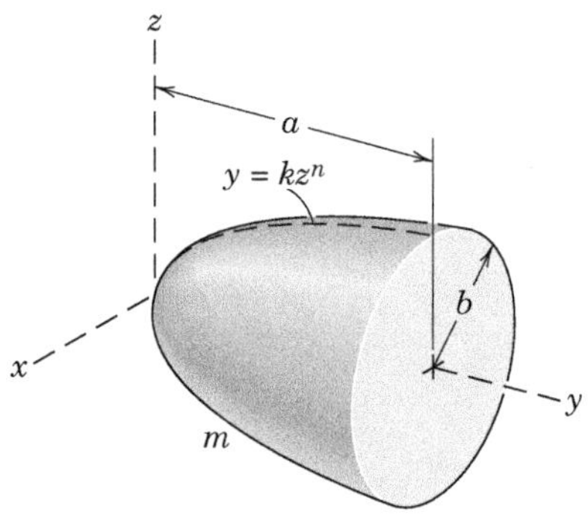

Problem B/15

B/16 Determine the mass moment of inertia about the x-axis of the solid spherical segment of mass m.

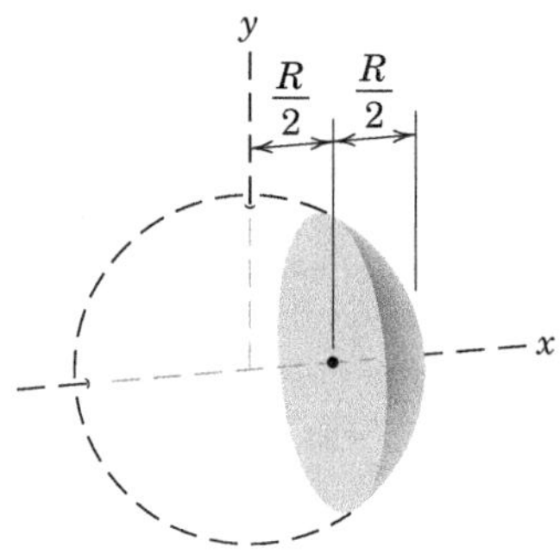

Problem B/16

B/17 Determine the moment of inertia about the generating axis of a complete ring (torus) of mass m having a circular section with the dimensions shown in the sectional view.

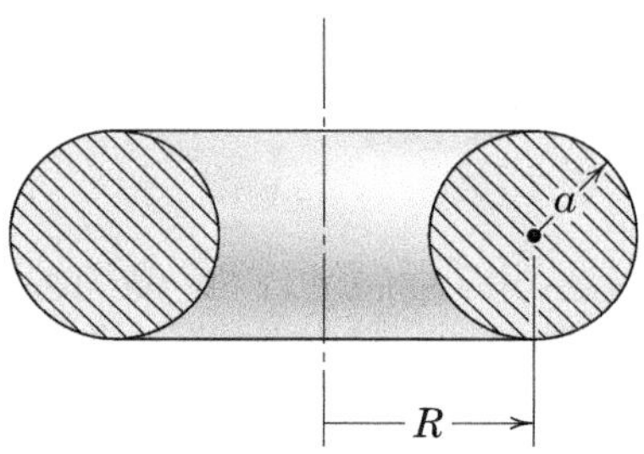

Problem B/17

B/18 The plane area shown in the top portion of the figure is rotated 180° about the x-axis to form the body of revolution of mass m shown in the lower portion of the figure. Determine the mass moment of inertia of the body about the x-axis.

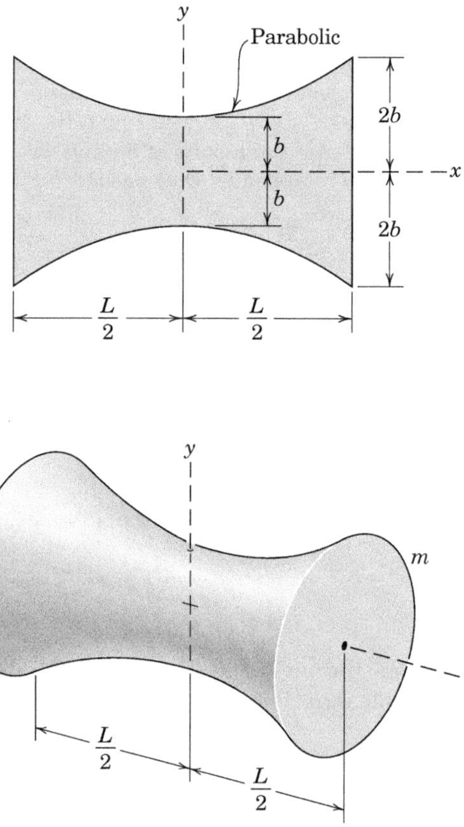

Problem B/18

B/19 Determine I_{yy} for the homogeneous body of revolution of the previous problem.

B/20 Determine the mass moment of inertia of the homogeneous square pyramid about the z-axis. The pyramid has mass m.

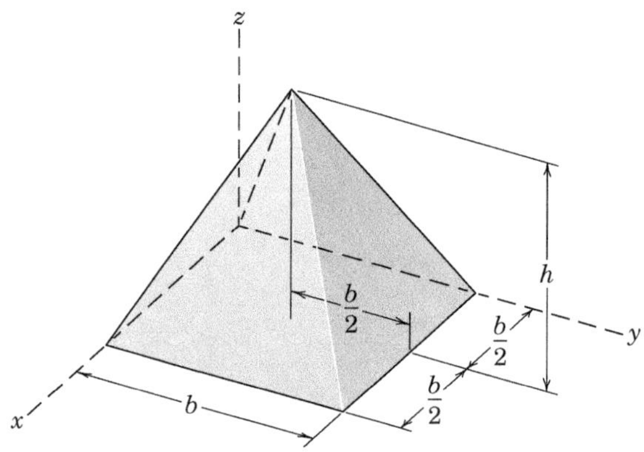

Problem B/20

B/21 For the square pyramid of Prob. B/20, determine the mass moment of inertia about the x-axis.

B/22 The thickness of the homogeneous triangular plate of mass m varies linearly with the distance from the vertex toward the base. The thickness a at the base is small compared with the other dimensions. Determine the moment of inertia of the plate about the y-axis along the centerline of the base.

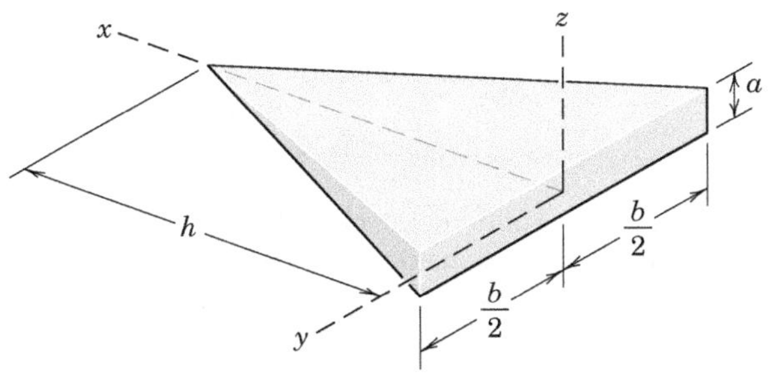

Problem B/22

B/23 Determine the moment of inertia, about the generating axis, of the hollow circular tube of mass m obtained by revolving the thin ring shown in the sectional view completely around the generating axis.

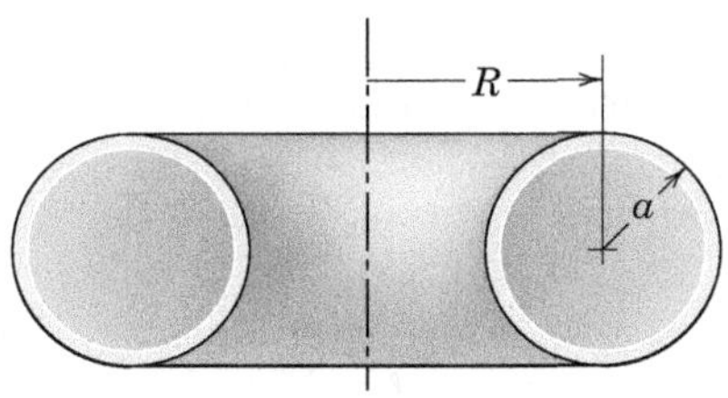

Problem B/23

B/24 Determine the moments of inertia of the hemispherical shell with respect to the x- and z-axes. The mass of the shell is m, and its thickness is negligible compared with the radius r.

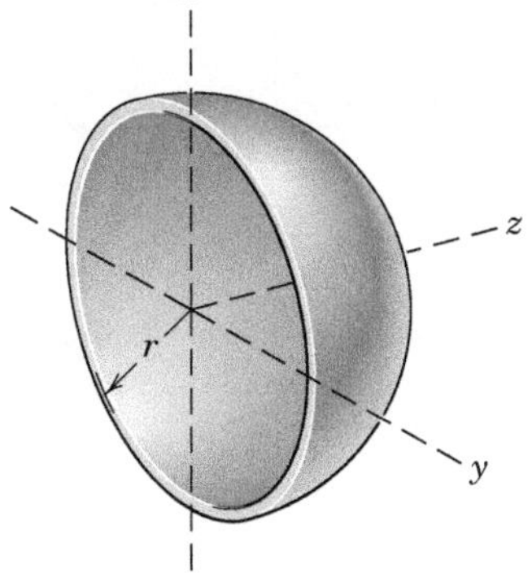

Problem B/24

B/25 The partial solid of revolution is formed by revolving the shaded area in the x-z plane 90° about the z-axis. If the mass of the solid is m, determine its mass moment of inertia about the z-axis.

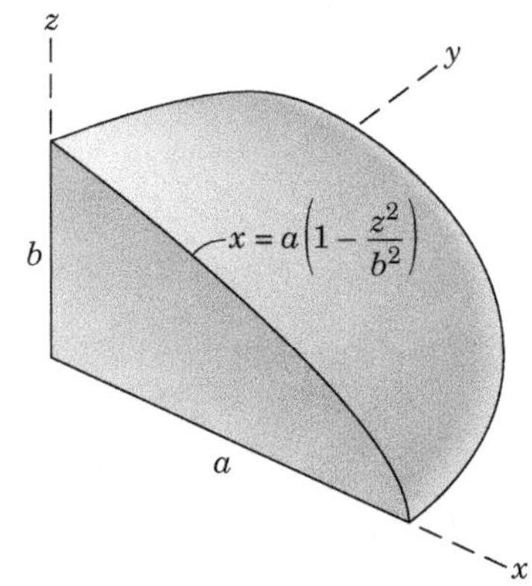

Problem B/25

B/26 For the partial solid of revolution in Prob. B/25, determine the mass moment of inertia about the x-axis.

▶B/27 A shell of mass m is obtained by revolving the quarter-circular section about the z-axis. If the thickness of the shell is small compared with a and if $r = a/3$, determine the radius of gyration of the shell about the z-axis.

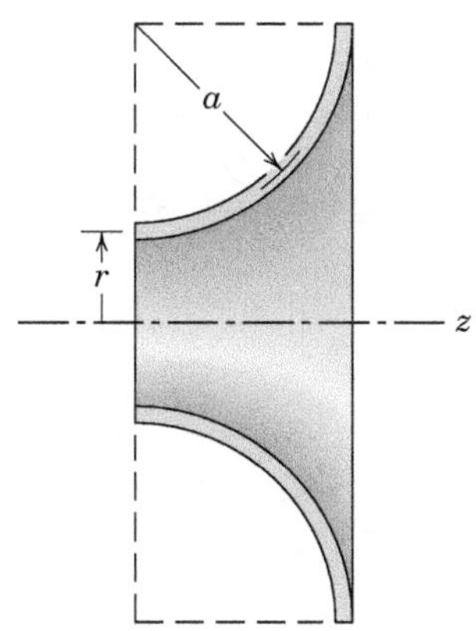

Problem B/27

►**B/28** Determine the mass moment of inertia and corresponding radius of gyration of the thin homogeneous parabolic shell about the y-axis. The shell has dimensions $r = 70$ mm and $h = 200$ mm, and is made of metal plate having a mass per unit area of 32 kg/m^2.

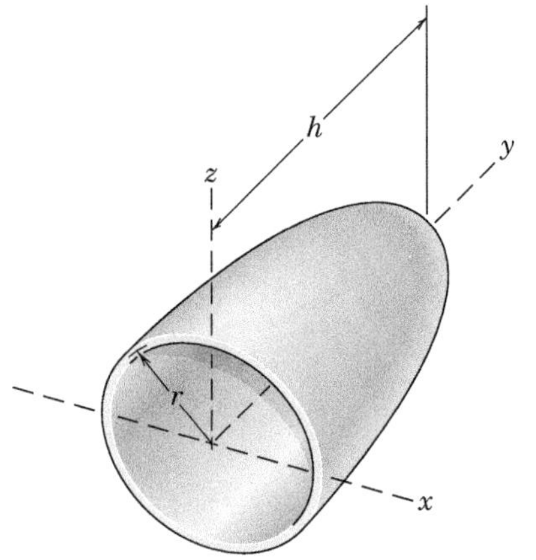

Problem B/28

►**B/29** For the parabolic shell of Prob. B/28, determine the mass moment of inertia and corresponding radius of gyration about the z-axis.

Composite and Parallel-Axis Exercises

B/30 Every "slender" rod has a finite radius r. Refer to Table D/4 and derive an expression for the percentage error e which results if one neglects the radius r of a homogeneous solid cylindrical rod of length l when calculating its moment of inertia I_{zz}. Evaluate your expression for the ratios $r/l = 0.01$, 0.1, and 0.5.

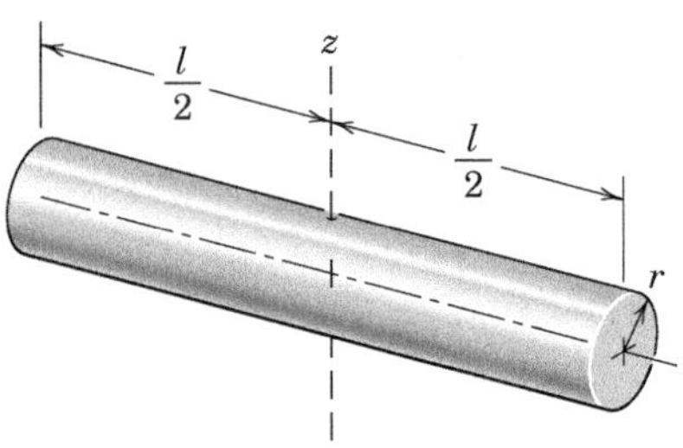

Problem B/30

B/31 The two small spheres of mass m each are connected by the light rigid rod which lies in the x-z plane. Determine the mass moments of inertia of the assembly about the x-, y-, and z-axes.

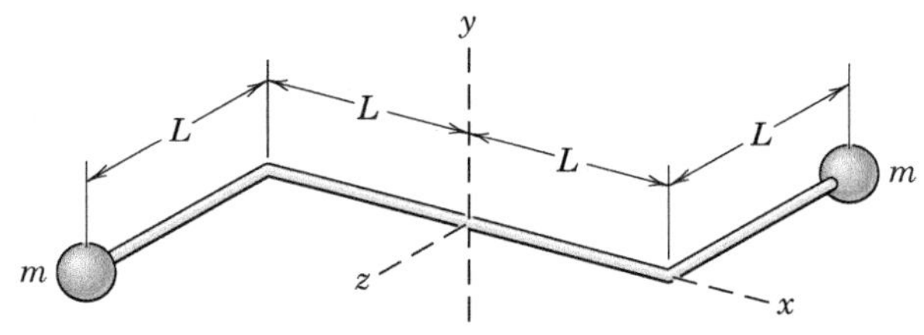

Problem B/31

B/32 The rectangular metal plate has a mass of 15 kg. Compute its moment of inertia about the y-axis. What is the magnitude of the percentage error e introduced by using the approximate relation $\frac{1}{3}ml^2$ for I_{xx}?

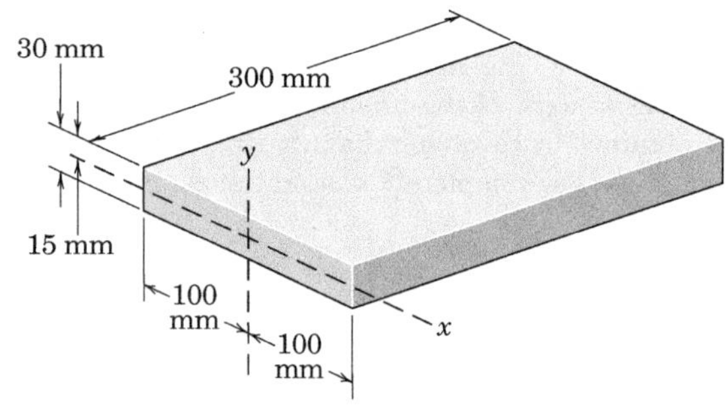

Problem B/32

B/33 Determine I_{xx} for the cylinder with a centered circular hole. The mass of the body is m.

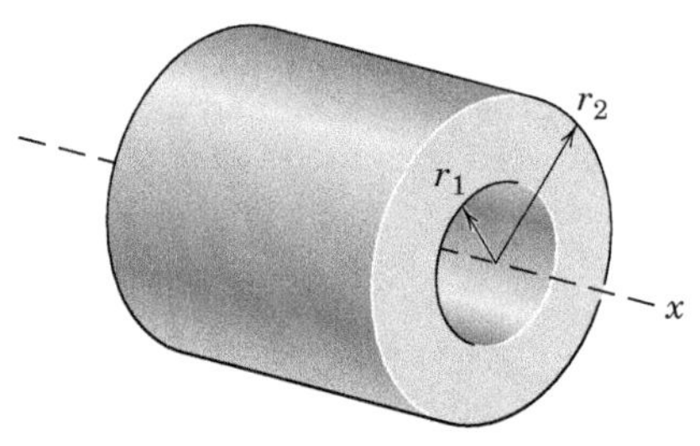

Problem B/33

B/34 Determine the mass moment of inertia about the z-axis for the right-circular cylinder with a central longitudinal hole.

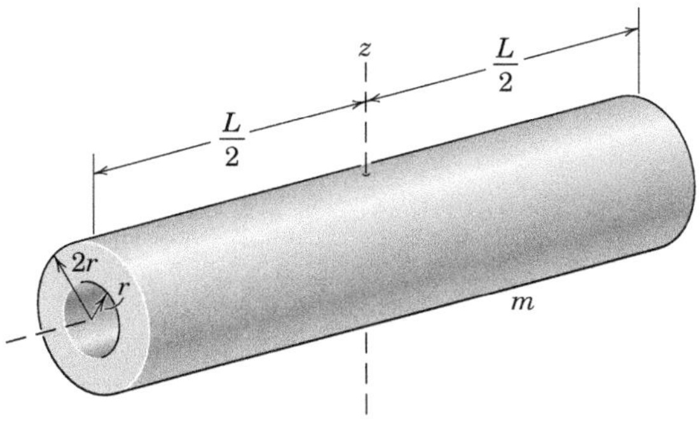

Problem B/34

B/35 Determine the moment of inertia of the half-ring of mass m about its diametral axis a-a and about axis b-b through the midpoint of the arc normal to the plane of the ring. The radius of the circular cross section is small compared with r.

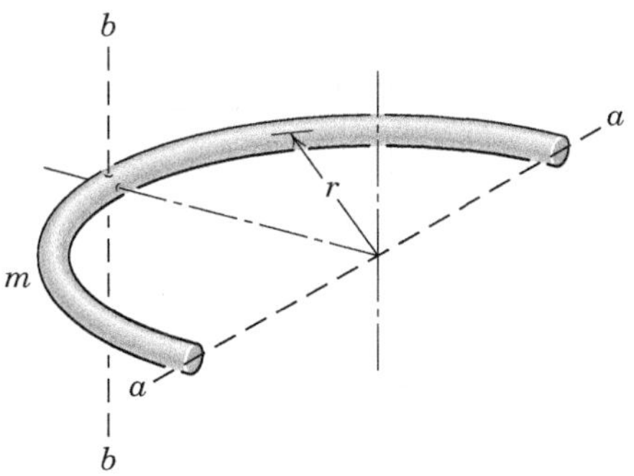

Problem B/35

B/36 A 6-in. steel cube is cut along its diagonal plane. Calculate the moment of inertia of the resulting prism about the edge x-x.

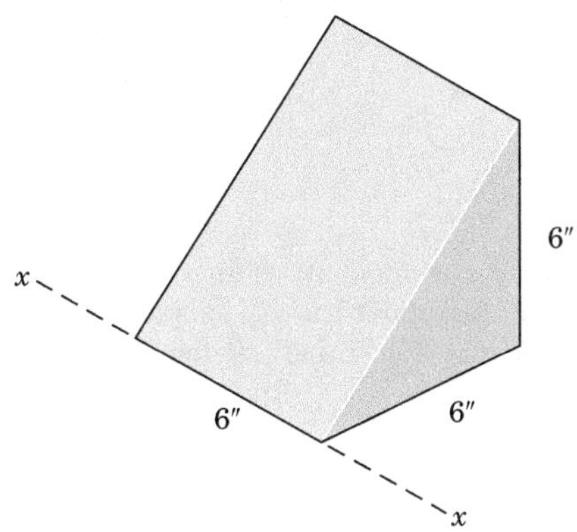

Problem B/36

B/37 The uniform coiled spring weighs 4 lb. Approximate its moments of inertia about the x-, y-, and z-axes from the analogy to the properties of a cylindrical shell.

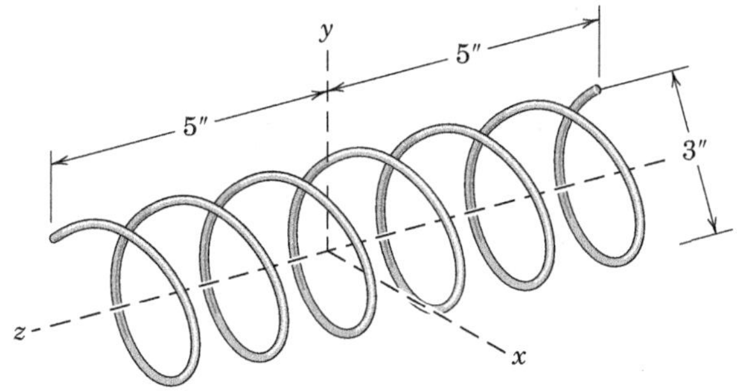

Problem B/37

B/38 Determine the length L of each of the slender rods of mass $m/2$ which must be centrally attached to the faces of the thin homogeneous disk of mass m in order to make the mass moments of inertia of the unit about the x- and z-axes equal.

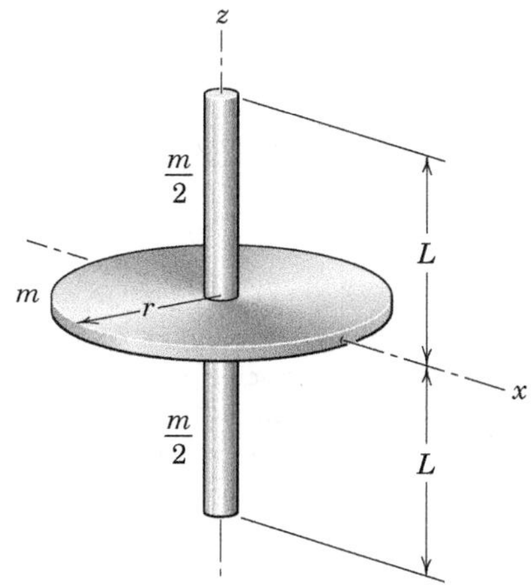

Problem B/38

B/39 A badminton racket is constructed of uniform slender rods bent into the shape shown. Neglect the strings and the built-up wooden grip and estimate the mass moment of inertia about the y-axis through O, which is the location of the player's hand. The mass per unit length of the rod material is ρ.

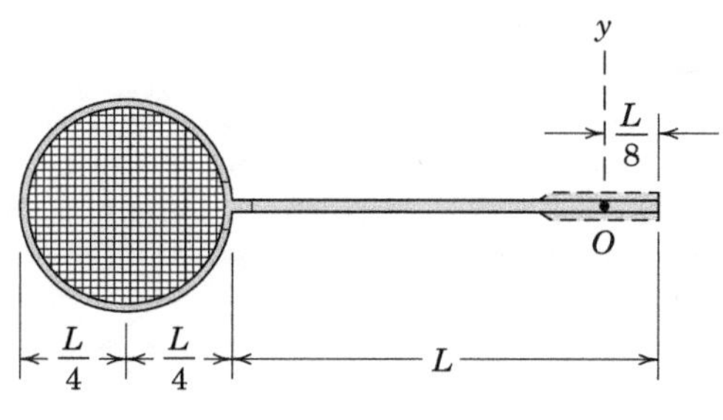

Problem B/39

B/40 Calculate the moment of inertia of the steel control wheel, shown in section, about its central axis. There are eight spokes, each of which has a constant cross-sectional area of 200 mm^2. What percent n of the total moment of inertia is contributed by the outer rim?

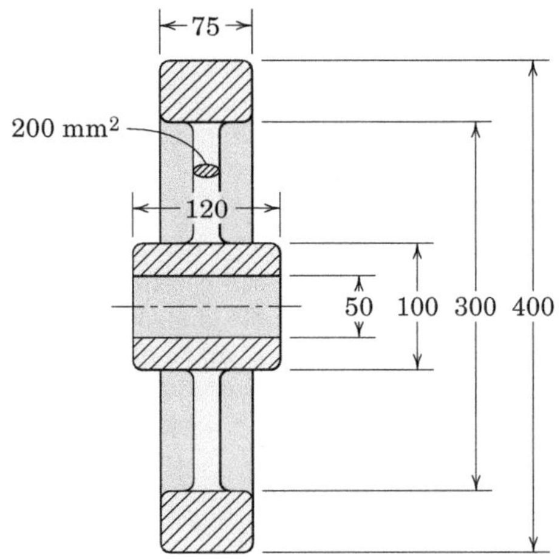

Dimensions in millimeters

Problem B/40

B/41 The welded assembly is made of a uniform rod which weighs 0.370 lb per foot of length and the semicircular plate which weighs 8 lb per square foot. Determine the mass moment of inertia of the assembly about the three coordinate axes shown.

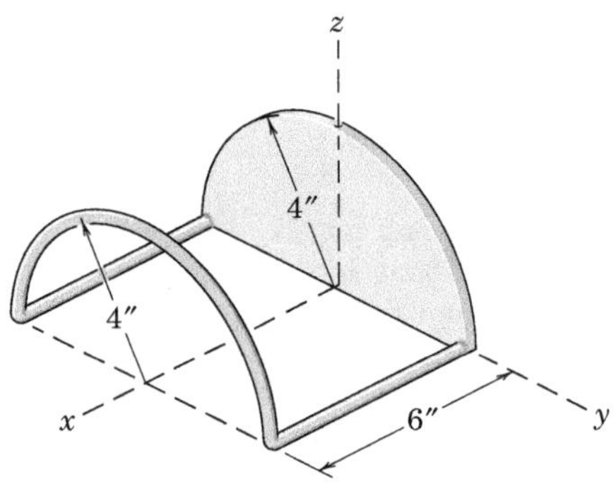

Problem B/41

B/42 The uniform rod of length $4b$ and mass m is bent into the shape shown. The diameter of the rod is small compared with its length. Determine the moments of inertia of the rod about the three coordinate axes.

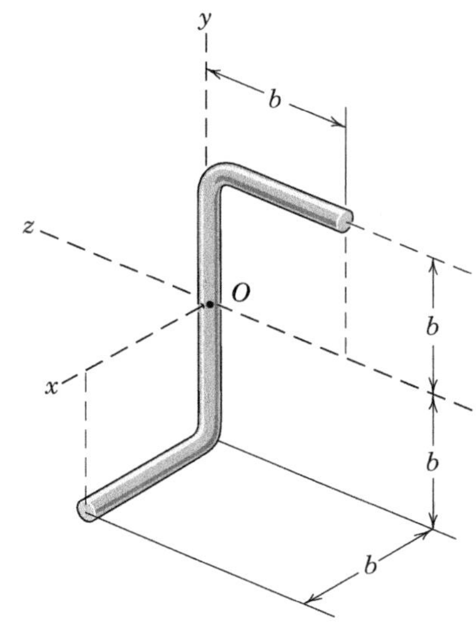

Problem B/42

B/43 The welded assembly shown in made from a steel rod which weighs 0.455 lb per foot of length. Determine the mass moment of inertia of the assembly (*a*) about the y-axis and (*b*) about the z-axis.

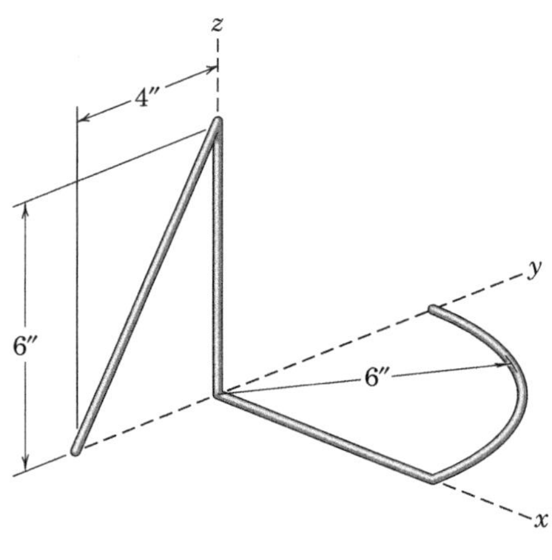

Problem B/43

B/44 Calculate the moment of inertia of the solid steel semicylinder about the x-x axis and about the parallel x_0-x_0 axis. (See Table D/1 for the density of steel.)

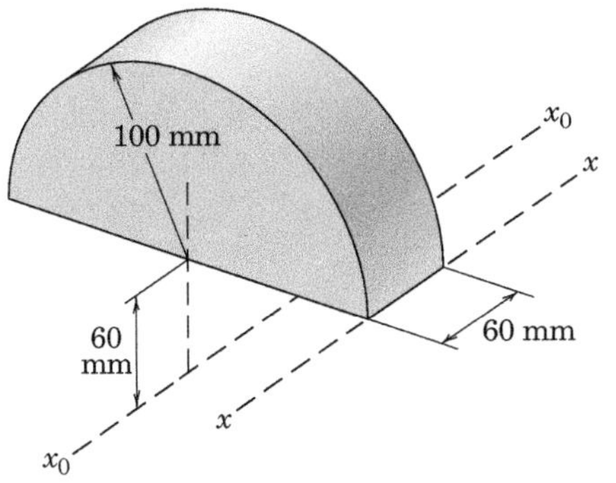

Problem B/44

B/45 The body is constructed of a uniform square plate, a uniform straight rod, a uniform quarter-circular rod, and a particle (negligible dimensions). If each part has the indicated mass, determine the mass moments of inertia of the body about the x-, y-, and z-axes.

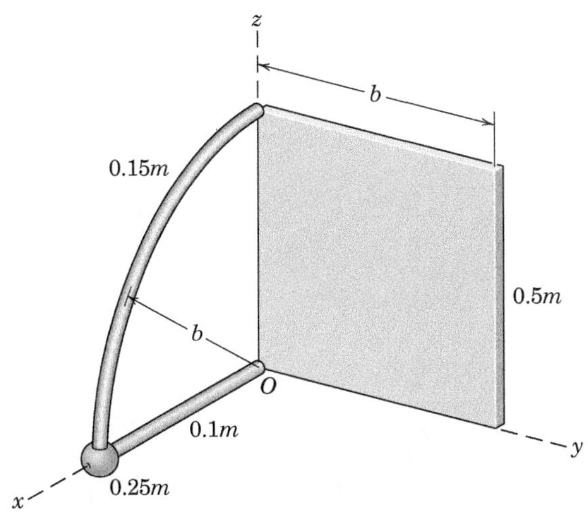

Problem B/45

B/46 The clock pendulum consists of the slender rod of length l and mass m and the bob of mass $7m$. Neglect the effects of the radius of the bob and determine I_O in terms of the bob position x. Calculate the ratio R of I_O evaluated for $x = \frac{3}{4}l$ to I_O evaluated for $x = l$.

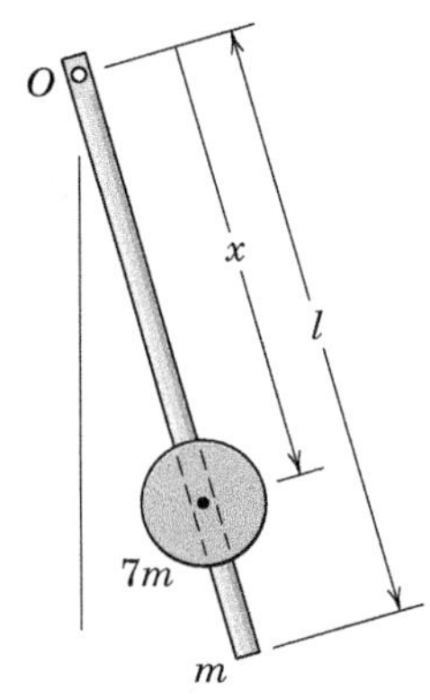

Problem B/46

B/47 Determine the mass moments of inertia of the bracket about the x- and x'-axes. The bracket is made from thin plate of uniform thickness and has a mass of 0.35 kg per square meter of area.

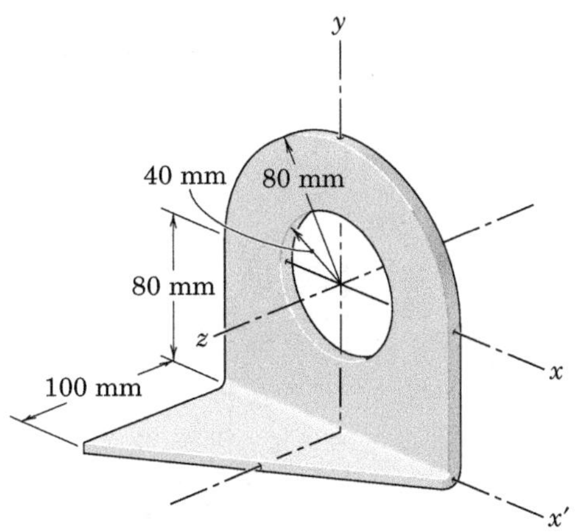

Problem B/47

B/48 A square plate with a quarter-circular sector removed has a net mass m. Determine its moment of inertia about axis A-A normal to the plane of the plate.

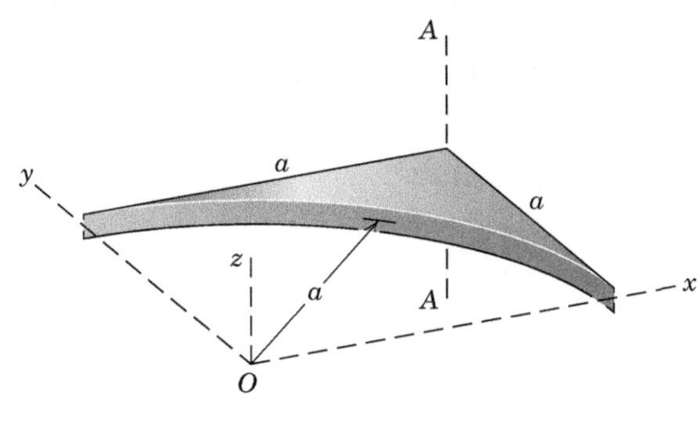

Problem B/48

B/49 The welded assembly consists of a cylindrical shell with a closed semicircular end. The shell is made from sheet metal with a mass of 24 kg/m^2, and the end is made from metal plate with a mass of 36 kg/m^2. Determine the mass moments of inertia of the assembly about the coordinate axes shown.

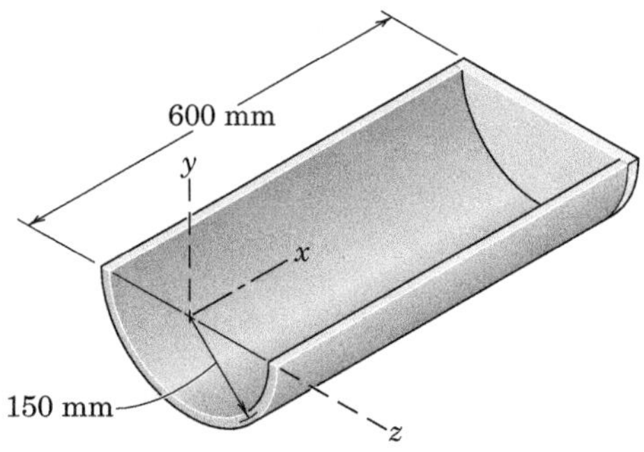

Problem B/49

B/50 Determine the mass moment of inertia of the steel bracket about the z-axis which passes through the midline of the base.

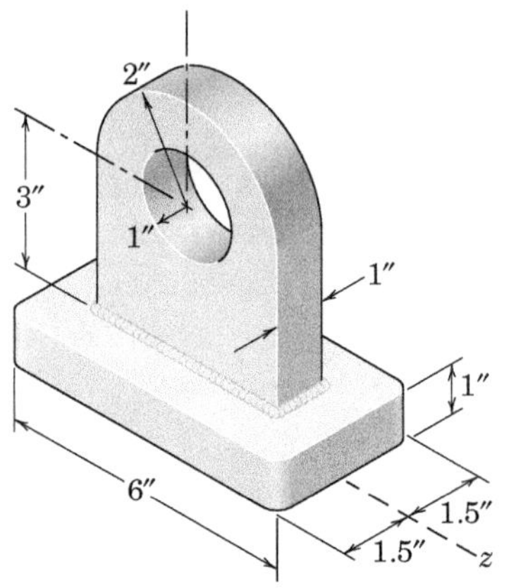

Problem B/50

B/51 The slender metal rods are welded together in the configuration shown. Each 6-in. segment weighs 0.30 lb. Compute the moment of inertia of the assembly about the y-axis.

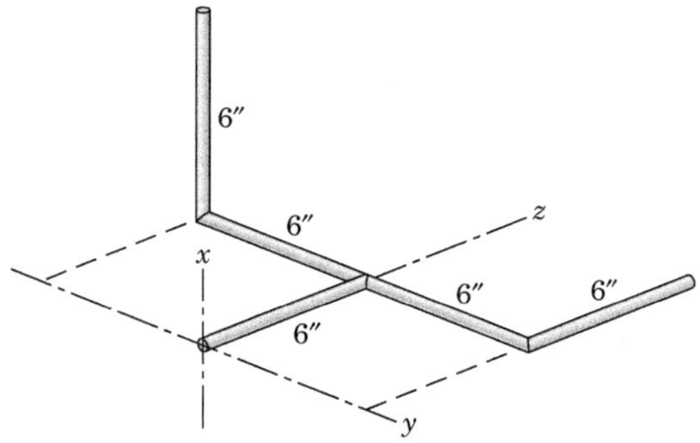

Problem B/51

B/52 The welded assembly is formed from thin sheet metal having a mass of 19 kg/m^2. Determine the mass moments of inertia for the assembly about the x- and y-axes.

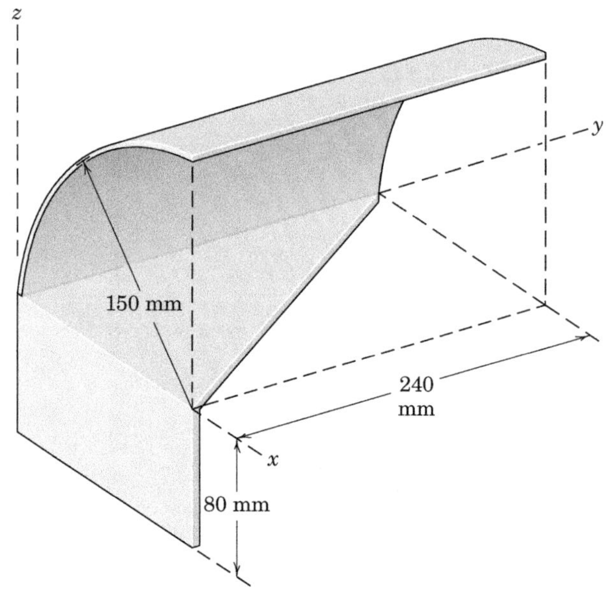

Problem B/52

B/53 Determine I_{xx} for the cone frustum, which has base radii r_1 and r_2 and mass m.

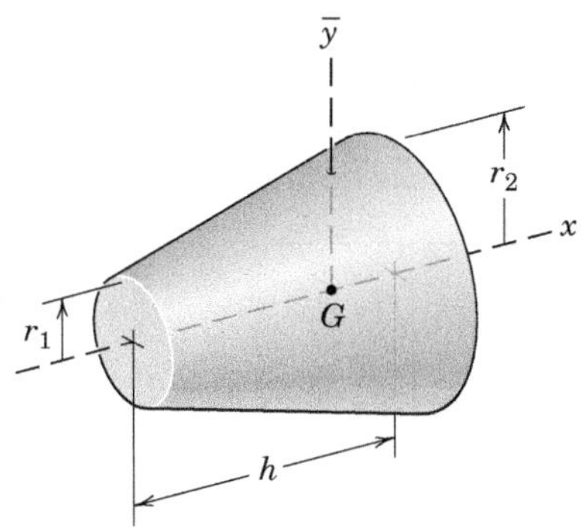

Problem B/53

***B/54** A preliminary design model to ensure rotational stability for a spacecraft consists of the cylindrical shell and the two square panels as shown. The shell and panels have the same thickness and density. It can be shown that rotational stability about the z-axis can be maintained if I_{zz} is less than I_{xx} and I_{yy}. For a given value of r, determine the limitation for L.

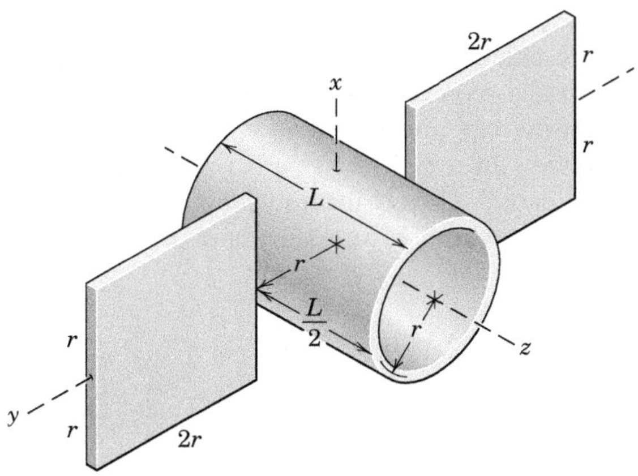

Problem B/54

▶**B/55** Determine the radius of gyration of the aluminum part about the z-axis. The hole in the upper surface is drilled completely through the part.

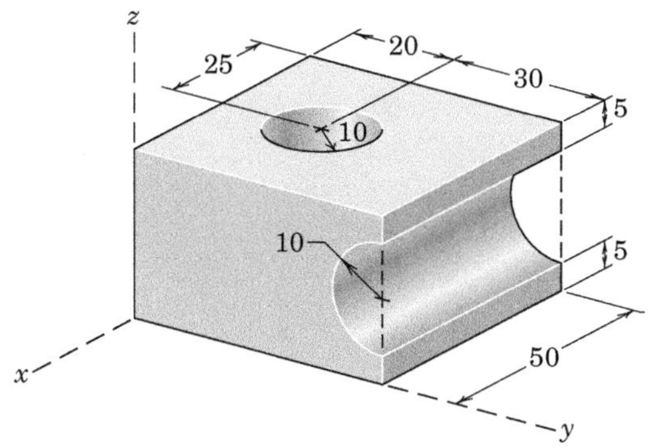

Dimensions in millimeters

Problem B/55

▶**B/56** Compute the moment of inertia of the mallet about the O-O axis. The mass of the head is 0.8 kg, and the mass of the handle is 0.5 kg.

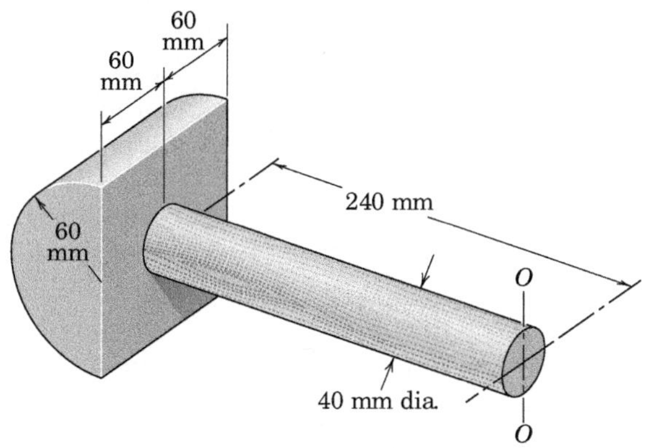

Problem B/56

B/2 Products of Inertia

For problems in the rotation of three-dimensional rigid bodies, the expression for angular momentum contains, in addition to the moment-of-inertia terms, *product-of-inertia* terms defined as

$$\begin{aligned} I_{xy} &= I_{yx} = \int xy\, dm \\ I_{xz} &= I_{zx} = \int xz\, dm \\ I_{yz} &= I_{zy} = \int yz\, dm \end{aligned} \qquad \textbf{(B/8)}$$

These expressions were cited in Eqs. 7/10 in the expansion of the expression for angular momentum, Eq. 7/9.

The calculation of products of inertia involves the same basic procedure which we have followed in calculating moments of inertia and in evaluating other volume integrals as far as the choice of element and the limits of integration are concerned. The only special precaution we need to observe is to be doubly watchful of the algebraic signs in the expressions. Whereas moments of inertia are always positive, products of inertia may be either positive or negative. The units of products of inertia are the same as those of moments of inertia.

We have seen that the calculation of moments of inertia is often simplified by using the parallel-axis theorem. A similar theorem exists for transferring products of inertia, and we prove it easily as follows. In Fig. B/6 is shown the x-y view of a rigid body with parallel axes x_0-y_0 passing through the mass center G and located from the x-y axes by the distances d_x and d_y. The product of inertia about the x-y axes by definition is

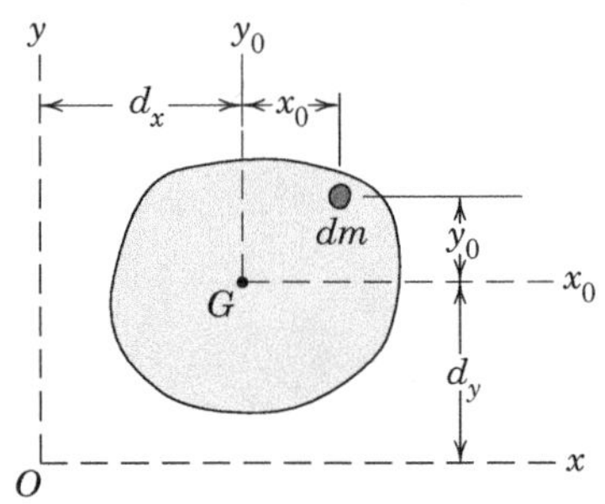

Figure B/6

$$\begin{aligned} I_{xy} &= \int xy\, dm = \int (x_0 + d_x)(y_0 + d_y)\, dm \\ &= \int x_0 y_0\, dm + d_x d_y \int dm + d_x \int y_0\, dm + d_y \int x_0\, dm \\ &= I_{x_0 y_0} + m d_x d_y \end{aligned}$$

The last two integrals vanish since the first moments of mass about the mass center are necessarily zero.

Similar relations exist for the remaining two product-of-inertia terms. Dropping the zero subscripts and using the bar to designate the mass-center quantity, we obtain

$$\begin{aligned} I_{xy} &= \bar{I}_{xy} + m d_x d_y \\ I_{xz} &= \bar{I}_{xz} + m d_x d_z \\ I_{yz} &= \bar{I}_{yz} + m d_y d_z \end{aligned} \qquad \textbf{(B/9)}$$

These transfer-of-axis relations are valid *only* for transfer to or from *parallel axes* through the *mass center*.

With the aid of the product-of-inertia terms, we can calculate the moment of inertia of a rigid body about any prescribed axis through the coordinate origin. For the rigid body of Fig. B/7, suppose we must determine the moment of inertia about axis *O-M*. The direction cosines of *O-M* are l, m, n, and a unit vector $\boldsymbol{\lambda}$ along *O-M* may be written $\boldsymbol{\lambda} = l\mathbf{i} + m\mathbf{j} + n\mathbf{k}$. The moment of inertia about *O-M* is

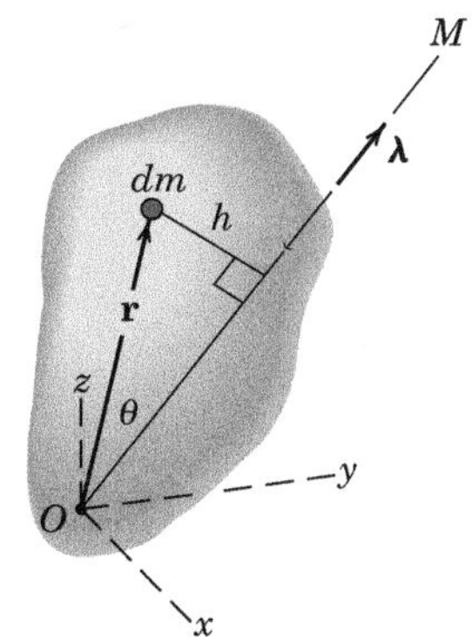

Figure B/7

$$I_M = \int h^2\, dm = \int (\mathbf{r} \times \boldsymbol{\lambda})\cdot(\mathbf{r} \times \boldsymbol{\lambda})\, dm$$

where $|\mathbf{r} \times \boldsymbol{\lambda}| = r \sin\theta = h$. The cross product is

$$(\mathbf{r} \times \boldsymbol{\lambda}) = (yn - zm)\mathbf{i} + (zl - xn)\mathbf{j} + (xm - yl)\mathbf{k}$$

and, after we collect terms, the dot-product expansion gives

$$(\mathbf{r} \times \boldsymbol{\lambda})\cdot(\mathbf{r} \times \boldsymbol{\lambda}) = h^2 = (y^2 + z^2)l^2 + (x^2 + z^2)m^2 + (x^2 + y^2)n^2 - 2xylm - 2xzln - 2yzmn$$

Thus, with the substitution of the expressions of Eqs. B/4 and B/8, we have

$$I_M = I_{xx}l^2 + I_{yy}m^2 + I_{zz}n^2 - 2I_{xy}lm - 2I_{xz}ln - 2I_{yz}mn \qquad \textbf{(B/10)}$$

This expression gives the moment of inertia about any axis *O-M* in terms of the direction cosines of the axis and the moments and products of inertia about the coordinate axes.

Principal Axes of Inertia

As noted in Art. 7/7, the array

$$\begin{bmatrix} I_{xx} & -I_{xy} & -I_{xz} \\ -I_{yx} & I_{yy} & -I_{yz} \\ -I_{zx} & -I_{zy} & I_{zz} \end{bmatrix}$$

whose elements appear in the expansion of the angular-momentum expression, Eq. 7/11, for a rigid body with attached axes, is called the *inertia matrix* or *inertia tensor*. If we examine the moment- and product-of-inertia terms for all possible orientations of the axes with respect to the body for a given origin, we will find in the general case an orientation of the x-y-z axes for which the product-of-inertia terms vanish and the array takes the diagonalized form

$$\begin{bmatrix} I_{xx} & 0 & 0 \\ 0 & I_{yy} & 0 \\ 0 & 0 & I_{zz} \end{bmatrix}$$

Such axes x-y-z are called the *principal axes of inertia*, and I_{xx}, I_{yy}, and I_{zz} are called the *principal moments of inertia* and represent the maximum, minimum, and intermediate values of the moments of inertia for the particular origin chosen.

It may be shown* that for any given orientation of axes x-y-z the solution of the determinant equation

$$\begin{vmatrix} I_{xx} - I & -I_{xy} & -I_{xz} \\ -I_{yx} & I_{yy} - I & -I_{yz} \\ -I_{zx} & -I_{zy} & I_{zz} - I \end{vmatrix} = 0 \qquad \textbf{(B/11)}$$

for I yields three roots I_1, I_2, and I_3 of the resulting cubic equation which are the three principal moments of inertia. Also, the direction cosines l, m, and n of a principal inertia axis are given by

$$\begin{aligned} (I_{xx} - I)l - I_{xy}m - I_{xz}n &= 0 \\ -I_{yx}l + (I_{yy} - I)m - I_{yz}n &= 0 \\ -I_{zx}l - I_{zy}m + (I_{zz} - I)n &= 0 \end{aligned} \qquad \textbf{(B/12)}$$

These equations along with $l^2 + m^2 + n^2 = 1$ will enable a solution for the direction cosines to be made for each of the three I's.

To assist with the visualization of these conclusions, consider the rectangular block, Fig. B/8, having an arbitrary orientation with respect to the x-y-z axes. For simplicity, the mass center G is located at the origin of the coordinates. If the moments and products of inertia for the block about the x-y-z axes are known, then solution of Eq. B/11 would give the three roots, I_1, I_2, and I_3, which are the principal moments of inertia. Solution of Eq. B/12 using each of the three I's, in turn, along with $l^2 + m^2 + n^2 = 1$, would give the direction cosines l, m, and n for each of the respective principal axes, which are always mutually perpendicular. From the proportions of the block as drawn, we see that I_1 is the maximum moment of inertia, I_2 is the intermediate value, and I_3 is the minimum value.

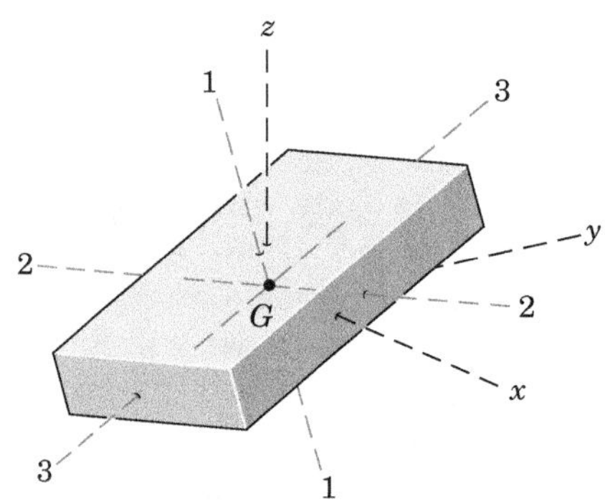

Figure B/8

*See, for example, the first author's *Dynamics, SI Version*, 1975, John Wiley & Sons, Art. 41.

Sample Problem B/6

The bent plate has a uniform thickness t which is negligible compared with its other dimensions. The density of the plate material is ρ. Determine the products of inertia of the plate with respect to the axes as chosen.

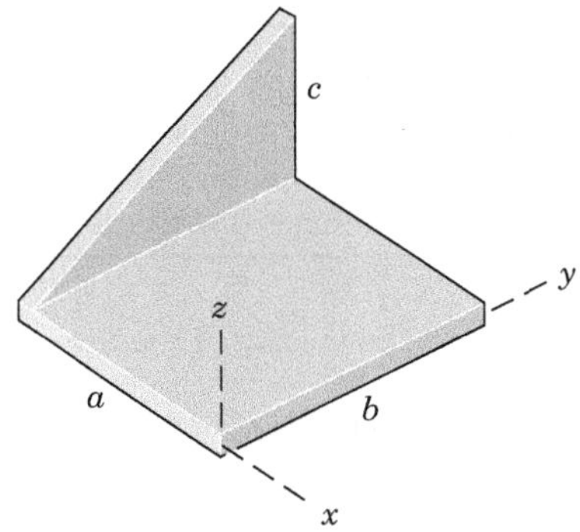

Solution. Each of the two parts is analyzed separately.

Rectangular part. In the separate view of this part, we introduce parallel ① axes x_0-y_0 through the mass center G and use the transfer-of-axis theorem. By symmetry, we see that $\bar{I}_{xy} = I_{x_0y_0} = 0$ so that

$$[I_{xy} = \bar{I}_{xy} + md_xd_y] \qquad I_{xy} = 0 + \rho tab\left(-\frac{a}{2}\right)\left(\frac{b}{2}\right) = -\frac{1}{4}\rho ta^2b^2$$

Because the z-coordinate of all elements of the plate is zero, it follows that $I_{xz} = I_{yz} = 0$.

Triangular part. In the separate view of this part, we locate the mass center G and construct x_0-, y_0-, and z_0-axes through G. Since the x_0-coordinate of all elements is zero, it follows that $\bar{I}_{xy} = I_{x_0y_0} = 0$ and $\bar{I}_{xz} = I_{x_0z_0} = 0$. The transfer-of-axis theorems then give us

$$[I_{xy} = \bar{I}_{xy} + md_xd_y] \qquad I_{xy} = 0 + \rho t\frac{b}{2}c(-a)\left(\frac{2b}{3}\right) = -\frac{1}{3}\rho tab^2c$$

$$[I_{xz} = \bar{I}_{xz} + md_xd_z] \qquad I_{xz} = 0 + \rho t\frac{b}{2}c(-a)\left(\frac{c}{3}\right) = -\frac{1}{6}\rho tabc^2$$

We obtain I_{yz} by direct integration, noting that the distance a of the plane of the triangle from the y-z plane in no way affects the y- and z-coordinates. With the mass element $dm = \rho t\, dy\, dz$, we have

② $$\left[I_{yz} = \int yz\, dm\right] \quad I_{yz} = \rho t\int_0^b\int_0^{cy/b} yz\, dz\, dy = \rho t\int_0^b y\left[\frac{z^2}{2}\right]_0^{cy/b} dy$$

$$= \frac{\rho tc^2}{2b^2}\int_0^b y^3\, dy = \frac{1}{8}\rho tb^2c^2$$

Adding the expressions for the two parts gives

$$I_{xy} = -\tfrac{1}{4}\rho ta^2b^2 - \tfrac{1}{3}\rho tab^2c = -\tfrac{1}{12}\rho tab^2(3a + 4c) \qquad \textit{Ans.}$$

$$I_{xz} = 0 \quad -\tfrac{1}{6}\rho tabc^2 = -\tfrac{1}{6}\rho tabc^2 \qquad \textit{Ans.}$$

$$I_{yz} = 0 \quad +\tfrac{1}{8}\rho tb^2c^2 = +\tfrac{1}{8}\rho tb^2c^2 \qquad \textit{Ans.}$$

Helpful Hints

① We must be careful to preserve the same sense of the coordinates. Thus, plus x_0 and y_0 must agree with plus x and y.

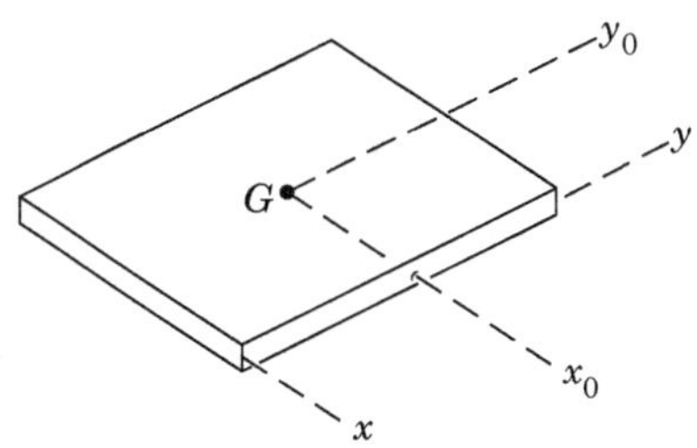

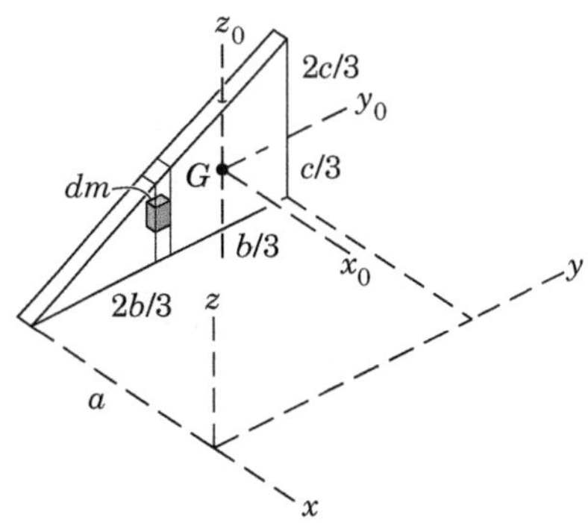

② We choose to integrate with respect to z first, where the upper limit is the variable height $z = cy/b$. If we were to integrate first with respect to y, the limits of the first integral would be from the variable $y = bz/c$ to b.

Sample Problem B/7

The angle bracket is made from aluminum plate with a mass of 13.45 kg per square meter. Calculate the principal moments of inertia about the origin O and the direction cosines of the principal axes of inertia. The thickness of the plate is small compared with the other dimensions.

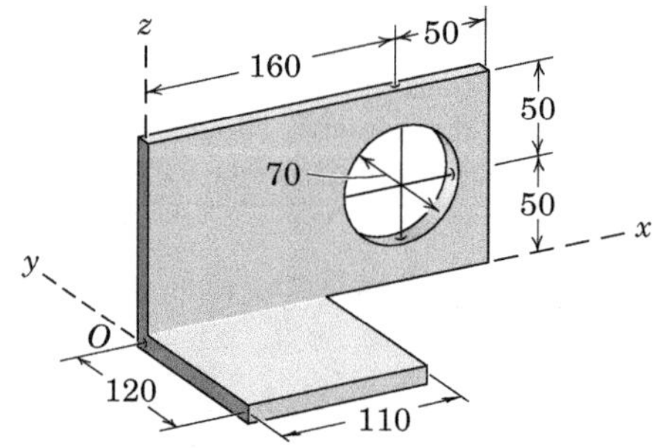

Dimensions in millimeters

Solution. The masses of the three parts are

$$m_1 = 13.45(0.21)(0.1) = 0.282 \text{ kg}$$

① $$m_2 = -13.45\pi(0.035)^2 = -0.0518 \text{ kg}$$

$$m_3 = 13.45(0.12)(0.11) = 0.1775 \text{ kg}$$

Part 1

$$I_{xx} = \tfrac{1}{3}mb^2 = \tfrac{1}{3}(0.282)(0.1)^2 = 9.42(10^{-4}) \text{ kg}\cdot\text{m}^2$$

② $$I_{yy} = \tfrac{1}{3}m(a^2 + b^2) = \tfrac{1}{3}(0.282)[(0.21)^2 + (0.1)^2] = 50.9(10^{-4}) \text{ kg}\cdot\text{m}^2$$

$$I_{zz} = \tfrac{1}{3}ma^2 = \tfrac{1}{3}(0.282)(0.21)^2 = 41.5(10^{-4}) \text{ kg}\cdot\text{m}^2$$

$$I_{xy} = 0 \qquad I_{yz} = 0$$

$$I_{xz} = \bar{I}_{xz} + md_xd_z$$

$$= 0 + m\frac{a}{2}\frac{b}{2} = 0.282(0.105)(0.05) = 14.83(10^{-4}) \text{ kg}\cdot\text{m}^2$$

Part 2

$$I_{xx} = \tfrac{1}{4}mr^2 + md_z{}^2 = -0.0518\left[\frac{(0.035)^2}{4} + (0.050)^2\right]$$

$$= -1.453(10^{-4}) \text{ kg}\cdot\text{m}^2$$

$$I_{yy} = \tfrac{1}{2}mr^2 + m(d_x{}^2 + d_z{}^2)$$

$$= -0.0518\left[\frac{(0.035)^2}{2} + (0.16)^2 + (0.05)^2\right]$$

$$= -14.86(10^{-4}) \text{ kg}\cdot\text{m}^2$$

$$I_{zz} = \tfrac{1}{4}mr^2 + md_x{}^2 = -0.0518\left[\frac{(0.035)^2}{4} + (0.16)^2\right]$$

$$= -13.41(10^{-4}) \text{ kg}\cdot\text{m}^2$$

$$I_{xy} = 0 \qquad I_{yz} = 0$$

$$I_{xz} = \bar{I}_{xz} + md_xd_z = 0 - 0.0518(0.16)(0.05) = -4.14(10^{-4}) \text{ kg}\cdot\text{m}^2$$

Helpful Hints

① Note that the mass of the hole is treated as a negative number.

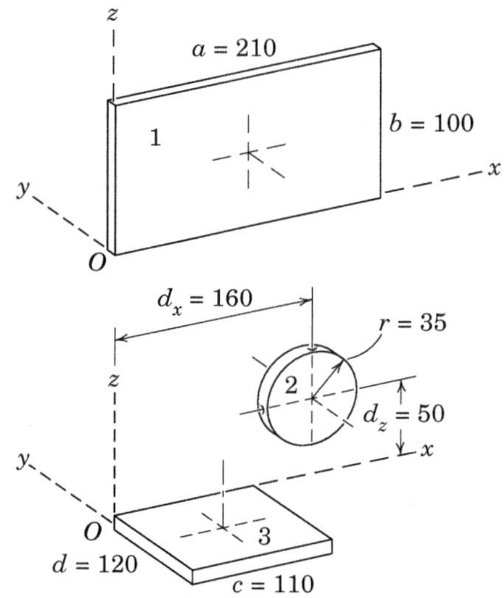

② You can easily derive this formula. Also check Table D/4.

Sample Problem B/7 (Continued)

Part 3

$$I_{xx} = \frac{1}{3}md^2 = \frac{1}{3}(0.1775)(0.12)^2 = 8.52(10^{-4})\ \text{kg}\cdot\text{m}^2$$

$$I_{yy} = \frac{1}{3}mc^2 = \frac{1}{3}(0.1775)(0.11)^2 = 7.16(10^{-4})\ \text{kg}\cdot\text{m}^2$$

$$I_{zz} = \frac{1}{3}m(c^2 + d^2) = \frac{1}{3}(0.1775)[(0.11)^2 + (0.12)^2]$$

$$= 15.68(10^{-4})\ \text{kg}\cdot\text{m}^2$$

$$I_{xy} = \bar{I}_{xy} + md_xd_y$$

$$= 0 + m\frac{c}{2}\left(\frac{-d}{2}\right) = 0.1775(0.055)(-0.06) = -5.86(10^{-4})\ \text{kg}\cdot\text{m}^2$$

$$I_{yz} = 0 \qquad I_{xz} = 0$$

Totals

$$I_{xx} = 16.48(10^{-4})\ \text{kg}\cdot\text{m}^2 \qquad I_{xy} = -5.86(10^{-4})\ \text{kg}\cdot\text{m}^2$$

$$I_{yy} = 43.2(10^{-4})\ \text{kg}\cdot\text{m}^2 \qquad I_{yz} = 0$$

$$I_{zz} = 43.8(10^{-4})\ \text{kg}\cdot\text{m}^2 \qquad I_{xz} = 10.69(10^{-4})\ \text{kg}\cdot\text{m}^2$$

Substitution into Eq. B/11, expansion of the determinant, and simplification yield

$$I^3 - 103.5(10^{-4})I^2 + 3180(10^{-8})I - 24\,800(10^{-12}) = 0$$

③ Solution of this cubic equation yields the following roots, which are the principal moments of inertia

$$I_1 = 48.3(10^{-4})\ \text{kg}\cdot\text{m}^2$$

$$I_2 = 11.82(10^{-4})\ \text{kg}\cdot\text{m}^2 \qquad \textit{Ans.}$$

$$I_3 = 43.4(10^{-4})\ \text{kg}\cdot\text{m}^2$$

The direction cosines of each principal axis are obtained by substituting each root, in turn, into Eq. B/12 and using $l^2 + m^2 + n^2 = 1$. The results are

$$l_1 = 0.357 \qquad l_2 = 0.934 \qquad l_3 = 0.01830$$

$$m_1 = 0.410 \qquad m_2 = -0.1742 \qquad m_3 = 0.895 \qquad \textit{Ans.}$$

$$n_1 = -0.839 \qquad n_2 = 0.312 \qquad n_3 = 0.445$$

The bottom figure shows a pictorial view of the bracket and the orientation of its principal axes of inertia.

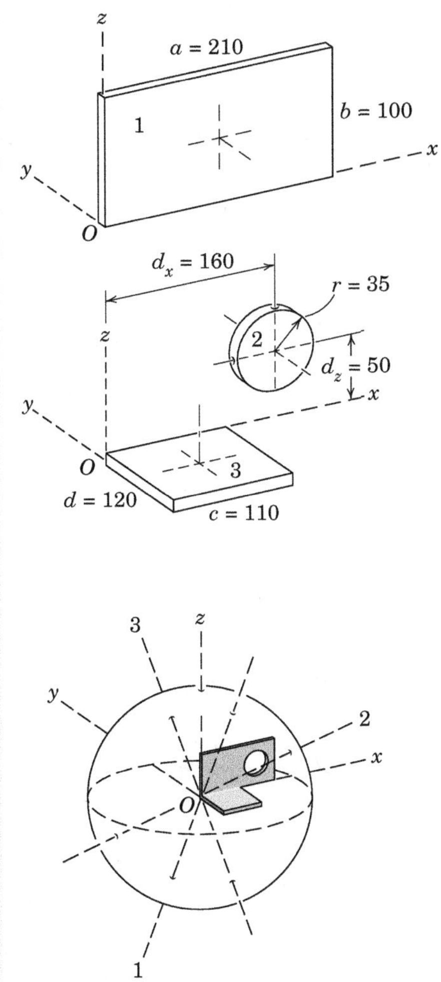

③ A computer program for the solution of a cubic equation may be used, or an algebraic solution using the formula cited in item 4 of Art. C/4, Appendix C, may be employed.

PROBLEMS

Introductory Problems

B/57 Determine the products of inertia about the coordinate axes for the unit which consists of four small particles, each of mass m, connected by the light but rigid slender rods.

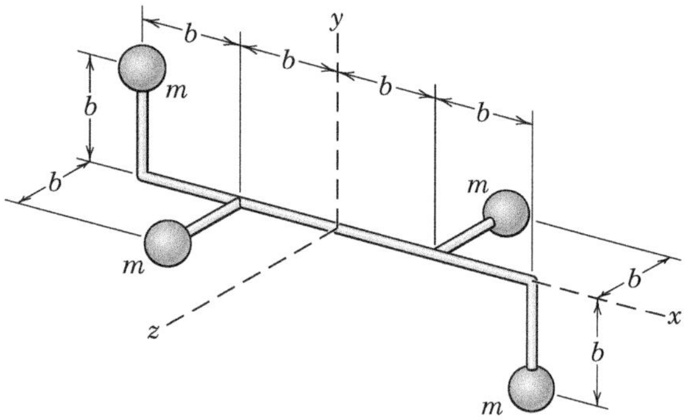

Problem B/57

B/58 Determine the products of inertia about the coordinate axes for the unit which consists of three small spheres, each of mass m, connected by the light but rigid slender rods.

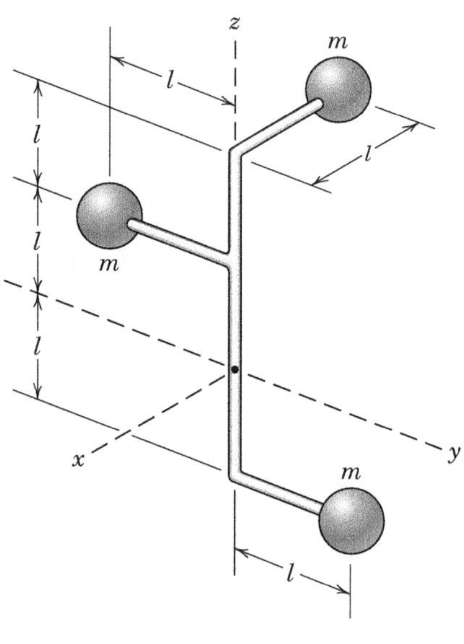

Problem B/58

B/59 Determine the product of inertia I_{xy} for the slender rod of mass m.

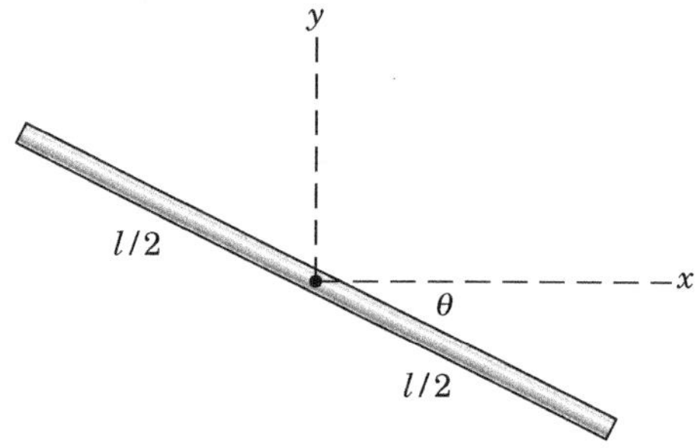

Problem B/59

B/60 The rod of Prob. B/1 is repeated here. Determine the product of inertia for the rod about the x-y axes. The rod has a mass ρ per unit length.

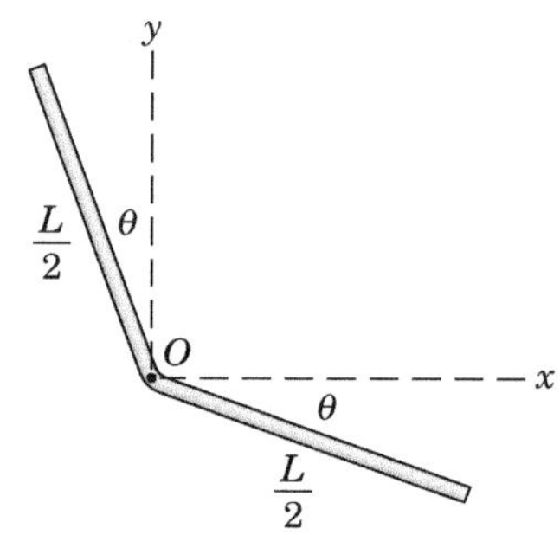

Problem B/60

B/61 Determine the products of inertia of the uniform slender rod of mass m about the coordinate axes shown.

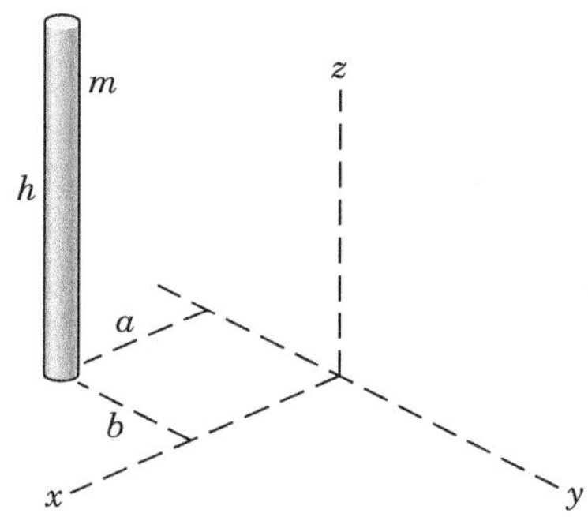

Problem B/61

B/62 Determine the products of inertia about the coordinate axes for the thin square plate with two circular holes. The mass of the plate material per unit area is ρ.

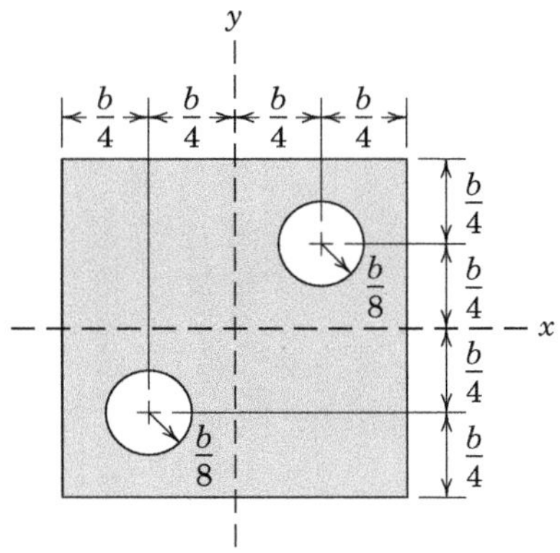

Problem B/62

B/63 Determine the products of inertia of the solid homogeneous half-cylinder of mass m for the axes shown.

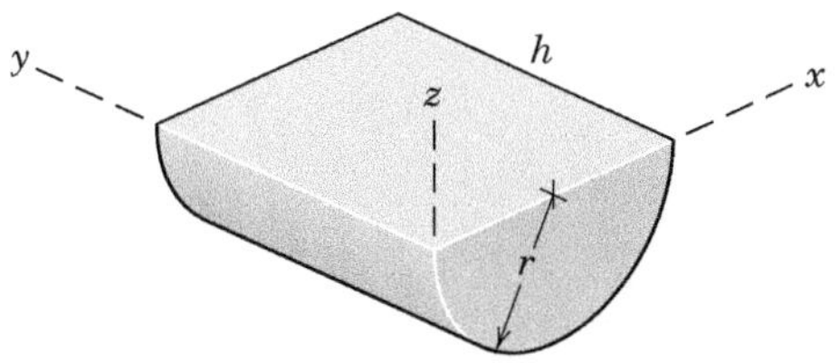

Problem B/63

B/64 Determine the products of inertia about the coordinate axes for the thin plate of mass m which has the shape of a circular sector of radius a and angle β as shown.

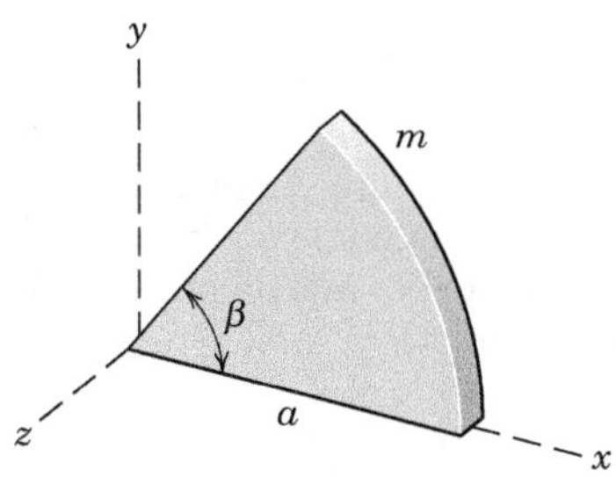

Problem B/64

B/65 The homogeneous plate of Prob. B/7 is repeated here. Determine the product of inertia for the plate about the x-y axes. The plate has mass m and uniform thickness t.

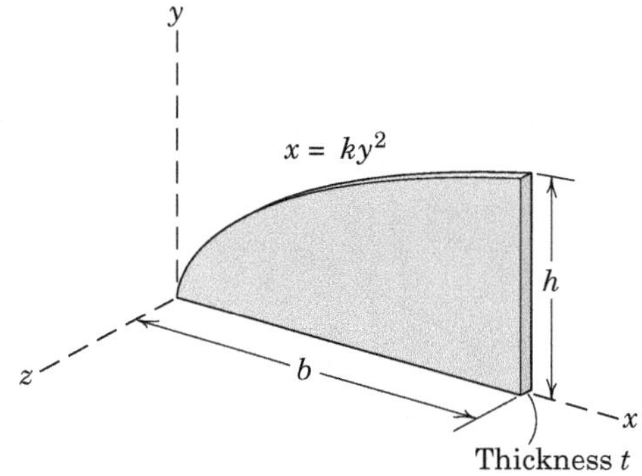

Problem B/65

B/66 Determine by direct integration the product of inertia of the thin homogenous triangular plate of mass m about the x-y axes. Then, use the parallel-axis theorem to determine the product of inertia for the plate about the x'-y' axes and the y''-y'' axes. What is the product of inertia about the centroid of the plate?

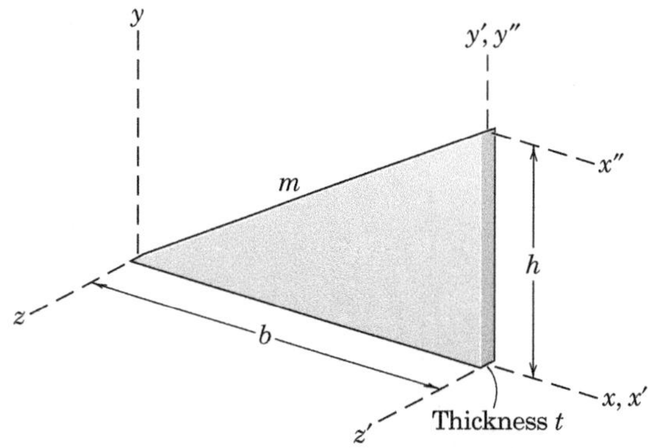

Problem B/66

Representative Problems

B/67 The aluminium casting consists of a 6-in. cube with a 4-in. cubical recess. Calculate the products of inertia of the casting about the axes shown.

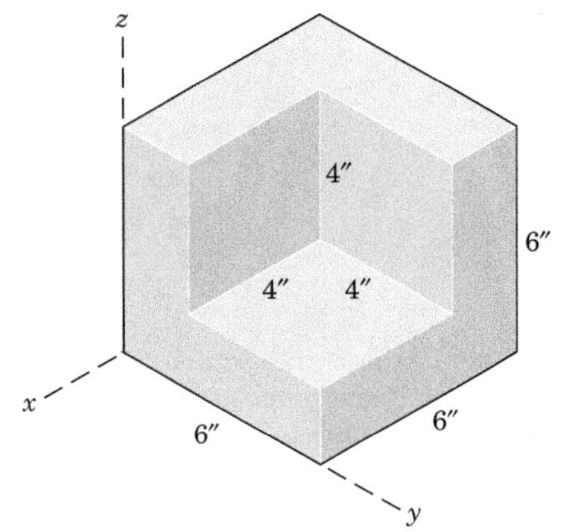

Problem B/67

B/68 Determine the products of inertia about the coordinate axes for the assembly which consists of uniform slender rods. Each rod has a mass ρ per unit length.

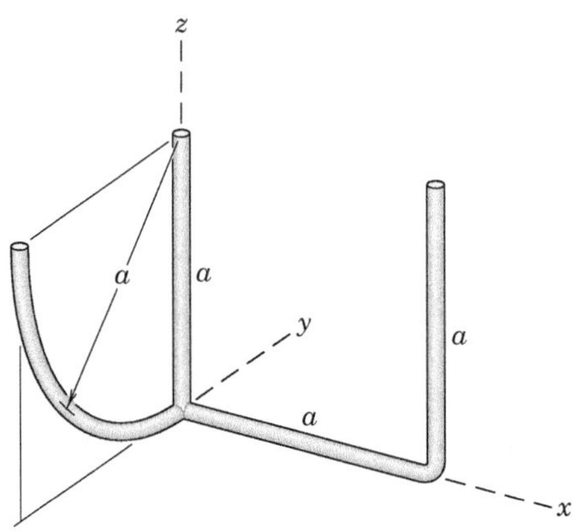

Problem B/68

B/69 The S-shaped piece is formed from a rod of diameter d and bent into the two semicircular shapes. Determine the products of inertia for the rod, for which d is small compared with r.

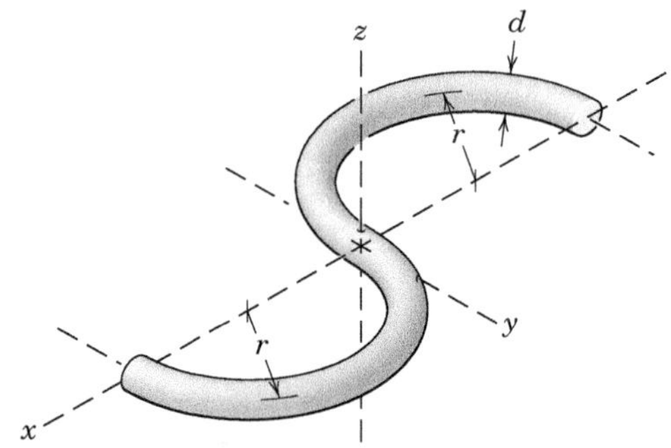

Problem B/69

B/70 Prove that the moment of inertia of the rigid assembly of three identical balls, each of mass m and radius r, has the same value for all axes through O. Neglect the mass of the connecting rods.

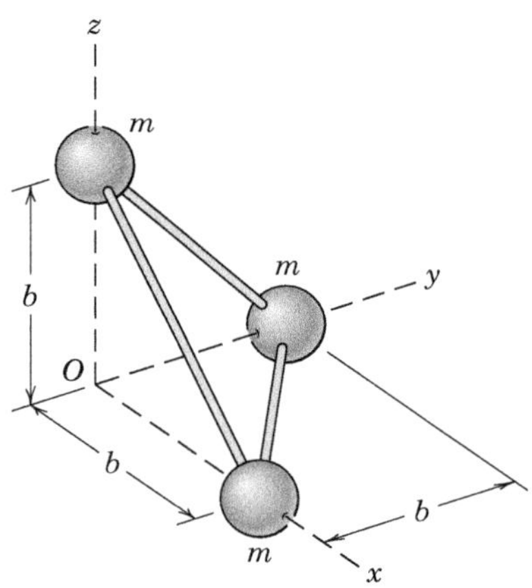

Problem B/70

►B/71 The plane of the thin circular disk of mass m and radius r makes an angle β with the x-z plane. Determine the product of inertia of the disk with respect to the y-z plane.

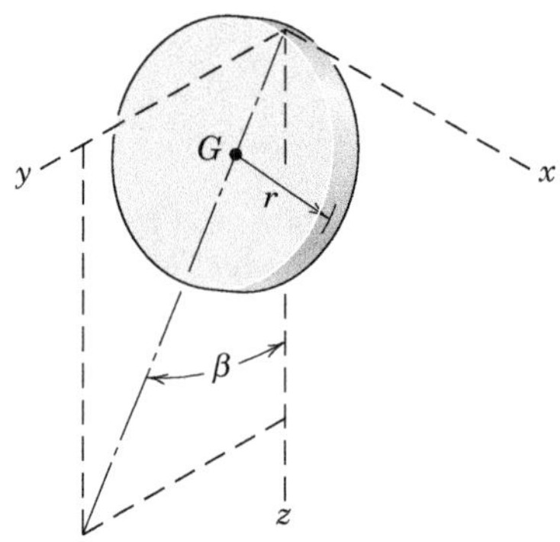

Problem B/71

Computer-Oriented Problems

***B/72** The L-shaped piece is cut from steel plate having a mass per unit area of 160 $\mathrm{kg/m^2}$. Determine and plot the moment of inertia of the piece about axis A-A as a function of θ from $\theta = 0$ to $\theta = 90°$ and find its minimum value.

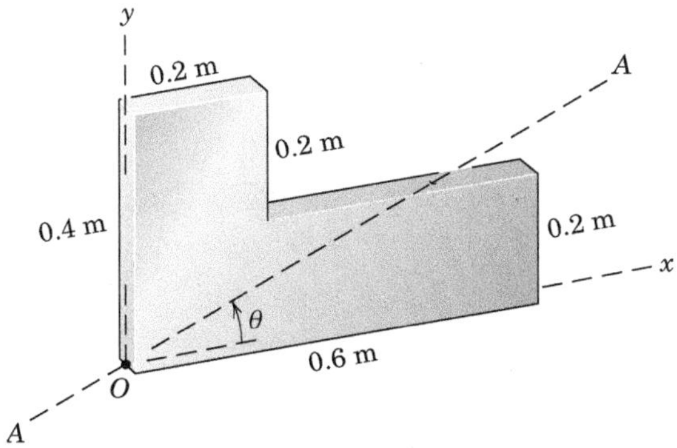

Problem B/72

***B/73** Determine the moment of inertia I about axis O-M for the uniform slender rod bent into the shape shown. Plot I versus θ from $\theta = 0$ to $\theta = 90°$ and determine the minimum value of I and the angle α which its axis makes with the x-direction. (*Note:* Because the analysis does not involve the z-coordinate, the expressions developed for area moments of inertia, Eqs. A/9, A/10, and A/11 in Appendix A of *Vol. 1 Statics*, may be utilized for this problem in place of the three-dimensional relations of Appendix B.) The rod has a mass ρ per unit length.

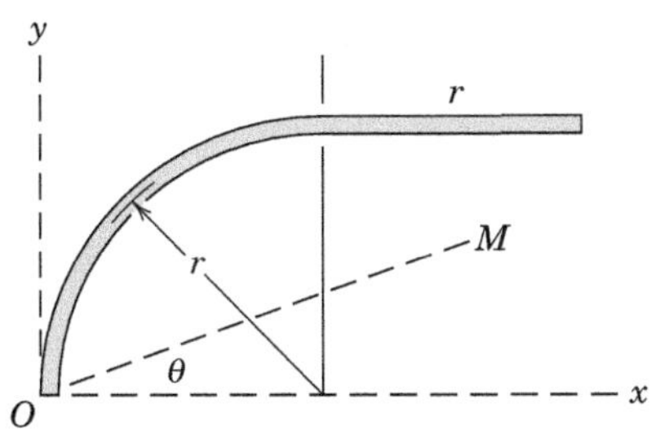

Problem B/73

***B/74** The assembly of three small spheres connected by light rigid bars of Prob. B/58 is repeated here. Determine the principal moments of inertia and the direction cosines associated with the axis of maximum moment of inertia.

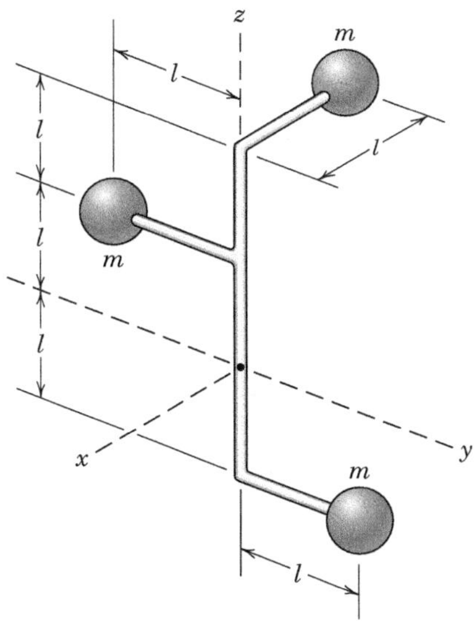

Problem B/74

***B/75** Determine the inertia tensor for the homogeneous thin plate about the x-, y-, and z-axes. The plate has a mass m and uniform thickness t. What is the minimum angle, measured from the x-axis, which will rotate the plate into principal directions?

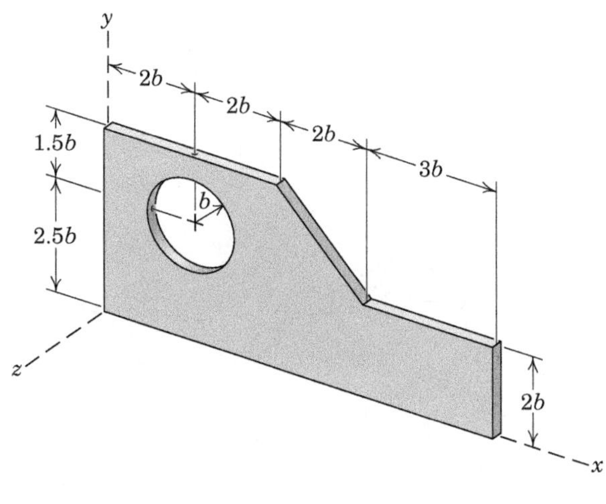

Problem B/75

*B/76 The thin plate has a mass ρ per unit area and is formed into the shape shown. Determine the principal moments of inertia of the plate about axes through O.

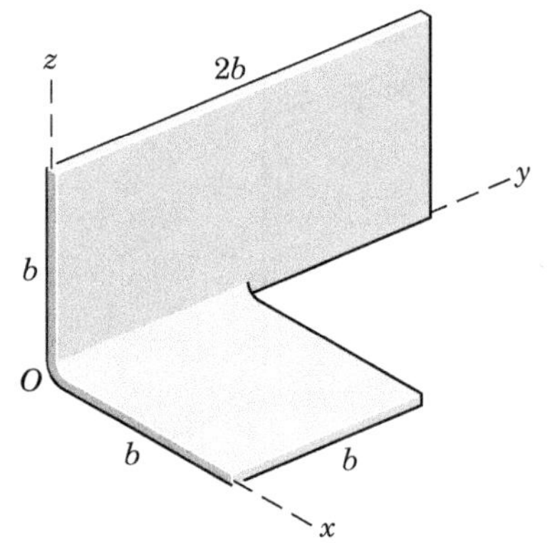

Problem B/76

*B/77 The slender rod has a mass ρ per unit length and is formed into the shape shown. Determine the principal moments of inertia about axes through O and calculate the direction cosines of the axis of minimum moment of inertia.

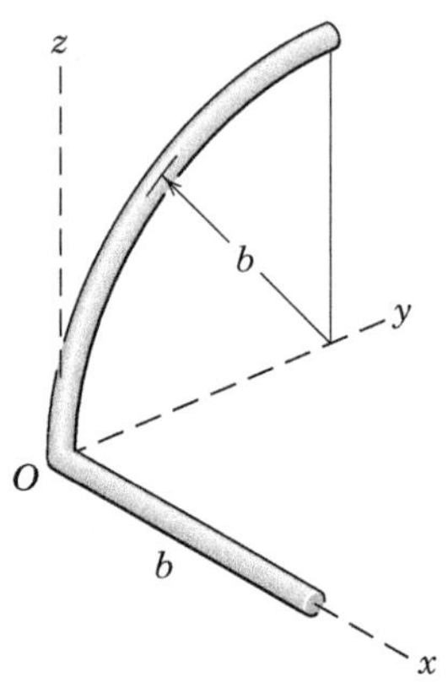

Problem B/77

*B/78 The welded assembly is formed from uniform sheet metal with a mass of 32 kg/m^2. Determine the principal mass moments of inertia for the assembly and the corresponding direction cosines for each principal axis.

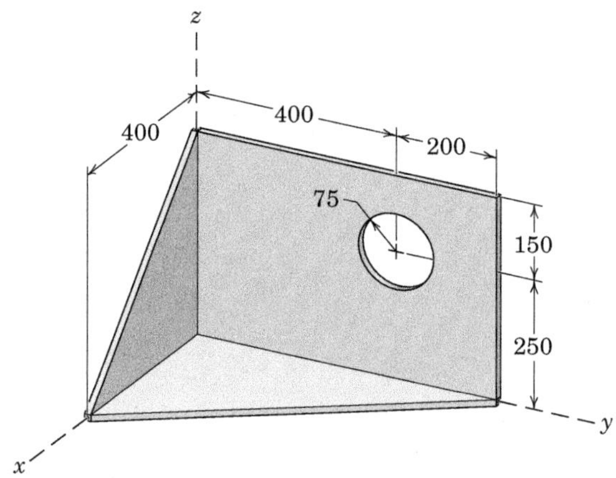

Dimensions in millimeters

Problem B/78

C SELECTED TOPICS OF MATHEMATICS

C/1 INTRODUCTION

Appendix C contains an abbreviated summary and reminder of selected topics in basic mathematics which find frequent use in mechanics. The relationships are cited without proof. The student of mechanics will have frequent occasion to use many of these relations, and he or she will be handicapped if they are not well in hand. Other topics not listed will also be needed from time to time.

As the reader reviews and applies mathematics, he or she should bear in mind that mechanics is an applied science descriptive of real bodies and actual motions. Therefore, the geometric and physical interpretation of the applicable mathematics should be kept clearly in mind during the development of theory and the formulation and solution of problems.

C/2 PLANE GEOMETRY

1. When two intersecting lines are, respectively, perpendicular to two other lines, the angles formed by the two pairs are equal.

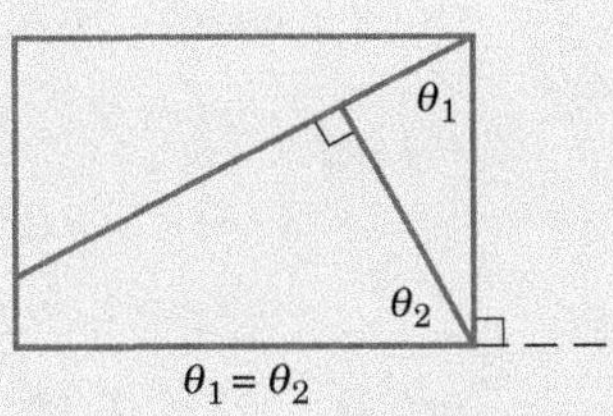

$\theta_1 = \theta_2$

2. Similar triangles

$$\frac{x}{b} = \frac{h - y}{h}$$

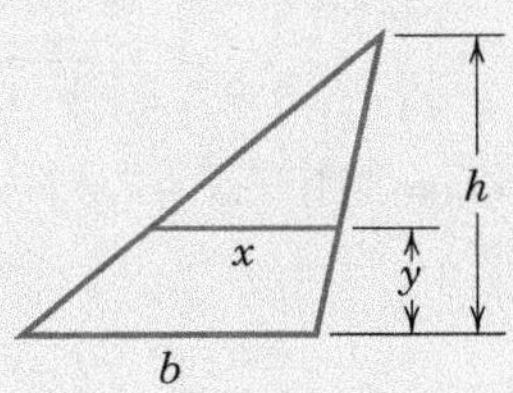

3. Any triangle

Area $= \frac{1}{2}bh$

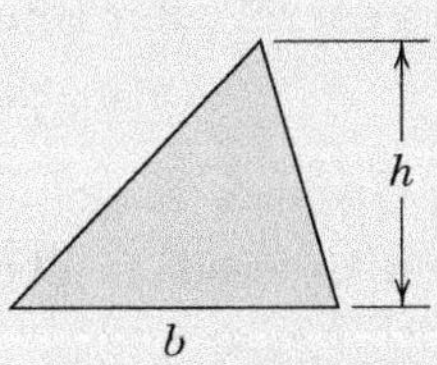

4. Circle

Circumference $= 2\pi r$
Area $= \pi r^2$
Arc length $s = r\theta$

Sector area $= \frac{1}{2}r^2\theta$

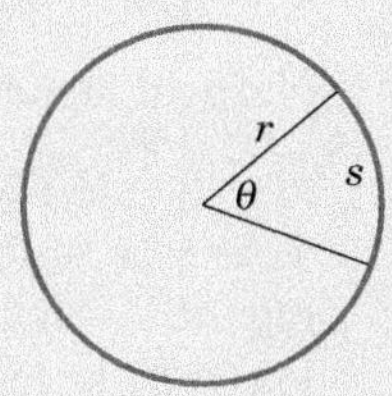

5. Every triangle inscribed within a semicircle is a right triangle.

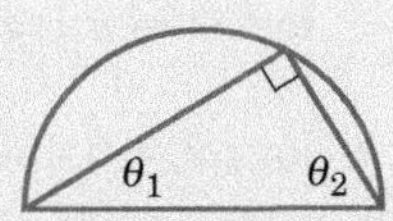

$\theta_1 + \theta_2 = \pi/2$

6. Angles of a triangle

$\theta_1 + \theta_2 + \theta_3 = 180°$
$\theta_4 = \theta_1 + \theta_2$

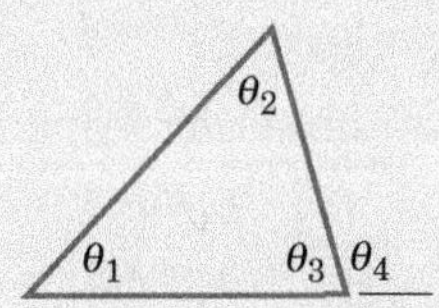

C/3 Solid Geometry

1. Sphere

Volume $= \frac{4}{3}\pi r^3$

Surface area $= 4\pi r^2$

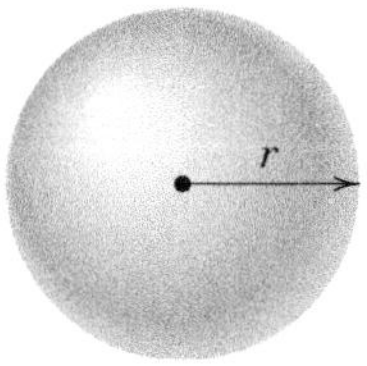

2. Spherical wedge

Volume $= \frac{2}{3}r^3\theta$

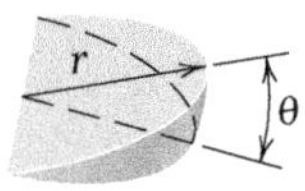

3. Right-circular cone

Volume $= \frac{1}{3}\pi r^2 h$

Lateral area $= \pi r L$

$L = \sqrt{r^2 + h^2}$

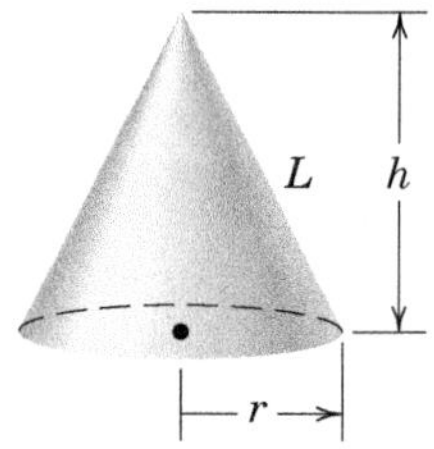

4. Any pyramid or cone

Volume $= \frac{1}{3}Bh$

where B = area of base

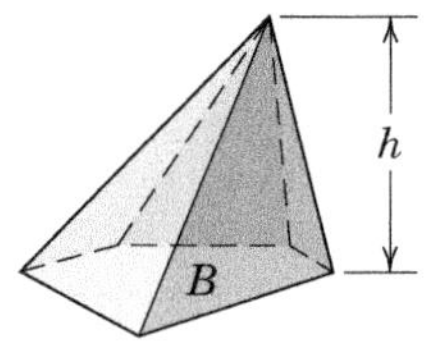

C/4 Algebra

1. Quadratic equation

$ax^2 + bx + c = 0$

$$x = \frac{-b \pm \sqrt{b^2 - 4ac}}{2a}, \; b^2 \geq 4ac \text{ for real roots}$$

2. Logarithms

$b^x = y, x = \log_b y$

Natural logarithms

$b = e = 2.718\,282$
$e^x = y, x = \log_e y = \ln y$
$\log(ab) = \log a + \log b$
$\log(a/b) = \log a - \log b$
$\log(1/n) = -\log n$
$\log a^n = n \log a$
$\log 1 = 0$
$\log_{10} x = 0.4343 \ln x$

3. Determinants

2nd order

$$\begin{vmatrix} a_1 & b_1 \\ a_2 & b_2 \end{vmatrix} = a_1 b_2 - a_2 b_1$$

3rd order

$$\begin{vmatrix} a_1 & b_1 & c_1 \\ a_2 & b_2 & c_2 \\ a_3 & b_3 & c_3 \end{vmatrix} = \begin{array}{l} +a_1b_2c_3 + a_2b_3c_1 + a_3b_1c_2 \\ -a_3b_2c_1 - a_2b_1c_3 - a_1b_3c_2 \end{array}$$

4. Cubic equation

$x^3 = Ax + B$

Let $p = A/3, q = B/2$.

Case I: $q^2 - p^3$ negative (three roots real and distinct)

$$\cos u = q/(p\sqrt{p}), \; 0 < u < 180^\circ$$
$$x_1 = 2\sqrt{p}\cos(u/3)$$
$$x_2 = 2\sqrt{p}\cos(u/3 + 120^\circ)$$
$$x_3 = 2\sqrt{p}\cos(u/3 + 240^\circ)$$

Case II: $q^2 - p^3$ positive (one root real, two roots imaginary)

$$x_1 = (q + \sqrt{q^2 - p^3})^{1/3} + (q - \sqrt{q^2 - p^3})^{1/3}$$

Case III: $q^2 - p^3 = 0$ (three roots real, two roots equal)

$$x_1 = 2q^{1/3}, x_2 = x_3 = -q^{1/3}$$

For general cubic equation

$$x^3 + ax^2 + bx + c = 0$$

Substitute $x = x_0 - a/3$ and get $x_0{}^3 = Ax_0 + B$. Then proceed as above to find values of x_0 from which $x = x_0 - a/3$.

C/5 Analytic Geometry

1. Straight line

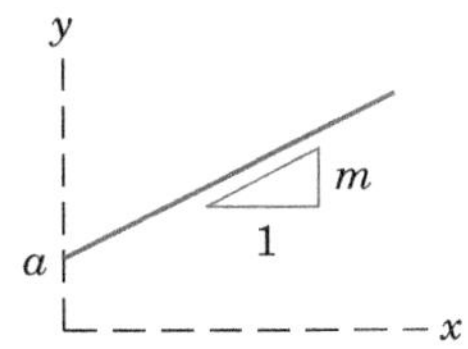

$y = a + mx$

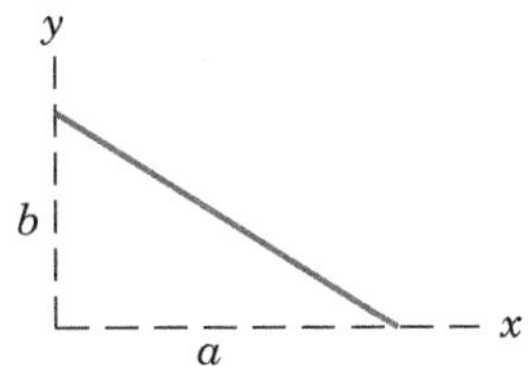

$\frac{x}{a} + \frac{y}{b} = 1$

2. Circle

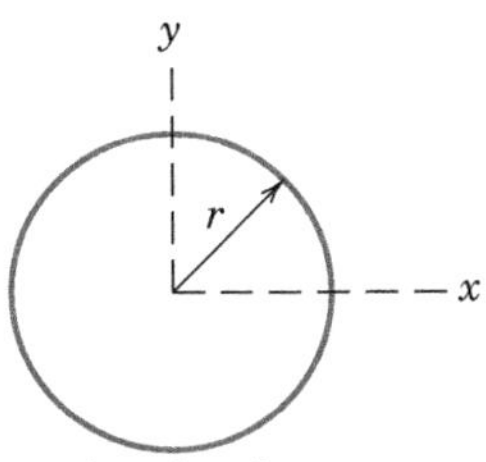

$x^2 + y^2 = r^2$

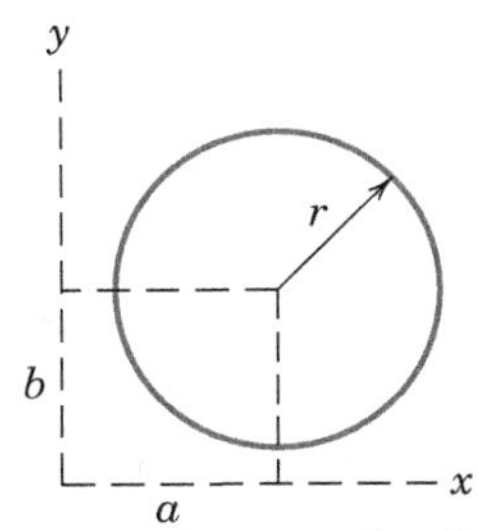

$(x - a)^2 + (y - b)^2 = r^2$

3. Parabola

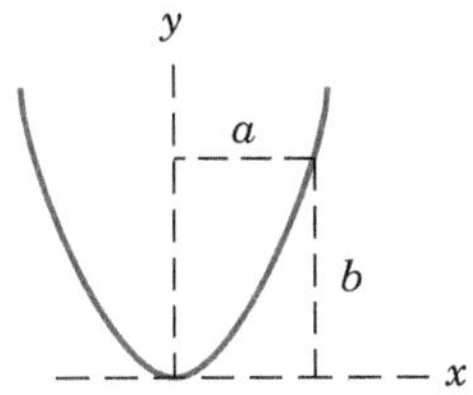

$y = b\frac{x^2}{a^2}$

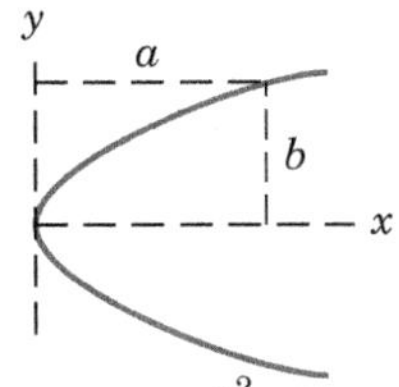

$x = a\frac{y^2}{b^2}$

4. Ellipse

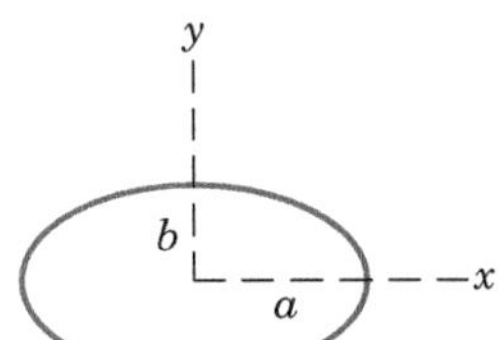

$\frac{x^2}{a^2} + \frac{y^2}{b^2} = 1$

5. Hyperbola

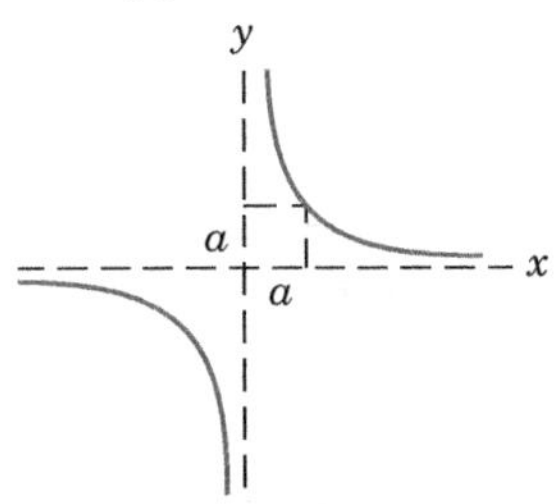

$xy = a^2$

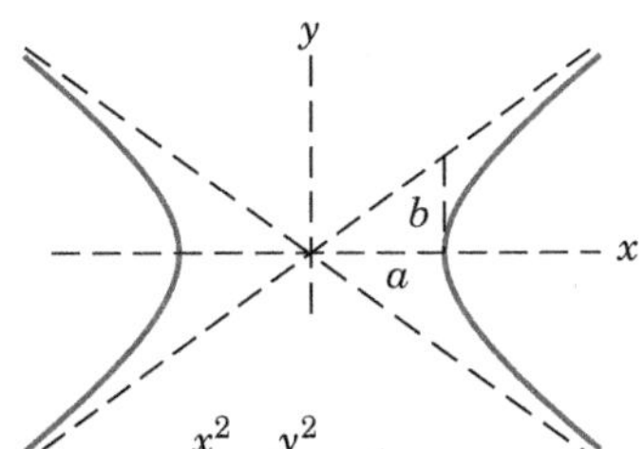

$\frac{x^2}{a^2} - \frac{y^2}{b^2} = 1$

C/6 Trigonometry

1. Definitions

$\sin\theta = a/c \quad \csc\theta = c/a$
$\cos\theta = b/c \quad \sec\theta = c/b$
$\tan\theta = a/b \quad \cot\theta = b/a$

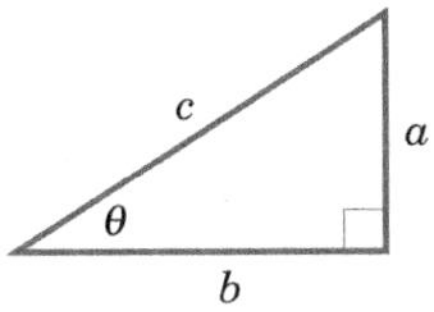

2. Signs in the four quadrants

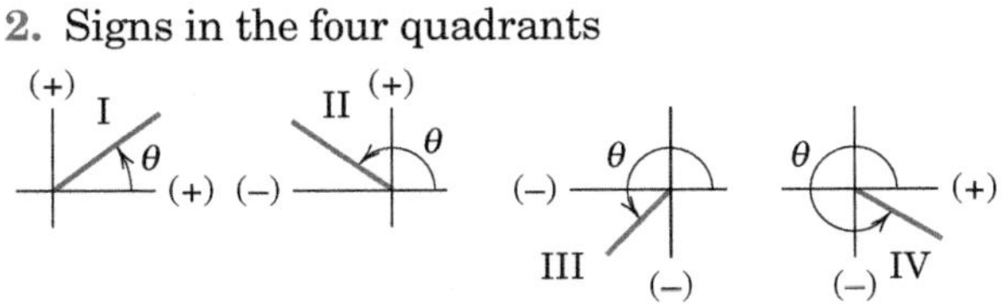

	I	II	III	IV
$\sin\theta$	+	+	−	−
$\cos\theta$	+	−	−	+
$\tan\theta$	+	−	+	−
$\csc\theta$	+	+	−	−
$\sec\theta$	+	−	−	+
$\cot\theta$	+	−	+	−

3. Miscellaneous relations

$$\sin^2\theta + \cos^2\theta = 1$$
$$1 + \tan^2\theta = \sec^2\theta$$
$$1 + \cot^2\theta = \csc^2\theta$$
$$\sin\frac{\theta}{2} = \sqrt{\tfrac{1}{2}(1 - \cos\theta)}$$
$$\cos\frac{\theta}{2} = \sqrt{\tfrac{1}{2}(1 - \cos\theta)}$$
$$\sin 2\theta = 2\sin\theta\cos\theta$$
$$\cos 2\theta = \cos^2\theta - \sin^2\theta$$
$$\sin(a \pm b) = \sin a\cos b \pm \cos a\sin b$$
$$\cos(a \pm b) = \cos a\cos b \mp \sin a\sin b$$

4. Law of sines

$$\frac{a}{b} = \frac{\sin A}{\sin B}$$

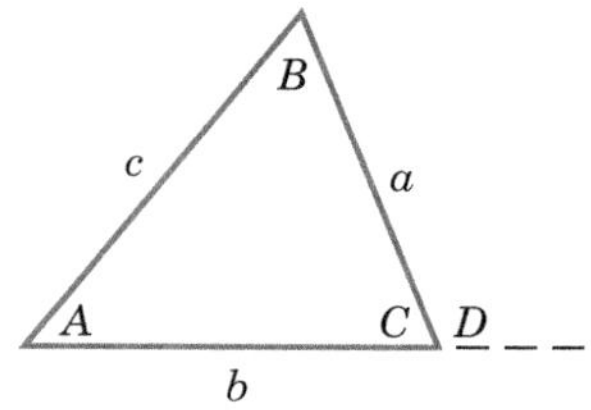

5. Law of cosines

$$c^2 = a^2 + b^2 - 2ab\cos C$$
$$c^2 = a^2 + b^2 + 2ab\cos D$$

C/7 Vector Operations

1. ***Notation.*** Vector quantities are printed in boldface type, and scalar quantities appear in lightface italic type. Thus, the vector quantity **V** has a scalar magnitude V. In longhand work vector quantities should always be consistently indicated by a symbol such as $\underline{V}$ or $\vec{V}$ to distinguish them from scalar quantities.

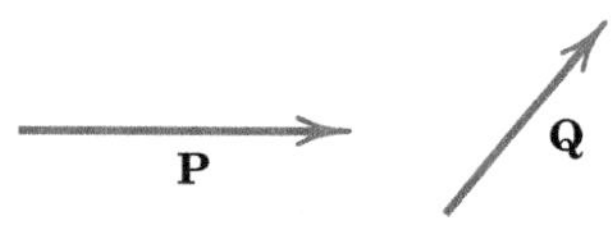

2. ***Addition***

Triangle addition $\mathbf{P} + \mathbf{Q} = \mathbf{R}$

Parallelogram addition $\mathbf{P} + \mathbf{Q} = \mathbf{R}$

Commutative law $\mathbf{P} + \mathbf{Q} = \mathbf{Q} + \mathbf{P}$

Associative law $\mathbf{P} + (\mathbf{Q} + \mathbf{R}) = (\mathbf{P} + \mathbf{Q}) + \mathbf{R}$

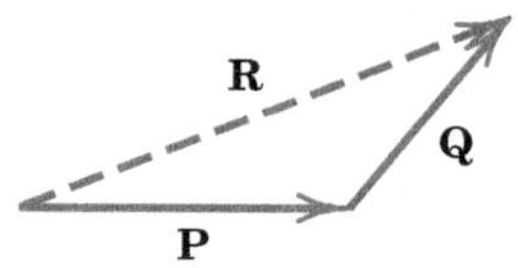

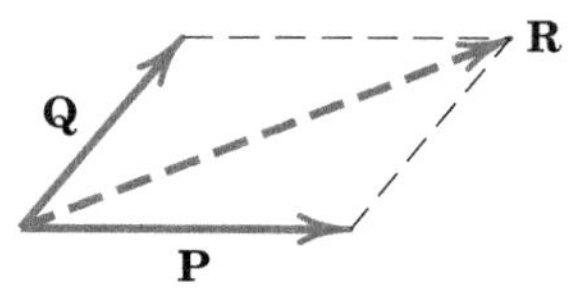

3. ***Subtraction***

$$\mathbf{P} - \mathbf{Q} = \mathbf{P} + (-\mathbf{Q})$$

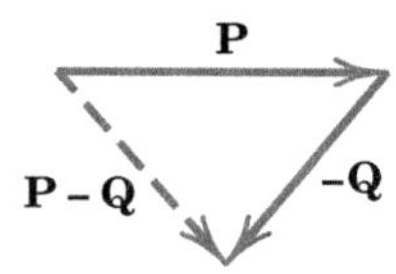

4. ***Unit vectors*** **i**, **j**, **k**

$$\mathbf{V} = V_x\mathbf{i} + V_y\mathbf{j} + V_z\mathbf{k}$$

where $|\mathbf{V}| = V = \sqrt{V_x^{\,2} + V_y^{\,2} + V_z^{\,2}}$

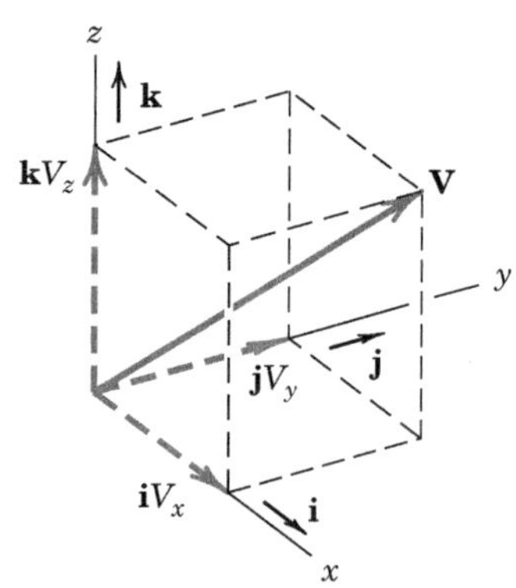

5. ***Direction cosines*** l, m, n are the cosines of the angles between **V** and the x-, y-, z-axes. Thus,

$$l = V_x/V \qquad m = V_y/V \qquad n = V_z/V$$

so that $\mathbf{V} = V(l\mathbf{i} + m\mathbf{j} + n\mathbf{k})$

and $l^2 + m^2 + n^2 = 1$

6. *Dot or scalar product*

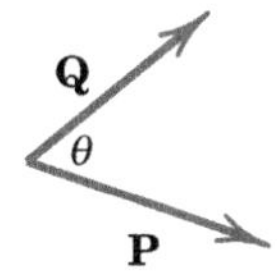

$$\mathbf{P}\cdot\mathbf{Q} = PQ \cos\theta$$

This product may be viewed as the magnitude of **P** multiplied by the component $Q \cos\theta$ of **Q** in the direction of **P**, or as the magnitude of **Q** multiplied by the component $P \cos\theta$ of **P** in the direction of **Q**.

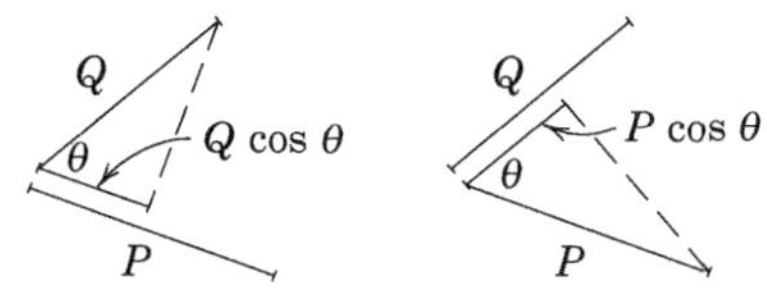

Commutative law $\quad \mathbf{P}\cdot\mathbf{Q} = \mathbf{Q}\cdot\mathbf{P}$

From the definition of the dot product

$$\begin{aligned}
\mathbf{i}\cdot\mathbf{i} &= \mathbf{j}\cdot\mathbf{j} = \mathbf{k}\cdot\mathbf{k} = 1 \\
\mathbf{i}\cdot\mathbf{j} &= \mathbf{j}\cdot\mathbf{i} = \mathbf{i}\cdot\mathbf{k} = \mathbf{k}\cdot\mathbf{i} = \mathbf{j}\cdot\mathbf{k} = \mathbf{k}\cdot\mathbf{j} = 0 \\
\mathbf{P}\cdot\mathbf{Q} &= (P_x\mathbf{i} + P_y\mathbf{j} + P_z\mathbf{k})\cdot(Q_x\mathbf{i} + Q_y\mathbf{j} + Q_z\mathbf{k}) \\
&= P_xQ_x + P_yQ_y + P_zQ_z \\
\mathbf{P}\cdot\mathbf{P} &= P_x^{\,2} + P_y^{\,2} + P_z^{\,2}
\end{aligned}$$

It follows from the definition of the dot product that two vectors **P** and **Q** are perpendicular when their dot product vanishes, $\mathbf{P}\cdot\mathbf{Q} = 0$.

The angle θ between two vectors $\mathbf{P}_1$ and $\mathbf{P}_2$ may be found from their dot product expression $\mathbf{P}_1\cdot\mathbf{P}_2 = P_1P_2 \cos\theta$, which gives

$$\cos\theta = \frac{\mathbf{P}_1\cdot\mathbf{P}_2}{P_1P_2} = \frac{P_{1_x}P_{2_x} + P_{1_y}P_{2_y} + P_{1_z}P_{2_z}}{P_1P_2} = l_1l_2 + m_1m_2 + n_1n_2$$

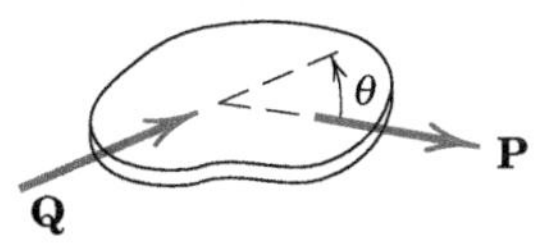

where l, m, n stand for the respective direction cosines of the vectors. It is also observed that two vectors are perpendicular to each other when their direction cosines obey the relation $l_1l_2 + m_1m_2 + n_1n_2 = 0$.

Distributive law $\quad \mathbf{P}\cdot(\mathbf{Q} + \mathbf{R}) = \mathbf{P}\cdot\mathbf{Q} + \mathbf{P}\cdot\mathbf{R}$

7. *Cross or vector product.* The cross product $\mathbf{P} \times \mathbf{Q}$ of the two vectors **P** and **Q** is defined as a vector with a magnitude

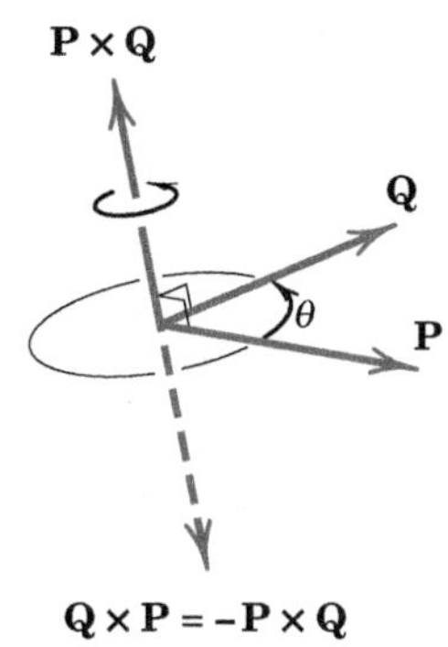

$$|\mathbf{P} \times \mathbf{Q}| = PQ \sin\theta$$

and a direction specified by the right-hand rule as shown. Reversing the vector order and using the right-hand rule give $\mathbf{Q} \times \mathbf{P} = -\mathbf{P} \times \mathbf{Q}$.

Distributive law $\quad \mathbf{P} \times (\mathbf{Q} + \mathbf{R}) = \mathbf{P} \times \mathbf{Q} + \mathbf{P} \times \mathbf{R}$

From the definition of the cross product, using a *right-handed coordinate system*, we get

$$\begin{array}{ccc}
\mathbf{i} \times \mathbf{j} = \mathbf{k} & \mathbf{j} \times \mathbf{k} = \mathbf{i} & \mathbf{k} \times \mathbf{i} = \mathbf{j} \\
\mathbf{j} \times \mathbf{i} = -\mathbf{k} & \mathbf{k} \times \mathbf{j} = -\mathbf{i} & \mathbf{i} \times \mathbf{k} = -\mathbf{j}
\end{array}$$

$$\mathbf{i} \times \mathbf{i} - \mathbf{j} \times \mathbf{j} = \mathbf{k} \times \mathbf{k} = \mathbf{0}$$

With the aid of these identities and the distributive law, the vector product may be written

$$\begin{aligned}\mathbf{P} \times \mathbf{Q} &= (P_x\mathbf{i} + P_y\mathbf{j} + P_z\mathbf{k}) \times (Q_x\mathbf{i} + Q_y\mathbf{j} + Q_z\mathbf{k}) \\ &= (P_yQ_z - P_zQ_y)\mathbf{i} + (P_zQ_x - P_xQ_z)\mathbf{j} + (P_xQ_y - P_yQ_x)\mathbf{k}\end{aligned}$$

The cross product may also be expressed by the determinant

$$\mathbf{P} \times \mathbf{Q} = \begin{vmatrix} \mathbf{i} & \mathbf{j} & \mathbf{k} \\ P_x & P_y & P_z \\ Q_x & Q_y & Q_z \end{vmatrix}$$

8. ***Additional relations***

Triple scalar product $(\mathbf{P} \times \mathbf{Q}) \cdot \mathbf{R} = \mathbf{R} \cdot (\mathbf{P} \times \mathbf{Q})$. The dot and cross may be interchanged as long as the order of the vectors is maintained. Parentheses are unnecessary since $\mathbf{P} \times (\mathbf{Q} \cdot \mathbf{R})$ is meaningless because a vector $\mathbf{P}$ cannot be crossed into a scalar $\mathbf{Q} \cdot \mathbf{R}$. Thus, the expression may be written

$$\mathbf{P} \times \mathbf{Q} \cdot \mathbf{R} = \mathbf{P} \cdot \mathbf{Q} \times \mathbf{R}$$

The triple scalar product has the determinant expansion

$$\mathbf{P} \times \mathbf{Q} \cdot \mathbf{R} = \begin{vmatrix} P_x & P_y & P_z \\ Q_x & Q_y & Q_z \\ R_x & R_y & R_z \end{vmatrix}$$

Triple vector product $(\mathbf{P} \times \mathbf{Q}) \times \mathbf{R} = -\mathbf{R} \times (\mathbf{P} \times \mathbf{Q}) = \mathbf{R} \times (\mathbf{Q} \times \mathbf{P})$. Here we note that the parentheses must be used since an expression $\mathbf{P} \times \mathbf{Q} \times \mathbf{R}$ would be ambiguous because it would not identify the vector to be crossed. It may be shown that the triple vector product is equivalent to

$$(\mathbf{P} \times \mathbf{Q}) \times \mathbf{R} = \mathbf{R} \cdot \mathbf{P}\mathbf{Q} - \mathbf{R} \cdot \mathbf{Q}\mathbf{P}$$

or

$$\mathbf{P} \times (\mathbf{Q} \times \mathbf{R}) = \mathbf{P} \cdot \mathbf{R}\mathbf{Q} - \mathbf{P} \cdot \mathbf{Q}\mathbf{R}$$

The first term in the first expression, for example, is the dot product $\mathbf{R} \cdot \mathbf{P}$, a scalar, multiplied by the vector $\mathbf{Q}$.

9. ***Derivatives of vectors*** obey the same rules as they do for scalars.

$$\frac{d\mathbf{P}}{dt} = \dot{\mathbf{P}} = \dot{P}_x\mathbf{i} + \dot{P}_y\mathbf{j} + \dot{P}_z\mathbf{k}$$

$$\frac{d(\mathbf{P}u)}{dt} = \mathbf{P}\dot{u} + \dot{\mathbf{P}}u$$

$$\frac{d(\mathbf{P} \cdot \mathbf{Q})}{dt} = \mathbf{P} \cdot \dot{\mathbf{Q}} + \dot{\mathbf{P}} \cdot \mathbf{Q}$$

$$\frac{d(\mathbf{P} \times \mathbf{Q})}{dt} = \mathbf{P} \times \dot{\mathbf{Q}} + \dot{\mathbf{P}} \times \mathbf{Q}$$

10. ***Integration of vectors.*** If $\mathbf{V}$ is a function of x, y, and z and an element of volume is $d\tau = dx\,dy\,dz$, the integral of $\mathbf{V}$ over the volume may be written as the vector sum of the three integrals of its components. Thus,

$$\int \mathbf{V}\,d\tau = \mathbf{i}\int V_x\,d\tau + \mathbf{j}\int V_y\,d\tau + \mathbf{k}\int V_z\,d\tau$$

C/8 Series

(The expression in brackets following a series indicates the range of convergence.)

$$(1 \pm x)^n = 1 \pm nx + \frac{n(n-1)}{2!}x^2 \pm \frac{n(n-1)(n-2)}{3!}x^3 + \cdots \qquad [x^2 < 1]$$

$$\sin x = x - \frac{x^3}{3!} + \frac{x^5}{5!} - \frac{x^7}{7!} + \cdots \qquad [x^2 < \infty]$$

$$\cos x = 1 - \frac{x^2}{2!} + \frac{x^4}{4!} - \frac{x^6}{6!} + \cdots \qquad [x^2 < \infty]$$

$$\sinh x = \frac{e^x - e^{-x}}{2} = x + \frac{x^3}{3!} + \frac{x^5}{5!} + \frac{x^7}{7!} + \cdots \qquad [x^2 < \infty]$$

$$\cosh x = \frac{e^x + e^{-x}}{2} = 1 + \frac{x^2}{2!} + \frac{x^4}{4!} + \frac{x^6}{6!} + \cdots \qquad [x^2 < \infty]$$

$$f(x) = \frac{a_0}{2} + \sum_{n=1}^{\infty} a_n \cos\frac{n\pi x}{l} + \sum_{n=1}^{\infty} b_n \sin\frac{n\pi x}{l}$$

$$\text{where } a_n = \frac{1}{l}\int_{-l}^{l} f(x)\cos\frac{n\pi x}{l}\,dx, \qquad b_n = \frac{1}{l}\int_{-l}^{l} f(x)\sin\frac{n\pi x}{l}\,dx$$

[Fourier expansion for $-l < x < l$]

C/9 Derivatives

$$\frac{dx^n}{dx} = nx^{n-1}, \qquad \frac{d(uv)}{dx} = u\frac{dv}{dx} + v\frac{du}{dx}, \qquad \frac{d\left(\frac{u}{v}\right)}{dx} = \frac{v\frac{du}{dx} - u\frac{dv}{dx}}{v^2}$$

$$\lim_{\Delta x \to 0} \sin \Delta x = \sin dx = \tan dx = dx$$

$$\lim_{\Delta x \to 0} \cos \Delta x = \cos dx = 1$$

$$\frac{d \sin x}{dx} = \cos x, \qquad \frac{d \cos x}{dx} = -\sin x, \qquad \frac{d \tan x}{dx} = \sec^2 x$$

$$\frac{d \sinh x}{dx} = \cosh x, \qquad \frac{d \cosh x}{dx} = \sinh x, \qquad \frac{d \tanh x}{dx} = \operatorname{sech}^2 x$$

C/10 Integrals

$$\int x^n\,dx = \frac{x^{n+1}}{n+1}$$

$$\int \frac{dx}{x} = \ln x$$

$$\int \sqrt{a+bx}\,dx = \frac{2}{3b}\sqrt{(a+bx)^3}$$

$$\int x\sqrt{a+bx}\,dx = \frac{2}{15b^2}(3bx-2a)\sqrt{(a+bx)^3}$$

$$\int x^2\sqrt{a+bx}\,dx = \frac{2}{105b^3}(8a^2-12abx+15b^2x^2)\sqrt{(a+bx)^3}$$

$$\int \frac{dx}{\sqrt{a+bx}} = \frac{2\sqrt{a+bx}}{b}$$

$$\int \frac{\sqrt{a+x}}{\sqrt{b-x}}\,dx = -\sqrt{a+x}\,\sqrt{b-x} + (a+b)\sin^{-1}\sqrt{\frac{a+x}{a+b}}$$

$$\int \frac{x\,dx}{a+bx} = \frac{1}{b^2}[a+bx-a\ln(a+bx)]$$

$$\int \frac{x\,dx}{(a+bx)^n} = \frac{(a+bx)^{1-n}}{b^2}\left(\frac{a+bx}{2-n}-\frac{a}{1-n}\right)$$

$$\int \frac{dx}{a+bx^2} = \frac{1}{\sqrt{ab}}\tan^{-1}\frac{x\sqrt{ab}}{a} \quad \text{or} \quad \frac{1}{\sqrt{-ab}}\tanh^{-1}\frac{x\sqrt{-ab}}{a}$$

$$\int \frac{x\,dx}{a+bx^2} = \frac{1}{2b}\ln(a+bx^2)$$

$$\int \sqrt{x^2\pm a^2}\,dx = \tfrac{1}{2}[x\sqrt{x^2\pm a^2}\pm a^2\ln(x+\sqrt{x^2\pm a^2})]$$

$$\int \sqrt{a^2-x^2}\,dx = \tfrac{1}{2}\left(x\sqrt{a^2-x^2}+a^2\sin^{-1}\frac{x}{a}\right)$$

$$\int x\sqrt{a^2-x^2}\,dx = -\tfrac{1}{3}\sqrt{(a^2-x^2)^3}$$

$$\int x^2\sqrt{a^2-x^2}\,dx = -\frac{x}{4}\sqrt{(a^2-x^2)^3}+\frac{a^2}{8}\left(x\sqrt{a^2-x^2}+a^2\sin^{-1}\frac{x}{a}\right)$$

$$\int x^3\sqrt{a^2-x^2}\,dx = -\tfrac{1}{5}(x^2+\tfrac{2}{3}a^2)\sqrt{(a^2-x^2)^3}$$

$$\int \frac{dx}{\sqrt{a+bx+cx^2}} = \frac{1}{\sqrt{c}}\ln\left(\sqrt{a+bx+cx^2}+x\sqrt{c}+\frac{b}{2\sqrt{c}}\right) \quad \text{or} \quad \frac{-1}{\sqrt{-c}}\sin^{-1}\left(\frac{b+2cx}{\sqrt{b^2-4ac}}\right)$$

$$\int \frac{dx}{\sqrt{x^2\pm a^2}} = \ln(x+\sqrt{x^2\pm a^2})$$

$$\int \frac{dx}{\sqrt{a^2 - x^2}} = \sin^{-1}\frac{x}{a}$$

$$\int \frac{x\,dx}{\sqrt{x^2 - a^2}} = \sqrt{x^2 - a^2}$$

$$\int \frac{x\,dx}{\sqrt{a^2 \pm x^2}} = \pm\sqrt{a^2 \pm x^2}$$

$$\int x\sqrt{x^2 \pm a^2}\,dx = \tfrac{1}{3}\sqrt{(x^2 \pm a^2)^3}$$

$$\int x^2\sqrt{x^2 \pm a^2}\,dx = \frac{x}{4}\sqrt{(x^2 \pm a^2)^3} \mp \frac{a^2}{8}x\sqrt{x^2 \pm a^2} - \frac{a^4}{8}\ln\,(x + \sqrt{x^2 \pm a^2})$$

$$\int \sin x\,dx = -\cos x$$

$$\int \cos x\,dx = \sin x$$

$$\int \sec x\,dx = \frac{1}{2}\ln\frac{1 + \sin x}{1 - \sin x}$$

$$\int \sin^2 x\,dx = \frac{x}{2} - \frac{\sin 2x}{4}$$

$$\int \cos^2 x\,dx = \frac{x}{2} + \frac{\sin 2x}{4}$$

$$\int \sin x\cos x\,dx = \frac{\sin^2 x}{2}$$

$$\int \sinh x\,dx = \cosh x$$

$$\int \cosh x\,dx = \sinh x$$

$$\int \tanh x\,dx = \ln\cosh x$$

$$\int \ln x\,dx = x\ln x - x$$

$$\int e^{ax}\,dx = \frac{e^{ax}}{a}$$

$$\int xe^{ax}\,dx = \frac{e^{ax}}{a^2}(ax - 1)$$

$$\int e^{ax}\sin px\,dx = \frac{e^{ax}\,(a\sin px - p\cos px)}{a^2 + p^2}$$

$$\int e^{ax} \cos px \, dx = \frac{e^{ax}(a \cos px + p \sin px)}{a^2 + p^2}$$

$$\int e^{ax} \sin^2 x \, dx = \frac{e^{ax}}{4 + a^2}\left(a \sin^2 x - \sin 2x + \frac{2}{a}\right)$$

$$\int e^{ax} \cos^2 x \, dx = \frac{e^{ax}}{4 + a^2}\left(a \cos^2 x + \sin 2x + \frac{2}{a}\right)$$

$$\int e^{ax} \sin x \cos x \, dx = \frac{e^{ax}}{4 + a^2}\left(\frac{a}{2} \sin 2x - \cos 2x\right)$$

$$\int \sin^3 x \, dx = -\frac{\cos x}{3}(2 + \sin^2 x)$$

$$\int \cos^3 x \, dx = \frac{\sin x}{3}(2 + \cos^2 x)$$

$$\int \cos^5 x \, dx = \sin x - \tfrac{2}{3}\sin^3 x + \tfrac{1}{5}\sin^5 x$$

$$\int x \sin x \, dx = \sin x - x \cos x$$

$$\int x \cos x \, dx = \cos x + x \sin x$$

$$\int x^2 \sin x \, dx = 2x \sin x - (x^2 - 2)\cos x$$

$$\int x^2 \cos x \, dx = 2x \cos x + (x^2 - 2)\sin x$$

Radius of curvature

$$\rho_{xy} = \frac{\left[1 + \left(\dfrac{dy}{dx}\right)^2\right]^{3/2}}{\dfrac{d^2y}{dx^2}}$$

$$\rho_{r\theta} = \frac{\left[r^2 + \left(\dfrac{dr}{d\theta}\right)^2\right]^{3/2}}{r^2 + 2\left(\dfrac{dr}{d\theta}\right)^2 - r\dfrac{d^2r}{d\theta^2}}$$

C/11 Newton's Method for Solving Intractable Equations

Frequently, the application of the fundamental principles of mechanics leads to an algebraic or transcendental equation which is not solvable (or easily solvable) in closed form. In such cases, an iterative technique, such as Newton's method, can be a powerful tool for obtaining a good estimate to the root or roots of the equation.

Let us place the equation to be solved in the form $f(x) = 0$. Part a of the accompanying figure depicts an arbitrary function $f(x)$ for values of x in the vicinity of the desired root x_r. Note that x_r is merely the value

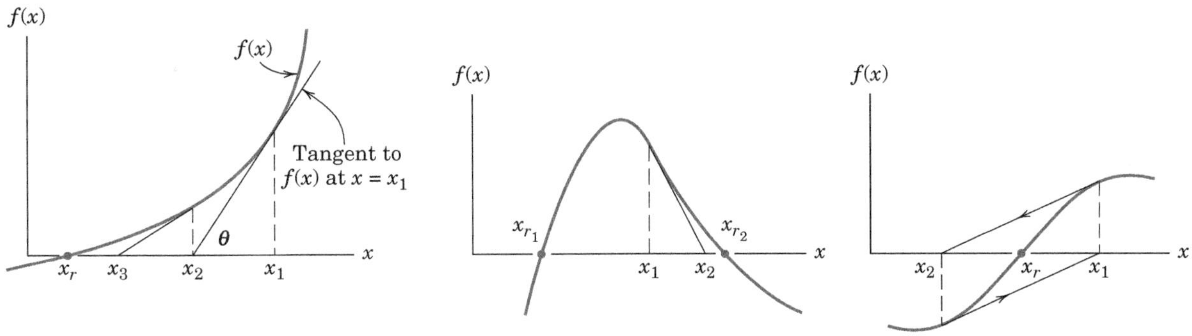

of x at which the function crosses the x-axis. Suppose that we have available (perhaps via a hand-drawn plot) a rough estimate x_1 of this root. Provided that x_1 does not closely correspond to a maximum or minimum value of the function $f(x)$, we may obtain a better estimate of the root x_r by extending the tangent to $f(x)$ at x_1 so that it intersects the x-axis at x_2. From the geometry of the figure, we may write

$$\tan \theta = f'(x_1) = \frac{f(x_1)}{x_1 - x_2}$$

where $f'(x_1)$ denotes the derivative of $f(x)$ with respect to x evaluated at $x = x_1$. Solving the above equation for x_2 results in

$$x_2 = x_1 - \frac{f(x_1)}{f'(x_1)}$$

The term $-f(x_1)/f'(x_1)$ is the correction to the initial root estimate x_1. Once x_2 is calculated, we may repeat the process to obtain x_3, and so forth.

Thus, we generalize the above equation to

$$x_{k+1} = x_k - \frac{f(x_k)}{f'(x_k)}$$

where

x_{k+1} = the $(k + 1)$th estimate of the desired root x_r

x_k = the kth estimate of the desired root x_r

$f(x_k)$ = the function $f(x)$ evaluated at $x = x_k$

$f'(x_k)$ = the function derivative evaluated at $x = x_k$

This equation is repeatedly applied until $f(x_{k+1})$ is sufficiently close to zero and $x_{k+1} \cong x_k$. The student should verify that the equation is valid for all possible sign combinations of x_k, $f(x_k)$, and $f'(x_k)$.

Several cautionary notes are in order:

1. Clearly, $f'(x_k)$ must not be zero or close to zero. This would mean, as restricted above, that x_k exactly or approximately corresponds to a minimum or maximum of $f(x)$. If the slope $f'(x_k)$ is zero, then the tangent to the curve never intersects the x-axis. If the slope $f'(x_k)$ is small, then the correction to x_k may be so large that x_{k+1} is a worse root estimate than x_k. For this reason, experienced engineers usually limit the size of the correction term; that is, if the absolute value of $f(x_k)/f'(x_k)$ is larger than a preselected maximum value, that maximum value is used.
2. If there are several roots of the equation $f(x) = 0$, we must be in the vicinity of the desired root x_r in order that the algorithm actually converges to that root. Part b of the figure depicts the condition when the initial estimate x_1 will result in convergence to x_{r_2} rather than x_{r_1}.
3. Oscillation from one side of the root to the other can occur if, for example, the function is antisymmetric about a root which is an inflection point. The use of one-half of the correction will usually prevent this behavior, which is depicted in part c of the accompanying figure.

Example: Beginning with an initial estimate of $x_1 = 5$, estimate the single root of the equation $e^x - 10 \cos x - 100 = 0$.

The table below summarizes the application of Newton's method to the given equation. The iterative process was terminated when the absolute value of the correction $-f(x_k)/f'(x_k)$ became less than 10^{-6}.

k	x_k	$f(x_k)$	$f'(x_k)$	$x_{k+1} - x_k = -\dfrac{f(x_k)}{f'(x_k)}$
1	5.000 000	45.576 537	138.823 916	−0.328 305
2	4.671 695	7.285 610	96.887 065	−0.075 197
3	4.596 498	0.292 886	89.203 650	−0.003 283
4	4.593 215	0.000 527	88.882 536	−0.000 006
5	4.593 209	$-2(10^{-8})$	88.881 956	$2.25(10^{-10})$

C/12 Selected Techniques for Numerical Integration

1. Area determination. Consider the problem of determining the shaded area under the curve $y = f(x)$ from $x = a$ to $x = b$, as depicted in part a of the figure, and suppose that analytical integration is not feasible. The function may be known in tabular form from experimental measurements, or it may be known in analytical form. The function is taken to be continuous within the interval $a < x < b$. We may divide the area into n vertical strips, each of width $\Delta x = (b - a)/n$, and then add the areas of all strips to obtain $A = \int y\, dx$. A representative strip of area A_i is shown with darker shading in the figure. Three useful numerical approximations are cited. In each case the greater the number of strips, the more accurate becomes the approximation geometrically. As a general rule, one can begin with a relatively small number of strips and increase the number until the resulting changes in the area approximation no longer improve the accuracy obtained.

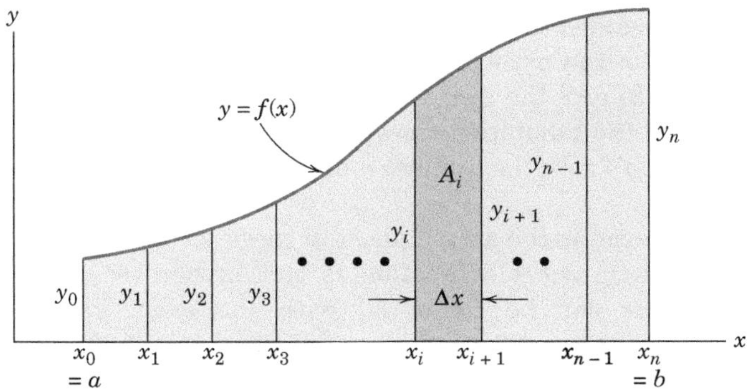

(a)

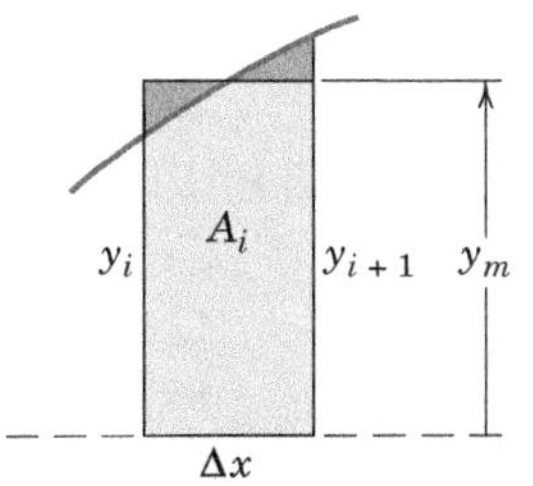

Rectangular

$A_i = y_m \Delta x$

$A = \int y \, dx \cong \Sigma y_m \Delta x$

(b)

I. *Rectangular* [Figure (*b*)] The areas of the strips are taken to be rectangles, as shown by the representative strip whose height y_m is chosen visually so that the small cross-hatched areas are as nearly equal as possible. Thus, we form the sum Σy_m of the effective heights and multiply by Δx. For a function known in analytical form, a value for y_m equal to that of the function at the midpoint $x_i + \Delta x/2$ may be calculated and used in the summation.

II. *Trapezoidal* [Figure (*c*)] The areas of the strips are taken to be trapezoids, as shown by the representative strip. The area A_i is the average height $(y_i + y_{i+1})/2$ times Δx. Adding the areas gives the area approximation as tabulated. For the example with the curvature shown, clearly the approximation will be on the low side. For the reverse curvature, the approximation will be on the high side.

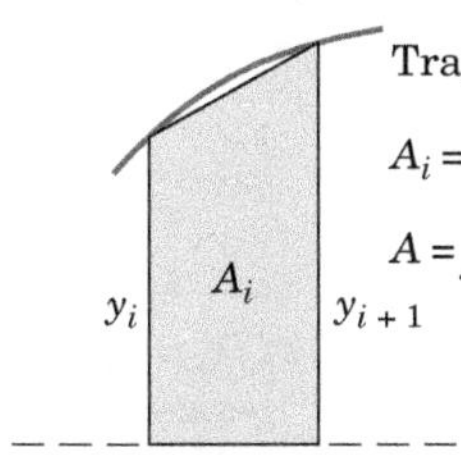

Trapezoidal

$$A_i = \frac{y_i + y_{i+1}}{2} \Delta x$$

$$A = \int y \, dx \cong \left(\frac{y_0}{2} + y_1 + y_2 + \cdots + y_{n-1} + \frac{y_n}{2} \right) \Delta x$$

(c)

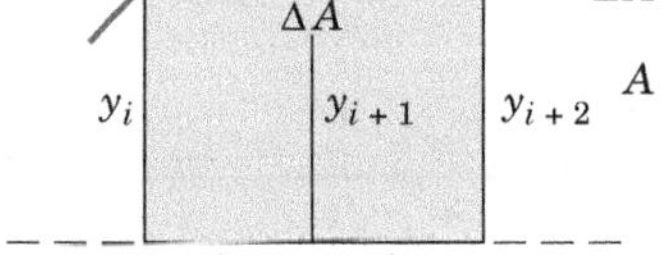

Parabolic

$$\Delta A = \frac{1}{3}(y_i + 4y_{i+1} + y_{i+2}) \Delta x$$

$$A = \int y \, dx \cong \frac{1}{3}(y_0 + 4y_1 + 2y_2 + 4y_3 + 2y_4 + \cdots + 2y_{n-2} + 4y_{n-1} + y_n) \Delta x$$

(d)

III. *Parabolic* [Figure (*d*)] The area between the chord and the curve (neglected in the trapezoidal solution) may be accounted for by approximating the function by a parabola passing through the points defined by three successive values of y. This area may be calculated from the geometry of the parabola and added to the trapezoidal area of the pair of strips to give the area ΔA of the pair as cited. Adding all of the ΔA's produces the tabulation shown, which is known as Simpson's rule. To use Simpson's rule, the number n of strips must be even.

Example: Determine the area under the curve $y = x\sqrt{1 + x^2}$ from $x = 0$ to $x = 2$. (An integrable function is chosen here so that the three approximations can be compared with the exact value, which is $A = \int_0^2 x\sqrt{1 + x^2}\,dx = \frac{1}{3}(1 + x^2)^{3/2}\big|_0^2 = \frac{1}{3}(5\sqrt{5} - 1) = 3.393\ 447$).

NUMBER OF SUBINTERVALS	AREA APPROXIMATIONS		
	RECTANGULAR	TRAPEZOIDAL	PARABOLIC
4	3.361 704	3.456 731	3.392 214
10	3.388 399	3.403 536	3.393 420
50	3.393 245	3.393 850	3.393 447
100	3.393 396	3.393 547	3.393 447
1000	3.393 446	3.393 448	3.393 447
2500	3.393 447	3.393 447	3.393 447

Note that the worst approximation error is less than 2 percent, even with only four strips.

2. Integration of first-order ordinary differential equations. The application of the fundamental principles of mechanics frequently results in differential relationships. Let us consider the first-order form $dy/dt = f(t)$, where the function $f(t)$ may not be readily integrable or may be known only in tabular form. We may numerically integrate by means of a simple slope-projection technique, known as Euler integration, which is illustrated in the figure.

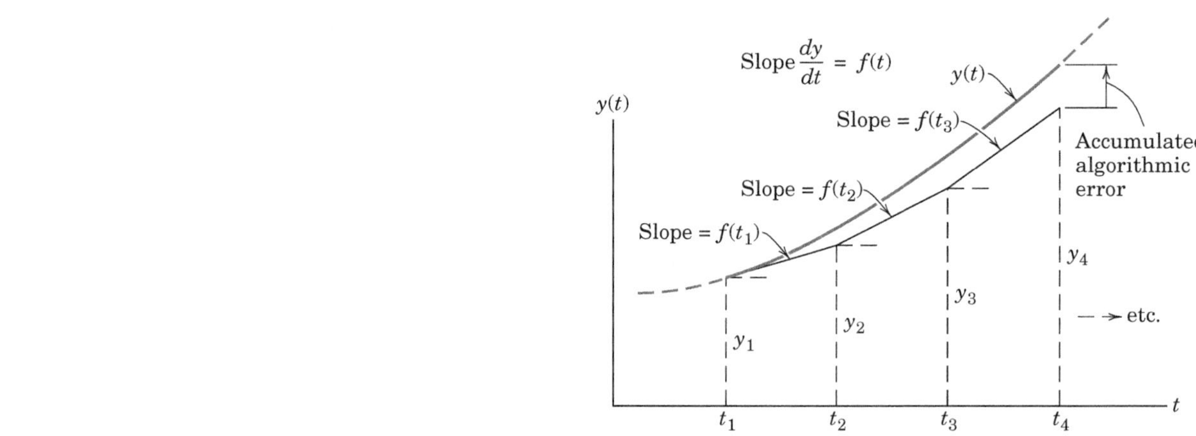

Beginning at t_1, at which the value y_1 is known, we project the slope over a horizontal subinterval or step $(t_2 - t_1)$ and see that $y_2 = y_1 + f(t_1)(t_2 - t_1)$. At t_2, the process may be repeated beginning at y_2, and so forth until the desired value of t is reached. Hence, the general expression is

$$y_{k+1} = y_k + f(t_k)(t_{k+1} - t_k)$$

If y versus t were linear, i.e., if $f(t)$ were constant, the method would be exact, and there would be no need for a numerical approach in that case. Changes in the slope over the subinterval introduce error. For the case shown in the figure, the estimate y_2 is clearly less than the true value of the function $y(t)$ at t_2. More accurate integration techniques (such as Runge-Kutta methods) take into account changes in the slope over the subinterval and thus provide better results.

As with the area-determination techniques, experience is helpful in the selection of a subinterval or step size when dealing with analytical functions. As a rough rule, one begins with a relatively large step size and then steadily decreases the step size until the corresponding changes in the integrated result are much smaller than the desired accuracy. A step size which is too small, however, can result in increased error due to a very large number of computer operations. This type of error is generally known as "round-off error," while the error which results from a large step size is known as algorithm error.

Example: For the differential equation $dy/dt = 5t$ with the initial condition $y = 2$ when $t = 0$, determine the value of y for $t = 4$.

Application of the Euler integration technique yields the following results:

NUMBER OF SUBINTERVALS	STEP SIZE	y at $t = 4$	PERCENT ERROR
10	0.4	38	9.5
100	0.04	41.6	0.95
500	0.008	41.92	0.19
1000	0.004	41.96	0.10

This simple example may be integrated analytically. The result is $y = 42$ (exactly).

D USEFUL TABLES

TABLE D/1 PHYSICAL PROPERTIES

Density **(kg/m^3) and *specific weight* (lb/ft^3)**

	kg/m^3	lb/ft^3		kg/m^3	lb/ft^3
Air*	1.2062	0.07530	Lead	11 370	710
Aluminum	2 690	168	Mercury	13 570	847
Concrete (av.)	2 400	150	Oil (av.)	900	56
Copper	8 910	556	Steel	7 830	489
Earth (wet, av.)	1 760	110	Titanium	4 510	281
(dry, av.)	1 280	80	Water (fresh)	1 000	62.4
Glass	2 590	162	(salt)	1 030	64
Gold	19 300	1205	Wood (soft pine)	480	30
Ice	900	56	(hard oak)	800	50
Iron (cast)	7 210	450			

* At 20°C (68°F) and atmospheric pressure

Coefficients of friction

(The coefficients in the following table represent typical values under normal working conditions. Actual coefficients for a given situation will depend on the exact nature of the contacting surfaces. A variation of 25 to 100 percent or more from these values could be expected in an actual application, depending on prevailing conditions of cleanliness, surface finish, pressure, lubrication, and velocity.)

CONTACTING SURFACE	TYPICAL VALUES OF COEFFICIENT OF FRICTION	
	STATIC, μ_s	KINETIC, μ_k
Steel on steel (dry)	0.6	0.4
Steel on steel (greasy)	0.1	0.05
Teflon on steel	0.04	0.04
Steel on babbitt (dry)	0.4	0.3
Steel on babbitt (greasy)	0.1	0.07
Brass on steel (dry)	0.5	0.4
Brake lining on cast iron	0.4	0.3
Rubber tires on smooth pavement (dry)	0.9	0.8
Wire rope on iron pulley (dry)	0.2	0.15
Hemp rope on metal	0.3	0.2
Metal on ice		0.02

TABLE D/2 SOLAR SYSTEM CONSTANTS

Universal gravitational constant	$G = 6.673(10^{-11})\ m^3/(kg{\cdot}s^2)$
	$= 3.439(10^{-8})\ ft^4/(lbf\text{-}s^4)$
Mass of Earth	$m_e = 5.976(10^{24})\ kg$
	$= 4.095(10^{23})\ lbf\text{-}s^2/ft$
Period of Earth's rotation (1 sidereal day)	= 23 h 56 min 4 s
	= 23.9344 h
Angular velocity of Earth	$\omega = 0.7292(10^{-4})$ rad/s
Mean angular velocity of Earth–Sun line	$\omega' = 0.1991(10^{-6})$ rad/s
Mean velocity of Earth's center about Sun	= 107 200 km/h
	= 66,610 mi/h

BODY	MEAN DISTANCE TO SUN km (mi)	ECCENTRICITY OF ORBIT e	PERIOD OF ORBIT solar days	MEAN DIAMETER km (mi)	MASS RELATIVE TO EARTH	SURFACE GRAVITATIONAL ACCELERATION m/s^2 (ft/s^2)	ESCAPE VELOCITY km/s (mi/s)
Sun	—	—	—	1 392 000 (865 000)	333 000	274 (898)	616 (383)
Moon	384 398[1] (238 854)[1]	0.055	27.32	3 476 (2 160)	0.0123	1.62 (5.32)	2.37 (1.47)
Mercury	57.3×10^6 (35.6×10^6)	0.206	87.97	5 000 (3 100)	0.054	3.47 (11.4)	4.17 (2.59)
Venus	108×10^6 (67.2×10^6)	0.0068	224.70	12 400 (7 700)	0.815	8.44 (27.7)	10.24 (6.36)
Earth	149.6×10^6 (92.96×10^6)	0.0167	365.26	12 742[2] (7 918)[2]	1.000	9.821[3] (32.22)[3]	11.18 (6.95)
Mars	227.9×10^6 (141.6×10^6)	0.093	686.98	6 788 (4 218)	0.107	3.73 (12.3)	5.03 (3.13)
Jupiter[4]	778×10^6 (483×10^6)	0.0489	4333	139 822 (86 884)	317.8	24.79 (81.3)	59.5 (36.8)

[1] Mean distance to Earth (center-to-center)

[2] Diameter of sphere of equal volume, based on a spheroidal Earth with a polar diameter of 12 714 km (7900 mi) and an equatorial diameter of 12 756 km (7926 mi)

[3] For nonrotating spherical Earth, equivalent to absolute value at sea level and latitude 37.5°

[4] Note that Jupiter is not a solid body.

TABLE D/3 PROPERTIES OF PLANE FIGURES

FIGURE	CENTROID	AREA MOMENTS OF INERTIA
Arc Segment 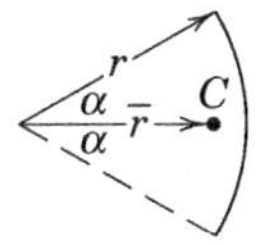	$\bar{r} = \dfrac{r \sin\alpha}{\alpha}$	—
Quarter and Semicircular Arcs 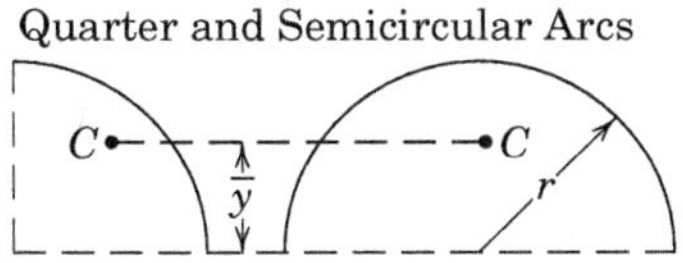	$\bar{y} = \dfrac{2r}{\pi}$	—
Circular Area 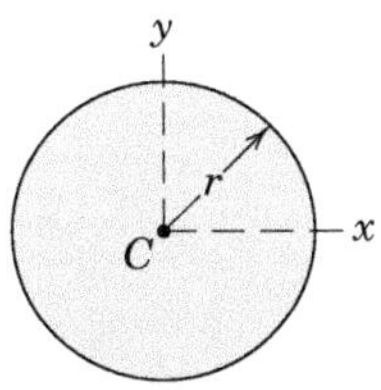	—	$I_x = I_y = \dfrac{\pi r^4}{4}$ $I_z = \dfrac{\pi r^4}{2}$
Semicircular Area 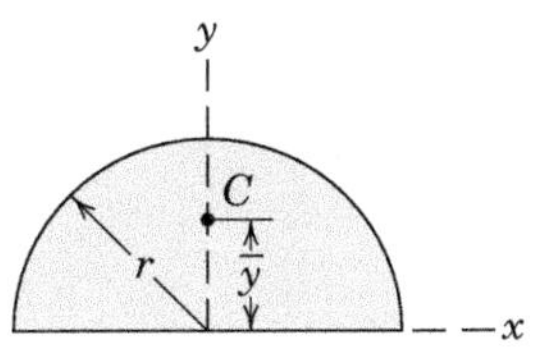	$\bar{y} = \dfrac{4r}{3\pi}$	$I_x = I_y = \dfrac{\pi r^4}{8}$ $\bar{I}_x = \left(\dfrac{\pi}{8} - \dfrac{8}{9\pi}\right) r^4$ $I_z = \dfrac{\pi r^4}{4}$
Quarter-Circular Area 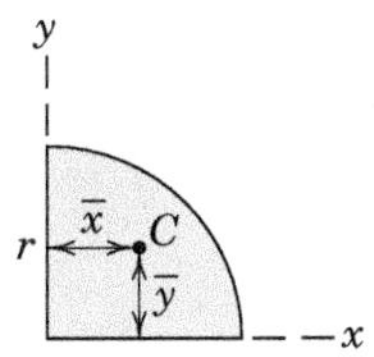	$\bar{x} = \bar{y} = \dfrac{4r}{3\pi}$	$I_x = I_y = \dfrac{\pi r^4}{16}$ $\bar{I}_x = \bar{I}_y = \left(\dfrac{\pi}{16} - \dfrac{4}{9\pi}\right) r^4$ $I_z = \dfrac{\pi r^4}{8}$
Area of Circular Sector 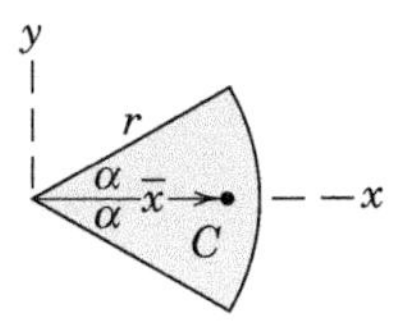	$\bar{x} = \dfrac{2}{3} \dfrac{r \sin\alpha}{\alpha}$	$I_x = \dfrac{r^4}{4}(\alpha - \dfrac{1}{2} \sin 2\alpha)$ $I_y = \dfrac{r^4}{4}(\alpha + \dfrac{1}{2} \sin 2\alpha)$ $I_z = \dfrac{1}{2} r^4 \alpha$

TABLE D/3 PROPERTIES OF PLANE FIGURES ***Continued***

FIGURE	CENTROID	AREA MOMENTS OF INERTIA
Rectangular Area 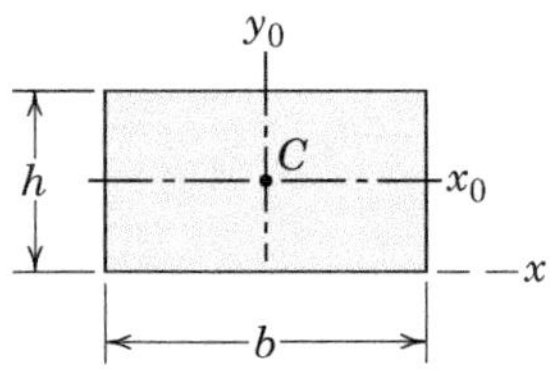	—	$I_x = \frac{bh^3}{3}$ $\bar{I}_x = \frac{bh^3}{12}$ $\bar{I}_z = \frac{bh}{12}(b^2+h^2)$
Triangular Area 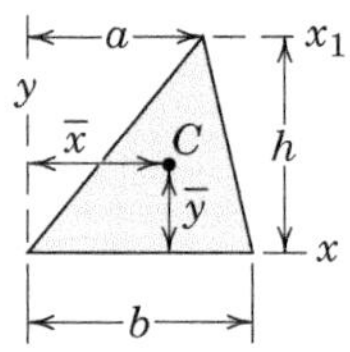	$\bar{x} = \frac{a+b}{3}$ $\bar{y} = \frac{h}{3}$	$I_x = \frac{bh^3}{12}$ $\bar{I}_x = \frac{bh^3}{36}$ $I_{x_1} = \frac{bh^3}{4}$
Area of Elliptical Quadrant 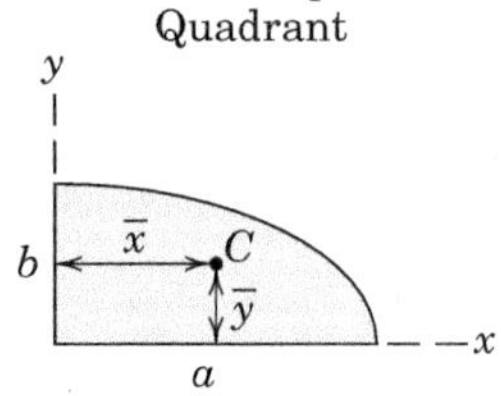	$\bar{x} = \frac{4a}{3\pi}$ $\bar{y} = \frac{4b}{3\pi}$	$I_x = \frac{\pi ab^3}{16}, \quad \bar{I}_x = \left(\frac{\pi}{16} - \frac{4}{9\pi}\right)ab^3$ $I_y = \frac{\pi a^3 b}{16}, \quad \bar{I}_y = \left(\frac{\pi}{16} - \frac{4}{9\pi}\right)a^3 b$ $I_z = \frac{\pi ab}{16}(a^2 + b^2)$
Subparabolic Area Area $A = \frac{ab}{3}$ 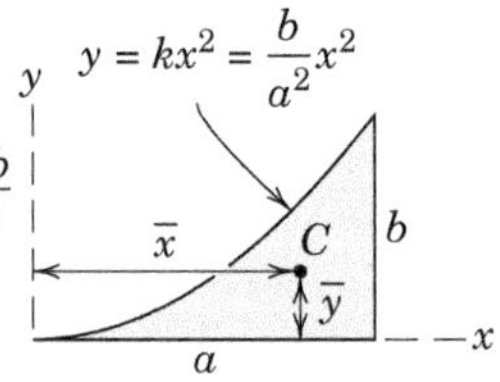	$\bar{x} = \frac{3a}{4}$ $\bar{y} = \frac{3b}{10}$	$I_x = \frac{ab^3}{21}$ $I_y = \frac{a^3 b}{5}$ $I_z = ab\left(\frac{a^2}{5} + \frac{b^2}{21}\right)$
Parabolic Area Area $A = \frac{2ab}{3}$ 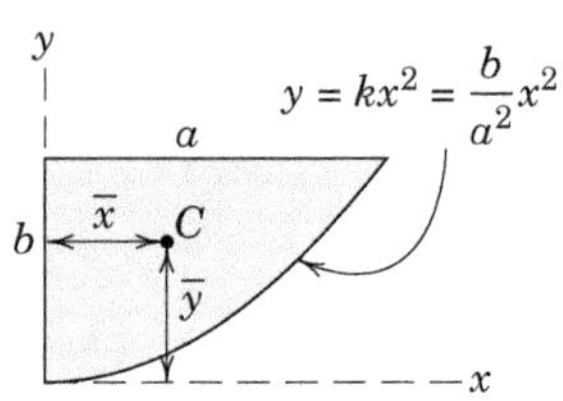	$\bar{x} = \frac{3a}{8}$ $\bar{y} = \frac{3b}{5}$	$I_x = \frac{2ab^3}{7}$ $I_y = \frac{2a^3 b}{15}$ $I_z = 2ab\left(\frac{a^2}{15} + \frac{b^2}{7}\right)$

TABLE D/4 PROPERTIES OF HOMOGENEOUS SOLIDS
(m = mass of body shown)

BODY	MASS CENTER	MASS MOMENTS OF INERTIA
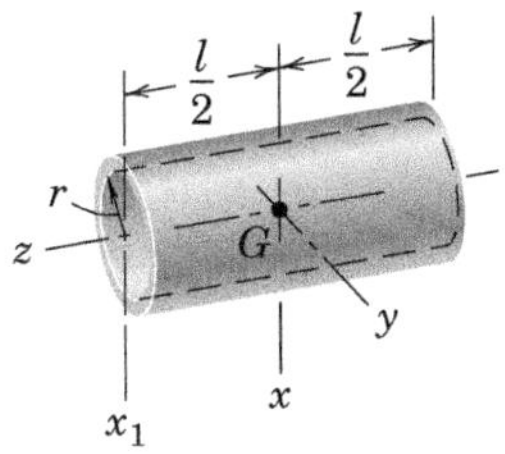 Circular Cylindrical Shell	—	$I_{xx} = \frac{1}{2}mr^2 + \frac{1}{12}ml^2$ $I_{x_1x_1} = \frac{1}{2}mr^2 + \frac{1}{3}ml^2$ $I_{zz} = mr^2$
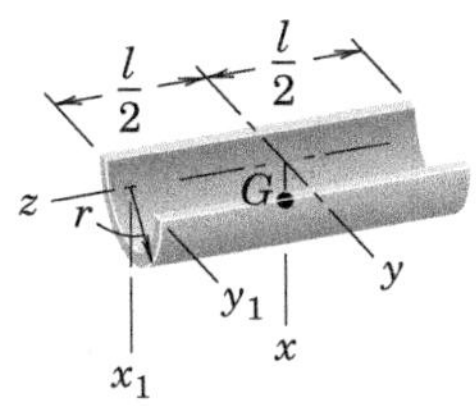 Half Cylindrical Shell	$\bar{x} = \frac{2r}{\pi}$	$I_{xx} = I_{yy}$ $= \frac{1}{2}mr^2 + \frac{1}{12}ml^2$ $I_{x_1x_1} = I_{y_1y_1}$ $= \frac{1}{2}mr^2 + \frac{1}{3}ml^2$ $I_{zz} = mr^2$ $\bar{I}_{zz} = \left(1 - \frac{4}{\pi^2}\right)mr^2$
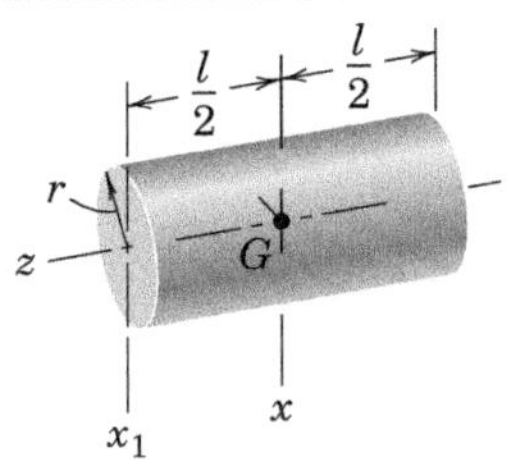 Circular Cylinder	—	$I_{xx} = \frac{1}{4}mr^2 + \frac{1}{12}ml^2$ $I_{x_1x_1} = \frac{1}{4}mr^2 + \frac{1}{3}ml^2$ $I_{zz} = \frac{1}{2}mr^2$
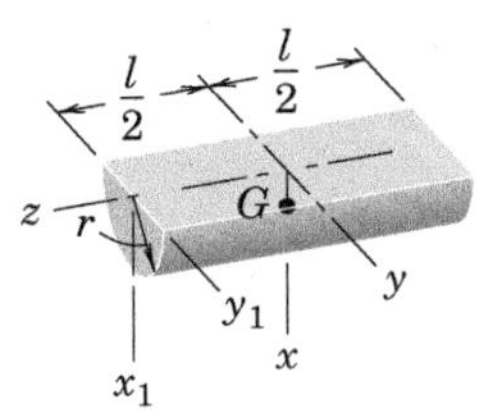 Semicylinder	$\bar{x} = \frac{4r}{3\pi}$	$I_{xx} = I_{yy}$ $= \frac{1}{4}mr^2 + \frac{1}{12}ml^2$ $I_{x_1x_1} = I_{y_1y_1}$ $= \frac{1}{4}mr^2 + \frac{1}{3}ml^2$ $I_{zz} = \frac{1}{2}mr^2$ $\bar{I}_{zz} = \left(\frac{1}{2} - \frac{16}{9\pi^2}\right)mr^2$
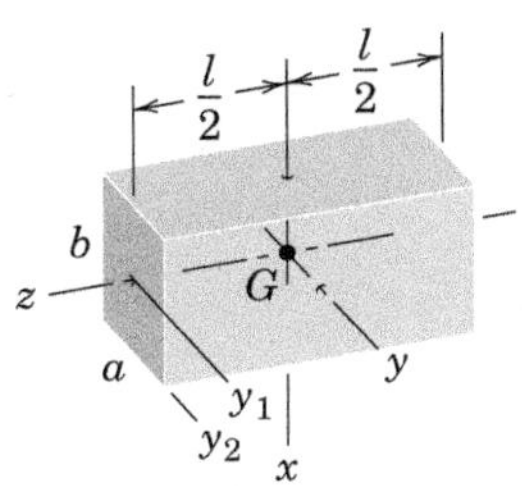 Rectangular Parallelepiped	—	$I_{xx} = \frac{1}{12}m(a^2 + l^2)$ $I_{yy} = \frac{1}{12}m(b^2 + l^2)$ $I_{zz} = \frac{1}{12}m(a^2 + b^2)$ $I_{y_1y_1} = \frac{1}{12}mb^2 + \frac{1}{3}ml^2$ $I_{y_2y_2} = \frac{1}{3}m(b^2 + l^2)$

TABLE D/4 PROPERTIES OF HOMOGENEOUS SOLIDS ***Continued***
(m = mass of body shown)

BODY		MASS CENTER	MASS MOMENTS OF INERTIA
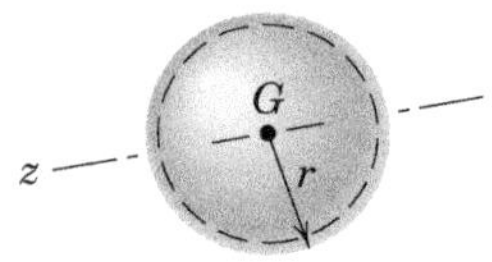	Spherical Shell	—	$I_{zz} = \frac{2}{3}mr^2$
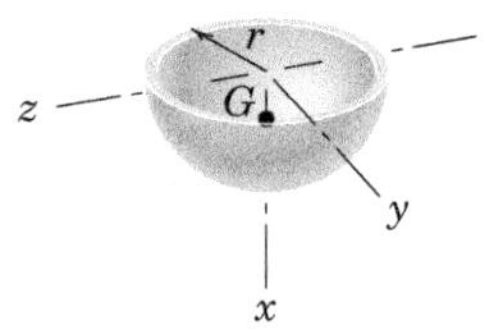	Hemispherical Shell	$\bar{x} = \frac{r}{2}$	$I_{xx} = I_{yy} = I_{zz} = \frac{2}{3}mr^2$ $\bar{I}_{yy} = \bar{I}_{zz} = \frac{5}{12}mr^2$
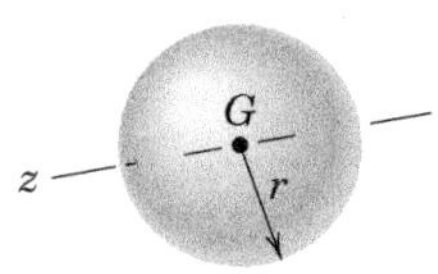	Sphere	—	$I_{zz} = \frac{2}{5}mr^2$
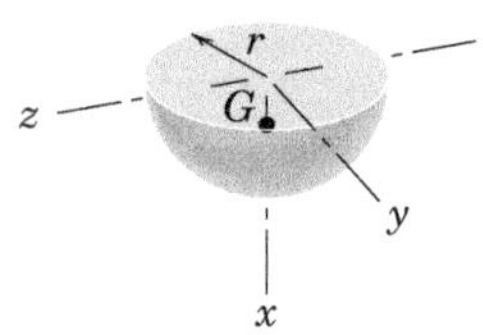	Hemisphere	$\bar{x} = \frac{3r}{8}$	$I_{xx} = I_{yy} = I_{zz} = \frac{2}{5}mr^2$ $\bar{I}_{yy} = \bar{I}_{zz} = \frac{83}{320}mr^2$
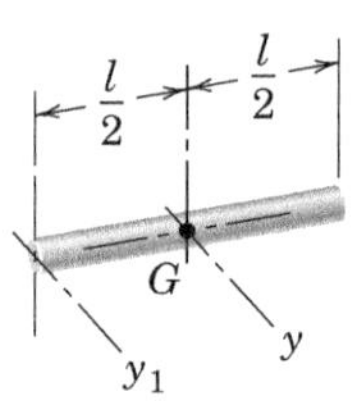	Uniform Slender Rod	—	$I_{yy} = \frac{1}{12}ml^2$ $I_{y_1y_1} = \frac{1}{3}ml^2$

TABLE D/4 PROPERTIES OF HOMOGENEOUS SOLIDS *Continued*
(m = mass of body shown)

BODY	MASS CENTER	MASS MOMENTS OF INERTIA
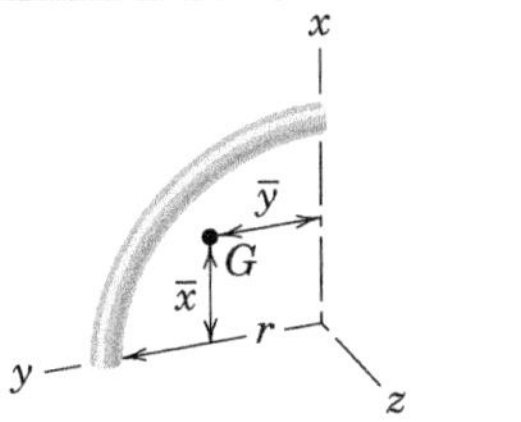 Quarter-Circular Rod	$\bar{x} = \bar{y}$ $= \frac{2r}{\pi}$	$I_{xx} = I_{yy} = \frac{1}{2}mr^2$ $I_{zz} = mr^2$
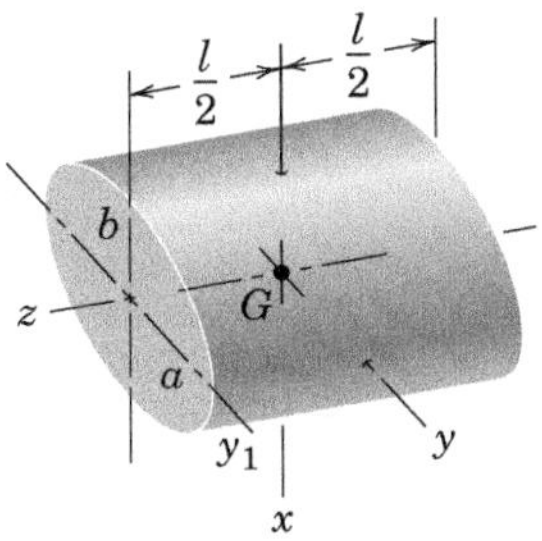 Elliptical Cylinder	—	$I_{xx} = \frac{1}{4}ma^2 + \frac{1}{12}ml^2$ $I_{yy} = \frac{1}{4}mb^2 + \frac{1}{12}ml^2$ $I_{zz} = \frac{1}{4}m(a^2 + b^2)$ $I_{y_1y_1} = \frac{1}{4}mb^2 + \frac{1}{3}ml^2$
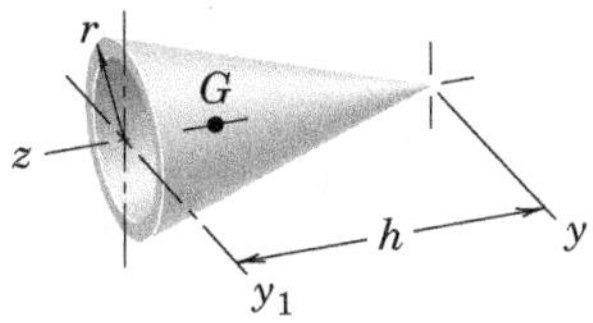 Conical Shell	$\bar{z} = \frac{2h}{3}$	$I_{yy} = \frac{1}{4}mr^2 + \frac{1}{2}mh^2$ $I_{y_1y_1} = \frac{1}{4}mr^2 + \frac{1}{6}mh^2$ $I_{zz} = \frac{1}{2}mr^2$ $\bar{I}_{yy} = \frac{1}{4}mr^2 + \frac{1}{18}mh^2$
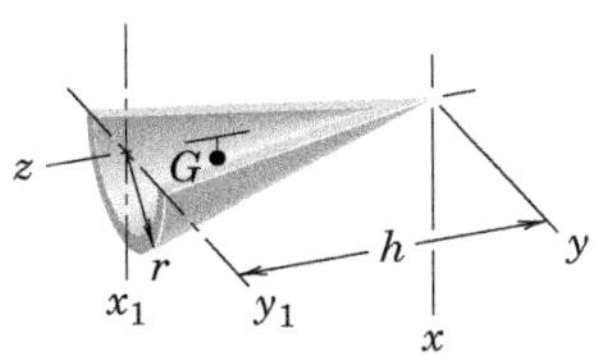 Half Conical Shell	$\bar{x} = \frac{4r}{3\pi}$ $\bar{z} = \frac{2h}{3}$	$I_{xx} = I_{yy}$ $= \frac{1}{4}mr^2 + \frac{1}{2}mh^2$ $I_{x_1x_1} = I_{y_1y_1}$ $= \frac{1}{4}mr^2 + \frac{1}{6}mh^2$ $I_{zz} = \frac{1}{2}mr^2$ $\bar{I}_{zz} = \left(\frac{1}{2} - \frac{16}{9\pi^2}\right)mr^2$
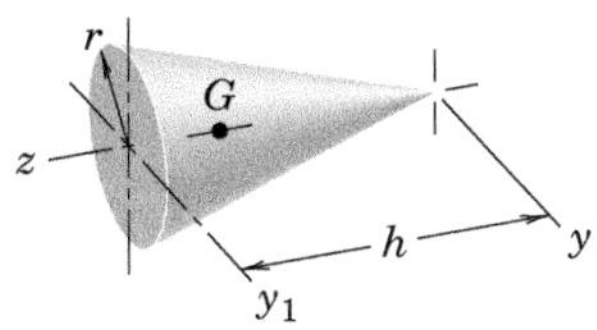 Right-Circular Cone	$\bar{z} = \frac{3h}{4}$	$I_{yy} = \frac{3}{20}mr^2 + \frac{3}{5}mh^2$ $I_{y_1y_1} = \frac{3}{20}mr^2 + \frac{1}{10}mh^2$ $I_{zz} = \frac{3}{10}mr^2$ $\bar{I}_{yy} = \frac{3}{20}mr^2 + \frac{3}{80}mh^2$

TABLE D/4 PROPERTIES OF HOMOGENEOUS SOLIDS *Continued*
(m = mass of body shown)

BODY	MASS CENTER	MASS MOMENTS OF INERTIA
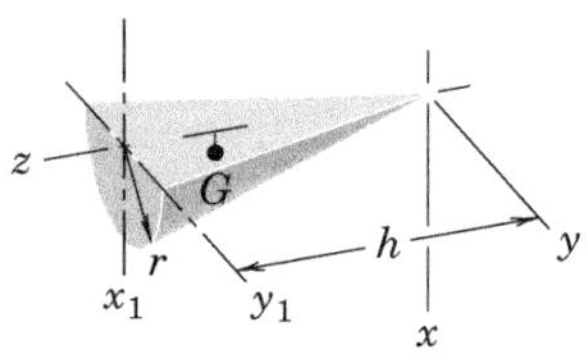 Half Cone	$\bar{x} = \frac{r}{\pi}$ $\bar{z} = \frac{3h}{4}$	$I_{xx} = I_{yy}$ $= \frac{3}{20}mr^2 + \frac{3}{5}mh^2$ $I_{x_1x_1} = I_{y_1y_1}$ $= \frac{3}{20}mr^2 + \frac{1}{10}mh^2$ $I_{zz} = \frac{3}{10}mr^2$ $\bar{I}_{zz} = \left(\frac{3}{10} - \frac{1}{\pi^2}\right)mr^2$
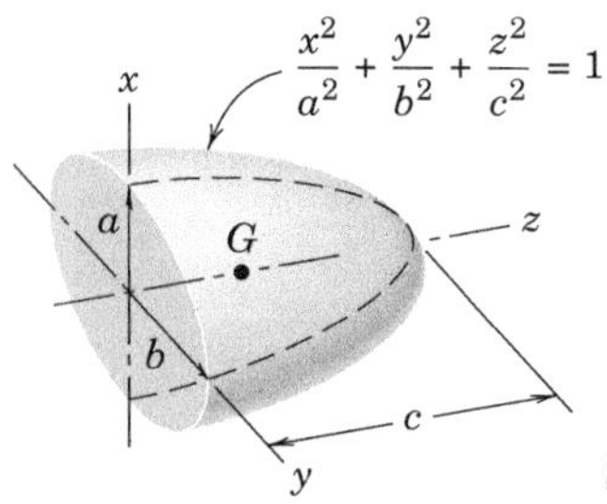 Semiellipsoid	$\bar{z} = \frac{3c}{8}$	$I_{xx} = \frac{1}{5}m(b^2 + c^2)$ $I_{yy} = \frac{1}{5}m(a^2 + c^2)$ $I_{zz} = \frac{1}{5}m(a^2 + b^2)$ $\bar{I}_{xx} = \frac{1}{5}m(b^2 + \frac{19}{64}c^2)$ $\bar{I}_{yy} = \frac{1}{5}m(a^2 + \frac{19}{64}c^2)$
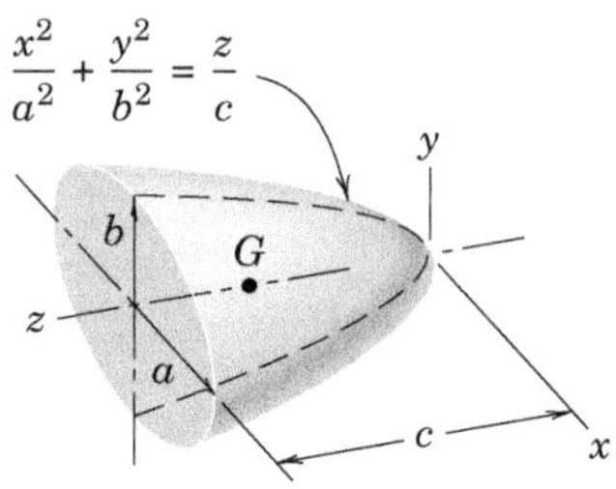 Elliptic Paraboloid	$\bar{z} = \frac{2c}{3}$	$I_{xx} = \frac{1}{6}mb^2 + \frac{1}{2}mc^2$ $I_{yy} = \frac{1}{6}ma^2 + \frac{1}{2}mc^2$ $I_{zz} = \frac{1}{6}m(a^2 + b^2)$ $\bar{I}_{xx} = \frac{1}{6}m(b^2 + \frac{1}{3}c^2)$ $\bar{I}_{yy} = \frac{1}{6}m(a^2 + \frac{1}{3}c^2)$
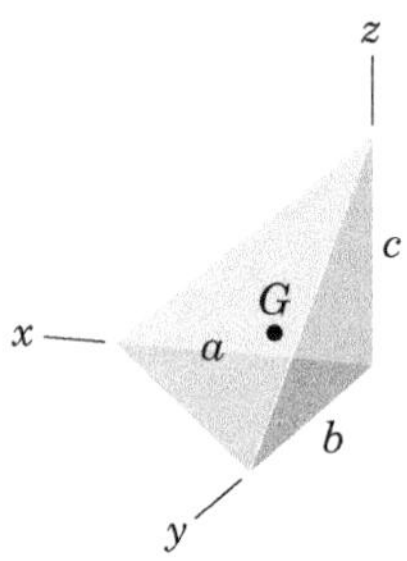 Rectangular Tetrahedron	$\bar{x} = \frac{a}{4}$ $\bar{y} = \frac{b}{4}$ $\bar{z} = \frac{c}{4}$	$I_{xx} = \frac{1}{10}m(b^2 + c^2)$ $I_{yy} = \frac{1}{10}m(a^2 + c^2)$ $I_{zz} = \frac{1}{10}m(a^2 + b^2)$ $\bar{I}_{xx} = \frac{3}{80}m(b^2 + c^2)$ $\bar{I}_{yy} = \frac{3}{80}m(a^2 + c^2)$ $\bar{I}_{zz} = \frac{3}{80}m(a^2 + b^2)$
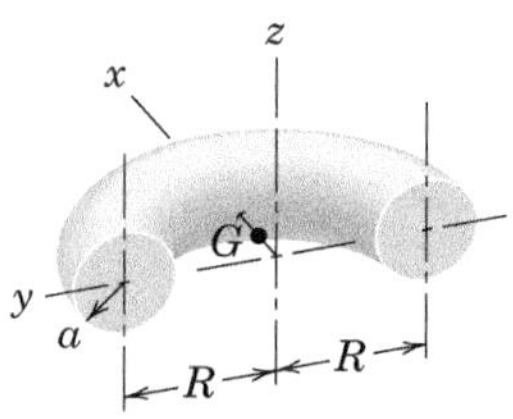 Half Torus	$\bar{x} = \frac{a^2 + 4R^2}{2\pi R}$	$I_{xx} = I_{yy} = \frac{1}{2}mR^2 + \frac{5}{8}ma^2$ $I_{zz} = mR^2 + \frac{3}{4}ma^2$

INDEX

Problem Answers

When a problem asks for both a general and a specific result, only the specific result might be listed below.

▶ Denotes that a problem is of increased difficulty
* Denotes that a problem is best solved utilizing a numerical solution

Chapter 1

1/1 (*a*) 108.7 slugs, (*b*) 15 570 N, (*c*) 1587 kg
1/2 For a 180-lb person: $m = 5.59$ slugs $= 81.6$ kg
$W = 801$ N
1/3 $V_1 + V_2 = 27$, $\mathbf{V}_1 + \mathbf{V}_2 = 6\mathbf{i} + 19.39\mathbf{j}$, $\mathbf{V}_1 - \mathbf{V}_2 = 18\mathbf{i} - 1.392\mathbf{j}$, $\mathbf{V}_1 \times \mathbf{V}_2 = 178.7\mathbf{k}$, $\mathbf{V}_2 \times \mathbf{V}_1 = -178.7\mathbf{k}$, $\mathbf{V}_1 \cdot \mathbf{V}_2 = 21.5$
1/4 $m = 0.01294$ slugs $= 0.1888$ kg, $W = 1.853$ N
1/5 $r = 1.258(10^8)$ km
1/6 $\mathbf{F} = (-2.19\mathbf{i} - 1.535\mathbf{j})10^{-8}$ N
1/7 $h = 0.732R$
1/8 $W_{abs} = 589$ N, $W_{rel} = 588$ N
1/9 $g = 29.9$ ft/sec^2, $W_h = 163.4$ lb
1/10 $h = 1.644(10^5)$ km
1/11 $d = 346\ 000$ km or $432\ 000$ km
1/12 $\theta = 1.770°$
1/13 $R_{es} = 1656$
1/14 $R_A = 2.19$, $R_B = 2.21$
1/15 SI: $[E] = \text{kg} \cdot \text{m}^2/\text{s}$, US: $[E]$ = lb-ft-sec
1/16 $[Q] = \text{ML}^{-1}\,\text{T}^{-2}$

Chapter 2

2/1 $v = -264$ ft/sec
2/2 $t = 2.69$ sec
2/3 $s = 22.2$ m, $v = 15.98$ m/s, $a = 8.99$ m/s^2
2/4 $t = 4.67$ s
2/5 $v = 3 - 10t + t^2$ m/s, $s = -4 + 3t - 5t^2 + \frac{1}{3}t^3$ m
2/6 $k = \frac{1}{6}$ m/s^4, $\Delta s = 54$ m
2/7 $v = \sqrt{{v_0}^2 - \frac{2}{3}k(s^3 - {s_0}^3)}$, $v = 9.67$ m/s
2/8 $s = s_0 + \frac{1}{c_2^2}\left[c_2(v - v_0) + c_1 \ln\left(\frac{c_1 + c_2 v_0}{c_1 + c_2 v}\right)\right]$
$s = s_0 + \frac{c_1 + c_2 v_0}{c_2^2}(e^{c_2 t} - 1) - \frac{c_1}{c_2}t$
2/9 $a = 3.61g$
2/10 $s = \frac{1}{4}(0.2t + 30)^2$ mm, $v = \frac{1}{10}(0.2t + 30)$ mm/s
$a = 0.02$ mm/s^2, For $v = 15$ mm/s: $t = 600$ s
$s = 5620$ mm
2/11 $v_2 = 139.0$ ft/sec, descending
2/12 $a = 1.2$ ft/sec^2
2/13 $t_{AC} = 2.46$ s
2/14 $h = 45.9$ m, $t = 7.70$ s
2/15 $s = 196.0$ ft
2/16 $s = 393$ m, $t = 14.16$ s
2/17 Child B wins by 1.988 ft
2/18 $\Delta a = 0.5$ m/s^2, $\Delta s = 64$ m
2/19 $v = 4.51$ m/s
2/20 $s = 3.26$ m, $t = 3.26$ s

2/21 $a = 16.28$ ft/sec^2, $v_1 = 39.5$ ft/sec, $t_1 = 2.43$ sec
2/22 $t = 59.8$ s
2/23 $v = 75\pi \cos(\pi t)$ mm/s, $a = -75\pi^2 \sin(\pi t)$ mm/s^2
$v_{max} = 75\pi$ or 236 mm/s, $a_{max} = 75\pi^2$ or 740 mm/s^2
2/24 $a = 1.168$ ft/sec^2, $v = 99.8$ mi/hr
2/25 $h = 2.61$ m
2/26 $s = 713$ m
2/27 $d = 21.9$ m
2/28 $v = 7.21$ ft/sec
2/29 $h = 125.4$ m, $t = 157.9$ s
2/30 $s = 12t^2 - \frac{1}{3}t^3 + \frac{10}{3}t^{3/2}$ ft, $s(10) = 972$ ft
2/31 $t = 108.9$ s
2/32 $a = -6.67$ m/s^2, $t \to \infty$
2/33 $\Delta x = 10.5$ in., $x_{total} = 26.5$ in.
2/34 $s = 5810$ ft
2/35 $s = 3710$ ft
2/36 $k_1 = 42.1$ s^{-2}, $k_2 = 42.0$ m^{-2}s^{-2}
2/37 (*a*) $v = 6.67$ mm/s, (*b*) $v = 29.3$ mm/s
(*c*) $v = 46.7$ mm/s
2/38 (*a*) $v = 9190$ ft/sec, (*b*) $v = 8033$ ft/sec
2/39 $v = 2\sqrt{\dfrac{K(L - D/2)}{LD}}$
***2/40** $t = 1.234$ sec
2/41 $t = 50.8$ s
2/42 $y = 8.05$ m, $v_{terminal} = 1.871$ m/s
2/43 $D = \dfrac{1}{2C_2}\ln\left(1 + \dfrac{C_2}{C_1}v_0^2\right)$
2/44 $v_B = 28.3$ ft/sec
2/45 $h = 36.5$ m, $v_f = 24.1$ m/s
2/46 $t_u = 2.63$ s, $t_d = 2.83$ s
2/47 $t_1 = 0.788$ sec, $t_{10} = 0.1279$ sec, $t_{100} = 0.0395$ sec
2/48 $t_1 = 0.795$ sec, $t_{10} = 0.1592$ sec, $t_{100} = 0.1246$ sec
2/49 (*a*) $s = 1206$ m, (*b*) $s = 1268$ m
2/50 $D = \dfrac{\ln 2}{k}$, $t = \dfrac{1}{kv_0}$
2/51 $x = 0.831$ ft
2/52 *A* leads *B* by 706 ft
2/53 $x_f = 61.6$ mm (stretch)
▶2/54 $x_f = -23.2$ mm (compression)
▶2/55 $v = \dfrac{g}{c}(e^{-ct} - 1) + \dfrac{k}{c - b}(e^{-bt} - e^{-ct})$
▶2/56 $t = 103.6$ s, $a_{max} = 3.61$ m/s^2
2/57 $t_1 = 0.900$ sec, $v_1 = 226$ ft/sec up
$t_2 = 14.94$ sec, $v_2 = 226$ ft/sec down
▶2/58 $t_1 = 1.460$ sec, $v_1 = 84.3$ ft/sec up
$t_2 = 5.97$ sec, $v_2 = 63.3$ ft/sec down
2/59 $a_{av} = 2.83$ m/s^2, $\theta = 225°$
2/60 $v_{av} = 20.6$ m/s, $\theta = 76.0°$
2/61 $(x, y) = (2.4, 3.24)$ in.

2/62 $\mathbf{v} = -6\mathbf{i} - 10.15\mathbf{j}$ mm/s, $\mathbf{a} = -10\mathbf{i} - 4.96\mathbf{j}$ mm/s^2
$t = 0.572$ s and 5.69 s
2/63 $v = 3\sqrt{5}$ in./sec, $a = 3$ in./sec^2
2/64 $u = 28.0$ m/s
2/65 $R_{max} = \dfrac{v_0^2}{g}$
2/66 $u = 81.8$ ft/sec
2/67 $v_0 = 45.6$ ft/sec, $\theta = 37.9°$
2/68 $v_0 = 131.0$ ft/sec, $d = 255$ ft, $t = 3.52$ sec
2/69 30.4 ft to the right
2/70 $v_0 = 5.04$ m/s, $\theta = 64.7°$
2/71 $E = \dfrac{mu^2 \sin^2\theta}{eb}$, $s = 2b \cot\theta$
2/72 $s = 3.10$ m, $\theta = 54.1°$
2/73 $L = 928$ ft
2/74 $d = 2.07$ m
2/75 $0.517° \leq \theta \leq 0.524°$
2/76 $\theta = 21.7°$
2/77 $\theta = 5.57°$
2/78 $d = 1.971b$, $0.986\sqrt{bg} < u < 1.394\sqrt{bg}$
2/79 $v = 21.2$ m/s, $s = 3.55$ m
2/80 42.2 ft up in the tree at *B*
2/81 $\theta = 22.0°$, 5.32 yd past pin
2/82 (*a*) $t = 1.443$ s, $h = 0.525$ m, (*b*) $t = 1.704$ s
$h = 0.813$ m
2/83 $d = 107.6$ m, $t_f = 2.92$ s
2/84 $R = 46.4$ m, $\theta = 23.3°$
2/85 $\theta = 2.33°$
2/86 (*a*) $(x_f, y_f) = (140, -31.9)$ ft
(*b*) $(x_f, y_f) = (163.4, -20)$ ft
2/87 $s = 455$ m, $t = 13.35$ s
2/88 $6.15 \leq v_0 \leq 6.68$ m/s
2/89 $d = 0.822$ ft, $h = 8.77$ ft
2/90 $h = 1.227$ m
2/91 $(x_f, y_f) = (32.5, 26.4)$ ft
2/92 $R = \dfrac{2u^2}{g}\tan\theta \sec\theta$
2/93 $\alpha = 50.7°$
2/94 $f_2 = \sqrt{1 - f_1}$, $f_2 = \dfrac{1}{2}$ if $f_1 = \dfrac{3}{4}$
▶2/95 $\theta = \dfrac{90° + \alpha}{2}$, $\alpha = 0: \theta = 45°$
$\alpha = 30°: \theta = 60°$, $\alpha = 45°: \theta = 67.5°$
▶2/96 $v_x = (v_0 \cos\theta)e^{-kt}$, $v_y = \left(v_0 \sin\theta + \dfrac{g}{k}\right)e^{-kt} - \dfrac{g}{k}$

$$x = \frac{v_0 \cos\theta}{k}(1 - e^{-kt})$$

$$y = \frac{1}{k}\left(v_0 \sin\theta + \frac{g}{k}\right)(1 - e^{-kt}) - \frac{g}{k}t$$

As $t \to \infty$, $v_x \to 0$ and $v_y \to -\dfrac{g}{k}$

2/97 $a = 6.77\text{ m/s}^2$
2/98 $a_A = 85.3\text{ m/s}^2, a_B = 31.3\text{ m/s}^2$
2/99 No answer
2/100 $v_A = 11.75\text{ m/s}, v_B = 13.46\text{ m/s}$
2/101 $\rho = 105.8\text{ m}$
2/102 $v = 19.81\text{ m/s}$
2/103 $a_n = 10.31\text{ ft/sec}^2, a_t = 9.09\text{ ft/sec}^2$
2/104 $a = 0.269\text{ m/s}^2$
2/105 $a = 3.26\text{ m/s}^2$ when the sprinter reaches the 60-m mark
2/106 $\rho = 266\text{ m}$
2/107 (a) $a_{av} = 4.94\text{ m/s}^2$, 1.138% difference
(b) $a_{av} = 4.99\text{ m/s}^2$, 0.285% difference
(c) $a_{av} = 4.998\text{ m/s}^2$, 0.0317% difference
$a_n = 5\text{ m/s}^2$ in each case
2/108 $a_n = 36.7\text{ ft/sec}^2, a_t = 20\text{ ft/sec}^2$
2/109 (a) $\mathbf{a} = -60\mathbf{i}\text{ m/s}^2$, (b) $\mathbf{a} = 52.0\mathbf{i} - 30\mathbf{j}\text{ m/s}^2$
2/110 (a) $\mathbf{a} = -31.9\mathbf{i} + 47.0\mathbf{j}\text{ m/s}^2$
(b) $\mathbf{a} = 49.6\mathbf{i} - 16.08\mathbf{j}\text{ m/s}^2$
2/111 $\rho_B = 163.0\text{ m}$
2/112 $a = 2.71\text{ m/s}^2$
2/113 $v = 356\text{ m/s}, a = 0.0260\text{ m/s}^2$
2/114 $v = 20\text{ m/s}$ or 72 km/h
2/115 $\rho = 1709\text{ m}$
2/116 Launch: $a_t = -6.69\text{ ft/sec}^2, \rho = 1859\text{ ft}$
Apex: $a_t = 0, \rho = 1740\text{ ft}$
***2/117** $t = 0.452$ and 2.67 sec
2/118 $N = 3.36\text{ rev/min}$
2/119 $v = 41.3\text{ km/s}$ or $148.9(10^3)\text{ km/h}$
2/120 Car B crosses first, $\delta = 25.9\text{ ft}$
2/121 $a = 11.47\text{ in./sec}^2$
2/122 $t = 1\text{ sec}: \dot{v} = -6.58\text{ ft/sec}^2, \rho = 142.2\text{ ft}$
$t = 2\text{ sec}: \dot{v} = 8.75\text{ ft/sec}^2, \rho = 149.7\text{ ft}$
2/123 $\rho = 2.60\text{ mm}$
2/124 (a) $a = 19.62\text{ m/s}^2, \theta_x = 0$
(b) $a = 38.9\text{ m/s}^2, \theta_x = -59.7°$
(c) $a = 97.3\text{ m/s}^2, \theta_x = -168.4°$
2/125 $a_n = 66.0\text{ mm/s}^2, a_t = 29.7\text{ mm/s}^2$
2/126 $\rho = 18\,480\text{ km}$
2/127 $a_n = 28.8\text{ in./sec}^2, a_t = 34.6\text{ in./sec}^2, \rho = 19.06\text{ in.}$
►2/128 $L = 46.1\text{ m}$
►2/129 $(x_C, y_C) = (22.5, -22.9)\text{ m}$
***2/130** $(a_n)_{max} = 11.01\text{ m/s}^2$ at $t = 9.62\text{ s}$
$|a_t|_{max} = 9.72\text{ m/s}^2$ at $t = 0$
$\rho_{min} = 288\text{ ft}$ at $t = 9.62\text{ s}$
2/131 $\dot{r} = 47.7\text{ ft/sec}, \dot{\theta} = -41.0\text{ deg/sec}$
2/132 $v = 10.91\text{ m/s}, \dot{r} = 9.76\text{ m/s}$
2/133 No answer
2/134 $v = 50\text{ mm/s}, a = 5\text{ mm/s}^2$
2/135 $\dot{r} = 15\text{ mm/s}, \dot{\theta} = 0.1\text{ rad/s}$
2/136 $\ddot{r} = -1\text{ mm/s}^2, \ddot{\theta} = -0.035\text{ rad/s}^2$
2/137 $\mathbf{v} = 1.5\mathbf{e}_r + 2.71\mathbf{e}_\theta\text{ ft/sec}$
$\mathbf{a} = -4.24\mathbf{e}_r + 1.344\mathbf{e}_\theta\text{ ft/sec}^2$
2/138 No answer
2/139 $\mathbf{v} = 6\mathbf{e}_r + 12.57\mathbf{e}_\theta\text{ in./sec}$
$\mathbf{a} = -4.19\mathbf{e}_r + 2.09\mathbf{e}_\theta\text{ in./sec}^2$
2/140 (a) $\mathbf{v}_A = v\mathbf{e}_r + l\Omega\mathbf{e}_\theta, \mathbf{a}_A = -l\Omega^2\mathbf{e}_r + 2v\Omega\mathbf{e}_\theta$
(b) $\mathbf{v}_B = 4v\mathbf{e}_r + 2l\Omega\mathbf{e}_\theta, \mathbf{a}_B = -2l\Omega^2\mathbf{e}_r + 8v\Omega\mathbf{e}_\theta$
2/141 $a = 2K^2\sqrt{R^2 - r_0^2}$
2/142 $\dot{r} = -14.17\text{ m/s}, \ddot{r} = 7.71\text{ m/s}^2$
$\dot{\theta} = 0.388\text{ rad/s}, \ddot{\theta} = 0.222\text{ rad/s}^2$
2/143 $v = 0.377\text{ m/s}, a = 0.272\text{ m/s}^2, \alpha = 19.44°$
2/144 $v_A' = 69.6\text{ mi/hr}, v_B' = 68.4\text{ mi/hr}$
2/145 $\dot{r} = 35.9\text{ ft/sec}, \dot{\theta} = 0.1834\text{ rad/sec}$
2/146 $\ddot{r} = -1.234\text{ ft/sec}^2, \ddot{\theta} = -0.1072\text{ rad/sec}^2$
2/147 $v = 360\text{ m/s}, a = 20.1\text{ m/s}^2$
2/148 $\dot{r} = -96.6\text{ m/s}, \dot{\theta} = 0.229\text{ rad/s}$
$\ddot{r} = 5.92\text{ m/s}^2, \ddot{\theta} = 0.391\text{ rad/s}^2$
2/149 $\dot{r} = -96.6\text{ m/s}, \dot{\theta} = 0.229\text{ rad/s}$
$\ddot{r} = 25.2\text{ m/s}^2, \ddot{\theta} = 0.345\text{ rad/s}^2$
***2/150** At entry: $\dot{\theta} = -2.13\text{ rad/s}, \ddot{\theta} = 1.565\text{ rad/s}^2$
$|\dot{\theta}|_{max} = 2.19\text{ rad/s}$ at $t = 1.633\text{ s}$
$|\ddot{\theta}|_{max} = 3.34\text{ rad/s}^2$ at $t = 1.373\text{ s}$
2/151 $v = 161.7\text{ ft/sec}, \dot{\theta} = -0.0808\text{ rad/sec}$
2/152 $a = 8.62\text{ ft/sec}^2, \ddot{\theta} = 0.01832\text{ rad/sec}^2$
2/153 $\ddot{r} = 12.15\text{ m/s}^2, \ddot{\theta} = 0.0365\text{ rad/s}^2$
2/154 $v_A = 1.190\text{ m/s}$ at $\theta_x = 125.2°$
$a_A = 7.54\text{ m/s}^2$ at $\theta_x = 225°$
2/155 $v = 25,470\text{ ft/sec}$
2/156 $\dot{r} = 17.68\text{ m/s}, \ddot{r} = 0.428\text{ m/s}^2$
$\dot{\theta} = -0.0442\text{ rad/s}, \ddot{\theta} = 0.00479\text{ rad/s}^2$
2/157 $\dot{r} = \dfrac{\frac{1}{2}at(2R\cos\alpha + at^2)}{\sqrt{R^2 + Rat^2\cos\alpha + \frac{1}{4}a^2t^4}}$
2/158 $\dot{\theta} = \dfrac{Rat\sin\alpha}{R^2 + Rat^2\cos\alpha + \frac{1}{4}a^2t^4}$
2/159 $\dot{r} = 8910\text{ ft/sec}, \ddot{r} = -1.790\text{ ft/sec}^2$
$\dot{\theta} = 3.48(10^{-4})\text{ rad/sec}, \ddot{\theta} = -1.398(10^{-7})\text{ rad/sec}^2$
2/160 (a) $v = 37.7\text{ km/s}, \beta = 32.0°$
(b) $v = 37.7\text{ km/s}, \beta = 17.01°$
2/161 $r = 21\,900\text{ m}, \dot{r} = -73.0\text{ m/s}, \ddot{r} = -2.07\text{ m/s}^2$
$\theta = 43.2°, \dot{\theta} = 0.00312\text{ rad/s}, \ddot{\theta} = -9.01(10^{-5})\text{ rad/s}^2$
►2/162 $r = 51.0\text{ ft}, \dot{r} = 91.4\text{ ft/sec}, \ddot{r} = -11.35\text{ ft/sec}^2$
$\theta = 31.9°, \dot{\theta} = -0.334\text{ rad/sec}, \ddot{\theta} = 0.660\text{ rad/sec}^2$
►2/163 $v = 50.2\text{ m/s}, \dot{v} = 6.01\text{ m/s}^2, \rho = 50.5\text{ m}, \beta = 12.00°$
►2/164 $r = 167.5\text{ ft}, \dot{r} = 151.0\text{ ft/sec}, \ddot{r} = -14.96\text{ ft/sec}^2$
$\theta = 23.0°, \dot{\theta} = -0.0880\text{ rad/sec}$
$\ddot{\theta} = -0.00897\text{ rad/sec}^2$
2/165 $\theta_1 = 60.8°, \theta_2 = 122.4°$

2/166 $x = -2050$ ft, $y = 5640$ ft, $z = 3950$ ft
$v_x = -102.6$ ft/sec, $v_y = 282$ ft/sec, $v_z = -124.4$ ft/sec
$a_x = 0, a_y = 0, a_z = -32.2$ ft/sec^2
2/167 $(x_f, y_f) = (390, 132.4)$ m
2/168 $\theta = 22.4°$ and $\phi = 21.5°$ or $\theta = 29.5°$ and $\phi = 62.5°$
2/169 $a = 27.5$ m/s^2
2/170 $v_\theta = -u \sin\theta, v_R = u\cos\theta\cos\phi, v_\phi = -u\cos\theta\sin\phi$
2/171 $a_{max} = \sqrt{r^2\omega^4 + 16n^4\pi^4 z_0^2}$
2/172 $\dot{R} = 12.26$ m/s, $\dot{\theta} = 0.1234$ rad/s, $\dot{\phi} = 0.0281$ rad/s
2/173 $\ddot{R} = 5.69$ m/s^2, $\ddot{\theta} = 9.52(10^{-4})$ rad/s^2
$\ddot{\phi} = -4.61(10^{-3})$ rad/s^2
2/174 $v_P = \sqrt{\dot{l}^2 + (l_0 + l)^2\omega^2 + \dot{h}^2}$
$a_P = \sqrt{(\ddot{l} - (l_0 + l)\omega^2)^2 + 4\dot{l}^2\omega^2 + \ddot{h}^2}$
2/175 $a = 219$ mm/s^2
2/176 $a_r = -19.82$ ft/sec^2, $a_\theta = -2.91$ ft/sec^2
$a_z = -0.386$ ft/sec^2
2/177 $v_B = 2.96$ m/s, $a_B = 0.672$ m/s^2
2/178 $a_P = 17.66$ m/s^2
►**2/179** $a_R = \ddot{R} - R\dot{\phi}^2 - R\dot{\theta}^2\cos^2\phi$
$a_\theta = \dfrac{\cos\phi}{R}\dfrac{d}{dt}(R^2\dot{\theta}) - 2R\dot{\theta}\dot{\phi}\sin\phi$
$a_\phi = \dfrac{1}{R}\dfrac{d}{dt}(R^2\dot{\phi}) + R\dot{\theta}^2\sin\phi\cos\phi$
2/180 $v_R = 0, v_\theta = R\omega\sqrt{1 - \left(\dfrac{h}{2R}\right)^2}, v_\phi = \dfrac{h\omega}{\sqrt{1 - \left(\dfrac{h}{2R}\right)^2}}$
►**2/181** $a_r = b\dot{\theta}^2(\tan^2\gamma\sin^2\beta - 1)e^{-\theta\tan\gamma\sin\beta}$
with $\beta = \tan^{-1}\left(\dfrac{b}{h}\right)$
►**2/182** $a_P = 0.904$ m/s^2
2/183 $\mathbf{v}_{B/A} = 120\mathbf{i}$ km/h, $\mathbf{a}_{B/A} = -2\mathbf{i}$ m/s^2
2/184 $v_B = 30.4$ mi/hr
2/185 $v_A = 1200$ km/h, $v_{A/B} = 1039$ km/h
2/186 $\mathbf{v}_{H/W} = -0.298\mathbf{i} + 1.928\mathbf{j}$ m/s
2/187 $\beta = 281°, t = 1.527$ hr
2/188 $\mathbf{v}_{A/B} = 70.9\mathbf{i} - 46.9\mathbf{j}$ km/h, $\mathbf{a}_{A/B} = 1.5\mathbf{i} + 2.60\mathbf{j}$ m/s^2
2/189 $\mathbf{v}_{A/B} = 3.00\mathbf{i} + 1.999\mathbf{j}$ m/s
$\mathbf{a}_{A/B} = 3.63\mathbf{i} + 0.628\mathbf{j}$ m/s^2
2/190 $a_A = 10.27$ ft/sec^2, $\dot{v}_B = 7.26$ ft/sec^2
2/191 $\theta = 28.7°$ below normal
2/192 At C: $\mathbf{v}_{A/B} = -597\mathbf{i} - 142.5\mathbf{j}$ km/h
At E: $\mathbf{v}_{A/B} = -492\mathbf{i} + 247\mathbf{j}$ km/h
2/193 At C: $\mathbf{v}_{A/B} = -614\mathbf{i} - 174.0\mathbf{j}$ km/h
$\mathbf{a}_{A/B} = -0.717\mathbf{i} - 1.280\mathbf{j}$ m/s^2
At E: $\mathbf{v}_{A/B} = -481\mathbf{i} + 359\mathbf{j}$ km/h
$\mathbf{a}_{A/B} = 38.6\mathbf{i} + 24.4\mathbf{j}$ m/s^2
2/194 $v_W = 14.40$ knots
2/195 $\dot{r} = -15.43$ m/s, $\dot{\theta} = 0.01446$ rad/s
2/196 $\ddot{r} = -1.668$ m/s^2, $\ddot{\theta} = 0.0352$ rad/s^2
2/197 $v_{B/A} = 9.55$ mi/hr
2/198 If $v_B = 30$ km/h: $\alpha = 31.3°$ or $74.3°$
If $v_B = 0$: $\alpha = 34.0°$ or $65.4°$
2/199 $\mathbf{a}_{B/A} = 0.733\mathbf{i} + 29.2\mathbf{j}$ ft/sec^2
2/200 $\alpha = 33.3°$, $\mathbf{v}_{A/B} = 73.1\mathbf{i} + 73.1\mathbf{j}$ ft/sec
2/201 $v_{A/B} = 36.0$ m/s, $\dot{r} = -15.71$ m/s, $\dot{\theta} = 0.1079$ rad/s
2/202 $\ddot{r} = 2.57$ m/s^2, $\ddot{\theta} = 0.01130$ rad/s^2
2/203 $\mathbf{v}_{A/B} = 71.5\mathbf{i} - 47.4\mathbf{j}$ ft/sec
2/204 $\mathbf{v}_{Ball/B} = -77.5\mathbf{i} - 50.8\mathbf{j}$ ft/sec
►**2/205** $\ddot{r} = -0.637$ m/s^2, $\ddot{\theta} = 0.1660(10^{-3})$ rad/s^2
►**2/206** *(a)* $\mathbf{v}_{A/B} = 50\mathbf{i} + 50\mathbf{j}$ m/s, $\mathbf{a}_{A/B} = 1.25\mathbf{j}$ m/s^2
(b) $\dot{v}_r = 0.884$ m/s^2, $\rho_r = 5660$ m
2/207 $a_B = 0.022$ m/s^2 down
2/208 $\mathbf{v}_A = -0.5\mathbf{j}$ m/s, $\mathbf{v}_B = 3\mathbf{j}$ m/s
2/209 $v_A = 1.8$ m/s up, $a_A = 3$ m/s^2 down
2/210 $v_A = 1.333$ ft/sec up incline
2/211 $t = 20$ sec
2/212 $2a_A + 2a_B + a_C = 0$, 2 degrees of freedom
2/213 $v_A = \dfrac{2\sqrt{x^2 + h^2}}{x}v_B$
2/214 $v_B = -\dfrac{3y}{2\sqrt{y^2 + b^2}}v_A$
2/215 $v_{B/A} = 1$ ft/sec, $a_{B/A} = 2$ ft/sec^2, $v_C = 4$ ft/sec
(all directed up incline)
2/216 $4v_A + 8v_B + 4v_C + v_D = 0$, 3 degrees of freedom
2/217 $a_x = -\dfrac{L^2 v_A^2}{(L^2 - y^2)^{3/2}}$
2/218 $v_A = 2.76$ m/s
2/219 $v_A = \dfrac{2y}{\sqrt{y^2 + b^2}}v_B$
2/220 $v_B = -\dfrac{3x_A v_A}{2\sqrt{x_A^2 + h^2}}$
$a_B = -\dfrac{3(v_A^2 + x_A a_A)}{2\sqrt{x_A^2 + h^2}} + \dfrac{3x_A^2 v_A^2}{2(x_A^2 + h^2)^{3/2}}$
(a minus value indicates a leftward direction)
2/221 $h = 300$ mm
2/222 $v = \dfrac{\dot{l}\sqrt{4y^2 + b^2}}{16y}, a = \dfrac{b^2\dot{l}^2}{256y^3}$
2/223 $v_B = \dfrac{1}{2^n}v_A$
2/224 $(v_A)_y = \dfrac{l\sqrt{2(1 + \cos\theta)}}{b\tan\theta}v_B$
2/225 $v_B = 62.9$ mm/s up
►**2/226** $a_B = 11.93$ mm/s^2 up
2/227 $v_B = 40.9$ mm/s
►**2/228** $\dot{v}_B = -15.98$ mm/s^2
2/229 $v_0 = 13.90$ ft/sec
2/230 $v = -7.27$ m/s

2/231 $t = 8.38$ s
2/232 From A to B: $a = 0$
From B to C: $a = a_n = 17.42$ ft/sec^2
2/233 $v_Q = 15$ ft/sec, $v_B = 61.8$ ft/sec
2/234 $\mathbf{v}_{B/A} = 355\mathbf{i} - 15.43\mathbf{j}$ mi/hr
$\mathbf{a}_{B/A} = 74.0\mathbf{i} + 70.5\mathbf{j}$ ft/sec^2
2/235 $t_1 = 2.27$ sec, $t_2 = 8.48$ sec
2/236 (*a*) $\mathbf{v}_{W/B} = -15.61\mathbf{i} - 0.928\mathbf{j}$ mi/hr
(*b*) $v_B = 12.86$ mi/hr, $v_{W/B} = 15.32$ mi/hr
2/237 $\dot{\theta} = 13.86$ rad/sec, $\ddot{\theta} = 215$ rad/sec^2
2/238 $t = 2.32$ s, $v_A = 1.439$ m/s down
2/239 $v = 414$ km/h
2/240 $a_r = 5\sqrt{3}$ ft/sec^2, $a_y = -5\sqrt{3}$ ft/sec^2, $a_t = 0$
$a_n = 10\sqrt{3}$ ft/sec^2, $\rho = \dfrac{6\sqrt{3}}{5}$ ft
2/241 $v = 7.51$ mm/s, $\mathbf{v} = 7.37\mathbf{i} - 1.470\mathbf{j}$ mm/s
$a = 7.24$ mm/s^2, $\mathbf{a} = -5.28\mathbf{i} - 4.96\mathbf{j}$ mm/s^2
$\mathbf{e}_t = 0.981\mathbf{i} - 0.1957\mathbf{j}$, $\mathbf{e}_n = -0.1957\mathbf{i} - 0.981\mathbf{j}$
$a_t = -4.20$ mm/s^2, $\mathbf{a}_t = -4.12\mathbf{i} + 0.822\mathbf{j}$ mm/s^2
$a_n = 5.90$ mm/s^2, $\mathbf{a}_n = -1.154\mathbf{i} - 5.78\mathbf{j}$ mm/s^2
$\rho = 9.57$ mm, $\dot{\beta} = 0.785$ rad/s
2/242 $v = 7.51$ mm/s, $\mathbf{v} = 7.37\mathbf{i} - 1.470\mathbf{j}$ mm/s
$a = 7.24$ mm/s^2, $\mathbf{a} = -5.28\mathbf{i} - 4.96\mathbf{j}$ mm/s^2
$\mathbf{e}_r = 0.795\mathbf{i} + 0.607\mathbf{j}$, $\mathbf{e}_\theta = -0.607\mathbf{i} + 0.795\mathbf{j}$
$v_r = 4.96$ mm/s, $\mathbf{v}_r = 3.95\mathbf{i} + 3.01\mathbf{j}$ mm/s
$v_\theta = -5.64$ mm/s, $\mathbf{v}_\theta = 3.42\mathbf{i} - 4.48\mathbf{j}$ mm/s
$a_r = -7.20$ mm/s^2, $\mathbf{a}_r = -5.73\mathbf{i} - 4.37\mathbf{j}$ mm/s^2
$a_\theta = -0.743$ mm/s^2, $\mathbf{a}_\theta = 0.451\mathbf{i} - 0.591\mathbf{j}$ mm/s^2
$r = 45.7$ mm, $\dot{r} = 4.96$ mm/s, $\ddot{r} = -6.51$ mm/s^2
$\theta = 37.3°$, $\dot{\theta} = -0.1233$ rad/s, $\ddot{\theta} = 0.01052$ rad/s^2
2/243 $\beta = 12.09°$
2/244 $(x, y, z) = (-1255, 1193, 0)$ ft
2/245 (*a*) $v_B = 0.293$ m/s, (*b*) $v_B = 0$, (c) $v_B = -0.250$ m/s
2/246 $h = 6048$ km
▶**2/247** (*a*) $a = b\sqrt{K^4 + \omega^4\theta_0^{\,2}\cos^2\phi}$
(*b*) $a = bK\sqrt{K^2 + 4\omega^2\theta_0^{\,2}}$
***2/248** At $t = 9$ s: $v_r = 90.6$ m/s, $v_\theta = -42.8$ m/s
$a_r = -2.39$ m/s^2, $a_\theta = -9.51$ m/s^2
***2/249** $t' = 0.349$ s
***2/250** $k = 0.00323$ ft^{-1}, $v_t = 99.8$ ft/sec, $v' = 113.5$ ft/sec
***2/251** $v_{1\text{ mi}} = 11.66$ knots, $v_{max} = 14.49$ knots
***2/252** $\theta_{max} = 110.4°$ at $t = 0.802$ sec
$\dot{\theta}_{max} = 3.79$ rad/sec at $t = 0.324$ sec
$\theta = 90°$ at $t = 0.526$ sec
***2/253** $\alpha = 48.5°$, $R = 81.1$ m
***2/254** $(v_{A/B})_{max} = 70$ m/s at $t = 47.1$ s and $s_B = 1264$ m
$(v_{A/B})_{min} = 10$ m/s at $t = 23.6$ s and $s_B = 557$ m
$(a_{A/B})_{max} = 6.12$ m/s^2 at $t = 0$ and $s_B = 0$
$(a_{A/B})_{min} = 2.52$ m/s^2 at $t = 10$ s and $s_B = 150$ m
***2/255** With drag: $R = 202$ m, Without drag: $R = 405$ m
***2/256** $a_{min} = 9.03$ m/s^2 at $\theta = 44.3°$ and $t = 0.237$ s
$a_{max} = 10.76$ m/s^2 at $\theta = 0$ and $t = 0$

Chapter 3

3/1 $t = 2.04$ s, $x = 8.15$ m
3/2 (*a*) $a = 1.118$ m/s^2 down incline, (*b*) $a = 0$
(*c*) $a = 2.04$ m/s^2 up incline
3/3 $t = 1.615$ sec, $d = 24.2$ ft, $d' = 18.17$ ft
3/4 $a = 4.46$ m/s^2 up incline
3/5 $T = 13.33$ kN, $a = 0.667$ m/s^2
3/6 $R = 846$ N, $L = 110.4$ N
3/7 $F = 2890$ N
3/8 $\mu_k = 0.0395$
3/9 (*a*) $a = 0.01990g$ up incline
(*b*) $a = 0.0498g$ down incline
3/10 $\alpha = \tan^{-1}\left[\dfrac{P}{(M + m)\,g\cos\theta}\right]$, For $P = 0$: $\alpha = 0$
3/11 Uphill: $s = 807$ m, Downhill: $s = 751$ m
3/12 $N_A = m\left[g + \dfrac{P}{\sqrt{3}(M + m)}\right]$
$N_B = m\left[g - \dfrac{P}{\sqrt{3}(M + m)}\right]$
3/13 $P = 1.673(M + m)g$
3/14 $\mathbf{v}_{A/B} = -87.7\mathbf{i} - 11\mathbf{j}$ ft/sec, $\mathbf{a}_{A/B} = -6.87\mathbf{i}$ ft/sec^2
3/15 $a = 0.098$ ft/sec^2, $R = 4870$ lb
3/16 $P = 43.8$ lb
3/17 $t = 6.16$ years, $s = 2.84(10^9)$ km
3/18 $P = 0.4$ lb, Coupler 1 fails first
3/19 (*a*) $a = 0.457$ m/s^2 down incline
(*b*) $a = 0.508$ m/s^2 up incline
3/20 (*a*) $a = 0$, (*b*) $a = 1.513$ m/s^2 up incline
3/21 $s = 1.710$ m

3/22 (*a*) $a = 0.0348g$ down incline
(*b*) $a = 0.0523g$ up incline
3/23 $F = 0.0206$ lb
3/24 $s = 64.3$ m
3/25 $v_{C/T} = 2.82$ m/s
3/26 $T_1 = 248$ N, $T_2 = 497$ N, $T_3 = 994$ N
3/27 (*a*) $a = 10.73$ ft/sec^2 up, (*b*) $a = 2.93$ ft/sec^2 up
3/28 (*a*) $W_B = 36.3$ lb, (*b*) $W_B = 96.7$ lb
3/29 $v = v_0 e^{-1.623(10^{-3})x}$, $v = 81.7$ mi/hr
3/30 $a = 0.532$ m/s^2
3/31 $M = 20.5$ N·m
3/32 $x = 201$ m
3/33 $R = 1.995$ MN, $L = 1.947$ MN, $D = 435$ kN
3/34 $v = \sqrt{\dfrac{2P}{\rho} - \mu_k gL}$
3/35 $a_A = 1.364$ m/s^2 right, $a_B = 9.32$ m/s^2 down
$T = 46.6$ N
3/36 $v = 0.490$ m/s, $x_{max} = 100$ mm
3/37 When $x = 0.5$ ft: $v = 14.47$ ft/sec
3/38 Case (*a*) is higher, $v = 7.43$ m/s at $t = 1.2$ s
3/39 $0.0577(m_1 + m_2)g \leq P \leq 0.745(m_1 + m_2)g$
3/40 (*a*) $h = 55.5$ m, (*b*) $h = 127.4$ m
3/41 $a_A = 7.95$ m/s^2 right, $a_B = 8.04$ m/s^2 down slot
$T = 25.0$ N
3/42 $v = 2100$ m/s
3/43 $t = 0.589$ s, $s_2 = 0.1824$ m
3/44 $t = 0.761$ s, $s_2 = 0.357$ m
▶3/45 (*a*) $T = 8.52$ N, (*b*) $T = 16.14$ N
▶3/46 $t = 13$ h 33 min, $v = 4.76(10^{-5})$ m/s
3/47 $N = 41.0$ N up, No
3/48 $N_B = 9.41$ N, $v = 4.52$ m/s
3/49 $T = m(L\omega^2 - g \sin \theta)$
3/50 $N = 394$ lb
3/51 (*a*) $R = 0.25$ lb, (*b*) $R = 0.271$ lb
3/52 $\rho = 24{,}000$ ft
3/53 $N = 8.63$ rev/min
3/54 $v = 29.1$ m/s, $N = 12.36$ kN
3/55 $N = 0.0504$ lb
3/56 (*a*) $v_B = 195.3$ km/h, (*b*) $N_A = 241$ N
3/57 $\Omega = 3.64$ rad/s
3/58 $N = 156.2$ lb, $a = 27.7$ ft/sec^2
3/59 $v = 3.13$ m/s, $T = 0.981$ N
3/60 $\omega = \sqrt{\dfrac{g}{\mu_s r}}$
3/61 $a_n = 0.818g$, $F = 2460$ lb
3/62 $a_t = -22.0$ ft/sec^2
3/63 $v = 18.05$ m/s or 65.0 km/h
3/64 $T = 1.901$ N, $v = 0.895$ m/s
3/65 $N = 11.53$ rev/min
3/66 $N = 2.89$ N toward O', $R = 1.599$ N
3/67 $N = 3.25$ N toward O', $R = 2.66$ N
3/68 $a_1 = 1.784$ m/s^2 up incline, $T = 2.68$ N
$(v_2)_{max} = 1.928$ m/s
3/69 $T = 1.76$ N, $F_\theta = 3.52$ N (contact on upper side)
3/70 $F_{OA} = 2.46$ N, $F_{slot} = 1.231$ N
3/71 $F_{OA} = 3.20$ N, $F_{slot} = 1.754$ N
3/72 $h = 35\,800$ km
3/73 $F = 0.0979$ lb
3/74 $P = 27.0$ N, $P_s = 19.62$ N
3/75 $N_A = 1643$ N, $N_B = 195.8$ N
3/76 $\rho = 3000$ km, $\dot{v} = 6.00$ m/s^2
3/77 Dynamic: $F_r = 4.79$ N, $F_\theta = 14.00$ N
Static: $F_r = 5.89$ N, $F_\theta = 10.19$ N
3/78 $\ddot{r} = -574$ ft/sec^2, $\ddot{\theta} = 0.802$ rad/sec^2
3/79 $v = 101.8$ mi/hr, $v_{min} = 0$, $v_{max} = 235$ mi/hr
3/80 $3.41 \leq \omega \leq 7.21$ rad/s
3/81 $\omega = 2.73$ rad/s
3/82 $F = 4.39$ N
3/83 $T = 2.52$ lb, $N = 0.326$ lb on side B
3/84 (*a*) $a_t = 8.50$ m/s^2, $v = 2.78$ m/s, $N = 280$ N
(*b*) $a_t = 0$, $v = 5.68$ m/s, $N = 795$ N
3/85 $N = \dfrac{1}{4\pi}\sqrt{\left(\dfrac{\mu_s g}{r\alpha}\right)^2 - 1}$
3/86 $\Omega = 4.77$ rad/s
3/87 Slips first at $\theta = 0$ and $\omega = 2.33$ rad/s
If $\alpha = 0$, $\omega = 3.84$ rad/s
3/88 $r = r_0 \cosh \omega_0 t$, $v_r = r_0\omega_0 \sinh \omega_0 t$, $v_\theta = r_0\omega_0 \cosh \omega_0 t$
3/89 $r = \dfrac{r_0}{2\sqrt{\mu_k{}^2 + 1}}\left[\left(\mu_k + \sqrt{\mu_k{}^2 + 1}\right)e^{\omega_0(-\mu_k + \sqrt{\mu_k{}^2+1})t} + \left(-\mu_k + \sqrt{\mu_k{}^2 + 1}\right)e^{\omega_0(-\mu_k - \sqrt{\mu_k{}^2+1})t}\right]$
3/90 $\beta = \cos^{-1}\left(\dfrac{2}{3} + \dfrac{v_0{}^2}{3gR}\right)$, $\beta = 48.2°$
3/91 $\dot{r} = 9620$ ft/sec, $\dot{\theta} = 1.133(10^{-4})$ rad/sec
$\ddot{r} = -1.153$ ft/sec^2, $\ddot{\theta} = -2.72(10^{-8})$ rad/sec^2
3/92 $\sigma = \dfrac{mL\omega^2}{2A}\left(\dfrac{1}{4} - \dfrac{r^2}{L^2}\right)$
▶3/93 $v = 5.52$ m/s
▶3/94 $N = 81.6$ N, $R = 38.7$ N
▶3/95 $s = 47.4$ m
▶3/96 $s = \dfrac{r}{2\mu_k} \ln\left(\dfrac{v_0{}^2 + \sqrt{v_0{}^4 + r^2g^2}}{rg}\right)$
3/97 (*a*) $(U_{1\text{-}2})_s = 3.38$ ft-lb, (*b*) $(U_{1\text{-}2})_W = -2.72$ ft-lb
3/98 $v_B = 7.16$ m/s
3/99 $k = 974$ lb/in.
3/100 (*a*) no motion, (*b*) $v_B = 5.62$ m/s
3/101 $v = 2.32$ m/s
3/102 $(U_{A\text{-}B})_f = -1.415$ J
3/103 $P = 0.400$ hp
3/104 $s_{AB} = 214$ ft, $s_{BA} = 262$ ft

3/105 (*a*) $(U_{A\text{-}B})_W = 1.570$ J, (*b*) $(U_{A\text{-}B})_s = -4.20$ J
3/106 (*a*) $v = 2.56$ m/s, (*b*) $x = 98.9$ mm
3/107 $k = 267$ lb/ft
3/108 (*a*) $v = 3.15$ ft/sec, (*b*) $v = 3.59$ ft/sec
3/109 (*a*) $v = 3.89$ ft/sec, (*b*) $v = 4.84$ ft/sec
3/110 $v_B = 4.25$ m/s
3/111 $P = 22.9$ N
3/112 $P = 0.393$ hp, $P = 293$ W
3/113 $v_B = 5.67$ m/s
3/114 $v = 1.881$ m/s
3/115 $P = 0.992$ kW
3/116 $e = 0.764$
3/117 (*a*) $P = 56.4$ hp, (*b*) $P = 87.6$ hp
3/118 $v = 7.08$ ft/sec
3/119 $\theta = 16.62°$
3/120 (*a*) $N_B = 48$ N right, (*b*) $N_B' = 29.4$ N right
(*c*) $N_C = 17.63$ N down, (*d*) $N_D = 29.4$ N left
3/121 With nonlinear term: $\delta = 0.377$ ft (4.52 in.)
Without nonlinear term: $\delta = 0.406$ ft (4.88 in.)
3/122 $e = 0.892$
3/123 $P = 0.1394$ hp
3/124 $v_0 = 5590$ m/s
3/125 (*a*) $F = 61{,}200$ lb, (*b*) $P = 3270$ hp
(*c*) $P = 6530$ hp, (*d*) $P = 2670$ hp
3/126 $P = 40.4$ kW
3/127 $v = 1309$ m/s
3/128 (*a*) $N_B = 4mg$, (*b*) $N_C = 7mg$
(*c*) $s = \dfrac{4R}{1 + \mu_k\sqrt{3}}$
3/129 $x = 53.2$ m
3/130 (*a*) $s = 0.221$ m down incline
(*b*) $s = 0.1713$ m up incline
3/131 (*a*) $s = 0.621$ m down incline
(*b*) $s = 0.425$ m down incline
3/132 $P = 20$ kW
3/133 $P_{in} = 36.8$ kW
3/134 $v = 7.80$ ft/sec
3/135 $\delta = 29.4$ mm
3/136 $s = 142.3$ mm
3/137 (*a*) $P_{30} = 5$ hp, $P_{60} = 16$ hp
(*b*) $P_{up} = 35.2$ hp, $P_{down} = -3.17$ hp
(*c*) $v = 70.9$ mi/hr
3/138 $v = 18.18$ ft/sec
3/139 $v = 6.95$ ft/sec
3/140 (*a*) $v_B = 9.40$ m/s, (*b*) $\delta = 54.2$ mm
3/141 $h = 95.6$ mm
3/142 $v = 1.248$ ft/sec
3/143 $v_B = 4.27$ ft/sec
3/144 $L = 49.0$ in.
3/145 $N_B = 14.42$ N
3/146 (*a*) $k = 393$ N/m, (*b*) $v = 1.370$ m/s, $\dot{\theta} = 2.28$ rad/s
3/147 (*a*) $v = 3.94$ ft/sec, (*b*) $x = 0.693$ in.
3/148 $\dot{\theta} = 4.22$ rad/s
3/149 (*a*) $k_T = 25.8$ N·m/rad, (*b*) $v = 1.255$ m/s
3/150 $v_2 = 2.26$ m/s
3/151 $v = 2.54$ ft/sec down, $d = 8.00$ in.
3/152 (*a*) $v = 3.06$ ft/sec, (*b*) $v = 1.641$ ft/sec
3/153 $\dot{\theta} = 9.27$ rad/sec
3/154 $h = \dfrac{3}{2}R$, $N_B = 6mg$ up
3/155 $v = 0.331$ m/s
3/156 $\rho = \rho_0\left(1 - \dfrac{2gy}{{v_0}^2}\right)$, $(v_0)_{min} = \sqrt{\rho_0 g}$
3/157 $P = 2.86$ lb
3/158 $v_0 = 5590$ m/s
3/159 $v_A = 2.30$ m/s
3/160 $v_B = 26\,300$ km/h
3/161 $v = 30.0$ ft/sec or 20.4 mi/hr
3/162 $v = 0.990$ m/s
3/163 $v_2 = 35.1$ km/h
3/164 $v_A = \sqrt{v_P{}^2 - 2gR^2\left(\dfrac{1}{r_P} - \dfrac{1}{r_A}\right)}$
3/165 $(v_B)_{max} = 0.962$ m/s
3/166 (*a*) $v = 0.635$ m/s, (*b*) $d_{max} = 0.1469$ m
3/167 $v = 6240$ ft/sec or 4250 mi/hr
3/168 $v = 3.06$ ft/sec
3/169 $\dot{\theta} = -1.045$ rad/s
3/170 (*a*) $k = 111.9$ N/m, (*b*) $v = 0.522$ m/s
3/171 $v_B = 5.92$ m/s, $N = 84.1$ N
▶**3/172** (*a*) $v = 0.865\sqrt{gr}$, (*b*) $v_{max} = 0.908\sqrt{gr}$
(*c*) $\theta_{max} = 126.9°$
3/173 $N = 33.3$ N
3/174 $t = 48.1$ s
3/175 $\Delta R = 568$ N
3/176 $\mathbf{G} = 14.40\mathbf{i} - 11.52\mathbf{j} + 6\mathbf{k}$ kg·m/s
$G = 19.39$ kg·m/s
$\mathbf{R} = 21.6\mathbf{i} - 14.4\mathbf{j}$ N
3/177 $|\Delta E| = 13\,470$ J, $n = 99.8\%$
3/178 $\mu_k = 0.302$
3/179 $v = 3.42$ m/s
3/180 $\mathbf{v}_2 = 82\mathbf{i} + 59.7\mathbf{j} + 39\mathbf{k}$ m/s
3/181 $v_C = 1.231$ m/s left
3/182 (*a*) $v = 2$ m/s, (*b*) $v = -2.5$ m/s
3/183 $d = 1.326$ m
3/184 $v = 2.10$ m/s down
3/185 $\mu_k = 0.030$
3/186 $v = 5.44$ m/s, $N = 2090$ N
3/187 $T = 2780$ N
3/188 $v = 13.83$ mi/hr, $\theta = 83.9°$ west of north
3/189 (*a*) and (*b*) $v' = \dfrac{v}{3}$, $n = \dfrac{2}{3}$

3/190 $N = 5960$ lb

3/191 $v_2 = v_1 + \frac{F}{mk}(1 + \mu_k) - \frac{\mu_k \pi g}{2k}, F = \frac{\mu_k \pi mg}{2(1 + \mu_k)}$

3/192 $v = 190.9$ m/s, $|\Delta E| = 17.18$ kJ (loss)

3/193 $v_3 = 17\,970$ km/h

3/194 $t_s = 3.46$ s

3/195 $t_s = 3.69$ s

3/196 $v_1 = 2.05$ m/s left, $v_2 = 0.878$ m/s right

3/197 $D = 9210$ lb

3/198 $v_f = 0.00264$ m/s, $F_{av} = 59.5$ N

3/199 $P = 704$ lb T

3/200 $v = 40.2$ ft/sec, $t = 18.57$ sec

3/201 (a) $v' = 19.56$ ft/sec or 13.33 mi/hr
(b) $a_A = 97.8$ ft/sec^2 left, $a_B = 195.6$ ft/sec^2 right
(c) $R = 12{,}150$ lb

3/202 $v = 8.06$ m/s, $\theta = 60.3°$

3/203 $R = 472$ lb, $a = 150{,}000$ ft/sec^2 ($4660g$)
$d = 0.075$ ft or 0.900 in.

3/204 $F = 147.8$ N, $\beta = 12.02°$

3/205 $R_x = 559$ lb, $R_y = 218$ lb

3/206 $n = 62.5\%, t' = \frac{3}{4}t$

3/207 $v' = 0.610$ m/s

3/208 $t = 1.863$ sec

3/209 $\theta = 20.7°, n = 99.8\%$

3/210 $R = 43.0$ N, $\beta = 8.68°$

3/211 $v = 12.32$ ft/sec, $N = 488$ lb

3/212 $d = 1.462$ m

3/213 (a) $v_B = 2.89$ ft/sec, (b) $v_C = 2$ ft/sec

►**3/214** $v_2 = 40.0$ mm/s right

3/215 (a) and (b) $H_O = 30.1$ kg·m^2/s

3/216 $\mathbf{H}_O = mv(c\mathbf{j} - b\mathbf{k}), \dot{\mathbf{H}}_O = F(-c\mathbf{i} + a\mathbf{k})$

3/217 (a) $\mathbf{G} = 8.49\mathbf{i} - 8.49\mathbf{j}$ kg·m/s
(b) $\mathbf{H}_O = -23.2\mathbf{k}$ kg·m^2/s, (c) $T = 24$ J

3/218 (a) $H_O = mb\sqrt{2ga}, \dot{H}_O = mgb$
(b) $H_O = 2ma\sqrt{ga}, \dot{H}_O = 0$

3/219 $t = 15.08$ s

3/220 $v_P = 17{,}723$ mi/hr

3/221 $\dot{\theta} = 26.0$ rad/s

3/222 $\dot{\theta} = 65.1$ rad/s

3/223 $\frac{d\omega}{dr} = -\frac{2\omega}{r}$

3/224 $\mathbf{H}_O = -8\mathbf{i} - 120\mathbf{j} + 600\mathbf{k}$ kg·m^2/s
$\mathbf{M}_O = -24\mathbf{j} + 240\mathbf{k}$ N·m

3/225 $\omega = 0.1721$ rad/s CW

3/226 $\mathbf{H}_O = 32.1\mathbf{i} - 9.40\mathbf{j} + 19.98\mathbf{k}$ lb-ft-sec

3/227 $\omega = \frac{\omega_0}{4}, n = \frac{3}{4}$

3/228 $v_B = 5.43$ m/s

3/229 $\omega = 0.824\frac{v}{L}$

3/230 $\mathbf{H}_O = -\frac{1}{2}mgut^2\mathbf{k}$

3/231 $\dot{\theta} = \left(\frac{6m_1}{4m_1 + 9m_2}\right)\frac{v_1}{L}$

3/232 (a) $H_A = 0, H_D = m\sqrt{g\rho^3}$ CCW
(b) $H_A = 0.714m\sqrt{g\rho^3}$ CCW
$H_D = 1.126m\sqrt{g\rho^3}$ CCW

3/233 $v_r = 88,870$ ft/sec, $v_\theta = 125{,}700$ ft/sec

3/234 $\theta = \cos^{-1}\left[\frac{v_0 h}{(h - z)\sqrt{v_0^2 + 2gz}}\right], v = \sqrt{v_0^2 + 2gz}$

3/235 $\omega' = 2.77$ rad/s CCW, $\theta = 52.1°$

3/236 $v_A = 11\,300$ km/h, $v_B = 19\,540$ km/h

3/237 $\theta = 52.9°$

3/238 $\omega = 9.68$ rad/sec, $U'_{1\text{-}2} = 0.227$ ft-lb

3/239 $(M_O)_{av} = 0.0302$ N·m

►**3/240** $\omega = 3.00$ rad/s, $U = 5.34$ J

3/241 $e = 0.724, n = 47.6\%$

3/242 $v_1' = 0.714$ m/s left, $v_2' = 4.09$ m/s right

3/243 $v_A' = \left(\frac{1 - 0.1p}{1 + p}\right)v, v_B' = \left(\frac{1.1}{1 + p}\right)v$

3/244 $\frac{m_1}{m_2} > \frac{1}{e}$

3/245 $\theta' = 31.0°, v' = 0.825v$

3/246 $e = \left(\frac{h_2}{h}\right)^{1/4}$

3/247 For $\theta = 40°$: $e = 0.434$

3/248 $h = 1.263$ ft, $s = 1.132$ ft

3/249 $\delta = 80.4$ mm

3/250 $h = 10.94$ in., $h_2 = 7.43$ in.

3/251 $\omega = 0.572$ rad/s CCW

3/252 $\frac{|\Delta T|}{T} = \frac{m_B}{m_A + m_B}$

3/253 $R = 11.63$ m

3/254 $x = 0.1088d$

3/255 $\delta = 52.2$ mm

3/256 $\theta = 2.92(10^{-4})$ deg

3/257 $\theta = 2.86°, n = 0.0475$

3/258 $e = \sqrt{\frac{h'}{h' + h}}, v_x = \frac{\sqrt{\frac{g}{2}}d}{\sqrt{h'} + \sqrt{h' + h}}$

3/259 $v_A' = 2.46$ m/s, $\theta_A = 40.3°$
$v_B' = 9.16$ m/s, $\theta_B = -88.7°$

3/260 $\theta = 82.3°$ or $-22.3°$

3/261 $v_A' = 5.12$ m/s, $\theta_A = 77.8°$
$v_B' = 9.67$ m/s, $\theta_B = 38.4°, n = 12.06\%$

3/262 $v_A' = 5.20$ m/s, $\theta_A = 106.1°$
$v_B' = 7.84$ m/s, $\theta_B = 49.9°, n = 12.72\%$

3/263 $h = 14.53$ mm, $e = 0.405$

3/264 $h_2 = 0.385$ m
▶3/265 $v_B' = 95.9$ ft/sec, $\beta = 46.5°$
▶3/266 $\alpha = 11.37°$ or $78.6°$
▶3/267 $v' = 6.04$ m/s, $\theta = 85.9°$, $\delta = 165.0$ mm
▶3/268 $2.04 \leq v_0 \leq 3.11$ m/s
3/269 $v = 1641$ m/s or 5910 km/h
3/270 $v = 7569$ m/s or 27 250 km/h
3/271 See Prob. 1/14 and its answer.
3/272 $d = 2520$ km
3/273 (*a*) $v_{rel} = 16,227$ mi/hr, (*b*) $v_{rel} = 18,306$ mi/hr
3/274 $\Delta h = 88.0$ km
3/275 $e = 0.01196$, $\tau = 1$ h 30 min 46 s
3/276 $\Delta E = 2.61(10^{12})$ J
3/277 $h_a = 32\,600$ km
3/278 $D = 0.01132$ N
3/279 (*a*) $v = 7544$ m/s, (*b*) $v = 7912$ m/s
(*c*) $v = 10\,398$ m/s, (*d*) $v = 10\,668$ m/s
3/280 $\Delta v = 3217$ m/s, $v = 7767$ m/s when $\theta = 90°$
3/281 $\tau_f = 21.76(10^6)$ s, $\tau_{nf} = 20.74(10^6)$ s
3/282 $t = 64.6$ days
3/283 $\Delta t = 71.6$ sec
3/284 $v_r = 284$ m/s, $\theta = 98.0°$
3/285 $h_{max} = 53.9$ km
3/286 $v_B = 7560$ m/s
3/287 $\alpha = 38.8°$
3/288 $\Delta v = 302$ ft/sec
3/289 $\Delta v = 148$ ft/sec
3/290 $\beta = 153.3°$
3/291 $\tau_f = 658.69$ h, $\tau_{nf} = 654.68$ h
3/292 $\Delta v_A = R\sqrt{\dfrac{g}{R+H}}\left(1 - \sqrt{\dfrac{R}{R+h}}\right)$
3/293 $\rho_p = 2\dfrac{(R+h_a)(R+h_p)}{2R+h_a+h_p}$
3/294 $p = 0.0514$ rad/s
3/295 $\Delta v = 2940$ m/s
***3/296** $h = 922,000$ mi
▶3/297 $\Delta v_A = 2370$ m/s, $\Delta v_B = 1447$ m/s
▶3/298 $a = 6572$ km (parallel to the *x*-axis), $\tau = 5301$ s
$e = 0.01284$, $v_a = 7690$ m/s, $v_p = 7890$ m/s
$r_{max} = 6.66(10^6)$ m, $r_{min} = 6.49(10^6)$ m
▶3/299 $a = 7462$ km, $\alpha = 72.8°$, No
▶3/300 $\beta = 109.1°$
3/301 $\mu_k = 0.382$
3/302 $a_{rel} = k\delta\left(\dfrac{1}{m_1} + \dfrac{1}{m_2}\right)$
3/303 $\mathbf{G} = 9\mathbf{i}$ kg·m/s, $\mathbf{G}_{rel} = 3\mathbf{i}$ kg·m/s
$T = 13.5$ J, $T_{rel} = 1.5$ J
$\mathbf{H}_O = -4.5\mathbf{k}$ kg·m²/s, $(\mathbf{H}_B)_{rel} = -1.5\mathbf{k}$ kg·m²/s
3/304 $P = 66.9$ kN
3/305 $F = 376$ lb
3/306 $F = 194.0$ kN
3/307 $x_{C/T} = 2.83$ m, $v_{rel} = 2.46$ m/s
3/308 No answer
3/309 $(v_{rel})_{max} = a_0\sqrt{\dfrac{m}{k}}$
3/310 $a_0 = 16.99$ m/s², $R = 0$
3/311 (*a*) and (*b*) $T = 112$ J
3/312 $T = 3ma_0 \sin\theta$, $T_{\pi/2} = 90$ N
3/313 $T_0 = m(g + a_0)(3 - 2\cos\theta_0)$, $T_{\pi/2} = 3m(g + a_0)$
3/314 $P_{rel} = 0.1206$ hp
▶3/315 (*a*) and (*b*) $h_2 = e^2 h_1$
▶3/316 $v_A = \sqrt{v_0^2 + 2gl\sin\theta + 2v_0\cos\theta\sqrt{2gl\sin\theta}}$
▶3/317 $g_{rel} = 9.825 - 0.03382\cos^2\gamma$ m/s²
▶3/318 No answer
3/319 $v_2 = 10.78$ ft/sec
3/320 $v = \sqrt{v_0^2 - 2gL\sin\theta}$
3/321 $N = 1177$ rev/min
3/322 $R = 46.7$ N
3/323 $v = 2.2$ m/s
3/324 $v_A = 7451$ m/s, $e = 0.0295$
3/325 $P = 652$ hp
3/326 $\delta = \sqrt{\dfrac{mgR(5 + 2\mu_k)}{k}}$
3/327 $v_B = 5.64$ ft/sec
3/328 No answer
3/329 $N_A = 23.2$ lb, $N_B = 148.3$ lb
3/330 $M = 18.96$ kN·m
3/331 $v = 39.0$ ft/sec
3/332 $T = 424$ N
3/333 $\delta = \dfrac{Pm}{k(M+m)}$, $P_{eq} = (M+m)g\sin\theta$, $\delta_{eq} = \dfrac{mg\sin\theta}{k}$
3/334 $L = 2540$ N, $D = 954$ N
3/335 $F = 327$ kN
3/336 $\delta = 2.55$ in.
3/337 $\mu_s = 0.292$
3/338 $(v_A)_0 = 72.6$ mi/hr
3/339 $s = 2.28$ m
3/340 (*a*) $k = 119.9$ N/m, (*b*) $v_{max} = 21.4$ m/s
(*c*) $a_{max} = 26.2$ m/s²
3/341 $t' = 8.50$ sec
3/342 $v = 0.382$ m/s, $d = 0.1325$ m
3/343 (*a*) $v = 2.15$ ft/sec, (*b*) $v' = 2.78$ ft/sec
3/344 $\Delta P = -4m\omega v_r$, $\Delta P = -24.3$ N
▶3/345 $v' = 182.9$ mi/hr
3/346 $\Delta v = 9000$ km/h
3/347 $x = 13.40$ ft or 47.3 ft
***3/348** (*a*) $\dot{\theta} = 1.414$ rad/s, (*b*) $\theta_{max} = 43.0°$
***3/349** (*a*) $\dot{\theta} = 3.98$ rad/s, (*b*) $\theta_{max} = 48.5°$
***3/350** $38.7° \leq \theta \leq 65.8°$

***3/351** $\theta_{max} = 53.1°$
***3/352** $e = 0.610, y = 1.396$ ft
***3/353** $t = 1.069$ s, $\theta = 30.6°$
***3/354** For $e = 0.5: R = 288$ ft, $v_B' = 95.9$ ft/sec
***3/355** $v_{max} = 0.691$ m/s at $x = 0.0696$ m
***3/356** $t = 0.1254$ s

Chapter 4

4/1 $\mathbf{r} = \frac{d}{7}(10\mathbf{i} + 6\mathbf{j}), \dot{\mathbf{r}} = \frac{v}{7}(4\mathbf{i} + 4\mathbf{j}), \ddot{\mathbf{r}} = \frac{F}{7m}\mathbf{i}$
$T = 6mv^2, \mathbf{H}_O = 4mvd\mathbf{k}, \dot{\mathbf{H}}_O = -Fd\mathbf{k}$
4/2 $\mathbf{H}_G = \frac{12}{7}mvd\mathbf{k}, \dot{\mathbf{H}}_G = -\frac{Fd}{7}\mathbf{k}$
4/3 $\mathbf{r} = b(\mathbf{i} + 3.61\mathbf{j} + 2.17\mathbf{k})$
$\dot{\mathbf{r}} = v(0.556\mathbf{i} + 0.748\mathbf{j} + 0.385\mathbf{k})$
$\ddot{\mathbf{r}} = \frac{F}{3m}\left(-\mathbf{j} + \frac{2}{3}\mathbf{k}\right), T = 16.5mv^2$
$\mathbf{H}_O = mvb(-15\mathbf{i} + 2.07\mathbf{j} - 1.536\mathbf{k})$
$\dot{\mathbf{H}}_O = Fb(14.50\mathbf{i} - 4\mathbf{j} + 3\mathbf{k})$
4/4 $\mathbf{H}_G = mvb(-12.92\mathbf{i} - 5.30\mathbf{j} + 9.79\mathbf{k})$
$\dot{\mathbf{H}}_G = Fb(0.778\mathbf{i} - 2\mathbf{j} + 6\mathbf{k})$
4/5 $a_C = 53.7$ ft/sec^2
4/6 $v = \sqrt{gR}, N = 2mg$
4/7 $F = 2.92$ N
4/8 $|M_O|_{av} = 0.7$ N·m
4/9 $T = 58.3$ lb
4/10 $D_x = 1.288$ lb left, $D_y = 35.5$ lb up, $N_E = 25.6$ lb up
4/11 $a_C = \frac{F}{2m} - g\sin\theta$ up incline
4/12 $\mathbf{G} = 8.66\mathbf{i} + 5\mathbf{j}$ kg·m/s, $\mathbf{H}_G = 0.225\mathbf{k}$ kg·m^2/s
$\mathbf{H}_O = -0.971\mathbf{k}$ kg·m^2/s
4/13 $\bar{a} = 15.19$ m/s^2
4/14 Mass-center accelerations are identical
4/15 (a) $\bar{a}_y = 2.69$ m/s^2, $a_{y_1} = a_{y_2} = 2.69$ m/s^2
(b) $\bar{a}_y = 2.69$ m/s^2, $a_{y_1} = \frac{25\tan\beta}{\tan\alpha + \tan\beta} - 9.81$ m/s^2
$a_{y_2} = \frac{25\tan\alpha}{\tan\alpha + \tan\beta} - 9.81$ m/s^2
(assumes $\tan\beta \leq 1.548\tan\alpha$)
4/16 $\mathbf{H}_O = 2m(r^2\omega - vy)\mathbf{k}$
4/17 $P = 3.24$ hp
4/18 $t = \frac{4mr^2\omega}{M}$
4/19 $\dot{\theta}' = 80.7$ rad/s
4/20 $v_A = 1.015$ m/s, $v_B = 1.556$ m/s
4/21 $v = 0.205$ m/s
4/22 $v = 0.355$ mi/hr, $n = 95.0\%$
4/23 $s = \frac{(m_1 + m_2)x_1 - m_2l}{m_0 + m_1 + m_2}$
4/24 $x = 0.316$ ft
4/25 $a_C = 10$ m/s^2
4/26 $\Delta Q = 2.52$ J, $I_x = 12.87$ N·s
4/27 (a) $\bar{a} = \frac{F}{2m}$, (b) $\ddot{\theta} = \frac{2Fb}{mL^2}$
4/28 $v = 0.877$ m/s
4/29 $v = 72.7$ km/h
4/30 $v_x = \sqrt{2gl}, \dot{\theta} = 2\sqrt{\frac{2g}{l}}$
▶4/31 $v = 3.92$ ft/sec
▶4/32 $(x_A, y_A) = (2270, -1350)$ m
▶4/33 $v_{b/c} = \sqrt{\left(1 + \frac{m_1}{m_2}\right)2gl(1 - \cos\theta)}$
$v_c = \sqrt{\frac{2gl(1 - \cos\theta)}{\frac{m_2}{m_1}\left(1 + \frac{m_2}{m_1}\right)}}$
▶4/34 $v_A = 103.6$ ft/sec
4/35 $u = 2070$ ft/sec
4/36 $\alpha = 17.22°$
4/37 $F = 750$ N
4/38 $F = 3.05$ lb
4/39 $R = 1885$ N
4/40 $T = 5.93$ N
4/41 $T = 2.85$ kN
4/42 $T = \rho u\frac{\pi d^2}{4}(u + v\cos\theta)$
4/43 $T = 32.6$ kN
4/44 $F_x = 442$ N, $F_y = 442$ N
4/45 $T = A(p + \rho v^2)$
4/46 $F = \rho Av^2\sin\theta, Q_1 = \frac{Q}{2}(1 + \cos\theta), Q_2 = \frac{Q}{2}(1 - \cos\theta)$
4/47 $h = 18.57$ ft
4/48 $n = 0.638$
4/49 $m = \rho\pi h(D^2 - d^2)$
4/50 $R = \frac{\pi d^2}{4}\left[\rho_B\left(1 - \frac{\rho_B}{\rho_A}\right)v^2 + (p_B - p_A)\right]$
4/51 $T = 6530$ lb
4/52 $T = 9.69$ kN, $V = 1.871$ kN, $M = 1.122$ kN·m
4/53 $R = 5980$ N
4/54 $p = 0.1556$ lb/in.^{2}
4/55 $M = 4.57$ lb-ft CCW
4/56 $R = mg + \rho\frac{\pi d^2}{4}u(u - v\cos\theta)$
4/57 $T = 12{,}610$ lb, $V = 162.3$ lb, $M = 1939$ lb-ft
4/58 $P = 5.56$ kN, $R = 8.49$ kN

4/59 At A: $V = 50.0$ lb, $P = 28{,}400$ lb, $M = 491$ lb-ft
At B: $V = 0$, $P = 28{,}200$ lb, $M = 0$

4/60 $P = 0.671$ kW

4/61 $m = 184.3$ kg

4/62 $v = \frac{1}{r}\sqrt{\frac{mg}{\pi\rho}}$, $P = \frac{mg}{2r}\sqrt{\frac{mg}{\pi\rho}}$

4/63 $M = \rho Q\left[\frac{Qr\cos\phi}{4A} - \omega(r^2 + b^2 + 2rb\sin\phi)\right]$

$\omega_0 = \frac{Qr\cos\phi}{4A(r^2 + b^2 + 2rb\sin\phi)}$

4/64 $\theta = 2.31°$, $a_y = 1.448$ m/s^2

4/65 $R_x = 311$ lb, $R_y = -539$ lb

▶4/66 $C = 100.3$ lb

▶4/67 $F = 391$ lb, $V = 156.2$ lb, $T = 27.3$ lb-ft
$M = 40.9$ lb-ft

▶4/68 $P = 1.197$ MW

4/69 $a_n = 8.31$ m/s^2, $a_t = 21.2$ m/s^2

4/70 $m = 10.86$ Mg

4/71 $a = 4.70$ m/s^2, $m' = 448$ kg/s

4/72 $m = m_0 e^{-\frac{a+g}{u}t}$

4/73 No answer

4/74 (*a*) $P = 1113$ lb, (*b*) $P = 1242$ lb

4/75 $a = 190.0$ ft/sec^2, $v_{max} = 178.0$ ft/sec

4/76 $F = \rho(x\ddot{x} + \dot{x}^2)$

4/77 $P = 209$ N

4/78 $R = \rho g x + \rho v^2$

4/79 1% mass reduction: 1.967% velocity increase
5% mass reduction: 10.04% velocity increase

4/80 $a = 0.498$ ft/sec^2

4/81 $P = 963$ lb

4/82 $P = 20.4$ N

4/83 $a = -1.603$ m/s^2

4/84 $v = 13.83$ ft/sec

4/85 $a = \frac{P}{\rho x} - \mu_k g - \frac{\dot{x}^2}{x}$

4/86 $x = 6.18$ m

4/87 $v = \sqrt{gL\sin\theta}$

4/88 $v = 7.90$ m/s

4/89 (*a*) $v_1 = \sqrt{2gh\ln\left(\frac{L}{h}\right)}$, (*b*) $v_2 = \sqrt{2gh\left[1 + \ln\left(\frac{L}{h}\right)\right]}$

$Q = \rho g h\left(L - \frac{h}{2}\right)$

4/90 $v = \frac{v_0}{1 + \frac{2\rho l}{m}}$, $x = \frac{m}{\rho}\left(\sqrt{1 + \frac{2v_0\rho t}{m}} - 1\right)$

▶4/91 $P = \frac{1}{2}\rho\left(\frac{v^2}{2} + gx\right)$, $R = \frac{\rho v^2}{4} + \rho g\left(L - \frac{x}{2}\right)$, $T_1 = \frac{\rho v^2}{4}$

▶4/92 $P = \frac{1}{2}\rho(v^2 + gx)$, $R = \rho g\left(L - \frac{x}{2}\right)$, $T_1 = \frac{\rho v^2}{2}$

$Q = \frac{\rho x}{4}(v^2 + gx)$

4/93 $\mathbf{r} = \frac{d}{7}(\mathbf{i} + 4\mathbf{j} + 6\mathbf{k})$, $\dot{\mathbf{r}} = \frac{2v}{7}(2\mathbf{i} + \mathbf{j} + 3\mathbf{k})$, $\ddot{\mathbf{r}} = \frac{F}{7m}\mathbf{k}$

$T = 13mv^2$, $\mathbf{H}_O = 2mvd\,(6\mathbf{i} + 3\mathbf{j} + \mathbf{k})$, $\dot{\mathbf{H}}_O = -Fd\mathbf{j}$

4/94 $\mathbf{H}_G = \frac{4mvd}{7}(18\mathbf{i} + 6\mathbf{j} + 7\mathbf{k})$, $\dot{\mathbf{H}}_G = -\frac{2Fd}{7}(2\mathbf{i} + 3\mathbf{j})$

4/95 $\bar{a} = 53.7$ ft/sec^2

4/96 $\omega = 12.37$ rad/sec

4/97 $a = 1.746$ m/s^2

4/98 $F = 812$ lb

4/99 $a_n = 4.67$ m/s^2, $a_t = 19.34$ m/s^2

4/100 $v = 133.3$ m/s

4/101 $t = 60$ s: $a = 5.64$ m/s^2, $a_{max} = 69.3$ m/s^2 at $t = 231$ s

4/102 $\mu = \frac{d}{2h}$, $v = \sqrt{\frac{mgd}{2\rho A h}}$

4/103 (*a*) $N_A = \frac{9b + 14h}{64h}mg$, $N_B = \frac{50h - 9b}{64h}mg$, $a = \frac{9dg}{32h}$

(*b*) $N_A = \frac{b + 6h}{16h}mg$, $N_B = \frac{10h - b}{16h}mg$, $a = \frac{dg}{8h}$

4/104 $F = 159.8$ lb

4/105 $\Delta Q = \rho g r^2$

4/106 $T = 21.1$ kN, $F = 12.55$ kN

4/107 $F = 3\rho g x$

4/108 (*a*) $a = \frac{g}{L}x$, (*b*) $T = \rho g x\left(1 - \frac{x}{L}\right)$, (*c*) $v = \sqrt{gL}$

4/109 $P = 151.8$ lb

▶4/110 $a = \frac{h}{H}g$, $v = h\sqrt{\frac{g}{H}}$, $R = 2\rho g\left(H - \frac{2h^2}{H}\right)$

▶4/111 $C = 4340$ N up, $D = 3840$ N down

▶4/112 $R = \rho g x\frac{4L - 3x}{2(L - x)}$

CHAPTER 5

5/1 $N = 60$ rev

5/2 $\mathbf{v}_A = 0.580\mathbf{i} + 0.1553\mathbf{j}$ m/s
$\mathbf{a}_A = -1.470\mathbf{i} + 0.849\mathbf{j}$ m/s^2

5/3 $\Delta\theta = 9$ rad, $N = 3.66$ rev

5/4 $r = 4.5$ in.

5/5 Startup: $\Delta\theta = 172.5$ rev, Shutdown: $\Delta\theta = 920$ rev

5/6 $\omega = 3.55$ rad/sec, $\omega_{av} = 1.231$ rad/sec

5/7 $\omega = 4$ rad/s, $\alpha = 6$ rad/s^2

5/8 $\left|\dot{\theta}\right|_{max} = \theta_0\omega_0$ when $\theta = 0$, $\left|\ddot{\theta}\right|_{max} = \theta_0\omega_0^2$ when $\theta = \theta_0$

5/9 $\boldsymbol{\omega} = 4\mathbf{k}$ rad/sec, $\boldsymbol{\alpha} = \pm 17.89\mathbf{k}$ rad/sec^2

5/10 $\Delta\theta = 10.99$ rad, $t = 1.667$ s
5/11 $\mathbf{v}_B = -11\mathbf{i}$ m/s, $\mathbf{a}_B = 22\mathbf{i} - 220\mathbf{j}$ m/s^2
5/12 $\mathbf{v}_P = 40(-20\mathbf{i} + 12\mathbf{j} - 9\mathbf{k})$ in./sec
$\mathbf{a}_P = 1600(-15\mathbf{i} - 16\mathbf{j} + 12\mathbf{k})$ in./sec^2
5/13 (*a*) $\mathbf{v}_A = 39.0\mathbf{i} - 22.5\mathbf{j}$ in./sec
$\mathbf{a}_A = 34.6\mathbf{i} + 240\mathbf{j}$ in./sec^2
(*b*) $\mathbf{v}_B = 46.5\mathbf{i} - 9.51\mathbf{j}$ in./sec
$\mathbf{a}_B = -45.4\mathbf{i} + 251\mathbf{j}$ in./sec^2
5/14 (*a*) $\mathbf{v}_A = 45\mathbf{e}_t$ in./sec, $\mathbf{a}_A = -90\mathbf{e}_t + 225\mathbf{e}_n$ in./sec^2
(*b*) $\mathbf{v}_B = 45\mathbf{e}_t + 15\mathbf{e}_n$ in./sec
$\mathbf{a}_B = -165\mathbf{e}_t + 195\mathbf{e}_n$ in./sec^2
5/15 $F = 4.33$ N, $\Delta\theta = 9$ rad
5/16 $\alpha = 15$ rad/sec^2, $a_B = 5$ ft/sec^2
5/17 $\mathbf{v}_A = 1.121\mathbf{i} + 0.838\mathbf{j}$ m/s
$\mathbf{a}_A = -4.48\mathbf{i} + 0.1465\mathbf{j}$ m/s^2
5/18 $t = 9.57$ s
5/19 $\omega_0 = 49.4$ rad/s
5/20 $\mathbf{v}_A = -0.382\mathbf{i} - 0.1025\mathbf{j}$ m/s
$\mathbf{a}_A = 0.540\mathbf{i} + 0.718\mathbf{j}$ m/s^2
5/21 (*a*) $\Delta\theta_B = 90$ rev, (*b*) $\omega_B = 70.7$ rad/s
(*c*) $\Delta\theta_B = 22.5$ rev
5/22 $\mathbf{v}_A = -0.374\mathbf{i} + 0.1905\mathbf{j}$ m/s
$\mathbf{a}_A = -0.757\mathbf{i} - 0.605\mathbf{j}$ m/s^2
5/23 $\mathbf{v}_A = -0.223\mathbf{i} - 0.789\mathbf{j}$ m/s
$\mathbf{a}_A = 3.02\mathbf{i} - 1.683\mathbf{j}$ m/s^2
5/24 $\mathbf{v}_A = -0.0464\mathbf{i} + 0.1403\mathbf{j}$ m/s
$\mathbf{a}_A = -0.1965\mathbf{i} + 0.246\mathbf{j}$ m/s^2
5/25 $\theta = 0.596$ rad
5/26 $t = 12.5$ s: $\Delta\theta = 187.5$ rev; $t = 25$ s: $\Delta\theta = 250$ rev
5/27 $N_B = 415$ rev/min
▶5/28 $\omega_C = \left(\dfrac{r_1}{r_2}\right)^2 \alpha_1 t,\ a_P = \dfrac{r_1^{\,2}}{r_2}\alpha_1\sqrt{1 + \left(\dfrac{r_1}{r_2}\right)^4 \alpha_1^{\,2} t^4}$
5/29 $\omega = \dfrac{0.643v}{\sqrt{L^2 - 0.413x_A^{\,2}}}$ CCW $\left(0 \le x_A \le \dfrac{L}{\sin 40°}\right)$
5/30 $\omega_{OA} = \dfrac{dv}{s^2 + d^2}$ CCW if $v > 0$
5/31 $v_P = 0.3$ m/s down, $v_C = 0.25$ m/s down
$\omega = 0.5$ rad/s CW
5/32 $v_B = 0.1760$ m/s down
5/33 $v_C = 2.38$ in./sec left
5/34 $v_P = -r\omega \sin\theta,\ a_P = -r\alpha \sin\theta - r\omega^2 \cos\theta$
(a negative value is directed leftward)
5/35 $v_P = -r\omega(\sin\theta + \cos\theta \tan\beta)$
$a_P = -r\alpha(\sin\theta + \cos\theta\tan\beta)$
$- r\omega^2(\cos\theta - \sin\theta\tan\beta)$
(a negative value is directed leftward)
5/36 $v = v_O\sqrt{2(1 + \sin\theta)},\ a = \dfrac{v_O^2}{r}$
5/37 $v_B = 0.755$ m/s
5/38 $a_B = 789$ mm/s^2 down
5/39 $\omega_{BC} = 1.903$ rad/s CW
5/40 $\omega_{OA} = -\dfrac{vh}{h^2 + s^2}$ CCW for $v > 0$
5/41 $\omega = 0.1258$ rad/sec CCW
$\alpha = 0.001684$ rad/sec^2 CW
5/42 $v_B = 0.0728$ m/s right, $a_B = 0.0964$ m/s^2 right
$\omega = 0.426$ rad/s CCW
5/43 $v = 2v_0 \cot\dfrac{\theta}{2}$
5/44 $\omega_{OB} = 0.572$ rad/s CCW
5/45 $h = 4g\left(\dfrac{b}{\pi v}\right)^2,\ h = 128.8$ mm
5/46 $\omega = 3$ rad/s CW, $v_A = 480$ mm/s, $v_O = 180$ mm/s
$a_C = 540$ mm/s^2
5/47 $v_C = 0.965$ ft/sec right
5/48 $a_C = 3.07$ ft/sec^2 left
5/49 $v = 2\dfrac{\sqrt{b^2 + L^2 - 2bL\cos\theta}}{L\tan\theta}\dot{s}$
5/50 $\dot{x} = \dfrac{L + x}{b}\dfrac{\tan\theta}{\cos\left(\dfrac{\theta + \delta}{2}\right)}\dot{c}$, where $\delta = \sin^{-1}\left(\dfrac{h}{b}\right)$
5/51 No answer
5/52 $\mathbf{a}_A = \dfrac{k^2 l}{b^2\cos^2\left(\dfrac{\theta}{2}\right)}\left(\mathbf{e}_n + \dfrac{1}{2}\tan\dfrac{\theta}{2}\,\mathbf{e}_t\right)$
5/53 $\alpha_2 = \dfrac{\dot{r}_1 r_2 - r_1\dot{r}_2}{r_2^{\,2}}\omega_1$
5/54 $\dot{\beta} = 6.28\dfrac{\cos\theta - 0.278}{1.939 - \cos\theta}$ rad/sec
5/55 $v_O = 26.2$ mm/s left
▶5/56 $\omega_{AB} = \dfrac{r\omega_0}{l}\dfrac{\cos\theta}{\sqrt{1 - \left(\dfrac{r}{l}\sin\theta\right)^2}}$

$$\alpha = \frac{r\omega_0^{\,2}}{l}\left(\frac{r^2}{l^2} - 1\right)\frac{\sin\theta}{\left[1 - \left(\dfrac{r}{l}\sin\theta\right)^2\right]^{3/2}}$$

▶5/57 $\omega_2 = 1.923$ rad/s
▶5/58 (*a*) $a = 0.909$ m/s^2 up, (*b*) $a = 0.918$ m/s^2 down
5/59 $\mathbf{v}_{B/A} = -15.78\mathbf{i} - 20.7\mathbf{j}$ in./sec
5/60 $\mathbf{v}_A = 3\mathbf{i} - \mathbf{j}$ m/s, $\mathbf{v}_B = 3.6\mathbf{i} + \mathbf{j}$ m/s
5/61 (*a*) $N = 91.7$ rev/min CCW
(*b*) $N = 45.8$ rev/min CCW
(*c*) $N = 45.8$ rev/min CW
5/62 $\omega = 4.54$ rad/sec CW
5/63 $\mathbf{v}_A = -1672\mathbf{i} + 107\,257\mathbf{j}$ km/h, $\mathbf{v}_B = 105\,585\mathbf{j}$ km/h
$\mathbf{v}_C = 1672\mathbf{i} + 107\,257\mathbf{j}$ km/h, $\mathbf{v}_D = 108\,929\mathbf{j}$ km/h
5/64 $v_D = 0.596$ m/s, $\dot{x} = 0.1333$ m/s
5/65 $\mathbf{v}_A = 27\mathbf{i} + 16\mathbf{j}$ in./sec, $\mathbf{v}_P = 11\mathbf{i}$ in./sec
5/66 $v_C = 1.528$ m/s

5/67 $\mathbf{v}_A = 0.5\mathbf{i} + 0.9\mathbf{j}$ m/s
5/68 $v_O = 6.93$ m/s, $\omega = 21.3$ rad/s CW
5/69 $v_O = 2$ ft/sec, $v_B = 2.83$ ft/sec
5/70 $(v_A)_{max} = 5.33$ m/s, $(v_A)_{min} = 3.56$ m/s
5/71 $\omega_{AB} = 0.268\dfrac{v_A}{L}$ CCW
5/72 $\boldsymbol{\omega}_{OA} = -3.33\mathbf{k}$ rad/s
5/73 $\mathbf{v}_A = \dfrac{v}{2}(\mathbf{i} + \mathbf{j})$, $\mathbf{v}_B = \dfrac{v}{2}(\mathbf{i} - \mathbf{j})$
5/74 $v_B = 4.38$ m/s, $\omega = 3.23$ rad/s CCW
5/75 (*a*) $\omega_{BC} = \omega$ CCW, (*b*) $\omega_{BC} = 2\omega$ CCW
5/76 $v_A = 4.57$ ft/sec
5/77 $\omega_{AB} = \omega_0$ CW, $v_B = 2.58r\omega_0$ down
5/78 $\omega_{AB} = 0.966\omega_0$ CCW, $v_B = \sqrt{2}r\omega_0$ at $\measuredangle 60°$
5/79 $\omega_0 = 1.452$ rad/s CW, $\omega_{AB} = 0.0968$ rad/s CCW
5/80 $\boldsymbol{\omega}_{AO} = 1.333\mathbf{k}$ rad/s, $\boldsymbol{\omega}_{BD} = 1.20\mathbf{k}$ rad/s
5/81 $\omega = 0.722$ rad/s
5/82 $\omega = 1.394$ rad/s CCW, $v_A = 0.408$ m/s down
5/83 $\dot{\theta} = 0.944$ rad/s CCW, $v_A = 0.298$ m/s left
5/84 $\omega = 16.39$ rad/s CCW, $v_B = 6.19$ m/s
5/85 $v_P = 0.2$ ft/sec
5/86 $\omega_{AB} = 1.725$ rad/s CCW, $\omega_{BC} = 4$ rad/s CCW
5/87 $\omega_{AOB} = 0.634$ rad/s CW
5/88 $\boldsymbol{\omega}_{AC} = 0.429\mathbf{k}$ rad/sec
5/89 $v_E = 0.514$ m/s
►5/90 $\omega = 18.22$ rad/s CCW
5/91 $v_A = 2.33$ m/s at $\measuredangle 31.0°$, $v_B = 2.33$ m/s at $\measuredangle 329°$
5/92 $v_A = 2.66$ m/s at $\measuredangle 335°$, $v_B = 1.949$ m/s at $\measuredangle 35.4°$
5/93 $v_B = 1.114$ m/s
5/94 $v_A = 2.24$ ft/sec at $\measuredangle 318°$
5/95 $v_A = 0.408$ m/s down
5/96 (*a*) $v_A = 6$ in./sec, $v_G = 6$ in./sec
(*b*) $v_A = 42$ in./sec, $v_G = 30$ in./sec
(*c*) $v_A = 30$ in./sec, $v_G = 21.6$ in./sec
5/97 (*a*) $\omega_{lower} = \dfrac{v}{r}$ CW, (*b*) $\omega_{upper} = \dfrac{v}{r}$ CCW
(*c*) $v_A = v$ right, $v_B = 2v$ right, $v_C = v$ right, $v_D = 0$
5/98 $v_D = 250$ mm/s
5/99 $v_A = 9.04$ in./sec, $v_C = 6.99$ in./sec
5/100 $\omega = 4.16$ rad/s CW, $v_B = 0.416$ m/s right
$v_G = 0.416$ m/s at $\measuredangle 60°$
5/101 $\omega = 4.37$ rad/s CW, $v_B = 0.971$ m/s right
5/102 $v_B = 269$ mm/s right
5/103 $v_A = 0.707$ m/s, $v_P = 1.581$ m/s
5/104 $v_A = 4.57$ ft/sec
5/105 $v_A = 18$ ft/sec right, $v_B = 12$ ft/sec left
$v_C = 15.30$ ft/sec at $\measuredangle 281°$, $v_D = 14.70$ ft/sec down
5/106 $\omega = 0.45$ rad/sec CCW, $v_A = 43.8$ ft/sec
$v_B = 46.5$ ft/sec
5/107 $v_O = 120$ mm/s, $v_P = 216$ mm/s
5/108 $\omega_{AB} = \omega_0$ CW, $v_B = 2.58r\omega_0$ down
5/109 $\omega_{AB} = 0.966\omega_0$ CCW, $v_B = \sqrt{2}r\omega_0$ at $\measuredangle 60°$
5/110 $\omega_{OB} = 11.79$ rad/s CCW
5/111 No answer
5/112 $\omega_{AB} = 0$, $\omega_{BD} = 8.33$ rad/s CCW
5/113 $v = 15.71$ ft/sec or 10.71 mi/hr right
$v_s = 6.98$ ft/sec left
5/114 $v_D = 0.596$ m/s
5/115 $\omega_{AB} = 1.414$ rad/sec CCW, $\omega_{BD} = 3.77$ rad/sec CW
5/116 $\omega_{AD} = 13.33$ rad/s CW, $v_A = 2.31$ m/s right
5/117 $v_A = 9.19$ ft/sec right
5/118 $v_A = 0.278$ m/s right
5/119 $v_A = 0.509$ m/s left
►5/120 $\omega = 1.10$ rad/sec CW
5/121 $a_A = 11.18$ m/s^2
5/122 (*a*) $a_A = 0.2$ m/s^2, (*b*) $a_A = 4.39$ m/s^2, (*c*) $a_A = 6.2$ m/s^2
5/123 $\alpha_{AB} = 0.1768$ rad/sec^2 CW
5/124 $\omega_{AB} = 0$, $\alpha_{AB} = \dfrac{v^2}{RL}$ CCW
5/125 $\theta = \sin^{-1}\left(\dfrac{r}{R}\right)$, $v_O = \sqrt{\dfrac{R}{r}a_O \sqrt[4]{R^2 - r^2}}$
5/126 $\alpha = 0.05$ rad/s^2 CW, $a_C = 0.05$ m/s^2 down
$b = 2$ m right of B
5/127 $\alpha = 0.286$ rad/s^2 CCW, $a_A = 0.653$ m/s^2 down
5/128 $\mathbf{a}_B = 0.0279\mathbf{i}$ m/s^2
5/129 $\mathbf{a}_{B/A} = -\dfrac{v^2}{2r}\mathbf{i} - a\mathbf{j}$
5/130 $\alpha_{AB} = 15.40$ rad/sec^2 CW
5/131 $\alpha_{AB} = 0.0268\dfrac{a_A}{L} - 0.01924\dfrac{v_A^2}{L^2}$ CCW if positive
$a_B = 1.035a_A - 0.0743\dfrac{v_A^2}{L}$ up incline if positive
5/132 $\alpha_{AB} = \omega^2$ CCW
5/133 $\boldsymbol{\alpha}_{AB} = -37.9\mathbf{k}$ rad/s^2, $\mathbf{a}_A = 17.30\mathbf{i}$ m/s^2
5/134 $a_A = 43.8$ ft/sec^2
5/135 $a_D = 1.388$ m/s^2
5/136 $(a_B)_t = 2.46$ m/s^2 left
5/137 (*a*) $\mathbf{a}_C = \dfrac{r\omega^2}{1 - \dfrac{r}{R}}\mathbf{j}$, (*b*) $\mathbf{a}_C = \dfrac{r\omega^2}{1 + \dfrac{r}{R}}\mathbf{j}$
5/138 $\alpha_{AB} = 9.98$ rad/s^2 CCW
5/139 $\alpha_{AB} = 1.578$ rad/s^2 CW
5/140 $\alpha_{ABD} = 0$, $\alpha_{BC} = 28$ rad/sec^2 CCW
5/141 $\alpha_{AB} = 3.64\omega_0^2$ CCW, $a_B = 6.82r\omega_0^2$ up
5/142 $\alpha = 696$ rad/s^2 CW, $a_B = 298$ m/s^2
5/143 $\mathbf{a}_A = -24\mathbf{i} - 270\mathbf{j}$ ft/sec^2
$\mathbf{a}_D = -265\mathbf{i} + 73.6\mathbf{j}$ ft/sec^2
5/144 $\alpha_{DB} = 1.234$ rad/s^2 CCW
5/145 $\alpha_{OB} = 628$ rad/s^2 CW
5/146 $\alpha = 8.00$ rad/s^2 CW, $(a_B)_t = 8.90$ m/s^2 at $\measuredangle 105°$
5/147 $\alpha_{AB} = 3.81$ rad/s^2 CCW, $\alpha_{CB} = 16.08$ rad/s^2 CCW

5/148 $\alpha_{AB} = 14.38$ rad/s² CW
5/149 $(a_B)_t = -23.9$ m/s², $\alpha = 36.2$ rad/s² CW
5/150 $\mathbf{a}_D = 0.786\mathbf{i}$ m/s²
5/151 $\mathbf{a}_A = -6.11\mathbf{j}$ ft/sec²
5/152 $\boldsymbol{\alpha}_{AC} = -0.0758\mathbf{k}$ rad/sec²
5/153 $\alpha_{AB} = 16.02$ rad/s² CW, $\alpha_{BC} = 13.31$ rad/s² CCW
5/154 $a_A = 4.89$ m/s² right, $a_{AB} = 0.467$ rad/s² CCW
►5/155 $a_E = 0.285$ m/s² right
►5/156 $a_D = 1997$ mm/s²
5/157 $\mathbf{v}_A = -225\mathbf{i} + 180\mathbf{j}$ mm/s
$\mathbf{a}_A = -675\mathbf{i} - 1733\mathbf{j}$ mm/s²
5/158 $\mathbf{v}_A = (r_1\Omega + u)\mathbf{i}$
$\mathbf{a}_A = r_1\dot{\Omega}\mathbf{i} + \left(r_1\Omega^2 + 2\Omega u - \dfrac{u^2}{r_2}\right)\mathbf{j}$, $\mathbf{a}_{\text{Cor}} = 2\Omega u\mathbf{j}$
5/159 $a_A = 10.06$ ft/sec²
5/160 $\mathbf{v}_A = -3.4\mathbf{i}$ m/s, $\mathbf{a}_A = 2\mathbf{i} - 0.667\mathbf{j}$ m/s²
5/161 $\mathbf{v}_{\text{rel}} = 81.2\mathbf{i} + 50\mathbf{j}$ ft/sec
5/162 $\mathbf{a}_{\text{Cor}} = 2\Omega v \sin\theta\mathbf{i}$, (*a*) $a_{\text{Cor}} = 0$, (*b*) $a_{\text{Cor}} = 0.0203$ m/s²
5/163 $\mathbf{v}_{\text{rel}} = 20\mathbf{i} - 9\mathbf{j}$ m/s, No
5/164 $\mathbf{a}_{\text{rel}} = -1.35\mathbf{i} - 6\mathbf{j}$ m/s
5/165 $\mathbf{a}_{\text{Cor}} = -2\omega u\mathbf{i}$
5/166 $\mathbf{v}_{\text{rel}} = -18.33\mathbf{j}$ ft/sec, $\mathbf{a}_{\text{rel}} = 3.33\mathbf{i} + 6.72\mathbf{j}$ ft/sec²
5/167 (*a*) $\mathbf{a}_{A/B} = -1.762\mathbf{i} + 0.698\mathbf{j}$ ft/sec²
(*b*) $\mathbf{a}_A = -10.42\mathbf{i} + 5.70\mathbf{j}$ ft/sec²
5/168 $v_{\text{rel}} = 837$ km/h
5/169 $\mathbf{a}_{\text{rel}} = -4.69\mathbf{k}$ m/s²
5/170 $\omega = 2.17$ rad/s CCW, $v_{\text{rel}} = 0.5$ m/s right
5/171 $\alpha = 12.24$ rad/s² CW, $a_{\text{rel}} = 2.61$ m/s² right
5/172 $\omega = 2.89$ rad/s CCW, $\alpha = 5.70$ rad/s² CCW
5/173 $\delta = \dfrac{\Omega L^2}{v}\sin\theta$, $\delta = 0.01125$ ft or 0.1350 in.
5/174 $\mathbf{v}_{\text{rel}} = -2.71\mathbf{i} - 0.259\mathbf{j}$ m/s
$\mathbf{a}_{\text{rel}} = 0.864\mathbf{i} + 0.0642\mathbf{j}$ m/s²
5/175 $\mathbf{v}_{\text{rel}} = -373\mathbf{i} - 686\mathbf{j}$ ft/sec
5/176 $\mathbf{a}_{\text{rel}} = 15.69\mathbf{i} - 11.05\mathbf{j}$ ft/sec²
5/177 $\mathbf{v}_{\text{rel}} = 22\mathbf{i} - 38.1\mathbf{j}$ ft/sec, $\mathbf{a}_{\text{rel}} = 25.1\mathbf{i} + 14.52\mathbf{j}$ ft/sec²
5/178 $\mathbf{v}_{\text{rel}} = 22\mathbf{i}' + 38.1\mathbf{j}'$ ft/sec
$\mathbf{a}_{\text{rel}} = -27.6\mathbf{i}' + 10.19\mathbf{j}'$ ft/sec²
5/179 $\omega = 4$ rad/s CW, $\alpha = 64$ rad/s² CCW
5/180 $v_{\text{rel}} = 3.93$ m/s at $\measuredangle 19.11°$
$a_{\text{rel}} = 15.22$ m/s² at $\measuredangle 19.11°$
$\omega_{BC} = 1.429$ rad/s CCW, $\alpha_{BC} = 170.0$ rad/s² CW
5/181 $v_{\text{rel}} = 7.71$ m/s at $\measuredangle 19.11°$
$a_{\text{rel}} = 13.77$ m/s² at $\measuredangle 19.11°$
$\omega_{BC} = 1.046$ rad/s CW, $\alpha_{BC} = 117.7$ rad/s² CW
►5/182 $\mathbf{v}_{\text{rel}} = -26\,220\mathbf{i}$ km/h, $\mathbf{a}_{\text{rel}} = -8.02\mathbf{j}$ m/s²
►5/183 $\alpha_{EC} = 12$ rad/sec² CCW
►5/184 $a_B = 27.3$ m/s² right
5/185 $\mathbf{v}_A = r\omega\left(\mathbf{i} + \dfrac{1}{3}\mathbf{j}\right)$, $\mathbf{a}_A = r\left[\left(\dfrac{\omega^2}{3} - \alpha\right)\mathbf{i} - \left(\omega^2 + \dfrac{\alpha}{3}\right)\mathbf{j}\right]$
5/186 $(x_P, y_P) = (-0.3, 0.4)$ m, $r = 0.5$ m
5/187 $t = \dfrac{1}{\sqrt{Kk}}\tan^{-1}\left(\omega_0\sqrt{\dfrac{k}{K}}\right)$
5/188 $\omega = \dfrac{2v}{L}$ CW, $v_C = v$ left
5/189 $\mathbf{v}_P = -0.45\mathbf{i} - 0.3\mathbf{j}$ m/s, $(\mathbf{a}_P)_t = 0.9\mathbf{i} + 0.6\mathbf{j}$ m/s²
$(\mathbf{a}_P)_n = 0.9\mathbf{i} - 1.35\mathbf{j}$ m/s²
5/190 $\omega_{AB} = 1.203$ rad/s CCW
5/191 $\omega = \dfrac{vh}{x^2 + h^2}$ CCW
5/192 $\alpha = -\dfrac{ah}{x^2 + h^2} - \dfrac{2xv^2h}{(x^2 + h^2)^2}$ CW
5/193 $a_C = 83.1$ ft/sec² up
5/194 $\omega_{AB} = 0.1133$ rad/s CW
5/195 $\mathbf{v}_A = 1249\mathbf{i} - 189.1\mathbf{k}$ ft/sec
$\mathbf{v}_B = 176\mathbf{i} + 1089\mathbf{j}$ ft/sec
5/196 $a_P = 11.14$ ft/sec²
5/197 $\theta = 70.5°$
5/198 $a_P = 0.625$ ft/sec²
►5/199 $\omega_{OA} = \dfrac{0.966dv}{d^2 + s^2 + 0.518ds}$ CCW if $v > 0$
5/200 $v_{\mathbf{A}} = 0.1732$ m/s down, $\omega = 1.992$ rad/s CCW
5/201 $v_B = 288$ mm/s
5/202 $\alpha_{DE} = 2.45$ rad/s² CCW
5/203 $\Delta\mathbf{v}_{\text{rel}} = -50.3\mathbf{i} + 87.1\mathbf{j}$ km/h
***5/204** $\omega_{OB} = 0$ when $\theta = 78.5°$ and 282°
***5/205** $(v_{\text{rel}})_{\min} = -2$ m/s at $\theta = 109.5°$
$(v_{\text{rel}})_{\max} = 2$ m/s at $\theta = 251°$
$(a_{\text{rel}})_{\min} = -15$ m/s² at $\theta = 0$ and 360°
$(a_{\text{rel}})_{\max} = 30$ m/s² at $\theta = 180°$
***5/206** $t = 0.0701$ s
***5/207** $|\omega_{AB}|_{\max} = 6.54$ rad/s at $\theta = 202°$
$|\omega_{BC}|_{\max} = 7.47$ rad/s at $\theta = 215°$
***5/208** $|\alpha_{AB}|_{\max} = 88.6$ rad/s² at $\theta = 234°$
$|\alpha_{BC}|_{\max} = 112.2$ rad/s² at $\theta = 182.1°$
***5/209** $|\omega_{AB}|_{\max} = 10.15$ rad/s at $\theta = 203°$
$|\omega_{BC}|_{\max} = 11.83$ rad/s at $\theta = 216°$
***5/210** $a_{\text{rel}} = 0.1067$ m/s² towards *B*
***5/211** $(v_A)_{\max} = 69.6$ ft/sec at $\theta = 72.3°$
***5/212** $a_A = 0$ when $\theta = 72.3°$

CHAPTER 6

6/1 $a = 3g$
6/2 $B = 1.5$ lb left
6/3 $N_A = 25.4$ lb
6/4 Wheel B, $a = 1.344$ m/s^2
6/5 $d = \dfrac{\mu_k h}{2}$
6/6 $(a)\ a = \dfrac{1}{2}g \cot\theta$, $(b)\ a = \dfrac{3}{4}g\cot\theta$, $(c)\ a = g\cot\theta$
6/7 $(a)\ P = \dfrac{c}{b}mg$, $(b)\ P = \dfrac{c}{b}mg$
6/8 $(a)\ N_A = 31.6$ N up, $N_B = 7.62$ N up
$(b)\ N_A = N_B = 19.62$ N up
$(c)\ N_A = 4.38$ N down, $N_B = 43.6$ N up
$a = 15$ m/s^2 right in all cases
6/9 $a = 5.66$ m/s^2
6/10 $P = \sqrt{3}(M + m)g$
6/11 $T = 24.5$ N, $A = 32.9$ N
6/12 $k_T = 6.01mgL$
***6/13** $\theta = 39.5°$
6/14 $F_A = 1.110$ kN at $\measuredangle 60°$, $O_x = 45$ N, $O_y = 667$ N
6/15 $a = 16.43$ ft/sec^2
6/16 $a = 1.443g$
6/17 $N_A = 6.85$ kN up, $N_B = 9.34$ kN up
6/18 $a = 3.92$ m/s^2
6/19 $a = 0.224g$
6/20 $P = \dfrac{4}{3\pi}(M + m)g$, $a = \dfrac{4}{3\pi}g$ right, $\mu_s \geq \dfrac{4}{3\pi}$
6/21 Dynamic: $P = 118.7$ lb, $\theta = 49.2°$, $N_B = 410$ lb up
Static: $N_B = 417$ lb up
6/22 $F_A = 11.37$ kN up, $F_B = 7.05$ kN up
6/23 $M = 196.0$ N·m
6/24 $C = 46.3$ lb
6/25 $N = 257$ kN up
6/26 $a = 16.10$ ft/sec^2, $\theta = 26.6°$
6/27 $(a)\ \theta = 51.3°$, $(b)\ \theta = 24.8°$; $a = \dfrac{5}{4}g$
▶6/28 $a = 16.74$ ft/sec^2
6/29 $W = 6460$ lb
6/30 $\theta = 0.964°$
6/31 $\mu = 1.167$
▶6/32 $R_B = 137.0$ lb, $R_F = 135.7$ lb, $F_f = 1.950$ lb right
6/33 $\alpha = 21.6$ rad/s^2 CCW
6/34 $R = 49.0$ N
6/35 $p = 125.5$ lb/in.^{2}
6/36 $\alpha = 1.193$ rad/s^2 CCW, $F_A = 769$ N
6/37 $d = 14.64$ in.
6/38 $\alpha = 9.12$ rad/sec^2
6/39 $(a)\ \alpha = 1.976$ rad/sec^2 CW, $(b)\ \alpha = 2.06$ rad/sec^2 CW
6/40 $(a)\ \alpha = \dfrac{g}{2r}$ CW, $O = \dfrac{1}{2}mg$, $(b)\ \alpha = \dfrac{2g}{3r}$ CW, $O = \dfrac{1}{3}mg$
6/41 $O = 109.1$ N
6/42 $R = 3.57$ lb
6/43 $\alpha = \dfrac{2M}{mr^2}$
6/44 $\alpha = \dfrac{3g}{5b}$ CW, $O = \dfrac{1}{2}mg$
6/45 $A = 56.3$ N
6/46 $b = 2.20$ in., $R = 14.62$ lb
6/47 $\alpha = \dfrac{8g}{3\pi b}$ CW, $O_x = 0.360mg$ left, $O_y = 0.640mg$ up
6/48 $A = B = 24.5$ N
6/49 $A\text{–}A$: $\alpha = \dfrac{3\sqrt{2}}{5}\dfrac{g}{b}$, $B\text{–}B$: $\alpha = \dfrac{3\sqrt{2}}{7}\dfrac{g}{b}$
6/50 Torus: $\alpha = \dfrac{4gR}{3a^2 + 4R^2}$ CW, $O = \dfrac{3a^2}{3a^2 + 4R^2}mg$
Ring: $\alpha = \dfrac{gR}{a^2 + R^2}$ CW, $O = \dfrac{a^2}{a^2 + R^2}mg$
6/51 $\alpha = 4.06$ rad/sec^2
6/52 $T = 91.1$ N
6/53 $M = \dfrac{\omega\rho d}{\tau}\left[\dfrac{1}{2}\pi r^4 + 4lt\left(\dfrac{1}{3}l^2 + rl + r^2\right)\right]$
6/54 $(a)\ k = 87.6$ mm, $(b)\ F = 2.35$ N, $(c)\ R = 531$ N
6/55 $(a)\ \alpha = 8.46$ rad/s^2, $(b)\ \alpha = 11.16$ rad/s^2
6/56 $\alpha = \dfrac{18g}{11L}$ CW, $O = 0.239mg$
6/57 $M_{\text{motor}} = 0.836$ lb-ft, $M_f = 0.1045$ lb-ft
6/58 $x = \dfrac{l}{2\sqrt{3}}$, $\alpha = \sqrt{3}\dfrac{g}{l}$
6/59 $\alpha = 0.893$ rad/s^2
6/60 $\alpha = \dfrac{3g}{10b}$, $O_y = \dfrac{9}{20}\rho bcg$, $O_z = \dfrac{37}{20}\rho bcg$
6/61 $F_n = 1.721mg\sin\theta$, $F_t = 0.640mg\cos\theta$
6/62 $R = 1341$ N
6/63 $a = \dfrac{2d_Ad_Bd_CM}{4I_Ad_B^{\,2} + 4(I_B + I_C)d_A^{\,2} + m_Dd_A^{\,2}d_C^{\,2}}$ down
6/64 $R = 17.60$ N
6/65 $\alpha = 62.6$ rad/sec^2, $O = 0.492$ lb
6/66 $a = 0.0758y$
6/67 $O = 1087$ lb, $\alpha = 0.0899$ rad/sec^2 CCW
6/68 $(a)\ M_A = 109.8$ N·m CCW
$(b)\ M_B = 5.51$ N·m CCW
▶6/69 $\omega_{\max} = 2.03$ rad/s CW, $R = 5.66$ kN
▶6/70 $\boldsymbol{\alpha} = -85.8\mathbf{i}$ rad/s^2
6/71 $b = 0.553$ m
6/72 $\omega = 7.2$ rad/sec, $v = 3$ ft/sec

6/73 $a = \frac{g}{2}$ down

6/74 $\mathbf{a}_A = \frac{F}{m}(1.681\mathbf{i} + 1.070\mathbf{j})$

6/75 $\mathbf{a}_A = \frac{F}{m}(-1.070\mathbf{i} - 0.681\mathbf{j})$

6/76 $\mathbf{a}_A = -\frac{Fr^2}{mk^2}\mathbf{i} - \left(\frac{F}{m} - r\omega^2\right)\mathbf{j}$

6/77 $\alpha = 48.8$ rad/s^2 CW from above (or $+\mathbf{k}$)
$\bar{a}_x = 0$, $\bar{a}_y = 5$ m/s^2

6/78 A: $\alpha_A = \frac{g}{r}\sin\theta$, $\mu_s = 0$

B: $\alpha_B = \frac{g}{2r}\sin\theta$, $\mu_s = \frac{1}{2}\tan\theta$

6/79 $\bar{a} = 13.80$ ft/sec^2 down incline
$F = 1.714$ lb up incline

6/80 $a_G = 3.88$ m/s^2 up incline, $F = 83.0$ N down incline

6/81 $a_G = 8.07$ m/s^2 up incline, $F = 42.9$ N up incline

6/82 $\omega = \frac{1}{k_O}\sqrt{2a\bar{r}}$

6/83 $A = \left(\frac{1}{2} + \frac{v^2}{3gr}\right)mg$

6/84 $N_A = 0.228mg$ at $\measuredangle 105°$, $a_A = 0.1130g$ down slot

6/85 (a) $R_f = \frac{\bar{I}a}{2dR}$, $R_r = -\frac{\bar{I}a}{2dR}$

(b) $R_f = \frac{mv^2d + \bar{I}a}{2dR}$, $R_r = \frac{mv^2d - \bar{I}a}{2dR}$

6/86 $a = \frac{8mg}{3\pi(m + 3M)}$ along $-\mathbf{i}$

$\alpha = \frac{8(m + M)g}{3\pi r(m + 3M)}$ along $-\mathbf{k}$

6/87 $T_A = 145.4$ lb, $T_B = 161.7$ lb

6/88 $\alpha = 22.8$ rad/s^2 CW, $\mu_s = 0.275$

6/89 $P = 20.2$ lb

6/90 $A_x = 5$ N, $A_y = 57.1$ N

6/91 $\alpha = \frac{12bg}{7b^2 + 3h^2}$ CW, $T_A = \frac{3(b^2 + h^2)}{7b^2 + 3h^2}mg$

6/92 $s = \frac{3d}{2}$

6/93 $N_B = 36.4$ N up

6/94 $T_B = 65.1$ lb

6/95 $T_B = 47.3$ lb, $a_T = 15.24$ ft/sec^2 right

6/96 $M_B = 3.55$ N·m CCW

6/97 $\mathbf{A} = \frac{3M}{2\sqrt{2}l}\mathbf{i}$, $\mathbf{B} = -\frac{3M}{2\sqrt{2}l}\mathbf{j}$

6/98 $\alpha = \frac{84a}{65L}$ CCW

6/99 $T = \frac{mg}{1 + \frac{r^2}{k^2}}$, $a = \frac{g}{1 + \frac{k^2}{r^2}}$, $v = \sqrt{v_O^2 + \frac{2gL}{1 + \frac{k^2}{r^2}}}$

6/100 $\omega = 2.97$ rad/sec CCW

6/101 $N_A = 53.6$ N, $N_B = 53.1$ N

6/102 $s = 18.66$ ft

6/103 $a_A = 8.50$ m/s^2 left, $T = 29.4$ kN

6/104 $v = 11.73$ m/s

6/105 $N_A = 15.93$ kN, $N_B = 18.33$ kN
$\mathbf{a}_G = 2.60\mathbf{i} - 0.782\mathbf{j}$ m/s^2

6/106 $A = 347$ lb

6/107 (a) $T = 1177$ N, (b) $N_A = 17\,980$ N, $N_B = 8500$ N

▶**6/108** $M = 2.58$ N·m CCW

▶**6/109** $M = 3.02$ N·m CCW

▶**6/110** $F = \left[mRn\left(1 - \frac{h^2}{2k^2}\right) + \frac{I_O}{R}\right]\alpha$

6/111 $\omega = 1.660\sqrt{\frac{g}{l}}$ CW

6/112 $\theta = 33.2°$

6/113 (a) $\omega = 0.541\sqrt{\frac{g}{b}}$, (b) $\omega = \sqrt{\frac{g}{b}}$, (c) $\omega = 1.414\sqrt{\frac{g}{b}}$

6/114 $v = 3.01$ m/s

6/115 $v = 2.97$ ft/sec

6/116 $k_T = 3.15$ N·m/rad

6/117 $\omega = 3.05$ rad/sec CCW

6/118 $\omega = 6.23$ rad/sec CW, $O = 86.2$ lb

6/119 $v_A = \sqrt{2gx\sin\theta}$, $v_B = \sqrt{gx\sin\theta}$

6/120 $h = 54.5$ mm

6/121 $v = 5.34$ ft/sec right

6/122 $N = 3240$ rev/min

6/123 $\Delta\omega = -1.236$ rad/sec

6/124 $\omega_{\text{max}} = 0.861\sqrt{\frac{g}{b}}$

6/125 $\omega = 3.31$ rad/s

6/126 $N_C = 124.7$ N

6/127 $k = 1.270$ kN/m

6/128 $\omega = 1.597$ rad/sec CW

6/129 $P = 140.4$ kW or 188.2 hp

6/130 $v_A = 2.45$ m/s right

6/131 $\Delta E = 0.435$ ft-lb

6/132 $\omega_B = 13.54$ rad/s CW

6/133 (a) $m = 1.196$ kg, (b) $\omega = 4.36$ rad/s CW

6/134 $\omega = 2.23$ rad/s CW

6/135 $v_{\text{diff}} = 2.18(10^{-3})\sqrt{xg\sin\theta}$, Torus leads

6/136 $\omega = 6.95$ rad/sec

6/137 (a) $P = 17.59$ kW, (b) $P = 11.06$ kW

6/138 $x = 0.211l$, $\omega_{\text{max}} = 1.861\sqrt{\frac{g}{l}}$ CW

6/139 $v_O = \sqrt{\frac{2r_o d}{m(\bar{k}^2 + r_o^2)}}\sqrt{P(r_o - r_i) - mgr_o\sin\theta}$

6/140 $v = 2.29$ m/s right

6/141 $v_O = 0.757$ m/s down

6/142 $N = 3720$ rev/min

6/143 $\omega = 4\sqrt{\dfrac{(1 - \cos\theta)}{(9\pi - 16)}\dfrac{g}{r}}$

6/144 (*a*) $k = 2280$ N/m, (*b*) $\omega = 1.613$ rad/s CW

***6/145** $\theta_{max} = 72.1°$

$\omega_{max} = 1.947$ rad/s CW when $\theta = 25.6°$

6/146 (*a*) $P = 17.76$ kW, (*b*) $P = 52.0$ kW

6/147 $v_A = \sqrt{3}\sqrt{\dfrac{M\theta}{m} - gb(1 - \cos\theta)}$ right

6/148 (*a*) $v_A = \sqrt{\dfrac{12}{5}gb}$, (*b*) $v_A = \sqrt{\dfrac{12}{5}gb}$

6/149 $\alpha = \dfrac{M}{mb^2} - \dfrac{g}{b}\sin\theta$ (CW if positive)

6/150 $\alpha = \dfrac{3P\sin\theta}{ml}$ CW

6/151 $\alpha = \dfrac{3g\cos\theta}{2b}$

6/152 $\theta = 7.08°$

6/153 $\alpha = 33.7$ rad/sec^2 CW

6/154 $a = \dfrac{2P}{5m}\tan\dfrac{\theta}{2} - g$ (up if positive)

6/155 $a = \dfrac{M}{2mb\sqrt{1 - \left(\dfrac{h}{2b}\right)^2}} - g$ (up if positive)

6/156 $k_T = \dfrac{m\bar{r}}{\theta}(a\cos\theta - g\sin\theta)$

6/157 $\theta = 64.3°$

6/158 $\alpha = \dfrac{P(2\cos^2\theta + 1)}{mb(8\cos^2\theta + 1)}$

6/159 $a = \dfrac{F}{2m} - g$ (up if positive)

6/160 $a = \dfrac{3}{8}\left(\dfrac{P}{m} - \dfrac{3g}{2}\right)$ (right if positive)

6/161 $N = 133.0$ rev/min

6/162 $\alpha = 135.3$ rad/s^2 CCW

6/163 $\alpha = 10.54$ rad/sec^2 CCW

6/164 $M = 0, M_B = 11.44$ kN·m

6/165 $\alpha = 27.3$ rad/s^2 CW

6/166 $a = 0.341$ m/s^2 down incline

6/167 $\omega = 1.811$ rad/s CCW

6/168 $\omega = 1.093$ rad/s

6/169 $\overline{H} = 2.66(10^{40})$ kg·m^2/s

6/170 $\omega = \dfrac{3}{5}\sqrt{\dfrac{2g}{L}}$ CW

6/171 $t = \dfrac{I}{\omega_0 k}$

6/172 $v = 0.5$ m/s right, $\omega = 7.85$ rad/s CW

6/173 $\omega = 1.202$ rad/s CW

6/174 (*a*) $t = 4.00$ s, (*b*) never stops

6/175 $\mathbf{v} = \dfrac{v_M}{M + m}(M\mathbf{i} + m\mathbf{j})$, $\omega = \dfrac{12v_M}{L}\left(\dfrac{m}{4M + 7m}\right)$ CCW

6/176 $H_G = \dfrac{11}{16}mRv_O$, $H_O = \dfrac{37}{32}mRv_O$

6/177 $v = 0.778$ m/s, $T_{av} = 100.7$ N

6/178 $\omega = \dfrac{3mv_1}{(M + m)L}$ CW, $\displaystyle\int_{t_1}^{t_2} O_x dt = \dfrac{M}{2(M + m)}mv_1$ right

6/179 $\mathbf{H}_G = -6.21\mathbf{k}$ lb-ft-sec

6/180 $\mathbf{G} = 3.12\mathbf{i} - 2.76\mathbf{j}$ lb-sec, $\mathbf{H}_O = -43.4\mathbf{k}$ lb-ft-sec

$\dot{\mathbf{H}}_O = -185.7\mathbf{k}$ lb-ft

6/181 $N = 2.04$ rev/s

6/182 $n = 1.405\%$

6/183 $\omega = 1.593$ rad/sec CCW, $n = 91.7\%$

6/184 $T = 0.750 + 0.01719t$ N

6/185 $\mathbf{v}_B = 0.267\mathbf{i} - 7.2\mathbf{j}$ m/s

6/186 $N = 32.0$ rev/min

6/187 $v_1 = 4.88$ m/s

6/188 $\omega = \dfrac{\left(\dfrac{l}{2} - x\right)\sqrt{2gh}}{\dfrac{1}{3}l^2 - lx + x^2}$

$x = 0$: $\omega = \dfrac{3}{2l}\sqrt{2gh}$ CW

$x = \dfrac{l}{2}$: $\omega = 0$

$x = l$: $\omega = \dfrac{3}{2l}\sqrt{2gh}$ CCW

6/189 (*a*) $\Omega' = 16.27$ rev/min CCW

(*b*) $\Omega' = 22.0$ rev/min CCW

6/190 (*a*) $\omega_2 = 6.57$ rad/s, (*b*) $\omega_3 = 1.757$ rad/s

(*c*) $\omega_4 = 1.757$ rad/s

6/191 $t = \dfrac{2v_0}{g(7\mu_k\cos\theta - 2\sin\theta)}$

$v = \dfrac{5v_0\mu_k}{7\mu_k - 2\tan\theta}$ down incline

$\omega = \dfrac{5v_0\mu_k}{r(7\mu_k - 2\tan\theta)}$ CW

6/192 $t = \dfrac{2r\omega_0}{g(2\sin\theta + 7\mu_k\cos\theta)}$

$v = \dfrac{2r\omega_0(\sin\theta + \mu_k\cos\theta)}{2\sin\theta + 7\mu_k\cos\theta}$ down incline

$\omega = \dfrac{2\omega_0(\sin\theta + \mu_k\cos\theta)}{2\sin\theta + 7\mu_k\cos\theta}$ CW

6/193 $T_{av} = 30.1$ lb

6/194 $v_1 = 2.41\sqrt{gl}$

6/195 $N = 4.78$ rev/sec

6/196 $N = 37.0$ rev/min

6/197 $M = 235\ \text{N}\cdot\text{m}$, $T_{upper} = 1252\ \text{N}$
$T_{lower} = 1194\ \text{N}$, $P = 3760\ \text{W}$
6/198 $\omega = 0.974$ rad/sec
6/199 $I = 3.45\ \text{kg}\cdot\text{m}^2$
6/200 $N_2 = 569$ rev/min, No
6/201 $P_{in} = 4500$ W
6/202 $v' = \frac{1}{3}[v + 2(v - v_0)\cos\theta]$, (*a*) $n = 0.543\%$
(*b*) $n = -3.88\%$
6/203 $\Omega = 1.135$ rad/s
▶6/204 $\mu_k = 0.1763$, $t = 1.037$ s, $v = 4.36$ m/s
6/205 $\theta = 35.5°$
6/206 $U_f = -0.336$ J
6/207 $\mathbf{G} = 6\mathbf{j}\ \text{N}\cdot\text{s}$, $\mathbf{H}_G = 0.45\mathbf{k}\ \text{N}\cdot\text{m}\cdot\text{s}$
6/208 $N = 504$ rev/min, $|\Delta E| = 98.1$ J
6/209 $M = 106.3$ lb-ft CCW, $A = 51.7$ lb
6/210 $a = 2.22\ \text{m/s}^2$ right
6/211 $P = 8.22$ kW
6/212 $(\mu_s)_{min} = \frac{2}{5}\tan\theta$
6/213 $v = 0.733$ ft/sec
6/214 $v_A = 11.35$ ft/sec down
6/215 $\frac{h}{b} \geq \frac{1}{\mu_s}$, $a = \frac{\mu_s\cos\theta - \sin\theta}{\cos\theta + \mu_s\sin\theta}g$
6/216 $\frac{h}{b} \geq \frac{1}{\mu_s}$, $a = \frac{\sin\theta + \mu_s\cos\theta}{\cos\theta - \mu_s\sin\theta}g$
6/217 $M = 2mr\dot{r}\omega$
6/218 $B = 2130$ lb left on fork
6/219 $t = 1206$ s
6/220 $(x_P, y_P) = (0.320r, 0.424r)$
6/221 $\omega_2 = \dfrac{m_2x + m_1\frac{L}{2}}{\frac{1}{3}m_1L^2 + m_2x^2}v_1$ CW
$m_2 = \frac{1}{2}m_1$: $(\omega_2)_{max} = 1.718\frac{v_1}{L}$ when $x = 0.291L$
6/222 $M_x = -46.0$ lb-in., $M_y = 16.11$ lb-in.
$M_z = -18.41$ lb-in.
▶6/223 $\bar{a}_x = -6.95\ \text{ft/sec}^2$, $\bar{a}_y = -4.01\ \text{ft/sec}^2$
▶6/224 $M = 0.482\ \text{N}\cdot\text{m}$ CCW
***6/225** $\omega_{max} = 2.35$ rad/s CW at $\theta = 18.88°$
***6/226** $\theta_{max} = 23.0°$, $\dot{\theta}_{max} = 0.389$ rad/s CW at $\theta = 11.52°$
***6/227** $O_x = \frac{3mg}{4}\sin\theta(3\cos\theta - 2)$, $O_y = \frac{mg}{4}(3\cos\theta - 1)^2$
***6/228** $t = 0.302$ s
***6/229** $\omega_{max} = 0.680$ rad/s CCW at $\theta = 22.4°$, $\theta_{max} = 45.9°$
***6/230** $\beta = 6.05°$, $\theta = 4.95°$
***6/231** $t = 2.85$ sec, $v_A = 75.7$ ft/sec
***6/232** $|M|_{max} = 3.93\ \text{N}\cdot\text{m}$ at $\theta = 145.6°$
$A_{max} = 88.2$ N at $\theta = 181.2°$
$B_{max} = C_{max} = 91.7$ N at $\theta = 184.2°$
***6/233** $|M|_{max} = 3.61\ \text{N}\cdot\text{m}$ at $\theta = 360°$
$A_{max} = 63.8$ N at $\theta = 287°$
$B_{max} = C_{max} = 73.7$ N at $\theta = 185.3°$
***6/234** $|M|_{max} = 4.64\ \text{N}\cdot\text{m}$ at $\theta = 144.6°$
$A_{max} = 105.0$ N at $\theta = 181.5°$
$B_{max} = 108.0$ N at $\theta = 184.1°$
$C_{max} = 121.7$ N at $\theta = 183.8°$
$O_{max} = 107.4$ N at $\theta = 180.7°$

Chapter 7

7/1 Finite rotations cannot be added as proper vectors
7/2 Infinitesimal rotations add as proper vectors
7/3 $v_z = -6$ ft/sec, $R = 2.81$ in., $a_P = 11{,}080\ \text{in./sec}^2$
7/4 $a_B = 1285\ \text{m/s}^2$
7/5 $\mathbf{v}_A = -0.8\mathbf{i} - 1.5\mathbf{j} + 2\mathbf{k}$ m/s, $v_B = 2.62$ m/s
7/6 No answer
7/7 $N_2 = 440$ rev/min
7/8 $\mathbf{v} = \omega[-l\cos\theta\mathbf{i} + (d\cos\theta - h\sin\theta)\mathbf{j} - l\sin\theta\mathbf{k}]$
$\mathbf{a} = \omega^2[(h\sin\theta\cos\theta - d\cos^2\theta)\mathbf{i} - l\mathbf{j} + (h\sin^2\theta - d\cos\theta\sin\theta)\mathbf{k}]$
7/9 $\boldsymbol{\omega} = p\mathbf{j} + \omega_0\mathbf{k}$, $\boldsymbol{\alpha} = -p\omega_0\mathbf{i}$
7/10 $\boldsymbol{\alpha} = -1.2\mathbf{i}\ \text{rad/sec}^2$, $\mathbf{a}_P = 35.8\mathbf{j} - 80\mathbf{k}\ \text{in./sec}^2$
7/11 $\boldsymbol{\alpha} = 12\pi^2\mathbf{j}\ \text{rad/sec}^2$, $\mathbf{v}_A = 5\pi(-4\mathbf{i} + 6\mathbf{j} - 3\mathbf{k})$ in./sec
$\mathbf{a}_A = -5\pi^2(25\mathbf{j} + 18\mathbf{k})\ \text{in./sec}^2$
7/12 $\boldsymbol{\alpha} = 50\pi\left(\frac{\pi}{2\sqrt{3}}\mathbf{i} + \mathbf{k}\right)\ \text{rad/sec}^2$
7/13 (*a*) $\omega = 26.5$ rad/s, (*b*) $\omega = 17.32$ rad/s
7/14 $\boldsymbol{\omega} = -0.4\mathbf{i} + 2.69\mathbf{k}$ rad/s, $\boldsymbol{\alpha} = 0.8\mathbf{j}\ \text{rad/s}^2$
7/15 $\boldsymbol{\alpha} = -1.5\mathbf{i} + 0.8\mathbf{j} + 2.60\mathbf{k}\ \text{rad/s}^2$
7/16 $\omega = 2.5$ rad/s, $\boldsymbol{\alpha} = 3\mathbf{j}\ \text{rad/s}^2$
7/17 $\boldsymbol{\omega} = 0.693\mathbf{j} + 2.40\mathbf{k}$ rad/s, $\boldsymbol{\alpha} = -1.386\mathbf{i}\ \text{rad/s}^2$
7/18 $\boldsymbol{\alpha} = -\left(\frac{2\pi}{\tau}\right)^2\frac{R}{r}\mathbf{i}$
7/19 $\mathbf{v}_A = \frac{2\pi R}{\tau}\left(\mathbf{i} - \mathbf{j} - \frac{r}{R}\mathbf{k}\right)$
$\mathbf{a}_A = -\left(\frac{2\pi}{\tau}\right)^2R\left[\left(\frac{R}{r} + \frac{r}{R}\right)\mathbf{i} + \mathbf{k}\right]$

7/20 $v_B = 3.95$ m/s, $a_B = 72.2$ m/s^2

7/21 $v_P = 3.48$ m/s, $a_P = 1.104$ m/s^2

7/22 (*a*) $\boldsymbol{\omega} = \pi\left(-\frac{2\pi}{3}\mathbf{j} + \mathbf{k}\right)$, $\boldsymbol{\alpha} = \frac{2\pi^3}{3}\mathbf{i}$

(*b*) $\boldsymbol{\omega} = \pi\mathbf{k}$, $\boldsymbol{\alpha} = \frac{8\pi^3}{3}\mathbf{j}$

7/23 $\mathbf{v}_A = 0.691\mathbf{j} + 1.448\mathbf{k}$ m/s, $\mathbf{a}_A = -11.70\mathbf{i}$ m/s^2

7/24 (*a*) $\boldsymbol{\alpha} = 6\mathbf{j} - 8\mathbf{k}$ rad/s^2, $a_A = 21.2$ m/s^2

(*b*) $\boldsymbol{\alpha} = 8\mathbf{i}$ rad/s^2, $a_A = 10.67$ m/s^2

7/25 $\boldsymbol{\omega} = -0.785\mathbf{i} - 2.60\mathbf{j} + 2.5\mathbf{k}$ rad/s, $\alpha = 11.44$ rad/s^2

▶7/26 $\alpha = 6.32$ rad/s^2

▶7/27 $\mathbf{v}_B = -14.35\mathbf{j}$ in./sec, $\mathbf{a}_B = 338\mathbf{i} + 194.8\mathbf{k}$ in./sec^2

$\boldsymbol{\alpha} = -31.0\mathbf{j}$ rad/sec^2

▶7/28 $\boldsymbol{\omega} = v\sqrt{\frac{1}{h^2} + \frac{1}{r^2}}\,\mathbf{i}$, $\boldsymbol{\alpha} = -\frac{v^2}{h^2}\left(\frac{r}{h} + \frac{h}{r}\right)\mathbf{j}$

7/29 $p = 28.2$ rad/s, $\mathbf{v}_{B/A} = 4.10\mathbf{i}$ m/s

7/30 $\boldsymbol{\alpha} = pq\mathbf{j}$

7/31 $\omega = 10.77$ rad/s, $\boldsymbol{\alpha} = -40\mathbf{j}$ rad/s^2

7/32 $\boldsymbol{\alpha} = -40\mathbf{j} + 6\mathbf{k}$ rad/s^2

7/33 $\mathbf{v}_A = -3\mathbf{i} - 1.6\mathbf{j} + 1.2\mathbf{k}$ m/s

$\mathbf{a}_A = -34.8\mathbf{j} - 6.4\mathbf{k}$ m/s^2

7/34 $\boldsymbol{\alpha} = -1.2\pi(\sqrt{3}\mathbf{i} + \mathbf{k})$ rad/s^2

7/35 $\boldsymbol{\omega}_n = \frac{1}{49}(-3\mathbf{i} + 20\mathbf{j} + 9\mathbf{k})$ rad/sec

7/36 $\boldsymbol{\alpha} = -\Omega p \sin\beta\mathbf{i} + \dot\beta(p\cos\beta - \Omega)\mathbf{j} - p\dot\beta\sin\beta\mathbf{k}$

7/37 $v_A = 6.8$ ft/sec, $a_A = 20.8$ ft/sec^2

7/38 $\mathbf{v}_A = -0.636\mathbf{i} - 4.87\mathbf{j} + 1.273\mathbf{k}$ m/s

7/39 $\mathbf{v}_A = -6\mathbf{j}$ m/s

7/40 (*a*) $\boldsymbol{\alpha} = -(3.88\mathbf{i} + 3.49\mathbf{j})10^{-3}$ rad/s^2

(*b*) $\boldsymbol{\alpha} = -3.49(10^{-3})\mathbf{j}$ rad/s^2

7/41 $\boldsymbol{\alpha} = p\omega_2\mathbf{i} - p\omega_1\mathbf{j} + \omega_1\omega_2\mathbf{k}$

7/42 $\boldsymbol{\alpha} = -3\mathbf{i} - 4\mathbf{j}$ rad/sec^2

7/43 $\mathbf{v}_A = \pi(0.1\mathbf{i} + 0.8\mathbf{j} + 0.08\mathbf{k})$ m/s

$\mathbf{a}_A = -\pi^2(6.32\mathbf{i} + 0.1\mathbf{k})$ m/s^2

7/44 $\mathbf{v}_A = \pi(0.1\mathbf{i} + 0.8\mathbf{j} + 0.08\mathbf{k})$ m/s

$\mathbf{a}_A = -\pi^2(6.32\mathbf{i} + 0.1\mathbf{k})$ m/s^2

7/45 $\mathbf{v}_A = -\Omega(R + b\sin\beta)\mathbf{i} + b\dot\beta\cos\beta\mathbf{j} - b\dot\beta\sin\beta\mathbf{k}$

$\mathbf{a}_A = -2b\Omega\dot\beta\cos\beta\mathbf{i}$

$- [\Omega^2(R + b\sin\beta) + b\dot\beta^2\sin\beta]\mathbf{j} - b\dot\beta^2\cos\beta\mathbf{k}$

7/46 $\boldsymbol{\alpha} = -40\pi^2\mathbf{i}$ rad/s^2, $\mathbf{a}_A = 2\pi^2(-2.4\mathbf{i} + 4\mathbf{j} - 5\mathbf{k})$ m/s^2

7/47 $\dot{\boldsymbol{\omega}} = \frac{1}{8}\mathbf{i}$ rad/sec^2

$\mathbf{a}_A = 0.313\mathbf{i} - 2.43\mathbf{j} - 0.1083\mathbf{k}$ ft/sec^2

7/48 $\mathbf{v}_A = -rp\mathbf{i} - (r\omega_1 + b\omega_2)\mathbf{k}$

$\mathbf{a}_A = -\omega_2(2r\omega_1 + b\omega_2)\mathbf{i} - r(\omega_1{}^2 + p^2)\mathbf{j} + 2rp\omega_2\mathbf{k}$

▶7/49 $\boldsymbol{\omega} = \pi(-3\mathbf{i} + \sqrt{3}\mathbf{j} + 5\mathbf{k})$ rad/sec

$\boldsymbol{\alpha} = \pi^2(4\sqrt{3}\mathbf{i} + 9\mathbf{j} + 3\sqrt{3}\mathbf{k})$ rad/sec^2

▶7/50 $\boldsymbol{\omega} = p\left[\cos\theta\mathbf{j} + \left(\sin\theta + \frac{R}{r}\right)\mathbf{k}\right]$, $\boldsymbol{\alpha} = \frac{p^2R}{r}\cos\theta\mathbf{i}$

▶7/51 $\alpha = 42.8$ rad/s^2

▶7/52 $\mathbf{v}_B = -\frac{2}{3}\mathbf{k}$ m/s, $\boldsymbol{\omega}_n = \frac{10}{49}\left(\frac{40}{3}\mathbf{i} - 2\mathbf{j} + 6\mathbf{k}\right)$ rad/s

7/53 $G = \sqrt{2}mb\omega$, $H_O = 3mb^2\omega$

7/54 $\mathbf{H}_O = mb^2\omega\left[\mathbf{i} + 2\mathbf{j} - \left(\frac{l^2}{6b^2} + 2\right)\mathbf{k}\right]$

7/55 $\mathbf{H}_G = -1.613\mathbf{j} - 744\mathbf{k}$ kg·m^2/s

$\mathbf{H}_A = -2.70\mathbf{j} - 744\mathbf{k}$ kg·m^2/s

7/56 $\mathbf{H}_O = \rho b^3\omega\left(-\frac{1}{2}\mathbf{i} - \frac{3}{2}\mathbf{j} + \frac{8}{3}\mathbf{k}\right)$, $T = \frac{4}{3}\rho b^3\omega^2$

7/57 $\mathbf{H}_G = \frac{1}{4}\rho b^3\omega(-\mathbf{i} - \mathbf{j} + 2\mathbf{k})$

7/58 $\mathbf{H}_O = \frac{1}{3}ml^2\omega\sin\theta(\sin\theta\mathbf{j} - \cos\theta\mathbf{k})$

7/59 $\mathbf{H}_O = mr\omega\left[-\frac{2}{3\pi}(b + 2c)\mathbf{j} + \frac{r}{2}\mathbf{k}\right]$

7/60 $\mathbf{H}_O = m\omega\left(\frac{b^2}{3} + \frac{r^2}{4} + h^2\right)\mathbf{i} + \frac{1}{2}mr^2p\mathbf{j}$

7/61 $\mathbf{H}_G = Ip(\mathbf{i} + \mathbf{j} + \mathbf{k}) + 2(I + mb^2)\boldsymbol{\Omega}$

where $\boldsymbol{\Omega} = \Omega_x\mathbf{i} + \Omega_y\mathbf{j} + \Omega_z\mathbf{k}$

7/62 $\mathbf{H}_O = -0.01\mathbf{i} + 0.0045\mathbf{j} + 0.0576\mathbf{k}$ lb-ft-sec

7/63 $\mathbf{H}_O = 2.38\mathbf{j} + 25.6\mathbf{k}$ lb-ft-sec, $T = 805$ ft-lb

7/64 $\mathbf{H}_O = \pi(-0.4\mathbf{j} + 0.6\mathbf{k})$ N·m·s, $T = 59.2$ J

7/65 $\mathbf{H}_O = \frac{1}{4}mr^2\omega[-\sin\alpha\cos\alpha\mathbf{i}$

$+ (\sin^2\alpha + 2\cos^2\alpha)\mathbf{k}]$, $\beta = 4.96°$

7/66 $\mathbf{H}_O = \frac{3}{10}mr^2\left[\left(\frac{1}{2} + \frac{2h^2}{r^2}\right)\Omega\mathbf{i} + p\mathbf{k}\right]$

$T = \frac{3}{10}mr^2\left[\left(\frac{1}{4} + \frac{h^2}{r^2}\right)\Omega^2 + \frac{p^2}{2}\right]$

7/67 $\mathbf{H}_O = 2m\omega\left(\frac{1}{3}l^2\sin^2\beta + b^2\right)\mathbf{k}$

7/68 $\mathbf{H}_G = -2\pi mfk'^2\sin\theta\mathbf{i} + mk^2(p + 2\pi f\cos\theta)\mathbf{k}$

7/69 $\mathbf{H}_O = \frac{1}{4}mr^2\left[-\omega_1\mathbf{i} + \left(1 + \frac{4b^2}{r^2}\right)\omega_2\mathbf{j} + 2p\mathbf{k}\right]$

$T = \frac{1}{8}mr^2\left[\omega_1{}^2 + \left(1 + \frac{4b^2}{r^2}\right)\omega_2{}^2 + 2p^2\right]$

7/70 $\mathbf{H}_{O'} = 0.1626(\mathbf{i} + 8\mathbf{j})$ lb-ft-sec, $T = 148.1$ ft-lb

7/71 $\mathbf{H}_O = m\omega\left[\frac{1}{6}c^2\sin 2\beta\mathbf{j} + \left(\frac{2}{5}r^2 + \frac{1}{3}c^2\cos^2\beta + 2b^2\right)\mathbf{k}\right]$

7/72 $\mathbf{H}_O = \frac{1}{6}mb^2\omega\sin 2\theta\mathbf{i}$

$+ m\omega\left(\frac{1}{3}c^2 + \frac{1}{3}b^2\cos^2\theta + a^2 + ac\right)\mathbf{k}$

$I_{\max} = m\left(\frac{c^2 + b^2}{3} + a^2 + ac\right)$, $I_{\min} = \frac{1}{3}mb^2$

$I_{\text{int}} = m\left(\frac{1}{3}c^2 + a^2 + ac\right)$

7/73 $M = \frac{1}{2}mbl\omega^2$

7/74 $A_x = B_x = 0, A_y = -\frac{1}{3}mR\omega^2, B_y = \frac{1}{3}mR\omega^2$

7/75 $\mathbf{B} = \frac{mbl\omega^2}{2c}(\cos\theta\mathbf{i} + \sin\theta\mathbf{j})$

7/76 $B_x = \frac{3Mb}{2lc}\sin\theta, B_y = -\frac{3Mb}{2lc}\cos\theta$

7/77 $\mathbf{M} = -2\rho b^3\omega^2\mathbf{i}$

7/78 $A = 576$ N, $B = 247$ N

7/79 $\mathbf{M} = -\frac{2}{\pi}mr\omega^2\mathbf{j}$

7/80 $\mathbf{M} = -\frac{4M_O}{3\pi}\mathbf{i}$

7/81 $\alpha = \frac{M}{I_O\cos^2\theta + I\sin^2\theta}$

7/82 $\mathbf{M} = -79.0\mathbf{i}$ N·m

7/83 $\mathbf{A} = 1608\mathbf{i}$ N, $\mathbf{B} = -1608\mathbf{i}$ N

7/84 $\mathbf{A} = -91.7\mathbf{j}$ N, $\mathbf{B} = 91.7\mathbf{j}$ N, $M = 10.8$ N·m

7/85 $\theta = \sin^{-1}\left(\frac{3g}{2\omega^2}\frac{b^2 - c^2}{b^3 + c^3}\right)$ if $\omega^2 \geq \frac{3g}{2}\frac{b^2 - c^2}{b^3 + c^3}$
else $\theta = 90°$

7/86 $\mathbf{M} = \frac{1}{8}mr^2\omega^2\sin 2\alpha\mathbf{j}$

7/87 $\beta = \cos^{-1}\left(\frac{4g}{5R\omega^2}\right)$ if $\omega^2 \geq \frac{4g}{5R}$, else $\beta = 0$

7/88 $N_A = \frac{mg}{2} - \frac{m_2v^2}{2\pi r}, N_B = \frac{mg}{2} + \frac{m_2v^2}{2\pi r}$

7/89 $M_x = \frac{1}{6}mb^2\omega^2\sin 2\beta$

7/90 $M_y = -\frac{1}{6}mb^2\alpha\sin 2\beta, M_z = \frac{1}{12}mb^2\alpha(1 + 4\sin^2\beta)$

7/91 $\omega = 2\sqrt{\frac{\sqrt{3}g}{l}}$

7/92 $M = \frac{2mr}{\pi}(g + 2r\omega^2)$

►7/93 $A = \frac{mg}{6}$

►7/94 $A = \frac{mg}{3}\left(\frac{7a + 2b}{2a + b}\right)$

7/95 CCW as viewed from above

7/96 Tendency to rotate to student's right

7/97 Decreased

7/98 $R = 712$ N

7/99 $b = 216$ mm

7/100 $M = M_1 = mk^2\Omega\frac{v}{r}$

7/101 Right-side normal forces are increased

7/102 $M = 1681$ N·m, Left rudder

7/103 $C = D = 948$ N

7/104 $\frac{h}{r} = \sqrt{\frac{3}{2}}$

7/105 $M = 3970$ lb-ft, CCW deflection

7/106 $\mathbf{\Omega} = 6.67\mathbf{k}$ rad/sec

7/107 $M = 5410$ kN·m, (*b*)

7/108 (*a*) No precession, $M_A = 12.56$ N·m
(*b*) $\Omega = 0.723$ rad/s, $M_A = 3.14$ N·m

7/109 $M_A = 30.9$ N·m, $M_O = 0$

7/110 $\Delta R_A = 98.1$ lb increase, $\Delta R_B = 98.1$ lb decrease

7/111 $\dot{\psi} = -6$ rev/min (retrograde precession)

7/112 $\dot{\psi} = \dot{\psi}_1 = 124.2$ rad/s (direct precession)

7/113 $R_A = 7.80$ lb up, $R_B = 12.20$ lb up

7/114 $\tau = 1.831$ s, Negative z-direction

7/115 $\mathbf{\Omega} = -1.230\mathbf{K}$ rad/sec, $\mathbf{M} = 67.7\mathbf{i}$ lb-in.

7/116 $\ddot{\psi} = \frac{M}{m({k_x}^2\cos^2\gamma + {k_z}^2\sin^2\gamma)}$

7/117 $f = 10$ Hz

7/118 $\tau = 0.0996$ s, Retrograde precession

7/119 (*a*) $p = 4\pi$ rad/s, $\theta = 0, \beta = 0, \dot{\psi} = 0$
(*b*) $p = 4\pi$ rad/s, $\theta = 10°, \beta = 3.03°$
$\dot{\psi} = 5.47$ rad/s
(*c*) $p = 0, \theta = 90°, \beta = 90°, \dot{\psi} = 4\pi$ rad/s

7/120 $\tau = 0.443$ sec

7/121 $\mathbf{M} = -196.3\mathbf{i}$ N·m, $M_O = 319$ N·m

7/122 $M = \frac{1}{12}ml^2p\omega_0\sin 2\phi$

►7/123 $A_z = -\frac{m\dot{\theta}}{2}\left(\frac{r^2p}{2b} + l\dot{\theta}\right), B_z = \frac{m\dot{\theta}}{2}\left(\frac{r^2p}{2b} - l\dot{\theta}\right)$
where $\dot{\theta} = 2\sqrt{\frac{2gl}{r^2 + 4l^2}}$

►7/124 $M_x = -I\omega\Omega_0\cos\omega t, M_y = 0, M_z = -I\omega\Omega_0\sin\omega t$

►7/125 (*a*) $\frac{h}{r} = \frac{1}{2}$

►7/126 $\mathbf{M} = 97.9\mathbf{i}$ lb-ft, $T = 73.8$ ft-lb

7/127 Direct precession: $\frac{l}{r} > \sqrt{6}$
Retrograde precession: $\frac{l}{r} < \sqrt{6}$

7/128 $\mathbf{H} = \frac{ma^2\omega}{6\sqrt{3}}(\mathbf{i} + \mathbf{j} + \mathbf{k})$

7/129 $p = \frac{mvh}{m_0k^2}$, Opposite to car wheels

7/130 $\boldsymbol{\alpha} = 77.9\mathbf{i}$ rad/s^2

7/131 $\boldsymbol{\alpha} = 8\sqrt{3}\mathbf{i} + 120\sqrt{3}\mathbf{j} + 52\mathbf{k}$ rad/sec^2
$\mathbf{a}_A = -2090\mathbf{i} - 369\mathbf{j} + 4810\mathbf{k}$ in./sec^2

7/132 $\boldsymbol{\omega}_n = \frac{9}{49}(2\mathbf{i} + \mathbf{k})$ rad/sec

7/133 $\mathbf{\Omega} = \Omega\mathbf{k}, \tau = \frac{4\pi r^2 p}{5gh}$

7/134 $\mathbf{H}_O = -0.707\mathbf{j} + 4.45\mathbf{k}$ kg·m^2/s, $T = 69.9$ J

7/135 $\boldsymbol{\omega} = \frac{2\pi}{\tau}\left[\left(\frac{r}{R} - \frac{R}{r}\right)\mathbf{j} + \frac{\sqrt{R^2 - r^2}}{R}\mathbf{k}\right]$

7/136 $\alpha = \left(\frac{2\pi}{\tau}\right)^2 \frac{\sqrt{R^2 - r^2}}{r}\mathbf{i}$

7/137 $\mathbf{v}_A = -\frac{4\pi}{\tau}\left(R - \frac{r^2}{R}\right)\mathbf{i}$

7/138 $\mathbf{a}_A = \left(\frac{2\pi}{\tau}\right)^2\left[\sqrt{R^2 - r^2}\left(\frac{2r^2}{R^2} - 3\right)\mathbf{j} + \left(3r - \frac{R^2}{r} - \frac{2r^3}{R^2}\right)\mathbf{k}\right]$

7/139 $N = 1.988$ cycles/min

7/140 $\mathbf{H}_O = 0.421\mathbf{j} + 1.281\mathbf{k}$ lb-ft-sec, $T = 11.30$ ft-lb

7/141 $\mathbf{H}_O = 0.0867\mathbf{i} + 0.421\mathbf{j} + 1.281\mathbf{k}$ lb-ft-sec
$T = 11.85$ ft-lb

7/142 $R_A = R_B = 159.3$ lb

7/143 $M = 271$ lb-ft

7/144 $M = 13.33$ N·m

7/145 $M = 2.70$ N·m

▶7/146 $\mathbf{H}_O = mr\omega\left(\frac{b}{2}\mathbf{i} + \frac{b}{\pi}\mathbf{j} + 2r\mathbf{k}\right)$

CHAPTER 8

8/1 $k = 701$ N/m, $k = 4.00$ lb/in., $k = 48.0$ lb/ft

8/2 $\omega_n = 18$ rad/sec, $f_n = 2.86$ Hz

8/3 $x = -2 \cos 18t$ in., $v_{max} = 36$ in./sec or 3 ft/sec
$a_{max} = 648$ in./sec^2 or 54 ft/sec^2

8/4 $x = 2.06 \sin (18t + 1.816)$ in., $C = 2.06$ in.
$\tau = 0.349$ sec

8/5 $\delta_{st} = 0.200$ m, $\tau = 0.898$ s, $v_{max} = 0.7$ m/s

8/6 $y = -0.0548$ m, $v = -0.586$ m/s, $a = 2.68$ m/s^2
$a_{max} = 4.9$ m/s^2

8/7 $f_n = \frac{1}{\pi}\sqrt{\frac{k}{m}}$

8/8 $f_n = 7.40$ Hz

8/9 $\tau = 2\pi\sqrt{\frac{m}{5k}}$

8/10 $f_n = 2.23$ Hz

8/11 $x = 25 \cos 14.01t$ mm

8/12 $\omega_n = 3\sqrt{\frac{k}{m}}$

8/13 $f_n = 1.332$ Hz

8/14 $x = -0.0371$ in., $v = -4.5$ in./sec, $\tau = 0.1305$ sec

8/15 (*a*) $k = 474$ kN/m, (*b*) $f_n = 0.905$ Hz

8/16 $f_n = 1.654$ Hz

8/17 (*a*) $k = k_1 + k_2$, (*b*) $k = \frac{k_1 k_2}{k_1 + k_2}$

8/18 $m = 2.55$ kg, $\mu_s = 0.358$

8/19 $\delta_{st} = \frac{mg \sin \theta}{3k}$, $\tau = 2\pi\sqrt{\frac{m}{3k}}$

8/20 (*a*) $v = 88.0 \cos 21.5t$ in./sec, (*b*) $x_{max} = 4.09$ in.

8/21 $\delta_{st} = 89.5$ mm

8/22 $\omega_n = 2\sqrt{\frac{6EI}{mL^3}}$

8/23 $\omega_n = \frac{1}{3}\sqrt{\frac{k}{m}}$

▶8/24 $\ddot{y} + \frac{k}{mL^2}y^3 = 0$

8/25 $\zeta = 0.418$

8/26 $c = 38.0$ N·s/m

8/27 $\zeta = 0.436$

8/28 $\zeta = 0.6$

8/29 $c = 2050$ N·s/m

8/30 $x = x_0(\cos 10.94t + 0.460 \sin 10.94t)e^{-5.03t}$

8/31 $\zeta = \frac{\delta_N}{\sqrt{(2\pi N)^2 + \delta_N{}^2}}$, where $\delta_N = \ln\left(\frac{x_0}{x_N}\right)$

8/32 $c = 1.721$ N·s/m

8/33 $\zeta = 0.559$

8/34 $c = 16.24(10^3)$ N·s/m

8/35 $\ddot{x}_1 + \frac{a^2c_1 + b^2c_2}{a^2m_1 + b^2m_2}\dot{x}_1 + \frac{a^2k_1 + b^2k_2}{a^2m_1 + b^2m_2}x_1 = 0$

8/36 $x_1 = -0.1630x_0$

8/37 $\ddot{x} + \frac{4c_1 + 9c_2}{4m_1 + 9m_2}\dot{x} + \frac{4k_1 + 9k_2}{4m_1 + 9m_2}x = 0$
$\omega_n = \sqrt{\frac{4k_1 + 9k_2}{4m_1 + 9m_2}}$, $\zeta = \frac{4c_1 + 9c_2}{2\sqrt{(4k_1 + 9k_2)(4m_1 + 9m_2)}}$

8/38 $(\dot{x}_0)_c = -\omega_n x_0$

8/39 (*a*) $x = 0.368$ ft or 4.42 in., (*b*) $x = 0.393$ ft or 4.72 in.

8/40 $\ddot{x} + \frac{a^2c_1 + b^2c_2 + c^2c_3}{a^2m_1 + b^2m_2 + c^2m_3}\dot{x} + \frac{a^2k_1 + b^2k_2 + c^2k_3}{a^2m_1 + b^2m_2 + c^2m_3}x = 0$
$\omega_n = 6.54$ rad/s, $\zeta = 0.254$

8/41 $\zeta = 0.274$, $c = 39.9$ lb-sec/ft

8/42 $x = x_0(1.171e^{-2.67t} - 0.1708e^{-18.33t})$

8/43 $\ddot{x} + \frac{b^2c}{a^2m}\dot{x} + \frac{k}{m}x = 0$, $\zeta = \frac{b^2c}{2a^2\sqrt{km}}$

▶8/44 $r = \frac{2\mu_k g}{\pi}\sqrt{\frac{m}{k}}$

8/45 $\zeta = 0.1936$

8/46 (*a*) $X = 13.44$ mm, (*b*) $X = 22.7$ mm

8/47 $\omega < 4.99$ rad/s and $\omega > 6.86$ rad/s

8/48 $\omega < 5.18$ rad/s and $\omega > 6.61$ rad/s

8/49 $c = 55.6\ \text{N}\cdot\text{s/m}$

8/50 $k = 5050\ \text{N/m}$

8/51 $(a)\ X = 0.251$ ft, $(b)\ X = 0.0791$ ft; $\delta_{st} = 0.25$ ft

8/52 $\omega_c = \sqrt{\dfrac{k}{6m}}$

8/53 $R_1 = 50\%$, $R_2 = 2.52\%$

8/54 $f = 3.83$ Hz

8/55 $\dfrac{\omega}{\omega_n} = \sqrt{1 - 2\zeta^2}$

8/56 $\omega < \sqrt{\dfrac{2}{3}}\omega_n$ and $\omega > \sqrt{2}\omega_n$

8/57 $N \leq 108.1$ rev/min and $N \geq 153.5$ rev/min

8/58 $M = 0.974$, $c = 208(10^3)\ \text{N}\cdot\text{s/m}$

8/59 $f = \dfrac{1}{2\pi}\sqrt{\dfrac{g}{\delta_{st}}}$

8/60 $2.38 < f_n < 5.32$ Hz

8/61 $\ddot{x}_i + 2\zeta\omega_n\dot{x}_i + \omega_n^2 x_i = \dfrac{k}{m}b\sin\omega t + \dfrac{c}{m}b\omega\cos\omega t$

8/62 $\ddot{x} + \dfrac{c}{m}\dot{x} + \dfrac{k_1 + k_2}{m}x = \dfrac{k_2}{m}b\cos\omega t,\ \omega_c = \sqrt{\dfrac{k_1 + k_2}{m}}$

8/63 $\ddot{x} + \dfrac{c_1 + c_2}{m}\dot{x} + \dfrac{k}{m}x = -\dfrac{c_2}{m}b\omega\sin\omega t,\ \zeta = \dfrac{c_1 + c_2}{2\sqrt{km}}$

8/64 $k = 227$ kN/m or 823 kN/m

8/65 $b = 1.886$ mm

8/66 $\ddot{y} + \dfrac{4k}{m}y = \dfrac{2k}{m}b\sin\omega t,\ \omega_c = 2\sqrt{\dfrac{k}{m}}$

8/67 $\ddot{x} + \dfrac{c}{m}\dot{x} + \dfrac{k_1 + \dfrac{b^2}{a^2}k_2}{m}x = \dfrac{k_2}{m}\dfrac{b}{a}b_0\cos\omega t$

8/68 $c = 44.6\ \text{N}\cdot\text{s/m}$

▶8/69 $E = \pi c\omega X^2$

▶8/70 $X = 4.19$ mm, $v_c = 15.16$ km/h

8/71 $\omega_n = \sqrt{\dfrac{2k}{m}},\ \tau = \pi\sqrt{\dfrac{2m}{k}}$

8/72 $\omega_n = \sqrt{\dfrac{3g}{2\sqrt{a^2 + b^2}}}$

8/73 $\tau = 2\pi\sqrt{\dfrac{2b}{3g}}$

8/74 $\tau = 2\pi\sqrt{\dfrac{5b}{6g}}$

8/75 $\Theta = 0.1271$ rad

8/76 $\tau = 2\pi\sqrt{\dfrac{ml^2L}{12JG}}$

8/77 $\ddot{\theta} + \dfrac{8g\sin\dfrac{\beta}{2}}{3r\beta}\sin\theta = 0,\ \tau = 1.003$ s

8/78 $\ddot{\theta} + \dfrac{3g\left(\cos\dfrac{\beta}{2} + 2\sin\dfrac{\beta}{2}\right)}{r(2 + 3\beta)}\sin\theta = 0,\ \tau = 1.060$ s

8/79 $\omega_n = \sqrt{\dfrac{3gh}{3r^2 + 2h^2}}$

8/80 $\ddot{\theta} + \dfrac{\dfrac{1}{2}m_1gL + \dfrac{4}{9}kL^2}{\dfrac{1}{3}m_1L^2 + \dfrac{4}{9}m_2L^2}\theta = \dfrac{\dfrac{2}{3}kLb\cos\omega t}{\dfrac{1}{3}m_1L^2 + \dfrac{4}{9}m_2L^2}$

8/81 $\zeta = \dfrac{cb}{2a}\sqrt{\dfrac{3}{km}},\ c_{cr} = \dfrac{2a}{b}\sqrt{\dfrac{km}{3}}$

8/82 $x = 0.558$ m

8/83 $\omega_n = 8.24$ rad/s

8/84 $\omega_n = 7.53$ rad/s

8/85 $\zeta = \dfrac{a^2c}{2b^2}\sqrt{\dfrac{3}{km}},\ c_{cr} = \dfrac{2b^2}{a^2}\sqrt{\dfrac{km}{3}}$

8/86 $\tau = \dfrac{2\pi k_O}{\sqrt{\dfrac{2kb^2}{m} + g\bar{r}}}$

8/87 $k = 3820$ N/m

8/88 $\ddot{x}_1 + \dfrac{a^2c_1 + b^2c_2}{a^2m_1 + b^2m_2 + k_O^2m_3}\dot{x}_1 + \dfrac{a^2k_1 + b^2k_2}{a^2m_1 + b^2m_2 + k_O^2m_3}x_1 = 0$

8/89 $m_{eff} = m_1 + \dfrac{m_2}{2}$

8/90 $\omega_n = \sqrt{\dfrac{k}{m_1 + \dfrac{k_O^2}{a^2}m_2}},\ \zeta = \dfrac{\dfrac{b^2}{a^2}c}{2\sqrt{k\left(m_1 + \dfrac{k_O^2}{a^2}m_2\right)}}$

8/91 $\omega_c = 3.92$ rad/s

8/92 $c = 198.4\ \text{N}\cdot\text{s/m}$

8/93 $\tau = \pi\sqrt{\dfrac{6(R - r)}{g}},\ \omega = \dfrac{\theta_0}{r}\sqrt{\dfrac{2g(R - r)}{3}}$

8/94 $\Theta = \dfrac{-\dfrac{3b}{2l}\omega^2}{\omega_n^2 - \omega^2}$, where $\omega_n = \sqrt{\dfrac{3k_T}{ml^2} - \dfrac{3g}{2l}}$

▶8/95 $\omega_c = 6.38$ rad/s, $\zeta = 0.247$

▶8/96 $\omega_n = 10.24$ rad/sec, $v = 11.95$ ft/sec, $\zeta = 1.707$

8/97 $f_n = 2.43$ Hz

8/98 $\tau = 6\pi\sqrt{\dfrac{m}{5k}}$

8/99 $\tau = 2\pi\sqrt{\dfrac{2l}{3g}}$

8/100 $\omega_n = \sqrt{\dfrac{k}{m\left(1 + \dfrac{\bar{k}^2}{r^2}\right)}}$, $\bar{k} = 0$: $\omega_n = \sqrt{\dfrac{k}{m}}$

$\bar{k} = r$: $\omega_n = \sqrt{\dfrac{k}{2m}}$

8/101 $\tau = 2\pi\sqrt{\dfrac{2r}{g}}$

8/102 $f_n = \dfrac{b}{2\pi l}\sqrt{\dfrac{k}{m}}$

8/103 $\omega_n = \sqrt{\dfrac{12k_T + 18\rho g l^2}{17\rho l^3}}$

8/104 $\tau = 0.326$ sec

8/105 $\tau = \pi\sqrt{\dfrac{6(R - r)}{g}}$

8/106 $x = \sqrt{\dfrac{\dfrac{\tau^2 k_T}{4\pi^2} - I_O}{2m}}$

8/107 $\omega_n = 3\sqrt{\dfrac{6k}{3m_1 + 26m_2}}$, $\omega = 3\theta_0\sqrt{\dfrac{6k}{3m_1 + 26m_2}}$

8/108 $\tau = 0.808$ s

8/109 $f_n = \dfrac{1}{2\pi}\sqrt{\dfrac{mgr_0}{3Mr^2 + m(r - r_0)^2}}$

8/110 $\omega_n = \sqrt{\dfrac{mgr \sin\alpha}{I + mr^2}}$

8/111 $\ddot{x} + \dfrac{a^2k_1 + b^2k_2 + c^2k_3}{a^2m_1 + b^2m_2 + c^2m_3 + k_O{}^2m_4}x = 0$

8/112 $\omega_n = \sqrt{\dfrac{6k}{m} - \dfrac{3g}{2l}}$, $k > \dfrac{mg}{4l}$

8/113 $f_n = 1.496$ Hz

8/114 $f_n = 1.142$ Hz

8/115 $\ddot{x} + 600x + 54x^3 = 0$, with x in in.

8/116 $\tau = 2\pi\sqrt{\dfrac{(\pi - 2)r}{g}}$

▸**8/117** $\tau = 41.4\sqrt{\dfrac{R}{g}}$

▸**8/118** $\tau = 4.44\sqrt{\dfrac{r}{g}}$

8/119 $f_n = \dfrac{1}{2\pi}\sqrt{\dfrac{2kb^2}{ml^2} - \dfrac{g}{l}}$, $k > \dfrac{mgl}{2b^2}$

8/120 $\tau = \dfrac{2\pi}{\sqrt{\dfrac{2k_T}{mb^2} + \dfrac{4g\sin\beta}{3b\beta}}}$

8/121 $X = 0.287$ m, $\tau = 0.365$ s

8/122 (*a*) $\omega_n = 2\sqrt{\dfrac{g}{5r}}$, (*b*) $\omega_n = \sqrt{\dfrac{2g}{3r}}$

8/123 (*a*) $f_n = 0.991$ Hz, (*b*) $f_n = 0.701$ Hz
(*c*) $f_n = 0.495$ Hz

8/124 $\omega_c = \omega_n = \sqrt{\dfrac{2g}{\sqrt{3}d} + \dfrac{4k}{m}}$

8/125 $Q = \dfrac{1}{2}kx_1{}^2(1 - e^{-2\delta})$, where $\delta = \dfrac{\pi}{\sqrt{\dfrac{km}{c^2} - \dfrac{1}{4}}}$

8/126 $\zeta = 0.0697$

8/127 $c = 69.3$ N·s/m

8/128 $\delta_0 = 0.712$ mm

8/129 $f_n = \dfrac{1}{\pi}\sqrt{\dfrac{3k}{m}}$

8/130 $(f_n)_y = 4.95$ Hz, $(f_n)_\theta = 10.75$ Hz, $N = 645$ rev/min

8/131 $\tau = 2\pi\sqrt{\dfrac{2mR^2 + MR^2 - \dfrac{1}{6}Ml^2}{g\sqrt{4R^2 - l^2}\left(m + \dfrac{M}{2}\right)}}$

▸**8/132** 28.9% increase in amplitude

***8/133** $t = 0.972$ s

***8/134** Large-angle: $\theta = 28.0°$ at $t = 1$ s
Small-angle: $\theta = 38.8°$ at $t = 1$ s

***8/135** $t = 0.0544$ s or 0.442 s

***8/136** $0 < k < 1.895$ lb/ft

***8/137** $y = 114.3(t - 0.1069 \sin 9.35t)$ mm

***8/138** $x_{\text{max}} = 0.1955$ ft at $t = 0.0462$ sec
$x_{\text{min}} = -0.0792$ ft at $t = 0.1923$ sec

***8/139** $x = 0.0926(t - 0.0913 \sin 10.95t)$ m

***8/140** (*a*) $y = 0.1722e^{-9.16t} - 0.0722e^{-21.8t}$ m
(*b*) $y = 0.1414e^{-10t}\sin(10t + 0.785)$ m

APPENDIX B

B/1 $I_{xx} = I_{yy} = \dfrac{1}{24}\rho L^3$, $I_{zz} = \dfrac{1}{12}\rho L^3$

B/2 $I_{xx} = \dfrac{1}{6}mh^2$, $k_x = \dfrac{h}{\sqrt{6}} = 0.408h$

$I_{yy} = \dfrac{1}{6}mb^2$, $k_y = \dfrac{b}{\sqrt{6}} = 0.408b$

$I_{zz} = \dfrac{1}{6}m(h^2 + b^2)$, $k_z = 0.408\sqrt{h^2 + b^2}$

B/3 $I_{yy} = \dfrac{1}{24}mb^2$, $k_y = 0.204b$

B/4 $I_{xx} = \dfrac{3}{10}mr^2$, $I_{yy} = \dfrac{3}{5}m\left(\dfrac{r^2}{4} + h^2\right)$

B/5 $I_{xx} = \dfrac{3}{7}mh^2$, $k_x = 0.655h$

B/6 $I_{yy} = \dfrac{1}{20}mb^2$, $k_y = 0.224b$

B/7 $I_{x'x'} = \frac{1}{5}mh^2, I_{y'y'} = \frac{8}{35}mb^2, I_{z'z'} = m\left(\frac{h^2}{5} + \frac{8b^2}{35}\right)$
B/8 $I_{xx} = \frac{1}{4}mb^2$
B/9 $I_{xx} = \frac{2}{7}mr^2$
B/10 $I_{yy} = I_{zz} = m\left(\frac{r^2}{7} + \frac{2h^2}{3}\right)$
B/11 $k_y = 39.6$ mm
B/12 $k_x = 39.0$ mm
B/13 $k_z = \frac{r}{\sqrt{3}}$
B/14 $I_{yy} = \frac{1}{2}m\left(h^2 + \frac{r^2}{3}\right)$
B/15 $I_{yy} = \frac{1}{2}\left(\frac{2+n}{4+n}\right)mb^2$
B/16 $I_{xx} = \frac{53}{200}mR^2$
B/17 $I = m\left(R^2 + \frac{3}{4}a^2\right)$
B/18 $I_{xx} = \frac{166}{147}mb^2$
B/19 $I_{yy} = m\left(\frac{83}{147}b^2 + \frac{23}{196}L^2\right)$
B/20 $I_{zz} = \frac{3}{5}mb^2$
B/21 $I_{xx} = m\left(\frac{3}{10}b^2 + \frac{1}{10}h^2\right)$
B/22 $I_{yy} = \frac{1}{10}mh^2$
B/23 $I = \frac{1}{2}m(2R^2 + 3a^2)$
B/24 $I_{xx} = I_{zz} = \frac{2}{3}mr^2$
B/25 $I_{zz} = \frac{8}{21}ma^2$
B/26 $I_{xx} = m\left(\frac{4}{21}a^2 + \frac{1}{7}b^2\right)$
►**B/27** $k_z = 0.890a$
►**B/28** $I_{yy} = 5.66(10^{-3})$ kg·m², $k_y = 53.8$ mm
►**B/29** $I_{zz} = 21.5(10^{-3})$ kg·m², $k_z = 104.9$ mm
B/30 $e = \frac{-100}{1 + \frac{l^2}{3r^2}}\%, \frac{r}{l} = 0.01\text{: } e = -0.030\%$
$\frac{r}{l} = 0.1\text{: } e = -2.91\%, \frac{r}{l} = 0.5\text{: } e = -42.9\%$
B/31 $I_{xx} = I_{zz} = 2mL^2, I_{yy} = 4mL^2$
B/32 $I_{yy} = 0.5$ kg·m², $e = 0.25\%$
B/33 $I_{xx} = \frac{1}{2}m({r_1}^2 + {r_2}^2)$
B/34 $I_{zz} = \frac{1}{12}m(15r^2 + L^2)$
B/35 $I_{aa} = \frac{1}{2}mr^2, I_{bb} = 2mr^2\left(1 - \frac{2}{\pi}\right)$
B/36 $I_{xx} = 1.898$ lb-in.-sec²
B/37 $I_{xx} = I_{yy} = 8.16(10^{-3})$ lb-ft-sec²
$I_{zz} = 1.941(10^{-3})$ lb-ft-sec²
B/38 $L = \frac{r\sqrt{3}}{2}$
B/39 $I_{yy} = \rho L^3\left(\frac{43}{192} + \frac{83\pi}{128}\right)$ or $2.26\rho L^3$
B/40 $I = 1.031$ kg·m², $n = 97.8\%$
B/41 $I_{xx} = 5.02(10^{-3})$ lb-ft-sec²
$I_{yy} = 5.84(10^{-3})$ lb-ft-sec²
$I_{zz} = 7.12(10^{-3})$ lb-ft-sec²
B/42 $I_{xx} = I_{zz} = \frac{3}{4}mb^2, I_{yy} = \frac{1}{6}mb^2$
B/43 (*a*) $I_{yy} = 3.27(10^{-3})$ lb-ft-sec²
(*b*) $I_{zz} = 3.68(10^{-3})$ lb-ft-sec²
B/44 $I_{xx} = 0.1107$ kg·m², $I_{x_0x_0} = 0.1010$ kg·m²
B/45 $I_{xx} = 0.408mb^2, I_{yy} = 0.6mb^2, I_{zz} = 0.525mb^2$
B/46 $I_O = m\left(7x^2 + \frac{1}{3}l^2\right), R = 0.582$
B/47 $I_{xx} = 3.71(10^{-5})$ kg·m², $I_{x'x'} = 4.95(10^{-5})$ kg·m²
B/48 $I_A = 0.1701ma^2$
B/49 $I_{xx} = 0.1670$ kg·m², $I_{yy} = I_{zz} = 1.356$ kg·m²
B/50 $I_{zz} = 12.40(10^{-3})$ lb-ft-sec²
B/51 $I_{yy} = 13.98(10^{-3})$ lb-ft-sec²
B/52 $I_{xx} = 3.65(10^{-2})$ kg·m², $I_{yy} = 2.10(10^{-2})$ kg·m²
B/53 $I_{xx} = \frac{3}{10}m\left(\frac{{r_2}^5 - {r_1}^5}{{r_2}^3 - {r_1}^3}\right)$
***B/54** $L > 4.54r$
►**B/55** $k_z = 39.9$ mm
►**B/56** $I_O = 0.0671$ kg·m²
B/57 $I_{xy} = -4mb^2, I_{xz} = -2mb^2, I_{yz} = 0$
B/58 $I_{xy} = 0, I_{xz} = I_{yz} = -2ml^2$
B/59 $I_{xy} = -\frac{1}{24}ml^2 \sin 2\theta$
B/60 $I_{xy} = -\frac{1}{24}\rho L^3 \sin 2\theta$
B/61 $I_{xy} = -mab, I_{xz} = \frac{1}{2}mah, I_{yz} = -\frac{1}{2}mbh$
B/62 $I_{xy} = -\frac{\rho\pi b^4}{512}, I_{xz} = I_{yz} = 0$
B/63 $I_{xy} = \frac{1}{2}mrh, I_{xz} = -\frac{4}{3\pi}mr^2, I_{yz} = -\frac{2}{3\pi}mrh$
B/64 $I_{xy} = \frac{ma^2 \sin^2 \beta}{4\beta}, I_{xz} = I_{yz} = 0$
B/65 $I_{xy} = \frac{1}{4}mbh$

B/66 $I_{xy} = \frac{1}{4}mbh, I_{x'y'} = -\frac{1}{12}mbh, I_{x''y''} = \frac{1}{4}mbh$

$\bar{I}_{xy} = \frac{1}{36}mbh$

B/67 $I_{xy} = I_{xz} = -0.0300$ lb-ft-sec^2

$I_{yz} = 0.01929$ lb-ft-sec^2

B/68 $I_{xy} = 0, I_{xz} = \frac{1}{2}\rho a^3, I_{yz} = -\frac{1}{2}\rho a^3$

B/69 $I_{xy} = \frac{2}{\pi}mr^2, I_{xz} = I_{yz} = 0$

B/70 $I_{xx} = I_{yy} = I_{zz} = m\left(\frac{6}{5}r^2 + 2b^2\right), I_{xy} = I_{xz} = I_{yz} = 0$

►**B/71** $I_{yz} = \frac{5}{8}mr^2 \sin 2\beta$

***B/72** $(I_{AA})_{min} = 0.535$ kg·m^2 at $\theta = 22.5°$

***B/73** $I_{min} = 0.1870\rho r^3$ at $\alpha = 38.6°$

***B/74** $I_1 = 9ml^2, I_2 = 7.37ml^2, I_3 = 1.628ml^2$

$l_1 = 0.816, m_1 = 0.408, n_1 = 0.408$

***B/75** $I_{xx} = 3.74mb^2, I_{yy} = 22.5mb^2, I_{zz} = 26.2mb^2$

$I_{xy} = 5.55mb^2, I_{xz} = I_{yz} = 0$

$\theta = 15.30°$ about $+z$-axis

***B/76** $I_1 = 3.78\rho b^4, I_2 = 0.612\rho b^4, I_3 = 3.61\rho b^4$

***B/77** $I_1 = 1.448\rho b^3, I_2 = 0.360\rho b^3, I_3 = 1.142\rho b^3$

$l_2 = 0, m_2 = 0.550, n_2 = 0.835$

***B/78** $I_1 = 1.509$ kg·m^2, $l_1 = 0.996$, $m_1 = -0.0876$

$n_1 = 0.00514$

$I_2 = 1.431$ kg·m^2, $l_2 = -0.0433$, $m_2 = -0.439$

$n_2 = 0.897$

$I_3 = 0.406$ kg·m^2, $l_3 = 0.0764$, $m_3 = 0.894$

$n_3 = 0.441$

Conversion Charts Between SI and U.S. Customary Units

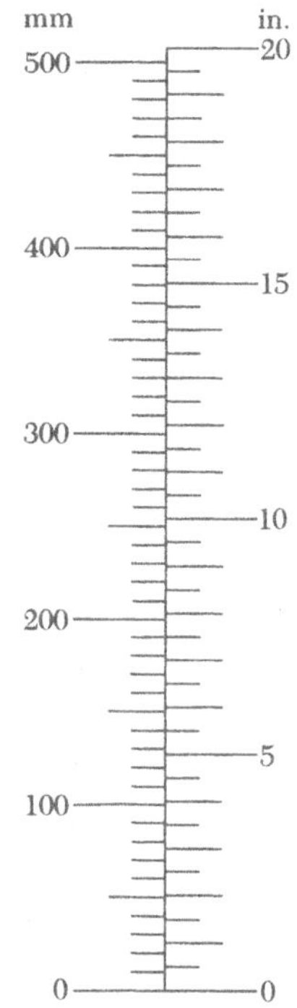

Length

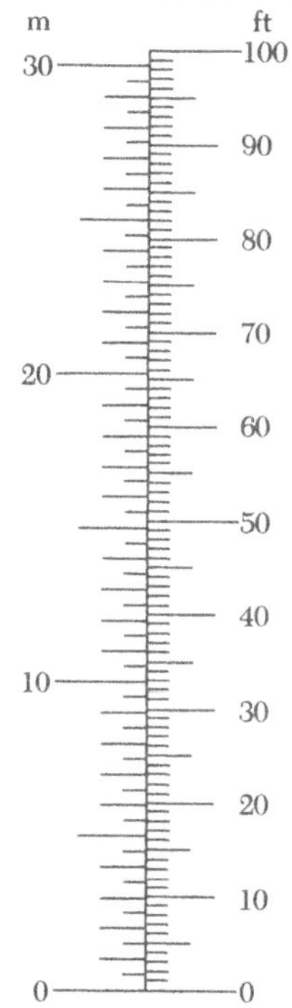

Length

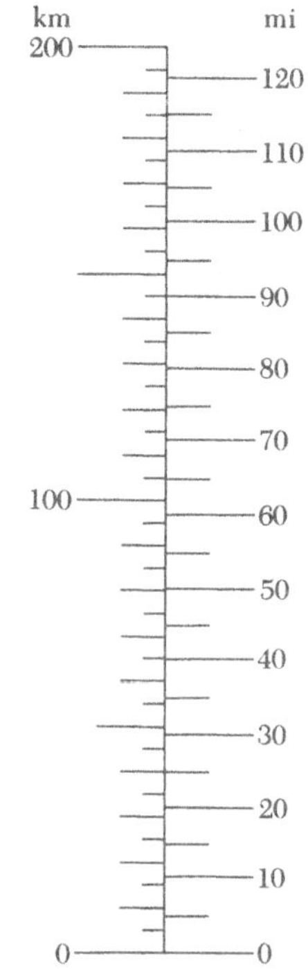

Length

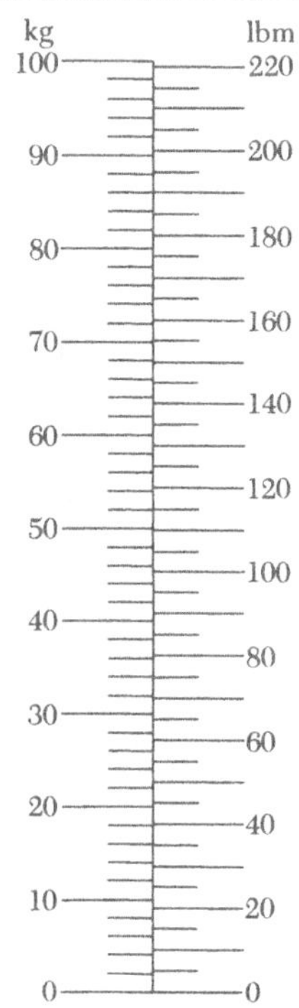

Mass

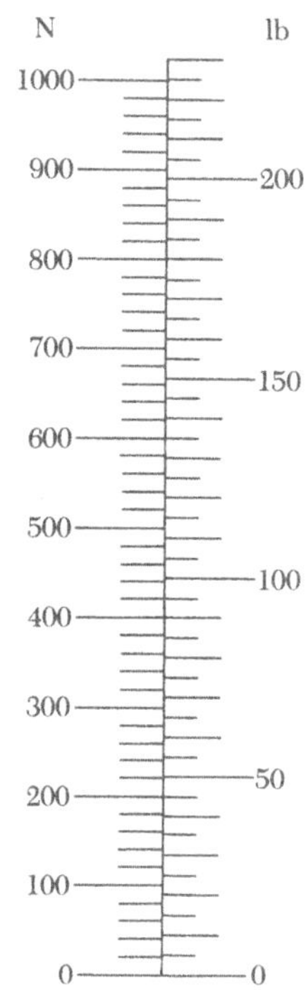

Force

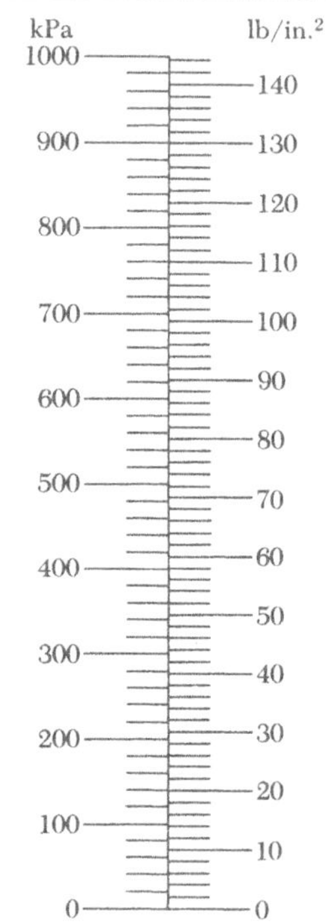

Pressure or Stress

Conversion Charts Between SI and U.S. Customary Units (cont.)

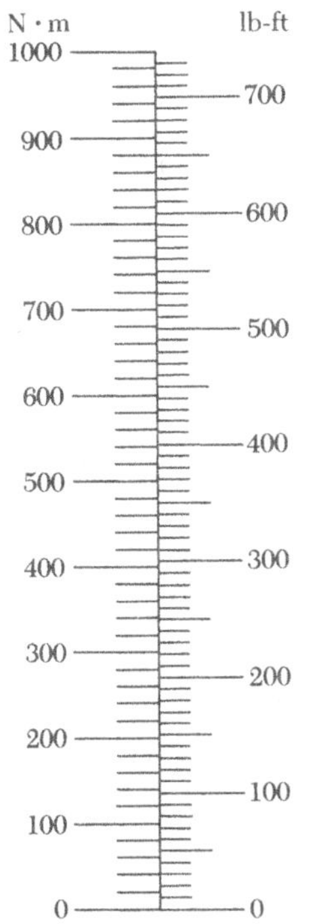

Moment or Torque

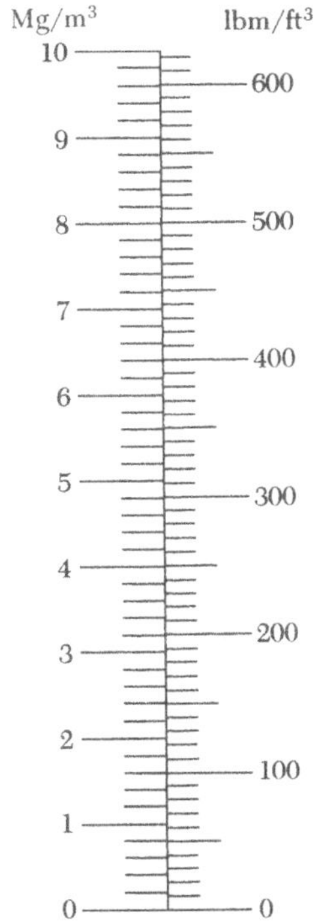

Density

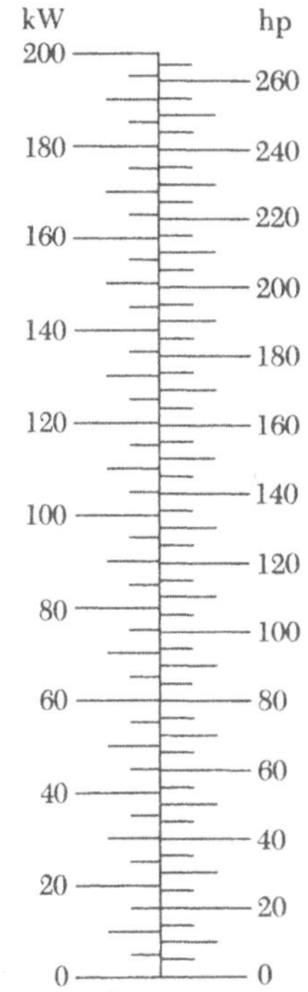

Power

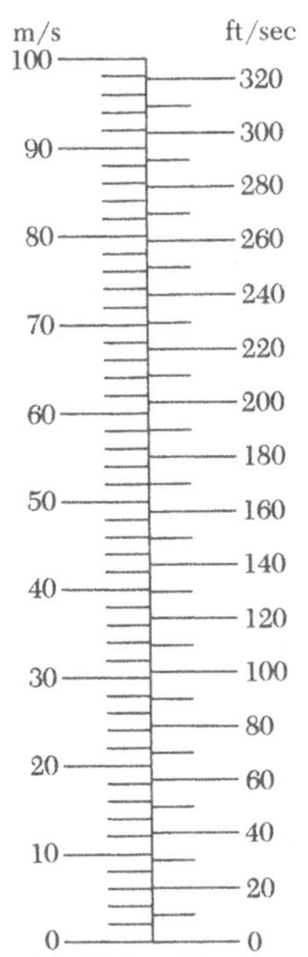

Velocity

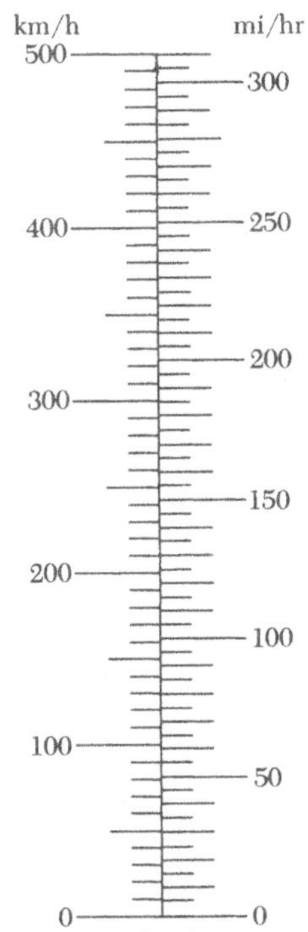

Velocity

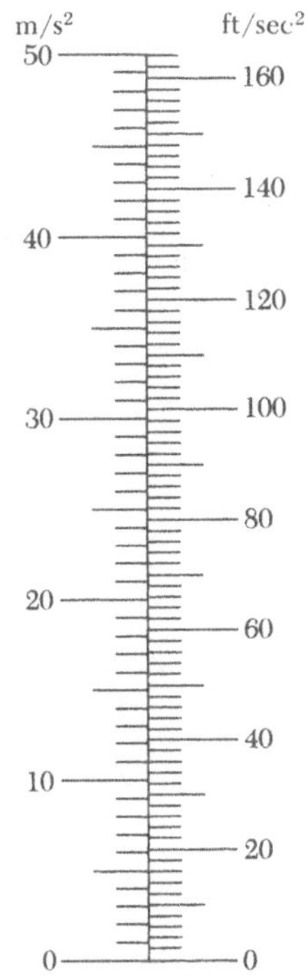

Acceleration